ANIMAL PHYSIOLOGY

ANIMAL PHYSIOLOGY

SECOND EDITION

RICHARD W. HILL

Michigan State University

GORDON A. WYSE

University of Massachusetts at Amherst

HarperCollinsPublishers

Sponsoring Editor: Claudia M. Wilson
Project Editor: Thomas R. Farrell
Cover Design: Wanda Lubelska
Text Art: Vantage Art, Inc.
Production: Kewal K. Sharma
Compositor: Ruttle, Shaw & Wetherill, Inc.
Printer and Binder: Murray Printing Co.

ANIMAL PHYSIOLOGY, Second Edition

Library of Congress Cataloging-in-Publication Data

Hill, Richard W.
  Animal physiology.

  Rev. ed. of: Comparative physiology of animals /
Richard W. Hill. 1976.
  Includes bibliographies and index.
  1. Physiology, Comparative.   I. Wyse, Gordon A.
II. Hill, Richard W. Comparative physiology of
animals.   III. Title.
QP33.H54 1988   591.1   87-14883
ISBN 0-06-042826-0

98 99 00 CRW 13 12 10 9

To
David and Christine
and to
Mary

CONTENTS

11 PRINCIPLES OF DIFFUSION AND DISSOLUTION OF GASES 244

12 EXCHANGES OF OXYGEN AND CARBON DIOXIDE 1: RESPIRATORY ENVIRONMENTS, INTRODUCTION TO RESPIRATORY EXCHANGE, AND EXTERNAL RESPIRATION 248

13 EXCHANGES OF OXYGEN AND CARBON DIOXIDE 2: TRANSPORT IN BODY FLUIDS 298

14 CIRCULATION 334

15   THE PHYSIOLOGY OF DIVING IN BIRDS AND MAMMALS   370

16   NEURONS AND NERVOUS SYSTEMS: THE BASIS OF EXCITABILITY   385

17   SYNAPSES   419

18   SENSORY PHYSIOLOGY   455

PREFACE

Our goal in this textbook is to provide a comprehensive introduction to comparative animal physiology at a level suitable for upper-division undergraduate and beginning graduate students. Our emphasis is deliberately conceptual, for we believe that the acquisition of proper conceptual frames of reference should be the primary objective of introductory study. We develop physiological concepts from fundamental principles of chemistry, biology, and physics; and we attempt throughout to step back from the welter of available biological detail to identify patterns of significance and importance.

We have tried to make this book equally suitable for a course in comparative physiology, animal physiology, or physiological ecology. Toward this goal, we have tried to balance the general and the comparative perspectives in physiology in a way that will enlighten students rather than mislead them with overgeneralizations or burden them with exceptions. We have employed the perspectives of comparative physiology—evolutionary, adaptational, and environmental—where it is practical and desirable to do so. We consider these perspectives to be important and valuable. Nevertheless, we have not found it practical or advantageous to maintain a uniformly comparative or ecological perspective for all topics. Thus, topics such as salt and water balance receive a more ecological and comparative treatment than do others such as neural transmission. We consider this balance between general and comparative viewpoints to be essential to our goal of providing a text of manageable size, which presents the perspective of comparative physiology without requiring a prerequisite course in general physiology.

We have chosen to emphasize the ecological relations of animals because in nature it is on an ecological stage that the physiological play is acted out. We have also emphasized evolutionary perspectives in recognition of the importance of history in determining the form and functional properties of modern species. Because comparative physiology draws its greatest strength as a discipline from broad study of the differences and similarities among animals, we have treated the entire animal kingdom, with the exception only of the protists and metazoan parasites. We believe that this broad scope provides a much more satisfactory approach than a treatment limited to only vertebrates or some other circumscribed phyletic assemblage. We have attempted to provide sufficient background information on anatomy and natural history for readers to appreciate the physiology of groups that may be otherwise unfamiliar. The intimate interrelations between morphology and function are repeatedly emphasized.

NEW TO THIS EDITION

Some will recognize that this text is the descendant of an earlier work by one of us (R.W.H.). In comparison to the previous book, this one is extensively rewritten, expanded, and reorganized. The greatest change is that we have added chapters on nervous and sensory functions, muscle and other effectors, and endocrines. Along with this enlarged scope, we have expanded coverage of control mechanisms throughout. We have also greatly expanded the introductory material (Chapters 1 and 2) to include discussions of control theory, homeostasis, the autonomic nervous system, biochemical control mechanisms, and the concept of evolutionary adaptation. We have divided this introductory material into two chapters with different overriding objectives. The material in Chapter 1 is a necessary introduction for all students. On the other hand, the various concepts of physiological regulation in Chapter 2 are more complex and could profitably be mastered either at the start of a course or later. We have tried to structure Chapter 2 so as to give instructors considerable flexibility in assigning its parts for reading.

Other chapters that are descended from the first edition of this book have been extensively rewritten, updated, and expanded in content. In the present book, for example, the physiology of hearts and pacemakers is discussed in detail; the principles of intracellular volume regulation in euryhaline animals are reviewed; the phenomenon of uncoupling of oxidative phosphorylation is discussed; the interpretive value of renal clearance studies is described; all information on renal tubular function is updated based on the explosion of data from microperfusion studies; passive models of mammalian urine concentration are discussed; the concept of the oxygen cascade is presented as an organizing principle for the treatment of oxygen delivery; the discussion of biophysical modeling of thermal relations has been expanded; and the material on water and salt relations of terrestrial animals has been synthesized by recognizing physiological (rather than merely taxonomic) groupings. A concerted effort has been made throughout to give increased prominence to synthetic principles and organizing concepts. Furthermore, the number of figures and tables has been substantially increased. Inevitably, expansions in some places require constrictions elsewhere. At first, we had intended to increase the number of chapters on special physiological problems, but we found that a decrease was necessary. We felt it important to retain at least one chapter in which an integrated approach to a physiological problem could be pursued in depth, and we chose diving as the subject, in part because it provides an instructive example of how inquisitiveness and new data can lead to a new world view (Chapter 15). Altitude physiology is now covered in the form of boxes within the chapters on oxygen delivery (Chapters 12 and 13). Biological rhythms are treated in Chapter 2.

ACKNOWLEDGMENTS

Where once people bemoaned the passage of an era when one or two individuals could together claim reasonable mastery of all of biology, the growth of knowledge has now reached a point where maintaining mastery of just a single subdiscipline—such as physiology—is a challenge. We are acutely aware of the magnitude of the challenge and here express our regret for any errors we have made and our plea that readers bring errors or other shortcomings to our attention. So that communications can be properly directed, we mention that Chapters 1–15 and 21 are mostly the work of Hill, whereas Chapters 16–20 and the material on endocrine biochemistry and invertebrate endocrinology in Chapter 21 are principally by Wyse.

The quality of the manuscript has been significantly improved by the comments and help of a number of biologists who have critically read part or all of it. Among them are Albert F. Bennett, University of California at Irvine; Eric Bittman, University of Massachusetts; Ronald L. Calabrese, Harvard University; Donald P. Christian, University of Minnesota, Duluth; Douglas A. Eagles, Georgetown University; Franz Engelman, University of California at Los Angeles; Robert E. Gatten, Jr., University of North Carolina, Greensboro; Richard J. Hoffman, Iowa State University; Theodore M. Hollis, The Pennsylvania State University; James L. Larimer, University of Texas, Austin; Leo E. Lipetz, The Ohio State University; L. M. Passano, University of Wisconsin, Madison; Henry D. Prange, Indiana University; Gregory Snyder, University of Colorado, Boulder; Donald H. Whitmore, University of Texas, Arlington; Leah H. Williams, West Virginia University. We also have profited immeasurably from our routine interactions with colleagues over the years. The contributions of those colleagues are often difficult to trace to their end effect on our thinking, and composing a list of all the individuals who have been influential and helpful would be a challenge. We wish to thank them all, however, and in the process acknowledge our awareness that this book is very much a product of the world community of comparative biologists. Where halftones appear in the text, we have nearly always received the generous cooperation of fellow scientists in obtaining original prints of the illustrations. We thank those individuals and all who have given us permission to use data or illustrations from their work. In particular, we thank the following men and women who have helped us obtain or have made available unpublished halftone illustrations: Bernd Heinrich, Dave Hinds, Michael Hlastala, Daniel Luchtel, Keith R. Porter, Frank L. Powell, Jane K. Townsend, and Walter S. Tyler. Kjell Johansen, whose data are featured prominently in several chapters, not only provided illustrative material but offered repeated encouragement over the years; his recent death is a source of sadness. Much of Hill's writing was done at the Marine Biological Laboratory in Woods Hole. Acknowledgment is made to the Laboratory and also to John W. H. Dacey (Woods Hole Oceanographic Institution), who was always eager to help in many ways. We also express our gratitude to Claudia M. Wilson, Thomas R. Farrell, and the staff of Harper & Row, without whose sympathetic editorial counsel and assistance this project could not have been completed. Finally, we thank our wives, Susan and Mary, for their steadfast support, and perhaps even more we thank our children, who have undoubtedly not always comprehended why we were off at the office so much, for their affection and forbearance.

Richard W. Hill
Gordon A. Wyse

NOTE TO THE STUDENT

We hope you will read the preface, which describes our goals and perspectives in this text. Our highest hope is that you will share the sense of excitement that we have felt in studying the grand patterns of animal function. If we have helped you in that quest, our efforts to create a readable and informative book will have been rewarded.

Writing a textbook involves a constant tension between the desire to recognize all available knowledge and the need to highlight general patterns that emerge from the background of detail. Inevitably, the identification of patterns entails the circumvention of some detail, the running roughshod over some data painstakingly gathered. The question for an author is always how to strike the proper balance. Too little attention to detail leads to abstractions so crude as to bear little resemblance to real nature. Yet, as one pundit has put it, the compression of too many details into the space available can create a black hole rather than a supernova. Another concern is that science is always changing; whole worldviews can be replaced with others. One of the finest of intellectual challenges is to preserve the capacity to make the intellectual leap from one worldview to another when justified by evidence. That is a lifelong challenge for you, as it is for us. We trust that the following pages do not reflect too many failures on our part in those respects.

We want to discuss here three additional issues we have confronted as authors. The first is the matter of giving references to our sources.

The knowledge described in this book is the product of decades of study by thousands of investigators. Scientists announce new knowledge by means of research reports in the scientific journals. In the formal scientific literature, when an author makes reference to a piece of such reported knowledge, it is customary to provide a citation to the journal and article in which the knowledge is documented; the author thus provides authority for the information and enables readers to "check out" the original research report for themselves. We debated whether we should do the same in this book. Our decision was that the text would become too cluttered if we provided a citation for every piece of knowledge, and we could devise no satisfactory formula for referencing some pieces of knowledge but not others. Thus, you will not find citations in the running text of our book. What we have done is to give you an ample reading list at the end of each chapter, and we have been certain to include in each list some individual research reports (idiosyncratically selected) as well as books and review articles. Should you wish to pursue a topic that you have learned about in this book, you will usually be able to find leads into the literature on the topic by using the reference lists we have provided. Sometimes, the lists deliberately include articles that express viewpoints at odds with our own presentation in this text. Two details concerning the reference lists are important. We have placed an asterisk (*) next to certain of the readings in each chapter that we believe would be especially worthwhile for your first explorations beyond this book. Also, we have clustered together in Appendix A at the end of the book some references that apply to a diversity of chapters. As a group, these references are extremely important, and most chapter reading lists include a reminder that you should consult Appendix A to find them.

Another topic we debated was whether to include a glossary. Because glossaries typically provide little more than verbatim quotes of definitions found in the text proper, we decided to omit a glossary and provide special directions to the definitions in the text. If you need a definition of a word or concept, turn to the page listed in bold type under that word or concept in the index. This system has the advantage that you will usually find explanatory material on the same page as you will find the definition.

The final matter we want to discuss is a conundrum: the question of what units of measure to use. There has been of late a concerted effort to adopt a single system of units in physiology. This system, called the Système International (SI, for short), recognizes seven base units of measure; all other units are to be derived from these seven. For instance, the base units of mass, length, and time are the kilogram (kg), meter (m), and second (s), respectively; the derived unit for energy is then the $m^2 \cdot kg/s^2$, which is known as a joule (J), and the derived unit for pressure is the $kg/(m \cdot s^2)$, which is called a pascal (Pa). In principle, we support the exclusive adoption of SI units; there is little to recommend having two, three, or four different units of measure for one parameter. Nonetheless, we believe also that there are legitimate practical concerns that must be weighed against the application of principle in a book of this sort. Most important, readers must be prepared to function with units that are in common usage, and those are not always the SI units; we believe we would do little to advance understanding by expressing blood pressures in kilopascals when the truth remains that for the immediate future at least, most of you who come to use such pressures in your work will be confronted with instruments calibrated in millimeters of mercury (mm Hg) or torr. Thus, for each parameter, we have chosen to emphasize units that are in reasonably common usage. If a unit is not part of the SI, we point that out and note its relation to the appropriate SI unit. Appendix B summarizes the SI units and their relations to traditional units. The following reference provides details on the SI: C. H. Page and P. Vigoureux, *The International System of Units*. National Bureau of Standards Special Publication 330, U.S. Government Printing Office, Washington, DC, 1974.

Richard Hill
Gordon A. Wyse

chapter *1*

Organism, Environment, and Adaptation

Physiology is the study of the function of organisms. In this text, our primary objective is to develop the basic concepts of animal physiology as presently understood. Our approach is *comparative*. That is, while we give substantial emphasis to mammals, we compare and contrast how functions are carried out in other vertebrates and invertebrates as well.

One advantage of the comparative approach to animal physiology is that it aids comprehension of our own functional attributes. By examining how particular functions are carried out by various animal groups that are built on diverse plans and occupy diverse environments, we gain a perspective on mammalian function that could not be obtained by looking at mammals alone. Indeed, advances in understanding of mammalian physiology in the past have often depended on studies of other organisms. The functional attributes of nerve fibers and vertebrate kidney tubules, for example, were first uncovered through investigations of squids and frogs, because the relevant tissues of the latter organisms are more amenable to study than those of mammals.

Another advantage of the comparative approach is that it permits patterns of physiological evolution to be identified. Comprehension of those patterns is often crucial to analyzing the adaptive value of physiological features (p. 5).

The comparative approach also has the inherent virtue of expanding physiological knowledge to include the full range of animals. An understanding of the functional attributes of fish, bees, and squids is of its own intrinsic interest to those who aspire to comprehend the natural world.

As we embark on our study of physiology, it is important that we focus at the outset on the concepts of organism and environment and on the process—adaptation—by which organisms have become matched to their environments. These are the concerns of this first chapter.

THE ORGANIZED AND DYNAMIC STATE OF BODY CONSTITUENTS

We are accustomed to seeing the world macroscopically, but this view is incomplete. Suppose for a moment that we could reduce our scale of vision to the atomic–molecular level and gaze at a woodland or pond. We would see atoms and molecules everywhere. In particular, here and there we would see self-sustaining physicochemical systems of high organization. Organization within these systems would be reflected both in the presence of large, complex molecules and in a patterned orientation of those molecules relative to one another. The organization of the systems would persist through time. These self-sustaining organized systems are what at a macroscopic level we call organisms.

We would see atoms and molecules moving into and out of each organized system. Some atoms and molecules would have a rapid passage; they would enter at a particular point, follow a particular path through the organized system, and exit at another particular point. These atoms and molecules would be undergoing ingestion and egestion. More importantly, we would see other atoms and molecules entering and exiting the *structure* of the organized system. They would be incorporated into or removed from the *organization itself*. In human beings, for example, iron atoms from foods are incorporated into hemoglobin in red blood cells, and later some of the atoms are excreted when the blood cells are broken down; the average longevity of a human red blood cell is only about 4 months. Calcium atoms enter the skeleton and later are withdrawn. As shown by the seminal investigations of Rudolf Schoenheimer and his colleagues in the 1930s, body fats and proteins are continually broken down and resynthesized at substantial rates. The resynthesis is carried out in part using molecules newly acquired from the environment, such as fatty acids and amino acids from foods. Human adults typically resynthesize about 3 percent of their body protein each day.

From this view of the world, we gain a number of important insights. (1) The material constituents of animals are in a state of dynamic exchange with the environment. In this regard, animals differ from objects such as telephones, which may be highly organized at an atomic–molecular level but do not exchange material with the outside world (except through surface wear). (2) Because atoms and molecules can change between being part of an organism and being part of its environment, it is not always easy to say where environment stops and organism begins. Consider, for example, a carbon atom in a piece of bread. When the bread is sitting on your table, the carbon atom is clearly part of your environment. Most of you would likely agree that it is still part of the environment when it is in your small intestine. But is it part of the environment or is it part of you when it is being transported in your bloodstream? And how should it be categorized when it has been absorbed by some cell? If finally the carbon atom becomes built into your own protein, we would firmly say that it is part of you. We recognize from this exercise that the material "boundaries" between an organism and its environment are not sharp. (3) A further and most important consequence of the exchange of atoms between organism and environment is that an organism is not a discrete material entity. Suppose you were to mark every atom in an adult animal's body at one point in time. If you then reexamined the animal 2 years later, many of the marked atoms would be gone, having been replaced with new, unmarked atoms taken in from the outside world. Thus, the precise material construction of an organism—unlike that of an inanimate object—does not persist through time. What then does persist? The *organization*. Part of the essential nature of organisms is that they are self-sustaining organizations.

The creation and maintenance of organization require not only material constituents but also energy (Chapter 3). Indeed, the acquisition and use of energy have loomed as major factors in animal evolution and are important themes of this book.

ORGANISM AND ENVIRONMENT

The environment is the theater in which the animal must function successfully. Physiological features that are appropriate in some environments may be woefully inappropriate in others. Thus, an accurate interpretation of an animal's physiology is dependent on detailed knowledge of the environment it occupies.

The *environment* of an animal is comprised of all the chemical, physical, and biotic components of its surroundings. As pointed out by the great physiologist Claude Bernard over a century ago, organism and environment are defined in terms of each other. Only by saying what "organism" is can we say what "environment" is. Consider, for example, a parasite inside a human being. From our usual perspective, the human is "organism," but from the perspective of the parasite, the human is "environment." All organisms, in fact, are part of the environment of other organisms.

An organism and its environment strongly interact with each other, further blurring the boundary between the two. All organisms are obviously influenced by aspects of their environment, such as temperature, light, and pollutants. Less obviously, the environment can be affected by the organism. Consider, for instance, a mouse in a small cavity in a tree. In winter, the mouse warms the cavity to temperatures above those in the outside world. Then the mouse responds physiologically to the elevated temperatures in the cavity. Clearly, mice are able to modify the thermal parameter of their own environment. In like manner, fish may deplete the water of oxygen and then must cope with low "environmental" oxygen levels.

We tend to divide the environment conceptually into different aspects or factors, a practice with advantages and disadvantages. The factors traditionally recognized have included temperature, humidity, light, wind, pH, salt concentration, oxygen concentration, food supply, competition, and predation. This factorial approach makes it easier for us to appreciate and analyze the multidimensional complexity of the environment, but it presents two dangers. First, we must recognize that some factors are more easily quantified than others. Temperature, for example, can be measured accurately and inexpensively. Light can be measured accurately but only with a more expensive instrument. Food supply can be quantified only approximately and with great difficulty. There is a temptation to study that which we can measure readily. The danger is that we may overemphasize the importance to animals of the factors we can easily quantify and underemphasize factors that we find more difficult to assess.

A second danger is that we may fail to take into account the *interaction* of environmental factors. Factors do not operate in isolation. Instead, the effect of any one factor may well depend on the simultaneous influence of others. Among carp, for example, the lowest tolerable environmental oxygen concentration depends on temperature. At 30°C, the carp require at least 1.3 mg of oxygen per liter of water, but about 0.8 mg O_2/L is sufficient at 1°C. Lowering the temperature decreases the metabolic rates of the fish, permitting them to survive in the face of lower oxygen levels. Interactions among factors can become very intricate, especially as the number of factors involved is increased. In turn, experiments to elucidate these interactions can become too complex to be practical. The danger is that simplifications introduced into experiments for the sake of practicality may cause us to overlook important interactive effects.

Microclimates

One important concern in evaluating animal environments is that we humans are relatively large organisms, and as we walk about, most of our sense organs are located several feet above the ground. Our conception of the climate and other conditions in an area may therefore bear little relation to the actual conditions experienced by many other animals. George Bartholomew, one of the founders of the modern ecological approach to comparative physiology, has expressed this important point especially well:

> Most vertebrates are much less than a hundredth of the size of man and his domestic animals, and the universe of these small creatures is one of cracks and crevices, holes in logs, dense underbrush, tunnels and nests—a world where distances are measured in yards rather than miles and where the difference between sunshine and shadow may be the difference between life and death. Climate in the usual sense of the word is, therefore, little more than a crude index to the physical conditions in which most terrestrial animals live.*

The actual climate in which a species lives is termed its *ecoclimate* or *microclimate*. Three examples will serve to illustrate the kind of essential information that is gained in studies of microclimatology. Figure 1.1 depicts temperature and humidity conditions in a habitat occupied by *Ligia oceanica,* a semiterrestrial marine isopod crustacean. Note that the entire width of the habitat depicted would fit well within the length of a human footprint. And yet the climate is far from uniform. On the slope to the left, the isopod experiences a hot substrate (34°C), cool air (20°C), direct sunlight, breezes, and a relatively low humidity (70 percent). Just a few centimeters away, at the base of the layer of pebbles, it finds moist, dark, and uniformly warm stillness.

Figure 1.2 shows the annual maximal and minimal temperatures at various depths below the ground surface in portions of the Arizona desert. The annual temperature excursion in the air at 1 m above the ground is about 40°C. The excursion on the desert surface is twice as great. The surface heats up more than the air during the daytime because of absorption of solar radiation, and it cools below air temperature at night because of radiative loss of heat to the cold

*From G. A. Bartholomew, *Symposia of the Society for Experimental Biology,* No. 18, pp. 7–29. Academic, New York, 1964.

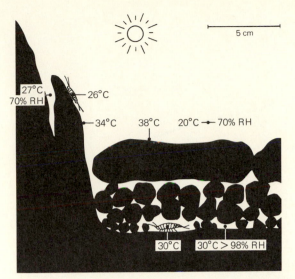

Figure 1.1 Diagrammatic section of the base of a red sandstone cliff and pebbles inhabited by *Ligia oceanica,* showing microclimatic temperatures and relative humidities (RH) and body temperatures of the animals. [From E. B. Edney, *J. Exp. Biol.* **30**:331–349 (1953).]

sky (see Chapter 6). In the soil, the total annual temperature excursion becomes rapidly smaller with increasing depth. Small rodents in their burrows find temperatures far below those on the surface during summer and much above those on the surface during winter. Although we humans think of deserts as hot places, the greatest thermal stress for nocturnal rodents may well be winter cold. The animals are in their burrows during the heat of the day in summer but in some cases venture out during the cold of night in winter.

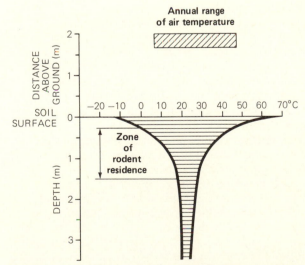

Figure 1.2 Annual range of temperatures in the soil of the Arizona desert near Tucson. Curve to the left depicts minimal temperatures recorded over the year; curve to the right depicts maximal temperatures recorded. The annual range of temperatures in the air is also shown. [From X. Misonne, *Mem. Inst. R. Sci. Natur. Belg., 2me Ser.* **59**:1–157 (1959).]

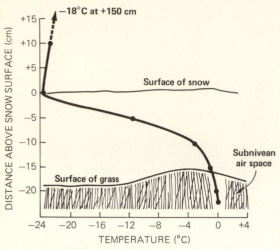

Figure 1.3 Temperature at points above, in, and under a 20-cm-thick snow cover in Sweden on 5 March, 1962, at 4 A.M. Various factors lead to development of a subnivean air space among the grass, between the ground and lower edge of the snow cover. It is postulated that small mammals can move about relatively freely in this air space. [After C.-C. Coulianos and A. G. Johnels, *Ark. Zool.* **15**:363–370 (1963).]

Finally, in Figure 1.3, we see that in cold northern regions, animals moving about under snow may find temperatures 20°C higher than in the air above.

Other Challenges to Understanding Animal Environments

The need to describe physical conditions in secluded places is only one of the hurdles that may be faced in describing the environments of animals. Another need is to understand the *behavior* of the animals sufficiently to know how they partition their time among the various microclimates available to them. Developing this behavioral understanding may be a major challenge if the species is small or secretive. Consider, for example, the difficulty of determining the lengths of time spent in various microclimates among the rocks by the isopods in Figure 1.1.

Just as our human experience of climate may bear little relation to the experience of other species, so also our sensory perceptions may be a poor guide to the perceptions of other species. This is another concern in understanding the relations of animals to their surroundings. We know, for example, that honeybees and certain other insects probably see flowers differently than we do because their eyes can sense reflected ultraviolet light to which we are blind. We also know that some birds can navigate using magnetic fields of which we are unaware. In these and other respects, we must be alert to the possibility that the sensory universe in which other animals live may be unlike our own.

ADAPTATION

From time immemorial, people have recognized that the properties of animals are often well matched to the environments they occupy. Animals living fully exposed to the heat of deserts often have exceptional capabilities of dealing with heat, for example; and ones exposed to arctic cold are often particularly well suited to the challenge. Several theories have been proposed to explain the match between animals and their environments. Evolution by natural selection is the one favored by most biologists today.

One way that people express the match between animals and environments is by saying that the properties of the animals are "adapted" to the conditions under which they live. This idea of "adaptation" is a central concept in animal physiology (as in most biological disciplines), and it bears some scrutiny not only because it is important but also because it is easily misused. Exactly what is meant by adaptation? And how might we determine whether a characteristic of an animal is in fact adaptive? There has been a call of late for increased rigor in answering these questions. One reason for the call is that all too often biologists in the past have seemed willing to assume, in essence, that every product of biological evolution is necessarily adaptive. Based on this assumption, the literature has become filled with impromptu rationalizations designed to explain the adaptive utility of each and every animal feature. To the extent that adaptation is simply assumed, it loses force as a scientific concept. There is a movement underway to establish an empirical science of adaptation. The function of such a science would be to provide definitions and standards whereby claims of adaptation could be evaluated by use of evidence rather than simple assertion.

To say that "an animal is adapted to its environment" is likely to mean little more than that the animal can survive and reproduce there. As already suggested, the really challenging question concerns not whether whole animals are adapted but whether particular animal characteristics are adaptive.

Suppose that a particular attribute of a species can exist in two genetically controlled states, A and B. For instance, let us consider as an attribute the affinity of the hemoglobin in red blood cells for oxygen, with state A being a relatively high affinity and state B being relatively low. Suppose now that there is a population of animals containing appreciable numbers of individuals with each type of hemoglobin. If individuals with type A enjoy a greater probability of surviving and reproducing successfully than ones with type B, then—by this very fact—the genes inducing state A will tend to increase in frequency in the population from one generation to the next (and those inducing state B will decrease). It could happen that after many years, virtually all individuals born into the population would be of type A. You will recognize this as the process of natural selection. If we were to come upon the population during the later phases of its evolutionary history and find it populated almost exclusively by individuals of type A, and if we were to know that state A had come to this position of prominence through the process of natural selection because it promoted superior survival and reproduction in comparison to alternative states (e.g., *B*), we would have little hesitancy in calling state *A* *adaptive*.

Suppose, however, that the population, during its early history—when individuals of type *B* are still reasonably abundant—goes through a low ebb in its size, containing perhaps only 10–20 individuals. And suppose also that during this low point, some catastrophe strikes, killing individuals *at random,* regardless of whether they are of type *A* or *B*. In a small population, it could happen that these random deaths would by sheer chance eliminate all the individuals possessing genes for state *A*. Then, when the population later grew back toward large size, it would consist solely of individuals of type *B*. If we were to come upon the population thus dominated by state *B* and yet were to know that state *B* had come to this position of prominence by chance—even though it provided a lower probability of survival and reproduction than an alternative state (*A*) that had once existed at high frequency in the population—we probably would not find it appropriate to term state *B* adaptive.

Processes such as that just described, in which chance assumes a preeminent role in altering gene frequencies, are termed *genetic drift*. There are also other ways in which character states can be brought to prominence in a population even though in the present environment they do not confer maximal probabilities of survival and reproduction by comparison to available alternative states. The issue of how important such processes have been in the course of evolution is intricate and controversial, and we do not pursue it. The important point for us is that not all traits that attain prominence in populations need to have done so by way of natural selection. If we then accept the view that the adaptive traits of organisms are those brought to prominence via natural selection by virtue of their advantages for survival and reproduction, we must conclude that not all evolution is adaptive evolution and not all products of evolution are adaptations. The conclusion is critical, for it means that we do indeed face a task of determining which contemporary characteristics of animals are and are not adaptive.

The definition of adaptation developed here may be stated as follows. A character state is *adaptive* if it has come to be present at high frequency in a population because it confers a maximal probability of survival and successful reproduction in comparison with available alternative states.* The environment must always be specified in evaluating adaptation; a character state adaptive in one environment may even be lethal in another. Also, the stipulation that a character state must be judged relative to *available* alternatives should not escape notice; a state

may be adaptive yet fall short of the optimum we can conceive because the optimum state may never have arisen by mutation or other sources of variation.

Occasionally, we are provided with a "natural experiment" that permits us to judge directly the adaptiveness of a character state. Industrial melanism provides a well-known example. Moths in the industrial regions of England have two genetically determined color states, light and dark. They were predominantly light colored prior to the industrial era, when light-colored lichens covered the tree trunks on which they rested during the day. With increasing industrialization, the lichens on the trees were killed by pollutants and soot from factories further darkened the tree trunks. Within 50 years the moth populations in industrial areas became predominantly dark colored. Experiments demonstrated that on dark tree trunks, the dark-colored moths were less susceptible to being sighted by avian predators than light-colored ones. In short, in the industrialized environment and in comparison with the available alternative coloration, dark coloration came to prominence in the moth populations via natural selection because it maximized survival. Dark coloration was indisputably adaptive to the sooty environment.

Often, we cannot judge directly the adaptiveness of character states in terms of the definition of adaptation. For instance, we cannot appraise directly whether the alveolar structure of the lung in mammals is adaptive (compare avian lungs, p. 288). The reason we cannot is that alternative types of lung structure are not to be found in modern populations. Thus, there can be no natural experiments in which the ecological success of individuals with the alveolar structure can be judged relative to that of members of the same species having another structure.

If the adaptiveness of a character state cannot be judged directly using the definition, indirect or surrogate criteria must be used (see p. 27). Several indirect criteria are believed useful in judging adaptiveness, although each in its own way is open to error. We mention several here:

1. Argument from correlation between habitat variables and animal variables. Data relevant to this kind of argument are one of the prominent fruits of the comparative approach to physiology. Suppose we discover, for example, that a particular character state has evolved independently in several animal groups that all occupy a particular environment. We might conclude from this evidence that the state is adaptive. As a particular case, witness that invaginated, lunglike respiratory organs are found in *air-breathing* representatives of several disparate groups of animals—including terrestrial vertebrates, insects, and pulmonate snails—*yet are almost unknown among water breathers*. This correlation between habitat and trait suggests that invaginated respiratory structures are adaptive to air breathing. If among related animals, the magnitude of a character state varies in

*This definition is in common use. However, professionals are not in universal agreement, and some would offer a somewhat different definition.

parallel with the magnitude of one or more environmental variables, this correlation might also be evidence of adaptation. For example, the ability of rodents to concentrate their urine varies inversely with the availability of water in their habitat, indicating that high concentrating ability is adaptive to aridity.

2. Argument by analogy to human design and practice. The vertebrate eye, for instance, may be judged to be an adaptation for image perception because many of its features closely match those of human devices (cameras) designed for processing images.

3. Argument from the performance of animals that have had a character state artificially altered, as by surgery or administration of pharmacological agents.

4. Argument from logic. This approach loses strength to the extent that the logic derives from hindsight rather than foresight.

EXOGENOUS CUES; PROXIMATE VERSUS ULTIMATE CAUSATION

Myriads of chemical and physical agents impinge on the sense organs of animals in their natural environments. Yet only some evoke active responses. A person, for example, might be exposed to cold and mild ionizing radiation in a particular environment. The cold would likely elicit active alterations of physiology and behavior, but the ionizing radiation would not elicit such responses because it is imperceptible to us. The agents or features of the environment that do elicit active physiological or behavioral responses are called *exogenous cues*. A cue is not merely any environmental agent. Instead, it is an agent that an animal—by evolution or learning—has come to use in controlling its behavior or physiology.

Clarity of understanding is often enhanced by recognizing that there are two levels of causation in animal responses to the environment, proximate and ultimate. The *proximate* environmental causes of a response are those agents of the environment that are *immediately* responsible for eliciting the response; exogenous cues are proximate causes. The *ultimate* causes of a response are the agents that *historically favored the evolution of the response through natural selection.*

Sometimes the proximate and ultimate causes are one and the same. When we look at an intensely bright light such as the sun, for instance, our response is to close our eyes quickly. The bright light is the proximate cause of this response. It is almost undoubtedly the ultimate cause as well, for intensely bright light poses risks of temporary or permanent blindness and we can reason that these risks were likely pivotal in favoring the evolution by natural selection of responses that protect our eyes.

Consider now another type of response to light. Specifically, consider responses to the *photoperiod,* the number of hours per day of daylight. In many species of birds and mammals, long photoperiods bring about reproductive readiness, including growth and activation of the gonads. Here the lengthy period of light per day is the proximate cause of the response, but logically it would seem unlikely to be the ultimate cause. *Considering light alone,* it is difficult to see why natural selection would have favored growth of the gonads in the presence of long photoperiods. We know, however, that in many parts of the world spring is a particularly favorable time of year for reproduction and long photoperiods occur in spring. Probably, such features of spring as favorable weather and lush plant growth were the true ultimate causes in the evolution of the response to long photoperiods. It is probable that long photoperiods became a cue for gonadal growth not because they hold intrinsic advantages for reproduction but because they signal spring's arrival.

The possibility that proximate causes may differ from ultimate causes is important in the analysis of adaptation to the environment. Animal responses are adaptive in respect to ultimate causes. Thus, when proximate and ultimate causes differ, it is necessary to look beyond the immediate to understand the process of adaptation. In the example we have just been discussing, the activation of gonadal growth is not an adaptation to long photoperiods but rather to breeding in a favorable season.

SELECTED READINGS

Bartholomew, G.A. 1986. The role of natural history in contemporary biology. *BioScience* **36**:324–329.

Endler, J.A. 1986. *Natural Selection in the Wild*. Princeton University Press, Princeton.

Florey, E. 1987. Fads and fallacies in contemporary physiology. In H. McLennan, J.R. Ledsome, C.H.S. McIntosh, and D.R. Jones (eds.), *Advances in Physiological Research*, pp. 91–110. Plenum, New York.

*Gould, S.J. and R.C. Lewontin. 1979. The spandrels of San Marco and the Panglossian paradigm: A critique of the adaptationist programme. *Proc. R. Soc. London Ser. B* **205**:581–598.

Gould, S.J. and E.S. Vrba. 1982. Exaptation: A missing term in the science of form. *Paleobiology* **8**:4–15.

Jorgensen, C.B. 1983. Ecological physiology: Background and perspectives. *Comp. Biochem. Physiol.* **75A**:5–7.

Lewontin, R.C. 1978. Adaptation. *Sci. Am.* **239**(3):212–230.

*Mayr, E. 1983. How to carry out the adaptationist program? *Am. Nat.* **121**:324–334.

Schoenheimer, R. 1942. *The Dynamic State of Body Constituents*. Harvard University Press, Cambridge.

Tracy, C.R. and J.S. Turner (eds.). 1982. What is physiological ecology? *Bull. Ecol. Soc. Am.* **63**:340–347.

Waterlow, J.C., P.J. Garlick, and D.J. Millward. 1978. *Protein Turnover in Mammalian Tissues and in the Whole Body*. North-Holland, New York.

Williams, G.C. 1966. *Adaptation and Natural Selection*. Princeton University Press, Princeton.

See also references in Appendix A.

chapter 2

The Internal Organization of the Animal: Basic Concepts

In Chapter 1, we emphasized that animals are in intimate interaction with their environment and persistently exchange material constituents with it. We also emphasized that they have an internal organization that is an essential aspect of life and that they maintain this organization despite the intimacy of their environmental interactions. In this chapter, we look more closely at the internal functional and structural organization of animals and how it is maintained.

BERNARD, CANNON, AND THE STABILITY OF THE INTERNAL ENVIRONMENT

Most cells of multicellular animals are not exposed directly to the outside world but instead are bathed by the body fluids. The famous nineteenth-century French physiologist Claude Bernard (1813–1878) termed the body fluids the *internal environment* (milieu intérieur) of the animal, emphasizing that they constitute the immediate environment of most cells of the body.

In studies of the glucose metabolism of mammals, Bernard had come to recognize that the liver takes up and releases glucose as necessary to maintain relatively stable concentrations in the blood and tissue fluids. He became impressed that, as a consequence, most cells in the body experience a relatively consistent environment with respect to glucose concentration. He also recognized that most cells in a mammal's body experience relative stability of temperature, oxygen level, osmotic pressure, pH, ion concentrations, and so on because various organs and tissues maintain all these parameters within narrow limits in the body fluids bathing the cells. Bernard then stated a hypothesis that is perhaps the most famous in animal physiology:

Constancy of the internal environment is the condition of free life.

His argument was that animals are able to lead lives of greater freedom and independence to the extent that they maintain a stable internal environment, sheltering their cells from the vagaries of the outside world.

In the twentieth century, Walter Cannon (1871–1945), a winner of the Nobel prize, further elaborated on the theme of internal consistency in animals. He emphasized not just the relatively stable composition of the body fluids in many organisms but also the relative consistency of organization and function within cells, tissues, and organs. He coined the term *homeostasis* to refer to the sum total of this internal structural and functional consistency. The term implies not only the condition of consistency itself but also the myriad physiological processes involved in maintaining it. For Cannon, homeostasis was a touchstone of highly evolved life. He viewed mammals as superior to frogs because of their greater extent of homeostasis, and he argued for using animal homeostasis as a model for human social organization.

Bernard's and Cannon's concepts of internal stability and homeostasis have been among the most influential in the history of biology. Despite the insights they provide, however, many biologists today believe that they have been overextended and overemphasized at times.

A Critique

We cannot be surprised in retrospect that both Bernard and Cannon spent their lives working mostly on mammals. Mammals and birds arguably exhibit the greatest internal stability of all animals. Bernard's and Cannon's ideas have been influential precisely because they aptly describe not only that stability but also its importance in endowing the birds and mammals with exceptional functional independence of the environment. The primary focus of concern is whether and how the ideas should be applied to other animal groups.

Consider, for example, that most fish and amphibians do not physiologically stabilize their internal temperature but instead allow it to vary with environmental temperature. If we insist—as Cannon did—on applying mammalian standards to these animals, their lack of internal thermal stability emerges as a defect. There is no reason, however, to hold them to mammalian standards. If we look at fish and amphibians in their own right, we cannot conclude with any confidence that their lack of internal thermal stability is a handicap. Indeed, in its own way it is a benefit because there are great energy savings to be realized by allowing internal temperature to equilibrate with external rather than opposing the natural physical tendency toward equilibration (see p. 133 for a further elaboration of these ideas).

There is no such thing as monolithic internal stability. Instead, animals are complex mixes of stability and instability. Whereas fish and amphibians, for example, do not stabilize their temperatures, they do ordinarily maintain great stability in the osmotic and ionic concentrations of their body fluids and in many other traits.

In the respects where we find internal stabilization in animals, we learn from Bernard and Cannon that it represents a means of attaining cellular independence of outside conditions. On the other hand, where we do not find internal stability, we should be slow to make adverse judgments. The reflex application of a mammal-centered dogma in such cases is more likely to thwart than promote understanding.

INTRODUCTION TO REGULATION

The rest of this chapter focuses on the fundamental mechanisms by which animals are able to maintain internal stability in the respects they do. We examine in this brief introductory section three elemental concepts: physiological balance, steady state, and regulation. We then proceed in the next section to look at the principles of organization of regulatory systems. After that, we turn to the properties of nervous and endocrine integrating systems and to mechanisms of biochemical regulation. Finally, the chapter ends with a look at the roles and properties of biological clocks.

The Concept of Balance

Consider the system diagrammed in Figure 2.1: a tub containing water with several inflows and outflows. If the rate of inflow of water by all routes combined

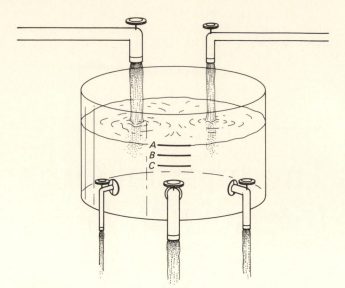

Figure 2.1 A tub of water with multiple inflows and outflows.

exactly equals the rate of outflow by all routes combined, the quantity of water in the tub will not change despite the continual flux of water through it. A system meeting these conditions is said to be *in balance.* By contrast, if inflows should exceed outflows, the quantity of water in the tub would gradually increase; the tub would then be said to be *in positive balance* with respect to water. Should outflows exceed inflows, the quantity in the tub would decrease, and the tub would be *in negative balance.*

These principles can be applied to many physiological systems (e.g., see p. 77 for their application to animal heat balance). Here we consider only the most obvious of applications: that to animal water balance. The quantity of water in an animal's body is analogous to that in the tub. The several avenues of water loss from an animal (e.g., urine, feces, pulmonary evaporation) correspond to the outflow pipes; and the animal's avenues of water gain (e.g., drinking water and water in food) correspond to the inflow pipes. In the animal as in the tub, water balance is attained if the rate of water loss by all avenues taken together is equaled by the rate of gain by all avenues. As we shall shortly see, balance is not all there is to the maintenance of internal stability. It is a key element, however.

The Concept of Steady State

If an animal's rate of gain of a commodity (e.g., water or heat) by each avenue of inflow is constant, if its rate of loss by each avenue of outflow is also constant, and if inflows balance outflows so the amount of the commodity in the body is constant, the animal is said to be *in steady state* with respect to the commodity in question. A steady state is a *dynamic* state and is conceptually quite distinct from a state of equilibrium. A system is at equilibrium when its potential energy—its potential to do work—is at a minimum. As you read this page, your body is probably near a steady state with respect to water and heat. Yet you certainly are not at equilibrium with your surroundings, for at equilibrium your temperature would be the same as that of the surrounding air and your tissues would be as dry as a mummy's.

The Concept of Regulation

Returning to the system in Figure 2.1, an additional concept must be added to complete the analogy between water exchange in the tub and the maintenance of internal stability in animals. The concept of balance (or that of steady state) says only that the quantity of water in the tub would be constant; it says nothing about the magnitude of that quantity. Thus, a state of balance could be established with the water at level *A* or *B* or *C*. Suppose though that we want the level to be always at a particular level, *B*. If at a given time it were not at *B*, we would then have to adjust some of the rates of inflow and outflow to create a transitory condition of imbalance so as to raise or lower the level to *B*, then reestablish and maintain a state of balance to keep it at *B*. The correction of deviations from *B* in this manner would achieve true internal stability of the water level in the tub and would constitute an example of regulation. **Regulation** is the active stabilization of a parameter—an amount, concentration, or intensity—at or near some particular level.

PRINCIPLES OF ORGANIZATION OF REGULATORY SYSTEMS

To explore how regulatory systems are organized, we shall look at a familiar problem, that of regulating the temperature of a body. In the process, we shall introduce a generalized terminology that will permit the concepts to be transferred readily to other situations.

Suppose in the water bath of Figure 2.2 we seek to regulate the temperature at the particular level of 40°C (this corresponds to regulating the heat content of the bath, for the temperature and heat content of such a body are simply proportional to one another). The parameter to be regulated—in this case temperature—can be termed the **controlled variable**. A bath at 40°C will ordinarily be warmer than room air and thus will lose heat to the air, the rate of loss depending on how cold the air is. This loss of heat will tend to cool the bath below 40°C and thus will represent a disturbance. In general, a **disturbance** is any interaction with the surrounding environment that tends to draw the controlled variable away from the level at which it is to be regulated. The bath is provided with a heater that opposes the tendency of the water to cool. Elements, such as the heater, that are actively employed to maintain or adjust the level of the controlled variable are termed **effectors**. The entity

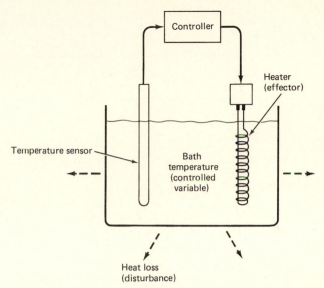

Figure 2.2 A thermoregulated water bath kept at a temperature, 40°C, above air temperature.

that is responsible for controlling the activity of effectors is known as a **controller;** in the particular case we are considering the controller sets the rate of heat production by the heater.

Clearly, the most straightforward way to regulate the bath temperature is to monitor that temperature and control the heater accordingly. This arrangement would constitute a feedback system. **Feedback** occurs when the effectors are controlled according to information about the controlled variable itself; that is, information on the effects being exerted by the effectors is "fed back" to control the behavior of the effectors. To achieve feedback, one or more sensors are required. A **sensor** (or **receptor**) is an element capable of measuring the current level of the controlled variable; in our particular example, it would take the form of a temperature-measuring device in the bath. For feedback, the sensor must have a means of communicating information about the level of the controlled variable to the controller.

Figure 2.3 is a generalized diagram of a control system. Study it to be certain you understand the meaning of each box and each link between boxes. Note that the boxes and their connecting links form a closed loop; symbolically, this is the essential prerequisite for feedback.

Two types of feedback are possible: negative and positive. In **negative feedback,** the effectors are controlled so that their activities *oppose* any deviation of the controlled variable from the level at which it is to be regulated. We have a negative-feedback system in Figure 2.2 if the heater produces heat when the bath temperature falls below 40°C but terminates heat production when the bath temperature rises above 40°C. *Negative feedback is a mechanism of regulating the controlled variable.* In **positive feedback,** the effectors are controlled so that their activ-

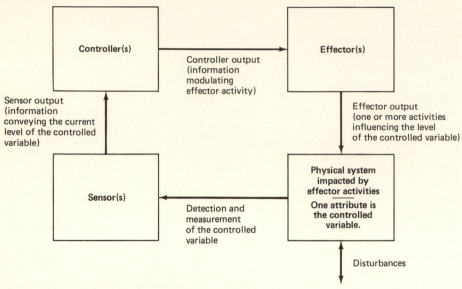

Figure 2.3 A generalized diagram of a control system.

ities *reinforce* deviations of the controlled variable from the level at which it is to be regulated. Positive feedback does not lead to regulation and is rare, but hardly unknown, in biological systems. The genesis of a nerve impulse, discussed in Chapter 16, is an example of positive feedback.

Negative-Feedback Control Based on a Reference Signal Now we need to consider more specifically how the effectors can be controlled in a negative-feedback fashion. Systems based on a reference signal are perhaps most easily understood inasmuch as they are common in devices of human design.

Suppose that within the controller there is a signal (or other piece of information) that corresponds exactly to the level at which the controlled variable is to be regulated, and suppose that this signal within the controller remains steady even as the controlled variable and other attributes of the control system and environment fluctuate. This signal would be termed a *reference signal* or *set point*. We have said earlier that we want the bath in Figure 2.2 to be held at 40°C. Thus, the reference signal or set point within the controller of Figure 2.2 would be set to correspond to 40°C.

Now the controller also receives a signal from the sensor, and this signal corresponds to the current *actual* level of the controlled variable. The essence of all systems based on a reference signal is this: The controller can determine the appropriate instructions to send to the effectors *by comparing the sensor signal and reference signal.*

One type of negative-feedback system based on a reference signal is *on–off control.* The defining feature of such a system is that the effectors are always instructed to be either fully on or fully off. If the bath in Figure 2.2 operated according to this princi-

ple, the heater would be activated to provide a fixed, invariant rate of heat output whenever bath temperature fell below reference temperature (40°C) and would be turned off completely whenever bath temperature rose above reference temperature. On–off control characteristically causes the controlled variable to oscillate above and below the reference level over time (witness the oscillations of air temperature in our homes in winter as the furnace is turned on and off).

Another type of negative-feedback system is *proportional control.* In this case, the activity of the effectors is increased *in proportion* to the difference between the current level of the controlled variable and the reference level (i.e., effector activity is graded). Applying this type of control to Figure 2.2, the rate of heat output from the heater would be zero if bath temperature were above the reference temperature (40°C), but at bath temperatures below the reference level the rate of heat output would be increased in proportion to the difference between bath temperature and reference temperature; thus, the rate of heat output would be twice as great if bath temperature were 36°C (4°C below reference temperature) than if it were 38°C (2°C below). Proportional controllers avoid the short-term oscillations characteristic of on–off controllers. However, they characteristically bring the controlled variable to a stable level that deviates from the reference level, this steady-state deviation being called a *load error.* Note that with proportional control in Figure 2.2, if the bath temperature were to be stable, the only way the heater would put out any heat to counteract the cooling of the bath would be if the bath temperature were below reference temperature; and if cooling intensified, the only way to get more heat output would be to increase the load error—that is, permit bath temperature to fall further below reference temperature. The body temperature of a person rises during vigorous exercise. Some investigators believe that this deviation from the normal body temperature fundamentally represents a load error required to activate heat-dissipating systems such as sweating.

An important general point to note is that although control systems do exert a stabilizing influence on their controlled variables, it is in their nature not to impart true constancy. The oscillations typical of on–off control and the load errors of proportional control exemplify this property.

Negative-Feedback Control Without a Reference Signal Negative-feedback devices of human design are so commonly based on reference signals that, by analogy, animal control systems have often been assumed to employ such signals. In certain quarters today, however, there is rising concern that this analogy has been misleading. Traditionally, for example, the systems that control body temperature in mammals and birds have been assumed to be dependent for their operation on a reference signal generated somewhere in the brain and corresponding to the "correct" temperature (e.g., 37°C in humans). But despite intensive searching, no one has yet found such a signal. A reference signal is by no means essential for negative-feedback control.

Examination of Figure 2.3 will reveal that there

are four fundamental variables within any control system. As listed at the top of Figure 2.4, these are the controlled variable and the outputs of the sensor, controller, and effector. To give these ideas greater specificity, let us again turn to the water bath in Figure 2.2 as an example. The controlled variable is the water temperature and, as indicated at the top of Figure 2.4, it may be symbolized by T. The effector output is the rate of heat production by the heater, H. The sensor output, S, is the intensity of the signal generated by the sensor, and the controller output, C, is the intensity of the signal generated by the controller.

Suppose now that a discrete functional relation exists between each successive pair of variables in the control loop. For instance, suppose that the heater linearly increases its rate of heat output as the intensity of the signal it receives from the controller is increased; this relation is shown by the graph relating H to C in the "effector" box of Figure 2.4. The graph of T as a function of H in the "controlled system" box indicates that under given ambient conditions, steady-state bath temperature increases as the rate of heat production is increased. As shown in the "sensor" box, we assume that the sensor increases the intensity of its output as bath temperature increases. Now we come to a key feature of our control system. As depicted in the "controller" box, let us assume that the controller *decreases* its

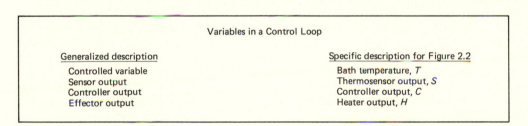

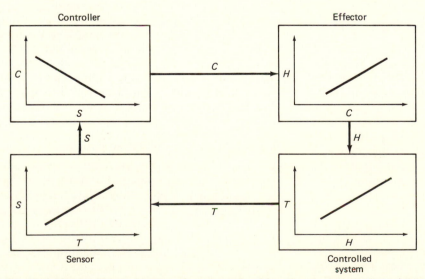

Figure 2.4 Variables in a control loop and an example of functional relations that might exist between successive variables.

output C as the intensity of the sensor signal S *increases*. Such an *inverse* relation between two variables somewhere in the control loop will itself cause the system to behave in a negative-feedback fashion, without need of an independent reference signal. To see this, note that a rise in bath temperature, causing a rise in sensor output, will lead to a fall in controller output and thus a fall in the rate of heat production by the heater; the fall in heat output will retard or reverse the rise in bath temperature (negative feedback). A system such as we have defined gradually "zeros in" on a stable bath temperature under given conditions. If air temperature is lowered, thus increasing cooling of the bath, the system will respond phenomenologically like a proportional control system: Bath temperature will fall, but because such a fall will provoke a rise in heater output, the fall will not be nearly as great as the decline in air temperature.

Changes in the Maintained Level of the Controlled Variable   An important property of all types of control system is that the level at which the controlled variable is maintained can be altered by changes in the operating characteristics of the control system. This principle is exemplified by the everyday experience of changing the "setting" of a home thermostat. Fever provides a familiar physiological example: An invasion of infectious microorganisms can cause a person's temperature-control system to raise the level at which body temperature is maintained (p. 117). In control systems based on a reference signal, the maintained level of the controlled variable can be altered by changing the reference signal or set point. In systems lacking a reference signal, a change in any of the functional relations between variables (e.g., in the slope of the C versus S function in Figure 2.4) will cause the controlled variable to be brought to a new level.

COORDINATION THROUGHOUT THE BODY: NERVES AND ENDOCRINES

When organisms were exclusively unicellular, not only the command-and-control function but also all other essential functions were carried out within the confines of single cells. With the evolution of multicellularity, however, cells became specialists. As some became specialized for movement, digestion, and reproduction, still others had to become specialized for the task of harmoniously integrating the functions of all the far-flung and disparate tissues in the body. Two types of cell evolved in this latter role: nerve and endocrine.

In the preceding section, we examined some of the abstract principles of physiological regulation. The actual *implementation* of many forms of regulation falls to the nervous and endocrine systems. In mammals, for example, parts of the central nervous system act as the controller for maintaining a stable body temperature. Sensory nerve cells provide the controller with information on the current body temperature, and the controller sends signals via other nerve cells to heat-generating tissues and other effectors.

Basic Features of Nervous Function

Nerve cells, known formally as **neurons,** are cells specialized for the repeated conduction of electrical signals from place to place in the body (see p. 391). Structurally, a neuron typically consists of a cell body (containing a cell nucleus) and many fine, often branching processes termed **nerve fibers,** which sometimes are highly elongate (see Figure 16.5). A **nerve** is a bundle of elongate nerve fibers, originating from many neurons and running together.

Neurons make specialized communicatory contacts, termed **synapses,** with other neurons and with effector cells such as muscle cells. When neuronal signals traveling along a neuron arrive at a synaptic contact, they exert specific physiological influences—excitatory or inhibitory—on the contacted cell.

Neurons perform a number of functions. Importantly, they are always key constituents of *integrative* centers such as brains. Some neurons perform *sensory* roles by initiating signals in response to physical or chemical agents such as light or touch. Such sensory signals are relayed to integrative centers by **afferent neurons** (*afferent* = "to bear toward"). Of particular interest for the analysis of regulatory systems are the **efferent neurons** (*efferent* = "to bear away"), which conduct control signals ("orders") from the integrative centers to effectors. In their performance of this latter role, neurons possess two noteworthy properties. First, they provide highly discrete lines of communication, because a particular terminal efferent neuron may make synaptic contact with just one or a relatively few effector cells. Second, signals carried by neurons travel rapidly and begin and end abruptly; for example, it would not be unusual for impulses to travel at 20–100 m/s along mammalian nerve fibers or for a fiber to be capable of transmitting 100 or more discrete impulses per second. In short, nervous lines of communication provide opportunities for fine control both spatially and temporally.

Basic Features of Endocrine Function

Endocrine cells are cells that release communicatory chemicals, termed **hormones,** directly into the blood (or sometimes into other body fluids). Carried throughout the body by the blood, these chemicals bathe the tissues and organs at large and elicit specific responses from some of them. Hormones are effective at low blood concentrations, on the order of 10^{-7}–10^{-12} M.

Two basic types of endocrine cell are recognized: neurosecretory cells and epithelial or nonneural endocrine cells. *Neurosecretory cells* are neuronlike cells that have assumed the role of producing and secreting communicatory chemicals into the blood. Their secretions are termed **neurosecretions** or **neurohormones.** Neurosecretory cells receive synaptic contact from typical neurons, and their secretion is thought to be controlled largely by inputs across these synapses. Unlike typical neurons, however, neurosecretory cells do not necessarily influence other cells synaptically. Their major communicatory output is their chemical secretion into the blood. Importantly, neurosecretory cells constitute a direct interface between the endocrine and nervous types of control system; they are endocrine cells immediately controlled by the nervous system. The **epithelial** or **nonneural endocrine cells** are hormone-producing cells that do not bear obvious affinities to nerve cells. Some glands composed of such cells (e.g., the vertebrate thyroid and anterior pituitary) are poorly innervated and yet are substantially under nervous-system control. The nervous system exerts this control by inducing neurosecretory cells to secrete neurohormones that in turn modulate the function of the poorly innervated epithelial glands. Hormones, it should also be stressed, are well known to affect the function of the nervous system. Interaction between the nervous and endocrine systems is in both directions.

Earlier, we discussed the spatial and temporal features of nervous lines of communication. Let us now examine endocrine systems from the same viewpoints. Once a hormone has been released into the blood, all tissues in the body are potentially bathed by it, and any given tissue can have evolved either not to respond or to respond in any of countless ways; the only limit on the number of tissues responding or on the number of their responses is that all responses need ideally to be compatible. In principle, hormones can exert either circumscribed or widespread effects; in practice, they commonly affect a whole tissue at least, and quite often they affect multiple tissues. Individual hormonal signals operate on much longer time scales than individual nervous signals. Initiation of hormonal effects requires at least seconds or minutes because the hormone, once released into the blood, must circulate to target tissues and diffuse to effective concentrations within them before responses can be elicited. After a hormone has entered the blood, a substantial time may pass before metabolic destruction and excretion reduce its concentration to ineffective levels. In the human bloodstream, for example, the three hormones vasopressin, cortisol, and thyroxine display half-lives of about 15 min, 1 h, and nearly 1 week, respectively. Thus, a single release of hormone can induce protracted stimulation of target tissues.

Nerves and Endocrines Compared

Comparing nervous and endocrine pathways as channels of communication between integrating centers and effector tissues, we see that nervous lines of communication are capable of an extreme fineness of control—both temporal and spatial—that is probably well outside the realm of possibility for endocrine systems. Thus, it is not surprising that the fine, rapid movements of locomotory muscles are always controlled neuronally.

At another extreme, if the task is to stimulate many far-flung tissues simultaneously and over a protracted time, clearly nervous systems can accomplish it also. However, to execute such a task, a nervous system requires large numbers of discrete neuronal connections between integrating centers and effector cells and typically must continue to send trains of impulses down all these connections for as long as stimulation is required. By contrast, an endocrine gland can accomplish the task with ostensibly greater economy—by secreting a single long-lasting chemical into the blood. Not unexpectedly, growth, development, and reproductive cycles—processes that involve diverse tissues and occur on time scales of days, months, or years—are often primarily controlled and coordinated by hormones.

The significant differences between nervous and endocrine channels of communication in their potential for spatial and temporal control help to explain why animals have evolved both types of communication.

The Autonomic Nervous System

Although we are deferring the detailed discussion of specific nervous and endocrine systems to later chapters (Chapters 16–21), we need to consider the autonomic nervous system here because it plays such key roles in regulating and coordinating many of the physiological processes to be discussed throughout the book. The concept of the autonomic nervous system has been developed in reference to vertebrates, and the present treatment is limited to them. Certain parts of some invertebrate nervous systems are sometimes described as autonomic, usually only by analogy to the vertebrate case.

The autonomic nervous system may be defined in terms of the effectors it controls. In this context, two classes of effector are recognized: (1) the striated (or skeletal) muscles, which are responsible for most outwardly visible movements such as locomotion, and (2) the internal effectors. The **internal effectors** include all the effectors besides striated skeletal muscles—such as the smooth muscles and digestive glands—and are so named because they exert most of their effects internally and unobtrusively. The **autonomic nervous system** is the part of the nervous system that controls the internal effectors. The part that controls the striated muscles is the **somatic nervous system.**

Originally, the autonomic nervous system was defined as the visceral motor part of the peripheral nervous system, that is, as the nerves that convey efferent (outgoing) messages from the central nervous system (CNS) to internal effectors. More recent studies, however, show that autonomic nerves may also contain significant numbers of sensory fibers that convey afferent signals from internal organs to the CNS. Therefore, in modern usage the autonomic nervous system is often considered to be more than merely a motor (efferent) system.

Specifically, the internal effectors controlled at least partly by the autonomic nervous system include the following:

1. smooth muscles such as those in the gut, blood vessels, eyes (iris muscles), urinary bladder, hair follicles, spleen, bronchi, and penis;
2. many exocrine glands such as digestive, sweat, and tear glands;
3. a few endocrine glands, notably the adrenal medullae (chromaffin tissue) and pineal;
4. the pacemaker region and other parts of the heart; and
5. certain other effectors such as the brown adipose tissue of mammals (a heat-producing tissue, see p. 106) and the integumentary chromatophores (color-change cells) and swimbladders of fish.

Obviously, the autonomic nervous system is pivotally involved in the *visceral* and *homeostatic* functions of the organism.

The autonomic nervous system has been subdivided in various ways. The subdivisions first specified by John Langley (1852–1925) in the early part of this century and most commonly acknowledged today are the **sympathetic, parasympathetic,** and **enteric.**

His scheme of subdivision is considered valid for mammals. It is usually used for nonmammalian vertebrates also, although its application to these other groups remains debatable. Langley's subdivisions are based on anatomy. They reflect differences in function, but especially in the nonmammalian groups, the lines of functional subdivision remain much less clear than those of anatomical subdivision. Our focus here shall be on mammals and on the sympathetic and parasympathetic divisions, which connect the CNS with peripheral effectors. The enteric division consists of a network of neurons in the walls of the gut, responsible in part for control of the gut musculature.

A characteristic of all vertebrate groups is that autonomic motor pathways leading from the CNS to the periphery have a synapse interposed between the CNS and the effector tissue (Figure 2.5). Thus, unlike somatic motor signals, autonomic efferent signals emanating from the CNS traverse a two-neuron relay in reaching their target. The synapses between these successive neurons are located within clusters of neuronal cell bodies called ganglia; neurons running from the CNS to the ganglia are termed **preganglionic,** whereas those running from the ganglia to the effectors are termed **postganglionic.** In the mammalian parasympathetic division, ganglia are located near or at the effectors, so the preganglionic parasympathetic neurons are long and the postganglionic ones are short (Figure 2.5). In contrast, sympathetic

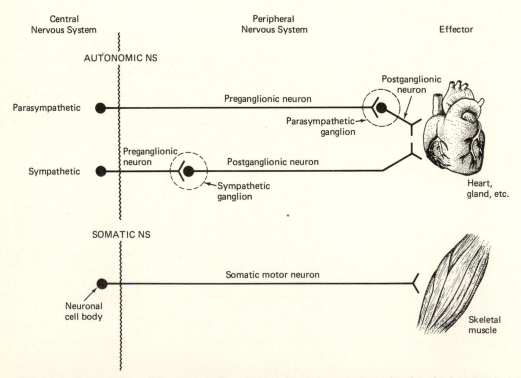

Figure 2.5 Organization of the mammalian autonomic nervous system, showing the heart as an example effector organ. Sympathetic ganglia are typically close to the spinal cord; parasympathetic ganglia are typically in the periphery, near target organs. In comparison, somatic motor neurons do not have a peripheral neuron-to-neuron synapse.

ganglia are mostly near the spinal cord, so that preganglionic sympathetic neurons are short and postganglionic ones are long.

The parasympathetic and sympathetic nerves are associated with different regions of the CNS in mammals. *Parasympathetic* preganglionic neurons exit the CNS in two groups of nerves: cranial and sacral. The cranial group consists of four of the pairs of cranial nerves: the oculomotor, facial, glossopharyngeal, and vagus; the sacral group of nerves emerges from the posterior, sacral part of the spine. The parasympathetic division is sometimes called the *craniosacral* division. In the *sympathetic* division (also called the *thoracolumbar* division), the pregan-

glionic neurons emerge from the thoracic and lumbar regions of the spine, and most of them terminate in sympathetic ganglia immediately lateral to the spine. As shown in Figure 2.6, these so-called paravertebral ganglia occur segmentally along the length of the spine and are interconnected by longitudinal nerve connectives, forming a *sympathetic chain* on each side of the vertebral column. Some preganglionic sympathetic neurons terminate in ganglia more distant from the spine, as in the celiac (solar) plexus; some directly innervate the adrenal medullae (p. 594).

The postganglionic neurons of the parasympathetic and sympathetic divisions differ overall in the chemical neurotransmitter substances that they release at their points of synapse with effector-tissue cells (see p. 443 for a discussion of neurotransmitters). Parasympathetic postganglionic neurons typically release primarily acetylcholine and thus are termed *cholinergic*. Sympathetic postganglionic neurons usually release mainly catecholamines—chiefly norepinephrine (noradrenaline) in mammals—and thus are called *adrenergic*.

Among effectors controlled by the autonomic nervous system, some are innervated by only one division. For example, most smooth muscles of blood vessels in mammals and the piloerector muscles of the hair follicles receive only sympathetic innervation. Figure 2.7 provides a graphic illustration of the effects of sympathetic activation of piloerector muscles. Many effectors, in contrast, receive both sympathetic and parasympathetic innervation. The re-

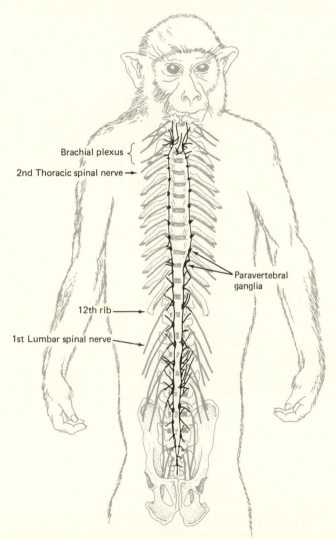

Figure 2.6 The sympathetic chains of the rhesus monkey (*Macaca mulatta*), drawn in black. The paravertebral ganglia, which appear as swellings along the chains, are interconnected by longitudinal connectives. Preganglionic sympathetic neurons enter the chains from the spinal cord. Postganglionic neurons travel from the chains to effectors innervated by the sympathetic system. Spinal nerves are shown in gray. [Reprinted by permission from J. Pick, *The Autonomic Nervous System*. Lippincott, Philadelphia, 1970.]

Figure 2.7 A cat in which the sympathetic chain on the right side of the body was surgically eliminated. When the cat was exposed to low air temperatures, only the fur on the left side of its body became fluffed. Fluffing increases the insulation provided by the fur and is effected by tiny pilomotor muscles in the hair follicles. When one of these muscles contracts, it causes erection of the hair to which it is attached. The pilomotor muscles are stimulated to contract by way of the sympathetic nervous system. Goosepimples are a manifestation of pilomotor contraction in humans. [Reprinted by permission from W. B. Cannon, H. F. Newton, E. M. Bright, V. Menkin, and R. M. Moore, *Am. J. Physiol.* **89**:84–107 (1929).]

sponses elicited by the two divisions in such cases are usually antagonistic. For instance, parasympathetic impulses delivered to the pacemaker of the heart slow the heart rate; sympathetic impulses accelerate it.

The *overall roles* of the parasympathetic and sympathetic divisions reflect their tendency to act antagonistically. The parasympathetic division (especially its cranial part) can be described as promoting restorative, restful processes. For example, it facilitates digestion by stimulating gastrointestinal motility and secretion by digestive glands; it slows the heart, induces constriction of lung passages, and promotes secretion of tears to moisten the eyes. In contrast, the sympathetic division promotes processes that draw on bodily energy reserves while inhibiting certain of the ones that restore reserves; it is particularly activated in the face of stress and readies the body to meet stress. The sympathetic division inhibits gastrointestinal motility; it stimulates the heart to beat more rapidly and strongly; by inducing constriction in certain blood vessels, it raises the blood pressure and reduces blood flow to some organs that are nonessential in immediate responses to stress, such as the kidneys and intestinal tract; it causes dilation of lung passages and of blood vessels in the skeletal muscles; and it stimulates the adrenal medullae to secrete epinephrine (adrenaline) and norepinephrine, which hormonally complement the activities of the sympathetic nervous system. Circumstances that might provoke strong, concerted activation of the sympathetic division include vigorous exertion, pain, threats to safety, and exposure to physical extremes such as severe heat or cold. Aptly, the sympathetic division has been said to prepare the animal for "fight or flight."

BIOCHEMICAL REGULATION

Enzymatically catalyzed chemical reactions taking place within the cells of the body are central to all aspects of animal function. In this section, we look briefly at a few of the salient mechanisms by which the rates of these reactions are regulated. These mechanisms are among the most ancient of regulatory processes; they operate within the confines of single cells, and their evolution probably preceded the evolution of multicellularity. In multicellular animals, the intracellular mechanisms of regulation have become functionally integrated with mechanisms operative at higher levels of organization. Specifically, the neural and endocrine regulatory systems just discussed, which serve to coordinate the intracellular activities of far-flung cells, operate by impinging on the intracellular regulatory mechanisms of those cells.

Discussions of enzymes often dwell almost exclusively on their ability to accelerate reactions. This is a very limited perspective, however, for enzymes play a crucial additional role as *regulators* of cell function. The rates of key reactions are commonly controlled by *modulation of the catalytic effectiveness* of the enzymes involved. The agents (e.g., cell metabolites) responsible for altering catalytic effectiveness are termed **enzyme modulators**.

One of the functional properties of an enzyme that can undergo modulation is the ease with which the enzyme can enter into combination with substrates (initial reactants) of the catalyzed reaction. For catalysis to occur, enzymes and their substrates must combine. Thus, when a modulator of an enzyme alters the ease of combination (as by modifying the enzyme's molecular shape), it potentially affects catalytic effectiveness. In the evolution of enzymes, responses to modulators have been subject to natural selection and often have become favorable to the harmonious regulation of cell function.

One important type of modulator is the allosteric type. On the surface of an enzyme molecule, there is a specialized molecular region, called the substrate-binding site, at which combination with substrate occurs. Some types of enzyme molecule have additional specialized regions at which they reversibly combine with specific metabolites that act as modulators. These latter regions are called **allosteric sites,** and the modulators that combine with them are called **allosteric modulators.**

Consider the following reaction sequence, in which the letters A, B, C, and D stand for chemical compounds and enz_1, enz_2, and enz_3 stand for enzymes:

$$A \xrightarrow{enz_1} B \xrightarrow{enz_2} C \xrightarrow{enz_3} D$$

One form of adaptive allosteric control, termed **feedback inhibition,** occurs when one of the products of a reaction sequence allosterically decreases the catalytic effectiveness of one of the early enzymes in the sequence. For example, in the sequence shown, product D could combine allosterically with enz_1, and the combination could decrease the catalytic effectiveness of enz_1 in converting A to B. An abundance of D in the cell, by promoting combination of enz_1 with D, would then inhibit, at its very beginning, the entire reaction sequence producing D. Conversely, if D were to become scarce, disassociation of D from enz_1 would promote entry of A into the reaction sequence forming D. The system would act to stabilize levels of D in the cell through negative feedback.

Sometimes, allosteric modulators *augment* the catalytic effectiveness of affected enzymes. One form of this phenomenon is termed **precursor activation.** For example, in the reaction sequence diagrammed earlier, precursor A might augment the catalytic effectiveness of enz_2 or enz_3, thereby promoting flow of material through the reaction sequence when precursor is readily available.

Alteration of the catalytic effectiveness of enzyme

molecules already extant in the cell provides a means for very rapid adjustment of cell function. Some of the other mechanisms of biochemical control involve the synthesis or destruction of enzyme molecules and thus operate on longer time scales. Two basic possibilities exist for exerting metabolic control by synthesis and destruction of enzyme: (1) The *concentrations* of extant enzymes might be altered, or (2) the *types* of enzyme molecule might be changed.

The first of these possibilities requires little discussion. Clearly, raising or lowering the concentration of an enzyme without otherwise altering its properties will tend, respectively, to promote or inhibit the catalyzed reaction. One way that some animals counter the depressing effects of low body temperatures on the rates of their metabolic processes is to increase the concentrations of key enzymes when they experience cold (p. 93).

Not uncommonly, animals are able to produce two or more enzymatic chemical compounds that catalyze the same reaction. Such compounds are called isozymic forms of an enzyme, or isozymes. That is, **isozymes,** defined broadly, are multiple molecular forms of a single enzyme produced by a single species. Isozymes commonly differ not only in chemical structure but also in catalytic properties. Thus, one way an organism can respond to altered circumstances is to abandon certain isozymes and turn to alternative ones that catalytically are especially effective under the new conditions. When transferred from warm to cold waters, for example, rainbow trout gradually replace one isozyme of the important brain enzyme acetylcholinesterase with another that more readily forms complexes with substrate at low temperatures (p. 95).

To summarize, cells possess some mechanisms of biochemical regulation, such as allosteric modulation, that operate on very short time scales and thus are well suited to the moment-by-moment control of cellular function. Other mechanisms, such as changing the types or concentrations of enzyme molecules, operate on comparatively long time scales and thus are particularly appropriate in responding to relatively persistent changes in circumstances.

BIOLOGICAL CLOCKS AND THE TEMPORAL ORGANIZATION OF ANIMALS

Organisms possess internal timing mechanisms termed **biological clocks.** By rhythmically modulating the physiological status of cells, tissues, and organs, these clocks endow the organism with an intrinsic **temporal organization,** meaning that many of its physiological and behavioral attributes change over time in rhythmic patterns quite apart from ostensible changes in the environment. Because the status of the organism is intrinsically changing over time, responses to environmental agents or stimuli often depend on the precise time at which they impact the organism; an environmental toxin may have a greater

probability of doing damage at one time of day than another, and an infectious agent may be most likely to gain a foothold at certain hours of the day. Biological clocks also activate overt physiological and behavioral processes of many kinds; a mouse, for example, may run about at a certain time each day and enter torpor at another time largely because of the action of its internal clocks. Clocks are interfaced with the other regulatory systems discussed in this chapter. For instance, biological clocks cause the body temperatures of humans and other mammals to rise and fall in a regular daily pattern by modulating the operating characteristics (possibly the set point) of the thermal control system (p. 12).

Circadian Rhythms

Phenomenology Certain plants are noted for raising their leaves during certain times of day and lowering them at other times.* They exhibit a **daily rhythm** of leaf movements, where the term *rhythm* means a regular, cyclical variation in function. In 1729, a Frenchman named M. de Mairan reported that certain of these plants continue to raise and lower their leaves in approximately a daily rhythm even when they are kept in constant darkness and at constant temperature. That is, their rhythm of leaf movements continues in more or less a daily pattern even when they are denied obvious environmental sources of information about the time of day. Experiments of a similar nature have since been performed on many plant and animal systems, and daily rhythms in many types of function in many types of organism—including even protists—have been shown to persist in a constant laboratory environment. Rhythms that continue in the absence of obvious environmental sources of information about time are termed **endogenous.** Those that fail to persist under such conditions are called **exogenous,** being dependent on outside cues for maintenance of their periodicity.

To determine if a rhythm is endogenous, the usual experimental approach is to test the organism under constant illumination and constant temperature; sometimes, other parameters such as humidity and barometric pressure are held constant as well. Cycles in light, temperature, and so on are the *obvious* environmental sources of information about time. If a biological rhythm persists without these information sources—that is, if it is endogenous—it is said perforce to be controlled by a biological clock. A few investigators argue that the primary timing mechanism of "biological" clocks is in fact not biological; they believe that organisms denied the obvious environmental sources of information about time are nonetheless obtaining time information from the environment, by detecting cycles in *subtle* parameters such as fields of gravity, geomagnetism, or ionizing

---

*Day and *daily* are used to refer to the *24-h* day throughout this discussion.

radiation. Currently, however, the majority view is that biological clocks are truly biological; that is, the primary timing mechanism of the clocks is believed to be a biochemical–biophysical attribute of cellular metabolism.

Let us now look in more detail at some of the properties of endogenous rhythms, using data on the activity of flying squirrels (*Glaucomys volans*) for illustration. Figure 2.8 provides a record of the activity of a flying squirrel kept in steady darkness at 20°C. It is apparent that the squirrel displayed a clear rhythm of activity, but the episode of activity shifted to an earlier time with each succeeding day. The time elapsing between the manifestation of a particular part of a rhythm on one day and the manifestation of that same part on the next day is defined to be the ***period*** of the rhythm. On average, the start of the squirrel's episode of activity on any one day followed the start on the preceding day by 23.6 h; thus, the period of the squirrel's rhythm was 23.6 h.

Figure 2.9 shows a record for another flying squirrel over a series of 53 days. For the first 7 days, the squirrel was in continuous darkness (except for one entire day when continuous light was provided). On day 8, the squirrel was placed on a photoperiodic cycle of 12 h of light and 12 h of dark per day (12L:12D). At first, the squirrel commenced its activity in the middle of the light phase, but gradually activity began at a later time each day until, on day 21, activity began at the start of the dark phase, the "natural" time for a nocturnal animal. Thereafter, activity continued to start at the beginning of the

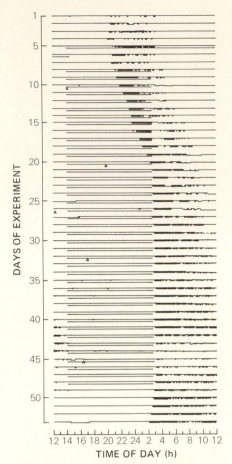

Figure 2.9　Activity of a flying squirrel (*Glaucomys volans*) as recorded on a running wheel over a period of 53 days at 20°C. See legend of Figure 2.8 for description of format. For each day the thin horizontal line running across the entire figure is the tracing of the unactivated recording pen attached to the running wheel. Horizontal lines just under the pen tracings indicate times that the lights were on. Note that the animal was in continuous darkness over days 1–4; lights were on continuously from hour 14 on day 5 to hour 14 on day 6; the animal was returned to continuous darkness for most of day 6 and all of day 7; from day 8 to day 48, the animal was placed on a light cycle of 12 h of light and 12 h of dark each day; and the animal was returned to continuous darkness on days 49–53. [From P. J DeCoursey, *Cold Spring Harbor Symp. Quant. Biol.* **25**:49–55 (1960). Copyright 1960 by the Cold Spring Harbor Laboratory.]

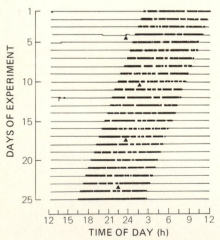

Figure 2.8　Activity of a flying squirrel (*Glaucomys volans*) as recorded on a running wheel over a period of 25 days in continuous darkness at 20°C. Each horizontal line corresponds to 1 day. Turning of the running wheel activated a pen to record a short vertical line for each rotation of the wheel; these vertical blips are often blended to give the appearance of a heavy, continuous line during periods of steady running. Note that activity started earlier with each passing day. Triangles indicate times at which food was added to the activity cage. [From P. J. DeCoursey, *Cold Spring Harbor Symp. Quant. Biol.* **25**:49–55 (1960). Copyright 1960 by the Cold Spring Harbor Laboratory.]

dark phase until, on day 49, the squirrel was returned to continuous darkness. Over days 49–53, the period of the activity rhythm reverted to being less than 24 h, and activity started at an earlier time with each passing day. This experiment illustrates a number of important and general principles. When the squirrel was first placed on the 12L:12D light cycle, its activity rhythm was *out of phase* with the environmental cycle, but gradually the activity rhythm shifted to be *in phase* and then it remained in phase as long as the environmental cycle was provided. The process by which a biological rhythm comes into phase with an environmental rhythm is termed ***entrainment,*** and the biological rhythm is said to be ***entrained*** by the environmental cues. An environmental cue that is capable of setting the phase of a biological rhythm is termed a **phasing factor** or **Zeitgeber** (German for "time-giver"). When obvious phasing factors or Zeitgeber are removed, as in Figure 2.8 or on days 49–53 of Figure 2.9, the biological rhythm that persists is said to **free-run** or to be a **free-running rhythm.** Because free-running rhythms usually have periods close to but not exactly equal to 24 h, they are termed **circadian rhythms,** meaning that they are *about* a day in length (*circa* = "about," *dies* = "a day").

Experiments such as those on the flying squirrels provide a great deal of information on the control of biological rhythms. We see that these squirrels have an endogenous rhythm of activity that persists even in the absence of obvious environmental information about the time of day. However, this endogenous circadian rhythm is not precisely timed to the 24-h day, and episodes of activity drift to occur at earlier and earlier times of day. When the squirrels are provided with a Zeitgeber in the form of a daily light–dark cycle, their activity is timed according to the external cues and does occur at 24-h intervals. The onset of darkness in nature serves to cue the beginning of activity, but we see that, teleologically speaking, the squirrels do not have to wait in a state of total ignorance each day to see when darkness will arrive. Rather, they have an endogenous sense of the time of day, and the onset of darkness simply serves as a cue that maintains a *precise* daily rhythm in a system that, in and of itself, will maintain an *approximate* daily rhythm.

Not all daily rhythms prove to be endogenous when tested, but rhythms that persist in the absence of obvious environmental information about the time of day are so widespread that endogenous circadian rhythmicity is believed to be an almost universal and probably ancient feature of life. Table 2.1 lists just some of the known endogenous circadian rhythms.

Table 2.1   SOME PROCESSES THAT SHOW CIRCADIAN RHYTHMICITY

Locomotor activity in many vertebrates and insects

Metabolic rate in many animals

Variations of body temperature (including torpor) in birds and mammals

Urine output and drinking in mammals

Adrenocortical hormone secretion and epidermal mitosis in mammals

Integumentary color change in fish and crabs

Oviposition, mating, and emergence of adults from pupae in insects

Female pheromone release and male pheromone sensitivity in insects

Mating in *Paramecium*

Bioluminescence and photosynthetic capacity in dinoflagellates

Several types of environmental stimulus have been found to be capable of acting as phasing factors or Zeitgeber for such rhythms. The great majority of rhythms can be entrained by daily cycles of light intensity. In addition, some can be entrained by cycles of temperature, sound, food availability, social interaction, or other parameters.

One of the truly remarkable features of endogenous circadian rhythms is that their free-running periodicities are not greatly affected by cellular temperature. The **frequency** of a rhythm is the inverse of its period and provides a measure of the *rate* at which the rhythm moves through an entire cycle. As will be seen in Chapter 3, the rates of most metabolic processes are typically quite sensitive to body temperature; heart rate, breathing rate, and metabolic rate, for example, are likely to double approximately if the body temperature of an animal is raised by 10°C. In sharp contrast, the frequencies of free-running circadian rhythms seldom increase or decrease by more than 20 percent when body temperature is elevated by 10°C, and often the frequencies of rhythms are nearly independent of temperature. A biological clock would obviously be of little use if it were highly sensitive to temperature; imagine the chaos if our wristwatches were to double their rate for every 10°C rise in temperature. Thus, the low thermal sensitivity of biological clocks is not surprising. Nonetheless, if the primary timing mechanisms of these clocks are attributes of living cells, we face the mystery of how they have been rendered rather immune to the thermal effects that so strongly influence most metabolic processes.

A brief focus on the word *circadian* is appropriate. Increasingly, it has been used as if synonymous with *daily*. This practice threatens to strip *circadian* of its special meaning and is pointless since we have no need of a new synonym for daily. *Any* rhythm with a period of 24 h is a daily rhythm. A rhythm is circadian only if it can persist with a period of approximately 24 h in the absence of obvious environmental sources of information about time.

The Uses of Circadian Clocks   Consider an animal that is nocturnally active—one that emerges from its nest at each dusk, spends the night in feeding and other activities, and retires to its nest at each dawn. There are at least two ways in which such an animal could keep its activity rhythm in phase with the daily environmental cycle. On the one hand, it could rely entirely on exogenous cues; that is, it could altogether lack any internal sense of time and thus emerge strictly when it witnessed the onset of dark each day and retire strictly when it witnessed the onset of light. On the other hand, it could possess an endogenous circadian rhythm of activity kept in phase with the environmental light–dark cycle by entrainment. What could be the advantages of the latter arrangement over the former? More generally, why have organisms evolved and retained circadian clocks? In this section, we briefly note some of the hypotheses offered in answer to these questions, acknowledging that in many specific cases, convincing and well-documented answers cannot be given.

One important advantage widely postulated for clocks is that they permit anticipation of and preparation for upcoming events. An animal strictly dependent on external cues must wait until appropriate cues appear before it knows to act, but the animal with an internal clock can anticipate when action will be necessary and thus prepare, even hours in advance.

Clocks also permit accurate timing of processes during periods of the day when environmental sources of time information are vague, difficult to use, or unreliable. For illustration, consider again the example of a nocturnally active animal. If it possesses circadian clocks, it can keep its clocks in phase with the daily environmental cycle by virtue of entrainment to dawn or dusk—cues that are particularly obvious and reliable—and then it can use the clocks to time its activities accurately throughout the long, dark part of the night, when environmental sources of time information (e.g., temperature fluctuations) are less obvious and less reliable.

Many animals depend on changes in daily photoperiod over the course of the year to time annual events in their life cycles; for example, the long period of daylight per day in spring may be used as a cue for reproduction, migration, or molting of the hair or feathers. Circadian clocks are often instrumental in such responses by virtue of being involved in measuring or detecting changes in photoperiod. In some instances, for example, animals are specially light sensitive at particular hours of the day, and photoperiodic responses are elicited only when the period of daylight in a day is long enough to encroach on the specially light-sensitive hours; in turn, timing of the light-sensitive hours is executed by a circadian clock.

Circadian clocks are also involved in some forms of directional orientation. Compass directions, for example, can be determined from the position of the sun if the time of day is known. In a number of animals, circadian clocks have been shown to provide the necessary information on the time of day (see p. 505).

Mechanistic Features of Clocks   Because many unicellular organisms evidence endogenous circadian rhythms, we know that circadian clocks can exist within the confines of a single cell. We also know that circadian clocks can continue to keep time even if the overt processes they control—such as locomotory activity—are experimentally blocked; that is, the overt processes are not essential parts of the timing mechanism. Clocks have been shown to be inherited and innate, not learned or imprinted. As noted earlier, they exhibit a low sensitivity to changes of temperature. Although they are impressively resistant to many chemical treatments—including exposure to certain potent metabolic poisons—their function can be affected substantially by other treatments, such as injection of heavy water, lithium, caffeine, or certain alcohols. All these observations throw light on the nature of circadian clocks, but as yet the biochemical–biophysical mechanisms of clocks remain obscure.

In multicellular animals, there is evidence that cells in many, if not all, individual tissues may possess circadian clocks. A *circadian organization* is one of the important features of such organisms. By this we mean that in the animal's body (1) subsets of cells, tissues, or organs can be identified *within* which circadian rhythms are normally synchronous; (2) *among* such subsets, rhythms are coordinated in harmonious phasic relations; and (3) in the presence of environmental entraining cues, whole subsets of cells, tissues, and organs—however far-flung in the body—are entrained. One of the major themes in recent work on biological clocks has been the search for circadian control centers—centers that play pivotal roles in maintaining the circadian organization of the body, as by imposing rhythms on other tissues and mediating entrainment. In primates, for example, there is now evidence for at least two major control centers, each of which synchronizes a diverse assemblage of circadian rhythms throughout the body. The anatomical location of control centers has in some cases been identified. For instance, the optic lobes in cockroaches and the pineal gland in certain birds are required for the maintenance of circadian locomotory rhythms. In mammals, certain control centers have been localized to two sets of cells in the hypothalamus known as the suprachiasmatic nuclei (or their homologs). Control centers may communicate with the rest of the body by neuronal or endocrine means.

There is evidence that disruption of the circadian organization of the body can have deleterious effects. The malaise known as jet lag, for instance, results when a person suddenly experiences a shift of many hours in the phase of the environmental cycle, thereby suffering a disturbance of the normal phasic relation between internal biological rhythms and the environmental cycle. Within an organism, the circadian rhythms of various tissues and organs ordinarily operate in particular phasic relations with one another. However, these internal phasic relations can be disrupted; this happens, for instance, in the aftermath of jet travel because some rhythms lag behind others in regaining synchronization with the cycles in the outside world. Evidence exists that disruption of normal internal phasic relations can be detrimental. In recent experiments, for example, when certain of the circadian rhythms of squirrel monkeys (*Saimiri sciureus*) were thrown out of their normal phasic relation with others, the animals became especially prone to undergoing a drop in their body temperature when exposed to moderately cold air.

Circatidal and Circannual Rhythms

Of all rhythms, the daily rhythms have received by far the greatest attention by students of biological clocks and thus have been emphasized here. Nonetheless, physiological and behavioral processes do undergo important cyclic variations on other time scales. Annual rhythms of reproduction, migration, fat accumulation, dormancy, and so on are well known. Along the seashore, animals living at the interface between the land and water display rhythms in synchrony with the tides, which usually rise and fall every 12.4 h (half a lunar day). For instance, fiddler crabs that scavenge for food on the sand or mud exposed by low tide become most active at low tide.

Some, but not all, annual and tidal rhythms persist even when animals are placed in a laboratory environment where they are denied access to any obvious environmental sources of information about the time of year or time of the tidal cycle. Typically, the periods of rhythms free-running under such conditions are only *approximately* a year or a tidal cycle in length. Thus, such endogenous rhythms have been termed **circannual** or **circatidal** in analogy with the circadian rhythms. Under natural conditions, some environmental parameters do vary in phase with the annual or tidal cycles; and the endogenous circannual and circatidal rhythms of animals evidently become entrained to cycles in such parameters, so that the biological rhythms are kept in phase with the actual tides and seasons. The annual photoperiodic cycle, for instance, is the putative Zeitgeber for certain circannual rhythms; and aspects of ebbing and flowing tidal water, such as mechanical agitation, have been postulated to serve as Zeitgeber for some of the circatidal rhythms. Concerning the question of whether the endogenous timing mechanisms for circannual and circatidal rhythms are dependent on circadian clocks, arguments have been presented on both sides. As yet, the nature of these timing mechanisms remains unresolved.

SELECTED READINGS

Aschoff, J., S. Daan, and G.A. Gross (eds.). 1982. *Vertebrate Circadian Systems*. Springer-Verlag, New York.

Bayliss, L.E. 1966. *Living Control Systems*. Freeman, San Francisco.

*Biological clocks. Special section in *BioScience*. 1983. *BioScience* 33:424–457.

*Brady, J. 1982. *Biological Timekeeping*. Cambridge University Press, New York.

Brown, F.A., Jr., J.W. Hastings, and J.D. Palmer. 1970. *The Biological Clock: Two Views*. Academic, New York.

Bünning, E. 1973. *The Physiological Clock*, 3rd ed. Springer-Verlag, New York.

Farner, D.S. 1985. Annual rhythms. *Annu. Rev. Physiol.* **47**:65–82.

Fuller, C.A., F.M. Sulzman, and M.C. Moore-Ede. 1979. Effective thermoregulation in primates depends upon internal circadian synchronization. *Comp. Biochem. Physiol.* **63A**:207–212.

*Gwinner, E. 1986. Internal rhythms in bird migration. *Sci. Am.* **254**(4):84–92.

Gwinner, E. 1986. *Circannual Rhythms*. Springer-Verlag, New York.

Hardy, J.D. and H.T. Hammel. 1963. Control system in physiological temperature regulation. In J.D. Hardy (ed.), *Temperature, Its Measurement and Control in Science and Industry,* Vol. 3, Part 3, pp. 613–625. Reinhold, New York.

Hochachka, P.W. and G.N. Somero. 1984. *Biochemical Adaptation*. Princeton University Press, Princeton.

*Johnson, C.H. and J.W. Hastings. 1986. The elusive mechanism of the circadian clock. *Am. Sci.* **74**:29–36.

*Mangum, C. and D. Towle. 1977. Physiological adaptation to unstable environments. *Am. Sci.* **65**:67–75.

Menaker, M., J.S. Takahashi, and A. Eskin. 1978. The physiology of circadian pacemakers. *Annu. Rev. Physiol.* **40**:501–526.

Moore-Ede, M.C., F.M. Sulzman, and C.A. Fuller. 1982. *The Clocks That Time Us*. Harvard University Press, Cambridge.

Moore-Ede, M.C. 1983. The circadian timing system in mammals: Two pacemakers preside over many secondary oscillators. *Fed. Proc.* **42**:2802–2808.

Naylor, E. 1985. Tidally rhythmic behaviour of marine animals. *Symp. Soc. Exp. Biol.* **39**:63–93.

Newsholme, E.A. and C. Start. 1973. *Regulation in Metabolism*. Wiley, New York.

Nilsson, S. 1983. *Autonomic Nerve Function in the Vertebrates*. Springer-Verlag, New York.

Page, T.L. 1982. Transplantation of the cockroach circadian pacemaker. *Science* **216**:73–75.

Pengelley, E.T. (ed.). 1974. *Circannual Clocks*. Academic, New York.

Pick, J. 1970. *The Autonomic Nervous System*. Lippincott, Philadelphia.

Pittendrigh, C. 1961. On temporal organization in living systems. *Harvey Lecture Ser.* **56**:93–125.

Saunders, D.S. 1982. *Insect Clocks,* 2nd ed. Pergamon, New York.

Werner, J. 1980. The concept of regulation for human body temperature. *J. Thermal Biol.* **5**:75–82.

Yamamoto, W.S. and J.R. Brobeck (eds.). 1965. *Physiological Controls and Regulations*. Saunders, Philadelphia.

*Yates, F.E. 1982. Outline of a physical theory of physiological systems. *Can. J. Physiol. Pharmacol.* **60**:217–248.

See also references in Appendix A.

chapter 3

Energy Metabolism

The subject of this chapter is *energy metabolism.* By this we mean the sum total of the processes by which animals acquire, interconvert, use, and dispose of energy.

We know that animals need inputs of energy to stay alive. An important first question, however, is, "Why?" At the root of the answer lies the *second law of thermodynamics.* One way of stating this law is that, in any isolated system, order tends spontaneously to change to disorder; nonrandom states move toward random ones. Order can be maintained in a system, however, by providing energy to it from the outside.

To explore these principles further, it is appropriate first to look at an inanimate system. The principles apply equally to inanimate and animate systems, and in some respects inanimate systems are more easily understood. Let us consider a circle of copper pipe with water flowing round and round in it, and initially let us consider the pipe and water to constitute an isolated system. A system is *isolated* if it exchanges neither matter nor energy with its surroundings. At first, there will be a highly nonrandom distribution of molecular motions in this system. The molecules in the walls of the pipe will exhibit only the incessant random motions that all molecules undergo on a molecular scale. The molecules of water will also display such random motions, but they will also exhibit highly oriented movements in the direction of their travel around the loop of pipe. Gradually, by way of intermolecular collisions, some of the energy of directional movement of water molecules will be transferred to molecules in the pipe walls and to other water molecules in such a way as to increase the intensity of random motions of the recipient molecules. Macroscopically, as this happens, the flow of

water around the pipe will slow, and the energy of the flow will become translated into heat. Finally, the flow of water will cease entirely. At that point, all the molecules in the system will merely be undergoing random molecular motions, and the original order in the system (the directed movements of water molecules) will have become entirely degraded to disorder. This is the second law of thermodynamics at work. How can we maintain nonrandomness in this system and keep the water flowing at its original rate? These ends cannot be achieved if the system remains isolated. However, they can be accomplished if we provide energy from the outside, as by delivering an electric current to a pump in the system. Here we see illustrated an important basic principle: Inputs of energy to a system can permit order to be maintained despite the system's inherent tendency to become ever more disordered.

As stressed in Chapters 1 and 2, a high degree of order or organization is one of the most essential characteristics of living organisms. Just as in the inanimate world, this order tends intrinsically to change toward greater disorder. For example, the blood coursing through the circulatory system tends to slow to a halt. Complex structural and physiological molecules tend to break down spontaneously into simpler parts; and the body fluids of a freshwater fish, normally having a far higher concentration of salts than the surrounding water, tend to become as dilute as fresh water. Animals require energy from the outside world precisely to create and maintain their vital internal organization against these tendencies toward greater disorganization.

Energy exists in a number of forms, and when animals use energy to maintain their internal organization, they alter its form. Like the very need for energy itself, this is one of the most fundamental principles of energetics. Energy is acquired in the form of chemical-bond energy in ingested foods. Much of it is converted to heat as it is used by the animal. This conversion has important implications, for whereas chemical-bond energy is an eminently useful form of energy for organisms, heat is not very useful, and heat cannot be converted by organisms back to chemical-bond energy. We need now to examine these ideas more closely. Already we can see, however, that the one-way conversion of energy from highly useful forms to heat imposes on animals the need to obtain more energy in the useful forms. This is why the search for food-bond energy goes on throughout life.

FUNDAMENTALS OF ANIMAL ENERGETICS

The Forms of Energy and the Capacity for Physiological Work

Energy is the capacity to do work, expressed physically as the product of force and distance. The forms of energy that are significant in biological systems include chemical energy, electrical energy, mechanical energy, and heat. *Chemical energy* is energy liberated or required when atoms are rearranged into new assemblages. *Electrical energy* is energy that a system possesses by virtue of separation of electrical charge. Mechanical energy and heat are both forms of kinetic energy. By *mechanical energy* we mean energy of organized motion, such as that possessed by a moving arm or the circulating blood. *Heat* is molecular kinetic energy, the energy of random molecular motion. As already mentioned, within any substance of uniform temperature above absolute zero, the atoms and molecules move randomly and continuously. Heat is the energy a substance possesses by virtue of these motions.

All forms of energy are by definition capable of doing work in one context or another. Here, however, we face a more exacting question: Are the forms capable of being used by *organisms* to do work? Can they be used, that is, for muscular contraction, nerve-impulse generation, protein synthesis, or other forms of *physiological* work?

Chemical-bond energy can be used directly or indirectly by animals to do all forms of physiological work, and both electrical and mechanical energy can be used by animals to accomplish certain forms of work. Heat, however, cannot be used by organisms to do any kind of work. Thermodynamics tells us that it is impossible to convert heat to work in a system of uniform temperature. In a steam turbine, large thermal differences exist, and heat is successfully converted to work. However, thermal differences within cells—the relevant functional units of organisms—are at most small and transient. Thus, we are not surprised to find that the only thing the

addition of heat can accomplish in living organisms is the increase of random molecular agitation, which is not a form of work.

By virtue of these considerations, the forms of energy of significance to living things are placed into two groups: (1) high-grade and (2) low-grade. The *high-grade* forms are capable of doing physiological work and include chemical, electrical, and mechanical energy. The *low-grade* form, heat, cannot be used to do physiological work. When energy is transformed from a high-grade form to heat, it is said to be "degraded."

The Inefficiency of High-Grade Energy Conversions and Its Implications

Having examined the forms of energy, we are now in a position to return to an important matter raised in the introduction, namely, the fate of energy when it is used physiologically. In this section, we approach the question in terms of basic principles. In the next section, we look in more detail at the actual categories of use of energy by animals.

When organisms transform high-grade energy from one form to another, the conversion is always incomplete, and some energy is degraded to heat. The *efficiency* of such a conversion is defined as the ratio

$$\frac{\text{output of high-grade energy or output of work}}{\text{input of high-grade energy}}$$

Typically, this efficiency is considerably less than 1. When a cell converts chemical-bond energy of glucose into chemical-bond energy of adenosine triphosphate (ATP), for example, only about 70 percent of the energy released from glucose is incorporated into bonds of ATP; the remaining 30 percent appears as heat (p. 40). When chemical-bond energy of ATP is used for contraction by a muscle cell, it is typical to find that at maximum only about 10–50 percent of the energy liberated from the ATP appears as energy of muscular motion; again, the remainder is lost as heat. These losses of high-grade energy as heat are partly a manifestation of the second law of thermodynamics.

Animals ingest high-grade chemical energy in the form of plant or animal tissue. They then transform the energy in myriad ways as they use it to carry out the tasks of being alive. Bond energy of an ingested sugar molecule, for example, may first be transferred to bond energy of ATP. Then the bond energy of ATP may be used to structure a protein (some of the energy appearing as protein-bond energy); or it may be used to establish a difference of electrical potential across a nerve-cell membrane (some of the energy appearing as electrical energy); or it may be used to drive the contraction of a muscle fiber. We recognize now that with each and every transformation, a toll is taken. Some energy is lost from the high-grade, work-capable forms to heat.

Major Paths of Animal Energy Use

The chemical-bond energy in the food of an animal is known as *ingested energy*. Some of this energy never really enters the animal but instead is merely egested in the feces; the egested chemical energy is known as *fecal energy* and is available to fuel the fires of life of those many organisms that consume fecal matter. The remainder of the ingested energy is taken up across the gut wall in the form of bond energy of assimilated foodstuffs and is known as the *absorbed* or *assimilated energy*. The absorbed energy is the energy available for the performance of work by the animal's cells.

The tasks that animals accomplish using their absorbed energy can be categorized in various ways. Here we recognize four categories. As we discuss them, our major focus will be on the fate of the energy used in each way. Figure 3.1 diagrams the major points made.

(1)   One category of tasks accomplished using absorbed energy is the *synthesis of organic compounds that are destined to be voided or sloughed off into the environment during the life of the individual*. Animals make organic compounds that are voided in their urine, for example. Gametes, milk, mucoid secretions, shed skin, and shed exoskeletons are just some of the other organic materials that are made and lost. The manufacture of such materials involves many biochemical steps in which degradation of energy can take place. Thus, much of the absorbed energy that is directed to the synthesis of the mate-

rials is converted to heat. Some of the energy, however, ends up as chemical-bond energy in the organic materials produced, and that energy is lost from the animal as chemical energy when the materials are voided. Other living things can then use that chemical energy as an energy source for their own metabolism.

(2)   *Synthesis of new tissue during growth*. The process of growth results in heat production. It also results in accumulation of chemical energy in the form of the animal's ever-increasing mass of tissue. That chemical energy can be used to meet the energy needs of the animal during times of food shortage and ultimately will meet the needs of predators or organisms of decay.

(3)   *Maintenance functions*. Much of an animal's absorbed energy is used in processes that maintain its integrity such as circulation, respiration, nervous coordination, activity of the gut, and tissue repair. Energy used in these functions is *virtually entirely degraded to heat within the body*. To see why, consider, for example, the circulation of the blood. As absorbed energy is directed to this function, it is converted to heat step by step until it is all heat. First, the chemical energy of absorbed food substances is converted into chemical energy of ATP and energy is lost as heat in the process. Second, when the energy of ATP is used to drive contraction of the heart muscle, some additional energy is lost as heat. Finally, a fraction of the chemical energy originally obtained from food molecules appears as mechanical energy of motion in the blood ejected from the heart, but even that energy is degraded to heat within the body in overcoming the viscous resistances that oppose motion of the blood through

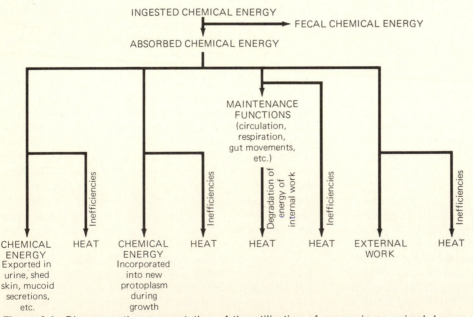

Figure 3.1 Diagrammatic representation of the utilization of energy in an animal. Lower line shows forms in which energy "leaves" the animal. Energy enters as chemical energy and leaves as heat, chemical energy, or external work.

the blood vessels. Work that takes place entirely in the body, such as the circulation of the blood, is termed **internal work**. The energy of internal work of all kinds is mostly or entirely degraded to heat within the body.

(4) *External work*. When animals apply forces to objects in the external world, they are said to perform **external work**. A mouse running across a field, for example, is doing external work using its leg muscles. Similarly, a person does external work when lifting boxes from the floor onto a shelf. Much of the energy directed to external work is degraded to heat within the body (e.g., in using ATP to set muscles in motion), but significantly *some energy leaves the animal as work*. What is the fate of that energy? The answer depends on the type of work. Many forms of external work, by their nature, fail to result in any long-lasting storage of the work energy as potential or mechanical energy; in these cases, the work energy is degraded to heat about as quickly as it exits the body. For example, if a mouse or other animal, starting from a standstill, travels a distance *horizontally* and stops, we recognize that none of the energy required for the trip has been stored as potential or mechanical energy once the trip has ended; although some energy left the animal's body as work, it was quickly and entirely degraded to heat in overcoming resistances to the animal's motion. On the other hand, some forms of external work do entail a significantly long-lasting storage of potential energy. For instance, if a mouse runs uphill or a bird flies to high altitude, a (small) part of the energy required for the trip is stored as potential energy of position by virtue of the net motion of the animal's body in opposition to gravity. Later, when the animal descends, that part will be converted to mechanical energy and then to heat.

Summary and Implications

A review of Figure 3.1 reveals that energy leaves an animal in three forms: (1) chemical energy, (2) work, and (3) heat. Of these, heat is quantitatively preeminent.

Sometimes, the mistaken idea is expressed that certain animals fail to produce heat. Heat production is in fact obligatory for living animals. If some, such as frogs, clams, or fish, are cool to the touch, it is not because they fail to produce heat. Rather, their rates of heat production are sufficiently low and their insulation sufficiently modest that they are not warmed appreciably by the heat they produce (p. 104).

As mentioned in the introduction to the chapter, organisms are unable to convert heat back into high-grade energy forms. Thus, *energy is not recycled* within individual organisms or, by extension, within the biosphere as a whole. The biosphere requires a continuing input of high-grade photic energy from the sun, and the individual animal requires a contin-uing input of chemical-bond energy, usually obtained ultimately from photosynthesis. The heat that organisms make is dissipated into their environment and in turn is radiated from earth into outer space.

The History of Views on Animal Heat Production

The steady emanation of heat from the human body and its cessation after death have probably stirred the curiosity of humans since the ancient time when the capacity for such curiosity first evolved. Yet the real significance of animal heat production has been appreciated only relatively recently. The maturation of concepts concerning animal heat is of no small interest, for an understanding of the origins of animal heat goes hand in hand with an understanding of some of the most fundamental attributes of life.

Now we recognize that heat is an inevitable *by-product* of the use of high-grade chemical energy in creating and maintaining the vital organization of living things. Interestingly, from the time of Aristotle until the nineteenth century, the significance of heat was generally viewed quite oppositely. Far from being a by-product, heat was usually seen as a primary source of life, a vital force that endowed many parts of the organism with their living attributes. This "vital heat" was thought to differ from the heat of a fire. It was believed to originate exclusively in the heart, lungs, or blood and to suffuse the rest of the body. When William Harvey described the circulation of the blood in the early seventeenth century, one of the principal roles attributed to the new-found circulation was transport of "vital heat."

At about the time of the American Revolution, Antoine Lavoisier in France showed that the ratio of heat production to carbon dioxide production was about the same for a guinea pig as it was for burning charcoal. From this and other evidence, he and the Englishman Adair Crawford argued convincingly that animal respiration is a slow form of combustion, and animal heat is the same as the heat thrown by fire. Still, all heat was believed to originate in the lungs, and the lungs were thought to be the exclusive site of oxygen utilization. Not until 1837 did Gustav Magnus show that the blood takes oxygen from the lungs to the rest of the body and returns carbon dioxide. Evidence for the all-important concept that tissues throughout the body make heat came a decade later when Hermann von Helmholtz demonstrated that muscular contraction liberates heat. In 1872, Eduard Pflüger showed that the tissues consume oxygen.

The discovery that all tissues use oxygen and produce heat was one of several lines of thought and investigation that came together in the nineteenth century to spawn our modern understanding of animal energetics and heat production. Other important developments were the flowering of the science of thermodynamics and revolutionary changes in the understanding of energy. In the 1840s, Julius Robert von Mayer and James Joule developed the seminal concept that heat, motion, electricity, and so on are all forms of one thing: energy. Mayer, a physician, conceptualized clearly for the first time the nature of animal energy transformations as we have described them in this chapter.

THE DEFINITION AND MEASUREMENT OF METABOLIC RATE

In the last section, we saw that animals assimilate chemical energy and in turn put out chemical energy, external work, and heat. The energy that is converted to heat or external work is said to be *consumed.* This term is somewhat unfortunate because it may imply that the animal has destroyed energy, which is impossible. The term, however, alludes to the fact that such energy has been removed by the animal from the chemical state that is the currency for physiologically useful energy transfer within the biosphere. In comparative physiology, the *metabolic rate* of an animal is defined to be its rate of energy consumption, that is, the rate at which it converts chemical energy to heat and external work.

One reason that the metabolic rate of an animal is significant is that the heat and external work exported from the animal reflect quantitatively the *overall* activity of its physiological machinery. Heat in particular is liberated by every energy-using process in the body; thus the rate of heat production is a reflection of the sum total of the rates of all such processes. Heat is always the principal component of the metabolic rate. From an ecological perspective, the metabolic rate of an animal is significant because, being the rate of consumption of chemical energy, it represents the drain placed by that animal on the physiologically useful energy supplies of the ecosystem.

Direct Measurement of Metabolic Rate (Direct Calorimetry)

An animal's metabolic rate can be measured in a device termed a *direct calorimeter,* which by definition assays the rate at which heat is dissipated from the animal's body. The name *calorimeter* is derived from the calorie, a unit of measure for heat (p. 29).

Direct calorimeters are technically complex when designed to be especially accurate, sensitive, and versatile. Their basic nature is illustrated nicely, however, by the relatively simple device that Antoine Lavoisier used in the first measurements of animal heat production. As shown in Figure 3.2, heat emanating from the experimental animal melted ice. Lavoisier collected the water produced over measured periods of time, and by knowing the amount of heat required to melt each gram, he was able to calculate the rate of heat output from the animal.

Thus far, we have not mentioned the measurement of external work, the second component of metabolic

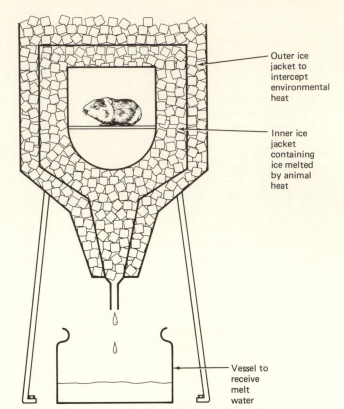

Figure 3.2 A schematic diagram of Lavoisier's calorimeter in cross section. The animal was surrounded by an ice-filled jacket. Ice melted by animal heat yielded water, which dripped out. A requirement for all direct calorimeters is that heat from the general environment must be excluded from the measurement of animal heat production. In Lavoisier's device, heat entering from the air surrounding the calorimeter was intercepted by a second ice-filled jacket that enclosed the jacket immediately surrounding the animal. [After A. Lavoisier and P.S. de Laplace, *Mém Acad. Sci.,* p. 355 (1780), as reprinted in *Oeuvres de Lavoisier*, Vol. 2. Imprimerie Impériale, Paris, 1862.]

Labels on figure:
- Outer ice jacket to intercept environmental heat
- Inner ice jacket containing ice melted by animal heat
- Vessel to receive melt water

rate. If the animal under study is at rest, no external work is being performed, so measurement of heat production alone will reflect its entire metabolic rate. In many cases where external work *is* being performed, the energy of the external work is rapidly degraded to heat, as in the case discussed earlier of an animal moving about horizontally. When this is true, a measure of heat production encompasses the energy of external work and thus again is sufficient in itself for determination of metabolic rate. If some energy of external work should be stored as potential energy of position, it must be measured independently and added to heat production to ascertain the metabolic rate.

Indirect Measurement of Metabolic Rate (Indirect Calorimetry)

Although direct calorimetry remains the method of choice for measuring metabolic rates in certain research applications, metabolic rates are usually measured by indirect methods that hold advantages of

---

BOX 3.1   DIRECT VERSUS INDIRECT MEASUREMENT

A method of measurement is **direct** if it measures a parameter in terms of the parameter's definition. A metabolic rate, for example, is defined to be a rate of production of heat and external work. Direct methods of measuring metabolic rate thus measure heat and work.

A method of measurement is **indirect** if it quantifies a parameter in ways not specified by the parameter's definition. Some methods of measuring metabolic rate, for instance, are based on quantifying an animal's rate of oxygen consumption. These are indirect, for the definition of metabolic rate mentions nothing about oxygen consumption.

The accuracy of an indirect method must be judged by determining how well its results agree with the results of direct methods. Often, indirect methods carry inherent risks of error, yet they are used in certain circumstances because they hold overriding advantages (e.g., practicality). When using indirect methods, cognizance of their limitations is essential.

---

relatively low cost or technical simplicity. The indirect methods fall into two classes, those based on respiratory gas exchange (principally oxygen consumption) and those based on material balance.

Methods Based on Respiratory Gas Exchange   If a mole of glucose is oxidized completely in the laboratory, stoichiometry dictates that 6 moles of molecular oxygen will be used and 6 moles of carbon dioxide and water will be produced. Furthermore, a certain amount of energy will be released as heat; for 1 mole of glucose this **heat of combustion** amounts to about 673,000 calories (cal), or 2800 kilojoules (kJ). We can therefore write the following equation for the oxidation of glucose.

$$C_6H_{12}O_6 + 6O_2 \rightarrow 6CO_2 + 6H_2O + 673,000 \text{ cal/mole}$$

The equation makes clear that, in the oxidation of glucose, a fixed, proportional relation prevails between the amount of heat produced and the amount of oxygen used: 673,000 cal of heat per 6 moles of O_2. Similarly, a proportional relation exists between heat production and CO_2 production: 673,000 cal of heat per 6 moles of CO_2. Knowing these relations, if we were to burn an *unknown* quantity of glucose in a test tube and we measured only the amount of O_2 used or the amount of CO_2 produced, we could calculate exactly the amount of heat produced.

If a chemical substance is oxidized metabolically by an animal, the intermediate steps of the oxidative reaction are bound to differ from those in a test tube. Energy that would be released immediately as heat in a test tube may well be temporarily stored in the animal in bonds of ATP before being degraded to heat as it is used in physiological work. Ultimately, however, as established by Max Rubner and Wilbur Atwater in the 1890s, if the chemical end products are the same in the animal as in the test tube, the stoichiometric relations that prevail among heat pro-

duction, O_2 consumption, and CO_2 production in the test tube will also prevail in the animal. Thus, if an animal were to be oxidizing glucose to CO_2 and H_2O and we were to measure either the O_2 consumed in the process or the CO_2 produced, we could calculate the animal's heat production from the proportionalities established in the previous paragraph. In essence, herein lies the rationale for estimating animal metabolic rates from data on rates of exchange of the respiratory gases, O_2 and CO_2.

There are three major classes of foodstuff—carbohydrates, lipids, and proteins—and during oxidation, the stoichiometric relations that prevail among heat production, O_2 consumption, and CO_2 production differ from one class to another. Table 3.1 gives the ratio of heat production to oxygen consumption and that of heat production to CO_2 production for each class of foodstuff during aerobic catabolism (*catabolism* = "destructive metabolism").

Suppose now that an investigator has measured an animal's rate of oxygen consumption. What does he or she then know about the animal's metabolic rate? In answering this question, we come face to face with the fact that as a measure of metabolic rate, oxygen consumption is indirect and carries certain inherent risks of error. To obtain the true metabolic rate, the rate of oxygen consumption must be translated into a rate of heat production, but this cannot be done unless the investigator knows what foodstuffs are being oxidized by the cells. From Table 3.1, we see that at an oxygen consumption of 10 mL/min, the metabolic rate is $10 \times 5.05 = 50.5$ cal/min if carbohydrates are oxidized but only $10 \times 4.74 = 47.4$ cal/min if lipids are oxidized. Generally, a mixture of foodstuffs is being used, and generally the investigator does not know the proportions of each. Accordingly, there is often uncertainty about the correct conversion factor to use in calculating metabolic rate from oxygen consumption. Commonly, a "rep-

Table 3.1   HEAT PRODUCTION PER UNIT WEIGHT, PER UNIT OXYGEN
CONSUMPTION, AND PER UNIT CO_2 PRODUCTION IN THE AEROBIC
CATABOLISM OF CARBOHYDRATES, LIPIDS, AND PROTEINS[a]

	Heat production (cal) per milligram oxidized	Heat production (cal) per milliliter O_2 consumed	Heat production (cal) per milliliter CO_2 produced
Carbohydrates	4.0–4.1 cal/mg	5.05 cal/mL	5.05 cal/mL
Lipids	9.3–9.5	4.74	6.67
Proteins	4.3–4.8	4.46	5.57

[a] Values given are for representative mixtures of each of the three foodstuffs. In the case of proteins, values depend on the metabolic disposition of nitrogen (see p. 34), and those given apply to mammalian catabolism. As is true throughout this text, gas volumes are expressed under standard conditions of temperature and pressure (0°C, 760 mm Hg of pressure). For more information, see M. Kleiber, *The Fire of Life*, 2nd ed. Krieger, Huntington, NY, 1975.

resentative'' factor of about 4.8 cal/mL O_2 is used; this is the approximate factor to be expected in an animal that is metabolizing a representative mixture of carbohydrates, lipids, and proteins. One of the strong points of using oxygen consumption to measure metabolism is that because the true conversion factor varies relatively little with foodstuff (Table 3.1), the potential errors incurred by using an approximate factor are not large. To illustrate, suppose that an investigator uses the conversion factor of 4.8 cal/mL O_2 but that the animal was oxidizing only carbohydrates with a true conversion factor of 5.05 cal/mL O_2. The investigator will then have underestimated the metabolic rate, but only by 5 percent. If, on the other hand, the animal was oxidizing only protein with a true conversion factor of 4.46 cal/mL O_2, the metabolic rate will have been overestimated, but again to only a relatively small extent, 7.5 percent. It is on these grounds that oxygen consumption is often used as a measure of metabolic rate despite uncertainties regarding foodstuffs.

Carbon dioxide production was commonly used to measure metabolic rate several decades ago but is seldom used any longer because of two considerations that render it distinctly inferior to oxygen consumption. The first of these is evident from Table 3.1, namely, that the conversion factor between heat production and CO_2 production depends very strongly on foodstuff. Accordingly, if the foodstuffs being oxidized are unknown and a ''representative'' conversion factor is used, metabolic rate can be misestimated by as much as 15–20 percent.

The second limitation of using CO_2 production to measure metabolic rate arises from a basic presumption of all methods based on respiratory gas exchange. Oxidation of foodstuffs occurs within the cells of the animal, and it is there that oxygen utilization and CO_2 production bear strict relations to heat production. We, of course, do not measure oxygen consumption and CO_2 production at the level of the cells. Rather, we must measure both the rate of oxygen consumption and the rate of CO_2 elimination at the respiratory organs (e.g., lungs). Accordingly, our measures of gas exchange relate strictly to metabolic rate only when a steady state exists, that is, only when gas exchange at the respiratory organs proceeds at the same rate as at the level of the cells. This consideration presents problems in certain situations even when oxygen consumption is being used. If a mammal, for example, suddenly starts to exercise, the muscle cells at first meet some of their increased oxygen demand by drawing oxygen from oxygen stores preexisting in the body. Respiratory oxygen uptake lags behind cellular oxygen uptake for a minute or more, and it is only after this period of adjustment that a steady state is again established such that the rate of respiratory oxygen consumption corresponds to the rate of cellular utilization. During the adjustment period, the respiratory rate of oxygen consumption does not provide an accurate measure of metabolic rate. Carbon dioxide production is subject to substantially greater and more-protracted deviations from steady state than oxygen consumption because of carbon dioxide's high solubility in the body fluids and the ease with which it reacts to form other compounds, such as bicarbonate (p. 323).

It is apparent that measures of oxygen consumption or CO_2 production could be converted entirely accurately to measures of metabolic rate if the foodstuffs being utilized by the cells were known (Table 3.1). There are methods for determining the nature of the foodstuffs, which we now briefly examine. First, it must clearly be recognized that the relations between gas exchange and metabolic rate depend on the foodstuffs being catabolized *in the cells*. Because animals interconvert and store food materials, the recent diet of the animal does not necessarily reveal the nature of the foodstuffs being oxidized at any given time. A respiratory parameter that is potentially diagnostic of foodstuff is obtained by *simultaneously* measuring *both* CO_2 production and O_2 consumption and taking their ratio:

$$\frac{\text{moles of } CO_2 \text{ produced per unit time}}{\text{moles of } O_2 \text{ consumed per unit time}}$$

When measured at the respiratory organs (e.g., lungs), this ratio is called the **respiratory exchange** *ratio, R.* When measured directly at the level of the cells, it is called the **respiratory quotient, RQ.** In steady state, R and RQ are equal, so RQ—the parameter of true biochemical significance—can be determined by measuring R; we shall assume such a steady state in our discussion here. The RQ needs to be interpreted with caution because it can be influenced by a variety of factors. Nonetheless, as shown in Table 3.2, the RQ values prevailing during oxidation of carbohydrates, lipids, and proteins are sufficiently different that the RQ can often be used to decipher the types of compound being oxidized. An RQ near 1.0, for example, would indicate that an animal's cells were catabolizing mostly carbohydrates, whereas an RQ near 0.7 would indicate predominantly lipid catabolism. Upon obtaining such RQ values, an investigator would have good insight into the particular conversion factor from Table 3.1 that should be used in calculating the metabolic rate from oxygen consumption. An RQ near 1.0, for instance, would suggest use of the carbohydrate factor, 5.05 cal/mL O_2. Unfortunately, RQ values that fail to approximate one of the extreme values (1.0 or 0.7) are not as easily interpreted. An RQ of 0.8, for example, could indicate catabolism of only protein, or of a particular mixture of lipid and carbohydrate, or of a mixture of all three foodstuffs. The problem is simple. There are three unknowns (the proportions of the three foodstuffs in the material being oxidized), but RQ encompasses only two knowns (CO_2 production and O_2 consumption). To solve for three unknowns, three knowns are needed. For certain applications, the problem can be solved by measuring a third known, namely, the rate of nitrogen elimination (which is reflective of protein catabolism).

In routine studies, by far the most common approach to measuring metabolic rates is to determine the rate of oxygen consumption alone and "live with" the relatively modest potential errors involved. Indeed, metabolic rates are often expressed simply as rates of oxygen consumption. We have already mentioned some pros and cons of using oxygen consumption as a measure of metabolic rate and need before closing to mention two additional features of the method—one a limitation, one an advantage. The limitation is that oxygen consumption reflects only *aerobic* metabolism; thus, if part or all of metabolism is anaerobic, alternative methods (e.g., direct calorimetry) must be used. An advantage is that because oxygen consumption is proportional to the ultimate yield of heat from foodstuffs, external work that results in storage of potential energy need not be measured independently; the heat equivalent of the potential energy is included in the metabolic rate computed from oxygen utilization.

Methods Based on Material Balance An alternative approach to the indirect determination of metabolic rate is based on measurement of the chemical-energy content of materials entering and leaving the body. In essence, the amounts of food eaten and of urine and feces eliminated are determined over a period of time. The chemical-energy content of the food is then determined by burning an equivalent amount of food in a bomb calorimeter (p. 34) and measuring the amount of heat evolved. Similarly, the chemical-energy content of the excreta is determined by burning. The metabolic rate is then estimated by subtracting the energy content of the excreta from that of the food. The logic of the method is straightforward: The animal must consume that chemical energy which it ingests but does not in turn void as chemical energy. Additional considerations are involved if the animal is increasing or decreasing its biomass. If, for example, the animal is growing and thus increasing the chemical-energy content of its body, then some of the chemical energy ingested but not voided is nonetheless not being consumed, and an estimate of this quantity must enter the calculation of metabolic rate. Furthermore, if chemical energy should enter or leave the body by routes other than the food, urine, and feces (e.g., by shedding of hair or feathers), these other routes need to be considered.

Measurements of ingestion, egestion, and other relevant processes must extend over an appreciable period of time—typically a day or more—if average, steady-state rates of input and output of chemical energy are to be estimated; the metabolic rate obtained is then the average rate over the entire test period. Thus, material-balance methods are suited only for long-term measurements of metabolism. To measure minute-to-minute variations in the metabolic rate, the methods of choice are those based on respiratory gas exchange or direct calorimetry.

Units of Measure

In the study of energy metabolism, the traditional unit of measure for energy is the *calorie* (also called the *gram-calorie* or *small calorie*), defined to be the amount of heat needed to raise the temperature of 1 g of water by 1°C, from 14.5 to 15.5°C. A *kilocalorie* (*kilogram-calorie* or *large calorie*) is 1000 cal. Sometimes, the kilocalorie is written *Calorie,* with a capital *C;* unfortunately, in some diet books this usage has become bastardized by the reversion of the *C* to

Table 3.2 RESPIRATORY QUOTIENTS FOR THE AEROBIC CATABOLISM OF THE THREE MAJOR CLASSES OF FOODSTUFFS

Foodstuff	Respiratory quotient
Carbohydrate	1.0
Protein	0.83[a]
Lipid	0.71

[a] The protein RQ is for animals such as mammals in which urea is the dominant nitrogenous end product.
Source: M. Kleiber, *The Fire of Life,* 2nd ed. Krieger, Huntington, NY, 1975.

lowercase, so kilocalories—confusingly and indefensibly—are being referred to as "calories."

If energy is expressed in calories or kilocalories, then *rates* of energy exchange or transformation—such as metabolic rates—are expressed in units of calories or kilocalories per unit time.

In the SI, the basic unit of measure for energy is the joule, named in honor of James Joule (p. 26). A *joule* (J) is defined to be a kilogram·meter²/second² (kg·m²/s²). A *watt* (W) is a joule/second (J/s) and is the basic SI unit for rates of energy exchange or transformation. These SI units are seeing ever-increasing use. An equivalency that permits interconversion of SI units and units in the calorie system is

$$1 \text{ cal} = 4.186 \text{ J}$$

To illustrate, as you sit quietly reading this page, your metabolic rate is likely to be near 23 cal/s, equivalent to $23 \times 4.186 = 96$ J/s, or 96 W. That is, you are producing heat about as rapidly as a 100-W light bulb.

SPECIFIC DYNAMIC ACTION OF FOOD

Under many circumstances, after a previously fasting animal consumes food, its metabolic rate increases *even though, in all respects other than the ingestion of food, conditions are kept constant.* This increase in metabolic rate, diagrammed schematically in Figure 3.3, is known as the *specific dynamic action (SDA)* or *calorigenic effect of food.* The *magnitude* of the SDA is defined to be the total *excess* metabolic heat production induced by the meal, integrated from the time metabolism first rises to the time that it falls back to the "background" level (Figure 3.3). To illustrate, suppose that a resting, fasting person has a metabolic rate of 1400 cal/min. If the person consumes some protein, his or her metabolic rate will rise within the first hour after the meal and then remain above 1400 cal/min for several hours. Suppose that it takes 5 h for the metabolic rate to return to 1400 cal/min, and suppose that the total metabolic heat production over the 5 h is 500,000 cal. If the person had not eaten the meal, his or her total heat production over 5 h would have been only $1400 \times 60 \times 5 = 420,000$ cal. Accordingly, the total *excess* heat production induced by the meal, the SDA, is $500,000 - 420,000 = 80,000$ cal.

The SDA tends to be equivalent to a certain *percentage* of the total physiological caloric value of the food ingested. For example, the SDA following a large protein meal tends to be proportionally greater than that following a small protein meal. Proteins not uncommonly exhibit SDA values near 25–30 percent (i.e., a protein meal with a caloric value of 100,000 cal will induce a SDA of 25,000–30,000 cal). Lipids and carbohydrates usually display lower SDA values. From this perspective, we see that one of the implications of the SDA is that a certain proportion

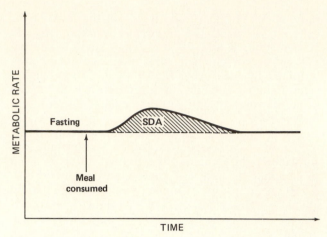

Figure 3.3 A schematic view of specific dynamic action. A resting, fasting animal is fed at the time marked and then remains at rest. The solid line shows the animal's actual metabolic rate. The dashed line depicts what the rate would have been, had the meal not been consumed. The hatched area is the magnitude of the SDA. There are circumstances in which the rise of metabolism attributable to SDA is obscured by other metabolic phenomena.

of the available energy from a meal is obligatorily dissipated as heat in the process of assimilating the meal; only the remaining proportion of the energy remains available for subsequent physiological uses.

The explanation for SDA remains unclear. The digestive process makes a contribution. However, the SDA often appears to arise principally in tissues such as the liver that receive nutrient molecules after assimilation from the digestive tract. One current hypothesis is that high levels of nutrients in tissues after a meal stimulate certain energy-demanding cycles of biochemical interconversion. Another long-standing idea is that the especially high SDA values of proteins may arise from processes involved in the disposition of unneeded protein nitrogen.

An animal is said to be *fasting* or *postabsorptive* when it has not eaten for a time sufficient to eliminate any SDA.

STANDARDIZED MEASURES OF METABOLIC RATE

The metabolic rate of an animal can be influenced by an immense variety of factors. Among these, the two that typically exert the greatest metabolic effects are the environmental temperature and the animal's level of physical (e.g., locomotory) activity—discussed in detail in Chapters 6 and 5, respectively. Other factors that can influence an organism's metabolic rate include age, sex, reproductive condition, hormonal balance, psychological stress, disease states, time of day, race, oxygen availability, and—as discussed in the previous section—ingestion of food.

For many purposes, standardized measures of

metabolic rate are useful. The goal in defining such measures is to standardize the influences of some of the major factors affecting metabolism. With these influences standardized, effects of other factors can be elucidated, and metabolic rates of species or other phyletic groups can be compared in a meaningful way.

Some animals (termed homeotherms, e.g., mammals and birds) regulate their body temperature physiologically. The effect of environmental temperature on these animals is such that, for each species, there is a range of environmental temperatures within which the metabolic rate is minimal. This range is called the *thermoneutral zone* (p. 100). The most common standardized measure of metabolism for these animals is called the **basal metabolic rate (BMR)**. The three conditions stipulated for measurement of a BMR are that the animal must be (1) in its thermoneutral zone, (2) resting, and (3) fasting. Thus, the effects of three of the more important factors affecting metabolic rate—environmental temperature, activity, and food ingestion—are standardized.

Other animals (termed poikilotherms, e.g., frogs and crayfish) allow their body temperature to fluctuate freely in concert with variations in environmental temperature. When such an animal is (1) resting and (2) fasting, its metabolic rate is termed its **standard metabolic rate (SMR)** for its prevailing body temperature.

It is hardly surprising that both of the standardized measures of metabolic rate we have discussed call for a resting condition. The level of activity is such a potent influence on metabolism that it must be specified, and, of all the levels of activity, the only one that is easily defined in reasonably quantitative terms for all species is rest. Unfortunately, however easy the experimenter may find it to rest, it is often no small challenge to get an experimental animal to rest completely. *Rest* thus has somewhat different meanings in different studies, and increasingly there has been an attempt at formal recognition of different gradations of the "resting condition." Some workers, for instance, apply the term *routine metabolic rate* to reasonably quiet animals exhibiting only relatively small, spontaneous movements and reserve *standard metabolic rate* for animals that have been coaxed to a truly minimal activity level.

When activity is truly minimal under standard or basal conditions, the metabolic rate approximates the rate necessary for simple physiological maintenance of life.

METABOLISM–WEIGHT RELATIONS

Figure 3.4*A* shows the general relation between basal metabolic rate (BMR) and body weight in species of placental mammals and illustrates a most significant characteristic: Although metabolic rate increases steadily with weight, it does not increase in propor-

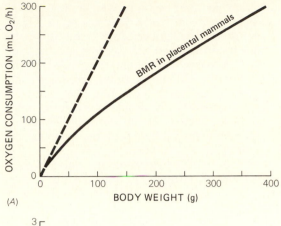

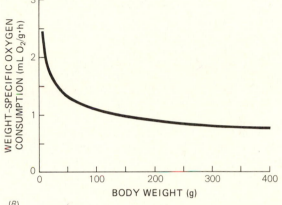

Figure 3.4 The general relation between basal metabolic rate and body weight in placental mammals. (*A*) Solid curve shows actual relation between rate of oxygen consumption and weight; dashed line shows how oxygen consumption would vary with weight if all mammals exhibited the same proportional relation between metabolism and weight as is seen in 10-g mammals. (*B*) Weight-specific rate of oxygen consumption as a function of weight. (Curves are based on equation given in the legend of Figure 3.5)

tion to weight. An average 10-g mammal, for example, exhibits a BMR near 20 mL O_2/h.* The dashed line in the figure shows how metabolic rate would increase with weight if larger mammals retained the same proportional relation between BMR and weight as the 10-g one. Under such circumstances, a 100-g mammal would have a BMR near 200 mL O_2/h. In actuality, however, the average BMR of a 100-g mammal is nearer 110 mL O_2/h. This same trend persists throughout the entire range of mammalian weights. Thus, the metabolic rate of a 400-g mammal is only about 2.7 times higher than that of a 100-g one, not four times higher.

Another way to see this relation is to examine the

*By convention, when expressing metabolic rates in terms of oxygen consumption, volumes of oxygen are expressed at standard conditions of temperature and pressure (STP): 0°C, 760 mm Hg.

metabolic rate per unit of body weight, termed the ***weight-specific metabolic rate***. The curve in Figure 3.4*A* is replotted in weight-specific terms in Figure 3.4*B*, where it can be seen that weight-specific metabolic rate decreases as weight increases. Under basal conditions a 3700-kg elephant produces only about 40 percent as much metabolic heat per gram as a 60-kg human and only about 5 percent as much as a 20-g mouse. It costs far less for an elephant to sustain a gram of elephant than for a mouse to sustain a gram of mouse.

Similar relations between metabolic rate and body weight are found in virtually all animal groups. Typically, within any given phyletic group, basal or standard metabolic rate is related to weight according to the following equation, which is termed an *allometric* equation:

$$M = aW^b$$

where *M* is the total metabolic rate, *W* is weight, and *a* and *b* are constants. If *b* were equal to 1.0, this equation would reduce to $M = aW$, indicating a proportional relation between metabolism and weight. However, *b* is usually less than 1.0 (commonly 0.6–0.9), meaning loosely that as *W* increases, *M* does not increase "as fast." Dividing both sides of the preceding equation by *W*, we get

$$\frac{M}{W} = aW^{(b-1)}$$

The expression on the left, *M/W*, is the weight-specific metabolic rate. Because *b* is usually between 0.6 and 0.9, the exponent $(b - 1)$ is usually between -0.4 and -0.1. The negative value of $(b - 1)$ signifies what we have already said, namely, that weight-specific metabolic rate decreases with increasing body weight.

The coefficients *a* and *b* depend on the animal group under consideration. Furthermore, in those organisms (the poikilotherms) that allow their body temperature to vary with environmental conditions, *a* in particular depends also on body temperature. Four examples of metabolism–weight equations are presented in Figure 3.5; note that besides being plotted, the equations themselves are given in the legend. The exponent *b* tends not to vary a great deal from one animal group to another; this consistency is exemplified by the four groups of vertebrates in the figure, which exhibit values of *b* ranging only from 0.65 to 0.73. Values of *a*, by contrast, vary widely among animal groups. In essence, *a* reflects the absolute level of metabolism; it is, in fact, equal to the theoretical metabolic rate of a 1-g animal. Of the groups shown in the figure, the birds have the highest metabolic rates for given weight and thus also have the highest value for *a* (7.54). The amphibians have the lowest metabolic rates and lowest value for *a* (0.4). The figure illustrates a fact that looms large in vertebrate energetics: Birds and mammals exhibit far

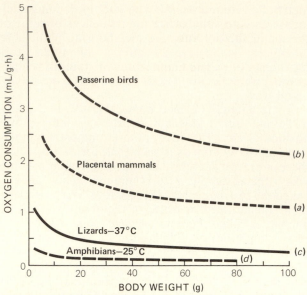

Figure 3.5 Weight-specific basal or standard metabolic rate as a function of body weight in four groups of vertebrates according to the following equations, where *M/W* is weight-specific metabolic rate expressed in mL O_2/g·h and *W* is body weight expressed in grams: (*a*) Basal metabolic rate in placental mammals: $M/W = 3.8W^{-0.27}$. (*b*) Basal metabolic rate in passerine birds: $M/W = 7.54W^{-0.28}$. (*c*) Standard metabolic rate in lizards at a body temperature of 37°C: $M/W = 1.33W^{-0.35}$. (*d*) Standard metaboic rate in temperate-zone amphibians (anurans and salamanders) at a body temperature of 25°C: $M/W = 0.355W^{-0.33}$. [Sources for equations—mammals: S. Brody, *Bioenergetics and Growth*. Reinhold, New York, 1945; birds: R.C. Lasiewski and W.R. Dawson, *Condor* **69**:13–23 (1967); lizards: J.R. Templeton, Reptiles. In G.C. Whittow (ed.), *Comparative Physiology of Thermoregulation*, Vol. I. Academic, New York, 1970; amphibians: W.G. Whitford, *Am. Zool.* **13**:505–512 (1973).]

higher resting metabolic rates than reptiles or amphibians (or fish) of equivalent body size. Among other things, this helps explain why the birds and mammals have much greater demands for food than the other groups.

If we take the logarithm of both sides of the equation $M = aW^b$, we get

$$\log M = \log a + b \log W$$

This indicates that a plot of log *M* against log *W* will be linear, for if we set $Y = \log M$ and $X = \log W$, we get $Y = \log a + bX$, a linear equation (*b* and log *a* are constants). Similarly, a plot of log(*M/W*) against log *W* will be linear. Data relating metabolic rate to weight are nearly always graphed on a log–log plot, where these linear relations pertain. Figure 3.6 presents two examples.

The values of *a* and *b* for an animal group are determined statistically and thus are subject to revision as new data accumulate. If the metabolic rate of a species is calculated from the equation $M = aW^b$, the result is a statistical prediction. Actual metabolic rates sometimes differ significantly from such predictions, and the differences themselves can be of considerable interest.

The Significance and Implications of Metabolism–Weight Relations

The tendency for weight-specific metabolic rate to decrease as body weight increases within a phyletic group has numerous practical implications. Small animals, for example, require food at a greater rate per unit of their body weight than larger, related animals. Some small shrews, to illustrate, require an amount of food (wet weight) equivalent to their body weight each day, but humans certainly do not demand 100–200 lb of food per day. The uninitiated might expect that the basal energy requirements of 3500 mice, each weighing 20 g (total weight: 70,000 g), would place no greater demands on a woodland ecosystem than a single 70,000-g deer. Because the weight-specific basal metabolic rate of a 20-g mouse is about eight times greater than that of a deer, however, the total basal metabolism of only about 440 mice would be equivalent to that of a deer. In these days of widespread pollution, the relation between metabolism and weight also has potential implications for the effects of environmental toxins. Because small animals must eat food and consume oxygen at greater rates per unit weight than large animals, they incur greater weight-specific doses of food-borne and airborne toxins per unit time.

The relation between metabolic rate and weight is often paralleled by similar relations between other physiological characteristics and weight; indeed, the metabolism–weight relation is just the tip of a physiological iceberg. Because small air-breathing animals, for example, require oxygen at greater rates per unit weight than larger, related animals, they typically need to breathe air at a greater rate per unit weight; this helps explain why small mammals inhale and exhale at greater frequencies than related large ones (p. 281). The circulatory system must transport oxygen at greater rates per unit weight in small animals; thus, the heart rates of small mammals are greater than those of their larger relatives (p. 345). Recent studies have extended knowledge of such relations all the way to the molecular level. In fish, the activity per unit weight of citrate synthase, an enzyme pivotal to aerobic catabolism, decreases with increasing body size.

Surface-to-Volume Ratios and the "Surface Law"

The surface area s of a sphere is proportional to the square of the sphere's radius r: $s \propto r^2$. The volume v, however, is proportional to the cube of the radius: $v \propto r^3$. From these relations we can write

$$s \propto (r^3)^{2/3}$$
$$s \propto v^{2/3}$$

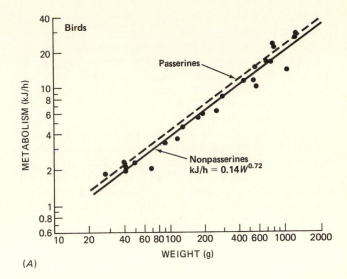

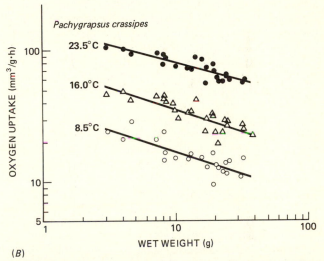

Figure 3.6 Relations between metabolic rate and body weight in two groups of animals plotted on double logarithmic coordinates. (A) Total basal metabolic rate as a function of body weight in birds. The points are measured values for 24 nonpasserine species; the solid line is statistically fitted to the points and conforms to the equation shown. The dashed line is the equation for passerines given in Figure 3.5. The data indicate that, contrary to long-standing belief, there is probably little or no difference between passerines and nonpasserines. (B) Weight-specific metabolic rate as a function of body weight at three different body temperatures in the common Pacific shore crab *Pachygrapsus crassipes*. [(A) after R. Prinzinger and I. Hänssler, *Experientia* **36**:1299–1300 (1980); (B) from J.L. Roberts, *Physiol. Zool.* **30**:232–242 (1957), used by permission of The University of Chicago Press, (B) © 1957 by The University of Chicago.]

In words, as spheres increase in volume, their surface area increases only as the 2/3-power of volume, signifying that the increase in area is less than fully proportional to that in volume. Thus, big spheres have less surface area per unit of volume (or weight) than little spheres.

The conclusion thus reached for spheres of various sizes applies equally to other sets of geometrically similar objects, whether they be cubes, cylinders, blood vessels, stomachs, hearts, lungs, or whole animals. As objects of a particular shape get larger, their surface area increases approximately as the 2/3-power of their volume and their surface-to-volume ratio falls.

Early students of metabolism–weight relations worked on mammals and developed a postulated explanation of those relations that is often referred to as the "surface law." The "law" stated that the basal metabolic rate of mammals is proportional to their body surface area. The quantitative accuracy of this statement has itself been subject to hot debate, but here our concern is with the reasoning behind it. That reasoning went as follows: (1) Mammals maintain high, relatively constant body temperatures (near 37°C) and thus tend to lose heat to their environment when living at thermoneutral environmental temperatures. (2) Because heat is lost across the outer body surfaces, the rate of heat loss is approximately proportional to body-surface area. (3) Small mammals have more surface area per unit weight than large mammals and thus lose heat more rapidly per unit weight. (4) Heat lost must be replaced metabolically. Thus, small mammals must produce heat at a greater rate per unit weight.

Is there truth in this reasoning? Unequivocally. However, we must consider a more exacting question. Does the reasoning provide the fundamental explanation of why weight-specific basal metabolic rate decreases with increasing weight? And here the answer is "probably not." We now recognize that inverse relations between weight-specific metabolic rate and weight are virtually universal among animal phyletic groups. In particular, such relations are seen in animals such as crabs, fish, and frogs that do *not* keep their bodies warm by virtue of metabolic heat production but instead allow their body temperatures to vary with ambient conditions (see Figs. 3.5 and 3.6). Clearly, the reasoning behind the "surface law" does not apply to such animals. Given that the physics of physiological thermoregulation cannot explain metabolism–weight relations in a majority of phyletic groups, we must seriously question whether it provides the fundamental explanation even in mammals. As yet, there are no definitively accepted mechanistic explanations of the metabolism–weight relations, despite many proposals.

THE ENERGY VALUES OF FOODSTUFFS

The amount of chemical energy that can be obtained by oxidation of foodstuffs is usually measured by burning the foodstuffs and measuring the heat of combustion evolved. Typically, the burning is carried out rapidly in an atmosphere of pure oxygen, in a device called a **bomb calorimeter.**

An important consideration is that if the amount of energy released by burning is to correspond to the amount liberated in animal catabolism, the end products of burning and catabolism must be the same. When the foodstuffs are carbohydrates and lipids, energy values from bomb calorimetry can be applied straightforwardly to *aerobic* catabolism, since both catabolism and burning yield the same fully oxidized products, CO_2 and H_2O. Proteins present a different picture because burning and catabolism do not equivalently oxidize the nitrogen atoms in proteins; the nitrogenous end products of burning are nitrogen oxides, whereas those of catabolism are more-reduced compounds. Not only must a correction be applied to the heat of combustion for protein, but the correction varies with the animal group because different groups dispose of nitrogen in different chemical forms (Chapter 9). Protein that yields 4.85 kcal/g in catabolism when ammonia is the nitrogenous end product will yield less energy, about 4.77 kcal/g, if urea is the product and still less, 4.34 kcal/g, if uric acid is. The ratio of heat production to oxygen consumption in protein catabolism and the RQ of protein also depend on the particular nitrogenous end products produced.

Table 3.1 (p. 28) gives the energy values of mixed carbohydrates, lipids, and proteins in mammalian catabolism. Note that, gram for gram, lipids yield about twice as much energy as the other foodstuffs. This explains why animals so commonly employ fats and oils as energy stores.

POSTSCRIPT

The reader who has arrived at this point will be interested to know that concerted mental effort is not very expensive energetically. Although the activity of our brains routinely accounts for a substantial portion—about 20 percent—of our total resting metabolism, the prominent physiologist Francis Benedict (1870–1957) found that the *increase* in energy consumption caused by an hour of hard study is slight: equivalent to the energy in a single oyster cracker or half a peanut.

SELECTED READINGS

Adkins, C.J. 1987. *An Introduction to Thermal Physics.* Cambridge University Press, New York.

Bertalanffy, L. Von. 1957. Quantitative laws in metabolism and growth. *Q. Rev. Biol.* **32**:217–231.

Brody. S. 1945. *Bioenergetics and Growth.* Reinhold, New York.

*Brown, A.C. and G. Brengelmann. 1965. Energy metabolism. In T.C. Ruch and H.D. Patton (eds.), *Physiology and Biophysics,* 19th ed. Saunders, Philadelphia.

Calder, W.A., III. 1981. Scaling of physiological processes in homeothermic animals. *Annu. Rev. Physiol.* **43**:301–322.

*Calder, W.A., III. 1984. *Size, Function, and Life History.* Harvard University Press, Cambridge, MA.

Costa, D.P. and G.L. Kooyman. 1984. Contribution of

specific dynamic action to heat balance and thermoregulation in the sea otter *Enhydra lutris*. *Physiol. Zool.* **57**:199–203.

Dyson, F.J. 1954. What is heat? *Sci. Am.* **191**(3):58–63.

Heglund, N.C. and G.A. Cavagna. 1985. Efficiency of vertebrate locomotory muscles. *J. Exp. Biol.* **115**:283–292.

Heller, H.C., R. Elsner, and N. Rao. 1987. Voluntary hypometabolism in an Indian yogi. *J. Thermal Biol.* **12**:171–173.

Heusner, A.A. 1985. Body size and energy metabolism. *Annu. Rev. Nutr.* **5**:267–293.

King, J.R. and M.E. Murphy. 1985. Periods of nutritional stress in the annual cycles of endotherms: Fact or fiction? *Am. Zool.* **25**:955–964.

*Kleiber, M. 1975. *The Fire of Life,* 2nd ed. Krieger, Huntington, NY..

Lasiewski, R. and W.R. Dawson. 1967. A re-examination of the relation between standard metabolic rate and body weight in birds. *Condor* **69**:13–23.

Lindstedt, S.L. and W.A. Calder III. 1981. Body size, physiological time, and longevity in homeothermic animals. *Q. Rev. Biol.* **56**:1–16.

Lomax, E. 1979. Historical development of concepts of thermoregulation. In P. Lomax and E. Schönbaum (eds.), *Body Temperature*, pp. 1–23. Dekker, New York.

*Mendelsohn, E. 1964. *Heat and Life. The Development of the Theory of Animal Heat*. Harvard University Press, Cambridge.

*Nagy, K.A. 1987. Field metabolic rate and food requirement scaling in mammals and birds. *Ecol. Monogr.* **57**:111–128.

Peters, R.H. 1983. *The Ecological Implications of Body Size*. Cambridge University Press, New York.

Prinzinger, R. and I. Hänssler. 1980. Metabolism–weight relationship in some small nonpasserine birds. *Experientia* **36**:1299–1300.

Ramsay, J.A. 1971. *A Guide to Thermodynamics*. Chapman and Hall, London.

*Schmidt-Nielsen, K. 1984. *Scaling: Why Is Animal Size So Important?* Cambridge University Press, New York.

Spanner, D.C. 1964. *Introduction to Thermodynamics*. Academic, New York.

Vleck, C.M. and D. Vleck. 1987. Metabolism and energetics of avian embryos. *J. Exp. Zool. Suppl.* **1**:111–125.

Webster, A.J.F. 1983. Energetics of maintenance and growth. In L. Girardier and M.J. Stock (eds.), *Mammalian Thermogenesis*. Chapman and Hall, London.

*Wieser, W. 1984. A distinction must be made between the ontogeny and the phylogeny of metabolism in order to understand the mass exponent of energy metabolism. *Respir. Physiol.* **55**:1–9.

Zeuthen, E. 1953. Oxygen uptake as related to body size in organisms. *Q. Rev. Biol.* **28**:1–12.

See also references in Appendix A.

chapter 4

Aerobic and Anaerobic Forms of Metabolism

In this chapter, we first examine certain of the biochemical pathways by which energy is released from foodstuffs and made available for performance of physiological work. Some of these pathways, termed **aerobic**, require oxygen; others, termed **anaerobic**, can function without oxygen. After examining mechanistic aspects of the pathways, we look at the interplay between aerobic and anaerobic modes of energy release in animals.

BASIC MECHANISTIC CONCEPTS

The energy-demanding processes in cells generally draw their energy from the high-energy compound adenosine triphosphate (ATP). Cleavage of ATP releases chemical energy, which can be used in activating a muscle, constructing a protein, transporting ions across a membrane, or any of many other energy-demanding events.

Animals receive the energy they require from the chemical bonds of ingested foodstuffs. Energy-demanding processes, however, cannot draw the energy they need directly from foodstuff bonds. Accordingly, cells possess intricate biochemical machinery to transfer energy from the bonds of foodstuffs to the bonds of ATP. The immediate chemical precursors of ATP (a triphosphate) are adenosine diphosphate (ADP) and inorganic phosphate. We see that the following reaction is crucial as an *energy shuttle* mechanism in cells:

$$\text{ADP} + \text{phosphate} + \text{energy} \rightleftharpoons \text{ATP}$$

Energy derived from foodstuff molecules is used to drive the reaction to the right, with some of the energy from the foodstuff bonds becoming stored in the bonds of the ATP produced. When the ATP is cleaved, reforming ADP, the energy is released for use in physiological work.

ATP is not transported from cell to cell in the body. Instead, each cell must possess a biochemical apparatus for making its own. A cell has several potential sources of foodstuff molecules for synthesis of ATP: (1) The cell may use molecules that have been freshly derived from digestion of foods and transported from the digestive tract by the bloodstream. (2) It may use foodstuff molecules that it previously stored. (3) Finally, it may use foodstuff molecules derived from stores in other cells. Certain cells are specialized for storage of foodstuffs; adipose-tissue cells, for instance, store lipids in abundance, and vertebrate liver cells store large quantities of glycogen. As appropriate, these cells release foodstuff molecules, such as fatty acids (from lipids) and glucose (from glycogen), into the bloodstream for transport to, and use by, other cells. The endocrine control of the release, uptake, and interconversion of foodstuff molecules is discussed in Chapter 21 (p. 613).

Biochemistry of Aerobic Catabolism

In the vast majority of animals, each cell possesses biochemical mechanisms that, using oxygen, completely oxidize foodstuff molecules to CO_2 and H_2O, capturing in bonds of ATP much of the chemical energy thus released. These **aerobic catabolic pathways** often can directly or indirectly oxidize all the major classes of foodstuffs. In outlining the pathways, we emphasize here just the catabolism of carbohydrates. Furthermore, our purpose is to provide an overview, not duplicate detailed treatments available in texts of cellular physiology or biochemistry.

The principal pathway of aerobic catabolism can be subdivided into four major sets of reactions: glycolysis, the Krebs citric acid cycle, the electron-transport chain, and oxidative phosphorylation.

Glycolysis *Glycolysis* is the name applied to the series of enzymatically catalyzed reactions shown in Figure 4.1*A,* by which glucose (or glycogen) is converted to pyruvic acid. The first step is that glucose is phosphorylated at the *cost* of an ATP molecule to form glucose-6-phosphate. Glucose-6-phosphate is then converted to fructose-6-phosphate, and the latter is phosphorylated to form fructose-1,6-diphosphate. This phosphorylation also is accomplished at the *cost* of an ATP molecule, so that at this point two ATP molecules have been invested in the reactions. Fructose-1,6-diphosphate is cleaved to form two three-carbon molecules, dihydroxyacetone phosphate and glyceraldehyde-3-phosphate. These compounds are interconvertible, and when glucose is being catabolized for release of energy, the former is converted to the latter, yielding two molecules of glyceraldehyde-3-phosphate. The reactions subsequent to glyceraldehyde-3-phosphate in Figure 4.1*A* are all multiplied by 2 to emphasize that two molecules follow these pathways for each glucose molecule that enters the system.

Glyceraldehyde-3-phosphate is next oxidized, with the addition of inorganic phosphate, to the three-carbon diphosphate 1,3-diphosphoglyceric acid. *This reaction is the only oxidative reaction in the entire glycolytic pathway* and will prove most significant in our subsequent discussion of anaerobic metabolism. The reaction does not in itself require oxygen; rather, it involves the concomitant reduction of one molecule of the enzyme cofactor, nicotinamide adenine dinucleotide (NAD), per molecule of glyceraldehyde-3-phosphate. The reduction of NAD is symbolized as NAD → $NADH_2$, and the fate of the $NADH_2$ will be discussed subsequently. The 1,3-diphosphoglyceric acid formed in the oxidation reaction is next converted to the monophosphate 3-phosphoglyceric acid, with the *formation* of one ATP per molecule. The 3-phosphoglyceric acid is converted in two steps to phosphoenolpyruvic acid, and the latter reacts to form pyruvic acid, again with the *formation* of one ATP per molecule. Three important consequences of glycolysis deserve note. First, each molecule of glucose is converted to two molecules of pyruvic acid. Second, two molecules of NAD are reduced to $NADH_2$ per molecule of glucose catabolized. Third, two molecules of ATP are used and four are formed for each glucose processed, yielding a net increase of two ATP molecules per glucose.

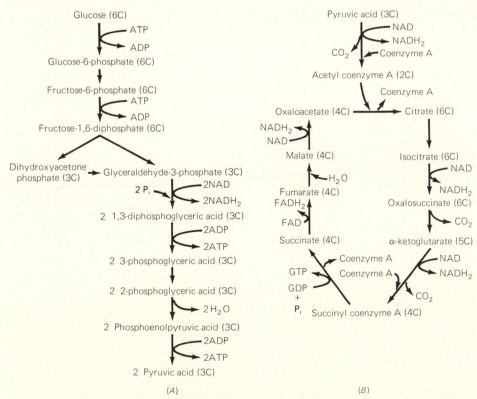

(A)   (B)

Figure 4.1 (*A*) Summary of the major reactions of glycolysis. (*B*) Summary of the major reactions of the Krebs citric acid cycle. The number of carbon atoms in each compound is indicated in parentheses. Some intermediate steps have been omitted for simplicity. P_i = inorganic phosphate. The expression for reduction of NAD, NAD → $NADH_2$, is shorthand; the actual reaction is $NAD^+ + 2H → NADH + H^+$.

The Krebs Citric Acid Cycle   During aerobic catabolism, the pyruvic acid formed in glycolysis is oxidized in the mitochondria by a cyclic series of enzymatically catalyzed reactions called the *Krebs citric acid cycle,* or *tricarboxylic acid cycle*. The cycle, named after Hans Krebs who in 1937 was the first to envision its features, is diagrammed in Figure 4.1*B*. For simplicity the reactions involved in the oxidation of just one molecule of pyruvic acid are shown, but remember that two molecules are processed for each molecule of glucose.

Pyruvic acid enters the Krebs cycle through a complex set of reactions in which it is oxidatively decarboxylated, forming CO_2 and a two-carbon acetyl group that emerges in combination with coenzyme A as acetyl coenzyme A. In the process, a molecule of NAD is reduced. Acetyl coenzyme A then reacts with oxaloacetate, the end result being that coenzyme A is released and the acetyl group is condensed with oxaloacetate (four-carbon) to form citrate (six-carbon). In the ensuing series of reactions, oxaloacetate is ultimately regenerated and then again can combine with acetyl coenzyme A. The reactions need not be reviewed stepwise here, though the following three points deserve emphasis. (1) Decarboxylations occur at two points, in the conversion of oxalosuccinate to α-ketoglutarate and in the conversion of α-ketoglutarate to succinyl coenzyme A. These two decarboxylations plus the one in the reaction of pyruvic acid to form acetyl coenzyme A account for the formation of three molecules of CO_2 for every molecule of pyruvic acid processed; in effect the six carbons of each glucose molecule thus emerge as six molecules of CO_2. (2) Oxidations occur at four points. At three of these NAD is reduced, whereas at one (the oxidation of succinate to form fumarate) another coenzyme, flavin adenine dinucleotide (FAD), is reduced. If we recall the formation of one $NADH_2$ in the reaction of pyruvic acid to form acetyl coenzyme A, we see that the processing of each pyruvic acid molecule results in formation of four $NADH_2$ and one $FADH_2$. (3) Finally, note that the reaction of succinyl coenzyme A to form succinate is accompanied by the formation of guanosine triphosphate (GTP) from guanosine diphosphate (GDP). The GTP is a high-energy compound, and it subsequently donates its terminal phosphate group to ADP, resulting in GDP and ATP. In essence, a molecule of ATP is generated.

Electron Transport, Oxidative Phosphorylation, and the Role of Oxygen   Thus far, it may appear paradoxical that in a discussion of aerobic catabolism we have not mentioned the involvement of molecular oxygen. Molecular oxygen is not a participant in any of the reactions of glycolysis or the Krebs cycle, and in a very narrow sense all the reactions can proceed without it. Oxygen nonetheless is essential; the reason for this lies in the disposition of the reduced cofactors, $NADH_2$ and $FADH_2$.

Oxidation reactions involve the removal of electrons, and when electrons are removed from one compound, they must be transferred to another. In the several oxidation reactions of glycolysis and the Krebs cycle, the immediate electron acceptors are the enzyme cofactors NAD and FAD. These cofactors, however, can never serve as the ultimate electron acceptors, for they are present in only limited quantities. If $NADH_2$ and $FADH_2$ were simply allowed to accumulate, the cell would soon run out of NAD and FAD, and the reactions of glycolysis and the Krebs cycle would come to a halt for lack of these vital, immediate electron acceptors. $NADH_2$ and $FADH_2$ are oxidized back to NAD and FAD in reactions with the electron-transport chain.

Located in the mitochondria, the *electron-transport chain (respiratory chain)* consists of a series of compounds (proteins or associated with proteins), each capable of undergoing reversible oxidation and reduction, which together serve to take electrons from $NADH_2$ and $FADH_2$ and pass them to oxygen. Oxygen acts as the *final electron acceptor*. As diagrammed schematically in Figure 4.2, the members of the chain function in a discrete order. Electrons removed from $NADH_2$ and $FADH_2$ are passed sequentially from one member of the chain to the next, in a series of reductions and oxidations, ultimately arriving at oxygen—along with "accompanying" protons (H^+ ions)—thereby reducing the oxygen to water. The special role played by oxygen is noteworthy. The cytochromes and other members of the electron-transport chain, like NAD and FAD, are present in limited quantities in the cell and therefore cannot act as terminal electron acceptors. Oxygen, on the other hand, is continuously supplied to the cell, and the product of its reduction, water, can be dissipated to the environment, thereby carrying electrons out of the cell.

The electron-transport chain, far from serving simply as a route for the oxidation of $NADH_2$ and $FADH_2$, is also pivotally involved in the transfer of energy from the bonds of foodstuffs to ATP. Molecular oxygen has a much higher affinity for electrons than the enzyme cofactors NAD and FAD, and there is a large decline in free energy as the electrons originally taken from nutrient molecules are passed through the electron-transport chain. A considerable portion of this energy is captured in bonds of ATP. The process of forming ATP from ADP using energy released in electron transport is called *respiratory-chain phosphorylation* or *oxidative phosphorylation*. As depicted in Figure 4.2, for every pair of electrons passed through the chain from $NADH_2$ to oxygen, three molecules of ATP are formed, in three distinct parts of the chain.

A common mode of expressing the yield of ATP in oxidative phosphorylation is as a *P/O ratio* (P = phosphate, O = oxygen), defined to be the number of ATP molecules formed per *atom* of oxygen reduced. For example, if a pair of electrons passes from

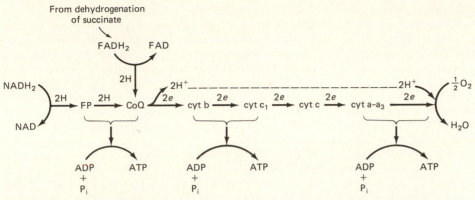

Figure 4.2 An abbreviated, conceptual summary of the electron-transport chain and oxidative phosphorylation. There is a large decline in free energy as electrons flow from $NADH_2$ to H_2O; the chain is believed to have evolved as a device for breaking up the total decline into a series of smaller energy drops, three of which are of sufficient size to support phosphorylation of ADP. As can be seen, formation of ATP by oxidative phosphorylation occurs at three locations along the chain. The biochemical mechanisms of harnessing energy released in electron transport to produce ATP (coupling mechanisms) are incompletely known; according to the chemiosmotic theory favored by most biochemists, electron transport pumps H^+ ions across mitochondrial membranes, and the subsequent back-diffusion of these ions occurs through an enzyme, ATP synthetase, which converts the energy inherent in the H^+ gradient into ATP bond energy. cyt = cytochrome; e = electron; P_i = inorganic phosphate; CoQ = coenzyme Q (also called ubiquinone); FP = the flavoprotein NADH dehydrogenase. Succinate dehydrogenase is also a flavoprotein, containing FAD as its flavin cofactor.

$NADH_2$ to oxygen (reducing the oxygen atom) and three ATP molecules are formed, the P/O ratio is 3.

Although we have been implying that the processes of electron transport and oxidative phosphorylation are inextricably linked, this is actually not the case. In fact, at an opposite extreme, it is possible for electrons to pass through the electron-transport chain without engendering the formation of any ATP at all. When this happens, all the energy liberated in the transport of the electrons appears immediately as heat.

Linkage of electron transport with oxidative phosphorylation is called **coupling.** When the two processes are *fully* ("tightly") *coupled,* the P/O ratio for electrons originating from $NADH_2$ is 3 and that for electrons from $FADH_2$ is 2 (see Figure 4.2). If the processes are completely *uncoupled,* the P/O ratio in either case is zero (0). Intermediate states of coupling are possible.

When coupling is tight, the rate of electron transport—and therefore the rate of aerobic catabolism as a whole—is controlled by availability of ADP. If ADP is unavailable and oxidative phosphorylation therefore cannot take place, electron transport must halt. Accordingly, if a cell has low demands for ATP and most of its ADP gets converted to ATP, the aerobic catabolic pathways responsible for producing ATP are slowed. By contrast, if the cell has high demands for ATP and therefore rapidly converts available ATP to ADP, causing ADP levels to rise, the pathways accelerate. By these and other negative-feedback effects, the rate of catabolism of foodstuffs is exquisitely tuned to the cell's needs for ATP.

If electron transport becomes uncoupled from oxidative phosphorylation, control by ADP is lost, and electrons can travel apace through the electron-transport chain even if ATP is present in abundance. The energy of foodstuff molecules can thereby be released—as heat—even when there is no need for it for the performance of physiological work. Uncoupling can be induced (directly or indirectly) in certain tissues by a variety of agents, including hormones, mitochondrial proteins, fatty acids, and Ca^{2+} ions. As might be guessed, one of the functions played by uncoupling in mammals is to increase heat production in cold environments (p. 106).

The Yield of ATP Let us now look at the total yield of ATP in the aerobic catabolism of glucose (assuming tight coupling). For each molecule of glucose, two molecules of $NADH_2$ are produced in glycolysis. Furthermore, eight $NADH_2$ molecules and two $FADH_2$ molecules are formed per glucose in the Krebs cycle and other reactions subsequent to formation of pyruvic acid (p. 38). When the ATP molecules produced from all these $NADH_2$ and $FADH_2$ molecules are added—taking the P/O for $NADH_2$ to be 3 and that for $FADH_2$ to be 2—it turns out that 34 ATP molecules are made per glucose by oxidative phosphorylation.*

---

*The two $NADH_2$ molecules produced in glycolysis, not being formed within the mitochondria, are processed somewhat differently from $NADH_2$ molecules produced in the Krebs cycle. In some tissues of some animals, the P/O for these $NADH_2$ molecules is probably just 2, making the total yield of ATP by oxidative phosphorylation 32 ATP molecules per glucose.

In addition to these ATP molecules, others are generated in the reactions of 1,3-diphosphoglyceric acid to form 3-phosphoglyceric acid (Figure 4.1*A*), phosphoenolpyruvic acid to form pyruvic acid (Figure 4.1*A*), and succinyl coenzyme A to form succinate (Figure 4.1*B*). These phosphorylations, in contrast to the oxidative phosphorylations, are known as *substrate-level phosphorylations* because they occur immediately in the reactions of substrates of glycolysis and the Krebs cycle. In all, six ATP molecules are formed by substrate-level phosphorylation per molecule of glucose.

In total, we see, 40 molecules of ATP are generated for each glucose molecule catabolized. However, because two ATP molecules are consumed in the phosphorylation of glucose and fructose-6-phosphate (see Figure 4.1*A*), the *net* yield is 38 molecules of ATP.

The energetic efficiency (p. 23) of the aerobic catabolic pathways depends on cell metabolite concentrations and is not yet certain, but in glucose catabolism with tight coupling, it is likely near 70 percent. That is, about 70 percent of the chemical energy released in oxidizing glucose to CO_2 and H_2O is captured as chemical-bond energy in ATP molecules.

The Biochemical Challenges Posed by Oxygen Deficiency

Biochemically, what are the implications if cells are denied oxygen or supplied with oxygen at an inadequate rate? In exploring this question, let us for simplicity take the case of cells entirely deprived of oxygen, even though for many animals this would be an unrealistically extreme state.

Without oxygen, the electron-transport chain is transformed from a route for the ultimate dissipation of electrons to a dead end, for electrons cannot be discharged by reducing oxygen at the end of the chain. For most animals, the consequences are easily summarized. In an immediate sense, once oxygen is denied, electrons entering the chain quickly bring the cytochromes and all other members of the chain to a statically reduced state, thereby rendering the chain unable to accept further electrons. The broader consequences are twofold. First, oxidative phosphorylation—the source of 34 out of every 38 ATP molecules produced in net fashion by aerobic glucose catabolism—is brought to a halt. Second, the electron-transport chain can no longer serve as a mechanism for reoxidizing the reduced cofactors $NADH_2$ and $FADH_2$.

As emphasized earlier (p. 38), NAD and FAD are present in cells in only limited quantities. Thus, although they act as immediate electron acceptors in a number of reactions of glycolysis and the Krebs cycle, they are wholly unsuited to serve as final repositories for electrons. If we consider NAD, FAD, or any other electron acceptor available in only limited quantity, a cell is said to be in *reduction–oxidation*

balance (*redox balance*) for the acceptor if it possesses the means to remove electrons from the electron acceptor as fast as they are added to it. During aerobic catabolism, oxygen and the electron-transport chain provide the means of maintaining redox balance with respect to NAD and FAD. To make ATP without oxygen, organisms must possess other mechanisms that will permit redox balance to be maintained while at least some ATP-generating reactions are sustained.

Not all tissues have evolved an ability to make ATP anaerobically. The mammalian brain, for example, succumbs rapidly when denied oxygen because it lacks anaerobic catabolic pathways.

Anaerobic Glycolysis in Vertebrate Skeletal Muscle: An Example of an Anaerobic Catabolic Pathway

In vertebrates, skeletal muscle and many other tissues are able to use the substrate-level phosphorylations of the glycolytic pathway to transfer energy from the bonds of glucose to ATP in the absence of oxygen. In the evolution of this ability, a key requirement was the acquisition of a means to reoxidize the $NADH_2$ molecules that are produced in the oxidation of glyceraldehyde-3-phosphate (Figure 4.1*A*). As shown in Figure 4.3, these $NADH_2$ molecules are in fact reoxidized by passing their hydrogen atoms to pyruvic acid, reducing the latter to lactic acid. In essence, pyruvic acid is used (temporarily) as a terminal electron acceptor by the anaerobic cell. The entire sequence of reactions from glucose to lactic acid is called *anaerobic glycolysis.*

Importantly, the anaerobic glycolytic pathway is in redox balance. Because one molecule of pyruvic acid is produced for each $NADH_2$ generated in the oxidation of glyceraldehyde-3-phosphate, the supply

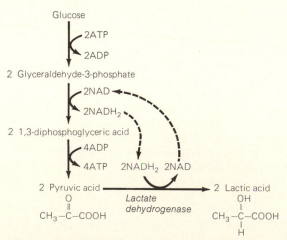

Figure 4.3 Principal features of the anaerobic glycolysis of glucose in vertebrate skeletal muscle. See Figure 4.1 for intermediate reactions of the glycolytic chain. The enzyme catalyzing reduction of pyruvic acid is lactate dehydrogenase.

of pyruvic acid keeps exact pace with the need for it as an electron acceptor.

Glucose and glycogen (a polymerized form of glucose) are the exclusive fuels of anaerobic glycolysis. For every molecule of glucose catabolized to lactic acid, a net yield of two ATP molecules is realized (p. 37); catabolism of glycogen yields three ATP molecules per glucose residue. These yields of ATP are vastly less than those from aerobic glucose catabolism, but as we shall see, the anaerobic pathway can nonetheless be a vitally important source of ATP in certain circumstances.

The major reason for the comparatively low yield of ATP per glucose molecule in anaerobic glycolysis is that only a small fraction (about 7 percent) of the free energy available from glucose is released in converting glucose to lactic acid. *Lactic acid is itself an energy-rich molecule.*

General Considerations in the Disposition of Catabolic End Products

Anaerobic glycolysis that yields lactic acid is by no means the only mechanism of anaerobic catabolism in animals, or even in vertebrates. We defer mention of other anaerobic pathways to future sections, however, and here take up the important matter of end-product disposition. *For an animal to make use of any catabolic pathway, it must ultimately possess means of satisfactorily disposing of the chemical end products generated.*

Insofar as living creatures are concerned, the principal products of *aerobic* catabolism, CO_2 and H_2O, are fully oxidized and not capable of being tapped for further energy. Thus, it is not surprising that animals fundamentally dispose of these end products by voiding them into the environment, although in some cases they transiently retain the products—particularly water—to perform chemical functions in the body (p. 154).

The end products of anaerobic catabolic pathways, in sharp contrast to CO_2 and H_2O, are always organic compounds that are far from fully oxidized and possess considerable further potential to yield energy. The high energy value of these products places a premium on retaining them in the body for future use as energy sources; excreting lactic acid, for instance, would be equivalent to voiding over 90 percent of the energy value of catabolized carbohydrate foods. On the other hand, unfettered retention is usually not possible either, because the end products typically exert direct or indirect harmful effects if allowed to accumulate to high concentrations. In the evolution of end-product disposition, these competing considerations seem to have been among the major operative factors. A third parameter of obvious importance is time. An animal that makes only short-term use of anaerobic pathways is more likely to be able to retain all the end-product molecules without adverse effect than one that continues reliance on anaerobic catabolism for days or weeks on end.

The Disposition of Lactic Acid in Vertebrates

When vertebrates resort to anaerobic glycolysis, they retain in their body the lactic acid produced and ultimately dissipate it metabolically. The major possible pathways of dissipation are depicted schematically in Figure 4.4. As shown in the figure, lactic acid itself is for the most part a metabolic cul-de-sac. That is, for lactic acid to undergo chemical utilization, it must first be converted back to pyruvic acid in a reversal of the very reaction that formed it in the first place. The conversion of lactic acid to pyruvic acid is an oxidative reaction, with NAD acting as the immediate electron acceptor. Once converted to pyruvic acid, the carbon chains of lactic acid can follow one of two major paths: oxidation in the Krebs cycle or reconversion to carbohydrate stores (glucose or glycogen). As we discuss these paths, it will be important to note that the dissipation of lactic acid *requires oxygen* regardless of the path followed.

After the carbon chains of lactic acid have been converted to pyruvic acid, one of their possible fates is full oxidation to CO_2 and H_2O via the Krebs cycle and electron-transport system. The process releases

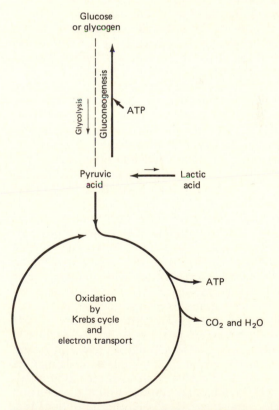

Figure 4.4 A brief biochemical map of major pathways of lactic-acid dissipation. It is also possible for an appreciable fraction of lactic-acid carbon chains to be incorporated into amino acids and proteins, by way of Krebs-cycle intermediates.

the energy available from the lactic acid and is capable of yielding 36 molecules of ATP per pair of lactic acid molecules catabolized. A functional electron-transport system is required to reoxidize and capture energy from all the $NADH_2$ and $FADH_2$ produced. Thus, molecular oxygen is essential.

The other possible major fate of the pyruvic acid formed from lactic acid is reaction to form glucose or glycogen. This conversion of lactic acid or pyruvic acid to carbohydrate is one form of *gluconeogenesis* (''new formation of glucose''). The principal chemical intermediates of the conversion are the same as those of glycolysis (Figure 4.1*A*) formed in reverse order. However, the gluconeogenic pathway involves some different enzymes than glycolysis and is not the stoichiometric opposite of glycolysis. Significantly, for instance, although the net yield of ATP in glycolysis is two molecules per glucose molecule catabolized, gluconeogenesis requires not two but six ATP molecules per glucose molecule formed. Gluconeogenesis cannot take place without oxygen because of its demand for ATP: The ATP must be made aerobically.

Just as tissues differ in their ability to carry out anaerobic glycolysis, they also differ in their capacity to utilize lactic acid for combustion or gluconeogenesis. Skeletal muscle itself, which is often one of the principal sources of lactic acid, has recently been discovered to be capable of not just lactic acid combustion but also extensive gluconeogenesis. Lactic acid made in skeletal muscle (or elsewhere) may also enter the blood and thus become available to tissues in general. The liver and kidneys may remove such lactic acid from the blood and are highly capable of carrying out gluconeogenesis with it. The liver, for example, uses some absorbed lactic acid to reconstitute its own glycogen depots and uses the rest to make glucose, which it releases into the blood for uptake by other tissues such as the skeletal muscles. The heart, lungs, and brain are among the tissues that can oxidize blood-borne lactic acid for production of ATP.

As we have stressed, the dissipation of lactic acid requires oxygen. Sometimes, the periods of production and dissipation of lactic acid are sequential: Tissues may go through a time of oxygen deprivation followed by one of oxygen availability and may make lactic acid in the first instance and dissipate it in the second. We have come to recognize in recent years, however, that it is also common for different parts of the body to be carrying out the synthesis and dissipation of lactic acid simultaneously. For example, even while certain skeletal-muscle cells may be turning to anaerobic glycolysis as a means of maximizing ATP production during strenuous exertion (see p. 48), other tissues—even other muscle cells—may be removing the lactic acid from the blood and using it as a fuel.

What fraction of lactic acid is oxidized, and what fraction is used in gluconeogenesis? The answer is not easy to come by and depends not only on the species but also on circumstances. Rats, for example, catabolize most lactic acid as fuel in the aftermath of exhausting exercise, but they also employ some of it for gluconeogenesis, and the fraction used in the latter way seems to increase with the deficit in carbohydrate stores.

Steady-State Versus Nonsteady-State Exercise; The Concept of Oxygen Deficit

From the viewpoint of ATP production, an animal is in steady state if (1) its rate of ATP use is matched by its rate of ATP production, (2) the materials required for ATP production are being brought into its body as fast as they are being used, and (3) the chemical by-products of ATP production are being removed from the body as fast as they are being generated. In steady state, even as ATP is being made and broken down, the body remains essentially constant in its levels of ATP and of the precursors and by-products of ATP production.

One of the virtues of aerobic catabolism is that it is capable of operating in steady state. Presuming that your reading habits are of the ordinary sort, you, for example, are in an aerobic steady state as you read this page. In a quietly reading person, needs for ATP are fully met, O_2 is taken in at the rate it is used, and CO_2 and H_2O are voided approximately as produced. The one respect in which such a person is not ostensibly in true steady state is in regard to fuels. If the individual is not eating, his or her fuel levels are obviously falling. However, the rate of decline during reading is small by comparison to the levels available; and averaged over the whole day (including meals), intakes of fuels probably closely match uses. Thus, even in respect to fuels, the departure from steady state is merely technical.

The capability of aerobic catabolism to operate in steady state is of great significance, for it means that aerobic ATP production can go on and on, without intrinsic self-limitations. Aerobic catabolism meets our ATP needs for a lifetime.

The *rate* of aerobic catabolism can be adjusted over a broad range; young men and women, for example, are generally capable of raising their rate of aerobic ATP production to at least 10 times the resting rate. This means that many forms of exercise can be aerobically fueled. Importantly, therefore, an approximate steady state can prevail during the exercise, permitting the activity in principle to go on and on. Consider, for example, a person walking or jogging. The first minutes of such exercise require special consideration, and we shall shortly need to examine them. The feature to be stressed here is that after a few minutes, the ATP needs of walking or jogging are fully met aerobically, and thus an approximate steady state prevails and the exercise can continue at length. As you might guess, the word ''approximate'' looms larger as the intensity of aerobically fueled exercise becomes greater; for example, the failure of fuels to be replenished exactly as fast as they are used—a mere technical matter at rest—may loom as a limiting factor during high exertion. The major point must not be lost, however: The ability of aerobic catabolism to operate in approximate steady state permits aerobically fueled exercise to be protracted. Such exercise is thus often termed *steady-state exercise.*

Consider now events during the abrupt onset of exercise. Suppose, for instance, that a person is initially at rest and begins suddenly to run at about 8 mph. Figure 4.5*A* shows how this individual's rate of oxygen consumption would change if his or her metabolism were supported fully aerobically at all times; there would be a stepwise increase in the rate of oxygen consumption simultaneous with the stepwise increment in ATP demand. When insects take to the air, their actual rate of oxygen consumption does in fact typically rise abruptly to a high level that is sufficient almost immediately to meet aerobically the full ATP demands of flight. Vertebrates do not function in this manner, however. The pulmonary

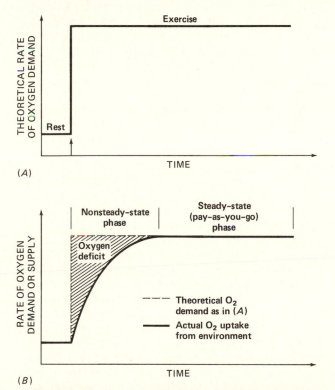

(A)

(B)

Figure 4.5 The concept of oxygen deficit. (A) The oxygen demand of a person who is initially at rest and abruptly starts running at about 8 mph, assuming unrealistically that all ATP is produced aerobically at all times. Running begins at the time marked by the arrow. (B) The actual rate of oxygen uptake from the environment (solid line) of the person in part (A), showing that there is an initial period of time when the full ATP demand is not in fact supported by oxygen uptake (nonsteady-state phase), followed by a period when oxygen uptake fully matches the oxygen demand (steady-state phase). The hatched area corresponds to the size of the oxygen deficit (see text).

and circulatory systems responsible for oxygen uptake and delivery in vertebrates do not instantly increase the rate at which they supply oxygen to the tissues but instead undergo a gradual increase. Figure 4.5*B* shows the *actual* pattern of oxygen consumption in a person undergoing sudden, vigorous exercise. The rate of oxygen consumption requires a matter of minutes to complete its increase during exercise of this sort, and only then is the ATP demand of the exercise entirely met by aerobic catabolism. Once the aerobic steady state has been established, the exercise is sometimes said to be in a *pay-as-you-go phase* because its oxygen cost is met by oxygen uptake on a moment-to-moment basis.

The first minutes of the exercise depicted in Figure 4.5*B* provide an example of nonsteady-state exercise. Before demonstrating this fact analytically, let us focus briefly on how the exercise is in fact being supported during the first minutes. Note that the (theoretical) oxygen *demand* of the exercise considerably exceeds the fresh *supply;* that is, aerobic ATP production using newly acquired oxygen is inadequate to meet the full ATP requirement. What are the sources of the additional ATP? Anaerobic glycolysis is one, and as we shall see, there are others.

To establish that the initial part of this exercise is an example of nonsteady-state exercise, let us consider both the aerobic and anaerobic components of ATP production. The initial part of the exercise is not in steady state with respect to aerobic catabolism because the aerobic ATP supply is less than the rate of ATP use. As to anaerobic glycolysis, a steady state does not exist for a number of reasons; one is that lactic acid, the major chemical end product of the reactions, accumulates in the body as the minutes pass.

Nonsteady-state conditions cannot persist for long by comparison to steady-state ones. In the case of the exercise we are discussing, the nonsteady-state circumstances are brought to an end by the ultimate rise of oxygen delivery to a rate sufficient to meet ATP costs fully aerobically. As we shall see later, if a nonsteady state is not thus supplanted by a steady state, it becomes self-terminating.

Whenever an animal's actual rate of oxygen uptake from the environment falls short of its oxygen requirement to make all ATP aerobically, we say that an *oxygen deficit* exists. The *magnitude* of the deficit is defined to be the sum total of extra oxygen that would have had to be taken up from the environment to provide all required ATP aerobically. Thus, in Figure 4.5*B,* the magnitude of the deficit incurred at the start of the exercise is the integrated area between the oxygen supply and demand curves.

Anaerobic glycolysis can permit the development of an oxygen deficit because by making ATP anaerobically, it alleviates some of the need for oxygen uptake. We now turn to two additional processes that can contribute to the development of oxygen deficits: use of phosphagens and of oxygen stores.

Phosphagens

The skeletal muscles of vertebrates and muscles of many invertebrates contain compounds termed *phosphagens* that serve as temporary stores of high bond energy. Creatine phosphate (Figure 4.6) is the phosphagen of vertebrate muscle and also occurs in some groups of invertebrates. The most widespread phosphagen of invertebrates is arginine phosphate (Figure 4.6), and other phosphagens are known, especially among annelids. The phosphagens are high-energy phosphate compounds. They are synthesized in reactions with ATP and subsequently can donate their phosphate groups to ADP to form ATP, as illustrated here for creatine phosphate:

creatine phosphate + ADP ⇌ creatine + ATP

During times of rest, when ATP supplies are relatively untaxed and cellular ATP concentrations are thus comparatively high, most of the creatine in the muscle cells of a human or other vertebrate comes to be phosphorylated. The concentration of creatine phosphate might then be three to six times the concentration of ATP. Subsequently, in times when ATP supplies are taxed and ATP concentrations fall, the reaction diagrammed earlier is shifted to the right, forming ATP without any contemporaneous need for oxygen. At the concentration ratios previously specified, each ADP molecule in the cell may in fact be rephosphorylated three to six times by this mechanism. To the extent that ATP is so made, an oxygen deficit is incurred.

Use of Internal Oxygen Depots

Earlier, although we did not call attention to it, we were careful to define an oxygen deficit to be the difference between an animal's oxygen requirement for fully aerobic function and its intake of oxygen *from the environment*. The concept is so defined because, in respect to oxygen intake, the parameter we can readily measure is precisely that: uptake from the outside world. An animal's body is likely to have *within* it at any given time a considerable quantity of molecular oxygen; you, for example, contain O_2 bound to the hemoglobin in your blood and the myoglobin in your muscle cells. When intake of O_2 from the environment lags behind O_2 demand, as at the start of exercise (Figure 4.5B), the cells of an animal can call upon the "stores" of O_2 already in the body to help make up the difference. This process contributes to the development of an oxygen deficit and leads to a (temporary) reduction in the size of the oxygen depots bound to hemoglobin and/or myoglobin within the body (pp. 305 and 318).

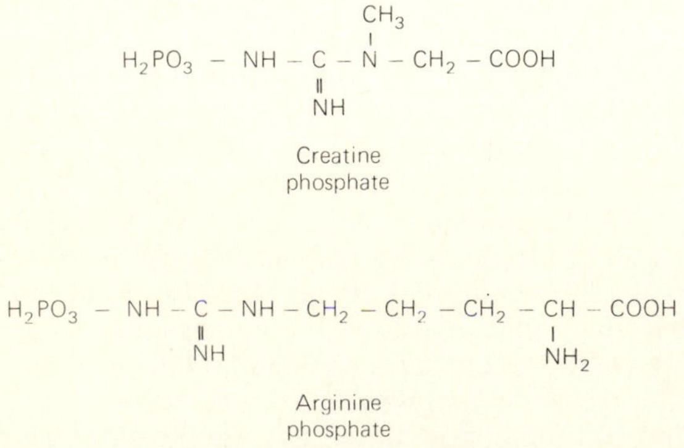

Comparative Properties of Mechanisms for ATP Production

We have now identified four physiological mechanisms of ATP production. The first is *aerobic catabolism using oxygen contemporaneously acquired from the environment*. The other three are mechanisms that enable animals to synthesize ATP at a greater rate than their current respiratory oxygen intake would permit. They are *anaerobic catabolism, anaerobic ATP production by use of phosphagens,* and *aerobic ATP production by use of internal oxygen depots*.

The latter three mechanisms permit animals to incur oxygen deficits and thus may be described collectively as the "mechanisms of oxygen deficit." In vertebrates, they are sometimes subdivided into two categories. Anaerobic glycolysis is known as the *lactacid* mechanism of oxygen deficit because it produces lactic acid. Phosphagen usage and the use of oxygen stores are known as the *alactacid* mechanisms of oxygen deficit.

In this section, we review some of the key properties of all four mechanisms of ATP production. Table 4.1 summarizes much of what will be said.

Steady State or Nonsteady State? Whereas aerobic catabolism using environmental oxygen frequently functions in approximate steady state, the other mechanisms often (or always) operate in nonsteady state and thus are potentially *self-limiting*. The use of phosphagens for net ATP production is always a nonsteady-state process because cell stores of phosphagen are drained and ultimately run out. Similarly, use of internal oxygen depots is nonsteady-state because the depots are depleted. Anaerobic catabolism *can* operate in steady state, as when its products are excreted or used by other tissues as fast as they are produced. However, it often operates in nonsteady state, and that mode of operation is our focus here. When a diving seal or a champion half-mile racer turns to anaerobic glycolysis for net ATP production and steadily accumulates lactic acid, a nonsteady

$$CH_3$$
$$H_2PO_3 - NH - C - N - CH_2 - COOH$$
$$NH$$

Creatine
phosphate

$$H_2PO_3 - NH - C - NH - CH_2 - CH_2 - CH_2 - CH - COOH$$
$$NH \qquad\qquad\qquad\qquad NH_2$$

Arginine
phosphate

Figure 4.6 Two important phosphagens. Creatine phosphate is also known as phosphocreatine or phosphorylcreatine. Arginine phosphate is also termed phosphoarginine or phosphorylarginine.

Table 4.1 SOME COMPARATIVE PROPERTIES OF MECHANISMS OF ATP PRODUCTION IN VERTEBRATES, INCLUDING NUMERICAL ESTIMATES FOR SOME OF THE PARAMETERS IN HUMAN BEINGS

Source of ATP	Mode of operation (mandatory or assumed)	Total potential ATP yield per episode of use[a] (unit: mole)	Rate of acceleration of ATP production at onset of use	Peak rate of ATP production[b] (unit: μmoles ATP/ g·min)	Rate of return to full potential for ATP production after use
Aerobic catabolism using O_2 from environment	Steady state	Very large (~200 in marathon, $>4 \times 10^6$ in lifetime)	Slow	Moderate (30 w/ glycogen fuel, 20 w/ fatty acid fuel)	—
Aerobic catabolism using O_2 preexisting in body	Nonsteady state	Small (0.2)	Fast	High	Fast
Phosphagen use	Nonsteady state	Small (0.4)	Fast	Very high (96–360)	Fast
Anaerobic glycolysis	Nonsteady state	Moderate (1.5)	Fast	High (60)	Slow

[a] Numerical estimates of total yields are computed from information in P.-O. Åstrand and K. Rodahl, *Textbook of Work Physiology*. McGraw-Hill, New York, 1970; a 75-kg person living 70 years is assumed.

[b] Peak rates of production are from P.W. Hochachka and G.N. Somero, *Biochemical Adaptation*. Princeton University Press, Princeton, 1984.

state prevails. As we shall discuss in more detail later, vertebrates become overwhelmed with fatigue when lactic acid accumulates to a high level in their body (p. 47). The buildup of lactic acid then effectively terminates further resort to anaerobic catabolism.

Total ATP Yield Because aerobic catabolism using environmental oxygen can operate in steady state, it is capable of supplying indefinite amounts of ATP (a lifetime's amount). By contrast, mechanisms operating in nonsteady state are limited in the amount of ATP they can produce in any one episode. For example, the amount of ATP that can be made from phosphagen is limited by the amount of phosphagen available. Table 4.1 shows some representative yields of ATP in mammals. Note that phosphagens and oxygen stores provide for only a relatively small production of ATP per episode. The yield from anaerobic glycolysis is larger but modest.

Rapidity of Onset of Increased ATP Production In vertebrates and many other animals (though not insects, p. 50), aerobic catabolism using environmental oxygen requires a relatively long time—minutes—to accelerate its rate of ATP production fully to a new high level (p. 43). This is because this mechanism of ATP production is *not self-contained;* oxygen from the environment is required, and the pulmonary and cardiovascular systems responsible for transporting it to the cells cannot increase their delivery markedly

in stepwise fashion. By contrast, mechanisms such as the use of phosphagens and anaerobic glycolysis *are self-contained* in the cells. At the start of a bout of exercise, for example, a muscle cell contains not only all the enzymes of glycolysis but also the sole required input: glycogen fuel. Nothing needs to be brought to the cell for glycolysis to take place, and in fact the cell is capable of nearly instantaneously increasing its glycolytic rate of ATP production to a high level. ATP production by use of phosphagens and oxygen stores is also capable of rapid acceleration.

Consider a human, initially at rest, who starts sprinting. If aerobic catabolism using environmental oxygen were the only source of ATP, the person would be able to accelerate only gradually, in parallel with the gradual rise in rate of ATP production. The fact that the person can in truth accelerate instantaneously is a manifestation of the capability of the self-contained ATP-generating mechanisms to accelerate their output of ATP instantaneously. When animals abruptly increase their intensity of exercise to a high level, they are said to engage in **burst exercise**. Among all vertebrates, the first stages of burst exercise receive their ATP supply predominantly from the mechanisms of oxygen deficit.

Peak Rate of ATP Production (Power) As shown in Table 4.1, the peak rate at which anaerobic glycolysis can make ATP is much greater than the peak rate for aerobic catabolism using environmental oxygen, and the rate for phosphagen usage is greater yet. Phosphagen usage cannot make a lot of ATP, but it can make its contribution very rapidly and thus can fleetingly support very intense exertion. Anaerobic catabolism can make a modest amount of ATP at a high rate, and aerobic catabolism using environmental oxygen can make an indefinite amount at a modest rate.

Resolution   Anytime ATP has been made by a nonsteady-state mechanism, cells are left in a state of imbalance that must be resolved before the mechanism can be used again to full effect. When internal oxygen depots have been drawn upon, they ultimately must be recharged. When phosphagens have been used, metabolism must eventually generate sufficient ATP (above and beyond that needed for other functions) to reconstitute them (p. 44). When lactic acid has been accumulated, it must be dissipated. After intense exercise or other activities employing the oxygen-deficit mechanisms, the time taken to dissipate lactic acid is much longer than that required to reconstitute oxygen stores or phosphagens. For example, in humans following exercise, a substantial accumulation of lactic acid may require about 15 min for half dissipation and even 1–2 h for full dissipation, whereas the half-time for reconstituting oxygen stores and phosphagens is just 30 s. Some fish display elevated lactic acid levels for many hours after intense exertion. The import of these comparisons is that the alactacid mechanisms are fully ready to be used again much sooner than anaerobic glycolysis.

Conclusions   A review of Table 4.1 and the preceding paragraphs reveals that each mode of ATP production has pros and cons; none is superior in all respects. Among most vertebrates, the two major sources of ATP are aerobic catabolism using environmental oxygen and anaerobic glycolysis. Because of this, a brief comparative summary of those two mechanisms is appropriate here.

Study of biochemical maps alone gives the impression that anaerobic glycolysis is inferior to aerobic catabolism. To begin with, the anaerobic mechanism unlocks only a small fraction of the energy value of foodstuffs, producing just two ATP molecules per glucose molecule rather than 38. Furthermore, only carbohydrate fuels can be tapped by anaerobic glycolysis, whereas the aerobic catabolic pathways can metabolize lipids and proteins as well. Anaerobic glycolysis also has the disadvantages that it is self-limiting, can produce only a limited net total amount of ATP in any one episode, and creates a product, lactic acid, that requires a long time for dissipation.

Anaerobic glycolysis also has its advantages, however. Because it does not require oxygen, it can both provide ATP when oxygen is unavailable and supplement aerobic ATP production in times of oxygen availability. Furthermore, because it can instanta-

neously achieve an exceptionally high rate of ATP production, it is well suited to the support of burst exercise.

Excess Postexercise Oxygen Consumption ("Oxygen Debt")

We are all aware that we breathe hard for a period following vigorous exercise. Figure 4.7 presents this phenomenon analytically. If we stop exercising suddenly, our oxygen need as assessed solely on the basis of our behavioral state falls in stepwise fashion to the resting level. Our actual rate of oxygen consumption, however, declines only gradually, remaining above the resting level for a long time. The oxygen consumed in excess of ordinary resting requirements in the aftermath of exercise has been termed *excess postexercise oxygen consumption (EPOC),* or *recovery oxygen.*

Another term for this oxygen consumption—one older, more familiar, and less appropriate—is *oxygen debt.* To see why the term is less appropriate, we need to understand its origins. Early investigators of exercise believed that dissipation of lactic acid is the major cause of the EPOC. They knew that the dissipation of lactic acid requires oxygen, and they reasoned that the presence of lactic acid after exercise thus creates an *obligation* to use oxygen *above and beyond the resting rate* to dissipate the acid. From this vantage they coined the terminology that an organism "incurs a debt to be paid in oxygen" when it resorts to anaerobic glycolysis, and it repays this "oxygen debt" with the extra oxygen taken in after exercise.

The problem with this terminology is that it implies a particular mechanistic interpretation of the EPOC, an interpretation that has recently become

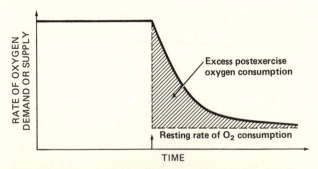

Figure 4.7   The concept of excess postexercise oxygen consumption (EPOC). Shown are the oxygen relations of a person who has been running in steady state at about 8 mph and who, at the arrow, suddenly comes to rest. The dashed line (obscured by the solid line on the plateau) shows the expected oxygen demand of the person under the unrealistic assumption that the only factor affecting oxygen demand is the individual's behavioral state (running or resting). The solid line shows the actual oxygen uptake from the environment (schematic). Compare with Figure 4.5.

increasingly suspect. We now know of many cases in which there is a nearly total lack of correspondence between the time course of the EPOC and that of lactate dissipation. In some instances, for example, lactic acid remains present at elevated levels long after the EPOC is over. The EPOC is in good measure not associated with lactate removal and thus does not meet the definition of an oxygen debt. What causes the EPOC is unknown, although one interesting hypothesis is that the coupling of electron transport and oxidative phosphorylation (p. 39) becomes loosened in the aftermath of exercise, so more electron transport—and hence oxygen consumption—is required to meet any given ATP need.

Fatigue

Fatigue is not well understood. There is no doubt, however, that it has multiple causes, depending partly on the type and duration of exercise. For example, during prolonged aerobic exercise demanding a substantial fraction of a person's maximal rate of oxygen consumption, fatigue is associated with (caused by?) depletion of muscle glycogen. One possible consequence of glycogen depletion would be that aerobic catabolism would have only lipid fuels available, and as can be seen in Table 4.1, the rate of ATP formation with lipid fuels is lower than with carbohydrates. Overheating can induce fatigue, and sometimes the cause is said to be psychological.

A particular variety of fatigue of interest here is that associated with lactic acid accumulation. Among vertebrates undergoing intense exercise involving sustained net lactate production, it is common to find that once lactic acid has accumulated to a certain level, the individual is profoundly overcome with fatigue. In some cases, virtual paralysis sets in. The level of lactic acid eliciting this state varies with the species and individual but is relatively consistent within any one individual (though see p. 49). At one time, the lactate ion was considered to be a specific "fatigue factor." More recently, however, the focus has shifted to the acid–base disturbances associated with lactic acid accumulation; the cells and body fluids are rendered more acidic, and this can cause numerous disruptions of normal function. There is reason to believe, for instance, that increases in acidity within muscle cells can interfere with force generation by causing disturbances of excitation-contraction coupling (see p. 514) or by interfering with the cyclic formation of actin-myosin cross-bridges (see p. 511). Some investigators believe that lactic acid is not in fact causative of fatigue in any form; rather, they believe that its accumulation to the "fatigue level" is merely a correlate of other changes that truly are responsible for the fatigue. Whatever the case, the important fact remains that the consequences of maximal resort to anaerobic glycolysis are of a very practical sort.

Muscle-Fiber Heterogeneity

Although we have spoken of vertebrate skeletal muscle as if it were a uniform tissue, in fact individual muscle fibers (muscle cells) may differ considerably in their functional properties, even within a single muscle. At one extreme are fibers that are poised especially to make ATP aerobically; these are relatively unsusceptible to fatigue but have relatively low peak power outputs. At the other extreme are fibers that are highly able to make ATP anaerobically; they can put out a lot of power but fatigue easily. This heterogeneity of fibers, which is of great functional importance, is discussed in detail in Chapter 19 and summarized in Table 19.2 (p. 524).

THE INTERPLAY OF AEROBIC AND ANAEROBIC CATABOLISM IN EXERCISE

The purpose of this section is to explore more systematically the interplay of aerobic and anaerobic catabolism in supplying the ATP requirements of various forms of muscular exercise. Initially, we limit our attention to vertebrates.

A given individual in a particular state of training is capable of a certain maximal rate of oxygen consumption. Exercise that in steady state requires exactly this maximum is called *maximal exercise,* whereas exercise requiring less than the maximal rate of oxygen consumption is called *submaximal.*

An oxygen deficit, albeit a small one, is typically incurred at the start of even light submaximal exercise, as shown in Figure 4.8A, because of the lags of the pulmonary and circulatory systems in heightening their rate of oxygen delivery. If the steady-state oxygen cost of exercise is less than 50–60 percent of maximal oxygen consumption, the initial oxygen deficit is often largely or entirely of an alactacid sort; that is, little or no net accumulation of lactic acid occurs, even though the turnover of lactic acid may increase sharply at the onset of exercise (indicating that some tissues are making it and others using it). In heavy submaximal exercise, such as shown in Figure 4.8B, the deficit incurred is larger and does entail net accumulation of lactic acid. Such buildup of lactic acid in submaximal exercise is nonetheless typically modest and nonfatiguing because before long the rate of oxygen intake becomes high enough to establish pay-as-you-go conditions, and reliance on net anaerobic glycolysis then ceases. Sometimes, when there has been a net accumulation of lactic acid at the start of submaximal exercise, the lactic acid is partly or even entirely dissipated during the exercise itself. However, in other cases, the accumulated lactic acid is retained until exercise has stopped.

Vertebrates are capable also of engaging in exercise that demands ATP at a greater rate than can be supplied aerobically at their maximal rate of oxygen

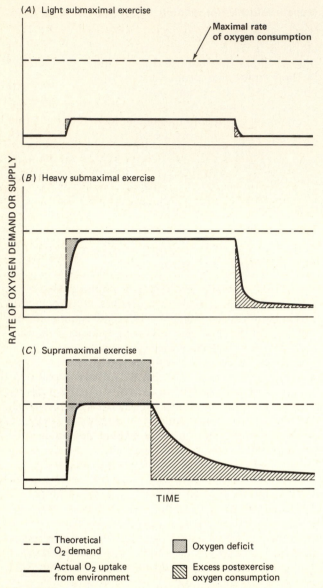

(A) Light submaximal exercise

Maximal rate of oxygen consumption

RATE OF OXYGEN DEMAND OR SUPPLY

(B) Heavy submaximal exercise

(C) Supramaximal exercise

TIME

- - - Theoretical O₂ demand

▭ Oxygen deficit

—— Actual O₂ uptake from environment

▨ Excess postexercise oxygen consumption

Figure 4.8 Stylized oxygen supply–demand diagrams for (A) light submaximal, (B) heavy submaximal, and (C) supramaximal exercise, as observed in mammals. The short-dashed line in each diagram shows what oxygen demand would be if all ATP were made aerobically. The solid line shows actual oxygen uptake from the environment. The long-dashed line is the maximal rate of oxygen uptake. Compare with Figures 4.5 and 4.7.

consumption. Such exercise is termed *supramaximal*. Its ATP demand must be met partly by anaerobic glycolysis even after oxygen delivery has reached its peak rate. Significantly, therefore, as illustrated in Figure 4.8*C*, a pay-as-you-go phase is never attained, and lactic acid continues to accumulate *throughout* the exercise. It is in supramaximal exercise that lactic acid levels can rise so high that the profound, often debilitating, "lactic acid" fatigue sets in.

Because an individual's limits on accumulation of lactic acid are absolute, an animal that has built up elevated lactic acid levels in one bout of exercise

displays a reduced capacity to accumulate lactic acid in future bouts until the original burden of lactic acid has been dissipated. This can be a factor in human athletic competition and in the life histories of animals. When vertebrate animals engage in sudden, peak-intensity running or swimming—such as human sprinting—the effort is typically a supramaximal one. Anaerobic glycolysis, by producing ATP, makes the effort possible, and the effort itself may have considerable survival value by enabling the animal to escape a predator or catch fleeing prey. However, lactic acid generated by the anaerobic glycolysis has its implications for survival too: It will induce profound fatigue if the exercise continues long enough, and once accumulated, it will require a long time for dissipation, during which the animal's capability for supramaximal exertion will be impaired. An individual that calls heavily on anaerobic glycolysis to escape from a predator must elude the attacker before exhaustion sets in and then must go through a period of dissipation of the lactic acid before it will be able to exert a similar effort in escape from another predator. In brief, behavioral capabilities are constrained by the biochemistry of ATP production.

Human Exercise

In some forms of exercise, most of the ATP is generated aerobically. In others, most is generated anaerobically. To illustrate, let us look briefly at several forms of human exertion.

In submaximal exercise such as walking or jogging that lasts an appreciable time, aerobic energy supply is overwhelmingly dominant, with mechanisms of oxygen deficit contributing just to the first minutes. The opposite extreme is represented by sprints such as the 100-yd dash. Running at 10 yd/s would require an oxygen consumption of over 0.3 L/s to be supported aerobically using environmental oxygen. The highest human oxygen consumption ever recorded is just 0.1 L/s, and it could by no means be attained in a span of 10 s. At least 90 percent of the ATP cost of a 100-yd dash is met during the dash by mechanisms of oxygen deficit.

In middle-distance races, such as the mile run, the pattern of energy expenditure is fundamentally like that in Figure 4.8*C*, with major contributions being made by both oxygen intake and oxygen deficit. Suppose, to illustrate, that a miler can average an oxygen consumption of 4.0 L/min, taking into account the initial lag in oxygen delivery. This would permit him or her to run at about 14.5 ft/s and complete the race in about 6 min on a strictly aerobic basis (using environmental oxygen). If, however, the racer can also contract an oxygen deficit of 17 L at an average rate of 4.3 L/min, then his or her total energy supply—aerobic plus anaerobic—will be equivalent to 8.3 L/min. This will permit a pace of about 22 ft/s, enabling completion of the race in about 4 min. In mile races, about one-third to one-half of the ATP required is generated by mechanisms of oxygen deficit, particularly anaerobic glycolysis. We see that this anaerobic contribution makes a big difference to the pace.

In long distance races, the capacity to exploit anaerobic glycolysis may be fully utilized, but the proportionate contribution of anaerobic catabolism to the support of the exercise falls off with the length of the race simply because the total cost increases while the energy available anaerobically remains fixed. Running the marathon might require energy equivalent to 650 L of oxygen. Only about 2 percent of this can conceivably be met by anaerobic glycolysis and other mechanisms of oxygen deficit. The marathon must thus be run largely aerobically on a pay-as-you-go basis.

Note that in the progression from sprint to mile to marathon, as the total ATP requirement of all-out exertion increases, the pace slows. By "total ATP requirement" we mean here the sum total of all ATP needed, integrated from start to finish. A brief effort having a low total ATP demand can be fueled largely by mechanisms such as phosphagen usage and anaerobic glycolysis that can sustain a very high *rate* of ATP production (Table 4.1). A long effort having a large ATP demand must be fueled by aerobic catabolism, which provides only a relatively low rate of ATP generation.

Two of the important determinants of athletic performance are an individual's maximal rate of oxygen consumption and maximal oxygen deficit. Significantly, both are increased by training, thereby increasing performance; for example, an appropriate training regime might be able to produce a 10–20 percent rise in a person's maximal rate of oxygen consumption. There are also large constitutional differences among individuals in their maximal rate of oxygen consumption and capacity to incur oxygen deficit; superior athletes often prove on testing to be exceptional in constitutional parameters of importance to their sport. Beyond young adulthood, an individual's maximal rate of oxygen consumption tends to decrease with age.

One reason the maximal rate of oxygen consumption is of interest is that it provides a measure of physical condition, for the strenuousness to an individual of any particular form of aerobic exercise depends roughly on how large a percentage of the individual's maximal oxygen consumption is required. If an activity demands 35 percent or less of an individual's maximal rate of oxygen consumption, it probably can be continued all day. If it requires 75 percent, however, it will likely be exhausting in 1–2 h.

Exercise in Other Vertebrates

The patterns evident in humans frequently have clear counterparts in other vertebrates. For example, migration in birds and salmon is of a similar nature to marathon running in that it is so prolonged as to require aerobic fueling, and the speed is thereby limited to what aerobic catabolism can support. We shall examine aerobic forms of exercise in some detail in Chapter 5. Here, our purpose is to take up the interesting discovery that related species often differ in their relative aerobic and anaerobic competence, and these differences can have important life-history consequences. Given that anaerobic glycolysis is invoked to a major extent only during comparatively intense exertion, our focus is to be mostly on forms of exercise analogous to long sprints and middle-distance races in humans.

As noted, striking relations are often evident between the energetics of intense exercise and attributes of life history. Consider, for example, two African predators, the wild hunting dog (*Lycaon*) and the cheetah (*Acinonyx*). Hunting dogs usually do not stalk their prey but simply approach and start chasing, and often the chases are protracted; the long chases of the dogs, being akin to marathons, must be supported aerobically. Cheetahs, by contrast, usually stalk prey and when they attack, do so at legendary speed; they give up the chase quickly if unsuccessful and then rest in a state of exhaustion for many minutes. Even if they catch their quarry, they often rest for a long time before starting to eat. The pattern of attack in cheetahs—with its advantage of extreme speed and its disadvantage of postexercise exhaustion—is symptomatic of anaerobic effort.

A particularly interesting comparison has been made between two large lizards, the spiny chuckwalla (*Sauromalus hispidus*) and a varanid (*Varanus gouldii*). The chuckwalla, a North American iguanid, is a relatively typical herbivorous lizard, usually slow moving but capable of brief bursts of speed. The varanid is an Australian monitor lizard, an active and wary predator capable of running down prey such as rodents and other lizards. One notable feature of varanids is that their lungs are the most complex found in reptiles (see Figure 12.27).

When the chuckwallas and varanids were induced to be vigorously active for 7 min at temperatures of 35–40°C, they exhibited considerable differences. The chuckwallas ultimately increased their rate of oxygen intake to about six times their standard rate of oxygen consumption (reaching an intake of 0.63 mL/g·h). The varanids, by contrast, increased their intake eightfold (attaining 0.87 mL/g·h). The varanids never gave evidence of exhaustion. The chuckwallas, on the other hand, often became exhausted; they accumulated considerably more lactic acid than the varanids and suffered a much larger drop in blood pH. Their drop in pH (acidosis) resulted partly from their lactic-acid accumulation and partly from their having a comparatively low blood-buffer capacity; the acidosis seriously degraded the oxygen-transport effectiveness of their blood hemoglobin, decreasing its affinity for oxygen and its total capacity to bind oxygen.

Clearly, the varanids and chuckwallas differ in

their reliance on aerobic and anaerobic catabolism for support of intense exertion. With their elaborate lungs, ability to take in oxygen relatively rapidly, and high blood-buffer capacity, the varanids are able to exercise vigorously without quickly exhausting: an important factor in their active, predatory way of life. The chuckwallas, on the other hand, rely heavily on anaerobic glycolysis to exercise vigorously and tend to become exhausted in short order. Their metabolic pattern, which is similar to that of most herbivorous lizards, nonetheless appears well matched to their way of life. Being plant eaters and not greatly besieged with enemies, the chuckwallas can ordinarily move about at the modest rate permitted by their relatively limited aerobic capacity; and yet because of their anaerobic capabilities, they are able to undertake bursts of rapid running when necessary.

When terrestrial amphibians are stimulated to be maximally active for a period of several minutes, it turns out that their peak rates of anaerobic and aerobic ATP production tend to be inversely correlated from one species to another. Toward one end of the spectrum are species such as certain ranid and hylid frogs that can attain only modest rates of oxygen consumption yet display rapid rates of anaerobic glycolysis. Their initial response to stimulation is to run or jump with great vigor, but within a few minutes they exhaust. Leopard frogs (*Rana pipiens*) illustrate this pattern. At the other end of the spectrum are species such as certain bufonid toads that attain relatively high rates of oxygen consumption yet have only modest anaerobic capabilities. They do not move as quickly as the first group when stimulated but can continue moving for a long time before tiring. The amphibians with high anaerobic competence escape danger by rapidly fleeing, whereas the species with high aerobic competence tend to rely on stationary defense mechanisms such as noxious skin secretions or defensive postures.

Among snakes that have received study, fast-moving species such as racers have relatively high capacities to produce ATP aerobically *and* anaerobically. On the other hand, both capacities are relatively low in slow-moving species such as boas.

Exercise in Invertebrates

The flight muscles of certain insects are, by a good measure, the most aerobically competent of all animal tissues as judged by the rate per gram at which they can aerobically synthesize ATP. These flight muscles contain very high levels of enzymes of the aerobic catabolic pathways, and in some cases, fully 40–50 percent of their tissue volume is mitochondria. Despite their high potential oxygen demand, insect flight muscles typically have almost no anaerobic competence; instead, as mentioned earlier, they remain fully aerobic even when suddenly called upon to increase greatly their power output, as at the outset of flight. A characteristic of insects that is un-

doubtedly important to their aerobic fueling of burst exercise is their tracheal respiratory system, which provides oxygen directly to each flight-muscle cell by way of gas-filled tubes and which often is ventilated by the action of the flight muscles themselves (p. 292).

In many cases, the locomotory muscles of invertebrates do possess well-developed anaerobic capabilities that are called into play to support sudden or intense exercise as in vertebrates. When a crayfish or lobster, for example, accelerates rapidly by flipping its tail, the tail muscle employs anaerobic glycolysis leading to accumulation of lactic acid. Interestingly, lactic acid cannot be called a common end product of anaerobic catabolism in invertebrates. Many groups produce other end products (p. 53). For example, when squids, octopuses, and scallops turn to anaerobic catabolism during burst swimming, their swimming muscles produce pyruvic acid in the usual way but then, instead of reducing it to lactic acid, reduce it to octopine by the following reaction, catalyzed by octopine dehydrogenase:

$$\text{pyruvic acid } + \text{ arginine } + \text{ NADH}_2 \rightarrow$$
$$\text{octopine } + \text{ NAD}$$

Why do this? One hypothesis revolves around the twin facts that the phosphagen of these animals is arginine phosphate and the product of its use, arginine, is potentially disruptive to cell function. Anaerobic catabolism is likely to be called into use at the same times as phosphagens, and its formation of octopine may be a way of detoxifying arginine.

ANAEROBIC AND AEROBIC CATABOLISM DURING HYPOXIA

When animals suffer an inadequacy of oxygen because the influx of oxygen from their environment is impaired, they are often said to be in a state of *hypoxia*. Besides vigorous exercise, hypoxia is the other major circumstance in which anaerobic catabolic pathways are often called forth to produce ATP. Hypoxia can arise in two major ways. First, animals sometimes enter environments in which they are unable to breathe; this is true of most terrestrial vertebrates when they dive, for example. Second, hypoxia can arise, even though the animal is in an appropriate environment for breathing, because the environment itself is low in oxygen.

Energy Metabolism in Vertebrate Diving

As a first case study of hypoxia, let us briefly consider vertebrate diving. When frogs or other amphibians dive, they are able to extract dissolved oxygen from the water across their skin (p. 274). Some reptiles also are able to obtain appreciable amounts of oxygen while under water; soft-shelled turtles (*Trionyx*), for example, acquire considerable oxygen across their external body surfaces and the membranes of their buccopharyngeal and cloacal cavities (which are ventilated with water). On the other hand, most diving reptiles and the diving birds and mammals lack means of obtaining any substantial amount of oxygen from the water. For them, the oxygen available during a dive is limited to that contained within their body at the time of submergence. Stored oxygen is found in their lung air. It is also found dissolved in their body fluids and chemically combined with the respiratory pigments in their blood (hemoglobin) and muscle (myoglobin). In many of these animals, research over the past decade has made it increasingly clear that their routine dives are typically short enough for their metabolism to remain largely or entirely aerobic. On the other hand, sophisticated capacities for support of metabolism using anaerobic glycolysis are often present and are known, or believed, to be invoked in certain circumstances. Here, as an example, we examine turtles. Diving in mammals and birds is the theme of Chapter 15.

As a group, freshwater and terrestrial turtles have an exceptional ability to survive oxygen deprivation. For example, in a study of reptiles placed in a pure nitrogen atmosphere at 22°C, the 45 investigated species of lizards, snakes, crocodilians, and sea turtles survived for 0.3–2 h, but 25 species of freshwater and terrestrial turtles endured for 6–33 h. In terms of diving, such data indicate that even the many species of turtles that cannot extract much oxygen from water should be capable of impressively long dives. One such species is the yellow-bellied or red-eared slider, *Pseudemys scripta*. Even when denied oxygen altogether by being submerged in oxygen-free water, it is reported to survive for about 20 h at 22°C. As shown in Figure 4.9, sliders in oxygen-free water first go through a period, termed phase I, during which they sustain their metabolic rate at the predive level while rapidly depleting their pulmonary and blood oxygen stores; their metabolism in this phase is probably fully aerobic. Later (phase II), they exhibit a sharp fall in their metabolic rate as their rate of oxygen extraction from stores also falls. Finally, the animals enter a truly remarkable period, phase III, in which they cease to obtain oxygen from their lungs or blood, and early on, their stores of oxygen bound to myoglobin in the muscles would seem surely to become similarly exhausted. In short, in phase III these vertebrate animals are either entirely without oxygen or virtually without it. Yet they live, metabolizing at a relatively low and gradually declining rate. Anaerobic glycolysis is their source of ATP, and lactic acid accumulates to high levels in their body fluids. In most vertebrates, however tolerant many tissues may be to anaerobic conditions, the brain and (often) the heart require oxygen. In these turtles, by contrast, even the brain endures anaerobically, although exhibiting reduced electrical activity.

Having noted the extraordinary capabilities of

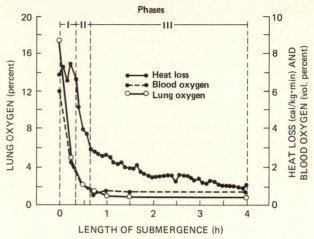

Figure 4.9 Rate of heat loss, blood oxygen content, and pulmonary oxygen content as functions of time in turtles, *Pseudemys scripta elegans,* during forced submergence for 4 h in oxygen-free water at 24°C. Values are averages for 5–10 turtles. Three phases of response are recognized, as discussed in the text. Because forced submergence, in comparison to voluntary submergence, can accelerate resort to anaerobic catabolism in turtles, experiments of this sort, although revealing patterns of response, do not necessarily reflect the exact timing of events during natural dives. The rate of heat loss to the surrounding water by the turtles was determined by direct calorimetry and provides a measure of the rate of heat production and thus metabolic rate. Blood oxygen content was determined on samples drawn from the heart. [From D.C. Jackson, *J. Appl. Physiol.* **24**:503–509 (1968).]

many turtles to survive anaerobically, we next inquire into the circumstances under which these capabilities are put to use. Evidence is mounting that the routine, voluntary dives of turtles and other diving reptiles (e.g., marine iguanas and sea snakes) are largely or entirely *aerobic*. For instance, a recent study of painted turtles (*Chrysemys picta*) undergoing voluntary dives in the summer revealed that the animals never had elevated concentrations of lactic acid in their blood. An important point to recognize is that the dives of turtles can be long by mammalian or avian standards, yet be supported entirely aerobically, in good part because the resting rates of oxygen consumption of turtles are well below those of mammals and birds of similar size (see also p. 356 for relevant circulatory physiology).

Painted turtles do make substantial use of anaerobic glycolysis in at least two situations: when forced by a predator to remain submerged and when hibernating. The turtles hibernate at the bottoms of lakes and ponds, remaining continually submerged for many weeks, and during this period, high levels of lactic acid accumulate in them. One reason hibernating turtles can remain submerged so much longer than turtles studied at warm temperatures is that their metabolic rates are depressed to a particularly profound extent by the low temperatures prevailing during hibernation. Among painted turtles placed in oxygen-*free* water, ones at 10°C survived 11–26 days.

In a set studied at 3°C, 40 percent survived for over 150 days.

A Classification of Responses to Hypoxia

Having looked in detail at the responses of a few vertebrates to hypoxia, we now take a broader but more cursory look at the other animals in this section and the next. Various species differ widely in their responses to oxygen deprivation. To categorize species meaningfully in this respect, it is essential that the temperature be standardized at least roughly, for as we have seen, anaerobic survival is often strongly dependent on temperature. Here we focus primarily on the temperatures most commonly studied, 10–25°C.

The following broad types of response to lack of oxygen are usually recognized. (1) Many animals, when denied oxygen, die within minutes or, at most, a few hours and—having an obligatory, steady demand for oxygen—are classed as *obligate aerobes*. Most vertebrates are in this category, as are many invertebrates, including most or all of the cephalopod mollusks, adult insects, and decapod crustaceans. (2) Certain other animals survive total oxygen deprivation for a substantial period—from a day to many weeks, depending on species—yet do not survive indefinitely. These are among the animals classed as *facultative anaerobes*. (3) Finally, a good number of Protozoa, some multicellular parasites (e.g., helminths of the mammalian intestine), and probably some multicellular free-living animals can survive without oxygen indefinitely. If animals of this type are able to live equally well with or without oxygen, they (like the preceding group) are called *facultative anaerobes*. Some are debilitated or killed in the presence of oxygen and are *obligate anaerobes*.

Anaerobiosis in Free-Living Facultative Anaerobes Exposed to Low-Oxygen Environments

In examining the free-living facultative anaerobes, it must be emphasized as an initial caveat that they do not necessarily form an entirely natural grouping physiologically. They are diverse in the length of time they can survive without oxygen and in their biochemistry of anaerobic catabolism. Furthermore, they probably vary considerably in their ability to sustain aerobic catabolism using oxygen stores during the initial stages of oxygen deprivation.

Species that can survive for at least a day without oxygen uptake are known from numerous invertebrate groups. Some of the bivalve mollusks that are facultative anaerobes are intertidal species, living between the limits of low and high tide along the seashore. Exposed to the air for many continuous hours at low tide, certain of them close their shells tightly, thus preventing desiccation but also preventing oxygen uptake from the air (p. 180). Many of the other invertebrate facultative anaerobes live in habitats

where they are likely on occasion to experience low *environmental* oxygen levels. The ribbed mussel (*Geukensia demissa*), for example, lives in the mud of salt marshes, often becoming buried; it survives in an atmosphere of nitrogen for a median of 5 days. Animals living in or near the bottom of lakes, ponds, or estuaries may experience severe and prolonged oxygen deprivation (pp. 253 and 254). Some are among the most tolerant to lack of oxygen of all free-living animals. Recently, for example, certain oligochaete worms of the genus *Tubifex* that burrow in anoxic (oxygen-free) sediments have been shown not only to survive but to feed, grow, and reproduce while deprived of oxygen for 7 months.

Among vertebrates, besides the turtles already discussed, a number of kinds of fish are known or reputed to be capable of surviving considerable periods without oxygen. The best studied species is the common goldfish (*Carassius auratus*), which, according to a recent study, can survive in oxygen-free water for 11–24 h at 20°C and 1.3–6 days at 10°C.

Only recently have methods of direct calorimetry been applied to facultative anaerobes, permitting their true metabolic rates to be measured during prolonged anaerobiosis. In one investigation, blue mussels (*Mytilus edulis*) were sealed in a vessel of aerated seawater, and then as they depleted the water of oxygen over the next 5 h, their rates of both total heat production and aerobic heat production were monitored. The results, shown in Figure 4.10, illustrate metabolic patterns that in basic form are shown also by certain other species of bivalves (e.g., oysters and ribbed mussels) when tested similarly. There is some dispute about the metabolism of *M. edulis* in well-aerated water; although the data shown (at the far left) indicate that about one-third of their total

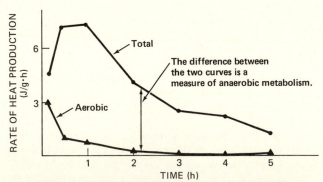

Figure 4.10 Average rates of total and aerobic heat production of blue mussels (*Mytilus edulis*) as functions of time after the animals were placed in a sealed vessel of seawater. The animals depleted the water of oxygen during the course of the experiment. Total heat production was measured directly. Aerobic heat production was determined by measuring oxygen uptake and calculating equivalent heat production using the oxycalorific coefficient for glucose (0.47 J/μmole O₂). Blue mussels are found attached to rocks and pilings in the intertidal zone. [After C.S. Hammen, *Comp. Biochem. Physiol.* **67A**:617–621 (1980).]

metabolic rate in aerated water is anaerobic, other data point to fully aerobic catabolism under such conditions. The subsequent pattern of change is not controversial, however, and has several attributes worthy of note. First, aerobic catabolism supported by oxygen uptake from the environment falls as environmental oxygen is depleted. Second, coincident with the drop in aerobic catabolism, anaerobic catabolism accounts for an increasing proportion of the total, such that metabolism becomes almost fully anaerobic after a few hours. Finally, the total metabolism falls as these events take place; after 24 h of anoxia, the total metabolic rate of *M. edulis* has declined to just 10 percent or less of the rate in aerated water.

Not just the data on mussels but also those on turtles (Figure 4.9) exemplify the important point that *a marked and deepening fall in metabolic rate is an almost universal response of facultative anaerobes to oxygen deprivation.* Anaerobic catabolic pathways never produce anywhere near as much ATP per foodstuff molecule as the aerobic ones. Given this reality, a drop in the body's ATP requirement is evidently one of the keys to long-term survival in low-oxygen environments.

What anaerobic catabolic pathways do facultative anaerobes use during prolonged exposure to low-oxygen environments? In many invertebrate anaerobes, such as the bivalve mollusks and parasitic helminths, lactic acid is (at most) only a minor product of anaerobic catabolism, and the principal products during long-term anaerobiosis are such compounds as succinic acid, propionic acid, and other volatile fatty acids. Clearly, these products signal that the biochemical pathways of anaerobic catabolism in these animals differ from simple anaerobic glycolysis. As currently understood, the pathways are elaborate, involving some of the reactions of the Krebs cycle and additional other reactions besides glycolytic ones. Importantly, some of these reactions are ATP generating. For instance, during protracted anaerobiosis in some animals, ATP-producing parts of the electron-transport chain are active, and ATP is generated in the formation of propionic acid from succinic acid. All the while, the critical requirement of maintaining redox balance (p. 40) is met without oxygen.

In comparison with simple anaerobic glycolysis, what advantages might these elaborate biochemical pathways have for animals that sustain all their vital functions anaerobically over ˙protracted periods? Among the answers that have been demonstrated or proposed are these. (1) The pathways seen in the invertebrate facultative anaerobes more fully tap the energy available from foodstuff molecules, yielding more ATP per foodstuff molecule than anaerobic glycolysis. (2) They often permit animals to catabolize anaerobically not only carbohydrates but also other classes of fuel molecules such as amino acids. (3) The pathways channel carbon chains from foodstuff molecules into intermediates of the Krebs cycle. This is important because some of these intermediates are needed for functions besides catabolism, notably biosynthesis. Lactic acid, by contrast, is a metabolic cul-de-sac (p. 41). (4) Pathways forming compounds such as succinic acid and propionic acid cause less of a drop in pH for equivalent ATP production than those forming lactic acid.

Commonly, invertebrate facultative anaerobes excrete considerable portions of their anaerobic end products during prolonged anaerobiosis. In a sense, this seems profligate because of the high energy value of the products (p. 41). On the other hand, ridding the body of the end products forestalls their accumulation to deleterious levels.

Among vertebrates, although turtles rely on simple anaerobic glycolysis even during prolonged anaerobiosis, goldfish and Crucian carp (*Carassius carassius*) use more-elaborate pathways (not fully understood) that yield CO_2 and ethanol, which are voided.

Most free-living facultative anaerobes are unable to survive indefinitely without oxygen, and an important unanswered question is why this is so. One possibility is that in some species anaerobic end products accumulate to deleterious levels. Another is that the relatively low rate of ATP production during anaerobiosis may be inadequate for long-term sustenance of all vital functions.

Interestingly, some of the invertebrate facultative anaerobes are now known to employ different anaerobic catabolic pathways during exercise than they employ during prolonged exposure to low-oxygen environments. The pathways used in exercise may provide for a greater *rate* of ATP production than those used during hypoxia.

The Rate of Aerobic Catabolism as a Function of Ambient Oxygen Level

A question that has long attracted the interest of physiologists is the effect of variations in environmental oxygen level on the rate of aerobic ATP production. Data on the subject are typically presented in the form shown in Figure 4.11*A*. To understand this sort of plot, the meaning of the "oxygen partial pressure" on the abscissa must be appreciated. Chapter 11 goes into the matter in detail. Here, suffice it to say three things. First, the oxygen partial pressure is a measure of the level (actually, chemical activity) of molecular oxygen in the air or water. Second, at sea level, the partial pressure is 150–160 mm Hg both in the open atmosphere and in well-aerated waters. Third, in any given body of water, the partial pressure is proportional to concentration; for example, water at sea level having a partial pressure of 75–80 mm Hg contains only half as much dissolved oxygen as well-aerated water.

The goldfish in Figure 4.11*A* exemplify a significant and widespread phenomenon: They are able to

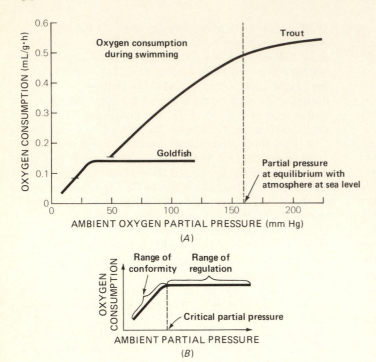

Figure 4.11 (*A*) Oxygen consumption as a function of ambient oxygen partial pressure in young, exercising goldfish (*Carassius auratus*) and speckled trout (*Salvelinus fontinalis*) at 20°C. Fish were compelled to swim near their maximal rate. Horizontal bars toward the left on each curve indicate the standard oxygen consumption of each species in well-aerated water at 20°C. The goldfish averaged 3.8 g in weight; the trout data are for 5-g individuals. The animals were maintained in aerated water except during experimentation. (*B*) Data for goldfish indicating the range of regulation, range of conformity, and critical partial pressure. [Data are from F.E.J. Fry and J.S. Hart, *Biol. Bull.* (*Woods Hole*) **94**:66–77 (1948); S.V. Job, *Univ. Toronto Biol. Ser. No. 61,* Publ. Ontario Fish. Res. Lab. No. 73, pp. 1–39 (1955).]

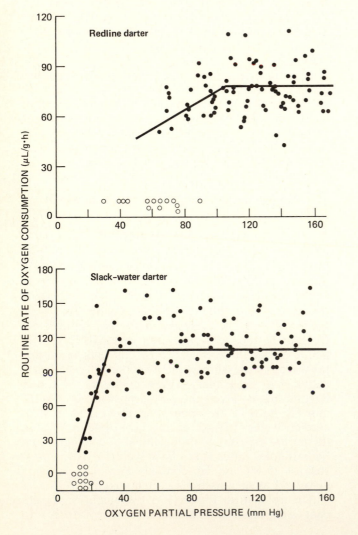

Figure 4.12 Rate of oxygen consumption during routine activity at 20°C as a function of ambient oxygen partial pressure in two fish of the genus *Etheostoma*: the redline darter (*E. rufilineatum*), an inhabitant of fast-flowing streams, and the slack-water darter (*E. boschungi*), a resident of slow-moving streams. Both species were relatively inactive during measurement: a state that is evidently typical for them in nature. Dots are individual data. The lines were fitted statistically and identify the P_c for the redline darter as 105 mm Hg and that for the slack-water darter as 30 mm Hg. Open circles mark ambient partial pressures at which deaths occurred. [Reprinted by permission from G. Ultsch, H. Boschung, and M.M. Ross, *Ecology* **59**:99–107 (1978).]

maintain a stable rate of oxygen consumption even as the environmental oxygen level is varied widely, from 34 mm Hg upward. When animals exhibit stability or near-stability of oxygen consumption in the face of environmental variation in the oxygen level, they are said to function as **oxygen regulators;** and the range of environmental oxygen partial pressures over which they so function is called the **range of oxygen regulation** or **oxygen independence** (see Figure 4.11*B*). If the ambient oxygen level is lowered far enough, inevitably a range of oxygen partial pressures is entered within which the animal's rate of oxygen consumption falls sharply as the ambient partial pressure is lowered. In this range, the animal is said to function as an **oxygen conformer,** and the range of partial pressures itself is called the **range of oxygen conformity** or **oxygen dependence** (again, note Figure 4.11*B*). The environmental partial pressure of oxygen at which regulation ceases and conformity begins is called the **critical partial pressure,** P_c (or critical tension, T_c). For the goldfish, P_c is about 34 mm Hg.

Figure 4.11*A* also shows data for trout of about the same body size as the goldfish, studied under similar conditions. One point illustrated by these data is that species can differ considerably in the response of their aerobic catabolic intensity to variations in environmental oxygen level: The trout data also show that P_c is not always as clearly defined as the results for goldfish would suggest; often, the best we can do is identify a *range* of partial pressures over which regulation gives way to conformity.

There was a time when each species was considered to exhibit a single species-specific P_c. Now we realize that in fact the response of a species to declining oxygen often depends strongly on the prevailing conditions. Each species displays a variety of P_c values and other response parameters, depending on circumstances. In general, factors that elevate a species' metabolic rate tend to raise its P_c; that is, when the demand for oxygen is increased, failure to meet it fully is likely to occur with less of a drop in the ambient oxygen level. Animals tend to exhibit higher P_c values when active than resting, for example, and an elevation of environmental temperature often raises the P_c of poikilothermic animals. Other factors that can affect P_c include the season, stage of the life cycle, body size, and oxygen level to which the animals have previously been acclimated. Considering all species and all conditions of study, a very wide range of P_c values is known: from as low as 5 mm Hg to as high as 250 mm Hg (the latter being well above oxygen partial pressures likely to occur in natural habitats, signifying that the range of regulation is unlikely to be entered).

There have been many attempts to relate differences in the P_c values of species to differences in the habitats they occupy. Some efforts have been befuddled by failure to take into account the variations of P_c that can occur within just a single species; for meaningful interspecific comparisons, the conditions of study of the several species must be as similar as possible. In some carefully executed studies, correlations between P_c and habitat have failed to emerge. On the other hand, there are instances in which species from relatively oxygen-poor environments tend to exhibit lower P_c values than related species from oxygen-rich environments. In other words, the species that are the more likely to confront low oxygen levels tend to be the more resistant to having their rates of oxygen consumption reduced by such levels (e.g., note the comparison of goldfish and trout, Figure 4.11). Figure 4.12 shows data from an interesting recent study of two species of darters living in a single Alabama watershed. One, the redline darter, occurs in fast-flowing streams where oxygen levels tend to be high because turbulence promotes aeration. The other, the slack-water darter, is found in slow-moving streams where oxygen partial pressures as low as 55 mm Hg have been measured. As the figure shows, the slack-water darter exhibits a much broader range of regulation than the redline. Indeed, lowering the oxygen partial pressure to 70–80 mm Hg—a level well above ones that occur in the habitats of slack-water darters—will cause a sharp depression of oxygen uptake and deaths in redline darters (while not affecting the oxygen uptake of slack-water darters at all).

When animals maintain a stable rate of oxygen uptake in the face of declining environmental oxygen levels, it is often because they undergo active protective responses. One such response is to breathe faster, thereby facilitating oxygen uptake across the gills or lungs (pp. 270 and 282). This response is shown by animals as diverse as clams, crabs, fish, and mammals. Other parameters that may be altered in ways promoting oxygen regulation include (1) the quantity and functional properties of blood oxygen-transport pigments (p. 321), (2) cardiovascular performance, and (3) tissue characteristics such as diffusion distances between blood capillaries and mitochondria.

When ambient oxygen levels fall below P_c and oxygen intake thus becomes constrained by oxygen availability, one important determinant of the consequences is the extent of the animal's anaerobic capabilities. In facultative anaerobes, when aerobic ATP production flags, anaerobic production can to some extent take its place (see p. 52). Animals that have a highly limited anaerobic capability (or none at all) are in a more tenuous position. The total possible anaerobic production of ATP in trout, crabs, squids, and mammals, for instance, is so modest that when such animals are unable to meet their oxygen needs, they almost immediately face an uncompromising need to curtail some of their functions for lack of ATP. The consequences may be of great potential ecological significance. For example, in salmon, perch, and trout—all highly aerobic fish—when oxygen levels are lowered progressively through the range of oxygen conformity during swimming, the peak sustained speeds the fish can maintain

---

BOX 4.1   HUMAN PEAK OXYGEN CONSUMPTION AND PHYSICAL PERFORMANCE AT HIGH ALTITUDES

When humans and other mammals are exposed to decreased atmospheric partial pressures of oxygen, they marshall vigorous defenses, discussed in Chapters 12 and 13. As a consequence, they enjoy a substantial measure of oxygen regulation as the ambient oxygen level falls with increasing altitude. Nonetheless, if we consider the most exacting of circumstances—the ability for *maximal* oxygen uptake at high montane altitudes—we find that humans function as conformers, as shown in Figure 4.13.

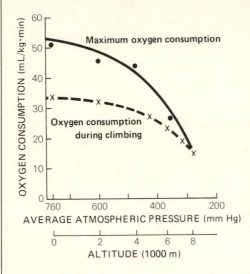

Figure 4.13 Oxygen consumption of mountaineers as a function of atmospheric pressure after about 2 months of acclimatization to high altitude. At all altitudes, the air consists of about 21 percent oxygen; thus, the oxygen partial pressure is about 21 percent of the total atmospheric pressure. The solid curve depicts maximum oxygen consumption; symbols (●) are mean experimental values. The dashed curve depicts oxygen consumption during normal mountain climbing; symbols (×) are mean experimental values—either measured directly or calculated from body weight and rate of climbing. [From L.G.C.E. Pugh, Animals in high altitudes: Man above 5,000 meters—mountain exploration. In D.B. Dill (ed.), *Handbook of Physiology, Section 4: Adaptation to the Environment.* American Physiological Society, Washington, DC, 1964.]

The figure also shows the rates of oxygen consumption of people climbing at various altitudes. The oxygen cost of any given rate of climbing is the same regardless of altitude. However, as people experience a decline in their peak capacity for oxygen intake at high altitude, they find that meeting any particular submaximal oxygen need becomes more and more arduous (p. 49), and thus they cut back on their rate of climbing. This is why their oxygen consumption during climbing falls with increasing altitude. At very high altitudes, even a very slow rate of climbing demands very close to the maximal possible oxygen consumption and thus is extremely taxing. Years ago, E.F. Norton, climbing without supplemental oxygen at 8500 m on Mt. Everest, reported his attempts to take 20 consecutive uphill steps. He never could. Recently, people have reached the peak of Mt. Everest (8848 m) without use of supplemental oxygen. Data gathered on them indicate that this feat is so close to the margin of what is possible that success or failure could well depend on whether the barometric pressure is relatively high or low on the day of the climb.

---

become lower and lower. Box 4.1 discusses another case of the relation between oxygen availability and physical performance.

SELECTED READINGS

*Åstrand, P.-O. and K. Rodahl. 1986. *Textbook of Work Physiology,* 3rd ed. McGraw-Hill, New York.

Belkin, D.A. 1964. Variations in heart rate during voluntary diving in the turtle *Pseudemys concinna. Copeia* **1964:**321–330.

Bennett, A.F. 1972. The effect of activity on oxygen consumption, oxygen debt, and heart rate in lizards *Varanus gouldii* and *Sauromalus hispidus. J. Comp. Physiol.* **79:**259–280.

*Bennett, A.F. 1978. Activity metabolism of the lower vertebrates. *Annu. Rev. Physiol.* **40:**447–469.

*Brooks, G.A. and T.D. Fahey. 1984. *Exercise Physiology.* Wiley, New York.

Chih, C.R. and W.R. Ellington. 1983. Energy metabolism during contractile activity and environmental hypoxia in the phasic adductor muscle of the bay scallop *Argopecten irradians concentricus. Physiol. Zool.* **56:**623–631.

*DeZwann, A. and V. Putzer. 1985. Metabolic adaptations

of intertidal invertebrates to environmental hypoxia (a comparison of environmental anoxia to exercise anoxia). *Symp. Soc. Exp. Biol.* **39**:33–62.

Dobson, G.P., W.S. Parkhouse, and P.W. Hochachka. 1987. Regulation of anaerobic ATP-generating pathways in trout fast-twitch skeletal muscle. *Am. J. Physiol.* **253**:R186–R194.

Gatten, R. E., Jr. 1981. Anaerobic metabolism in freely diving painted turtles (*Chrysemys picta*). *J. Exp. Zool.* **216**:377–385.

*Gatten, R.E., Jr. 1985. The uses of anaerobiosis by amphibians and reptiles. *Am. Zool.* **25**:945–954.

Gleeson, T.T. 1980. Lactic acid production during field activity in the Galapagos marine iguana, *Amblyrhynchus cristatus*. *Physiol. Zool.* **53**:157–162.

Hammen, C.S. 1980. Total energy metabolism of marine bivalve mollusks in anaerobic and aerobic states. *Comp. Biochem. Physiol.* **67A**:617–621.

Heisler, N. 1982. Transepithelial ion transfer processes as mechanisms for fish acid–base regulation in hypercapnia and lactacidosis. *Can. J. Zool.* **60**:1108–1122.

Herbert, C.V. and D.C. Jackson. 1985. Temperature effects on the responses to prolonged submergence in the turtle *Chrysemys picta bellii*. *Physiol. Zool.* **58**:655–681.

Herreid, C.F., II. 1980. Hypoxia in invertebrates. *Comp. Biochem. Physiol.* **67A**:311–320.

Hochachka, P.W. 1980. *Living Without Oxygen*. Harvard University Press, Cambridge.

*Hochachka, P.W. 1986. Defense strategies against hypoxia and hypothermia. *Science* **231**:234–241.

Hochachka, P.W. and M. Guppy. 1987. *Metabolic Arrest and the Control of Biological Time*. Harvard University Press, Cambridge.

*Hochachka, P.W. and G.N. Somero. 1984. *Biochemical Adaptation*. Princeton University Press, Princeton.

Jones, D.R. 1982. Anaerobic exercise in fish. *Can. J. Zool.* **60**:1131–1134.

Lehninger, A.L. 1986. *Biochemistry,* 4th ed. Worth, New York.

*Macleod, D., R. Maughan, M. Nimmo, T. Reilly, and C. Williams (eds.). 1987. *Exercise. Benefits, Limits and Adaptations*. E. & F.N. Spon, London.

Mangum, C. and W. Van Winkle. 1973. Responses of aquatic invertebrates to declining oxygen conditions. *Am. Zool.* **13**:529–541.

Nadel, E.R. 1985. Physiological adaptations to aerobic training. *Am. Sci.* **73**:334–343.

Nicholls, D. and R. Locke. 1983. Cellular mechanisms of heat dissipation. In L. Girardier and M.J. Stock (eds.), *Mammalian Thermogenesis*. Chapman and Hall, London.

Ott, M.E., N. Heisler, and G.R. Ultsch. 1980. A re-evaluation of the relationship between temperature and the critical oxygen tension in freshwater fishes. *Comp. Biochem. Physiol.* **67A**:337–340.

Putnam, R.W. 1979. The basis for differences in lactic acid content after activity in different species of anuran amphibians. *Physiol. Zool.* **52**:509–519.

Rome, L.C., P.T. Loughna, and G. Goldspink. 1985. Temperature acclimation: Improved sustained swimming performance in carp at low temperatures. *Science* **228**:194–196.

Saz, H.J. 1981. Energy metabolism of parasitic helminths: Adaptations to parasitism. *Annu. Rev. Physiol.* **43**:323–341.

Shick, J.M., A. DeZwaan, and A.M.Th. DeBont. 1983. Anoxic metabolic rate in the mussel *Mytilus edulis* L. estimated by simultaneous direct calorimetry and biochemical analysis. *Physiol. Zool.* **56**:56–63.

Smellie, R.M.S. and J.F. Pennock (eds.). 1976. *Biochemical Adaptation to Environmental Change*. Biochemical Society Symposia, No. 41. The Biochemical Society, London.

Smith, E.L. and R.L. Hill. 1983. *Principles of Biochemistry,* 7th ed. McGraw-Hill, New York.

*Somero, G.N. and J.J. Childress. 1980. A violation of the metabolism–size scaling paradigm: Activities of glycolytic enzymes in muscle increase in larger-size fish. *Physiol. Zool.* **53**:322–337.

Sutton, J.R. and N.L. Jones. 1983. Exercise at altitude. *Annu. Rev. Physiol.* **45**:427–437.

Taigen, T.L. and F.H. Pough. 1985. Metabolic correlates of anuran behavior. *Am. Zool.* **25**:987–997.

*Ultsch, G.R., H. Borschung, and M.J. Ross. 1978. Metabolism, critical oxygen tension, and habitat selection in darters (*Etheostoma*). *Ecology* **59**:99–107.

Von Brand, T. 1946. *Anaerobiosis in Invertebrates*. Biodynamica, Normandy, MO.

See also references in Appendix A.

chapter 5

The Active Animal

The principal aim of this chapter is to examine the energetic costs of exercise and the significance of those costs in the life histories of animals. In a sense, we are completing the discussion of exercise started in Chapter 4. There, we emphasized the interplay of aerobic and anaerobic catabolism in the support of exercise. Here, we focus on forms of exercise that are mostly aerobically fueled (i.e., submaximal forms), because although highly anaerobic forms of exertion (e.g., sprints) sometimes spell the difference between life and death, it is aerobic types of exercise that predominate in the lives of animals in regard to energetic cost.

As noted in Chapter 3, exercise is one of the factors that most potently affect the metabolic rates of animals. Table 5.1 illustrates, for example, that during sustained activities, the metabolic rates of humans range over more than an order of magnitude, depending on the type of activity.

METHODOLOGY IN THE STUDY OF ACTIVE ANIMALS

The study of animals that are actively moving presents challenging methodological problems. A brief look at current techniques of study is worthwhile for a number of reasons. One particularly important reason is that our understanding of active animals continues to be constrained because of limits in available techniques.

Laboratory Studies

One of the principal topics of interest in the study of exercise is the relation between speed of locomotion and metabolic rate. To investigate this relation, a means of controlling a moving animal's speed must be available. For running or walking animals, the device most commonly used is a motor-driven treadmill, as diagrammed in Figure 5.1. The animal stands on a belt, which is driven round and round by a motor, and to keep its position, it must walk or run at exactly the same speed as the belt is passing beneath its feet. The treadmill can be tilted at an angle relative to the horizontal to study uphill or downhill running.

Flying animals such as birds have been trained to fly against the air current in a wind tunnel; then their speed of flight has been controlled by varying the speed of the current. For swimming animals such as fish, a device analogous to a wind tunnel—but filled with water instead of air—can be used.

One of the greatest challenges in employing devices of these sorts lies in training animals to use them. Vance Tucker, who pioneered the use of wind tunnels in the study of avian flight, initially tried to prevent birds from landing during experiments simply by placing an electrical grid on the floor of the test chamber. He reports on one parakeet that evaded being shocked by turning upsidedown as it fell to the floor; by landing on its back rather than its feet, it was insulated from the grid by feathers!

Once an animal has been induced to run, fly, or swim at a steady speed, its metabolic rate is usually determined by measuring its rate of oxygen consumption.

Field Studies

One of the central goals for the physiological study of animal activity is a better understanding of the natural biology of animals. To understand how animals function in their natural habitats, studies on individuals actually living in those habitats are par-

Table 5.1 REPRESENTATIVE METABOLIC RATES OF YOUNG ADULT HUMANS OF AVERAGE BUILD DURING SUSTAINED, AEROBICALLY SUPPORTED TYPES OF EXERCISE

Type of activity	Metabolic rate[a] (kcal/min)
Lying down	1.5
Sitting	1.7
Standing	2.1
Walking at 2 miles per hour (mph)	2.9
Walking at 4 mph	5.1
Bicycling at 13 mph	7.6
Jogging at 7 mph	14
Crawl swimming at 2 mph	14
Running at 10 mph	20

[a] In interpreting these values, recognize that variability among individuals can be considerable. Walking, bicycling, and running are assumed to be on the level.

Source: P.-O. Åstrand and K. Rodahl, *Textbook of Work Physiology.* McGraw-Hill, New York, 1970.

ticularly desirable. The techniques presently available for the study of free-roaming animals have limits, however. For example, neither heat production nor oxygen consumption can ordinarily be measured on a free-ranging individual. Thus, the metabolic rates of such individuals must be assessed by techniques that are more open to error. Here, we look at a few of the methods that are currently available for physiological study of animals loose in the wild.

A technique that is useful in the laboratory as well as in field situations is *radio telemetry*. The animal is outfitted externally or internally with a small, battery-powered radio transmitter (''telemeter'') that is provided with one or more sensors of physiological variables. Information on the variables is then transmitted by radio, permitting the animal freedom of movement while under study. Telemeters can be used to monitor any of the numerous physiological parameters that either are intrinsically electrical (e.g., the electrocardiogram) or can be transduced into electrical signals with available technology (e.g., blood pressure or body temperature). Thus, as shown in Figure 5.2, the blood pressure and heart rate of a giraffe can be studied even as the individual gallops across the plains of Africa.

Most of our knowledge of the energetic costs of *particular types of exercise* in nature has been obtained by simulating the activities in the laboratory and using standard laboratory methods to measure metabolism (e.g., oxygen consumption). The confidence that can be placed in such laboratory measures as indicative of the natural situation clearly depends on the fidelity of the simulations. An important constraint in this type of research is that animals must be made to continue activities for long enough that a *steady-state* measure of their metabolism can be obtained. The flight of a migrating bird can be simulated well. But simulation of the activity of a bird flitting about in the trees is more challenging.

Often, it is of interest to determine the *total daily energy expenditure* or *average daily metabolic rate* of a free-roaming animal. The two principal techniques used for this purpose at present are (1) time–energy budgets and (2) doubly labeled water methods.

Time–Energy Budgets Probably the most common way in which the total daily energy expenditure of animals in nature has been estimated is through integration of laboratory data on the energetic costs of activities and field data on the time spent in each type of activity. To illustrate, if we wanted to know the daily energy expenditure of a bird, we might estimate the amount of time spent resting, singing, foraging, defending territorial boundaries, and flying by observing the bird in the wild. Such a summary of an animal's time expenditures is termed a *time budget*. We would estimate the energetic cost of each activity by simulating the activity in the laboratory. Then we would compute the bird's total daily energy demand in the wild by multiplying the cost per unit time of each activity by the time spent in the activity and summing the products. Such a computational integration of energy data with the time budget is called a *time–energy budget*.

Each step in the construction of a time–energy budget presents challenges. We have already pointed out the possible difficulties in simulating activities in the laboratory. Observing and categorizing an animal's activities in nature over the full 24-h day can also be difficult.

Doubly Labeled Water Methods These methods of determining total daily energy expenditure derive their name from the fact that the animal under study is administered water composed of unusual isotopes of *both* hydrogen and oxygen. The most common version of the doubly labeled water method is called

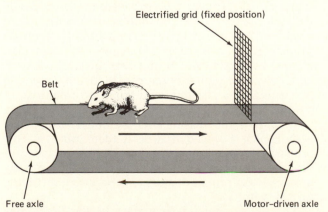

Figure 5.1 Diagram of a treadmill. Turning of the motor-driven axle causes the belt to move continually as indicated by arrows. If the animal fails to run as fast as the belt moves, it is swept backward into the electrified grid and receives a mild shock. Animals learn to avoid being shocked.

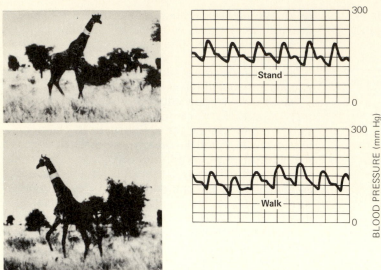

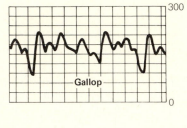

1 s

Figure 5.2 A giraffe equipped with a blood pressure telemeter. The transmitter was taped externally to the neck and the pressure transducer inserted surgically in the right carotid artery. Records were then obtained while the giraffe was free in the wild. The arrangement permitted measurement of both blood pressure and heart rate in a variety of exercise states. The sharp drops in pressure during galloping coincided with the animal's front-hoof beats. [From R.L. Van Citters, W.S. Kemper, and D.L. Franklin, *Science* **152**:384–386 (1966). Copyright 1966 by the American Association for the Advancement of Science.]

the $D_2{}^{18}O$ *method* because the isotope used to label hydrogen is deuterium and that used for oxygen is oxygen-18.

In the $D_2{}^{18}O$ method, the animal is simultaneously administered measured amounts of D_2O and $H_2{}^{18}O$. These mix with the animal's body water, and subsequently the rates of elimination of the deuterium and oxygen-18 from the body water are followed. From these rates, the animal's rate of carbon dioxide production can be calculated, thus providing a measure of metabolic rate. Once the D_2O and $H_2{}^{18}O$ have been administered, the animal can be released in its natural habitat, subsequently to be recaptured for determination of the amounts of deuterium and oxygen-18 eliminated during the experimental period. The metabolic rate computed is then the *average* rate over the entire period.

The initial observation that led to the development of the $D_2{}^{18}O$ method was that the oxygen of expired carbon dioxide is in isotopic equilibrium with the oxygen of body water. Thus, if the body water consists of given proportions of $H_2{}^{16}O$ (ordinary water) and $H_2{}^{18}O$, the CO_2 expired by the animal will contain both oxygen-16 and oxygen-18 in approximately the same proportions. Accordingly, when the concentration of $H_2{}^{18}O$ is experimentally elevated, the excess oxygen-18 will gradually be voided to the environment in expired CO_2, and the rate at which

this occurs will depend on the rate of CO_2 production and elimination.

There is a second major route by which excess oxygen-18 is voided from the body water. Water lost through evaporation, urination, and other mechanisms consists of both $H_2{}^{16}O$ and $H_2{}^{18}O$, meaning that some of the excess $H_2{}^{18}O$ will be carried away in general water losses. Accordingly, the total rate of dissipation of excess oxygen-18 is in fact a function of *both* the rate of CO_2 production *and* the rate of water loss. If an investigator were to measure only the rate of oxygen-18 elimination, he or she would not know how much this reflected dissipation of oxygen-18 in either water or CO_2 alone. However, if an independent measure of the rate of water loss is obtained, the fraction of oxygen-18 lost in water can be calculated and subtracted from the total loss of oxygen-18 to obtain the amount of oxygen-18 lost in CO_2. This is the reason D_2O is administered along with $H_2{}^{18}O$. Labeled hydrogen (deuterium) is lost primarily in the form of water. Thus knowledge of the rate of elimination of deuterium from the body provides the required independent measure of the rate of water loss.

The $D_2{}^{18}O$ method is subject to the limitations inherent in any method that estimates metabolic rates by measurement of CO_2 elimination (p. 28). However, it appears to be the most intrinsically reliable

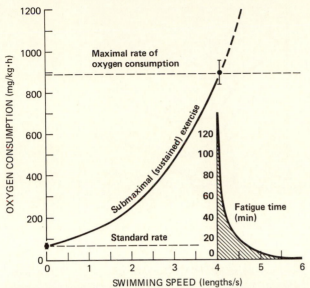

Figure 5.3 Oxygen consumption (solid line) as a function of swimming speed in yearling sockeye salmon (*Oncorhynchus nerka*) acclimated and tested at 15°C. Fish weighed about 50 g, were about 18 cm long, and were tested in a "water tunnel" in aerated fresh water. Speeds requiring less than the maximal rate of oxygen consumption could be sustained for long periods. On the other hand, speeds requiring a (theoretical) rate of oxygen consumption greater than the maximal rate (note rising dashed line) demand a steady net production of ATP by anaerobic glycolysis and are relatively rapidly fatiguing. [After J.R. Brett, *J. Fish. Res. Board Can.* **21**:1183–1226 (1964).]

of the methods now available for the estimation of the metabolic rates of free-ranginig animals.

ENERGETIC COSTS OF EXERCISE

Let us begin examination of the energetics of exercise by looking at the relations between metabolic rate and speed in three forms of vertebrate locomo-

tion: swimming in fish, running in mammals, and flying in birds.

Cost per Unit Time as a Function of Speed

Figure 5.3 depicts data on the rate of energy consumption of yearling sockeye salmon during swimming in a water tunnel. Our particular concern here is with submaximal effort: exercise that requires less than the maximal rate of oxygen consumption (p. 47). Speeds that meet this description are often called *sustained speeds,* for inasmuch as they can be supported entirely by aerobic ATP production, they can be sustained for long times (p. 43). For the yearling salmon, speeds of up to 4 body lengths/s constituted submaximal effort. The energy costs of such exercise can be assessed accurately by measuring steady-state oxygen consumption, and that was the approach taken. As the data show, the cost of swimming increased *exponentially* with swimming speed.* This is a common pattern in fish.

Exponential relations plot as straight lines when graphed on semilogarithmic coordinates (p. 84). Thus, it is convenient to use such coordinates in presenting data on the exercise energetics of fish. The data for sockeye salmon swimming at 15°C, which we have just seen on rectangular coordinates in Figure 5.3, are replotted semilogarithmically in Figure 5.4 (see middle dashed line). Figure 5.4 also presents results for sockeye salmon at two other temperatures and data for several other species. In each case, only sustained swimming speeds are depicted, and the highest rate of oxygen consumption shown is in fact the maximal possible rate. As can be seen, both (1) the relation between oxygen consumption and speed and (2) the maximal rate of oxygen consumption depend on the species and the

---

*The characteristics of exponential functions are discussed in detail on p. 83.

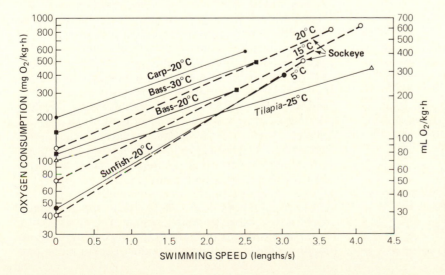

Figure 5.4 Oxygen consumption as a function of swimming speed in small (50 ± 15 g) fish of five species: carp (*Cyprinus carpio*) at 20°C; largemouth black bass (*Micropterus salmoides*) at 20 and 30°C; sockeye salmon (*Oncorhynchus nerka*) at 5, 15, and 20°C; pumpkinseed sunfish (*Lepomis gibbosus*) at 20°C; and *Tilapia nilotica* at 25°C. Fish were exercised up to maximum sustained speeds. Note that oxygen consumption is expressed on a logarithmic scale. Data for sockeyes at 15°C are the same as in Figure 5.3 [From J.R. Brett, *Respir. Physiol.* **14**:151–170 (1972).]

temperature. Bass at 20°C, for example, exhibit a maximal oxygen consumption only 37 percent as great as sockeye salmon at the same temperature, and they can attain speeds of only 2.4 lengths/s, as compared to 3.7 lengths/s in the salmon. The data on bass and salmon also exemplify thermal effects. In each species, raising the temperature not only increases the resting metabolic rate (zero speed) but also elevates the metabolic rate associated with any given speed.

Data on the rate of oxygen consumption as a function of running speed in mammals are presented in Figure 5.5. Note that the coordinates are rectangular and thus comparable to those in Figure 5.3. In contrast to fish, mammals generally exhibit a *linear* relation between energy expenditure and speed. Because the animals in Figure 5.5 were not exercised to peak sustained speeds, their highest recorded rates of oxygen consumption are not indicative of their maximal rates. However, the data do permit interspecific comparison of the quantitative relation between metabolism and speed. Significantly, the slope of this relation tends to increase with decreasing body size—indicating that small mammals must increase their weight-specific metabolic rates to a considerably greater extent than large mammals to attain

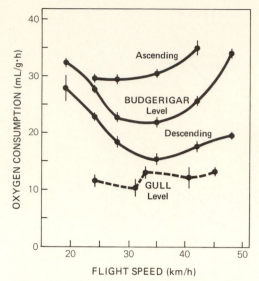

Figure 5.6 Oxygen consumption as a function of flight speed in budgerigars, *Melopsittacus undulatus* (solid curves), and laughing gulls, *Larus atricilla* (dashed curve). The birds were trained to fly in a wind tunnel and wore transparent masks, through which air was circulated, so that their respiratory gas exchange could be monitored. The gulls were studied during horizontal (level) flight only. The budgerigars flew horizontally and also at ascending and descending angles of 5°. Vertical bars delimit twice the standard error on either side of the mean response. The cost of flight was elevated to some extent by the extra drag contributed by the mask and tubing attached to the mask. [From V.A. Tucker, *J. Exp. Biol.* **48**:67–87 (1968); data for gulls from V.A. Tucker, *Sci. Am.* **220**(5):70–78 (1969).]

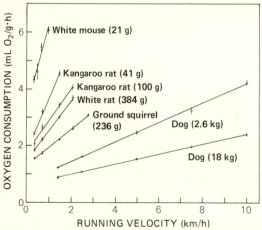

Figure 5.5 Oxygen consumption as a function of running speed at 22–27°C in six species of mammals: laboratory mice, laboratory rats, Merriam's kangaroo rats (*Dipodomys merriami,* average weight: 41 g), bannertailed kangaroo rats (*D. spectabilis,* average weight: 100 g), round-tail ground squirrels (*Citellus tereticaudus*), and domestic dogs (mongrels weighing 2.6 kg and Walker foxhounds weighing 18 kg). Animals were studied during horizontal running on treadmills. They were not necessarily exercised to their maximal rates of oxygen consumption. Horizontal bars indicate mean oxygen consumption, and vertical bars delimit twice the standard error on either side of the mean. By extrapolation, the oxygen consumption at zero speed is obtained as the *Y*-intercept. *Y*-intercepts exceed basal metabolic rates for the various species by a factor of 1.7 on the average. The greater-than-basal oxygen consumption at zero speed probably at least partly reflects the energetic cost of maintaining a running posture. [From C.R. Taylor, K. Schmidt-Nielsen, and J. L. Raab, *Am. J. Physiol.* **219**:1104–1107 (1970).]

any particular absolute speed (km/h). A similar trend is evident in fish. A 150-g salmon that increases its weight-specific oxygen consumption by the same amount as a 50-g salmon will be able to swim at about the same number of body lengths per second—meaning that because it has a longer body, it can swim at a greater absolute speed than the 50-g fish for the same weight-specific metabolic effort.

Finally, let us turn to data on budgerigars (parakeets) and laughing gulls flying at sustained speeds in a wind tunnel. As seen in Figure 5.6, the relations between oxygen consumption and speed in these birds are of still another nature than those seen in fish or mammals. Although oxygen consumption during horizontal swimming in fish increases steadily and exponentially with speed and although that during horizontal running in mammals typically increases linearly with speed, oxygen consumption during horizontal flight in budgerigars passes through a minimum, decreasing as speed is increased from 20 to 35 km/h and increasing as speed is elevated above 35 km/h. In the laughing gulls, the relation between metabolic rate and speed during horizontal flight is still more complex. Again we see evidence that larger animals (gulls in this case) can make more rapid progress for a given weight-specific metabolic rate than smaller ones.

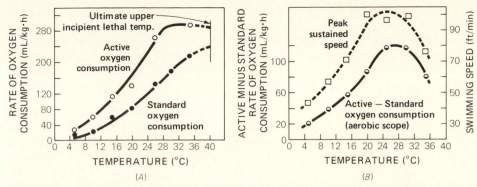

Figure 5.7 (*A*) Oxygen consumption as a function of temperature in 3.8-g goldfish (*Carassius auratus*) acclimated to test temperatures. The lower curve shows the standard rate of oxygen consumption, and the upper curve shows the rate during maximal sustained swimming (active rate). (*B*) The solid line depicts the difference between active and standard oxygen consumption at each temperature, termed the aerobic scope for activity; compare with part (*A*). The dashed line depicts maximal sustained swimming speed as a function of temperature. The left ordinate gives units for aerobic scope, whereas the right ordinate gives units for swimming speed. [From F.E.J. Fry and J.S. Hart, *Biol. Bull.* (*Woods Hole*) **94**:66–77 (1948).]

Aerobic Scope for Activity and Aerobic Expansibility

An animal's *aerobic scope for activity* at any particular temperature is usually defined to be the difference between its maximal exercise-induced rate of oxygen consumption at that temperature and its resting rate of oxygen consumption at the same temperature. To illustrate, let us consider the results of experiments on goldfish in Figure 5.7*A*. First, the standard, resting rate of oxygen consumption of the fish was measured as a function of temperature. Then, the fish were exercised to peak sustained speeds to determine the maximal rate of oxygen consumption that could be elicited by activity at each temperature (the

so-called "active" rate of oxygen consumption). The standard rate was then subtracted from the maximal rate at each temperature to obtain the aerobic scope for activity, which is plotted as the solid curve in Figure 5.7*B*. The scope represents the extent to which the rate of oxygen consumption can be increased above the resting level in support of activity at each temperature. It varies both from temperature to temperature, as shown already in Figure 5.7, and from species to species, as exemplified in Figure 5.8.

In addition to showing the aerobic scope of goldfish, Figure 5.7*B* also depicts their peak sustained swimming speed as a function of temperature. Analysis of the two curves reveals two major points.

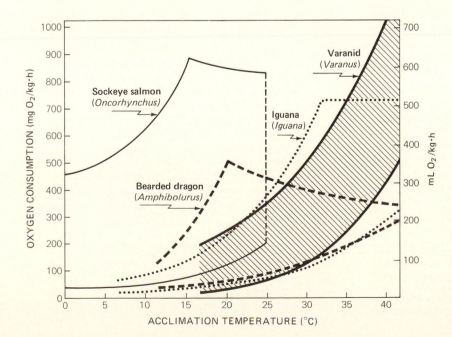

Figure 5.8 The standard rate of oxygen consumption (lower line) and the maximal rate elicited by activity (upper line) in sockeye salmon and three species of lizards. Aerobic scopes can readily be visualized. More-recent studies indicate that the maximal oxygen consumption in some varanids is higher than shown here: over 1000 mL/kg·h at 42°C. [From J.R. Brett, *Respir. Physiol.* **14**:151–170 (1972).]

First, the aerobic scope and peak swimming speed of goldfish tend to vary together as temperature is raised and lowered; at temperatures where the fish are capable of increasing their oxygen consumption to a particularly marked extent, they are also able to swim especially rapidly. The second point, however, is that this relation is not quantitatively simple. Note, for instance, that although the scope is six times greater at 25°C than at 5°C, the maximal swimming speed is just over two times greater. The lesson is that one must be cautious in extrapolating from scope to exercise performance. Fundamentally, the scope tells us about *oxygen consumption*. It is a measure of the flexibility and capability of the systems responsible for the uptake, transport, and use of oxygen during exercise. Differences in scope may not faithfully mirror differences in locomotory performance, especially when the comparison is being made across species lines.

One interesting point illustrated by Figure 5.8 is that fish and lizards, despite living in such different environments as water and air, have basically similar peak rates of oxygen consumption and aerobic scopes. This similarity extends to other poikilothermic vertebrates as well.

Increases in temperature tend within limits to expand aerobic scope in poikilothermic animals. In some species, such as many turtles and lizards (e.g., *Varanus,* Figure 5.8), aerobic scope increases with temperature all the way up to the maximal healthful temperature. In other species (e.g., goldfish, Figure 5.7*B*), the scope goes through a maximum at intermediate temperatures. The reasons for the latter phenomenon are not always clear. One factor for aquatic animals is that the concentration of dissolved oxygen in water typically declines as temperature increases (p. 246), making oxygen more difficult to obtain at high temperatures than low.

Another way of looking at the flexibility of aerobic catabolism is as the *ratio* of active oxygen consumption over resting oxygen consumption: known as **aerobic expansibility.** Suppose that a fish were to have a resting rate of oxygen consumption of 0.05 and a maximal rate of 0.30 mL/g·h; and suppose the comparable figures for a mammal were 2 and 12 mL/g·h. The fish would have a far lower aerobic scope than the mammal, and yet the expansibilities of the two would be identical: 6. That is, in *proportion* to their own resting oxygen consumption, both species would be identical in their ability to increase oxygen consumption. This example illustrates the insight that expansibilities can provide.

There can be a lot of variation in expansibility within a phyletic group. The expansibilities of young adult men and women, for example, range from about 8 to 20, depending on constitution and training.* Other large mammals tend to be similar to hu-

*In mammals and birds, expansibility and scope are usually calculated by comparing the peak rate of oxygen consumption with the *basal* rate.

mans in their expansibilities. Small running mammals tend to have relatively low expansibilities near 5–10. Bats and small birds in flight, on the other hand, may display expansibilities of at least 15–25. A rough average for all mammals is 10. *A value of 10 is also the rough average for all other classes of vertebrates as well.*

As noted in Chapter 3 and discussed further in Chapter 6 (p. 104), the standard metabolic rates of fish, amphibians, and reptiles are typically no more than 1/10 to 1/4 as high as the basal rates of mammals and birds of similar body size. Given that expansibility averages about 10, we see that the *peak* rates of oxygen consumption of fish, amphibians, and reptiles are of the same order of magnitude as the *basal* rates of mammals and birds. Table 5.2 provides data on two pairs of vertebrates: (1) a salmon and rat of similiar size and (2) a monitor lizard and guinea pig of similar size. Salmon and monitor lizards are among the most aerobically competent of fish and reptiles, and yet still it is clear that the mammals operate on an entirely different aerobic plane, having active rates and scopes that are well above those of the other animals.

The highest aerobic expansibilities occur in insects, a number of which consume oxygen 50–200 times faster when flying than when at rest. Some of the values for insects, however, are not entirely comparable to those for other animal groups because many insects increase their body temperature markedly during flight (p. 128) and thus a substantial thermal effect may be present in addition to the effect of activity.

Cost of Transport (Cost per Unit Distance)

Cost of transport is defined to be the metabolic expenditure required to cover a unit of distance. For an animal traveling at any given speed, it is calculated by dividing the animal's metabolic rate (cost per unit time) by its speed (distance covered per unit time). Suppose, for instance, that a 50-g salmon displays an oxygen consumption of 30 mL/h when swimming at 2400 m/h; then its cost of transport when traveling at that speed is

$$\left(30\,\frac{\text{mL O}_2}{\text{h}}\right) \Big/ \left(2400\,\frac{\text{m}}{\text{h}}\right) = 0.0125\,\frac{\text{mL O}_2}{\text{m}}$$

Cost of transport is a measure of the effectiveness of the animal in using its metabolic resources to cover distance, animals with a high cost being less effective than those with a low cost. The measure is familiar to all of us in its inverse form for expressing the performance of automobiles. We often are concerned about our car's miles per gallon. If we spoke of gallons per mile instead, we would be using an expression of cost of transport, and clearly we would want to minimize the cost. Often cost of transport is expressed in weight-specific terms. Returning to the salmon mentioned earlier, we see that its weight-

Table 5.2 RATES OF AEROBIC CATABOLISM DURING REST AND DURING EXERCISE OF PEAK INTENSITY IN TWO PAIRS OF VERTEBRATES: A FISH AND MAMMAL OF SIMILAR BODY WEIGHT, AND A LIZARD AND MAMMAL OF SIMILAR WEIGHT

| Species | Body weight (g) | Test temperature[a] (°C) | Oxygen consumption | | |
			Basal or standard (mL/g·h)	Active (mL/g·h)	Aerobic scope (mL/g·h)
Rat (*Rattus*)	230	30	0.9	4.6	3.7
Salmon (*Oncorhynchus*)	230	15	0.05	0.49	0.44
Guinea pig (*Cavia*)	880	30	0.6	3.7	3.1
Monitor lizard (*Varanus*)	670	40	0.11	1.0	0.89

[a] Test temperatures for the fish and lizard were those at which aerobic scope was maximal.

Sources: Mammals—P. Pasquis, A. Lacaisse, and P. Dejours, *Respir. Physiol.* **9**:298–309 (1970), basal rates calculated according to Kleiber equation therein; salmon—J.R. Brett, *J. Fish. Res. Board Can.* **22**:1491–1501 (1965); varanid—A.F. Bennett, *J. Comp. Physiol.* **79**:259–280 (1972).

specific metabolic rate while covering 2400 m/h is 0.6 mL O_2/g·h. Thus, its weight-specific cost of transport is 0.6/2400 = 0.00025 mL O_2/g·m. This is the cost of moving one gram of body weight over a distance of one meter.

In Figure 5.6 we have seen the relation between rate of oxygen consumption and horizontal flight speed in budgerigars. The *cost per hour* (mL O_2/g·h) is minimal at a speed of 35 km/h. Thus, a budgerigar with given nutrient reserves could stay aloft for the *longest time* by flying at this speed. The *cost of transport* (cost/g·km) of budgerigars flying horizontally is graphed as a function of speed in Figure 5.9. Here, we see that the *cost of covering a given distance* is minimal at a speed of 42 km/h. Thus, a budgerigar with given nutrient reserves could cover the *longest distance* by flying at this latter speed. At 42 km/h, the bird uses more than the minimal amount of energy *per unit time,* but by flying at 42 km/h rather than 35 km/h, it also covers more distance per unit time. It is the interaction of these two functional relations that determines the speed at which cost per unit of distance is minimal.

Having examined how cost of transport and speed are related in a flying bird, let us now briefly examine

their relations in representative swimming fish and running mammals. Figure 5.10 shows cost of transport as a function of speed in the young salmon we examined earlier in Figure 5.3. Note that at low speeds the cost is very high. Although the increment in the rate of oxygen consumption caused by activity at such speeds is quite small (see Figure 5.3), the animal must also sustain its standard maintenance oxygen consumption, and the latter does not contribute to locomotion; all told then, the rate of metabolism per unit distance covered is large. There is a broad range of speeds, from about 1 to 2.5 lengths/ s, where the cost of transport is close to minimal.

Before looking at mammals, it is appropriate to introduce the distinction between *net* and *gross* cost of transport. The gross cost is the total cost per unit of distance, including the resting oxygen consumption as well as the increment caused by activity. All the costs discussed thus far are gross costs. The net cost is obtained by subtracting the resting oxygen consumption from the gross cost. It expresses the *increment* in oxygen consumption required to traverse a unit of distance. To illustrate, if a mammal has a resting rate of oxygen consumption of 2 mL/ g·h and a total consumption of 4 mL/g·h when run-

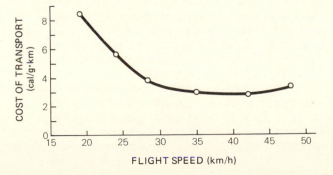

Figure 5.9 Cost of transport as a function of speed during horizontal flight in still air in budgerigars. Curve was computed from data for horizontal flight in Figure 5.6 [From V.A. Tucker, *J. Exp. Biol.* **48**:67–87 (1968).]

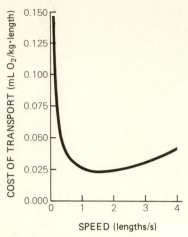

Figure 5.10 Cost of transport as a function of speed in year-ling sockeye salmon at 15°C. Both parameters are expressed relative to the body length of the fish. Data are the same as those in Figure 5.3.

ning at 4 km/h, the gross cost is 4/4 = 1 mL/g·km, but the net cost is (4–2)/4 = 0.5 mL/g·km.

As seen in Figure 5.5, mammals often display a linear relation between rate of oxygen consumption and running speed. For any given species, the Y-intercept of the line relating oxygen consumption to speed represents the oxygen consumption of an animal in the running posture at zero speed. If we treat the Y-intercept as the resting rate of oxygen consumption, then it turns out that, in mammals, the net cost of transport is a constant regardless of speed. Figure 5.11 depicts both the gross and net costs of transport in one of the species shown in Figure 5.5, the ground squirrel. At low speeds, the gross cost is very high, for the same reason it is high in salmon: The resting oxygen consumption, which in an im-

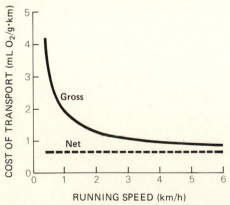

Figure 5.11 Net and gross cost of transport as functions of speed in ground squirrels, *Citellus tereticaudus*. Data are the same as those in Figure 5.5. Net cost was calculated by subtracting the oxygen consumption at zero speed (extrapolated Y-intercept, Figure 5.5) from the total oxygen consumption while running, then dividing by speed.

mediate sense contributes nothing to locomotion, is included in the cost and in fact dominates because the increment caused by running at low speeds is small. As running speed increases, the gross cost approaches the net cost asymptotically.

Minimal Cost of Transport as a Function of Body Size
Suppose that for each species studied we identify the *minimal* cost of transport displayed by the species during horizontal locomotion, regardless of the speed at which the minimum occurs. (As seen earlier, flying and swimming animals often display discrete minima; in the case of running animals an approach often used is to treat the net cost as the minimal cost.) If we then plot minimal cost of transport as a function of body weight, we come upon some of the major recent revelations in the study of exercise energetics.

As shown in Figure 5.12, animals tend to fall into relatively discrete groups depending on their primary mode of locomotion, whether it be running, flying, or swimming. Furthermore, within each group, the minimal cost of transport is a reasonably regular function of body size.

Let us elaborate. The three solid lines in Figure 5.12 show the relation between minimal cost of transport and body weight for runners, fliers, and swimmers. Consider, first, animals for which the primary mode of locomotion is flying; even though they represent such diverse taxonomic groups as insects, bats, and birds, note that values for their minimal cost of transport during flight all tend to fall along a *single* line. Similarly, if we consider lizards and species of mammals that move primarily by walking or running, we find that their values for cost of transport during running tend to fall along one line, and recently in fact it has been shown as well that running ants, cockroaches, and land crabs fall on the same line as the lizards and mammals. Remarkably, in short, among animals engaged in their primary form of locomotion, the minimal cost of transport evidenced by a species of given body size *depends principally on the species' form of locomotion, not its taxonomic position.*

Note that, for animals of a particular size, running is the most costly form of locomotion, while swimming is the least costly. The differences in cost among the three forms of locomotion are more substantial than they may appear from visual inspection of Figure 5.12 because cost of transport is plotted there on a logarithmic scale. For a 100-g animal, the cost of running a unit of distance is about 3.6 times as great as that of flying the same distance and about 14 times greater than the cost of swimming, according to the analysis in Figure 5.12.

Among animals of any one locomotory type, as suggested earlier (p. 62), species of large body size cover distance at less weight-specific cost than ones of small body size. During flight, for instance, the minimal weight-specific cost of transport for a 100-g

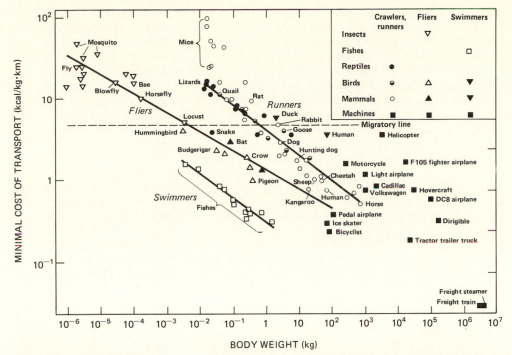

Figure 5.12 Minimal cost of transport as a function of body weight for animals and machines, plotted on log–log coordinates. Data were compiled from a diversity of sources; in each instance, the cost of transport plotted is that pertaining at the speed that was found to minimize, or approximately minimize, the cost. As specified by the key in the upper right corner, symbols indicate both the type of organism and the type of locomotion. Note, for example, that data for running mammals are plotted as open circles, whereas those for swimming mammals are plotted as closed, inverted triangles; accordingly, the values for running and swimming humans are distinguished as a circle and triangle, respectively. Representative species and machines are identified at random. The three solid lines indicate the relation between cost of transport and body weight for runners, fliers, and swimmers. The data reviewed earlier in Figures 5.9, 5.10, and 5.11 are included in this figure. [Reprinted by permission from V.A. Tucker, *Am. Sci.* **63**:413–419 (1975).]

bird is only about 12 percent as great as that for a 0.01-g insect. Within a locomotory type, cost of transport and body weight are related allometrically:

minimal weight-specific cost of transport = aW^b

where W is body weight, a and b are constants, and b is typically about −0.2 to −0.4. Given that cost of transport and weight are related in this manner, their relation plots as a straight line on log–log coordinates, as in Figure 5.12 (see p. 32). One way that the relation between cost of transport and body weight has been interpreted is this: Suppose that we define a "step" to be one cycle of movement of the locomotory appendages—be they the legs of a running animal, the wings of a flying animal, or the tail of a swimming fish. Within a locomotory type, the energy cost per unit body weight to take one step is approximately the same regardless of body size. Small animals, however, need to take many more steps than large animals do to cover a given absolute distance (e.g., 1 km) and thus experience a higher weight-specific cost per unit distance.

Having now identified the major trends that come

to light in the analysis of cost of transport, let us look briefly at a number of more-specific concerns.

(1) When animals engage in a form of locomotion that for them *is not primary,* their cost of transport often proves to be quite different from the cost incurred by animals for which the form of locomotion *is primary.* Compare, for instance, the costs for swimming in ducks and humans with those for swimming in fish (Figure 5.12). The cost for a human to swim a given distance is fully 30–40 times greater than the cost that would be incurred by a fish of similar body size, based on extrapolation of the line for fish. Among other things, this comparison indicates that our own experience of the effort involved in swimming a distance provides no gauge whatsoever of the effort demanded of animals adapted to move primarily by swimming.

(2) A long-standing question has been whether bipedal running differs in cost from quadrupedal running. The question has been of interest in studies of human evolution, for example, because the advent of bipedalism could possibly have raised or lowered the energetic costs of moving about. Now that data

have been collected on many species, it appears that the costs of transport for bipedal and quadrupedal running are in fact not fundamentally different. As shown in Figure 5.13, lizards and mammals running quadrupedally do not differ systematically from birds or mammals running bipedally. As also shown, the range of costs at any one body size is substantial. Relatively ungainly runners such as penguins and geese tend to incur higher costs, for their size, than other runners that appear to be more graceful or adept, such as roadrunners, ostriches, horses, or dogs.

(3) In quadrupedal mammals, the legs are positioned directly under the body, but in lizards the legs project sideways and then down, giving the lizards a sprawling gait. For years it has been argued that in the evolution of vertebrates, the abandonment of the "push-up" posture seen in lizards improved the energetic efficiency of locomotion. However, as shown in Figures 5.12 and 5.13, recent studies have revealed that the minimal energetic cost of covering distance is in fact no different in lizards than it is in mammals.

(4) In humans (Figure 5.12), swimming is a much more costly mode of covering distance than running, but ice skating and bicycling are much less costly than running. Bicycling ranks as one of the least costly of all animal-powered forms of locomotion.

(5) As also shown in Figure 5.12, when effects

of size are taken into account, animals compare favorably with machines such as airplanes, ships, and automobiles in the efficiency with which they can cover distance.

(6) When we consider animals that undertake long migrations, we are struck by the fact that although certain small and medium-sized fish and flying animals—insects, bats, and birds—are migrants, long migrations are rare among small or medium-sized running animals. Some *large* running animals, such as caribou and reindeer, are noted for their migrations; their minimal costs of transport are similar to those of relatively small fish. Vance Tucker has pointed out that if we draw a line (the so-called "migratory line") across Figure 5.12 at about 4.7 kcal/kg·km, most migratory species will be found below the line, and species above the line will be unlikely to be migrants. Apparently, for species that must expend in excess of about 4.7 kcal/kg·km, locomotion is usually too costly for the evolution of long-distance travels to be favored by natural selection.

(7) Now that we have discussed some of the interesting insights to be gained from sets of data such as Figure 5.12, it is perhaps best to end on a cautionary note by pointing out that an animal's minimal cost of transport is far from being the only parameter of relevance to its locomotory performance. One point to remember is that animals do not always move at speeds that minimize their cost of transport. Another point is that the peak rates of aerobic catabolism of animals are also of importance to performance. Consider mammals and lizards of a particular body size, for example. The two types of animal are similar in their net costs of transport and therefore are capable of covering a given distance with similar amounts of energy. Mammals, however, can attain far higher rates of oxygen consumption than lizards, and thus the mammals can achieve higher sustained *speeds* than lizards.

Energetics of Migration

Many animals undertake lengthy migrations that are energetically demanding. Those that do not feed along the way have attracted particular interest, for they must accomplish the entire feat using stored nutrient reserves. It may seem a bit fatuous to ask if, given their known reserves and their measured cost of locomotion, they can accomplish their migrations; obviously at least a portion of the individuals get where they are going. By asking (and attempting to answer) the question, however, we can gain insight into the energetic challenge that migration presents and delineate some of the parameters that affect success or failure.

Studies of migrating insects and vertebrates have shown that fat is the preeminent fuel, and it is instructive first to examine the advantages that fat holds. In Chapter 3 (p. 28), we noted that fat yields

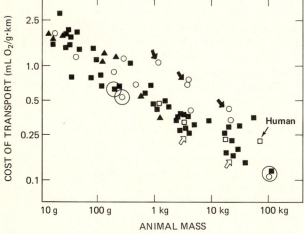

Figure 5.13 Net cost of transport of running in lizards (▲), birds (○), and mammals (□, ■). Animals running bipedally are represented by open symbols; those running quadrupedally, by closed symbols. The large circles identify data for birds and mammals that were compared directly using identical experimental techniques, showing that the cost of bipedal running in the birds was not different from the cost of quadrupedal running in mammals of similar body size. The two large circles at the left encircle data for roadrunners (*Geococcyx californianus*) and ground squirrels (*Citellus tridecemlineatus*); the single large circle at the right encircles data for ostriches (*Struthio camelus*) and Shetland ponies. Solid arrows point to data for geese and penguins; open arrows, to data for dogs. (Reprinted by permission from *Nature,* vol. 282, pp. 713–716, Copyright © 1979 Macmillan Journals Limited.)

about 9.4 kcal of energy per gram when completely oxidized, whereas carbohydrates and proteins yield only half as much, 4–5 kcal/g. The high energy density of fat gives it an advantage for migrating animals, for they must transport each gram of fuel at an energetic *cost* until it is used and they gain the most in return from the fuel that provides the most energy per unit weight. The advantage of fat over the other major form of stored fuel, glycogen, is actually much greater in this respect than the data already presented suggest. Whereas lipids are stored in pure form, glycogen is stored with an appreciable amount of water. This water increases the weight of the glycogen stores without increasing their energy value and gives them an effective energy density (in vertebrates and insects) of only about 1.1 kcal/g: just 12 percent of the value for lipids. Glycogen is not without its advantages as a fuel; it can be catabolized relatively quickly by comparison to lipids (Table 4.1) and, unlike lipids, can be used by vertebrates to support anaerobic glycolysis. Glycogen or glucose is the preeminent fuel of short-term exertion in many animals. Storage of glycogen, however, is clearly an uneconomical use of body weight for the migrant.

At the start of migration in birds that undertake uninterrupted long-distance trips, fat often accounts for 25–50 percent of body weight. Commonly, this fat content is reduced to near 5 percent of body weight by the end of migration. Vance Tucker has estimated the nonstop distance that birds can travel by using these values for fat accumulation and depletion in tandem with measures of minimal cost of transport gathered mostly in laboratory studies. The basic rationale of his computations may be illustrated by example. If a 40-g bird uses fat equivalent to 25 percent of its initial body weight during migration, then its total energy available is about 94 kcal (10 g of fat catabolized, 9.4 kcal/g of fat). Based on studies of budgerigars in wind tunnels, the minimal cost of transport for a 40-g bird is about 0.116 kcal/km (Figure 5.9). Thus, a bird flying through still air at the speed that minimizes cost of transport could travel 94/0.116 = 810 km. Using a relation between cost of transport and body weight like that in Figure 5.12, Tucker calculated flight range as a function of weight as depicted in Figure 5.14. Note that for a given percentage of fat depletion, flight range increases with body weight. The reason is that the weight-specific cost of transport is lower in large birds than in small ones (Figure 5.12).

How do the actual flight ranges of birds compare with those estimated in Figure 5.14? Tucker cites three remarkable examples of birds that migrate over water and cannot stop to feed along the way. Ruby-throated hummingbirds fly 800 km (500 mi) across the Gulf of Mexico. Starting at a body weight of about 4.7 g, their flight range estimated from Figure 5.14 for 50 percent fat is about 1100 km. Hummingbirds are known to commence migration with 40 per-

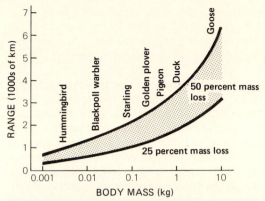

Figure 5.14 Nonstop flight range in still air as a function of body weight at takeoff in birds. Fat is presumed to be the fuel of flight, and range is calculated assuming that fat equivalent to 25 percent (lower line) or 50 percent (upper line) of initial body weight is utilized. It is assumed that birds fly at speeds that minimize their cost of transport. [From V.A. Tucker, *Am. Zool.* **11**:115–124 (1971).]

cent or more of fat. Thus, we see that the estimate of Figure 5.14 is compatible with the actual performance of the birds, though it appears that an 800-km flight represents a substantial fraction of their capability. Blackpoll warblers weighing about 20 g fly 1300 km from New England to Bermuda and have an estimated range (Figure 5.14) of about 1500 km for 50 percent fat. Golden plovers, which weigh near 200 g, migrate 3900 km from the Aleutian to the Hawaiian islands. This distance far exceeds their estimated range of 2600 km for 50 percent fat. In all three of these species, ranging in weight from 4.7 to 200 g, we see that actual flight distances are relatively close to or exceed the estimated range. In interpreting this result, we recognize that Tucker's analysis does not take account of some important but poorly understood factors such as the effects of headwinds or tailwinds and the possibility that flight speed might not always minimize the cost of transport. Also, alternative approaches to estimating migratory range sometimes yield significantly different results. One of Tucker's main conclusions bears repeating, nonetheless. A land bird migrating over water has no second chance if it fails. Thus, if indeed the margin of safety is sometimes as narrow as appears, we must be impressed with the high evolutionary premium that would have been placed on developing abilities for accurate navigation and for utilizing winds to advantage.

Birds migrating over land, of course, can stop to replenish their fuel reserves. The white-crowned sparrow, for example, migrates as far as 4000 km along the west coast of the Northern Hemisphere. The migration is accomplished in a series of nocturnal flights covering 100–600 km. Between flights the birds feed for one to several days, restoring their fat supplies. Weighing about 30 g and carrying about 20 percent fat on takeoff, their maximum range for a single night's flight is estimated to be about 700 km by Tucker's method of calculation—a figure that agrees favorably with the peak distances the birds are known to cover.

Salmon undertake some of the most spectacular of migrations. Spawned in freshwater lakes and streams, they travel to the ocean when young and later return to fresh water to breed as adults. Their migrations against the current in rivers and streams as adults are accomplished without eating and sometimes are of remarkable length. For instance, chinook salmon, after entering the mouths of rivers, are known to travel upstream as far as 3500 km (2200 mi) to spawning grounds in the Yukon.

Sockeye salmon have been studied in some detail physiologically, and the results indicate that they, like some birds, very nearly reach their energetic limits during migration. The body composition of sockeyes was determined before and after a 1000-km migration up the Fraser River from Vancouver to Stuart Lake in British Columbia. The trip depleted their fat reserves by over 90 percent, and in the females protein was depleted by 50–60 percent. Taking into account that the females invest considerable energy in the development of their ovaries, it was calculated that the daily energy demand of swimming in males and females was near 75–80 percent of the peak demand that could be met aerobically. The maximal rate of aerobic catabolism is relatively high in salmon (Figure 5.4), and operating at 75–80 percent of maximum for days on end is an exceptional feat of endurance. Without both of these unusual attributes, salmon could not migrate as they do. Laboratory studies indicate that adult salmon realize a minimal cost of transport at swimming speeds of about 1.8 km/h. In the Fraser River, however, they appear from energetic considerations to average about 4.2 km/h—a speed that hastens completion of their journey but evidently imposes a cost of transport that is nearly twice the minimum.

As a final example of the energetics of migration, we may briefly consider the migratory locusts of Africa, *Schistocerca gregaria*. At takeoff, the 2-g locust carries an average of about 0.2 g of fat. Utilization of all this fuel would yield about 1900 cal of energy, and the cost of flying is about 130 cal/h. Thus, flights would appear to be limited energetically to durations of about 14 h. Actual flights last 6–8 h and are interspersed with periods of fuel replenishment that are a scourge to humans (a swarm can weigh up to 10,000–20,000 tons and consume as much food in a day as over a million people).

ENERGY BUDGETS: CASE STUDIES

In nature, animals typically engage in diverse activities, and for ecological analysis it is often desirable to devise a budget showing the energetic cost in-curred by each separate type of activity. Such a budget can be used both to estimate the animal's total energy expenditure and to calculate which particular activities are responsible for imposing large demands on energy resources. Because energy budgets vary enormously not only between species but also within a species from one time of year to another, the information available on budgets cannot readily be generalized or summarized. Thus, emphasis is placed here on illustrative examples.

A Time–Energy Budget for a Hummingbird

To illustrate the approach of developing a time–energy budget (p. 59), let us look at one of the classic, early studies in which the approach was used: Oliver Pearson's analysis of the energy expenditure of a wild Anna hummingbird, summarized in Table 5.3. Pearson determined the metabolic rate associated with three types of behavior—hovering, perching, and roosting—using measures of oxygen consumption in the laboratory; he took care to adjust these measures to temperatures comparable to those in the field. He estimated the time that a single wild hummingbird spent in each type of behavior by observing the animal on 2 days in early September. With these data, he estimated the daily energy demand of each type of behavior by multiplying the time devoted to the behavior by the hourly cost of the behavior. Pearson did not know whether the hummingbird spent the night simply sleeping or whether it entered torpor. Accordingly, he estimated the energy demand during roosting under both assumptions. The total daily energy demand was calculated by summing the daily costs of all three behaviors and turned out to be about 10.3 kcal under the assumption of simple sleep overnight, or about 7.6 kcal under the assumption that the animal entered torpor for most of the night. It is noteworthy that flight, though it occupied just 10 percent of the day, accounted for nearly as much energy expenditure as either perching or sleep. Subsequent studies of hummingbird hovering have suggested that the cost per hour is perhaps only 60 percent as great as Pearson measured; this would lower the estimates of daily cost of hovering and of total daily energy expenditure, but still, hovering accounts for a large portion of the total daily expenditure in relation to the fraction of time it occupies.

Some Results of D$_2$^{18}O Analysis and Their Use in Evaluating Time–Energy Budgets

The technique of time–energy analysis was the earliest analytical method applied to the estimation of energy expenditures of wild animals, and to this day the technique has marked virtues (e.g., low expense). For a long time, however, no rigorous knowledge of its accuracy was available. One of the important uses

Table 5.3 A TIME–ENERGY BUDGET FOR A MALE ANNA HUMMINGBIRD (*Calypte anna*)

Behavior	A Hours per day de-voted to behavior[a]	B Hourly cost of behavior[b] (mL O_2/h)	C Daily cost of behavior: $A \times B$ (mL O_2)
Perching	10.53	75.4	794
Flying (hovering)	2.35	272	639
Roosting	11.13	64.5 ------------- simple sleep ------------------- 718	
		12.6 ---------------- torpor ----------------------- 140	

Totals (total daily energy expenditure):
simple sleep: 2151 mL O_2 = 10.3 kcal
torpor: 1573 mL O_2 = 7.6 kcal

[a] Time devoted to each behavior was determined for a single wild male observed on the campus of the University of California over 2 days in September 1953.
[b] The hourly energetic cost of each behavior was determined in studies of several animals in the laboratory.
Source: Data from O.P. Pearson, *Condor* **56**:317–322 (1954). For an update see W.A. Calder, *Auk* **92**:81–97 (1975).

of the doubly labeled water method (p. 59) has been to check the validity of time–energy budget techniques.

In a number of instances, the time–energy technique has indeed been validated. As an example, let us consider an approach developed for time–energy analysis of longtail pocket mice (*Perognathus formosus*) living in the Nevada desert. Field data were obtained on burrow temperatures, desert surface temperatures, and the lengths of time the animals spent in their burrows and on the desert surface. Laboratory data were obtained on resting metabolic rate as a function of environmental temperature and on the increment in metabolic rate caused by ordinary nocturnal activity. Assuming that the animals were at rest in their nest when in their burrows, the metabolic expenditure over the period in the burrows was estimated by first calculating nest temperature from the measured burrow temperature, then determining resting metabolic rate at the nest temperature, and finally multiplying this metabolic rate by the number of hours spent in the burrow. The metabolic expenditure for the period on the desert surface was calculated by first determining the resting metabolic rate at the temperature of the desert surface, then adding the laboratory-measured activity increment to get total metabolic rate, and finally multiplying this metabolic rate by the number of hours spent on the desert surface. The expenditures for the periods in the burrow and on the desert surface were then added to get total daily expenditure. This method of time–energy analysis was applied during 8 months of the year, and simultaneously the total daily energy expenditure was measured using $D_2^{18}O$. Significantly, the greatest monthly discrepancy between the two methods was about 22 percent, and the average discrepancy was just 12 percent. Thus, for these mice, a carefully applied time–energy analysis provides reasonably accurate estimates of energy expenditure in the wild. Evidently, environmental tem-

perature and activity are the major modulators of metabolic rate in these animals, and because the time–energy analysis takes both factors into account, it is realistic in its results. Both the time–energy and $D_2^{18}O$ analyses showed that the energy demand of the pocket mice is far greater in winter than summer, the daily demand in February and March being about twice as great as in July and August.

A $D_2^{18}O$ study of another rodent of the Nevada desert, Merriam's kangaroo rat (*Dipodomys merriami*), not only provided data on the energetics of the species but also pointed to ways that errors can be incurred in using time–energy analysis. From spring to fall (March–October), daily energy expenditure measured by $D_2^{18}O$ varied systematically and inversely with mean ambient temperature, being over three times higher in March (mean temperature about 10°C) than August (about 30°C). This is the pattern expected of an animal that maintains a high and relatively constant body temperature, since to keep its body temperature high, it must increase its rate of metabolic heat production as the environment gets colder and colder (p. 101). In winter, however, daily energy expenditures that were far below such expectations were found; for instance, even though the mean ambient temperature in November was 12°C, animals studied in November displayed expenditures similar to those studied in September, when the temperature was 25°C. It was concluded that the animals in winter were undergoing daily torpor or some other process that radically reduced their energy costs. Notably, prior data on these kangaroo rats had failed to indicate the routine occurrence in winter of any such process, and thus, a time–energy analysis carried out prior to the $D_2^{18}O$ studies could not possibly have taken such a process into account. Time–energy analyses can be remarkably accurate, but only if all processes important to the daily energy budget are known. The $D_2^{18}O$ method, on the other hand, has the advantage that it simply measures energy expenditure and thus needs to make no prior assumptions about the processes determining the expenditure.

A Daily Energy Budget for Young Sockeye Salmon

Figure 5.15A depicts aspects of the behavior and temperature relations of young sockeye salmon in Babine Lake, British Columbia. In the summer, this lake is thermally stratified (p. 253), as may be seen by comparing the left (temperature) and right (depth) ordinates. The young fish spend most of the daylight hours in the cold, deep waters (hypolimnion) but ascend to the warm surface at dusk to feed on plankton. Later in the night, they descend to 10–12 m, and then they return to the surface waters at dawn for

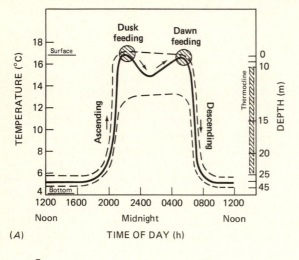

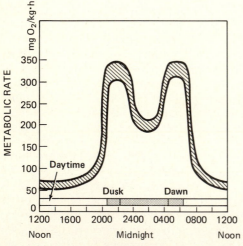

Figure 5.15 (A) The pattern of daily vertical migration and feeding in young sockeye salmon *Oncorhynchus nerka* in Babine Lake, British Columbia, during midsummer. Dashed lines indicate the general range of temperatures experienced by the fish at various times of day, and the heavy line indicates the approximate mean response. The lake is thermally stratified, as reflected by comparison of the right and left ordinates. (B) The estimated daily cycle of energy demand in young, 2-g sockeyes. [From J.R. Brett, *Am. Zool.* **11:**99–113 (1971).]

another bout of feeding. Afterward, they descend into the hypolimnion. This pattern of behavior affects energy demand in at least two important ways. First, the fish experience large changes in temperature, the mean diurnal temperature being about 5°C and the nocturnal temperature being 14–17°C; as we have seen earlier, temperature affects both the standard metabolic rate and the cost of activity. Second, the fish undergo changes in their level of activity; presumably they are more active during their feeding periods than at other times. From laboratory studies on the energetic effects of temperature and feeding activity, the cycle of metabolic demand in Figure 5.15B has been constructed. Demand is lowest during the period of relative rest at cold temperatures during the day. Peaks of demand are reached during the bouts of feeding in the warm surface waters at dawn and dusk. Over the middle of the night, when the fish are not feeding but remain in warm waters, their metabolic demand is intermediate. Integrating mean demand over the entire 24 h, the estimated daily energetic cost comes out to be about 2800 mL O_2/kg = 13.5 kcal/kg (or 27 cal for a 2-g fish). This is the rate at which energy is consumed (degraded to heat); the fish must actually assimilate more energy because they are growing and thus directing some food to the production of new biomass.

J.R. Brett, who worked out this daily energy cycle for young salmon, pointed out the most interesting feature that by descending into cold waters for much of the day, these animals realize energetic savings analogous to those enjoyed by homeotherms that undergo daily torpor (compare Figure 6.54). Whereas the homeotherm lowers its body temperature and metabolic rate by physiological adjustments in thermoregulatory control, the young salmon does so by the behavioral expedient of entering cold waters. Experiments have shown that when young salmon consume submaximal amounts of food per day, their growth rate is maximized at relatively low temperatures; the reduced catabolic demand at such temperatures allows more of the ingested food to be directed to growth. Brett has suggested that the descent to cold waters during the day might permit the fish to maximize the growth potential of their morning meal.

An Annual Energy Budget

The construction of annual energy budgets presents exceptional challenges, which have yet to be met fully. Figure 5.16 presents an early attempt: a partial estimated budget for tree swallows, based on measures on captive birds. The first step was to determine the rate of energy utilization of caged birds at seasonally appropriate temperatures (using material balance methods); this so-called "existence energy" utilization included the basal energy requirement plus components attributable to thermoregulation, specific dynamic action, and the activity of the caged

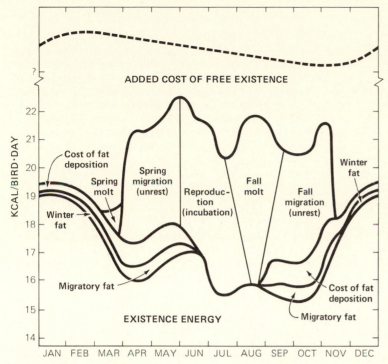

Figure 5.16 An estimated partial annual energy budget for the tree sparrow (*Spizella arborea*). Lowermost curve depicts existence energy utilization for birds overwintering in Illinois and spending the summer in northern Manitoba. Other energetic costs have been added to the existence energy curve, as discussed in the text. Existence energy utilization was determined in laboratory cages and is probably greater in the wild, especially in winter. The upper dashed line, although purely hypothetical, reflects the original investigator's belief that, with the additional costs of free existence added in, total daily energy demand might well be fairly constant over the year. [From G.C. West, *Auk* **77**:306–329 (1960). Used by permission of the American Ornithologists' Union.]

birds. Then several additional energy demands were added to the existence energy demand: (1) the energy value of fat that was laid down and the estimated metabolic cost of depositing it, (2) the estimated energy cost of molting, (3) the cost of nocturnal behavioral restlessness shown by captive birds during the spring and fall migratory periods, and (4) an estimate (now questionable) of the cost of keeping the eggs and young warm (incubation). The investigator who developed this annual budget frankly noted that some of the cost estimates are preliminary or even hypothetical. Furthermore, there are a number of known omissions; the full cost of migration is not included, for instance. These considerations, if nothing else, emphasize the magnitude of the challenge that can be faced in constructing an annual energy budget.

An important point illustrated by Figure 5.16 is that the major energy-demanding events in the life cycle of tree swallows occur more or less sequentially during the year. First there is the spring migration, then reproduction, then the major molt, and then fall migration. Sequencing of a similar sort occurs in many other birds and assures that the animals do not have to meet the full-blown costs of two or more major energy-demanding processes at the same time.

Costs and Rewards in Bumblebee Foraging

From the viewpoint of energy, collecting food has both costs and rewards. By assessing both, we can hope to gain understanding of the evolved and learned foraging behavior of animals.

When foraging, bumblebees (*Bombus*) fly from one flower (or flower cluster) to another, landing on each long enough to probe the flower with their tongue and collect available nectar. Two major costs of foraging must be considered. (1) Flight is itself very costly; it can easily elevate the metabolic rate of the bee to 20–100 times the resting rate. The cost of flight per unit time is essentially independent of air temperature. (2) To fly, bumblebees require that the temperature of the flight muscles in their thorax be 29°C or higher (p. 129). During flight, heat produced by the muscular activity of flying keeps the muscles at such temperatures. However, when bees alight on flowers, they potentially can cool quickly to below the necessary flight temperature if the surrounding air is cool; should this happen, they would be unable to take to the air again in search of other flowers. To keep their flight muscles warm while they are alighted, bees produce heat by a process analogous to human shivering (p. 129). The intensity and *energetic cost* of this shivering become greater as the air temperature becomes lower. Although no shivering may be necessary at an air temperature of 25°C, shivering at 0–5°C may be so intense as to raise the bee's metabolic rate to as high a level as prevails during flight.

Considering the costs of *both* flying and shivering, the *average* metabolic expenditure per unit time demanded by foraging tends to increase as the air becomes cooler. If the air is warm enough to preclude any need for shivering when bees are alighted, the bees have high metabolic rates when they are flying but low rates when they are not. If the air is cold, the bees have high metabolic rates all the time, whether flying or alighted.

The energy *reward* that can be obtained per unit time from any particular species of flower depends on (1) the volume of nectar obtained per flower, (2) the sugar concentration of the nectar, and (3) the number of flowers from which a bee can extract nectar per unit time. The latter parameter, in turn, depends on the spacing of the flowers and the difficulty of penetrating flowers to obtain their nectar.

Some species of plants yield sufficient sugar per flower that bumblebees can realize a net energy profit when foraging from them regardless of air temperature. The rhodora (*Rhodora canadensis*), for example, can be expected to yield sugar equivalent to about 0.4 cal/flower. At 0°C, a queen bee might expend energy at a rate of 3.0 cal/min while foraging. Accordingly, the bee could break even energetically by taking the nectar from about seven or eight rhodora flowers per minute. In fact, bees can tap almost 20 such flowers per minute. Thus, even at 0°C, they not only can meet their costs of foraging but also can accumulate a surplus of nectar to contribute to the hive.

In contrast, some plants yield so little sugar per flower that they are profitable sources of nectar only when relatively high temperatures prevail and the bees' costs of foraging are thereby reduced. For example, bees typically visit blossoms of wild cherry (*Prunus pennsylvanica*) only at high air temperatures. The blossoms yield sugar equivalent to only about 0.05 cal apiece. At 0°C, queens would have to tap an unrealistically high number of about 60 blossoms/min just to meet their costs of foraging.

SELECTED READINGS

Alexander, R.M. 1984. Walking and running. *Am. Sci.* **72**:348–354.

Alexander, R.M. and G. Goldspink (eds.). 1977. *Mechanics and Energetics of Animal Locomotion*. Chapman and Hall, London.

*Åstrand, P.-O. and K. Rodahl. 1986. *Textbook of Work Physiology,* 3rd ed. McGraw-Hill, New York.

Bennett, A.F. 1978. Activity metabolism of the lower vertebrates. *Annu. Rev. Physiol.* **40**:447–469.

*Bennett, A.F. and K.A. Nagy. 1977. Energy expenditure in free-ranging lizards. *Ecology* **58**:697–700.

Bill, R.G. and W.F. Herrnkind. 1976. Drag reduction by formation movement in spiny lobsters. *Science* **193**:1146–1148.

Brett, J.R. 1965. The swimming energetics of salmon. *Sci. Am.* **213**(2):80–85.

*Brett, J.R. 1972. The metabolic demand for oxygen in fish, particularly salmonids, and a comparison with other vertebrates. *Respir. Physiol.* **14**:151–170.

*Brooks, G.A. and T.D. Fahey. 1984. *Exercise Physiology.* Wiley, New York.

Chassin, P.S., C.R. Taylor, N.C. Heglund, and H.J. Seeherman. 1976. Locomotion in lions: Energetic cost and maximum aerobic capacity. *Physiol. Zool.* **49**:1–10.

Chew, R.M. and A.E. Chew. 1970. Energy relationship of the mammals of a desert shrub community. *Ecol. Monogr.* **40**:1–21.

Davis, R.W., T.M. Williams, and G.L. Kooyman. 1985. Swimming metabolism of yearling and adult harbor seals *Phoca vitulina. Physiol. Zool.* **58**:590–596.

Fedak, M.A. and H.J. Seeherman. 1979. Reappraisal of energetics of locomotion shows identical cost in bipeds and quadrupeds including ostrich and horse. *Nature* **282**:713–716.

Feldkamp, S.D. 1987. Swimming in the Californa sea lion: Morphometrics, drag and energetics. *J. Exp. Biol.* **131**:117–135.

Gessaman, J.A. (ed.). 1973. *Ecological Energetics of Homeotherms*. Utah State University Press, Logan.

Gold, A. 1973. Energy expenditure in animal locomotion. *Science* **181**:275–276.

Hagan, R.D. and S.M. Horvath. 1978. Effect of diurnal rhythm of body temperature on muscular work. *J. Thermal Biol.* **3**:235–239.

Hainsworth, F.R. 1981. Energy regulation in hummingbirds. *Am. Sci.* **69**:420–428.

Heglund, N.C., C.R. Taylor, and T.A. McMahon. 1974. Scaling stride frequency and gait to animal size: Mice to horses. *Science* **186**:1112–1113.

*Heinrich, B. 1979. *Bumblebee Economics*. Harvard University Press, Cambridge.

Herreid, C.F., II and C.R. Fourtner (eds.). 1981. *Locomotion and Energetics in Arthropods*. Plenum, New York.

Macleod, D., R. Maughan, M. Nimmo, T. Reilly, and C. Williams (eds.). 1987. *Exercise. Benefits, Limits and Adaptations*. E. & F.N. Spon, London.

Mullen, R.K. and R.M. Chew. 1973. Estimating the energy metabolism of free-living *Perognathus formosus. Ecology* **54**:633–637.

*Nagy, K.A. 1987. Field metabolic rate and food requirement scaling in mammals and birds. *Ecol. Monogr.* **57**:111–128.

Parker, K.L., C.T. Robbins, and T.A. Hanley. 1984. Energy expenditures for locomotion by mule deer and elk. *J. Wildl. Manage.* **48**:474–488.

*Pedley, T.J. (ed.). 1977. *Scale Effects in Animal Locomotion*. Academic, New York.

Robinson, S. 1980. Physiology of muscular exercise. In V.B. Mountcastle (ed.), *Medical Physiology,* 14th ed., Vol. II. Mosby, St. Louis, MO.

Rothe, H.-J., W. Biesel, and W. Nachtigall. 1987. Pigeon flight in a wind tunnel. II. Gas exchange and power requirements. *J. Comp. Physiol.* **157B**:99–109.

*Schmidt-Nielsen, K. 1972. Locomotion: Energy cost of swimming, flying, and running. *Science* **177**:222–228.

Taylor, C.R. 1985. Force development during sustained locomotion: A determinant of gait, speed and metabolic power. *J. Exp. Biol.* **115**:253–262.

*Taylor, C.R. 1987. Structural and functional limits to oxidative metabolism: Insights from scaling. *Annu. Rev. Physiol.* **49**:135–146.

Taylor, C.R., S.L. Caldwell, and V.J. Rowntree. 1972. Running up and down hills: Some consequences of size. *Science* **178**:1096–1097.

Taylor, C.R., G.M.O. Maloiy, E.R. Weibel, V.A. Langman, J.M.Z. Kamau, H.J. Seeherman, and N.C. Heglund. 1980. Design of the mammalian respiratory system. III. Scaling maximal aerobic capacity to body mass: Wild and domestic mammals. *Respir. Physiol.* **44**:25–37.

Taylor, C.R., A. Shkolnik, R. Dmi'el, D. Baharav, and A. Borut. 1974. Running in cheetahs, gazelles, and goats: Energy cost and limb configuration. *Am. J. Physiol.* **227**:848–850.

*Tucker, V.A. 1969. The energetics of bird flight. *Sci. Am.* **220**(5):70–78.

Tucker, V.A. 1971. Flight energetics in birds. *Am. Zool.* **11**:115–124.

*Tucker, V.A. 1975. The energetic cost of moving about. *Am. Sci.* **63**:413–419.

Weis-Fogh, T. 1968. Metabolism and weight economy in migrating animals, particularly birds and insects. In J.W.L. Beament and J.E. Treherne (eds.), *Insects and Physiology*. Elsevier, New York.

See also references in Appendix A.

chapter 6

Thermal Relations

The temperature of an animal's body generally has profound effects on function. The cells, tissues, and organs of all organisms have upper and lower lethal temperatures, and within the thermal range compatible with life, rates of function are typically highly dependent on temperature. Animals display several different types of relation with their thermal environment. These relations and their physiological and ecological implications are the subject of this chapter.

Two particularly prevalent types of thermal relation are *homeothermy* and *poikilothermy*. Homeothermy is the physiological maintenance of a relatively constant internal temperature regardless of external temperature. The tissues of the homeothermic animal are permitted to function in a more or less stable thermal milieu, but the organism must expend energy to maintain a state of thermal disequilibrium between itself and its environment. In poikilothermy, the body temperature is allowed to vary with ambient conditions; thus, the body temperature will be low in a cold environment and high in a warm environment. The poikilothermic animal realizes a considerable energy saving compared to the homeothermic animal by not maintaining a thermal disequilibrium between its body and the environment, but its cells, tissues, and organs must cope with a changing internal temperature. Organisms that predominantly or exclusively display one or the other of these thermal relations are termed **homeotherms** and **poikilotherms**, respectively. The birds and mammals are classed as homeotherms, and many of the lower vertebrates and invertebrates are classed as poikilotherms.

The effects of temperature on individual organisms have much ecological significance. We mentioned in Chapter 3 that metabolic rate in both ho-meotherms and poikilotherms is often strongly dependent on ambient temperature. Thus, the energetic demands placed on the environment by an organism increase and decrease with thermal circumstances. Temperature is also significant in that the animal must be able to survive the various thermal challenges imposed on it throughout the year. The distribution and habitat of species may thus be influenced by thermal effects. Temperature per se may exceed viable limits at certain times or places, or temperature may interact with other factors to produce a physiologically or ecologically stressful situation. Mammals and birds exposed to heat in deserts, for example, may face critical problems of dehydration if required to expend large amounts of water in evaporative cooling to keep their body temperatures from rising to threatening extremes. Fish in summer may have high metabolic rates because their body temperatures are elevated in the warm water, and yet, at the same time, they may be faced with relatively low oxygen availability because warm water tends to hold less dissolved oxygen than cold water. The interaction of these factors may prove critical.

The profound ecological impact of temperature may nowhere be more strikingly illustrated than in our temperate woodlands. In the warm seasons of the year, we witness vigorous photosynthesis by plants and sustained activity of birds, mammals, reptiles, insects, amphibians, and other animals. In the winter, plants and poikilothermic animals such as amphibians and insects enter a state of quiescence; and activity in the woodland becomes largely limited to those birds and mammals that, by virtue of high internal temperatures, can continue to move about and forage for food despite the low ambient temperatures. We see the profound influence that temperature may exert on the nature of the whole ecological community, an influence that is also displayed by the changes in community types evident as one travels from the equator to the poles.

MECHANISMS OF HEAT TRANSFER

At the outset of a study of thermal relations, it is important to appreciate the difference between *heat* and *temperature*. One fundamental distinction between the two is that heat is an *extensive* property, whereas temperature is an *intensive* one. To explore what this means, let us consider a uniform solid substance in which the individual constituent molecules possess, on average, a certain known kinetic energy. *Heat* is a form of energy: the energy that this substance—or any other—possesses by virtue of the random motions of its molecules (p. 23). The *quantity* of heat in the substance depends on the average kinetic energy of its molecules, but it also depends on other parameters, one being the size or *extent* of the piece of substance under consideration. A piece that is twice the size of another also contains twice the heat. Temperature, by contrast, would not differ in two such pieces but is a property that all parts of a uniform substance possess in common. *Temperature* is a direct, immediate measure of the average kinetic energy possessed by individual molecules and expresses the substance's *intensity* of hotness (or coldness). Succinctly, temperature is a measure of the tendency of a substance to give up heat.

Figure 6.1 presents an overview of the thermal exchanges between an animal and its environment. Metabolism contributes heat to the animal's body. Evaporation of body water, on the other hand, carries heat away. Heat also moves between the animal and its surroundings by virtue of convection, conduction, and exchange of electromagnetic radiation; the net effect of each of these three avenues of heat exchange may be to carry heat away from the animal's body or add heat to the body, depending on conditions. For the temperature of the animal's body to remain constant, gains of heat from metabolism and the environment must be counterbalanced by losses of equal magnitude (note the analogy with Figure 2.1). If gains and losses are unequal, heat will either accumulate in the body or be subtracted in net fashion, and the body temperature will rise or fall accordingly.

To analyze animal heat exchanges, it is necessary to understand the principles governing each of the four physical mechanisms of heat transfer: conduction, convection, evaporation, and radiation. In pursuit of this understanding, the simple thermal model of an animal depicted in Figure 6.2 will be useful. The core of the animal's body is considered to be at a uniform *body temperature,* symbolized by T_B. The temperature of the environment is called *ambient temperature,* T_A. The temperature of the body surface often differs from T_B and T_A and thus is distinguished as T_S. Separating the body core from the

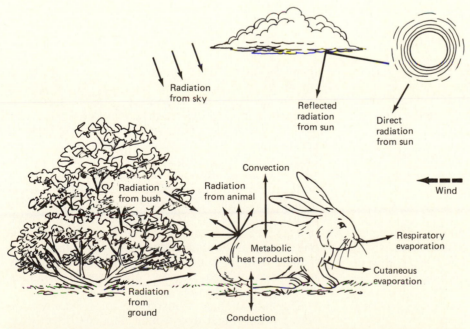

Figure 6.1 Some avenues of heat exchange between an animal and its environment. The animal receives visible electromagnetic radiation from the sun and receives infrared (thermal) radiation from not only the sun but also the ground, bush, and atmosphere. It also emits infrared radiation toward all these objects. It exchanges heat convectively with the wind and conductively with the ground. Furthermore, it gains heat from its own metabolism and loses heat by respiratory and cutaneous evaporation.

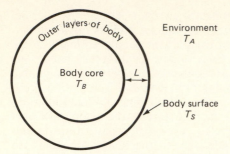

Figure 6.2 A schematic representation of the animal body. The outer layers of the body have thickness L. The body core is at temperature T_B, and the surrounding environment is at temperature T_A. The temperature of the body surface is T_S.

surface are the outer layers of the organism, wherein temperature gradually changes from T_B on the inside to T_S on the outside.

What are conduction and convection? **Conduction** is diffusional transfer of heat. That is, when heat moves through a medium by conduction, no macroscopic movement of the medium is involved; instead, energy is transferred from place to place by molecular interactions, such as intermolecular collisions. **Convection** is transfer of heat by a moving fluid. We are familiar with these modes of heat transfer from everyday experience. If you sit on a hot rock, for example, you receive heat from the rock through the seat of your pants largely by conduction. If you stand in a hot wind, on the other hand, you receive heat from the air by convection. It is important to recognize that, for a given temperature difference, heat transfer by convection is much more rapid than that by conduction. Consider, for example, a horizontal surface that is 10°C warmer than the surrounding air; if the air is moving at just 10 mph, the consequent convection will carry heat away from the surface about 70 times faster than if the air were perfectly still and conduction were the sole mode of heat transfer. The hastening of heat transfer by fluid movement is again a matter of everyday experience. If, for example, you dip your fingers in ice water, you will note that the sensation of cold is more severe if you swirl them around than if you keep them still.

Conduction

Suppose we have a flat layer of material of thickness d, bounded on either side by parallel faces. If the temperature on one side of the layer is T_1, that on the other side is T_2, and heat is moving through the layer by conduction, then the rate of heat transfer from one side to the other per unit of cross-sectional area, H, can be expressed as follows

$$H \text{ (in units of cal/cm}^2\text{·s)} = k\,\frac{T_1 - T_2}{d}$$

where k is a constant.[*]

This equation expresses several important con-

*The calorie (cal) and other units of measure for heat are discussed on p. 29.

cepts. First, the rate at which heat traverses the layer by conduction increases as the temperature difference prevailing across the layer ($T_1 - T_2$) is increased. Second, the rate of heat transfer from one side of the layer to the other decreases as the thickness of the layer (d) is increased. The ratio of the temperature difference divided by the thickness [$(T_1 - T_2)/d$] is known as the **thermal gradient**. Note that the rate of heat transfer is proportional to this gradient. The proportionality coefficient, k, depends on the material of which the layer is composed and is known as the material's **conductivity**. Some materials, such as air, have a low conductivity, signifying that they conduct heat relatively poorly. Others, such as water, have higher conductivity and thus are less insulative (the conductivity of water is about 20 times that of air at physiological temperatures).

One function of the fur of mammals (or the winter overcoats of people) is to trap a layer of relatively motionless air around the body. As a first approximation, heat transfer through this layer may be considered to occur by conduction. Indeed, from the viewpoint of heat transfer, the benefit of fur (or heavy clothing) in a cold environment is that, by limiting flow of air near the skin, it favors an intrinsically slow mechanism of heat loss from the body, conduction, over a more-rapid one, convection. In Figure 6.2, the "outer layer" of the body might be taken for present purposes to represent the fur. In line with the concepts developed earlier, increasing the thickness L of the fur will tend to slow heat loss from the animal to the environment.

Convection

There are two types of convection, free and forced. If the surface of an animal or other object is warmer than the surrounding air, currents will be set up in the air even if there are no extrinsic forces moving the air. In brief, air in contact with the surface is warmed, and since warm air is less dense than cool air, the warmed air will rise away from the surface, creating an air flow that removes heat from the surface convectively. Convection that thus results strictly from a difference in temperature between a surface and surrounding fluid is termed **free convection**. By contrast, when a fluid, such as air, is driven across a surface by some extrinsic force, the resultant convection is termed **forced convection;** wind is one example. Forced convection is not only more familiar than free but also can produce considerably higher rates of heat transfer. Thus, we shall emphasize forced convection.

The rate of heat transfer by forced convection between an object and a fluid depends directly on the difference in temperature between the *surface* of the object and the fluid. Suppose, for instance, that the animal in Figure 6.2 is exposed to a wind. Then the rate of convective heat transfer per unit of surface area will be

$$H \text{ (in units of cal/cm}^2\text{·s)} = h_c(T_S - T_A)$$

The proportionality coefficient, h_c, is called a **convection coefficient.** It is influenced by many factors, including wind speed, shape, orientation to the wind, and type of flow (laminar or turbulent). Often, parts of an animal can approximately be modeled as cylinders; the fingers, arms, and torso of a person, for instance, can be considered to represent cylinders of various sizes. When a cylinder is oriented with its long axis perpendicular to the wind and flow is laminar, then

$$h_c \propto \frac{\sqrt{V}}{\sqrt{D}}$$

where V is the wind velocity and D is the diameter of the cylinder. This tells us that if a warm body part is exposed to a cold wind and the temperature difference $(T_S - T_A)$ is held constant, the rate of heat loss per unit of surface area will increase with the square root of the wind speed. Heat loss per unit area also increases as the square root of the diameter is decreased, this being one reason why body parts of small diameter (e.g., fingers) are more susceptible to being cooled than ones of large diameter.

We commonly think only of external media in considering convective heat transport. Yet, one of the important functions of the circulatory system in many animals is to move heat convectively within the body. Mammals in hot environments, for example, quickly die of overheating if their circulatory systems are impaired in their ability to carry metabolically produced heat from the depths of the body to the body surfaces where the heat can be voided (see p. 157).

Evaporation

For water to change from its liquid to vaporous state, it must absorb a substantial amount of heat, known as the **latent heat of vaporization.** This heat is absorbed *from the surface at which the change of state occurs* and is carried away with the water vapor produced. Accordingly, evaporation of body water from the respiratory passages or integument of an animal ab-

stracts heat from the animal's body. The latent heat of vaporization of water depends on the prevailing temperature and ranges from about 570 to 595 cal per gram of water at physiological temperatures. For reasons not fully clear, mammalian sweat is measured to have a somewhat higher latent heat, near 620 cal/g. The physical laws governing the rate of evaporation are discussed at the beginning of Chapter 7 (pp. 136–138).

Electromagnetic Radiation

To appreciate the principles of radiant heat exchange between animals and their environments, it is first appropriate to look at the electromagnetic spectrum, depicted in Figure 6.3. Out of the entire spectrum, our eyes perceive as light just a narrow range of wavelengths, from 0.4 to 0.72 μm. Radiation at longer wavelengths is known as **infrared** or **thermal radiation.** It is not visible but, if strong enough, may be sensed as heat.

All objects *emit* electromagnetic radiation, provided only that their temperature is above absolute zero. The higher the surface temperature of an object, the shorter the wavelengths at which it will emit (Figure 6.3). Organisms and most objects in their environments, being fairly cool, radiate only at infrared wavelengths, which are too long to be visibly perceptible. The embers of a fire are hot enough to radiate at the longer wavelengths of the visible spectrum and thus appear red-orange. The sun, being still hotter, radiates at the shorter as well as the longer wavelengths of the visible spectrum and therefore produces a nearly white light. It is to be emphasized that the radiative emissions from organisms are of the same basic nature as those from the sun or a fire but are at such long wavelengths that our eyes do not perceive them and are of sufficiently low intensity that our thermal sensors generally do not signal their immediate presence. (When we *see* plants or animals outdoors, we perceive *solar* radiation of visible wavelengths that has been *reflected* from them; we do not see the radiation that the organisms are *emitting* as a function of their surface temperatures.)

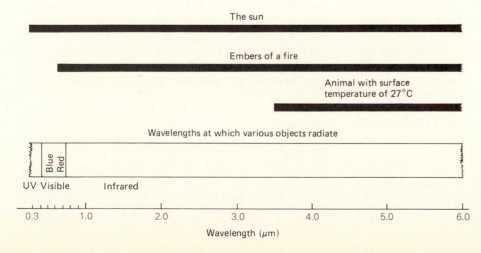

Figure **6.3** The electromagnetic spectrum, showing the approximate range of wavelengths at which various objects radiate by virtue of their surface temperatures. All three objects emit energy at long wavelengths in the infrared (not shown here).

When electromagnetic radiation of whatever wavelength, visible or infrared, impinges on an object, some of the radiant energy is absorbed, some is reflected, and if the object is transparent to the wavelength of concern, some is transmitted. Radiant energy that is absorbed is converted into heat at the surface of the absorbing object; we are familiar, for example, with the way that radiation from a fire warms our skin. Radiation thus provides a route of heat transfer among organisms and objects in the biosphere. Heat energy radiated by one organism or object travels in a straight line, suffering little absorption by the air, until it strikes, and can be absorbed by, other organisms or objects in the surrounding environment. Any animal, in fact, experiences a veritable multitude of radiative exchanges with objects in its environment (Figure 6.1). It receives radiation from the sun, sky, trees, rocks, soil, grass, and so on. It also emits radiant energy to them. The net result may be either a gain or loss of heat.

Radiant exchanges of heat integrate with conductive, convective, and evaporative exchanges to determine the overall thermal flux between the animal and environment. We quickly appreciate how a lizard emerging in the cool desert morning may warm to well above air temperature by basking in the sun. We are perhaps less familiar with situations where radiant heat exchange tilts the balance toward cooling an organism. To cite an example from common experience, consider that during winter in a well-insulated house people usually feel comfortable if the air temperature is near 22°C (72°F). Yet if you have ever spent an evening in a cabin or other poorly insulated building in winter, you will probably have experienced a sense of chill even though the air temperature was at least as high as 22°C. The difference between the poorly insulated and well-insulated buildings is the temperature of the interior walls. Because the interior wall surfaces are colder in a poorly insulated building, they act as a more powerful radiative "heat sink." You emit radiation toward the walls of whatever building. The cold walls of a poorly insulated structure emit less toward you in return, however, than the warmer walls of a well-insulated building. Thus, in the poorly insulated building your net radiative loss of heat to the walls is greater and the environment as a whole feels colder even though conductive, convective, and evaporative relations may resemble those in a well-insulated structure.

What determines the rate at which a surface (animal or otherwise) emits radiant energy? Two factors are involved: (1) the temperature of the surface and (2) certain other surface properties quantified as *emissivity*.

The temperature of a surface completely determines the surface's *maximum possible* intensity of radiant emission not only at each particular wavelength but also (as a corollary) at all wavelengths

combined. In fact, the maximum possible intensity at all wavelengths combined is a mathematically simple function of surface temperature, given by the *Stefan–Boltzmann equation:*

$$H_{max} \text{ (in units of cal/cm}^2\cdot\text{s)} = \sigma T_S^4$$

In this equation, T_S must be expressed in absolute degrees (°K), and σ is a constant called the Stefan–Boltzmann constant (1.35×10^{-12} cal/cm$^2\cdot$s$\cdot$°K^4). Figure 6.4 depicts the maximum possible intensity of radiation as a function of wavelength for two surface temperatures in the physiological range; the area under each curve indicates the total, integrated intensity at all wavelengths and thus corresponds to the value that would be computed from the Stefan–Boltzmann equation at each temperature. Note that the maximum possible intensity of radiation increases markedly as surface temperature increases.

The *emissivity* (emittance) of a surface is an expression of the extent to which the surface attains its maximum possible intensity of radiation. Emissivity, symbolized ϵ, varies from 0 to 1 and represents the fraction of the maximum that a surface actually emits; for example, a surface with an emissivity of 0.9 emits at a rate equal to 90 percent of the maximal rate set by its temperature. By introducing emissivity into the Stefan–Boltzmann equation, we get an equation for the *actual* intensity at which a surface emits radiation:

$$H \text{ (in units of cal/cm}^2\cdot\text{s)} = \epsilon\sigma T_S^4$$

The *absorptivity* of a surface is a value between 0 and 1 that represents the fraction of incident radiation that the surface absorbs; for instance, a surface with an absorptivity of 0.9 absorbs 90 percent of radiation incident on it. For a given surface, both emissivity and absorptivity typically vary from one wavelength to another. Significantly, however, *they are always equal to each other at any particular wavelength.*

As noted earlier, except for the sun, natural objects in the biosphere (including animals) emit radiation only at relatively long infrared wavelengths (about 4 μm and longer). Many of the radiant exchanges of physiological and ecological significance thus occur at such wavelengths, and it is appropriate to focus briefly on the properties of surfaces in this part of the electromagnetic spectrum. Remarkably, virtually all natural objects, *regardless of their visible color,* have high average emissivities and correspondingly high absorptivities at the infrared wavelengths in question. Soil, snow, plant leaves, the white hair of a snowshoe hare, the skin of white and black humans—all these objects, when considered at wavelengths of 4 μm and beyond, have integrated emissivities and absorptivities of 0.9 or higher. Animals and the objects in their surroundings are thus "good" emitters at the wavelengths concerned, meaning that they emit at intensities that approach the maximum possible for their prevailing surface

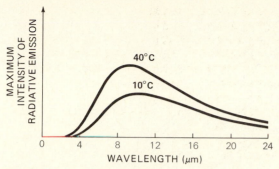

Figure 6.4 Maximum intensity of radiative emission as a function of wavelength for an object at two different surface temperatures, 10 and 40°C. Curves such as these are derived from Planck's spectral-distribution law. The wavelength at which each curve peaks, which gets shorter as the temperature is raised, is predictable from Wien's displacement law.

temperatures, and they are "good" absorbers, meaning that they absorb a high fraction of the radiant energy they receive from other animals and objects. When two surfaces of temperatures T_1 and T_2 are in exchange, the *net* radiant transfer of heat is typically from the warmer surface to the cooler at a rate proportional to $(T_1^4 - T_2^4)$.

The Sky as a Radiant Object One of the objects in the biosphere that deserves special note is the sky. In the atmosphere above us, each molecule, from the earth's surface to the limits of outer space, emits radiation as a function of its temperature. Thus, the surface of the earth is steadily showered with radiation *emitted* by the sky above. One way to express the intensity of this radiation is to pretend that the sky is a solid surface and ask what the temperature of this surface would have to be for it to emit at the intensity observed (assuming its emissivity to be 1). This temperature is called the ***radiant sky temperature***. At least at night, the radiant temperature of the clear sky is typically much below the air temperature at ground level. To exemplify, during a summer night in the Arizona desert, the air temperature near the ground was 30°C, but the radiant temperature of the clear sky was simultaneously below freezing, −3°C. When animals and other objects are exposed to the clear nocturnal sky, they emit radiation toward it but, in comparison to what they emit, tend to receive little in return. The sky thus tends to act as a radiant "heat sink," soaking up radiant heat in net fashion. When small mammals live in burrows under the snow in winter, they do not merely find protection against rapid convective loss of heat to the cold air above; they also avoid radiative loss of heat to the exceedingly cold nighttime sky by interposing the snow as a barrier to radiant exchange between themselves and the atmosphere. The radiant temperature of the sky is increased by cloudiness; frosts are more likely on clear than cloudy nights because clear skies act as stronger heat sinks than cloudy ones.

Absorption of Solar Radiation Although animals and other natural objects uniformly exhibit high integrated absorptivities at wavelengths beyond 4 μm, they vary considerably in their absorptivities at shorter wavelengths, as is evident from their differences in visible color. Absorptivities at these shorter wavelengths are important in analyzing the thermal impact of direct and reflected solar radiation, for most solar radiation arriving at the surface of the earth is at wavelengths below 3 μm.

Light-colored animals have substantially lower absorptivities to the wavelengths of solar radiation than related dark-colored ones, and in many circumstances, this means that the light-colored animals absorb less heat when exposed to the sun, as common sense would suggest. Common sense is not always an infallible guide, however, and in this case data have accumulated in recent years showing that sometimes, among *feathered* or *furred* animals, light-colored forms absorb *more* heat from the sun than dark-colored ones. The root of the explanation is that light-colored feathers and hairs not only are less absorptive to solar radiation than their dark-colored counterparts but also are typically more transparent. One circumstance in which this latter difference is known to be important is when animals are exposed to a wind. A wind tends to blow away heat that is absorbed right at the outer surface of the pelage or plumage, thus preventing that heat from penetrating to the living flesh of the animal. In such circumstances, as shown in Figure 6.5, the living tissues may gain more heat if the body covering is light than dark, because as the transparency of the body covering is increased, an increasing fraction of incident solar radiation penetrates into the depths of the plumage or pelage (away from the wind-blown surface) before being absorbed and converted into heat.

Aquatic Versus Terrestrial Environments We have seen that because air has a high transparency to both visible and infrared wavelengths, animals on land can exchange heat radiatively with distant objects. These objects themselves can vary considerably, one to another, in their radiant temperatures. Water, by contrast, is virtually opaque to infrared wavelengths, and thus aquatic animals exchange infrared radiation just with the water that is very near their bodies. Aquatic environments tend in consequence to be far less thermally complex than terrestrial ones. Water *is* modestly transparent to visible radiation, meaning that an aquatic animal near the surface can be warmed by the sun.

Operative Temperatures

In terrestrial animals, the effects of the thermal environment on metabolic rate are usually studied in laboratory enclosures ("metabolic chambers") having at least two significant peculiarities. First, the wind velocity within the enclosures is typically very

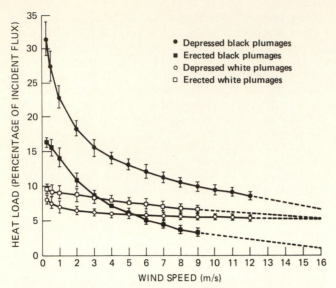

Figure 6.5 Percentage of incident (simulated) solar radiation absorbed as heat through the plumage of white and black pigeons, as a function of wind speed. Experiments were performed on the plumage both when erected (fluffed out) and depressed (flattened). Note that black plumage absorbs more solar radiation as heat than white plumage when the wind speed is near zero. However, once the wind speed exceeds 3 m/s (6.7 mph), erected white plumage absorbs more heat than erected black plumage. Dashed lines are extrapolations. Squares and circles show mean measured values; vertical bars delimit ±1 standard deviation. [Reprinted by permission from G.E. Walsberg, G.S. Campbell, and J.R. King, *J. Comp. Physiol.* **126B:**211–222 (1978).]

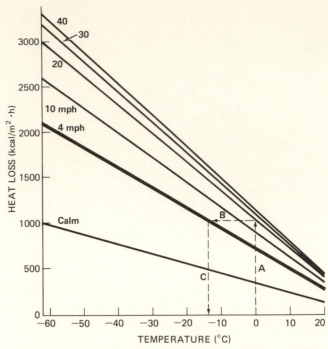

Figure 6.6 A nomogram that adjusts for convective differences between a natural and standard environment according to the assumptions of the wind-chill index. Each line depicts the rate of heat loss from a person in the shade as a function of air temperature, at a given wind speed. To determine the wind-chill index for a given prevailing combination of temperature and wind speed, first read up from the prevailing temperature to the line for the prevailing wind speed (step *A*). Then read horizontally to the 4-mph line (step *B*) and down (step *C*) to determine the temperature that at 4 mph imposes the same heat loss as the prevailing temperature at the prevailing wind speed. The example shows that when the air temperature is 0°C and the wind speed is 20 mph, the wind-chill index is −14°C.

low, and second, the walls of the enclosures usually have radiant temperatures approximating air temperature.

Consider now a terrestrial animal in a natural environment, where the wind speed may be substantial and some surrounding objects may have radiant temperatures very different from air temperature. How might we use laboratory data to predict the metabolic rate of the animal in the natural environment? One approach to this important question is to use the physical laws of heat transfer to calculate what air temperature in a standard laboratory enclosure would elicit the same metabolic response as the complex set of thermal conditions in the natural environment. That laboratory temperature is then said to be the ***operative temperature*** (or equivalent temperature) of the natural environment.

An elementary yet familiar version of the operative-temperature concept is the ***wind-chill index,*** which adjusts only for convective differences between the natural environment and a standard laboratory environment. For any given combination of air temperature and wind speed in the natural environment, the wind-chill index is the air temperature that in the presence of a standard, low wind speed (4 mph) elicits the same metabolic response from a human being as the natural environment (Figure 6.6).

POIKILOTHERMY (ECTOTHERMY)

In this and the following sections, we review some of the properties of poikilothermy, homeothermy, and other thermal relations with the environment. It is to be emphasized strongly from the outset that poikilothermy and homeothermy are best viewed as *types of thermal relations* displayed by certain animals under certain conditions. They are not necessarily deterministic properties of particular animal groups; an animal, for example, may display poikilothermy under some circumstances and homeothermy under others. Saying that a *species* is a poikilotherm implies that it always, or at least nearly always, displays poikilothermy. Classically, all the animals except the birds and mammals were viewed as poikilotherms. Whereas most groups in this vast assemblage do appear to respond largely poikilothermically, we now recognize more and more that many do not on certain occasions, and some never do so. These exceptional cases are discussed mainly toward the end of this chapter.

The defining characteristic of *poikilothermy* is that the animal's body temperature is determined by equilibration with the thermal conditions of the environment and varies as environmental conditions vary. In aquatic poikilotherms such as clams, starfish, crayfish, and fish, body temperature is largely determined by conductive and convective equilibration with the surrounding water, and the body temperature therefore typically approximates water temperature, as illustrated in Figure 6.7*A*. The animal, of course, produces internal heat metabolically, and this may result in some elevation of body temperature over water temperature. Water, however, absorbs heat so effectively and these animals are so poorly insulated that the difference between body and water temperature is usually slight. In terrestrial poikilotherms, such as frogs, snails, and many insects, body temperature does not necessarily approximate the temperature of the surrounding air, for it is determined by the *totality* of environmental thermal conditions. Radiant input of heat from the sun, for example, may elevate body temperature well above air (ambient) temperature, and evaporation of water from the animal may depress body temperature a few degrees below ambient temperature. In a laboratory setting where environmental surfaces have radiant temperatures close to air temperature, the relations of Figure 6.7*A* often apply to terrestrial poikilotherms as approximations.

Poikilothermic animals have often been called "cold-blooded," in reference to their coolness to the touch under certain conditions. In point of fact, this term is quite inappropriate because body temperature may rise to high levels in many species, provided only that there is adequate input of heat from the environment. Lizards in the desert, for example, may have body temperatures higher than those of humans. The term *poikilotherm* (*poikilos* = "manifold" or "variegated") refers to the lack of regulated constancy of body temperature and is more descriptive of the actual physiological status of these organisms. Recently, these animals have increasingly been termed *ectotherms,* in reference to the fact that their body temperatures are determined primarily by external thermal conditions. The term *ectotherm* emphasizes the mechanism by which body temperature is determined, whereas *poikilotherm* emphasizes the variation of body temperature with environmental conditions. Both terms have value. Although poikilotherms mostly lack physiological mechanisms of controlling their body temperature, they are not necessarily devoid of means of control. They often exert exquisite control behaviorally, by selecting their thermal environment (see p. 124).

Resting Metabolic Rates and Other Rates as Functions of Body Temperature; Q_{10}

For analyzing thermal effects on metabolism in poikilotherms, the temperature of immediate functional relevance is that of the tissues: the *body* temperature. As shown in Figure 6.7*B*, the resting metabolic rate of poikilotherms in general *rises approximately exponentially* with body temperature. An exponential relation signifies that the metabolic rate increases by a given *multiplicative factor* for given *additive increment* in the temperature. For example, the metabolic rate might increase by a *factor of 2* for each *increment* of 10°C in temperature. Then if the metabolic rate were 0.5 cal/h at 0°C, it would be 1.0 cal/h at 10°C, 2.0 cal/h at 20°C, and 4.0 cal/h at 30°C. The relation between metabolic rate and body temperature is usually, in fact, only approximately exponential. That is, the factor by which metabolism increases for a given increment in temperature is usually not precisely constant from one temperature range to the next but might, for example, be 2.5 between 0 and 10°C but only 1.8 between 20 and 30°C. An approximately exponential relation is often termed *quasiexponential*.

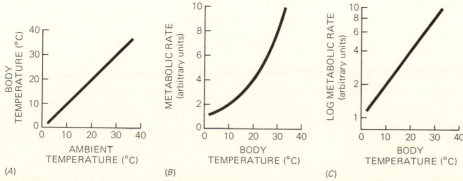

Figure 6.7 Common relations between body temperature, resting metabolic rate, and ambient temperature in poikilotherms. (*A*) Body temperature as a function of ambient temperature. (*B*) Resting metabolic rate as a function of body temperature. (*C*) The logarithm of resting metabolic rate as a function of body temperature. Metabolic rate is expressed in the same arbitrary units in parts (*B*) and (*C*).

If metabolic rate M is truly exponentially related to temperature T, the relation can be described by an equation of the form

$$M = a \cdot 10^{bT}$$

where a and b are constants. Taking the common logarithm of both sides of this equation, we get

$$\log M = \log a + bT$$

This latter equation demonstrates that if M is an exponential function of T, then $\log M$ is a linear function of T (for $\log a$ and b are constants). This result provides the rationale for the common practice of plotting metabolism–temperature data on semi-logarithmic coordinates. The logarithm of metabolic rate is plotted on the ordinate, and temperature itself is plotted on the abscissa; and a linear relation is expected if metabolic rate is a truly exponential function of temperature. The curve of Figure 6.7B is replotted on semilogarithmic coordinates in Figure 6.7C, illustrating this "linearizing" effect. A similar comparison is provided in Figure 6.8, using data on tiger moth caterpillars. As emphasized earlier, metabolic rate commonly proves to be related to temperature in only an approximately exponential fashion. When this is the case, the semilogarithmic plot is not precisely linear, an effect that is evident in Figure 6.8B.

One simple way to describe an exponential relation—and compare it with other such relations—is to specify the factor by which the dependent variable increases when the independent variable is raised by some *standardized* increment. To analyze relations between biological rates and body temperature in this way, the standardized increment in temperature is taken to be 10°C, and the factorial increase in rate over this increment of temperature is termed the **temperature coefficient** or **Q_{10}**. That is, Q_{10} is defined as

$$Q_{10} = \frac{R_T}{R_{(T-10)}}$$

where R_T is the rate at any given body temperature T and $R_{(T-10)}$ is the rate at body temperature T minus 10°C. To illustrate, if the metabolic rate of an animal were 2.2 cal/h at a body temperature of 25°C and 1.0 cal/h at 15°C, Q_{10} would be 2.2. Clearly, the magnitude of Q_{10} reflects thermal *sensitivity*. A Q_{10} of 3, indicating that metabolic rate triples for a 10°C increase in temperature, would reflect far greater thermal sensitivity than a Q_{10} of only 2, indicating just a doubling of metabolic rate for the same increment in temperature. Q_{10} values for resting metabolic rate in poikilotherms tend usually to be in the neighborhood of 2, though substantially lower or higher values are sometimes found. If metabolic rate is a truly exponential function of temperature, then Q_{10} is a constant over all temperature ranges. When, as is usual, metabolic rate is not a strictly exponential function of temperature, then Q_{10} varies with the particular

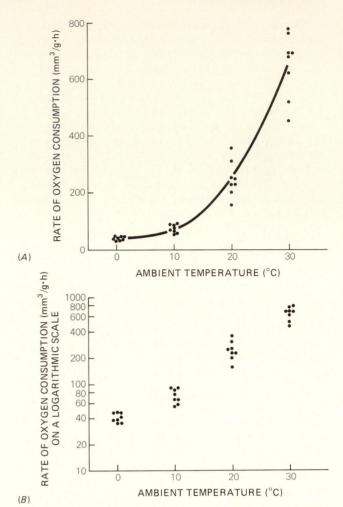

Figure 6.8 The rate of oxygen consumption of tiger moth caterpillars (Arctiidae) as a function of temperature. The same data are plotted on rectangular coordinates (A) and on semi-logarithmic coordinates (B). Body temperature approximated ambient temperature during these measurements. [Data from P.F. Scholander, W. Flagg, V. Walters, and L. Irving, *Physiol. Zool.* **26**:67–92 (1953).]

range of temperature considered. The Q_{10} between 0 and 10°C, for example, might be 2.3, but the Q_{10} for the same animal between 20 and 30°C might be only 1.9.*

---

*When thermal sensitivity changes with temperature, a detailed description may require analysis of Q_{10} over temperature spans of less than 10°C. In fact, Q_{10} can be calculated for a temperature span of any width, using the van't Hoff equation:

$$Q_{10} = \left(\frac{R_2}{R_1}\right)^{10/(T_2 - T_1)}$$

where R_2 is the rate at any temperature T_2 (measured in °C) and R_1 is the rate at *any* lower temperature T_1 (°C). The Q_{10} calculated is the factor by which the rate would increase over a *10°C increment* in temperature if the thermal sensitivity evident between T_1 and T_2 were to prevail over a full span of 10°C. That is, the Q_{10} expresses the thermal sensitivity between T_1 and T_2, adjusted to the standard 10°C increment.

One value of the semilogarithmic plot of metabolic rate against temperature is that its *slope* reflects changes in Q_{10}. (The slope is not, however, equal to Q_{10}.) A constant slope indicates a constant Q_{10}. If the slope is increasing or decreasing as temperature is raised, Q_{10} (and thermal sensitivity) is also increasing or decreasing, respectively. The slope in Figure 6.8B, for example, is greater between 10 and 20°C than between 0 and 10°C, indicating a higher Q_{10} in the former thermal range than in the latter.

Just as resting metabolic rate increases with body temperature in poikilotherms, other physiological rates, such as heart rate and breathing rate, also generally show a strong positive thermal dependence. Behavioral rates are also affected; frogs, insects, and fish, for instance, often move strikingly more slowly at low than high body temperatures. Many rates bear exponential or approximately exponential relations to temperature, although some exhibit fundamentally different (e.g., linear) relations.

The Arrhenius Relation

Briefly now, it is appropriate to consider the kinetics of *individual chemical reactions* of the sorts that comprise metabolism. When reactions are studied "in a test tube" at physiological temperatures, their rates typically increase exponentially or quasiexponentially as temperature is increased. The Q_{10} values of the reactions tend to be 2 to 3, signifying that the reaction rates increase by *100–200 percent* for every increment of 10°C. Chemical bonds within the reacting molecules must be strained and contorted to the breaking point for the reactions to occur. The degree of strain and contortion of the bonds in a reacting molecule is a function of the kinetic energy possessed by the molecule: The higher the energy level, the greater the intramolecular strains and the greater the probability of bond breakage. Recognizing that an increase in temperature reflects an increase in molecular kinetic energy, these principles explain in a general sense why reaction rates increase with temperature. In a specific sense, however, a substantial paradox seemingly remains. When temperature is raised by 10°C in the physiological range, the *average* kinetic energy of molecules is raised by just *3–4 percent*. How is it possible then for reaction rates to increase by 100–200 percent? Svante Arrhenius is famed for his solution to this paradox, formulated in the 1880s.

Within a population of molecules at given temperature, there is a wide range of variation in the kinetic energies possessed by individual molecules. During intermolecular collisions, molecules interchange energy; thus, a particular molecule may be high in energy at one moment but low at another. At any specific moment, nonetheless, the fraction of all molecules possessing any given kinetic energy is fixed and predictable, depending on temperature, as

shown in Figure 6.9. Arrhenius reasoned that for a molecule to undergo a particular reaction, its energy level must equal or exceed some threshold level, which he termed the *energy of activation.* In Figure 6.9, the shaded and cross-hatched parts under the curves represent the fractions of molecules possessing at least the energy of activation (for some particular reaction) at each respective temperature. Arrhenius pointed out that this fraction increases several-fold when the temperature is raised by 10°C, even though the average kinetic energy of all molecules increases only slightly. This is the explanation for why reactions accelerate as much as they do when the temperature is elevated. Based on these principles, Arrhenius devised the following equation, now named in his honor, to predict the relation between reaction rate and temperature:

$$\ln \frac{k_2}{k_1} = \frac{E}{R}\left(\frac{1}{T_1} - \frac{1}{T_2}\right)$$

where k_1 and k_2 are, respectively, the reaction rate constants at two absolute temperatures T_1 and T_2, R is the universal gas constant, and E is the energy of activation.

Metabolism, to a large extent, represents the sum total of a complex array of chemical reactions. Thus, the dependence of metabolic rates on tissue temper-

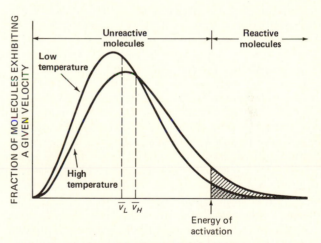

Figure 6.9 A schematic representation of the effects of temperature on molecular velocities in a population of molecules. Molecular kinetic energy is directly dependent on molecular velocity. Each curve depicts the fraction of molecules possessing a given velocity at a given time (or the probability that any particular molecule will attain that velocity). One curve is for a relatively low temperature, the other, for a relatively high temperature. On the abscissa, $\bar{v}_L$ and $\bar{v}_H$ identify the *average* velocities at the low and high temperatures, respectively. The velocity corresponding to the energy of activation for some particular reaction is also identified, and the shaded and cross-hatched regions indicate the fractions of molecules having at least that energy of activation at the low and high temperatures, respectively. Raising the temperature has a greater proportional effect on the fraction of molecules that are reactive than on the average molecular velocity. The curves are derived from the Maxwell–Boltzmann distribution equations.

ature can be explained substantially as a manifestation of the thermal sensitivity inherent in individual reactions.

Escape from the Tyranny of the Arrhenius Equation

The strong dependence of metabolic rates and other functional rates on body temperature led Sir Joseph Barcroft (1872–1947) to the apt remark that poikilotherms are subject to a certain "tyranny of the Arrhenius equation." Organisms, as chemicophysical systems, must simply cope, to some extent, with the thermal sensitivity inherent in such systems. Life is, however, an adaptively evolving kind of chemicophysical system, and we have come to expect it not to show the static inflexibility of a test-tube reaction. We can presume that there would sometimes be selective advantages for poikilotherms to escape the tyranny of the Arrhenius equation, and forms of "escape" are now known in both the acute and chronic time frames.

Before examining the forms of escape, we need to clarify the distinction between acute and chronic exposure to temperature. *Acute* means short-term. *Chronic* means long-term. Thus, an animal is chronically exposed to a temperature if it has been at that same temperature for a long time. It is acutely exposed if it has only just recently been placed at the temperature. When exposure to a temperature is acute, the animal cannot fully bring into play any forms of physiological adjustment that require substantial time for their manifestation (e.g., synthesis of new enzymes). On the other hand, chronic exposure to a temperature allows the full development of long-term responses. Suppose that an individual that has been living at 20°C is suddenly (within a few hours' time) exposed sequentially to 15, 10, and 5°C. Then suppose that it is left at 5°C for a number of weeks. We might well find that its responses to 5°C after the passage of time are different from those it initially displayed upon acute exposure to 5°C. Furthermore, if the animal—having chronically adjusted to 5°C—were then acutely exposed once more to 10 and 15°C, we might well find that its responses at 10 and 15°C are different from those displayed earlier when it had been living at 20°C. An animal's response to acute exposure can depend on the previous chronic exposure.

In our earlier discussion of metabolic rates as functions of temperature (see Figures 6.7 and 6.8), we were looking at responses to acute temperature exposures.

Cases of Low Sensitivity to Acute Changes in Temperature

Some poikilotherms display very low metabolic sensitivity to acute changes of temperature within certain ranges. Such instances of escape from the tyr-

anny of the Arrhenius equation, although the exception and not the rule, are now known to be reasonably common.

Narrow ranges of low thermal sensitivity are observed in a number of lizard species, for instance, and are exemplified by the data on desert iguanas (*Dipsosaurus dorsalis*) in Figure 6.10. Between 35 and 40°C, the Q_{10} for the standard metabolic rate of these lizards is not significantly different from 1.0, indicating no change in resting metabolic rate as body temperature is acutely shifted up or down in this range. By contrast, the Q_{10} is 2.1–3.5 in the other temperature ranges depicted. Another case of low thermal sensitivity in a vertebrate is shown in Figure 6.11. Embryos of birds in the egg function poikilothermically: Their body temperatures approximate the environmental temperature. As shown by the figure, however, the metabolic rates of the embryos of Heermann's gulls are unaffected by acute changes of temperature between 30 and 40°C. The embryos' heart rates also are thermally independent at about 33°C and above. The full significance of such ranges of low metabolic sensitivity to temperature has yet to be understood. It seems instructive, nonetheless, that in the instances we have examined, these ranges fall within the span of body temperatures that are

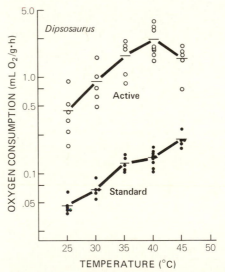

Figure 6.10 Standard (resting) and active rates of oxygen consumption during acute exposure to various ambient temperatures in the desert iguana (*Dipsosaurus dorsalis*). Under the conditions of the experiments, body temperature can be presumed approximately equal to ambient temperature. Horizontal bars depict means. Active rates were determined during 2 min of maximal activity induced by electrical stimulation. It is important to note that anaerobic processes accounted for 60–80 percent of ATP production during activity; thus, the rate of oxygen consumption during activity, indicating only the rate of aerobic ATP production, reflects less than half of energy expenditure during activity (see Chapter 4). Total energy expenditure during activity varied with temperature more or less in parallel with the curve for oxygen consumption that is shown. During rest, metabolism is aerobic and thus accurately reflected by oxygen consumption alone. [From A.F. Bennett and W.R. Dawson, *J. Comp. Physiol.* **81**:289–299 (1972).]

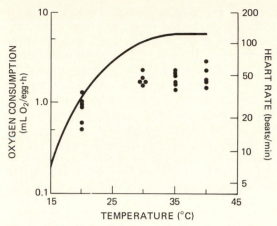

Figure 6.11 Rate of oxygen consumption (dots) and average heart rate (line) during acute exposure to various body temperatures in 1-week-old embryos of Heermann's gull (*Larus heermanni*). Oxygen consumption was measured on intact eggs, heart rate on embryos removed from eggs. Incubation in this species lasts 25 days; thus, the studied embryos would have to complete over 2 weeks of additional development before hatching. [After A.F. Bennett and W.R. Dawson, *Physiol. Zool.* **52**:413–421 (1979).]

common among invertebrates that occupy the intertidal zone, the region between the high and low tide lines along the seashore. These animals often experience particularly large and rapid changes of temperature by virtue of being alternately immersed in the ocean and exposed to the air and sun above. In one instance, for example, barnacles along the English shore had body temperatures of 30°C when exposed to warm air and sunshine at low tide, but the temperature of the ocean was 14°C and thus their body temperatures plunged rapidly by 16°C upon immersion during the next incoming tide. Perhaps the regular recurrence of such sharp temperature fluctuations has placed a particular premium on the evolution of low thermal sensitivity. In any case, many intertidal invertebrates—including certain anemones, worms, barnacles, mussels, snails, and sea urchins—exhibit substantial ranges in which their *resting* rates of aerobic metabolism show little or no thermal sensitivity, as illustrated in Figure 6.12. The range of temperatures in which resting metabolism is thermally insensitive typically corresponds fairly well with the temperatures actually experienced by the animals in their native habitat, even shifting with the seasons in some cases.

Responses to Chronic Environmental Changes: Acclimation and Acclimatization

In the responses of animals to chronic changes of temperature, forms of "escape from the tyranny of the Arrhenius equation" are widespread. To start our examination of these phenomena, let us introduce certain basic concepts by using the data for fence lizards, *Sceloporus occidentalis,* presented in Figure 6.13. Two groups of lizards were maintained

particularly likely to prevail. Desert iguanas regulate their body temperatures behaviorally to be in the high 30s or low 40s when they are active during the day (p. 125). Parental Heermann's gulls maintain their eggs at temperatures that fluctuate around 37°C. Thus, both the lizards and the gull embryos have evolved capacities for metabolic stabilization that are effective within specific thermal ranges that they commonly experience.

Ranges of low acute thermal sensitivity—sometimes as broad as 10°C or more—are particularly

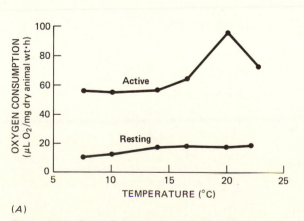

(A)

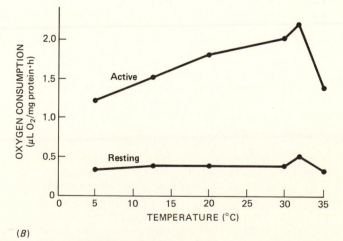

(B)

Figure 6.12 Rate of oxygen consumption during acute exposure to various ambient water temperatures in resting and active intertidal invertebrates: (A) Barnacles (*Semibalanus balanoides*). (B) Periwinkles (*Littorina littorea*). Body temperature approximates ambient temperature. Active barnacles were undergoing normal cirral beating (feeding movements); resting ones maintained a small opening between their opercular valves but showed no cirral beating. Active periwinkles were crawling, whereas resting ones were quiescent. [Part (A), for animals of 1.5 mg dry weight, after R.C. Newell and H.R. Northcroft, *J. Mar. Biol. Assoc. U.K.* **45**:387–403 (1965); (B), for animals of 30 mg protein content collected in May, after R.C. Newell and V.I. Pye, *Comp. Biochem. Physiol.* **38B**:635–650 (1971).]

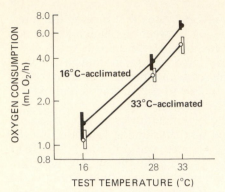

Figure 6.13 Metabolic responses to acute changes in temperature in two groups of resting, fasting fence lizards (*Sceloporus occidentalis*). One group (upper curve and solid symbols) was acclimated for 5 weeks to 16°C prior to testing; the other group (lower curve and open symbols) was acclimated to 33°C for 5 weeks prior to testing. Circles indicate means; vertical bars indicate ±2 standard deviations of the mean. Note that oxygen consumption is plotted on a logarithmic scale. [From W.R. Dawson and G.A. Bartholomew, *Physiol. Zool.* **29**:40–51 (1956). Used by permission of The University of Chicago Press. Original figure © 1956 by the University of Chicago.]

for 5 weeks at 16 and 33°C, respectively. Their resting metabolic rates were then determined during short-term (acute) exposure to 16, 28, and 33°C, and it can be seen that the responses of the two groups were different. Those that had been living at the cooler temperature, 16°C, had higher metabolic rates at any given test temperature than those that had been living at the warmer temperature, 33°C. The only known difference between the two groups was the temperature at which they had been living. When differences in physiological state appear after exposure to environments differing in only one or two well-defined parameters (e.g., temperature), we say that *acclimation* has occurred.

It is important to appreciate certain implications of this type of acclimation. Figure 6.14 presents a hypothetical example resembling the data for the lizards in basic outline. The upper solid line shows the *acute* response of cold-acclimated individuals, whereas the lower solid line shows the *acute* response of warm-acclimated animals. The dashed line connects the point for warm-acclimated animals tested at *their* (warm) temperature of acclimation with the point for cold-acclimated animals tested at *their* (cold) temperature of acclimation. The dashed line thus shows the responses of the animals to *chronic* temperature changes; it is the metabolism–temperature curve for animals that are allowed to live at each temperature for a long period of time before being tested. On a semilogarithmic plot, as in Figure 6.14, the slope of the metabolism–temperature curve reflects Q_{10} or thermal sensitivity (p. 85). Thus, we see the very important point that the chronic response (dashed line) is less thermally sensitive than the acute response of *either* acclimation group. Put another way, the difference in metabolism

between warm and cold test temperatures is less if the animals are permitted to acclimate to each test temperature than if they are changed suddenly from one temperature to the other. This same principle is evident in the data for lizards in Figure 6.13. In essence, we see that the type of acclimation shown by the lizards represents a mechanism for relative stabilization of metabolic rate—a type of escape from the tyranny of the Arrhenius equation.

An alternative way to look at these results is provided in Figure 6.15*A*. Suppose we start with 33°C-acclimated lizards at 33°C and within a few hours lower their temperature to 16°C. Metabolism will fall, as shown, according to the acute response curve for 33°C-acclimated animals and will be much reduced. If the lizards are then left at 16°C for several weeks so that acclimation can occur, metabolism will *rise* back toward its original level. *Note that this rise occurring during acclimation acts to reduce the effect of the change in temperature on metabolism.*

After metabolic rate has been raised or lowered by an abrupt change in temperature, any subsequent, long-term tendency for it to return toward its original level even though the new temperature continues to prevail is called metabolic *compensation.* The rise from points *y* to *z* in Figure 6.15*A*, tending to return the metabolic rate toward its original level *x*, is an example. Compensation is *partial* if the metabolic rate fails to return all the way to its original level, as in Figure 6.15*A*. A full return, as in Figure 6.15*B*, would represent *complete* compensation. Usually, compensation is only partial.

Earlier, we defined acclimation to represent changes in physiological state resulting from long-term adjustment to environments differing in only one or two well-defined parameters, such as temperature. As such, acclimation is a laboratory phenomenon. In nature, environments probably never differ in only one or two well-defined parameters. Winter

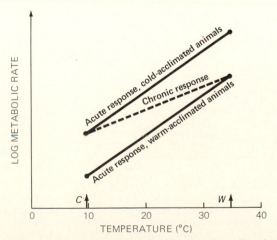

Figure 6.14 A hypothetical example of acclimation to temperature. The arrows labeled *C* and *W* mark, respectively, the temperatures to which the "cold" and "warm" groups of animals were acclimated.

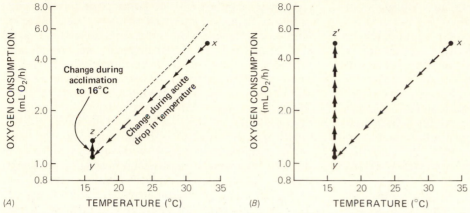

Figure 6.15 (*A*) Average changes in resting metabolism of lizards that have been living at 33°C when the temperature is lowered acutely to 16°C and then held at 16°C for 5 weeks, based on Figure 6.13. Point *x* is the metabolic rate of 33°C-acclimated lizards at 33°C. During the acute drop in temperature, metabolism falls to point *y*, the metabolic rate for 33°C-acclimated lizards at 16°C. Then metabolism rises over the period of acclimation to point *z*, the rate for 16°C-acclimated lizards at 16°C. Thin arrows show change in metabolism during acute temperature change and follow the curve for 33°C-acclimated animals in Figure 6.13. Thick arrow shows change in metabolism during acclimation to 16°C. Dashed curve shows the acute response of 16°C-acclimated lizards (see Figure 6.13). (*B*) Average changes in metabolism that would occur if animals were treated as in part (*A*) but displayed complete metabolic compensation during acclimation to 16°C. Acclimation (thick arrows) would then lead to a metabolic rate (*z'*) identical to that initially shown at 33°C (*x*).

in temperate woodlands, for example, implies not only lower temperatures than those in summer, but also shorter days, lowered atmospheric humidities, altered food sources, and a good many other changes. Differences in physiological state that appear after long-term exposure to different natural environments are said to represent *acclimatization*. A classic example is illustrated in Figure 6.16. Mussels from relatively cold, high-latitude waters pump water across their gills faster at any test temperature (temperature of acute exposure) than mussels from warmer, lower-latitude waters. This acclimatization is thermally compensatory; because of it, the popu-

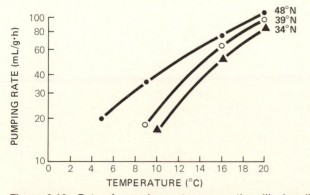

Figure 6.16 Rate of pumping water across the gills, in milliliters of water per gram of body weight per hour, as a function of acute test temperature for 50-g mussels (*Mytilus californianus*) gathered at three latitudes along the West Coast of the United States. Pumping rate is significant because mussels extract both food and oxygen from the water pumped. [From T.H. Bullock, *Biol. Rev. (Cambridge)* **30**:311–342 (1955).]

lations of mussels living chronically in the cold, northern and warmer, southern waters of our West Coast are more similar to each other in pumping rates than they otherwise would be.

The distinction between acclimatization and acclimation is important to make because, although the phenomena are closely related, they present significantly different challenges when we seek to analyze their causation. We cannot immediately pinpoint the environmental factors responsible for differences observed between populations in different natural environments, whereas in laboratory environments we can control, and thus readily identify, the factors of interest. In the case of the two groups of fence lizards acclimated to different temperatures in the laboratory, for example, there can be little doubt that the differences in temperature were responsible for the physiological differences observed between the groups. In the case of the populations of mussels acclimatized to several latitudes, however, although differences in temperature between northern and southern waters may have been primarily responsible for inducing the observed physiological differences among the populations, we cannot be sure without carrying out controlled experiments. Perhaps temperature differences were inconsequential for the acclimatization of the mussels, and differences in day-length or food supplies among latitudes were critical. Despite these complexities, it remains true that poikilotherms exhibiting thermal acclimation in the laboratory usually show *qualitatively* similar responses in natural habitats that differ in temperature. If, for example, metabolic rate at a given test temperature is elevated by acclimation to low temperatures in the laboratory, it usually is also elevated during cool seasons in nature.

Sometimes animals fail to display acclimation or acclimatization of their metabolic rates. In the many instances where they do exhibit the phenomena, diverse patterns are found in the details. As already suggested by the examples we have cited (Figures 6.13 and 6.16), the most common pattern is for animals from cold environments to exhibit higher metabolic rates at a given test temperature than ones from warm environments. In such cases, living chronically in the cold is said to *translate* (shift) *metabolism upward*.

When a physiological *rate* or *capacity* to perform a function is altered by long-term exposure to changed conditions, the process is termed **capacity acclimation** or **capacity acclimatization**. The upper or lower temperatures that are lethal or incapacitating can also be changed, and then we speak of **resistance acclimation** or **resistance acclimatization**. During these latter processes, the range of temperatures compatible with adequate function is typically shifted downward by long-term exposure to cold conditions and upward by exposure to warm conditions. To cite one example, cockroaches (*Blatta*) acclimated to 30°C are unable to move if their temperature is lowered to about 7.5°C; but if they are acclimated to 14–17°C, they are not immobilized until their temperature has fallen to about 2.0°C.

Acclimation and acclimatization, whether of the capacity or the resistance type, require appreciable amounts of time to become manifest, but it is difficult to generalize about the actual amount of time required inasmuch as it varies considerably with the species and function under study. Sometimes at least a partial expression of acclimation is evident within the first day after animals have been placed under new environmental conditions, but full expression usually requires days, weeks, or even months.

The Third Time Frame of Response; Genotypic Versus Phenotypic Differences Between Populations

Up to now, we have discussed two time courses of response to alterations in the environment: responses to acute environmental changes and the longer-term responses known as acclimation and acclimatization. Both of these types of response, by definition, occur within the life span of the individual and, although conditioned by the genotype, are phenotypic. A third significant time frame is that of evolutionary time, in which natural selection operating on alternative genic alleles can result in genotypic physiological differences between populations.

Whenever physiological differences are found between populations living in different environments, it is important to consider whether the differences are attributable to genetic divergence or to acclimatization in genetically similar populations. Genetic differences can be presumed only if the physiological differences persist after steps have been taken to eliminate differences in acclimatization. What might such steps be? An illustration of one approach is provided by a classic study of limpets (*Acmaea limatula*) living in the intertidal zone. Adult limpets attach firmly to rocks, moving about only slightly. Thus, the adults occupying the lowest reaches of the intertidal zone at the study site could be considered not to mingle with those found higher on the shore near the midpoint of the zone. The limpets living in the low intertidal zone were immersed in the sea at least 90 percent of the time, whereas those from higher up, near the midpoint of the zone, were immersed only about 50 percent of the time. It was postulated that the "high" animals, being more often exposed to the air and sun, would experience a higher mean body temperature than the "low" animals. When tested at temperatures from 9 to 29°C, the heart rates of "low" animals proved to be consistently higher than those of "high" animals at a given test temperature. The question then was: Were the differences in heart rate attributable to acclimatization, or were the "low" and "high" animals distinct microgeographic physiological races? To answer the question, reciprocal transplants were performed; "high" animals were transferred to the low habitat, and vice versa. Within 2–4 weeks the transplanted animals assumed heart rates similar to those of the natives in their new habitat. The transplants thus revealed that the "high" and "low" animals were not genetically different in the respect under study. Instead, their physiological differences represented acclimatization, a phenotypic adjustment to prevailing environmental conditions.

Another approach to determining the basis for physiological differences between two populations is to acclimate animals from both populations to a single laboratory environment and ascertain whether the differences persist. If in fact they disappear, genetic divergence between the populations is contraindicated. On the other hand, what if the two groups of wild-caught animals continue to differ physiologically even in the laboratory? This result would suggest genetic divergence, but it would not truly demonstrate it, for animals that grew up in different natural environments may thereby have acquired phenotypic differences that are partly or fully irreversible. To test conclusively for genotypic divergence, we would need to compare animals, descended from the two groups, that have been reared from conception in a single laboratory environment. Unfortunately, this ideal experiment often proves difficult in practice.

Climatic Adaptation

When related animals living in different climates exhibit physiological differences that are known or believed to be caused by evolutionary, genotypic di-

---

BOX 6.1   ACCLIMATION AND ACCLIMATIZATION IN GENERAL

The principle that *the current physiological status of an animal depends on its environmental history* is one of great generality, transcending the study of temperature and metabolic rates. Animals may show long-term physiological adjustments in response to diverse environmental agents, including—in addition to temperature—humidity, salinity, oxygen supply, photoperiod, and food supply, to name just some. Furthermore, acclimation or acclimatization can potentially be exhibited in virtually any physiological property (and sometimes in behavioral and morphological properties as well). The basic concepts developed here in the limited context of thermal effects thus have applicability in other diverse contexts.

---

vergence, those differences are commonly called *climatic adaptations.* Historically, the mere presence of genetically based physiological differences between populations has often been deemed sufficient justification for calling the differences "adaptations." Unfortunately, therefore, the term *climatic adaptation* has tended to be applied to all genetically based differences, regardless of whether they have met more exacting standards for being judged adaptive (p. 5). Another challenge for the future is that many of the differences that are termed climatic adaptations have not yet been tested adequately to establish conclusively their genetic basis.

A favorite approach in the study of climatic adaptation has been to compare animals along latitudinal gradients. Many instances are now known in which animals from relatively cold, high latitudes are less tolerant of high temperatures (or more tolerant of low temperatures) than related animals from warmer, lower latitudes. The upper lethal temperatures of high-latitude species of fish, for example, tend to be low by comparison to those of tropical fish; indeed, some species from the perpetually cold antarctic seas die when the water temperature has risen to only 5–10°C.

Looking at rate functions, it is a common field observation that related poikilotherms living at cold and warm latitudes seem to function behaviorally at much the same level despite large differences in body temperature. Thus, for example, fish, crabs, and starfish in the frigid waters of northern Maine are not strikingly more lethargic than their relatives in the warm waters of Bermuda. Such observations, although obviously subjective, strongly suggest the presence of physiological compensations to prevailing thermal conditions. Investigators seeking to evaluate quantitatively the presence of such compensations have frequently compared the standard (or routine) metabolic rates maintained at low, polar temperatures by animals native to high and low latitudes. Some of the first studies of this type indicated that, among fish, crustaceans, and some other taxa, the standard (or routine) metabolic rates exhibited by polar species at polar ambient temperatures are substantially higher than the rates that related temperate-zone or tropical species would be expected to maintain at such temperatures. Some more-recent studies have obtained similar results, but others, even on similar animals, have concluded that related high- and low-latitude species differ little, if at all, in their standard (or routine) metabolic rates at low ambient temperatures. Indeed, a significant theoretical question has been raised: Even if polar species might profit by being able to maintain higher levels of *activity* than low-latitude species when temperatures are low, of what advantage would it be for them to maintain a higher *resting* rate of metabolism as well? Currently, uncertainty remains on these issues.

Data on some other rates seem clearer. For example, rates of development at low ambient temperatures are higher in some high-latitude species of insects and amphibians than in related low-latitude species, thus aiding completion of development by the high-latitude forms within the relatively short, cool growing seasons that they experience. At near-freezing temperatures, protein synthesis by the liver occurs much more rapidly in polar fish than in temperate-zone ones, and the skeletal muscles of the polar fish are capable of greater power output. Indeed, temperate-zone fish need to be as warm as 15–25°C to develop the same muscle power output as some antarctic fish display at 0°C.

Biochemical Aspects of Escape from the Tyranny of the Arrhenius Equation: Enzymes

Changes in such parameters as metabolic rate and heart rate, which we observe in studies of whole animals, must ultimately reflect changes at the biochemical and biophysical level within cells of the organism. Thus, we are led to such questions as: What biochemical and biophysical alterations underlie the processes of acclimation and acclimatization? What attributes of biochemical organization permit many of the intertidal invertebrates to have relatively stable resting metabolic rates over broad ranges of temperature? What are the biochemical differences between an antarctic fish that is killed by heat at 5°C and a tropical fish that is killed by cold at 10°C?

A decrease in body temperature means that the molecules in each cell undergo a drop in kinetic energy. One of the fundamental problems that thus arises is that molecules have a lesser probability of attaining the energy of activation required for any given reaction and reaction rates accordingly tend to slow (p. 85). What can a cell do to ameliorate these effects? A key concept in attempting to answer this question is that the rate-limiting reactions in cellular metabolism are typically catalyzed by enzymes. By modifying the amounts of enzymes or their catalytic properties, a cell can prevent a drop in temperature from exerting as large an effect on reaction rates as it otherwise would, thus escaping the tyranny of the Arrhenius equation.

One experimental approach used extensively in studying modulation of enzymes is to search for changes in enzymatic activity. To illustrate, let us consider succinic dehydrogenase, the enzyme that oxidizes succinate to fumarate in the Krebs cycle (see Figure 4.1). Suppose the tissue of interest is muscle. We could homogenize a sample of muscle, establish a controlled concentration of succinate in the homogenate, and measure the rate at which fumarate is formed from succinate; this rate would measure the overall catalytic ability—that is, the *activity*—of the enzyme present in the tissue under the test conditions. By comparing muscle samples from fish that have been acclimated to two different temperatures, we could determine whether acclimation has affected the activity of succinic dehydrogenase. This comparison has actually been performed on goldfish, and the activity at given test temperature proved to be higher in cold-acclimated than in warm-acclimated individuals, indicating changes at the enzymatic level in the cold-acclimated animals that would tend to compensate for the reduced kinetic energies prevailing at cold temperatures. Similar comparisons have been performed on a considerable variety of enzymes, mostly in fish but also in some invertebrates. In general, enzymes associated with glycolysis, the Krebs cycle, and the electron transport chain—that is, enzymes involved in the generation of ATP—have proved to show increased activities in cold-acclimated individuals. This is entirely in line with the whole-animal data showing that metabolic rate is frequently elevated at given test temperature in cold-acclimated animals.

A shortcoming of information on enzyme activity is that it does not necessarily tell us what properties of the enzyme have changed. An increase in activity during cold acclimation, for example, could reflect an increase in enzyme concentration, a change in the cellular milieu effecting increased enzymatic efficiency, or a change in the chemical structure of the enzyme. Recently, attention has increasingly been directed to identifying the specific properties of enzyme systems that are altered under various circumstances.

Basic Enzyme Kinetics and Potential Sites of Modulation For an enzyme molecule to catalyze a reaction, it must first combine with a molecule of substrate (initial reactant) to form an *enzyme–substrate complex*. The substrate is then converted to product while united with the enzyme, forming an *enzyme–product complex*, which subsequently dissociates to yield free product and free enzyme (the latter ready to combine with new substrate). Symbolically, if E, S, and P represent molecules of enzyme, substrate, and product, respectively, the major steps are:

$$E + S \rightarrow E\text{–}S \rightarrow E\text{–}P \rightarrow E + P$$
$$\text{complex} \quad \text{complex}$$

In a cell, a collision between enzyme and substrate molecules does not necessarily result in an enzyme–substrate complex; the two molecules may instead simply bounce apart. The outcome of a collision depends in good part on properties of the enzyme quantified as *enzyme–substrate affinity.* By definition, an enzyme has high affinity for substrate if it exhibits a high probability of complexing with substrate molecules it encounters. Conversely, an enzyme that forms complexes with difficulty is said to have low affinity. Given that formation of an enzyme–substrate complex is necessary for conversion of substrate to product, the affinity of enzyme for substrate can play a major role in determining how many substrate molecules are converted to product by each enzyme molecule per unit time. If a cell has suffered a reduction in reaction rates because of a drop in its temperature, one way for it to return those rates toward their original levels would be to increase the affinities for substrate of the enzymes involved.

Once an enzyme–substrate complex has been formed, conversion of substrate to product must await attainment of an energy level at least as high as the energy of activation (p. 85). Thus, the magnitude of the energy of activation is also an important determinant of reaction rate; if the activation energy is relatively low, for instance, substrate molecules will have a relatively high probability of reaching or exceeding it in any given length of time, and the rate of formation of product will therefore be relatively high. The magnitude of the energy of activation is set by the molecular properties of the enzyme and accordingly can potentially be modified in advantageous ways by the cell.

Because both the enzyme–substrate affinity and energy of activation of an enzyme depend on the molecular nature of the enzyme, they are termed *qualitative* enzymatic parameters. A third parameter that potentially can be modified advantageously is *quantitative,* namely, the enzyme concentration. Raising the concentration, for instance, accelerates the catalyzed reaction by increasing the availability

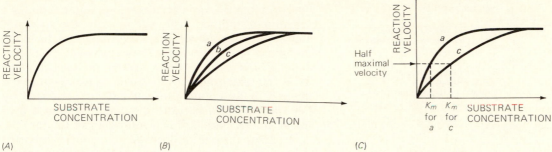

Figure 6.17 (*A*) The velocity–concentration curve for an enzyme exhibiting hyperbolic reaction kinetics. (*B*) Velocity–concentration curves for enzymes having high (*a*), medium (*b*), and low (*c*) affinity for substrate; the amount of enzyme is presumed constant regardless of affinity. (*C*) An illustration of the calculation of the apparent Michaelis constant.

of catalytic sites. For a cell to alter one of the qualitative enzymatic parameters, it often needs to synthesize a new type of enzyme molecule, but to alter concentration, it needs only to synthesize more or less of the type of molecule it has already been making.

Thus far, we have identified several possibilities by which a cell might compensate for changes in its temperature by modification of enzyme parameters. Before examining the extent to which animals actually capitalize on these possibilities, it will be useful to look at enzyme kinetics from one additional angle. Figure 6.17*A* illustrates how reaction rate is affected by substrate concentration in many enzymatically catalyzed reactions. Note that at relatively low substrate concentrations, the velocity of the reaction increases as substrate concentration increases, but a point is reached where further increases in substrate concentration no longer alter velocity. This behavior follows from the necessity that substrate be combined with enzyme before it can react to form product. At low substrate concentrations, the amount of substrate available is the limiting factor in determining velocity; all available enzyme molecules are not "occupied" at any one time, and raising the substrate concentration increases the rate of formation of product by allowing fuller utilization of available enzyme. At some substrate concentration, however, substrate becomes sufficiently abundant that available enzyme molecules are utilized fully or **saturated**. Then the amount of enzyme present becomes limiting, and further increases in substrate concentration cannot enhance velocity.

Figure 6.17*B* shows how an enzyme's affinity for substrate affects the velocity–concentration curve. Line *a* represents an enzyme of relatively high affinity. Line *c,* by contrast, shows how the curve would be changed if the *affinity* were considerably lower than in *a* but the *amount* of enzyme were unaltered. The diagram brings out the important point that affinity affects reaction rates only at subsaturating con-

centrations of substrate. Usually concentrations in cells are indeed subsaturating. A convenient numerical expression of affinity is the **apparent Michaelis constant, K_m,** defined to be the substrate concentration required to attain one-half of the maximal velocity. K_m is derived in Figure 6.17*C* for lines *a* and *c* from Figure 6.17*B*. Note that the low-affinity enzyme (line *c*) has the greater K_m. *Thus, K_m and affinity are related inversely. A high K_m means low affinity, and a low K_m means high affinity.*

Figure 6.18 illustrates two mechanisms of compensation for a reduction in temperature. In both parts, the upper solid line depicts reaction velocity as a function of substrate concentration for a hypothetical enzyme at high temperature, and the lower solid line shows the extent to which reaction velocities would be depressed by a lowering of temperature with no change in either the concentration or affinity of the enzyme. The dashed line in part (*A*) shows reaction rates at the low temperature if the lowering of temperature is accompanied by a rise in enzyme concentration, whereas the dashed line in (*B*) shows the effect of an increase in enzyme–substrate affinity at the low temperature. Note that at the substrate concentrations prevailing in cells, either a rise in enzyme concentration or a rise in affinity tends to elevate the reaction rate at low temperatures back toward the rate seen at high temperatures, thus partly compensating for the drop in temperature.

Patterns of Enzyme Modulation in Animals Changes in enzyme concentration seem to play an important part in acclimation and acclimatization. We indicated earlier (p. 92), for example, that acclimation to cold commonly evokes increases in the activities of enzymes involved in the production of ATP. There is reason to believe that, in many instances, these increases in activity are attributable to increases in enzyme concentration.

Although species living in different climates some-

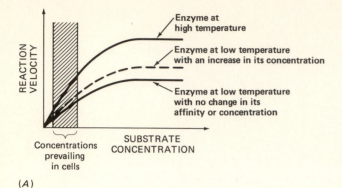

(A)

(B)

Figure 6.18 Illustrations of how increases in enzyme concentration (A) or enzyme–substrate affinity (B) can compensate for a reduction in temperature. See text for discussion. The shaded vertical bars show a range of substrate concentrations that might prevail in cells. In (A), all three curves have the same affinity. In (B), enzyme concentration is assumed constant for all three curves. It is evident that if substrate concentrations change as temperature is altered, those changes will interact with changes in enzyme parameters in determining the overall outcome; the behavior of substrate concentrations is a focus of current interest.

times differ adaptively in enzyme concentrations, more is known at present about their differences in activation energy. Related species that ordinarily operate at different tissue temperatures have often evolved different forms of particular enzymes, such that the enzyme variants possessed by the cold-tissue species (e.g., polar fish) provide lower energies of activation than the homologous variants in warm-tissue species (e.g., tropical fish or mammals). Such differences may be great enough to give the cold-tissue species at least 10-fold greater reaction rates when temperatures are low. Acclimation and acclimatization (*within* a species) seem only rarely to involve changes in activation energy.

In examining enzyme–substrate affinities, it is appropriate to start by considering the effects of *acute changes in temperature* on the affinity of extant enzyme molecules. The affinity for substrate of an enzyme protein, like other catalytic properties, is dependent on the three-dimensional structure of the protein molecule. Many of the bonds that stabilize

three-dimensional structure are of weak types (e.g., hydrogen bonds, van der Waals interactions) that are sensitive to temperature. Accordingly, the structure and affinity for substrate of an enzyme depend on the immediately prevailing temperature. Commonly, animals have evolved enzymes that increase in affinity as the temperature is dropped, at least within the range of body temperatures usually experienced by the species. This pattern is evident in Figure 6.19 (recall that K_m and affinity are inversely related).

An acute drop in temperature, of course, tends to slow a reaction by decreasing the kinetic energies of reacting molecules. If that selfsame drop in temperature causes the enzyme catalyzing the reaction to increase its affinity for substrate, however, then the reduction in reaction rate will be less than if affinity were to remain unchanged (presuming subsaturating substrate concentrations). That is, the increase in affinity will tend to compensate for the reduction in kinetic energies. Possibly, an extreme form of this process is what permits certain animals (e.g., some intertidal invertebrates) to display ranges of metabolic temperature insensitivity (p. 86). Interpretation of the significance of affinity changes is presently difficult because substrate concentrations may also shift with temperature and we know little about such shifts.

Early in the comparative study of enzyme–substrate affinity, it was hypothesized that species inhabiting cold climates might have evolved enzymes of higher affinity than related species inhabiting warm climates. In fact, however, various species, *when tested at their respective physiological ranges of temperature,* generally resemble each other in affinity. As an illustration, consider the data on the lactate dehydrogenases of fish in Figure 6.20. The

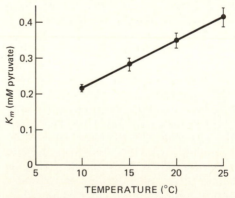

Figure 6.19 The effect of acute changes in temperature on the apparent Michaelis constant, K_m, of lactate dehydrogenase from muscle of bluefin tuna (*Thunnus thynnus*). This enzyme catalyzes the interconversion of pyruvic acid and lactic acid (Figure 4.3). The K_m shown is the concentration of pyruvic acid (mM) required at the prevailing temperature to attain a half-maximal rate of lactate formation. The vertical lines depict 95 percent confidence limits. Assays were run in a buffered solution permitted to undergo usual variations in pH with temperature. [After P.H. Yancey and G.N. Somero, *J. Comp. Physiol.* **125:**129–134 (1978).]

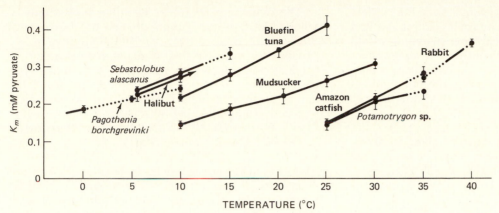

Figure 6.20 The apparent Michaelis constant, K_m, as a function of test temperature for muscle lactate dehydrogenases isolated from seven species of fish and from rabbits. See legend of Figure 6.19 for further description. Parts of curves drawn as solid lines pertain to body temperatures normally experienced by each species in its natural habitat. Dotted portions of curves refer to temperatures not normally experienced. [Reprinted by permission from P.H. Yancey and G. N. Somero, *J. Comp. Physiol.* **125**:129–134 (1978).]

enzymes possessed by the various species are indeed different, as is evident from their different K_m–temperature plots. Note, however, that all the species are similar in the K_m values that they exhibit at their respective physiological temperatures; polar fish tested at polar temperatures (e.g., *Pagothenia*) have about the same affinity as tropical fish tested at tropical temperatures (e.g., *Potamotrygon*). Evidently, affinity needs to be kept in a certain range for enzymes to carry out their catalytic *and* regulatory roles, and *conservation* of affinity has been one of the achievements of the evolution of interspecific enzyme variants.

For animals exposed to unusual temperatures, an important property of enzyme variants that has already been implied but not stressed is that enzyme molecules may well take on unsatisfactory affinity properties at temperatures outside the normal range for the species. Note, for example, that the lactate dehydrogenase of *Sebastolobus* (Figure 6.20) would have an exceptionally low affinity at tropical temperatures, and therefore the conversion of pyruvate to lactate would be impaired. A more dramatic decline of affinity at elevated temperatures is shown by the acetylcholinesterase of the antarctic fish *Pagothenia* in Figure 6.21. Perhaps one reason that *Pagothenia* cannot live in warm waters is the deterioration of this enzyme's ability to bind to its substrate.

A final question of interest is whether individual animals modify the types of enzyme they possess in response to the thermal regime they experience. The results of one important experiment on this question are presented in Figure 6.21. Acetylcholinesterase is a vital enzyme in the brain and other parts of the vertebrate nervous system (p. 440). As shown in the figure, rainbow trout acclimated to 17°C possess a form of acetylcholinesterase that increases its affinity for substrate in the usual way as temperature is dropped acutely from 30 to 17°C. However, as the

temperature is lowered below 17°C, this acetylcholinesterase progressively loses affinity for substrate. This decrease of affinity, occurring *in conjunction* with decreases in the kinetic energies possessed by reacting molecules, could confront fish possessing this acetylcholinesterase with reaction rates that are so retarded at low temperatures as to be debilitating. If trout are acclimated to 2°C, they switch to a new form of acetylcholinesterase which, as shown in Figure 6.21, has affinity properties that are much more suitable to life at low temperatures. In brief, trout have the capacity to switch between two isozymes, or molecular forms (p. 17), of acetylcholinesterase, each of which is well suited to a particular thermal range. Similar switching of isozymes with changes of acclimation temperature is seen also in many other key enzymes of trout.

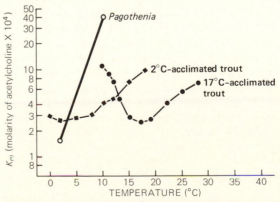

Figure 6.21 The apparent Michaelis constant, K_m, as a function of test temperature for acetylcholinesterase from the brains of three groups of fish: (1) rainbow trout (*Salmo gairdneri*) acclimated to 17°C—closed circles; (2) rainbow trout acclimated to 2°C—closed squares; and (3) an antarctic fish, *Pagothenia borchgrevinki*, collected from waters at −2°C—open circles. [From G.N. Somero and P.W. Hochachka, *Am. Zool.* **11**:159–167 (1971).]

The experiments on trout were among the earliest on the role of isozymes in thermal acclimation and gave rise to the notion that isozyme switching would prove common. Quite to the contrary, however, most species have turned out not to exhibit much, if any, isozyme switching. That is, each tissue of an animal seems generally to be able to make just one form of an enzyme; and while the concentration of enzyme is commonly altered during acclimation (p. 93), the *type* is not. Why then is isozyme switching so prominent in trout? Possibly, the answer lies in another unusual feature of salmonid fish, their tetraploidy. With double the number of gene copies of most animals, the salmonids are likely to be unusually able to carry and use the genetic information for multiple enzyme forms.

Lipid Composition and Membrane Fluidity

Another biochemical attribute that deserves discussion is lipid composition and its implications for membrane fluidity. The temperature at which a lipid solidifies depends on its degree of chemical unsaturation and on other aspects of its composition. Highly saturated lipids solidify at higher temperatures than relatively unsaturated lipids having equivalent fatty-acid chain lengths. Many instances are known, among not only poikilothermic animals but also bacteria and plants, in which individuals living at low temperatures deposit lipids having lower solidification temperatures than individuals of the same species living at high temperatures; sometimes, in particular, the lipids of the cold-acclimated individuals are comparatively low in saturation. By virtue of these changes in lipid composition, deleterious "hardening" of lipids at low temperatures is avoided, and the cold- and warm-acclimated animals enjoy similar lipid viscosities despite the difference in the temperatures at which they are living (viscosity is conserved).

It seems particularly important that suitable "fluidity" be maintained in the lipids of the cellular and intracellular membranes. Stiffening of these membranes at low temperatures can adversely alter their important biophysical properties. Furthermore, many enzymes are bound to membranes, and solidification of membrane lipids can adversely affect the function of such enzymes, as by deforming the enzyme proteins or interfering with proper molecular flexibility during catalysis.

The Threat of Freezing

It is well to start a discussion of freezing with two points that are elaborated in detail in Chapter 7. First, the addition of dissolved matter to water lowers its freezing point. Second, in general, the extent to which the freezing point of a solution is depressed below 0°C increases in proportion to the total concentration of dissolved matter present in the solution (see p. 148).

The extracellular and intracellular body fluids of animals have freezing points below 0°C because of their solute content. However, in the absence of any special protections against freezing, the extent of this freezing-point depression is slight: Body fluids typically freeze at −0.1 to −1.9°C, depending on the animal group considered. Thus, if poikilotherms are exposed to ambient conditions only a bit more severe than those necessary to freeze water, they themselves face a threat of freezing.

The Freezing Process and Its Implications The implications of ice formation depend on its location. Under natural conditions, intracellular freezing, if widespread in the body, seems to be almost always fatal; among other things, intracellular ice formation probably physically damages the cellular ultrastructure on which function depends. Widespread ice formation limited to the extracellular body fluids is fatal to many animals, but strikingly, it is tolerated by others. This tolerance is significant because, for reasons only starting to be well understood, ice often begins its formation in the extracellular fluids and then tends to remain limited to those fluids.

Let us look in detail at the process of extracellular ice formation in a system exposed to gradually declining temperatures. An important attribute of the slow freezing of a solution is that water tends to freeze out in relatively pure form. Solutes, that is, tend to be excluded from the ice crystals. As depicted in Figure 6.22*A,B,* these excluded solutes enter the unfrozen portion of the extracellular fluid, raising its total solute concentration. This elevation of the concentration of the unfrozen extracellular fluid, of course, lowers the freezing point of the fluid. Thus, if the temperature is temporarily held constant, the formation of ice is a self-limiting process; water freezes out only until the freezing point of the unfrozen fluid becomes low enough to equal the prevailing temperature, and then the ice mass and concentrated fluid will exist side-by-side at equilibrium.

In the unfrozen animal, the intra- and extracellular fluids have similar total solute concentrations (osmotic pressures); accordingly, there is little or no tendency for water to enter or leave the cells by osmosis. This circumstance is disrupted by freezing in the extracellular fluids. Because freezing causes the concentration of the extracellular fluids to rise above that of the intracellular fluids, it induces osmotic exit of water from the cells, as shown in Figure 6.22*C.* At any given temperature, this osmotic loss of intracellular water is itself self-limiting; it ceases once the intracellular concentration has risen to equal the extracellular concentration and both concentrations have become high enough for the freezing point of the intra- and extracellular fluids to equal the prevailing temperature. To a degree, the osmotic loss of water from the cells is protective; by lowering the intracellular freezing point as the temperature falls, it helps to prevent intracellular freezing, which usually is fatal. On the other hand, taken to sufficient

cool by 3–7°C is known in many reptiles, and certain bats supercool.

For many freezing-intolerant species of insects, extensive supercooling is the principal mechanism by which the overwintering life stage survives, and it is among them that the greatest known capacities to supercool are found. In winter, an ability to supercool to 20–25°C below the freezing point of the body fluids is about average, and supercooling by 30–35°C is not uncommon. At the extreme, there are now several known examples of insects that, by virtue of extensive antifreeze depression of their freezing points and extensive supercooling, have supercooling points of −50 to −65°C. They can overwinter in exposed microhabitats (e.g., plant stems) in some of the most severe climates on earth.

Freezing Tolerance An ability to survive extracellular freezing, it is becoming clear, is far more widespread than appreciated even 10–15 years ago. In the intertidal zone along northern shores, sessile or slow-moving invertebrates clinging to rocks or pilings frequently experience freezing conditions when exposed to the air during winter low tides. Many of these animals—including certain mussels, barnacles, and snails—actually freeze and survive; some tolerate solidification of 60–80 percent of their body water as ice. Increasing numbers of insects are known to tolerate bodily freezing; in their frozen state, some survive temperatures below −50°C. Recently, certain amphibians that overwinter on land, such as spring peepers (*Hyla crucifer*), have been discovered to survive freezing at −2 to −9°C. Among the animals that survive freezing in winter, tolerance to freezing is typically much diminished, or lost, in summer.

A fascinating discovery has been made recently in studies of freezing-tolerant insects. With the approach of winter, many species (e.g., bald-faced hornets, *Vespula maculata*) synthesize solutes that *promote* freezing and add them to their extracellular fluids. These solutes act as nucleating agents, thus limiting the extent to which the extracellular fluids can supercool. The benefit of promoting extracellular freezing is believed to lie in preventing intracellular freezing (which would be lethal). Having relatively high supercooling points, the extracellular fluids of these animals freeze at relatively high temperatures, *before* temperatures have fallen low enough to make intracellular freezing likely. Then, as temperatures fall, the frozen state of the extracellular fluids protects the intracellular fluids from freezing, as described earlier (p. 96).

Hibernation and Estivation

When poikilotherms pass the winter in a resting state, they are frequently said to be in **hibernation.***

*An alternative, but less commonly used, term is *brumation*.

When they enter a resting state in response to heat or drought, the condition is called **estivation** (see also pp. 179 and 188). These states bear only superficial resemblance to the like-named states in birds and mammals. Thus, the use of the same terms to describe seasonal dormant conditions in all animals is to some degree unfortunate. Having special terms of some sort for the resting states of poikilotherms is appropriate, nonetheless, because when the animals are in these states, they often are in a special physiological condition. Hibernating poikilotherms are not merely summer animals that happen to be cold, for example. They may display altered blood composition (e.g., high antifreeze levels); their skin color or state of hydration may be modified (p. 164); and so on. In general, these are matters that deserve more research attention than they have received.

In a very general sense, poikilotherms may reap two types of benefit from the quiescent states that many enter during times of environmental stress. First, by virtue of their special physiological state, the animals may enjoy an enhanced *physiological* ability to cope with extreme conditions (e.g., increased freezing resistance). Second, they are often permitted to remain continuously in favorable microhabitats. Their metabolism is depressed, often to a special extent that is more profound than accounted for by mere inactivity at the prevailing conditions of temperature; and frequently their stores of nutrients (e.g., body fat) have been augmented. Thus, for long periods, they need not feed. They can function as energetically self-contained units, and having found a microhabitat sheltered from the full harshness of the outside world, they need not venture from it.

HOMEOTHERMY IN BIRDS AND MAMMALS

Homeothermy is regulation of body temperature by physiological means. Although now known to occur in several animal groups, it has been studied most exhaustively in the birds and mammals, and an examination of these forms will serve to introduce many basic principles before we examine the phenomenon as it occurs elsewhere in the animal kingdom. All birds and mammals are not homeothermic all the time. The alternative thermal relations sometimes displayed by these groups are discussed toward the end of the chapter.

Sometimes birds and mammals have been termed endotherms rather than homeotherms. **Endothermy** is the maintenance of an appreciable difference between body temperature and ambient temperature by virtue of internal (metabolic) production of heat (*endo* = "inner"). When birds and mammals occupy cool environments, their metabolic heat production is responsible for keeping them warm. Thus, indisputably, they are endothermic. However, there are two good reasons for calling them homeotherms in preference to endotherms. First, in common usage, endothermy implies only the elevation of body temperature through metabolic heat production; it does not necessarily imply maintenance of a *stable* body temperature by this means. Thus, birds and mammals, which maintain stable temperatures in cool environments through adjustment of metabolic heat production, display a particular *type* of endothermy. Second, birds and mammals often have mechanisms for keeping their body temperature from rising in hot environments. In short, avian and mammalian homeothermy, although including endothermy, goes well beyond the austere implications of endothermy alone.

Body Temperatures in Birds and Mammals

When in thermally unstressful environments, placental mammals typically maintain deep-body temperatures averaging about 37°C. Temperatures of thermally unstressed birds are usually 3–4°C higher. On the other hand, temperatures tend to be lower in marsupials, some of the primitive placental mammals, and especially monotremes; the platypus, for example, exhibits a deep-body temperature of 30–33°C.

An important characteristic of birds and mammals is that their deep-body temperatures are not held absolutely constant but are allowed to vary within limits characteristic of each species. Under thermally nonstressful circumstances, there is commonly a daily cycle in temperature; people, for example, exhibit daily variations in rectal temperature having an amplitude of about 1.5°C, the low point occurring during sleep. Exercise is often accompanied by some elevation of deep temperature; for instance, a man or woman with a resting temperature near 37°C might well undergo an increase to 39–41°C during vigorous exercise. Exposure to hot or cold environments is also often accompanied, respectively, by an elevation or depression of deep temperature. The variations routinely observed in the deep-body temperatures of birds and mammals should not be taken to signify an absence or failure of temperature control. On the contrary, body temperature typically is not permitted to cross well-defined upper and lower limits, and there is every reason to believe that most, if not all, normal variations within the typical range for a species are subject to control.

The Basic Form of the Relation Between Metabolic Rate and Ambient Temperature

Figure 6.24 illustrates how the resting metabolic rate of a bird or mammal typically varies in response to changes in ambient temperature. Within a certain range of ambient temperatures known as the ***thermoneutral zone,*** resting metabolic rate is independent of ambient temperature; the metabolic rate also is lower in this range than at other ambient temperatures. The upper and lower limits of the thermoneutral zone are termed the ***upper*** and ***lower critical temperatures,*** respectively, and depend on not only the

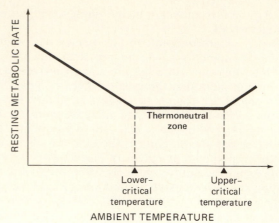

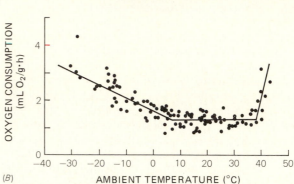

Figure 6.24 Metabolic responses of birds and mammals to changes in ambient temperature. The thermal environment is assumed to be uniform; that is, the radiant temperatures of environmental surfaces are assumed equal to air temperature. (*A*) General type of relation between resting metabolic rate and ambient temperature, indicating terminology used in description of response curve. (*B*) Resting metabolic rate of fasting white-tailed ptarmigan (*Lagopus leucurus*) as a function of ambient temperature. [Reprinted with permission from R.E. Johnson, *Comp. Biochem. Physiol.* **24**:1003–1014 (1968). Copyright 1968, Pergamon Press.]

species but also other attributes of the animal, such as its thermal history. Metabolic rate increases as the ambient temperature falls below the lower critical temperature or rises above the upper critical temperature, reflecting the animal's need to perform physiological work to maintain its stable internal temperature as the environmental temperature becomes relatively cold or warm. The metabolic rate of a resting and postabsorptive bird or mammal in the thermoneutral zone is termed its ***basal metabolic rate*** (see p. 31).

Fundamental Explanations for the Form of the Metabolism–Temperature Curve

How is the physiological regulation of body temperature achieved in birds and mammals? We shall look at the mechanistic details later. But first, in this and several following sections, we take an extensive look at basic principles. To a large extent, this analysis is an exploration into why resting metabolic rate varies as it does with ambient temperature.

For simplicity, let us assume in this section that

the thermal environment of the animal is uniform, meaning that all environmental surfaces have radiant temperatures equal to air temperature and it is thus appropriate to speak of a single ambient temperature, T_A. Suppose, too, that wind speed is held constant.

If the environment is cooler than the bird or mammal, the animal tends to lose heat passively by conduction, convection, radiation, and evaporation. Let us focus for a moment on just the avenues of heat transfer—conduction, convection, and radiation—that do not entail water transfer. These are termed collectively the mechanisms of **dry (nonevaporative) heat transfer.** The *rate* of heat loss by each of these mechanisms tends to increase as the difference in temperature between the body and environment $(T_B - T_A)$ increases. Indeed, we can say as a first approximation that *if factors other than temperature are held constant,* then the *overall* rate of heat loss by *all* the mechanisms of dry heat transfer combined is proportional to the difference between body and ambient temperature.

$$\text{overall rate of dry heat loss} \propto T_B - T_A$$

In essence, the difference in temperature between an animal's body and its surroundings provides the "driving force" for dry heat exchange.

An analysis limited to just dry heat exchange can provide useful insights into animal thermal relations at ambient temperatures within and below the thermoneutral zone. Above thermoneutrality, evaporation is too significant to be disregarded even as a first approximation.

The Thermoneutral Zone　Typically, even the highest ambient temperature encompassed by the thermoneutral zone is below the body temperature of the bird or mammal. At all ambient temperatures within the thermoneutral zone, therefore, animals lose heat by conduction, convection, and radiation—as well as evaporation. As discussed earlier (p. 77), for an animal to maintain a constant body temperature in an environment where it is losing heat, its rate of metabolic heat production must equal its rate of heat loss. When we consider that the metabolic rate of a bird or mammal in the thermoneutral zone remains *constant* even as ambient temperature is *varied,* a paradox seemingly arises. Suppose, to illustrate, that a mammal maintaining a body temperature of 37°C has a thermoneutral zone extending from 30 to 20°C. At an ambient temperature of 30°C, $(T_B - T_A)$ equals 7°C, but at an ambient temperature of 20°C, $(T_B - T_A)$ is much greater, 17°C. In short, the driving force for dry heat loss increases markedly as the ambient temperature is lowered in the thermoneutral zone; thus, based on a consideration of the driving force alone, we would conclude that the animal must be losing heat more rapidly at 20 than at 30°C. How can this conclusion be reconciled with the fact that the animal's metabolic heat *production* is held constant and yet its body temperature also remains constant? The answer to the paradox is that within the ther-

moneutral zone, birds and mammals vary their insulation—their resistance to dry heat loss. As the ambient temperature is lowered and the driving force for dry heat loss becomes greater, insulation is increased. This increase in insulation counterbalances the increase in the driving force so that *the actual rate of dry heat loss in fact remains relatively constant,* and, accordingly, metabolic heat production need not be increased for the body temperature to remain constant. *Modulation of insulation against a background of constant metabolic heat production is the principal mechanism of thermoregulation in the thermoneutral zone.*

Other factors may also be involved. As the ambient temperature is raised in the thermoneutral zone, evaporative heat loss often increases—thus in fact permitting some decline in dry heat loss—and the body temperature may start to rise, helping to maintain the driving force for dry heat loss (p. 102).

The width of the thermoneutral zone depends on the species and varies enormously. Small animals tend to have narrower zones than large animals. Among mice, a thermoneutral zone extending from about 30 to 35°C would not be unusual. At another extreme, eskimo dogs have a thermoneutral zone encompassing a span of about 55°C: from −25 to +30°C. Given that thermoregulation is accomplished over this extremely wide range of ambient temperatures largely by modulation of insulation, one can only marvel at the insulatory flexibility of these animals.

Temperatures Below Thermoneutrality　Inevitably, there are limits to the increase of insulation. These come to the fore at the lower critical temperature. Consider an animal that is subjected to a steadily declining ambient temperature. As the temperature falls, the driving force for dry heat loss $(T_B - T_A)$ steadily increases. Within the thermoneutral zone, however, the rate of heat loss itself is prevented from rising by the animal's increases in insulation, and, accordingly, heat production need not be augmented. The lower critical temperature represents the ambient temperature below which the animal's insulatory adjustments become inadequate to counterbalance fully the increase in the driving force favoring heat loss. That is, as the ambient temperature falls below the lower critical temperature and $(T_B - T_A)$ increases commensurately, *the rate of heat loss itself increases* and therefore the animal must counter with an increase in its rate of heat production. The lower the ambient temperature falls, the greater the rate of heat loss becomes, and thus the more the rate of heat production must be raised. *Below thermoneutrality, modulation of the rate of metabolic heat production becomes the salient mechanism of thermoregulation.*

Sometimes, insulation is maximized at the lower critical temperature and thus remains constant (at its maximal level) below thermoneutrality. In other cases, birds and mammals continue to increase their insulation to a modest extent even as ambient temperature falls below the lower critical temperature.

The increase in metabolic rate that is associated with decreases in ambient temperature below thermoneutrality is one of the most striking ways in which homeotherms differ from poikilotherms. For the poikilotherm, low ambient temperatures imply low body temperatures and correspondingly depressed metabolic rates. For the homeotherm, however, low ambient temperatures imply increased metabolic effort to maintain a high, stable body temperature.

Temperatures Above Thermoneutrality To appreciate the processes at work above thermoneutrality, it is important first to recognize that except in special circumstances (see p. 130), the basal metabolic rate of a bird or mammal represents essentially an unavoidable minimum rate of internal heat production. In the long run, at least, this heat must be lost to the environment as rapidly as it is produced or the animal's body temperature will rise and keep rising to deleterious extremes.

In the thermoneutral zone, metabolic heat is carried away as fast as it is produced by a combination of passive evaporation and dry heat transfer. This state of affairs is challenged by increases in ambient temperature because such increases tend to reduce the driving force for dry heat transfer ($T_B - T_A$), but within thermoneutrality counterbalancing decreases in insulation—sometimes aided by increases in body temperature—serve to keep the actual rate of dry heat loss at an adequate level. Near the upper critical temperature, insulation either reaches its minimum or, at least, becomes incapable of sufficient further reduction to offset additional decreases in ($T_B - T_A$). Thus, as the ambient temperature rises further, the rate of dry heat loss falls so low that the combination of dry heat transfer and passive evaporation becomes inadequate to void metabolic heat. Birds and mammals commonly respond by sweating, panting, or otherwise *actively* augmenting their rate of evaporative water—and heat—loss. These responses demand metabolic effort and thereby frequently account (partly or wholly) for the increase in metabolic rate that commences at the upper critical temperature. Dependence on active evaporative cooling becomes particularly acute once the ambient temperature has risen high enough to exceed body temperature because then conduction, convection, and radiation carry heat *into* the body and evaporation must carry away this exogenous heat as well as metabolically produced heat.

It may seem paradoxical that an animal under heat stress would increase its metabolic rate: its rate of internal heat production. This increase, however, often represents the price that must be paid to augment evaporative cooling. The amount of heat carried away by the evaporation of a gram of water is large enough to exceed greatly the increase in metabolic heat production required to promote the evaporation. Thus, the *net* effect of active evaporative cooling is to increase the rate of heat loss.

Active evaporative cooling is not the only response of birds and mammals to high ambient temperatures. We defer a full discussion of other responses to later sections (pp. 110–112), but here we briefly consider one response: hyperthermia. Many birds and mammals allow their body temperature to rise by several degrees (or more) as they are exposed to high ambient temperatures. One benefit of such hyperthermia is that it helps prevent a drop in the driving force for dry heat loss ($T_B - T_A$). Suppose, to take an extreme example, that when the ambient temperature is elevated from 38 to 40°C, a bird allows its body temperature to increase from 41 to 43°C. In this case, the driving force, being 3°C at either ambient temperature, would not be diminished at all by the increase in environmental temperature. To the extent that decreases in the driving force for dry heat loss can be prevented by development of hyperthermia, higher ambient temperatures can be tolerated before the animal is forced to make active use of its body water in service of heat dissipation. Interestingly, the elevations of body temperature that occur in response to high ambient temperatures evidently do not induce increases in metabolic rate in some species. In other instances, however, metabolism is elevated by the hyperthermia, and this can account for some of the rise in resting metabolic rate observed at high ambient temperatures.

The Concepts of Insulation and Thermal Conductance

We have already made a number of allusions to insulation, referring in particular to the *insulation of the animal as a whole*. Now we need to look more closely at this holistic or "whole-body" concept of insulation. **Insulation,** thus conceived, is defined to be the overall resistance to dry heat transfer between an animal's body core and environment.

In principle, insulation is measured identically whether we are considering a living animal or an inanimate object such as an overcoat or piece of fiberglass batting. Because of this, insight into the measurement of animal insulation may be gained by taking a moment to consider first an inanimate system. The diagram in Figure 6.25 depicts a simple apparatus for measurement of insulation. It consists of a central, metal sphere containing an electric heating element. The sphere is entirely enclosed by a uniform layer of the material whose insulative value is to be measured. In turn, the entire array is placed in a controlled-temperature environment. When the heating element is turned on, thereafter producing heat at a known and constant rate, the temperature inside the sphere will at first gradually rise. Ultimately, however, the sphere will become warm enough relative to the environment that heat will be

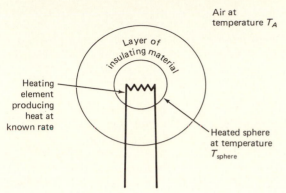

Figure 6.25 A simple apparatus for measuring insulation.

lost from the sphere as fast as it is being produced. Then the temperature of the sphere will stabilize. Suppose we call this steady-state sphere temperature T_{sphere}.

Consider now that the thermal difference between the sphere and environment ($T_{sphere} - T_A$) is the driving force for heat loss across the layer of insulating material in steady state, and by the procedure we have described we have determined how large a driving force is required to cause heat to be lost at a particular known rate (i.e., a rate equal to the rate of heat production). It is precisely this knowledge that gives us a quantitative measure of the insulation provided by the material. If insulation is symbolized by I, then

$$I = \frac{\text{thermal difference required to drive heat loss}}{\text{rate of heat loss}} \quad (1)$$

$$I = \frac{T_{sphere} - T_A}{\text{rate of internal heat production}} \quad (2)$$

It is not difficult to see that this is an intuitively appropriate way to measure insulation. The higher the insulation I, the more the inside of the sphere will be warmed above environmental temperature by a given rate of internal heat production.

Returning now to animals, we recognize that *under conditions wherein their heat losses are mostly dry,* their insulation can be calculated analogously. If the animal, in steady state, has a core body temperature T_B and metabolic rate M, then in parallel with Equation (2), its insulation is

$$I = \frac{T_B - T_A}{M} \quad (3)$$

An important attribute of this measure of insulation is that it is a *composite* index of *all* factors that affect the ease of heat transfer between body core and environment. It is not, for example, a measure of the insulation of just the pelage or plumage alone.

The inverse of insulation is termed ***thermal conductance,*** *C: C* = 1/*I.* Conductance expresses the overall *facility* with which heat can move between

the body core and environment. A high value of C signifies that dry heat transfer occurs with ease.

Further Consideration of the Form of the Metabolism–Temperature Curve

Equation (3) can be rearranged to yield an equation that has occupied a prominent place in the analysis of avian and mammalian thermal relations:

$$M = \frac{1}{I}(T_B - T_A) = C(T_B - T_A) \quad (4)$$

In a slightly different form, this equation has been called the ***linear heat transfer equation.*** It has also been described as a version of ***Newton's law of cooling*** or ***Fourier's law of heat flow.*** Like Equation (3) from which it is derived, Equation (4) should be applied only where heat losses can be considered to be dry to a first approximation: that is, in the thermoneutral zone and especially below.

We noted previously that if the ambient temperature is gradually lowered within an animal's thermoneutral zone, the driving force for heat loss from the animal ($T_B - T_A$) is increased but so also is the animal's insulation (conductance is decreased). Recognizing that ($T_B - T_A$) and I occur, respectively, in the numerator and denominator of Equation (4), the equation makes clear how an increase in ($T_B - T_A$) can be counterbalanced exactly by an increase in I, thereby enabling M to remain constant even as T_A is lowered. We also noted earlier that oftentimes the lower critical temperature represents the ambient temperature at which insulation becomes maximized (conductance becomes minimized). When this is the case, once the ambient temperature has fallen below lower critical, I and C become constants. Equation (4) makes the consequence quantitatively explicit: Below thermoneutrality, M increases in *proportion* to increases in ($T_B - T_A$).

Consider an animal maintaining a constant body temperature as well as constant insulation below thermoneutrality, so that T_B, I, and C in Equation (4) are all constants. If we plot the equation as simply a mathematical formulation, we obtain a straight line having two particular properties, illustrated in Figure 6.26*A:* First, the slope of the line is $-C;$ and second, the line intersects the abscissa at the ambient temperature that equals subthermoneutral T_B. As a model of an animal's metabolism–temperature curve, Figure 6.26*A* is flawed, of course, because it ignores the fact that metabolism does not truly fall below the basal level. Figure 6.26*B* is thus a more realistic model. The key insight to be gained from this analysis is that if an animal meets the stipulations of constant T_B, C, and I at subthermoneutral temperatures, the absolute value of the slope of its metabolism–temperature curve below thermoneutrality may be expected to equal the conductance maintained below thermoneutrality. Figure 6.26*C* depicts curves for two animals differing in subthermoneutral con-

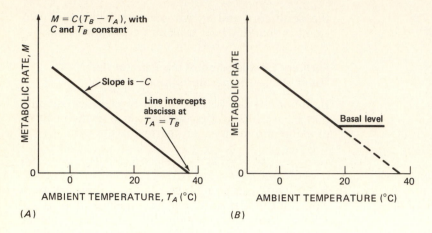

(A)

(B)

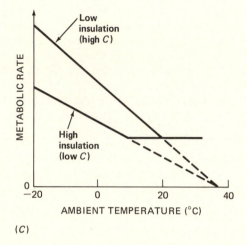

(C)

Figure 6.26 Illustration of a model for the relation between metabolic rate and ambient temperature in and below the thermoneutral zone. For this particular illustration, body temperature is taken to be 37°C. Part (A) depicts the basic equation relating metabolic rate to ambient temperature below thermoneutrality. Part (B) is derived from (A) by truncating the equation at the level of the basal metabolic rate. Part (C) depicts plots for two animals that adhere to the model developed in (B); although both are assumed to have the same basal metabolic rate and body temperature, one exhibits relatively high insulation (low conductance) below thermoneutrality, whereas the other displays relatively low insulation (high conductance).

ductance. Note that the one with lower conductance (higher insulation) exhibits a shallower slope. Figure 6.26C also illustrates that if two animals have the same basal metabolic rate and body temperature, the one that can attain higher insulation will enjoy a lower lower-critical temperature. Overall, the figure makes clear the considerable energetic advantages provided by high insulation at low ambient temperatures.

The main use of the principles of graphical analysis just described is to facilitate qualitative, visual interpretation of metabolism–temperature plots. In Figure 6.42 (p. 121), for example, given that the data basically adhere to the model of Figure 6.26, one can tell at a glance that the winter fox is able to attain higher insulation than the summer fox.

In concluding this section, it is important to be explicit about a fundamental matter, namely, that such properties of the metabolism–temperature curve as the critical temperatures and measured insulation are not merely attributes of the animal but depend strongly on environmental conditions such as wind speed and radiant conditions. For this reason, although metabolism–temperature curves determined under standard laboratory conditions are indispensable for interspecific comparisons, they are often not directly applicable to natural circumstances

(p. 81). At air temperatures below thermoneutrality, for example, high winds in nature tend to raise the metabolic costs of thermoregulation above those in the laboratory, and solar radiant inputs tend to lower the costs.

Birds and Mammals Versus Other Vertebrates: Metabolism and Insulation Compared

The thermoneutral zones of birds and mammals commonly encompass ambient temperatures that are 10, 20, 30, or even more degrees Celsius below body temperature (e.g., see Figure 6.24B). Put another way, birds and mammals can maintain their body temperatures substantially above ambient temperature with only their *minimal* rate of heat production. If this is so, why do not reptiles and amphibians likewise keep warm by virtue of their internal heat production? The answer is twofold:

1. The resting metabolic rates of reptiles and amphibians are inherently much lower than those of birds and mammals. If lizards, for example, are warmed to avian or mammalian body temperatures—so that the cells of the lizards are operating at the selfsame temperatures as the cells of the mammals or birds—the lizards

prove to have standard metabolic rates that are only one-fourth to one-tenth as high as the basal metabolic rates of mammals or birds of equivalent body size (e.g., see Figure 3.5). A dramatic increase in the intensity of metabolism seems to have been a key factor in the evolution of avian and mammalian homeothermy.

2. The bodies of reptiles and amphibians are much less insulated than those of birds and mammals.

In brief, unlike birds and mammals, reptiles and amphibians do not typically produce heat rapidly enough or retain it well enough to warm their bodies much above ambient temperature.

It is worth noting that the low insulation of reptiles and amphibians is actually often a benefit in their ectothermic way of life. They warm themselves using external sources of heat such as the sun, and high insulation would interfere.

Factors that Affect Insulation

We now turn to a more-detailed and mechanistic look at many of the features of homeothermy that we have been discussing heretofore at a phenomenological level. We first consider insulation.

Birds and mammals employ several mechanisms to modify their insulation. As we have seen, these are the predominant mechanisms of thermoregulation within the thermoneutral zone. One means of varying insulation is elevation or flattening of the hairs or feathers; these responses are termed *pilomotor responses* in the case of mammals and *ptilomotor responses* in the case of birds. The hairs or feathers are raised or fluffed out as the ambient temperature falls (Figure 2.7), thus trapping a thicker layer of stagnant air around the animal (p. 78). Another mechanism is alteration of peripheral or superficial blood flow (*vasomotor responses,* see p. 345). Constriction of the peripheral vessels at cool temperatures results in retarded convective movement of heat to the body surface via the blood. Vasodilation at warm temperatures results in enhanced heat loss. Insulation may also be modified by changes in *posture* that alter the amount of body area directly exposed to ambient conditions. Many birds, for example, hold their wings away from their body when temperatures are high. At low temperatures, mammals often curl up, and some birds tuck their heads under their body feathers or squat so as to enclose their legs in the ventral plumage. Besides postural adjustments, other behavioral means may also be employed to modulate insulation; some species, for example, build nests or huddle with conspecifics when ambient temperatures are low.

In addition to the variable insulatory parameters, there are parameters that affect insulation but are more or less fixed for any given animal. Outstanding among these is body size. In explaining why insulation depends on body size, the relation between surface-to-weight ratio and size (see p. 33) seems to be significant at temperatures below thermoneutrality, even if its explanatory power within the thermoneutral zone is suspect: Because small birds and mammals have more body surface per unit of weight than large ones, they tend to suffer comparatively high weight-specific rates of heat loss when exposed to a cold environment. A second reason that insulation depends on body size is that small species tend to be limited to a thinner coat of hair or feathers than large species. Hair thicknesses of 5–6 cm, for example, are common among large mammals, but a mouse with such a thick pelt would be ensconced in a ball of fur, unable even to walk effectively. To quantify the relation between insulation and body size, animals of various sizes have been studied under standardized conditions of environmental convection and radiation (those prevailing in laboratory "metabolism chambers," p. 81) and at temperatures low enough to maximize their insulation (minimize their conductance). Under these circumstances, minimal weight-specific conductance tends to be a regular function of body weight within sets of related animals: C(in units of heat loss/g·h·°C) $= aW^{-0.5}$, where a is a constant depending on the taxonomic group and specific conditions (e.g., time of day). Based on this equation, a 10-g rodent might be expected to require a weight-specific metabolic rate about 10 times as high as that required by a 1000-g rodent to maintain a given body temperature at a given subthermoneutral ambient temperature.

Modes of Increasing Heat Production Below Thermoneutrality

Below the lower critical temperature, heat production must be elevated as ambient temperature falls. Although all metabolic processes result indirectly in production of heat, birds and mammals have evolved processes that have the specific function of generating heat for thermoregulation. These *thermogenic processes* accomplish little or no meaningful physiological work in the strict sense of work but instead emphasize the conversion of chemical energy to heat.

Shivering   The mechanism of thermogenesis with which we are most familiar is shivering, and it appears that all adult mammals and birds use this mechanism. *Shivering* is a high-frequency, relatively uncoordinated contraction of skeletal-muscle motor units mediated via the nervous system. All muscular contraction liberates heat, and here the conversion of chemical energy to thermal energy becomes the primary function of the contraction.

Nonshivering Thermogenesis (NST)   If rats that have been living at warm temperatures are transferred to

6°C, they shiver violently at first; but over the next few weeks of life at 6°C, although they continue to maintain metabolic rates well above basal, they gradually cease overt shivering. If the cold-acclimated rats are then administered curare, a drug that completely blocks contraction of skeletal muscle, they continue to maintain an elevated metabolic rate when exposed to cold, confirming that they have developed mechanisms of *nonshivering thermogenesis (NST):* mechanisms by which they can augment their heat production in service of thermoregulation without shivering. NST is widespread in mammals. It was long thought not to occur in birds, but in the past decade the issue has been reopened by scattered reports of avian NST (and even brown fat).

Of all the possible sites of NST in mammals, the one that is best understood and evidently dominant is *brown adipose tissue (BAT),* also called *brown fat.* This is a type of lipid tissue that differs greatly from the "white" fat with which we are more familiar. It is distinguished by great numbers of relatively large mitochondria and by other cytological characteristics. It receives a rich supply of blood vessels and is well innervated by the sympathetic nervous system. The rich blood supply and the yellowish cytochrome pigments of its dense supply of mitochondria are in large part responsible for its characteristic brownish-red color. The function of brown adipose tissue was obscure to early anatomists and physiologists, and it was often considered to be a gland (the "hibernation gland"). Only since 1961 have we appreciated its function as a site of thermogenesis. Release of norepinephrine into the tissue by the sympathetic nervous system results in a great increase in oxidation of lipid, with consequent liberation of heat.

Brown fat—like NST in general—is particularly prominent in three types of mammal: (1) cold-acclimated or winter-acclimatized adults (particularly in species of small to modest body size), (2) hibernators, and (3) newborn individuals. The brown fat tends to occur in discrete masses, located in such parts of the body as the interscapular region, neck, axillae, and abdomen (e.g., Figure 6.27). Some of the masses of brown fat are so positioned that the heat they produce is delivered directly by the vascular system or by conduction to vital organs such as the heart, brain, or spinal cord. Among hibernators, brown fat (often constituting 1–3 percent of total body weight) has usually been thought to play an important role in rewarming the body during emergence from hibernation (p. 132); recently, however, this view has been called into question. Newborn mammals use brown fat in routine thermogenesis; some are unable to shiver, and some (e.g., guinea pigs) usually do not shiver even though able. Why newborns tend to be more reliant on NST than adults is not fully clear. Among humans, brown fat occurs as prominent masses in newborns; its occurrence in adults has been controversial, but recent evidence indicates that it is indistinctly present. Brown fat has

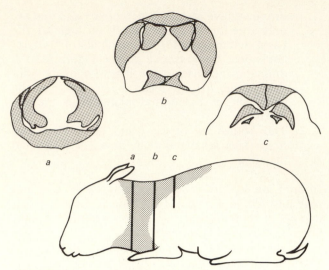

Figure 6.27 The gross morphology of brown fat in newborn domestic rabbits. The side view shows, in gray, the general location of brown-fat deposits. Images *a*, *b*, and *c* are cross sections at the positions indicated on the side view. [Reprinted by permission from M.J.R. Dawkins and D. Hull, *J. Physiol. (London)* **172**:216–238 (1964).]

been implicated in body-weight control as well as thermoregulation. It is claimed to help oxidize excess foodstuffs.

Brown fat may not be an animal's only substantial site of NST. Controversial evidence exists, for example, that the skeletal muscles of some mammals are capable of NST.

Biochemists have been intensely interested in how tissues might increase their heat production during NST. One mechanism proposed is increased ion pumping by the Na^+–K^+ active-transport pump situated in cell membranes (p. 142). Because this pump derives its energy from ATP, an increase in its rate of pumping—like the increase in myofibrillar contraction during shivering—will raise a cell's rate of ATP utilization; and given that the energy released from ATP appears in short order as heat, the effect is to speed conversion of foodstuff energy to heat. There is evidence for the acceleration of the pump during NST, but its quantitative contribution to heat production may well be small.

A consensus has emerged that the dominant mechanism of heat production by brown fat is uncoupling of oxidative phosphorylation from electron transport, a process that frees catabolism from limitations imposed by availability of ADP and permits unbridled oxidation of foodstuffs, with immediate release of their bond energy as heat (p. 39). Evidence has recently accumulated that a mitochondrial protein of molecular weight 32,000 ("thermogenin") is instrumental in promoting uncoupling [by facilitating dissipation of the mitochondrial proton gradient that, according to the chemiosmotic theory (Figure 4.2), is created by electron transport]. When rats are exposed to cold, not only does their number of brown fat cells increase, but also the cellular content of the uncoupling protein increases.

Exercise The prime function of exercise, unlike that of shivering and nonshivering thermogenesis, is ordinarily not production of heat. Nonetheless, as seen in Chapter 5, exercise can greatly increase an animal's rate of heat production, and we must consider whether the heat generated by exercise can meet thermoregulatory needs. From our human perspective the answer seems obvious. Probably, everyone has had the experience on a cold day of feeling warm without shivering while exercising and yet shivering uncomfortably after a long period of rest. Clearly, during physical activity, the heat produced by our exercise can substitute for heat generated by shivering. When we broaden our perspective to include other animals, however, we find that our human experience cannot necessarily be extrapolated to other species. Analytically, the reason is that exercise not only tends to elevate heat *production* but also tends to degrade body insulation and thereby facilitate heat *loss*. Its net effect depends on the relative magnitudes of these two effects. How does exercise facilitate heat loss? Some ways are (1) it entails fully extending the appendages, which are potentially major sites of heat loss; (2) the flexing of the body during activity can disrupt the cover of plumage or pelage; and (3) movement can effectively create a wind across body surfaces.

From a comparative perspective, we in fact know relatively little about the net effect of exercise. Probably, other large animals generally resemble us in gaining a net thermal advantage from exercise and thus being able to curtail shivering or NST when active. At another extreme, we know that in some small mammals and birds under some conditions, heat loss is facilitated sufficiently during exercise to offset completely any potential thermal gain from the activity of the exercising muscles; thus, the exercise heat production does not substitute for shivering or NST, and the latter must continue unabated during physical activity if the animal is to maintain its usual high body temperature. Some small rodents in cold environments actually experience a drop in body temperature when they undertake running.

Some Pros and Cons of Shivering and NST A potential disadvantage of shivering thermogenesis is that it is reduced or even eliminated by exercise, for shivering (uncoordinated contraction) ceases in muscles directly involved in coordinated movements (e.g., running). NST, on the other hand, is not inhibited by activity. At least in small rodents, these distinctions are important, for as we have said, the heat production of exercise alone may not be sufficient to keep the body warm during exertion. Rats (*Rattus*) acclimated to 30°C have little capacity for NST and rely mainly on shivering to keep warm when at rest in a cold environment. During running, their shivering is suppressed, and they are able to maintain a stable body temperature only at air temperatures of about 10°C and above. Cold-acclimated rats, by contrast, have a well-developed capability for NST and thus when running are better able than warm-acclimated rats to supplement the heat production of exercise with additional thermogenesis. Running cold-acclimated rats can maintain a stable body temperature at air temperatures as low as −20°C.

Regional Heterothermy

Structures such as legs, tails, and ears are potentially major sites of heat loss. Keeping the appendages warm in cold surroundings presents much the same problem we have heretofore recognized in comparisons of large and small species, namely, that the surface area of small structures is so great relative to their size that their rate of heat loss per unit of weight tends to be high. Appendages also are often more thinly covered with fur or feathers than other body parts (see also p. 79). In a bird or mammal that keeps its appendages at the same temperature as the body core, the appendages contribute disproportionately, for their weight, to the overall weight-specific metabolic demands of homeothermy.

One way that a bird or mammal can limit heat losses across its appendages is to eschew keeping the appendages as warm as the rest of its body. The difference between appendage temperature and ambient temperature is the driving force for heat loss from the appendages. Allowing the appendages to cool toward ambient temperature reduces this driving force, in effect compensating for the appendages' relatively low weight-specific resistance to heat loss. Not uncommonly, the deep tissues of appendages—especially their distal parts—are 10, 20, 30, or even more degrees Celsius cooler than the head, thorax, or abdomen. The phenomenon of maintaining different temperatures in different living parts of the body is termed ***regional heterothermy***. Because maintenance of lowered temperatures in the appendages in cold surroundings reduces the total metabolic cost of maintaining a given core body temperature, it effectively increases the animal's overall insulation (*I*).

Figures 6.28 and 6.29 provide illustrations of regional heterothermy, showing in particular that sled dogs allow their nose and feet to cool, and Virginia opossums permit profound cooling of their ear pinnae. The opossums have only recently expanded their range into northern climates and when exposed to subfreezing temperatures are prone to letting their ear pinnae (and tail) become frostbitten. In contrast, many species that trace a long ancestry in frigid climates control their regional heterothermy so that tissue temperatures do not fall below freezing even if the environment becomes substantially colder. For example, in a variety of arctic canids—including foxes and wolves as well as sled dogs (Figure 6.28)—the foot pads are routinely allowed to cool to near

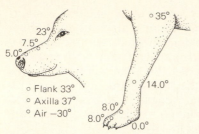

Figure 6.28 Subcutaneous temperatures (°C) on the head and foreleg of an arctic sled dog at an air temperature of −30°C. [From L. Irving and J. Krog, *J. Appl. Physiol.* **7**:355–364 (1955).]

0°C but, even when in contact with much colder substrates (e.g., −30 to −50°C), are prevented from cooling further by increases in the circulatory delivery of heat.

Appendages, especially their distal extremities, often consist largely of bone, tendon, cartilage, skin, and other tissues that metabolically are relatively inactive. As a consequence, the appendages often do not have sufficient endogenous heat production to

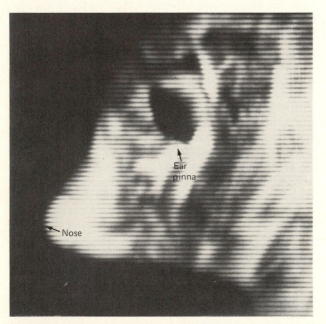

Figure 6.29 An infrared radiograph, or thermal map, of the head and neck of a Virginia opossum (*Didelphis marsupialis*) at an ambient temperature of 10°C. In this image, which was produced using infrared detection equipment, shades of gray are used to indicate temperatures on the animal's body surface. Surfaces that were close to ambient temperature are colored almost black, whereas those that were much warmer than ambient are colored almost white; intermediate temperatures are represented by intermediate gray tones. Note that the surface temperature of the ear pinna, a naked flap of skin, was virtually the same as ambient. By contrast, the surfaces of the snout were much warmer than ambient. Over the thick fur of the posterior head and neck, surface temperatures were variegated but generally relatively low; regions of high surface temperature within these thickly furred parts (e.g., below the ear) represent places where the fur had parted, allowing relatively free flow of heat to the environment.

keep themselves warm in a cold environment, and thus their temperature is largely dependent on how rapidly heat is brought to them from the thorax, abdomen, or head by the circulating blood. Curtailment of circulatory heat delivery is the mechanism by which the appendages are rendered hypothermic. This curtailment may be achieved by simply restricting blood flow to the appendage or by other means (see below).

When a high rate of circulatory heat delivery is provided to an appendage, the temperature of the appendage may be maintained at a high level even in a cold environment; and as we have already noted, a warm appendage acts as a site of especially rapid heat loss. Many animals adaptively modulate heat delivery to their appendages. Whereas they curtail heat delivery and allow appendage temperatures to fall when heat conservation is of advantage, they augment heat delivery when heat dissipation is to be favored. Black-tailed jackrabbits, for example, when at rest in a cool environment limit blood flow to their huge ear pinnae sufficiently that their pinna temperature falls virtually to ambient, but when they exercise, as shown in Figure 6.30, they augment blood flow and pinna temperature considerably. Evidently, the exercise produces a surfeit of metabolic heat and the pinnae are put to use to dissipate the heat by convection and radiation; indeed, one reason the pinnae may be so large is that they provide a non-evaporative means of voiding heat and thus may spare body water from use in cooling these animals, which occupy water-poor environments. Opossums, rats, and muskrats sometimes warm their tails when they exercise. Seals heat up their flippers, and goats warm their horns.

We might expect tissues that sometimes function at low temperatures to show special adaptations to this condition, and in fact a number of such adaptations have been identified. For example, segments of nerve axons from the naked lower legs of cold-acclimated herring gulls fail to respond to stimulation only when their temperature has fallen below about 4°C; but segments of the *same* axons taken from the thickly feathered upper legs fail at about 12°C. Along a different line, we noted earlier that solidification of lipids can have deleterious effects (p. 96). In a number of mammals from both cold and warm climates—including caribou, deer, wolves, and foxes—lipids (e.g., marrow lipids) from the distal extremities of the legs exhibit considerably lower solidification temperatures than lipids from the abdomen or proximal portions of the legs.

Countercurrent Heat Exchange

As we have noted earlier, one mechanism by which an animal can curtail flow of body heat into an appendage is to reduce the rate of blood flow into the appendage. This mechanism, however, is highly non-specific: It does not merely limit heat flow to the

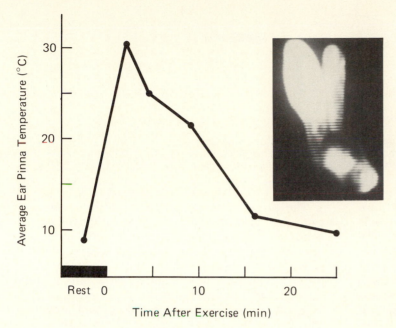

Figure 6.30 The graph shows the average surface temperature of the ear pinnae of a jackrabbit (*Lepus californicus*) while at rest and at various times after a 6-min bout of running. Air temperature was about 8°C. Ear temperature was only 9°C when the jackrabbit was at rest before exercising but rose to 30°C immediately after exercise; it then fell back toward air temperature. Inset is an infrared radiograph of a jackrabbit exhibiting elevated ear temperature after exercise; as described in the legend to Figure 6.29, light-colored regions have relatively high surface temperatures. [Graph after R.W. Hill, D.P. Christian, and J.H. Veghte, *J. Mamm.* **61**:30–38 (1980).]

appendage. It also subjects the appendage to reduced rates of inflow and outflow of all commodities transported by the blood, including oxygen, nutrients, and wastes. Does a mechanism exist for *selectively* curtailing heat flow? One does. It rests on the principle of countercurrent exchange and depends on a special morphological arrangement of the blood vessels servicing the appendage.

Figure 6.31 schematically depicts two possible arrangements of the arteries and veins in a limb. In both cases, the arteries are located relatively deep within the appendage. In (*A*) the veins are superficial, but in (*B*) they are closely juxtaposed to the arteries. For simplicity, let us assume that the rate of blood flow is identical in both systems. The vascular arrangement in (*A*) does nothing to conserve heat; blood flowing through such a system loses heat steadily to a cold environment and, returning to the body core much cooler than when it entered the limb, must be rewarmed considerably by metabolic heat production. In contrast, the vascular arrangement in (*B*) promotes heat conservation because the venous blood flowing in such close proximity to the arteries picks up heat lost from the arterial blood and carries it back to the body core, thus impeding loss of heat to the environment; if the area of contact between veins and arteries is sufficiently extensive, blood may be little cooler on its return to the body core than it was upon entry into the limb. The system shown in Figure 6.31*B* is known as a **countercurrent heat exchange system** because it depends on heat exchange between two closely juxtaposed fluid streams flowing in opposite directions.

Short-Circuiting of Heat Flow A useful way to conceptualize the effect of countercurrent heat exchange in an appendage is to think of it as short-circuiting flow of heat into the appendage. Figure 6.32 illustrates that in the presence of countercurrent exchange, although *blood* flows all the way to the end of an appendage before returning to the body core, *heat*—to a substantial extent—flows only part of the length of the appendage before starting its return. This short-circuiting impedes access of heat to the outer extremities of the appendage, causing the outer extremities to be cooler than they otherwise would be [compare (*A*) and (*B*) in Figure 6.31]. This is an important consequence because the outer extremities are typically where heat is most readily lost to the environment.

A vascular countercurrent exchanger short-cir-

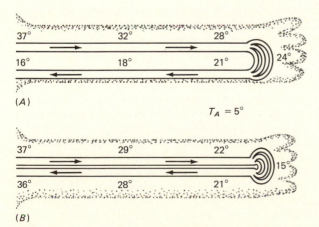

Figure 6.31 A diagrammatic representation of circulation in a limb of a mammal showing hypothetical temperature changes of the blood in the absence (*A*) and presence (*B*) of countercurrent heat exchange. Arrows indicate direction of blood flow. In (*B*), the venous blood takes up heat (thus cooling the arterial blood) all along its path of return because even as it becomes warmer and warmer, it steadily encounters arterial blood that is warmer yet.

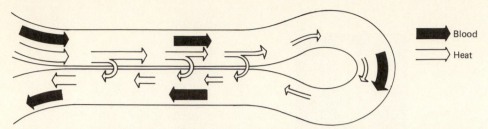

Figure 6.32 Short-circuiting of the flow of heat into the outer reaches of an appendage by countercurrent exchange. A vascular exchanger short-circuits only those commodities that can pass through the walls of the vessels involved. Other commodities follow the path of the blood.

cuits the flow of only those commodities that are able to pass through the walls of the blood vessels involved. Heat, you will note, is short-circuited by the exchangers we have been discussing precisely because it can pass through the walls of arteries and veins. If oxygen and nutrients could diffuse through the walls of these vessels, they too would be short-circuited. However, they cannot diffuse through such thick-walled vessels and thus travel to the outer limits of the appendage with the blood.

Countercurrent Exchangers and Regional Heterothermy Vascular arrangements meeting the prerequisites for countercurrent exchange (close juxtaposition of arteries and veins) are widely reported in appendages that display regional heterothermy. Such arrangements are known, for example, in the arms of people; in the legs of many mammals and birds, including some tropical species such as sloths and armadillos; in the flippers and flukes (tail fins) of dolphins; in the tails of numerous rodents; and in the ears of rabbits. Anatomically they vary from the relatively simple to the complex. There may simply be a close intermingling of rather ordinary veins and arteries, as is found in the human arm. In the flippers and tail fins of dolphins, on the other hand, the major arteries are *surrounded* by venous channels (Figure 6.33); with this arrangement, heat leaving the arteries can hardly help but pass to venous blood. In some animals (such as armadillos and sloths), the main arteries and veins of the limb split up to form many fine vessels that intermingle in a complex network termed a *rete mirabile* ("wonderful net").

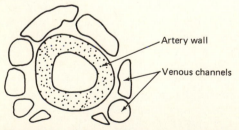

Figure 6.33 A cross section through an artery in the tail fin of a dolphin (*Tursiops truncatus*). Surrounding the thick-walled artery are numerous thin-walled venous channels. [After P.F. Scholander and W.E. Schevill, *J. Appl. Physiol.* **8:**279–282 (1955).]

The extent to which a countercurrent exchanger can cool the outer extremities of an appendage depends on several properties of the exchanger. For example, cooling of the outgoing arterial blood is promoted by a high degree of contact between arteries and veins and by a relatively slow rate of blood flow through the exchanger.

In the arms of humans, the flippers of dolphins, and many other appendages, there are two sets of veins, one superficial and not in close proximity to the major arteries, the other deep and part of a countercurrent exchange system. By modulating the return of blood along these two venous systems, the animal can adaptively emphasize heat dissipation or conservation in its limb according to its thermal status (see Figure 6.31). As we are aware from common experience, return of blood via the superficial veins of our arms is augmented in warm surroundings and reduced in the cold.

Responses to High Heat Loads: Introduction

Sweating, panting, and other modes of actively augmenting evaporative heat loss are well recognized as mechanisms by which birds and mammals can increase their rate of heat dissipation when confronted with high environmental or metabolic heat loads. Importantly, however, evaporation carries heat away only at the expense of body water. Thus, whereas recourse to evaporative cooling may solve problems of body temperature regulation, it may simultaneously create problems of water balance. Conflict between the demands of temperature and water regulation can be especially acute for residents of deserts and other similar habitats where environmental heat loads can be high and yet water is in short supply.

It would seem correct to say that for many mammals and birds, especially ones from hot-arid climates, active evaporative cooling is in fact a last line of defense against heat loading. Other defenses are marshaled preferentially, and only when these other defenses prove inadequate is body water used actively to void heat. We discuss these "front-line" defenses first. They include behavioral maneuvers and manipulations of body temperature and insulation (see also pp. 191–193).

Behavioral defenses are often of crucial importance for animals living in hot climates. Many desert rodents, for example, are nocturnal; they occupy burrows during the day and thus entirely evade the extreme heat we humans associate with the desert surface (Figure 1.2). Diurnal birds and mammals often seek shade during the heat of the day and restrict their activity, thus minimizing their metabolic heat load. Camels orient their body to present a minimum of surface area to the sun on hot days.

Insulation in Hot Environments

We noted earlier (p. 102) that in a uniform thermal environment, as the ambient temperature rises through the thermoneutral zone toward body temperature, birds and mammals reduce and eventually minimize their insulation: their resistance to dry heat exchange with the environment. By decreasing insulation *in an environment that is cooler than their body,* they facilitate loss of metabolic heat by the dry mechanisms of heat transfer (e.g., convection) and thus minimize their need for evaporative cooling. Mechanistically, they typically compress their pelage or plumage, vasodilate cutaneous vascular beds, augment heat flow to appendages, and assume postures that maximize their exposed body-surface area.

What happens, however, in an environment that is so hot that the net combined effect of conduction, convection, and radiation is to carry heat *into* the animal? Theory predicts that animals should increase their insulation in such an environment, for just as high insulation can retard outflux of heat from the body in a cold environment, it can slow influx of heat into the body in a hot one. When conditions favor heat influx by conduction, convection, and radiation, evaporative cooling becomes the only possible means of heat loss, signifying that evaporation must carry away all the heat incoming from the environment. To the extent that heat gains from the environment are slowed by maintaining high insulation, demands for evaporative cooling are reduced. Thus, increasing insulation in a hot environment provides a means of water conservation.

In fact, not all birds and mammals respond as theory suggests; some do not increase their insulation when conditions favor dry heat gain. Others do, however. A notable example is provided by jackrabbits, *Lepus alleni.* In environments that are warm but nonetheless cooler than their body, the jackrabbits minimize their insulation; but when conditions become so hot as to favor dry heat gain, they raise their insulation. Indeed, their insulation in a hot environment can be just as high as in a very cold one.

Some birds and mammals of hot-arid regions have evolved strikingly thick insulatory coverings. The dorsal pelage of dromedary camels in summer, for example, can be at least 5–6 cm thick, and the plumage of ostriches can be 10 cm thick when erected. In line with what we have been saying, such thick pelage or plumage may have evolved as an aid to water

conservation because it can act as a barrier to heat influx. The surface of the dorsal pelage can become as hot as 70–85°C in animals like camels and sheep when exposed to the sun in hot climates. The pelage insulates the body from these enormous heat loads. Much of the solar radiant heat absorbed by the pelage is passively lost to the environment by reradiation and convection before it can reach living tissues.

Body Temperature in Hot Environments

Cycling   Some birds and mammals exploit cycling of their body temperature as a mechanism of reducing reliance on evaporative cooling in a hot environment. To see the potential benefits of cycling, let us consider a classic example of the phenomenon, the dromedary camel. When dehydrated in summer, dromedaries permit their body temperature to fall to 34–35°C overnight and increase to over 40°C during the day. By allowing their temperature to rise each day—rather than rigidly maintaining it at a particular level—the animals *absorb and store* a considerable amount of the heat incident on them during the day, and thus they avoid having to dissipate that heat evaporatively. Knowing that it takes about 0.8 cal to warm 1 g of camel by 1°C, we can readily calculate that a 400-kg camel will absorb about 1.9×10^6 cal by allowing its body temperature to rise by 6°C; dissipation of that heat by evaporation would require over 3 L of water. When night falls, conditions become favorable for cooling by convection and radiation, and these mechanisms (plus passive respiratory and integumentary water loss) provide for the nocturnal fall in body temperature. Thus, in the end, cycling of body temperature permits much of the heat load of the day to be dissipated *nonevaporatively* at night.

In essence, any species that allows its body temperature to rise and fall substantially with the passage of day and night will partake of the advantages of cycling just described. However, the saving of water by cycling requires that heat be temporarily stored in the body, and for a particular rise in body temperature, small-bodied species in a given environment store only a much smaller fraction of their total diurnal heat load than large-bodied ones. Thus, among species that cycle on a daily basis, it is in the large-bodied ones (such as camels) that cycling yields truly great benefits.

Some small-bodied desert animals substantially curtail water losses by undergoing *multiple* cycles per day. Antelope ground squirrels, for example, forage during daylight hours by undertaking numerous forays from their burrows. Their body temperature rises during each foray and then falls by nonevaporative heat loss during each return underground.

Hyperthermia   When exposed to hot conditions, many birds and mammals display **controlled hyper-**

thermia: a controlled elevation of their body temperature to exceptional levels. This is a virtually universal response in birds; whereas resting avian body temperatures are typically 40–41°C in the absence of heat stress, increases to 44–45°C are common in hot environments. Some mammals, such as humans, normally allow little rise in their resting body temperature under hot conditions, but others permit large increases; to cite two extreme examples, the oryx (*Oryx beisa*) and Grant's gazelle (*Gazella granti*), both native to hot environments, permit their rectal temperatures to reach 45.5–47°C (114–116°F!) without ill effect.

Of course, a rise in body temperature entails heat storage, and thus to some extent the benefits of the rise are the very ones we have already described under the rubric of "cycling." However, a high body temperature *in and of itself* holds advantages for water conservation. Already (p. 102) we have seen that in a uniform thermal environment characterized by temperature T_A, if T_A, though high, is *below T_B*, then an elevation of T_B is of advantage because it enhances the driving force ($T_B - T_A$) for *dry* heat loss and thus spares body water by reducing dependence on evaporative heat loss. What happens, however, if the environment is so hot that conduction, convection, and radiation carry heat *into* the animal? Hyperthermia still holds advantages: The rise in T_B will reduce the rate of heat influx by decreasing the driving force ($T_A - T_B$) favoring influx and thus will reduce the rate of active evaporative cooling required to void incoming heat.

Keeping a Cool Brain   Recently, it has become increasingly clear that in many species of mammals and birds the brain is kept cooler than the thorax and abdomen when the animals are occupying a warm or hot environment, especially during exercise. To cite an extreme example, when gazelles (*Gazella thomsonii*) run vigorously in a warm environment, their thoracic temperature may rise to 43–44°C but their brain may be as much as 2.7°C cooler. The brain evidently tolerates less of an elevation of temperature than most organs. Thus, the temperature of the thorax and abdomen—the bulk of the body—can be allowed to rise to higher levels with brain cooling than without. Brain cooling permits enhanced exploitation of the advantages of hyperthermia and body-temperature cycling.

How is the brain kept cooler than the thorax? Frequently, the key process is countercurrent cooling of arterial blood supplying the brain. The arteries carrying blood toward the brain come into intimate contact with veins or venous blood draining the upper respiratory passages (e.g., nasal passages). In many mammals, the site of this contact is the cavernous sinus located at the base of the skull; there, the arteries in some species divide into a plexus of small vessels (the carotid rete mirabile) that is immersed in a lake of venous blood. As noted, the venous blood closely juxtaposed to the arteries is returning toward the heart from the upper respiratory passages. Evaporation from the respiratory passage walls has cooled this blood, and as the venous blood flows by the arteries, it cools the arterial blood traveling toward the brain (Figure 6.34).

Conclusion   Consider a day in the life of a dromedary camel: a day so hot that convection and radiation carry heat into the animal's body for hours. During the morning, the camel will deal with its metabolic and environmental influxes of heat by simply storing the heat, allowing its body temperature to rise. Ultimately, however, on such a hot day, the body temperature will reach the highest level the camel can tolerate: 40–41°C rectal. Then storage has to stop, and until the cool of evening comes, the animal will need to resort to sweating to void heat as rapidly as it is gained. The rate of gain from the environment will be slowed during this period, nonetheless, by the camel's insulatory pelt and by hyperthermia—thus slowing the rate of sweating needed. During both the heat-storage and sweating phases of the day, we see here a vivid illustration of the key roles that insulation and modulation of body temperature can play in sparing body water.

Active Evaporative Cooling

Active facilitation of evaporation is the ultimate line of defense for birds and mammals faced with high environmental or metabolic (e.g., exercise-induced) heat loads. When animals cool themselves evaporatively, they take advantage of a salient physical property of water: its absorption of a lot of heat when it changes state from a liquid to a gas. Details of the physics of evaporation are discussed earlier in this chapter (p. 79) and especially in Chapter 7 (p. 136). Here, we focus on the four major mechanisms of actively enhancing evaporative cooling that are

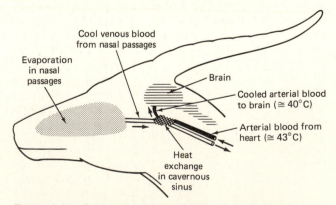

Figure 6.34   Schematic representation of the anatomical arrangement promoting cooling of the brain in certain mammals. Dark vessels symbolize arterial blood flow and light vessels, venous blood flow. In the cavernous sinus, the two flows are in intimate juxtaposition. [Reprinted by permission from C.R. Taylor and C.P. Lyman, *Am. J. Physiol.* **222**:114–117 (1972).]

known to be employed by birds and mammals: sweating, panting, gular fluttering, and saliva spreading.

Transpirational Water Loss and Sweating During sweating, fluid is secreted, by way of the sweat-gland ducts, through the epidermis onto the skin surface. But let us consider first what happens when there is no sweating, when the skin forms a barrier between the body fluids and air. The skin of birds and mammals is poorly permeable to water and thus greatly impedes water loss (see p. 181). Nonetheless, skin is not entirely impermeable, and water constantly diffuses outward through it, evaporating when it reaches the surface. This steady diffusive loss of water through the skin is known as *transpirational* or *insensible* ("unperceived") *water loss*. There is recent evidence that some birds and mammals enhance transpiration when they are under heat stress. Modulation of transpirational evaporation is thus implicated as a mechanism of thermoregulation.

When the sweat glands are activated, they breech the barrier posed by the skin by secreting water directly onto the skin surface. Consequently, the rate of cutaneous evaporation may come to exceed the background, transpirational rate by 50 or more times. Vigorous sweating occurs in many mammalian species, such as humans, horses, camels, and some kangaroos. Rodents and lagomorphs (rabbits and hares), however, lack integumentary sweat glands, and in some mammals that possess such glands—such as dogs and pigs—secretion rates are so low that sweating appears to play little or no role in thermoregulation. Sweating does not occur in birds. Among mammals that employ sweating in thermoregulation, sweat production can be profuse; humans working strenuously in the desert, for example, can attain sweating rates of 2 L/h. Sweat is a saline solution. Concentrations of sodium and chloride in sweat are lower than in the blood plasma, but prolonged sweating can nonetheless impose a significant drain on the body's pool of Na^+ and Cl^-.

For furred mammals that sweat, the insulative barrier provided by the pelage takes on special significance in a hot environment because it prevents unbridled access of environmental heat to the evaporative surface on the skin. We have seen, for instance, that the sun beating down on a camel can heat its fur surface to 70–80°C. Were it not for the fur, this heat would impinge without limit on the skin, vaporizing sweat. A constraint on the pelage in sweating animals is that it cannot be so thick or dense as to interfere with adequate dissipation of vaporized water from the skin to the atmosphere. Peoples endemic to deserts have long been cognizant of these matters. They *do* cover their bodies with clothing (i.e., they place a barrier between themselves and the environmental heat), and the clothing is loose-fitting, thus allowing water vapor from their skin surface to be carried readily away.

Panting *Panting* is an increase in the rate of breathing in response to heat stress and occurs widely in birds as well as mammals. By virtue of evaporation from the warm, moist membranes lining the respiratory tract, air inhaled becomes saturated with water vapor. The net amount of water vapor it can carry away when exhaled depends on not only its initial humidity but also its temperature at the time of exhalation; these are matters discussed in Chapter 8 (see especially p. 184). Here, we emphasize simply that respired air does typically carry away water vapor, and panting increases the rate of this process by increasing the flow of air in and out of the respiratory tract.

In some species, the breathing rate during panting increases progressively as the extent of heat stress increases. In others, there is an abrupt change in respiratory frequency at the onset of panting, and within a wide range of thermal stress, the rate of panting is independent of the degree of stress. Dogs exemplify this pattern; whereas in cool air they breathe around 10–40 times per minute, their breathing rate jumps abruptly to 200 or more breaths per minute when panting begins. Analysis indicates that animals with such a stepwise change in respiratory rate often pant at the resonant frequency of their thoracic respiratory structures. This is of advantage because breathing requires considerable muscular work and therefore entails substantial heat production. Panting at the resonant frequency reduces the muscular effort required and thus reduces the amount of heat that must be produced in the process of augmenting heat dissipation.

By comparison to sweating, panting holds certain advantages, two of which we note here. First, no loss of salts need occur during panting because evaporation occurs within the body and only water vapor is carried away in the exhalant air. Second, the breathing activities of panting assure that air saturated with water vapor is driven forcibly away from the evaporative surfaces, whereas during sweating the removal of water-laden air is often dependent on less predictable, controllable, or vigorous forces, such as free convection or external winds.

Panting also has certain liabilities in comparison to sweating. Because of the muscular effort involved in panting, evaporation of a given quantity of water is likely to require more energy—and entail more heat production—with panting than with sweating, even when resonant frequencies are used. Another potential liability of panting is that it can induce *respiratory alkalosis:* an elevation of the pH of the body fluids attributable to excess removal of carbon dioxide (p. 332). Ordinarily, when animals are not panting, ventilation of the respiratory exchange membranes deep in the lungs—the alveolar membranes of mammals or the air-capillary membranes of birds—is closely regulated so that the rate of dissipation of CO_2 is equal to the rate of metabolic production of CO_2 (see p. 284). This ventilatory regulation acts to maintain stable concentrations of CO_2 and bicarbonate in the body fluids. During panting, the potential exists for breathing to carry CO_2 away faster than it is produced, because the rate of breathing is increased for thermoregulation rather than being governed simply by the need for exchange of oxygen and CO_2. Excessive dissipation of CO_2 will lower the concentration of CO_2 in the body fluids and consequently shift the following reactions to the left:

$$CO_2 + H_2O \rightleftharpoons H_2CO_3 \rightleftharpoons H^+ + HCO_3^-$$

The result will be a lowering of the concentration of H^+ in the body fluids and a corresponding increase in pH. Such excessive alkalinity—or alkalosis—can have major deleterious effects because many enzymes and cellular processes are acutely sensitive to pH. (In junior high school, we probably all witnessed children render themselves dizzy by deliberately breathing too rapidly.)

It is now established that when heat stress is light to moderate, little or no alkalosis develops during panting in many species of mammals and birds. A key concept in understanding this result is that the respiratory exchange membranes themselves—the alveolar or air-capillary membranes—must be hyperventilated for alkalosis to develop. Merely increasing the rate of air flow in and out of the upper airways (or air sacs) will not induce alkalosis because the membranes lining these airways, while moist, are too thick or poorly vascularized to permit any substantial exchange of CO_2 between the airstream and blood. When resting birds or mammals are panting under light to moderate heat stress, alkalosis is typically avoided by limiting most of the increase in ventilatory volume to the nonrespiratory upper airways; the breaths of panting, as shown in Figure 6.35, are predominantly too shallow to carry air to the respiratory exchange membranes. By contrast, if heat stress becomes extreme, resting but panting animals often develop severe alkalosis. This alkalosis may be associated with a transition to deeper, more labored breathing. Evidently, in the face of a thermoregulatory emergency, the increased evaporation that can be achieved by this altered pattern of breathing takes precedence, and the pattern is adopted despite its tendency to induce alkalosis by heightening the influx of air to the exchange membranes. Some panting species have superior tolerance to alkalosis.

Gular Fluttering Many birds, as exemplified in Figure 6.36, augment evaporative cooling by rapidly vibrating their gular area (the floor of their mouth)

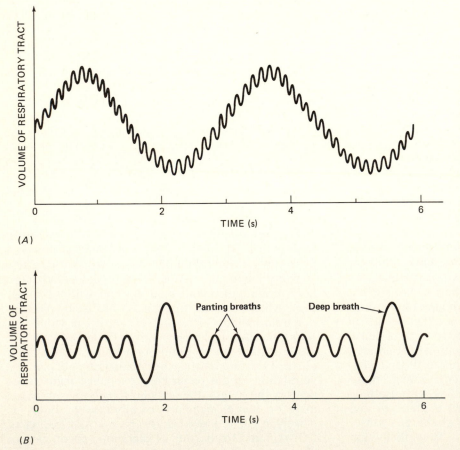

Figure 6.35 Schematic representations of two breathing patterns that serve to limit hyperventilation of the respiratory exchange membranes during panting. (A) Breathing consists of simultaneous high-frequency and low-frequency components in pattern (A), which is known as compound ventilation and has been reported in some mammals (e.g., sheep) as well as birds (e.g., pigeons). The low-frequency component (two cycles of which are shown) resembles ordinary breathing; the volume of air moved in and out during each inhalation and exhalation is relatively large, and these breaths serve to ventilate the respiratory exchange membranes deep within the lungs. The high-frequency component represents panting per se and is superimposed on the low-frequency component. The high-frequency breaths move relatively small volumes of air in and out and are too shallow to carry air to the respiratory exchange membranes. (B) In this pattern, which has been reported in flamingoes, low-amplitude and high-amplitude breaths occur sequentially rather than simultaneously. [Based on J.M. Ramirez and M.H. Bernstein, *Fed. Proc.* **35**:2562–2565 (1976); C. Bech et al., *Physiol. Zool.* **52**:313–328 (1979).]

Figure 6.36 Young (2- to 6-day-old) cattle egrets (*Bubulcus ibis*) undergoing gular fluttering in their nest. The arrow points to the gular area, which in these animals is vibrated up and down at a rate of 800–1000 cycles/min. Cattle egrets nest in exposed locations, and in hot climates the sun beating down during the day can quickly overheat nestlings. Parents protect their young from overheating by shading them. However, if the parents are forced off the nest, the young must combat overheating physiologically; they are able to gular flutter even at 1 day of age, when their overall motor abilities are so meager that they can barely hold their heads up. [Reprinted with permission from J.W. Hudson, W.R. Dawson, and R.W. Hill, *Comp. Biochem. Physiol.* **49A**:717–741 (1974). Copyright 1974, Pergamon Press.]

while holding their mouth open. This ***gular fluttering,*** driven by flexing of the hyoid apparatus, promotes evaporation by increasing flow of air over the bird's moist and highly vascular oral membranes. Flutter rates of 70–1000 cycles/min have been observed in various species. Usually, fluttering occurs at a consistent frequency, which apparently matches the resonant frequency of the structures involved. Fluttering is commonly employed synchronously with panting.

Gular fluttering shares certain positive attributes with panting: It creates a vigorous, forced flow of air across evaporative surfaces and does not entail salt losses. Unlike panting, gular fluttering cannot induce alkalosis because it enhances only oral air flow. Gular fluttering involves movement of less-massive structures than panting; consequently, the muscular work—and metabolic heat production—required to achieve a given increment in evaporation can be considerably less with gular fluttering than with panting.

Saliva Spreading   When exposed to heat stress, many rodents and marsupials spread saliva copiously on their limbs, tail, chest, or other body parts. Spreading of saliva on furred regions of the body is a relatively inefficient use of body water for evaporative cooling because the evaporative surface thus created—on the outer surface of the fur—not only suffers from the potential disadvantages of being directly exposed to the environment (p. 113) but also is insulated from the living tissues of the body by the underlying pelage. Nonetheless, saliva spreading is useful for cooling in certain circumstances. In many rodents, for example, saliva spreading is the sole means of markedly increasing evaporative cooling,

and it is employed in emergency situations to prevent excessive hyperthermia while a cool environment is sought.

Control Mechanisms

We saw in Chapter 2 (p. 9) that a key component in any control system is a controller that receives sensory input and uses that input to generate appropriate commands to effector tissues. As portrayed in Chapter 2, the controller was a discrete, localized entity: a box. Realistically, physiological controllers are seldom so tidy.

The hypothalamus and its associated preoptic tissues are the principal sites at which controller functions for thermoregulation are carried out. This does not mean, however, that all sensory input of thermoregulatory significance is fed raw into these regions of the brain, nor that these regions are the exclusive fount of commands to effector tissues. In fact, many other parts of the central nervous system, including parts of the spinal cord as well as brain, play roles in modifying or integrating sensory input and in originating or modulating commands.

Messages governing thermoregulatory effector processes are communicated by controller centers to effector tissues largely via the nervous system, although some glands, such as the thyroid and adrenals are also known to be involved or implicated (e.g., see p. 597). Shivering and behavioral responses to heat and cold (e.g., curling up) are activated via somatic motor neurons. Piloerection (Figure 2.7), peripheral vasoconstriction, thermogenesis by brown fat, and sweating are elicited by the sympathetic division of the autonomic nervous system.

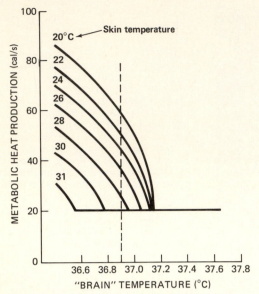

(A)

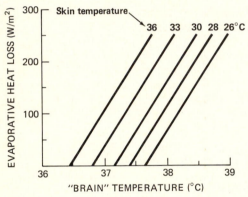

(B)

Figure 6.37 Metabolic rate (*A*) and sweating intensity (*B*) as functions of brain and skin temperatures in humans. Each solid line shows the relation between effector activity and brain temperature at a particular skin temperature. Brain temperature was measured as the temperature of the auditory tympanum; tympanic temperature closely parallels hypothalamic temperature. [Part (*A*) after T.H. Benzinger et al., *Proc. Natl. Acad. Sci. USA* **47**:730–739 (1961); (*B*) reprinted by permission from J.A.J. Stolwijk and J.D. Hardy, in D.H.K. Lee (ed.), *Handbook of Physiology, Section 9: Reactions to Environmental Agents*, pp. 45–68. American Physiological Society. Bethesda, 1977.]

One of the most interesting challenges in the study of thermoregulation has been to identify the sensory inputs to the controller and their interactions. Significantly, the hypothalamus and preoptic region of the brain, which you will recall are primary seats of controller function, are themselves thermally sensitive. Controller action is exquisitely sensitive to changes in hypothalamic–preoptic thermosensory input; this is a key factor in the defense of a stable brain temperature. The controller also receives sensory inputs from other parts of the central nervous system (e.g., spinal cord and medulla oblongata), from the abdomen, and from warm and cold receptors distributed widely in the skin, including the scrotum and udder in some species. The cutaneous receptors provide an early-warning system: Alterations in the outside thermal environment that might ultimately induce changes in deep-body temperature are typically heralded first as changes in skin temperature.

With so many sensory inputs to the controller, a continuing question has been how they are integrated in determining controller output. In general, considering any particular pair of inputs, if both inputs signal changes of temperature in the same direction, the effects on controller output are mutually reinforcing; if the inputs signal changes in opposite directions, the effects are mutually antagonistic. Figure 6.37 illustrates the interactions between brain and skin temperature in the control of human shivering and sweating. Let us look first at part (*A*). Any one solid line shows the relation between shivering intensity (as reflected by metabolic rate) and brain temperature at a constant skin temperature. If skin temperature has fallen to 20°C, brain temperature needs just to fall to below about 37.2°C to elicit shivering; if the skin temperature is 30°C, on the other hand, only brain temperatures below 36.8°C induce shivering. Whatever the skin temperature, once brain temperature has fallen low enough to initiate shivering, the intensity of shivering increases markedly as brain temperature falls further. The intercepts of the solid lines with any given vertical line indicate the relation between shivering intensity and skin temperature at constant brain temperature; for instance, the intercepts along the dashed line depicted in the figure reveal the extent to which shivering is stimulated by a fall in skin temperature from 28 to 20°C, when brain temperature is constant at 36.9°C. Some salient features of the overall pattern are these: (1) Variations in skin temperature affect metabolic rate in a negative-feedback fashion even if brain temperature is stable; indeed, these effects of skin temperature *help assure* a stable brain temperature. (2) If brain temperature itself rises or falls by even tenths of a degree Celsius in the range near 37°C (its ordinary level), metabolism is modified powerfully in negative-feedback fashion, tending to reverse the change in brain temperature. The control of sweating follows analogous principles, as seen in Figure 6.37*B*.

A major question in the study of temperature regulation is: What exactly is regulated? Is it brain temperature, skin temperature, rectal temperature, or what? Taking cognizance of the multiplicity of thermal sensory inputs to the controller, one popular answer is that the regulated variable is a holistic measure of temperature throughout the body, the exact measure being a weighted average of all temperatures used by the controller, each weighted according to its influence on controller function.

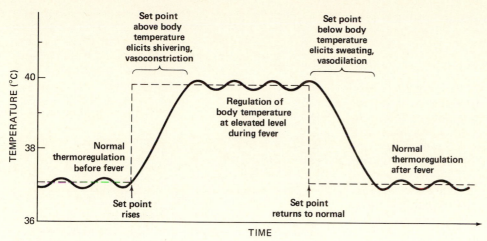

Figure 6.38 Phenomenology of fever. The dashed line depicts the "set point" for body temperature. The solid line shows the actual body temperature.

Fever and Exercise Controller function is affected by nonthermal factors in some instances. During fever, for example, body temperature is regulated at an elevated level because of a chemically mediated alteration of controller function. Phenomenologically, the analogy between this change in the physiological controller and the resetting of a home thermostat is so close that fever is often described as a "resetting of the body's temperature set point," even though the controller may not literally operate on the basis of set points (see pp. 10–12). At the outset of fever, as diagrammed in Figure 6.38, elevation of the "set point" elicits shivering, peripheral vasoconstriction, and other processes serving to raise the body temperature to the new "set point." Later, when the "set point" falls back to normal, the body

temperature is brought into correspondence with its revised "set point" by such processes as sweating and vasodilation. The rise in "set point" during fever is the culmination of a series of events set in motion by the presence in the body of chemicals known as **pyrogens,** among which are lipopolysaccharides in bacterial cell walls.

Exercise often provokes large increases in deep-body temperature. Among humans, for example, rectal temperature is elevated increasingly as the intensity of exercise increases, with levels of 39.0–41.9°C being observed during vigorous exertion (interestingly, and for reasons largely unknown, these high temperatures do not cause adverse clinical symptoms in this context). The control of body temperature during exercise is not yet well understood, although

---

BOX 6.2 CONTROL MECHANISMS IN LOWER VERTEBRATES

Studies on fish and lizards have revealed remarkable parallels between control of their *behavioral* thermoregulation (p. 124) and the control of physiological thermoregulation in mammals and birds. The controller for behavioral thermoregulation in fish and lizards is located in the brain and receives thermal sensory inputs both from the forebrain and from one or more peripheral parts of the body (e.g., skin or abdomen). Furthermore, the inputs from the brain and periphery are integrated in a manner fundamentally similar to the integration in mammals and birds. Evidently, important features of the sensory and controller mechanisms in vertebrate homeotherms evolved before homeothermy itself.

More evidence for this last conclusion comes from recent studies on the effects of bacterial infection in poikilothermic vertebrates. Among fish, amphibians, and reptiles, animals given a choice of thermal environments often choose a warmer environment when infected than when not. By choosing a warmer environment, they elevate their body temperature. In brief, that is, infected animals develop a *behavioral fever.* Animals prevented from behaviorally elevating their body temperature when infected often suffer greater morbidity and mortality than ones not prevented, suggesting that fever evolved as a defense against disease.

to some extent the rise in temperature probably represents a load error, as discussed earlier (p. 11).

Body Size and Thermal Relations

From a strictly physiological view, large body size typically conveys advantages in coping with the exigencies of heat or cold. The advantages follow in good part from the relation between body size and surface-to-weight ratio. Because the rate of heat exchange with the environment by conduction, convection, and radiation is substantially dependent on surface area and because large animals have less surface area per unit of weight than small, it follows that—other things being equal—large animals lose or gain heat at lower rates per unit of body mass than small ones. In cold surroundings the large animal can offset heat losses with a lower weight-specific metabolic rate, whereas in hot surroundings it can offset heat gains with a lower weight-specific rate of loss of evaporated water.

We have earlier discussed the effects of body size on metabolism and insulation in cold environments (p. 105). Looking at hot environments, we find that small animals are in a less advantageous physiological position than large animals not only because of their higher surface-to-weight ratios but also because of their higher weight-specific metabolic rates. That is, the weight-specific heat load of the small animal under hot conditions is enhanced relative to that of the large animal both by a greater weight-specific rate of input of heat from the outside world and by a greater weight-specific rate of internal production of heat. By quantifying these factors, Schmidt-Nielsen identified the relation shown in Figure 6.39 between body size and the weight-specific rate of evaporative water loss needed to dissipate total heat load while undergoing moderate activity under exposed conditions during a summer day in the Nevada desert. In interpreting this relation, it is important to note that many mammals die under hot conditions when they have lost water equivalent to about 15 percent of their body weight, though camels, donkeys, Merino sheep, and many other inhabitants of hot climates can tolerate dehydration of over 25 percent. Under the conditions stipulated in Figure 6.39, donkeys and humans must expend water equivalent to 1.2–1.4 percent of their body weight each hour to maintain a stable body temperature near 37°C. Merely because of greater size, animals as large as camels have expenditures only about half as great: 0.7 percent/h [dromedaries in fact enjoy even lower expenditures, largely because they exploit the advantages of *not* keeping their body temperature stable (p. 111)]. At the other extreme, small desert rodents weighing 10–100 g—such as kangaroo mice and kangaroo rats—are predicted to require water expenditures of 12–30 percent/h to keep cool evaporatively while moving about in the desert sun. These enormous predicted expenditures offer insight into why

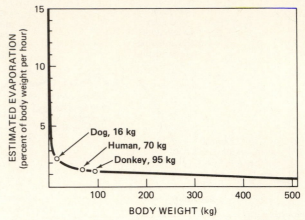

Figure 6.39 Relation between body size and evaporative water loss needed for mammals to maintain a body temperature near 37°C while undergoing moderate activity (e.g., walking) during exposure to the sun in the Nevada desert in summer. The line is based on theory and on data for dogs, humans, and donkeys. Note that the unit of measure for evaporation is a weight-specific expression of water loss: g H_2O/100 g body weight · h. [After K. Schmidt-Nielsen, *Desert Animals.* Clarendon, Oxford, 1964; see also L.F. Soholt et al., *Comp. Biochem. Physiol.* **57A**:369–371 (1977).]

most desert rodents are nocturnal: To wander freely during the day, a small mammal would at times have to confront rates of desiccation so great as to be lethal in an hour or two. Small desert birds are also subject to the physiological disadvantages inherent in small size, but most are diurnal. One advantage that birds have is that they routinely allow their body temperatures to rise to 44–45°C, and thus they typically enjoy lower water expenditures than mammals of equal size under given conditions (p. 112). As a group, small desert birds are more tied to sources of water—succulent foods or water holes—than small mammals; apparently, the water losses of the birds in the summer desert are great enough that steady access to replacement water is essential (p. 193).

Whereas large birds and mammals frequently meet the stresses of heat and cold from a more advantageous physiological position than small forms, the small species have the outstanding behavioral advantage of being more readily able to avoid thermal extremes by moving into burrows, cavities, shade, and other relatively equitable microhabitats. The large animal is often required to confront the full severity of the thermal environment. Furthermore, the large animal must find and consume more food and water than the small animal, and unlike its smaller relatives, it cannot readily establish food stores adequate to meet its needs over stressful periods of summer or winter. An ameliorating factor for large animals is their greater mobility by comparison to small species: Large animals can travel more extensively in search of the food and water they need. Large desert mammals undoubtedly benefit from their ability to seek out water holes that may be scattered widely; this is an important factor in their capacity to rely on evaporative cooling as a protection against overheating. The potential advantages of small size, on the other hand, are illustrated by some animals living in areas of severe cold. Deep snow cover may be life threatening to large herbivores, which cannot simply burrow down to the plants that they depend on for food. In the Rocky Mountains, elk, bighorn sheep, and other large species routinely leave the highest altitudes during winter, moving to areas below timberline where the trees protect them from high winds and food can be found above the surface of the snow. Small pikas and some small rodents, on the other hand, remain active all winter above timberline, burrowing under the snow to avoid the intense cold above and consuming food stored from the previous summer.

Climatic Adaptation

Abundant evidence exists that among birds or mammals of given size, thermoregulatory properties can depend considerably on the climate in which the animals have evolved.

Interspecific Comparisons Mammals of the arctic typically have thicker and better insulating pelage than similarly sized mammals of the warm tropics. Differences in plumage insulation between arctic and tropical birds are not so well studied and may well tend to be smaller than the differences in pelage insulation between mammals. A possible constraint on the development of the plumage as an insulatory covering is that its conformation must remain suitable aerodynamically.

Metabolic relations to temperature are of particular interest because of their importance in animal energetics. As shown by the classic set of data in Figure 6.40, tropical mammals—by comparison to arctic—have higher lower-critical temperatures and must increase their metabolic rate proportionately more above the basal level for any given drop in temperature below lower-critical temperature. Because the metabolic rates in Figure 6.40 are expressed relative to basal rates rather than in absolute units, differences among species in the slope of the metabolism–temperature curve below thermoneutrality do not quantitatively reflect differences in maximal overall insulation. It is clear though that some of the arctic forms—by virtue of their relatively thick pelage and other properties—attain very high levels of insulation and, as a consequence, enjoy enormous energetic advantages. In the extreme cases of the eskimo dog and arctic fox, metabolism need not be elevated above the basal level until ambient temperature has fallen to $-20°C$ or below, and the resting metabolic rate would apparently need to exceed basal by less than a factor of 2 even at the lowest temperature ever recorded in the Arctic, around $-70°C$. Of course, not all arctic animals enjoy such low lower-critical temperatures. The collared lemming, for example, has a lower-critical temperature near $15°C$ and must sustain a metabolic rate well above basal when abroad during cold winter months. Nonetheless, the lemming is far better protected against cold than tropical mammals.

There is increasing evidence that basal metabolic rate (BMR) is sometimes correlated with climate. Polar species of birds, for example, tend to have

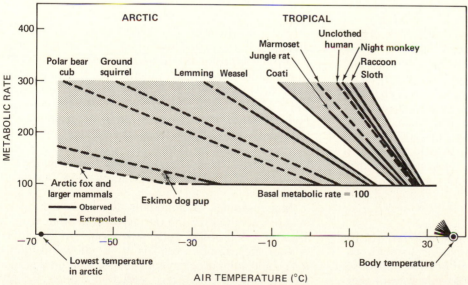

Figure 6.40 Resting metabolic rate of arctic and tropical mammals as a function of ambient temperature. All curves are adjusted to a common standard, such that basal metabolic rate equals an arbitrary value of 100. Arctic species were studied at Point Barrow, Alaska; tropical species, in Panama. [From P.F. Scholander, R. Hock, V. Walters, F. Johnson, and L. Irving, *Biol. Bull. (Woods Hole)* **99**:237–258 (1950).]

BMRs higher than those of temperate species of similar body weight, whereas tropical or desert species tend to have comparatively low BMRs for their size. Desert and semidesert rodents also tend to have relatively low BMRs; an extreme example is provided by the naked mole rat of equatorial Africa (*Heterocephalus*), which has a BMR only 20–40 percent as high as the average for placental mammals of its weight. The benefit of a high *basal* rate in cold climates is not obvious, since the basal rate prevails only when temperatures are thermoneutral. The adaptive advantage of a low basal rate in warm climates is clearer, however: An animal that can allow its metabolic heat production to fall to unusually low basal levels is able to enjoy a reduced overall heat load under hot conditions.

In the abstract, it might be expected that arctic species of mammals and birds would have evolved lower body temperatures than related tropical ones. However, for reasons unknown, body temperature turns out to be basically a conserved character within each taxonomic group: The temperature maintained in the absence of heat or cold stress does not differ from one climate to another. One bit of evidence for adaptation in body temperature is that some mammals of hot climates are able to tolerate higher degrees of hyperthermia than related mammals of temperate or cold climates. In birds, by contrast, the ability to tolerate hyperthermia seems to be about the same regardless of climate.

Intraspecific Divergence   Increasing attention has been paid of late to the possibility that populations of a single species living in different climates may differ genetically in their physiological features. As yet, not many examples of such "physiological race" formation have been documented, but some interesting cases have come to light. In the Near East, the golden spiny mouse (*Acomys russatus*) ranges from a hot climate along the Dead Sea to a cooler climate in the mountains of southern Sinai. Mice from the Dead Sea and mountains do not differ in body temperature or insulation, but they do differ dramatically in their capacity for nonshivering thermogenesis. The mountain mice can attain at least twice the rate of heat production of the Dead Sea mice and are far more resistant to cold. These differences persist among animals born and reared in the laboratory and thus are genetic.

Acclimation and Acclimatization

Figure 6.41 provides a basic frame of reference for interpreting results on acclimatization to cold seasons in birds and mammals. One possible response, shown in part (A), is for the maximal rate of thermogenesis to increase, with little or no change in insulatory characteristics. In this case, the resting metabolic rate required for thermoregulation at any given temperature is unaltered, but the cold-accli-

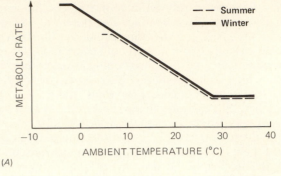

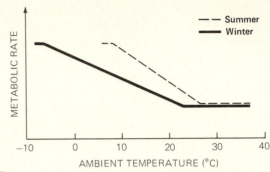

Figure 6.41   Simple metabolic acclimatization (A) and simple insulatory acclimatization (B). The plateau of each curve at the left indicates where metabolic rate has been maximized.

matized animal can maintain its body temperature at lower ambient temperatures than the warm-acclimatized animal by virtue of an increased ability to augment heat production. This type of response is termed simple ***metabolic acclimatization***. The enhanced ability to increase metabolic rate might, for example, result from an increase in capacity for nonshivering thermogenesis. Another possible response, shown in part (B), is for maximal insulation to increase, with little or no change in the peak rate of metabolic heat production. Here, the resting metabolic rate required for thermoregulation at any given temperature below thermoneutrality is reduced, and even though peak metabolism remains unchanged, the cold-acclimatized animal can again maintain its body temperature at lower ambient temperatures than the warm-acclimatized animal, because of its increased ability to retard heat loss to the environment. This is termed simple ***insulatory acclimatization***. Metabolic and insulatory acclimatization can occur together.

Many species of mammals and birds exhibit insulatory acclimatization during winter, either with or without metabolic acclimatization. In northern Alaska, for example, red foxes—as shown in Figure 6.42—and such other mammals as collared lemmings and varying hares undergo sufficient changes in their insulation with the passing of the seasons that their metabolic rate in winter at −30°C, the average winter air temperature, is in fact little higher than their metabolic rate in summer at 5°C, the average summer

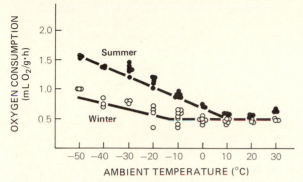

Figure 6.42 Resting metabolism of a single arctic red fox (*Vulpes vulpes*) during summer and winter. [From J.S. Hart, *Rev. Can. Biol.* **16**:133–174 (1957); based on data of L. Irving, H. Krog, and M. Monson, *Physiol. Zool.* **28**:173–185 (1955).]

air temperature. Birds and mammals often molt to a better-insulating pelage or plumage as winter approaches, and this constitutes a prime mechanism of their insulative acclimatization. The molt is largely under photoperiodic control. Large and medium-sized mammals tend to enjoy a greater proportional increase in the insulative value of their fur in winter than small mammals.

There seems to be a tendency among northern mammals for small species to undergo a greater proportional increase in their peak metabolic rate in winter than large ones; in white-footed mice, for example, the maximal rate of thermogenesis rises by 96 percent in winter, but in snowshoe hares it increases by only 24 percent. Presumably, part of the reason that large species show less metabolic acclimatization is that they are in a better position to rely on insulatory defenses than small species. Sometimes, as shown in Figure 6.43, small mammals, while undergoing metabolic acclimatization, exhibit no significant insulatory acclimatization. Some birds also fail to show insulatory acclimatization.

As winter approaches, animals in nature are exposed to gradually declining temperatures, to shortening daylengths, and often to changes in nutritional regime and other factors. Any or all of these changes can exert effects on acclimatization. By contrast, cold acclimation in the laboratory generally involves just a change in temperature, and the responses of cold-acclimated birds and mammals are commonly different from those of winter-acclimatized animals of the same species. Frequently, as in acclimatization, there is an increase in peak metabolism. However, whereas overall insulation below thermoneutrality is generally either not changed or increased by winter acclimatization, it is usually either not changed or *decreased* by cold acclimation. This latter type of response, which is quite common, is illustrated in Figure 6.44; note that the cold-acclimated individual expends *more* energy at a given temperature below thermoneutrality than the warm-acclimated one. The reasons why insulation might decrease during cold acclimation are not fully clear. One important consideration is that, because of the absence of appropriate photoperiodic stimuli for molting, the insulatory value of an animal's pelage or plumage typically undergoes little or no change during cold acclimation of several weeks' duration. A process that promotes an actual decrease in insulation in at least some mammals is that blood flow to the ears, legs, and tail is enhanced during cold acclimation; this keeps the appendages warmer but facilitates heat loss to the environment.

Birds and mammals undergo acclimatization and acclimation to warm conditions as well as cold ones. When people, for example, repeatedly experience high heat loads by virtue of exercise or intermittent exposure to hot environments, their sweating physiology undergoes dramatic changes. They start to sweat at lower body temperatures (an example of an

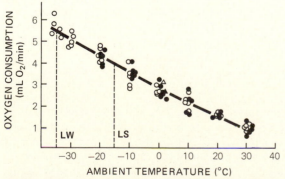

Figure 6.43 Resting metabolism of deer mice (*Peromyscus maniculatus*) during summer (closed symbols) and winter (open sumbols). Vertical dashed lines indicate temperatures at which survival is limited to 200 min in summer (LS) and winter (LW); note that winter animals survive for this time at a temperature 20°C lower than summer animals. [From J.S. Hart, *Rev. Can. Biol.* **16**:133–174 (1957).]

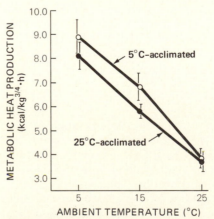

Figure 6.44 Metabolic heat production as a function of ambient temperature in laboratory rats (*Rattus norvegicus*) acclimated for 60 days to 5°C (upper line) or 25°C (lower line). Points represent mean values; vertical bars depict ±1 standard deviation around the mean. [From W. Cottle and L.D. Carlson, *Am. J. Physiol.* **178**:305–308 (1954).]

acclimation-induced change in controller function); their rate of sweating under any given level of heat stress is markedly increased; and the salt concentration of their sweat is considerably decreased, thus aiding salt conservation. By virtue of such changes, the people enjoy an enhanced capacity to work under hot conditions and experience less discomfort.

Ontogeny of Homeothermy

At birth, birds and mammals lack the full thermoregulatory capabilities of adults of their species. Thus, their thermal and energetic relations to the environment change—often markedly—during postnatal development. Study of the physiological ontogeny of animals is important from several viewpoints. One is that to understand the ecological and evolutionary forces at work on a species, we must understand the physiology of each life stage through which individuals must pass.

There is a great range of thermoregulatory competence in mammals and birds at birth. Caribou exemplify one extreme; born under windy and near-freezing conditions, they can keep themselves warm even while still wet with amniotic fluid. At the other extreme, opossums (*Didelphis*) are unable to thermoregulate even at room temperature until over 80 days old. There are many species that are intermediate. They have relatively little or no thermoregulatory competence at birth but develop advanced competence within a week to a month. Included are passerine birds, small rodents, domestic rabbits, and others. Let us start by looking at one of these species in detail, the white-footed mouse.

As shown in Figure 6.45A, when mice studied as isolated individuals are 2 days old, they can keep their body temperature above 34°C only if their environment is warm: 30°C (86°F) or higher. Gradually, however, as they mature, they develop the ability to maintain a body temperature of at least 34°C at lower and lower ambient temperatures, until at 18–20 days of age they are able to thermoregulate even when the air is freezing cold. Figure 6.45B reveals that even 2-day-old mice exhibit modest homeothermic responses: They often increase their metabolic rate when the air temperature is lowered from 35 to 30°C. Such young animals, however, being limited in their ability to keep themselves warm, lapse into a poikilothermic type of relation between metabolism and ambient temperature at lower temperatures; as the air is cooled from 20 to 0°C, their metabolic rate declines quasiexponentially, along with the decline in their body temperature. In contrast, 18-day-old mice consistently respond to decreases in air temperature with increases in metabolic heat production, even as the temperature approaches 0°C. Figure 6.46 summarizes some of the developmental features of individual nestlings. Parts (*A*) and (*B*) reveal that as nestlings become older, their insulation and capacity to produce heat increase markedly. Their improvement in thermoregulatory ability is a reflection of these changes. At about 14 days, they first gain the ability to maintain a high body temperature at low air temperatures while isolated from their nest, parents, and siblings; and this sets the stage for them to leave the protective confines of their natal nest, which they start to do at about 16 days. Weaning occurs at about 20 days.

As might be expected, young animals that have only meager abilities to thermoregulate generally display a high tolerance to reduced body temperatures. Young white-footed mice, for example, that are cooled to body temperatures of 1–3°C for 2 h generally survive, even though they cease breathing. Such tolerance to hypothermia often diminishes with age.

Within related groups of animals, sharp differences in the pace of development are often observed. For example, the 1-day-old chicks of laughing gulls (*Larus*) can run about and thermoregulate at air temperatures of 20–25°C. The similarly sized 1-day-old chicks of cattle egrets (*Bubulcus*), on the other hand, are physically immature and poorly coordinated, destined not to be able to thermoregulate at 20–25°C until at least 8 days old or leave the nest until 2 weeks old. Young of the former type, that are relatively mature at hatching or birth and gain comparatively early independence of their parents, are termed *precocial*. Those of the latter type, that are hatched or born in a relatively immature, helpless state and thus require protracted parental care, are called *altricial*. What are the implications of these two forms of development? More specifically, granting that precocial development conveys certain obvious advantages (e.g., early independence), does altriciality also have virtues?

In fact, altriciality is believed to have certain potential advantages. In arriving at this conclusion, one major theme is that altricial young, by comparison to precocial ones, may be able to channel a greater fraction of their available energy into growth and maturation because they enjoy reductions in some of the alternative demands on their energy resources. A point to recognize is that it is of advantage for young birds and mammals to be warm during at least a substantial part of their postnatal development; low body temperatures suppress growth and other maturational processes. Now one of the features of altricial development is extensive parental care, including extensive incubation: Over the course of postnatal development, altricial young tend to be incubated more than precocial ones. To the extent that parents keep young warm, the young do not have to divert their own energy resources into production of metabolic heat for thermoregulation. Thus, the relatively great amount of incubation associated with altriciality is one mechanism by which the energy resources of altricial young are spared for use in growth and maturation.

Passerine birds typically exhibit no thermoregulatory responses whatever for a number of days after hatching. This particular form of physiological altri-

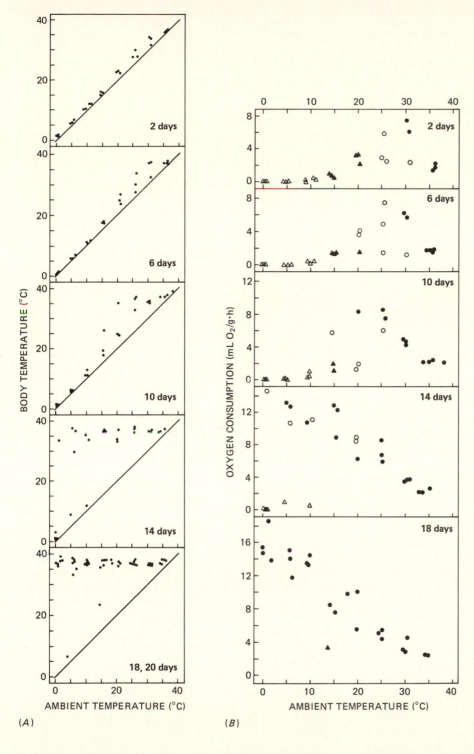

Figure 6.45 Body temperature (*A*) and metabolic rate (*B*) as functions of ambient temperature in young white-footed mice (*Peromyscus leucopus*) at 2, 6, 10, 14, and 18–20 days of age. Age is indicated at the lower right of each block in (*A*) and the upper right of each block in (*B*). Young mice were studied individually—without nest, siblings, or parents—and were exposed to test temperatures for 2.5–3.0 h. In (*A*), the diagonal line in each block is a line of equality between body temperature and ambient temperature. In (*B*), symbols indicate approximate body temperatures: ●, 34°C or higher; ○, 24.0–33.9°C; ▲, 14.0–23.9°C; △, 13.9°C or lower. [After R.W. Hill, *Physiol. Zool.* **49**:292–306 (1976).]

ciality itself may constitute a mechanism of preferentially directing available energy resources into growth and maturation: The absence of physiological thermoregulation in the early-age young *guarantees* that they will not use their energy resources for thermoregulatory heat production. On the other hand, this physiological immaturity leaves the young vulnerable to cooling when their parents are absent.

Among rodents that undergo altricial development, the young commonly exhibit homeothermic metabolic responses promptly after birth. These early responses may be barely detectable when individual young are studied in isolation, as we have seen in white-footed mice (Figure 6.45). The young do not live in isolation in nature, however. Rather, they live as groups of littermates huddled within an insulating nest. At least in white-footed mice, the homeothermic abilities of the early-age young *are* adequate to keep them warm as a huddled group in a nest (without the parents) even at low air temper-

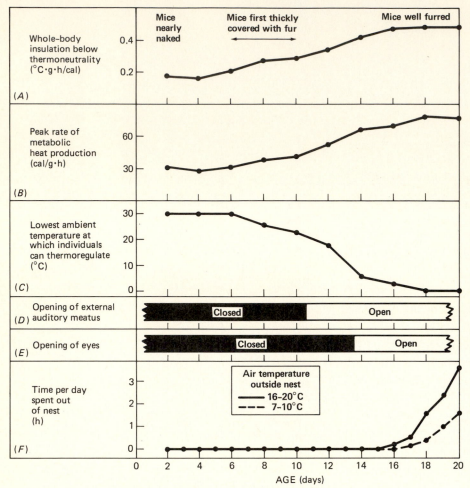

Figure 6.46 A summary of some major developmental changes occurring between birth and weaning in white-footed mice (*Peromyscus leucopus*). Weaning occurs at about 20 days of age. Parts (*A*)–(*C*) are based on the same studies as in Figure 6.45; data are for individual animals, isolated from their nest, siblings, and parents. In part (*C*), animals were rated as capable of thermoregulation if they could keep their body temperature at 34°C or higher. Hearing undergoes dramatic improvement when the external auditory meatus (ear canal) opens (*D*). As shown in (*F*), nestlings undertake temporary trips away from the nest starting several days before weaning; the time they spend away each day depends on the air temperature outside the nest.

atures for several hours. Probably, the huddled young allow their parents to keep them warm when the parents are present. When the parents are absent, however, the young divert some of their own energy resources into thermoregulation. Presumably, their reward for using energy in this way is that they avoid hypothermia, which in itself would slow their development.

THERMOREGULATORY AND ENDOTHERMIC PHENOMENA IN LOWER VERTEBRATES AND INVERTEBRATES

As stressed at the beginning of this chapter, homeothermy and poikilothermy are *types of thermal relation* between the animal and its environment. Some animals so predominantly show one type of relation or the other that we call them poikilotherms or homeotherms. It has become increasingly clear, however, that many animals exhibit both types of relation at different times or exhibit thermal relations that are not adequately described as either poikilothermy or homeothermy. In this and the next section, we examine some of these phenomena. Before proceeding, we need to distinguish clearly between the

two concepts mentioned in the title of this section: thermoregulation and endothermy. **Thermoregulation** is the maintenance of a more or less stable body temperature, whereas **endothermy** is the maintenance of a difference between body temperature and ambient temperature through metabolic heat production. In birds and mammals, these go hand in hand. However, the two are not necessarily coupled. Some animals thermoregulate without depending on elevation of their body temperature by internal heat production, and some maintain their body temperature above ambient by virtue of metabolic heat but do not control their rates of heat production and loss in such a way as to maintain a stable body temperature.

Behavioral Thermoregulation

Maintenance of body temperature at a relatively stable level by behavioral means is termed **behavioral thermoregulation**. In certain forms, it occurs in homeotherms. Our focus here, however, is on poikilotherms, among which it is very widespread. Sometimes, behavioral thermoregulation is outwardly simple, involving no more than the selection of one environmental temperature over another when a choice is possible. In other instances, however, the processes of behavioral thermoregulation are elaborate.

For illustration here, let us look at the lizards, some of which exhibit behavioral thermoregulation in particularly sophisticated forms. Placed in a uniform thermal environment in a laboratory, lizards typically have body temperatures close to ambient—the characteristic response of poikilothermy. In nature, however, they typically exploit possibilities for heating and cooling during the day to maintain relatively stable body temperatures that may be considerably higher than air temperature. Desert lizards, for example, emerge in the morning and bask in the sun until their body temperature has risen to within the usual range maintained during daily activity. Thereafter, they maintain their temperature within this range until nightfall by a variety of mechanisms. Frequently, they are observed to shuttle back and forth between sun and shade. They also can modify the amount of their body surface exposed to the direct rays of the sun by changing their posture and orientation to the sun. They may flatten themselves against the substrate to lose or gain heat, depending on substrate temperature, and when the substrate has become very hot during midday, they may minimize body contact by elevating all but their feet off the ground or even climbing on bushes. If circumstances aboveground become so uniformly hot that their body temperature cannot be held below threatening levels, they retreat underground. By thus exploiting the numerous opportunities for heating and cooling in their environment, many species can maintain body temperatures that vary by only several degrees Celsius over long periods during the day. The mean temperature depends on the species. To cite one quantitative example, desert iguanas (*Dipsosaurus dorsalis*) typically maintain average abdominal temperatures of 38–42°C during the day, and they often keep their temperature within 2–3°C of the mean for hours on end.

One of the chief limitations of behavioral thermoregulation is that it is dependent on thermal opportunities available in the environment and thus susceptible to being thwarted by vagaries of the environment. Confronted with a cool, cloudy day, for example, a desert iguana may be unable to warm itself to temperatures anywhere near as high as those it "prefers." Physiological mechanisms of thermoregulation—although they may be costly in energy, water, or other terms—confer increased independence.

Thermal Relations of Reptiles

Some reptiles do not thermoregulate. However, thermoregulation is common not only in lizards but also in other reptilian groups, and it is our focus here. Nearly always, the chief mechanisms of thermoregulation are behavioral. Nonetheless, many reptiles have evolved physiological mechanisms of thermoregulation that are used to supplement the dominant behavioral mechanisms; and there are a few known or postulated instances in which thermoregulation has become chiefly physiological. We described behavioral thermoregulation in lizards in the preceding section. Now we examine additional aspects of reptilian thermoregulation, again emphasizing lizards.

Physiological Capabilities Neither lizards nor the vast majority of other modern-day reptiles have evolved thermogenic or insulatory capabilities adequate for their bodies to be warmed metabolically (see p. 104). That is, they have not evolved endothermy; and in this respect an enormous gulf separates them from birds and mammals. Lizards, however, have evolved at least three basic physiological mechanisms of exerting control over their temperature, two of which parallel avian and mammalian mechanisms closely enough as to be potentially homologous. (1) When exposed to heat stress, many lizards actively augment evaporation by panting or gular pumping. (2) Many medium-sized and large lizards physiologically modify the ease of heat transfer between themselves and the environment so that they are able to warm more quickly than they cool; specifically, their rate of warming when T_A exceeds T_B by a given amount is greater than their rate of cooling when T_B exceeds T_A by the selfsame amount. In the marine iguana of the Galápagos Islands, to cite a particularly marked example, cool animals in a warm environment warm up about twice as fast as warm animals in a cool environment cool down. Such responses allow lizards to remain at high body temperatures for more time each day than would otherwise be possible; for example, the animals heat up relatively rapidly in the morning but cool down relatively slowly in the evening. Vasomotor and cardiac adjustments are responsible for the differential heating and cooling rates. During heating, the heart circulates blood to the periphery more rapidly and the cutaneous vascular beds are more dilated than during cooling; in this way heat transfer from the environment to the animal is facilitated. (3) A third physiological mechanism of thermoregulation widely seen in lizards is alteration of skin color. Commonly, the skin is darkened when it is of advantage to increase absorption of solar radiant energy and lightened when absorption is to be limited.

Other groups of reptiles also possess physiological mechanisms of exerting control over their temperature. For example, some crocodilians and turtles possess physiological (cardiovascular) abilities to warm up faster than they cool down; and under heat stress, some turtles spread saliva or urine on themselves.

As noted earlier, endothermy is rare in extant reptiles. Some instances of it are known or suspected, however. The best-documented case is that of the female Indian python (*Python molurus*). During brooding of eggs (but not otherwise), she in fact thermoregulates by way of controlled endothermy, using heat produced by spasmodic body contractions. In some of the huge sea turtles, partly by virtue of their size, at least some body regions are warmed to a few degrees Celsius (or more) above water temperature by routine metabolic production of heat, especially during swimming.

Effects of Body Temperature on Functional Capabilities   When cellular temperatures are stabilized, the possibility exists for cells to become specialized to function most effectively at those temperatures. Such thermal specialization is often cited as one of the chief potential benefits of thermoregulation. Recognizing that many lizards and other reptiles regulate their body temperatures near certain "preferred" levels for many hours each day using behavioral and—to some extent—physiological means, a key question is whether in fact each species is able to function "best" within the range of body temperatures it maintains. The answer, as now understood, is not altogether simple.

Many functions do indeed seem to be carried out best when body temperature is within the species-specific preferred range, indicating cellular thermal specialization. For example, among certain species of lizards that have preferred temperatures near 40°C, testicular development at the onset of the breeding season is most rapid and complete at similarly high temperatures; among other species that prefer temperatures near 30°C, the testicles develop optimally near 30°C and are damaged by 40°C. Hearing, digestion, and the response of the immune system to bacterial invasion are just some of the other processes known to take place optimally in at least some reptiles when body temperatures are in or near the preferred range. There are exceptions to these patterns, however, and there are traits that seem in general not to be optimized at preferred temperatures. Speed and endurance during bursts of anaerobically fueled escape running in lizards, for example, are relatively independent of temperature and not necessarily maximal at the preferred temperature.

Concluding Comment   Study of reptiles probably provides a roughly accurate glimpse into the nature of vertebrate thermal relations prior to the evolution of the full-blown homeothermy seen in birds and mammals. From our discussion here and from Box 6.2 (p. 117), we see that modern reptiles possess a partial set of the key thermoregulatory attributes of birds and mammals. From this we may conclude with substantial confidence that avian and mammalian homeothermy did not evolve as an entirely novel suite of processes but as an elaboration on preexisting behavioral and rudimentary physiological mechanisms of thermoregulation. We do not know when

homeothermy first appeared. Certain authorities believe that it was already present in some dinosaurs.

Warm-Bodied Fish

Whereas the body temperatures of most fish closely approximate water temperature, we now know that in species of tunas and lamnid sharks, temperatures within the deep swimming muscles and certain other regions of the body exceed water temperature, sometimes substantially. These fish are large, streamlined, fast-swimming animals that lead a wide-ranging predatory existence, including capture of such speedy prey as squid and herring. The vigorous activity of their swimming muscles produces the heat that warms the muscles. However, high heat production is not in itself adequate to explain any substantial elevation of tissue temperature in gill-breathing animals: If metabolic heat is carried freely to the gills in the venous blood, it will be lost so readily to the surrounding water across the vast respiratory surfaces that no appreciable elevation of the animal's body temperature will occur. Crucially, the tunas and lamnid sharks have evolved vascular countercurrent exchange networks that short-circuit circulatory outflow of heat from their swimming muscles. In many species, as diagrammed in Figure 6.47, the major longitudinal arteries and veins that carry blood to and from the swimming muscles run along either side of the body just under the skin. Smaller arteries branch off from the longitudinal arteries and penetrate inward through the muscles. In turn, blood is brought outward from the muscles in veins that discharge into the longitudinal veins leading back to the heart. The arteries and veins carrying blood inward and outward from the muscles are closely juxta-

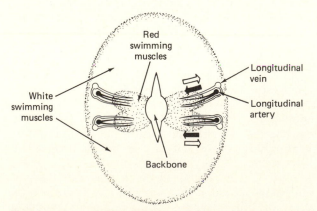

Figure 6.47 A diagrammatic cross section of a bluefin tuna showing the nature of the blood supply to the red swimming muscles in highly schematic form. The longitudinal arteries, carrying blood along the length of the body, give off small arteries that penetrate inward through the muscles toward the backbone. Small veins running in close juxtaposition to the small arteries return blood peripherally to the longitudinal veins, which lead back to the heart. Dark vessels and dark arrows refer to arterial flow; light vessels and light arrows refer to venous flow.

posed, forming countercurrent exchange networks. By virtue of branching of the vessels, these networks become extraordinarily elaborate in the vasculature that supplies the red (dark) muscles which are chiefly responsible for power generation during cruising. In these elaborate vascular beds, huge numbers of minute arteries and veins, each measuring only about 0.1 mm in diameter, are closely intermingled in thick layers of vascular tissue—a true *rete mirabile*. With these countercurrent exchange arrangements, much of the heat picked up by the venous blood in the red muscles is transferred to the ingoing arterial blood rather than being carried to the periphery of the body and gills, where it would readily be lost to the environment. Thus, heat produced by the red muscles tends to be retained within them.

Figure 6.48 depicts the relation between red-muscle temperature and water temperature in bluefin tuna, the largest of all tunas. As can be seen, bluefins maintain fairly constant muscle temperatures even as the ambient temperature varies from 7 to 30°C. Evidently, these fish are physiological thermoregulators. Their mechanisms of thermoregulation are not known. One postulate is that they reduce the efficiency of heat retention by their countercurrent exchange networks as ambient temperature rises, thus maintaining a smaller elevation of their muscle temperature at high water temperatures than at low water temperatures. Figure 6.49 illustrates that in yellowfin and skipjack tunas, unlike bluefins, red-muscle temperature is elevated over water tempera-

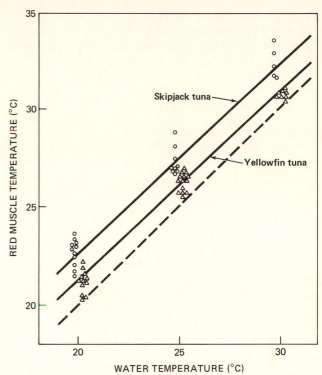

Figure 6.49 Red-muscle temperature as a function of water temperature in small skipjack tuna (*Katsuwonis pelamis*) and yellowfin tuna (*Thunnus albacares*) while swimming in an aquarium. The dashed line is a line of equality between muscle and water temperature. Circles and triangles are data for skipjack and yellowfins, respectively. Solid lines statistically summarize the relations between muscle and water temperature in the two species. Larger, wild fish of these species are sometimes observed to exhibit greater temperature differentials between muscles and water (e.g., 5–10°C in skipjacks). [Reprinted by permission from A.E. Dizon and R.W. Brill, *Am. Zool.* **19:**249–265 (1979).]

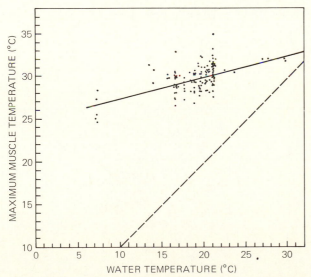

Figure 6.48 Maximum temperatures recorded in the swimming muscles of bluefin tuna (*Thunnus thynnus*) captured in waters of various temperatures. Maximum muscle temperatures were characteristically found within the red muscles. The dashed line is a line of equality between muscle temperature and ambient temperature; the vertical distance between it and the solid line at any given water temperature indicates the extent to which muscle temperature is elevated above the water temperature. [Reprinted with permission from F.G. Carey and J.M. Teal, *Comp. Biochem. Physiol.* **28:**205–213 (1969). Copyright 1969, Pergamon Press.]

ture by a relatively constant amount regardless of the water temperature; if water temperature falls by, say, 10°C, muscle temperature falls by about the same amount. Phenomenologically, these species present cases of endothermy without thermoregulation. Recent evidence indicates, nonetheless, that in certain respects they do actively regulate their tendency to retain heat; for example, they decrease heat retention when they are highly active in warm waters, thus preventing their activity from driving their muscle temperatures too high.

The swimming muscles are not the only tissues kept warm. Among certain tunas, for example, the brain, eyes, and stomach (during digestion) are warmed. In each case, the arteries and veins serving the organ form a *rete*, which causes organ temperature to be elevated by short-circuiting outflow of locally produced heat.

One of the biggest questions regarding tunas and lamnids is why they have evolved endothermy and thermoregulation. Unfortunately, there are no definitive answers as yet. Conjecture concerning the swimming muscles centers around the idea that lo-comotory performance may be aided by elevation of muscle temperature. Any such aid to performance would be significant for animals so dependent on high-intensity exertion for their livelihood. One postulate is that a rise in muscle temperature permits greater power output by a direct, facilitating effect on the contractile process. Another is that elevated temperatures aid oxygen delivery to the muscle mitochondria by accelerating the facilitated diffusion of oxygen that is mediated by myoglobin (see p. 318).

Thermoregulation in Colonies of Social Insects

Physiological regulation of colony temperature is widespread in colonies of bees and wasps. Honeybees (*Apis*) provide the best-studied and possibly most-elaborate example. The brood of honeybees requires temperatures within a narrow range, about 32–36°C, for proper development. When a hive is rearing brood, it maintains the temperature of the brood combs within this range even though the air temperature outside the hive may fall to −30°C or rise to 50°C. When ambient temperatures are low, worker bees cluster together within the hive and elevate their individual rates of metabolic heat production (see the next section). When ambient temperatures become high, the workers disperse within the hive and fan with their wings in a cooperative pattern that moves fresh, relatively cool air from outside the hive across the brood combs. Should dispersal and fanning not in themselves keep the brood adequately cool, the bees collect water and spread it within their nest, whereupon it evaporates into the airstream produced by fanning.

Endothermy and Physiological Thermoregulation in Solitary Insects

The solitary insect at rest characteristically metabolizes at a sufficiently low rate that, given its small body size, its body temperature is not appreciably elevated over ambient temperature. Insects in flight, however, may exhibit very high metabolic rates; strong flyers may actually release more heat per gram than active birds or mammals. Thus, it is not surprising to find that many insects have body temperatures that are several to many degrees higher than air temperature when in flight. Some of these species, although endothermic, do not thermoregulate; for example, during flight in certain small American geometrid moths, thoracic temperature is elevated above air temperature by about the same amount, 6°C, whether the air temperature is 10 or 20°C. The last two decades have witnessed, however, a rapidly escalating appreciation of the remarkable fact that many solitary insects *physiologically thermoregulate* during flight and some other active states. Usually, these insects regulate the temperature of their thorax but not that of their abdomen. The thorax, of course,

is the location of the flight muscles and thus the site where most heat is generated.

Sphinx moths are some of the insects in which thermoregulation during flight has been most thoroughly studied; they are particularly large insects, weighing as much as several grams and thus being equivalent in weight to some of the smallest birds and mammals. One example of thermoregulation by a species of sphinx moth during flight is shown in Figure 6.50. Note that the flying moths keep their thoracic temperature within a narrow range, 38–43°C, even as the air temperature varies between 12 and 35°C. Thermoregulation is not limited to just such large forms. Worker bumblebees (*Bombus vagans*), averaging 0.12 g in weight, for instance, stabilize their thoracic temperature at about 32–33°C when air temperatures are 9–24°C during foraging. Certain other lepidopterans and bees as well as some dragonflies and beetles thermoregulate in flight.

Some instances of endothermy and physiological thermoregulation are also known during terrestrial activity in insects. In nearly all such cases, the primary source of heat is again the flight muscles, activated to undergo "shivering" of one form or another (see later). Dung beetles sometimes become markedly endothermic while working in dung piles. Some katydids and crickets thermoregulate while singing.

Flight Temperatures, Preflight Warm-up, and Shivering The flight muscles of insects must be able to produce a certain minimal power output (depending on species) before flight is possible. Within a broad range of temperatures, the power output that flight muscles can attain increases with their temperature. Thus, muscle temperature is potentially an important determinant of whether flight is possible. Small in-

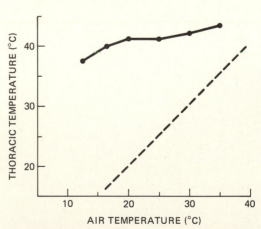

Figure 6.50 Mean thoracic temperature of freely flying sphinx moths (*Manduca sexta*) as a function of air temperature. The dashed line is the line of equality between thoracic and air temperatures; thus, at any given air temperature, the vertical distance between it and the solid line indicates the extent to which the thorax is warmer than the air. [Data from B. Heinrich, *J. Exp. Biol.* **54**:141–152 (1971).]

sects such as fruit flies, mosquitoes, and midges have such high surface-to-volume ratios that the activity of their flight muscles cannot warm them appreciably above the temperature dictated by ambient conditions. It is noteworthy then that small insects commonly are able to fly over a broad range of thoracic temperatures, including, in some cases, temperatures as low as 0–5°C. An important facet of the flight physiology of small, poikilothermic flyers is that they apparently require only a modest fraction of their maximal power output to stay aloft; thus, they can fly at relatively low thoracic temperatures, at which their power output is substantially submaximal. In contrast, many large insects, including those known to thermoregulate, require a near-maximal power output to take off and remain airborne. Accordingly, they also require high thoracic temperatures. Sphinx moths, *M. sexta,* for example, are unable to fly unless their thorax is at least as warm as 35–38°C, and worker bumblebees, *B. vagans,* require about 29°C.

Even large insects typically cool to the temperature dictated by ambient conditions when at rest; a sphinx moth resting in cool night air at 20°C, for example, has a thoracic temperature little different from air temperature. Accordingly, insects often find themselves too cool to take off. Diurnal species may be able to warm to flight temperature by basking in the sun. Often, however, insects must invoke physiological heat production to warm their thorax sufficiently to take to the air: a process known as ***preflight warm-up.***

The mechanisms of preflight warm-up involve high-frequency activation of the flight muscles and are often termed ***shivering.*** In many insects, such as moths and butterflies, the muscles responsible for the upstroke and downstroke of the wings are contracted synchronously during shivering (rather than alternately as in flight), thus working against each other. Heat is evolved, and the wings are often seen to go through low-amplitude vibrations. In some forms, such as certain bees and flies, on the other hand, the upstroke and downstroke muscles may contract substantially or wholly asynchronously during shivering, but motions of the wings are prevented nonetheless, possibly by mechanical uncoupling of the wings from the thoracic flight apparatus.

When sphinx moths warm from a low temperature, their flight muscles shiver at a higher and higher intensity as their thoracic temperature increases to the flight level. Then suddenly the pattern of muscular contraction changes, the wings are driven through the flapping motions of flight, and the animal takes to the air.

Thermoregulatory Mechanisms The mechanisms of thermoregulation during flight and other activities are not fully understood. One possible mechanism is to increase the rate of metabolic heat production as ambient temperature is lowered. It is well accepted that the rate of heat production by *shivering* can be modulated according to thermoregulatory needs, but the flight muscles can be used for shivering only when the insect is not flying. Honeybees and bumblebees often maintain high and stable body temperatures for long periods when working in the hive. Under these circumstances, the flight muscles are available for shivering, and their rate of heat production is modulated according to demands for thermoregulation. Solitary queen bumblebees, *Bombus vosnesenskii,* for example, incubate their first brood in the spring by keeping their abdomen at an elevated temperature and pressing it against the brood, as shown in Figure 6.51. Heat is brought to the abdomen from the thorax, where it is produced by the flight muscles. As shown in Figure 6.52, the queen bee thermoregulates by increasing her metabolic

Figure 6.51 A queen bumblebee, *Bombus vosnesenskii,* incubating her brood by pressing her abdomen against it. (Photograph courtesy of Bernd Heinrich.)

rate—evidently by increasing her shivering intensity—as the air temperature falls, a response analogous to that in birds and mammals.

Modulation of shivering can also be used to thermoregulate during intermittent flight. For example, when a bumblebee forages for nectar—flying from flower to flower—it can shiver or not shiver *while it is alighted* on each flower. At air temperatures near 24°C, the heat generated just from intervals of flight seems adequate to keep the thorax at flight temperature, 32–33°C. At lower air temperatures, however, additional heat is needed to keep the thorax warm, and it is produced by shivering while the bee is alighted. Overall, therefore, the bee's metabolic rate is higher at low air temperatures than high ones (p. 73).

Available evidence indicates that when insects fly continuously, the intensity of work by their flight muscles is typically set by the requirements of flight and not modulated in service of thermoregulation. When sphinx moths fly at a certain speed, for example, their rate of metabolic heat production is virtually identical whether the air temperature is 15 or 30°C. Furthermore, they remain aloft continuously for long periods, even feeding while airborne (hovering). How can an insect in continuous flight maintain a stable thoracic temperature over a range of air temperatures even though its metabolic rate is constant? The answer in many cases is that insects in flight regulate their thoracic insulation; this is done by modulating the rate of heat transfer into the abdomen, at least in part via adjustments in the rate of blood flow between the thorax and abdomen. To the extent that heat produced in the thorax is carried into the abdomen, its loss to the environment is facilitated, both because the surface area across which it can be lost is increased and because abdominal body surfaces are often less well insulated than thoracic ones. In sphinx moths—to take one example—the thorax is covered with a dense, insulating layer of furry scales; and of course, during flight, it has a prodigious rate of heat production within. By virtue of these properties, if the thorax were to function as an entirely isolated system, its temperature during flight would be elevated by 20°C or more above ambient. Now when ambient temperatures are low, the heart of a flying sphinx moth beats weakly, and blood circulates only slowly between the thorax and abdomen; thus, heat produced in the thorax tends to be retained there, warming the thorax markedly. By contrast, when ambient temperatures are high, the heart beats more vigorously; and because heat is thereby lost to a greater degree across the abdomen, the thorax is not warmed as much.

In a few instances, solitary insects employ evaporative cooling in thermoregulation. Worker honeybees, for instance, prevent overheating of their head and thorax during flight under high heat loads by repeatedly regurgitating fluid, manipulating it on their tongue, and sucking it back into their mouth.

CONTROLLED HYPOTHERMIA IN BIRDS AND MAMMALS

Although birds and mammals frequently function in the homeothermic mode and maintain high, relatively stable body temperatures, many mammals and some birds have the ability to relax their homeothermic responses and allow their body temperatures to fall, even to close approximation with low ambient temperatures. There are a number of forms of such *controlled hypothermia*.

States in Which Body Temperature Is Maintained Well Above Ambient Temperature

In many instances, although the body temperature of a bird or mammal is permitted to fall, it nonetheless is regulated at a level well above ambient temperature. For example, black-capped chickadees (*Parus atricapillus*) sometimes allow their core temperature to fall by about 10°C, to 31–34°C, while sleeping overnight in freezing-cold winter weather. Bears of some species—and possibly raccoons, skunks, and other related animals—become modestly hypothermic during extended periods of sleep in winter; black bears in their winter dens, for instance, have temperatures of 31–33°C. There is no well-accepted, general nomenclature for states such as this, although the specific condition seen in overwintering bears has been termed "winter sleep" or "carnivorean lethargy."

Even a modest drop in body temperature during cold weather reduces an animal's rate of heat loss by reducing the difference between body and ambient temperature. Accordingly, it extends the length of time the animal can survive on body fat and other food reserves.

Hibernation, Estivation, and Daily Torpor

Hibernation, estivation, and daily torpor are states in which the animal relaxes its homeothermic processes virtually completely within a certain range of ambient temperatures and, like a poikilotherm, *al-*

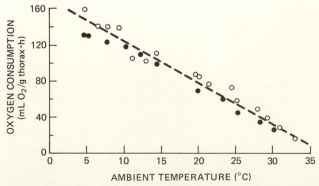

Figure 6.52 The rate of oxygen consumption of two queen bumblebees (*Bombus vosnesenskii*) as a function of air temperature while incubating brood. Closed and open symbols refer to the two individuals. [Reprinted by permission from B. Heinrich, *J. Comp. Physiol.* **88**:129–140 (1974).]

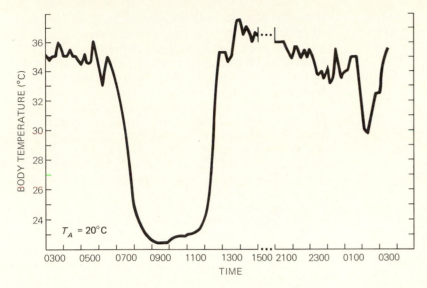

Figure 6.53 Body temperature over 24 h in a pigmy mouse (*Baiomys taylori*) exposed without food to an air temperature of 20°C. Note the profound drop in body temperature to levels approaching ambient temperature during the episode of torpor (0700–1230). [Reprinted from J.W. Hudson, *Physiol. Zool.* **38**:243–254 (1965), by permission of The University of Chicago Press. © 1965 by the University of Chicago.]

lows its body temperature to approximate ambient temperature.

When body temperature is freed to approximate ambient temperature for periods of several days or longer during winter, the phenomenon is termed *hibernation*. When this occurs during summer, it is called *estivation*. If body temperature is freed to approximate ambient temperature for only part of each day, generally on many consecutive days, the phenomenon is termed *daily torpor,* regardless of season. These three forms of controlled hypothermia are thus distinguished largely, if not completely, on the basis of their duration and time of occurrence. They appear to be different manifestations of a single fundamental physiological process.

The types of change in body temperature and metabolism that occur during these forms of hypothermia are illustrated in Figures 6.53 and 6.54 using data for species of mice undergoing daily torpor. During episodes of hypothermia, the heart rate and breathing rate decline along with metabolism. The animal may retain some ability to move and respond behaviorally to its environment at body temperatures well below normal, but there is increasing lethargy as body temperature falls, and at low temperatures the lethargy becomes extreme.

The chief benefit of these hypothermic states is a reduction in the animal's energy demands. This reduction is predicated, in a sense, on two different principles: (1) The animal no longer elevates its rate of metabolism to keep itself warm, and (2) the resultant decline in body temperature in itself lowers metabolism because the metabolic rates of tissues typically drop quasiexponentially as tissue temperatures fall ("Q_{10} effect," see p. 83). The dramatic impact of these changes in metabolic physiology is illustrated in Figure 6.55. The length of time spent in hypothermia is also a factor in overall energy savings. A kangaroo mouse that allows its temperature to drop to 17°C for about 10 h/day reduces its total

daily energy costs by about 30 percent. When hibernators remain at body temperatures near 5°C almost continuously for weeks or months on end, their savings are enormous.

Entry into controlled hypothermia reduces an animal's water expenditures as well as its energy expenditures; in fact, for individuals suffering water shortage, the savings of water may be of greater significance than those of energy. Respiratory water

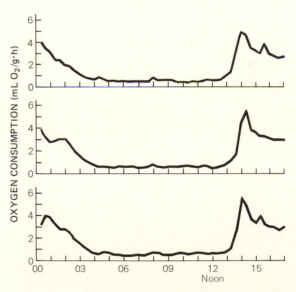

FASTERN STANDARD TIME (h)

Figure 6.54 Oxygen consumption over three consecutive days (from top to bottom) in a white-footed mouse (*Peromyscus leucopus*) studied at an air temperature of 13–15°C with an abundance of food. The animal required a resting metabolic rate of about 3.0 mL O_2/g·h for maintenance of high body temperatures. It underwent a prolonged episode of torpor on each day, as indicated by the drop in metabolic rate. Body temperature determined during an episode of torpor similar to those shown was 16.8°C. [Reprinted with permission from R.W. Hill, *Comp. Biochem. Physiol.* **51A**:413–423 (1975). Copyright 1975, Pergamon Press.]

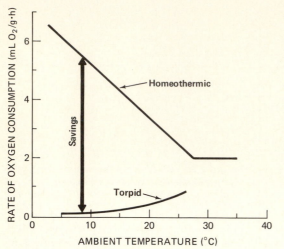

Figure 6.55 Metabolic rate as a function of ambient temperature in homeothermic and torpid kangaroo mice, *Microdipodops pallidus*. At ambient temperatures of 5–25°C, the body temperatures of the mice are 32–37°C when they are homeothermic, but during torpor their body temperatures are just 1–3.5°C above ambient. As illustrated, the vertical distance between the "homeothermic" and "torpid" lines at a given ambient temperature depicts the energetic savings per unit time of entering torpor at that temperature. Torpor is provoked by insufficiency of food. Single episodes of torpor can last from just a few hours to 2 days or more. Thus, torpor in kangaroo mice represents, in a way, an intergrade between daily torpor and hibernation: one bit of evidence that daily torpor and hibernation are in fact different manifestations of a single basic process. [After J.H. Brown and G.A. Bartholomew, *Ecology* **50**:705–709 (1969).]

losses are reduced during hypothermia for two reasons: (1) Because oxygen requirements are reduced, the rate of ventilation of the lungs is decreased; and (2) because body temperature is lowered, exhaled air is cooler than during homeothermy and thus carries less water vapor per unit volume. Transcutaneous water losses are also reduced because the drop in body temperature lowers the vapor pressure of the body fluids (p. 137).

Birds and mammals that are capable of hibernation, estivation, or daily torpor are often termed heterotherms. A *heterotherm* is an animal that sometimes regulates its body temperature physiologically and sometimes does not. The heterotherm can enjoy, in a sense, the best of both the homeothermic and poikilothermic worlds. When thermoregulating at high body temperatures, the animal is able to move about with the independence of external thermal conditions that we have emphasized as a prime advantage of homeothermy. When in hypothermia, on the other hand, it enjoys the comparatively low requirements for energy and water characteristic of poikilothermy. Insects that thermoregulate physiologically during flight are also heterotherms and partake of similar benefits.

In What Respects Is "Controlled" Hypothermia Controlled?   Birds and mammals sometimes suffer a lowered body temperature merely because their physiological efforts to keep warm are overwhelmed, as when people become hypothermic after falling into icy waters. During hibernation, estivation, and daily torpor, by contrast, the drop in body temperature is brought about by a change in the operation of the animal's thermoregulatory control centers, resulting in a directed curtailment of efforts to keep warm. Perhaps the most striking evidence of the controlled nature of hibernation, estivation, and daily torpor is the ability of animals to arouse from these conditions: Hypothermic individuals are able to warm themselves back to a high body temperature using their own metabolic heat production. Arousal is accomplished by intense shivering and, in the case of mammals, nonshivering thermogenesis (p. 106). Species differ widely in the minimal body temperature from which they are able to arouse. Many hibernators can arouse from temperatures of 5°C or less.

We have stressed that animals in controlled hypothermia allow their body temperatures to approximate ambient temperature. An important corollary is that they often permit their body temperatures to rise and fall in tandem with fluctuations in the environmental temperature. We must thus wonder what happens if the environment becomes cold enough that body temperature is in danger of falling below the minimal level from which arousal is possible. Sometimes, animals simply let their temperatures fall below that level, a lethal situation. Frequently, though, hypothermic animals undergo one of two other responses, both of which provide striking evidence that although the animals may behave poikilothermically in certain respects, they in truth monitor their temperatures and exert control if needed. Sometimes, animals immediately arouse if their body temperatures start to fall too low. More remarkably, sometimes they start to thermoregulate at a reduced body temperature. An animal that requires a body temperature of at least 4°C to arouse, for example, might keep its temperature at that level even if the ambient temperature drops to 0 or −20°C, increasing its metabolism as the ambient temperature falls so as to offset the increasing cooling effect of the air.

Distribution and Natural History   Hibernation is known in many mammals, including hamsters, many ground squirrels, dormice, woodchucks, some bats, some monotremes, and some marsupials. Such animals typically store considerable quantities of body fat during the months preceding entry into hibernation and utilize this stored nutrient material over the winter. Hibernators arouse periodically (often frequently), and at such times they may void urine and feces and consume food they have stored in their burrow or den.

Estivation has received much less attention than hibernation, partly because it is not as easy to detect. It has been reported mostly in species of desert ground squirrels.

Daily torpor is found in a great many mammals and birds in both warm and cold situations. It occurs, for example, in numerous species of bats and rodents and in certain hummingbirds, swallows, swifts, and caprimulgids (nightjars, poorwills, nighthawks). A characteristic of daily torpor is that the animal is hypothermic for part of each day but maintains an elevated body temperature during the rest of the day. Feeding and other activities are carried out during the latter phase. When bats are undergoing daily torpor, they become hypothermic during daylight hours and emerge to forage at night; hummingbirds become torpid at night and feed in daylight.

Control of Timing and Duration The ability to enter controlled hypothermia seems clearly to have evolved several to many different times, in different evolutionary lines of mammals and birds. These polyphyletic origins are nowhere more evident than in the controls of timing and duration, for these controls can vary considerably from one taxonomic group to another. Furthermore, the controls are complex enough to make analysis difficult. Thus, many gaps remain in our understanding of them. All in all, generalizations about these controls do not come easily.

There is good evidence that hibernation is substantially under the control of an annual biological clock (p. 21) in certain animals, most notably the golden-mantled ground squirrel (*Spermophilus lateralis*). There are other hibernators, however, in which such a clock appears to be either absent or weak in its effects; in these, hibernation is evidently cued immediately by seasonal environmental changes (e.g., decreasing amounts of light per day). In some instances, environmental cueing of hibernation may be remote in time; witness the suggestion that in some mammals the warm days of *spring* set in motion a series of physiological events culminating in hibernation. One generality that emerges is that entry into hibernation is usually governed by factors operative well in advance of its occurrence (whether the factors be clocks, decreasing photoperiod, or whatever). This dependency on long-range controls enables animals to prepare extensively, as by fattening or caching food. It also reflects the role of hibernation as a preprogrammed defense against a predictably difficult season. Notably, although hibernation is an adaptation for energy conservation, its onset often depends little, if at all, on actual experience of energy hardship.

Sometimes, daily torpor seems to play a role similar to hibernation. In some species of mice, for example, daily torpor occurs in some individuals without actual experience of energy hardship; the incidence of this "spontaneous" torpor is markedly greater in winter than other seasons; and one factor precipitating the winter increase in torpor is shortening daylengths. More generally, however, daily torpor seems to be employed as a response to immediate

hardship. Often, animals undergo daily torpor only when suffering food shortage. The length of time spent in torpor per day may be increased as the food shortage becomes more severe.

Effects of daylength on hibernation or daily torpor evidently are mediated in at least some species by the pineal gland and its hormone melatonin. Secretion of melatonin is increased as days shorten.

CONCLUDING COMMENT

In ending this chapter, it is perhaps well to offer a rejoinder to the common belief that homeothermy is superior to poikilothermy. Although we may be led to this belief by our own homeothermy, this is hardly a valid basis for objective judgment. In nature's benign indifference (Camus' phrase), the only criterion of success is perpetuation of the species, involving the subsidiary elements of survival and reproduction at the individual level. The living world provides the incontrovertible evidence that, by this criterion, poikilothermy can be as successful a mode of life as homeothermy; many poikilothermic taxa have persisted for hundreds of millions of years. In fact, poikilothermy and homeothermy are simply extremes in a continuum of thermal relations that have been exploited in the evolution of different niches within which success is possible, and our scientific explorations into thermal biology are directed to understanding the various roles that the thermal relations may play in the lives of species. Although we may speak of advantages and disadvantages of various thermal relations, these are usually highly relative judgments made in hindsight. Homeothermy, for example, is not only an "advantage" but an essential for terrestrial animals that carry out their life functions unabated in the dead of winter in cold climates. Extending the argument a step further, however, we recognize that only some animals remain fully active in winter, and the others persist perfectly well. The types of thermal relations take on merits and detriments only in the context of particular niche specializations.

SELECTED READINGS

Aschoff, J. 1981. Thermal conductance in mammals and birds: Its dependence on body size and circadian phase. *Comp. Biochem. Physiol.* **69A**:611–619.

Bakken, G.S., W.R. Santee, and D.J. Erskine. 1985. Operative and standard operative temperature: Tools for thermal energetics studies. *Am. Zool.* **25**:933–943.

Bayne B.L., C.J. Bayne, T.C. Carefoot, and R.J. Thompson. 1976. The physiological ecology of *Mytilus californianus* Conrad. 1. Metabolism and energy balance. *Oecologia* **22**:211–228.

Bligh, J. 1973. *Temperature Regulation in Mammals and Other Vertebrates*. North-Holland, Amsterdam.

Bligh, J. 1978. Thermal regulation: What is regulated and how? In Y. Houdas and J.D. Guieu (eds.), *New Trends in Thermal Physiology*, pp. 1–10. Masson, New York.

*Brück, K. 1986. Basic mechanisms in thermal long-term and short-term adaptation. *J. Thermal Biol.* **11**:73–77.

Bullock, T.H. 1955. Compensation for temperature in the metabolism and activity of poikilotherms. *Biol. Rev.* **30**:311–342.

Burggren, W.W. and B.R. McMahon. 1981. Oxygen uptake during environmental temperature change in hermit crabs: Adaptation to subtidal, intertidal, and supratidal habitats. *Physiol. Zool.* **54**:325–333.

Carey, F.G. 1982. A brain heater in swordfish. *Science* **216**:1327–1329.

Chappell, M.A. 1980. Thermal energetics and thermoregulatory costs of small arctic mammals. *J. Mamm.* **61**:278–291.

*Clarke, A. 1983. Life in cold water: The physiological ecology of polar marine ectotherms. *Oceanogr. Mar. Biol. Annu. Rev.* **21**:341–453.

Cold-hardiness in poikilothermic animals. A collection of papers. 1982. *Comp. Biochem. Physiol.* **73A**:517–640.

Dawson, W.R. 1975. On the physiological significance of the preferred body temperatures of reptiles. In D.M. Gates and R.B. Schmerl (eds.), *Perspectives of Biophysical Ecology,* pp. 443–473. Springer-Verlag, New York.

*Dawson, W.R., R.L. Marsh, and M.E. Yacoe. 1983. Metabolic adjustments of small passerine birds for migration and cold. *Am. J. Physiol.* **245**:R755–R767.

DeVries, A.L. 1983. Antifreeze peptides and glycopeptides in cold-water fish. *Annu. Rev. Physiol.* **45**:245–260.

Dill, D.B. (ed.). 1964. *Handbook of Physiology. Section 4: Adaptation to the Environment.* American Physiological Society, Washington, DC.

Duman, J. and K. Horwath. 1983. The role of hemolymph proteins in the cold tolerance of insects. *Annu. Rev. Physiol.* **45**:261–270.

Gates, D.M. 1980. *Biophysical Ecology.* Springer-Verlag, New York.

*Girardier, L. and M.J. Stock (eds.). 1983. *Mammalian Thermogenesis.* Chapman and Hall, London.

*Graves, J.E. and G.N. Somero. 1982. Electrophoretic and functional enzymic evolution in four species of eastern Pacific barracudas from different thermal environments. *Evolution* **36**:97–106.

Greenleaf, J.E. 1979. Hyperthermia and exercise. *Int. Rev. Physiol.* **20**:157–208.

*Grout, B.W.W. and G.J. Morris (eds.). 1987. *The Effects of Low Temperatures on Biological Systems.* Edward Arnold, London.

Hagan, R.D. and S.M. Horvath. 1978. Effect of diurnal rhythm of body temperature on muscular work. *J. Thermal Biol.* **3**:235–239.

Hammel, H.T., F.T. Caldwell, Jr., and R.M. Abrams. 1967. Regulation of body temperature in the blue-tongued lizard. *Science* **156**:1260–1262.

*Heinrich, B. (ed.). 1981. *Insect Thermoregulation.* Wiley, New York.

Heinrich, B. 1987. Thermoregulation in winter moths. *Sci. Am.* **256**(3):104–111.

Heinrich, B. and T.P. Mommsen. 1985. Flight of winter moths near 0°C. *Science* **228**:177–179.

Heller, H.C., X.J. Musacchia, and L.C.H. Wang (eds.). 1986. *Living in the Cold.* Elsevier, New York.

Hellstrøm, B. and H.T. Hammel. 1967. Some characteristics of temperature regulation in the unanesthetized dog. *Am. J. Physiol.* **213**:547–556.

Hensel, H. 1981. *Thermoreception and Temperature Regulation.* Academic, New York.

Hill, R.W. and D.L. Beaver. 1982. Inertial thermostability and thermoregulation in broods of redwing blackbirds. *Physiol. Zool.* **55**:250–266.

Himms-Hagen, J. 1976. Cellular thermogenesis. *Annu. Rev. Physiol.* **38**:315–351.

*Hochachka, P.W. 1986. Defense strategies against hypoxia and hypothermia. *Science* **231**:234–241.

Hochachka, P.W. and M. Guppy. 1987. *Metabolic Arrest and the Control of Biological Time.* Harvard University Press, Cambridge.

*Hochachka, P.W. and G.N. Somero. 1984. *Biochemical Adaptation.* Princeton University Press, Princeton.

*Holeton, G.F. 1974. Metabolic cold adaptation of polar fish: Fact or artefact? *Physiol. Zool.* **47**:137–152.

Hulbert, A.J. 1987. Thyroid hormones, membranes, and the evolution of endothermy. In H. McLennan, J.R. Ledsome, C.H.S. McIntosh, and D.R. Jones (eds.), *Advances in Physiological Research,* pp. 305–319. Plenum, New York.

*Irving, L. 1972. *Arctic Life of Birds and Mammals.* Springer-Verlag, New York.

Johnsen, H.K., A.S. Blix, J.B. Mercer, and K.-D. Bolz. 1987. Selective cooling of the brain in reindeer. *Am. J. Physiol.* **253**:R848–R853.

Johnson, I.A. and P. Harrison. 1985. Contractile and metabolic characteristics of muscle fibers from antarctic fish. *J. Exp. Biol.* **116**:223–236.

Karasov, W.H. and J.M. Diamond. 1985. Digestive adaptations for fueling the cost of endothermy. *Science* **228**:202–204.

*Kluger, M.J. 1979. *Fever: Its Biology, Evolution, and Function.* Princeton University Press, Princeton.

*Lyman, C.P., J.S. Willis, A. Malan, and L.C.H. Wang. 1982. *Hibernation and Torpor in Mammals and Birds.* Academic, New York.

Manis, M.L. and D.L. Claussen. 1986. Environmental and genetic influences on the thermal physiology of *Rana sylvatica. J. Thermal Biol.* **11**:31–36.

McNab, B.K. and W. Auffenberg. 1976. The effect of large body size on the temperature regulation of the Komodo dragon, *Varanus komodoensis. Comp. Biochem. Physiol.* **55A**:345–350.

Murphy, D.J. 1983. Freezing resistance in intertidal invertebrates. *Annu. Rev. Physiol.* **45**:289–299.

Nadel, E.R. (ed.). 1977. *Problems with Temperature Regulation during Exercise.* Academic, New York.

Newell, R.C. 1979. *Biology of Intertidal Animals.* Marine Ecological Surveys Ltd., Faversham.

Nicholls, D. and R. Locke. 1983. Cellular mechanisms of heat dissipation. In L. Girardier and M.J. Stock (eds.), *Mammalian Thermogenesis,* pp. 8–49. Chapman and Hall, London.

Pough, F.H. and R.M. Andrews. 1984. Individual and sibling-group variation in metabolism of lizards: The aerobic capacity model for the origin of endothermy. *Comp. Biochem. Physiol.* **79A**:415–419.

Prosser, C.L. and D.O. Nelson. 1981. The role of nervous systems in temperature acclimation of poikilotherms. *Annu. Rev. Physiol.* **43**:281–300.

Rothwell, N.J. and M.J. Stock. 1985. Biological distribution and significance of brown adipose tissue. *Comp. Biochem. Physiol.* **82A**: 745–751.

Schmid, W.D. 1982. Survival of frogs in low temperature. *Science* **215**:697–698.

*Schmidt-Nielsen, K. 1964. *Desert Animals*. Oxford University Press, Oxford.

Segal, E. 1956. Microgeographic variation as thermal acclimation in an intertidal mollusc. *Biol. Bull.* **111**:129–152.

Smith, M.A.K. and A.E.V. Haschemeyer. 1980. Protein metabolism and cold adaptation in antarctic fishes. *Physiol. Zool.* **53**:373–382.

*Somero, G.N. 1978. Temperature adaptation of enzymes: Biological optimization through structure–function compromises. *Annu. Rev. Ecol. Syst.* **9**:1–29.

Stevens, E.D. and W.H. Neill. 1978. Body temperature relations of tunas, especially skipjack. In W.S. Hoar and D.J. Randall (eds.), *Fish Physiology,* Vol. 7, pp. 315–359. Academic, New York.

Taylor, C.R. 1977. Exercise and environmental heat loads: Different mechanisms for solving different problems? *Int. Rev. Physiol.* **15**:119–146.

Thiessen, D.D. 1983. Thermal constraints and influences on communication. *Adv. Study Behav.* **13**:147–189.

Trayhurn, P. and D.G. Nicholls (eds.). 1986. *Brown Adipose Tissue*. Edward Arnold, London.

Vogt, F.D. 1986. Thermoregulation in bumblebee colonies. I. Thermoregulatory versus brood-maintenance behaviors during acute changes in ambient temperature. *Physiol. Zool.* **59**:55–59.

Walsberg, G.E. 1983. Coat color and solar heat gain in animals. *BioScience* **33**:88–91.

Weathers, W.W. 1981. Physiological thermoregulation in heat-stressed birds: Consequences of body size. *Physiol. Zool.* **54**:345–361.

Werner, J. 1980. The concept of regulation for human body temperature. *J. Thermal Biol.* **5**:75–82.

*Whittow, G.C. (ed.). 1970–1973. *Comparative Physiology of Thermoregulation,* 3 vols. Academic, New York.

*Zachariassen, K.E. 1985. Physiology of cold tolerance in insects. *Physiol. Rev.* **65**:799–832.

See also references in Appendix A.

chapter 7

Exchanges of Salts and Water: Mechanisms

Animals are composed in large part of solutions. The universal biological solvent is water, and among the important solutes are a variety of ions such as sodium, potassium, calcium, chloride, sulfate, and phosphate. This and the following chapter are concerned with the exchanges of these vital materials between the organism and its outside world.

Animals occupy all the major environments of the earth—land, oceans, fresh waters, and the interfaces among them. The body fluids of all animals differ in salt and water composition from the environmental medium. The differences may be relatively small, as in many marine invertebrates. Or they may be large. The concentrations of the major physiological ions in the body fluids of freshwater animals are much higher than those in fresh water. The concentrations in marine bony fish are much lower than those in seawater. Terrestrial animals are not even surrounded with an aqueous solution. We see that in this dimension, as in so many others, animals are at disequilibrium with their surroundings. From experience and theory, we know that there are always forces tending to bring disequilibria to equilibrium. In particular, there are forces tending to bring animal body fluids to the same composition as the surrounding environmental medium. The body fluids of freshwater animals tend to be diluted. Those of marine bony fish tend to be concentrated. The water resources of terrestrial animals tend to be vaporized. The animal must combat these tendencies, must maintain the disequilibria, by active, energy-demanding intervention.

This chapter is a review of the mechanisms by which water and salts are exchanged. These mechanisms are so diverse that the chapter is perforce a hybrid one. Evaporation is a mechanism by which animals lose both water and heat. We discuss it first

because the physical laws of evaporation are equally relevant to the material we have just covered in Chapter 6 and the material we cover next in Chapter 8. After looking at the logical converse of evaporation—uptake of water vapor from the air—we examine a number of mechanisms by which water and ions are exchanged in the aqueous phase: diffusion, active transport, and osmosis. Then we look at aspects of food, drinking water, metabolism, and kidney function as they apply to water and salt balance. Chapter 8 will bring the principles mastered in this chapter to bear on an integrated treatment of salt–water balance in the major groups of animals.

EVAPORATION

Humidity and the Water Vapor Pressure of Air

The rate of evaporation of water from an animal into the air depends partly on the humidity of the air. Thus, a logical starting point for understanding the physical laws of evaporation is a look at how humidity is expressed.

The atmosphere consists of a mixture of gases: oxygen, nitrogen, argon, and—most importantly for our present perspective—water vapor. According to Dalton's law of partial pressures, the total pressure exerted by a mixture of gases is in fact the sum of the *partial pressures* exerted by the individual constituents of the mixture. The partial pressure exerted by water vapor is known as the *water vapor pressure* or simply the *vapor pressure*. To illustrate, suppose some air exerts a total pressure of 760 mm Hg* and

*Gas pressures, as well as blood pressures, can be expressed in a variety of units, and at present there is little unanimity among physiologists on the units to be used. The unit used preferentially in this text is the *millimeter of mercury,* abbreviated *mm Hg;* it is the pressure exerted by a column of mercury 1 mm high under

is composed, in molar terms, of 20 percent oxygen, 75 percent nitrogen, and 4 percent water vapor. Then the partial pressures exerted by oxygen and nitrogen would be, respectively, 20 percent and 75 percent of the total pressure, and the water vapor pressure would be 4 percent of the total pressure. That is, the water vapor pressure would be 0.04 × 760 mm Hg = 30 mm Hg (see also Chapter 11).

The water vapor pressure of air is one expression of the humidity, or water content, of the air. As the concentration of water vapor in a body of air is increased, the vapor pressure increases in proportion (assuming constant temperature).

The water vapor pressure can be raised only so high in air of a particular temperature. Air that has reached its maximum vapor pressure is said to be *saturated,* and its vapor pressure is termed the *saturation vapor pressure*. As shown in Table 7.1, the saturation vapor pressure depends strongly on temperature; in the physiological range, it approximately doubles for every 10–12°C increase in temperature. If a body of air has reached saturation with water vapor, it cannot hold any more water in the vapor state; so if water vapor is injected into such air from one source or another, the excess vapor promptly condenses out in the form of liquid water droplets (e.g., fog).

To illustrate the significance of these concepts, consider that humans and other terrestrial vertebrates humidify air in their lungs until it is approximately saturated at body temperature. If a person inhales saturated air at 20°C, warms and saturates it at body temperature (37°C), and breathes the air out saturated at 37°C, every liter of air has its vapor pressure raised from 17.5 mm Hg to 47.1 mm Hg between inhalation and exhalation—representing about a 2.5-fold increase in the water content of that air. Warming of respired air greatly promotes respiratory loss of water by raising the saturation vapor pressure of the air.

The humidity of air is sometimes expressed *relative* to the air's saturation vapor pressure (which depends strongly on the temperature of the air). One expression of this sort is the *saturation deficit,* which is the *difference* between the actual, prevailing vapor pressure and the saturation vapor pressure. Another relative expression of humidity is called, quite simply, the *relative humidity.* It is the *ratio* of the actual vapor pressure over the saturation vapor pressure. For instance, if some air at 18.6°C—for which the saturation pressure is 16 mm Hg—has an actual vapor pressure of 4 mm Hg, its saturation deficit is 16 − 4 = 12 mm Hg and its relative humidity is 4/16 = 25 percent.

Humidity can also be expressed in absolute terms,

Table 7.1 THE SATURATION WATER VAPOR PRESSURE AT SELECTED TEMPERATURES IN THE PHYSIOLOGICAL RANGE.

Temperature (°C)	Saturation vapor pressure[a] (mm Hg)
0	4.6
10	9.2
20	17.5
30	31.8
37	47.1
40	55.3

[a] The saturation pressure is independent of the composition of the air or other gas.

that is, in terms independent of saturation vapor pressure. One absolute expression of humidity is the *concentration* of water vapor (e.g., grams of water per liter of air), also called the *vapor density.* The prevailing *vapor pressure* is also an absolute expression of humidity and is in fact the most useful expression for the analysis of evaporation.

Within a body of air, water always diffuses from regions of high vapor pressure to regions of low vapor pressure. Suppose though that we are looking at an interface between air and liquid water. How can we predict the direction and rate of water movement between the gas and liquid phases? Again, water always tends to diffuse from higher to lower vapor pressure. To apply this important principle, we must first understand what the vapor pressure of a liquid is.

The Vapor Pressure of Aqueous Solutions

Any particular aqueous solution, if placed in contact with air in a closed system, will tend to establish a characteristic equilibrium vapor pressure in the air. That vapor pressure is termed the *vapor pressure of the solution.* The vapor pressure of pure water at a particular temperature is the same as the saturation vapor pressure of air at the same temperature. Thus, Table 7.1 can be used to look up the vapor pressure of pure water. A useful way to view the vapor pressure of a solution is that it is a measure of the *tendency of the solution to inject water vapor into the air.* Water at 30°C has a much greater tendency to inject water vapor into air than water at 10°C (Table 7.1).

Suppose we have a closed container into which we introduce pure water and dry air, and suppose we keep the air temperature at 20°C. Once the system is sealed, water will evaporate into the air, and the vapor pressure of the air will gradually rise. If the water itself is at 20°C, it will tend ultimately to establish a vapor pressure in the air of 17.5 mm Hg, which is identical to the saturation vapor pressure of the air; thus, when pure water and air are of the same temperature, an equilibrium is reached wherein the air is exactly saturated. Suppose instead that the water is at 10°C while the air remains at 20°C. Evaporation will then raise the vapor pressure in the air

standard gravitational acceleration. Another unit in common usage by physiologists is the *torr,* which is essentially identical to a millimeter of mercury. An *atmosphere* is 760 mm Hg, or 760 torr. The basic SI unit of pressure—the newton per square meter (N/m²), or *pascal* (Pa)—is just starting to be used widely by physiologists (1 kPa ≃ 7.5 mm Hg).

only to 9.2 mm Hg; the tendency of cool water to inject water vapor into air is insufficient to saturate warmer air. If the water is at 30°C, it will tend to inject water vapor into the air until a vapor pressure of 31.8 mm Hg is established in the air, but the air, being at 20°C, will hold only 17.5 mm Hg. Once the cool air in contact with warmer water has become saturated, water vapor will continue to enter it from the water but will quickly condense out. This phenomenon is at work when condensation occurs above a hot cup of coffee or when a fog is formed over a warm lake early on a summer morning.

The vapor pressure of an aqueous solution is dependent on not only its temperature but also its solute concentration (see p. 148 for a refined discussion of what is meant here by concentration). Raising the concentration of salts or other solutes tends to depress the vapor pressure. This effect is relatively small at the concentrations of most animal body fluids; even body fluids as concentrated as seawater have vapor pressures only about 2 percent lower than pure water. Sometimes, the effect of solutes can be of greater consequence. For instance, if salt left behind from evaporated sweat were allowed to accumulate indefinitely on the skin of a sweating person so that sweat newly secreted onto the skin surface became highly concentrated with salt, the vapor pressure of the sweat could be depressed by as much as 25 percent—markedly reducing its tendency to inject water vapor (and void heat) into the air.

Physical Laws of Evaporation

At the interface of an aqueous solution and air, water moves in net fashion from higher to lower vapor pressure. Thus, condensation occurs if the prevailing vapor pressure of the air exceeds the vapor pressure of the solution. Our chief concern here is with evaporation. Evaporation occurs if the vapor pressure of the solution exceeds the prevailing vapor pressure in the air. Furthermore, the rate of evaporation is proportional to the difference between the two vapor pressures.

Symbolically, if VP_s is the vapor pressure of an animal body fluid from which evaporation is occurring, VP_a is the prevailing vapor pressure in the air, and E is the rate of evaporative water loss per unit of body surface area

$$E = h_e (VP_s - VP_a)$$

where h_e is a constant for a given animal under given conditions. Other things being equal, lowering the ambient vapor pressure hastens evaporation; as we know, people are more effectively cooled by sweat when the ambient humidity is low. Evaporative water loss is also hastened if the vapor pressure of the body fluids is raised; thus, other things being equal, animals with high body temperatures lose water more rapidly than those with low body temperatures.

One important determinant of the coefficient h_e in the above equation is the permeability to water of the skin or other membrane that separates the body fluids from the air. The skin of amphibians, for example, is generally more permeable than that of reptiles. Thus, amphibians have a higher h_e than reptiles and experience more-rapid cutaneous evaporation at any given vapor pressure difference ($VP_s - VP_a$).

The rate of evaporation from an animal body surface also depends on the rate of air movement across the surface. If we consider an animal losing water evaporatively across its skin, the evaporation itself will tend to raise the vapor pressure of the air next to the skin if the air is still; this elevation of VP_a will decrease the vapor pressure difference ($VP_s - VP_a$) across the skin and thus impede evaporation. A wind will blow water-laden air away from the skin surface, replacing it with drier air from the atmosphere at large, thereby preventing the diminution of ($VP_s - VP_a$) and facilitating evaporative water loss. In physical terms, this is why a breeze is so welcome when we are sweating.

UPTAKE OF WATER FROM THE AIR

When in thermal steady state, most animals cannot gain water from the air even when the air is *saturated* with water vapor, for reasons that seem clear from the principles developed in the preceding section. If frogs, for example, are placed in saturated or near-saturated air, evaporative loss of metabolic heat is so impeded that their body temperatures rise appreciably above air temperature. This elevation of body-fluid temperature causes the body-fluid vapor pressure to be greater than the vapor pressure of the air, even though the air may be saturated and even taking into account the effects of solutes in the body fluids. Thus, the gradient of vapor pressure favors *evaporative loss* of water from frogs in saturated air, just as it favors loss in subsaturated air. Similar arguments apply to many other poikilotherms and, of course, to homeotherms. Elevation of body temperature above air temperature preempts the possibility of a vapor-pressure gradient favoring water uptake.

If an animal could be cooler than the air, the vapor-pressure gradient might favor uptake of water, just as water is drawn out of the air to condense on the sides of an iced-tea glass on a humid summer day. In the case of the iced-tea glass, droplets of water on the side of the glass—because of their low temperature—have a vapor pressure below the vapor pressure of the air; thus, water moves from the air into the droplets, and the droplets grow. Water has been observed to condense on the skin of lizards emerging from cool burrows into warm air, and possibly the lizards lick or absorb the condensed water. Note though that this is a transitory situation, for the warm air and the condensation itself will warm the lizards and ultimately reverse the vapor-pressure gradient to favor evaporation. Calculations suggest that frogs might experience *steady-state* condensation of water onto their skin when cooled by radiant loss of heat to clear skies (p. 81) on humid nights.

Some arthropods—including certain insects, ticks, and mites—are remarkable for their ability to gain body water persistently from the air even though the vapor pressure of their blood and general body fluids is *greater* than that of the air from which the water is being absorbed. A desert cockroach, *Arenivaga investigata*, for example, can gain water when at steady state in a thermally uniform laboratory environment even when the ambient relative humidity is as low as 79–83 percent. Mealworms (larval *Tenebrio*) can gain water down to 88 percent relative humidity, and firebrats (*Thermobia*) down to 45 percent. Dehydrated arthropods capable of this type of water uptake gain water from the air steadily until fully rehydrated; the process is not merely transitory. At high ambient humidities, gain of about 10 percent of body weight per day is the rule. Much recent research has focused on the location of uptake. *Arenivaga* and a number of ixodid ticks, it turns out, take up water at the mouth. Mealworms and firebrats, on the other hand, absorb water rectally. The mechanisms by which these animals absorb water even though their blood vapor pressures exceed ambient vapor pressures are often unclear and are currently subject to debate. Some investigators believe that true, primary active transport of water sometimes occurs (p. 141). Others disagree, arguing—for example—that the animals produce—at their mouths or rectums—*localized* pockets of body fluids sufficiently concentrated with solutes to have vapor pressures below ambient vapor pressures. Such pockets are known to occur in the rectal walls of some insects, where they are more usually thought of as helping to concentrate and dry the excrement (p. 240).

PASSIVE SOLUTE MOVEMENTS

With this section we turn to several mechanisms of solute and water movement across membranes when both sides of the membrane are bathed by aqueous solutions. These mechanisms apply to direct transfers between the animal and environment in aquatic organisms and are ubiquitously involved in solute and water movements across membranes within the animal.

Solutes move across biological membranes by many mechanisms. Various classifications of the mechanisms have been proposed, but none has received universal acceptance. We shall subdivide the mechanisms into *passive* and *active,* and in this section we focus in particular on passive solute transfer. The distinction between passive and active solute movements will be elaborated (p. 141) after some background information has been developed.

The ionic relations between a freshwater fish and its environment are diagrammed in Figure 7.1. In all aquatic animals, the internal body fluids are separated from the environmental fluids by a cellular

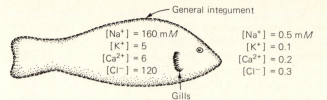

$[Na^+] = 160 \, mM$ $[Na^+] = 0.5 \, mM$
$[K^+] = 5$ $[K^+] = 0.1$
$[Ca^{2+}] = 6$ $[Ca^{2+}] = 0.2$
$[Cl^-] = 120$ $[Cl^-] = 0.3$

Figure 7.1 A schematic representation of a freshwater fish showing, inside the fish, the concentrations (mM) of certain ions that might prevail in the blood plasma and, outside the fish, the concentrations (mM) that might prevail in the environment.

body covering. Also, in all animals that have been studied, at least some solutes differ in concentration between the inside and outside of the organism, as in the fish. Some solutes are able to pass through the body covering, and to them the covering is said to be *permeable*.

When there is a concentration difference for a particular permeating solute across the body covering of an animal, the solute tends to diffuse in net fashion from its area of greater concentration to its area of lesser concentration. In the example of the freshwater fish in Figure 7.1, concentration differences tend to cause all four ions to diffuse from the body fluids into the surrounding water; to the extent such diffusion occurs, it dilutes the body fluids, thereby presenting the fish with a challenge to homeostasis. We shall later look in detail at this challenge and how it is met. For the present, let us examine more thoroughly the process of diffusion. It is a form of passive solute movement.

Simple Diffusion in Response to a Concentration Gradient

Consider a membrane that is permeable to glucose and that separates two solutions of differing glucose concentration, as in Figure 7.2. Glucose will diffuse in net fashion from the side of higher concentration to the side of lower concentration until the concentration on both sides is equal, that is, until a state of ***concentration equilibrium*** is reached.

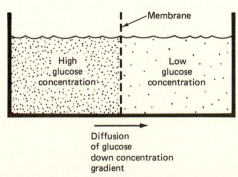

Figure 7.2 A vessel divided into two halves by a glucose-permeable membrane and filled with solutions differing in glucose concentration.

The motive force for this diffusion is the continuous random movement displayed by all atoms and molecules (molecular kinetic energy). When a concentration difference exists for a solute, random molecular motions tend to carry more molecules of the solute away from areas of high concentration than are carried into such areas. Thus, at a macroscopic level, a net flux of solute molecules away from the region of high concentration is observed. When the concentration becomes equal on both sides of a membrane, random molecular motions continue to carry solute molecules across the membrane in both directions, but the fluxes in both directions are equal and the concentrations on the two sides of the membrane therefore do not change any further. *Diffusion* may be defined, very simply, as a movement of material caused by the inexorable random motions of atoms and molecules.

The basic laws of solute diffusion in response to a concentration difference will seem familiar because they are conceptually similar to the laws of thermal diffusion discussed in Chapter 6. Let M represent the rate of net movement of a solute per unit of cross-sectional area from a region of high concentration C_1 to a region of low concentration C_2. Then

$$M = D \frac{C_1 - C_2}{X}$$

where X is the distance between C_1 and C_2 and D is a proportionality factor termed the diffusion coefficient. The change in concentration per unit distance $[(C_1 - C_2)/X]$ is called the **concentration gradient**. Note that the rate at which solute diffuses from the region of concentration C_1 to the region of concentration C_2 increases as the difference between the concentrations $(C_1 - C_2)$ increases. The rate also increases as the distance separating the two concentrations (X) decreases; diffusion is a notoriously slow process for moving substances from one place to another in the macroscopic world, but when only the thickness of a cell membrane separates regions of differing concentration, diffusion can transfer substances rapidly. One factor affecting the diffusion coefficient D is the ease with which solute moves through the medium; when diffusion is occurring across a membrane, this factor is the **permeability** of the membrane to the solute in question. Another influence on D is the temperature. Higher temperatures mean higher molecular kinetic energies and thus more-rapid diffusion (Q_{10} is typically less than 1.5 but can be as great as 2 or 3).

Convective flow over the surfaces of a cell or animal can affect rates of diffusion. Suppose the concentration of a permeating solute is 100 mM in the body fluids of an animal and 2 mM in the ambient water. Figure 7.3 illustrates that outward diffusion of the solute will tend to create a boundary layer of elevated solute concentration next to the animal's body surface, in effect increasing X, the distance between the two extreme concentrations, and thus

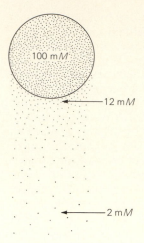

Figure 7.3 Diffusional concentration of a permeating solute in a boundary layer next to an organism, presented schematically. The boundary layer may be very thin yet still have a significant impact on the rate of diffusion.

decreasing the rate of diffusion. Increasing the flow of water over the animal will tend to carry solute away, decrease the thickness of the boundary layer, and increase the rate of diffusive solute loss from the body fluids. Investigators have debated for years whether the mucous films secreted by many aquatic animals might act to trap boundary layers and thus slow diffusional solute exchange.

Electrical Effects in Simple Diffusion; the Electrochemical Gradient

If there should be a difference of electrical potential across a membrane, it will affect the diffusion of all electrically charged solutes (e.g., ions). Positively charged solutes will be attracted toward the negative side of the membrane, and negatively charged solutes will be repelled from that side. There are thus two major influences on simple diffusion: differences of solute concentration and differences of electrical potential.

The electrical difference across a membrane sometimes promotes diffusion of a solute in the same direction as the concentration difference for the solute, thereby causing diffusion to proceed faster than accounted for by the concentration gradient alone. On the other hand, as illustrated in Figure 7.4, the concentration and electrical gradients may act in opposite directions. If the electrical effect in the figure is small relative to the concentration effect, Na$^+$ ions will diffuse in net fashion from left to right—as dictated by the concentration difference—but slower than if the electrical difference were absent. On the other hand, if the electrical effect is larger than the concentration effect, Na$^+$ ions will diffuse in net fashion from right to left—as dictated by the electrical difference—even though that direction of diffusion is opposite to what the concentration gradient would indictate. *A large electrical difference may*

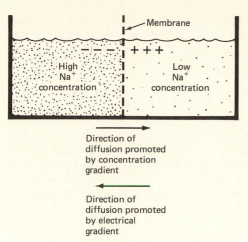

Direction of
diffusion promoted
by concentration
gradient

Direction of
diffusion promoted
by electrical
gradient

Figure 7.4 A vessel divided into two halves by a sodium-permeable membrane. The membrane is charged, such that the right side is positive with respect to the left.

cause solute to move in such a direction as to increase a concentration difference.

How do we decide, in a particular situation, whether the electrical or the concentration effect is "larger"? The *Nernst equation,* described in Chapter 16 (p. 398), enables the calculation of the electrical potential difference that is exactly equal in its effect to any given concentration difference. Furthermore, equations that are close relatives of the Nernst equation can be used to compute a value called the *electrochemical gradient* that expresses quantitatively the *net* effect on diffusion of both concentration and electrical differences taken together.

Passive Solute Movements Defined

Passive solute movements are those that can be accounted for by the differences of concentration and electrical potential existing across a membrane. Net passive flux of solute is in the direction of the electrochemical gradient, and net passive flux in either direction ceases once electrochemical equilibrium is reached. If the solute in question is not charged, electrochemical equilibrium is identical to concentration equilibrium (equal dissolved concentration throughout the system). If the solute is charged, electrochemical equilibrium prevails when the concentration difference and potential difference across a membrane are related as in the Nernst equation (i.e., equal but opposite in their effects).

Facilitated Diffusion

When solutes cross living membranes passively, they sometimes act as if the process of traversing the membrane is fundamentally the same as the process by which molecules of sugar might diffuse to uniform concentration within a beaker of water. That is, crossing the membrane is a case of simple diffusion, as implied throughout the preceding discussion. Re-

search has indicated, however, that passive movement of solutes across living membranes commonly is aided by the association of the solutes with specific structural constituents of the membrane, a process called *facilitated diffusion.* One postulate is that during facilitated diffusion, solute molecules combine with carrier molecules in the membrane and cross the membrane in the form of solute–carrier complexes.

The kinetics of crossing a membrane by facilitated diffusion differ from those of simple diffusion (exhibiting saturation, for instance), and the rate of crossing the membrane is enhanced ("facilitated"). Facilitated diffusion is a passive type of solute movement, nonetheless, because its direction can always be accounted for by the differences of concentration and electrical potential existing across the membrane, and it leads toward electrochemical equilibrium.

ACTIVE TRANSPORT

Animals can transport many solutes across membranes against their electrochemical gradients by using metabolic energy. Mechanisms capable of such transport are called *active-transport mechanisms* or *pumps.* Pumps do not necessarily move solute against the electrochemical gradient, but they are capable of doing so.

We have seen in the case of the freshwater fish (Figure 7.1) that Na^+ and Cl^- are so much more concentrated in the fish's body fluids than in the ambient water that they tend passively to diffuse out of the fish, just as a rock tends passively to roll downhill. One major mechanism by which fish replace the ions lost by diffusion is by actively transporting Na^+ and Cl^- from the dilute water into their concentrated body fluids. Many fish, for instance, can pump chloride ions from waters as dilute as 0.02–0.03 mM Cl^- into body fluids containing over 100 mM Cl^-. Such inward pumping is analogous to making a rock roll uphill in that it requires energy, which is supplied from ATP. Active transport is sometimes called "uphill transport."

Although active-transport mechanisms have evolved for a great variety of solutes—including inorganic ions, amino acids, and sugars—there has yet been no conclusive demonstration of pumps for either of two important commodities: water and oxygen. Interest in the possibility of active water transport remains high, and there are certain situations where it is regarded by some investigators as an attractive hypothesis (e.g., uptake of water vapor from the atmosphere by certain arthropods).

Properties of Pumps and Pumping

Active-transport mechanisms are always highly specific for certain solutes; a particular mechanism is likely to transport no more than one or two ordinary physiological solutes. The rate of pumping is sensitive to temperature, Q_{10} values of 2–3 being common. Pumping is also sensitive to factors affecting the cell's supply of ATP and other energy intermediates. Poisons and inhibitors of catabolic metabolism interfere with active transport, and oxygen deprivation of tissues often decreases rates of active transport.

The molecular mechanism of active transport has been subject to intense research for several decades but has proved to be one of the more recalcitrant problems in cellular physiology and is not as yet fully understood. Progress has been notable, however, to the point that some components of some pumps have been purified chemically. Important among these is an enzyme called sodium-, potassium-activated ATPase, or—for short—*Na*$^+$*–K*$^+$ *ATPase*. Most animal cells pump Na$^+$ outward across the cell membrane and K$^+$ inward, thus maintaining a relatively low internal concentration of Na$^+$ and a relatively high internal concentration of K$^+$ (see pp. 396 and 399). Na$^+$–K$^+$ ATPase is intimately involved. It cleaves ATP for release of energy and, for each ATP, moves, in some manner, three Na$^+$ ions outward and two K$^+$ ions inward.

Solutes being pumped across a membrane associate with solute-specific structural elements in the membrane. Some theories postulate energy-driven carrier molecules, which pick up solute on one side of the membrane and carry it through to be released on the other side; others postulate energy-driven gates. Because transport sites—whether carrier molecules or gates—are necessarily limited in number, active transport exhibits saturation kinetics. As illustrated in Figure 7.5, for example, the rate at which a crayfish actively absorbs Na$^+$ from its freshwater environment increases with the environmental Na$^+$ concentration only up to a certain concentration. At *very* low environmental Na$^+$ concentrations, all transport sites are not functioning at any one time because of limited availability of Na$^+$. As the Na$^+$ concentration is raised, the sites are brought more fully into play and the rate of transport accordingly accelerates. At what is still quite a low Na$^+$ concentration, however, Na$^+$ becomes sufficiently available

that all sites function at their peak transport rate all the time; the system is then *saturated,* and further elevation of Na$^+$ concentration does not accelerate transport.

Tandem Ion Movements

Characteristically, active transport of an ion across a membrane in one direction is accompanied by movement of an ion of the opposite charge in the same direction or by movement of an ion of the same charge in the opposite direction. Such tandem ion movements preclude the unbridled magnification of differences of electrical potential across the membrane.

Both of a pair of tandem ion movements may be active. This is thought to be the case in the uptake of Na$^+$ and Cl$^-$ by the gills of freshwater fish and crayfish, for instance, as illustrated in Figure 7.6. There are two independent transport mechanisms, one for Na$^+$ and one for Cl$^-$. Each mechanism extrudes ions of the same charge as the ions it is taking up from the water. The chloride pump exchanges bicarbonate ions (HCO$_3^-$) for Cl$^-$ ions; the bicarbonate ions are formed from metabolically produced carbon dioxide and represent a metabolic waste product. The sodium–transport mechanism exchanges either hydrogen (H$^+$) or ammonium (NH$_4^+$) ions for Na$^+$ ions; the hydrogen ions are formed when bicarbonate is formed and may react with the nitrogenous waste product ammonia (NH$_3$) to form ammonium ions. In total, the gills are not only actively absorbing Na$^+$ and Cl$^-$ but also extruding waste ions.

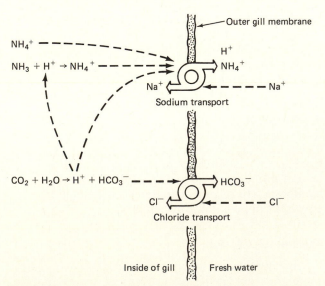

Figure 7.6 Current view of ion transport across the gills of freshwater fish and crayfish. This diagram depicts the phenomenology of transport but omits details at the cellular level. Similar mechanisms are seen in other freshwater animals: frogs, clams, and worms.

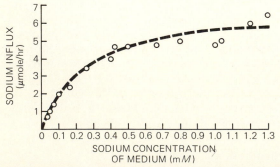

Figure 7.5 Rate of sodium influx in a sodium-depleted crayfish (*Astacus pallipes*) as a function of environmental sodium concentration at 12–13°C [From J. Shaw, *J. Exp. Biol.* **36**:126–144 (1959).]

Sometimes, movements of ions in tandem are not both active; instead, one of a pair of tandem movements is active and the other is passive. Suppose, for example, that the Na^+ concentration is lower on the outside of a membrane than on the inside and the membrane is pumping Na^+ (exclusively) from outside to inside. The active influx of the positive Na^+ ions will tend to render the inside of the membrane positive with respect to the outside. This electrical potential difference created by the Na^+ pump may become large enough to cause inward *diffusion* of Cl^- ions even if Cl^-, like Na^+, is more dilute on the outside than inside. The Cl^- ions would then be said to *follow passively* the active movement of Na^+ ions. The result of such a situation would be that both Na^+ and Cl^- would move from the side of the membrane where their concentrations are lower to the side where they are higher even though the membrane, in a strict sense, is actively pumping only Na^+. A pump, such as the Na^+ pump in this example, that creates a difference of electrical potential across a membrane is termed an **electrogenic pump** (p. 400).

When Is Transport Truly Active?

The essence of an *active* transport mechanism is that it is capable of transporting a solute against its electrochemical gradient, that is, in the direction opposite to that dictated by concentration and electrical gradients. Obviously, such transport demands an input of free energy. There has traditionally been an inclination to stipulate that the immediate source of this energy must be ATP or some other chemical intermediate in the cell's apparatus for releasing energy from carbon compounds.

Consider now the case of a carrier molecule that is *not* energy driven but that facilitates the *linked* movement of two solutes, A and B, in opposite directions across a membrane. Suppose that solute A has been concentrated on the inside of the membrane by an active-transport mechanism using ATP. Solute A will then tend to diffuse across the membrane from inside to outside in response to its concentration gradient. Importantly, because the movements of A and B are linked by the carrier molecule, the diffusion of A from inside to outside may induce movement of B from outside to inside *even if this movement of B is against B's electrochemical gradient.* Would such movement of B be "active" transport or not? Some taxonomists of membrane phenomena would view B's movement as an unorthodox case of facilitated diffusion. Others emphasize that B's movement is energy driven (the energy is supplied by the diffusion of A) and that the source of the energy is only a short step removed from ATP (ATP energy was used to establish the concentration gradient of A). They thus term B's movement *secondary active transport* ("secondary" because the immediate source of energy is the potential energy inherent in A's concentration gradient rather than ATP or another source of chemical-bond energy).*

While recognizing the importance and fascination of these concerns, whole-animal physiologists usually emphasize phenomenology, calling a transport mechanism simply "active" if it can move solute against the electrochemical gradient using metabolic energy, whether directly or indirectly. In this book, we usually take that approach.

ADDED ASPECTS OF SOLUTE MOVEMENT

Whereas active transport is one of the principal means by which animals manipulate solute concentrations in their body fluids, the active involvement of animals in solute exchange is not limited just to active transport, and in this sense the word *passive* is unfortunate as applied to many solute movements. The rates of passive, diffusive movements depend, for example, on the permeability properties of membranes, and because animals construct and maintain membranes with particular properties, they exert "active" influences on these "passive" movements. Another good example of how animals can be actively involved in passive solute movements was discussed earlier: An ion may be caused to diffuse against its concentration gradient without any assistance of carriers by following an electrical gradient set up by the active transport of another ion.

The maintenance of a concentration difference across a membrane does not always signify even an indirect involvement of active transport. Animals have in their body fluids a variety of high-molecular-weight solutes, such as respiratory pigments and other proteins, to which the body covering is typically impermeable. Concentration gradients between the organism and outside world for these substances are established and persist because the organism maintains the impermeable body covering and manufactures the solutes within the covering, both of these being energy-requiring activities.

Donnan Equilibria

Mention of nonpermeating solutes, such as proteins, sets the stage for an analysis of the diffusive phenomenon known as **Donnan equilibrium.** This type of equilibrium is a mechanism that very commonly accounts for "passive" maintenance of concentration gradients of permeating charged solutes.

Molecules of proteins and other nonpermeating solutes are often charged, and it frequently happens that the *net* charge of all such solutes in a body fluid

---

*Lest this issue seem remote, linked movements of two solutes of this type are believed to be common. Active uptake of sugars and amino acids in the mammalian intestine seems, for instance, to be of the secondary type, driven by Na^+ diffusion, the gradient of Na^+ being maintained by primary active transport.

is not zero. To see the implications for the distributions of charged permeating solutes, let us turn to the simplified example shown in Figure 7.7. At the top we have two fluid compartments separated by a membrane that is permeable to Na^+ and Cl^- but impermeable to protein. The membrane separates two equimolar solutions (see "initial concentrations"), one composed of Na^+Cl^- and the other of Na^+ and protein$^-$ (the protein is presumed to be monovalently anionic). The situation on side 1 simulates that already described for body fluids in that the net charge on the nonpermeating solutes (protein) is not zero but negative. Now the system as shown at the top of the figure is not at equilibrium. The concentrations of protein and Cl^- are strongly unequal on the two sides of the membrane. However, diffusion cannot distribute both of these solutes evenly on both sides because the protein cannot cross to side 2. Thus, the equilibrium that develops is not one in which all solutes are equal in concentration on both sides but, instead, is a Donnan equilibrium.

We can best understand the Donnan equilibrium by analyzing its development. Specifically, we consider how the system at the top of Figure 7.7 will behave as it moves toward equilibrium. Cl^-, being able to permeate the membrane and initially being far more concentrated on side 2 than side 1, will tend to diffuse toward side 1. This movement of Cl^- in response to its concentration gradient will cause side

1 to become negatively charged relative to side 2, thus establishing an electrical potential gradient across the membrane. This gradient will attract Na^+ to side 1. Significantly, the movement of Na^+ to side 1 in response to the electrical gradient will create a concentration gradient for Na^+, with the concentration on side 1 becoming increasingly greater than that on side 2. As the Na^+ concentration gradient increases, the tendency for Na^+ to return to side 2 in response to its concentration gradient will increase. Ultimately, this tendency will become great enough to oppose exactly the tendency for Na^+ to move toward side 1 in response to the electrical gradient established by Cl^- movement. Net Na^+ movement will then cease. Net Cl^- movement will also cease because the accumulation on side 1 of Cl^- ions that are not neutralized by accompanying Na^+ ions will establish an electrical gradient (side 1 negative) large enough to oppose fully the movement of Cl^- in response to the Cl^- concentration gradient. At this point, the system will have arrived at a stable Donnan equilibrium, with concentrations on either side of the membrane being as shown at the bottom of Figure 7.7. The concentrations of Na^+ and Cl^- at equilibrium are described by the equation

$$\frac{[Na^+]_1}{[Na^+]_2} = \frac{[Cl^-]_2}{[Cl^-]_1}$$

The essence of the Donnan equilibrium is that *the presence of a nonpermeating charged solute on one side of a membrane*—because it constitutes a reservoir of electrical charge that cannot be distributed equally on the two sides of the membrane—*results in an equilibrium in which each permeating charged solute is more concentrated on one side than the other.* There is also a difference of electrical potential across the membrane at equilibrium, with the side containing the nonpermeating solute having the charge of that solute. Furthermore, at equilibrium, the total molar concentration of all solutes taken together is greater on the side with the nonpermeating solute, meaning that the osmotic pressure is greater on that side; this effect is evident from the "final" concentration histograms in Figure 7.7.

Because nonpermeating charged solutes such as proteins are ubiquitous in living systems, the suite of Donnan effects—involving solute concentrations, osmotic pressures, and transmembrane differences of electrical charge—must be considered in many circumstances.

OSMOSIS

Consider a membrane separating two glucose solutions of different concentrations. If the membrane is permeable to glucose, we have seen that glucose will diffuse from the side with the higher glucose concentration to the side with the lower glucose concentration. It is also true that if the membrane is permeable

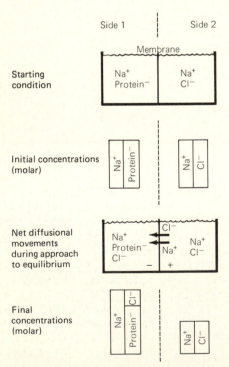

Figure 7.7 Development of a simple Donnan equilibrium. The membrane is permeable to Na^+ and Cl^- but is impermeable to the protein solute. The height of the concentration histograms is proportional to concentration. See text for explanation.

to water, water will move passively from the side with the higher water concentration (lower glucose concentration) to the side with the lower water concentration (higher glucose concentration). Passive movement of water across a membrane is termed *osmosis*. Although osmosis is indeed passive, it is not synonymous with diffusion of water because, for reasons as yet not fully clear, the rate of water movement by osmosis often exceeds that accounted for by principles of diffusion.

Referring back to the freshwater fish in Figure 7.1, we see that because solutes are more concentrated in the animal's body fluids than in the environment, water will tend to enter the fish by osmosis. Accordingly, the fish faces a dual problem of diffusional loss of solutes to its environment and osmotic uptake of water. Both of these processes tend to dilute its body fluids.

The Concept of Osmotic Pressure; Direct Measurement

In the basic analysis of osmosis, it is often useful to stipulate a membrane that is permeable only to water because confounding effects of solute movement across the membrane are thereby eliminated from consideration. Such a membrane is described as *semipermeable*. Although there are artificial semipermeable membranes, biological membranes are always permeable to at least some solutes. Thus, after seeing what is to be learned from analysis of water movements across semipermeable membranes, we shall have to return to the complexities of simultaneous solute and water movements.

Suppose we have two solutions, *A* and *B,* and we wish to predict the direction of osmosis between them when they are separated by a semipermeable membrane. What measurement could we make on each solution independently so as to make this prediction? Suppose we separate solution *A* from pure water by a semipermeable membrane and measure the *tendency of water to enter* solution *A* by osmosis. Suppose we also separate solution *B* from pure water by a semipermeable membrane and measure the tendency of water to enter *B*. If it turned out that water tended to enter *B* more strongly than it tended to enter *A,* we would expect that when we separated *A* and *B* by a semipermeable membrane, the direction of osmosis would be from *A* to *B* (Figure 7.8).

How are we to measure these tendencies for water entry we have just discussed? Suppose we separate a solution—it could be *A* or *B*—from pure water by a semipermeable membrane using the apparatus in Figure 7.9, where the membrane is mounted as the end of an ideal, frictionless piston within a cylinder. If the piston is free to move, water will travel by osmosis from the side filled with pure water to the side filled with solution, and the resulting changes in volume of the two fluid compartments will cause the piston to move to the right as in Figure 7.9*B*. Now

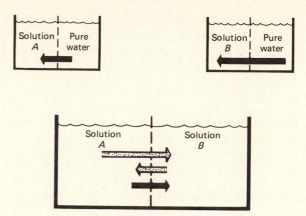

Figure 7.8 Dark arrows symbolize by their length the tendency for water to cross the membrane by osmosis. When solutions *A* and *B* are each separated from pure water, the tendency for water entry is greater for *B*. Thus, when *A* and *B* are separated from each other, water moves toward *B*. (Light arrows are repeated from upper parts; dark arrow is the difference between them.)

if we return to the starting condition and exert a force on the piston as in Figure 7.9*C*, an increased hydrostatic pressure will be produced in the solution compartment. The difference in hydrostatic pressure induced between the two compartments will tend to drive water molecules through the pores of the membrane by ultrafiltration in the direction opposite to

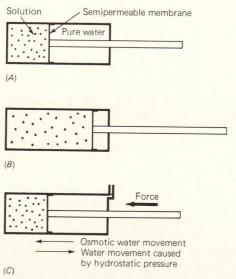

Figure 7.9 (*A*) A cylinder containing an aqueous solution separated from pure water by a semipermeable membrane. The membrane is mounted as the end of an ideal, frictionless piston. (*B*) A later condition of the system if the piston is left free to move. (*C*) Stable condition of the system if the increased hydrostatic pressure produced in the solution compartment by force on the piston is equal to the osmotic pressure of the solution. The open side arm on the pure-water compartment permits hydrostatic pressure in that compartment to remain constant as the hydrostatic pressure in the solution compartment is increased.

osmotic water movement. Clearly, we can adjust the hydrostatic pressure so that there will be no net movement of water across the membrane, that is, so that osmotic movements will be exactly counterbalanced by movements resulting from the hydrostatic pressure difference. The difference of hydrostatic pressure across the membrane required to achieve this condition is termed the osmotic pressure of the solution. Succinctly, the *osmotic pressure of a solution is the difference of hydrostatic pressure that must be created between that solution and pure water to prevent exactly any net osmotic movement of water when the solution and pure water are separated by a semipermeable membrane.*

The osmotic pressure of a solution is a measure of the tendency of water to enter the solution by osmosis from pure water. The greater is this tendency, the greater is the difference of hydrostatic pressure needed to oppose it. Returning to solutions A and B discussed earlier (Figure 7.8), if solution A were a 0.5-M glucose solution and were placed in the apparatus of Figure 7.9, a difference in hydrostatic pressure of about 11.2 atmospheres (atm) would be required to oppose exactly osmosis into the solution. If solution B were a 1.0-M glucose solution, its measured osmotic pressure would be about 22.4 atm. From these figures, we would know that water tends to enter B more readily by osmosis. Thus, if A and B were separated by a semipermeable membrane, the direction of osmosis would be from A to B—that is, *from lower to higher osmotic pressure.*

Solutions by Themselves Do Not Exert Hydrostatic Pressures as a Consequence of Their Osmotic Pressures

We have seen that the osmotic pressure of a solution is measured as a difference of hydrostatic pressure and can be expressed in units of hydrostatic pressure such as atmospheres. It is vital to recognize, however, that in this context hydrostatic pressure is being used simply as a measurement device: a means of assessing the tendency of a solution to take on water osmotically. Solutions by themselves do not exert hydrostatic pressures because of their osmotic pressures. As noted earlier, the osmotic pressure of a 1.0-M glucose solution is about 22.4 atm—a prodigious figure. If a 1.0-M glucose solution actually exerted a hydrostatic pressure of 22.4 atm, it would explode glass bottles. In fact, the hydrostatic pressure within a bottle of glucose solution differs only negligibly from that in a like bottle of pure water.

There are circumstances when hydrostatic pressures are generated by osmotic pressures. These circumstances involve *interaction between solutions.* If two solutions of differing osmotic pressure are separated by a water-permeable membrane and there is resistance to expansion of the solution into which water moves by osmosis, an increased hydrostatic pressure will be produced in that solution. To illus-

trate, suppose we have a rigid, completely sealed container divided into two compartments by a rigid semipermeable membrane and we place a 1.0-M glucose solution in one compartment and pure water in the other. Osmosis into the glucose solution will raise the hydrostatic pressure there because it entails the addition of matter (water) to the inexpansible compartment that contains the glucose solution. Indeed, the hydrostatic pressure in the glucose compartment will continue to rise until it exceeds that in the water compartment by about 22.4 atm because only then will the hydrostatic pressure difference prevent any further net osmotic entry of water into the glucose compartment. To mention a more-biological example, if a red blood cell is dropped into distilled water, osmotic water entry will raise the hydrostatic pressure in the cell, stretching the cell membrane outward like a balloon until it bursts.

The Direction and Rate of Osmosis

Earlier, we defined osmosis as passive movement of water across a membrane. We can refine that definition by noting that water movement is passive if it is in the direction dictated by osmotic pressures and if it leads toward equalization of osmotic pressure on both sides of a membrane.

The *direction* of osmosis between two solutions is from the solution of lower osmotic pressure toward the solution of higher osmotic pressure, and the net *rate* of osmotic water flux per unit area of membrane is proportional to the gradient of osmotic pressure across the membrane. This gradient is defined as $(\pi_1 - \pi_2)/X$, where π_1 and π_2 are the osmotic pressures of the solutions and X is the distance separating π_1 and π_2. The proportionality coefficient between the rate of osmotic water flux and the osmotic gradient depends in part on the permeability of the membrane to water and the temperature.

If we have two solutions, A and B, and they have the same osmotic pressure, they are said to be *isosmotic.* If A has a lower osmotic pressure than B, A is said to be *hyposmotic* to B, and B is said to be *hyperosmotic* to A; in this terminology, water moves osmotically from the hyposmotic solution into the hyperosmotic one.

A simple point that should be kept clearly in mind because it helps greatly in the analysis of salt–water exchange is that osmotic pressures have one major claim to fame: They govern movements of *water* across membranes. That is why they are measured and why they are significant.

Determinants of the Osmotic Pressure of a Solution

The osmotic pressure of a solution is determined by the total concentration of dissolved solutes, but this is a superficial statement that requires refinement. A point to be emphasized from the start is that only *dissolved* solutes contribute to the osmotic pressure; red blood cells or other suspended materials do not. We can consider a solution to consist of solute "particles" dissolved in water, with each separate dissolved entity constituting a particle. If a glucose molecule is placed in solution, it constitutes a single dissolved particle. If a NaCl molecule is placed in solution, it dissociates into two ions, Na^+ and Cl^-, and each of these constitutes a separate dissolved particle. With this introductory background, we can state that the *osmotic pressure of a solution is proportional to the effective concentration of dissolved particles, regardless of the size or chemical nature of the particles.* The meaning of *effective concentration* will be elaborated below. We emphasize here that each dissolved particle—be it a glucose molecule, a large protein molecule, or a Na^+ ion—makes a roughly equal contribution to the osmotic pressure of a solution.

When nonelectrolytes are dissolved in water, the individual molecules go into solution separately, and each constitutes a dissolved particle. Because a mole of solute contains a fixed number of molecules (Avogadro's number) regardless of the solute in question, equimolar solutions of various nonelectrolytes have equivalent particle concentrations and equivalent osmotic pressures. Thus, a 0.1-*M* solution of glucose has the same dissolved particle concentration as a 0.1-*M* solution of urea, and both have virtually the same osmotic pressures. Similarly, a 0.1-*M* solution of a large, nondissociating protein also has essentially the same osmotic pressure.*

When electrolytes are dissolved in water, the individual molecules dissociate into two or more ions, each of which constitutes a dissolved particle. Ideally, then, a 0.1-*M* solution of NaCl would be expected to exhibit twice the osmotic pressure of a 0.1-*M* glucose solution, and a 0.1-*M* solution of Na_2SO_4 would exhibit three times the osmotic pressure of the glucose solution. Electrolytes, however, do not show this ideal behavior. Dissociation may not be complete; and even when it is complete, as with strong electrolytes, the dissolved anions and cations tend to be attracted to each other, and the solution behaves as if it had a lower concentration of dissolved particles than would be surmised from simple dissociation kinetics. A 0.1-*M* solution of NaCl, for example, exhibits about 1.9 times the osmotic pressure of a 0.1-*M* glucose solution, not twice the osmotic pressure. The particle concentration of a so-

lution as judged from its actual osmotic behavior is variously termed the **effective particle concentration,** the **osmotic concentration,** or the **osmotic activity.** An important point to recognize is that the effective particle concentration is, in reality, a fictional concentration for strong electrolytes. A mole of NaCl in solution does dissociate into 2 moles of particles, but the solution behaves as if there were fewer free particles because of electrostatic attractions among the ions.

The physical chemist van't Hoff showed that the gas laws have application to the osmotic pressures of dilute solutions. The fundamental gas law may be written $PV = nRT$, where P is the pressure of the gas in atmospheres, V is the volume in liters, n is the number of moles of gas, R is the gas constant [0.082 L·atm/(°K·mole)], and T is the absolute temperature. By rearrangement, we get

$$P = \frac{n}{V} RT$$

and it should be clear that n/V is the concentration. As applied to simple solutions of one solute, n/V is taken to represent the molar concentration C, and, substituting the osmotic pressure π for P, we get for solutions of *nonelectrolytes*

$$\pi = CRT$$

From this equation, one can calculate, for instance, that the predicted osmotic pressure of a 1-*M* solution of nonelectrolyte at 0°C (273°K) is 22.4 atm. This value, 22.4 atm at 0°C, is considered to be the ideal osmotic pressure of a 1-*M* solution of nonelectrolyte. Actual solutions typically deviate from ideal behavior unless very dilute. Thus, to compute correct actual osmotic pressures, one must multiply results of the above equation by an "osmotic coefficient" G: $\pi = GCRT$. The coefficient is 1.024 for a 0.1-*M* solution of sucrose, for example, indicating a slight but significant departure from ideal behavior. For simple solutions of *electrolytes*, G is the effective number of particles yielded per solute molecule.

Units of Measure

As we have seen, osmotic pressures can be expressed in units of hydrostatic pressure, such as atmospheres or mm Hg. An alternative system of units is that based on **osmolarity.** A 1-osmolar solution is defined to be one having an effective particle concentration of one Avogadro's number of particles per liter. That is, it has the same osmotic pressure as a 1-*M* solution of ideal nonelectrolyte. In hydrostatic-pressure units, such a solution has an osmotic pressure of 22.4 atm at 0°C; thus, the equivalency between the two sets of units is

$$1 \text{ osmolar} = 22.4 \text{ atm} \quad (\text{at } 0°C)$$

An **osmole** is an Avogadro's number of effective particles.

---

*In point of fact, solutions of various nonelectrolytes have the same ratio of dissolved particles to water molecules when they are of the same molality, not molarity. Thus, it is solutions of identical molality that exhibit the same osmotic properties. In biological work we are routinely confronted with "preconstructed" solutions such as body fluids for which it is easier to determine chemical concentrations in molar terms than in molal terms. At physiological concentrations, the errors involved in working with molar expressions of osmotically effective concentrations are small, and thus molar expressions (as well as molal ones) are in common use.

A third set of units in common usage, that based on freezing-point depression, is introduced in the next section.

Colligative Properties of Solutions and the Indirect Measurement of Osmotic Pressure

Properties of solutions that depend on the effective particle concentration, regardless of the chemical nature of the particles, are termed *colligative properties.* Four of these properties deserve note: osmotic pressure, freezing-point depression, boiling-point elevation, and vapor pressure.* An increase in the effective particle concentration increases the osmotic pressure, increases the freezing-point depression (decreases the freezing point), increases the boiling-point elevation (increases the boiling point), and decreases the vapor pressure of a solution. The latter three effects are easily remembered by recognizing that increases in the effective particle concentration of a solution tend to impair any change in state of the water. Thus, as the particle concentration is increased, lower temperatures are required for water to change from a liquid to a solid (the freezing point is lowered), higher temperatures are required for boiling, and the equilibrium vapor pressure established in air through evaporation in a closed system is reduced.

All four colligative properties normally vary predictably with each other as the effective particle concentration of a solution is varied. Accordingly, if any one colligative property of a solution has been measured, all the others can usually be computed. This is the fundamental basis for the indirect measurement of osmotic pressure. For various technical and practical reasons, physiologists interested in osmotic pressures usually measure either the freezing-point depression or the vapor pressure of a solution and then calculate its osmotic pressure, rather than determining osmotic pressure directly. A 1-osmolar solution has a freezing-point depression of 1.86°C.

What Causes Osmosis?

The prevailing view is that osmosis occurs as a direct consequence of differences in water *concentration,* very much in the way that diffusion of solutes occurs because of a difference in solute concentration. Water moves from *its* region of higher concentration toward its region of lower concentration.

An alternative view that prevailed early in the study of osmosis and that has recently been championed anew by P. Scholander and H. Hammel holds that solutions behave much like gases (this is why, they claim, the gas laws apply to the calculation of osmotic pressure). Solute molecules, like gas mole-

---

*When we speak of "depression" and "elevation," the reference is pure water. A freezing-point depression of 1°C, for example, signifies that freezing occurs at 1°C below zero.

cules, are viewed as exerting an outward force (positive pressure) on the walls of a container, and the outward thrust of the solute molecules is viewed as stretching the solvent matrix of the solution, inducing, quite literally, a negative hydrostatic pressure in the solvent (water). Within a single solution, the positive solute pressure and negative solvent pressure cancel. Nonetheless, it is claimed, water moves from a solution of lower osmotic pressure to one of higher osmotic pressure because the hydrostatic pressure within the solvent matrix is lower (more negative) in the latter.

INTERRELATIONS BETWEEN SOLUTES AND OSMOSIS

Because the osmotic pressures of solutions depend on solute concentrations, osmotic water movements are interrelated with solute behavior. The interrelations can be complex, for in the real world, membranes are not semipermeable and passive but instead permit or induce some solutes to cross from side to side. Here, we examine several concepts of use in understanding the relations between osmosis and solute movement.

A Uniformly Distributed Solute Does Not Contribute to the Osmotic Gradient

A solute that is initially equal in its osmotically effective concentration on the two sides of a membrane—or that diffuses to concentration equilibrium—does not contribute to the difference of osmotic pressure across the membrane. For example, if a membrane separates two complex solutions containing glucose and if the glucose concentration is equal on the two sides, the contribution of glucose to the osmotic pressure of each solution is the same, and any difference in osmotic pressure between the solutions will have to be due to other solutes.

Nonpermeating Solutes and Colloid Osmotic Pressure

If a solute is *nonpermeating* and more concentrated on one side of a membrane than the other, it contributes a persistent component to the osmotic pressure differential across the membrane: a component that cannot be eliminated by the solute's own diffusion to concentration equilibrium. This is an important principle in many situations. For example, it is important in the analysis of water flux across the walls of blood vessels because blood plasma typically contains relatively high concentrations of proteins that do not readily cross vessel walls.

The elevation of the osmotic pressure of a solution caused by nonpermeating proteins in the solution is termed the *colloid osmotic pressure* (or oncotic pressure) of the solution. Colloid osmotic pressure is a relative concept; that is, it is the extent that the nonpermeating proteins cause elevation of the osmotic pressure of a solution relative to some particular other solution. When speaking of blood plasma, the comparison solution is typically tissue fluid. The colloid osmotic pressure of human arterial blood plasma is about 25 mm Hg. This means that the plasma proteins impart to the plasma an osmotic pressure that is 25 mm Hg higher than that of tissue fluid. The presence of the proteins creates a persistent tendency for the plasma to take on water by osmosis from the tissue fluids (p. 346).

Interactions of Passive Solute and Water Fluxes

When permeating solutes are not at electrochemical equilibrium and thus are diffusing across a membrane, their flux tends to alter the osmotic-pressure gradient across the membrane by removing osmotically effective solute from one side and transferring it to the other. On the other hand, osmotic flux of water across a membrane tends to alter electrochemical gradients of solutes because solutes tend to become concentrated on the side losing water and diluted on the opposite side. In addition to these interactive effects of diffusion and osmosis on os-

motic and electrochemical gradients, the rates of solute and water flux frequently interact *dynamically*. Flux of water in one direction can, for example, increase rates of solute movement in the same direction or lower rates of solute movement in the opposite direction.

Active Solute Fluxes and Osmosis

Active transport of solutes often *creates* gradients of osmotic pressure, thus setting up conditions favorable to osmosis in one direction or another. Organisms appear to lack mechanisms for active water transport. Thus, active solute transport is in fact one of the preeminent means by which organisms control directions of water flux.

Consider the simplified diagram of an epithelial membrane separating two solutions, A and B, in Figure 7.10. If the cells in the membrane were to transport a solute such as Na^+ actively from solution A to solution B, it could happen that B would thereby become hyperosmotic to A; and if the epithelium were water permeable, water would then move across the epithelium (through the cells or through the intercellular spaces) by osmosis from A to B.

BOX 7.1 OSMOTICITY VERSUS TONICITY

Solutions can be classified according to their osmoticity or tonicity. The generally accepted distinction between these systems of classification is this: (1) The osmoticity system refers to relative osmotic pressures, as discussed earlier (p. 146). Isosmotic solutions have identical osmotic pressures. (2) The tonicity system refers to effects on cellular volume. An *isotonic* solution is one into which cells can be placed without affecting cellular volume. If cells are placed in a *hypotonic* solution, they swell because of an osmotic influx of water. In a *hypertonic* solution, they shrink because of osmotic water loss.

If cellular membranes were semipermeable, the two systems of classification would be synonymous in that solutions that were isosmotic, hyposmotic, or hyperosmotic to the contents of a cell would be—respectively—isotonic, hypotonic, or hypertonic. The systems of classification are not in fact synonymous because solutes may be capable of crossing the cellular membrane.

To illustrate, consider the practical problem of devising a solution that will preserve the volume of human red blood cells. Let us assume—not that inaccurately—that the cell membrane is totally impermeable to the significant osmotically effective solutes (e.g., salts) within the cell. The membrane is impermeable to Na^+ and Cl^-, so if we place cells in a NaCl solution that is *isosmotic* to the cell contents (about 0.15-*M* NaCl = 0.28 osmolar), that NaCl solution will be *isotonic* as well; solutes will not cross the cell membrane in either direction, so the cell contents and bathing solution will remain perpetually isosmotic, and water will neither enter nor leave the cell. Red blood cells are permeable to urea. Thus, if we immerse cells in an initially isosmotic urea solution, urea will tend to diffuse to concentration equilibrium across the cell membranes rather than remaining strictly on the outside and counterbalancing the osmotic effect of the normal intracellular solutes. The cells will consequently become hyperosmotic to the bathing solution, take up water osmotically, swell, and burst. A urea solution *isosmotic* to normal red cells is *not isotonic*.

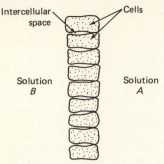

Figure 7.10 A schematic diagram of an epithelial membrane separating two solutions, *A* and *B*.

This type of coupling of water flux with active solute transport is relatively straightforward.

Often, water movement across an epithelial membrane is more perplexing. In a number of tissues, water has been found to move steadily across an epithelium even though there is no measurable difference in osmotic pressure between the solutions on the two sides of the epithelium, a phenomenon sometimes termed *isosmotic water flux*. There are also cases in which the solutions on the two sides of an epithelium do differ measurably in osmotic pressure, but water moves from the side of higher osmotic pressure toward the side of lower osmotic pressure. Isosmotic water flux and water flux against the osmotic gradient are commonly believed to be explained by the principles of osmosis, *with the critical osmotic-pressure gradients existing within the epithelial membrane,* not across it. Osmosis in response

to such highly circumscribed osmotic gradients is sometimes termed *local osmosis*. The osmotic gradients within the epithelium are believed to be created by active solute transport. The water flux is thus coupled with and effected by active solute flux.

A classical model of how local osmosis could cause isosmotic water flux across an epithelium is shown in Figure 7.11. On the side facing solution *A,* adjacent cells of the epithelium are connected by *tight junctions,* which have traditionally been thought to be virtually impermeable to water and solutes, thus sealing one end of each intercellular space. As shown, NaCl is actively transported from the cells into the one end of each intercellular space, the cellular supply of NaCl being replenished by uptake (not diagrammed) of NaCl from solution *A* into the cells. Importantly, the extrusion of NaCl into the intercellular space raises the osmotic pressure there to be well above the osmotic pressure of solution *A,* causing osmosis to carry water from *A* into the space. This osmotic entry of water induces bulk flow (streaming) of intercellular solution toward *B*. As the solution flows the length of the intercellular space, it is diluted more and more by osmotic influx of water, so that by the time it is discharged into *B*, its osmotic pressure has been reduced approximately to the level seen in *B* or *A*. Thus, the end result is movement of both NaCl and water from *A* to *B* even though there is no osmotic gradient across the epithelial membrane as a whole.

The particular concept of isosmotic water flux diagrammed in Figure 7.11 is called the *standing-gradient model* in reference to the presence of a steady-state

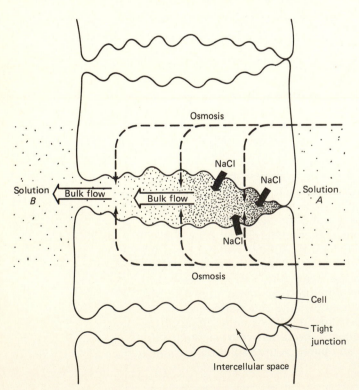

Figure 7.11 Schematic representation of the standing-gradient model of isosmotic water flux. Heavy dark arrows indicate active transport of NaCl. The density of dots indicates relative osmotic pressure.

gradient of osmotic pressure along the length of each intercellular space. There are a number of other models of isosmotic water flux that have been postulated. What most of the models have in common is the concept that active solute transport can raise or lower the osmotic pressure in microscopically fine channels within an epithelium—intercellular spaces or cellular infoldings—and thereby induce water flux by osmosis. Strikingly, secretory and absorptive epithelia of all kinds (intestinal, renal, salivary, gall bladder, etc.) are distinguished by the presence of such fine channels.

Conceptually, it should be clear that elevation of the osmotic pressure in the intercellular spaces of Figure 7.11 could cause osmosis from solution *A* into the intercellular spaces even if solution *A* were hyperosmotic to solution *B*. This would then constitute the first step in a process of *osmotically* moving water across an epithelium from a side of higher osmotic pressure to a side of lower osmotic pressure (see p. 240).

In recent years, it has come to light that the tight junctions in some epithelia are not tight but leaky, permitting flux of water and solutes at appreciable rates. This discovery and others are currently causing much debate over the standing-gradient model and other models.

DETERMINANTS OF WATER FLUX

Taking now a broader perspective on biological water flux, there are, overall, three types of force known to induce water flux in animals: gradients of osmotic pressure, hydrostatic pressure, and matric potential. Let us look briefly at the nature of the latter two types of gradient before considering how different types of gradient can interact.

(1) *Hydrostatic-pressure gradients*. When the hydrostatic pressure differs on the two sides of a membrane, water tends literally to be forced through the membrane from the high-pressure side to the low-pressure side. Water movement thus induced by a gradient of hydrostatic pressure is termed **filtration** or **ultrafiltration** because as the water flows through pores in the membrane, some solutes may be swept through the pores in the water stream whereas others may be left behind.

(2) *Matric-potential gradients*. Soil, skin, wood, paper, seeds, proteins, and many other substances take in water and hold it within their structural matrix by virtue of adhesive and cohesive forces. Note, for example, how paper or hair becomes softer on a humid day because of such water imbibition. Such materials attract water more avidly the drier they are. The tendency of a substance to attract and retain adhesively and cohesively bound water is quantified as its **matric potential**. If two substances in contact differ in matric potential, water tends to move from the one of lower potential into the one of higher potential.

Gradients of osmotic pressure, hydrostatic pressure, and matric potential can be present simultaneously. Their combined effect on the direction and rate of water flux is then assessed by expressing all the gradients in a single system of units of measure and adding or subtracting them as appropriate. Let us look at a case study in which gradients of osmotic and hydrostatic pressure are simultaneously present.

A Case Study: Filtration Pressure

In most vertebrates and some other groups of animals, fluid is introduced into the kidney tubules from the blood by filtration. The anatomical relationship between blood capillaries and kidney tubules in vertebrates is diagrammed schematically in Figure 7.12. Each blood capillary receives blood at relatively high hydrostatic pressure from the arteries, and this hydrostatic pressure tends to force water out of the blood plasma into the kidney tubule. The process is true filtration, for while small solutes such as ions and amino acids are carried along with the water into the kidney tubule, large solutes such as proteins are left behind in the plasma. Accordingly, the plasma has a colloid osmotic pressure with respect to the fluid in the kidney tubule, and this tends to cause osmosis of water *out* of the kidney tubule into the plasma. Another force opposing entry of water into the kidney tubule is the hydrostatic pressure in the tubule itself. The three forces are illustrated in Figure 7.12.

The *net* force promoting entry of fluid into the kidney tubule is termed the **filtration pressure** and is calculated as follows:

filtration pressure = hydrostatic pressure of plasma
    − hydrostatic pressure of
    renal tubular fluid − colloid
    osmotic pressure of plasma

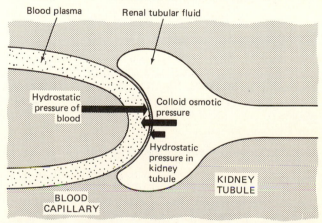

Figure 7.12 Schematic diagram of a filtration site in a vertebrate kidney, showing by the heavy arrows the direction of water flux promoted by relevant osmotic and hydrostatic pressures. The relative lengths of the arrows represent the relative average magnitudes of the three forces favoring water flux. (See Figure 10.1 for greater anatomical detail.)

In Munich-Wistar rats, for illustration, the blood pressure is about 50 mm Hg, the hydrostatic pressure in the tubular fluid is about 13 mm Hg, and the colloid osmotic pressure averages about 28 mm Hg. Using these values we get that the filtration pressure = 50 − 13 − 28 = 9 mm Hg.

Here, we see how the net effect of hydrostatic and osmotic forces acting on a single system can be calculated. The end result demonstrates that net water flux is from the plasma into the kidney tubule.

TYPES OF REGULATION

We have been looking at a number of salt–water exchange mechanisms at the physicochemical level. Now we turn to higher levels of organization—organs or whole organisms—and examine some of the fundamental principles operative there. A good starting point is to identify three types of regulation carried out by animals.

Volume regulation is the regulation of the body's total water content. *Ionic regulation* is the regulation of the concentrations of ions in the body fluids. *Osmotic regulation* (osmoregulation) is regulation of the osmotic pressure of the body fluids, that is, regulation of the ratio between all dissolved particles (regardless of chemical nature) and water.

Challenges to all three types of regulation often go hand in hand. When water is carried into a freshwater fish by osmosis, for example, the entry of the water tends to bloat the body (a challenge to volume regulation), dilute ions in the body fluids (a challenge to ionic regulation), and lower the osmotic pressure of the body fluids (a challenge to osmotic regulation). Similarly, the homeostatic physiological responses of animals to all three types of challenge are often intimately interconnected. Nonetheless, the three forms of regulation are distinct, and it is usually helpful for analysis of salt–water relations to keep the distinctions clearly in mind.

URINE

The kidneys help regulate the composition of the body fluids by removing and excreting water, salts, and other solutes. Here, we examine some of the fundamental regulatory roles that the kidneys can play.

Urinary Osmotic Pressure and Osmoregulation

Urine may be isosmotic, hyperosmotic, or hyposmotic relative to the blood plasma and other body fluids. The relative osmotic pressure of the urine is expressed as the **osmotic U/P ratio,** defined to be the ratio of urinary osmotic pressure over plasma osmotic pressure. This ratio reveals a great deal about the role of the urine in osmoregulation.

Suppose that a freshwater fish, being hyperosmotic to its environment, takes up a quantity of water into its body fluids by osmosis from the ambient medium. This water will act to dilute the fish's body fluids and reduce their osmotic pressure. Now an important question is, can the fish restore its original osmotic pressure (osmotically regulate) through production of urine? A bit of reflection will reveal that the answer is ''yes,'' but *only* if the fish is able to produce a urine more dilute than its body fluids, that is, a hyposmotic urine (U/P < 1.0). Such urine will preferentially void water over solutes, because it will contain more water in relation to its osmotically effective solute content than the body fluids. By this preferential disposal of water—and the converse, preferential retention of solutes—voiding the urine will act to raise the osmotic pressure of the body fluids back toward the original level. Contrast this with the situation if the fish's urine were always isosmotic to its body fluids (U/P = 1.0). Then the fish, having suffered dilution of its body fluids, would be capable merely of voiding water in the same relation to solutes as pertained in the diluted body fluids; and even as the urine was voided, the body fluids would remain just as dilute as they were prior to urine production.

We see from this example that *production of isosmotic urine (U/P = 1.0) cannot serve directly to bring about osmotic regulation.* On the other hand, *production of hyposmotic urine (U/P < 1.0) has an osmoregulatory effect in an animal suffering from dilution of its body fluids.* As one might expect, most freshwater animals have evolved the capacity to render their urine hyposmotic to their body fluids. We humans as well as many other terrestrial animals also have that capacity, which serves us well after an evening of too much iced tea or beer.

If an animal has suffered an *increase* in its internal osmotic pressure to abnormally high levels, production of an isosmotic urine would again fail to bring about osmoregulation. Urine more concentrated than the body fluids, however, would preferentially dispose of solutes in relation to water, and thus it would favorably alter the ratio of solutes to water in the body fluids. We see that *production of a hyperosmotic urine (U/P > 1.0) has an osmoregulatory effect in an animal suffering from concentration of its body fluids.* The ability to produce hyperosmotic urine is not nearly as widespread as that to produce hyposmotic urine. The greatest capacities to concentrate the urine are found in mammals, birds, and insects—all primarily terrestrial groups that face risks of dehydration.

Urine and Volume Regulation

The renal organs typically help regulate the body's water content by voiding greater or lesser amounts of water as required. The urine, in fact, can play a critical role in volume regulation even when not playing any direct role in osmoregulation. Freshwater crabs, for example, are strongly hyperosmotic to their environment and thus experience an osmotic flux of water into their body fluids. They meet this challenge to volume regulation by producing an ample flow of urine. However, in those species that have been investigated, the urine is isosmotic to the body fluids, meaning that direct responsibility for maintaining a high internal osmotic pressure must fall elsewhere. This example illustrates nicely that there is a distinction between volume regulation and osmoregulation.

Urine and Ionic Regulation

The part played by the urine in ionic regulation can be analyzed in terms analogous to those introduced for analysis of osmoregulation. For each ion, an *ionic U/P ratio* can be computed; this ratio is the concentration of the ion in the urine divided by its concentration in the blood plasma. Taking Na^+ as an example, if the U/P Na^+ ratio is greater than 1.0, the urine contains more Na^+ in relation to water than the plasma, and thus urine production is acting to lower the concentration of Na^+ in the plasma. On the other hand, if the U/P Na^+ ratio is less than 1.0, urine production is acting to conserve Na^+ preferentially and raise the plasma Na^+ concentration.

The distinction between ionic and osmotic regulation is illustrated well by examining the urine of marine teleost fish. These fish are hyposmotic to seawater. They thus lose water osmotically to their environment and gain ions by diffusion from the environment; both of these processes act to raise *ionic and osmotic* concentrations in their body fluids. The marine teleosts produce a urine that is isosmotic to their body fluids (osmotic U/P = 1.0); thus, the urine can play no direct role in solving their osmoregulatory problem. Yet the urine differs substantially from the body fluids in its solute *composition*. In particular, the U/P ratios for Mg^{2+}, Ca^{2+}, and SO_4^{2-} are well above 1.0. Therefore, the production of urine is serving the important ionic regulatory role of keeping down the internal concentrations of these solutes, which the fish tend to gain from the seawater.

FOOD AND DRINKING WATER

The *overall* composition of available food and drinking water is an important consideration in water and salt relations. In this section, we consider a series of examples that illustrate this principle.

Relative Osmoticity of Predator and Prey

Each species of aquatic animal maintains characteristic osmotic and ionic concentrations in its tissues and body fluids. When an animal is consumed, its salt and water composition, as well as its nutrient content, may have significant implications for the predator. Marine teleost fish and mammals, for example, are markedly hyposmotic to seawater, whereas most marine plants and invertebrates are approximately isosmotic with seawater. Consider a fish or mammal that consumes a meal of invertebrates. The tissues and body fluids of the invertebrate prey, being markedly hyperosmotic to those of the predator, impose a salt load on the predator—a salt load that tends to raise the osmotic pressure of the predator's body fluids. The predator must eliminate the salt load to maintain osmotic homeostasis. By contrast, consider a fish or mammal that consumes a meal of fish. In this case, the tissues and body fluids of the prey are much closer to isosmoticity with those of the predator, and any salt load that may be incurred is substantially less than that imposed by the invertebrate meal. In a very real sense, the fish-eating predator benefits from the osmotic work the prey performed in maintaining its body fluids hyposmotic to seawater—an intriguing lesson in ecological energetics.

The Water Content of Air-Dried Foods

Desert animals are often in a precarious state of water balance. Many consume air-dried seeds and other dry plant matter, and the water content of the food, as well as its nutrient content, may be critical to survival. It thus is significant to recognize that these foods, even though ostensibly dry, equilibrate with air moisture and vary in their water content as humidity varies. At 25°C, "dry" pearled barley, for example, contains 3.7 g of water per 100 g dry weight at 10 percent relative humidity and 18.1 g per 100 g at 76 percent relative humidity. Hygroscopic plant material equilibrates rapidly with air humidity. Furthermore, humidity tends to be higher nocturnally than during the day, and higher below ground than above. Animals can increase their water intake by feeding at night or by storing food in burrows prior to ingestion.

Animals and Saline Drinking Water

Some animals can gain water from drinking saline waters such as seawater, whereas others cannot. A critical factor is whether the salts in the water can be eliminated in less water than was taken in with them.

We have all heard Coleridge's famous line from "The Rime of the Ancient Mariner," "Water, water, everywhere, nor any drop to drink." Sailors discovered long ago that drinking seawater was worse than drinking no water at all: Drinking the seawater paradoxically dehydrated them. We now know that a key consideration is that the maximum Cl^- concentration that humans can achieve in their urine is less than the concentration of Cl^- in seawater. Thus, the predominant anion in seawater can be eliminated only at the cost of more water than was taken in, meaning that to excrete the Cl^- in ingested seawater, a person not only must use all the water ingested, but also must draw on other bodily reserves of water. (Another problem is that the $MgSO_4$ in seawater induces diarrhea, thus increasing fecal water loss. $MgSO_4$ under the guise of Epsom salts is a well-known home laxative.) Some desert mammals are capable of concentrating their urine much more than people can and thus *are* able to gain water from seawater by excreting the salts in less water than was consumed with them.

Succulent Foods in the Desert

Some desert herbivores live on succulent plants. Here the amount of water in the food is high, but other problems may be presented. To illustrate, consider that some succulent desert plants, known as *halophytes* or salt-loving species, live in saline soils and often have very high salt concentrations in their tissues. The total salt concentration in their tissues may exceed that of seawater by as much as 50 percent, and some species contain high amounts of toxic oxalic acid as well. These plants constitute a major part of the diet of the sand rat (*Psammomys obesus*), for example, and are consumed at times by dromedary camels. Analytically, the salt levels of these plants pose much the same problems as those just discussed in regard to seawater. To take the case of just one ion in particular, the Na^+ concentration in the plant tissues may be five times that in mammalian plasma. Thus, a herbivore must be able to achieve a U/P Na^+ ratio greater than 5 if it is to excrete the Na^+ and enjoy any net gain of water from the plants. Many mammals cannot achieve such a high U/P Na^+ ratio. Furthermore, because the fleshy parts of the plants may be 80–90 percent water, an animal such as the sand rat that depends on them for nutrition must eat large quantities and deal with a correspondingly high salt load—illustrating again that the total composition of a food needs consideration in appraising osmotic–ionic implications.

Desert animals that eat animals or succulent plants are consuming packets of water that have been accumulated from the arid environment through the physiological efforts of the consumed organisms. The predator may thus sometimes take advantage of specialized abilities of the prey. Animals of relatively low mobility may consume ones of greater mobility, which can collect water over a larger area. Animals with modest capabilities to conserve water may prey on ones with highly developed capabilities. The adaptation of certain species to exploit the adaptations of other species, graphically illustrated in the desert, is a principle of the most widespread application.

The Special Significance of Protein Foods

Carbohydrates and lipids are composed primarily of carbon, hydrogen, and oxygen, and their oxidative catabolism thus results mostly in formation of CO_2 and water. The CO_2 is voided by the respiratory

organs, and the water contributes to the organism's water resources. In contrast, proteins contain large amounts of nitrogen as well as carbon, hydrogen, and oxygen, and their catabolism results in nitrogenous wastes as well as CO_2 and H_2O. These nitrogenous wastes may have implications for water balance, especially when excreted in solution. The principal nitrogenous waste of mammals, for example, is urea, a highly soluble compound. Because urea is necessarily voided in solution in the urine, a given protein intake, implying production of a given amount of urea, obligates the animal to a certain loss of urinary water, depending on the concentrating ability of the kidney. If a mammal is concentrating urea maximally in its urine, a high-protein meal may well demand greater urinary water loss than a low-protein meal. Thus, the high-protein meal may place a greater strain on water balance.

Concluding Comment

One of the major points of this discussion of food and water has been that total composition must be considered and that the animal, in taking in needed nutrients or water, may take in excesses of other materials. It is well to remember that although the physiological machinery can operate to correct such excesses, deficits must inevitably be replaced by uptake from the outside world, and the animal must adapt to handle the total composition of those foodstuffs and waters that it must consume to meet its deficits.

METABOLIC WATER

When foodstuffs are aerobically catabolized, water and CO_2 are produced, as illustrated by the equation for glucose catabolism:

$$C_6H_{12}O_6 + 6O_2 \rightarrow 6CO_2 + 6H_2O$$

The water produced in catabolism is known as *metabolic water* or *oxidation water,* in contrast to *preformed water,* which is water taken in as such from the environment. Metabolic water is produced in all animals. Table 7.2 gives the amount of water formed per gram of foodstuff oxidized.

The Calculation of Net Gain or Loss of Water in Catabolism

That which gives also takes away. The catabolism of foodstuffs not only yields water but also imposes certain *obligatory water losses,* which may be classed as respiratory, urinary, and fecal. To assess the net impact of catabolism on water balance, the obligatory water losses must be subtracted from the gains of metabolic water. This important principle will be exemplified after briefly discussing the nature of the obligated losses. Throughout we assume a terrestrial, air-breathing animal.

Table 7.2 AVERAGE AMOUNT OF METABOLIC WATER
FORMED IN THE OXIDATION OF THE BASIC
FOODSTUFFS

	Grams of water formed per gram of food
Starch	0.56
Lipid	1.07
Protein when urea is the end product of nitrogen metabolism	0.40
Protein when uric acid is the end product of nitrogen metabolism	0.50

Source: K. Schmidt-Nielsen, Terrestrial animals in dry heat: Desert rodents. In D.B. Dill (ed.), *Handbook of Physiology. Section 4: Adaptation to the Environment.* American Physiological Society, Washington, DC, 1964.

1. *Respiratory losses* of water are obligated by the oxidation of all foodstuffs because the oxidation is impossible without oxygen and the oxygen must be acquired from the atmosphere by breathing. Breathing entails losses of water by evaporation. The magnitude of the losses depends partly on the respiratory physiology of the species and partly on ambient conditions (e.g., humidity).

2. *Urinary losses* of water are obligated by the oxidation of proteins but not by that of carbohydrates or lipids. The catabolism of proteins yields nitrogenous wastes that often must be excreted in the urine at a cost in water.

3. *Fecal losses* must be taken into account in the case of all foodstuffs when considering the catabolism of ingested, rather than stored, foods. Ingested foods usually contain *preformed* water. If the animal loses more water in its feces than it took in with its ingested food, it then suffers a net fecal loss in processing the food, and the fecal loss must be subtracted from metabolic water production in assessing the overall effect of catabolizing the ingested food. On the other hand, if the feces contain less water than was taken in with the food, then the animal realizes a net gain of *preformed* water, which is ignored in calculating the net gain of metabolic water.

The use of these concepts is exemplified in Box 7.2, which you should study closely. The kangaroo

---

BOX 7.2 METABOLIC WATER GAIN IN KANGAROO RATS

Desert kangaroo rats (*Dipodomys*) were studied at an air temperature of 25°C and a relative humidity of 33 percent. They were fed pearled barley but given no drinking water. This box illustrates how water losses obligated by catabolism are taken into account in calculating the rats' net gain of metabolic water.

METABOLIC WATER PRODUCTION (0.54 g H₂O/g barley)
By knowing the nutrient composition of pearled barley and the amount of water produced in the catabolism of the major foodstuffs (Table 7.2), it can be calculated that the nutrients derived by a kangaroo rat from each gram of barley (dry weight) yield about 0.54 g of metabolic water during cellular oxidation.

OBLIGATED WATER LOSSES (total: 0.47 g H₂O/g barley)

1. *Respiratory* (0.33 g H₂O/g barley). Oxidation of the nutrients derived from a gram of barley requires consumption of about 810 mL of oxygen; this entails an evaporative loss of about 0.33 g of water across the lungs at the conditions of the experiments.
2. *Urinary* (0.14 g H₂O/g barley). The protein in a gram of barley yields about 0.03 g of urea. Urea can be concentrated sufficiently in the urine that this amount can be voided in about 0.14 g of water.
3. *Fecal* (0 g H₂O/g barley). Each gram (dry weight) of barley ingested contains about 0.1 g of preformed water, and the feces resulting from the digestion of a gram of barley contain only about 0.03 g of water. Thus, there is a net gain of *preformed* water in the digestion of the barley; this is ignored in calculating the net gain of metabolic water.

NET GAIN OF METABOLIC WATER (0.07 g H₂O/g barley)
Because about 0.54 g of water is produced in the oxidation of a gram of barley and a total of only about 0.33 + 0.14 = 0.47 g is obligatorily lost, the kangaroo rat realizes a net gain of about 0.07 g of water in the catabolism of each gram of barley. This can be used to offset other water losses, or, if it represents an excess, can be excreted.

rats discussed in the box were fed pearled barley exclusively, and even under the relatively arid conditions prevailing during experimentation, they realized a net gain of water in catabolizing the barley. Note that the net gain is computed by first determining the immediate yield of metabolic water in the cellular oxidation of a gram of barley and then subtracting the water losses obligated by the catabolism of a gram of barley.

The Myth of Exceptional Metabolic-Water Production in Desert Rodents

Formation of water de novo in the body at first can seem almost magical. Perhaps this is why the subject of metabolic-water production has attracted more than its share of mythology. At moderate temperatures and even at low relative humidities (as low as about 20 percent), kangaroo rats and some other desert rodents can live indefinitely on air-dried seeds without drinking water, meeting most of their water needs with metabolic water and the remainder with the small amounts of preformed water in the seeds. Under similar conditions, most mammals would quickly die. These striking contrasts have given rise to the myth that desert rodents produce especially large amounts of metabolic water.

In responding to this myth, the first and foremost thing to recognize is that *the amount of metabolic water produced per gram of foodstuff oxidized is fixed by the stoichiometry of the aerobic catabolic pathways* and thus, for a given foodstuff, is the same for all animals, except that in protein metabolism the amount is dependent on the particular nitrogenous end product made. Suppose we place kangaroo rats and laboratory rats at an air temperature of 25°C and relative humidity of 33 percent and we feed both species the same food—pearled barley—so both are metabolizing similar foodstuffs. Presuming that their overall rates of metabolism are similar, we are led by the principles of stoichiometry to conclude that the two species will produce *similar* amounts of metabolic water per unit time. Yet if we provide no drinking water, the kangaroo rats will thrive and the laboratory rats will perish.

Table 7.3 looks at the *full* impact of catabolism on water balance in these species. Both species experience about the same obligatory respiratory water loss per gram of barley metabolized. However, laboratory rats cannot concentrate urea in their urine as much as kangaroo rats and thus are obligated to a greater urinary water loss. Furthermore, because laboratory rats produce substantially moister feces and void greater quantities of feces they, unlike the kangaroo rats, are obligated to an overall fecal loss. When all the losses obligated by catabolism are subtracted from metabolic water production, the kangaroo rats enjoy a net *gain* of water per gram of barley catabolized and the laboratory rats suffer a net *loss*. The kangaroo rats can use their excess

Table 7.3 APPROXIMATE CATABOLIC GAINS AND LOSSES OF WATER IN CAGED KANGAROO RATS (*DIPODOMYS*) AND LABORATORY RATS (*RATTUS*) WHEN EATING PEARLED BARLEY AND DENIED DRINKING WATER AT 25°C AND 33 PERCENT RELATIVE HUMIDITY[a]

	Kangaroo rat	Laboratory rat
Metabolic water produced	0.54 g/g	0.54 g/g
Losses obligated by catabolism		
Respiratory	0.33	0.33
Urinary	0.14	0.24
Fecal	0.00	0.03
TOTAL	0.47	0.60
Net gain of metabolic water	0.07	−0.06

[a] Values are expressed as grams of water per gram (dry weight) of pearled barley metabolized.

catabolic water to offset other water losses. The laboratory rats incur a larger water deficit with every gram of barley they use.

In retrospect, we see that the critical difference between these species is not in how abundantly they *produce* metabolic water but in how effectively they *conserve* what they make. It is the exceptional water-conservation capability of kangaroo rats and other desert rodents that permits some metabolic water to be left over once all the water debts incurred by catabolism have been paid.

The Myth of Water in the Camel's Hump

Cutting into the hump of a dromedary camel reveals quickly the error of the old myth that the hump is filled with water. However, the discovery of what the hump does contain—mostly fat—has repeatedly sparked a new myth that the hump is a water store in disguise. As shown in Table 7.2, catabolism of a gram of lipid yields more than a gram of metabolic water. Thus, it has been argued, the lipid in the hump provides in a remarkable way for two of the needs of an animal that undertakes long treks in an arid environment: It is a source of energy, and it is a lightweight form in which to carry water.

Again, the Achilles heel of the myth lies in a full accounting of water gains and losses in catabolism. This type of accounting is greatly simplified when looking at stored lipids or carbohydrates for two reasons: First, oxidation of lipids and carbohydrates does not obligate urinary water losses, and second, in considering oxidation of stored nutrients we need not be concerned with fecal losses. Thus, the problem resolves to considering metabolic water production in relation to obligatory respiratory losses. At temperatures and humidities prevailing in the desert, camels lose more water across their lungs in obtaining oxygen for the oxidation of lipid than they gain from the oxidation. Therefore, in fact, their catabolism of lipid (for energy) imposes a water deficit.

DISTRIBUTION AND MOVEMENTS OF WATER AND SOLUTES WITHIN THE BODY

The water and salts within an animal's body can be viewed as occupying a number of body compartments. The plasma compartment is one of these; by definition, it includes all water and salts in the blood plasma. Other major compartments include the intracellular compartment (water and salts within cells) and the intercellular (tissue-fluid) compartment. When an animal, in relating to its environment, experiences a shortage or overload of water or salts, the distribution of these materials among its body compartments may prove to be critical in understanding its overall response.

Consider, for example, the case of mammals subjected to dehydration while exposed to the heat stress of the summer desert. Under such conditions, dromedary camels can tolerate a water loss of over 25 percent of their body weight, whereas many other large mammals can survive only about 13 percent dehydration. A significant reason for the difference lies in the distribution of water losses among the fluid compartments of the body. When mammals are dehydrated under heat stress, a frequent cause of death is a sudden, rapid rise in body temperature. This "explosive" hyperthermia is the ultimate consequence of a series of events initiated by dehydration. During dehydration, water is lost in part from the plasma compartment. This loss increases the viscosity of the blood and makes the heart work harder to pump the blood. The circulation of the blood plays a vital thermoregulatory role, carrying metabolic heat from the body core to the skin where the heat can be dissipated to the environment. When dehydration has so elevated blood viscosity that the heart cannot maintain an appropriate circulatory rate, metabolic heat accumulates in the body core and rapidly drives the body temperature to lethal extremes. Now, dehydration of the body does not dehydrate all fluid compartments equally. In human beings, during profound dehydration, the plasma compartment bears a disproportionately large share of the burden of water loss; men, for instance, who had lost about 10 percent of their total body water had lost a much-greater fraction—about 25 percent—of their plasma volume. In dromedary camels, by contrast, the blood compartment is relatively protected; camels that had lost about 30 percent of their total body water had lost only about 20 percent of their plasma volume. Apparently, an important reason for the high tolerance of dehydration in camels lies in the distribution of the water loss among their fluid compartments: Protection of their plasma compartment permits an exceptional degree of total dehydration before blood viscosity rises to threatening levels.

In tetrapod vertebrates faced with deficiencies of calcium in their diet, adequate calcium concentrations are maintained in the blood, muscles, and other tissues by removal of calcium from the skeleton under mediation of parathyroid hormone, a hormone of the parathyroid glands (p. 612). The animal thus copes with an environmental stress by redistributing material within its body. When a terrestrial crustacean molts, calcium is removed from the old exoskeleton under hormonal control, temporarily stored in the body (e.g., in the hepatopancreas), and later used to calcify the new exoskeleton. These transfers within the body reduce the animal's need for calcium from the environment.

SELECTED READINGS

Barnhart, M.C. 1983. Gas permeability of the epiphragm of a terrestrial snail, *Otala lactea*. *Physiol. Zool.* **56**:436–444.

Berridge, M.J. and J.L. Oschman. 1972. *Transporting Epithelia*. Academic, New York.

*Boulpaep, E.L. (ed.). 1977. Symposium on isotonic water movement. *Yale J. Biol. Med.* **50**:97–163.

Burton, R.F. 1983. Inorganic ion pairs in physiology: Significance and quantitation. *Comp. Biochem. Physiol.* **74A**:781–785.

Campbell, G.S. 1977. *An Introduction to Environmental Biophysics*. Springer-Verlag, New York.

Christensen, H.N. 1975. *Biological Transport,* 2nd ed. Benjamin, Reading, MA.

Davson, H. and M.B. Segal. 1975. *Introduction to Physiology,* Vol 1. Academic, New York.

Evans, D.H. 1975. Ionic exchange mechanisms in fish gills. *Comp. Biochem. Physiol.* **51A**:491–495.

Friedman, M.H. 1986. *Principles and Models of Biological Transport*. Springer-Verlag, New York.

Gerencser, G.A. (ed.). 1984. *Chloride Transport Coupling in Biological Membranes and Epithelia*. Elsevier, New York.

*Giese, A.C. 1979. *Cell Physiology,* 5th ed. Saunders, Philadelphia.

Greenaway, P. 1985. Calcium balance and moulting in the Crustacea. *Biol. Rev.* **60**:425–454.

*Kirschner, L.B. 1983. Sodium chloride absorption across the body surface: Frog skins and other epithelia. *Am. J. Physiol.* **244**:R429–R443.

Kylstra, J.A., I.S. Longmuir, and M. Grace. 1968. Dysbarism: Osmosis caused by dissolved gas? *Science* **161**:289.

Lowry, W.P. 1969. *Weather and Life*. Academic, New York.

Machin, J. 1979. Atmospheric water absorption in arthropods. *Adv. Insect Physiol.* **14**:1–48.

Maloney, P.C. and T.H. Wilson. 1985. The evolution of ion pumps. *BioScience* **35**:43–48.

McClanahan, L., Jr. 1972. Changes in body fluids of burrowed spadefoot toads as a function of soil water potential. *Copeia* **1972**:209–216.

Murrish, D.E. and K. Schmidt-Nielsen. 1970. Water transport in the cloaca of lizards: Active or passive? *Science* **170**:324–326.

Noble-Nesbitt, J. 1977. Active transport of water vapor. In B.L. Gupta, R.B. Moreton, J.L. Oschman, and B.J. Wall (eds.), *Transport of Ions and Water in Animals*, pp. 571–597. Academic, New York.

*Oschman, J.L. 1980. Water transport, cell junctions, and "structured water." In E.E. Bittar (ed.), *Membrane Structure and Function,* Vol. 2, pp. 141–170. Wiley, New York.

*Potts, W.T.W. and G. Parry. 1964. *Osmotic and Ionic Regulation in Animals.* Macmillan, New York.

Schmidt-Nielsen, K. 1964. *Desert Animals.* Oxford University Press, London.

Scholander, P.F. and H.T. Hammel. 1976. *Osmosis and Tensile Solvent.* Springer-Verlag, New York.

Sidell, B.D. and J.R. Hazel. 1987. Temperature affects the diffusion of small molecules through cytosol of fish muscle. *J. Exp. Biol.* **129:**191–203.

Ussing, H.H., N. Bindslev, N.A. Lassen, and O. Sten-Knudsen (eds.). 1981. *Water Transport Across Epithelia.* Munksgaard, Copenhagen.

See also references in Appendix A.

chapter 8

Exchanges of Salts and Water: Integration

The goal of this chapter is to review the salt and water relations of the animal groups in an integrated fashion. To a great extent, we shall be examining how gains and losses of salts and water are balanced by animals to achieve stability in the volume and composition of their body fluids (see p. 8). The chapter builds on the mechanistic concepts developed in Chapter 7. It is organized primarily by habitat rather than phylogeny in the belief that a habitat orientation provides for the most straightforward synthesis.

PROPERTIES OF NATURAL WATERS

The total concentration of salts in a body of water is usually expressed in terms of *salinity*. The salinity is the number of grams of inorganic matter dissolved per kilogram of water; the units of salinity are thus g/kg, or parts per thousand (ppt, sometimes written ‰). Natural waters are categorized by their salinity. Waters with a salinity of less than 0.5 ppt are generally classed as fresh (the average for lakes and rivers is 0.1–0.2 ppt). The salinity of seawater varies from place to place but is usually 34–37 ppt. Where ocean water mixes with fresh water along the continental coasts, waters of intermediate salinity, termed *brackish waters,* are formed. Waters with salinities of 0.5–30 ppt are generally considered brackish, whereas those above 30 ppt are classed as seawater. Salinities much greater than those of seawater, even exceeding 200 ppt, can be found in some of the inland saline lakes.

Representative data on the salt *composition* of fresh water and seawater are given in Table 8.1. Note that the major monovalent ions in seawater are Na^+ and Cl^-, and the major divalent ions are Mg^{2+} and SO_4^{2-}; the *relative* proportions of these ions remain

consistent even as the total salinity varies a bit from place to place. The relative proportions of ions in fresh waters vary considerably with the source (e.g., "hard" waters are high in Ca^{2+}). Seawater is so much more concentrated than fresh water that where the two mix to form brackish waters, the relative proportions of ions typically resemble those in seawater, each ion being diluted equally.

As to osmotic pressure, average seawater has the convenient property (convenient for memory, that is) of being almost precisely a 1-osmolar solution. Fresh waters have very low osmotic pressures of about 0.0005–0.01 osmolar.

The open ocean tends to provide a more stable physicochemical environment than bodies of fresh or brackish water. In comparison to a marine animal, a freshwater animal is likely to face greater temporal and spatial variation not only in the relative proportions of ions but also in temperature, pH, and oxygen concentration.

Table 8.1 CONCENTRATIONS OF MAJOR IONS IN FRESH WATER AND SEAWATER

Ion	Concentration (mM)	
	Fresh water[a]	Seawater[b]
Sodium	0.35	470
Chloride	0.23	548
Magnesium	0.21	54
Sulfate	0.19	28
Calcium	0.75	10
Potassium	0.08	10
Bicarbonate	1.72	2

[a] Freshwater values are worldwide averages for rivers.
[b] Seawater values prevail at a salinity of 34.3 ppt.
Source: Fresh water: I.A.E. Bayly and W.D. Williams, *Inland Waters and Their Ecology.* Longman, Camberwell, Australia, 1973. Seawater: H. Barnes, *J. Exp. Biol.* **31:**582–588 (1954).

FRESHWATER ANIMALS

The Composition of the Body Fluids

All freshwater animals are hyperosmotic to their environment and thus are known as **hyperosmotic regulators**. As shown in Table 8.2, the internal osmotic pressures of various species span at least an order of magnitude, but even the blood of the freshwater mussel *Anodonta*, which exhibits one of the lowest osmotic pressures known (about 0.04 osmolar), is substantially more concentrated than fresh water. Apparently, body fluids as dilute as fresh water are incompatible with life. As shown also in Table 8.2, the major ionic solutes in the blood of freshwater animals are Na^+ and Cl^-; these two ions characteristically account for much of the blood osmotic pressure.

Passive Salt and Water Exchange

Being hyperosmotic to their surroundings, freshwater animals face the problem of a continuous osmotic uptake of water, which, unopposed, would lead to dilution of their body fluids. Ions—Na^+ and Cl^- in particular—are more concentrated in their body fluids than in the environment, suggesting that the ions would tend to diffuse out. Measurements of the electrical gradient between blood and water across the outer body membranes of freshwater animals have consistently confirmed that, at least for Na^+ and Cl^-, the electrochemical gradient indeed favors outward diffusion. This diffusional loss of major ions tends, like the osmotic influx of water, to dilute the body fluids.

In a broad sense, we expect the energetic demands of osmotic and ionic regulation to vary directly with rates of passive water and salt exchange. The more rapidly water is taken up by osmosis and the more rapidly salts are lost by diffusion, the more metabolic energy the animal is forced to expend in counteracting these processes so as to maintain homeostasis. Three factors are important in determining passive rates of exchange: (1) the osmotic and ionic gradients between the body fluids and surrounding medium, (2) the permeability of the body wall to water and ions, and (3) the surface area for exchange. Each of these deserves brief discussion.

Gradients   Freshwater animals typically have less-concentrated body fluids than related marine forms. Most marine decapod crustaceans, for example, are virtually isosmotic with seawater (about 1.0 osmolar), but freshwater decapods, which are believed to be descended from marine forms, have lower osmotic pressures (e.g., 0.4 osmolar for the crayfish *Astacus* in Table 8.2). Similarly, freshwater mussels have very low osmotic pressures but are derived from marine groups that are nearly isosmotic with seawater. The lower body fluid concentrations seen in freshwater forms result in lower osmotic and ionic gradients between their body fluids and their freshwater environment than would otherwise be the case and thus can be viewed as adaptive to reducing the energy demands of osmotic–ionic regulation. W. T. W. Potts has estimated on thermodynamic grounds that the energetic cost of osmotic regulation is about one-tenth as great in crayfish (*Astacus*) as it would be if they retained the blood osmotic pressure of their marine ancestors, and the cost in freshwater mussels (*Anodonta*) is less than one-thousandth as great.

In considering evolutionary changes in the internal solute concentrations of animals, it is important to recognize the full functional implications. Because intracellular and extracellular osmotic pressures are

---

Table 8.2   OSMOTIC CONCENTRATION AND CONCENTRATIONS OF SOME SOLUTES IN THE BLOOD PLASMA OF SOME FRESHWATER ANIMALS

	Osmotic concentration (milliosmole/kg)	Solute concentrations (mM)					
		Na^+	K^+	Ca^{2+}	Mg^{2+}	Cl^-	HCO_3^-
Chlorohydra viridissima, hydrozoan coelenterate, tissue fluids	45	—	—	—	—	—	—
Anodonta cygnaea, freshwater mussel	44	15.6	0.5	6	0.2	11.7	12
Viviparus viviparus, snail	76	34	1.2	5.7	<0.5	31	11
Astacus fluviatilis, crayfish	436	212	4.1	15.8	1.5	199	15
Aedes aegypti, mosquito larva	266	100	4.2	—	—	51.3	—
Salmo trutta, brown trout	326	161	5.3	6.3	0.9	119	—
Rana esculenta, frog	237	109	2.6	2.1	1.3	78	26.6

Source: W.T.W. Potts and G. Parry, *Osmotic and Ionic Regulation in Animals.* Pergamon, Oxford, 1964; C. Little, *J. Exp. Biol.* **43**:23–37 (1965); J. Shaw and R.H. Stobbart, *Adv. Insect Physiol.* **1**:315–399 (1963).

similar in an animal, any alteration in the osmotic concentration of the extracellular body fluids implies a change in that of the cells, the functional units of the organism. A reduction, for example, in extracellular osmotic pressure requires a reduction in the concentration of at least some intracellular solutes, and the biochemical machinery must adapt successfully to such changes. One cannot help being impressed with some groups, such as the bivalve mollusks, which having evolved in isosmoticity with the sea, have adjusted to the low internal osmotic concentrations seen in freshwater forms.

Permeabilities The permeability of the outer body surface is generally relatively low in freshwater animals. Crayfish, for example, are at least 10 times less permeable to water, Na^+, and Cl^- than certain strictly marine decapod crustaceans of similar body size. The low permeabilities of freshwater animals are an important factor in reducing their rates of passive solute and water exchange and thus reducing the energy costs of maintaining homeostasis. Among freshwater animals, an appreciable range of permeabilities is to be found, with species that expose large areas of unprotected soft tissue to the water (e.g., snails and mussels) often being especially permeable overall. Available evidence indicates that the gill surfaces of freshwater animals are often more permeable to water and salts than the nonrespiratory body surfaces. This high permeability of gill surfaces is not unexpected, given that the gills must allow free diffusion of oxygen into the body fluids. It raises the significant consideration, however, that differences in overall salt/water permeability among freshwater animals may at times be secondary effects of differences in their demands for respiratory gas exchange. Gills not only exhibit high permeabilities per unit surface area but also have large surface areas and thus often are the major sites of diffusive solute loss and osmotic water gain.

In some instances, compensatory interrelations appear to exist between body-fluid osmotic pressure and the overall permeability of the animal. Freshwater mollusks have relatively high overall permeabilities, for example, and we can speculate that they evolved exceptionally low body-fluid solute concentrations because maintenance of high concentrations would have imposed too high a metabolic cost, given the tendency toward rapid passive salt and water exchange imposed by their high permeabilities. Those freshwater animals that have relatively high internal concentrations also tend to have relatively low overall permeabilities.

Surface Area The surface-to-volume ratio is as significant in passive salt and water exchange with the environment as it is in thermal exchange (see p. 118). Small animals, when compared to large ones of similar body form, have more body surface area per unit

weight; thus, if other things are equal, they will experience greater rates of diffusive salt loss and osmotic water gain per unit of weight. This principle implies a greater weight-specific energy cost for maintenance of homeostasis in the small forms unless they have lower permeabilities, lower internal solute concentrations, or other compensations.

Common Regulatory Mechanisms

Most freshwater animals share fundamentally similar mechanisms of osmotic–ionic regulation. This suite of mechanisms, to be discussed in this section, has been found in such diverse phyletic groups as the freshwater mussels, crayfish, earthworms, leeches, mosquito larvae, lampreys, teleost (bony) fish, frogs, toads, and soft-shelled turtles.

Urine As we have seen, freshwater animals are faced with a continuous osmotic influx of water. The water is voided in a copious urine. Table 8.3 presents data on the rate of urine production in a number of invertebrates and vertebrates. Recognizing that the rate of urinary water output is a measure of the rate of osmotic water influx, we see that it would not be unusual for these animals to face an osmotic water load equivalent to one-third of their body weight each day.

As also shown in Table 8.3, the urine of freshwater animals is typically strongly hyposmotic to their blood plasma and contains much lower concentrations of Na^+ and Cl^- than the plasma. Thus, the renal organs not only solve the animal's volume-regulation problem but conserve the major blood solutes and aid osmotic regulation as well (pp. 152 and 153).

Whereas the broad picture is that the urine of freshwater animals is copious and dilute, it is vital to recognize that the renal organs characteristically respond to conditions in the body fluids, and therefore the exact volume and composition of the urine vary with circumstances. For example, if an animal were to experience an increase in its osmotic water load, an increase in its urine output would ordinarily occur. In experiments on crayfish (*Astacus*), the Na^+ concentration in the urine of starved animals averaged 4.4 mM; but when the crayfish were fed, increasing their supply of Na^+, urinary Na^+ rose to 11.9 mM.

Although freshwater animals can typically limit their urinary concentrations of Na^+ and Cl^- to low levels, there is always some loss of these ions in the urine and this loss may be critical when the maintenance of ion balance is under challenge. The rate of loss of ions in the urine depends in part on the rate of urine production and therefore on the rate of osmotic water flux into the animal. Thus, factors that affect the rate of osmotic influx, such as the permeability to water of the integument and gills, can indirectly affect the animal's ability to deal with prob-

Table 8.3 REPRESENTATIVE URINARY PARAMETERS FOR SOME FRESHWATER ANIMALS[a]

Species	Urine volume (mL/100 g body wt · day)	Urine/plasma ratio Na$^+$	Urine/plasma ratio Osmotic pressure	Temperature (°C)	Reference
Viviparus viviparus, snail	36–131	0.28	0.20	19	2
Astacus fluviatilis, crayfish	8	0.006– 0.06	0.10	20	1, 5
Aedes aegypti, mosquito larva	≤20	0.05	0.12	?	6, 7
Rana clamitans, frog	{ 32 8	— —	— —	20–25 5	8 8
Xenopus laevis, clawed toad	58	0.10	0.16	25	4
Carassius auratus, goldfish	33	0.10	0.14	?	3

[a] Data for different species were gathered under a diversity of conditions and should be interpreted in that light.
Source: (1) G.W. Bryan, *J. Exp. Biol.* **37**:83–112 (1960). (2) C. Little, *J. Exp. Biol.* **43**:39–54 (1965). (3) J. Maetz, *Symp. Zool. Soc. London* **9**:107–140 (1963). (4) R.L. McBean and L. Goldstein, *Am. J. Physiol.* **219**:1115–1123 (1970). (5) L.E.R. Picken, *J. Exp. Biol.* **13**:309–328 (1936) (lower of two average osmotic U/P ratios reported here). (6) J.A. Ramsay, *J. Exp. Biol.* **27**:145–157 (1950). (7) J.A. Ramsay, *J. Exp. Biol.* **30**:79–89 (1953). (8) B. Schmidt-Nielsen and R.P. Forster, *J. Cell. Comp. Physiol.* **44**:233–246 (1954).

lems of ion balance. Volume regulation and ionic regulation are intimately related.

Active Ion Uptake Besides losing ions in their urine, freshwater animals also lose them by direct outward diffusion across their body surfaces. Foods replace some of the ions. In addition, at least the major ions, Na$^+$ and Cl$^-$, are actively transported into the blood directly from the ambient water.

The capabilities of most freshwater organisms for such active uptake of Na$^+$ and Cl$^-$ are remarkable. For example, some crayfish, fish, and frogs having plasma concentrations of 100–200 mM can extract Na$^+$ and Cl$^-$ in net fashion from waters as dilute as 0.01 mM. The site of active transport varies among the animal groups and in some cases has yet to be positively identified. Teleost fish and decapod crustaceans absorb Na$^+$ and Cl$^-$ across the gill epithelium. Among frogs, transport occurs across the gills in tadpoles but across the general integument in adults. Earthworms and leeches also use the general integument. Midge and mosquito larvae absorb ions across their anal papillae; interestingly, when larvae are reared in 0.9 percent NaCl and their need for active uptake is thus essentially abolished, their anal papillae are small (Figure 8.1).

Active uptake of Na$^+$ and Cl$^-$ is usually linked to extrusion of such ions as NH$_4^+$, HCO$_3^-$, and H$^+$ (see p. 142). Regulation of Na$^+$ and Cl$^-$ is thus tied to other vital functions: excretion of nitrogenous and respiratory wastes and acid–base regulation. As alluded to earlier, the Na$^+$ and Cl$^-$ pumps of freshwater animals can extract ions from waters where the ions are present at only very low concentrations; this indicates that the pumps have a high affinity for ions. The affinity of the pumps is much higher in freshwater crustaceans and fish than in their marine

or estuarine relatives, suggesting that the evolution of high-affinity pumps was one of the significant steps in the colonization of dilute waters. Inexplicably, however, the freshwater forms have lower maximal transport rates than the marine or estuarine forms.

Although the evidence is overwhelming that the Na$^+$ and Cl$^-$ pumps of freshwater animals typically play a major role in meeting needs for ions, evidence has come to light that certain aquatic amphibians at times exhibit little or no discernible drop in body Na$^+$ over a number of weeks when fasted in Na$^+$-free water or treated with a drug that depresses activity of the Na$^+$ pump. These observations suggest a remarkable ability to limit diffusive and urinary Na$^+$ losses under such circumstances. They also point to a need for fuller understanding of the role of the Na$^+$ pump in the lives of these amphibians in their native environments.

Food and Drinking Water Because most studies have been performed on fasting animals, the role of food in meeting ion needs is not well understood, although the prevailing view is that inputs of ions via active transport exceed those from foods in a majority of freshwater animals. These animals typically have to excrete copious amounts of water just to deal with their osmotic water influxes and thus would not be expected to drink. This expectation has usually been confirmed when examined experimentally, but recent studies on teleost fish in fresh water have revealed that some do drink sufficiently to raise their total water influx by 4–50 percent over the osmotic influx alone. Reasons for and consequences of this drinking are unclear.

Summary and Quantitative Example The usual pattern of salt–water balance in freshwater animals is

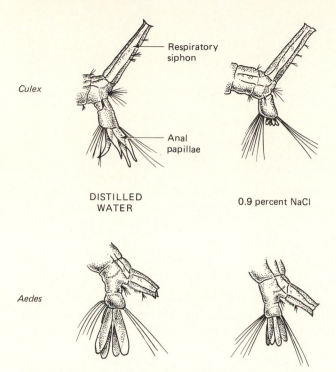

Figure 8.1 Posterior ends of larvae of *Culex pipiens* and *Aedes aegypti*, showing the difference in size of the anal papillae in animals reared in distilled water and in 0.9 percent NaCl. [From V.B. Wigglesworth, *J. Exp. Biol.* **15**:235–247 (1938).]

summarized diagrammatically in Figures 8.2 and 8.3. To review the pattern in words, let us look quantitatively at the gains and losses of water and Na$^+$ in a crayfish (*Astacus*). When fasting at 20°C in water containing 0.5-mM Na$^+$, a 29-g crayfish excretes about 0.1 mL of urine per hour. This provides a measure of the animal's rate of uptake of water by osmosis, probably principally across the gills. The urine is very dilute (1-mM Na$^+$) relative to the body fluids (204-mM Na$^+$). Only about 0.1 μmole of sodium is therefore lost per hour in the urine. Diffu-

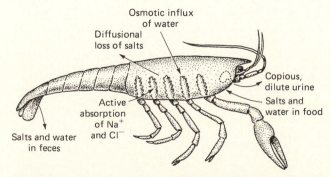

Figure 8.2 Summary of salt and water exchanges in crayfish. The renal organ, or green gland, is shown as a dashed, saccular structure opening at the base of the second antenna. The gills, which are covered by the carapace and not actually visible externally, are shown at the bases of the legs.

sional loss of sodium is much greater, approximately 10 μmole/h. The urinary and diffusional losses of sodium are replaced by active uptake across the gills at a rate near 10 μmole/h.

Why Is the Urine Hyposmotic?   In principle, production of a dilute urine is not a necessity for freshwater existence. Urine volume must be adequate to void the osmotic water load, and if the urine were isosmotic to the body fluids rather than hyposmotic, salt losses would be increased. Nonetheless, osmotic and ionic balance could be maintained if the increased salt losses were replaced by heightened active uptake of ions from the environment. Why then have the vast majority of freshwater animals evolved the capability of voiding a urine that is strongly hyposmotic to their blood?

In producing a hyposmotic urine, the renal organs start off with a fluid that is isosmotic to the blood and actively extract NaCl from it, lowering its osmotic pressure and returning the ions to the blood. Every Na$^+$ or Cl$^-$ ion thus removed from the urine prior to excretion is an ion that does not have to be replaced by active uptake from the ambient water. In the urine as it is formed in the renal organs, ion concentrations are initially as high as in the blood; only gradually—as ions are resorbed—do urinary ion concentrations fall to low levels. By contrast, ion concentrations in the ambient water are always very low. To remove an ion from the urine as it is formed, the active-transport mechanisms of the renal organs, on average, must operate against only a modest concentration gradient between the urine and blood. To remove an ion from the ambient water, however, the transport mechanisms on the body surface must inevitably operate against a large concentration gradient between the ambient water and blood. If we assume that when ions are being transported into the blood against a gradient, the net cost of adding each ion to the blood increases as the magnitude of the gradient increases, then it would be energetically less costly for the animal to extract an ion from its urine prior to excretion than to replace that ion from the ambient water. Thus, the widespread evolution of an ability to dilute the urine by ion resorption would make sense. Potts has calculated on thermodynamic grounds that the cost of osmoregulation would be 2.7 times higher in crayfish (*Astacus*) if they produced urine isosmotic to their blood instead of hyposmotic urine.

Effects of Temperature   Temperature exerts important effects on many of the individual processes involved in water and salt relations: rates of diffusion, osmosis, active transport, and so on. Not surprisingly, therefore, the composition and even the volume of the body fluids of animals can change when the temperature changes. Lowering the temperature from summer to winter levels has been observed, for example, to result in lower osmotic pressures and

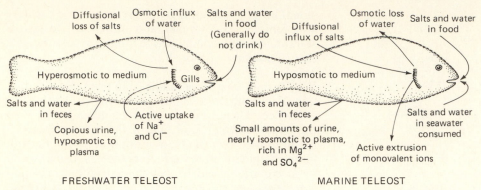

Figure 8.3 Summary of salt and water exchanges in freshwater and marine teleosts.

ion concentrations in the body fluids of quite a few freshwater animals. In some cases, this may be because rates of active ion uptake from the water are depressed by low temperatures. Frogs at winter temperatures decrease their urine output (see Table 8.3) and tend to retain water. Consequently, their body weight increases by about 6–9 percent, they become edematous, and—at least for a long time—their body fluids are diluted.

Exceptional Patterns of Regulation

The general pattern of osmotic–ionic regulation that we have discussed does not apply to all freshwater animals. As usual, a look at the exceptions can be as conceptually revealing as a look at the rule.

A number of freshwater animals produce urine that is very nearly isosmotic to their blood plasma, demonstrating that life in fresh water without production of dilute urine is more than a theoretical possibility. Among these animals is a toadfish, *Opsanus tau*, which is primarily marine but invades freshwater creeks. Other known examples include certain crabs that spend much or all of their lives in fresh waters. Two of these are *Potamon niloticus*, an African species restricted to fresh water, and *Eriocheir sinensis*, the wool-handed crab of Asia and northern Europe, which spends much of its adult life in rivers but migrates to the sea for breeding. Freshwater crabs typically maintain high internal osmotic pressures in comparison to other freshwater animals: about 510 milliosmolar in *P. niloticus* and 640 milliosmolar in *Eriocheir*, for example (cf. Table 8.2). As usual, the major blood solutes are Na^+ and Cl^-. Not only is the urine of the crabs virtually isosmotic with their blood, but also its ionic composition is very similar to that of blood; thus, the loss of Na^+ and Cl^- per unit volume of urine excreted is appreciable. A critical feature of these crabs is that the permeabilities of their body surfaces to water are exceptionally low. As a consequence, the crabs are not faced with high rates of osmotic water influx, and their rates of urine production are correspondingly limited. *Eriocheir*, for instance, excretes water equivalent to only about 3.6 percent of its body

weight per day, and *P. niloticus* excretes less than 0.6 percent per day (cf. Table 8.3). Importantly, the low urine output of the crabs—which results from their low water permeability—restricts the loss of salts in their urine. The quantities of Na^+ and Cl^- that are lost in the urine and by diffusion, although large by comparison to those lost by a crayfish, can be replaced by active uptake from the ambient water. There is reason to think that the toadfish and freshwater crabs are relatively recent immigrants to fresh water; this may help explain why they have not evolved the ability to make dilute urine.

Lacewing larvae, *Sialis lutaria*, do not appear to take up ions from fresh water under normal physiological conditions. Their urinary losses of Na^+ and Cl^- are low for two reasons: First, concentrations of the ions in their urine are low; and second, the larvae are among the most impermeable of freshwater animals known and thus need produce only small volumes of urine. With low losses of Na^+ and Cl^-, the ions evidently can be replaced from the diet alone.

ANIMALS IN THE OCEAN AND MORE-SALINE WATERS

Life is believed to have originated in the sea, and to this day the fauna of the oceans is more phylogenetically diverse than that in the other major habitats. All phyla and most classes of animals have marine representatives. Many groups have invaded fresh water and the land. In turn, there have been many reinvasions of the oceans. Thus, whereas some of our modern marine animals have a continuously marine ancestry, others trace their history to forms that occupied other habitats. As we shall see in this section, phylogenetic history appears sometimes to have had a major impact on the pattern of salt–water balance in contemporary species.

The Condition of Most Marine Invertebrates

Most marine invertebrates are isosmotic, or nearly so, to seawater. Included are such diverse forms as sponges, coelenterates, annelids, mollusks, echinoderms, and most crustaceans. These animals do not face problems of osmotic regulation. Sometimes, their internal osmotic pressure is somewhat higher than that of the water, but this represents a passive osmotic equilibrium resulting from Donnan effects (p. 144). The body fluids of all species that have received study differ in ionic composition from seawater, as illustrated in Table 8.4. The differences may be minor, as in the echinoderms (note *Echinus*), or may be more substantial, as in the squid *Loligo* and the crab *Carcinus*. Any given ion can be found relatively concentrated in some forms but relatively dilute in others (e.g., note magnesium in *Loligo* and in *Carcinus*). The adaptive significance of the differences in ionic composition among these animals remains generally obscure.

The differences in ionic composition between the extracellular body fluids and seawater in isosmotic marine invertebrates imply ionic regulation. Donnan effects may produce significant differences in ionic composition between blood and water in animals, such as the crustaceans, that have relatively high blood protein concentrations. These effects do not account for more that a small part of the total difference between internal and external salt composition, however. The animals under discussion are typically relatively permeable to both water and many ions, and we must look to active processes of salt uptake or excretion to explain the regulation of their internal ionic composition. Active uptake of ions from the environment at the body surface or from ingested seawater is widely indicated. Furthermore, in the crustaceans, mollusks, and some other groups, although the renal organs make a urine that is approximately isosmotic to the body fluids, they alter the ionic composition of the urine from that in the body fluids, thus contributing to ionic regulation (p. 153). In most decapod crustaceans, for example, the urine contains Mg^{2+} and SO_4^{2-} at higher concentrations that the blood (U/P = 1.1–4.2 for a number of species). This is important in maintaining reduced blood concentrations of these ions (see *Carcinus*, Table 8.4). Rates of urine production are modest but not trivial—up to 2–8 percent of body weight per day in a number of crabs, lobsters, and cephalopod mollusks.

Hagfish: The Only Vertebrates with Blood Salt Concentrations that Render Them Isosmotic to Seawater

The hagfish, an exclusively marine group of primitive jawless fishes, resemble the mass of marine invertebrates in having body fluids that are approximately isosmotic with seawater. Most of the blood osmotic pressure is due to dissolved salts, principally Na^+ and Cl^- (Table 8.4). Urine composition is known to differ from plasma composition, implicating the kidneys in ionic regulation. It seems likely that active uptake mechanisms are also involved. These fish appear to be the only modern vertebrates that might possibly claim a continuously marine ancestry.

Teleost Fish: Animals Hyposmotic to Seawater

The body fluids of the marine teleost fish are osmotically more dilute than the environment. These fish are therefore known as *hyposmotic regulators*. Although some of the marine invertebrates are also hyposmotic regulators, we look first at the fish because they have been studied exceptionally thoroughly and illustrate many of the broad patterns of such regulation.

The osmotic concentrations of marine teleosts are typically in the range of 0.3–0.5 osmolar—values that are higher than those of freshwater teleosts (about 0.25–0.35 osmolar) but not exceptionally so (which, incidently, explains why the meat of marine fish does not taste especially more salty than that of freshwater fish). It is currently believed that the ancestral jawed fish evolved in fresh waters and therefore that our modern-day marine teleosts are descended from freshwater fish. The low concentrations of the body fluids of the marine teleosts are thought to represent a carry-over from their freshwater ancestry.

The evolution of higher osmotic pressures in the

Table 8.4 CONCENTRATIONS OF CERTAIN IONS IN THE BODY FLUIDS OF SOME MARINE ANIMALS

	Concentration (mmole/kg)[a]					
	Na$^+$	K$^+$	Ca^{2+}	Mg^{2+}	Cl$^-$	SO$_4{}^{2-}$
Seawater	478	10.13	10.48	54.5	558	28.77
Aurelia mesogleal fluid, coelenterate	474	10.72	10.03	53.0	580	15.77
Echinus coelomic fluid, echinoderm	474	10.13	10.62	53.5	557	28.70
Mytilus blood, mussel	474	12.00	11.90	52.6	553	28.90
Loligo blood, squid	456	22.20	10.60	55.4	578	8.14
Carcinus blood, crab	531	12.26	13.32	19.5	557	16.46
Myxine blood, hagfish	537	9.12	5.87	18.0	542	6.33

[a] Figures are recalculated to apply to animals living in seawater of the composition given.
Source: W.T.W. Potts and G. Parry, *Osmotic and Ionic Regulation in Animals*. Pergamon, Oxford, 1964. Used by permission of Pergamon Press Ltd.

marine teleosts than the freshwater teleosts has presumably been adaptive energetically because by having heightened osmotic pressures the marine fish experience a reduced osmotic differential between their body fluids and the ocean. Nonetheless, given that the internal osmotic pressures of marine teleosts are near 0.4 osmolar and the ocean's osmotic pressure is near 1 osmolar, the osmotic difference between the blood and environment is around 0.6 osmolar, which is approximately twice as great as the difference maintained between blood and environment by freshwater teleosts. This consideration in itself would tend to saddle the marine fish with a relatively high rate of osmotic water flux. Importantly, however, the marine fish are typically less permeable to water than freshwater fish, so in fact the osmotic fluxes experienced by the two groups are similar (for given body size). Of course, in the case of marine fish, water tends to leave the body osmotically rather than enter. For a hyposmotic animal, the sea is a desiccating environment.

Because the concentrations of Na^+, Cl^-, and many other ions are lower in the body fluids of marine teleosts than in the sea, these fish are also faced with a tendency to gain salts by diffusion. The concentration gradients between the plasma and environment for the two major plasma ions, Na^+ and Cl^-, are large by comparison to the (oppositely directed) gradients seen in freshwater teleosts. Nonetheless, the picture that emerges in many, but not all, marine teleosts is that (1) the rate of Cl^- influx is limited by what is apparently a low gill permeability to Cl^-, and (2) the rate of Na^+ influx is limited, despite a relatively high gill permeability to Na^+, because in the gills the plasma is so electrically positive with respect to the environment that the electrochemical gradient for Na^+, which is critical, is near zero.

Replacement of Water Losses   Osmotic losses of water tend to deplete the body fluids and raise their concentration. Urinary losses also occur, although they tend to be small by comparison to the osmotic losses. To replace their water losses, marine teleosts drink seawater. For some, the intake is less than 1 percent of their body weight per day, but others consume over 50 percent/day, and the average is probably 10–20 percent/day. Given that the major losses are osmotic, these data on water ingestion give an idea of the magnitude of the daily osmotic outflux.

In marine teleosts, as in freshwater teleosts, the process of volume regulation exacerbates problems of ionic regulation (the two forms of regulation are inextricably linked because the major solutes of the body fluids are ions). At first sight, drinking seawater seems to be a straightforward way to obtain water. Consider, however, that when seawater is first taken into the gut, it is strongly hyperosmotic to the body fluids, and water (H_2O) would therefore be expected to travel by osmosis from the body fluids into the gut fluids, not vice versa. In fact, this happens. Stud-

ies on a number of species indicate that as seawater travels through the esophagus, stomach, and (in at least some instances) the anterior intestine, not only do Na^+ and Cl^- diffuse into the blood across the gut wall but also water *enters* the gut fluids by osmosis. Thus, gradually, the ingested seawater is diluted. Uptake of water from the gut fluids does not take place until the fluids reach parts of the intestine in which Na^+ and Cl^- are actively transported from the gut contents into the blood. *The uptake of water is driven by—and dependent on—this active uptake of salts.* The fluids in the posterior gut have sometimes been found to be hyposmotic to the plasma because of the active ion transport, but they may also be isosmotic or even somewhat hyperosmotic to the plasma. Indeed, in eel intestine studied experimentally, water uptake proved possible even when the osmotic pressure of the gut contents exceeded that of the plasma by 0.15–0.2 osmolar. Local osmosis (p. 150) driven by active solute uptake is thus implicated as the mechanism of water uptake.

The end result of the processes in the gut is absorption of 50–80 percent of the water from ingested seawater. Importantly, however, because of the active absorption of salts and also their diffusive uptake, the proportion of NaCl absorbed from the ingested water is much greater. In some species, the amount of NaCl absorbed from the gut per day is so great as to equal or exceed the total steady-state quantity in the tissues and body fluids of the fish. This influx of NaCl is *obligated by the process of obtaining water* and increases problems of Na^+ and Cl^- regulation.

The gut is poorly permeable to the major divalent ions in seawater, Mg^{2+} and SO_4^{2-}. Thus, although these diffuse into the body fluids to a small extent, they are mostly left behind in the gut and excreted in the feces.

Urinary Excretion   We now turn to the question of how marine teleosts eliminate excess ions that enter their body fluids from the gut or that diffuse in from the ocean across their gills or other body surfaces.

The kidneys of marine teleosts have not evolved the capability of producing urine hyperosmotic to the blood plasma and in fact excrete urine that is essentially isosmotic to the plasma. Considering all solutes taken together, therefore, the kidneys are unable to void solutes in disproportion to water and accordingly are unable to keep the osmotic pressure of the body fluids from rising toward that of the environment. On the other hand, the *composition* of the urine differs considerably from that of the plasma, and the *kidneys are the principal organs of Mg^{2+} and SO_4^{2-} regulation.* Whereas U/P ratios for Na^+, Cl^-, and K^+ are below 1.0, those for Mg^{2+} and SO_4^{2-} (and Ca^{2+} as well) are substantially above 1.0, signifying that the kidneys void the major divalent ions preferentially in relation to water and thus keep plasma concentrations of these ions from increasing.

For every milliliter of water ingested, absorbed,

and renally excreted, a marine teleost is left with an excess of solutes, because although the water enters hyperosmotic to the body fluids, it leaves isosmotic to them. From the viewpoint of osmoregulation, therefore, production of urine is an outright liability, and we would expect the volume of urine to be limited to the minimum necessary for excretion of solutes that are not voided by other routes. Nitrogenous wastes and the principal ions, Na^+ and Cl^-, are voided across the gills. Thus, the role of the kidneys is largely limited to excretion of Mg^{2+} and SO_4^{2-}, and the rate of urine production can be low. Urine volumes of several species have been measured to be just 0.5–3.5 percent of body weight per day, and even these may be inflated because laboratory stresses can induce diuresis (high urine outflow).

Extrarenal Salt Excretion   The gills assume prime responsibility for excreting the major ions, Na^+ and Cl^-. Chloride ions are extruded from the blood into the seawater by active transport against both concentration and electrical gradients. The gills also extrude Na^+, but whether this process is active or passive remains unclear. The voiding of these ions by the gills of marine teleosts provides an example of *extrarenal salt excretion*: excretion of ions by structures other than the kidneys.

Present evidence indicates that excretion of the monovalent ions by the gills is accomplished without concomitant excretion of water. The process thus admirably meets the needs of the fish not only for the regulation of plasma NaCl concentration but also for the regulation of osmotic pressure.

Cells of distinctive morphology found in profusion in the gill membranes and called *chloride cells* are believed to be responsible for the active extrusion of Cl^-. Recently, chloride cells have been discovered on the inner opercular membranes, jaw, and certain other surfaces of some species, implicating these other body parts in Cl^- excretion.

Summary   The pattern of water–salt regulation in marine teleost fish is summarized diagrammatically in Figure 8.3 and is reviewed here using quantitative data for southern flounder (*Paralichthys lethostigma*). In a study of individuals weighing about 1 kg, C.P. Hickman estimated that the fish suffered an osmotic water loss of 7.9 percent of their body weight per day. To replace this water and meet urinary water losses of 0.4 percent of body weight per day and fecal losses of 2.7 percent/day, the fish had to drink seawater equivalent to 11 percent of their body weight per day. From the ingested seawater, the fish absorbed 76 percent of the water (H_2O) but in doing so absorbed much higher percentages of the Na^+ (99 percent) and Cl^- (96 percent). Looking at divalent ions, only 16 percent of the Mg^{2+} and 11 percent of the SO_4^{2-} ingested were absorbed from the gut, the remainder being voided in the feces; 68 percent of the Ca^{2+} was absorbed. The scanty urine had a U/P ratio for Mg^{2+} of 99 and a U/P for SO_4^{2-} of 331; the

urine was believed to carry away the total absorbed quantities of these ions. The urine also had a fairly high U/P for Ca^{2+} (7.2) but did not excrete more than a minor fraction of the Ca^{2+} absorbed. The U/P ratios for Na^+ and Cl^- were low—0.08 and 0.79, respectively. Given the large quantities of Na^+ and Cl^- absorbed from ingested seawater, the low U/P ratios for these ions, and the small volume of urine produced, the urine served to excrete only 0.1 percent of the Na^+ and 1.1 percent of the Cl^- absorbed from the gut. The gills excreted the remainder, plus whatever NaCl entered the fish by diffusion from the ocean. Excretion of much of the Ca^{2+} absorbed could not be accounted for by known mechanisms.

Hyposmotic Regulation in Some Arthropods of Saline Waters

Numerous arthropods living in the ocean or more-saline waters maintain their body fluids hyposmotic to their environment. Among these are many grapsid and ocypodid crabs, some isopods, penaeid shrimp, and palaemonid prawns; when in seawater, for example, the fiddler crab *Uca pugnax* has an internal osmotic pressure of about 0.86 osmolar, and prawns such as *Palaemonetes varians* are more dilute yet, near 0.7 osmolar. Additional crustacean hyposmotic regulators include the small branchiopods known as brine shrimp, found in inland saline waters; one species is *Artemia salina* (see Figure 8.8), which can survive an exceedingly broad range of salinities, from waters about one-tenth as concentrated as seawater to saturated solutions of NaCl (about 300 ppt). Finally, larvae of a number of insect species, as well as adults of some (e.g., some water-boatmen), inhabit coastal seawaters or inland saline waters and regulate hyposmotically. Larvae of the mosquito *Aedes detritus* and the fly *Ephydra riparia*, for example, maintain internal osmotic pressures near 0.4 osmolar when in seawater.

Mechanisms of Hyporegulation   The mechanisms of hyposmotic regulation have been well studied in two species of brine shrimp, *A. salina* and *Parartemia zietziana*, and it seems clear that their pattern of regulation strikingly parallels that of teleost fish. For such small animals, their rate of osmotic water loss is relatively low, indicating low water permeability. To replace water losses they drink their saline medium (*Artemia*, at least, drinks both orally and anally). Water is then absorbed from the gut; this absorption in all likelihood occurs by local osmosis and is driven by active uptake of NaCl from the gut fluids (whereas the gut fluids become isosmotic to the blood in *Parartemia*, they remain hyperosmotic in *Artemia*). NaCl is actively extruded into the environment across the gills.

The crabs, prawns, and shrimp are not as well understood. The current trend is to see these groups as functioning largely along the same lines as teleosts, although available data do not preclude some radically different interpretations in some cases. Among certain crabs (e.g., *Uca pugnax*), unlike fish, the urine can be modestly hyperosmotic to the blood, but urine volume is so low that other parts of the body—probably the gills—must nonetheless assume most of the burden of NaCl extrusion and osmoregulation.

Air-breathing insect larvae typically enjoy especially low overall permeabilities, in part because they need not expose gas-permeable membranes to the water. Although much remains uncertain about their mechanisms of osmoregulation, a central factor that sets them apart from the other hyposmotic invertebrates is an ability to produce urine and feces that are strongly hyperosmotic to their blood plasma. Larvae of the mosquito *Aedes detritus*, for instance, can achieve an osmotic U/P ratio of about 3, and those of the fly *Ephydra riparia* can attain 10. The excretions of even the mosquito are sufficiently concentrated that the animal can drink seawater, excrete the salts in the urine, and have water left over to meet osmotic losses or other water needs. Furthermore, the U/P ratios for *all* the major ions can simultaneously be sufficiently high to permit regulation of their blood concentrations. Some saline-water mosquito larvae have well-developed anal papillae and seem to extrude ions actively across them.

Why Are These Arthropods Hyposmotic to Their Environments? The blood osmotic pressures of some of these arthropods are believed to be strongly influenced by their heritage. *Artemia*, for example, seems likely to be descended from freshwater branchiopods. The insect larvae not only trace a terrestrial ancestry but develop into terrestrial adults. Their blood osmotic pressures when in seawater are within the range typical of insects generally, and their ability to produce urine strongly hyperosmotic to their blood is a characteristic they share with terrestrial insects, which likewise face a desiccating environment.

The hyposmotic prawns and crabs occur in estuarine and other brackish-water settings, where they may experience not just full-strength seawater but a range of lower salinities. Possibly, the lowering of their blood osmotic pressures from levels typical of their strictly marine relatives represents a compromise, minimizing the cost of regulation in an environment that on average is more dilute than seawater.

Marine Reptiles, Birds, and Mammals: Additional Hyposmotic Vertebrates

Marine reptiles, birds, and mammals, in common with teleosts, are hyposmotic to the ocean. All are descended from terrestrial forms. Their internal osmotic pressures, although often slightly higher than those of terrestrial and freshwater relatives, reflect this heritage, being near 0.4 osmolar.

Permeabilities in these animals are relatively low, in part because they have inherited integuments adapted to limiting water loss in the dehydrating terrestrial environment and in part because, being air breathers, they do not expose relatively permeable respiratory membranes to the seawater. They nonetheless confront problems of water loss and salt loading. Water is lost, for example, by evaporation across the respiratory tract; it also is lost, to some extent, across the skin not only when the animals are immersed in seawater but also when they are exposed to the air. A salt load is imposed by the diet. Organisms taken as food are often isosmotic with seawater (e.g., most invertebrates) and therefore strongly hyperosmotic to the body fluids of the predator (p. 153); although the overall concentration of salts in the bodies of such prey is typically below the concentration of salts in seawater because some of the solutes in the prey are amino acids and other organic compounds, the prey are still much more concentrated in salts than the reptile, bird, or mammal. In addition, quantities of seawater itself are probably often taken in with the food. Note that dietary salt loading is a problem for animals such as sea gulls and herons when they are consuming marine plants and animals, even though they do not immerse their bodies in the ocean to any great extent. For the most part, marine reptiles, birds, and mammals have been thought to abstain from drinking seawater; at least among mammals, however, this thinking is currently being re-evaluated.

Reptiles and Birds There are relatively few marine reptiles, but they include representatives of all the major groups: snakes, lizards, turtles, and crocodiles. The numerous marine birds include representatives of several orders. Reptiles are generally unable to produce urine hyperosmotic to their blood. The capabilities of marine birds to concentrate their urine are incompletely understood, but for most the maximum urinary concentration appears to be isosmotic to the blood or only modestly hyperosmotic (up to about twice the blood osmotic pressure).

In most or all of the truly marine reptiles and birds, *salt glands* assume a central role in osmotic–ionic regulation. These are glands located in the head that produce concentrated salt solutions (Figure 8.4). They empty into the nasal passages in birds and lizards; on the other hand, they empty into the orbits in turtles, and into the mouth in sea snakes. Not only are the secretions of the salt glands strongly hyperosmotic to the blood (by a factor of 4–5 in many species), but also, as indicated in Table 8.5, they contain concentrations of Na^+ and Cl^- (and K^+ as well) that exceed those in seawater. Thus, reptiles and birds with salt glands could drink seawater and void the major monovalent ions while still realizing a net gain in water. Their kidneys alone do not provide such a capability. In an experiment on a double-crested cormorant fed fish and NaCl, about half the total NaCl load was voided by the salt glands and about half was voided by way of the cloaca in urine

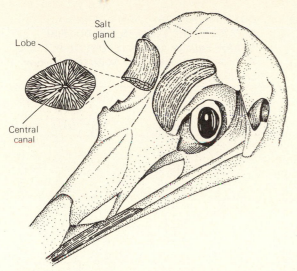

Figure 8.4 Salt glands of the herring gull. Each gland lies in a shallow, crescent-shaped depression in the skull above the eye. The cross section shows that the gland consists of many longitudinal lobes. Each lobe contains a great many branching, radially arranged secretory tubules that discharge into a central canal. The glands empty into the nasal passages via ducts, and their secretion flows out through the external nares. [From K. Schmidt-Nielsen, *Circulation* **21**:955–967 (1960).]

and feces; importantly, the accompanying loss of water in the salt-gland secretions was only half the cloacal loss because of the concentrating ability of the salt glands.

The rate of secretion by salt glands is varied in response to needs to excrete excess salt. The ingestion of a salt load is followed promptly by an increase in the rate of secretion. In addition, the salt glands undergo long-term changes (acclimatization) dependent on the conditions faced by the animal. For instance, if a bird experiences a chronic increase in salt ingestion (e.g., after migrating from a freshwater

Table 8.5 SODIUM CONCENTRATIONS IN THE SALT-GLAND SECRETIONS OF VARIOUS MARINE REPTILES AND BIRDS[a]

	Sodium concentration (mM)
Standard seawater	470
Marine iguana (*Amblyrhynchus cristatus*), nasal secretion	840
Loggerhead turtle (*Caretta caretta*), orbital secretion	732–878
Leach's petrel (*Oceanodroma leucorhoa*)	900–1100
Herring gull (*Larus argentatus*)	600–800
Brown pelican (*Pelecanus occidentalis*)	600–750
Humboldt penguin (*Spheniscus humboldti*)	725–850

[a] Chloride concentrations are typically about the same.
Source: K. Schmidt-Nielsen, *Circulation* **21**:955–967 (1960); K. Schmidt-Nielsen and R. Fange, *Nature* **182**:783–785 (1958).

to a marine habitat), its salt glands will characteristically increase in size, concentrating ability, and peak secretory rate. These changes are reversible. Studies on birds indicate that the short-term and long-term control of salt-gland function is mediated in part by hormones (e.g., adrenal cortical steroids). Furthermore, major or primary control is exerted by the autonomic nervous system; this control is probably guided by information on blood osmotic pressure provided by osmoreceptors located in or near the heart and brain.

The tears observed in marine turtles when they emerge onto land to lay eggs are of some renown. We now understand that they are secretions of orbital salt glands. If you carefully watch a herring gull standing by the ocean, you will see salt-gland secretions dripping slowly from the end of its bill, or perhaps the bird will flick the drops away with a shake of its head.

Although salt glands have been reported in 14 orders of birds, they have not been reported in passerines. Some passerines live in salt marshes, and one of these, a subspecies of the savannah sparrow, *Passerculus sandwichensis beldingi*, has attracted considerable interest. It can drink 75 percent seawater or 1.1 osmolar NaCl with impunity and may well consume water from the salt marshes. It does not possess the extrarenal route of voiding salts provided by salt glands but, compared to most birds, has kidneys with extraordinary concentrating capabilities. The mean maximum Cl^- concentration in the urine is reported to be 960 mM, and the mean maximum osmotic pressure is about 2 osmolar. Another subspecies of this sparrow, *P. s. brooksi*, that has less contact with salt marshes has a maximum urinary concentrating ability—about 1 osmolar—that is still high by comparison to most birds and yet lower than that of *beldingi*, providing an interesting case of physiological divergence at the subspecies level.

Mammals As a group, mammals have evolved the capability of producing urine that is strongly hyperosmotic to their blood plasma. This ability has been important to species occupying arid terrestrial environments and appears to be central to hyposmotic regulation in marine species. Salt glands or other routes for extrarenal salt excretion have not been identified.

As important as the kidneys probably are to seals and whales, urinary concentrating abilities in these animals are not exceptional by mammalian standards, according to available data. In harbor seals (*Phoca vitulina*), for example, the maximum osmotic U/P ratio is about 6, and the maximum urinary Na^+ and Cl^- concentrations that have been measured are slightly above 500 mM. These renal concentrating abilities are high by comparison to those of reptiles and most birds; but they would not, for example, permit a gain of water by drinking seawater, given that the maximum recorded Cl^- concentration is a bit less than the average Cl^- concentration in seawater, 548 mM (p. 153).

Not a lot is known about overall patterns of salt and water balance in seals and whales. Subsistence on a diet of fish—even fish as salty as herring—seems to pose no insurmountable challenges. To date, however, it remains unclear whether or how most species could subsist entirely on invertebrates, which are substantially more salty yet.

Another issue under debate is the extent to which seals and whales drink seawater. The debate has been rekindled of late by data showing that at least some species of both groups do drink, contrary to long-standing belief. Studies have been made, for instance, of species of seals that haul out on land for weeks on end in hot climates during the breeding season; they then not only fast but also expend water in thermoregulation by sweating or panting. Some of these, we now know, drink seawater. Evidently, a key consideration in understanding such behavior is that mammalian urine can simultaneously carry high concentrations of both urea and salts. Thus, even if an animal's urine were only approximately as concentrated in salts as seawater, drinking could be of advantage, for the flux of *ingested* seawater through the body could carry off urea and thereby (in effect) permit *metabolic* water to be used largely for nonexcretory functions, such as thermoregulation.

Perhaps the most complete account of water and salt balance in a marine mammal is not for a seal or whale but for a bat, *Pizonyx vivesi*. This bat lives on desert islands in the Gulf of California where it subsists on fish and crustaceans. It gets most of its water as preformed and metabolic water from its food. But it can concentrate NaCl sufficiently in its urine (about 615 mM) to gain water from seawater, and it meets a fraction of its water needs by drinking. This fraction increases as flying time increases because flight greatly increases evaporation and therefore water needs.

Elasmobranch Fish: Animals That Are Hyperosmotic but Hypoionic to Seawater

The elasmobranch fish (sharks, skates, and rays) have adopted a novel solution to the osmotic problems of living in the sea. As shown in Table 8.6, their blood salt concentrations are similar to those of marine teleosts and well below those in seawater. The osmotic pressure of their blood, however, is slightly higher than that of seawater because of high concentrations of urea and, to a lesser extent, trimethylamine oxide (TMAO). The hyperosmoticity established by the maintenance of high urea and TMAO concentrations means that these fish experience a small osmotic influx of water rather then being faced with the potential problems of osmotic desiccation seen in most other marine vertebrates.

Urea is the principal nitrogenous end product of protein catabolism in elasmobranchs. The gills of marine elasmobranchs are reportedly less permeable

Table 8.6 COMPOSITION OF PLASMA, URINE, AND RECTAL-GLAND FLUID IN THE DOGFISH SHARK, *Squalus acanthias*, AND OF PLASMA AND URINE IN THE COELACANTH *Latimeria chalumnae*[a]

	Milli-osmolarity	Solute concentrations (mM)			
		Na$^+$	Cl$^-$	Urea	TMAO
Squalus acanthias					
Seawater	930	440	496	0	0
Plasma	1018	286	246	351	71
Urine	780	337	203	72	6
Rectal secretion	1018	540	533	15	—
Latimeria chalumnae					
Seawater	1035	470	548	0	0
Plasma	931	197	187	377	122
Urine	961	184	15	388	94

[a] The composition of the seawater in which fish were living is also indicated.

Source: J.W. Burger and W.N. Hess, *Science* **131**:670–671 (1960); J.J. Cohen et al., *Am. J. Physiol.* **194**:229–235 (1958); R.W. Griffith, *Proc. R. Soc. London Ser. B* **208**:329–347 (1980); B. Schmidt-Nielsen et al., *Comp. Biochem. Physiol.* **42A**:13–25 (1972).

to urea than those of marine teleosts, and urea is resorbed from the urine of elasmobranchs as the urine is formed in the kidneys. These are important adaptations to maintaining the high plasma levels. Urea in high concentrations alters the structure of many proteins, and its concentration is kept fairly low in most vertebrates (around 2–7 mM in human plasma). In comparison, the levels in marine elasmobranchs are extraordinary (Table 8.6). Some enzymes and other macromolecules in elasmobranchs have evolved exceptional resistance to urea's deleterious effects. Some organs, such as the heart, have in fact become dependent on urea for proper function. Recent studies have revealed, however, that many elasmobranch proteins are just as sensitive to urea's perturbing effects as homologous proteins in other vertebrates. How can this be? One answer seems to be that TMAO counteracts the effects of urea. Evidently, the retention of *both* urea and TMAO (in a particular ratio) is an adaptation for minimizing enzyme perturbation.

In most aquatic animals, the blood osmotic pressure is due largely to dissolved salts. Thus, problems of osmotic and ionic regulation are related in particular ways. If the animal is hyperosmotic to its medium, it confronts passive losses of salts and gains of water. If hyposmotic, the converse is true. In the elasmobranchs these relations are to some extent uncoupled. Being slightly hyperosmotic to seawater, they tend to gain water by osmosis, but because their blood salt concentrations are below those in seawater, they also tend to gain excess salts by diffusion and by consuming foods isosmotic with their medium. Owing to their osmotic inputs of water, the marine elasmobranchs, unlike teleosts, do not need to drink seawater to meet water needs. Recall that teleosts incur a heavy salt load in extracting needed water from ingested seawater. The fact that elasmobranchs avoid this added salt load must be seen as one of the chief advantages of maintaining hyperosmoticity through retention of urea and TMAO.

Excess salts are removed in elasmobranchs by the kidneys and, extrarenally, by rectal glands. These glands void into the rectum a secretion (Table 8.6) that is isosmotic with the body fluids but that contains only traces of urea and approximates or slightly exceeds the concentrations of Na^+ and Cl^- present in seawater. Excess divalent ions are voided largely by the kidneys along with significant quantities of NaCl (Table 8.6). The rate of urine production is low, and the urine is moderately hyposmotic to the blood. The relative roles of the rectal glands and kidneys in excreting NaCl are not as yet clear. Nor is it clear whether active ion elimination occurs across the gills.

The modern marine elasmobranchs are believed to be descended from freshwater ancestors. The moderate concentration of salts in their blood supports this viewpoint, and their retention of urea and TMAO is thought to be an adaptation to colonization of the sea. Many elasmobranch species are today found in fresh waters. Some do not show any elevation of urea concentration by comparison to other freshwater animals and may trace a continuously freshwater ancestry. Most, however, have urea levels that are 25–30 percent as high as those of marine elasmobranchs. Such levels are far higher than those in most freshwater animals and have the disadvantage of raising blood osmotic pressure, thereby aggravating problems of osmotic flooding in fresh water. Evidently, these species have invaded fresh water from the sea, and their elevated urea levels are a reflection of this heritage.

Latimeria

The crossopterygian fish, presumed ancestors of the terrestrial vertebrates (p. 179), were believed extinct until a coelacanth, *Latimeria chalumnae*, was captured off Africa in 1939. *Latimeria* resembles the marine elasmobranchs in its osmotic–ionic regulation. As shown in Table 8.6, about half of its blood osmotic pressure is caused by urea and TMAO. Although the blood remains hyposmotic to seawater, the osmotic gradient is small, suggesting that drinking, if it occurs, is limited. *Latimeria* has a rectal gland, but its function is not yet established.

RESPONSES OF ANIMALS TO CHANGES IN SALINITY; BRACKISH WATERS

Brackish waters occur in estuaries, salt marshes, mangrove swamps, lagoons, rain-washed tide pools, and in some of the inland seas (e.g., the Caspian Sea, salinity 13 ppt). Although the brackish waters cover less than 1 percent of the earth's surface, they are of enormous commercial importance and biological interest. They are of commercial importance because estuaries and salt marshes are among the most productive of the earth's ecosystems. They are of special physiological interest because animals in brackish habitats frequently face extraordinary challenges of temporal and spatial variation in their environment. When the tide goes out, for example, the salinity at a point in an estuary may fall dramatically. Animals in a tide pool may be immersed in seawater one hour, diluted seawater after a heavy rain the next hour, and seawater again once the incoming tide swirls around their habitat. Although the main focus of this section is on animals in brackish habitats, we shall also look at some other circumstances in which animals confront large changes in salinity, as when migratory fish travel between the oceans and fresh waters.

In the taxonomy of animal relations to variable salinity, one distinction that is made is between *stenohaline* and *euryhaline* species. The former are able to survive only over narrow ranges of salinity, whereas the latter survive broad ranges.

Osmoregulators and *osmoconformers* are also distinguished. A species is described as *osmoregulating* or *homeosmotic* if it maintains a relatively constant internal osmotic pressure over a range of environmental salinities. It is described as *osmoconforming* or *poikilosmotic* if its internal osmotic pressure approximately equals the external osmotic pressure over a range of the latter.

Figure 8.5 illustrates the responses of three invertebrates to changes in external osmotic pressure. The dashed line is called the *line of isosmoticity* (equivalency of internal and external osmotic pressure) and is generally entered on graphs of this sort as a reference; the extent of departure of the response of an animal from the isosmotic line is a reflection of the animal's degree of osmoregulation. Note that *Mytilus* is a strict osmoconformer. *Palaemonetes* is an osmoregulator over a wide range of external osmotic pressures. It maintains its internal osmotic pressure below that of the medium at ambient osmotic concentrations above 0.6 osmolar but maintains its internal osmotic pressure above that of the medium at lower ambient concentrations. The data show that *Palaemonetes* possesses mechanisms for both hyposmotic and hyperosmotic regulation. *Carcinus* displays an intermediate response. It conforms at ambient osmotic concentrations above about 1 osmolar, being incapable of hyposmotic regulation. At ambient concentrations below 1 osmolar, its internal osmotic pressure falls with external osmotic pressure but remains well above isosmoticity, indicating a substantial capability for hyperosmotic regulation.

Some of the physiological implications of regulation and conformity are analogous to those we discussed in regard to temperature relations. The conformer does not incur energy demands for osmoregulation, but to survive over a range of external osmotic pressures, its tissues must be able to function over a similar range of internal solute concentrations. The regulator shields its tissues from

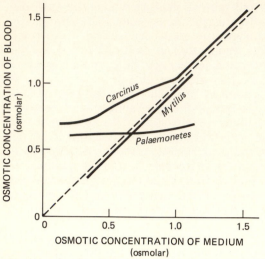

Figure 8.5 Osmotic concentration of the blood as a function of the osmotic concentration of the medium in three invertebrates exposed to various concentrations and dilutions of seawater. *Mytilus edulis,* a common mussel, is a strict osmoconformer. The green crab, *Carcinus maenas,* regulates hyperosmotically in brackish waters but is an osmoconformer at higher salinities. The shrimp *Palaemonetes varians* shows both hyperosmotic and hyposmotic regulation. The dashed line is the line of isosmoticity. For clarity, the curve for *Mytilus* has been displaced slightly below the isosmotic line, and the upper part of the curve for *Carcinus* has been displaced slightly above the line. In fact, these curves lie directly on the isosmotic line. [Sources of data—*Carcinus:* M. Duval, *Ann. Inst. Oceanogr.* **2:**232–407 (1925); *Mytilus:* W.T.W. Potts and G. Parry, *Osmotic and Ionic Regulation in Animals.* Pergamon, Oxford, 1964; *Palaemonetes:* N.K. Panikkar, *Nature* **144:**866–867 (1939).]

large changes in solute concentration but does so at a metabolic cost.

The story of the serious oyster pathogen known as MSX provides just one illustration of the importance that can attach to the physiology of regulation and conformity. MSX, a protistan parasite, has at times devastated the oyster fishery of the Chesapeake Bay. The parasite cannot survive at tissue osmotic pressures below about 0.4 osmolar, and oysters (*Crassostrea*) are osmoconformers. Thus, Bay waters having salinities below 0.4 osmolar are a safe haven: The tissue osmotic pressures of the osmoconforming oysters are low enough there to preclude parasitization. Serious spread of the parasite has occurred during periods of drought because the low rainfall during such periods has allowed the salinity to rise in certain parts of the Bay that ordinarily are more dilute than 0.4 osmolar.

Responses of Most Marine Invertebrates: The Osmoconformers

Most invertebrates characteristic of the oceans are osmoconformers and when placed in diluted seawater soon become approximately isosmotic with the environment. As discussed earlier, most of the ma-

rine invertebrates are relatively permeable creatures adapted to ionic but not osmotic regulation in seawater. Thus, it is not surprising to find them unable to osmoregulate in brackish waters. The hagfish also are osmoconformers.

Recognizing that animal cells assume the same osmotic pressure as the extracellular body fluids, the inability of these animals to regulate the concentration of their blood when exposed to dilute waters means that their cells confront the challenges of intracellular dilution. The cells of most marine osmoconformers have quite limited tolerance to such challenges. The great majority of species exhibit decreased vigor at salinities below about 30 ppt, and the marine fauna diminishes markedly as the salinity in brackish habitats falls below that level. Relatively few of the marine coelenterates, annelids, mollusks, or echinoderms, for example, are to be found in distinctly brackish situations.

Some marine osmoconformers, by contrast, exhibit extensive tolerance of cellular dilution and are found over broad ranges of salinity. The oysters and mussels (Figure 8.5) already discussed, for example, can occur at salinities ranging from about 5 ppt to full-strength seawater, and the starfish *Asterias rubens* lives in waters more dilute than 15 ppt in the Baltic Sea. Not uncommonly, adults of such species are found at lower salinities than are compatible with reproduction and early development; thus, populations at the extremes of environmental dilution are maintained by reproductive input from more-concentrated waters.

Numerous factors affect the range of salinity tolerance of a species, including temperature, time of exposure, and the rapidity of salinity changes. Species interactions are also important; a species that might otherwise live in dilute waters may be excluded if its prey cannot tolerate the degree of dilution, and conversely a species may prosper in dilute waters that its predators cannot tolerate.

As noted earlier, the salinity can change rapidly in some brackish habitats such as estuaries and tide pools. Certain of the invertebrates are able to escape exposure to the extremes of salinity in such habitats by sealing themselves off temporarily or invading protective microhabitats during salinity fluctuations; thus, they can avoid stresses that the extremes of salinity would pose. Snails and bivalve mollusks, for instance, can seal themselves in their shells for many hours or even days. The potential consequences are graphically illustrated by tests on certain mussels, *Mytilus edulis,* that were living in an estuary where they were regularly washed with *fresh* water at low tide; the salinity in the mantle cavity (the cavity within the shells) of these animals never fell below 19.5 ppt. Animals that burrow in the bottom sediments also enjoy protection because the interstitial water within the sediments does not come to equilibrium with transitory extremes of salinity in the water above. In one estuary, for example, the salinity was found to stay near 21 ppt at a depth of 5 cm in the sediments while the salinity in the waters above varied from 2 to 29 ppt over the tidal cycle.

If osmoconformers are shifted from seawater to dilute waters, their body fluids are initially more concentrated than their new environment, and accordingly (unless they seal themselves off) they tend to take on water by osmosis and lose solutes by diffusion. The osmotic influx of water causes an increase in their body weight, and in soft-bodied animals like worms it causes overt swelling. Many stenohaline species die in this bloated state. The euryhaline osmoconformers, by contrast, typically manifest powers of volume regulation; as illustrated in Figure 8.6, they gradually return their weight and volume toward the normal level. Their excess water is believed generally to be excreted in urine isosmotic to their blood.

Invertebrate Osmoregulators

Insofar as we now know, the marine and brackish-water species of some phyletic groups, such as the mollusks and echinoderms, are uniformly osmoconformers. In these groups, the critical differences between euryhaline and stenohaline species reside squarely at the level of *cellular* tolerance to changes in solute concentration. Some other phyletic groups present a different picture: Among their marine and brackish-water representatives, they include osmoregulators as well as osmoconformers, and the species that live in distinctly brackish waters tend to be osmoregulators. In these phyletic groups, survival in dilute waters is associated with *shielding* the cells from changes in concentration.

The crustaceans display this latter pattern outstandingly. So do nereid polychaete worms. Here, we focus on the crustaceans as an example.

Osmoconforming marine crustaceans are generally limited in their distribution to the oceans and only slightly dilute waters. The spider crab *Maia squinado*, for example, is poikilosmotic, dies in a few hours at salinities much below 28 ppt, and has an exclusively marine distribution. The crustaceans that extend into dilute waters are osmoregulators.

Examples of Crustacean Osmoregulators   Two physiological groups of osmoregulating crustaceans are recognized:

1. The *hyper-isosmotic regulators* that regulate hyperosmotically in dilute waters but lack powers of hyposmotic regulation and thus conform at higher salinities.
2. The *hyper-hyposmotic regulators* that regulate hyperosmotically at low salinities and hyposmotically at high salinities.

The green crab, *Carcinus maenas*, shown in Figure 8.5, is an example of the first group. It is a common shore crab, often found in tide pools, and penetrates estuaries to salinities as low as about 10 ppt. Another hyper-isosmotic regulator is the blue crab (*Callinectes sapidus*) of gastronomic fame, which extends all the way into potable (essentially fresh) waters. Hyper-isosmotic regulation is seen also in brackish-water amphipod crustaceans, as illustrated in Figure 8.7.

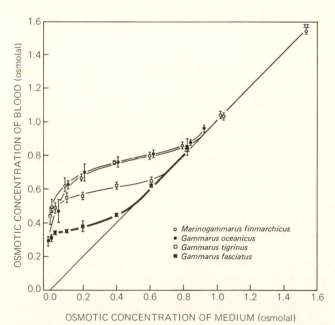

Figure 8.7 Hyper-isosmotic regulation in four species of gammarid amphipods exposed to various concentrations and dilutions of seawater. The species form a graded series in their physiology and ecological relations. The two species that maintain the highest blood osmotic pressures in dilute media occur in marine and brackish habitats but not in fresh waters. *Gammarus tigrinus,* which maintains intermediate internal concentrations when in dilute media, is largely brackish in distribution. *G. fasciatus,* which has evolved the most dilute body fluids, is a freshwater species. [From H.O. Werntz, *Biol. Bull. (Woods Hole)* **124**:225–239 (1963).]

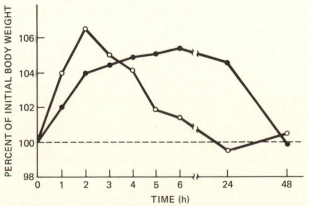

Figure 8.6 Changes in body weight of mussels, *Geukensia demissa,* after sudden transfer from 36 ppt seawater to either 27 ppt (open symbols) or 3 ppt (closed symbols). Transfer occurred at 0 h. The shells were propped open so the animals could not avoid exposure to the diluted environment. [After S.K. Pierce, *Comp. Biochem. Physiol.* **39A**:103–117 (1971).]

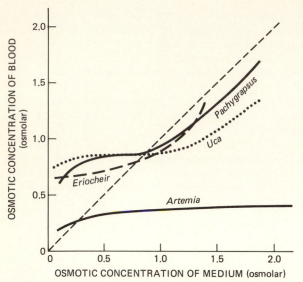

Figure 8.8 Osmotic concentration of the blood as a function of the osmotic concentration of the medium in the Pacific shore crab *Pachygrapsus crassipes;* the fiddler crab *Uca crenulata;* the wool-handed crab, *Eriocheir sinensis;* and the brine shrimp, *Artemia salina.* [Sources of data—*Pachygrapsus* and *Uca:* L.L. Jones, *J. Cell. Comp. Physiol.* **18:**79–92 (1941); *Artemia:* P.C. Croghan, *J. Exp. Biol.* **35:**219–233 (1958); *Eriocheir:* W. Scholles, *Z. Vergl. Physiol.* **19:**522–554 (1933), and R. Conklin and A. Krogh, *Z. Vergl. Physiol.* **26:**239–241 (1938).]

Figures 8.5 and 8.8 depict several of the hyper-hyposmotically regulating crustaceans. The members of this group include many fiddler crabs (genus *Uca*), some shore crabs such as *Pachygrapsus,* and many palaemonid prawns found in shore and estuarine waters. A good number of these animals maintain their blood osmolarity below environmental osmolarity when in seawater; possible reasons have been discussed earlier (p. 168). Furthermore, many are capable of hyposmotic regulation in waters substantially more concentrated than seawater. The adaptive advantage of such an extensive range of hyposmotic regulation is not always apparent. The advantage seems clear, however, in species such as *Artemia salina* and *Palaemonetes varians,* which can be found in inland saline waters of high concentration.

Young stages of crustaceans frequently are more stenohaline than adults. Eggs of blue crabs, for example, require salinities above 23 ppt to develop, whereas adults can live in virtually fresh water. The properties of the young have obvious implications for where spawning can take place.

Mechanisms of Osmoregulation in Euryhaline Crustaceans The crustaceans that osmoregulate are substantially less permeable to water and salts than stenohaline marine osmoconformers; furthermore, among the osmoregulating forms, there is a tendency for species that enter fresh or near-fresh waters to be less permeable than ones that are limited to more-concentrated brackish waters. The mechanisms of

hyposmotic regulation in crustaceans have been discussed earlier (p. 167). Hyperosmotic regulation involves active uptake of NaCl from the environment (across the gills, at least in crabs) and urinary elimination of excess water gained by osmosis. In euryhaline decapods (crabs, shrimp), the urine is approximately isosmotic to the blood (see p. 164), but some of the gammarid amphipods can produce hyposmotic urine. An important point to recall is that the rate of osmotic water influx into a hyperosmotically regulating animal has significant implications for salt balance because salts are carried away in the urine flow necessitated by the osmotic inflow. From this perspective one can see that the low permeability to water of the osmoregulating crustaceans lessens their problems of ion regulation as well as those of volume regulation. Another factor that lessens regulatory problems is that many of these animals permit their blood osmotic and salt concentrations to fall to some extent as they enter more and more dilute waters (see Figures 8.5, 8.7, and 8.8). The decline in blood osmotic pressure helps to limit the magnitude of the osmotic gradient favoring water influx, and the decline in blood ion concentrations tends to limit electrochemical gradients favoring diffusive ion loss.

The responses of the green crab, *Carcinus maenas,* to a declining environmental salinity provide a case study of hyperosmotic regulation. As the ambient osmolarity declines, although the blood osmolarity of *Carcinus* also declines, the difference in osmotic pressure between the blood and medium nonetheless increases (Figure 8.5). Consequently, the rate of osmotic water flux into the blood also increases. *Carcinus* responds by increasing its rate of urine production as the ambient water becomes more dilute; its urine output rises from 4 to 30 percent of body weight per day when the ambient salinity is reduced from 35 to 14 ppt. Sensation of the ambient salinity by the antennules is involved in this response because urine production increases even if only the antennules are bathed with dilute water. Another major reaction of *Carcinus* to lowering of the environmental salinity is an increase in its rate of active Na$^+$ and Cl$^-$ uptake across the gills. The increase in Na$^+$ uptake is triggered by a fall in blood Na$^+$. In seawater, the blood Na$^+$ concentration is near 470 mM; a fall to 360–390 mM causes the rate of active Na$^+$ uptake to increase by 15-fold to its maximum.

In a hyper-hyposmotically regulating crustacean, the *direction* of active transport of Na$^+$ and Cl$^-$ is expected to reverse from inward to outward when the animal moves from an environment in which hyperosmotic regulation occurs to one in which hyposmotic regulation pertains. Such a reversal has now been documented in a number of species. When moved from an environment where they are strongly hyperosmotic (4 ppt) to one where they are strongly hyposmotic (60 ppt), the fiddler crabs *Uca pugnax* and *U. pugilator* not only reverse their direction of

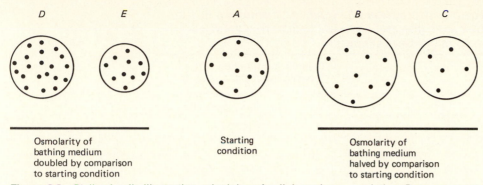

Osmolarity of
bathing medium
doubled by comparison
to starting condition

Starting
condition

Osmolarity of
bathing medium
halved by comparison
to starting condition

Figure 8.9 Stylized cells illustrating principles of cellular volume regulation. Dots represent osmotically effective particles. The cell membranes are presumed permeable to water, and all cells are at equilibrium with their bathing medium as depicted. See text for discussion.

active NaCl transport and decrease their urine flow (from 5 to 0.5 percent of body weight per day) but also increase their drinking (from 1.4 to 11.8 percent of body weight per day) and increase their permeability to water.

Cellular Volume Regulation

Suppose that an animal's blood and other extracellular body fluids are suddenly diluted. In the period immediately following this dilution, the extracellular fluids are hyposmotic to the intracellular fluids. Therefore, the cells will take on water by osmosis and swell until, by processes of water gain and solute loss, they are again isosmotic with the body fluids. This type of problem obviously can occur in osmoconforming animals, but it is not limited to them. As we have just seen, many osmoregulating animals experience significant dilution of their extracellular fluids when they move into more-dilute environments. Conversely, both osmoconforming and osmoregulating animals may experience concentration of their extracellular fluids when they enter environments of increasing concentration, and the process of osmotic equilibration between their intracellular and extracellular fluids may then cause the cells to lose water and shrink. How can an animal maintain or reestablish a constant cell volume in the face of changes in the osmolarity of its extracelluar fluids?

In Figure 8.9, consider cell *A*, which contains 10 osmotically effective particles and is presumed to be isosmotic with the ambient medium in which it is immersed. Suppose that we halve the osmotic pressure of the ambient medium. If the number of particles in the cell were to remain constant, then for the cell to come to osmotic equilibrium with the diluted medium, it would have to take on water until its volume doubled, thus reducing the concentration of particles inside the cell by half. The cell would then look like *B*. On the other hand, if the number of particles in the cell were to be halved, as in *C*, the cell would be at osmotic equilibrium with the diluted medium when it had the same volume as in *A*. This

illustrates that *if the extracellular fluids of an animal become diluted, the number of osmotically effective particles in each cell must be reduced if the cells are to remain isosmotic with the extracellular fluids and keep (or regain) their original volume.* Suppose now that we return to cell *A* and double the osmotic pressure of the bathing medium. If the number of particles in the cell were to stay constant, the volume of the cell ultimately would be halved as in *E*; only by doubling the number of particles as in *D* could the cell retain its original volume. Thus, if the extracellular fluids of an animal increase in osmolarity, cells must increase their content of osmotically effective particles if they are to remain isosmotic with the extracellular fluids and retain (or regain) their original volume.

Inorganic ions are by far the dominant solutes in the blood and other *extracellular* body fluids of marine invertebrates. However, in the *intracellular* fluids, organic compounds account for a substantial part of the total osmotic pressure: often 40–60 percent in ocean-dwelling individuals. The principal types of dissolved organic compounds to be found in the cells are free amino acids, other amino compounds such as taurine, glycine-betaine, trimethylamine oxide, and organic phosphates.

In euryhaline invertebrates of all phyla that have received study, cellular volume regulation is achieved principally by modifying the amounts of organic solutes in the cells (changes in the quantities of inorganic solutes are also involved but typically to a much lesser extent). The solutes, organic or inorganic, whose intracellular quantities are changed so as to assure relative constancy of cell volume are termed **osmotic effectors**.

Table 8.7 illustrates the type of response seen in the tissues of euryhaline invertebrates. Note that when blue crabs were transferred from full-strength seawater to 50 percent seawater and permitted time to acclimate to the new environment, their muscle cells did not become bloated (tissue water content remained almost unchanged) despite a considerable drop in extracellular osmotic pressure because the

Table 8.7 **INTRACELLULAR OSMOTIC CONTRIBUTIONS OF VARIOUS COMPOUNDS IN THE LEG MUSCLES OF BLUE CRABS (*Callinectes sapidus*) ACCLIMATED FIRST TO SEAWATER AND LATER TO 50 PERCENT SEAWATER**[a]

Compounds	Osmotic contribution (milliosmole/kg intracellular water)	
	Acclimated to seawater	Acclimated to 50 percent seawater
Amino acids		
Glycine	362	282
Arginine	136	95
Proline	74	49
Serine	52	5
Alanine	38	23
Glutamate	10	3
Aspartate	8	3
Methionine	6	1
Valine	5	6
Other	19	9
Taurine	69	37
TOTAL of measured organic compounds	780	514
Inorganic ions		
Na^+	40	28
K^+	186	162
Cl^-	46	26
TOTAL of measured inorganic ions	271	217
TOTAL intracellular osmotic pressure accounted for by listed compounds	1051	731
Blood osmotic pressure (milliosmole/L)	1100	850
Tissue water content (percent water)	77.1	77.8

[a] Although the data are expressed as concentrations (milliosmole/kg intracellular water), changes in the intracellular *quantities* of solutes are indicated because the water content of the tissues did not change (note data on tissue water content). Changes in the content of measured amino compounds are also large in the nervous tissue of these crabs, but they are minor in gill tissue. Blood concentrations of amino compounds are very low and virtually invariant.

Source: J.F. Gerard and R. Gilles, *J. Exp. Mar. Biol. Ecol.* **10:**125–136 (1972).

cellular content of osmotic effectors decreased. About 80 percent of the decrease in osmotic effectors was attributable to amino acids and taurine.

In the euryhaline invertebrates as a group (as in blue crabs), certain amino acids (e.g., glycine, arginine, proline, and serine) tend to be more involved as osmotic effectors than others. Certain biochemists believe that these particular amino acids and taurine have been favored as agents of cellular volume regulation because changes in their concentrations intrinsically have relatively small effects on enzyme function.

The mechanisms involved in lowering the intracellular content of free amino acids when euryhaline invertebrates enter dilute waters can be summarized as follows, insofar as they are presently understood. Cells increase their rate of amino acid oxidation, releasing the products of catabolism, CO_2 and NH_3, into the blood. Additionally, cells increase their permeability to amino acids, permitting accelerated diffusion of amino acids out into the blood. Amino acids thus entering the blood may be incorporated into blood proteins (thereby being stored within the body), or they may be excreted, or they also may be catabolized to form CO_2 and NH_3, which are voided into the environment. Indeed, an increased rate of nitrogen excretion has proved to be readily demonstrable in a number of invertebrates after transfer to dilute waters. Part of the excreted NH_3 may be voided as NH_4^+. At least in certain osmoregulating crabs, the excretion of NH_4^+ is coupled (see p. 142) with the active uptake of Na^+, which itself is accelerated in dilute waters.

Processes involved in elevating the intracellular content of amino acids when euryhaline invertebrates are transferred from dilute to concentrated waters may include a decrease in the rate of amino acid catabolism, an increase in protein breakdown, and possibly an increase in the rate at which amino acids are actively transported from the blood into the cells. In mussels (*Mytilus*), breakdown of intracellular protein to produce amino acids is important, and populations exposed to high salinities have especially high frequencies of a gene which codes for an enzyme that is particularly effective in the catalysis of protein breakdown.

By comparison to ocean-dwelling invertebrates, the teleost fish and most other vertebrates maintain low extracellular and intracellular osmotic pressures. Intracellular concentrations of inorganic ions are rather similar in the vertebrates and invertebrates, however; in parallel, intracellular concentrations of free amino acids and other organic compounds are very much lower in most of the vertebrates than in ocean-dwelling invertebrates. It turns out that the osmotic effectors of greatest importance to cellular volume regulation in vertebrates are usually inorganic ions. Potassium commonly plays a preeminent role.

Euryhaline Marine and Migratory Fish

Many fish that live in the oceans are limited, physiologically or ecologically, to marine salinities. Many others have marine and brackish distributions, and some, such as the killifish *Fundulus heteroclitus,* occur routinely in fresh, brackish, and marine waters. Of particular interest are the fish that migrate between fresh waters and the oceans, typically breeding in one habitat and undergoing much of their growth and maturation in the other. Those that ascend rivers and streams from the sea to breed are termed ***anadromous*** ("running upward"), whereas those that grow in fresh waters and descend to the sea for breeding are termed ***catadromous*** ("running downward"). The former group includes species of salmon, smelt, shad, and lampreys. Freshwater eels of Europe (*Anguilla anguilla*) and North America (*A. rostrata*) belong to the latter group.

The migratory and other euryhaline fish are characteristically excellent osmoregulators; they generally show only a small drop in their internal osmotic pressure when in fresh water by comparison to seawater. The chinook salmon (*Oncorhynchus tshawytscha*), for example, has an internal osmotic concentration of about 0.41 osmolar when in the ocean and 0.36 osmolar when at its freshwater spawning grounds.

In many of the species that migrate between the sea and fresh water, physiological capacities for living in the two environments are gained or lost during the life cycle: The animals are not fully euryhaline at all stages of development. Salmon eggs, for example, do not develop normally in saline waters. Young coho salmon (*O. kisutch*) are several months old before they develop full tolerance to marine salinities, and they do not migrate to the oceans until a year or more of age; chum salmon fry (*O. keta*) develop salinity tolerance more quickly and migrate sooner. As adults, some species remain euryhaline during their migrations; individual Atlantic salmon (*Salmo salar*), for instance, make multiple trips up rivers for spawning. Lampreys, *Petromyzon marinus,* illustrate another pattern: They become stenohaline freshwater fish as they make their single spawning run into rivers.

Migratory and other euryhaline teleosts and lampreys regulate hyperosmotically when in dilute waters and hyposmotically when at oceanic salinities. Their mechanisms of regulation in dilute and concentrated waters are like those earlier discussed for freshwater and marine teleosts, respectively. Thus, when they move from seawater to fresh water, their direction of active NaCl transport across the gills reverses from outward to inward, drinking of the medium decreases, urine production increases, and their urine becomes hyposmotic to the blood. In addition, at least in eels, the activity of the NaCl-uptake mechanisms in the intestine decreases when the animals enter fresh water. Experiments on several species demonstrate that the fish exhibit a considerably lower permeability to water when in seawater than when in fresh water; the lower permeability in seawater helps offset the effect of a steeper osmotic gradient between the blood and medium (see p. 166). Paradoxically, permeabilities to Na^+ and Cl^- appear to be lower when fish are in fresh water rather than seawater.

Many of the physiological changes that occur when fish migrate between the sea and fresh water are believed to be under hormonal control. Numerous hormones have been implicated, but great areas of uncertainty remain. In many species, the adenohypophysial hormone prolactin is vital for existence in fresh water; it reduces the permeability of the gills to Na^+, and it augments urine flow by effects on the kidneys and urinary bladder. A role in osmotic–ionic regulation seems also to be firmly established for certain steroid hormones of the interrenal gland (homolog of the adrenal cortex). At least in eels, for example, aldosterone promotes active uptake of Na^+ across the gills when fish are in fresh water; upon transfer of the fish to seawater, cortisol and corticosterone promote extrusion of Na^+ across the gills and active uptake of Na^+ from the gut. Other hormones implicated in the control of gill or kidney function include adrenaline, neurohypophysial hormones such as vasotocin, and hormones of the urophysis. Effects of hormones on the kidneys are discussed further in Chapter 10 (p. 220).

Common Responses of Freshwater Animals in Brackish Waters

Most freshwater animals are capable only of hyperosmotic regulation and do not tolerate a substantial elevation of their internal solute concentration. If a leopard frog, crayfish, carp, or freshwater mussel is placed in progressively more-concentrated brackish water, its internal osmotic pressure rises and ultimately becomes equal to the osmotic pressure of the medium (Figure 8.10). As a rule of thumb, freshwater animals do not survive well at ambient osmotic pressures exceeding the blood osmotic pressures that

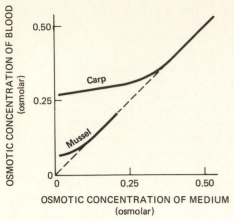

Figure 8.10 Osmotic concentration of the blood as a function of the osmotic concentration of the medium in a carp and freshwater mussel (*Anodonta*). [Data from M. Duval, *Ann. Inst. Oceanogr.* **2:**232–407 (1925).]

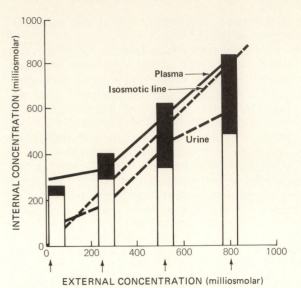

Figure 8.11 Properties of blood and urine in *Rana cancrivora* acclimated to fresh water and various dilutions of seawater for at least 48 h. The solid and long-dashed lines portray the osmotic concentrations of blood plasma and urine, respectively, as functions of environmental osmotic concentration. The bars show the osmotic pressure caused by ions (open part) and urea (solid part) in the plasma at the external concentrations indicated by arrows. Most of the ionic osmotic pressure was due to NaCl. [From M.S. Gordon, K. Schmidt–Nielsen, and H.M. Kelly, *J. Exp. Biol.* **38:**659–678 (1961).]

they normally maintain when occupying fresh water. A few species venture well into estuaries, but the fauna characteristic of fresh water has virtually disappeared at salinities above 10 ppt.

Euryhaline Amphibians: Some Special Cases

Most amphibians are virtually limited to fresh waters, but there are some exceptional species that extend into concentrated brackish waters. *Rana cancrivora,* a crab-eating frog of southeast Asia, occurs in coastal mangrove swamps up to salinities of 29 ppt. The green toad, *Bufo viridis,* of Europe, the Middle East, and western Asia may be found at nearly as high concentrations, though its salinity tolerance varies seasonally and between populations.

In both *R. cancrivora* and *B. viridis,* the osmotic concentration of the plasma rises only slightly in waters up to 10 ppt, but then it increases in parallel with external concentration, remaining somewhat hyperosmotic (Figure 8.11). The adaptation of *R. cancrivora* to high salinities is reminiscent of marine elasmobranchs. As depicted in Figure 8.11, about 40 percent of the frog's increase in internal osmotic pressure at high ambient salinities is attributable to inorganic ions (mostly NaCl), but 60 percent is due to retention of urea. The frog in high salinities, being hyperosmotic to its environment, takes up water by osmosis. Because its internal ion concentrations are below those of the medium, salts are also taken up passively—the reverse of the situation in waters of low salinity. Excess water is voided in the urine, which always remains hyposmotic to the blood. Urinary excretion also seems adequate to void the salt load, though extrarenal excretion remains a possibility. In *Bufo viridis,* most (85 percent) of the increase in internal osmotic pressure at high salinities is due to NaCl, though there is again some increase in blood urea concentration.

The tadpoles of *R. cancrivora* live at even higher salinities than the adults and exhibit a very different pattern of osmoregulation. They remain hyposmotic to the medium when at salinities above about 10 ppt; in full seawater, their internal concentration is around 0.5 osmolar. Mechanisms of regulation include drinking of the medium and extrarenal salt excretion. This species presents a particularly striking example of a change in osmotic–ionic relations during development.

Factors that Affect Responses to Variable Salinity

The exact response of a species to a change in environmental salinity can be affected by many factors, and the complexities thus introduced into the study of salt–water balance are of considerable current interest. Some of the factors of known significance are the following:

(1) *Temperature.* American lobsters (*Homarus*), for instance, can live at lower salinities when exposed to 15°C than 25°C. Cod (*Gadus morhua*) in winter cannot maintain normal blood ion concentrations at oceanic salinities if the temperature falls below 2°C; where a thermal gradient exists, they cluster along the warm side of the 2°C line, a fact used to advantage by fishermen.

(2) *Water composition.* Calcium, for instance, is important for maintenance of the integrity of membranes. Many tropical marine teleosts that do not ordinarily penetrate very far into estuaries enter fresh waters that receive drainage from limestone

regions, possibly because high Ca^{2+} concentrations in the latter waters permit the fish to maintain relatively low permeabilities.

(3) *Direction of salinity change*. When switched acutely between two salinities, for example, euryhaline invertebrates typically recover their normal volume much more rapidly when the transfer is from the high to the low salinity than when it is from low to high.

(4) *Stage of the life cycle* (e.g., p. 174).

(5) *Acclimatization and acclimation*. As shown at the top in Figure 8.12, freshly collected mussels from the North Sea (30 ppt) and the brackish Baltic Sea (15 ppt) differ substantially in the range of salinities over which their gill cilia maintain normal activity. When mussels from each habitat were acclimated to the salinity characteristic of the other habitat, however, their salinity limits for ciliary activity gradually shifted. After 30 days, the North Sea animals living at Baltic salinity assumed limits like those originally seen in Baltic Sea animals, and vice versa—demonstrating that the original differences between the two groups resulted largely from acclimatization.

(6) *Racial differences*. Lampreys (*Petromyzon marinus*) ordinarily migrate to the sea as adults but have become landlocked in some North American freshwater lakes. Adults from the landlocked populations exhibit osmoregulatory difficulties at salinities above 16 ppt, whereas adults from migratory populations, if tested, can osmoregulate well at salinities at least as high as 34 ppt even before they have left rivers and had actual experience with saline waters. The lampreys provide one of a number of examples in which genetic divergence between populations seems likely (although not definitively proved).

RESPONSES TO DRYING OF THE HABITAT IN AQUATIC ANIMALS

Lungfish

Residents of puddles, ditches, small ponds, intermittent streams, and the like are often confronted with drying of their habitat. It is hypothesized that this was an important factor in the emergence of vertebrates onto land. In the Devonian, two groups of related bony fish with lungs and fleshy fins appeared: the Crossopterygii and the Dipnoi. It is believed that these fish occupied transient bodies of fresh water and that their lungs evolved under a selective pressure to resort to air breathing when waters became stagnant. The lungs and lobed fins also probably permitted some species to migrate across the land from drying streams or ponds to seek other bodies of water. The crossopterygians—represented today only by one marine species, *Latimeria chalumnae*—are believed to have given rise to the amphibians. The Dipnoi are currently represented by three genera of lungfish, which are found in transient bodies of fresh water in Africa (genus *Protopterus*), South America (*Lepidosiren*), and Australia (*Neoceratodus*).

If *Protopterus* or *Lepidosiren* are confronted with drying of their habitat, they are able to survive by entering a resting condition within a chamber of mud dug in the bed of a lake or stream. The phenomenon has been well studied in *P. aethiopicus*, which has only rudimentary gills and relies on air breathing even if occupying well-aerated waters. When the water level falls to a low point, the animal digs into the mud and positions itself head upward. It makes occasional trips to the surface for breathing as long as water remains. However, once the water has dried, the fish curls up, and mucus it has secreted hardens into a cocoon that opens only into its mouth. The mouth hole is used for breathing, and the cocoon serves to decrease evaporative water losses. Metabolism ultimately drops to about 10 percent of the ordinary resting level; this reduces respiratory water losses. The fish does not feed, and urine production virtually ceases. Urea becomes the major nitroge-

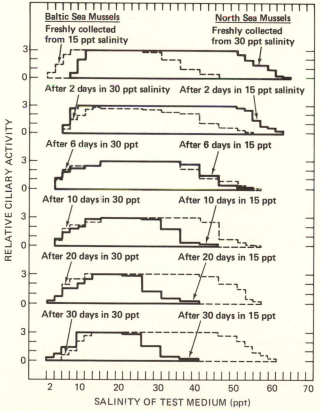

Figure 8.12 Relative ciliary activity of the gills (on a scale of 0 to 3) as a function of test salinity in mussels (*Mytilus edulis*) from the North and Baltic Seas. Data for North Sea animals are drawn with a solid line; those for Baltic Sea animals are drawn with a dashed line. At the top are data for freshly collected animals living at their native salinities. Below are data at various times after the North Sea animals were transferred to 15 ppt and the Baltic Sea animals were transferred to 30 ppt. [Reprinted by permission from H. Theede, *Kieler Meeresforschungen* **21**:153–166 (1965).]

nous end product and builds up in the blood to levels approaching those of marine elasmobranchs. The fish can survive in this condition for more than a year.

Other Freshwater Animals

Many other freshwater animals burrow into the substrate and enter a resting condition during times of drought. This has been reported, for example, in some leeches, snails, water mites, and amphibians. Mucoid coverings are common.

Many forms enter remarkable dormant life stages that are tolerant of becoming fully desiccated in air, and often are tolerant to other environmental extremes as well. Examples of such desiccation-tolerant stages include the eggs of some nematodes, rotifers, water fleas (cladocerans), and other crustaceans and the embryonic cysts of coelenterates and gemmules of sponges. Certain fish (mostly cyprinodonts) also produce desiccation-tolerant eggs. Tardigrades ("water bears") and bdelloid rotifers are able to enter a desiccated dormant condition as adults. Desiccation-tolerant stages often can endure drought for long periods (even several years), and frequently they can be carried about by wind, thus dispersing the species across land.

Intertidal Animals

Marine animals that live along the seashore may be exposed to the air once or twice a day during the ebb of the tide. Many of these intertidal animals remain in protective microhabitats during low tide; small crustaceans and starfish, for instance, are often found under clumps of seaweed. On the other hand, some barnacles, mussels, limpets, and snails remain fully exposed to the air at low tide and face potential problems not only of desiccation but also of overheating.

These exposed animals have shells and can seal themselves off from the atmosphere. Full sealing, however, minimizes not only their water loss but also their exchange of oxygen and carbon dioxide. Capacities for anaerobic metabolism are well developed in many intertidal animals (p. 52). Some species (e.g., certain bivalves) appear to seal themselves and resort to anaerobiosis routinely during low tide. Many species ordinarily do not, however. Many intertidal barnacles maintain a small opening between their opercular valves when the tide is out, permitting some evaporation of water but also providing for oxygen uptake; only if they become considerably desiccated do they seal the opening and turn to anaerobiosis. Some intertidal bivalves gape intermittently or leave their valves partly open continuously. Many types of intertidal animals are able to feed only during those limited parts of the day when they are immersed in water; this restriction on food intake may place a premium on using aerobic metabolism

whenever possible, because aerobic catabolism yields more ATP from a quantity of foodstuff than anaerobic catabolism.

Tolerance to desiccation is often high in intertidal animals, and within a phyletic group it is typically greatest in species or populations that occupy the highest reaches of the intertidal zone where exposure to the air is most protracted. Some intertidal species rank with the most desiccation-tolerant of animals; the limpets *Acmaea scabra* and *Patella vulgata,* for example, can recover from water loss equivalent to 60–65 percent of the weight of their soft body parts. High desiccation tolerance is important not merely because intertidal animals are exposed to the air but because an elevated rate of evaporation may be a price that they must pay to continue aerobiosis or prevent overheating.

Some snails and limpets hold water in their external body cavities when the tide is out, thereby buffering their body fluids against evaporative concentration. Large individuals of a species have lower surface-to-volume ratios than small, and in certain snails larger individuals survive longer in dry air. This factor has been related to distribution in several species: By comparison to large individuals, small ones tend to be found lower in the intertidal zone, where air exposure is shorter.

ANIMALS ON LAND: FUNDAMENTAL PHYSIOLOGICAL CONSIDERATIONS

Animal life originated and spent much of its evolutionary history in water. The land and its plant life represented a vast ecological reservoir within which animals would ultimately establish themselves in an immense variety of ecological niches. The earliest animals that ventured to spend time on land, to consume the productivity of the land, and, ultimately, to develop on land were able to escape competitors and predators in their primordial habitat. In these regards, positive selective pressures for terrestriality must have been great. But early animal life was adapted to living in an abundance of water, and evaporative losses of water on land constituted a physiological problem of paramount importance for all stages of the life cycle.

In studies of the salt–water relations of terrestrial animals, the problems of maintaining water balance on land have deservedly been the focal point, and the present discussion will reflect that emphasis. As in aquatic animals, nonetheless, processes of salt and water balance are significantly intertwined and cannot be divorced from one another; the excretion of excess salts, for example, inevitably entails water loss in terrestrial animals since salts are voided in solution (e.g., urine). Obtaining adequate salts seems usually to present no great problem for terrestrial animals (dietary inputs are adequate); exceptions occur, however, and in such cases problems of salt acquisition and conservation, rather than ones of water balance, may be preeminent.

In discussing the water balance of terrestrial animals, we shall often emphasize how they avoid becoming excessively desiccated, for in terms of water relations, dehydration constitutes the major threat to well-being on land. We must not forget, however, that water is often plentiful enough that animals need not exercise their water-conserving mechanisms maximally. Sometimes, overhydration may even be a problem.

In later sections, we shall examine case studies to see how the discrete elements of water exchange fit into integrated patterns of water balance in a number of animal groups. The present section is devoted to the fundamentals of physiology and behavior relevant to water economy on land. For organizing purposes, we distinguish humidic and xeric animals. The *humidic animals* are ones restricted to humid, water-rich environments, whereas the *xeric animals* are those capable of living in dry, water-poor environments.* These types represent ends of a continuum; so not all animals can readily be classed as being of one type or the other.

Microclimates

Humidic animals, as we have said, must remain for most of their lives in moist, water-rich microenvironments. Some, such as earthworms, are found in soil. Others, such as slugs and centipedes, live in leaf litter or under rocks or rotting logs. The majority of frogs and toads stay in or near bodies of water, and when they venture away from water, they remain in protected microenvironments like the tall grass frequented by leopard frogs. Some humidic animals, such as most amphibians and all terrestrial crabs, have never broken the primordial tie to standing water for breeding.

Humidic animals at times venture briefly from their protected microhabitats, but if they are prevented from returning, the results are quickly devastating. If exposed continually to dry air and wind, they dehydrate rapidly to death.

The xeric animals, by contrast, can live successfully in the open air and expose themselves routinely to the full drying power of the terrestrial environment. Some have succeeded in deserts and other equally dry environments such as grain stores. The major xeric groups are the insects, arachnids (ticks, spiders), reptiles, birds, and mammals. Although xeric animals often seek protected, humid microenvironments, they are not stringently tied to them as humidic animals are.

---

* The term "xeric" is widely used. However, there is no standardized term to describe the animals restricted to moist habitats; "humidic" is used here because it seems best to capture some of the essence of such creatures. The term "mesic" is sometimes applied to animals intermediate between humidic and xeric forms.

Integumentary Permeability to Water

Of all the modes by which terrestrial animals lose water, evaporation—respiratory and integumentary—is one of the most significant. As discussed in Chapter 7 (p. 138), the rate of integumentary evaporative water loss depends on the permeability of the integument to water, as well as on the difference in water vapor pressure between the body fluids and surrounding air.

One of the factors that looms largest in restricting humidic animals to their protective microhabitats is their high integumentary permeability to water. The integument of earthworms, the skin of most amphibians, and most of the fleshy surfaces of snails and slugs, for example, provide little barrier to evaporative water loss; in fact, rates of cutaneous water loss from snails (*Helix*), leopard frogs, and many other amphibians are reported to be 50–100 percent as great as rates of evaporation from open dishes of water of equivalent surface area. With their high permeability, humidic animals can restrict integumentary water losses only by limiting the vapor pressure gradient across their integument—that is, by occupying air with a high vapor pressure.

The evolution of low integumentary permeability must be seen as one of the single most important steps toward a xeric existence. For a long time, the physical basis for the low integumentary permeability of the vertebrate xeric groups was not known. Recent research has started to resolve the issue, however, so that now it seems correct to say that in all the major xeric groups—vertebrate and invertebrate alike—*very thin layers of lipids* are responsible for low integumentary permeability. In mammals and probably birds and reptiles, these lipids (e.g., glycolipids) are laid down in the spaces between cells in the thin, superficial horny layer of the skin (e.g., stratum corneum). In insects and arachnids, the lipids (e.g., hydrocarbons and waxy esters) are contained in the outermost layers of the cuticle; these layers, termed the *epicuticle,* are only 1–2 μm thick. Interestingly, because most of the resistance to water loss across an insect's integument resides in the thin, lipoid epicuticle and not in the thicker, chitinous part of the exoskeleton, soft-bodied insects can have permeabilities as low as hard-bodied ones. The importance of the epicuticular lipids is emphasized also by the fact that arthropods such as many millipedes and centipedes that either lack epicuticular lipids or possess lipids of different types than insects often are quite permeable to water despite the apparent similarity of their exoskeletons to those of insects.

In the insects and arachnids with protective epicuticular lipids, the permeability of the integument to water typically increases as the cuticular temperature is raised, and frequently a *transition temperature* can be identified, above which the increase in permeability becomes particularly marked. Some data support the view that the transition temperature is literally a discrete point on the temperature scale (e.g., Figure 8.13*A*), whereas other results indicate that the marked increase in permeability is gradual, not stepwise (e.g., Figure 8.13*B*). In any case, the changes of permeability with temperature are generally believed to reflect changes in the physical structure of the protective lipoid layers. Often the transition temperature is so high that it would not be experienced by insects in nature. However, this is not always the case. The cockroach *Periplaneta americana,* for example, experiences a marked increase in its perme-

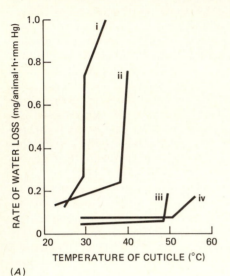

Figure 8.13 Rate of evaporative water loss of dead insects as a function of temperature, expressed per unit of saturation deficit in the air. Measured and expressed in this way, changes in the rate of water loss within a species reflect changes in integumentary permeability. (*A*) Results on (i) large nymphal cockroaches, *Periplaneta americana;* (ii) larval butterflies, *Pieris brassicae;* (iii) adult locusts, *Schistocerca gregaria;* and (iv) mealworms, larval *Tenebrio molitor.* (*B*) Results on the African migratory locust, *Locusta migratoria.* [Data for (*A*) from J.W.L. Beament, *J. Exp. Biol.* **35:**494–519 (1958) and J.W.L. Beament, *J. Exp. Biol.* **36:**391–422 (1959); (*B*) reprinted by permission from J.P. Loveridge, *J. Exp. Biol.* **49:**1–13 (1968).]

ability starting at 25–30°C; thus, it might naturally encounter temperatures high enough to degrade significantly its protection against evaporative water loss. In general, insect species that live in warm, dry habitats tend to have higher transition temperatures than those from cool, moist habitats.

Not only in insects but also in other phyletic groups, if integumentary permeability is measured in a variety of species under similar conditions, it is not uncommon to discover that species from relatively arid habitats have relatively low permeabilities. Look, for example, at the cutaneous water losses of two iguanid lizards, *Iguana* and *Sauromalus,* in Table 8.8. The *Sauromalus,* which occurs in deserts, exhibits an integumentary permeability only about 30 percent as great as that of the *Iguana,* which is a tropical, semiaquatic form. In a compilation by E.B. Edney of data on adult and larval insects (see Selected Readings), the average permeability of 10 arid-habitat species was found to be only one-quarter that of 7 species from relatively moist environments. Arid-habitat insects tend to possess thicker lipid layers and more-protective types of lipids than moist-habitat ones.

Based on appearances, the integuments of the xeric animals are often thought to be so impervious that the animals' cutaneous water losses are utterly dwarfed by their respiratory ones. Frequently, this is not at all true. Looking again at Table 8.8, for example, we see that even though the lizards enjoy relatively low cutaneous losses, their respiratory losses are also low when at rest, and in fact their cutaneous losses constitute *well over one-half* of their total resting evaporative water losses. In humans, some other mammals, some birds, and some insects, integumentary losses also account for approximately one-half or more of total evaporative losses during rest under moderate conditions. There are some birds and mammals, on the other hand, in which the contribution of integumentary losses is much less than one-half.

Respiratory Evaporation

Membranes that are freely permeable to the respiratory gases seem inevitably to be highly permeable to water. Thus, in terrestrial animals, as in many aquatic animals, the need for respiratory gas exchange has special implications for water balance (see also Chapter 12).

Among the humidic animals, some groups have respiratory surfaces that are directly exposed to the air. Earthworms, some isopods, and amphibians, for instance, respire appreciably—or in some cases entirely—across their general integument. The arrangement is a detriment from the viewpoint of evaporative water loss because movement of air across exposed respiratory surfaces can greatly exceed that necessary for exchange of oxygen and carbon dioxide, so water loss can be much greater than that to

Table 8.8 MEAN RESPIRATORY AND CUTANEOUS WATER LOSS IN FOUR REPTILES RESTING AT 23°C IN DRY AIR[a]

	Respiratory water loss (mg/mL O_2)	Cutaneous water loss [mg/(cm² · day)]	Total evaporative water loss (percent body weight/day)	Cutaneous loss as percent of total
Pseudemys scripta, slider turtle	4.2	12.2	2.0	78
Terrapene carolina, Eastern box turtle	4.2	5.3	0.9	76
Iguana iguana, iguana	0.9	4.8	0.8	72
Sauromalus obesus, chuckwalla	0.5	1.3	0.3	66

[a] The species are listed, from top to bottom, in order of increasing aridity of habitat, the slider turtle being amphibious and the chuckwalla being desert-dwelling. Note that resistance to evaporative water loss increases in the same order.
Source: P.J. Bentley and K. Schmidt-Nielsen, *Science* **151**:1547–1549 (1966). Copyright 1966 by the American Association for the Advancement of Science.

which the animal is obligated to support its metabolism. For the animals involved, cutaneous respiration can also have advantages. Frogs, for instance, obtain oxygen from the water across their skin when they are submerged; lungs do not provide this capability.

Most terrestrial animals have evolved invaginated respiratory structures. In the truly xeric groups, reliance is placed exclusively on such invaginated structures for respiration; the integument is virtually impermeable to oxygen and carbon dioxide. The enormous advantage of this arrangement is that access of air to the permeable respiratory membranes can be controlled and thus can be limited to levels required for exchange of oxygen and carbon dioxide; in turn, respiratory evaporation is limited. The reptiles, birds, and mammals control access of air to their lungs by regulating the amplitude and frequency of their breathing movements. Insects and some spiders have a system of respiratory tubes, the tracheae and tracheoles, which branch throughout their body (Figure 12.35). The openings to this system on the body surface, called spiracles, are closed and opened, as needed, to limit evaporation yet permit adequate influx of oxygen. The importance of keeping the spiracles closed much of the time is illustrated by this observation: The rate of evaporative water loss from resting insects is increased by a factor of 2–12 when they are induced experimentally to keep their spiracles steadily open.

Cooling of Exhalant Air in Vertebrates As discussed in Chapter 7 (p. 137), the saturation vapor pressure of air approximately doubles for every 11°C increase in air temperature. This physical law bears particular import for warm-bodied air breathers. When a bird or mammal inhales air, the temperature of the air is raised to the deep-body temperature of the animal, and the air becomes saturated with water vapor at that temperature. Depending on conditions, a substantial addition of water to the air may be involved. For example, consider a mammal inhaling saturated air at 20°C. Such air has a vapor pressure of 17.5 mm Hg and contains 17.3 mg H_2O/L. By the time the air reaches the lungs, it will be saturated at 37°C and thus will have a vapor pressure of 47.1 mm Hg and contain 44.0 mg H_2O/L. Thus, even though the air was saturated to begin with, it will contain 26.7 mg/L more water once it is in the lungs, all of this added water being drawn from the animal's body. If the air were then to be exhaled at body temperature and saturated with water vapor, it would carry all the added water away into the environment.

Usually, air is indeed saturated when it is exhaled, but in many birds and mammals, the temperature of exhaled air is well below the deep-body temperature. Suppose the mammal we have been discussing were to reduce the temperature of air from its lungs to 25°C before exhaling it. On leaving the body, the air would then have a vapor pressure (presuming saturation) of 23.8 mm Hg and contain 23.1 mg H_2O/L. The air would still be carrying away some body water (5.8 mg/L to be exact). Importantly, however, the reduction in the temperature of the exhalant air would have caused 78 percent of the water added during inhalation to be recovered before exhalation.

In those birds and mammals that reduce the temperature of air before it is exhaled, the cooling of the air occurs in the nasal passages and is mediated by a countercurrent phenomenon. To understand the process, let us first look in more detail at what happens during inhalation, using our example of a mammal breathing 20°C air. As inhaled ambient air travels up the nasal passages, it is progressively warmed to about 37°C, and it takes up additional water vapor as its temperature is elevated. The heat that warms the air and the latent heat of vaporization for the added water vapor are drawn from the walls and other surfaces within the nasal passages. At the end of inspiration, the outer ends of the nasal passages may be left as cool as 20°C if the ambient air is saturated; they may be even cooler than 20°C if the ambient air is not saturated, because of evaporative cooling. Inspiration will leave a gradient of increasing temperature along the walls of the nasal passages from the nostrils to the upper passageways. During the ensuing expiration, air will arrive at the upper nasal passages from the lungs at 37°C and saturated. However, as the air moves down the nasal passages, it will encounter the increasingly cooler surfaces established during inspiration and lose heat to them. Cooling of the exhalant air will lower its saturation vapor pressure, causing water to condense out on the passage walls, liberating the heat of vaporization. Thus, by the time the exhalant air exits the nostrils, it will have released a portion of *both the water and heat* that it took up during inhalation. The countercurrent process in the nasal passages differs from that discussed earlier (p. 108) in that the opposing flows (inhalation and exhalation) are separated in time instead of space. Thus, the process has been termed **intermittent countercurrent exchange.**

If this process seems unfamiliar, it may be because cooling of the nasal exhalant air in us humans is comparatively modest; when breathing air at 20°C, for example, we might exhale at 32–34°C. The cooling of nasal exhalant air in small mammals is far greater, as illustrated by the data points in Figure 8.14; it would not be unusual for a small mammal to exhale air at 22–23°C when inhaling at 20°C. Even in medium-sized and large mammals, the cooling can be appreciable in species with long snouts (and nasal passages), such as dogs (Figure 8.14). Importantly, even a 10°C reduction in the temperature of exhalant

air will effect large water savings because the water content of saturated air is such a sensitive function of its temperature.

Experiments on dogs have revealed that when air is exhaled *orally,* it remains nearly at deep-body temperature. Although dogs inhale and exhale exclusively through their nose when not under heat stress, they exhale orally to some (variable) extent when panting. In comparison to nasal exhalation, oral exhalation increases the amount of water vapor carried away in the exhalant air (because of the increase in exhalant temperature). Thus, a dog can vary its evaporative cooling—even while panting at a steady, resonant frequency (p. 113)—by altering the extent of oral and nasal exhalation. This type of system may prove common in panting species.

Besides occurring in mammals and birds, cooling of nasal exhalant air also takes place in some lizards when they are maintaining high, behaviorally regulated body temperatures.

An Algebraic Expression of Respiratory Evaporative Water Loss   As a general principle, the rate of respiratory evaporative water loss from animals depends on both their rate of oxygen consumption and the amount of water lost per unit of oxygen consumed:

water loss (mg/h) = oxygen consumption (mL/h)
    × water loss per unit oxygen
        consumed (mg/mL)

Cooling of nasal exhalant air is one mode of reducing the amount of water lost per unit of oxygen consumed. Another mode of reducing that parameter would be to increase the amount of oxygen extracted from each volume of respired air; marine mammals, for instance, breathe relatively slowly and deeply and are noted for extracting a comparatively large fraction of the oxygen from air they breathe, consequently reducing their evaporative water loss. Under thermally unstressful conditions, terrestrial reptiles, birds, and mammals breathing relatively dry air exhibit water losses in the range of 0.4–4.0 mg/mL O_2 (e.g., see Table 8.8).

An animal's rate of oxygen consumption is the second major determinant of its rate of respiratory evaporative water loss. Birds and mammals suffer exceptional respiratory water losses by comparison to reptiles because of their high metabolic rates and oxygen demands; differences in metabolic rate help explain, for instance, why the total evaporative water losses of white mice in thermoneutrality are about five times those of similarly sized collared lizards (*Crotaphytus collaris*) even when the lizards are at mammalian body temperature. The *net* effect of differences in metabolic rate on water balance is not so dramatic, however, because a relatively high metabolic rate also causes a high rate of metabolic water *production* (see Table 8.10, p. 194).

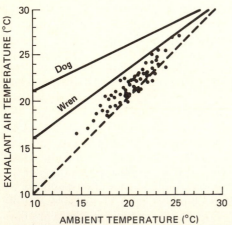

Figure 8.14 Nasal exhalant temperature as a function of ambient air temperature. The dashed line is a line of equality between exhalant and ambient temperature. The plotted symbols (•) are individual data points for 18 species of small mammals, including shrews, bats, mice, squirrels, and rabbits. The solid lines are for mongrel domestic dogs and cactus wrens (*Campylorhynchus brunneicapillum*). The data for dogs and wrens were gathered at low humidities; those for the small mammals were gathered at unspecified humidities. [Data from M.B. Goldberg, V.A. Langman, and C.R. Taylor, *Respir. Physiol.* **43**:327–338 (1981) (dogs); W.D. Schmid, *Comp. Biochem. Physiol.* **54A**:305–308 (1976) (small mammals); K. Schmidt–Nielsen, F.R. Hainsworth, and D.E. Murrish, *Respir. Physiol.* **9**:263–273 (1970) (wrens).]

Table 8.9 TOTAL RATE OF EVAPORATIVE WATER LOSS IN SELECTED ANIMALS, MEASURED
IN DRY OR RELATIVELY DRY AIR AT 25–32°C

Species	Evaporative water loss (percent body wt/h)	Ambient temperature (°C)	Body weight (g)	Reference
Ghost crab (*Ocypode quadrata*)	1.1	25	44	5
Fiddler crab (*Uca annulipes*)	1.7	25	~5	4
Isopod (intertidal) (*Ligia oceanica*)	10.6	32	0.9	4
Isopod pillbug (terrestrial) (*Armadillidium vulgare*)	2.7	32	0.2	4
Migratory locust (*Locusta migratoria*)	0.3–0.5	30	1.9	6
Tsetse fly (*Glossina morsitans*)	0.5	25	0.03	2
Frog (semiaquatic) (*Rana temporaria*)	1.7	26	29	7
Desert spadefoot toad (*Scaphiopus couchii*)	1.4	26	26	7
Desert iguana (*Dipsosaurus dorsalis*)	0.035	26	26	7
Sand lizard (*Uma notata*)	0.036	27	16	3
Desert kangaroo rat (*Dipodomys merriami*)	0.15	27	30–37	3
Lab mouse (*Mus musculus*)	0.39	27	9–24	3
House finch (*Carpodacus mexicanus*)	0.72	25	19	1
Brown towhee (*Pipilo fuscus*)	0.25	25	39	1

Source: (1) G.A. Bartholomew and W.R. Dawson, *Physiol. Zool.* **26**:162–166 (1953). (2) E. Bursell, *Trans. R. Entomol. Soc. London* **111**:205–235 (1959). (3) R.M. Chew and A.E. Dammann, *Science* **133**:384–385 (1961). (4) E.B. Edney, *Trans. R. Soc. South Africa* **36**:71–91 (1961). (5) C.F. Herreid, *Comp. Biochem. Physiol.* **28**:829–839 (1969). (6) J.P. Loveridge, *J. Exp. Biol.* **49**:15–29 (1968). (7) V.H. Shoemaker et al., *Science* **175**:1018–1020 (1972).

Evaporative Water Loss: Overview and Quantification

Table 8.9 lists measured total rates of evaporative water loss in a number of invertebrates and vertebrates under roughly similar ambient conditions. Expressed as percentage loss of body weight per hour, the rates of evaporation in various species span over two orders of magnitude. Note that animals such as the land crabs, pillbug, and amphibians would lose water at rates equivalent to 26–65 percent of their body weight per day if exposed at moderate temperatures to dry air; the necessity of their staying in protective microhabitats is obvious. Insects have relatively low evaporative losses. Among vertebrates, the lizards, birds, and mammals have substantially lower rates of loss than amphibians, and lizards have lower rates than birds or mammals.

A factor that has enormous influence on the weight-specific rate of evaporative water loss is body size. Small animals have a greater integumentary surface area per unit weight than related large animals, and they also tend to have a greater respiratory oxygen demand per unit weight than their larger relatives (p. 32). These considerations tend to impart relatively high weight-specific cutaneous *and* respiratory evaporative losses to the small animals within a phyletic group. Figure 8.15 shows the relation between the total weight-specific rate of evaporation and body weight in birds when exposed to moderate temperatures. Similar relations have been found in virtually every phyletic group examined.

In Table 8.9, all the vertebrate examples are roughly equivalent in body size, so size is not a major factor in the comparisons. However, most of the invertebrate examples are at a disadvantage relative to the vertebrates because of size alone. The evaporative rate of the locust is about one-third of that to be expected for a bird of its size (Figure 8.15). When the minute size of the tsetse fly is taken into account, the effectiveness of insect water-conservation mechanisms becomes particularly impressive.

Let us now look briefly at a number of additional factors that affect rates of evaporative water loss.

(1) *Ambient humidity.* Many animals, as illustrated in Figure 8.16, experience a linear increase in their rate of evaporative water loss as the ambient vapor pressure decreases. On the other hand, many other animals—either in the short term or long term—are capable of increasing their resistance to evaporative water loss as the humidity declines; for them, although evaporative water loss increases as the ambient vapor pressure decreases, the increase is nonlinear and not as great as would be expected from a simple linear relation.

(2) *Activity.* Exercise usually increases substantially the rate of evaporative water loss. This increase occurs at least partly because exercise raises the need for respiratory gas exchange. The increase in metabolism during exercise also augments metabolic

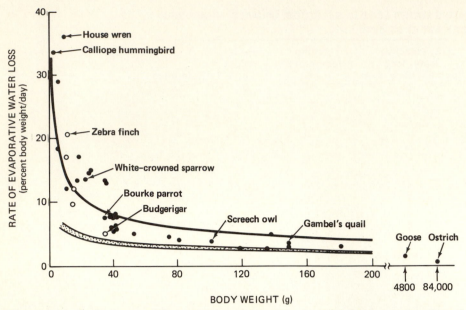

Figure 8.15 Rate of evaporative water loss (percent body weight/day) as a function of body weight in birds resting at 23–25°C in relatively dry air. Open circles identify seed-eating species from arid habitats that are especially tolerant of water deprivation. The heavy curve is a regression line fitted to the data points. The shaded area shows rates of metabolic water production (percent body weight/day) to be expected on the basis of standard relations between resting metabolic rate and weight; the upper limit of the shaded area assumes metabolism of pure carbohydrate, and the lower limit assumes pure lipid. [Data points from E.J. Willoughby and M. Peaker, in G.M.O. Maloiy (ed.), *Comparative Physiology of Osmoregulation in Animals,* Vol. 2. Academic, New York, 1979; format and curves for metabolic water production from G.A. Bartholomew and T.J. Cade, *Auk* **80:**504–539 (1963).]

water *production,* so exercise may or may not induce a substantial *net* increase in the rate of water depletion of the animal. In fact, if conditions are such (e.g., high humidity) that the water produced per unit oxygen consumed exceeds the respiratory water loss per unit consumed, an increase in metabolism, as during exercise, may actually have a net positive effect on water balance.

(3) *Effects of water stress.* Numerous animals have been observed to increase their resistance to evaporative water loss when exposed to water de-

privation, low humidities, or other forms of water stress. Sometimes, the integumentary permeability to water is reduced. Animals may also decrease their respiratory water losses; a number of insects, for example, restrict opening of their tracheal spiracles more closely to the minimum necessary for oxygen uptake when exposed to dehydrating circumstances (p. 292).

(4) *Stage of the life cycle.* Not only evaporative water loss but also other aspects of water balance change with the stage of the life cycle. In general,

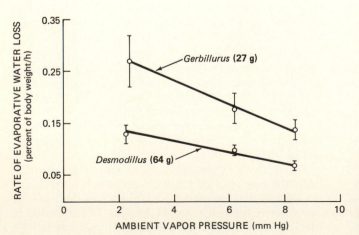

Figure 8.16 Minimal resting evaporative water loss at 23°C as a function of ambient vapor pressure in two rodents from the Namib Desert. Note that the larger species, *Desmodillus auricularis,* has a lower weight-specific rate of evaporation than the smaller species, *Gerbillurus paeba.* [Data from D.P. Christian, *Comp. Biochem. Physiol.* **60A:**425–430 (1978).]

TERRESTRIAL ANIMALS: CASE

typically far outweigh the
water production. Migrator
ing in dry air, for example
fold increase in their *net* rat
temperature is raised from

In birds, mammals, and
ards) that have mechanism
evaporation in service of th
atures high enough to evoke
ing can mean enormous i
Dogs, for example, sometim
respiratory water loss by 10-
mans can maintain sweat ou
hours. Recognizing the magn
that can be incurred, we she
discussed in Chapter 6 (p. 1
mals often exploit behavi
means of avoiding active ev
environments.

**High Humidities Often Limit
ture Tolerance** High amb
evaporative water loss and t
orative cooling of the body
may not be able to tolerate
atures when the humidity is
We humans quickly underst
applies to organisms, like
pant, or otherwise actively
keep cool at high ambient te
ple is also of importance to
merely cooled passively. Fi
example, sometimes have
proaching their lethal tempe
exposed to the sun on beach
bit of passive evaporative c
ference between life and deat
temperatures that are 2–3°C
midity is 80 percent than if i

TERRESTRIAL ANIMALS: CA

In this section, we look at a
briefly summarizing what ha
and discussing some feature
amphibians have been select
part because they so beautif
sity of macroenvironments t
organisms having only meage
to prevent water loss. We al
reptiles, birds, and mammals
ing in deserts and other dema
animals have achieved the e
water constraints.

Amphibians

The terrestrial amphibians h
riety of habitats, from the sh
forests, to deserts. Despite th

this area of knowledge has been disappointingly neglected. An interesting example is provided by the hatchlings of painted quail (*Excalfactoria chinensis*). Their weight-specific evaporative losses are two to three times those of adults because (being smaller) they have higher surface-to-weight ratios and higher weight-specific metabolic rates than adults and also because (for unknown reasons) they have a greater respiratory loss of water per unit of oxygen consumed. The high evaporative losses of the hatchlings not only present a challenge to water balance but also cool the animals substantially and thus are a factor in restricting their capabilities for homeothermy (p. 122).

Excretory Water Loss

Terrestrial animals characteristically modulate the concentration, composition, and volume of their urine in service of osmotic, ionic, and volume regulation. We humans, for example, when dehydrated produce a limited volume of urine that is hyperosmotic to our blood (U/P can reach 4), but after we have consumed large amounts of water, we void a copious urine that is hyposmotic to our blood (U/P can be as low as 0.1–0.2). The bloodsucking bug *Rhodnius prolixus* (a vector of Chagas' disease), after taking a blood meal, produces an abundance of crystal-clear urine that is isosmotic to its body fluids. If it goes without feeding for a period, however, its urine reaches twice the osmotic concentration of its body fluids and is much reduced in volume.

While recognizing the fundamental *regulatory* role of urine, we also need to consider in particular the capabilities of animals to *minimize* the loss of water in their urine. Such capabilities are important when animals face dehydrating circumstances. Urinary water losses can be reduced both by increasing the concentration of the urine and by decreasing the urinary solute load.

Urinary Concentrating Ability Earthworms (*Lumbricus*), amphibians, and most of the rest of the humidic groups are unable to raise the osmotic concentration of their urine above that of their blood. Reptiles too, although xeric, are generally incapable of making hyperosmotic urine.

Certain of the major xeric groups—notably birds, mammals, and insects—possess the ability to make urine that is hyperosmotic to their blood. Among terrestrial insects, maximum osmotic U/P ratios of 2–3.5 have been observed in blowflies (*Calliphora erythrocephala*), desert locusts (*Schistocerca gregaria*), and stick insects (*Carausius morosus*); and mealworms (*Tenebrio*) can produce a virtually dry excrement, equivalent to a U/P ratio of 8. The mechanisms that render the urine of insects hyperosmotic to the blood reside in the rectum and thus operate on the feces as well as the urine.

For most birds, including many that live in arid

regions, the maximum U/P ratio is 1.5–2.5. However, values approaching 6 have been observed in a few forms such as the black-throated sparrow (*Amphispiza bilineata*), which lives in deserts, and a salt-marsh race of the savannah sparrow, *Passerculus sandwichensis beldingi* (p. 169). In mammals, an enormous range of urinary concentrating abilities is to be found, and the U/P ratios achieved by many species are well above those seen in other phyletic groups. To illustrate, the maximum reported osmotic U/P ratio is about 3 for muskrats, 4 for humans, 8 for dromedary camels, 9 for laboratory rats, 14 for Merriam's kangaroo rats and gerbils (*Meriones*), and—at the highest extreme—25–27 for two species of Australian desert mice (*Notomys alexis* and *Leggadina hermannburgensis*). Among related mammals, concentrating ability tends to be correlated with life history, being greatest in species that confront relatively severe problems of desiccation or dietary salt loading. Large desert mammals (e.g., camels) do not attain the concentrating abilities of small species, perhaps because their size places them in a relatively advantageous position in other physiological respects (pp. 118 and 185).

The abilities of insects, birds, and mammals to concentrate their urine permit them to limit the amount of water they must lose in excreting *soluble* wastes.

Salt Glands Nasal salt glands (p. 168) are found in a good number of terrestrial birds (e.g., ostriches, roadrunners) and lizards, where they permit inorganic ions to be excreted at higher concentrations than in the urine. Unlike marine birds and reptiles, many of the terrestrial species produce secretions that are richer in K^+ than Na^+. Emphasis on K^+ excretion is especially appropriate in herbivores because plant tissues are rich in K^+.

Nitrogenous Wastes and Excretory Water Loss (See also Chapter 9.) Nitrogenous wastes are usually voided in the urine by terrestrial animals. When the wastes take the form of highly soluble compounds such as urea or ammonia, this excretion in the urine may entail the loss of substantial amounts of water. Many of the humidic animals—earthworms, isopods, some terrestrial mollusks, and most amphibians—excrete nitrogen principally in highly soluble chemical forms. Mammals also do (they produce mostly urea). Some of the groups that produce highly soluble nitrogenous wastes have evolved means of reducing the water demands of nitrogen excretion. Some isopods and snails void ammonia as a gas, for example, and mammals possess exceptional abilities to concentrate urea in their urine (p. 208).

Many of the xeric groups of animals excrete nitrogen mostly in the form of poorly soluble compounds such as uric acid, urates, allantoin, or guanine. It is a testimony to the advantages of this type of nitrogen excretion for land-dwelling animals that it has

evolved independently
nitrogenous wastes are
nids, snails, some xeric
incorporating waste ni
precipitate, these anim
their urine and thus r
some species, precipit
excreted with so little w
in the urine as a solid p
Birds, reptiles, and sor
nificant amounts of ura
potassium urate; these
trogen but also cationi
form.

Cessation of Urine Excre
iccation, some terrestri
toads, cease altogethe
known as *anuria*. They
terials such as nitrogen
in their bodies.

Tolerance to Desiccatio

If an animal is in deh
length of time it can sur
rate at which it loses boo
extent of loss it can tol
midic and xeric groups
in their tolerance of
brates, for example, var
erate 25–50 percent los
for reptiles, which are
Within a humidic or xer
dration is sometimes (bu
habitat. The semiaquati
ample, tolerates loss of
weight, whereas the ter
tolerates 43 percent, and
Scaphiopus holbrooki t
worms (*Lumbricus*) and
desmus) are among the
animals to desiccation.
cent loss of weight.

Dehydration concent
tually all terrestrial anim
amphibians and isopods.
rises with dehydration
pected on the basis of si
vailing body-fluid solute
volume. Other animals,
mammals, limit their ris
(i.e., they osmoregulate
content of their body flui
Solutes may be removed
on the other hand, seem
solutes in their bodies;
solution during dehydrat
tential return to the boo
(see p. 206).

they find wet soil or standing water. In fact, dehydrated animals typically take up water faster than normally hydrated ones. One reason water uptake is particularly rapid in dehydrated animals is that dehydration elevates the blood osmotic pressure, thus favoring water entry. A second reason is that during dehydration many amphibians increase the apparent permeability of their skin to water influx (whether skin permeability itself increases or there is some other change that mimics an increase in permeability is currently under debate). As a result of these changes, leopard frogs, for example, take up water about five times faster when 75 percent hydrated than when 95 percent hydrated.

Many terrestrial amphibians are able to ward off dehydration for a time when they are away from water sources by using their bladder as a canteen. At the onset of dehydration, the walls of the bladder are rendered permeable to water. Given that the urine inside is typically hyposmotic to the blood, water then leaves the bladder contents by osmosis. In addition, NaCl is actively transported out of the bladder, thus further motivating osmotic outflux of water. The capacity of the bladder to hold fluid in terrestrial frogs and toads is remarkable; the water contained in the filled bladder is equal to 20–50 percent of the animal's bladder-empty weight. By contrast, in strictly aquatic amphibians the bladder is usually tiny.

Control of Responses to Dehydration In terrestrial amphibians, the neurohypophysial hormone *arginine vasotocin,* commonly known as *antidiuretic hormone (ADH),* activates a suite of coordinated responses that retard or reverse the process of dehydration. Release of ADH is stimulated if the volume of the body fluids is decreased or if the osmotic pressure of the body fluids is increased.

The animal's overall reaction to ADH has appropriately been called the "water-balance response." In its complete form (not shown by all amphibians), this response involves changes at three sites in the body: the kidneys, bladder, and skin. (1) ADH causes the kidneys to reduce their rate of urine production and elevate the urine concentration toward isosmoticity with the blood (see p. 216). (2) ADH stimulates the bladder to increase its permeability to water and its rate of Na^+ resorption; both of these responses augment return of water from the bladder to the blood. (3) Finally, ADH causes the skin to increase its apparent permeability to water influx, thereby facilitating rehydration when the animal encounters a suitable water source.

Responses to ADH tend to be most pronounced in species that are particularly terrestrial. The responses are essentially absent in some strictly aquatic amphibians, such as the clawed toad, *Xenopus.*

The Suite of Characteristics in Fossorial Desert Forms
A good number of frogs and toads, such as *Bufo*

cognatus and the spadefoot *Scaphiopus couchii* in North America, live out their lives in deserts or other arid habitats even though their integuments provide no more protection against evaporative water loss than those of semiaquatic frogs like *Rana pipens* (see Table 8.9). These species are indeed remarkably similar to the majority of terrestrial amphibians in all physiological respects, although they sometimes show modest, quantitative improvements over their more-mesic relatives, such as by having a larger bladder, a somewhat greater tolerance of dehydration, or an accelerated pace of rehydration. For all amphibians, behavior looms large in the maintenance of water balance. For the forms living in deserts, where fatal dehydration can take merely a few hours, stringent behavioral control of water loss is one of the prime requirements of life. These desert amphibians are largely nocturnal and spend much of their time in protective microhabitats, especially in burrows underground. One can imagine that they go through repeated ups and downs in their water reserves. There are times when they exhaust their bladder water and dehydrate, suffering ever more concentrated body fluids despite exposing themselves as little as possible to the full brunt of their desert habitat; there are other times when they find moist sands or pools of rainwater and quickly restock their water supplies. They have the critical ability to become dormant deep in the ground during the driest seasons, in some cases sloughing off a protective cocoon (p. 188). Spadefoot toads, *S. couchii,* are known to spend many months of the year in dormancy, during which urea accumulates to very high levels in their body fluids. In their burrows, they are in intimate contact with soil, and the soil initially tends to draw water out of the toads by virtue of its matric potential (p. 151). However, the accumulation of urea in the toads helps ultimately to impede the integumentary water losses to the soil by raising the osmotic pressure of the body fluids. Indeed, the accumulation of urea may even create a gradient of water potential favoring water influx from the soil into the animals. In these respects, the retention of urea aids water balance.

The "Radical" Suite of Characteristics in Certain Arboreal, Arid-Zone Frogs Recently, it has come to light that a number of species of frogs in two genera—*Chiromantis* of Africa and *Phyllomedusa* of South and Central America—have physiological abilities to conserve water that are extraordinarily different from those of most amphibians. These abilities permit them to lead relatively exposed, arboreal lives in arid or semiarid regions. One unusual trait of these frogs is that they excrete much of their nitrogenous waste in the form of poorly soluble uric acid or urates; *P. sauvagei,* for example, excretes 80 percent of nitrogen in these forms. A second unusual trait is that the integuments of these frogs are exceptionally poorly permeable to water, so their evaporative water losses are little different from those of lizards of similar size. In *Phyllomedusa,* the low permeability of the integument results from lipids (mainly waxy esters) that are spread on the skin surface (Figure 8.17).

Xeric Groups: Insects and Arachnids

The suite of characteristics that permits insects and arachnids to prosper in many of the world's dry habitats includes the following attributes: low integumentary water permeability provided chiefly by epicuticular lipids, spiracular limitation of respiratory water losses, excretion of waste nitrogen in poorly soluble form, an ability (at least in many insects) to concentrate solutes in the urine to such an extent that the urinary osmotic pressure substantially exceeds blood osmotic pressure, an ability (in some) to extract water from feces sufficiently that the excrement emerges in relatively dry form, and finally, an ability (in some) to absorb water vapor from the atmosphere. Few water budgets have been worked out for insects, but it is apparent that certain species are so effective in limiting water losses that they can maintain water balance at moderate temperatures and comparatively low humidities while denied drinking water and eating only air-dried foods; examples include grain beetles and clothes moths. Mealworms (larval *Tenebrio*) can maintain their body fluids at nearly constant levels over a month's fast in dry air, apparently replacing their small water losses through metabolic water production.

Xeric Groups: Desert Mammals

Some small mammals, by virtue of well-developed behavioral and physiological abilities to conserve water, are able to subsist in deserts on only metabolic water and the small amounts of preformed water contained in air-dried plant matter. The kangaroo rats (*Dipodomys*) of North America provide a classic example. They moderate their environment by being active on the desert surface only at night and spending the day in protective burrows (Figure 1.2). Physiologically, they are able to reduce their rate of water loss to a very low level by virtue of a number of traits: They exhibit exceptionally low cutaneous water permeability; they cool their exhalant air by nasal countercurrent exchange (p. 183); they can produce very concentrated urine (U/P = 14); and they can restrict their fecal water losses exceptionally by digesting and absorbing a high fraction of the food they ingest and avidly resorbing water in the gut. To clarify how these remarkable animals can survive on air-dried foods without drinking water, let us look at their water budget in detail.

Figure 8.18 depicts, as solid lines, the *minimal* water losses attainable by *D. merriami* over a range of humidities at 25°C when eating pearled barley. Note that the evaporative, urinary, and fecal losses are superimposed on each other so that the heavy solid line represents total losses. Evaporative losses

Figure 8.17 Frogs of the genus *Phyllomedusa* spread lipids secreted by integumentary glands over their skin surface using a series of stereotyped limb movements, some of which are shown here. Arrows show direction of motion; (*A*) and (*B*) are in sequence. The lipids sharply reduce cutaneous evaporative water loss. [Reprinted by permission from L.A. Blaylock, R. Ruibal, and K. Platt-Aloia, *Copeia* **1976**:283–295 (1976).]

decrease with increasing humidity, but the minimal attainable fecal and urinary losses are independent of humidity. Total water losses reflect the decreasing evaporative losses as humidity increases. Water inputs are also shown in Figure 8.18 as dashed lines. Metabolic water production per 100 g of metabolized pearled barley is, of course, independent of humidity. The preformed water contained in air-dried barley increases with humidity because the barley comes to equilibrium with the water vapor in the air. It is assumed in Figure 8.18 that no drinking water is

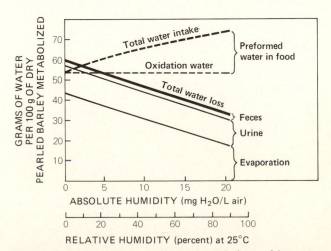

Figure 8.18 Parameters in the water relations of kangaroo rats (*Dipodomys merriami*) eating pearled barley at 25°C. Animals normally consume 100 g of pearled barley in about a month. [From B. Schmidt-Nielsen and K. Schmidt-Nielsen, *J. Cell. Comp. Physiol.* **38**:165–181 (1951).]

available. Metabolic and preformed water inputs are superimposed so that the heavy dashed line represents total water intake. The animal will be in water balance when water inputs equal water losses, and the figure indicates that balance is possible at relative humidities of around 10 percent and higher.

Figure 8.18 was constructed from information on the individual parameters of the water-balance equation. Experiments on kangaroo rats have shown that they can maintain body weight on a diet of pearled barley with no drinking water at 25°C and 24 percent relative humidity. Some tested individuals could also maintain their weight at 10 or 15 percent relative humidity, though others showed a small decrease. The tested animals could not maintain water balance at 5 percent relative humidity. These results basically confirm the abilities indicated by the analysis in the figure.

In the kangaroo rats, we see animals in which metabolic water is the dominant water source. As emphasized in Chapter 7, metabolic water assumes a position of prominence in such animals not because they produce unusual amounts of it but because their exceptional conservation of water reduces their water *needs* to low levels.

Up to now, we have looked at inputs and outputs of water in kangaroo rats as functions of humidity at a single temperature. How is water balance affected by changes in temperature? Metabolic water production and evaporation are, respectively, the dominant modes of water input and loss in these animals. The rates of water flux by both of these processes are regular functions of ambient temperature. Specifically, the rate of metabolic water production, although independent of temperature in the thermoneutral zone, *increases with decreasing temperature* below thermoneutrality (i.e., the rate of water production varies in concert with the metabolic rate). The rate of evaporative water loss by rodents characteristically *decreases with decreasing temperature* in and below the thermoneutral zone. The opposite effects of temperature on the rates of metabolic water production and evaporation yield the type of relation illustrated in Figure 8.19. The figure shows that the *net* effect of metabolism and evaporation on water balance becomes dramatically more favorable as temperature falls: Metabolism yields much more water than is lost by evaporation at low temperatures, but evaporation dominates at high temperatures.

Figure 8.19 helps to clarify recent data on *D. merriami* living wild in the desert. Regardless of season, the animals maintained a virtually constant blood osmotic pressure of about 0.35 osmolar. Their urinary osmotic pressure, however, went through pronounced seasonal changes, being around 1 osmolar (U/P ≃ 3) in midwinter but rising to about 4 osmolar (U/P ≃ 11) in midsummer. Low winter temperatures apparently place the animals in such a favorable situation with respect to metabolic water production

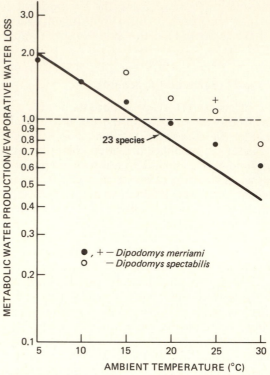

Figure 8.19 The ratio of the rate of metabolic water production to the rate of evaporative water loss as a function of ambient temperature for resting rodents in dry or nearly dry air. The solid line shows the average relation for 23 desert and nondesert species. The open circles are data for *Dipodomys spectabilis*. The closed circles are from one study of *D. merriami;* the cross is from another study of the species, that depicted in Figure 8.18. Above the dashed line, metabolic water production exceeds evaporative water loss. Note that all points would be shifted upward by increasing the humidity since evaporative water loss would decrease as humidity increased. A diet of barley is assumed. [After R.E. MacMillen and D.E. Grubbs, in D.H. Johnson (ed.), *Progress in Animal Biometeorology,* Vol. 1, pp. 63–69. Swetz & Zeitlinger, Lisse, The Netherlands, 1976.]

and evaporation that they enjoy a water surplus, which they excrete renally (lowering the concentration of solutes in the urine). Summer, however, demands nearly maximal frugality in the use of water for urinary excretion.

A good number of rodents besides the kangaroo rats also survive in arid regions on air-dried foods without drinking. Included are some of the gerbils and jerboas of the Old World, hopping mice of Australia, and pocket mice and kangaroo mice of the New World. Some (p. 131) reap major water savings by undergoing daily torpor (*Dipodomys* enter torpor, but to an uncertain extent).

Many mammals of arid regions, although unable to survive on a dry diet without drink, have sufficient abilities to conserve water that succulent plant or animal foods can supply their needs for preformed water. Like the mammals discussed so far, they have achieved independence of standing water for drinking. Included in this category are pack rats (*Neo-toma*), which eat succulent plant foods, including considerable amounts of cacti; certain grasshopper mice (*Onychomys*), which consume large quantities of insects; and several species of desert ground squirrels, which eat growing plants, insects, and dead vertebrates.

Unlike rodents and other small mammals, large desert mammals such as camels are limited in their ability to escape to protective microhabitats; shade is about the only possibility and may not be easy to find. The large animals enjoy physiological advantages, however, as discussed in Chapter 6 (p. 118) and earlier in this chapter (p. 185). Dromedary camels, besides having the advantages of their size, achieve water economy in several ways: (1) They are able to produce relatively dry feces and concentrated urine (U/P = 8); (2) they curtail their urine production relatively rapidly and profoundly when faced with dehydration (see p. 233), allowing solutes (e.g., Na$^+$, urea) to accumulate to some extent in their body fluids rather than excreting them; (3) they have thick fur that provides an effective heat shield (p. 111); (4) they exploit the benefits of large daily changes in body temperature (p. 111); and (5) they minimize heat influx behaviorally, as by orienting a minimal body surface area toward the direct rays of the sun if possible. They also have an exceptional tolerance to dehydration (p. 157). Camels can travel for days or weeks, depending on conditions, without water but do have to drink occasionally, at least in the summer months. Their drinking feats are of some renown, for unlike people they can make up for their full water deficit in one rapid drinking bout even when substantially dehydrated. A 400-kg camel dehydrated by 20 percent would require 80 L of water to rehydrate fully and would likely drink it all in just 10 min (equivalent to a 70-kg person drinking a case and a half of 12-oz cans of water in that time). Such prodigious feats of drinking helped give rise to the notion that camels store excess water prior to long desert journeys. In fact, camels drink only to replace prior losses.

Among the other large desert-dwelling mammals are the oryx (*Oryx beisa*) and Grant's gazelle (*Gazella granti*) of East Africa. Although these species do not seek shade even in the hottest times of day, both are able to live entirely without drinking water for much or all of the year under the conditions they experience. They concentrate their urine to about the same extent as camels. In explaining their independence of drinking water, a major factor is that they enjoy an unusual tolerance of hyperthermia. When exposed to air at 45°C, moderately dehydrated animals allow their core temperature to rise to 0.5–2.0°C *above* air temperature (their brain is cooler, p. 112). Critically, such extreme hyperthermia, coupled with heat storage during the warm-up phase, reduces or eliminates any need to expend body water in active evaporative cooling (p. 112).

Xeric Groups: Desert Birds and Lizards

The birds and lizards are mostly small and thus are readily able to exploit protective microenvironments. However, in contrast to most small mammals, they are generally diurnal and thus confront the heat and aridity of the desert relatively directly. Birds and lizards share the advantages of excreting most of their waste nitrogen, along with appreciable quantities of cations such as K$^+$, in precipitated form as uric acid and urates. In at least some species, water is conserved by nasal countercurrent cooling of exhalant air, and some possess salt glands. Birds under heat stress seek shade and often become very still, thus limiting their metabolic heat production and respiratory evaporation. They also are noted for allowing their body temperatures to rise to high levels (around 45°C), thus avoiding or minimizing recourse to active evaporative cooling (p. 112). Birds have the advantage over both lizards and small mammals of being able to fly to distant watering places.

Desert lizards characteristically eat animal foods (e.g., insects) or living plant tissues of reasonably high water content. The water in their foods must often suffice to meet all preformed water needs since often there is no drinking water in their vicinity. Water budgets for two herbivorous species are shown in Table 8.10. Note that water in the food is the predominant water source for these animals, as it probably is generally for desert reptiles.

Some desert birds also get along without drinking water by eating water-rich animal or plant foods. A lot of attention has centered on the species of desert birds that subsist on seeds and other air-dried plant parts. Are they also independent of drinking water? A number of seed-eating species have proved capable of maintaining water balance without drinking water *under moderate conditions of temperature* in the laboratory. Water budgets for budgerigars and zebra finches under such conditions are presented in Table 8.10, for example. These budgets are notable for their similarity to the budget of kangaroo rats under similar conditions (Figure 8.18), metabolic water serving as the major water source. Note, however, that whereas kangaroo rats can spend most of their lives in deserts under moderate thermal conditions, these birds, being neither nocturnal nor fossorial, cannot. The difference is important, and it seems clear that during the warm months in deserts, the seed-eating birds must have access to drinking water. Indeed, both zebra finches and budgerigars in the Australian deserts visit watering holes in large flocks, and desert travelers have used this behavior to find water.

CONTROL OF WATER AND SALT BALANCE

The mechanisms of water and salt regulation are generally under hormonal control in both vertebrates and invertebrates.

Table 8.10   WATER BUDGETS FOR DESERT LIZARDS (*Dipsosaurus dorsalis* and *Sauromalus obesus*) AND BIRDS (*Melopsittacus undulatus* and *Poephila guttata*)[a]

| | Lizards, free-living in desert, presumed not to have drinking water, but eating water-rich foods | | Birds, given seed but no drink, in laboratory | |
	Desert iguana in summer	Chuckwalla in spring	Budgerigar, 23°C, 30–55 percent RH	Zebra finch, 24°C, 34 percent RH
Gains (g/100 g·day)				
Food	2.7	2.1	1.7	2.3
Metabolic	0.4	0.3	7.7	9.2
TOTAL	3.1	2.5	9.4	11.5
Losses (g/100 g·day)				
Evaporation	0.9	1.2	6.2	6.7
Salt gland	0.3	0.3	—	—
Feces	1.9	0.8	} 3.2	} 3.3
Urine	0.1	0.2		
TOTAL	3.1	2.5	9.4	10.0

[a] The lizard budgets are for animals living in the desert at times of year when they are in water balance. The bird budgets are for animals living under moderate laboratory conditions. All values are expressed as percent of body weight per day. Rounding of original data has caused some small discrepancies between sums of components and "totals." The birds lack salt glands. Fecal and urinary losses were measured together for the zebra finches. Only urinary losses were measured for the budgerigars, and the total for urine and feces was estimated on the basis that the birds were in water balance; most of the combined urinary–fecal loss would appear to be urinary. RH = relative humidity.
Source: B. Krag and E. Skadhauge, *Comp. Biochem. Physiol.* **41A**:667–683 (1972) (budgerigar); P. Lee and K. Schmidt-Nielsen, *Am. J. Physiol.* **220**:1598–1605 (1971) (zebra finch); J.E. Minnich, *Comp. Biochem. Physiol.* **35**:921–933 (1970) (iguana); K.A. Nagy, *J. Comp. Physiol.* **79**:39–62 (1972) (chuckwalla).

Vertebrates

Two of the main groups of hormones known to control water and salt balance in the terrestrial vertebrates are (1) hormones of the neurohypophysis (posterior pituitary) and (2) hormones of the adrenal cortex or its homologs. The neurohypophysial hormones of particular importance are neurosecretory octapeptides known as ***antidiuretic hormones (ADH)***. Each species typically secretes just one ADH, but the chemical nature of the ADH varies somewhat from group to group (p. 592). The adrenal hormones of prime interest are steroids, called ***mineralocorticoids,*** produced by the adrenal cortex or homologous interrenal tissue (p. 594). ***Aldosterone*** is the mineralocorticoid of chief importance. Corticosterone and deoxycorticosterone are also of note.

Effects of Aldosterone and ADH in Mammals   Despite intricate (and incompletely understood) interactions in hormone effects, it is an instructive first approximation to state that *aldosterone is a chief hormone of volume regulation* in mammals and *ADH is the primary hormone of osmoregulation.* Both hormones affect ion regulation.

Aldosterone stimulates the renal tubules to increase both their resorption of Na^+ from the urine and their secretion of K^+ into the urine. Most obviously, this action affects the quantities of these ions in the body fluids. Less obviously, the action of aldosterone is one of the most important elements in the routine regulation of extracellular fluid volume (including blood volume). To understand this latter role of aldosterone, consider that although Na^+ is largely excluded from the intracellular water, Na^+ and its accompanying anions (mostly Cl^-) constitute most of the osmotically active particles in the extracellular fluids. Presuming that the osmotic pressure of the extracellular fluids is regulated at a more or less fixed level, the volume of the extracellular fluids therefore depends pivotally on the body's content of Na^+. If, for example, excess NaCl were to accumulate in the body (as occurs, for instance, in certain disease states), most of it would be in the extracellular fluids, and maintenance of a normal extracellular osmotic pressure would require retention of enough excess water to dilute the excess NaCl, thereby expanding extracellular volume. Variations in aldosterone secretion, by controlling renal excretion of Na^+, thus play a central role in the regulation of extracellular volume. Actually, although aldosterone has its major effects on the kidneys, it is an all-purpose Na^+-retention hormone, stimulating the salivary glands, sweat glands, intestine, and possibly the mammary glands to increase resorption of Na^+ (and typically augment release of K^+). Furthermore, aldosterone is one stimulus for salt appetite.

The pivotally important function of ADH in mammals is to control the excretion of *pure water* (*osmotically free water*) relatively independently of solute excretion. Consider that an animal has a certain *quantity* of urea, salts, and other solutes that it must excrete per day. If the solutes collectively are excreted at the maximal concentration the species can achieve, the accompanying water loss can be considered to be strictly obligated by solute excretion.

However, if the solutes are excreted at less than maximal concentration, then the urine contains some additional water that is not obligated by solute excretion; in essence, the urine has been diluted by the addition of pure water above and beyond the amount needed to void solutes, and the additional water represents a specific excretion of water itself. The urine can therefore be considered to consist of two components: (1) the solutes and their associated water and (2) a quantity of additional pure water. The magnitude of the latter component is controlled by ADH in service of osmotic and volume regulation. If a person, for instance, has a constant daily solute output but consumes little water on one day and a modest amount the following day, the osmotic U/P ratio might drop from 3 to 2 between the first and second days (along with an increase in urine volume) because the kidneys, under control of ADH, would excrete the extra water consumed on the second day, thus preventing it from diluting the body fluids. On another day, if the person were to drink great quantities of water, the U/P ratio might drop to 0.5 along with a further increase in urine volume, reflecting again the renal excretion of water present in excess. As might be guessed from the name "antidiuretic hormone," high levels of ADH in the blood induce production of a relatively concentrated and sparse urine ("antidiuresis"). Low levels of ADH permit production of a relatively dilute and abundant urine ("diuresis"). Aldosterone also affects urinary volume, by helping to control the amount of solute in the urine.

Control of Aldosterone and ADH Secretion in Mammals We now come to the question of how mammals control the secretion of aldosterone and ADH. The mechanisms are both complex and incompletely understood.

For the volume of the body fluids to be regulated, either the volume itself or reliable correlates of volume must be sensed, for otherwise the regulatory systems would not "know" whether to promote an increase or decrease in volume at any particular moment. Probably volume itself is not sensed. However, there is good evidence that certain correlates of volume are; for example, both the blood pressure and the extent to which blood-vessel walls are stretched are functions of blood volume, and pressure and stretch receptors have been identified (e.g., in and around the heart).

In like vein, if the osmotic pressure of the body fluids is to be regulated, either it or close correlates must be sensed. There is no doubt that appropriate receptors for osmoregulation are present (e.g., in the hypothalamus), but whether they respond to osmotic pressure itself, Na$^+$ concentration, and/or other correlated parameters remains uncertain.

Aldosterone secretion is controlled to a major extent by a second hormonal system, called the renin–angiotensin system (p. 611), which itself is partly un-der control of pressure receptors and other detectors of blood volume. A decrease in blood pressure, signifying reduced blood volume, activates secretion by the kidneys of renin, which in turn causes formation in the blood of angiotensin II. The angiotensin stimulates the adrenal glands to secrete aldosterone, which in turn induces increased Na$^+$ resorption from the urine, tending to expand extracellular fluid volume. The angiotensin also stimulates thirst (motivating addition of water to the body fluids) and release of ADH.*

Secretion of ADH, as just suggested, is partly responsive to changes in blood volume; pressure sensors and other sensors of volume affect ADH secretion not only via the renin–angiotensin system but also by way of nervous inputs to the hypothalamus. Decreases in pressure activate ADH secretion, a response favoring fluid retention. Importantly, secretion of ADH is also under control of osmoreceptors or other detectors of the concentration of the body fluids. Increases in the osmotic pressure of the body fluids induce increased ADH secretion; in turn, the ADH favors the specific retention of water by the renal tubules, thus tending to lower the osmotic pressure of the body fluids. Thirst also is partly under control of osmoreceptors or detectors of Na$^+$ concentration.

Natriuretic Hormones After years of speculation and debate, the existence of at least one type of natriuretic hormone seems to be established. The name *natriuretic* signifies that the hormone promotes addition of Na$^+$ to the urine. The known natriuretic hormone is produced by the heart and is a polypeptide or mix of polypeptides. In nonmammalian vertebrates, it is evidently manufactured in both the atria and ventricles. However, in mammals it is synthesized by the atria alone and thus is known as ***atrial natriuretic peptide (ANP)*** or ***atrial natriuretic factor (ANF)***. The effects of the hormone are currently being elucidated. Its actions in mammals are, in many ways, antagonistic to the renin–angiotensin–aldosterone system. It inhibits secretion of the hormones in the system; it directly affects the kidneys to promote Na$^+$ excretion; and in opposition to angiotensin (p. 611), it dilates blood vessels. It also has effects on brain centers involved in the control of water–salt balance and blood pressure. Secretion of ANP is stimulated by expansion of extracellular fluid volume (detected by heart-wall stretching), and the hormone induces a loss of extracellular fluid. Secretion is also stimulated by stress. There is some evidence that other natriuretic hormones or factors are secreted by the brain; these differ markedly from ANP in their mode of action.

---

* Plasma K$^+$ is another important factor controlling aldosterone secretion. High K$^+$ levels promote secretion of aldosterone, which in turn promotes urinary K$^+$ excretion.

Overview of Other Vertebrates and Hormones

Among amphibians, reptiles, and birds, aldosterone and ADH appear generally to assume the same basic regulatory roles that they have in mammals, aldosterone being important in controlling Na^+ retention and ADH in controlling water retention. From one group of animals to another, however, there exist numerous differences of detail in hormone action. As noted, ANP occurs in other vertebrates besides the mammals; in elasmobranchs, for example, it has recently been shown to help control secretion of Na^+ by the rectal gland. Other hormones having demonstrated or putative roles in water–salt regulation in mammals and/or other groups of tetrapods include prolactin, oxytocin, kallikrein, adrenaline, and prostaglandins. Calcitonin, parathormone, and certain other hormones regulate calcium metabolism. Hormonal controls of salt–water balance in fish have been studied mostly in euryhaline species and were discussed earlier (p. 177).

Invertebrates

Among invertebrates, control mechanisms have been studied most thoroughly in certain of the insects. When the blood-sucking bug *Rhodnius prolixus* feeds, it may consume up to 10 times its own weight in blood. Then, over the next few hours, it excretes most of the water from the meal in a virtual flood of urine, which also carries away excess ions; the process concentrates the nutritious part of the meal (e.g., proteins) in the gut. The diuresis following a meal is activated by a neurosecretory hormone produced principally by about a dozen neurons whose cell bodies are located in the mesothoracic ganglionic mass. These neurons are stimulated to release diuretic hormone by nerve impulses arising from stretch receptors in the abdominal wall.

Diuretic hormones have been demonstrated in a number of other insects too, including tsetse flies (*Glossina*), which feed on blood, and stick insects (*Carausius*), which obtain large quantities of water in their succulent plant food. The stick insects also have an antidiuretic hormone, as do such other species as cockroaches (*Periplaneta*) and locusts (*Schistocerca*). The insect diuretic and antidiuretic hormones are all neurosecretions. An interesting recent development is the apparent discovery of a brain hormone in cockroaches (*Periplaneta*) that governs cuticular water permeability.

Control of water–salt balance is also executed at least partly by neurosecretory hormones in a number of other invertebrate groups. In crustaceans, for example, hormones produced by cells in the optic ganglia, brain, and thoracic ganglia affect the movements of salts and water across the gills, renal-organ membranes, and gastrointestinal wall. Most details of these systems remain to be eludicated.

SELECTED READINGS

Ar, A. and H. Rahn. 1980. Water in the avian egg: Overall budget of incubation. *Am. Zool.* **20:**373–384.

Baldwin, G.F. and L.B. Kirschner. 1976. Sodium and chloride regulation in *Uca. Physiol. Zool.* **49:**158–171, 172–180.

Belovsky, G.E. 1981. Sodium dynamics and adaptations of a moose population. *J. Mammal.* **62:**613–621.

Bentley, P.J. 1966. Adaptations of Amphibia to arid environments. *Science* **152:**619–623.

Bentley, P.J. 1982. *Comparative Vertebrate Endocrinology.* Cambridge University Press, New York.

Bentley, P.J. and T. Yorio. 1979. Evaporative water loss in anuran Amphibia: A comparative study. *Comp. Biochem. Physiol.* **62A:**1005–1009.

Bliss, D.E. 1968. Transition from water to land in decapod crustaceans. *Am. Zool.* **8:**355–392.

Bradley, T.J. 1987. Physiology of osmoregulation in mosquitoes. *Annu. Rev. Entomol.* **32:**439–462.

*Bryan, G.W. 1960. Sodium regulation in the crayfish *Astacus fluviatilis. J. Exp. Biol.* **37:**83–112.

Calder, W.A., III and E.J. Braun. 1983. Scaling of osmotic regulation in mammals and birds. *Am. J. Physiol.* **244:**R601–R606.

Cantin, M. and J. Genest. 1986. The heart as an endocrine gland. *Sci. Am.* **254**(2):76–81.

Carey, C. 1986. Tolerance of variation in eggshell conductance, water loss, and water content by red-winged blackbird embryos. *Physiol. Zool.* **59:**109–122.

Carpenter, R.E. 1968. Salt and water metabolism in the marine fish-eating bat, *Pizonyx vivesi. Comp. Biochem. Physiol.* **24:**951–964.

Che Mat, C.R.B. and W.T.W. Potts. 1985. Water balance in *Crangon vulgaris. Comp. Biochem. Physiol.* **82A:**705–710.

Davenport, J. 1985. Osmotic control in marine animals. *Symp. Soc. Exp. Biol.* **39:**207–244.

Diehl, W.J. 1986. Osmoregulation in echinoderms. *Comp. Biochem. Physiol.* **84A:**199–205.

Dill, D.B. (ed.). 1964. *Handbook of Physiology. Section 4: Adaptation to the Environment.* American Physiological Society, Washington, DC.

*Edney, E.B. 1977. *Water Balance in Land Arthropods.* Springer-Verlag, New York.

*Gilles, R. (ed.). 1979. *Mechanisms of Osmoregulation in Animals.* Wiley, New York.

Greenwald, L. 1971. Sodium balance in the leopard frog (*Rana pipiens*). *Physiol. Zool.* **44:**149–161.

Griffith, R.W. 1980. Chemistry of the body fluids of the coelacanth, *Latimeria chalumnae. Proc. R. Soc. London Ser. B* **208:**329–347.

*Gupta, B.L., R.B. Moreton, J.L. Oschman, and B.J. Wall (eds.). 1977. *Transport of Ions and Water in Animals.* Academic, New York.

Hadley, N.F. 1980. Surface waxes and integumentary permeability. *Am. Sci.* **68:**546–553.

*Hickman, C.P., Jr. 1968. Ingestion, intestinal absorption, and elimination of seawater and salts in the southern flounder, *Paralichthys lethostigma. Can. J. Zool.* **46:**457–466.

Hilbish, T.J., L.E. Deaton, and R.K. Koehn. 1982. Effect of an allozyme polymorphism on regulation of cell volume. *Nature* **298:**688–689.

Holmes, W.N. and J.G. Phillips. 1985. The avian salt gland. *Biol. Rev.* **60:**213–256.

Holmes, W.N. and R.B. Pearce. 1979. Hormones and osmoregulation in the vertebrates. In R. Gilles (ed.), *Mechanisms of Osmoregulation in Animals,* pp. 413–533. Wiley, New York.

Kirschner, L.B. 1961. Thermodynamics and osmoregulation. *Nature* **191**:815–816.

*Kirschner, L.B. 1979. Control mechanisms in crustaceans and fishes. In R. Gilles (ed.), *Mechanisms of Osmoregulation in Animals,* pp. 157–222. Wiley, New York.

*Koehn, R.K. and T.J. Hilbish. 1987. The adaptive importance of genetic variation. *Am. Sci.* **75**:134–141.

Krogh, A. 1939. *Osmotic Regulation in Aquatic Animals.* Cambridge University Press, London.

Leader, J.P. 1972. Osmoregulation in the larva of the marine caddis fly, *Philanisus plebeius* (Walk.) (Trichoptera). *J. Exp. Biol.* **57**:821–838.

Lockwood, A.P.M. 1976. Physiological adaptation to life in estuaries. In R.C. Newell (ed.), *Adaptation to Environment,* pp. 315–392. Butterworths, Boston.

Loveridge, J.P. 1968. The control of water loss in *Locusta migratoria migratorioides* R and F. *J. Exp. Biol.* **49**:1–13, 15–29.

MacMillen, R.E. and E.A. Christopher. 1975. The water relations of two populations of noncaptive desert rodents. In N.F. Hadley (ed.), *Environmental Physiology of Desert Organisms,* pp. 117–137. Dowden, Hutchinson, and Ross, Stroudsburg, PA.

MacMillen, R.E. and D.S. Hinds. 1983. Water regulatory efficiency in heteromyid rodents: A model and its applications. *Ecology* **64**:152–164.

Maddrell, S.H.P. 1976. Functional design of the neurosecretory system controlling diuresis in *Rhodnius prolixus. Am. Zool.* **16**:131–139.

*Maetz, J. 1974. Aspects of adaptation to hypo-osmotic and hyper-osmotic environments. *Biochem. Biophys. Perspectives Mar. Biol.* **1**:1–167.

Maloiy, G.M.O. (ed.). 1972. *Comparative Physiology of Desert Animals.* Symposia of the Zoological Society of London, No. 31. Academic, New York.

*Maloiy, G.M.O. (ed.). 1979. *Comparative Physiology of Osmoregulation in Animals.* Academic, New York.

McAfee, R.D. 1972. Survival of *Rana pipiens* in deionized water. *Science* **178**:183–185.

Nagy, K.A. and C.C. Peterson. 1987. Water flux scaling. In P. Dejours, L. Bolis, C.R. Taylor, and E.R. Weibel (eds.), *Comparative Physiology: Life in Water and on Land,* pp. 131–140. Liviana Press, Padova, Italy.

Newell, R.C. 1976. Adaptations to intertidal life. In R.C. Newell (ed.), *Adaptation to Environment,* pp. 1–82. Butterworths, Boston.

Nielsen, B. 1984. The effect of dehydration on circulation and temperature regulation during exercise. *J. Thermal Biol.* **9**:107–112.

Oglesby, L.C. 1981. Volume regulation in aquatic invertebrates. *J. Exp. Zool.* **215**:289–301.

Oglesby, L.C. 1982. Salt and water balance in the sipunculan *Phascolopsis gouldi:* Is any animal a "simple osmometer"? *Comp. Biochem. Physiol.* **71A**:363–368.

Peaker, M. and J.L. Linzell. 1975. *Salt Glands in Birds and Reptiles.* Cambridge University Press, Cambridge.

Potts, W.T.W. 1954. The energetics of osmotic regulation in brackish- and fresh-water animals. *J. Exp. Biol.* **31**:618–630.

Potts, W.T.W. and G. Parry. 1964. *Osmotic and Ionic Regulation in Animals.* Macmillan, New York.

Rahn, H., C.V. Paganelli, I.C.T. Nisbet, and G.C. Whittow. 1976. Regulation of incubation water loss in eggs of seven species of terns. *Physiol. Zool.* **49**:245–259.

Rankin, J.C. and J. Davenport. 1981. *Animal Osmoregulation.* Blackie, Glasgow.

Remane, A. and C. Schlieper. 1971. Biology of brackish water. *Die Binnengewässer* **25**:1–372.

*Schmidt-Nielsen, K. 1964. *Desert Animals.* Oxford University Press, London.

Shaw, J. 1959. Salt and water balance in the East African fresh-water crab, *Potamon niloticus* (M. Edw.). *J. Exp. Biol.* **36**:157–176.

*Shoemaker, V.H. and K.A. Nagy. 1977. Osmoregulation in amphibians and reptiles. *Annu. Rev. Physiol.* **39**:449–471.

*Skadhauge, E. 1981. *Osmoregulation in Birds.* Springer-Verlag, New York.

Taylor, C.R. 1969. The eland and the oryx. *Sci. Am.* **220**(1):88–95.

Tracy, C.R. 1976. A model of the dynamic exchanges of water and energy between a terrestrial amphibian and its environment. *Ecol. Monogr.* **46**:293–326. [See also C.R. Tracy and W.L. Rubink, *Comp. Biochem. Physiol.* **61A**:559–562 (1978).]

Tucker, L.E. 1977. Regulation of ions in the haemolymph of the cockroach *Periplaneta americana* during dehydration and rehydration. *J. Exp. Biol.* **71**:95–110.

*Yancey, P.H., M.E. Clark, S.C. Hand, R.D. Bowlus, and G.N. Somero. 1982. Living with water stress: Evolution of osmolyte systems. *Science* **217**:1214–1222.

See also references in Appendix A.

chapter 9

Nitrogen Excretion and Metabolism

When animals catabolize organic molecules for release of chemical energy, the atoms of the molecules appear in a variety of catabolic end products. Carbon atoms typically appear in carbon dioxide, hydrogen atoms in water, and oxygen atoms in both carbon dioxide and water. The fourth most plentiful atom is nitrogen, which is a characteristic constituent of amino acids and proteins. The disposition of nitrogen atoms during catabolism is the principal topic of this chapter.

Animals receive a steady supply of amino acids in their diet and use them to synthesize a variety of functional nitrogenous compounds, including nucleic acids and many types of protein such as enzymes and muscle filaments. Proteins and other nitrogenous compounds are not usually used in the role of metabolically available storage compounds, however. Thus, amino acids obtained in the diet in excess of those needed for synthesis of functional nitrogenous compounds are in general used promptly in one of two ways. They may be catabolized as energy sources. Or alternatively, they may be used as sources of carbon chains for synthesis of nonnitrogenous compounds—including glycogen, fat, and other metabolically useful storage compounds, which themselves are often destined to be catabolized for release of energy. When amino acids are used as energy sources or for synthesis of nonnitrogenous compounds, reactions that remove their nitrogen atoms are among the first steps.

To give these concepts more specific meaning, let us look at some possible fates of a molecule of serine that is not needed for synthesis of protein or other nitrogenous compounds. As shown in Figure 9.1, the nitrogen-containing amino group is first removed by an enzymatically catalyzed deamination reaction, forming ammonia (NH_3) and pyruvic acid. The py-

ruvic acid can then be directed into the citric acid cycle (p. 38) and oxidized to carbon dioxide and water, with release of energy that is partially incorporated into ATP. Alternatively, the pyruvic acid can be used in the synthesis of the storage compound glycogen via gluconeogenesis (p. 42). Although the paths followed here by serine are especially direct, all deaminated amino acids can be channeled into the usual anabolic and catabolic pathways.

In the case of most amino acids, removal of the amino group is commonly accomplished not by direct deamination, as in the case of serine, but by transfer of the amino group to another compound—that is, by a *transamination* reaction. As shown at the top of Figure 9.2, for example, the amino group of aspartic acid can be transferred under the catalysis of a transaminating enzyme to α-ketoglutaric acid, forming oxaloacetic acid and glutamic acid; or, similarly, alanine can undergo transamination with α-ketoglutaric acid to form pyruvic acid and glutamic acid. Note that the effect of transamination is to remove the amino groups from the initial amino acids. The pyruvic acid arising from alanine can react as indicated in Figure 9.1. The oxaloacetic acid arising from aspartic acid (Figure 9.2) also can enter the citric acid cycle (see Figure 4.1) or can participate in glycogen formation through gluconeogenic reactions. Unlike deamination reactions, transamination reactions do not result in formation of ammonia but, as illustrated, generate other amino acids—in this case glutamic acid. In mammals and a number of other taxa, amino groups removed from most amino acids tend to be channeled, directly or indirectly, to the formation of glutamic acid by transamination reactions. The glutamic acid can then be deaminated to yield ammonia, as shown at the bottom of Figure 9.2. When this occurs, the ultimate effect of transamination to form glutamic

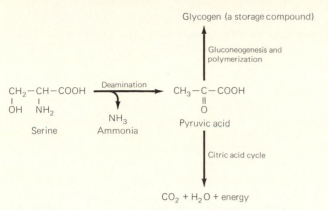

Figure 9.1 Some routes of serine catabolism showing the initial deamination reaction and two possible fates of the resulting pyruvic acid. The disposition of the nitrogen liberated in such reactions is the principal subject of this chapter.

acid and the subsequent deamination of glutamic acid is to remove amino groups as ammonia. The entire process has thus been termed *transdeamination*.

BASIC ASPECTS OF THE DISPOSITION OF NITROGEN

Ammonia, as just suggested, is a common end product of amino acid catabolism, and many animals excrete waste nitrogen in the form of ammonia. However, ammonia is highly toxic, and this and other considerations militate against the final disposition of nitrogen as ammonia in many animals. Often, the nitrogen removed from amino acids is incorporated into less-toxic end products such as urea or uric acid (Figure 9.3).

Animals consume and are composed of a great variety of nitrogenous compounds. When different types of compound are catabolized, their nitrogen

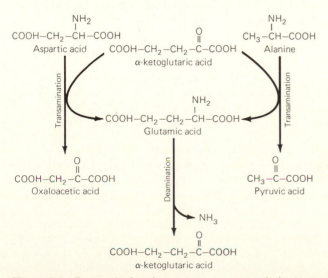

Figure 9.2 Removal of amino groups as ammonia by transdeamination. See text for explanation.

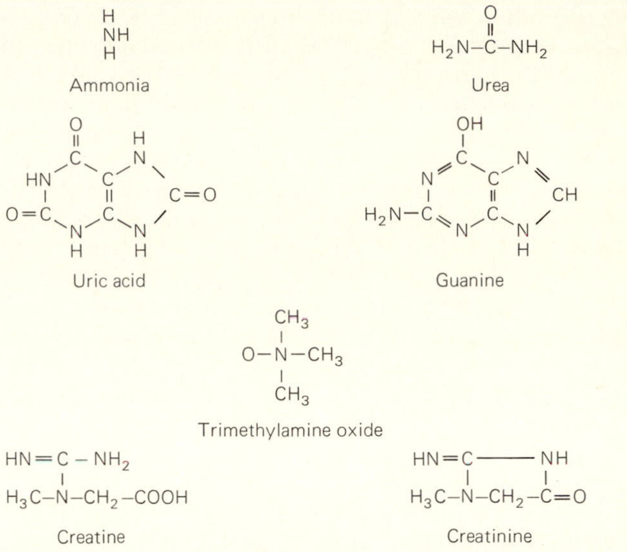

Figure 9.3 Some nitrogenous compounds excreted by animals.

may appear in different end products. Nitrogen from one type of compound may also be incorporated into more than one end product. It is often important to know the ultimate source of the nitrogen atoms found in the various end products produced. Humans, for example, produce large amounts of urea and small amounts of ammonia and uric acid. The urea and ammonia arise largely from catabolism of amino acids and proteins. The uric acid, on the other hand, comes from catabolism of the nucleic acid purines, adenine and guanine.

Because the first step in protein catabolism is typically the depolymerization of the proteins into amino acids, protein nitrogen and amino acid nitrogen enter a common pool and appear in the same end products. In turn, because proteins and amino acids are the dominant nitrogenous compounds of animals, their nitrogenous end products predominate among the various end products produced. Nucleic acid catabolism accounts for roughly 5 percent of catabolic nitrogen.

In the nitrogenous excretions of an animal, the relative dominance of any given end product is generally expressed as the percentage of total catabolic nitrogen represented by that end product. Suppose, to illustrate, that an animal produces 1 mole of ammonia and 0.08 mole of urea over a period of time. The mole of ammonia will carry a mole of nitrogen, whereas the 0.08 mole of urea will carry 0.16 mole of nitrogen (two atoms of nitrogen per urea molecule). There is a total of 1.16 moles of catabolic nitrogen. Of this, 86 percent (1/1.16) is carried in ammonia, and 14 percent is carried in urea. Generally, one or a few end products account for a large proportion of the total nitrogen (and these are characteristically the products of protein and amino acid catabolism). In many animals, the major nitrogenous end product is ammonia, urea, or uric acid. Organisms excreting a preponderance of their nitrogen in one of these forms are termed, respectively, **ammonotelic, ureotelic,** and **uricotelic.** It is important to recognize that these terms are relative. Mammals, for example, are classed as ureotelic. This does not mean that they produce only urea, nor does it necessarily mean that urea constitutes an overwhelming fraction of their catabolic nitrogen. The percentage of excreted nitrogen that is carried in urea is as low as 64 percent in some mammals and as high as 88 percent in others. The terms *ammonotelic, ureotelic,* and *uricotelic* are very useful in facilitating discussion, but we must stress that no one term can fully characterize nitrogen excretion in an animal when, in fact, the excretion may involve many compounds in varying proportions. As a broad generalization, the predominant nitrogenous end product, whatever it may be, rarely represents over 90 percent of the total catabolic nitrogen, often accounts for 60–90 percent, and commonly accounts for only 40–60 percent.

We have been describing the products of nitrogen catabolism as end products, rather than "wastes," for a biologically significant reason. It is true that catabolic end products—be they carbon dioxide, water, or nitrogenous compounds—are generally not used as sources of elements for anabolic (synthetic) processes. It is also true that catabolic end products generally cannot be, or are not, oxidized further for release of energy. These considerations, however, do not mean, in and of themselves, that these end products are wastes. End products often perform important functions. We need only recall, for example, the great significance of metabolic water in the water relations of many animals (p. 154). We have seen in Chapter 8 that urea plays an important role in the osmotic relations of animals such as marine elasmobranchs (p. 170), and other functional roles played by nitrogenous end products will be mentioned in this chapter. Substances become wastes only insofar as they have no useful function or are present in excess of needs. They are then generally excreted, though some animals store excess nitrogenous end products in their bodies for short or long periods of time.

THE MAJOR NITROGENOUS END PRODUCTS IN DETAIL

The list of compounds known to play some role in nitrogen excretion in various species is long, including allantoin, allantoic acid, amino acids, ammonia, creatinine, guanine, hippuric acid, pyrimidines, trimethylamine oxide, urea, uric acid, and urates. Some of these compounds are illustrated in Figure 9.3.

Ammonia

Deamination of amino acids yields ammonia, and ammonia can also arise in the degradation of purines and pyrimidines. Ammonia (NH_3) reacts with hydrogen ions to form ammonium ions (NH_4^+):

$$NH_3 + H^+ \rightleftharpoons NH_4^+$$

At the ordinary pH values of animal body fluids, this reaction is shifted strongly to the right; thus, ammonia typically exists in body fluids primarily as the ammonium ion. For simplicity of discussion, we shall use the term "ammonia" to refer to NH_3 and NH_4^+ taken together; when it is important to distinguish the two forms, we shall use their chemical symbols. Both NH_3 and the salts of NH_4^+ are highly soluble.

Ammonia is quite toxic to animals. Thus, blood concentrations are kept low: under 0.5 mg/100 mL in vertebrates and under 10 mg/100 mL even in the more tolerant of invertebrates. Because of ammonia's toxicity, ammonotelic animals must unfailingly remove ammonia from their bodies as rapidly as it is formed.

Most ammonotelic animals are aquatic. Although they may void some ammonia in their urine, they are believed generally to lose most of it directly to the ambient water across their body surfaces, especially the gills. The mechanisms of loss are currently a matter of uncertainty and debate. As discussed in Chapter 7 (p. 142), in certain groups of freshwater animals, the active-transport mechanisms that take up Na^+ from the water commonly are believed to exchange NH_4^+ (or H^+) for Na^+; according to this view, the animals in part carry out the excretory function of voiding NH_4^+ using active transport in tandem with one of their vital mechanisms of salt regulation. Recently, evidence has been presented that some marine invertebrates and fish also possess active-transport mechanisms that extrude NH_4^+ in tandem with uptake of Na^+, and the mechanisms apparently can account for a significant fraction of their ammonia excretion (e.g., 20–60 percent in several species). This finding, if confirmed, is particularly unexpected in the marine teleost fish because, having body fluids more dilute than their environment, they suffer substantial influxes of excess Na^+ from drinking and diffusion; and the Na^+–NH_4^+ exchange, although it aids ammonia excretion, increases their load of excess Na^+ by at least a modest amount. Besides being lost by active Na^+–NH_4^+ exchange, ammonia can also diffuse out of an aquatic animal's body in the form of NH_3 or NH_4^+. Free ammonia, NH_3, passes through membranes far easier than NH_4^+, but concentration gradients favorable to outward diffusion are typically greater for NH_4^+ than NH_3. A recent appraisal indicates that in fish about 30 percent of the diffusive loss is in the form of NH_4^+ and 70 percent is as NH_3. In general, the relative roles of diffusion, active Na^+–NH_4^+ exchange, and other possible modes of ammonia loss are not well understood. Diffusion of NH_3 is probably the typical dominant mode of loss in aquatic animals.

The reaction of NH_3 with H^+ to form NH_4^+ is a buffer reaction (p. 324). An examination of the reaction reveals that changes in the concentrations of NH_3 and NH_4^+ can affect the pH of body fluids. The reaction plays a significant role in acid–base balance in a number of situations, being important, for example, in the urinary excretion of acid by mammals.

Urea

Urea is synthesized from amino acid nitrogen in many animals. It is a highly soluble compound that generally diffuses readily across membranes. Importantly, it is much less toxic than ammonia. In humans, blood concentrations are normally in the range of 15–40 mg/100 mL, and much higher concentrations, although abnormal, can be tolerated. As noted in Chapter 8, very high concentrations can be found in marine elasmobranchs (around 2000 mg/100 mL), estivating lungfish, and *Rana cancrivora* in saline water.

The most thoroughly known biochemical mechanism for synthesis of urea, the one used by ureotelic vertebrates, is the ***ornithine–urea cycle***. Other routes for synthesis from general amino acid nitrogen have been proposed and may be used by some vertebrates or invertebrates. The ornithine–urea cycle is diagrammed in Figure 9.4. Note that one of the nitrogens appearing in urea originates from free ammonia; in turn, it is derived from deamination reactions, especially deamination of glutamic acid. The second nitrogen appearing in urea comes from the amino group of aspartic acid. Amino groups from most amino acids can make their way to aspartic acid or glutamic acid by transamination reactions. Those amino groups that enter urea by way of aspartic acid need never appear as free ammonia. Urea synthesis occurs primarily in the liver in mammals and apparently in most other ureotelic vertebrates as well.

A noteworthy feature of the ornithine–urea cycle is that the compound from which urea is immediately derived is the amino acid arginine. Cleavage of arginine by the enzyme *arginase* yields urea and ornithine, the latter being recycled. It is currently hypothesized that some of the enzymes of the ornithine–urea cycle originally evolved to serve a nutritive function, that of synthesizing arginine. This nutritive pathway, it is suggested, was then appropriated to the function of producing urea by the evolutionary addition of arginase and certain other enzymes. The ornithine–urea cycle is known today to supply arginine for protein synthesis in a number of ureotelic groups.

The synthesis of urea requires energy; in the ornithine–urea cycle, four high-energy phosphate bonds are used in the structuring of each molecule of urea (two high-energy bonds per nitrogen atom). Production of ammonia does not require this investment of energy. Thus, ureotelic animals gain the advantage of producing an end product of relatively low toxicity but pay an energetic price in comparison to ammonotelic animals.

Some animals lack the ornithine–urea cycle as a whole but possess the enzyme arginase. They cannot channel nitrogen from general amino acid catabolism into urea but can produce relatively small amounts of urea by hydrolyzing dietary arginine. Urea is also formed in the degradation of purines and pyrimidines in various animals.

Purines and Purine Derivatives

Uric Acid, Urates, and Their Formation from Amino Acid Nitrogen   The most important purines from the viewpoint of nitrogen excretion are uric acid, the dihydrate of uric acid, and urate salts. Animals excreting a preponderance of their nitrogen in any of these forms, or a mixture of them, are properly termed uricotelic. Uric acid and urates are of low toxicity.

Uric acid is poorly soluble in water (6.5 mg dissolves in 100 mL at 37°C). The urate salts, although more soluble than uric acid, are still of very low solubility by comparison to urea or ammonia (e.g., about 130 mg/100 mL for monosodium urate and 250 mg/100 mL for monopotassium urate). Because of their low solubility, uric acid and urates can be excreted as a semisolid paste or even as a dry pellet or powder. Thus, excretion of nitrogen in these forms can potentially be accomplished at a considerable savings in water by comparison to the excretion of highly soluble end products that must necessarily be voided in solution.

Experiments have shown that a variety of cations—including Na^+, K^+, NH_4^+, Ca^{2+}, and Mg^{2+}—

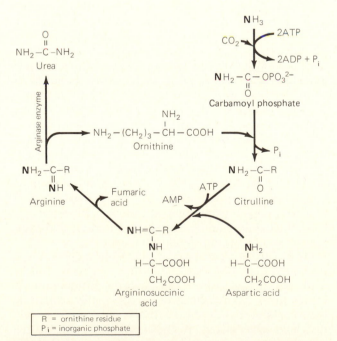

Figure 9.4 An outline of the ornithine–urea cycle. The nitrogen atoms that end up in urea are written bold. Ornithine is recycled over and over.

can be incorporated into uric acid excrement in a poorly soluble state. There is uncertainty over the chemical form taken by these ions; they can be present as urate salts, but apparently they also can be bound in some manner to precipitated uric acid. Regardless, the cations are removed from solution, and this can appreciably reduce the water demands of cation excretion. Calculations indicate, for example, that because of precipitation of K^+ in the excrement, desert iguanas (*Dipsosaurus*) can void as much as 5000 milliequivalents of K^+ per liter of water—an effective concentration that is well above the highest K^+ concentration achieved by reptilian salt glands.

Uricotelic animals synthesize uric acid and urates from the nitrogen of amino acids and proteins. The route of synthesis is complex and is described in biochemistry texts. Suffice it to say here that the nitrogen atoms are derived directly from certain amino acids (glycine, aspartic acid, and glutamine) but can be derived ultimately from other amino acids (by transamination) or from free ammonia. Like urea, uric acid requires energy for its synthesis. Whereas the energy content per milligram of nitrogen is about 0.4 cal higher in the urea molecule than in ammonia, uric acid contains about 3.2 cal more energy per milligram of nitrogen than ammonia.

Pathways of purine synthesis are phylogenetically ancient and ubiquitous in modern animals, serving the nutritive function of producing purines (e.g., adenine) for such compounds as ATP and the nucleic acids. There is little question that these nutritive pathways were appropriated to the function of nitrogen excretion in the evolution of uricotelic animals.

Uricolysis and the Excretion of Nitrogen from Nucleic Acid Purines   When the purines of nucleic acids, adenine and guanine, are catabolized, their nitrogen—so-called "purine nitrogen"—can appear in a variety of end products. Many animals, regardless of their disposition of protein nitrogen, first convert adenine and guanine to uric acid. Species that excrete protein nitrogen predominantly as uric acid generally excrete purine nitrogen in this form as well. Humans and other primates also excrete purine nitrogen as uric acid even though they are ureotelic (and if their uric acid metabolism becomes abnormal, gout can result). Most animals that excrete protein nitrogen in another form than uric acid convert the uric acid arising from purine catabolism to some other compound prior to excretion. The end product is dependent on which of a series of *uricolytic enzymes* are present. Figure 9.5 summarizes the enzymes and the reactions they catalyze. Most mammals, some turtles, and some gastropods possess a uricase enzyme but none of the other uricolytic enzymes. The uricase oxidizes uric acid, splitting the pyrimidine ring to form *allantoin*. Purine nitrogen is then excreted in this form. Allantoin is a slightly more soluble compound than uric acid. Some teleost fish possess one enzyme in addition to uricase: allantoinase. Allan-

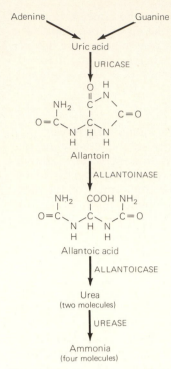

Figure 9.5   Uricolytic reactions and the enzymes that catalyze them. The structures of guanine, uric acid, urea, and ammonia are given in Figure 9.3. The formation of two molecules of urea from allantoic acid may in fact require two enzymes, allantoicase and ureidoglycollate lysase.

toinase converts allantoin to *allantoic acid,* splitting the imidazole ring. Allantoic acid is still only a slightly soluble compound. In addition to the preceding enzymes, elasmobranchs, terrestrial amphibians, and most teleosts possess an enzyme or enzymes that degrade allantoic acid to urea. Many invertebrates, either in their own tissues or in their intestinal microorganisms, have, in addition to all the above enzymes, a urease that degrades urea to ammonia.

As shown in Figure 9.6, animals often have evolved biochemical pathways that permit purine and amino acid nitrogen to be excreted in a common chemical form (examples include elasmobranchs, birds, and many invertebrates). However, there are also prominent examples of a lack of such convergence in the end products of purine and amino acid catabolism (e.g., mammals and most teleosts).

Allantoin and allantoic acid usually arise primarily from the catabolism of nucleic acid purines and thus appear in only small quantities in the excreta. This is not always so. Terrestrial insects generally convert both amino acid nitrogen and purine nitrogen to uric acid, and many excrete the nitrogen predominantly in this form. Many others (e.g., certain heteropteran bugs) exhibit high uricase activity and convert much uric acid to allantoin. In these, allantoin can be the primary nitrogenous end product, for it includes the amino acid nitrogen as well as the purine nitrogen. Still other insects (e.g., certain lepidopterans) pos-

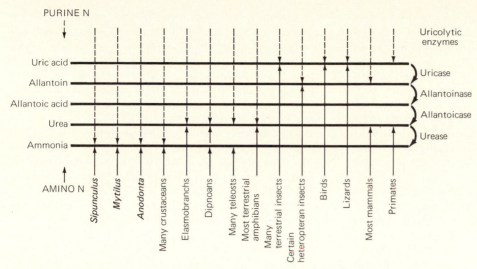

Figure 9.6 Principal nitrogenous end products of purine and amino acid catabolism in certain groups of animals. *Dashed lines:* purine catabolism; *solid lines:* amino acid catabolism. [After M. Florkin and G. Duchateau, *Arch. Int. Physiol.* **53**:267–307 (1943).]

sess both uricase and allantoinase and excrete the preponderance of their nitrogen as allantoic acid.

Guanine Excretion The purine guanine is the principal nitrogenous end product of spiders, scorpions, certain ticks, and some other arthropods. It is even less soluble in water than uric acid and contains more nitrogen per unit weight than uric acid.

Other End Products

Trimethylamine (TMA) and trimethylamine oxide (TMAO) are found in the tissues of many marine invertebrates and fish, and the oxide especially appears as a nitrogenous excretory product in some animals. Both compounds are highly soluble and diffusible and of low toxicity. The elasmobranchs and some marine teleosts are examples of animals that excrete substantial amounts of nitrogen as TMAO [in elasmobranchs, the blood concentration of TMAO is elevated along with that of urea (p. 170)]. Uncertainty remains over the extent to which fish synthesize TMAO de novo; often, their TMAO appears to be derived largely from TMA and TMAO in invertebrate foods. Inexplicably, TMA and TMAO do not occur to any substantial extent in freshwater animals.

Creatine and its internal anhydride, creatinine, occur as relatively minor excretory compounds in many vertebrates and a few invertebrate taxa. Mammals, for example, may excrete from less than 1 percent to over 10 percent of their nitrogen in these forms, mostly creatinine.

In some animals—such as certain crustaceans, bivalve mollusks, insects, echinoderms, and turtles—amino acids are lost to such an extent that they can represent 10 percent or more of total nitrogen loss. It is unclear whether the amino acids should in general be considered true excretory compounds. In

some instances, they may be lost rather accidentally because they can diffuse across the integument or are incompletely resorbed from the urine.

INTERRELATIONS BETWEEN HABITAT, WATER RELATIONS, AND THE FORM OF NITROGEN EXCRETION: BASIC CONCEPTS

Early in this century, it had become clear that the form of nitrogen excretion often varies systematically with habitat among phylogenetically related animals. This observation gave rise to certain classic concepts of the interrelations between habitat and nitrogen excretion. The purpose of this section is to outline those concepts and other related ideas. To a substantial extent, the classic concepts have stood the test of time, although we shall see in subsequent sections that they have been subject to challenge and refinement.

Ammonotelism is the rule today in most aquatic groups but is uncommon in terrestrial animals. The correlation between ammonotelism and the aquatic habitat is understandable in terms of the chemical properties of ammonia. A property of key importance is that ammonia is toxic enough that it must be kept at low levels in the body fluids by being promptly excreted after its formation. As we shall see shortly, maintenance of low levels of ammonia in the body fluids would generally demand a high rate of urine production if renal excretion were necessary; it could thus tax the animal's water resources. Aquatic animals, however, need not excrete ammonia renally and in fact possess means of voiding ammonia across their body surfaces directly into the surrounding water at no cost in body water whatsoever. These means include diffusion and Na^+–NH_4^+ exchange, as already described (p. 200). Given that aquatic animals have the means to void ammonia innocuously, their ammonotelic state is of no surprise, for ammonotelism is probably the primordial form of nitrogen excretion and formation of ammonia is metabolically inexpensive.

The fact that terrestrial animals are typically ureotelic or uricotelic suggests the validity of either or both of two propositions: first, that ammonotelism has disadvantages in terrestrial life and, second, that the other modes of nitrogen excretion have advantages. Both propositions can be rationalized in terms of the chemical properties of the nitrogenous end products.

The contrasting implications of ureotelism and ammonotelism can be explored by looking at vertebrates. Although many fish are ammonotelic, terrestrial amphibians are mostly ureotelic, suggesting that a basic transition occurred in the mode of nitrogen excretion when vertebrates emerged onto land. A critical factor for the terrestrial amphibian is that unlike its aquatic forebears, it cannot lose its nitrogenous end products across its body surfaces into surrounding water but must excrete the end products in its urine at a cost in body water. To see the implications, consider first that, in molar terms, steady-state blood concentrations of urea in ureotelic vertebrates are *at least* 50 times those of ammonia in ammonotelic vertebrates. This difference in the blood concentrations of urea and ammonia reflects the relative toxicities of the two compounds: *Urea, being far less toxic than ammonia, can be allowed to accumulate in the body fluids to far higher levels.* Remembering that one molecule of urea carries twice the nitrogen of a molecule of ammonia, we see that an animal excreting only ammonia would have to achieve a urine:plasma concentration ratio at least 100 times that of an animal excreting only urea in order to void the same amount of nitrogen in the same amount of water. Put another way, *if both animals had the same U/P ratio for their respective end products, the ammonotelic animal would have to void at least 100 times more water to excrete the same amount of nitrogen.* Given the problems of water conservation on land, the advantages for a terrestrial animal of making a low-toxicity end product, rather than ammonia, are evident.

Another factor that would have favored the evolution of ureotelism in terrestrial animals is that animals on land can sometimes face water scarcity so profound that depletion of body water by urine production becomes a threat to life. Because of its low toxicity, urea can be allowed to accumulate in the body fluids to a substantial extent. Thus, a ureotelic animal faced with dehydration can curtail urine production even to the extent of failing to excrete nitrogenous wastes as fast as they are produced. Water balance can take transitory supremacy over nitrogen excretion. If ammonia were the nitrogenous end product of a terrestrial animal, by contrast, the need for its elimination would be of such urgency that a considerable urine output would ordinarily have to

be maintained regardless of the implications for water balance. Terrestrial amphibians faced with dehydration commonly cease to void urine altogether. In some instances, they then accumulate urea to high enough concentrations in their body fluids that it becomes a significant factor in impeding cutaneous water loss or even inducing water influx across the skin from surrounding soil (p. 190). Such effects on transcutaneous water exchange may themselves have played a part in the evolution of ureotelism, for only a highly soluble and relatively nontoxic compound could build up to requisite concentrations in the body fluids.

Uricotelism is more common than ureotelism among terrestrial animals. The distinctive advantages of uric acid and urates for terrestrial life rest on their low solubility. These compounds can carry both nitrogen and cations (e.g., K^+) out of the body in precipitated form, permitting the associated water losses to be sharply limited (see p. 201). The low toxicity and poor solubility of uric acid and urates also give them advantages over urea when circumstances favor curtailment of excretion. Because urea is highly soluble, its concentration in the body fluids increases steadily whenever it is not being excreted as fast as it is produced metabolically; this buildup of the concentration of urea cannot continue indefinitely, for urea does become toxic at high concentrations. Storage of urea is characteristically temporary. By contrast, uric acid and urates, when accumulated, are deposited as precipitates within the body; because their solubilities are low, their concentrations in the body fluids cannot increase above low levels regardless of the amounts stored. Uric acid and urates are suited to indefinite storage. Some uricotelic animals indeed accumulate a portion of their nitrogenous end products for prolonged periods, possibly throughout life in certain cases. Stores of uric acid and urates sometimes may assume dynamic physiological roles (p. 205).

NITROGEN EXCRETION IN THE ANIMAL GROUPS

Invertebrates

Aquatic Groups Aquatic invertebrates are usually ammonotelic and are believed typically to void most ammonia across their outer body surfaces, especially their gills. An appreciable amount of urinary ammonia excretion occurs in some, such as cephalopod mollusks (p. 235) and many of the freshwater animals. At least in the latter, however, evidence indicates that the volume of urine produced is typically governed by needs for body-fluid volume regulation rather than nitrogen excretion; the abundant osmotic flux of water through the body appears to act as a rather passive reservoir for dissipation of ammonia, just as the ambient water serves this purpose in a more general sense. A good number of aquatic insects are ammonotelic, and some of these are unusual in that they excrete most ammonia in their urine and at very high concentrations relative to their blood.

Production of urea has been observed in some aquatic invertebrates (e.g., crabs, snails), but urea is typically only a relatively minor end product in such animals. Few, if any, aquatic invertebrates are routinely ureotelic. However, some freshwater snails accumulate high levels of urea in their blood when estivating, and some snails are evidently ureotelic in certain seasons or when starved.

Nitrogen excretion in the aquatic gastropod and bivalve mollusks is not generally well understood, and one of the more perplexing observations in this regard is that some marine and freshwater snails possess appreciable internal stores of uric acid, stores that sometimes are as large as those typical of terrestrial snails. The stores suggest that these aquatic snails synthesize uric acid, but the fraction of their total catabolic nitrogen that ends up in uric acid has not been assessed. At one extreme, uric acid could prove to be a minor end product, perhaps derived largely from purines ingested in food. At the other extreme, certain aquatic snails might prove to be uricotelic.

Ammonia Excretion in Terrestrial Invertebrates Although ammonia is usually just a minor end product in terrestrial animals, certain humidic invertebrates provide interesting exceptions. Earthworms (*Lumbricus*) are ammonotelic when well fed. Terrestrial isopod crustaceans are also ammonotelic. Significantly, the isopods void much of their ammonia into the atmosphere as NH_3 gas across their integuments, a process that obviates any specific need to use water for the ammonia excretion (although it probably does require a relatively permeable integument). Isopods have high levels of the enzyme glutaminase in their body wall. Evidently, NH_3 destined to be voided across the body surface is largely formed within the body wall itself, by release from circulating glutamine—an arrangement that helps to minimize ammonia levels in the general body fluids. High NH_3 levels in the body wall and integument may contribute to formation of the exoskeleton by creating an alkaline environment favorable to deposition of $CaCO_3$.

Many terrestrial pulmonate snails and at least one terrestrial prosobranch snail also are known to dissipate appreciable quantities of ammonia as NH_3 gas, although volatilization of ammonia is not their principal mode of nitrogen excretion. Enzymes that release NH_3 are found at high levels in the mantle, the shell-forming tissue. The NH_3 almost certainly plays a role in the deposition of $CaCO_3$ in the shell.

Research over the past 15 years has yielded the unexpected revelation that many species of cockroaches not only fail to excrete uric acid but void substantial quantities of ammonia in their urine and feces. In the common American cockroach *Peripla-neta americana,* for example, ammonia has been observed to account for about 40 percent of excreted nitrogen. Clearly, ammonia is an important end product in such animals, but its true fractional contribution to the disposition of catabolic nitrogen is difficult to appraise, in part because these cockroaches also store uric acid (i.e., they produce more nitrogenous end product than they excrete).

Ureotelism in Terrestrial Invertebrates Ureotelism, like ammonotelism, is unusual in terrestrial invertebrates, but a few cases are known or suspected. Earthworms (*Lumbricus*), for instance, become ureotelic when starved, and at least one common terrestrial planarian, *Bipalium kewense,* excretes as much as 70 percent of its catabolic nitrogen as urea. Strong evidence exists that urea is generated by the ornithine–urea cycle in these animals, suggesting that this biochemical pathway is phylogenetically ancient. Some slugs may be ureotelic, and some land snails—although evidently not ureotelic when active—accumulate urea to high concentrations when estivating (a possible advantage in retaining body water, p. 190).

Purines and Purine Derivatives in Terrestrial Invertebrates Poorly soluble compounds are the dominant nitrogenous end products in most terrestrial invertebrates. In the majority of terrestrial insects, for example, the principal nitrogenous excretion is uric acid, allantoin, or allantoic acid (the question of why some insect species excrete mainly one of these compounds and other species excrete mainly another is interesting and unanswered). Spiders, scorpions, and a number of other groups of arachnids excrete mostly guanine. Among most terrestrial gastropod mollusks, uric acid is the predominant nitrogenous end product, but other purines—guanine and xanthine—are often produced in quantity as well. The fact that terrestrial arthropods and mollusks and certain terrestrial vertebrates (e.g., reptiles and birds) have independently evolved an emphasis on poorly soluble purines and purine derivatives as nitrogenous end products speaks to the distinct advantages of such compounds in terrestrial life.

Temporary or permanent storage of purines has been observed in certain land crabs and in many insects and snails. A significant series of studies on cockroaches (*Periplaneta*) has recently revealed that stores of uric acid and urates in their fat body are not inert and probably play significant physiological roles. When cockroaches are placed on a low-protein diet, they lose uric acid from their fat body; the hypothesis has been made that the uric acid is used as a source of nitrogen for synthesis of proteins or for other anabolic processes. Symbiotic bacteria would probably be instrumental in such uses. A similar role for stored uric acid has also been hypothesized in certain other insects, such as boll weevils. The evidence for use of uric acid in anabolism is strongest for termites (*Reticulitermes*). They do not store uric acid. However, they have a nitrogen-poor diet, and with the aid of bacterial symbionts, they clearly recycle nitrogen in their bodies, using uric acid as a source of nitrogen for protein synthesis.

When faced with dehydration and a decrease in blood volume, cockroaches remove certain inorganic ions from their blood and sequester them within their bodies. In this way, increases in the concentrations of the ions in the blood are limited, and the ions remain available for release during rehydration, thus at that time helping to limit dilution of the blood by the influx of water. Evidently, uric acid and urates in the fat body are important sites for the storage and release of Na^+.

Changes in Nitrogen Excretion During the Life Cycle

Dragonflies (*Aeschna*), caddis flies (*Phryganea*), and lacewings (*Sialis*) are among the insects that are terrestrial as adults but aquatic as larvae or nymphs. Significantly, when these insects change habitats at the time of metamorphosis, they also undergo a transition in their pattern of nitrogen excretion. The aquatic stages are ammonotelic, but the adults are uricotelic. Among blowflies, the meat-eating larvae, which occupy a moist habitat, are ammonotelic, whereas the pupae and aerial adults are uricotelic.

Vertebrates

Fish   Freshwater and marine teleosts are ammonotelic, although they sometimes also excrete appreciable quantities of other end products, especially urea and, in the case of marine species, trimethylamine oxide. The liver and, to a lesser extent, the kidneys are the chief internal sites of ammonia production in teleosts. Ammonia is lost mainly across the gills. Principally, the ammonia exiting via the gills is removed as such from the blood; but in some cases a substantial fraction of the ammonia (e.g., 40 percent in sculpins, *Myoxocephalus*) is formed in the gills themselves, by enzymatic deamination of circulating amino acids. The chemical release of ammonia in situ in the gills may help keep down the level of ammonia in the general body fluids.

In contrast to teleosts, marine elasmobranchs, many freshwater elasmobranchs, holocephalan fish (chimaeras), and the coelacanth *Latimeria* possess an active ornithine–urea cycle and are ureotelic. The importance of urea in the salt–water balance of marine forms has been discussed in Chapter 8.

Lungfish have attracted considerable interest because of the insight they might provide in understanding the transition of vertebrate life onto land. As discussed in Chapter 8 (p. 179), certain lungfish, such as *Protopterus,* enter a dormant state in the mud when the bodies of water in which they live dry up. They then undergo most significant changes in nitrogen excretion. *Protopterus* in water is ammonotelic, although it does produce considerable amounts of urea. When *Protopterus* is dormant in its cocoon, its metabolism becomes depressed, urine production ceases, and although considerable catabolism of protein occurs, nitrogen is not excreted. The fish produces urea virtually exclusively, and urea accumulates to high levels in its body fluids. Clearly, the change from ammonotelism to ureotelism is adaptive; ammonia could not be allowed to accumulate in the body during dormancy, and its excretion in the urine would quickly drain the water resources of the dormant animal. Research has shown that urea is synthesized via the ornithine–urea cycle in *Protopterus* and furthermore that the activities of the cycle enzymes do not increase during dormancy. When the fish are dormant, their rate of urea production is about the same as when they are not dormant, but their release of free ammonia is suppressed almost completely. Interestingly, the activities of enzymes of the ornithine–urea cycle are much lower in *Neoceratodus,* a lungfish that does not enter dormancy, than in *Protopterus*. Apparently, *Protopterus* continually maintains a relatively active ornithine–urea cycle as protection against a nonrainy day.

The presence of an active ornithine–urea cycle in *Protopterus* suggests that an active cycle was evolved (or retained from ancestral fish) as a response to the exigencies of living in intermittent bodies of water. The progenitors of amphibians are believed to have occupied such bodies of water, and it seems likely that they were at least facultatively ureotelic. Ureotelism may well have originally become a feature of the line leading to amphibians owing to the advantages of urea as a storage compound during periods of water stress.

Amphibians   Adult amphibians occupy a diversity of habitats and thus provide interesting comparisons. Adult semiaquatic and terrestrial frogs and toads (anurans) are mostly ureotelic. Recognizing that mammals and some semiaquatic turtles are also ureotelic, we see that ureotelism, despite its rarity among invertebrates, is a prominent feature in the adjustment of vertebrates to terrestrial life. Even many of the anurans that live in arid regions, including the fossorial desert species of North America (e.g., spadefoot toads), are ureotelic. Recently, it has come to light, however, that a few species of arboreal frogs from arid regions are uricotelic; these frogs, members of the African genus *Chiromantis* and the South American genus *Phyllomedusa,* face particular problems of water balance because of their exposed habitats (p. 190). Turning to opposite extremes of habitat, at least some of the amphibians that are principally aquatic as adults are ammonotelic. An example is the mudpuppy (*Necturus*).

The clawed frog, *Xenopus laevis,* is an aquatic species that is ammonotelic when living in water. Despite being principally aquatic, it can live for long periods in humid conditions out of water. Then it becomes ureotelic, and urea accumulates in its body fluids. In contrast to *Protopterus, Xenopus* increases its rate of urea synthesis under such conditions, a response that is correlated with heightened activity of certain of the enzymes of the ornithine–urea cycle in the liver.

In parts of the world where water is abundant, the tadpoles of semiterrestrial and terrestrial amphibians typically develop in permanent or semipermanent bodies of water (e.g., ponds and lakes), which provide a reliably aquatic environment. Interestingly, such species are ammonotelic as tadpoles and become ureotelic at the time they metamorphose into adults—thus providing a particularly striking example of a correlation between nitrogen excretion and habitat. Ontogenetic data on bullfrogs are presented in Figure 9.7. The numbered stages along the abscissa refer to morphological steps in development. Whereas animals of stages X to XVII are tadpoles, those of stages XX to XXIV are undergoing metamorphosis, and metamorphosis is largely complete at stage XXV. Note the great increase in urea excretion during metamorphosis and also the sudden increases in the activities of enzymes of the ornithine–urea cycle in the liver.

In arid regions, the larvae of terrestrial amphibians often develop in highly temporary ponds, which may become severely reduced in volume or even dry up before metamorphosis is complete. In some cases, the larvae even spend part of their development out of water—for instance, in a foam nest provided by the parents. Recent studies of certain arid-zone spe-cies that are ureotelic as adults (e.g., *Leptodactylus, Scaphiopus*) have revealed that their larvae are either ureotelic throughout development or become ureo-telic unusually early. *Phyllomedusa sauvagei,* which as an adult is arboreal and uricotelic, is ureotelic as a tadpole.

Reptiles and Birds Birds, lizards, and snakes are uricotelic. The turtles that have received study ex-crete ammonia, urea, and uric acid in various pro-portions. Recognizing that these compounds collec-tively may represent from 85 percent to as little as 40 percent of total nitrogen excretion in turtles, we may note that their proportions exhibit notable cor-relations with habitat. As exemplified in Table 9.1, semiaquatic turtles tend to be distinctly ureotelic; turtles common to dry terrestrial habitats, on the other hand, tend toward uricotelism. The American alligator can be ammonotelic or uricotelic, depending on conditions. Normally, the alligator excretes about half of its total nitrogen as ammonia and half as uric acid, but when overly hydrated it excretes predom-inantly ammonia, and when dehydrated it excretes mostly uric acid. Ammonia is voided in the urine as NH_4^+. It has been argued that in alligators (and also in some turtles and insects) a substantial production of NH_4^+ under circumstances of unrestricted water availability may represent a mechanism for Na^+ con-servation, for ammonium salts take the place of so-dium ones in the urine.

Mammals We have seen repeatedly in this chapter that modes of nitrogen excretion are often not only adaptive but also phylogenetically labile. From this viewpoint, the condition in mammals is surprising, for all mammals, regardless of habitat, are ureotelic. An informed biologist knowing the patterns of nitro-gen excretion in other animal groups but knowing nothing about this character in mammals would prob-ably predict the appearance of uricotelism, at least in species facing severe problems of water balance. All the other terrestrial classes of vertebrates have evolved uricotelism, and the mammals themselves have the capacity to synthesize purines from amino acid nitrogen and can form uric acid from other pur-

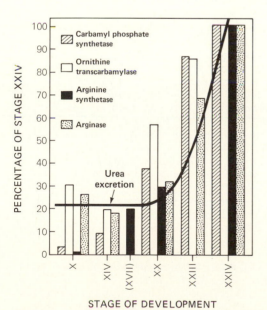

STAGE OF DEVELOPMENT

Figure 9.7 Excretion of urea and activities of enzymes of the ornithine–urea cycle in the liver as functions of stage of de-velopment in bullfrog tadpoles (*Rana catesbeiana*). Urea ex-cretion and enzyme activities are expressed relative to their levels at stage XXIV. [From G.W. Brown, Jr. and P.P. Cohen, Biosynthesis of urea in metamorphosing tadpoles. In W.D. McElroy and B. Glass (eds.), *A Symposium on the Chemical Basis of Development.* © The Johns Hopkins University Press, Baltimore, 1958.]

Table 9.1 PERCENTAGE OF TOTAL URINARY NITROGEN IN THE FORM OF AMMONIA, UREA, AND URIC ACID IN TWO TURTLES

Species	Percentage of total nitrogen in the form of		
	Ammonia	Urea	Uric acid
Semiaquatic freshwater turtle (*Pseudemys scripta*)	4–44	45–95	1–24
Desert tortoise (*Gopherus agassizii*)	3–8	15–50	20–50

Source: W. H. Dantzler and B. Schmidt-Nielsen, *Am. J. Physiol.* **210**: 198–210 (1966).

ines. Nonetheless, uric acid typically appears in only small quantities in the urine as a specific product of purine catabolism.

Mammals are able to restrict their urinary water losses to relatively low levels despite excreting a highly soluble nitrogenous end product because they have kidneys that are uniquely capable of concentrating urea in the urine. The urea U/P ratio is typically higher than that for any other solute and can greatly exceed the total osmotic U/P ratio—meaning that urea can be concentrated to a much greater extent than solutes as a whole. In humans, for example, the maximal osmotic U/P ratio is about 4.2, and the maximal chloride ratio is about 3.5, but the maximal urea ratio is about 170. Many desert rodents, which possess some of the most effective of mammalian kidneys, can achieve urinary urea concentrations of 2.5–5.0 M, corresponding to 7–14 g of nitrogen per 100 mL. Because urinary urea concentrations achieved by mammals are much greater than those attained by other animals, the water losses obligated by nitrogen excretion in mammals are exceptionally low in comparison to other ureotelic groups. In fact, urinary nitrogen : water ratios attainable by desert rodents can equal or exceed those observed in some of the *uricotelic* vertebrates that void their uric acid in a relatively fluid mix (e.g., certain birds). On the other hand, some birds and reptiles that void uric acid in the form of relatively dry pellets achieve nitrogen : water ratios that are several times higher than the highest mammalian values.

It has been argued that the basic design of the mammalian kidney, as described in Chapter 10, provided such great potential for concentrating urea in the urine that even when mammals invaded deserts, there was little selective pressure for the evolution of uricotelism. The reverse argument is that mammals for some reason have not been able to make the biochemical and physiological adjustments required for uricotelism and have thus remained tied to ureotelism. Such a tie to ureotelism would then provide a basis for great selective pressures for the evolution of a kidney with pronounced concentrating abilities. These hypotheses concerning the evolution of kidney function and universal ureotelism in mammals provide interesting food for thought.

Mammals are not known to store urea in the sense of accumulating it. Through the activities of bacteria in the gut, however, ruminants, macropod marsupials, and some other mammals are able to derive nitrogen for protein synthesis from urea; in at least some such species, urea excretion is curtailed during periods of low dietary protein intake. In ruminants, urea is introduced into the rumen fluid from the blood, and there bacteria accomplish what the mammal cannot—synthesis of protein using urea as a nitrogen source. When the bacteria are carried through the rest of the gut, the bacterial protein is digested, yielding amino acids that are absorbed across the intestine and used in protein synthesis by the mammal. In this way, nitrogen can be recycled within the confines of the ruminant's body.

Embryonic Nitrogen Catabolism and the Needham Hypothesis Reptiles and birds, as well as insects and terrestrial gastropod mollusks, produce eggs encased in tough shells that limit evaporative water loss. In these groups, embryonic nitrogenous wastes are stored within the egg, and it has been logically satisfying to find that the embryos are predominantly uricotelic (although evidence exists that embryos of some reptiles and birds pass through stages of ammonotelism and ureotelism early in development). The embryos of placental mammals are ureotelic; the urea they produce is passed into the maternal circulation and voided by the mother. Around 1930, Joseph Needham made the influential suggestion that ureotelism in adult mammals is a carryover from the fetal situation. This suggestion seems not entirely satisfying in view of the general lability of nitrogen excretion in the animal kingdom and the numerous examples of changes in nitrogenous end products during the life cycles of other animals.

SELECTED READINGS

Becker, W. and H. Schmale. 1978. The ammonia and urea excretion of *Biomphalaria glabrata* under different physiological conditions: Starvation, infection with *Schistosoma mansoni,* dry keeping. *Comp. Biochem. Physiol.* **59B**:75–79.

Bidigare, R.R. 1983. Nitrogen excretion by marine zooplankton. In E.J. Carpenter and D.G. Capone (eds.), *Nitrogen in the Marine Environment,* pp. 385–409. Academic, New York.

Brown, G.W., Jr. and P.P. Cohen. 1958. Biosynthesis of urea in metamorphosing tadpoles. In W.D. McElroy and B. Glass (eds.), *A Symposium on the Chemical Basis of Development,* pp. 495–513. Johns Hopkins Press, Baltimore.

Bursell, E. 1967. The excretion of nitrogen in insects. *Adv. Insect Physiol.* **4**:33–67.

Campbell, J.W. (ed.). 1970. *Comparative Biochemistry of Nitrogen Metabolism.* Academic, New York.

*Campbell, J.W. 1973. Nitrogen excretion. In C.L. Prosser (ed.), *Comparative Animal Physiology*, 3rd ed., Vol. 1. Saunders, Philadelphia.

*Campbell, J.W. and L. Goldstein (eds.). 1972. *Nitrogen Metabolism and the Environment.* Academic, New York.

*Dantzler, W.H. 1982. Renal adaptations of desert vertebrates. *BioScience* **32**: 108–113.

Evans, D.H. 1986. The role of branchial and dermal epithelia in acid-base regulation in aquatic vertebrates. In N. Heisler (ed.), *Acid-Base Regulation in Animals*, pp. 139–172. Elsevier, New York.

Evans, D.H. and J.N. Cameron. 1986. Gill ammonia transport. *J. Exp. Zool.* **239**:17–23.

Janssens, P.A. and P.P. Cohen. 1968. Biosynthesis of urea in the estivating African lungfish and in *Xenopus laevis* under conditions of water-shortage. *Comp. Biochem. Physiol.* **24**:887–898.

Jones, R.M. 1980. Nitrogen excretion by *Scaphiopus* tadpoles in ephemeral ponds. *Physiol. Zool.* **53**:26–31.

Kinnear, J.E. and A.R. Main. 1975. The recycling of urea nitrogen by the wild tammar wallaby (*Macropus eugenii*) — a "ruminant-like" marsupial. *Comp. Biochem. Physiol.* **51A**:793–810.

Kormanik, G.A. and J.N. Cameron. 1981. Ammonia excretion in the seawater blue crab (*Callinectes sapidus*) occurs by diffusion, and not Na^+/NH_4^+ exchange. *J. Comp. Physiol.* **141**:457–462.

*Mangum, C. and D. Towle. 1977. Physiological adaptation to unstable environments. *Am. Sci.* **65**:67–75.

Mullins, D.E. and D.G. Cochran. 1975. Nitrogen metabolism in the American cockroach. I and II. *Comp. Biochem. Physiol.* **50A**:489–500.

Potrikus, C.J. and J.A. Breznak. 1981. Gut bacteria recycle uric acid nitrogen in termites: A strategy for nutrient conservation. *Proc. Natl. Acad. Sci. USA* **78**:4601–4605.

Potts, W.T.W. 1967. Excretion in the molluscs. *Biol. Rev.* **42**:1–41.

Regnault, M. 1987. Nitrogen excretion in marine and freshwater Crustacea. *Biol. Rev.* **62**:1-24.

Shoemaker, V.H. and L.L. McClanahan, Jr. 1975. Evaporative water loss, nitrogen excretion and osmoregulation in Phyllomedusine frogs. *J. Comp. Physiol.* **100**:331–345.

Tucker, L.E. 1977. Regulation of ions in the haemolymph of the cockroach *Periplaneta americana* during dehydration and rehydration. *J. Exp. Biol.* **71**:95–110.

Wieser, W. 1972. A glutaminase in the body wall of terrestrial isopods. *Nature* **239**:288–290.

See also references in Appendix A.

chapter 10

Renal Organs

Earlier, in Chapters 7 and 8, we frequently described the vital roles played by kidneys and other renal organs in maintaining salt and water balance. In this chapter, we go beyond the phenomenological approach taken heretofore and examine the mechanisms by which the urine is fashioned. Knowledge of these mechanisms helps us understand why some feats are within the competence of a species' renal organs but others are not.

The renal organs of all the various phyletic groups are far from being universally homologous, and they differ considerably in details of their mechanisms and capacities. Yet they do bear certain features in common. All renal organs are tubular structures that communicate directly or indirectly with the outside world. All produce and eliminate aqueous solutions derived from the blood or other body fluids. Most importantly, *their function is the regulation of the composition and volume of the body fluids through controlled excretion of solutes and water.*

The mammalian kidney—to focus on just one example—performs many particular roles. It rids the body of urea. It excretes Na^+, Cl^-, K^+, and numerous other ions in greater or lesser amounts day by day, closely regulating the concentration of each in the body fluids. It is one of the principal organs of acid–base regulation, excreting H^+ as appropriate to keep the pH of the body fluids within narrow limits. It also regulates the osmotic pressure of the body fluids through the controlled excretion of water. When we consider that it performs all these roles simultaneously by structuring the composition of a single fluid output, we see that the kidney must be classed as one of the marvels of life.

BASIC MECHANISMS OF RENAL FUNCTION

Often, it is appropriate to view urine formation as occurring in two steps, though these "steps" may sometimes be partly contemporaneous. First, an aqueous solution, called *primary urine,* is introduced into the renal tubules. Second, this solution is modified as it moves through the renal tubules and other excretory passages, ultimately becoming the *definitive urine* that is eliminated.

Modes of Introducing Fluid into Renal Tubules

Ultrafiltration One widespread mechanism by which fluids are introduced into the renal tubules is ultrafiltration under force of blood pressure (p. 151). This is the mechanism operative in most vertebrates and in certain invertebrates such as mollusks and decapod crustaceans. We shall illustrate here with the vertebrate example.

The vertebrate kidney consists of many tubules, called *nephrons*, each of which begins blindly with a hemispherical, invaginated *Bowman's capsule* (Figures 10.1 and 10.3). The Bowman's capsule surrounds a small, anastomosing cluster of capillaries, the *glomerulus* (Figure 10.1), which is supplied with blood at relatively high pressure by branches of the renal artery. The glomerulus and Bowman's capsule together constitute a *renal corpuscle.* (Another name is Malpighian corpuscle, and sometimes the entire renal corpuscle is called a glomerulus.) The glomerular capillaries are closely applied to the walls of the Bowman's capsule, and the lumen of the capillaries is separated from the lumen of the Bowman's capsule

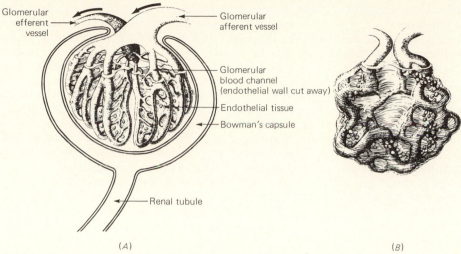

Glomerular efferent vessel

Glomerular afferent vessel

Glomerular blood channel (endothelial wall cut away)

Endothelial tissue

Bowman's capsule

Renal tubule

(A)

(B)

Figure 10.1 (*A*) A human renal corpuscle. The endothelial cells that constitute the walls of the glomerular blood channels form continuous sheets of endothelial tissue. The tissue is cut away at the top of the glomerulus so that its laminar orientation can be seen. Arrows show direction of blood flow. The Bowman's capsule is depicted diagrammatically. The inner membrane of the capsule, constructed of specialized cells termed *podocytes,* actually interdigitates with the sheets of endothelial tissue so that there is intimate juxtaposition of all blood channels and the capsular membrane. The lumen of the blood channels is separated from the lumen of the Bowman's capsule by only a single layer of endothelial cells, a thin basement membrane, and a single layer of podocytes. (*B*) Glomerulus of a frog (*Rana pipiens*) Note that it also consists of blood channels running through continuous sheets of endothelial tissue but is of simpler construction than the human glomerulus. [Glomeruli from H. Elias, A. Hossman, I.B. Barth, and A. Solmor, *J. Urol.* **83**:790–798 (1960). © 1960 The Williams & Wilkins Company, Baltimore.]

by only two cell layers and a thin basement membrane. The latter structures act as a filter. Fluid is driven through this filter from the blood plasma into the lumen of the Bowman's capsule by the hydrostatic pressure of the blood. In a simplified way, the fluid may be envisioned as moving through pores in the filtering membranes in the manner of small streams (bulk flow). The process is actually far more complex. One highly important determinant of whether a solute will pass through in the fluid streams is its molecular size (as simple pore theory would suggest), but molecular charge and shape can also be significant. Inorganic ions and such small organic molecules as glucose, urea, and amino acids pass through freely. Thus, their concentrations are virtually the same within the Bowman's capsule as in the blood plasma. At the other extreme, solutes with molecular weights of several thousand or more—such as the plasma proteins—are almost entirely excluded from passage. In sum, the filtrate introduced into the Bowman's capsules closely resembles the blood plasma in its composition of low-molecular-weight solutes but differs from plasma in being almost devoid of high-molecular-weight solutes such as proteins.

As discussed and illustrated in Chapter 7 (p. 151, Figure 7.12), entry of fluid into renal tubules under the force of blood pressure is opposed both by the colloid osmotic pressure of the blood plasma and by the hydrostatic pressure prevailing in the tubules themselves. The blood pressure at the sites of filtration must be high enough to overcome these opposing forces and create a net positive force, a *filtration pressure,* favoring the entry of primary urine into the renal tubules. We estimated earlier (p. 152) that in certain mammals (e.g., rats), with the glomerular capillary pressure near 50 mm Hg, the filtration pressure averages about 9 mm Hg. The blood pressure in the glomerular capillaries of mammals is significantly higher than that in most capillary beds, partly because the arterioles leading to the glomeruli are relatively large in diameter and thus offer a relatively low resistance to flow. The high capillary blood pressures resulting from such vascular peculiarities are vitally important because, given the colloid osmotic pressure of the blood, the filtration pressure and rate of filtration would be meager if hydrostatic pressures in the glomeruli were no higher than those in capillaries generally.

The rate of primary urine formation by all of an animal's renal tubules taken together is called the *filtration rate.* In vertebrates, it is termed more specifically the *glomerular filtration rate* or *GFR.* Humans, for example, have a GFR of about 120 mL/min. At this rate, the equivalent of all the plasma water in the body is filtered about every half hour. Most of the filtered water is ultimately resorbed back into the blood rather than being excreted, but the sheer magnitude of the rate of filtration signifies that the nephrons enjoy very intimate access to the plasma in carrying out their functions of waste removal and regulation of plasma composition.

To some degree in mammals and to a greater degree in certain other groups, the volume of urine flow is controlled by variations in GFR. In principle, the GFR can be reduced either by curtailing filtration into all nephrons more or less equally or by selectively reducing filtration into some nephrons while leaving others unaffected. The rate of filtration into a nephron depends on its glomerular blood pressure. This pressure can be modulated by vasomotor changes in the diameter (and hence flow resistance) of the afferent glomerular blood vessel. Such vasomotor changes are under control of the autonomic nervous system and circulating hormones.

Variation in GFR is not the only mechanism of controlling the rate of urine production. As we shall see later, variation in the rate of resorption of fluid by the nephrons is also important and often (as in mammals) is preeminent.

Secretion   Another method by which water and solutes can be moved into renal tubules is based on active solute secretion. This is the mechanism in insects and some marine fish, for example. Certain essentials of a secretory system are illustrated in artificial, stepwise fashion in Figure 10.2. We assume just two uncharged solutes, and we start, in Figure 10.2A, with a system in which the osmotic pressure and concentrations of both solutes are equal on the inside and outside of the renal tubule. We also assume for simplicity that the fluid outside the tubule is sufficiently abundant that movements of solutes and water into the tubule do not appreciably alter the outside concentrations. In Figure 10.2B, an active-transport mechanism secretes a quantity of solute X into the lumen of the renal tubule, increasing the inside concentration of X and also increasing the inside osmotic pressure. In Figure 10.2C, water moves inward by osmosis, following the osmotic gradient set up by secretion, and the volume of fluid in

the tubule thus increases. Because of this increase in volume, the inside concentration of solute Y, initially the same as the outside concentration, is reduced. In Figure 10.2D, solute Y diffuses inward such as to neutralize its concentration gradient. The system is still not at equilibrium, and the analysis presented here is simplified. Nonetheless, this treatment is sufficient to illustrate the important principle that active secretion of even just one solute into a renal tubule can lead to passive influx of water and other solutes. As in ultrafiltration, a complex solution of many body solutes can be introduced into the renal tubule. The animal must expend energy to accomplish this task. In ultrafiltration systems, energy is expended in maintaining a suitably high blood pressure, whereas in secretory systems, energy is required for active transport. In secretory systems, the membranes of the renal tubules again act as something of a filter. Their permeability to the various solutes that *might* enter passively determines which solutes *do*, in fact, enter the renal tubule.

Alteration of Fluid Before Elimination

As fluid introduced into a renal tubule moves down the tubule and through other parts of the excretory system, it is characteristically altered extensively in volume and composition before elimination as definitive urine. Some of the water in the fluid is usually resorbed back into the blood. Solutes may be resorbed or added. These processes are the predominant *regulatory* ones in renal function. Consider, for instance, that ultrafiltration produces a primary urine that is little different from blood plasma (except for the virtual absence of proteins). It is the processes taking place *after* filtration that adjust the absolute and relative quantities of the components in the excreted urine so as to promote plasma homeostasis.

Processes adjusting the amount and composition of the urine are not necessarily limited to the renal tubules. Although in mammals the form of the urine

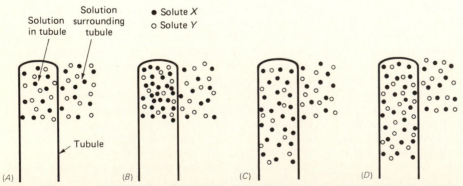

Figure 10.2   A simplified illustration of a secretory system. There are two uncharged solutes, symbolized by open and closed circles. The renal tubule is assumed to be completely surrounded by the outside solution, but only a small sample of the outside solution is shown at the upper right of the tubule in each step. Movement of water into the tubule is represented by an increase in the length of the tubule filled with solution. See text for explanation.

is essentially finalized once the urine leaves the kidneys, in other groups the urinary bladder, cloaca, or other postrenal structures sometimes play pivotal roles in modifying the urine.

Along the renal tubules and other excretory passages, the urine is separated from blood vessels or blood spaces by the membranes forming the passage walls; the processes of modification of the urine are in reality processes of exchange between the urine and blood. Some solutes are actively transported in or out of the urine. These active solute movements, in turn, can promote ("drive") a variety of passive solute or water movements by virtue of the interactive effects discussed in Chapter 7. For example, active transport of one ion may create electrical gradients affecting the diffusion of other ions; and active transport of certain solutes may alter osmotic gradients, thereby inducing osmotic water fluxes which in turn can alter concentration gradients of other solutes. In all of this, the permeability properties of the walls of the excretory passages also play a major role by affecting the facility with which water and various solutes can cross the barrier between blood and urine. One way that a solute can become concentrated in the urine, for example, is if the renal tubular walls permit an osmotic outflux of water yet oppose diffusive outflux of the solute in question. Trapped in the renal tubules, the solute will become ever more concentrated as water is extracted from the urine.

URINE FORMATION IN AMPHIBIANS

We shall look at amphibians first because knowledge of the function of their nephrons provides a good base of comparison for study of all other vertebrates. In fact, amphibians were the subjects for many of the very first investigations of vertebrate nephron function (in the 1920s and 1930s). The purpose of this section is not only to describe how amphibians form urine but also to bring out many additional general principles by example.

Each nephron of amphibians (Figure 10.3) consists of (1) a Bowman's capsule; (2) a convoluted segment known as the *proximal convoluted tubule;* (3) a short, relatively straight segment of small diameter, the *intermediate segment;* (4) a second convoluted segment known as the *distal convoluted tubule;* and (5) a relatively straight segment, the *initial collecting duct.* The nephrons are microscopically small, and there are hundreds or thousands in each kidney. The initial collecting ducts of the nephrons feed into *collecting ducts* (Figure 10.3C), and all the collecting ducts of each kidney connect to a single *ureter,* which carries fluid from the kidney to the bladder.

Ultrafiltration and Events in the Proximal Tubule

Measurements of glomerular blood pressure have demonstrated that it is adequate for ultrafiltration, and analysis of fluid taken from the Bowman's capsules of amphibians by micropuncture has confirmed that the fluid is an ultrafiltrate of the plasma. Thus, the fluid entering the proximal convoluted tubule resembles plasma in composition except that it lacks the plasma proteins. It is approximately isosmotic with the plasma. (The difference in osmotic pressure between the tubular fluid and plasma, owing to the plasma proteins, is very important in analyzing ultrafiltration but represents less than 1 percent of the total osmotic pressure of either the plasma or tubular fluid.)

One important function of the proximal convoluted tubule is the active resorption of glucose. Glucose is a valuable metabolite that, because of its small molecular size, cannot be withheld from the urine at the site of ultrafiltration. However, in amphibians, as in many other animals, it is promptly reclaimed and returned to the blood. Amino acids are other valuable metabolites that are freely carried into the nephrons in the ultrafiltrate. They are known to be recovered from the urine by active resorption in the proximal tubules of some other vertebrates and likely are treated similarly in amphibians.

The amount of water filtered into the Bowman's capsules each day typically far exceeds the amount needing to be excreted. The amounts of Na^+ and Cl^- filtered are also well in excess of the amounts to be excreted: Na^+ and Cl^- are overwhelmingly the dominant solutes in the filtrate (as in the plasma), yet considering the magnitude of dietary gains and possible cutaneous exchanges of NaCl, only modest urinary excretion of NaCl (if any at all) is demanded for maintenance of salt balance. As noted before, a high rate of filtration assures that the nephrons will have intimate access to the plasma solution in performing their regulatory functions. It also necessitates resorption of much of the water and NaCl filtered. This resorption begins in the proximal convoluted tubule.

Sodium is actively resorbed across the walls of the proximal tubule. Chloride may also be resorbed actively in some species, but in general its resorption is passive, induced by the electrical gradient set up by active Na^+ resorption. Sodium and chloride, as noted, are the dominant solutes in the filtrate, and the quantities resorbed in the proximal tubule are large. Yet the osmotic pressure of the tubular fluid does not fall in the proximal tubule because there is also a large loss of water from the fluid, induced by the resorption of NaCl (p. 149). Importantly, the walls of the proximal tubule are freely permeable to water. Thus, as solute is resorbed from the tubular fluid, water exits osmotically across the walls of the tubule in such a manner that the tubular fluid remains isosmotic to the blood plasma and other surrounding body fluids (Figure 10.4). In species that have received study, 20–40 percent of the filtered water and NaCl have been found to be resorbed in the proximal tubule.

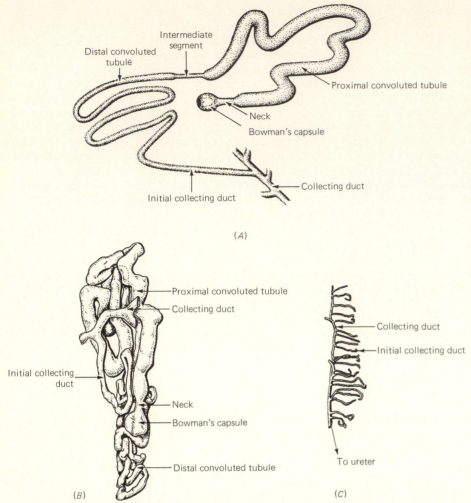

(A)

(B)   (C)

Figure 10.3 (A) Schematic representation of an amphibian nephron. (B) Nephron of a bullfrog (*Rana catesbeiana*) in its natural configuration, magnified 54 times. The transition from proximal convoluted tubule to intermediate segment and distal convoluted tubule occurs where the proximal convoluted tubule has finally doubled back to about the level of the Bowman's capsule and is hidden from view. (C) A single collecting duct of *Rana catesbeiana* showing the connections of the initial collecting ducts of many nephrons. The many collecting ducts of the kidney connect to a common ureter. [Parts (B) and (C) from G.C. Huber, Renal tubules. In E.V. Cowdry (ed.), *Special Cytology,* Vol. I Medical Department of Harper & Row, New York, 1928.]

Events in the Distal Convoluted Tubule

Active resorption of NaCl from the renal tubular fluid continues in the distal convoluted tubule, gradually lowering the quantity of NaCl destined for excretion toward the level appropriate for maintenance of salt balance. (Ideally, if the animal has dietary inputs of NaCl that exceed its requirements, the total resorption of NaCl by the time the urine is excreted will be just sufficient for the urinary output of NaCl to offset exactly the excess gains. If the animal is in negative NaCl balance, resorption will likely be maximized, but since urine totally devoid of NaCl cannot be made, the ideal of *no* NaCl excretion will not be attained.)

A major function of the distal convoluted tubule in many amphibians—a function that may be shared by the urinary bladder and collecting ducts—is control of the excretion of pure ("osmotically free") water. On the one hand, the renal tubular fluid may remain approximately isosmotic with the blood plasma during its passage through the distal tubule, collecting duct, and bladder—meaning that the urine will contain no more water than demanded by its content of solutes (amphibians cannot render their urine any more concentrated than their blood plasma). On the other hand, the tubular fluid may become substantially more dilute than the plasma during passage through the distal tubule, collecting duct, and bladder—signifying that the urine will then carry out of the body an "extra" quantity of water, an amount not merely necessitated by solute excre-

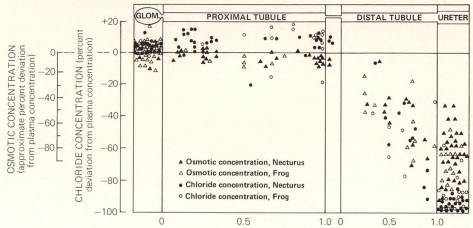

Figure 10.4 Chloride and osmotic concentrations in the renal tubular fluid of *Necturus maculosus* and the frog *Rana pipiens* during states of relative diuresis. Fluid was sampled by micropuncture. The scales on the abscissa indicate the point of sampling as a fraction of the total length of the proximal tubule and distal tubule, respectively. Chloride concentration is expressed as percent deviation from the chloride concentration in the plasma. (For example, a value of -40 in the distal tubule of a frog would indicate that the chloride concentration in the tubule was below the plasma concentration by an amount equal to 40 percent of the plasma concentration in the same frog.) Osmotic concentration is also expressed *approximately* as percent deviation from the osmotic concentration in the plasma; for the exact meaning of measured osmotic concentrations, see original research report. [From A.M. Walker, C.L. Hudson, T. Findley, Jr., and A.N. Richards, *Am. J. Physiol.* **118**:121–129 (1937); percent deviation of urinary osmotic concentration from plasma osmotic concentration estimated from calibration data by linear regression.]

tion. The concept of the specific excretion of pure water was discussed in Chapter 8 (p. 194), and you should be certain to review that material if the ideas are not still fresh. The extent of pure-water excretion is controlled in the distal convoluted tubule, collecting duct, and bladder by varying the degree to which osmotic water resorption keeps pace with solute re-

sorption. The extent of water resorption is in itself controlled by modulating the permeability to water of the walls of these excretory passages. The control of permeability is exercised at least partly by the antidiuretic hormone (ADH) of amphibians, arginine vasotocin (p. 190).

When ADH levels are low, the permeability of the

BOX 10.1   QUANTITY VERSUS CONCENTRATION

In analyzing renal function, it is always useful to maintain a clear distinction between measures of quantity (or mass) and those of concentration. The importance of the distinction is illustrated nicely by the events in the proximal tubule of amphibians. As shown in Figure 10.4, the *concentrations* of Na^+, Cl^-, and water in the tubular fluid remain, on average, unchanged. Yet, the *quantities* of these substances exiting the proximal tubule are much lower than those entering.

Measures of quantity and concentration are each informative, although in different ways. Quantity is an absolute measure, whereas concentration is a relative measure, expressing the quantity of solute in relation to the quantity of water. As a general principle, the *quantity* of a solute (e.g., Na^+) voided per day by an organism in its definitive urine is to be compared with the organism's daily gain of that solute (also taking account of extrarenal losses, if any) to determine if the organism is in balance for that solute or gaining or losing the solute in net fashion. The *concentration* of a solute in the urine, by contrast, is not directly useful in calculations of solute balance. However, when the urinary concentration of a solute is incorporated into a U/P ratio, it helps to indicate in what direction the *blood concentration* of the solute is being impelled by the actions of the renal organs or their subparts (p. 153).

distal convoluted tubule to water is low, and consequently ion and water resorption from the tubular fluid are significantly uncoupled. The active resorption of NaCl tends to dilute the tubular fluid and thus create an osmotic gradient favorable to water resorption, but the low permeability to water of the tubule walls impedes the water resorption, with two important, complementary consequences. First, as illustrated in Figure 10.4, the ion resorption renders the tubular fluid steadily more dilute than the blood, both in osmotic pressure and ion concentrations. Second, although significant amounts of water may be resorbed across the walls of the distal tubule, a relatively high proportion of the water entering the tubule passes through to be excreted in the urine. These processes contribute to the production of a copious, dilute urine—a urine appropriate to excreting large quantities of excess water while still retaining useful solutes. The production of a copious urine is called *diuresis*.

When ADH is secreted, it induces the walls of at least some parts of the distal convoluted tubule to become more permeable to water, and the distal tubule then functions more like the proximal tubule. Osmotic water resorption is promoted. Thus, in the context of ongoing NaCl resorption, a greater frac-

tion of the water entering the distal tubule is resorbed than when ADH levels are low, and the development of large differences in osmotic pressure between the tubular fluid and plasma is opposed. These processes contribute to the formation of a relatively scanty, relatively concentrated urine—a urine that contains little, if any, more water than that demanded for solute excretion. The state of producing a scant urine is called *antidiuresis*.

The effects of changes in tubular water permeability during diuresis and antidiuresis are diagrammed conceptually in Figure 10.5. Note that, in essence, the ability to modulate permeability provides a mechanism for controlling the amount of water excretion independently of solute excretion (p. 194).

Another function localized in the distal tubule is active H^+ secretion into the tubular fluid. The amount of H^+ added is adjusted to maintain a normal pH in the body fluids.

Integrated Effects of ADH on Nephron Function

In amphibians as well as in reptiles and birds, ADH (arginine vasotocin) not only increases the permeability of the distal convoluted tubules to water but

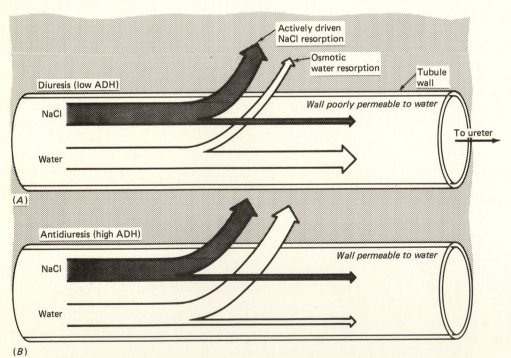

Figure 10.5 Conceptual view of events in the distal convoluted tubule during diuresis and antidiuresis. (*A*) In diuresis, low permeability of the tubule walls impedes osmotic water resorption; thus, a relatively large fraction of the water remains in the tubule and the ratio of solute to water (osmotic pressure) in the tubule is dramatically reduced. (*B*) In antidiuresis, the tubule walls are more permeable to water; thus, a larger fraction of the water is resorbed and the ratio of solute to water is affected relatively little. An important simplification in this diagram is that solutes other than NaCl, such as urea, are ignored. Because water moves toward osmotic equilibrium, water resorption in (*B*) does not proceed in amphibians beyond the point at which the total osmotically effective concentration of all solutes taken together in the tubular fluid is similar to that in the blood plasma. It is also a simplification to assume that NaCl and water resorption necessarily occur in the same parts of the tubule.

BOX 10.2   PRODUCTION OF HYPOSMOTIC URINE

To produce urine that is hyposmotic to their blood plasma, amphibians actively resorb solutes from the renal tubular fluid while the fluid is separated from the blood by membranes poorly permeable to water. Among all the animal groups that are capable of making hyposmotic urine, this mechanism, in its fundamentals, appears to be universally used.

---

also decreases the glomerular filtration rate by inducing vasoconstriction in afferent glomerular blood vessels; evidently, the GFR is decreased largely by reducing the numbers of nephrons filtering rather than by reducing the filtering rates of all individual nephrons. The decrease in GFR acts to reduce urine flow and in this respect complements the increase in water resorption induced by ADH in the distal tubules.

In some frogs and toads, ADH has been shown to increase the rate of active NaCl resorption from the renal tubules. Like the effects on GFR and distal-tubule permeability, this effect also tends to reduce urine volume because it promotes solute-driven water resorption and decreases the solute load of the urine.

Clearly, an elaborate *pattern* of control over nephron function is exerted by way of ADH. If the animal confronts excesses of water influx (e.g., when living in water), secretion of ADH is restricted. Then the GFR is relatively high and distal-tubule resorption of water is relatively low, and a voluminous, dilute urine results. If dehydration sets in, ADH is secreted from the neurohypophysis, apparently under control of osmoreceptors (which detect an increase in body-fluid osmolarity) and of pressure or stretch receptors (which signal a decrease in blood volume) (p. 195). The ADH then induces a reduction in GFR, an increase in distal-tubule water resorption, and an increase in NaCl resorption, thus promoting production of a scanty, concentrated urine.

The renal responses to ADH are not as well developed in some amphibians from consistently moist or wet habitats as they are in species that are more terrestrial and thus more likely to experience dehydration (p. 190).

Other Aspects of Urine Formation

Bladder Function   Interspecific differences exist in the properties of the urinary bladder, but in many species of amphibians the bladder closely resembles the distal convoluted tubules in respect to basic features of water and salt exchange. In such species, the bladder wall is poorly permeable to water when ADH levels are low but becomes quite permeable when ADH levels are high. NaCl is actively resorbed across the bladder wall, and the resorption is stimulated by ADH. As discussed in Chapter 8 (p. 190), it is well established that during times of dehydration, many semiterrestrial and terrestrial amphibians resorb urine stored in their bladder (via action of ADH) to supplement body water resources.

Urea Excretion   The nephrons, bladder, and other excretory passages of amphibians seem generally to be poorly permeable to urea. Thus, as water is resorbed from the renal fluid, urea in the fluid tends to be concentrated. Filtration is one source of urinary urea, and in many amphibians it is probably the sole source. However, in at least some ranid frogs (e.g., bullfrogs), urea is also actively secreted into the renal tubular fluid across the nephron walls.

The Extent of Water and NaCl Resorption   It is difficult to generalize about the fraction of filtered water that is resorbed by the time the urine is excreted because the fraction varies widely depending on species and conditions. Among hydrated amphibians, 20–60 percent resorption is common. During dehydration, the percentage of water resorbed increases. For example, green frogs (*Rana clamitans*) resorb about 62 percent when living in water but 84 percent after a few hours in air. NaCl is also resorbed to a variable degree; resorption of 95–99 percent of the amount filtered would not be unusual.

Hormonal Controls   The effects of ADH on renal function are generally emphasized not only because they are important but also because they are relatively well understood. Clearly, however, much remains to be learned about endocrine controls, and other hormones are involved. Aldosterone, for example, may well prove to promote Na^+ resorption across the nephrons and has rather clearly been shown to affect the bladder (p. 194).

METHODS OF STUDYING RENAL FUNCTION

Having examined the function of the renal organs in one group of animals, let us now briefly discuss some of the methods by which renal function is elucidated.

Clearance Studies

Suppose we have measured the quantity of a substance excreted per unit time. We can then ask: What volume of blood plasma would have to be completely depleted of this substance—that is, completely *cleared* of the substance—to yield the excreted quantity? This volume is termed the **plasma clearance** of the substance.

To illustrate, suppose a toad excretes 100 mL of urine per day and the urine contains 2 μmole Na$^+$/mL. The daily excretion of Na$^+$ would then be 2 × 100 = 200 μmole. Suppose we measure the plasma concentration of Na$^+$ and find it to be 100 μmole/mL. We would then know that to obtain the quantity of Na$^+$ excreted per day, 2 mL of plasma would have to be completely cleared of Na$^+$. Thus, the plasma clearance of Na$^+$ would be 2 mL/day. In algebraic form, if V is the volume of urine excreted per unit time, U is the urinary concentration of the substance in question, and P is the plasma concentration, then the clearance of the substance per unit time, C, is computed as

$$C = \frac{UV}{P}$$

The calculation of the plasma clearance for a substance should not be taken to imply that the kidneys obtain the excreted quantity by completely clearing part of the plasma while leaving the rest of the plasma untouched. Actually, of course, the excreted quantity is obtained by incompletely clearing the plasma at large. Although the plasma clearance is in this sense an artificial concept, it is also a powerful concept in the analysis of renal function, as we shall shortly see.

Suppose there is a plasma solute that meets the following three criteria: First, it enters the nephrons by glomerular filtration and is not introduced into the urine by any other mechanism; second, it is freely filtered, so its concentration in the filtrate is the same as its concentration in the plasma; and third, it is not resorbed in the nephrons or elsewhere, so that once it has entered the nephrons, it is destined to be fully excreted. Measurement of the plasma clearance of such a substance permits the quantification of glomerular filtration rate, as may be shown readily by algebra. If F is the filtration rate (volume of filtrate produced per unit time), then the *amount* of substance filtered per unit time is FP, because the concentration of substance in the filtrate is the same as the concentration in the plasma, P. The amount of substance *excreted* per unit time is UV, as indicated earlier. For a substance that is neither added to the urine nor resorbed after filtration, the amount filtered and the amount excreted must be the same. Thus, $FP = UV$. By rearrangement, $F = UV/P$. Therefore, *for a substance meeting the criteria listed at the start of this paragraph, the plasma clearance is precisely equal to the glomerular filtration rate.* Substances used to measure GFR in this manner are artificially infused into the bloodstream. The most commonly used substance is **inulin,** a fructose polysaccharide

derived from Jerusalem artichokes. It is known to meet all the necessary criteria in mammals, amphibians, and some (but not all) other vertebrates.

Most of the native plasma constituents that enter the nephrons by filtration (e.g., Na$^+$, glucose) undergo postfiltration exchange with the blood, being either resorbed from the urine prior to excretion or added to the urine, as by active secretion. Whether a substance is added or removed in net fashion following filtration can be determined by comparing its clearance with the GFR. Studies of bullfrogs, for example, revealed that the urea clearance could be several times higher than the GFR; in other words, the volume of plasma being cleared of urea per unit time could substantially exceed the volume filtered. This finding pointed to the important conclusion that in bullfrogs urea is transferred from the plasma into the urine by secretion across the nephron walls in addition to being introduced by filtration at the glomeruli. Many substances exhibit clearances lower than the GFR, indicating that they are resorbed in net fashion after filtration. Consider, for example, results for toads (*Bufo marinus*) living in distilled water: Their GFR (inulin clearance) was about 6 mL/h, and their Na$^+$ clearance was about 0.04 mL/h. Because we know that Na$^+$ is freely filtered, these data tell us that, every hour, the Na$^+$ contained in 6 mL of plasma entered the nephrons by filtration, yet only the amount contained in 0.04 mL of plasma was ultimately excreted. Most of the filtered Na$^+$ must have been resorbed.

A useful quantitative parameter is the **relative clearance** of a substance, defined to be the ratio of that substance's plasma clearance to the GFR (e.g., inulin clearance). For a freely filtered substance, this ratio expresses the fraction of the amount filtered that is excreted. To illustrate, the relative Na$^+$ clearance for the toads just discussed was

$$\frac{\text{Na}^+ \text{ clearance}}{\text{GFR (inulin clearance)}} = \frac{0.04 \text{ mL/h}}{6 \text{ mL/h}}$$

$$= 0.007 = 0.7 \text{ percent}$$

Thus, these toads were excreting only 0.7 percent of the amount of Na$^+$ filtered. When the toads were transferred to a saline-water environment, their relative Na$^+$ clearance rose to 53 percent, indicating increased fractional excretion (diminished resorption) of filtered Na$^+$. Clearly, if the relative clearance of a substance exceeds 100 percent, some of the excreted amount must have entered the urine by means other than filtration.

In conclusion, studies of plasma clearance permit us to measure the filtration rate and to determine whether, and to what extent, particular solutes are resorbed or added to the urine in net fashion following filtration. These types of clearance analysis can be applied to all animal groups that form their primary urine by filtration but not, however, to groups using secretion to form the primary urine.

Other Methods of Analysis

We know from clearance studies that Na^+ is resorbed from the urine between filtration and excretion in amphibians. But where is it resorbed? And how? Is the resorption active or passive? To answer such questions, techniques for the direct study of individual nephrons must be used. Many of these techniques have proved to be applicable whether organisms form their primary urine by filtration or by secretion.

A technique of major importance is *micropuncture*. Fine micropipets are inserted into individual nephrons at identified points, permitting samples of tubular fluid to be drawn for analysis of composition. One of the reasons amphibians assumed a pivotal position in the pioneering studies of nephron function is that their nephrons are relatively large, visible, and accessible within the kidneys, thus making them amenable to micropuncture. When samples from amphibian nephrons were analyzed for glucose, for example, the glucose concentration proved to fall virtually to zero by the end of the proximal convoluted tubule; thus the proximal tubule was identified as the site of glucose resorption. When inulin is provided, micropuncture reveals that its concentration increases as fluid flows through the amphibian proximal tubule. Recognizing that inulin enters the nephrons only by glomerular filtration, the increase of its concentration from one end to the other of the proximal tubule must result not from addition of solute but from removal of water; thus, resorption of water in the proximal tubule is demonstrated. In bullfrogs, the concentration of urea increases even more than the concentration of inulin as fluid flows through the proximal tubule; this indicates that the proximal tubule is at least one of the places where urea is secreted into the tubular fluid.

In the distal convoluted tubule of amphibians, as shown in Figure 10.4, the concentrations of Cl^- and Na^+ decline, indicating ion resorption. *Studies of the difference of electrical potential* across the nephron wall can help to clarify the nature of this resorption. In the late part of the distal tubule, for instance, the tubular fluid proves to be electrically negative relative to fluid bathing the outside of the nephron, demonstrating that the outflux of Na^+ is opposed not only by the concentration gradient for Na^+ but also by the prevailing electrical gradient. Thus, the Na^+ outflux is evidently active.

It is not necessary that the fluid passing through a nephron be the native fluid produced by filtration. Techniques of *microperfusion* permit solutions of an investigator's choosing to be introduced into a nephron to ascertain how the nephron will alter the solution composition. The solutions, of course, can be constructed with particular analytical objectives in mind. A revolutionary recent innovation has been the development of techniques for removing nephron *segments* from the kidney and perfusing them in vitro. Importantly, such techniques are permitting functional analysis of parts of the nephron that in the kidney are too inaccessible for study.

FISH

Renal function in fish holds particular interest because the renal organs assume such very different tasks in the two major environments that fish occupy. Teleosts living in seawater produce a relatively small volume of urine that is virtually isosmotic with their blood plasma but rich in divalent ions (Mg^{2+}, SO_4^{2-}, Ca^{2+}). In fresh water, by contrast, teleosts characteristically void an abundant urine that is much more dilute than the blood.

Here, we discuss only the teleosts and elasmobranchs. In these groups, the overall renal design is similar to that seen in amphibians in that each kidney consists of many nephrons that discharge into collecting ducts. The nephron universally includes a proximal tubule, at least part of which (in virtually all species) is believed to be homologous with the proximal convoluted tubule of amphibians. In most freshwater and some marine fish, the proximal tubule is believed to resorb NaCl and to be sufficiently permeable to water that isosmoticity between the tubular fluid and blood plasma is maintained by osmosis.

Introduction of Fluid into the Nephrons

Elasmobranchs, virtually all freshwater teleosts, and many marine teleosts use glomerular filtration to form the primary urine. Variations in their GFRs can be large and constitute one of the major controls on urine flow.

Among the teleosts with glomerular nephrons, the total filtration surface of the kidneys tends to be substantially smaller in marine species than in freshwater species, and there is evidence (currently subject to controversy) that the GFR also tends to be substantially lower in the marine forms. These structural and functional correlations make sense from the viewpoint that marine teleosts need to produce only a small volume of urine by comparison to freshwater ones.

The glomeruli of many marine teleosts are degenerate by comparison to those of freshwater teleosts. Even more remarkably, in some marine species, which are termed *aglomerular,* many or all of the nephrons lack glomeruli altogether. The aglomerular fish form their primary urine by secretory mechanisms, and such mechanisms are probably used in combination with filtration even in some of the marine fish with glomerular nephrons. The marine teleosts are believed to be descended from freshwater teleost ancestors that had glomerular nephrons. Because ultrafiltration has seemed to be a mechanism well suited to the production of primary urine at a high rate, the degeneration or loss of the glomerular ultrafiltration system in some marine teleosts has usually been thought to be a consequence of the low demands for urine production they face. A case can be made against this interpretation, however, and possibly the selective advantages for the evolution of aglomerularism in some marine fish were more subtle than merely curtailing primary-urine production. It has been suggested, for instance, that the aglomerularism of a number of Antarctic teleosts serves to prevent urinary loss of their glycoprotein antifreezes (p. 98); as blood proteins go, these glycoproteins are relatively small molecules, small enough to be filtered appreciably in a glomerular kidney.

A few aglomerular fish, such as the toadfish *Opsanus tau* and the Malaysian pipefish *Microphis boaja,* occur in fresh water. These species, all immigrants from the sea, indicate that a freshwater existence is possible without ultrafiltration as a means of introducing large quantities of fluid into the nephrons.

The Distal Tubule

In elasmobranchs, virtually all freshwater teleosts, most euryhaline teleosts, and some marine teleosts, the nephron includes a segment that is homologous to the distal convoluted tubule of amphibians. This segment is believed usually to be the principal site where the urine can be rendered hyposmotic to the blood. Possibly, the collecting ducts also participate in producing dilute urine.

Significantly, marine teleosts commonly lack the distal convoluted tubule. Inasmuch as they are believed to be descended from freshwater ancestors, the absence of the distal tubule probably represents a secondary loss rather than a primitive condition. This loss can be understood in terms of the osmotic relations of marine teleosts and the function of the distal tubule in producing hyposmotic urine. Marine teleosts are hyposmotic to their environment and thereby face continuous osmotic desiccation. They have no need of a nephron segment specialized for the elimination of excess water.

A few species of teleosts that lack the distal tubule are found living in fresh water. Included are the aglomerular toadfish (*O. tau*) and pipefish (*M. boaja*) mentioned earlier and some euryhaline glomerular species such as killifish (*Fundulus heteroclitus*) and three-spined sticklebacks (*Gasterosteus aculeatus*). Even in the absence of the distal tubule, production of urine hyposmotic to plasma might be possible, because other structures (e.g., the collecting ducts) could potentially carry out the diluting function. In toadfish, although the osmotic U/P ratio of voided urine is 0.92, that of uretal urine direct from the kidneys is lower, 0.85; still, even the latter degree of dilution is meager compared to that attained by most freshwater animals (Table 8.3; see also p. 164).

Other Aspects of Nephron and Bladder Function

Fluid Resorption   Fish generally filter considerably more water than they excrete. Resorption of water, driven largely by NaCl resorption, occurs not only in their nephrons but also, in many species, in the urinary bladder. Recent studies reveal that bladder resorption can be a major factor in limiting the urine volumes of marine teleosts.

Concentration of Divalent Ions by Marine Teleosts   The urine of marine teleosts contains Mg^{2+}, SO_4^{2-}, and Ca^{2+} at concentrations much higher than those prevailing in the plasma. These ions are actively secreted into the urine, probably in the proximal nephron tubules. The bladder also plays a crucial role in concentrating divalent ions, for the bladder wall does not permit escape of such ions from the urine. Thus, as NaCl is actively removed from the urine in the bladder and as water follows the NaCl osmotically (keeping the urine isosmotic to the plasma), the divalent ions in the urine become ever more concentrated.

Resorption of Urea and TMAO by Elasmobranchs   Elasmobranchs have glomerular nephrons, so urea and trimethylamine oxide (TMAO) freely enter their primary urine. Maintenance of their high plasma levels of these compounds is dependent on extensive resorption, which takes place across the nephrons and is probably active.

Control of Renal Function   There is considerable evidence, mostly from studies of euryhaline species, that fish exert sensitive control over their excretory function. For example, transfer from fresh water to seawater may bring about one or more of the following changes: a decrease in GFR mediated largely by a reduction in the number of filtering nephrons, an increase in the active secretion of Mg^{2+} and SO_4^{2-} into the urine, and (probably) an increase in the water permeability of the nephrons or other excretory structures (e.g., bladder). To give a quantitative example, when rainbow trout (*Salmo*) were moved from fresh water to seawater, their GFR fell from about 8.6 to 1.2 mL/kg·h, and concomitantly the percentage of their nephrons filtering dropped from about 45 to 5 percent.

Although changes in the excretory function of fish are believed to be under hormonal control, the endocrinology remains largely unclear. Whether fish possess a functional ADH of the same chemical class as tetrapods is uncertain. Prolactin, an anterior pituitary hormone, is instrumental in the survival of many species in fresh water. It is believed to promote production of a copious, dilute urine by increasing the GFR and decreasing the osmotic permeability of the bladder and nephrons.

REPTILES AND BIRDS

Certain features are consistent in the kidneys of reptiles, birds, and mammals. The primary urine is always formed by glomerular ultrafiltration in these groups, and the nephrons always include proximal and distal convoluted segments and discharge into collecting ducts, which in turn lead to the ureters. The nephrons of reptiles are broadly similar to those of amphibians; each consists of a Bowman's capsule, proximal convoluted tubule, distal convoluted tubule, and certain short interconnecting segments. Birds, by contrast, have two types of nephron. Some avian nephrons structurally resemble their reptilian counterparts and are called *reptilian-type nephrons*. Other avian nephrons, as shown in Figure 10.6, have a long segment interposed between the proximal and distal convoluted tubules; this segment is arranged in the shape of a hairpin loop and is called the *loop of Henle*. Because mammalian nephrons also have loops of Henle, avian nephrons with loops are called *mammalian-type nephrons*.

In the avian kidney, the mammalian-type nephrons are organized into sets. Among the nephrons of a set, as depicted in Figure 10.6, the Bowman's capsules and proximal and distal convoluted tubules are all positioned toward the same part of the kidney that houses the reptilian-type nephrons, but the loops of Henle all project in a compact parallel array toward the direction of the ureter. Such a parallel array of loops of Henle is called a *medullary cone*. Each kidney includes many cones. Collecting ducts carrying the outflow from both the reptilian-type and mammalian-type nephrons run through the medullary cones on their way to the ureter. By virtue of the arrangement of loops of Henle into cones, the substance of the avian kidney has a clear macroscopic structure, consisting of two histologically distinct types of tissue: (1) the cones and (2) a mass of tissue known as the *renal cortex*, which contains the reptilian-type nephrons and the convoluted tubules of mammalian-type nephrons. In the kidneys of fish, amphibians, and reptiles, the nephrons are not arrayed into patterns giving rise to such distinct zones of tissue.

Renal Function

Reptilian kidneys cannot produce urine that is more concentrated than the blood plasma. Avian kidneys can, however, and the evidence is strong that the urine is rendered hyperosmotic to the blood by a process of countercurrent multiplication that involves the highly ordered arrangement of the loops of Henle and collecting ducts in the medullary cones. The principles of operation of the mammalian countercurrent multiplier are discussed in the next section. The avian multiplier is believed to function sim-

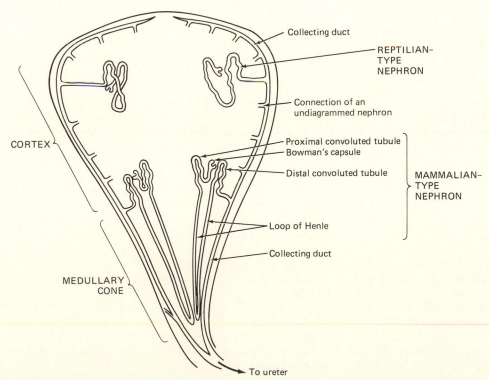

Figure 10.6 The organization of a lobe of an avian kidney, in cross section. [After E.J. Willoughby and M. Peaker, in G.M.O. Maloiy (ed.), *Comparative Physiology of Osmoregulation in Animals.* Academic, New York, 1979.]

ilarly, although urea cannot play a role in birds as it does in mammals.

One piece of evidence that points to a pivotal role for the medullary cones in production of hyperosmotic urine is this: Among related species of birds, there tends to be a direct correlation between the maximal concentrating ability of the kidneys and the numbers of mammalian-type nephrons and cones per unit volume of renal tissue.

The functions of the various segments of the nephrons have not been extensively documented by micropuncture studies in reptiles and birds. As in other vertebrates, NaCl and water are resorbed from the tubular fluid as it passes through the nephrons. Insofar as is now known, the proximal convoluted tubule is sufficiently permeable to water that the tubular fluid remains isosmotic to the blood plasma while in that part of the nephron; if urine hyposmotic to the plasma is produced, hyposmoticity is established in the distal convoluted tubule and perhaps the collecting duct.

In both reptiles and birds, uric acid is actively secreted into the tubular fluid, as well as being filtered. Secretion accounts for the main share of the excreted amount.

Variations in the GFR, partly under control of ADH (arginine vasotocin), play a substantial role in adjusting the rate of urine production in some reptiles and birds. Among at least some birds, reduction of the GFR is achieved selectively—by curtailing glomerular blood flow, and thus filtration, at the reptilian-type nephrons while permitting the mammalian-type nephrons to continue relatively unimpeded filtration. This mode of reducing the GFR has two direct effects, both of which are adaptive to situations (e.g., dehydration) that induce antidiuresis. First, the rate of primary urine formation is reduced; and second, the mammalian-type nephrons, which can produce an especially concentrated urine, are given increased functional prominence. In some reptiles and birds, ADH appears to induce an increase in the osmotic permeability of the distal tubules and/or collecting ducts. However, many terrestrial reptiles exhibit little or no evidence of such an ADH effect.

Postrenal Modification of Urine

In both birds and reptiles, the ureters discharge into the cloaca, and from there the urine may be moved retrogradely into the lower intestine. The composition and volume of the urine may be modified in the cloaca and intestine prior to excretion. Resorption of NaCl and water has been documented in the cloaca–intestine of a number of birds and lizards; depending on the species and conditions, the consequent conservation of water and NaCl may be appreciable. Often, uric acid and urates are carried into the cloaca largely in the form of supersaturated colloidal suspensions and then are precipitated in the

cloaca–intestine; this arrangement minimizes risks of occluding the ureters or other small-bore renal tubules. Evidence from lizards indicates that precipitation is induced in the cloaca by a whole suite of processes that tend to destabilize the colloidal state; these processes include water resorption from the urine and secretion of K^+, H^+, and urate into the urine. Recently, the surprising discovery has been made that an Australian lizard (*Amphibolurus*) and a sea turtle (*Chelonia*) can make urine that is modestly hyperosmotic to their blood plasma (U/P = 1.5 in the lizard). The mechanism is unknown, but at least in the lizard it resides in the cloaca–rectum, not the kidneys.

MAMMALS

The kidneys of mammals differ morphologically from those of amphibians and reptiles in two prominent respects. First, the nephrons have loops of Henle interposed between their proximal and distal convoluted segments, and second, the loops of Henle and collecting ducts are organized into parallel arrays. As we have seen, birds share these features to a degree. Discussion of the full physiological significance of the features has been postponed to this point, however, because the original impetus for understanding the function of the loops of Henle came from studies of mammals, and to this day it is in mammals that theories of the function of the loops are most thoroughly informed by data.

Anatomy of the Kidney

All nephrons in the kidneys of mammals have a loop of Henle. As shown in Figure 10.7, each loop consists of two parallel *limbs* connected by a *hairpin bend;* the **descending limb** leads from the proximal convoluted tubule to the bend, and the **ascending limb** runs from the bend to the distal convoluted tubule. The descending limb begins with a **thick segment,** and the ascending limb terminates with a thick segment. Interposed between these thick segments at various positions and for various lengths is a segment of very small diameter, the **thin segment**. The thin segment differs cytologically from the intermediate segment discussed earlier (Figure 10.3) and occurs only in mammals and birds. The loop of Henle varies considerably in length among species of mammals and among the nephrons within the kidneys of one species.

As shown in Figure 10.7, the glomeruli and convoluted tubules of the nephrons in each kidney are aggregated toward the outer surface of the kidney, whereas the loops of Henle and collecting ducts project inward, toward the renal pelvis (the renal pelvis, a tubular structure, represents the expanded inner end of the ureter that drains the kidney). Because of this highly ordered arrangement of the renal tubules, histologically distinct zones are apparent in the kid-

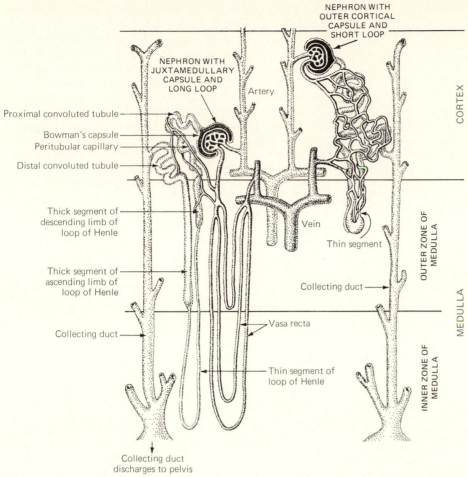

NEPHRON WITH
OUTER CORTICAL
CAPSULE AND
SHORT LOOP

NEPHRON WITH
JUXTAMEDULLARY
CAPSULE AND
LONG LOOP

Artery

CORTEX

Proximal convoluted tubule

Bowman's capsule
Peritubular capillary

Distal convoluted tubule

Vein

Thin segment

Thick segment of
descending limb of
loop of Henle

OUTER ZONE OF MEDULLA

Thick segment of
ascending limb of
loop of Henle

Collecting duct

MEDULLA

Collecting duct

Vasa recta

INNER ZONE OF MEDULLA

Thin segment of
loop of Henle

Collecting duct
discharges to pelvis

Figure 10.7 Diagram of human nephrons and their circulatory supply. See Figure 10.8 for relation of the cortex, medulla, and pelvis in whole kidneys. The collecting ducts discharge into the renal pelvis at the inner surface of the medulla. [After H.W. Smith, *The Kidney. Structure and Function in Health and Disease.* Oxford University Press, New York, 1951.]

ney tissue. In cross section (see Figures 10.8 and 10.10), each kidney is seen to consist of an outer layer, the *cortex,* which surrounds an inner body of tissue, the *medulla;* the medulla in turn surrounds and sometimes projects into the renal pelvis. The cortex (see Figure 10.7) consists of Bowman's capsules, convoluted tubules, the beginnings of collecting ducts, and associated vasculature; whereas the medulla consists of loops of Henle and collecting ducts as well as their associated vasculature. Within the medulla, the loops of Henle and collecting ducts run in parallel to one another. In rodents and insectivores, the medullary tissue of each kidney typically forms a single coherent structure. By contrast, in many other mammals, including humans, subdivisions of the medullary tissue (termed medullary pyramids) are evident.

Let us now briefly trace the path of fluid through the renal tubules of mammals, using the nephron to the left in Figure 10.7 as a reference. After filtration into the Bowman's capsule, fluid moves first through the proximal convoluted tubule and then descends into the medulla in the loop of Henle. After rounding the bend of the loop, the fluid returns to the cortex, passes through the distal convoluted tubule, and enters a collecting duct. The fluid then again passes through the medulla, ultimately being discharged from the collecting duct into the renal pelvis. From there the fluid flows down the ureter to be excreted. A convention worthy of note is that when fluid flows from the cortex toward the medulla, it is said to move *deeper* into the kidney.

We have already mentioned that various nephrons in the kidney of a species may have loops of Henle of dramatically different lengths. A look at Figure 10.7 (left side) reveals that there is a region of the medulla wherein the only *loop* elements are *thin* descending and ascending segments of relatively long loops of Henle. This region is called the *inner zone* of the medulla. Loops that project into the inner zone are termed *long loops*. Loops that turn back within the outer medulla or cortex are called *short loops* (e.g., the nephron to the right in Figure 10.7 has a short loop). Bowman's capsules may be positioned

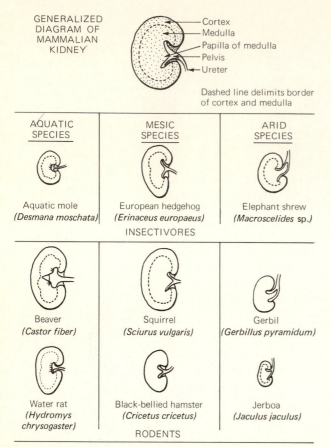

Figure 10.8 Diagrams of the kidneys of nine mammals. Aquatic species show little or no development of the papilla. (*Desmana* and *Hydromys* lack the papilla. *Castor* has two very shallow papillae.) Mesic species have papillae. The papilla is especially developed in arid species, so much so that it often penetrates well into the ureter. [From I. Sperber, *Zool. Bidr. Upps.* **22**:249–432 (1944).]

near the outer cortical surface, at mid-depth in the cortex, or within the cortical tissue next to the medulla (the *juxtamedullary* position). As depicted in Figure 10.7, nephrons with short loops tend to have their capsules positioned toward the outer cortex, whereas ones having long loops tend to have mid-cortical or juxtamedullary capsules. Importantly, differences in anatomy among nephrons within a kidney are often paralleled by differences of function.

The total number of nephrons has been estimated for a variety of species and varies widely. The laboratory rat has about 30,000 nephrons in each kidney, the dog about 400,000, and humans about 1.1 million.

Comparative Anatomy of the Kidney in Relation to Concentrating Ability and Habitat

Before the physiology of the loops of Henle began to be understood, anatomical evidence suggested that the loops were intimately involved in the production of urine hyperosmotic to the blood plasma. This evidence centered attention on the physiology of the loops.

The mere fact that the mammalian kidney differs from the reptilian kidney both in being able to produce hyperosmotic urine and in having loops of Henle arranged in parallel arrays gives the impression that the loops are likely to play an important role in producing concentrated urine. Several types of comparative morphological data on species of mammals strongly bolster this impression.

In certain mammals of moist environments, such as the mountain beaver (*Aplodontia*) and muskrat (*Ondatra*), all loops of Henle are short, and there is no inner medulla. These species are noted for having only meager urinary concentrating abilities. In mammals that are able to achieve high urinary concentrations, at least a modest proportion (15–20 percent or more) of the nephrons have long loops.

Commonly, the renal medulla has a roughly pyramidal shape and forms a projection into the pelvis known as the **renal papilla** (Figure 10.8). Because the papilla is composed in good part of long loops of Henle, its prominence gives an indication of the number and length of such loops. In 1944, Sperber reported observations on the medullary structure of about 140 species of mammals living in diverse environments. He found that the papilla was uniformly lacking or poorly developed in species inhabiting wet or aquatic environments. It was present, however, in species from mesic environments and was most developed in species from arid environments, as shown in Figure 10.8. Insofar as habitat may be taken as an indicator of demands for urinary concentration, these results indicated that there is a greater development of the long loops of Henle in species that produce relatively concentrated urine.

Following Sperber's work, a number of investigators took pains to quantify medullary size, usually expressing it relative to total kidney size. When the ability of species to concentrate their urine was examined as a function of their relative medullary size, a positive correlation was often found, as shown by the results of one classic study in Figure 10.9. Again, the arrays of loops of Henle in the medulla were implicated in the production of concentrated urine.

Figure 10.10 shows the renal papillae of two famous rodents from arid regions: a kangaroo rat (*Dipodomys*) and the sand rat, *Psammomys obesus*. Although both rodents have a sizable papilla, that of *Psammomys* is clearly the more prominent. Recent studies of *Psammomys* have revealed that its renal medulla is particularly elaborately organized; and among its long loops, an especially high proportion extend far into the papilla rather than turning back only a fraction of the way toward the tip (Figure 10.10*C*). *Psammomys* experiences far higher dietary salt loads than *Dipodomys*, for it subsists largely on succulent plants of high salt content (p. 154), whereas *Dipodomys* is a grain eater. Renally, *Psammomys* is currently known to have a bit greater ability to concentrate its urine relative to plasma than *Dipodomys* (maximum osmotic U/P = 17 versus 14). What is more striking, however, is that *Psammomys* pro-

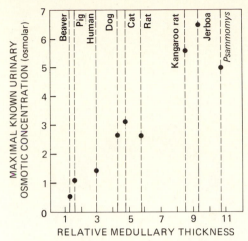

Figure 10.9 Maximal urinary osmotic concentration measured on nine mammals as a function of relative medullary thickness. Relative medullary thickness was determined as 10 times the ratio of medullary thickness to kidney size, kidney size being expressed as the cube root of the product of the three primary dimensions of the whole kidney. [From B. Schmidt-Nielsen and R. O'Dell, *Am. J. Physiol.* **200**:1119–1124 (1961).]

duces far greater volumes of concentrated urine than *Dipodomys* (perhaps 20 times as much volume per unit body weight). The organization and size of the papilla in *Psammomys* are seen to correlate with aspects of renal function that are adaptive to voiding the species' particularly high salt loads.

Theories of How the Urine Is Concentrated

In considering how mammals manufacture urine that is more concentrated in total solutes than their blood plasma, a distinction must be made between the mechanisms concentrating *urea* and those concentrating *other major solutes*. We start by examining some aspects of how nonurea solutes are concentrated; then we look at how urea is concentrated; and finally we discuss some intriguing recent ideas on interrelations between the mechanisms concentrating urea and nonurea solutes. Despite enormous advances over the past 30 years, these topics remain subject to both uncertainty and intense research. There is reason to believe that mechanisms may be different in some mammals (e.g., *Psammomys*) than in others. Here, we emphasize the processes of urinary concentration as they are understood from research on laboratory rabbits and rats.

Introductory Concepts of How Nonurea Solutes Are Concentrated
The urine contains a complex mix of nonurea solutes, predominantly inorganic ions such as Na^+, K^+, NH_4^+, Ca^{2+}, Cl^-, and SO_4^{2-}. An important operational parameter is the osmotically effective concentration of all these solutes *taken together,* termed the **total concentration of nonurea solutes**.

On its way out of the kidney, urine is discharged

from the nephrons into the collecting ducts and flows down the ducts, traversing the renal cortex and then the medulla (see Figure 10.7). At the point where urine enters the collecting ducts, the total concentration of nonurea solutes in the urine is typically lower than that in the blood plasma. However, when a state of antidiuresis prevails, as the urine passes in the collecting ducts through successively deeper layers of the medulla, its total concentration of nonurea solutes becomes progressively elevated, ultimately reaching levels well above the plasma concentration. The immediate mechanism of concentrating the nonurea solutes is abstraction of water from the urinary fluid. Urinary nonurea solutes are substantially trapped within the collecting ducts because the duct walls are poorly permeable to such solutes. Thus, as water is withdrawn from the urine, the nonurea solutes in the urine become more concentrated. How is water withdrawn? The fluids that *surround* the collecting ducts in the medulla, known as the **medullary interstitial fluids,** have a high NaCl concentration. In fact, their NaCl concentration rises steadily with increasing depth in the medulla, so that near the inner medullary surface the osmotic pressure attributable to NaCl is much above plasma osmotic pressure. As urine in the collecting ducts flows through the medulla and encounters the interstitial fluids having higher and higher NaCl-induced osmotic pressures, it progressively loses water by osmosis, for in antidiuresis the duct walls are freely permeable to water.

An important attribute of these processes is that *a high NaCl concentration on the outside of the collecting ducts serves to concentrate not only NaCl but also many other nonurea solutes on the inside.* This happens because the solutes involved are substantially barred from crossing the walls of the collecting ducts, yet the ducts are permeable to water. Thus, when high interstitial NaCl concentrations are encountered deep in the medulla, the primary process of equilibration between the urine and interstitial fluid is osmosis, and urinary nonurea solutes are concentrated *indiscriminantly* until their total osmotic concentration reaches the osmotic concentration of NaCl in the interstitium.

Now we must consider how the gradient of NaCl concentration is created in the medullary interstitial fluids. The loops of Henle are responsible.

A Single Effect Based on Active NaCl Transport   A well-documented part of the mechanism that creates the interstitial NaCl gradient resides in the thick segments of the ascending limbs of the loops of Henle. The cells in the walls of each ascending thick tubule actively transport NaCl out of the tubular fluid into the medullary interstitial fluid. The precise mechanisms involved have been subject to controversy for over a decade and are yet to be fully resolved. Suffice it to say here that both ions—Na^+ and Cl^-—are transported out of the ascending limb, and *at least* the Na^+ is moved by true, primary active transport. The consequences of the transport of NaCl, illus-

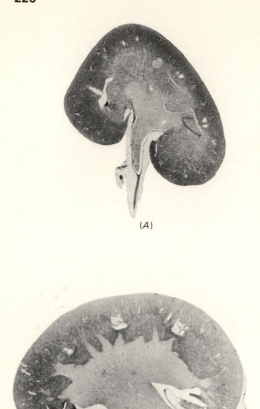

Figure 10.10 Cross sections of the kidneys of (*A*) a kangaroo rat (*Dipodomys*) and (*B*) the sand rat, *Psammomys obesus*. Part (*C*) shows a nephron with long loop of Henle in the *Psammomys* kidney. [Parts (*A*) and (*B*) from B. Schmidt-Nielsen, Organ systems in adaptation: The excretory system. In D.B. Dill (ed.), *Handbook of Physiology. Section 4: Adaptation to the Environment.* American Physiological Society, Washington, DC (1964); part (*C*) after B. Kaissling, C. de-Rouffignac, J.M. Barrett, and W. Kriz, *Anat. Embryol.* **148**:121–143 (1975).]

trated in Figure 10.11, depend on the permeability characteristics of the ascending limb and the adjacent descending limb of the loop of Henle. The walls of the ascending limb are poorly permeable to water. Thus, not only does the transport of NaCl lower the concentration of NaCl in the tubular fluid of the ascending limb and raise the concentration of NaCl in the adjacent interstitial fluid, but it also creates a difference of osmotic pressure between the ascending-limb fluid and interstitial fluid. The permeability characteristics of the descending limb appear to vary from species to species; nonetheless, by some combination of passive exchanges (osmosis or solute diffusion), the fluid in the descending limb is permitted to come readily to approximate osmotic equilibrium with the interstitial fluid. Therefore, as transport of NaCl out of the ascending limb increases the NaCl concentration and osmotic pressure of the adjacent interstitial fluid, it similarly raises the NaCl concen-

tration and osmotic pressure of the tubular fluid in the adjacent descending limb. In rabbits, for example, the descending limb is freely permeable to water. Thus, elevation of the concentration of the interstitial fluid induces an osmotic loss of water from the descending limb that brings the descending-limb fluid up to the same osmotic pressure as the interstitial fluid. Since NaCl is the predominant solute in the descending-limb fluid, the osmotic exit of water from that fluid appreciably raises the NaCl concentration of the fluid as well.

To summarize, at least along the thick ascending segment, active transport of NaCl out of the ascending limb lowers the NaCl concentration and osmotic pressure of the ascending-limb fluid and raises the NaCl concentration and osmotic pressure of the adjacent interstitial and descending-limb fluids. The *difference* in concentration thus created between the ascending-limb fluid and *adjacent* interstitial and de-

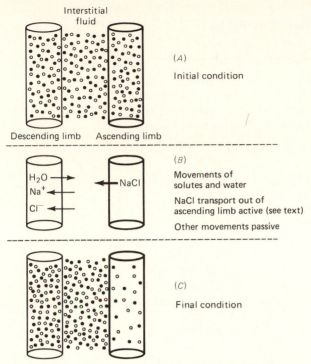

Figure 10.11   Schematic diagram of the development of gradients of osmotic pressure and NaCl concentration between adjacent parts of the ascending and descending limbs in the loop of Henle. The walls of the ascending limb are drawn thick to represent their relative impermeability to water. For simplicity, Na^+ and Cl^- are the only solutes shown—Na^+ as closed circles and Cl^- as open circles. In (A), the Na^+ and Cl^- concentrations are uniform throughout. In (B), NaCl is actively pumped, accompanied by passive movements discussed in the text. The consequences of the processes in (B) are shown in (C). The osmotic and ionic concentrations of the ascending-limb fluid are lowered from their original levels, whereas the concentrations in the interstitial and descending-limb fluids are raised. These processes may well be limited to the thick part of the ascending limb.

scending-limb fluids is termed the **single effect** of the active-transport mechanism.

Countercurrent Multiplication   The major hurdle in understanding how mammals produce concentrated urine was crossed in the 1940s and 1950s when Kuhn, Hargitay, Ramel, Martin, and Ryffel applied the concept of countercurrent multiplication to the loops of Henle. In the classic model of countercurrent multiplication generated by their work, it was assumed that all parts of each ascending limb actively transport NaCl as just discussed. Here, we develop that classic model. Later, we examine revisions suggested by recent research.

One basic property of the countercurrent multiplier is provided by the hairpin shape of the loop of Henle: This shape establishes two fluid streams that are oppositely directed (countercurrent), intimately juxtaposed, and connected. A second basic property of the multiplier is that energy is used within it (in

the form of active NaCl transport) to create a difference of concentration between adjacent parts of the oppositely directed fluid streams. This difference is the single effect already discussed and amounts to roughly 200 milliosmolar. As shown in Figure 10.12, the single effect is oriented from *side to side* in the loop. The pivotal contribution of the multiplier is to translate, or multiply, this single effect into a much larger difference of concentration from *end to end* in the loop. As shown again in Figure 10.12, an end-to-end difference of 600 milliosmolar would not be unusual; many mammals can create substantially greater differences.

The mechanism of countercurrent multiplication is diagrammed in Figure 10.13. Although osmotic pressures are shown in the figure and the following discussion is phrased in those terms, it will be important to remember that differences in osmotic pressure in the loop are paralleled by differences in NaCl concentration.

In part (A) of Figure 10.13, the entire loop of Henle and the interstitial space are filled with fluid of the same osmotic pressure as that exiting the proximal convoluted tubule—approximately isosmotic with the blood plasma. In part (B), active transport establishes a single-effect osmotic gradient of 200 milliosmolar all along the loop. In part (C), fluid moves through the loop in countercurrent fashion. Fluid that was concentrated in the descending limb during step (B) is thus brought around opposite to the descending limb in the ascending limb so that both limbs and the interstitial space are filled with concentrated fluid at the inner end of the loop. Now when, in (D), the single-effect osmotic gradient is

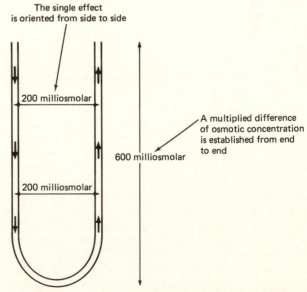

Figure 10.12   The distinction between the longitudinally and transversely oriented gradients of osmotic concentration in the loop of Henle. The longitudinal gradient is created from the transverse gradient by countercurrent multiplication. Gradients are expressed as differences in osmotic concentration from one place to another. Note the countercurrent flow in the two limbs of the loop of Henle.

BOX 10.3 COUNTERCURRENT MULTIPLIERS VERSUS COUNTERCURRENT EXCHANGERS

When oppositely directed fluid streams are closely juxtaposed and commodities are actively or passively exchanged between them, the effect of the countercurrent arrangement is to magnify or preserve longitudinal gradients in the levels of those commodities (longitudinal gradients being those oriented along the axis of fluid flow). The countercurrent arrangement has this effect because it impedes longitudinal flux of exchangeable commodities.

Two functional types of countercurrent system are recognized: *active* and *passive*. The active systems are the *countercurrent multipliers,* exemplified by the loops of Henle. The passive systems are called *countercurrent exchangers* (or *countercurrent diffusion exchangers*) and are exemplified by the heat exchangers in the appendages of birds and mammals, illustrated in Figures 6.31 and 6.32.

In an active system, metabolic energy is used *in the countercurrent system itself* to induce flux of commodities in or out of the fluid streams; within the loop of Henle, for example, energy is used to transport NaCl out of the ascending limb. In a passive system, fluxes of commodities in or out of the fluid streams occur without expenditure of metabolic energy *in the countercurrent system itself*. In the heat exchanger of Figures 6.31 and 6.32, for example, heat does not move out of one blood vessel and into another because of any metabolic energy expenditure within the countercurrent system; instead, it follows thermal gradients that exist because energy expenditure *elsewhere* in the body has caused the body core to be warmer than the environment.

Active countercurrent systems *create* gradients along their longitudinal axes. Note, for instance, that if the loops of Henle were turned off, the gradient of NaCl concentration from the outer to the inner side of the medulla would entirely disappear. Passive systems, by contrast, act only to *accentuate or preserve* longitudinal gradients that already exist for other reasons (e.g., the thermal gradient between the body core and end of the appendage in Figure 6.31).

again established between the ascending limb and interstitial fluid, the interstitial fluid is elevated to 500 milliosmolar at the inner end, rather than the 400 milliosmolar developed in step (B), and the fluid in the descending limb also reaches this higher osmotic concentration. Steps (E) and (F), and (G) and (H), repeat this process. *Fluid concentrated in the descending limb moves around into the ascending limb, providing a basis on which the single effect can produce an ever-increasing osmotic concentration in the interstitial fluid and descending limb at the inner end of the loop.* Meanwhile, the steady influx of 300-milliosmolar fluid into the beginning of the descending limb and the dilution of the tubular fluid that occurs during its flow from deep in the medulla to the top of the ascending limb combine to keep the osmotic pressure of the interstitial fluid at the outer (cortical) end of the loop near 300 milliosmolar. Thus, the difference in osmotic pressure between the two ends of the loop becomes greater and greater, to the extent that it substantially exceeds the single effect.

In the renal medulla, there are thousands of loops of Henle, all aligned in parallel. We would expect that by their combined action they would create in the medullary tissue as a whole a gradient of increasing osmotic pressure from its outer to inner side. The

classic data that originally confirmed this expectation are shown in Figure 10.14.

Further Notes on the Mechanisms of Concentrating Nonurea Solutes Figure 10.15 summarizes changes in the concentration of nonurea solutes in the tubular fluid when the kidney is producing concentrated urine. Note that the concentration of nonurea solutes in the tubular fluid rises to a high level at the hairpin bend of the loop of Henle, but by the time fluid exits the loop, it is actually *more dilute* than the blood plasma. The fluid then makes one last pass through the medulla, traveling down a collecting duct to be discharged into the renal pelvis. On this pass, final concentration of the nonurea solutes occurs.

The concentration of nonurea solutes in the definitive urine depends on the magnitude of the NaCl concentration in the interstitial fluids of the innermost medulla, for the urine osmotically equilibrates with those fluids just before leaving the kidney. In turn, the inner medullary NaCl concentration itself depends on a number of properties of the countercurrent multiplier system, including the size of the single effect, the rate of fluid flow through the loops of Henle, and the lengths of the loops. Lengthening of the loops tends to increase the end-to-end gradient

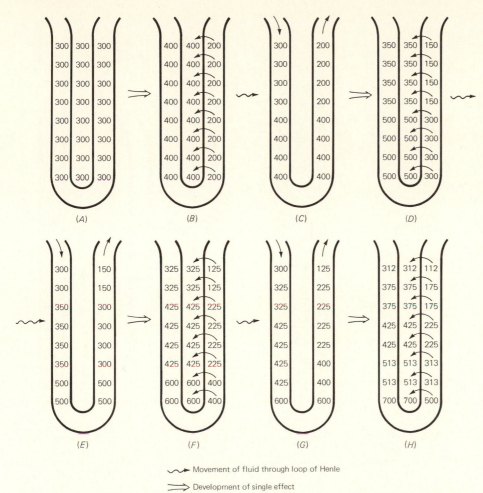

Movement of fluid through loop of Henle

Development of single effect

Figure 10.13 Simplified illustration of the operation of the countercurrent multiplier in the loop of Henle. Numbers are osmotic concentrations in units of milliosmolarity (mOsm). The operation of the multiplier is represented in a series of alternating steps. In (A), the entire system is at 300 mOsm. In (B), a single-effect osmotic gradient of 200 mOsm is developed all along the loop, and in (C) fluid moves through the loop. These steps are repeated in (D) through (H). The amount of fluid movement through the loop decreases progressively in steps (C), (E), and (G). Note that fluid entering the descending limb is always at 300 mOsm. This produces a tendency for the osmotic concentration at the cortical end of the descending limb and interstitial space to remain near 300 mOsm. For simplicity, osmotic pressures developed by the single effect have been computed by raising the osmotic pressure in the descending limb and lowering that in the adjacent ascending limb by equivalent amounts so as to produce a difference of 200 mOsm between the two limbs. [Based on the concept of R.F. Pitts, *Physiology of the Kidney and Body Fluids,* 3rd ed. Year Book Medical Publishers, Inc., Chicago, 1974.]

of NaCl concentration that can be maintained by the loops and thus tends to raise the inner-medullary NaCl concentration. This explains why, among related animals, species with relatively thick medullas and prominent papillas tend to be capable of producing relatively concentrated urine (p. 224).

Most investigators are inclined to believe that countercurrent multiplication holds energetic advantages. Specifically, it is thought to maintain a high inner-medullary NaCl concentration at relatively low energetic cost. This is so because with countercurrent multiplication, active transport does not have to maintain directly the full difference in concentration between the inner-medullary interstitial fluids and

blood plasma; instead, active transport has only to generate a smaller difference of concentration (the single effect), which is then magnified.

Concentration of Urea Urea is typically highly concentrated in the urine of antidiuretic mammals and can contribute half or more of the total urinary osmotic pressure. The walls of the collecting ducts prevent most solutes in the urine and medullary interstitial fluid from diffusing to electrochemical equilibrium. Urea, however, is a notable exception, and urea is concentrated in the urine according to quite different principles than other solutes. Urea is present at high concentrations in the medullary interstitial

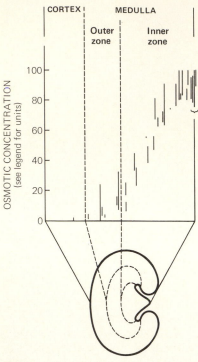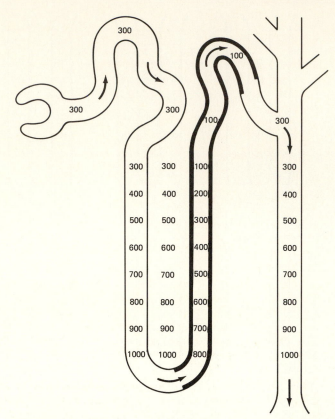

Figure 10.14 Osmotic concentrations determined along the axis of the papilla in kidneys of dehydrated rats. The value 0 corresponds to the osmotic concentration of plasma. The highest concentration measured (about 1000 milliosmolar greater than plasma at the tip of the papilla) is set equal to 100, and other increases of concentration above the plasma concentration are expressed as a percentage of this maximal deviation from the plasma concentration. [That is, the ordinate is equal to: 100 (measured concentration − plasma concentration)/(maximal concentration − plasma concentration).] Throughout the cortex, the osmotic concentration is equivalent to the plasma concentration. The osmotic concentration increases steadily with depth in the medulla; this increase is attributable both to an increase in NaCl concentration with depth and to an increase in urea concentration. [From H. Wirz, B. Hargitay, and W. Kuhn, *Helv. Physiol. Pharmacol. Acta* **9:**196–207 (1951).]

Figure 10.15 Osmotic pressures attributable to nonurea solutes in the nephrons and collecting ducts of the *concentrating* kidney. Values, expressed in milliosmolarity, are approximate and meant only to illustrate general trends as now understood. For simplicity, effects of urea are ignored. Parts of the tubules that are poorly permeable to water are drawn thick; those permeable to water are drawn thin. The location of the transition from poorly permeable to freely permeable walls at the end of the distal convoluted tubule is subject to debate.

fluid, and the walls of the collecting ducts *in the inner medulla* are permeable to urea. Basically then, high urea concentrations in the urine reflect the diffusion of urea to concentration equilibrium across the walls of the inner-medullary collecting ducts.

How does urea come to be present at high concentration in the medulla? Put simply, much more urea is filtered than is excreted, and some of the urea resorbed along the nephrons accumulates in the medulla. The thick ascending tubule in the loop of Henle, the distal convoluted tubule, and the cortical and outer-medullary parts of the collecting duct are poorly permeable to urea. In these same tubular regions, NaCl is actively transported out of the tubular fluid, diluting the fluid and thus setting up conditions favorable to osmotic outflux of water. Although the ascending tubule and (at least sometimes) the distal convoluted tubule are poorly permeable to water, the collecting duct is permeable to water during antidiuresis. Thus, when tubular fluid diluted by active

NaCl resorption enters the cortical and outer-medullary collecting duct, it loses water osmotically; and since the permeability to urea of these tubules is low, urea—trapped inside the tubules—becomes concentrated in the tubular fluid as water is lost. The important net result is that the tubular fluid bears urea at high concentration by the time it enters the inner-medullary collecting duct, which is permeable to urea. Urea therefore diffuses from the tubular fluid into the inner-medullary interstitial fluid, a process that charges the interstitium with urea. According to present thinking, this whole sequence of events is self-reinforcing because urea also *enters the tubular fluid* from the medullary interstitial fluid *in the loops of Henle.* By such recycling, the urea concentration in tubular fluid arriving in the inner-medullary collecting ducts (after passage through the loops of Henle) tends automatically to rise in parallel with the urea concentration of the interstitial fluid. Thus, with a steady influx of new urea from filtration, a gradient favorable for diffusion of urea *into* the interstitial fluid from the collecting ducts is maintained even though the interstitial concentration rises to a high level. High concentrations of urea in the interstitial fluid promote high urinary concentrations because diffusive outflux of urea from the collecting-duct fluid continues only to the point of concentration equilibrium with the interstitium.

It is clear in principle that the generation of a high urea concentration in the medullary interstitium must require energy. One of the major investments of energy is made by way of active NaCl resorption, for the resorption of NaCl sets the stage for concentrating urea sufficiently in the collecting-duct fluid that urea will diffuse out of the collecting ducts into the medullary interstitial fluid.

Urea contributes powerfully to the *absolute* osmotic pressure of both the urine and the medullary interstitial fluid. However, because urea diffuses to concentration equilibrium across the walls of the inner-medullary collecting ducts, it does not, except transiently, make a direct contribution to gradients of osmotic pressure between the collecting-duct fluid and interstitial fluid (p. 148). An important consequence is that resorption of *water* from the urine in the collecting ducts (which occurs osmotically) is controlled by *nonurea* solutes.

"Passive" Models of NaCl Concentration in the Inner Medulla   In the classic concept of how a gradient of interstitial NaCl concentration is created from the outer to the inner side of the medulla, the single effect in the countercurrent multiplier is presumed to be created along the entire length of the loop of Henle by active NaCl transport out of the ascending limb. As we now look briefly at current debates over this concept, it will be useful to recognize two regions of the loop of Henle, which are evident from examination of the left-hand loop in Figure 10.7. The *outer region* of each loop consists of the thick segment of the ascending limb and the part of the descending limb adjacent to it. The *inner region* is composed of the thin segment of the ascending limb and its adjacent descending limb.

Active resorption of NaCl has been clearly demonstrated in the thick ascending segment. Thus, a consensus exists that countercurrent multiplication occurs according to the classic model in the outer regions of the loops of Henle.

For a long time, the applicability of the classic model to the inner regions of the loops has been open to question because the cells of the thin ascending segment are not morphologically typical of active-transport cells. Lately, functional studies using the technique of in vitro microperfusion have indicated that in at least some species (e.g., rabbit) ions are probably *not* actively resorbed from the tubular fluid in the thin ascending segment. New models spurred by this discovery hold that countercurrent multiplication still does occur in the inner regions of the loops of Henle of these species and accounts for the gradient of increasing NaCl concentration from the outer to the inner margin of the inner-medullary

zone. According to these models, however, generation of the single effect in the inner regions of the loops of these animals does not involve active solute transport by either the ascending or descending limb. The models are called "passive" models because of this purported origin of the single effect, but they are passive only in an immediate sense. The high urea concentration in the inner medulla is of key functional importance in these models. As we have seen, production of this high concentration is dependent on active NaCl transport elsewhere in the kidney (e.g., in the thick ascending segments).

The single effect is postulated to be generated in the inner regions of the loops of the animals concerned by a combination of (1) the high medullary urea concentration and (2) special permeability characteristics of the descending and ascending limbs. Let us outline here one particular view of how this occurs (that of J. Kokko and F. Rector). Evidence indicates that the descending limb is poorly permeable to both urea and NaCl but permeable to water. Thus, high urea concentrations in the medullary interstitial fluid induce an osmotic outflux of water from the descending-limb fluid and thereby concentrate NaCl within the descending limb. The ascending limb is evidently poorly permeable to water, modestly permeable to urea, and quite permeable to NaCl. Fluid entering the ascending limb has just had its NaCl concentration raised by passage through the descending limb. Thus, NaCl diffuses out across the ascending limb, charging the interstitial fluid with NaCl; and even though this loss of solute renders the fluid in the ascending limb hyposmotic to the interstitial fluid, water outflux from the ascending limb is limited by the limb's low water permeability. Phenomenologically, the results of all these processes are argued to be not greatly different from those created by active NaCl transport (Figure 10.11). The interstitial and descending-limb fluids are endowed with higher NaCl concentrations than adjacent ascending-limb fluid, and this single effect is multiplied countercurrently.

Further exploration of such passive models of inner-medullary NaCl concentration is a focus of much current research.

The Vasa Recta   The blood capillaries of the medulla schematically take the form of hairpin loops paralleling the loops of Henle, as shown in Figure 10.7. These vascular loops are termed *vasa recta*.

By virtue of their looped shape, the vasa recta can have water-permeable and solute-permeable walls like other capillaries—thus permitting free exchange between the blood and the perfused tissue—yet the circulation of the blood is prevented from destroying the concentration gradients of NaCl and urea in the medullary interstitium. Consider what would happen if blood, after flowing into the medulla from the cortex, simply exited the medulla on the pelvic side. On its way into the medulla, as it encountered ever more concentrated interstitial fluids, the blood would lose water to the interstitium osmotically and take up NaCl and urea by diffusion. Exiting on the pelvic side, it would leave all that water behind and take the solutes away, diluting the medulla on both accounts. Instead, after entering the deep medulla, the blood reverses direction and flows back toward the cortex; and on its way out, as it encounters more and more dilute interstitial fluids, it reabsorbs water and yields NaCl and urea. The familiar tendency of countercurrent flow to preserve gradients oriented along the axis of flow is apparent. The vasa recta exchange solutes and water with the interstitium only passively and thus are described as countercurrent diffusion exchangers.

Recall that the final process of concentrating the urine involves osmotic loss of water from the collecting ducts into the medullary interstitium. This water, if allowed to accumulate in the interstitium, would itself dilute the medullary concentration gradients. An important function of the blood flow through the vasa recta is to carry away the resorbed water. Evidently, as the blood in the vasa recta dynamically loses and regains water during its passage in and out of the medulla, the colloid osmotic pressure caused by the blood proteins introduces a bias for the gains of water by the blood to exceed losses.

An Overview of Events in the Concentrating Kidney

The fluid introduced into the Bowman's capsule by ultrafiltration is approximately isosmotic to the blood plasma and contains similar concentrations of inorganic ions, glucose, and amino acids. A major function of the proximal convoluted tubule is resorption of NaCl and water; in fact, measures on various species indicate that 60–80 percent of the filtered amounts of NaCl and water are resorbed by the time the tubular fluid reaches the beginning of the loop of Henle. In addition, glucose, many amino acids, and HCO_3^- are almost completely resorbed in the proximal tubule, and well over half of filtered K^+ is resorbed. The mechanisms involved are complex and not yet fully understood. Sodium is actively resorbed, and this is a process of principal importance. The resorption of HCO_3^-, Cl^-, glucose, amino acids, and certain other solutes is coupled in one way or another with this active Na^+ transport. The walls of the proximal tubule are freely permeable to water, and water is resorbed osmotically at a sufficient pace that, despite all the solute resorption, the osmotic pressure of the tubular fluid does not fall measurably below plasma osmotic pressure. The mechanism of this isosmotic resorption of water may be local osmosis.

Although the tubular fluid enters the loop of Henle isosmotic to plasma (~300 milliosmolar), it exits the loop hyposmotic (perhaps 100–150 milliosmolar) for reasons already discussed (p. 228). In the distal convoluted tubule, Na^+ is actively resorbed, with Cl^- following passively (usually) in the electrical field generated by Na^+ transport. The walls of much of the distal tubule are poorly permeable to water; thus, osmotic water resorption is sufficiently limited that, given the removal of NaCl, the tubular fluid remains strongly hyposmotic to plasma (Figure 10.15). Potassium is added to the tubular fluid (partly passively, partly actively) in the distal convoluted tubule and cortical collecting tubule. This addition of K^+ controls the amount of K^+ eliminated in the urine, for by virtue of events in the proximal tubule and loop, the tubular fluid contains little K^+ when it first enters the distal tubule.

Perhaps 5 percent or less of the originally filtered volume reaches the collecting duct. In the concentrating kidney, the collecting duct is permeable to water (the terminal distal tubule may also be). Thus, dilute tubular fluid arriving there from the main distal tubule promptly comes to isosmoticity with the cortical interstitium (~300 milliosmolar) by osmotic outflux of water (Figure 10.15). This tubular fluid then descends from the cortical to the pelvic end of the collecting duct. As it does so, it encounters ever higher interstitial NaCl concentrations and attains higher concentrations of urea and nonurea solutes by the mechanisms heretofore discussed. Water is resorbed osmotically. Especially in the cortical part of the collecting duct but also in the inner-medullary part, active Na^+ resorption occurs (Cl^- following); this process both determines the final amount of NaCl to be excreted and, by reducing the amount of nonurea solute, enhances osmotic reduction of urinary volume. In the end, antidiuretic mammals typically excrete only 1 percent or less of the filtered NaCl and water.

Regulation of Urinary Volume and Concentration; ADH

Mammals are typically capable of adjusting the volume and concentration of their urine over a broad range. Humans in antidiuresis, for example, can limit the volume of their urine to less than 1 percent of the filtered amount and raise its concentration to about 1200 milliosmolar (U/P = 4); in diuresis, by contrast, they can increase the volume to about 15 percent of the filtered amount and lower the concentration to about 50 milliosmolar (U/P = 0.2).

Antidiuretic hormone (arginine vasopressin in most mammals) plays a major role in controlling the volume and concentration of the urine. The principal demonstrated effect of ADH is to cause a dramatic increase in the permeability to water of the collecting ducts and perhaps, as well, the terminal parts of the distal convoluted tubules.

In the concentrating kidney, although we have not said so heretofore, high blood levels of ADH are responsible for the high permeability of the collecting ducts to water. This high permeability is responsible for the relatively unfettered resorption of water from fluid in the ducts, a process that reduces the volume of the urine and concentrates nonurea solutes.

When blood levels of ADH are low, the collecting ducts and entire distal convoluted tubules are poorly permeable to water. Thus, from the time the tubular fluid exits the loops of Henle to the time it is discharged into the renal pelvis, it is barred from freely coming to osmotic equilibrium with the surrounding cortical and medullary interstitial fluids. Given that the tubular fluid is already hyposmotic to plasma when it exits the loops, its tendency is to lose water osmotically. The low permeability of the walls of the distal tubules and collecting ducts impedes such water loss, however. Furthermore, because of the low permeability, as NaCl is actively resorbed in the distal tubules and collecting ducts, the tubular fluid becomes ever more hyposmotic to the plasma. An abundant, dilute urine results.

Note again that ADH fundamentally controls *water* excretion (see pp. 194 and 214). The amount of each particular nonurea solute excreted is adjusted by specific tubular mechanisms (e.g., active resorption and secretion) according to needs for solute homeostasis. The level of ADH helps determine the amount of water excreted with the nonurea solutes.

When the kidney converts from producing concentrated urine to producing dilute urine, not only do the collecting ducts become less permeable to water, but also the magnitude of the osmotic gradient in the interstitial fluids of the medulla is reduced. In dogs shifted from antidiuresis to profound diuresis, for example, the osmotic pressure of the inner-medullary interstitium fell from 2400 milliosmolar to 500 milliosmolar. An extensive interplay of processes, not fully understood, is responsible for this change. An important consequence of the drop in medullary interstitial osmotic pressures is that osmotic gradients across the walls of the collecting ducts in the diluting kidney are reduced; thus, the driving force for water resorption from the urine is decreased.

Comparisons of rodents from arid and mesic habitats have revealed that the arid species tend to maintain larger pituitary stores of ADH per unit body weight and probably can synthesize ADH faster. These properties make sense, given that the arid species maintain their kidneys in an antidiuretic state more of the time.

Adrenal Control of Sodium Excretion

Aldosterone stimulates resorption of Na^+ in the cortical collecting tubule and possibly also in the distal tubule and some other parts of the collecting tubule. The implications of this effect and the modes of control of aldosterone secretion are discussed in Chapter 8 (p. 194). Here and there around the world, soils low in Na^+ are found, and herbivores in such regions may need to conserve Na^+ assiduously to avoid entering negative Na^+ balance. Aldosterone is involved in this conservation; rabbits in a Na^+-poor region of Australia, for example, have especially large adrenals, greatly elevated circulating aldosterone levels, and low urinary Na^+ levels. Aldosterone also stimulates K^+ secretion into the renal tubular fluid.

Other hormones that affect renal Na^+ excretion include additional adrenal steroids, calcitonin, and the natriuretic hormones (p. 195).

Glomerular Filtration Rate

In controlling the rate of urine production, variations in the GFR do not generally assume the importance in mammals that they do in many other vertebrates. Instead, the GFR is kept relatively stable, and the principal mode of increasing or decreasing the rate of urine flow is adjustment of the fraction of filtered water that is resorbed.

Like most generalizations, this one should not be interpreted too absolutely. Ups and downs in the GFR do occur and sometimes assume major importance. Dromedary camels, for example, have been observed to undergo a threefold drop in GFR when denied drinking water for 10 days; their rate of urine flow dropped virtually in proportion.

The GFR is an allometric function of body weight in mammals, increasing in proportion to the 0.7 power of weight. Thus, large mammals produce fil-

---

BOX 10.4   THE SIMILARITY OF EFFECTS OF ADH IN AMPHIBIANS AND MAMMALS

ADH has retained a single basic effect on the renal tubules all the way from amphibians to mammals; namely, it increases the permeability to water of tubular membranes that otherwise are poorly permeable. This increase in the permeability to water has the important result that the tubular fluids come to osmotic equilibrium with the fluids surrounding the tubules. In amphibians, the fluids surrounding the distal tubules and collecting ducts are osmotically similar to the general body fluids, and in the presence of ADH, the urine approaches isosmoticity with the body fluids. Mammals have surrounded the collecting ducts with fluids that are hyperosmotic to the general body fluids. Thus, the urine is rendered hyperosmotic in the presence of ADH.

trate at a lower rate per unit weight than small mammals, just as they have lower metabolic rates per unit weight.

CRUSTACEANS

In moving on now to consider some of the invertebrates, we first examine two groups—the crustaceans and mollusks—that are believed to form their primary urine by ultrafiltration. We emphasize the decapods (e.g., crayfish and crabs) in discussing the crustaceans.

The renal organs of adult decapod crustaceans are a pair of **antennary glands** or **green glands** located in the head and opening to the outside independently near the bases of the second antennae (see Figure 8.2). The basic structure of the antennary gland in freshwater crayfish is illustrated at the top in Figure 10.16. Each gland begins with a closed **end sac** or **coelomosac** lying to the side of the esophagus. Following the end sac is the **labyrinth,** or green body, a sheet of spongy tissue consisting of anastomosing channels. The **nephridial canal,** which also has a spongy internal morphology, leads from the labyrinth to the expanded **bladder,** and the bladder empties to the outside. The nephridial canal is found in only certain freshwater decapods. The modern marine forms, lacking the nephridial canal, are believed to represent the primitive decapod condition, and it is thought that the canal made its appearance when decapod crustaceans migrated from the sea into fresh

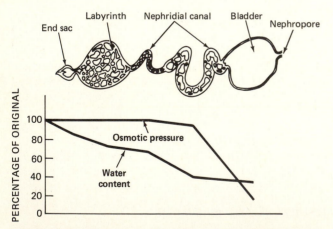

Figure 10.16 Changes in the osmotic pressure and water content of the urine as it passes through the antennary gland and bladder of crayfish (*Australopotamobius pallipes* and *Orconectes virilis*) living in fresh water. Values are expressed as percentages of the osmotic pressure and water content (volume) of the filtrate introduced into the end sac. Values are plotted immediately below the parts of the renal apparatus to which they apply. The diagram of the renal structures is distorted. In life, the nephridial canal is tightly convoluted and partly enveloped by the sheetlike labyrinth. [After J.A. Riegel, in B.L. Gupta, R.B. Moreton, J.L. Oschman, and B.J. Wall (eds.), *Transport of Ions and Water in Animals*. Academic, New York, 1977.]

water. The various parts of the renal organ vary morphologically among species. The labyrinth is lacking in some shrimps, for example, and in hermit crabs the bladder ramifies into a complex tubular network that penetrates much of the thorax and sends diverticula into the abdomen. The physiological significance of much of this morphological diversity is unknown.

The walls of the end sac are thin. Arteries coming fairly directly from the heart supply a network of small vessels or lacunae on the outer surface of the end sac (though a true capillary bed has not been demonstrated). Such morphological evidence has long suggested that fluid enters the end sac by filtration under the force of blood pressure. Physiological studies have supported this hypothesis.

In crayfish, minute vesicular structures called **formed bodies** are found suspended in the fluid within the end sac and elsewhere in the renal passages. Such bodies, which are believed to be introduced by the cells lining the passages, have also been reported in the renal tubular fluids of a number of other animals, including snails, frogs, and insects; but little is known of their functional properties except in crayfish. They seem to have a marked tendency to attract water, and one postulate is that they help to induce water flux—filtration—into the end sac.

The labyrinth, although it may alter the volume and composition of the urine, is by all accounts incapable of rendering the urine appreciably hyposmotic to the blood. Pioneering micropuncture studies indicated, furthermore, that in crayfish the nephridial canal is chiefly responsible for diluting the urine. These studies revealed that the nephridial canal is in certain respects functionally analogous to the distal convoluted tubule of amphibians: a segment of the renal tubule, positioned relatively distally, that renders the urine hyposmotic to the blood by actively resorbing NaCl (and perhaps other solutes) across membranes poorly permeable to water. Elaborating on this discovery, some investigators voiced the hypothesis that for a decapod to produce dilute urine, it must possess a nephridial canal. This view received support from the morphological evidence, noted earlier, that the nephridial canal is a freshwater "invention." Crayfish, which have the canal, can produce hyposmotic urine; but marine decapods and such freshwater forms as the crab *Eriocheir sinensis* (p. 164), which lack the canal, are unable to manufacture a dilute urine.

Restudy of crayfish, again using micropuncture, has provided additional information. The results of the reinvestigation, as depicted in Figure 10.16, confirm that the nephridial canal can appreciably dilute the urine. However, the principal site of dilution in these studies proved to be the bladder. At this time, a full appraisal of the relative roles of the nephridial canal and bladder in producing urine hyposmotic to the blood must await further study. Clearly, however, the nephridial canal is only one element in the overall adaptation of the crayfish renal organ to produce dilute urine.

Active resorption (conservation) of NaCl has been demonstrated to be one of the major processes by which crayfish dilute their urine. Another known attribute of renal function in crayfish is that the volume of urine excreted is controlled in good measure by adjustments in the amount of water resorbed during passage of the urine through the renal organ. As shown in Figure 10.16, crayfish (*Austropotamobius*) may resorb about 70 percent of the volume they filter when living in fresh water. These crayfish voluntarily emerge into the air, and then their resorption increases markedly.

In various marine crabs, there is evidence for active resorption of Ca^{2+} and glucose and active secretion of Mg^{2+} and SO_4^{2-} into the urine. Interestingly, one important site of Mg^{2+} secretion and glucose resorption is the bladder. In some land crabs, urine is passed to the gill chambers after excretion and there is altered, as by salt resorption.

MOLLUSKS

The renal organs of mollusks are termed *kidneys* or *nephridia*. They are tubular or saccular structures that empty into the mantle cavity or directly to the outside. Each nephridium generally connects to the pericardial cavity via a *renopericardial canal*. The morphology of the nephridia is highly variable among species. In the bivalves, most cephalopods, and some gastropods, there are two nephridia, but in most gastropods there is only one.

Excretion in *Octopus*

One of the most thoroughly understood molluscan kidneys is that of the octopus *Octopus dofleini*. Thus, although the renal complex of cephalopods includes several unusual features, we shall discuss *Octopus* first.

In common with other squids and octopuses, *O. dofleini* possesses two branchial hearts in addition to its systemic heart. These receive venous blood from the body and pump it through the gills (see p. 367). The nephridia are associated with the branchial hearts, as diagrammed schematically in Figure 10.17. Each branchial heart bears a relatively thin-walled protuberance, the branchial heart appendage, which communicates with the lumen of the heart. In *Octopus* the pericardium encloses only the side of the branchial heart bearing the heart appendage. A long renopericardial canal leads from the pericardial cavity to the large *renal sac*. The sac, in turn, empties into the mantle cavity.

As noted, the connection of the nephridium with the pericardial cavity is a common molluscan feature. It has long been thought that fluid is generally intro-

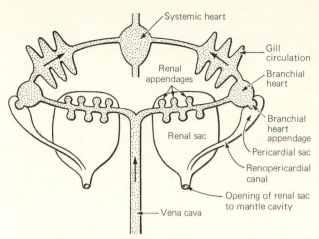

Figure 10.17 Schematic representation of the renal complex and associated circulatory system in *Octopus*. The parts of the circulatory system are stippled grey. Arrows show the direction of blood flow. Fluid is driven into the pericardial sacs by ultrafiltration across the branchial heart appendages and flows to the renal sacs through the renopericardial canals. The renal sacs discharge into the mantle cavity. [From A.W. Martin and F.M. Harrison, Excretion. In K.M. Wilbur and C.M. Yonge (eds.), *Physiology of Mollusca*, Vol. II. Academic, New York, 1966.]

duced into the molluscan nephridium by ultrafiltration from the blood across the heart wall into the pericardial cavity. This hypothesis has been particularly appealing in view of the close association of the nephridia with the very source of blood pressure. For *Octopus* and certain other cephalopods (e.g., squids, *Loligo*), the evidence is compelling that the pericardial fluid (i.e., primary urine) is an ultrafiltrate of the blood forced into the pericardial sacs across the branchial heart appendages under the force of pressure developed in the branchial hearts. The filtrate flows to the renal sacs via the renopericardial canals. Glucose and amino acids are promptly reclaimed by being actively resorbed along the canals.

As depicted in Figure 10.17, blood approaches the branchial hearts in a large vein, the vena cava, which divides to form two lateral branches that enter the hearts. The lateral branches of the vena cava pass by the renal sacs and there bear many glandular diverticula, the so-called renal appendages, which are closely applied to the walls of the sacs. These appendages have long been thought to be sites of exchange between the blood and the fluid in the renal sacs. Cephalopods excrete considerable quantities of ammonia in their urine. Evidence on *Octopus* indicates that ammonia diffuses into the renal sacs from the blood across the renal appendages and effectively becomes trapped in the urine by reacting to form ammonium ion. Interestingly, much of this ammonia is chemically released in the renal appendages themselves, at least in part through enzymatic breakdown of circulating glutamine by glutaminase. As a consequence, blood leaving the region of the renal appendages often has a higher ammonia concentration than blood entering, despite considerable diffusion of ammonia into the urine. It is significant that the blood travels directly to the gills after passing through the branchial hearts, and there blood ammonia is lost by diffusion into the water in the mantle cavity. More ammonia is lost across the gills than in the urine.

Excretion in Bivalves and Gastropods

Bivalves and gastropods do not possess branchial hearts. Their nephridia open into the pericardial cavity surrounding the systemic heart, as shown in Figure 10.18. Not uncommonly, structures called pericardial glands are associated with the wall of the heart or pericardium. These structures are known to be involved in excretion. They have been thought to function in secretion; recent evidence indicates that they are sites of ultrafiltration.

In comparison with cephalopods, the pressures developed by the hearts of bivalves and most gastropods are very low (see p. 366), and reservations about the adequacy of those pressures to induce ultrafiltration have repeatedly been expressed. However, it now seems clear that cardiac pressures are generally adequate, and numerous other pieces of evidence point to the occurrence of ultrafiltration. Thus, today's consensus is that, in most gastropods and bivalves, (1) the primary urine is formed by filtration, (2) this filtration generally occurs across the heart wall (e.g., atrial wall), and (3) the filtrate is introduced into the pericardial cavity.

At least some of the terrestrial pulmonate snails constitute exceptions. In them, primary urine is formed in the renal sac, the enlarged main chamber of the kidney into which the renopericardial canal leads. Their primary urine has the composition expected of an ultrafiltrate of the blood, but whether it is formed by filtration under force of blood pressure remains unclear.

The urine of bivalves and gastropods undeniably is altered as it flows through the kidney, although our knowledge of the processes taking place is limited. Resorption of glucose and amino acids has been observed in certain species, as has addition of nitrogenous wastes. Freshwater and terrestrial forms can render their urine hyposmotic to the blood, and the sites where this occurs have been pinpointed in some species. In the freshwater mussel *Anodonta cygnea,* for example, the osmotic pressure of the urine has been reduced to about half that of the blood by the time the urine reaches the bladder (Figure 10.18).

ANNELIDS

In polychaete and oligochaete annelids, the usual arrangement is two renal tubules, or *nephridia,* per body segment—either in most or all of the body segments or in the segments of a limited part of the body. The morphology of the nephridia is highly variable among species. Typically, a nephridium is an unbranched, convoluted tubule. The tubule may be differentiated, to a greater or lesser extent, into morphologically distinct regions. The nephridia project into the coelomic cavity, and usually each nephridium opens directly to the coelom at its coelomic end via a ciliated funnel (*nephrostome*). At the other end, the nephridium opens directly or indirectly to the outside. Most commonly, each nephridium discharges via a pore through the body wall (*nephridiopore*), but in some earthworms the nephridia discharge into the gut. The latter arrangement has been postulated (without direct evidence) to permit water resorption from the urine in the gut, thereby aiding water conservation.

Coelomic fluid is by all accounts swept into the nephridia of polychaetes and oligochaetes at their nephrostomes; in fact, this influx is believed in general to be the primary or exclusive source of nephridial tubular fluid. A favorite hypothesis—supported by only a small amount of indirect evidence—is that coelomic fluid itself is formed at least partly by ultrafiltration across coelomic blood vessels. Should this hypothesis be correct, the coelomic circulation, coelomic cavity, and nephrostomes would together constitute a system for introducing into the renal tubules a primary urine that is an ultrafiltrate of the blood.

One of the most thoroughly studied of the oligochaetes and polychaetes is the common nightcrawler *Lumbricus terrestris.* As shown in Figure 10.19, each of its nephridia consists of several morphologically distinct segments. The osmotic pressure of the tubular fluid does not change from the time the fluid enters at the nephrostome until it exits the narrow tubule, but in the middle and wide tubules, the urine can be markedly diluted by NaCl resorption across membranes poorly permeable to water.

Another annelid to have received close study is the amphibious blood-sucking leech *Hirudo medicinalis.* This animal's 17 pairs of nephridia are in certain respects quite different in structure from the ne-

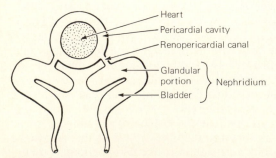

Figure 10.18 Schematic representation of the renal complex of bivalve mollusks. [From A.W. Martin and F.M. Harrison, Excretion. In K.M. Wilbur and C.M. Yonge (eds.), *Physiology of Mollusca,* Vol. II. Academic, New York, 1966; in turn, after E.S. Goodrich, *Q. J. Microsc. Sci.* **86:**113–392 (1945).]

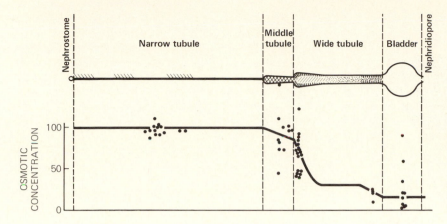

Figure 10.19 Osmotic concentration of the nephridial fluid in *Lumbricus terrestris*. Nephridia were dissected out in Ringer's solution and samples of nephridial fluid taken by micropuncture. Osmotic concentrations are expressed as percentages of the osmotic concentration of the surrounding Ringer's solution. The line drawn through the data represents the investigator's interpretation of the results. The diagonal dashed lines along the narrow tubule indicate ciliated portions of the tubule. [From J.A. Ramsay, *J. Exp. Biol.* **26**:65–75 (1949).]

phridia of oligochaetes and polychaetes. In each nephridium is a long tortuous *central canal* imbedded in connective tissue. Also imbedded in the connective tissue, surrounding the central canal and intertwining with each other, are a bed of blood capillaries and a bed of anastomosing fine tubules called *canaliculi*. The canalicular system discharges into the proximal end of the central canal. Perhaps the most notable morphological feature of the leech nephridium is that it has no opening to the coelomic cavity, so coelomic fluid cannot serve as primary urine.

Recent research has revealed that *Hirudo* forms its primary urine by secretion, in a manner that seems remarkably similar to that in insects. The site of formation is the canaliculi. The canalicular fluid is hyperosmotic to the blood and contains much higher concentrations of K^+ and Cl^- than the blood, indicating that active secretion of certain ions into the canaliculi establishes conditions favorable for the osmotic entry of water and the diffusive entry of other solutes (p. 212). Once canalicular fluid has entered the central canal, the nephridium starts avidly to reclaim salts; and by the time the urine is excreted from the distal end of the canal, it can be strongly hyposmotic to the blood.

ANIMALS WITH PROTONEPHRIDIA

In certain primitive groups of polychaete annelids, the nephridia do not open into the coelom at their inner ends but terminate blindly in clusters of specialized cells called *solenocytes* (Figure 10.20). Each solenocyte consists of a rounded cell body bearing a long flagellum that projects into a fine tubule. The tubule connects with the lumen of the nephridium. Nephridia ending blindly in solenocytes are also reported in the larvae of many annelids and mollusks. Among flatworms, rotifers, gastrotrichs, nemertines, and several other groups, the nephridia end blindly in cells bearing *tufts* of cilia or flagella (Figure 10.20). Two types of such tufted end structure are recognized, *flame cells* and *flame bulbs*—so named because under the microscope, their beating tufts of cilia or flagella are reminiscent of a flickering candle. Ne-

phridia that at their inner ends terminate blindly in solenocytes, flame cells, or flame bulbs are collectively called *protonephridia*. Although some authorities have concluded that all protonephridia trace a common evolutionary origin, others feel that there are several independently evolved types. To the extent that the latter view is correct, attempts to generalize about the functions of these organs would be especially fraught with pitfalls.

Because protonephridia range in size from small to extremely small, they have been difficult to study, and knowledge of their physiology is limited. Currently, the favored hypothesis is that primary urine is introduced into them by ultrafiltration, at least in those nephridia bearing flame cells or bulbs. According to this view, beating of the cilia or flagella creates a region of negative pressure (suction) inside the blind end of the nephridium by driving tubular fluid down the nephridial tubule, away from the blind end. This internal suction, it is postulated, induces new fluid (primary urine) to enter the nephridium by filtration across regions of the flame cell or flame bulb specialized to permit such inflow.

It is argued by some investigators that in certain marine animals, protonephridia are not truly regulatory organs; instead, it is said that the protonephridia merely hasten flux of water through the animal's body by virtue of steady excretory activity, thereby helping to wash metabolic wastes out of the body fluids. On the other hand, protonephridia have assumed clear regulatory roles in some animals, such as the freshwater rotifer *Asplanchna priodonta* (remarkably, this species has been studied in some detail despite a total *body* length of ~ 1 mm). As in most freshwater animals, the urine of *Asplanchna* has a substantially lower osmotic pressure and Na^+ concentration than the body fluids. These low urinary concentrations seem clearly to result from active ion resorption along the length of each protonephridium. If *Asplanchna* that have been feeding in lake water are fasted and placed in distilled water, they dramatically reduce the osmotic pressure and Na^+ concentration of their urine. Furthermore, when animals are transferred from relatively concen-

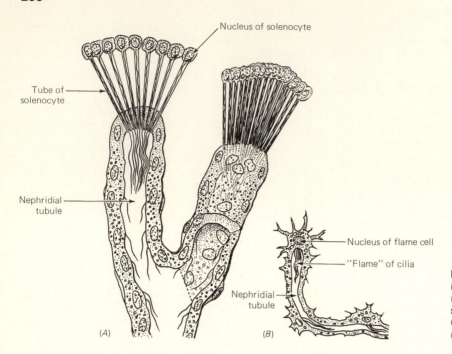

Nucleus of solenocyte

Tube of
solenocyte

Nephridial
tubule

Nucleus of flame cell

"Flame" of cilia

Nephridial
tubule

(A)   (B)

Figure 10.20 Inner ends of protonephridia. (A) Solenocytes as seen in a polychaete annelid (*Phyllodoce paretti*). (B) Flame cell as seen in a polyclad flatworm. [From E.S. Goodrich, *Q. J. Microsc. Sci.* **86**:113–392 (1945).]

trated to relatively dilute waters, they increase both the rate of beating of their flame bulbs and the frequency of emptying of their urinary bladders, indicating that they hasten water excretion.

INSECTS

Anatomy of the Renal Complex

Most insects possess Malpighian tubules, and these are often referred to as the "excretory tubules." We must emphasize from the outset, however, that the lower gut plays such a central role in the formation of urine that it must be included in any definition of the renal complex.

The *Malpighian tubules* are long, slender, blind-ended tubules that, as shown in Figure 10.21, typically arise from the junction of the midgut and hindgut. They number from 2 to over 200, depending on species. Projecting into the hemocoel, they are bathed by the blood (hemolymph). The blind ends of the tubules are usually free in the hemocoel, but sometimes they are attached to the rectum. The walls of the tubules consist of a single layer of epithelial cells, surrounded on the outside by a thin fibrous sheet, the basal lamella. Although the tubules exhibit little histological differentiation along their length in some species, they are differentiated into two to four distinct regions in numerous others. In many species, the various tubules within an individual are morphologically similar, but in others two or more types of tubule are present. We know that anatomical heterogeneity along or among the tubules is often paralleled by functional heterogeneity, but much remains to be learned about the subject.

The hindgut typically consists of a relatively small-diameter *anterior hindgut* (intestine) and an expanded posterior part, the *rectum* (Figure 10.21). The walls of the anterior hindgut usually consist of a single layer of cuboidal epithelial cells. In some insects, the walls of the rectum are similar; but in many species, parts of the rectal wall consist of thick, columnar epithelial cells, sometimes associated with secondary cell layers (Figure 10.21). The thickened parts of the rectal wall have traditionally been termed *rectal glands,* and the term is still in use although we now recognize that the function of these parts is not typically glandular. The rectal glands in many insects look like simple thickenings of the rectal wall and are termed *rectal pads.* By contrast, adult Diptera (e.g., flies) have columnar epithelial cells arrayed in cones or papillae that project into the rectal lumen; these rectal glands are called *rectal papillae.* For many years, investigators have recognized that species possessing rectal pads or papillae tend to be especially capable of resorbing water from their excreta.

Function of the Malpighian Tubules

The basic function of the Malpighian tubules is to form the primary urine. The tubules are not supplied with blood vessels; and for this and a number of other reasons, ultrafiltration has long seemed unlikely to be the mechanism. The evidence for a secretory mechanism is now strong.

Active secretion of K^+ into the tubules is the key process driving the formation of primary urine in many, if not most, insects. Although all the events involved in introducing primary urine into the tubules are not yet understood, current thinking can be summarized as follows (see also Figure 10.2). Potassium is secreted vigorously from the blood into the Mal-

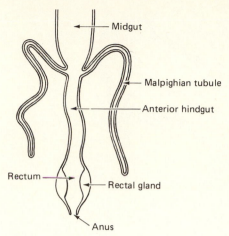

Midgut

Malpighian tubule

Anterior hindgut

Rectum

Rectal gland

Anus

Figure 10.21 Generalized schematic representation of the posterior gut and Malpighian tubules in insects. The rectal glands are depicted as longitudinal thickenings of the wall of the rectum.

pighian tubules—so vigorously that the K^+ concentration in the tubular fluid is 6–30 times higher than that in the blood. Often, Na^+ is also actively secreted into the tubules, but usually Na^+ secretion is subsidiary to K^+ secretion and the Na^+ concentration within the tubules remains below that in the blood. Chloride is the predominant anion accompanying the K^+ and Na^+; the Cl^- influx is apparently often passive (following the electrical gradient set up by active cation secretion), but evidence exists for active Cl^- secretion in some species. The active flux of ions into the tubules motivates osmotic entry of water, which typically occurs at a sufficient pace that the tubular fluid remains approximately isosmotic to the blood. Local osmotic processes (p. 150) are implicated in water entry because although the tubular fluid is sometimes slightly hyperosmotic to the blood, it sometimes is isosmotic and in unusual cases is hyposmotic during water entry. Many solutes are believed to enter the primary urine passively, as by following concentration gradients set up by water influx; these solutes include amino acids, sugars, diverse organic wastes and toxins, and a number of inorganic ions. Proteins are largely excluded because of their size. Certain organic compounds—notably some detoxification products and plant toxins (e.g., alkaloids)—are actively secreted into the primary urine by some insects.

In some species, the Malpighian tubules have unusual functional features that are attuned to meeting special challenges peculiar to the life histories of the species involved. Mosquito larvae that live in coastal saline waters, for instance, actively secrete Mg^{2+} and SO_4^{2-} into their tubular fluid.

The Malpighian tubules may also be involved in resorption. Glucose, for example, appears to be actively reclaimed by the tubules in many insects, and K^+ is resorbed by the tubules in some.

In the end, the fluid passed into the hindgut by the Malpighian tubules contains numerous solutes. It has a solute composition quite unlike that of the blood, but it characteristically resembles the blood in osmotic concentration. Usually, the fluid contains a lot of K^+; in the stick insect, *Dixippus morosus,* for example, the Malpighian tubules sometimes empty the equivalent of all the body's K^+ into the hindgut *every 3 h*. It would not be unusual either for them to void the equivalent of all the body's water every 10–48 h.

Considering the inputs to the hindgut from the tubules, it is clear that ordinarily the hindgut must extensively resorb water and certain solutes. The need for K^+ resorption is particularly striking. Large quantities of K^+ typically appear in the outflow of the Malpighian tubules because of the central role ordinarily played by K^+ secretion in forming the primary urine, but clearly much of the K^+ must ultimately be returned to the blood if the insect is to survive. This resorption of K^+ usually occurs primarily in the rectum, and the K^+ returned to the blood becomes available to participate again in the secretory production of primary urine. In short, K^+ is recycled.

Function of the Hindgut

The function of the anterior hindgut has been relatively neglected, although by now there is evidence that it often modifies the urinary fluid and in some species is the principal architect of the definitive urine. The rectum occasionally serves as little more than a storage site, but in most species it is responsible for extensive alterations of the volume and composition of the urine in service of body-fluid homeostasis. Later, we shall examine the overall regulatory potential of the rectum. Here, we turn to one of the focal questions in research on insect excretion, namely, how the urine is rendered hyperosmotic to the blood.

Production of Concentrated Urine   When insects produce urine that is hyperosmotic to their blood (p. 187), the process of concentration usually occurs in the rectum. At least three basically different mechanisms of concentrating the urine have evolved. We first look at the processes at work in species possessing rectal glands.

In cockroaches (*Periplaneta*), desert locusts (*Schistocerca*), blowflies (*Calliphora*), and presumably other insects having rectal pads or papillae, the urine is concentrated by *water resorption in excess of solute resorption.* This water resorption raises the osmotic concentration of the solutes remaining in the rectum, and it continues even when the osmotic pressure of the rectal contents has risen to be two or more times higher than the osmotic pressure of the blood bathing the rectum. Perhaps the most compelling evidence for concentration of the urine by this type of process comes from experiments in which the rectum has been filled with a pure solution of a solute (e.g., trehalose) that is neither resorbed nor secreted across the rectal wall. The *amount* of such a solute in the rectum is fixed during the course of an experiment. When locusts (*Schistocerca*) were treated in this way, water was resorbed from the rectal contents to such an extent that the rectal osmotic pressure rose to be nearly three times the blood osmotic pressure. These results show that the rectal wall can move water against large, opposing osmotic gradients between the rectal fluid on the inside and blood on the outside. Furthermore, in the short term, this water resorption can occur even in the absence of simultaneous solute resorption.

At first, experimental observations of this type seemed to point toward active transport of water by the rectum. However, a consensus has emerged over the past decade that the resorption of water in fact occurs by osmosis. The mechanism of resorption rests in part on a complex microarchitecture in the rectal pads or papillae. The details of structure and possibly of function vary from species to species. However, the mechanism in its essentials is believed to be similar in a variety of insects. Here, we review the hypothesized process of water resorption using the blowfly *Calliphora* as a specific example. The relevant anatomy is shown in Figure 10.22, and the process of water resorption is diagrammed in Figure 10.23.

In the rectal papilla of the blowfly, adjacent cells of the columnar epithelium are tightly joined on the side facing the rectal lumen and on the opposite (basal) side, but in between the cells are separated by an elaborate network of minute intercellular channels and spaces. As depicted in Figure 10.22, the intercellular spaces communicate at the apex of the papilla with infundibular channels that run along the basal side of the epithelial cells and connect with general blood spaces. It is believed that the epithelial cells actively secrete solutes into the intercellular spaces, thereby rendering the fluid in the spaces strongly hyperosmotic both to the blood and to the fluid in the rectal lumen. Osmosis then carries water out of the rectal fluid into the intercellular spaces; that is, because of the *locally* high osmotic pressure in the intercellular spaces, water is osmotically withdrawn from the rectal fluid even though the latter is thereby rendered increasingly hyperosmotic to the blood. Entry of water into the intercellular spaces causes fluid to flow in streams through the spaces toward the apex of the papilla and then through the infundibular channels toward the blood cavity (hemocoel). The fluid exiting the intercellular spaces is highly concentrated, but as it flows past the epithelial cells in the infundibular channels, solutes are believed to be actively or passively resorbed from the fluid into the cells across membranes poorly permeable to water, with two highly significant consequences. First, the fluid flowing through the infundibular channels is diluted, so in the end a highly

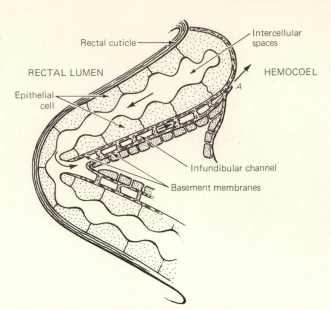

Figure 10.22 Highly diagrammatic represenation of a rectal papilla of the blowfly *Calliphora erythrocephala*. The papilla is about 1 mm long and about 0.5 mm in diameter at its base. Each fly has four papillae. The epithelial cells run between the cuticle and outer basement membrane. The intercellular space, depicted for simplicity as a single broad cavity running through the epithelial cells, actually exists as a complex, interconnecting network of small channels and spaces *between* the epithelial cells. These channels and spaces ultimately communicate with the infundibular channels near the apex of the papilla. The infundibular channels, bounded by basement membranes (connective tissue), are traversed by bar-shaped bridges of basement material; these bridges do not form complete septa and do not prevent flow of fluid through the channels. Arrows depict the direction of hypothesized fluid flow. Fluid exits to general blood spaces (hemocoel) at the point marked *A*. [After B.L. Gupta and M.J. Berridge, *J. Morphol.* **120**:23–82 (1966).]

concentrated fluid is not introduced into the blood. Second, solutes are returned to the epithelial cells and thus can again be secreted into the intercellular spaces, permitting continued osmotic water resorption from the rectal fluid without great need for new solutes from any source. The nature of the solutes involved remains under investigation. The inorganic ions Na^+, K^+, and Cl^- are strongly implicated; possibly, some organic solutes also play roles.

A different type of concentrating mechanism has been described in mealworms (larval *Tenebrio molitor*). These animals are able to live in stores of dry grain without access to drinking water. They can produce pellets of excrement (feces and urine combined) that are ostensibly dry, having an effective osmotic U/P ratio near 8—substantially higher than the U/P ratios achieved by locusts or blowflies (near 3). In common with some other larval and adult coleopterans, *Tenebrio* has a cryptonephridial complex with leptophragmata. The **cryptonephridial complex** is an association between the Malpighian tubules and rectum; in brief, the distal parts of the tubules (parts nearest the blind ends) are closely applied to the

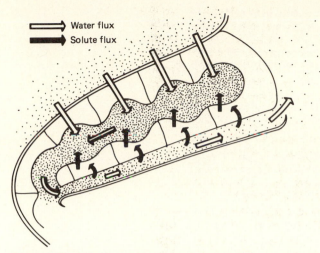

Water flux
Solute flux

Figure 10.23 Outline of proposed mechanism of water resorption in the rectal papillae of blowflies. Light and dark arrows show water and solute fluxes, respectively. The density of stippling in fluid-filled spaces indicates relative osmotic pressures. See Figure 10.22 for anatomy. In the short term, water resorption can occur totally without solute resorption from the rectum, as illustrated. However, solutes are in fact continually lost to the hemocoel, thus gradually depleting the epithelial cells of solute. Accordingly, long-term water resorption demands that epithelial-cell solutes be supplemented, at least partly by solute resorption from the rectum. Even then, however, water is resorbed in excess of solutes.

outer rectal wall, and these parts of the tubules and the rectum are together enclosed by a **perinephric membrane,** which separates them from the hemocoel. The **leptophragmata** are tiny "windows" of very thin tissue formed at points of fusion between the walls of the Malpighian tubules and overlying perinephric membrane. The mechanism of concentrating and drying the excrement in *Tenebrio* is not fully understood, but an important part of it resides in the cryptonephric Malpighian tubules. Potassium, followed by chloride, is actively transported from the blood into the lumen of each tubule, purportedly across the leptophragmata. In a marked departure from the usual condition in insects, however, water is barred (probably by the perinephric membrane) from entering the tubular fluid from the blood. The fluid in the tubules thus is rendered strongly hyperosmotic to the blood, and a gradient favoring osmotic resorption of water from the closely juxtaposed rectal lumen is created.

A third mechanism of concentrating the urine has been reported in certain saline-water mosquito larvae (e.g., *Aedes campestris* and *A. taeniorhynchus*). These larvae can live in waters that are markedly hyperosmotic to their blood, and under such circumstances they must drink to replace water lost osmotically across their body surfaces. Unlike the insects discussed heretofore, their main challenge is not water conservation; instead, it is to excrete ions at higher concentrations than in the drinking water, so

that a net gain of water (H_2O) will be realized from drinking and excretion. With these thoughts in mind, investigators reasoned that the larvae might concentrate their urine not by resorption of water from the rectal fluid but by secretion of solutes into the fluid. Experiments confirmed this expectation. At least five ions are actively secreted into the rectal fluid: Na^+, K^+, Mg^{2+}, Cl^-, and HCO_3^-.

Regulatory Activities of the Rectum In terrestrial insects, the rectum is a resorptive structure in its overall mode of operation; it characteristically resorbs, for example, most of the water, K^+, Na^+, and Cl^- introduced into the hindgut by the Malpighian tubules. In locusts at least, all three of these ions are resorbed by active-transport pumps, which can operate against 10- to 50-fold concentration gradients. The locust rectum also actively resorbs amino acids, acetate, and phosphate. Acidification of the urine by active secretion of H^+ occurs in the rectum and contributes to the precipitation of uric acid and urates.

Increasingly it becomes clear that the insect rectum possesses impressive abilities to adjust the composition of the urine so as to promote homeostasis in the blood and other body fluids. A number of insects, for example, are known to be able to render their urine either hyposmotic or hyperosmotic to the blood, depending on osmoregulatory requirements. In the terrestrial insects, hyperosmoticity is created by water resorption in excess of total solute resorption; hyposmoticity results from solute resorption in excess of water resorption.

Several studies have demonstrated that the ionic composition of the urine is also adjusted in the rectum in service of homeostasis. In one set of experiments, for example, fasting locusts were permitted to drink either tap water or a saline solution containing Na^+, K^+, Cl^-, and other ions. As shown in Table 10.1, the saline-fed locusts had high ion concentrations in the urine in their rectum; in fact, the ion concentrations in their urine were somewhat higher than the concentrations in the saline drinking water—a necessity if the locusts were to void excess ions from their drink without depleting body water. In contrast, the rectal fluids of the water-fed locusts were virtually depleted of ions; these animals had almost no ion input, and they conserved ions avidly. The volume of urine accumulated in the rectum by saline-fed locusts was many times greater than that accumulated by water-fed locusts. Taking account of both volume and concentration, therefore, the quantities of ions excreted in the urine of the saline-fed locusts exceeded those excreted by the water-fed animals by hundreds of times.

As discussed in Chapter 8 (p. 196), diuretic and antidiuretic neurohormones have been identified in insects. These hormones are known to affect both rates of resorption in the rectum and rates of primary-urine formation in the Malpighian tubules. Al-

Table 10.1 COMPOSITION OF THE BLOOD, ANTERIOR-HINDGUT FLUID, AND RECTAL FLUID IN *Schistocerca gregaria* DEPRIVED OF FOOD BUT PROVIDED WITH TAP WATER OR A SALINE SOLUTION[a]

Experimental treatment	Fluid	Osmotic pressure (osmolar)	Ion concentrations (mM)		
			Cl^-	Na^+	K^+
Water-fed	Blood	0.40	115	108	11
	Anterior-hindgut fluid	0.42	93	20	139
	Rectal fluid	0.82	5	1	22
Saline-fed	Blood	0.52	163	158	19
	Anterior-hindgut fluid	—	192	67	186
	Rectal fluid	1.87	569	405	241

[a]All values are means. The high osmotic pressure in the scanty rectal fluid of water-fed animals is presumed to be caused by organic solutes.

Source: J.E. Phillips, *J. Exp. Biol.* **41**:69–80 (1964).

though the hormones are believed to act on membrane permeabilities and rates of active solute transport, details of their mode of action remain generally unknown.

Other Groups with Malpighian Tubules

Malpighian tubules are found in some other terrestrial arthropods: centipedes, millipedes, and spiders. There has been little study of excretory function in these animals.

SELECTED READINGS

Abramow, M. 1979. Control mechanisms in mammals. In R. Gilles (ed.), *Mechanisms of Osmoregulation in Animals*, pp. 349–412. Wiley, New York.

Bentley, P.J. 1969. Neurohypophysial function in Amphibia: Hormone activity in the plasma. *J. Endocrinol.* **43**:359–369.

Bradley, T.J. 1987. Physiology of osmoregulation in mosquitoes. *Annu. Rev. Entomol.* **32**:439–462.

Braun, E.J. and W.H. Dantzler. 1972. Function of mammalian-type and reptilian-type nephrons in kidney of desert quail. *Am. J. Physiol.* **222**:617–629.

Braun, G., G. Kümmel, and J.A. Mangoes. 1966. Studies on the ultrastructure and function of a primitive excretory organ, the protonephridium of the rotifer *Asplanchna priodonta. Pflügers Arch.* **289**:141–154.

Brenner, B.M. and F.C. Rector, Jr. (eds.). 1981. *The Kidney*, 2nd ed. Saunders, Philadelphia.

Dantzler, W.H. 1985. Comparative aspects of renal function. In D.W. Seldon and G. Giebisch (eds.), *The Kidney. Physiology and Pathophysiology*, pp. 333–364. Raven, New York.

Dobbs, G.H., III, Y. Lin, and A.L. DeVries. 1974. Aglomerularism in antarctic fish. *Science* **185**:793–794.

Edney, E.B. 1977. *Water Balance in Land Arthropods.* Springer-Verlag, New York.

Garland, H.O. and I.W. Henderson. 1975. Influence of environmental salinity on renal and adrenocortical function in the toad, *Bufo marinus. Gen. Comp. Endocrinol.* **27**:136–143.

*Gilles, R. (ed.). 1979. *Mechanisms of Osmoregulation in Animals.* Wiley, New York.

*Gupta, B.L., R.B. Moreton, J.L. Oschman, and B.J. Wall

(eds.). 1977. *Transport of Ions and Water in Animals.* Academic, New York.

Hevert, F. 1984. Urine formation in the lamellibranchs: Evidence for ultrafiltration and quantitative description. *J. Exp. Biol.* **111**:1–12.

*Jamison, R.L. and R.H. Maffly. 1976. The urinary concentrating mechanism. *N. Engl. J. Med.* **295**:1059–1067.

Jard, S. and F. Morel. 1963. Actions of vasotocin and some of its analogues on salt and water excretion by the frog. *Am. J. Physiol.* **204**:222–226.

*Jones, H.D. and D. Peggs. 1983. Hydrostatic and osmotic pressures in the heart and pericardium of *Mya arenaria* and *Anodonta cygnea. Comp. Biochem. Physiol.* **76A**:381–385.

Kirschner, L.B. 1967. Comparative physiology: Invertebrate excretory organs. *Annu. Rev. Physiol.* **29**:169–196.

Kokko, J.P. 1974. Membrane characteristics governing salt and water transport in the loop of Henle. *Fed. Proc.* **33**:25–30.

Kümmel, G. 1973. Filtration structures in excretory systems. A comparison. In L. Bolis, K. Schmidt-Nielsen, and S.H.P. Maddrell (eds.), *Comparative Physiology*, pp. 221–240. North-Holland, Amsterdam.

Little, C. 1965. The formation of urine by the prosobranch gastropod mollusc *Viviparus viviparus* Linn. *J. Exp. Biol.* **43**:39–54.

Long, W.S. 1973. Renal handling of urea in *Rana catesbeiana. Am. J. Physiol.* **224**:482–490.

Lote, C.J. (ed.). 1986. *Advances in Renal Physiology.* Croom Helm, London.

*Maddrell, S.H.P. 1981. Functional design of the insect excretory system. *J. Exp. Biol.* **90**:1–15.

*Maloiy, G.M.O. (ed.). 1979. *Comparative Physiology of Osmoregulation in Animals.* Academic, New York.

Phillips, J.E. 1977. Excretion in insects: Function of gut and rectum in concentrating and diluting the urine. *Fed. Proc.* **36**:2480-2486.

*Pitts, R.F. 1974. *Physiology of the Kidney and Body Fluids*, 3rd ed. Year Book Medical, Chicago.

*Riegel, J.A. 1972. *Comparative Physiology of Renal Excretion.* Oliver and Boyd, Edinburgh.

*Schipp, R. and F. Hevert. 1981. Ultrafiltration in the branchial heart appendage of dibranchiate cephalopods: A comparative ultrastructural and physiological study. *J. Exp. Biol.* **92**:23–35.

Schmidt-Nielsen, B. 1964. Organ systems in adaptation: The excretory system. In D.B. Dill (ed.), *Handbook of*

Physiology. Section 4: Adaptation to the Environment. American Physiological Society, Washington, DC.

Scoggins, B.A., J.R. Blair-West, J.P. Coghlan, D.A. Denton, K. Myers, J.F. Nelson, E. Orchard, and R.D. Wright. 1970. The physiological and morphological response of mammals to changes in their sodium status. In G.K. Benson and J.G. Phillips (eds.), *Hormones and the Environment*, pp. 577–602. Cambridge University Press, Cambridge.

Smith, H.W. 1956. *Principles of Renal Physiology*. Oxford, New York.

Stolte, H., B. Schmidt-Nielsen, and L. Davis. 1977. Single nephron function in the kidney of the lizard, *Sceloporus cyanogenys. Zool. Jahrbuch. Physiol.* **81**:219–244.

Stoner, L.C. 1977. Isolated, perfused amphibian renal tubules: The diluting segment. *Am. J. Physiol.* **233**:F438–F444.

Tyler-Jones, R. and E.W. Taylor. 1986. Urine flow and the role of the antennal glands in water balance during aerial exposure in the crayfish, *Austropotamobius pallipes* (Lereboullet). *J. Comp. Physiol.* **156B**:529–535.

Ussing, H.H., N. Bindslev, N.A. Lassen, and O. Sten-Knudsen (eds.). 1981. *Water Transport Across Epithelia*. Munksgaard, Copenhagen.

*Valtin, H. 1983. *Renal Function*. Little, Brown and Co., Boston.

Wessing, A. (ed.). 1975. Excretion. *Fortschr. Zool.* **23**:1–362. (Includes articles on many animal groups.)

Wilson, R.A. and L.A. Webster. 1974. Protonephridia. *Biol. Rev.* **49**:127–160.

See also references in Appendix A.

chapter *11*

Principles of Diffusion and Dissolution of Gases

In this brief chapter, we examine the physical behavior of gases in preparation for the discussion of respiratory gas exchange and circulatory gas transport in succeeding chapters. Although the respiratory gases, oxygen and carbon dioxide, are emphasized, the principles developed apply equally to other substances that exist as gases at physiological temperatures, such as nitrogen, argon, and gaseous anesthetics. Water vapor is to some extent a special case because water exists as both a gas and liquid at physiological temperatures; it has been discussed separately (p. 136).

GASES IN THE GAS PHASE

In a mixture of gases such as air, the *mole fractional concentration* of any particular component gas is the fraction of the total moles of gas present represented by the component in question. To illustrate, in a volume of dry atmospheric air near sea level, the number of moles of oxygen is 20.95 percent of the total number of moles of all atmospheric gases taken together; thus, the mole fractional concentration of oxygen in dry air is 0.2095 (20.95 percent). The *volume fractional concentration* of a component gas in a mixture is the fraction of the total volume represented by the component in question and, according to Avogadro's principle, is essentially equal to its mole fractional concentration. Thus, returning to the example of oxygen in dry air, if we were to remove the oxygen from a volume of air at given temperature and pressure and subsequently restore the residual gas mixture to the same temperature and pressure, we would find that the original volume had been reduced by about 20.95 percent.

John Dalton (1766–1844), in his law of partial pressures, enunciated the important concept that the total pressure exerted by a mixture of gases is the sum of individual pressures—so-called *partial pressures*—exerted by each of the several component gases in the mixture. The partial pressure exerted by each component gas is independent of the other gases present. Furthermore, in a volume of mixed gases, each component gas behaves in terms of its partial pressure as if it alone occupied the entire volume. Thus, the partial pressure of each component in a mixture can be calculated from the universal gas law by knowing just the temperature, the molar quantity of that component, and the volume occupied by the mixture as a whole.

An important corollary of these principles is that in a mixture of gases, the fraction of the total pressure exerted by each component gas is identical to the mole (or volume) fractional concentration of the component. That is, each gas in a mixture contributes to the total pressure in proportion to its molar representation in the mixture. Algebraically, if P_{tot} is the total pressure of a gas mixture, P_x is the partial pressure of a particular component gas (x), and F_x is the mole or volume fractional concentration of that component expressed as a percentage, then

$$P_x = \frac{F_x}{100} P_{tot}$$

To illustrate, let us look again at dry atmospheric air, which near sea level consists of approximately 20.95 percent oxygen, 78.1 percent nitrogen, 0.03 percent carbon dioxide, and small percentages of other gases. If the total pressure of a body of dry air is 760 mm Hg (1 atm), then by the above formula, the partial pressure of oxygen in the air is $(20.95/100)(760) =$

159 mm Hg, and the partial pressures of nitrogen and carbon dioxide are 594 mm Hg and 0.23 mm Hg, respectively.*

A final consideration requiring attention in this discussion of gases in the gas phase is the expression of *concentrations*. The concentration of a gas is the amount of gas per unit volume and can be expressed in a number of ways. Again taking oxygen in the atmosphere as an example, it would obviously be straightforward to express its concentration as the weight of oxygen per liter of air or as the number of moles of oxygen per liter. Suppose, however, that we wish to express the concentration as the volume of oxygen per liter of air. We would then have to recognize that the volume of oxygen measured depends strongly on the conditions of temperature and pressure prevailing during measurement. To avoid confounding true variations in concentration with variations in the temperature and pressure of measurement, a convention is adopted: Namely, when concentration is expressed as the volume of a specific pure gas (e.g., oxygen) per liter of air, the volume of the specific gas is corrected to standard conditions of temperature and pressure (STP): 0°C and 760 mm Hg. Consider a body of dry atmospheric air at 20°C and 740 mm Hg. Because this air consists of 20.95 percent oxygen, at the *prevailing* conditions it contains about 210 mL of oxygen per liter. Correcting the volume of oxygen to *standard* conditions, we get (273°K/293°K)(740 mm/760 mm)(210 mL)= 190 mL. Thus, the air contains 190 mL O_2 at STP per L. The two volumes, 210 mL and 190 mL, must be carefully distinguished. The 210 mL is the actual volume of oxygen that would be measured under prevailing conditions. However, if we are to compare the oxygen concentrations of various gas mixtures at various temperatures and pressures, we must express their oxygen contents in comparable terms; and it is in this light that the expression at STP is indispensable. Compare, for example, the body of dry air at 20°C and 740 mm Hg with another body of dry air at 20°C and 700 mm Hg. Under their respective prevailing conditions, both bodies of air contain 210 mL O_2/L. However, correcting volumes of oxygen to STP, we find that the former body of air contains 190 mL O_2 at STP/L, whereas the latter (because it is under less pressure and is thus less dense) contains just 180 mL O_2 at STP/L. These figures, unlike those expressed at prevailing conditions, reveal that the former body of air contains more molecules of oxygen per liter than the latter. Concentrations expressed in weight/liter, moles/liter, and volume at STP/liter are interconvertible by simple proportionalities.

It is convenient to note that for gases in the gas phase, if temperature is held constant, the concentration of a gas and its partial pressure are proportional. If temperature varies, the concentration prevailing at any particular partial pressure becomes lower as the temperature becomes higher.

GASES IN SOLUTION

Gases dissolve in aqueous solutions. Molecules of oxygen, for example, become distributed among water molecules in much the same way that the molecules of glucose or the ions of sodium chloride are incorporated among water molecules during dissolution of solids. The molecules of oxygen do not appear in the solution as tiny bubbles any more than sodium chloride in solution appears as tiny crystals. Bubbles represent gas that is not in solution.

The concentration of a gas in a solution can be expressed in the usual units: weight/liter, moles/liter, and volume at STP/liter. A frequent form of expressing concentration is in units of milliliters of gas at STP per liter of solution; for instance, we might state that the oxygen concentration of a solution is 2 mL O_2/L. What is meant by such an expression is that *if we were to extract all the oxygen from solution into the gas phase*, we would obtain 2 mL of pure gaseous oxygen (measured at STP) from each liter of liquid.

If oxygen-depleted water is brought into contact with air containing oxygen at a partial pressure of 159 mm Hg, oxygen will dissolve in the water until a certain equilibrium concentration is achieved. The *solution* is then said to have a partial pressure of oxygen of 159 mm Hg. If this solution is brought into contact with air containing oxygen at 140 mm Hg, the solution will lose oxygen to the air until a new, lower equilibrium concentration is realized. The partial pressure of oxygen in the solution will then be 140 mm Hg. In general, the **partial pressure of any given gas in solution** is precisely equal to the partial pressure of the same gas in a gas phase with which the solution is at equilibrium. The partial pressure of a gas in solution is often called the gas **tension**. Sometimes, the term *tension* is also applied to partial pressures of gases in gas mixtures.

As just indicated, there are limits to the solubility of gases in solutions. An aqueous solution exposed to a given gas-phase partial pressure will dissolve only so much gas. The *solubilities* of gases in solutions are expressed in a standardized way, namely, as the concentration of gas (volume of dissolved gas at STP/liter) when the solution is at equilibrium with a gas phase in which the partial pressure of the gas is 760 mm Hg (i.e., when the partial pressure of the gas in solution is 760 mm Hg). This quantity is known as the **absorption coefficient**. The absorption coefficient of oxygen in distilled water at 0°C, for example, is about 49 mL/L. This means that such water will dissolve 49 mL of oxygen (at STP) per liter if allowed to equilibrate with an atmosphere in which the partial pressure of oxygen is 760 mm Hg.

There are several important characteristics of gas solubility that can be elucidated by examining absorption coefficients. The absorption coefficients of oxygen, nitrogen, and carbon dioxide in distilled water at 0°C are, respectively, 49, 24, and 1713 mL/L. These values illustrate that the solubilities of various gases are different and, in particular, the solubility of carbon dioxide is far greater than the solubilities of oxygen and nitrogen in aqueous solutions. The absorption coefficients of oxygen in distilled water at 0, 20, and 40°C are, respectively, 49, 31, and 23 mL/L. These values exemplify the important point that *solubilities of gases decrease strongly with increasing water temperature*. The absorption coefficients of oxygen at 0°C in waters of 0, 29, and 36 ppt salinity are, respectively, 49, 40, and 38 mL/L. Here, we see that *increasing salinity decreases gas solubilities*.

The concentration of a gas in an aqueous solution at any given partial pressure can readily be computed from the absorption coefficient using the formula

$$C = \frac{A}{760} P$$

where C is the concentration in mL of gas at STP/L, P is the partial pressure of the gas in solution in mm Hg, and A is the absorption coefficient in mL/L. Clearly, the concentration of any given gas is proportional to its partial pressure in any given body of water. The absorption coefficient, however, varies with the gas under consideration and with the temperature and salinity of the water. Thus, the constant of proportionality, $A/760$, also varies with these parameters.

To illustrate the use of the preceding formula, consider distilled water at 0°C in equilibrium with dry atmospheric air at 1 atm of pressure. As given earlier, the absorption coefficient of oxygen is 49 mL/L, and the partial pressure of oxygen in the air and water is 159 mm Hg. We get that

$$\text{mL of } O_2 \text{ (at STP)/L} = \frac{49}{760}(159) = 10.2$$

By similar computations, we can determine that the concentration of carbon dioxide in such water is only about 0.5 mL/L; although carbon dioxide is much more soluble than oxygen, it is present at only low partial pressure in water equilibrated with the atmosphere.

Inasmuch as the absorption coefficient for a given gas varies with the temperature and salinity of the water, it should be clear that the concentration of a gas at a given partial pressure also varies with temperature and salinity. Consider, for example, solutions of oxygen in distilled water at 40°C, distilled water at 0°C, and seawater (36 ppt) at 0°C — all at oxygen tensions of 159 mm Hg. These solutions, respectively, will contain 4.8, 10.2, and 8.0 mL O_2/L even though their oxygen tensions are identical. This example serves to emphasize that the partial pressure of a gas in solution does not in itself tell much about the concentration of the gas. It is possible for a solution that has a higher partial pressure than another solution to have a lower concentration. For example, distilled water at 40°C and an oxygen tension of 159 mm Hg contains 4.8 mL O_2/L; but distilled water at 0°C and an oxygen tension of 130 mm Hg—a lower tension—contains 8.4 mL O_2/L—a much higher concentration (this is a reflection of the higher solubility of oxygen in colder water).

DIFFUSION OF GASES

Gases diffuse from areas of higher partial pressure to areas of lower partial pressure. This is true within aqueous solutions, within mixtures of gases, and across gas–water interfaces. Sometimes, but *only* sometimes, diffusion in the direction of the partial-pressure gradient also means diffusion in the direction of the concentration gradient. In a body of water of uniform temperature and salinity, if the oxygen tension is greater in one region than another, the oxygen concentration is also greater. Diffusion will occur from the region of higher tension to the region of lower tension, which in this case also happens to be from the region of higher concentration to that of lower concentration. By contrast, as just shown, it is possible for cold water to have a higher oxygen concentration but lower oxygen tension than warm water. In this situation, if the cold and warm bodies of water are in contact, diffusion will again occur in the direction of the tension gradient, but this means it will be against the concentration gradient (oxygen will diffuse into the cold water from the warm). In any system, equilibrium is attained when the partial pressure is uniform throughout.

To analyze the factors affecting the *rate* of gas diffusion, consider two bodies of fluid separated by some intervening material of thickness d. One body, for example, could be the water in contact with the gills of a fish, and the other could be the blood within the gills; in this instance, the intervening material would be the outer gill membranes. Alternatively, in a tube filled with still air, we could consider the air at one end to be one body and that at the far end to be the other body; the intervening material would then be the air in between. Regardless of the particular situation, suppose that we number the bodies 1 and 2, and suppose that the partial pressure of a given gas (e.g., oxygen) in body 1 is P_1 and that of the same gas in body 2 is P_2. Suppose also that P_1 exceeds P_2, signifying that the gas will diffuse in net fashion from body 1 to 2. In steady state, the rate of this diffusion is given by the following formula:

$$\begin{array}{l}\text{amount of gas reaching} \\ \text{body 2 from body 1 by} \\ \text{diffusion per unit time}\end{array} = KA(P_1 - P_2)\frac{1}{d}$$

Note that the rate of diffusion increases in proportion to the difference in partial pressure between the bodies. The rate also increases in proportion to A, the cross-sectional area of the diffusion path; but it is inversely related to *d*, the length of the diffusion path. The coefficient *K* in the equation symbolizes what is often called **Krogh's coefficient of diffusion**. It depends on the particular diffusing gas and reflects, in part, the ease with which that gas is able to pass through the material separating the two bodies of concern. Note the similarity of the diffusion equation for gases to the equations presented earlier for diffusion of heat (p. 78) and solutes (p. 140).

IMPLICATIONS OF CHEMICAL REACTION IN SOLUTION

In water or other liquids, only those gas molecules that are in solution as gas molecules contribute to the partial pressure of gas. Thus, according to the formula we have just discussed, only gas molecules in solution as gas molecules immediately affect the direction and rate of diffusion. Also, if the partial pressure of a gas in a solution is known and the concentration is computed from the partial pressure using the absorption coefficient, the concentration obtained is the concentration of gas in solution as gas molecules. These considerations must be taken into account whenever gases undergo chemical reaction in solution.

In natural waters, the present considerations are chiefly relevant to the analysis of carbon dioxide, for unlike oxygen, nitrogen, and other atmospheric gases, carbon dioxide reacts with the water—to an extent that ranges from minor to major—forming (principally) bicarbonate ions. Molecules of carbon dioxide that so react do not contribute to the partial pressure of carbon dioxide in solution.

Not only carbon dioxide but also oxygen can undergo extensive chemical reaction in animal bloods. Many animals have in their blood an oxygen-transport pigment, such as hemoglobin, which combines chemically with oxygen. Oxygen molecules that become bound to hemoglobin molecules do not contribute to the partial pressure of oxygen dissolved in the blood solution and thus—importantly—do not interfere with the capacity of the blood solution to gain more oxygen by diffusion. Net diffusive flux of oxygen into the blood ceases only when the concentration of oxygen dissolved as oxygen has risen to the point that the partial pressure of oxygen in the blood solution equals the partial pressure of the oxygen source (e.g., lung air).

SELECTED READINGS

Fenn, W.O. and H. Rahn (eds.). 1964. *Handbook of Physiology. Section 3: Respiration*, Vol. I. American Physiological Society, Washington, DC.

Rao, F.R. 1974. *An Introduction to Respiratory Physiology*. American Elsevier, New York.

chapter 12

Exchanges of Oxygen and Carbon Dioxide 1: Respiratory Environments, Introduction to Respiratory Exchange, and External Respiration

Of all the exchanges of materials between an animal and its environment, those of the respiratory gases are often the most urgent. A person, for example, can live for many hours or days without exchanging nutrients, nitrogenous wastes, or water but will die within minutes if denied oxygen. The urgency of the need for oxygen arises from the role oxygen plays as final electron acceptor in the aerobic biochemical apparatus responsible for energy transfer from foodstuffs to adenosine triphosphate, ATP (Chapter 4). Dissipation of carbon dioxide is often a pressing concern as well because accumulation of carbon dioxide in the body can rapidly acidify the body fluids and exert other deleterious effects.

In this chapter, after looking at the environments of animals from a respiratory perspective and taking an overview of the gas-exchange process as a whole, we examine in detail the mechanisms by which oxygen is acquired from the surrounding air or water and by which carbon dioxide is released. Later, in Chapters 13 and 14, we shall look in depth at circulatory transport of gases from place to place within the body.

RESPIRATORY ENVIRONMENTS

The Contrasting Properties of Air and Water

The two major environments of animals, air and water, differ in several key physical properties. An appreciation of the differences is crucial for a comparative understanding of respiratory physiology.

Oxygen Concentrations Because oxygen is not very soluble in water, concentrations of oxygen are regularly much lower in aquatic environments than air. To illustrate, Table 12.1 lists the oxygen concentrations of bodies of air and water when the partial pressure of oxygen (oxygen tension) is 159 mm Hg. As established in Chapter 11, 159 mm Hg is the partial pressure of oxygen in dry air when the total pressure is 1 atm. Thus, the concentrations tabulated for air approximate those in the atmosphere near sea level, and the values given for water are representative of the upper limits of concentration attained in natural waters when in equilibrium with the atmosphere. Clearly, air provides a much richer source of oxygen than water. The quantitative importance of this difference is illustrated by considering a hypothetical animal attempting to obtain a liter of oxygen by completely extracting the oxygen from a volume of its environmental medium. At 0°C, this animal would have to pass over its respiratory surfaces 4.8 L of air, 90 L of fresh water, or 125 L of seawater (assuming an oxygen tension of 159 mm Hg). The dramatic difference between the oxygen concentrations in air and water helps explain why the highest known metabolic rates are found in air-breathing animals: in insects among invertebrates and in birds and mammals among vertebrates. Probably, the in-

Table 12.1 CONCENTRATION OF OXYGEN IN AIR AND WATER AT A PARTIAL PRESSURE OF 159 mm Hg AT THREE TEMPERATURES[a]

Medium	Concentration of oxygen (mL/L) at specified temperature		
	0°C	12°C	24°C
Air	209	200	192
Fresh water	10.2	7.7	6.2
Seawater (36 ppt)	8.0	6.1	4.9

[a] Concentrations are expressed as mL O_2 at STP/L.

herently low oxygen concentrations in natural waters have imposed limits on the metabolic rates of aquatic animals.

The percentage by which the oxygen concentration is reduced by a rise in temperature is much greater in water than air. As shown, for example, in Table 12.1, if the oxygen tension is kept constant and the temperature is raised from 0 to 24°C, the oxygen concentration in air declines by about 8 percent (because of the decrease in air density), but the oxygen concentration in fresh water or seawater declines by about 40 percent (because of the fall in oxygen solubility). Recognizing that air is characteristically rich in oxygen at whatever temperature, the small percentage drop in the oxygen concentration of air caused by a rise in temperature is probably not of much consequence for terrestrial animals. On the other hand, a rise in temperature can seriously threaten the oxygen supplies of aquatic animals. In Chapter 6, we saw that the metabolic rates of fish and most other poikilotherms increase quasiexponentially as the temperature increases. At elevated temperatures, aquatic animals can be caught in a respiratory trap of increased demands for oxygen coupled with much reduced availability of oxygen in their environment.

Relative Changes in Oxygen and Carbon Dioxide Tensions   In analyzing changes in the gas content of the medium, an important parameter is the change in gas concentration as a ratio of the change in gas tension:

$$\frac{\Delta \text{ concentration}}{\Delta \text{ partial pressure}}$$

This ratio is termed the **capacitance coefficient** of the medium for the gas in question. In air of given temperature, one corollary of the universal gas law is that the capacitance coefficients for oxygen and carbon dioxide (and other gases as well) are equal. In water, however, the capacitance coefficient for carbon dioxide is at least 23 times that for oxygen at physiological temperatures; this is partly because carbon dioxide is much more soluble than oxygen (p. 246) and partly because, in some waters, carbon dioxide reacts to form bicarbonate.

To see the implications of these properties, let us consider for simplicity an animal with discrete respiratory organs—lungs or gills—that obtains all its ox-

ygen and voids all its carbon dioxide across those organs. Let us also assume that the animal is in steady state and has a respiratory exchange ratio of 1.0, meaning that when it consumes a given volume (at STP) of oxygen, it produces a like volume (at STP) of carbon dioxide (p. 28). In passing air through its lungs—if it is an air breather—or in passing water across its gills—if it is a water breather, this animal will raise the carbon dioxide *concentration* of the respired medium by the same amount as it lowers the oxygen *concentration*. For example, if it removes 80 mL of oxygen from each liter of air—reducing the oxygen concentration of the respired air by 80 mL/L, it will add 80 mL of carbon dioxide to each liter—thus raising the carbon dioxide concentration by 80 mL/L. Now, if the respiratory medium is indeed air, the capacitance coefficients for oxygen and carbon dioxide will be identical. Thus, given that the carbon dioxide *concentration* is raised by the same amount as the oxygen *concentration* is lowered, the carbon dioxide *tension* will be raised by the same amount as the oxygen *tension* is lowered. For instance, when the air leaves the lungs, if its oxygen tension has been reduced by 60 mm Hg, its carbon dioxide tension will have been elevated by 60 mm Hg. If the respiratory medium is water, by contrast, the capacitance coefficient for carbon dioxide will greatly exceed that for oxygen. Again, the carbon dioxide *concentration* of the respired medium will be raised by the same amount as the oxygen *concentration* is lowered, but because of the difference in capacitance coefficients, the carbon dioxide *tension* will be elevated much less than the oxygen *tension* is lowered. For example, if the capacitance coefficient for carbon dioxide is 30 times greater than that for oxygen and if when water leaves the gills its oxygen tension has been reduced by 60 mm Hg, its carbon dioxide tension will have been raised by just 2 mm Hg.

To summarize, the elevation of the carbon dioxide tension of the respired medium tends to be roughly similar to the decrease in its oxygen tension in air breathers but is much less than the decrease in oxygen tension in water breathers. Whereas air breathers commonly raise the carbon dioxide tension in the respired medium by several tens of mm Hg, water breathers typically raise it just slightly (never more than 5 mm Hg). These properties follow largely from contrasting *physical* attributes of air and water.

Rates of Diffusion   In the face of any given partial-pressure gradient, gases diffuse at vastly greater rates through air than water. This is a difference of crucial importance for analyzing gas transfer not only in the environment but also within the respiratory passages of animals, as we shall see repeatedly. For oxygen, the Krogh diffusion coefficient (p. 247) is at least 200,000 times greater in air than water at physiological temperatures. The coefficient for carbon dioxide in air exceeds that in water by 6000 times or more.

Density, Viscosity, and Buoyancy Water is much more dense and viscous than air. The density of air at 1 atm of pressure and 17°C is 0.0012 g/mL, whereas that of fresh water at the same temperature is essentially 1 g/mL—over 800 times higher. The viscosity of water is 35 times higher than that of air at 40°C and over 100 times higher at 0°C.

The greater density and viscosity of water dictate that, within broad limits, aquatically respiring animals must expend more energy to move a given volume of medium through their respiratory passages than air-breathing animals. This problem faced by water breathers is compounded by the fact that each volume of water carries less oxygen than each volume of air. The aquatically respiring animal thus must often work much harder than the air-breathing animal to obtain a given volume of oxygen, which means that in the water breather, a greater percentage of the oxygen taken up must be directed to the metabolic effort of obtaining more oxygen. In resting people, 1–3 percent of metabolism is involved in ventilating the lungs, whereas in resting fish 10–20 percent is probably directed to ventilating the gills.

According to Archimedes principle, objects immersed in a fluid are buoyed up by a force equal to the weight of the fluid displaced. The density of water is close to that of protoplasm. Gill filaments of aquatic animals are thus supported near neutral buoyancy by the water, and this is often vitally important in keeping the individual filaments suspended so that their entire surface area is in contact with the water. The buoyant force provided by air is negligible in comparison to the weight of protoplasm. Thus, respiratory structures of terrestrial animals that project into the medium must have sufficient structural rigidity to be self-supporting.

Evaporation As described in Chapter 8 (p. 182), the problem of water loss by evaporation into air has clearly been a major factor influencing the form and function of air-breathing organs.

Oxygen and Carbon Dioxide Tensions in the Environment

Animals can be influenced greatly by the oxygen and carbon dioxide tensions prevailing in their habitat (e.g., p. 53). Here, we first take a brief, general look at the biological and physical factors that affect ambient gas tensions and then examine the state of affairs in a variety of particular environments.

Figure 12.1 depicts the basic processes at work in a segment of an aquatic or terrestrial environment. The animals and plants living there exert strong influences on oxygen and carbon dioxide tensions. During the day, under adequate illumination, the net effect of photosynthetic plants is to add oxygen to the medium and extract carbon dioxide. Opposite effects are exerted by animals, by saprotrophic bacteria and fungi, and by photosynthetic plants at night.

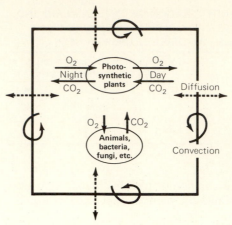

Figure 12.1 Schematic representation of the processes that affect partial pressures of oxygen and carbon dioxide in a segment of an aquatic or terrestrial environment. Oxygen and carbon dioxide are exchanged with adjacent segments of the environment by diffusion and convection.

An important question is whether the oxygen produced in photosynthesis is less than, equal to, or greater than the oxygen demands of all organisms in the system. This relationship will determine the net effect of the resident organisms on the oxygen tension. Similarly, the consumption of carbon dioxide in photosynthesis must be compared to total carbon dioxide production to determine the net effect of the resident organisms on the carbon dioxide tension. As illustrated in Figure 12.1, any given segment of the environment exchanges oxygen and carbon dioxide with neighboring segments through the physical processes of diffusion and convection. Diffusion always tends to equalize gas tensions throughout, and convection commonly exerts the same effect. The oxygen and carbon dioxide tensions actually prevailing in a segment of the environment depend on the interplay of the biotic and physical processes. The net effect of the resident organisms, for example, may be to raise or lower the oxygen tension by comparison to that in neighboring parts of the environment. Diffusion and convection, on the other hand, will exert an equalizing effect. If the rate of exchange by diffusion and convection is high, very little tension difference may develop. But if the rate of exchange is low, the oxygen tension may rise well above or fall well below that in surrounding segments of the environment.

As mechanisms of transporting gases, diffusion and convection differ markedly. A key attribute of diffusion for our present purposes is that the rate of diffusive gas transfer from place to place is an inverse function of the intervening distance (p. 246). Although diffusion can effect very rapid gas transfer over microscopically small distances, transfer over distances of inches or feet is relatively slow in air and extremely slow in water. In sharp contrast, convection—in the form of wind or water currents—can move gases rapidly over even great distances.

Open-Air Habitats According to all available measures, the percentage composition of dry air in reasonably open habitats is remarkably uniform from place to place on the earth's surface both at sea level and over the altitudinal range occupied by animals. This uniformity of composition is a consequence of the continual convective mixing of winds. Oxygen represents about 20.95 percent of the volume of dry air, and carbon dioxide averages about 0.03 percent. The level of carbon dioxide displays more variation than that of the other major components of dry air but generally does not exceed 0.06 percent even on city streets—or in forests when dissipation of carbon dioxide produced by resident animals and plants is limited by atmospheric thermal inversion.

Given that the percentage composition of dry air is so uniform in open terrestrial habitats, variations in the partial pressures of oxygen and carbon dioxide arise chiefly from changes in barometric pressure and water vapor pressure. A key factor is the decrease in barometric pressure with altitude. At 14,900 ft in the Peruvian Andes, for example, dry air still consists of 20.95 percent oxygen, but the total pressure is only about 0.59 atm and thus the oxygen tension is only about 59 percent as high as the tension at sea level (i.e., it is 93 mm Hg in dry air).

Secluded Terrestrial Habitats The exchange of gases between secluded terrestrial habitats and the open atmosphere may be restricted sufficiently for the metabolic activities of resident organisms to lower the oxygen tension and raise the carbon dioxide tension of the air. Oxygen tensions well below atmospheric, and carbon dioxide tensions well above, have sometimes been observed, for example, within decaying tree trunks, anthills, and bird nests.

Many terrestrial animals, such as earthworms, live within the soil. The interstitial air spaces among the soil particles provide paths for diffusional and microconvective gas exchange with the open atmosphere, and in porous soils the composition of air in the interstitial spaces may remain close to atmospheric down to a substantial depth provided the exchange paths with the atmosphere remain air-filled. During rains, the interstitial spaces can become filled with water, however. Because diffusion of gases is vastly slower in water than in air and because microconvective exchange is diminished by water's greater viscosity and density, gas tensions at depth in wet soil can deviate profoundly from those in the atmosphere. In one study, the oxygen tension at a depth of 30 cm in a field fell from 153 to 46 mm Hg after a rain, and the carbon dioxide tension rose from 1.5 to 46 mm Hg. Such changes may be responsible for the emergence of earthworms during heavy rain.

Burrows constructed by rodents and other relatively large animals are one interesting form of soil microhabitat. Within occupied rodent burrows, oxygen tensions have sometimes been measured to be as high as atmospheric but in other cases have proved to be 50 mm Hg or more below atmospheric. Carbon dioxide tensions ranging from near atmospheric to at least 45 mm Hg above have been observed. Recently, the factors that affect gas tensions in rodent burrows have been analyzed. The metabolic rate of the occupant and the size and depth of the nest chamber are all potentially important. In many cases, the tunnel connecting the subterranean nest chamber with the ground surface turns out to be of little consequence as a conduit for gas exchange; instead, most oxygen that reaches the nest chamber arrives by way of diffusion through the soil, and carbon dioxide leaves principally by the same route. This explains why an increase in soil wetness can substantially lower the oxygen tension in the nest chamber. It also means that when rodents plug their burrow entrances as a defensive measure, they may pay little respiratory price.

In some instances, the air within burrow systems is renewed by mass flow induced by ambient air currents. The physical principles accounting for such flow have such widespread application that they are developed separately in a box, Box 12.1, which you should read at this time.

The burrow systems of prairie dogs are one situation where the principles developed in Box 12.1 apply. In those burrow systems, the mounds built at various openings differ in height and shape, creating a situation like that in Figure 12.2C. Tests indicate that even light winds along the surface of the ground induce a significant mass flow of fresh air through the burrows, thus helping to keep burrow oxygen tensions high and carbon dioxide tensions low. Termite mounds may also be ventilated by the principles in Box 12.1.

Bodies of Water The atmosphere, as we have seen, is so thoroughly mixed by convective currents that oxygen and carbon dioxide tensions in terrestrial habitats generally do not deviate from those in the open air unless there is some physical barrier to gas exchange. Circumstances in bodies of water can be very different. Convective mixing can be much less active in water than air, partly because of water's greater density and viscosity; and diffusion through water is slow. Consequently, large differences in gas tension often exist from one region of a body of water to another even without intervening physical barriers. The air above the ground in a forest or field is generally of almost uniform composition. In a lake of equivalent volume, gas tensions can vary substantially from place to place.

Gas tensions in fast-flowing streams tend to be maintained near equilibrium with the atmosphere by turbulence. As already suggested, it is in deeper or more-stable bodies of water that tensions deviating from atmospheric are likely to be found. Waters in which organisms consume oxygen more rapidly than it is produced in photosynthesis may well become depleted of oxygen if exchange with the atmosphere

---

BOX 12.1   INDUCTION OF FLOW THROUGH TUBES BY AMBIENT CURRENTS

If the opening of a tube faces partly or fully *into* the path of an ambient air or water current—as in Figure 12.2*A*—it is easy to see that the ambient current will induce mass flow of air or water through the tube. But what if the tube is oriented at a right angle to the current, as in Figure 12.2*B*? Is it then possible for the ambient current to induce flow through the tube? Depending on the geometry of the system, the answer can well be "yes."

Figure 12.2*C* illustrates one principle by which an ambient current can induce flow through a tube at right angles. We envision a current flowing along a hard substrate and a burrow or other tube opening at two spots on the substrate surface. Furthermore, one of the openings is situated on an elevation of the substrate. In this type of circumstance, the fluid stream flowing along the substrate will accelerate as it passes over the elevation, and *Bernoulli's principle* (familiar in the study of airfoils) will come into play. According to this principle, the lateral pressure exerted by a fluid stream decreases as the velocity of the stream increases. Thus, the ambient pressure at the elevated orifice of the tube is reduced in comparison to that at the lower orifice, and fluid is made to flow through the tube as indicated in Figure 12.2*C*. One of the remarkable features of this induction of fluid flow through the tube is that, unlike the situation in (*A*), it does not depend on the direction of the ambient current. Provided the current is homogeneous, the acceleration of the fluid stream as it passes over the elevation will cause the pressure at the upper orifice to be lower than that at the lower one, whatever the direction.

Another possibility for setting up flow through a tube depends on the principle of *viscous entrainment* or *viscous suction*. When a current flows across the opening of a tube, there is a tendency for fluid to be drawn from the tube into the current because of the viscosity, or resistance to shear forces, of the fluid. Other things being equal, this force of suction is greater, the greater the velocity of the current. Thus, if a tube is open at two places and the two orifices are exposed to different ambient current velocities, as in Figure 12.2*D*, fluid will tend to flow through the tube toward the end exposed to the greater velocity. This principle can lead to ventilation of a tube in situations where the Bernoulli principle does not apply. When a fluid flows in laminar

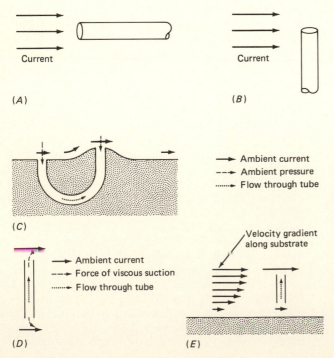

Figure 12.2 Mechanisms by which ambient currents can induce flow through tubular structures. Solid arrows indicate qualitatively, by their length, the velocity of the ambient current. Dashed arrows indicate qualitatively, by their length, the magnitude of the ambient pressure (*C*) or the magnitude of the force of viscous suction (*D*). Part (*E*) illustrates that fluid flowing in laminar fashion along a substrate exhibits a velocity gradient near the surface; thus, higher and lower orifices of a tubular structure positioned near the substrate are exposed to different fluid velocities. [For further detail see S. Vogel, *Life in Moving Fluids.* Willard Grant Press, Boston, 1981.]

fashion along a substrate, as illustrated in Figure 12.2*E,* there is a gradient of fluid velocity within a short distance of the substrate. Fluid relatively far from the substrate flows more rapidly than that near the substrate. In this commonplace situation, there is no difference in pressure associated with the velocity gradient. Bernoulli's principle does not say that *any* rapidly moving stream will exert less pressure than *any* more slowly moving stream; rather, it applies only when a *particular* fluid stream is accelerated or decelerated. As illustrated, however, the velocity gradient itself can induce flow through a tube by virtue of differential viscous entrainment at orifices located different distances from the substrate. Again, the direction of the ambient current does not matter.

is slow. On the other hand, oxygen tensions of up to twice the equilibrium value with air have been reported in waters in which there is intense photosynthetic activity. Carbon dioxide tensions ranging from 0 to over 55 mm Hg have been reported, the latter in acid bogs and lakes.

For a deep body of water such as a deep lake or the ocean, the *sources* of oxygen are predominantly near the surface. The surface waters are the ones that can dissolve oxygen from the air. Furthermore, only waters relatively near the surface receive enough light for photosynthesis to supply oxygen. The depths—being dark and removed from the atmosphere—are largely dependent on influxes of surface waters for renewal of their oxygen supplies. These considerations have important implications for gas tensions in the deep waters.

In the open oceans, oxygen tensions generally remain reasonably high even at thousands of meters below the surface. This is because water movements on a global scale circulate oxygen-laden surface waters into the depths at a sufficient rate to replace oxygen used by organisms.

By contrast, depletion of oxygen is common in the depths of lakes in summer, owing to consequences of **thermal stratification** (thermal layering of the water). Increasing the temperature of water decreases its density. Consequently, when the surface waters of a lake are warmed by the sun in the summer, they tend to float on top of the colder (more dense) deep waters, and two thermally distinct layers (or strata) are formed. There may well be active circulation *within* each of these layers; but waters of different densities are reluctant to mix, and typically there is little convective exchange across the interface between the warm surface waters and cold deep waters. Now it frequently happens that light penetration into the cold, deep layer is inadequate for photosynthesis *within* that layer to meet the oxygen demands of all the organisms living there. Thus, when thermal stratification curtails mixing of oxygen-rich surface waters into the deep waters, a severe state of oxygen depletion can develop in the deep waters, as illustrated in Figure 12.3. Sessile inhabitants of the deep waters are unable to migrate away when such conditions develop. Also, some of the mobile animals prefer or require low temperatures

(e.g., some trout); these creatures may be caught between the conflicting demands of needing to swim to upper waters to find satisfactory oxygen tensions, yet needing to avoid the upper waters owing to their elevated temperatures. Sometimes, massive die-offs of certain species ("summer kills") result.

In the fall, the upper waters of thermally stratified lakes cool off, and as the density of the upper waters approaches that of the lower waters, convective mixing is restored between the lower waters and surface (fall overturn). High oxygen tensions are then reestablished throughout the lake. Heavy accumulations of snow on ice-covered lakes in winter, however, can again cause oxygen levels to plummet, by blocking light penetration to the water—and thus impairing photosynthesis—while turbulent mixing with the atmosphere is simultaneously prohibited by the ice.

Estuaries can become stratified by salinity effects (as well as thermal ones) because less-saline waters (such as formed by river outflows) tend to float on top of more-saline ones. The consequent interruption of free mixing between the surface and depths can then allow oxygen levels in the deep waters to fall. In large parts of the estuarine Chesapeake Bay, oxygen levels descend virtually to zero in the deep waters during summer. Oysters are thereby killed, and blue crabs are forced to migrate into the shallower waters. Oxygen depletion can also occur in some fully marine situations.

Human activities have profoundly altered oxygen levels in many bodies of water. A fundamental problem is that, by one of two major processes, people commonly increase the levels of dissolved or suspended organic matter undergoing decay in bodies of water. One of these processes is the straightforward addition of organic material, such as sewage. The other is the addition of inorganic chemicals such as nitrates and phosphates that fertilize the waters, promoting increased algal growth and eventually producing an increased amount of dead organic material when the algae and organisms that feed on the algae die. A buildup of dead organic matter, however arrived at, tends to promote increased activity of fungi, bacteria, and other decay organisms that sap oxygen resources. Consequently, oxygen tensions often are diminished. Rivers loaded with increased amounts of decaying matter often exhibit reduced oxygen levels.

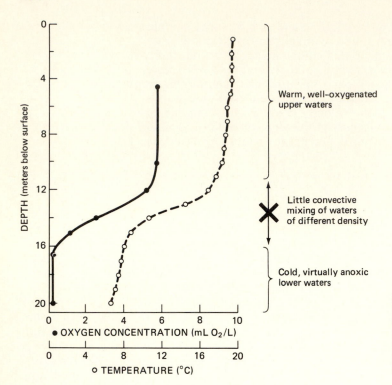

Figure 12.3 Dissolved oxygen concentration (●) and temperature (○) as functions of depth in a thermally stratified lake (Douglas Lake, Cheboygan County, Michigan) during July 1969. Oxygen levels in the upper waters are near equilibrium with the atmosphere, but those at depth are near zero. (Data gathered by a class of physiology students.)

In lakes or estuaries that become stratified, an increase in the amount of decaying matter often causes oxygen depletion of the deep waters to develop more rapidly and proceed to greater extremes. Animals that require relatively high oxygen tensions, such as many game fish, may consequently be replaced by others, such as carp and catfish, that are tolerant of low oxygen levels.

Sediments Problems of gas exchange similar to those in wet soils prevail in the bottom sediments of bodies of water. Movement of oxygen by diffusion and convection from the open water through the interstitial spaces of sediments occurs slowly and can meet only a small oxygen demand. When organic matter in the sediment supports any substantial degree of saprotrophic activity, completely anaerobic conditions may prevail within millimeters of the substrate surface. Most animals that live in bottom sediments have adopted the expedient of drawing their oxygen supplies directly from the open water above, rather than depending on interstitial supplies. Buried clams, for example, pump water over their respiratory surfaces through siphons that project to the substrate surface. Polychaete worms that live in tubes in the substrate often pump water through their tubes. Some build U-shaped tubes resembling the one in Figure 12.2C and conceivably could enjoy a flow of water through their tubes induced by ambient currents flowing along the substrate surface.

FUNDAMENTALS OF GAS EXCHANGE

The gas exchange systems of all animals can be sche-

matized as in Figure 12.4. There are three fundamental steps: (1) Oxygen and carbon dioxide must move between the ambient medium and the respiratory exchange membranes; (2) they must traverse the exchange membranes; and (3) they must move between the exchange membranes and the internal tissues of the animal. Each of these steps must occur at an adequate rate; the chain is only as strong as its weakest link. Here, we discuss the steps individually.

The Exchange Membrane

There is always a membrane separating the internal tissues of the animal from the environmental medium. Flux of oxygen across this exchange membrane is exclusively by diffusion; instances of active oxygen transport are unknown in the animal kingdom. Diffusion is the exclusive mechanism of carbon dioxide flux across the exchange membrane in some animals; in others, however, while diffusion is the

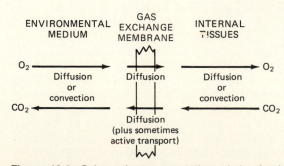

Figure 12.4 Schematic representation of the fundamental processes in animal gas exchange.

principal mechanism, flux by active transport also occurs. Active outward transport of carbon dioxide is best established among freshwater animals (p. 142). It may also occur, however, in other groups, even being postulated in mammals by some investigators.

Recognizing that diffusion is the sole means by which oxygen enters animals and the sole or principal mode of carbon dioxide efflux, the principles of diffusion reviewed in Chapter 11 deserve recall. Diffusion occurs strictly in the direction of the partial-pressure gradient; oxygen, for example, will not enter an animal unless the oxygen tension on the outside of the exchange membrane exceeds that on the inside. Physics tells us too that the rate of diffusion depends directly on a membrane's area and inversely on its thickness. Thus, we readily understand why respiratory exchange membranes are characteristically thin and often are thrown into elaborate patterns of invagination or evagination that increase their surface area.

Movement of Oxygen and Carbon Dioxide Between the Medium and Respiratory Exchange Membrane: External Respiration

Oxygen in the ambient medium may be brought to the environmental surface of the exchange membrane, and carbon dioxide carried away, by diffusion or convection or a combination of both. The movement of gases between the medium and exchange membrane is termed *external respiration* in any case.

Convective movement of the ambient medium across the respiratory exchange membranes is called *ventilation*. If the animal creates ventilatory currents using forces of suction or positive pressure developed at the expense of metabolic energy, the ventilation is termed *active*. On the other hand, if ambient air or water currents directly or indirectly induce flow of the medium across the exchange membranes, the ventilation is termed *passive* (e.g., Figure 12.2). Whereas active ventilation draws on the animal's energy resources, it is potentially more reliable, controllable, and vigorous than passive ventilation.

Movement of Oxygen and Carbon Dioxide Between the Respiratory Exchange Membrane and Internal Tissues

It is instructive from the outset in discussing this topic to review some calculations performed by the great physiologist August Krogh (1874–1949). Krogh wanted to define the maximum distance over which diffusion through an organism could meet the gas exchange requirements of tissues. He centered his attention on oxygen because it diffuses more slowly than carbon dioxide in aqueous media. Oxygen diffuses less rapidly through animal tissues than pure water, and Krogh assumed, on the basis of some evidence, a rate of one-third the rate in water (at 20°C). He considered a spherical cell and, postulating a metabolic rate of 0.1 mL O_2/g·h and an oxygen tension of about 159 mm Hg at the cell surface, computed that diffusion of oxygen through such a cell will suffice to meet the oxygen demands of all parts of the cell only if the cell radius does not exceed 0.9 mm. Convective movement of oxygen through the cell is necessary if the cell has a larger radius, if the metabolic rate is higher, or if the oxygen tension at the surface is lower. These calculations, though somewhat theoretical, indicate in any case that whether we are considering spherical cells or tissues of other shapes, *movement of oxygen through tissues by diffusion alone will generally suffice to meet oxygen demands only over short distances even when the oxygen tension of the source is high.*

Two general possibilities for gas movement between the respiratory exchange membranes and tissues become evident. Either (1) the exchange membranes may be very near all cells, in which case gas movement by diffusion might be able to meet the requirements of all tissues; or (2) the exchange membranes may be distant from some cells, in which case convective gas transport within the animal will generally be essential. One of the major roles that circulatory systems often play is the convective transport of gases; in an animal with a circulatory system, oxygen picked up by the circulating blood at the respiratory exchange membranes can be carried rapidly—by mass flow—over great distances. In many animals, the requirements of gas transport have exerted a major influence on the attributes of the circulatory system and the composition of the blood.

The Oxygen Cascade

For aerobic catabolism to occur at an adequate rate, the mitochondria must be supplied with adequate oxygen. In an organism such as a crayfish, salmon, or human, the mitochondria in the depths of the body are far removed from the environmental source of oxygen. Elaborate systems for external respiration and circulation are involved in transporting oxygen from the environment to the mitochondria. With each successive step in this transport, the oxygen tension falls. Thus, a plot of changes in oxygen tension along the path from environment to mitochondrion has the appearance of a cascade: an *oxygen cascade*.

The oxygen cascade in healthy, resting people at sea level is shown in Figure 12.5. In the alveoli of our lungs—which are the sites of gas exchange between our lung air and blood—the oxygen tension is already well below the oxygen tension in the atmosphere, because we do not fully empty our lungs when we exhale and thus the fresh air we inhale mixes with a large residuum of "stale" air. A small further drop in oxygen tension occurs between our alveolar air and arterial blood because the air and blood do not fully equilibrate in the lungs. Then, as our arterial blood flows through the capillary beds in our sys-

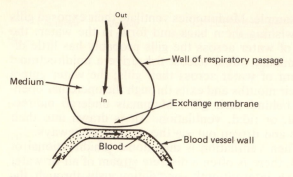

Figure 12.7 Tidal ventilation of the respiratory exchange membrane. Heavy arrows show the directions of flow of medium and blood.

over 30 percent below in resting humans, for example (Figure 12.5). The oxygen tension of the blood cannot become higher than the oxygen tension of the medium next to the exchange membranes. Thus, tidal ventilation imposes limits on peak blood tensions. An attribute worth remembering for the discussion about to come is this: With tidal ventilation of the exchange membranes, the oxygen tension of the blood characteristically remains at least slightly below the oxygen tension of *exhaled* medium.

Figure 12.8 shows two possible patterns of oxygen exchange when ventilation of the exchange membranes is unidirectional. Part (*A*) supposes that the blood flows along the exchange membrane in the same direction as the medium. The pattern of gas exchange in this case is known as ***concurrent exchange***. Part (*B*) depicts ***countercurrent exchange*** in

which the blood and medium flow in opposite directions along the exchange membrane.

With concurrent exchange, oxygen-depleted afferent blood initially meets fresh, incoming medium, as shown at the left in part (*A*). Then, as blood and medium flow along together, they gradually approach equilibrium at some oxygen tension intermediate between their respective starting tensions. On leaving the exchange membrane, blood undergoes its final exchange with medium that has a tension considerably below ambient. Concurrent exchange shares with tidal exchange the feature that the oxygen tension of blood cannot rise above the tension of exhaled medium.

With countercurrent exchange, oxygen-depleted afferent blood initially meets medium that has already been substantially deoxygenated, as shown at the right in part (*B*). However, as the blood flows along the exchange surface, it steadily encounters medium of higher and higher oxygen tension. Thus, even as the blood picks up oxygen and its tension rises, a tension gradient favoring further uptake of oxygen is maintained. The final exchange of the blood is with fresh, incoming medium of high oxygen tension. Countercurrent exchange is intrinsically a more effective mode of exchange than either tidal or concurrent exchange. This is reflected in Figure 12.8 in two complementary ways. First, with countercurrent exchange the oxygen tension of blood can rise well above that of exhalant medium and, in principle, may approach that of inhalant medium. Second, comparing parts (*A*) and (*B*) of the figure, it is evident that the oxygen tension of the medium falls more—and that of the blood rises more—in part (*B*):

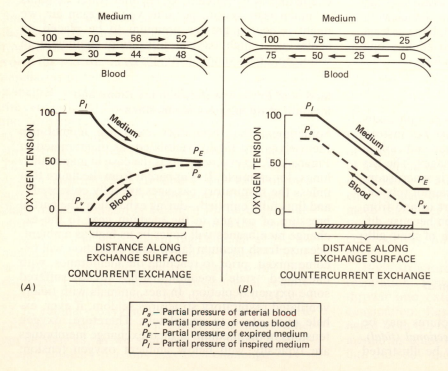

P_a — Partial pressure of arterial blood
P_v — Partial pressure of venous blood
P_E — Partial pressure of expired medium
P_I — Partial pressure of inspired medium

Figure 12.8 Schematic representation of oxygen transfer between blood and environmental medium (*A*) when the blood and medium flow in the same direction along the respiratory exchange surface (concurrent exchange) and (*B*) when they flow in opposite directions (countercurrent exchange). Numerical values are oxygen tensions in arbitrary units. It has been assumed for simplicity that the flow rates of blood and medium are the same and that oxygen content and tension are related proportionally and similarly in the blood and medium.

Countercurrent exchange permits a more complete transfer of oxygen from medium to blood than concurrent (or tidal) exchange under comparable conditions.

A third mode of exchange sometimes associated with unidirectional ventilation, known as **cross-current exchange,** is illustrated in Figure 12.9. In this mode, some blood exchanges exclusively with oxygen-rich medium and some exclusively with oxygen-poor medium. Consequently, cross-current exchange permits the tension of efferent blood as a whole to be above that of exhalant medium but does not enable it to rise as high as the tension attained by countercurrent exchange under comparable circumstances.

The modes of exchange that we have discussed can be ranked as follows in terms of their *intrinsic* efficiency in transferring oxygen and carbon dioxide between the medium and blood: Countercurrent exchange is superior to cross-current, and cross-current is superior to concurrent or tidal. It is important to recognize though that the *actual magnitude* of the differences in efficiency among modes depends on conditions (e.g., the diffusion coefficient of the exchange membrane) and may be either substantial or vanishingly small.

EXTERNAL RESPIRATION IN AQUATIC INVERTEBRATES AND ALLIED GROUPS

We now proceed to review the mechanisms of external respiration, treating first the aquatic invertebrates and their allied groups, then the vertebrates, and finally the insects and arachnids.

Sponges, Coelenterates, and Flatworms

Sponges, coelenterates, and flatworms (such as *Planaria*) lack a circulatory system. Once oxygen has penetrated the body proper of these animals, it probably travels only by diffusion or via highly localized convective movements, such as movements of protoplasm within cells or of pockets of body fluid between cells. This suggests that the rate of oxygen movement within the body proper would not be sufficient to supply the needs of cells far removed from an exchange surface with the environment. We are not surprised therefore to find that the bodies of these animals are so constructed that the medium can get close to most cells.

The bodies of sponges are penetrated by many pores, and it is significant that most of the cells in a sponge lie near the outer surface of the animal or along the lining of a pore or other water channel within the body. Ventilation is unidirectional, as shown in Figure 12.10. Water is drawn in through the pores from the environment by flagellary action and exits via larger openings, the oscula, which receive water from many pores. Additionally, because sponges have body openings both near to and far from their point of attachment with the substrate, many species may experience water flow through their bodies that is induced passively by ambient currents flowing along the substrate (see Box 12.1).

Coelenterates possess an internal body cavity, the gastrovascular cavity, which opens to the outside by way of a single orifice, or mouth (Figure 12.10). It is notable that most of the cells of the body, including nerve and muscle cells, are located near the outer surface of the animal or near the gastrovascular cavity. The region between the outer surface and gastro-

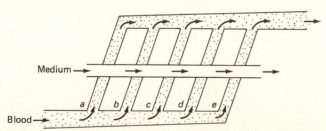

Figure 12.9 Schematic diagram of cross-current exchange. The afferent blood vessel breaks up into smaller vessels, each of which makes exchange contact with just a limited segment of the tube carrying medium. The smaller vessels then coalesce to form a single efferent vessel. The medium is gradually depleted of oxygen during its passage through the tube; thus, blood traveling in the first vessel to make exchange contact with the medium (a) is exposed to a high oxygen tension, but that in the final vessel (e) is exposed to a low, nearly exhalant oxygen tension. One of the characteristics of cross-current exchange is that the efficiency of gas transfer between medium and blood is independent of the direction of flow of the medium.

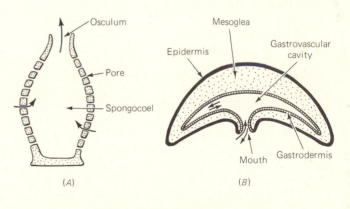

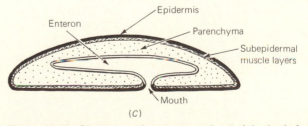

Figure 12.10 Diagrammatic representations of the body form of a sponge, coelenterate, and flatworm: (A) radial section of an asconoid sponge, (B) radial section of a coelenterate medusa, and (C) longitudinal section of a turbellarian flatworm. Heavy arrows in (A) and (B) indicate water currents through body cavity.

vascular cavity can be quite thick, as in many jelly-fish, but it is filled largely with a nonliving membranous or gelatinous mesoglea. Little direct physiological information has been gathered on respiration in coelenterates. Probably, the oxygen needs of cells near the outer surface of the animal are met by diffusion directly from the ambient water. Water currents are established in the gastrovascular cavity by the action of cilia and flagella and at least sometimes are highly directional. These are thought to bring oxygen-laden water from the environment to all parts of the cavity at a sufficient rate to assure oxygen tensions adequate for the diffusional supply of the cells near the cavity walls; this notion requires further testing, however, especially in large species, which sometimes reach diameters of several feet.

The flatworms, as their name suggests, are highly flattened dorsoventrally. In the free-living forms, or Turbellaria, the body is covered on the outside by an epidermis, with the major muscle layers lying just beneath; usually, also, there is an internal body cavity, or enteron, opening to the outside via a single orifice, the mouth (Figure 12.10). The space between the gastrodermal cells lining the enteron and the subepidermal muscle layers is largely occupied by a population of cells termed the mesenchyme or parenchyma. Fluid-filled spaces are much in evidence among these parenchymal cells. We are again faced with a marked paucity of information on respiration. Circulation of water in the enteron probably plays a minor role, if any, in supplying oxygen. Thus, most oxygen enters the flatworm across the epidermis, which is generally ciliated. There are several features of turbellarians that are probably significant in gas exchange. First, the major muscle layers, which are probably sites of substantial oxygen uptake, are located just beneath the exchange surface with the medium. Second, the pockets of fluid among the parenchymal cells are squeezed during body movements. Convection within these pockets is probably important to gas transfer through the animal. Third, the highly flattened shape of turbellarians assures a relatively short distance between all cells and an exchange surface with the environment. It is generally thought that the limited capacities of these animals for internal gas transfer have been a significant factor in limiting their body thickness.

Annelids

A fairly sophisticated circulatory system has made its appearance by the phylogenetic level of annelids. The presence of a circulatory system not only permits cells to be further from the body surface but *permits the meaningful development of specialized respiratory structures* because oxygen taken up in a specialized region of the body can be carried to cells throughout the body.

The oligochaete annelids (e.g., earthworms) are typically smooth bodied and respire entirely across their general body surface, which is well vascular-

ized. Among the marine polychaete annelids, the general integument typically is also well suited to a role in respiratory gas exchange, and in some species it must assume the entire task. In many polychaetes, however, there occur localized evaginations that amplify the surface area for gas exchange and often are particularly heavily vascularized. Such evaginations are properly termed gills. The gills of the various species are extremely diverse in morphology and location. Frequently, they are developments of the lateral appendages, or parapodia, which extend on either side of each body segment. They vary (Figure 12.11) from simple, flattened plates of tissue (as in *Nereis*) to elaborate, branching trees of filaments (as in *Arenicola*).

Epidermal cilia are found in many groups of polychaetes and often create water currents over the respiratory surfaces. Dorsal cilia commonly propel water along the dorsal integument; often the direction of this flow is anterior to posterior, and inasmuch as blood in certain dorsal vessels flows in the opposite direction, a potential exists for countercurrent exchange. Muscular movements of the parapodia (e.g., during swimming) also may create ventilatory currents.

Many polychaetes live in tubes constructed of sand grains, hardened viscous secretions, or calcareous material. Parapodial gills are often reduced in these forms, and anterior gills that are presented to the water at the mouth of the tube are common. The most spectacular of these are the beautiful fans of pinnately divided tentacles found in sabellids and serpulids: the fanworms (Figure 12.12). Some tube-dwellers (e.g., *Arenicola;* Figure 12.11*E*) have well-developed parapodial gills. Furthermore, the general integument still presents possibilities for respiratory exchange. Tube-dwellers commonly circulate water through their tubes by ciliary activity, parapodial paddling, or undulatory or peristaltic body contractions; up to 50–70 percent of the oxygen in the water may be removed in its passage.

Mollusks

Gill-Breathing Forms   In mollusks, outfolding of the dorsal body wall produces a sheet of tissue, the mantle, that commonly overhangs part or all of the rest of the body and is responsible for generating the shell. Where the mantle overhangs the rest of the body, it encloses an external body cavity, the mantle cavity. The gills of mollusks typically are suspended in this cavity and thus provide our first example of internal gills (p. 257). The arrangement of the mantle, mantle cavity, and gills is diverse in various molluscan groups, as illustrated in Figure 12.13*A–C*. In many snails there is but a single gill, whereas in certain chitons there are over 20 pairs. The more primitive gills (as seen in chitons and certain snails) are plumose, consisting of many pinnately arranged filaments along a central axis. This structure has been modified, sometimes extensively, in many mollusks,

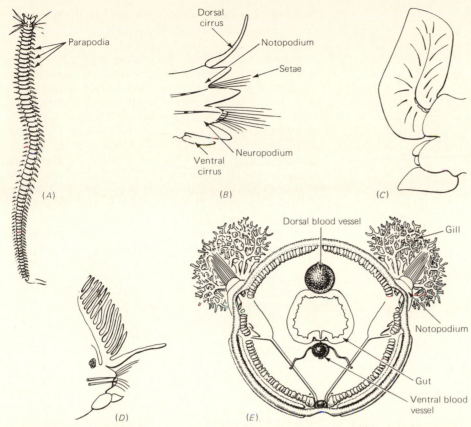

Figure 12.11 (*A*) Dorsal view of a nereid polychaete showing the position of the parapodia. (*B*) Lateral view of a parapodium of the nereid *Nereis pelagica*. The parapodium consists of a dorsal division, the notopodium, and a ventral division, the neuropodium. Each division bears a cluster of chitinous bristles, the setae, and a fleshy tentacular process, the cirrus. The parapodia consist of broad, well-vascularized lobes and constitute important sites of respiratory gas exchange.

The parapodia of *Nereis* are gills in the general sense of the word. However, when applied to polychaetes, the word *gill* is often used in a more restrictive sense to refer to specialized developments of the parapodia (or other parts of the body). Specialized branchial developments are seen in (*C*), (*D*), and (*E*). (*C*) Parapodium of *Phyllodoce groenlandica*. The dorsal and ventral cirri are developed into large, broad lamellae. (*D*) Parapodium of *Eunice harassii*. A filamentous branchial structure arises from the base of the dorsal cirrus. (*E*) Transverse section of a gill-bearing body segment in *Arenicola marina,* viewed from the anterior end. The gill consists of hollow, branching outgrowths of the body wall and is attached just behind the notopodium. Gills are found on only 13 of the body segments. [Parts (*A*)–(*D*) from P. Fauvel, *Faune de France,* Vol. 5, *Polychètes errantes*. Fédération Française des Sociétés de Sciences Naturelles, Office Central de Faunistique, Paris, 1923; part (*E*) from G.P. Wells, *J. Mar. Biol. Assoc. U.K.* **29**:1–44 (1950).]

as we shall see. The gills may not be the only sites of gas exchange; indeed, the mantle and other body surfaces often contribute substantially.

The surfaces of both the mantle and the gills are generally heavily ciliated, and ventilation of the mantle cavity and gills is most commonly accomplished by ciliary action. Usually, flow through the mantle cavity is unidirectional; water enters and leaves via well-defined incurrent and excurrent openings. Some mollusks have replaced ciliary ventilation to a greater or lesser extent with muscular ventilation. Muscular ventilation reaches its zenith in the cephalopods. Most cephalopods swim by using muscular contractions of the mantle. Water is alternately sucked into the mantle cavity through incurrent openings and then driven forceably outward through a ventral funnel, producing a propulsive force. The gills in the mantle cavity are ventilated during inhalation.

Countercurrent exchange between the water and blood is common in mollusks. As shown in Figure 12.13C, for example, the cilia on the gill leaflets of many prosobranch snails drive water across the leaflets from left to right. Blood flows through the exchange vessels of the leaflets in the opposite direction. Countercurrent exchange also occurs in the gills of squids and octopuses.

The lamellibranch mollusks, which include the clams and mussels, are the focus of an intriguing debate. Their gills, radically modified from the primitive molluscan condition, take the form of broad,

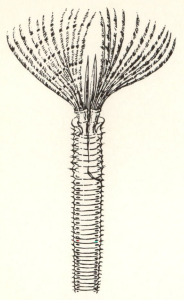

Figure 12.12 Dorsal view of the anterior end of the tube-dwelling sabellid polychaete *Sabella pavonina*. The array of pinnately divided tentacles is projected into the water at the mouth of the tube in which the worm lives. Water is driven along the tentacles by cilia, and the tentacles function in both feeding and respiration. [From P. Fauvel, *Faune de France*, Vol. 16, *Polychètes sédentaires.* Fédération Française des Sociétés de Sciences Naturelles, Office Central de Faunistique, Paris, 1927.]

thin sheets—or lamellae—of fused or semifused filaments. The basic pattern of water flow through such gills is shown in Figure 12.14, using the most extensively fused type of gill (the eulamellibranch gill) for illustration. Cilia drive incoming water through pores (ostia) on the gill surface into water channels that run *within* the gill; the water channels then ultimately carry the water to exhalant passages. Interestingly, the direction of water flow within the water channels is opposite to the direction of blood flow in the major blood vessels of the gill; thus, countercurrent gas exchange can occur. The lamellibranch gill represents, at least in part, an adaptation for feeding: As the abundant flow of incoming water passes through the array of pores leading to the interior water channels, food particles suspended in the water are removed, ultimately to be delivered to the mouth. Now, the question arises: Do the gills still play any role as special organs of external respiration? Or have they become feeding organs exclusively? Undoubtedly, in some cases the gills play a special respiratory role. However, in others, they may not. Most species of lamellibranchs lack oxygen-transport pigments in their blood. Furthermore, recent experiments on mussels indicate that the circulation of their blood plays little or no role in oxygen delivery (p. 366). *If the circulation is not transporting appreciable amounts of oxygen from place to place, then the gills cannot be taking up much oxygen for use by other tissues.* The tissue masses of mussels are generally thin, and it may be that all tissues obtain

their oxygen directly by diffusion from the water flowing through the mantle cavity.

Pulmonate Gastropods   Molluscan gills are typically delicate structures that collapse without the buoyant support of water. Thus, when we turn to the dominant group of terrestrial mollusks, the snails and slugs of the subclass Pulmonata, we are not surprised to find that gills have been abandoned. In these animals, the mantle cavity has become a lung; gas exchange occurs across the well-vascularized mantle surface. As shown in Figure 12.13*D,* this lung opens to the outside via a single, closable, porelike opening, the pneumostome, formed by the mantle edge.

Calculations indicate that the pulmonate lung can meet gas exchange requirements while functioning entirely as a diffusion lung. Nonetheless, some species ventilate the lung by raising and lowering the floor of the mantle cavity. The pneumostome is alternately closed and opened in both ventilating and nonventilating forms. Closing of the pneumostome helps limit evaporative water losses (p. 183) and occurs more frequently in dehydrated individuals than in well hydrated ones.

The pulmonates have radiated back into the aquatic environment (chiefly fresh water), and there three respiratory patterns are evident. (1) Some aquatic species have retained the air-breathing habit and periodically come to the surface to refresh the air in their lung. These are in fact dual breathers. In well-aerated water, they probably obtain about as much oxygen from the water across general body surfaces as they obtain from the air. If the oxygen tension of the water becomes reduced, however, they depend more on lung breathing and surface more frequently. (2) Some aquatic pulmonates are believed to ventilate their lung with water, probably tidally. (3) Finally, some have reverted to gill breathing. Notably, their gills are not homologous to the primitive molluscan gill (lost in the original evolution of pulmonates) but are novel developments; the gills of freshwater limpets, for example, are evaginations of the foot. Such cases emphasize the strong evolutionary bias toward evaginated respiratory structures in aquatically respiring animals.

Crustaceans

Although some of the very small crustaceans, such as copepods, lack specialized respiratory structures and evidently respire entirely across general body surfaces, gills are present in most larger crustaceans and many of the smaller ones. Crustacean gills are nearly always closely associated with the thoracic or abdominal appendages, arising from the appendages themselves or close to their bases. The entire body surface of crustaceans is covered with a chitinous cuticle. The gills are no exception, but the covering over them is thin and permeable. External cilia are lacking. Thus, ventilation is always accomplished by

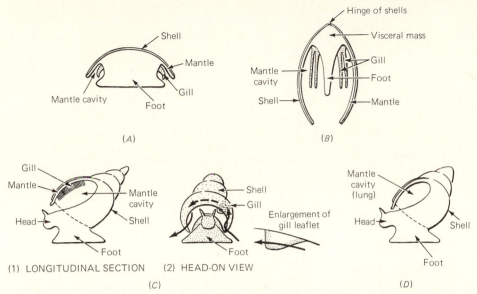

Figure 12.13 Schematic representations of the arrangement of the mantle, mantle cavity, and gills in several groups of mollusks. (*A*) Transverse section of a chiton. The mantle, extending laterally away from the foot, delimits a mantle trough (mantle cavity) running the length of the animal on each side. Many pairs of gills project into the mantle troughs in a serial arrangement from anterior to posterior. In life, the lateral edges of the mantle are appressed to the substrate except in localized regions at the anterior and posterior ends of the animal (inhalant and exhalant openings). Ciliary action on the gills produces a ventilatory current. Water enters anteriorly, follows the channel between the gills and mantle, passes across the gills to enter the channel between the gills and foot, and then flows posteriorly in that channel to exit. (*B*) Transverse section of a lamellibranch clam. The mantle cavity is relatively capacious, and the gills are suspended in the cavity. There are just two gills, each being folded to produce two half-gills or demibranchs. Ventilation in eulamellibranchs is discussed in the text. (*C*) Diagrams of the condition in many prosobranch gastropods. There is only one gill, the left, and it has become modified and fused to the mantle, assuming the form of many triangular gill leaflets that hang into the mantle cavity something like the pages of a book. There is a broad anterior orifice into the mantle cavity (see *C2*). Ciliary action on the gill leaflets produces a ventilatory stream, water entering at the left of the animal, passing across the gill leaflets, and exiting at the right (indicated by heavy arrows in *C2*). (*D*) Longitudinal section of a terrestrial pulmonate gastropod. Gills are lacking, and the walls of the mantle cavity have become richly vascularized, transforming the mantle cavity into an air-breathing lung. The mantle cavity opens to the outside only through a small porelike orifice. When the mantle cavity is ventilated, air passes both in and out through this pore. Sometimes the surface area of the cavity is increased by ridges or tubular outpocketings.

muscular contraction, typically by beating of certain appendages. Respiration has been studied most extensively in the decapod crustaceans, and they are emphasized here.

Aquatic Decapods   A well-defined carapace covers the head and thorax of decapods dorsally and overhangs the thorax laterally, fitting more or less closely around the bases of the thoracic legs (Figure 12.15*A*). The carapace delimits two lateral branchial chambers in which the gills lie. The gills are all thoracic and number from 3 to 26 on each side. Each gill consists of a central axis to which are attached a great many lamellar plates, filaments, or dendritically branching tufts. These structures are richly vascularized. Each branchial chamber is ventilated by a specialized appendage located toward its anterior end and known as the scaphognathite or gill bailer (Figure 12.15*B*). The gill bailer beats back and forth, generally driving

water outward through an anterior exhalant opening; a negative pressure is thus created within the branchial chamber, causing water to be drawn in at other openings, the location of which varies with species (note Figure 12.15*B*). Ventilation is unidirectional, and countercurrent exchange may occur.

Many decapods, although not all, markedly increase their rate of ventilation if the oxygen tension of the ambient water falls. In this respect they resemble many other aquatic animals, including fish and some clams. An increase in the rate of water flow across the gills can help sustain a steady rate of aerobic catabolism in the face of a drop in the oxygen available from each unit volume of water.

Semiterrestrial and Terrestrial Decapods   All the semiterrestrial and terrestrial crabs retain gills. The cuticular covering of crustacean gills is probably significant in this regard because it stiffens the gills by

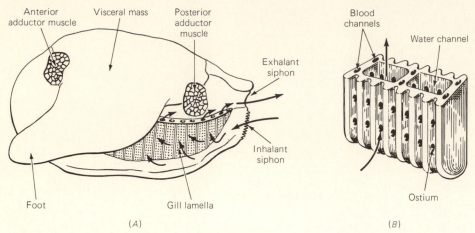

(A)   (B)

Figure 12.14 Aspects of gill structure and ventilation in eulamellibranch clams as exemplified by the freshwater mussel *Anodonta*. (*A*) A specimen of *Anodonta* with the left shell removed (semidiagrammatic). One of the four gill lamellae suspended in the mantle cavity is shown. It has been sectioned longitudinally near its dorsal attachment with the body to reveal the water channels within. (*B*) The structure of the lamella in greater detail (semidiagrammatic). Ostia on the surfaces of the lamella lead into the water channels, which run dorsoventrally within the lamella.

Ventilatory currents are indicated by heavy arrows. Water from the mantle cavity is forced into the water channels through the ostia by the action of cilia surrounding the ostial openings. The water then passes dorsally through the water channels to enter a suprabranchial chamber above the gill (not shown). The water flows posteriorly in the suprabranchial chamber to exit through the exhalant siphon. These animals and many other bivalve mollusks bury themselves in the substrate, with only their siphons projecting to the water above. The action of the ostial cilia pulls water into the mantle cavity through the inhalant siphon. [Part (*A*) after and part (*B*) from T.J. Parker and W.A. Haswell, *A Text-book of Zoology,* Vol. I, 6th ed. (as revised by O. Lowenstein). Macmillan and Company, Ltd., London, 1940. Part (*B*) used by permission of Macmillan, London and Basingstoke.]

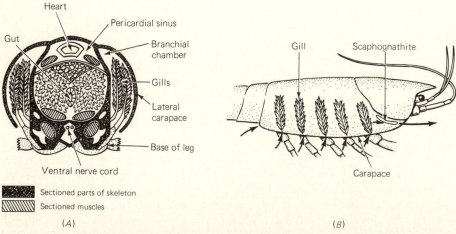

Sectioned parts of skeleton
Sectioned muscles

(A)   (B)

Figure 12.15 (*A*) Transverse section of a crayfish at the level of the heart (semidiagrammatic). The carapace overhangs the thorax laterally, delimiting an external body cavity, the branchial chamber, on each side. The gills arise near the bases of the legs and lie in the branchial chambers. (*B*) Water flow through the branchial chambers of a crayfish. The gills and scaphognathite are enclosed by the carapace and not visible externally. The beating of the scaphognathite drives water out of the branchial chamber anteriorly. Water is drawn in around the bases of the legs and the posterior margin of the carapace according to the pressure gradient established by the scaphognathite. Heavy arrows indicate water flow. [Part (*A*) after F. Plateau, *Arch. Biol.* **1**:595–695 (1880).]

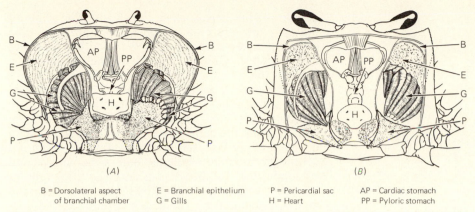

B = Dorsolateral aspect of branchial chamber	E = Branchial epithelium	P = Pericardial sac	AP = Cardiac stomach
	G = Gills	H = Heart	PP = Pyloric stomach

Figure 12.16 Internal anatomy of two terrestrial crabs viewed dorsally: (*A*) *Gecarcinus lateralis* and (*B*) *Ocypode quadrata*. The dorsal carapace and certain internal organs have been removed. Note the large branchial chambers (especially in *Gecarcinus*) and the thin, vascularized branchial epithelium lining the walls of the branchial chambers. The branchial epithelium of *Ocypode* (*B*) bears vascularized tufts or papillae, whereas that of *Gecarcinus* (*A*) does not. [From D.E. Bliss, *Am. Zool.* **8:**355–392 (1968).]

comparison, for example, to molluscan gills. The gills of terrestrial crabs tend to be reduced in size and number by comparison to those of marine crabs. The branchial chambers tend to be enlarged, and frequently some part of the epithelial lining of each chamber has become highly vascularized and thus well suited to respiratory exchange. In some species, the surface area of the epithelium of the branchial chambers is increased by evaginated folds or papillae (Figure 12.16). The trends toward the reduction of the gills and the development of lunglike branchial chambers in terrestrial crabs provide a striking parallel to the situation in pulmonate gastropods. The branchial chambers typically are ventilated with air, driven by beating of the scaphognathites.

Recently, an elaborate lung of unexpected morphology has been discovered in a crab (*Pseudothelphusa*) that inhabits montane regions in Trinidad. In this lung, the surface area for exchange is increased by the *invagination* of fine, anastomosing air passages into blood sinuses from the branchial chambers. The air passages are ventilated by a bellowslike mechanism, and gas exchange between the air and blood is particularly effective.

The Pseudotracheae of Terrestrial Isopods The specialized respiratory structures of isopod crustaceans are derived from the abdominal appendages (pleopods), each of which consists of two lamellar branches lying flat against the underside of the abdomen. Among terrestrial isopods as well as many marine forms, the inner (dorsal) branch of each abdominal appendage is delicate and vascular and acts as a gill, whereas the outer (ventral) branch serves as a gill cover, or operculum. A striking and novel feature of respiratory morphology in many terrestrial isopods is the development of invaginations of various types on the inner surfaces of the opercula (Figure 12.17). These ***pseudotracheae*** act as diffusion

lungs. Evidently, they are particularly important in dry air. The relatively exposed surfaces of the gills and the general integument may be unable to function adequately in respiration when the air is dry; thus, dry air can asphyxiate species that lack the invaginated respiratory surfaces of pseudotracheae.

The Book Gills of Horseshoe Crabs

The horseshoe crabs possess novel respiratory structures known as ***book gills***. As shown in Figure 12.18, the underside of the abdomen of a horseshoe crab

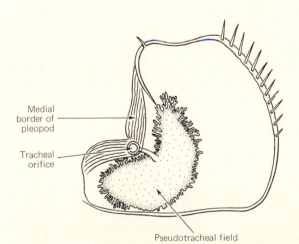

Figure 12.17 The outer branch (operculum or exopodite) of the first pleopod in the terrestrial isopod *Porcellio scaber*, showing the pseudotracheae. A branching, air-filled hollow within the operculum (pseudotracheal field) communicates with the outside through the tracheal orifice. The pseudotracheae are well supplied by the circulatory system and play an important role in respiratory gas exchange. [After a drawing by Verhoeff as represented in A. Vandel, *Faune de France*, Vol. 64, *Isopodes terrestres*. Fédération Française des Sociétés de Sciences Naturelles, Office Central de Faunistique, Paris, 1960.]

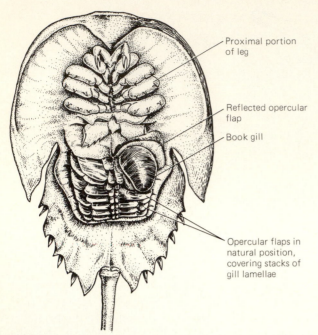

Proximal portion
of leg

Reflected opercular
flap

Book gill

Opercular flaps in
natural position,
covering stacks of
gill lamellae

Figure 12.18 A ventral view of a horseshoe crab (*Limulus poly-phemus*). There are five pairs of overlapping gill opercula, each operculum covering a book gill. A sixth pair of opercula lies anterior to the gill opercula; these are the genital opercula and do not bear gills. One gill operculum has been reflected anteriorly, spreading the numerous, very thin lamellae of the book gill out like an accordion. In life, the lamellae are stacked closely, like pages of a book, under the operculum. Only the proximal segments of the legs are shown. [From a drawing by Ralph Russell, Jr.]

bears six overlapping pairs of heavy, flaplike appendages, termed opercula. The underside of each operculum (excepting the members of the most anterior pair) gives rise to about 100 broad, thin gill lamellae, stacked on top of each other dorsoventrally. These lamellae look much like pages of a book, thus their name. The gills are ventilated by rhythmic flapping motions of the opercula.

Echinoderms

Starfish The starfish have a limited hemal circulatory system; most gas transport within the body of a starfish is accomplished by circulation of fluids in the perivisceral coelom and water vascular system. As shown in Figure 12.19, the water vascular system sends a tube, the radial canal, along the length of each arm. Attached laterally to the radial canal are many tube feet that project outward on the oral side of the animal. Associated with each tube foot is a muscular bulb, the ampulla, which by contracting can extend the tube foot hydraulically. The tube feet end in suckers with which they can grip the substrate, and their most obvious function is locomotion. In addition though, they are important sites of external respiration, for collectively they present a very large surface area to the environment, and gases

diffuse readily across their thin walls. The walls of the tube feet are ciliated internally, and these cilia circulate water vascular fluid between the tube feet and ampullae. Cilia or flagella on the external surfaces of the tube feet circulate ambient water across those surfaces. Convective movement of the internal and external fluids is also produced by ordinary locomotory activity. As shown in Figure 12.19, the ampullae and the canals of the water vascular system lie in the perivisceral coelom. Oxygen taken up across the tube feet probably enters the perivisceral coelomic fluid by diffusion across the thin walls of the water vascular system. The perivisceral coelomic fluid is circulated by the action of cilia on the coelomic walls, thus carrying oxygen throughout the animal and supplying the digestive organs, gonads, and other structures lying in the coelomic cavity.

The *branchial papulae* of starfish are fine, fingerlike evaginations of the coelomic wall found on the aboral and sometimes the oral surfaces of the animal. The papulae are ciliated internally and externally and, being thin-walled, provide sites for direct exchange of gases between the coelomic fluid and external medium (Figure 12.19). As their name suggests, they play a significant role in external respiration, but the tube feet play at least as great a role, if not greater, in many species.

The respiratory system of starfish seems simple by comparison to that of many of the other relatively large invertebrates. The oxygen exchange requirements of these sluggish animals are not great, however.

Holothuroideans Various respiratory modifications occur in the groups of echinoderms other than starfish, but here we look at just one other group, the holothuroideans or sea cucumbers, because in them a truly unusual respiratory apparatus makes its full appearance. The water vascular system of sea cucumbers, including the anterior tentacles developed from it, probably still plays a significant respiratory role. The remarkable feature of most sea cucumbers is their possession of elaborate, tidally ventilated *water lungs,* which account for half of oxygen uptake in at least some species. These lungs are termed *respiratory trees* (Figure 12.20) and arise as invaginations of the lower gut, or cloaca. They extend far up into the coelomic cavity and branch into many fine tubules. The trees are filled with water by a series of cloacal contractions: The anal sphincter closes, and water in the cloaca is driven up into the trees; the sphincter then opens, the cloaca dilates and refills from the environment, the sphincter closes, and water is again propelled into the trees. Six to 10 such cycles are required for one inhalation. The trees are then emptied in a single step. Oxygen diffuses across the walls of the trees into the perivisceral coelom where it is carried about in the currents of coelomic fluid. The hemal circulatory system may also play a role in distributing oxygen from the trees.

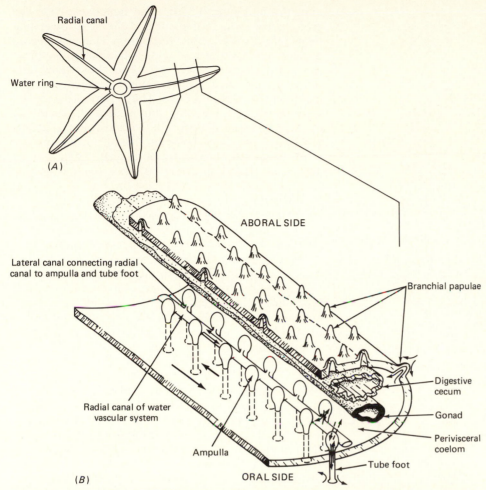

Radial canal

Water ring

(A)

ABORAL SIDE

Lateral canal connecting radial
canal to ampulla and tube foot

Branchial papulae

Radial canal of water
vascular system

Digestive
cecum

Gonad

Perivisceral
coelom

Ampulla

Tube foot

(B)

ORAL SIDE

Figure 12.19 Semidiagrammatic representation of major structures involved in gas exchange in a starfish. (A) The general plan of the water vascular system. A circular canal, the water ring, sends out a radial canal along the length of each arm. The water ring and radial canals are situated on the oral side of the body cavity. (B) Diagram of part of an arm, with the aboral–lateral integument cut away on one side. The ampullae lie in the perivisceral coelom and connect with tube feet that project through the integument on the oral side. Each ampulla connects with the radial canal of the arm through a valved lateral canal. Two digestive (pyloric) ceca and two gonadal branches run along each arm in the perivisceral coelom; only one of each of these is shown. The branchial papulae are thin-walled evaginations of the perivisceral coelomic wall and appear externally as minute, fingerlike projections. Solid arrows show movements of ambient water, water vascular fluid, and coelomic fluid. Dashed arrows indicate diffusion of gases between the ampullar fluid and perivisceral coelomic fluid.

Why Are Water Lungs Rare? The elaborate respiratory trees of sea cucumbers are remarkable because tidally ventilated lungs of any sort are rare among water breathers. In rationalizing that rarity, a key concept is that water is relatively dense and viscous, and thus it is relatively expensive energetically to set in motion. Whereas unidirectional ventilation requires just a single acceleration of each volume of medium passed over the respiratory surfaces, tidal ventilation requires two accelerations: one to inhale and one to exhale. Tidal ventilation thus compounds the energetic problems presented by water's density and viscosity, and it increases the cost of bringing a volume of water into contact with the respiratory surfaces. Tidal ventilation probably also increases the time required to exchange a volume of water. Add to these considerations that the oxygen available from a volume of water is relatively low, and the case against tidal ventilation in water breathers becomes apparent (see also p. 279). Why sea cucumbers have evolved such ventilation we do not know. Air is a far more favorable medium for tidal ventilation because it can be moved back and forth rapidly at relatively low metabolic cost.

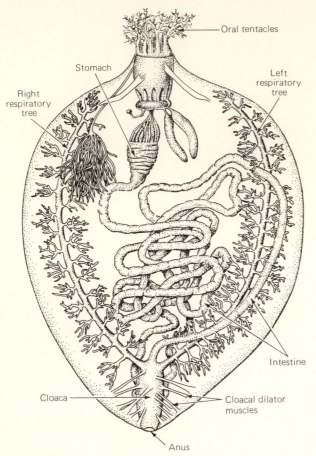

Oral tentacles

Stomach

Left respiratory tree

Right respiratory tree

Intestine

Cloaca

Cloacal dilator muscles

Anus

Figure 12.20 Internal anatomy of a sea cucumber (*Thyone*), viewed from the ventral side. The two respiratory trees, which actually branch more extensively than shown, arise from the anterior end of the cloaca and lie within the coelomic cavity. [After W.R. Coe, *Echinoderms of Connecticut. Connecticut State Geol. Nat. Hist. Surv. Bull.* **19**:1–152 (1912). Used by permission of the Connecticut State Geological and Natural History Survey.]

EXTERNAL RESPIRATION IN FISH

Gill Breathing in Teleosts

The buccal cavity of teleost fish communicates with the environment not only via the mouth but by lateral pharyngeal openings, the gill slits. The gills are arrayed across these lateral openings and are covered by protective external flaps, the opercula. The structure of the gills is illustrated in Figure 12.21. On each side of the fish are four branchial arches that run dorsoventrally between the gill slits. Each arch bears two rows of **gill filaments** splayed out laterally in a V-shaped arrangement; collectively, the filaments form a corrugated array separating the buccal cavity on the inside from the opercular cavity on the outside (Figure 12.21*A*). Each filament bears a series of folds, called **secondary lamellae,** on its upper and lower surfaces (Figure 12.21 *B*); the lamellae run perpendicular to the long axis of the filament and number 10–40 per *millimeter* of filament length on each side. As

depicted in Figure 12.21*C,* the lamellae divide the space between one filament and the next lower filament into rows of elongated pores. The entire array of filaments and their secondary lamellae thus forms a sievelike arrangement between the buccal and opercular cavities. The secondary lamellae are the major sites of gas exchange. They are richly perfused with blood and are very thin walled; in them the average distance between blood and water is generally only 0.6–6 μm. Probably, the gills of fish are the most elaborately organized of all gills. Their surface area is estimated to be 10–60 times that of the rest of the body.

Water flows through the gill sieve from the buccal to the opercular side. Blood flows through the secondary lamellae in the opposite direction. Thus, as shown in Figure 12.21*D,* there is countercurrent exchange along the respiratory exchange membranes. Traditionally, this has been viewed as an adaptation to water breathing: By permitting extraction of a relatively high percentage of the oxygen from water, countercurrent exchange could help to compensate for the low oxygen concentration of water. There is some solid evidence for this line of thought. Nonetheless, other recent studies have determined that in some instances, under actual gas exchange conditions, the advantage of countercurrent exchange over concurrent exchange in teleosts is much less than theoretically possible and sometimes insignificant. The role of countercurrent exchange thus needs further investigation. The percentage of oxygen extracted from ventilated water by resting teleosts has been reported to range from very high values of 80–85 percent down to 30 percent or less.

Active fish tend to have more gill surface area per unit of body weight than sluggish fish. They tend to have thinner lamellae, more closely spaced lamellae, and a shorter diffusion distance between the blood and water across the lamellar membranes. These features are conducive to meeting the greater gas exchange requirements of an active way of life.

Recent studies have revealed that in some teleosts the skin can account for up to 30 percent of gas exchange in well-aerated water.

Ventilatory Mechanics: Buccal–Opercular Pumping
In general, water flow across the gills of teleosts is maintained almost without interruption by the synchronization of two pumps: a **buccal pressure pump,** which forces water from the buccal cavity through the gills into the opercular cavity, and an **opercular suction pump,** which sucks water from the buccal cavity into the opercular cavity. The relative dominance of the two pumps varies from species to species. Here, we take a generalized view. We look first at the action of each pump separately and then at the integration of the pumps over the respiratory cycle. It will be important to remember throughout that water flows from regions of higher pressure to ones of lower pressure.

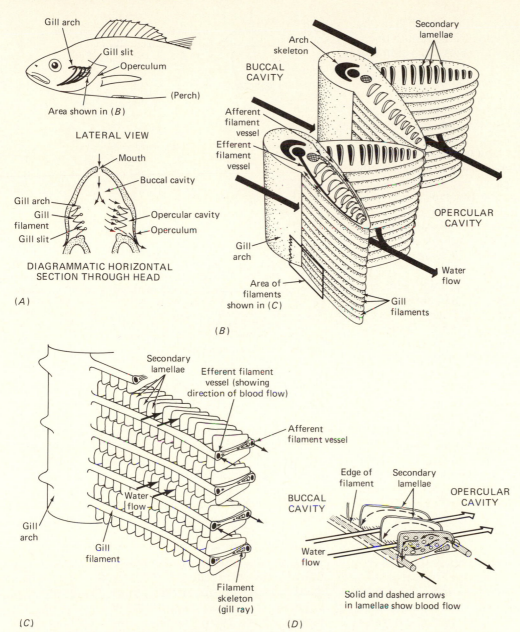

Figure 12.21 Major features of the branchial respiratory system in teleost fish. (*A*) The general arrangement of the gills. The lateral view shows the orientation of the gill arches under the operculum. The horizontal section shows the orientation of the buccal cavity, gill filaments, opercular cavity, and operculum. Arrows indicate direction of water flow. (*B*) Enlarged, diagrammatic view of segments of two gill arches. Note the two rows of gill filaments on each arch and the secondary lamellae that project dorsally (and ventrally) from each gill filament. Heavy arrows show direction of water flow. Light arrows within the upper left gill filament show direction of blood flow within the filament. (*C*) An enlarged and more-realistic view of four filaments and their secondary lamellae, showing that the lamellae on the upper and lower surfaces of the filaments divide the spaces between adjacent filaments into arrays of fine pores like a sieve, through which water is pumped. In tench, for example, both the dorsal and ventral filament surfaces bear over 20 lamellae per *millimeter* of filament length, assuring intimate contact between the water current and respiratory surfaces. Small arrows next to blood vessels show direction of blood flow in longitudinal filament vessels; note that blood returns toward the body along the edge of the filament facing the respiratory water current. (*D*) Enlarged view of a filament and three lamellae, showing that blood flow in the secondary lamellae is counter to water flow across the lamellae. The foremost lamella has been sectioned to show that blood flows in a sheet through the lamella. [Parts (*A*) and (*B*) after J.H. Bijtel, *Arch. Neer. Zool.* **8**:267–288 (1951); parts (*C*) and (*D*) after M. Morgan and P.W.A. Tovell, *Z. Zellforsch. Mikroskop. Anat.* **142**:147–162 (1973).]

The buccal cavity is filled with water when the floor of the cavity is depressed with the mouth open. The depression of the floor increases the volume of the cavity, thus decreasing buccal pressure below ambient pressure and resulting in influx of water. The mouth is then closed and the floor of the cavity raised. This increases buccal pressure and drives water from the buccal cavity through the gills into the opercular cavities. Thin flaps of tissue, which act as passive valves, project across the inside of the oral opening from the upper and lower jaws. During the refilling phase of the buccal cycle, when buccal pressure is below ambient, these valves are pushed inward and open by the influx of water through the mouth. During the positive-pressure phase, however, the valves are forced against the oral opening on the inside and help to prevent reflux of water from the buccal cavity through the mouth.

The opercular cavities can be expanded and contracted by lateral movements of the opercula and other muscular activities. Running around the rim of each operculum is a thin sheet of tissue that acts as a passive valve, capable of sealing the slitlike opening between the opercular cavity and the ambient water. When the opercular cavity is expanded, the pressure in the cavity falls below the pressures in the buccal cavity and ambient water. Water is thus sucked into the opercular cavity from the buccal cavity through the gill sieve and would be sucked in readily from the environment were it not for the action of the rim valve; because the pressure in the opercular cavity is lower than ambient during the sucking phase, the rim valve is pulled medially against the body wall at that time, substantially sealing the opercular opening and preventing influx of ambient water. After the sucking phase, the opercular pump enters its discharge phase. Contraction of the opercular cavity raises the opercular pressure above ambient pressure, forcing the rim valve open and discharging water from the opercular cavity through the opercular opening.

The temporal integration of the buccal and opercular pumps is illustrated schematically in Figure 12.22. In step (A) the buccal cavity is being refilled. Expansion of the cavity produces a pressure below ambient; if the buccal pump were the only pump, flow through the gills from the buccal side would not occur at this point, and, in fact, there would be backflow through the gills into the buccal cavity because of the lowered pressure in the cavity. It is at this time, however, that the opercular pump is in its sucking phase. Opercular pressure is reduced well below buccal pressure, and water is drawn through the gills from the buccal cavity. Step (B) is a short transition stage in which the opercular pump is completing its sucking phase and the buccal pump is beginning its pressure phase. In step (C), the opercular pump is in its discharge phase. Pressure is elevated in the opercular cavity, but because the buccal pump is in its pressure phase, buccal pressure is elevated to an

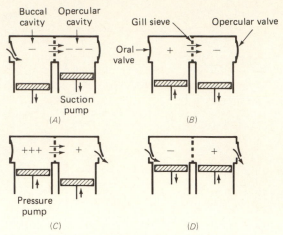

Figure 12.22 Schematic diagram of phases of the respiratory cycle in teleost fish. Plus (+) and minus (−) symbols indicate pressures in the buccal and opercular cavities relative to *ambient* pressure. The buccal and opercular pumps are represented by pistons. The gill sieve is interposed between the buccal and opercular cavities; arrows through the gill sieve indicate water flow. Phases (B) and (D) are transitional and short in duration. See text for discussion. [Reprinted by permission from G. M. Hughes, *New Sci.* **11**(247):346–348 (1961).]

even greater degree, and water again flows through the gills from the buccal cavity. Only in step (D), which occupies just a short part of the respiratory cycle, is the pressure gradient such as to cause backflow of water through the gills. In all, the two pumps are beautifully integrated to produce almost continuous, unidirectional flow across the gills: The opercular pump sucks while the buccal pump is being refilled, and the buccal pump develops positive pressure while the opercular pump is being emptied.

Ventilatory Mechanics: Ram Ventilation When swimming fish attain speeds of 50–80 cm/s, their motion through the water, if they open their mouths, can itself elevate the buccal pressure sufficiently to ventilate the gills at rates adequate to meet oxygen requirements. Many fast-swimming fish, in fact, cease buccal–opercular pumping when they reach such speeds and, holding their mouths open, allow their motion to ventilate their gills, a process known as *ram ventilation*. Such ventilation transfers the muscular effort of ventilation from the buccal and opercular muscles to the swimming muscles and conceivably reduces the cost of ventilation. Tunas, mackerel, and bonito are active fish that swim continuously and use ram ventilation all the time; they have long been claimed in fact to be obligate ram ventilators, but recent experiments have brought the claim under dispute.

Ventilatory Control As a rule, resting teleosts markedly increase their rate of ventilation by buccal–opercular pumping if subjected to decreases in ambient oxygen tension. Sometimes, they increase ventilation in response to increases in the ambient tension of carbon dioxide as well. But overall, whereas lowered oxygen is a potent ventilatory stimulus for teleosts, elevated carbon dioxide is a weak one. The reasons for this pattern of ventilatory control attract interest, particularly when we realize that in mammals and many other air breathers, elevated blood carbon dioxide is an extremely potent ventilatory stimulus (p. 284). Sensitivity to ambient oxygen tension is not difficult to understand; lowered oxygen tensions are common in bodies of water, and increasing the flow of water across the gills obviously can help compensate for a reduction in the oxygen concentration of the water. As to carbon dioxide, a key consideration is that water breathers *never* raise the tension of carbon dioxide by very much in water passing over their gills, provided the rate of ventilation is adequate to meet oxygen needs (p. 249). Thus, at any given ambient carbon dioxide tension—whether high or low—increases or decreases in ventilatory rate will not greatly affect the carbon dioxide tension to which the gills are exposed.

Ventilation by buccal–opercular pumping is also modulated in response to exercise. The rate of ventilation is increased as swimming speed increases.

Besides adjusting their ventilation rate, fish may also exert control over their branchial oxygen exchange by modulating the ease with which oxygen can traverse their gill membranes to enter their blood. In rainbow trout (*Salmo gairdneri*), for example, the ease of oxygen transfer is modulated by adjusting the proportion of secondary lamellae that are actively perfused with blood. Only about 60 percent of the lamellae are perfused at rest, but the proportion increases during exposure to low ambient oxygen tensions and probably during exercise.

Air Breathing in Fish

Low oxygen tensions prevail commonly in sluggish bodies of fresh water and sometimes in sluggish coastal marine waters. Many species of bony fish that live in such habitats have evolved mechanisms of tapping the rich oxygen resources of the air. For the most part, the air-breathing fish retain functional gills and, with respect to oxygen acquisition, are in fact dual breathers. The extent to which they rely on the air depends on a number of factors. They increase their use of air breathing as the oxygen tension in their aquatic habitat falls. They also tend to resort increasingly to air breathing as the temperature rises, because high temperatures elevate their oxygen needs. Notably, air-breathing fish typically void most of their carbon dioxide into the water—across their gills or skin—even when relying on the air for most of their oxygen.

Some fish that exploit the air lack marked anatomical specializations for doing so. American eels (*Anguilla rostrata*) provide a case in point. At low air temperatures, when their oxygen demands are low, eels sometimes come out onto land in moist situa-

tions. They then meet about 60 percent of their oxygen requirement by uptake across their skin and 40 percent by buccal air gulping. Their gills, which are of quite ordinary structure, are probably the primary site of oxygen uptake from the air they gulp. American eels do not ventilate their swimbladder, but the gas in their swimbladder is rich in oxygen, and when they first emerge onto land, they meet some of their oxygen requirement from that source; after a few hours, however, they exhaust the supply and become fully dependent on the atmosphere.

In most air-breathing fish, some part or branch of the alimentary canal has become specialized as an air-breathing organ. Such specialization always includes the development of a high degree of vascularization and may include evagination or invagination of the walls of the structure. Sometimes, the buccal cavity has become adapted to air breathing, as in electric eels (*Electrophorus electricus*), which have the walls of the cavity (and pharynx) thrown into highly vascular papillae. Or the opercular cavity may be involved; in mudskippers (*Periophthalmus*), for example, the inner walls of the opercula and adjacent parts of the gill chambers are vascularized and folded. In some catfish (*Saccobranchus*), diverticula of the gill chambers that are employed in aerial respiration extend all the way to the tail region. Sometimes, the stomach or intestines are used for air breathing, as in armored catfish (*Plecostomus*). The swimbladder is used by many fish; exceedingly elaborate patterns of folding of the swimbladder wall may then be present.

The air-breathing organ is nearly always inflated with air by buccal pumping. After air has been taken into the buccal cavity, the mouth is closed and the cavity compressed, thus driving the air into the stomach, swimbladder, or other relevant structure.

A potential problem for air-breathing fish is that oxygen taken up from the air may be *lost* to oxygen-poor waters across their gills. This probably helps explain why the gills of these animals are often reduced in their gas exchange capability in comparison to those of other fish (e.g., the gill surface area may be reduced). In extreme cases, the gills have become so atrophied as to make air breathing obligatory; this is the case in electric eels, for instance. Air-breathing fish have often also evolved circulatory shunts that enable some blood to bypass the gas exchange surfaces of their gills as it flows from the heart to the rest of the body. Modifications of the gills and circulation that impair oxygen loss to the water simultaneously impair carbon dioxide loss; this is an important consideration since carbon dioxide is generally voided into the water, not the air. Air-breathing fish often have elevated blood tensions of carbon dioxide in comparison to other fish.

The Lungfish The three genera of lungfish (dipnoans) are of particular interest because their air-breathing organs, which are enlarged diverticula of the pharynx, are believed to be homologous to the lungs of higher vertebrates. These diverticula are indeed called *lungs* in the restrictive sense of indicating this homology [according to the general definition of a lung (i.e., a lung is an invaginated respiratory structure), the diverticula are obviously lungs, as, for example, are swimbladders used for aerial respiration].

As shown in Figure 12.23, the walls of dipnoan lungs are thrown into a complex pattern of ridges and septa, not unlike the lining of many amphibian lungs. The Australian lungfish (*Neoceratodus*) is a dual breather with a single lung and well-developed gills. The African and South American genera (*Protopterus* and *Lepidosiren*), by contrast, have bilobed lungs and in general have much-reduced gills lacking secondary lamellae; they are obligate air breathers and drown if denied air.

EXTERNAL RESPIRATION IN AMPHIBIANS

Gills: Anatomy and Ventilation

The gills of aquatic amphibian larvae (tadpoles) are of different origin and structure than the gills of adult fish. They develop as outgrowths of the integument of the pharyngeal region and project into the medium from the body wall (Figure 12.24). The gills are ex-

ternal in all young amphibian larvae; in salamander larvae, they remain so, but in the frogs and toads (Anura) an outgrowth of the integument, termed the operculum, soon encloses the gills in a chamber that opens to the outside posteriorly, usually via a single aperture. The gills are generally lost at metamorphosis, but external gills remain throughout life in certain aquatic salamanders, such as the mudpuppy (*Necturus*).

Mudpuppies ventilate their gills by waving them back and forth in the water; the frequency of these movements is increased with declining oxygen tension and increasing temperature. Muscular movements of the gills are also reported in many amphibian larvae that have external gills. In anuran larvae having the gills enclosed in an operculum, ventilation is accomplished by buccal pumping. Water is taken in through the mouth and nares and driven back through the pharyngeal gill slits into the opercular cavity containing the gills. It then exits via the opercular aperture.

Lungs: Anatomy and Ventilation

Paired lungs develop from the ventral wall of the pharynx in most amphibians near the time of metamorphosis. The lungs of many adult amphibians are simple, well-vascularized sacs; their internal surface

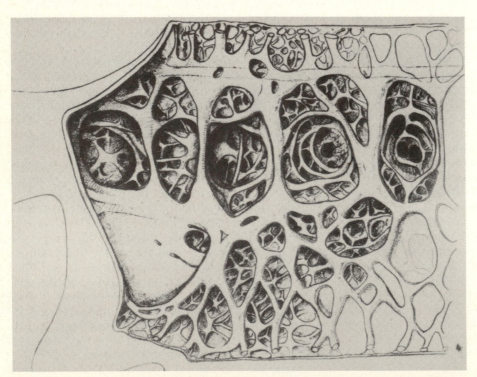

Figure 12.23 The ventral internal surface of the anterior portion of a lung of *Protopterus aethiopicus*. Respiratory surface area is greatly enhanced by a complex pattern of vascularized ridges and septa. Compartmentalization in other parts of the lung is similar but less elaborate. The side compartments in the wall of the lung open to a central cavity that runs the length of the lung and communicates anteriorly with a short pulmonary canal leading to the esophagus. [From M. Poll, *Ann. Mus. R. Afr. Centr. (Ser. 8)* **108**:129–172 (1962).]

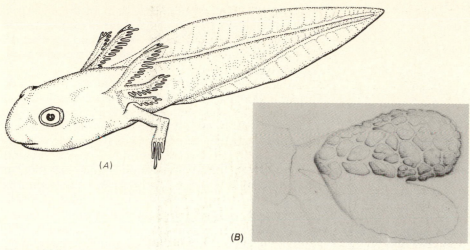

Figure 12.24 Respiratory structures of amphibians. (*A*) A 3-week-old larva of the salamander *Ambystoma maculatum,* showing the external gills. (*B*) A lung of the frog *Rana temporaria,* showing the compartmentalization of the wall by ridges and septa. The lung has been sectioned and the dorsal aspect reflected to reveal the inner surface of the ventral wall. [Part (*A*) from a drawing by Ralph Russell, Jr.; part (*B*) from M. Poll, *Ann. Mus. R. Afr. Centr.* (*Ser. 8*) **108:**129–172 (1962).]

area is increased little, if at all, by folding, and in this respect the lungs are less well developed than those of the modern lungfish. Particularly among frogs and toads, the walls of the lungs may be more elaborate; they may be thrown into a complex pattern of folds and septa, giving them something of a honeycombed appearance (Figure 12.24). Still, the lungs retain their basic saclike form. The central cavity of each lung remains open and provides access to the various side compartments formed by the folding of the walls.

It is instructive to compare the actual vascularized surface area in a lung with the area that would be provided by complete vascularization of the walls of a simple sac of the same gross dimensions as the lung. The ratio of these two quantities is less than 1 in amphibians with simple, unfolded lungs because the entire internal surface area is not vascularized. In the frog *Rana esculenta,* which has a highly divided lung, the ratio is about 8—illustrating the extent to which folding can increase the surface for respiratory exchange. The highest surface areas observed in amphibians are much lower, however, than those attained in mammals for equivalent lung volume. People have about 15 times the respiratory surface per unit of lung volume as *R. esculenta.*

Amphibians fill their lungs by buccopharyngeal pressure. This basic mechanism is presumably carried over from their piscine ancestors and, as mentioned earlier, is often employed by amphibian larvae to ventilate their gills. Most studies of pulmonary ventilation have been performed on frogs, and though several patterns differing in detail have been reported, the essentials of the buccopharyngeal pressure pump are quite uniform. Air is taken into the buccal cavity through the nares or mouth when the pressure in the cavity is reduced by lowering the floor of the cavity. When the floor of the cavity is then raised with the mouth closed and the nares at least partially sealed by valves, the increase in pressure forces air down into the lungs. This stretches the lungs and elevates pulmonary pressure. The lungs would discharge upon opening of the mouth or nares were it not for the fact that the glottis, the slitlike opening of the lung passage into the pharynx, is closable. The glottis is closed by muscular contraction after inhalation. The nares are then opened, and the animal often pumps air in and out of its buccal cavity through the nares by lowering and raising the floor of the cavity. This so-called *buccopharyngeal pumping* is easily observed in common frogs and is to be distinguished from pulmonary ventilation. After a period of time, the glottis is opened, and air from the lungs is exhaled. Exhalation results in part from the elastic recoil of the expanded lungs and to that extent is passive, that is, does not involve muscular contraction. It may also involve contraction of muscles in the walls of the lungs and body wall.

As noted earlier, differences in detail in the sequence of ventilatory events have been reported. Observations on the bullfrog (*Rana catesbeiana*) illustrate one possibility (Figure 12.25). In this species, filling of the buccal cavity in preparation for inflation of the lungs occurs before pulmonary exhalation. With the glottis closed, air is drawn into the buccal cavity through the nares (step 1, Figure 12.25). Much of this air comes to lie in a posterior depression of the buccal floor, situated ventrally to the opening of the glottis. Next, the glottis is opened, and pulmo-

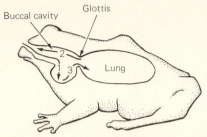

Figure 12.25 Diagrammatic representation of a bullfrog, showing three major steps in the pulmonary ventilatory cycle. See text for explanation. [From C. Gans, *Evolution* **24**:723–734 (1970).]

nary exhalant air passes in a coherent stream across the dorsal part of the buccopharyngeal cavity to exit through the nares (step 2). There is little admixture of the exhalant air with the fresh air in the depression of the buccal floor. This fresh air is then driven into the lungs when, in step 3, the buccal floor is raised with the nares closed. After inflation of the lungs, the frog commences buccopharyngeal pumping with the glottis closed; one important effect of this pumping is that it washes out residual pulmonary exhalant air from the buccal cavity, so that when the next pulmonary ventilatory cycle begins, the cavity is filled with a relatively fresh mixture.

The Relative Roles of Respiratory Sites

Collectively, amphibians present us with four gas exchange sites of potential importance: lungs, gills, the buccal cavity, and the skin. Usually, at least three of these are present in a given individual. One of the central questions in amphibian respiratory physiology is how total gas exchange is partitioned among the available gas exchange sites. In general, the buccal cavity makes just a relatively minor contribution to overall exchange. Thus, the major sites are the lungs, gills, and skin. Besides varying with species, the partitioning of exchange among sites can depend on the animal's level of activity, body size, developmental stage, and other factors.

If we consider adult frogs, toads, and lunged salamanders while they are at rest on land, we find that the relative contributions of their lungs and skin to oxygen uptake depend strongly on temperature. At 5°C, these amphibians, on average, acquire about two-thirds of their oxygen across their skin. As the temperature rises and their total oxygen requirements increase, however, their lungs assume more and more responsibility for oxygen uptake, as evidenced by increases in the amplitude and frequency of pulmonary ventilatory cycles. At 25°C, roughly two-thirds of oxygen uptake is across the lungs, on average. Carbon dioxide, in contrast to oxygen, is lost principally across the skin whatever the temperature; at 5 or 25°C, about 75–80 percent of carbon dioxide efflux is cutaneous. There is an important

similarity here to air-breathing fish. In them also, it will be recalled, the specialized air-breathing organ typically plays only a minor role in carbon dioxide elimination even when functioning as the dominant site of oxygen uptake.

In winter, frogs often hibernate at the bottoms of ponds and lakes. Their exchange of oxygen and carbon dioxide is then believed to be entirely cutaneous. A factor that is important in permitting cutaneous exchange with the water to meet the entire respiratory requirement for long periods is the depressed metabolic rate of the hibernating animal; metabolism is depressed in part because of the low temperatures and also, in at least some species, because of a seasonal effect. In some frogs, the ability to meet resting gas exchange requirements cutaneously while submerged is not limited just to very low temperatures; *Rana esculenta*, for example, can survive for 2 or 3 weeks while submerged at 15°C, provided the water is well aerated. We emphasized in Chapter 8 that the high permeability of the skin of frogs presents problems of rapid evaporative water loss on land and rapid passive exchange of salts and water during submergence. In terms of gas exchange, we see here, however, that the permeability and vascularity of the skin have advantages.

An interesting topic that has only recently been approached quantitatively is the ontogeny of gas exchange partitioning in amphibians that metamorphose from aquatic, gill-bearing tadpoles to amphibious, lunged adults. Figure 12.26 shows the course of events in bullfrogs. During metamorphosis, the importance of both the skin and gills in oxygen uptake declines, as the lungs take over most responsibility for oxygen acquisition. In accord with what we have said earlier, however, the lungs never assume more than a minor role in carbon dioxide elimination; rather, as the gills wane, the role of the skin increases.

EXTERNAL RESPIRATION IN REPTILES

There are reptiles such as soft-shelled turtles (*Trionyx*) that can breathe without use of their lungs under certain circumstances (p. 50). However, for most modern reptiles, the lungs are the virtually exclusive site of gas exchange. In this majority of reptiles, a significant development is that the lungs bear full responsibility for elimination of carbon dioxide as well as uptake of oxygen; the skin, an important site of carbon dioxide elimination in amphibians, has become poorly permeable not only to water but also to the respiratory gases.

Lung Morphology

The simplest type of reptilian lung—seen in many lizards and snakes—resembles the amphibian lung in being a saccular structure with an open central cavity (Figure 12.27A); this type of lung is termed unicam-

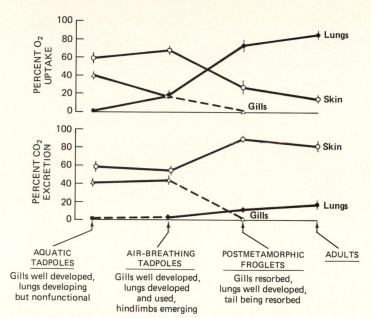

Figure 12.26 The percentages of oxygen uptake and carbon dioxide excretion occurring across lungs, gills, and skin in bullfrogs (*Rana catesbeiana*) at four stages of development. All animals were immersed in well-aerated water and had access to air during study; the temperature was 20°C. If animals at the "air-breathing tadpole" stage are placed in water of reduced oxygen tension, they increase reliance on the lungs to obtain oxygen and decrease ventilation of their gills; the latter response helps assure that oxygen acquired by way of the lungs from the air will not be lost across the gills into the oxygen-poor water. [After W.W. Burggren and N.H. West, *Respir. Physiol.* **47**:151–164 (1982).]

eral ("one-chambered"). The walls of the lung are thrown into a honeycomblike pattern of vascularized partitions, greatly increasing their surface area (Figure 12.27*B*). Air flows in and out of the central cavity, but gas exchange between the cavity and the depths of the honeycomblike cells on the walls is probably largely by diffusion.

A major development observed in several groups of reptiles is the subdivision of the main lung cavity into numerous smaller cavities, forming a so-called multicameral lung. As shown in Figure 12.27*C*, this type of lung occurs in monitor lizards, animals noted for their active way of life and relatively high aerobic competence (p. 49). It is also found in crocodilians and turtles. One noteworthy development in the multicameral lung is the appearance of a cartilage-reinforced tube (bronchus) running lengthwise *within* the lung (Figure 12.27*C*); such an arrangement is necessary to provide air flow to all the chambers. Clearly, the multicameral type of lung can provide for a great deal more surface area per unit of lung volume than the unicameral type because the septa between chambers as well as the outer walls can develop elaborate gas exchange surfaces.

Lung Ventilation

In reptiles we find a major transition in the mode of ventilation that is carried over into the birds and mammals: The lungs are filled by suction rather than by buccal pressure. That is, air is drawn into the lungs by an expansion of pulmonary volume that creates a subatmospheric pressure within the lungs.

As will be discussed in more detail subsequently (p. 281), the lungs elastically assume a certain volume, termed their relaxation volume, if there are no active muscular forces tending to expand or contract them.

In some reptiles, such as lizards and crocodilians, the lungs are expanded beyond their relaxation volume by muscular activity during inspiration, and during expiration they contract at least in part under forces of elastic rebound. In other reptiles, such as at least some snakes, the lungs are compressed below their relaxation volume by muscular activity during expiration and expand elastically on inspiration. In either case, the basic process of inspiration is suction. Responsibility for ventilating the lungs is transferred from muscles of the buccal cavity to muscles of the thorax and abdomen. An important consequence is that the buccal cavity—in both form and function—is freed from an important set of constraints and permitted to become more exclusively specialized for feeding and other nonrespiratory activities.

The mechanics of ventilation vary considerably from one group of reptiles to another. Here, we look just at the example of lizards. Running over and between the ribs on each side of the body are sheets of muscles, which can expand or contract the volume enclosed by the rib cage. These muscles, termed ***costal muscles*** (*costa* = "rib"), are believed to play a central role in lizard ventilation. During inhalation, the glottis is opened, and the lizard's lungs are filled by expansion of the thoracic cavity. Certain costal muscles are thought to expand the rib cage at this time, concomitantly expanding the lungs. This mechanism has been referred to as a *costal suction pump*. After inflation of the lungs, the glottis is closed, and the inspiratory muscles relax. During the ensuing respiratory pause, the buccal cavity is often ventilated by buccopharyngeal pumping. After several seconds to several minutes, expiration occurs, followed quickly by another inspiration. Both passive and active forces are involved in expiration. The

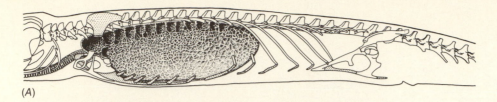

(A)

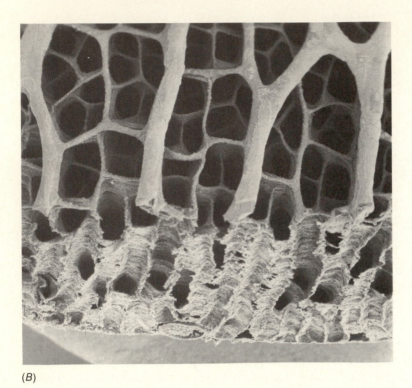

(B)

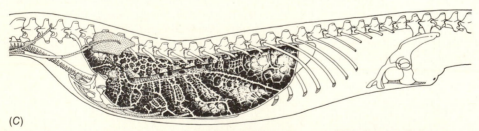

(C)

Figure 12.27   (*A*) A longitudinal section of a green lizard (*Lacerta viridis*) showing the gross internal structure of a unicameral lung. (*B*) A scanning electron micrograph of the lung wall of a tegu lizard (*Tupinambis nigropunctatus*) showing the honeycomblike pattern of vascularized partitions. Magnification: 35×. (*C*) A longitudinal section of a monitor lizard (*Varanus exanthematicus*) showing the gross internal structure of a multicameral lung. [Parts (*A*) and (*C*) reprinted by permission from H.-R. Duncker, In J. Piiper (ed.), *Respiratory Function in Birds, Adult and Embryonic,* pp. 2–15. Springer-Verlag, New York, 1978; (*B*) kindly provided by Daniel Luchtel and Michael Hlastala.]

glottis is opened, elastic forces drive air out, and, in addition, contraction of thoracic and abdominal muscles compresses the thoracic cavity. Certain costal muscles are active in this process.

The overall pattern of breathing in lizards is one of pulmonary ventilatory cycles, performed with the glottis open, separated by periods of *apnea* (no breathing) during which the glottis is closed and the lungs are filled. This pattern, known as *intermittent* or *periodic breathing,* is seen also in other reptiles as well as amphibians and air-breathing fish. The buccopharyngeal pumping often observed in reptiles during the apneic phase is probably, in general, an aid to olfaction.

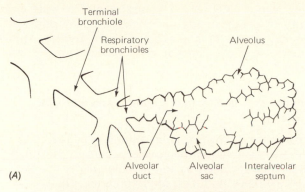

(A)

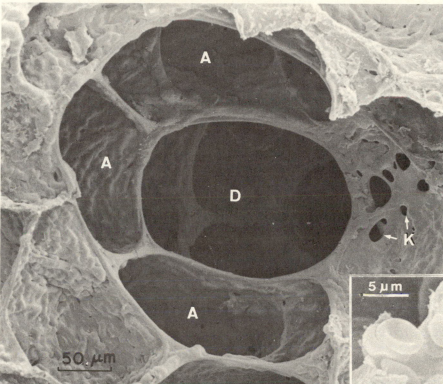

(B)

Figure 12.28 (A) Diagrammatic cross section of the terminal airways in the mammalian lung. Terminal bronchioles, having diameters of 0.5 mm or less in humans, represent the final branches of the purely conducting (nonrespiratory) bronchial–bronchiolar tree (see discussion of the anatomical dead space in the text). Each terminal bronchiole branches into two or more respiratory bronchioles. Respiratory bronchioles lead to alveolar ducts, which in turn terminate in alveolar sacs. The alveoli are minute outpocketings of the alveolar ducts and sacs. Alveoli first appear along the respiratory bronchioles and form a continuous lining along the alveolar ducts and alveolar sacs. Adjacent alveoli are separated by thin interalveolar septa. The walls of the alveoli are richly invested with capillaries. (B) A scanning electron micrograph of an alveolar duct (D) lined with alveoli (A). Inset shows red blood cells at about 10-fold greater magnification. K, pore of Kohn. [Part (A) from J. Hildebrandt and A.C. Young, Anatomy and physics of respiration. In T.C. Ruch and H.D. Patton (eds.), *Physiology and Biophysics,* 19th ed. Saunders, Philadelphia, 1965; (B) reprinted from E.R. Weibel, in S.C. Wood and C. Lenfant (eds.), *Evolution of Respiratory Processes,* pp. 289–316. Dekker, New York, 1979. Reprinted by courtesy of Marcel Dekker, Inc.]

EXTERNAL RESPIRATION IN MAMMALS

Lung Morphology

The lungs attain an extreme degree of internal sub-division in mammals. The primary bronchus that en-ters each lung branches and rebranches dendritically within the lung, giving rise to secondary and higher-order bronchi of smaller and smaller diameter and then to ever-smaller fine tubes known as bronchioles. The final bronchioles end blindly in ***alveolar sacs*** (Figure 12.28). The bronchial–bronchiolar tree is so elaborate that in humans the alveolar clusters are separated from the trachea by an average of 23 branches. As shown in Figure 12.28, the walls of the alveolar sacs are composed of numerous semi-spherical outpocketings called *alveoli*. There are about 300 million alveoli in the lungs of human adults, each measuring about 0.25 mm across. The collective surface area of these alveoli is an astounding 120–140 m^2: about 60–70 times the total external body surface area.

The trachea and bronchi and all but the last few branches of bronchioles are not much involved in gas exchange; accordingly, they are known as the **conducting** (as opposed to *respiratory) airways,* and collectively they are said to comprise the **anatomical dead space.** They are lined with a relatively thick epithelium and do not receive a particularly rich vascular supply. The last few branches of bronchioles and the alveoli are lined with thin, highly flattened epithelial cells and are richly supplied with blood capillaries. It is in these parts of the lung that gas exchange occurs. The alveoli constitute most of the exchange surface. To move between air and blood, gases must diffuse across the alveolar epithelium, the capillary endothelium, and a basement membrane separating the two. These structures are all so thin (Figure 12.29) that their total average diffusive thickness is just 0.3–0.6 μm.

Functional Components of Lung Volume

In this and the next two sections, we look at a number of analytical concepts useful in understanding pulmonary function. Although we have postponed discussing these concepts until they could be considered in the context of mammalian respiration, it will be apparent that many of them have applicability to other vertebrate lungs.

Figure 12.30 depicts a useful conceptual subdivision of the total lung volume. The amount of air inhaled and exhaled per breath is known, in general, as the **tidal volume** and, in resting individuals, as the **resting tidal volume.** The latter is about 500 mL in adult humans. As we are well aware, we neither exhale nor inhale maximally when at rest. The maximal amount of air that an individual can expel beyond the resting expiratory level is termed the resting **expiratory reserve volume.** On the other hand, the maximal amount that can be inhaled beyond the resting inspiratory level is the resting **inspiratory reserve volume.** As shown in the figure, the average resting expiratory reserve in healthy young men is about 1200 mL; the average resting inspiratory reserve is about 3100 mL. Tidal volume, according to this conception, is increased above the resting level by using parts of the inspiratory and expiratory reserves. The maximal possible tidal volume, which is termed the **vital capacity,** is attained by fully using both reserves and thus is the sum of the resting tidal volume and the resting inspiratory and expiratory reserve vol-

umes: about 4800 mL in young men. The vital capacity of people tends to be increased by physical training, but advancing age and some diseases tend to decrease it. It is smaller in females than males, on average.

Patterns of Gas Flux in the Lungs; Alveolar Gas Tensions

Most of the change in lung volume over the course of the resting ventilatory cycle occurs in the *respiratory* airways such as the alveolar ducts and sacs (Figure 12.28). As a first approximation, the volume of the anatomical dead space may be considered constant. With these points in mind, let us consider the pattern of air flow in the lungs during resting ventilation, using quantitative values for adult humans.

The volume of the anatomical dead space in humans is about 170 mL. Therefore, when a person exhales while at rest, 170 mL of the 500 mL expired is air from the dead spaces. A bit of reflection will reveal that this 170 mL is in fact relatively *fresh* air. It is air that lodged in the dead spaces at the end of the previous inhalation; never drawn deeply enough into the lungs to reach the respiratory exchange membranes, it has suffered virtually no depletion of oxygen or enrichment with carbon dioxide. On the other hand, 330 mL of the 500 mL expired is gas emptied from the respiratory exchange parts of the lungs; this gas has a substantially lowered oxygen tension and elevated carbon dioxide tension. At the end of a resting exhalation, 2400 mL of gas remains in the lungs (Figure 12.30). Importantly, all this remaining gas has been in the respiratory exchange regions, although some of it—for the moment—is pushed up into the dead spaces. During the ensuing inhalation, the gas left in the dead spaces at the end of exhalation is the first to be drawn into the respiratory spaces. Following it into the respiratory spaces is 330 mL of fresh air. Finally, of the total of 500 mL of fresh air inhaled, 170 mL fills the dead spaces at the end of inhalation.

We see that the 330 mL of fresh air that reaches the respiratory exchange regions on each inhalation mixes there with a much larger volume—2400 mL—of gas that has already been in the respiratory spaces. On this basis alone, it is evident that oxygen tensions prevailing near the respiratory exchange membranes do not swing radically up and down as air is inhaled and exhaled. Rather, they remain relatively stable.

An additional important factor that stabilizes gas tensions at the respiratory exchange membranes is that fresh, incoming gases are drawn by mass flow (convection) only to the depth of the first alveolar ducts (or possibly the alveolar sacs). From that depth to the alveolar membranes, gas flux occurs largely or entirely by diffusion. Thus, streams of fresh air cannot possibly wash across the alveolar membranes. It is evident that the effective cessation of mass flow slightly upstream from the principal respi-

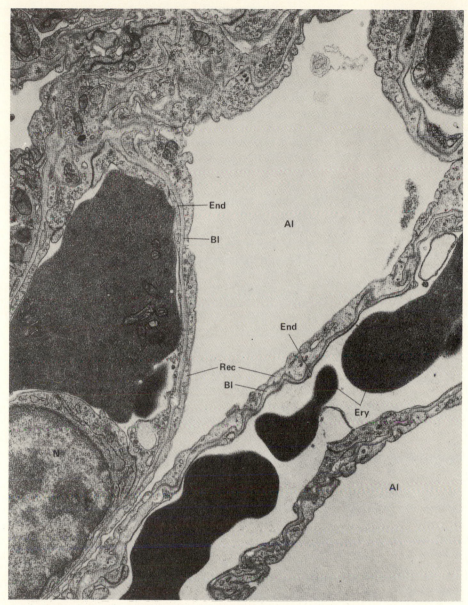

Figure 12.29 Electron micrograph of the lung of a mouse, magnified 32,000 times. Two alveoli (Al) are separated by an interalveolar septum containing a blood capillary in which erythrocytes (Ery) can be seen. The lumen of each alveolus is separated from the lumen of the capillary by (1) respiratory epithelial cells (Rec) lining the alveolus, (2) a basement membrane (Bl), and (3) capillary endothelial cells (End) lining the capillary. The capillary endothelial cells are very attenuated except in the region of the cell nucleus (N). The respiratory epithelial cells are similarly attenuated. (Electron micrograph kindly supplied by K.R. Porter and M.A. Bonneville.)

ratory exchange membranes is a logical outcome of having the exchange membranes positioned at the ends of tidally ventilated, blind-ended tubes. The distances over which oxygen and carbon dioxide must move by diffusion are short. Thus, given that the diffusion is occurring through a gaseous medium, the rates of diffusion pose no impediment to adequate oxygen and carbon dioxide exchange. If water were the medium filling the lung passages, however, rates of diffusion could impose a serious limitation on

overall rates of gas exchange; this undoubtedly has been a factor militating against the evolution of tidally ventilated water lungs.

Gas tensions prevailing in the alveoli of mammalian lungs are a dynamic function of both (1) the rate that fresh air is brought into the depths of the lungs and (2) the rate of gas exchange with the blood. The alveolar oxygen tension in resting humans near sea level is about 100 mm Hg. The alveolar carbon dioxide tension is about 40 mm Hg. These are also the

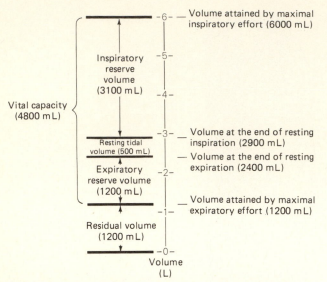

Figure 12.30 Average lung volumes in healthy young adult men. Volumes include the anatomical dead space as well as respiratory spaces. The inspiratory and expiratory reserve volumes depicted are the reserves available during resting ventilation. The residual volume is the volume remaining in the lungs after maximal expiratory effort.

approximate tensions prevailing in air exhaled from the respiratory spaces of the lungs.

Ventilation Rates and the Use of Inspired Oxygen in Rest and Exercise

The overall rate of ventilation is commonly expressed as the **respiratory minute volume:** the volume of new air moved into the lungs per minute. This in turn is partitioned into two components: the tidal volume V_T and the frequency of breaths f:

minute volume (mL/min)
$= V_T$ (mL/breath) $\times f$ (breaths/min)

Given that resting people breathe about 12 times per minute, their minute volume is about 6 L/min (500 mL/breath $\times$ 12 breaths/min).

We have seen that the oxygen and carbon dioxide tensions in the alveoli of people at rest are about 100 and 40 mm Hg, respectively. What would happen to these tensions if a person were to elevate his or her metabolic rate above the resting level without increasing respiratory minute volume? Clearly, the increase in the rate of oxygen extraction from an unaltered rate of supply of new air would lower the alveolar oxygen tension, and similarly the alveolar carbon dioxide tension would be raised. In fact, of course, the respiratory minute volume is increased when metabolism is elevated, as during exercise. Indeed, increases in minute volume are matched to increases in metabolic demand so that *over a broad range of rates of oxygen consumption, alveolar gas tensions remain remarkably stable.* Only during

heavy exertion do gas tensions in the alveoli deviate more than slightly from resting values.

The minute volume is increased during exercise by increases in both the tidal volume and the ventilatory frequency. Typically, most of the increase in tidal volume in humans is achieved by inspiring beyond the resting inspiratory limit—that is, by using the inspiratory reserve volume; the expiratory reserve volume is also used to some extent. People cannot maximize their tidal volume and their ventilatory frequency simultaneously because the time needed for one ventilatory cycle tends to increase as the tidal volume increases. At peak sustained performance, trained athletes can achieve minute volumes of over 100 L/min, combining tidal volumes of 3000 mL or more with respiratory frequencies in excess of 30 breaths/min.

The **alveolar ventilation rate** is defined to be the rate that new air is brought into the respiratory spaces of the lungs (the respiratory bronchioles, alveolar ducts and sacs, and alveoli). It is calculated by subtracting the volume of the anatomical dead space V_D from the tidal volume and multiplying by the respiratory frequency:

$$\text{alveolar ventilation rate} = (V_T - V_D) \times f$$

By dividing the preceding equation by the earlier equation for overall respiratory minute volume, it becomes evident that the fraction of inhaled air that reaches the respiratory spaces is $(V_T - V_D)/V_T$. Recall now that, to a first approximation, V_D is a constant. With this in mind, we see from the expression just derived that *a greater fraction of the inspired air reaches the respiratory spaces when the tidal volume is high than when it is low.*

These considerations set the stage for an analysis of the oxygen utilization coefficient in humans. The air entering and leaving the lungs consists of two components: (1) Air that passes strictly in and out of the dead spaces; this component suffers virtually *no* oxygen depletion. (2) Air that enters the respiratory spaces; this component consistently suffers about a 33 percent reduction in its oxygen content at sea level. Now, we know that the latter component comes to represent a greater and greater fraction of all respired air as the tidal volume increases. *An important consequence is that the oxygen utilization coefficient for respired air as a whole increases as the tidal volume increases.* In humans, for example, the oxygen utilization coefficient at sea level increases from about 20 percent when the tidal volume is 500 mL to about 30 percent when the tidal volume is 2000 mL. Exercising individuals, we see, make more extensive use of the oxygen in inhaled air *even though the gas in their respiratory spaces becomes no more depleted of oxygen than at rest.*

The oxygen utilization coefficients of mammals, being generally 20–30 percent, are below the coefficients seen in water-breathing fish, about 30–80 percent, as might be expected (p. 257).

Pulmonary Ventilation and Body Size

In mammals ranging in size from shrews to whales, lung volume constitutes a rather steady proportion of total body volume. Specifically, lung volume in liters averages about 6 percent of body weight in kilograms. The resting tidal volume tends consistently to be about a tenth of lung volume, meaning that mammals of all sizes have a resting tidal volume of roughly 6 mL/kg body weight. The resting oxygen utilization coefficient is also about the same in all mammals regardless of weight (near 20 percent). These latter two facts signify that, for mammals of all sizes, the amount of oxygen delivered per resting breath is approximately a constant proportion of body weight. Now, in Chapter 3 we saw that the resting weight-specific rate of oxygen consumption tends to increase with decreasing body size. If the weight-specific oxygen demands of a small mammal are greater than those of a big mammal and yet the small animal receives only about the same amount of oxygen per breath per unit weight, then the small animal must breathe more frequently. This is in fact the case. A 20-g mouse might breathe 100 times per minute at rest, whereas resting humans take about 12 breaths/min, for example.

In contrast to prior understandings, several extensive recent studies have found that total alveolar surface area varies approximately in *proportion* to body *weight* among mammals. These studies have also confirmed that in the species examined not only the resting but also the peak weight-specific rate of oxygen consumption increases with decreasing body size. Thus, the studies lead to the unexplained conclusion that small mammals are able to attain a substantially higher rate of oxygen uptake per unit of alveolar surface area than large mammals. Another, more readily rationalized, conclusion of the studies is that alveolar surface area tends to be greater in active species of a given body size than sluggish ones.

Mechanics of Ventilation

Unlike other vertebrates, mammals have a true *diaphragm* that completely separates their thoracic and abdominal cavities and plays a central role in ventilation. This sheet of muscular and connective tissue is dome-shaped, projecting further into the thorax at its center than at its edges. The edges are attached to the body wall. Contraction of the diaphragm muscles tends to flatten the diaphragm, pulling the center away from the thorax toward the abdomen. This increases the volume of the thoracic cavity, resulting in expansion of the lungs and inflow of air by suction. It also reduces abdominal volume, as witnessed by the bulging of the abdominal wall on inspiration.

Among the other muscles important to ventilation are the external and internal intercostals that run obliquely between each pair of adjacent ribs. Contraction of the external intercostals rotates the ribs anteriorly and outward, expanding the thoracic cavity. The fibers of the internal intercostals run crossways to those of the externals, and, in general, their contraction rotates the ribs posteriorly and inward, decreasing the volume of the thoracic cavity. You can easily demonstrate the action of these muscles on yourself by consciously expanding and contracting your rib cage while feeling the ribs with your fingers.

A significant development in mammals (also seen in birds) is that the glottis is no longer used to close off the lungs between inspiration and expiration as in reptiles and amphibians. Thus, the lungs remain filled only as long as contraction of the inspiratory muscles maintains the thorax in an expanded condition. Furthermore, breathing is continuous rather than interrupted by periods of apnea.

The lungs and thoracic wall together form an elastic system. Much like a hollow rubber ball with a hole on one side, they will elastically assume a certain equilibrium volume, known as their *relaxation volume,* if they are freed from the influence of any external forces. For the volume of the lungs to deviate from the relaxation volume, muscular effort must be exerted. In adult human males, when the thoracopulmonary system assumes its relaxation volume, the volume of gas in the lungs is about 2400 mL. Note that this is identical to the volume that remains in the lungs after a resting expiration (Figure 12.30).

As might be surmised from what we have just said, while inspiration in resting humans is active—involving muscular expansion of the lungs to greater than their relaxation volume—expiration is largely or completely passive. The thoracic cavity is enlarged during inspiration by contraction of the diaphragm, external intercostal muscles, and anterior internal intercostals. Relaxation of these muscles at expiration permits an elastic return of the lung volume to the relaxation volume.

During exercise in humans, not only is tidal volume increased, but respiratory frequency is also. Thus, additional muscular activity is required both to amplify changes in lung volume during each breath and to hasten the inspiratory and expiratory processes. The external intercostals assume a greater role in inspiration than during rest. Expansion of the rib cage by these muscles during quiet breathing is of relatively minor importance by comparison to the action of the diaphragm in effecting expansion of the lungs, but during heavy exertion, expansion of the rib cage comes to account for about half of the change in lung volume during inspiration. In addition, active forces contribute to expiration during exercise. The most important muscles in expiration are the internal intercostals, which actively contract the rib cage, and muscles of the abdominal wall, which contract the abdominal cavity, forcing the diaphragm upward into the thoracic cavity. These mus-

BOX 12.2   MAMMALS AT HIGH ALTITUDE

The environment at high montane altitudes is challenging in many respects. It can be cold, windy, dehydrating, and high in ultraviolet radiation. The most immediate challenge for a mammal at high elevation, however, is to meet the oxygen demands of its cells, because its source of oxygen—the atmosphere—is rarefied. Here, we discuss several dimensions of this challenge. Other aspects are reviewed elsewhere (pp. 56 and 320).

We noted earlier (p. 251) that the oxygen tension in dry air at 4540 m (14,900 ft) is about 93 mm Hg and thus is just 59 percent as high as the tension at sea level. Permanent human settlements occur at this and moderately higher elevations, nonetheless; and the people there are far from sedentary. Peruvians at such altitudes, for example, are miners and enjoy soccer as a pastime. At the top of Mt. Everest—8848 m—the oxygen tension in dry air is only about one-third that at sea level: near 53 mm Hg. Recently, a few people have reached the peak without supplementary oxygen. Their experiences show that such altitudes are at the very fringes of human endurance and are accessible only to individuals of exceptional constitution and physical condition (p. 56).

When a person born and reared at low altitude ascends the high mountains, his or her functional traits change over time as acclimatization occurs. There remains debate, however, over whether such an individual will ever reach the physiological status of people born and reared at high altitude. So, in analyzing people at high altitudes, at least three classes need to be distinguished: the newly arrived lowlander, the acclimatized lowlander, and the native highlander. Among other mammals, many species resemble humans in being predominantly of lowland distribution; but at another extreme, there are species that are limited to high altitudes.

Figure 12.31*A* depicts the oxygen cascade for Peruvians at sea level and 4500 m. It is immediately evident that despite the large drop in ambient oxygen tension at 4500 m, the venous tension is reduced only modestly. This *conservation* of venous tension results from significant reductions at altitude in two of the tension drops of the oxygen cascade. The drop in tension between inspired air and alveolar air is about 32 mm Hg at altitude, as compared to 43 mm Hg at sea level; and the drop between arterial and mixed venous blood is about 11 versus 55 mm Hg. The reasons for the lessening of the arteriovenous drop in tension at altitude relate to blood gas transport and thus are discussed in Chapter 13 (p. 320). Here, we examine pulmonary function and systemic tissue responses.

One of the most important defenses marshaled by humans at altitude is an increase in their pulmonary ventilation. For example, in meeting any particular oxygen demand, the ventilation of native highlanders at 4500 m is perhaps 40 percent higher than that of people residing at sea level. When lowlanders first ascend to high altitude, an increase in their rate of breathing occurs in short order; this increase is probably principally activated by the reduction in their arterial oxygen tension (p. 284). As lowlanders pass their first days at altitude, their breathing becomes even more rapid, evidently because of an increasing sensitivity of their respiratory control centers to hypoxic stimulation. The final level of ventilation reached by acclimatized lowlanders tends to be modestly higher than that shown by some groups of native highlanders (hypoxic sensitivity in those highlanders is said to be "blunted").

The increased ventilation rate of people at altitude elevates the flux of fresh air to their alveoli and thus raises their alveolar oxygen tension above what it would otherwise be. That is, the increased ventilation is responsible for lessening the drop in tension between the ambient air and alveolar air, offsetting some of the decrement in the ambient tension. Oxygenation of the blood is thereby aided.

The heightened ventilation at altitude tends to lower the blood CO_2 concentration. Thus, it induces a rise in blood pH: respiratory alkalosis (p. 113 and 332). The rise in pH is quite marked soon after an ascent to higher elevation. During acclimatization

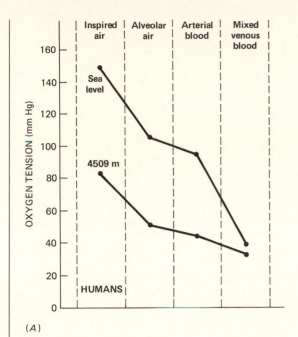

(A)

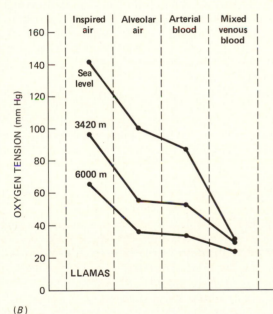

(B)

Figure 12.31 The oxygen cascade at high altitude compared with that at sea level in two species: humans and llamas (*Lama glama*). Shown are average oxygen tensions in inspired (tracheal) air, alveolar air, arterial blood, and mixed venous blood during rest. The data in (A) are for two groups of native male Peruvians studied at their altitude of residence: students and scientists of Lima (altitude: 10 m) and miners of Morococha (4509 m). The data in (B) are for three male llamas born and reared at sea level and studied under three conditions: (1) at sea level, (2) at 3420 m after 10 weeks of acclimatization, and (3) during acute exposure to a simulated altitude of 6000 m after acclimatization to 3420 m. The "inspired" tensions are for tracheal air; they are lower than tensions in the open atmosphere because the air has become humidified by the time it reaches the trachea. [Data for part (A) from J.D. Torrance, C. Lenfant, J. Cruz, and E. Marticorena, *Respir. Physiol.* **11**:1–15 (1970); data for part (B) from N. Banchero, R.F. Grover, and J.A. Will, *Respir. Physiol.* **13**:102–115 (1971).]

the elevation of pH is substantially reversed by renal acid–base regulatory mechanisms. Nonetheless, acclimatized lowlanders often remain somewhat alkalotic, and even native highlanders may be slightly alkalotic by comparison to people at sea level.

As we have stressed, oxygen tensions in the systemic blood capillaries are prevented from falling excessively by processes such as hyperventilation at altitude. Nonetheless, capillary oxygen tensions *do* in fact decline. At sea level, blood enters the capillaries at about 94 mm Hg and exits at about 39 mm Hg (Figure 12.31A). When the altitude is 4500 m, it enters at a much lower tension, 44 mm Hg, and exits at a modestly lower one, 33 mm Hg. Thus, the "head of pressure" for diffusion of oxygen into tissue cells is reduced at altitude. Recently, there has been a lot of interest in how the tissues of mammals might accommodate, and investigations of various species have suggested that several types of *tissue-level acclimatizations* sometimes occur. The existence of these acclimatizations remains controversial, in part because of methodological criti-

Continued on next page.

BOX 12.2 *(Continued)*

cisms leveled against the investigations. Nonetheless, we note that the investigations point to the following types of acclimatizations at high altitudes: (1) increases in the density of functioning capillaries, serving to reduce the average *diffusion distance* between capillaries and mitochondria; (2) increases in myoglobin, a compound that facilitates diffusion of oxygen (p. 318); (3) increases in numbers of mitochondria; and (4) increases in the catalytic activities of certain enzymes of aerobic catabolism.

Figure 12.31*B* depicts oxygen cascades for the llama, one of the few native highland species to be thoroughly studied. The llama hyperventilates at high altitude; this seems to be a near universal mammalian response. However, llamas are less reliant on this defense than humans, perhaps because their hemoglobin has a higher affinity for oxygen than ours and thus does not require as high an alveolar oxygen tension to become loaded with oxygen. Unlike humans, llamas do not hyperventilate at all at 3400 m (note that the inspired–alveolar drop is the same as at sea level); at 6000 m they hyperventilate less than humans at the same altitude. At all altitudes, llamas have a lower mixed venous tension than humans, suggesting that they may place greater reliance on tissue-level adaptations that permit function in the face of reduced capillary oxygen tensions.

cles hasten expiration and may also compress the lungs beyond their relaxation volume, thereby enhancing tidal volume through use of some of the expiratory reserve volume.

The same basic groups of muscles are used for ventilation in other mammals, but their relative importance varies. Among large quadrupeds, for example, movements of the rib cage tend to be constrained, and the diaphragm bears especially great responsibility for ventilation.

The Control of Ventilation

The respiratory control centers are located in the medulla of the brain and communicate with the respiratory muscles via motor neurons. If the medulla is isolated experimentally from all neural inputs, an animal will continue to breathe rhythmically. This type of evidence demonstrates that there is an endogenous respiratory rhythm in the medulla, but it is abundantly clear that this rhythm is modulated by a variety of neural inputs in the intact animal. Another portion of the brain, the pons, is routinely involved in controlling the rhythm, and we all know that higher, conscious centers can intervene to modify the pattern of ventilation. (We can stop breathing, for example, when we want to.) Sensory neurons from the lungs relay information to the control centers concerning the degree of pulmonary expansion. Some of these neurons modulate the respiratory rhythm by inhibiting inspiration when the lungs are expanded, whereas others act to excite inspiration when the lungs are compressed; these responses are known as the Hering–Breuer reflexes.

Because the primary function of the lungs is the supply of oxygen and removal of carbon dioxide, and because we know that ventilation is regulated so as to maintain stable alveolar tensions of these gases,

it is clear that the parameters of ventilation must ultimately be controlled according to information on gas exchange. This information is provided by chemoreceptors and chemosensitive zones that sense aspects of body fluid composition that are indicative of respiratory status.

When the concentration of carbon dioxide in the blood or other body fluids rises or falls, the concentration of hydrogen ions increases or decreases in tandem (p. 323). In some combination—and by pathways not yet fully understood—these concentrations are sensed by chemosensitive zones near the ventrolateral surface of the medulla, and deviations from normal levels exert a potent influence on respiration. Ventilation increases or decreases in such a way as to bring the concentrations back to normal—a negative-feedback system. For example, if the carbon dioxide concentration of the blood is elevated, ventilation is increased, resulting in a greater rate of removal of carbon dioxide. The potency of these effects is illustrated by the observation that an increase in the arterial carbon dioxide tension of just 4 mm Hg—from 40 to 44 mm Hg—will cause about a *doubling* of the respiratory minute volume in people.

Oxygen tension is sensed not in the medulla but in peripheral chemoreceptive bodies known as the **carotid** and **aortic bodies**. The former are positioned along the two common carotid arteries (near where each branches to form internal and external carotids); they receive blood flow from the carotids and relay sensory information to the medulla via the glossopharyngeal nerves. The latter are located along the aortic arch and relay information via the vagi. Both the carotid and aortic bodies, being richly perfused with blood emanating from major arteries, are in an excellent position to monitor arterial oxygen tension.

Whereas ventilation is markedly stimulated by

even small elevations of arterial *carbon dioxide* tension above normal, the arterial *oxygen* tension must fall to 50–60 mm Hg—much below normal (~ 95 mm Hg)—before ventilation is reliably stimulated. Thus, sensation of CO_2 tension and/or H^+ concentration is paramount in regulating ventilation under usual resting conditions. If the arterial oxygen tension does fall below 50–60 mm Hg, marked stimulation of ventilation occurs. Thus, sensation of oxygen tension is important in ventilatory regulation in certain circumstances, as at high altitude. The carotid and aortic bodies become more sensitive to lowered oxygen tension when the CO_2 tension is elevated.

The question of how the ventilation rate is controlled during exercise is one of the major unresolved issues in respiratory physiology. Arterial oxygen and carbon dioxide tensions remain little changed from resting values except during vigorous exertion. This stability results *because* the ventilation rate increases in tandem with metabolic rate. However, arterial gas tensions are far too stable to *account* for the increases in ventilation on the basis of the simple negative-feedback systems discussed heretofore; for example, an increase of about 4 mm Hg in the arterial carbon dioxide tension is required to provoke a doubling of ventilation rate, yet during exercise the measured tension may not be elevated to that extent even when ventilation has reached 10–15 times the resting level. There is considerable conviction that gas tensions—particularly the carbon dioxide tension and associated H^+ concentration—*are* involved in the control of ventilation during exercise; but how they are involved remains enigmatic. Besides controls mediated by gas tensions, there is increasing evidence for controls initiated in the nervous system: so-called "neural" controls. These are postulated to take two forms. (1) Parts of the brain, such as the hypothalamus, that initiate motor signals to the locomotory muscles involved in exercise might simultaneously initiate stimulatory neural signals to the ventilatory centers. (2) Sensors of movement or pressure in the limbs might stimulate the ventilatory centers in accord with the vigor of limb activity they detect. One persuasive piece of evidence for "neural" controls is that when people suddenly begin to exercise at a moderately heavy level, a marked increase in their ventilation rate is evident within just one or two breaths; this response seems far too rapid to be mediated by changes in the chemical composition of body fluids.

Lung Surfactant

The tendency of a bubble to collapse shut is an inverse function of its radius. Thus, pulmonary alveoli, which are bubblelike and of very small radius, might be expected to exhibit a high inherent tendency to collapse. Furthermore, if two bubbles of identical composition but different radius are connected together, the smaller will develop a higher internal pressure and thus tend to collapse by emptying into the larger. We must wonder why some alveoli do not collapse while others expand.

A bubble's tendency to collapse is a direct function of its surface tension as well as being an inverse function of its radius. The inherent tendency of alveoli to collapse is reduced in a very important way by the presence on their inner walls of an exceedingly thin layer of material that reduces their surface tension. This material, known as **lung surfactant** or **surpellic material,** consists of phospholipids. It exhibits the remarkable property that its surface tension decreases dramatically if alveolar size decreases; this decrease in surface tension helps prevent small alveoli from emptying into large by opposing the effect of decreasing radius. Surfactants have been reported from the lungs of all groups of terrestrial vertebrates and lungfish.

EXTERNAL RESPIRATION IN BIRDS

Anatomy

The structure of the avian lung, which is very different from that of other vertebrate lungs and yet a logical derivative of reptilian lungs, is illustrated schematically in Figure 12.32A. Bifurcation of the trachea gives rise to two primary bronchi, one of which enters each lung. The primary bronchus passes through the lung and within the lung is termed the **mesobronchus.** Two groups of branching **secondary bronchi** arise from the mesobronchus. One group, which arises at the anterior end of the mesobronchus, ramifies over the ventral surface of the lung; a second group originates toward the posterior end of the mesobronchus and spreads over the dorsolateral lung surface. For simplicity, we term these the *anterior* and *posterior* groups of secondary bronchi, although they formally are termed the *medioventral* and *mediodorsal* groups, respectively. Also in the spirit of simplicity, each group is represented as just a single passageway in Figure 12.32A. The anterior and posterior secondary bronchi are connected by a great many small tubes, 0.5–2.0 mm in internal diameter, termed **tertiary bronchi** or **parabronchi.** As depicted in Figure 12.33, each parabronchus gives off radially along its length an immense number of finely branching **air capillaries.** These are profusely surrounded by blood capillaries and are the sites of gas exchange. The air capillaries are only 3–10 μm in diameter, and the exchange surface been air capillaries and blood capillaries is enormous, 200–300 mm^2/mm^3 of tissue in the parabronchial walls. Air flows through the parabronchi, but exchange between the parabronchial lumen and the surfaces of the air capillaries is probably largely by diffusion. The parabronchi, air capillaries, and associated vasculature constitute the bulk of the lung tissue.

Figure 12.34 depicts the **air sacs,** which—located outside the lungs—occupy a considerable portion of

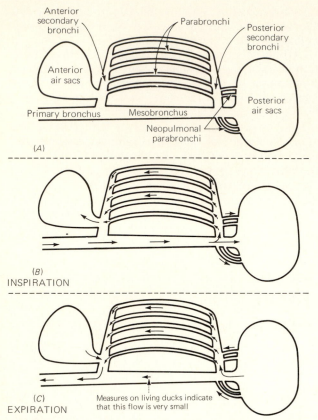

Figure 12.32 Aspects of anatomy and air flow in the respiratory system of birds. (A) A schematic representation of the anatomy of the bird lung and its connections with the air sacs. [For a detailed description, see H.-R. Duncker, *Respir. Physiol.* **14**:44–63 (1972).] (B and C) Known and presumptive directions of air flow in the lung and air sacs during inspiration (B) and expiration (C).

the thoracic and abdominal body cavities. Usually, there are nine air sacs, divisible into two groups. The *anterior sacs* (two cervical sacs, two anterior thoracic sacs, and a single interclavicular sac) open to various anterior secondary bronchi. The *posterior sacs* (two abdominal sacs and two posterior thoracic sacs) open to the posterior mesobronchi. (Each mesobronchus terminates at its connection with an abdominal air sac.) The air sacs are thin-walled, poorly vascularized structures that play little role in gas exchange between the air and blood but nonetheless, as we shall see, are essential components of the respiratory system.

The structures of the lung described thus far (mesobronchus and anterior and posterior sets of secondary bronchi connected by parabronchi) are present in all birds, and their connections with the air sacs are similar in all birds. These pulmonary structures are collectively termed the ***paleopulmonal system,*** or simply ***paleopulmo.*** Most birds, in addition, have a more or less extensively developed system of respiratory, parabronchial tubes running directly between the posterior air sacs, on the one hand, and the posterior parts of the mesobronchi and posterior

secondary bronchi, on the other hand (Figure 12.32A)—and sometimes being even more elaborate. This system is called the ***neopulmonal system,*** or simply ***neopulmo.*** The paleopulmonal system is always dominant. The neopulmo never represents more than 20–25 percent of the total lung volume even in its most highly developed form.

Ventilatory Mechanics

Avian lungs are compact, rigid structures in comparison to mammalian lungs. They also contrast sharply with mammalian lungs in that they undergo little change in volume over the course of the ventilatory cycle. The air sacs, on the other hand, expand and contract substantially during ventilation. Like bellows, the air sacs suck and push gases through the relatively rigid tubing of the lungs. In comparison to mammalian ventilation, this avian process turns out to be an energetically inexpensive way to move air.

The rib cage surrounding the lungs themselves is relatively rigid. During inspiration, contraction of internal intercostal muscles and several other thoracic muscles expands other parts of the rib cage (especially those posterior to the lungs), and the sternum swings downward and forward. These movements enlarge the air sacs by expanding the thoracoabdominal cavity. Some of the external intercostals and abdominal muscles compress the thoracoabdominal cavity and air sacs at expiration.

The Pattern of Air Flow

The pattern of air flow through the avian respiratory system remained one of the great unresolved mysteries in respiratory physiology until just 10 or 15 years ago. The basic pattern now seems to be firmly understood, although many details require further study.

Let us look first at flow through the paleopulmo, the part of the lung that is dominant in all birds. During inspiration, both the anterior and posterior sets of air sacs are expanded and therefore must receive gas. As depicted in Figure 12.32B, air from the outside flows through the mesobronchus to enter the posterior air sacs and posterior secondary bronchi; in turn, the air entering the posterior secondary bronchi is drawn anteriorly through the parabronchi under the force of suction developed in the anterior air sacs. Air exiting the parabronchi enters the anterior sacs. Three aspects of the events at inhalation deserve emphasis: (1) The posterior air sacs are filled with relatively fresh air coming directly from the environment. (2) The anterior air sacs are filled substantially with gas that has passed across the respiratory exchange surfaces in the parabronchi; this gas has suffered a drop in its oxygen content and rise in its carbon dioxide content. (3) The direction of ventilation of the parabronchi in the paleopulmo is from posterior to anterior during inspiration.

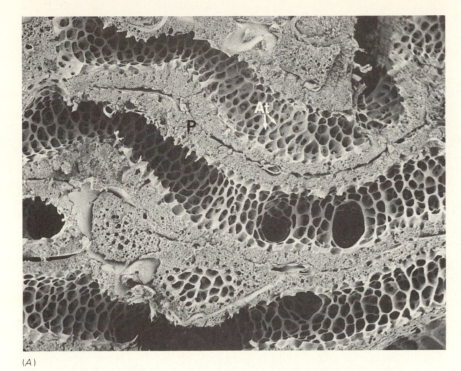

(A)

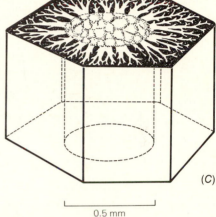

(C)

├──────────┤
0.5 mm

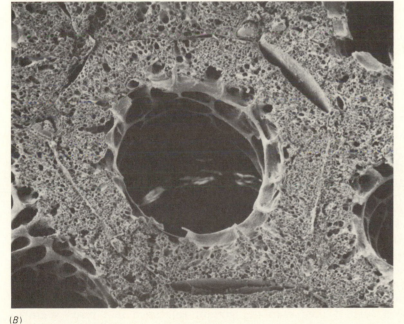

(B)

Figure 12.33 (*A*) A longitudinal section through three parabronchi in the lung of a chicken (*Gallus*) as viewed by scanning electron microscopy (magnification: 20×). Holes marked At are openings (atria) leading to arrays of air capillaries. Tissue marked P, known as parenchyma, is an intermingling of air and blood capillaries. (*B*) A cross section through a parabronchus in the chicken lung, as viewed by scanning electron microscopy (magnification: 65×). Compare with *C*. (*C*) A diagrammatic cross section through a parabronchus, depicting the branching air capillaries in the parenchyma. The parenchyma surrounding the lumen of each parabronchus assumes a hexagonal shape. [*A* and *B* kindly provided by Walter S. Tyler and Dave Hinds; *C* from E.H. Hazelhoff, *Verslag van de gewone vergaderingen der Afdeling Natuurkunde van de Nederlanse Akademie van Wetenschappen* **52**:391–400 (1943); English translation available in *Poultry Sci.* **30**:3–10 (1951).]

During expiration, the air sacs are compressed and discharge gas. As shown in Figure 12.32*C*, air exiting the posterior air sacs predominantly enters the posterior secondary bronchi to pass anteriorly through the parabronchi. This air is relatively fresh, having entered the posterior sacs more or less directly from the environment during inspiration. Gas exiting the parabronchi anteriorly, combined with gas exiting the anterior air sacs, is directed into the mesobronchus

via the anterior secondary bronchi and is exhaled. Recall that the anterior air sacs were filled with gas from the parabronchi during inhalation. Thus, the exhaled gas is mostly gas that has passed across the respiratory exchange surfaces. Three aspects of the expiratory events deserve emphasis: (1) The relatively fresh air of the posterior air sacs is directed mostly to the parabronchi; outflow toward the environment along the length of the mesobronchus is

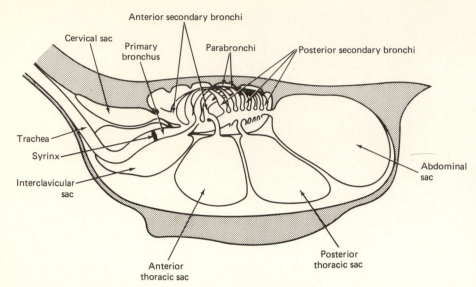

Figure 12.34 Semirealistic diagram of the respiratory system of a domestic goose, showing the position of the lungs and the positions and connections of the air sacs. [Reprinted with permission from J.H. Brackenbury, *Comp. Biochem. Physiol.* **68A:**1–8 (1981). Copyright 1981, Pergamon Press.]

minimal, according to present evidence. (2) The exhaled gas has largely passed across the respiratory surfaces. (3) Air flows through the paleopulmonal parabronchi from posterior to anterior.

In all, *flow through the paleopulmonal parabronchi occurs during both inhalation and exhalation and is in the same direction in both instances*. Shortly, we shall consider the implications of the unidirectional parabronchial flow for gas exchange.

One of the biggest unresolved questions in the study of avian lungs is how air is directed along its elaborate and, in some ways, unexpected path through the air sacs and paleopulmo. Passive, flap-like valves appear to be entirely absent. Active, muscular valving could occur, but evidence for its presence is at best circumstantial. One long-standing concept is that the complex architecture of the lung passages may itself create aerodynamic conditions that direct air along its inspiratory and expiratory paths without need of either passive or active valves. This concept has received interesting support from studies of excised duck lungs. The lungs, removed from the animal and disconnected from the air sacs, were ventilated by an alternating pump attached to the anterior mesobronchi; significantly, air entered the posterior secondary bronchi to pass anteriorly through the paleopulmonal parabronchi on both "inhalation" and "exhalation." In this situation, active valving was impossible.

Ventilation of the neopulmo is incompletely understood. Probably, however, as shown in Figure 12.32, flow through many of the neopulmonal parabronchi is bidirectional.

The Mode of Gas Exchange in the Paleopulmo   When the unidirectional flow of air through the paleopulmonal parabronchi first became well established, countercurrent exchange between the blood and air was quickly postulated. Soon, however, the postu-

late was disproved by clever experiments, which showed that the efficiency of gas exchange between the air and blood is not diminished if the direction of air flow in the parabronchi is artificially reversed. Morphological and functional studies have now shown convincingly that blood flow in the respiratory exchange vessels occurs in a *cross-current* pattern (Figure 12.9). With countercurrent exchange ruled out, attempts have been made to determine if the observed unidirectionality of air flow might aid gas exchange in other ways. At present, however, advantages for the unidirectional air flow have not been established.

Comparisons with Mammals

In comparing the disparate respiratory systems of birds and mammals, certainly one of the quickest questions to come to mind is whether one or the other system is superior. As yet, no compelling evidence exists that either type of system, in actual practice, provides distinct advantages in the overall result: oxygen uptake from the environment and carbon dioxide elimination. Advantages have been claimed for the avian system. For example, the cross-current mode of gas exchange has been credited with the greater tolerance that birds seem to have to lowered ambient oxygen tensions (an advantage during high-altitude flight). However, the tolerance of birds to low oxygen levels cannot at present be ascribed with confidence to their type of lung. Both birds and mammals have evolved high-performance respiratory systems. Perhaps they have arrived at similar end points in their capabilities for gas exchange. Perhaps the differences in the paths they have followed to those end points relate not so much to gas exchange per se as to other factors, such as the need for the avian respiratory system to be compatible with a body plan attuned to flight.

Numerous differences of functional detail are to be found between birds and mammals. Resting birds, for example, typically breathe at only about one-half or one-third the frequency of resting mammals of equivalent body size, but the birds have greater tidal volumes.

An intriguing difference between avian and mammalian lungs is manifest near the time of hatching or birth. Mammals *suddenly* inflate their lungs with air as they emerge from the birth canal; they make a quick transition from no air breathing (placental gas exchange) to complete reliance on air breathing. This seems not to be possible for birds: The gas exchange elements of avian lungs, the air capillaries, cannot suddenly be inflated out of a collapsed state because the lungs are relatively rigid and the air capillaries themselves are of such extremely small diameter. As a bird egg matures, an air space develops within it, and 2 or 3 days before hatching the young bird starts to breathe from the air space. During the ensuing hours, until hatching occurs, the bird is then able to make an *essential gradual transition* from respiration across its chorioallantoic membranes to full-fledged pulmonary respiration. It is in this period, in fact, that the air capillaries undergo most of their prehatching development. Room for the required air space within the egg is created by evaporative loss of water over the course of egg development, and it appears that egg permeability and features of the egg environment have become finely attuned to assuring that an appropriate amount of evaporation will occur.

INSECTS

Early in this chapter, we noted that oxygen and carbon dioxide must move between all cells of the body and the surfaces of exchange with the environment and that transport by diffusion and microconvection within the cells and intercellular fluids will generally suffice only over short distances. We have seen that most of the reasonably large and active animals possess localized respiratory exchange surfaces, remote from many cells; and thus they depend on circulation of blood or other body fluids to move gases rapidly within their bodies. Among the insects, we find a truly remarkable respiratory system, one that brings the exchange surface itself close to all cells.

Anatomy

Basic features of the anatomy of the insect respiratory system are illustrated in Figure 12.35. Access to the outside is provided by pores, termed *spiracles,* along the lateral body wall. Air-filled tubes, or *tracheae,* penetrate the body from each spiracle and branch repeatedly, reaching all parts of the animal. Figure 12.35 shows only the major branches. In a few insects, the tracheal trees arising from different spiracles remain independent, but usually they anastomose via large longitudinal and transverse connectives to form a fully interconnected tracheal system.

The spiracles number from a single functional pair to as many as 10 or 11 pairs. They are segmentally arranged and may occur on the thorax, abdomen, or both. In most insects they can be closed by spiracular muscles.

The tracheae represent invaginations of the epidermis and are lined with a thin cuticle. Typically, the cuticle is thrown into folds that run a spiral course along the tracheae, providing resistance against collapse. The tracheae become finer with increasing distance from the spiracles and finally give rise to very fine, thin-walled end tubules termed *tracheoles,* believed to be the major sites of gas exchange with the tissues. Tracheoles are perhaps 200–350 μm long and are believed to end blindly. They generally taper from a diameter approximating 1 μm at their origin to a diameter of 0.1–0.2 μm at the end.

The layout of the tracheal system varies im-

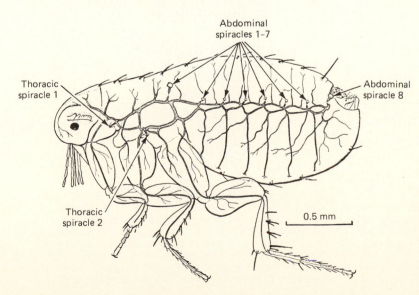

Abdominal spiracles 1–7

Thoracic spiracle 1

Abdominal spiracle 8

Thoracic spiracle 2

0.5 mm

Figure 12.35 The main tracheae in a flea, *Xenopsylla*. There are 10 pairs of spiracles, two thoracic pairs and eight abdominal pairs, as labeled. The tracheae arising from the abdominal spiracles are narrowed for a short distance near the spiracular openings. All spiracular openings are closable. In the case of the abdominal spiracles, closing is accomplished by muscular compression of the narrow tracheal segment near the spiracular opening. [From V.B. Wigglesworth, *Proc. R. Soc. London Ser. B* **118**:397–419 (1935). Used by permission of The Royal Society.]

mensely among species, but the usual result is that all organs and tissues are thoroughly invested with fine tracheae and tracheoles. The degree of tracheation of various organs and tissues appears generally to vary directly with their metabolic requirements. In most tissues, the tracheoles usually run between the cells. But in the flight muscles of many species, the tracheoles penetrate the muscle cells, indenting the cell membrane inward, and run among the individual myofibrils, in close proximity to the arrays of mitochondria. The average distance between adjacent tracheoles within the flight muscles of strong fliers is commonly on the order of just 3 μm. Other muscles, the nervous system, the rectal glands, and other active tissues also tend to be richly supplied by the tracheal system, although intracellular penetration is not nearly as common as in flight muscles. In the epidermis of the bug *Rhodnius,* tracheoles are much less densely distributed than in active flight muscles, but, still, cells are usually within 30 μm of a tracheole. In other words, no cell is separated from a tracheole by more than two or three other cells.

Tracheolar Liquid   In some tissues of some species, the terminal ends of the tracheoles are filled with liquid when the animals are at rest. Then, during exercise or when the insects are exposed to oxygen-deficient environments, the amount of liquid decreases and gas penetrates further into the tracheoles, as illustrated in Figure 12.36. This increased filling of the tracheoles with gas facilitates the exchange of oxygen and carbon dioxide between the tissues and environment because oxygen and carbon dioxide move through the tracheoles by diffusion, and a gaseous medium, of course, is far more favorable for rapid diffusion than a liquid medium.

In general, the extent of liquid filling of tracheoles is believed to represent a balance between capillary forces, tending to draw water into the tracheoles, and the colloid osmotic pressure of the surrounding cells and body fluids, tending to restrain entry of

water. The mechanism of liquid withdrawal during exercise or exposure to oxygen deficiency is incompletely understood. There is considerable support for the hypothesis that locomotor activity and hypoxia can lead to the accumulation of metabolites which increase the osmotic pressure of the cells and body fluids surrounding the tracheoles. This elevation of osmotic pressure would lead to the osmotic withdrawal of water.

Air Sacs   Air sacs, generally associated with major tracheae, are a common feature of the insect respiratory system. Some occur as swellings along tracheae, and air can flow through them. Others occur as blind endings of tracheae or as blind, lateral diverticula of tracheae. Air sacs tend to be particularly well developed in active insects and in them may occupy a considerable fraction of the body volume (Figure 12.37).

Diffusion as a Mechanism of Gas Flux in the Tracheal System

Many insects, at least when relatively inactive, display no regular ventilatory movements. This observation suggests that gas exchange through the tra-

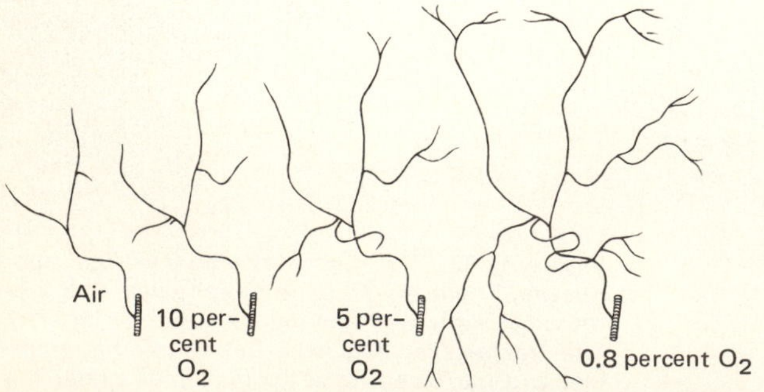

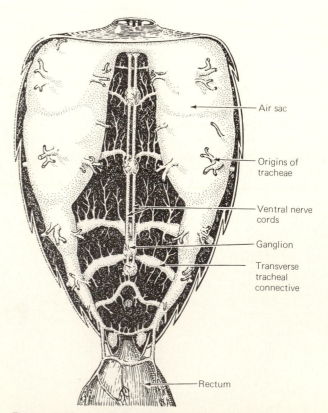

Figure 12.36   The extent of penetration of gas in a tracheolar bed of a resting flea (*Xenopsylla*) when exposed to air and to diminishing ambient concentrations of oxygen. Gas-filled tracheoles stand out from surrounding tissues and are readily visualized. [From V.B. Wigglesworth, *Proc. R. Soc. London Ser. B* **118**:397–419 (1935). Used by permission of The Royal Society.]

Figure 12.37   A dorsal view of the abdomen of a worker honeybee (*Apis*), showing the large air sacs and associated tracheae (including transverse connectives). The gut is reflected posteriorly, and the tracheal supply to the rectum is shown. Major air sacs are also found in the head and thorax. [From K. Dreher, *Z. Morphol. Oekol. Tiere* **31**:608–672 (1936).]

cheal system might occur entirely by diffusion, and numerous analyses have confirmed that diffusion between the spiracles and tracheoles is often adequate to meet respiratory requirements. The tracheoles are usually sufficiently densely distributed in all tissues that diffusion can also account for exchange between the tracheolar membranes and all cells. Thus, the circulatory system need play little role in gas transport.

Whether diffusion through the tracheal system will be adequate to meet an insect's requirements for gas exchange depends on several factors, among which body size and metabolic rate deserve discussion. If we consider two insects that are of identical body proportions but different size, one factor immediately stands out as placing the small insect in a more favorable position than the large to have its requirements for gas exchange through the tracheal system met by diffusion alone: The rate of diffusive gas transfer is an inverse function of the length of the diffusion path (p. 246), and the smaller insect will enjoy a shorter average distance between its body surfaces and tissues. Other considerations, in and of themselves, place the smaller insect at a comparative disadvantage: The cross-sectional area of the tracheal–tracheolar airways will be smaller in the smaller insect, and each unit of tissue will tend to have greater gas exchange requirements in the smaller insect (p. 32). Nonetheless, when all these factors are considered together, the effect of the shorter path length in the smaller insect dominates, and small size indeed proves to be more conducive than large to having diffusion suffice as the sole means of tracheal gas flux.

In a given insect, how can the rate of oxygen diffusion through the tracheal system be increased to meet elevated metabolic oxygen requirements? One key mechanism is that the tracheolar oxygen tension may be allowed to fall. Assuming the ambient oxygen tension to be constant, such a drop in tracheolar tension will hasten the arrival of oxygen into the tracheoles by increasing the driving force for diffusion through the tracheal system: the difference in tension between the two ends of the system. This mechanism of accelerating tracheal diffusion can be carried only so far, however, because the tracheolar tension is also a key component of the driving force for diffusion of oxygen from the tracheoles into the tissues; decreases in the tracheolar tension tend to reduce this latter driving force. If oxygen demands in the tissues should rise to high levels, it may become impossible for the two diffusional systems—the tracheal system and the tracheolar–tissue system—to operate fast enough simultaneously to meet oxygen needs (Figure 12.38). Short tracheal diffusion paths, such as found in small insects, are an asset in avoiding this eventuality because any given rate of diffusion through the tracheal system can be attained at a higher tracheolar tension if the diffusion path is short rather than long, other things being equal.

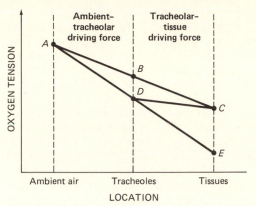

Figure 12.38 The interplay between the driving forces for ambient–tracheolar and tracheolar–tissue diffusion: a thought exercise using fictional values. The slope of each line is proportional to the drop in oxygen tension: the driving force for diffusion. In the upper line (*ABC*), the ambient–tracheolar and tracheolar–tissue drops in tension are equal. Reducing the tracheolar tension from *B* to *D* increases the driving force (*AD*) for ambient–tracheolar diffusion by 70 percent but decreases the driving force (*DC*) for tracheolar–tissue diffusion. The tracheolar–tissue driving force can be increased by 70 percent from its original value—thus accelerating tracheolar–tissue diffusion to the same extent as ambient–tracheolar diffusion—by allowing tissue tension to drop to *E*. This process of raising both driving forces in concert by dropping tissue tension in parallel with tracheolar tension can be carried only to the point that tissue tension is at the minimum necessary for function.

These principles, largely derived from physics, suggest when quantified that gas transport through the tracheal system by diffusion alone might occur in even relatively large insects when metabolic rates are low but is likely only in small insects when metabolic rates are raised. The biological evidence generally supports these conclusions. Diffusion appears to be the principal or only mode of transport in most larvae, all pupae, and most resting adults. During flight, many insects are known to ventilate the tracheal airways actively. Calculations have demonstrated, however, that tracheal diffusion should be adequate to meet oxygen demands during flight in small insects, such as fruit flies, and diffusion may be the principal mechanism of transport in these forms even during flight.

Ventilation of the Tracheal System

Many adult insects ventilate the tracheal system. This is seen in some large species at rest and, as noted already, is common during activity, especially flight. It is important to recognize from the outset that only the major tracheae are ventilated. From these, exchange with the tracheoles via the smaller tracheae is still largely or completely by diffusion. Essentially then, ventilation serves to reduce the path length for diffusion by moving air convectively to a certain depth in the tracheal system.

The mechanisms of ventilation are diverse. In the

broadest terms, the airways are alternately compressed and expanded by muscular activity. The air sacs commonly act as bellows, forcing or sucking air through the tracheae. This role of the air sacs is reminiscent of the situation in birds and helps to explain the especial development of air sacs in active insects. Ventilation is sometimes tidal and sometimes unidirectional. Patterns of unidirectional flow are established by opening and closing of spiracles in synchrony with ventilatory movements. In resting locusts (*Schistocerca gregaria*), for example, only certain thoracic spiracles are open during inspiration and only certain abdominal ones are open during expiration; thus, air flows posteriorly in the tracheae from the thorax to the abdomen. In resting but ventilating insects, pumping movements of the abdomen commonly provide the chief propulsive force for ventilation. Such pumping is often evident during flight as well.

A mechanism of ventilation that seems to be important in many insects during flight is *autoventilation:* gas flow in and out of the tracheae supplying the flight muscles induced by flight movements. As the thorax is deformed in synchrony with the wingbeats, tracheae and air sacs within are compressed and expanded, driving gas tidally in and out of the thoracic spiracles and tracheae. The phenomenon has been well documented in locusts (*Schistocerca*), which hold their second and third most anterior pairs of spiracles continuously open during flight; about 80 percent of the influx of fresh air into the tracheae of their flight muscles is then by autoventilation.

Control of Diffusional and Ventilatory Exchange

When at rest, insects respiring by diffusion are commonly observed to restrict access to their tracheal system by maintaining their spiracles partly closed or by periodically opening and closing them. This behavior has evolved, we believe, as a mechanism of limiting evaporative water losses. Indeed, experiments show that forcing the spiracles to remain fully open increases the rate of evaporative water loss by 2–12 times.

To understand the physical basis for the restriction of evaporative water losses ordinarily seen, consider these arguments: If the tracheal air is assumed to be saturated with water vapor, then the water vapor tension gradient across the spiracular openings will be constant in a given atmosphere whether the spiracles are fully open or constricted. The rate of diffusion varies directly with the cross-sectional area of the spiracular openings, however, and thus constriction reduces water loss.

It is important to recognize that partial closing of the spiracles need not interfere with proper exchange of oxygen and carbon dioxide despite its effect on the rate of water vapor diffusion. Suppose, for example, that when the spiracles of an insect are wide open, diffusion of oxygen to the tracheoles occurs at an adequate rate with only a trivial drop in oxygen

tension across the spiracles themselves and a drop of about 10 mm Hg through the remainder of the tracheal system—as shown by the solid line in Figure 12.39. Partial constriction of the spiracles will impair inward diffusion of oxygen across the spiracles and thus cause tracheal oxygen tensions to fall. However, as they fall, the gradient of oxygen tension across the spiracles will increase, thus tending to override the reduction in the diffusion rate caused by spiracular constriction. Suppose that a tension drop of 20 mm Hg across the spiracles proves adequate to restore the rate of diffusion through the spiracles to its original level. Then diffusion to the tracheoles will occur at the original rate with a total tension drop from the atmosphere to the tracheoles of 30 mm Hg—as shown by the dashed line in Figure 12.39. We see by this example that, within limits, water loss can be reduced without any reduction in the rate of influx of oxygen, by permitting a reduction in the tracheolar oxygen tension.

Insects actively modulate the ease of gas transfer between their tracheal system and the atmosphere. For example, they typically open their spiracles more fully or more frequently when they are active rather than resting (opening the spiracles reduces the drop in oxygen tension across them and thus helps maintain higher tracheolar oxygen tensions, as evident in Figure 12.39). The modulation of spiracular opening is under strong control of parameters that reflect the adequacy of respiratory gas exchange. The most potent stimulus for opening of the spiracles is an increase in the carbon dioxide tension and/or acidity of the body fluids; decreased oxygen tension may stimulate spiracular opening but typically is far less influential. In these respects, the control of the spiracles resembles the control of pulmonary ventilation in mammals. There is evidence also that spiracular opening is modulated in response to immediately prevailing evaporative conditions. In resting tsetse flies

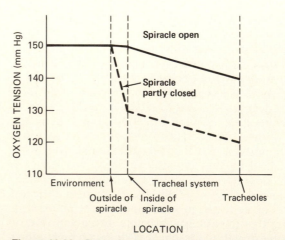

Figure 12.39 Drops in oxygen tension in the respiratory system of an insect with spiracles fully open (solid line) or partly closed (long-dashed line). The rate of oxygen diffusion from the atmosphere to the tracheoles is the same in both cases.

(*Glossina*), for example, the spiracles remain 30 percent open at 60 percent relative humidity but only 10 percent open in dry air; as a result, the rate of water loss is about the same in the two environments. In insects that ventilate the tracheal system, the ventilation rate is controlled in response to respiratory gas tensions and sometimes ambient humidity. These controls function analogously to the spiracular controls just discussed.

Discontinuous Respiration

In many insects at rest, both oxygen and carbon dioxide are exchanged at more or less continuous rates with the atmosphere. A different pattern, termed *discontinuous respiration,* has been observed in various pupae and quiescent larvae and adults. Whereas oxygen consumption proceeds continuously in these insects, their rate of carbon dioxide elimination displays strong intermittent bursts (Figure 12.40).

Discontinuous respiration represents a special case of spiracular control of diffusional gas exchange. Its chief benefit is believed to be exceptional water conservation, although the mechanisms by which conservation is enhanced relative to more usual patterns of respiration are not entirely understood. Pupae, in which discontinuous respiration has been most thoroughly studied, do not ingest water, sometimes for months. They thus have special requirements for conserving water.

In the large, diapausing pupae of cecropia moths, bursts of carbon dioxide release last around 15–30 min and may be separated by interburst periods of 4–17 h at 20°C. During a burst, the spiracles are fully open, but between bursts they are mostly closed. Once the spiracles have been closed at the end of a burst, the tracheal oxygen tension quickly falls to around 30–40 mm Hg, creating a large gradient of oxygen tension across the spiracles; this gradient evidently induces some inward diffusion of oxygen through the mostly closed spiracles and thus contributes to the continuous oxygen influx. A key attribute of discontinuous respiration is that during the interburst period, oxygen removed from the tracheal gas by the metabolizing tissues is not replaced by a like volume of carbon dioxide. One reason is that the respiratory quotient is only 0.7 (see p. 29); another is that about two-thirds of the carbon dioxide produced during the interburst period does not enter the tracheal gas but instead is stored in solution in the body fluids, as bicarbonate. Because oxygen is steadily removed from the tracheal gas but not replaced by an equivalent volume of carbon dioxide, a negative pressure (partial vacuum) is created in the tracheal airways, and air accordingly is *sucked* in across the spiracles during a substantial part of the interburst period. This bulk inflow of air—known as passive suction ventilation—contributes to oxygen uptake. It is thought also to play a key role in reducing water losses by blowing outward-bound water vapor inward, away from the spiracles where it might diffuse into the atmosphere. When the spiracles are opened during a burst, the carbon dioxide dissolved in the body fluids during the preceding interburst comes out of solution and is lost, setting the stage for another long period of carbon dioxide storage.

Respiration in Aquatic Insects

Nearly all aquatic insects retain an air-filled tracheal system. They have evolved a number of intriguing ways of interfacing this system with the ambient water or air.

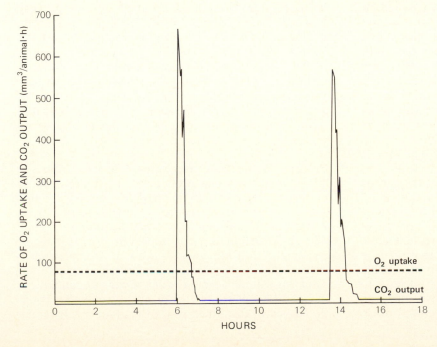

Figure 12.40 Rates of oxygen uptake and carbon dioxide release from a 5-g diapausing cecropia pupa, illustrating discontinuous respiration. [From H.A. Schneiderman and C.M. Williams, *Biol. Bull.* (*Woods Hole*) **109**:123–145 (1955).]

Breathing from the Atmosphere   Perhaps the simplest respiratory expedient for aquatic insects is to breathe directly from the atmosphere. Typically, this entails localization of the functional spiracles to one end of the body, which then is projected to the air at the water's surface. A commonplace example is provided by mosquito larvae, which hang at the surface from a posterior abdominal tube, at the end of which the spiracles open (Figure 8.1). The spiracular openings of insects such as this are rendered hydrophobic either by oily secretions or by being surrounded by arrays of *hydrofuge* ("water-repelling") *hairs*. In this way, when the animals are at the water's surface, the water pulls away from the spiracular openings rather than adhering, thus granting free access to the air.

Breathing from Compressible Bubbles (Compressible Gas Gills)   Many insects carry packets of air on the outside of their bodies when under water. These packets communicate with the tracheal system via spiracles. Probably everyone has observed bubbles being carried under the wings of water beetles. Bubbles are also carried at the posterior tip of the abdomen or among the legs in various species. These bubbles become established and persist because hydrofuge structures on the body surface render the affinity of the surface to air greater than that to water.

It is obvious that bubbles contain a store of oxygen when they are established at the water's surface. It is less obvious that this store can be replenished while the insect is under water. As oxygen is withdrawn from a bubble, the oxygen tension within the bubble falls. When it has fallen below the oxygen tension of the surrounding water, dissolved oxygen from the water will diffuse into the bubble, renewing its oxygen supply. The bubble thus can act as a *gas gill*.

A complete analysis of the dynamics of a bubble requires attention to carbon dioxide and nitrogen as well as oxygen. Before undertaking that analysis, we must state that we are specifically concerned here with bubbles classed as *compressible*. By definition, such bubbles are ones in which the hydrostatic pressure of the bubble gas remains always the same as the ambient water pressure. Because of this property, compressible bubbles shrink in proportion to any drop in their total gas content.

Carbon dioxide is so soluble in water that when it is added to a bubble from metabolism, it diffuses away rapidly into the environment. Thus, the carbon dioxide tension in a bubble remains low, and, effectively, oxygen withdrawn from a bubble by metabolism is not replaced with an equivalent volume of carbon dioxide. This means that the percent composition of nitrogen and the nitrogen tension in a bubble will rise as the oxygen tension in the bubble falls. Thus, there will be a tendency for nitrogen to diffuse out of a bubble into water of lower nitrogen tension, as well as a tendency for oxygen to diffuse

into the bubble. The loss of nitrogen will gradually reduce the size of the bubble, with two detrimental effects on the functioning of the bubble as an oxygen supply. First, as the size decreases, the surface area of the gas gill will decrease, and second, the smaller bubble will contain a smaller reserve of oxygen. Eventually, the insect will have to return to the surface and renew the bubble from the atmosphere. Bubbles last longer under water and are more effective in extracting oxygen than might at first be expected because of the different properties of oxygen and nitrogen. At given tension difference, oxygen diffuses across an air–water interface about three times as rapidly as nitrogen. This means that when the composition of a bubble is altered by metabolic oxygen uptake, there is a physical bias toward reestablishing the original composition by inward diffusion of oxygen from the water rather than by outward diffusion of nitrogen.

It is instructive to consider the outcome of an intriguing experiment on back-swimmer bugs (*Notonecta*), which carry a bubble on their ventral surface. Bugs in a system equilibrated with atmospheric air lived under water for 6 or 7 h. Surprisingly, however, if the atmosphere was pure oxygen and the water was equilibrated with pure oxygen, the bugs survived under water for only 35 min. How could this be? Underwater survival of the bugs in pure oxygen was shortened because their gas gill was nonfunctional. As oxygen was withdrawn from their bubble, only oxygen remained, and little or no tension gradient was established between the bubble and water. Thus, the animals could use the oxygen carried in their bubble from the water's surface but could not gain appreciable amounts of oxygen by diffusion when submerged. This experiment illustrates the essential role played by the inert gas nitrogen in the functioning of a compressible gas gill. It is the presence of nitrogen that permits the oxygen in a bubble to be diluted by metabolism, thus establishing a favorable tension gradient for inward diffusion of oxygen. We also see that because the bugs in the air-equilibrated system survived 10–12 times as long as the bugs in pure oxygen even though they carried less oxygen in their bubbles at the time of submergence, oxygen derived from the gill action of a bubble during submergence far exceeds that obtained at the water's surface.

Plastron Respiration: Incompressible Gas Gills   In some aquatic insects, parts of the body are covered extremely densely with fine hydrofuge hairs; the bug *Aphelocheirus aestivalis,* for example, has between 2 and 2.5 million hairs per *square millimeter*. Such densely distributed hairs represent one means of trapping a thin film of gas—known as a *plastron*— that *cannot be displaced*. A plastron remains constant in volume over a wide range of conditions, including exposure of the insect to water that is entirely free of dissolved gases (wherein a substantial

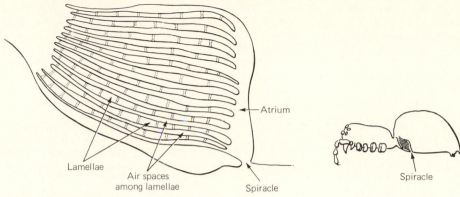

Figure 12.41 Section of a book lung of a spider (semidiagrammatic). Inset shows the position of the book lungs in the anteroventral abdomen of a two-lunged spider. The lamellae of the book lung are attached not only at the anterior wall of the lung cavity but also along both lateral walls of the cavity. [From J.H. Comstock, *The Spider Book*. Doubleday, Garden City, NY, 1912.]

vacuum must develop in the plastron). Because the gas space is incompressible and permanent in plastron breathers, it can serve indefinitely as a gas gill, unlike the compressible type of gas gill already discussed. Some insects with plastron respiration remain submerged for months in well-aerated water.

Breathing Directly from the Water; Tracheal Gills   In many aquatic insects—unlike those discussed heretofore—the spiracles are obliterated or otherwise nonfunctional. Usually, the tracheal system remains gas filled and thus continues to function as the path of least resistance for diffusion of gases within the body. However, all the body surfaces are bathed with water, and the superficial branches of the tracheal system undergo gas exchange with the water by way of diffusion across the integument.

Commonly, there is a dense proliferation of fine tracheae under the general integument of these insects. In addition—and in a remarkable parallel with numerous other groups of aquatic animals—many of these insects have evolved evaginations of their body surface that are densely supplied with superficial tracheae and covered with just a thin cuticle. Such evaginations, termed *tracheal gills,* may occur on the outer body surface or in the rectum. The larvae of caddis flies and nymphs of dragonflies are examples of insects possessing tracheal gills.

ARACHNIDS

Scorpions, many spiders, and some other arachnids possess a novel type of respiratory structure, the ***book lung*** (Figure 12.41), which may or may not bear any homology with the book gills of horseshoe crabs. The number of book lungs varies from a single pair (as in certain spiders) to four pairs (in scorpions). Book lungs represent invaginations of the ventral abdomen and are lined with a thin cuticle. Each consists of a chamber, the atrium, opening to the outside by a closable ventral pore, or spiracle. The dorsal or anterior surface of the atrium is thrown into a great number of lamellar folds. Blood circulates within the lamellae, whereas the spaces among them are filled with gas. Lamellae commonly number into the hundreds, and the blood-to-gas distance is often just 1 μm or less.

Calculations indicate that diffusion should be adequate to meet requirements for gas exchange between the lamellar surfaces and ambient air. In some spiders, the book lungs in fact function as diffusion lungs. Other species of spiders are reported to ventilate the atrium. Transport of oxygen and carbon dioxide between the book lungs and the rest of the body is by circulation of the blood.

The scorpions have only book lungs, but tracheal systems are found in many spiders. Basically, spiders have two pairs of respiratory structures: either two pairs of book lungs, or two pairs of tracheal trees, or one pair of each. The tracheal systems take two forms. Sometimes, a tracheal tree consists of a tuft of many relatively short tubes and functions analogously to a book lung; the tracheae do not ramify throughout the body but simply rest in abdominal blood spaces, and the blood must carry gases between the tracheae and the rest of the animal. This form of tracheal system has been termed a ***tracheal lung***. In other cases, the tracheae ramify throughout the body.

SELECTED READINGS

Anderson, J.F. and K.N. Prestwich. 1982. Respiratory gas exchange in spiders. *Physiol. Zool.* **55**:72–90.

Barnhart, M.C. and B.R. McMahon. 1987. Discontinuous carbon dioxide release and metabolic depression in dormant land snails. *J. Exp. Biol.* **128**:123–138.

Booth, C.E. and C.P. Mangum. 1978. Oxygen uptake and transport in the lamellibranch mollusc *Modiolus demissus. Physiol. Zool.* **51**:17–32.

Bouverot, P. 1985. *Adaptation to Altitude–Hypoxia in Vertebrates.* Springer-Verlag, New York.

*Brackenbury, J.H. 1981. Airflow and respired gases within the lung–air-sac system of birds. *Comp. Biochem. Physiol.* **68A**:1–8.

Bramble, D.M. and D.R. Carrier. 1983. Running and breathing in mammals. *Science* **219**:251–256.

Brendel, W. and R.A. Zink (eds.). 1982. *High Altitude Physiology and Medicine.* Springer-Verlag, New York.

Brockway, A.P. and H.A. Schneiderman. 1967. Strain-gauge transducer studies on intratracheal pressure and pupal length during discontinuous respiration in diapausing silkworm pupae. *J. Insect Physiol.* **13**:1413–1451.

*Burggren, W.W., K. Johansen, and B. McMahon. 1985. Respiration in phyletically ancient fishes. In R.E. Forman, A. Gorbman, J.M. Dodd, and R. Olsson (eds.), *Evolutionary Biology of Primitive Fishes,* pp. 217–252. Plenum, New York.

Burggren, W.W. and N.H. West. 1982. Changing respiratory importance of gills, lungs and skin during metamorphosis in the bullfrog *Rana catesbeiana. Respir. Physiol.* **47**:151–164.

Burton, F.G. and S.G. Tullett. 1985. Respiration of avian embryos. *Comp. Biochem. Physiol.* **82A**:735–744.

Comparative Physiology of Respiration in Vertebrates (A Symposium). 1972. *Respir. Physiol.* **14**:1–236.

Comroe, J.H., Jr. 1974. *The Physiology of Respiration,* 2nd ed. Year Book Medical, Chicago.

*Dejours, P. 1981. *Principles of Comparative Respiratory Physiology,* 2nd ed. Elsevier/North Holland, New York.

Dempsey, J.A. and H.V. Forster. 1982. Mediation of ventilatory adaptations. *Physiol. Rev.* **62**:262–346.

Dickinson, P.S., D.J. Prior, and C. Avery. 1988. The pneumostome rhythm in slugs: A response to dehydration controlled by hemolymph osmolality and peptide hormones. *Comp. Biochem. Physiol.* **89A**:579–585.

Duncker, H.-R. 1974. Structure of the avian respiratory tract. *Respir. Physiol.* **22**:1–19.

*Duncker, H.-R. 1978. General morphological principles of amniotic lungs. In J. Piiper (ed.), *Respiratory Function in Birds, Adult and Embryonic,* pp. 2–15. Springer-Verlag, New York.

Eldridge, F.L., D.E. Millhorn, and T.G. Waldrop. 1981. Exercise hyperpnea and locomotion: Parallel activation from the hypothalamus. *Science* **211**:844–846.

Gans, C. 1970. Respiration in early tetrapods—The frog is a red herring. *Evolution* **24**:723–734.

Heath, D. and D.R. Williams. 1981. *Man at High Altitude,* 2nd ed. Churchill Livingstone, London.

*Hornbein, T.H. (ed.). 1981. *Regulation of Breathing.* Dekker, New York.

*Houlihan, D.F., J.C. Rankin, and T.J. Shuttleworth (eds.). 1982. *Gills.* Cambridge University Press, Cambridge.

Hughes, G.M. (ed.). 1976. *Respiration of Amphibious Vertebrates.* Academic, New York.

Hughes, G.M. and M. Morgan. 1973. The structure of fish gills in relation to their respiratory function. *Biol. Rev.* **48**:419–475.

Innes, A.J. and E.W. Taylor. 1986. The evolution of air-breathing in crustaceans: a functional analysis of branchial, cutaneous and pulmonary gas exchange. *Comp. Biochem. Physiol.* **85A**:621–637.

Johansen, K. 1970. Air breathing in fishes. In W.S. Hoar and D.J. Randall (eds.), *Fish Physiology,* Vol. 4. Academic, New York.

Kiceniuk, J.W. and D.R. Jones. 1977. The oxygen transport system in trout (*Salmo gairdneri*) during sustained exercise. *J. Exp. Biol.* **69**:247–260.

Krogh, A. 1941. *The Comparative Physiology of Respiratory Mechanisms.* University of Pennsylvania Press, Philadelphia.

Lenfant, C. 1973. High altitude adaptation in mammals. *Am. Zool.* **13**:447–456.

Lenfant, C. and K. Johansen. 1968. Respiration in the African lungfish, *Protopterus aethiopicus. J. Exp. Biol.* **49**:437–452.

Liem, K.F. 1981. Larvae of air-breathing fishes as countercurrent flow devices in hypoxic environments. *Science* **211**:1177–1179.

Maclean, G.S. 1981. Factors influencing the composition of respiratory gases in mammal burrows. *Comp. Biochem. Physiol.* **69A**:373–380.

Metcalfe, J., M.K. Stock, and R.L. Ingermann (eds.). 1987. *Development of the Avian Embryo.* Liss, New York.

*Mill, P.J. 1972. *Respiration in the Invertebrates.* Macmillan, London.

Paul, R., T. Finke, and B. Linzen. 1987. Respiration in the tarantula *Eurypelma californicum:* evidence for diffusion lungs. *J. Comp. Physiol.* **157B**:209–217.

Perry, S.F. 1983. Reptilian lungs. Functional anatomy and evolution. *Adv. Anat. Embryol. Cell Biol.* **79**:1–81.

*Piiper, J. (ed.). 1978. *Respiratory Function in Birds, Adult and Embryonic.* Springer-Verlag, New York.

*Piiper, J. 1982. Respiratory gas exchange at lungs, gills and tissues: Mechanisms and adjustments. *J. Exp. Biol.* **100**:5–22.

Piiper, J. and P. Scheid. 1975. Gas transport efficacy of gills, lungs, and skin: Theory and experimental data. *Respir. Physiol.* **23**:209–221.

Rahn, H. and B.J. Howell. 1976. Bimodal gas exchange. In G.M. Hughes (ed.), *Respiration of Amphibious Vertebrates,* pp. 271–285. Academic, New York.

Rahn, H. and C.V. Paganelli. 1968. Gas exchange in gas gills of diving insects. *Respir. Physiol.* **5**:145–164.

Randall, D. 1982. The control of respiration and circulation in fish during exercise and hypoxia. *J. Exp. Biol.* **100**:275–288.

*Randall, D.J., W.W. Burggren, A.P. Farrell, and M.S. Haswell. 1981. *The Evolution of Air-Breathing in Vertebrates.* Cambridge University Press, New York.

Scheid, P., C. Hook, and C.R. Bridges. 1982. Diffusion in gas exchange of insects. *Fed. Proc.* **41**:2143–2145.

Scheid, P. and J. Piiper. 1972. Cross-current gas exchange in avian lungs: Effects of reversed parabronchial air flow in ducks. *Respir. Physiol.* **16**:304–312.

Schmidt-Nielsen, K. 1972. *How Animals Work.* Cambridge University Press, London.

Seller, T.J. (ed.). 1987. *Bird Respiration.* CRC Press, Boca Raton, Florida.

Sinha, N.P. and P. Dejours. 1980. Ventilation and blood acid–base balance of the crayfish as functions of water oxygenation (40–1500 torr). *Comp. Biochem. Physiol.* **65A**:427–432.

Snyder, G.K. 1987. Capillary growth and diffusion distances in muscle. *Comp. Biochem. Physiol.* **87A**:859–861.

*Steen, J.B. 1971. *Comparative Physiology of Respiratory Mechanisms*. Academic, New York.

Sutton, J.R. and N.L. Jones. 1983. Exercise at altitude. *Annu. Rev. Physiol.* **45**:427–437.

Taylor, E.W. 1982. Control and co-ordination of ventilation and circulation in crustaceans: Responses to hypoxia and exercise. *J. Exp. Biol.* **100**:289–319.

Tenny, S.M. and D. Bartlett, Jr. 1981. Some comparative aspects of the control of breathing. In T.H. Hornbein (ed.), *Regulation of Breathing,* pp. 67–101. Dekker, New York.

Vogel, S. 1981. *Life in Moving Fluids*. Willard Grant Press, Boston.

*Weibel, E.R. 1984. *The Pathway for Oxygen*. Harvard University Press, Cambridge.

*Weibel, E.R. and C.R. Taylor (eds.). 1981. Design of the mammalian respiratory system. *Respir. Physiol.* **44**:1–86, 151–164.

Weis-Fogh, T. 1967. Respiration and tracheal ventilation in locusts and other flying insects. *J. Exp. Biol.* **47**:561–587.

Wells, M.J. and J. Wells. 1982. Ventilatory currents in the mantle of cephalopods. *J. Exp. Biol.* **99**:315–330.

West, J.B. 1982. Respiratory and circulatory control at high altitudes. *J. Exp. Biol.* **100**:147–157.

*West, J.B. 1984. Human physiology at extreme altitudes on Mount Everest. *Science* **223**:784–788.

West, J.B. and S. Lahiri. 1984. *High Altitude and Man*. American Physiological Society, Bethesda.

Whitford, W.G. 1973. The effects of temperature on respiration in the Amphibia. *Am. Zool.* **13**:505–512.

*Wood, S.C. and C. Lenfant (eds.). 1979. *Evolution of Respiratory Processes. A Comparative Approach*. Dekker, New York.

See also references in Appendix A.

chapter 13

Exchanges of Oxygen and Carbon Dioxide 2: Transport in Body Fluids

With an Introduction to Acid–Base Physiology

In many animals, the respiratory gases must be swept from one region of the body to another by circulation of the blood or other body fluids if adequate rates of internal gas transport are to be achieved (p. 255). In this chapter, we focus on the respiratory functions of flowing body fluids. We look first at oxygen transport and later at carbon dioxide transport. As part of the discussion of carbon dioxide, we also consider the general principles of acid–base physiology.

RESPIRATORY PIGMENTS: INTRODUCTION

If we consider an animal that employs circulation of blood to transport oxygen to its tissues, we quickly recognize that the rate of blood flow required to meet a given oxygen demand will tend to be an inverse function of the amount of oxygen carried by each unit volume of blood. The required rate of blood flow, for example, would be reduced by a factor of 10 if the amount of oxygen carried per unit volume of blood were increased 10-fold, assuming a consistent fraction of the oxygen carried by the blood to be yielded to the tissues. Although the circulation of the blood serves many functions, the demand for oxygen transport often overrides other factors in setting the rate of the circulation (p. 334). Thus, increasing the capacity of each unit volume of blood to carry oxygen can greatly reduce the circulatory effort required of an animal.

Oxygen dissolves in blood and other body fluids in accord with the physical laws outlined in Chapter 11. The amount of oxygen carried in solution per unit volume increases in proportion to the prevailing oxygen tension. However, at physiological tensions,

the dissolved oxygen concentration of body fluids is never very great. In animals that depend on the circulation of body fluids for oxygen transport, the capacity of the fluids to carry oxygen commonly has been increased to well above that permitted by simple solution by the evolution of specialized oxygen-transport compounds. These are compounds that are dissolved or suspended in the circulating body fluid and that *undergo reversible chemical combination with oxygen*. Human hemoglobin is one example. Human blood contains enough hemoglobin that chemical combination of oxygen molecules with hemoglobin accounts for over 98 percent of the oxygen carried by the blood when leaving the lungs; less than 2 percent is in simple solution.

When we say that a compound undergoes reversible combination with oxygen, we mean that it tends to take up oxygen when exposed to high oxygen tensions and release oxygen when oxygen tensions are low. The compounds in body fluids that function in this way fall into four groups: the *hemoglobins, hemocyanins, hemerythrins,* and *chlorocruorins*. All these compounds are pigments and proteins. Thus, they are collectively called the *respiratory pigments* or *respiratory proteins*.

It is to be stressed that the hemoglobins, hemocyanins, hemerythrins, and chlorocruorins are *groups* of compounds. There are many different hemoglobins, for example, all classed together because they possess heme groups in their molecular structure.

Although respiratory pigments commonly serve to aid the routine transport of oxygen by circulating body fluids, they may also play other roles. Hemoglobins, for example, are often found within muscle or nerve cells, where they function (in part) to facilitate diffusion of oxygen through the cells. Respiratory pigments may also serve as storage depots for oxygen.

RESPIRATORY PIGMENTS: DISTRIBUTION AND CHEMICAL PROPERTIES

Circulating respiratory pigments are virtually universal among the relatively large and active members of the major higher phyla—with the notable exception of the insects and other tracheate arthropods, which generally do not depend on circulation for internal gas transport (p. 289). Respiratory pigments are typically absent in the animals that lack a circulatory system or other system in which mass internal circulation can occur (e.g., a coelomic cavity); sponges, coelenterates, and most flatworms lack the pigments, for example. Respiratory pigments are also absent in some animals that do depend on circulation of body fluids for internal gas transport; these creatures are typically ones that have relatively low oxygen demands, such as many of the echinoderms.

All the known respiratory pigments are metalloproteins. The metal in the pigment molecules is iron or copper, bound in an organic complex. Combination with oxygen occurs at the site of the metal complex. The pigments combine with and release molecular oxygen: O_2.

The process of combining with oxygen is termed *oxygenation,* whereas that of releasing oxygen is called *deoxygenation.* The prefixes *oxy-* and *deoxy-,* when appended to the name of a respiratory pigment, refer, respectively, to pigment that is combined, or not combined, with oxygen; for example, hemoglobin combined with oxygen is called *oxyhemoglobin,* whereas hemoglobin devoid of bound oxygen is termed *deoxyhemoglobin.* Oxygenation is not synonymous with oxidation. When a respiratory pigment combines with oxygen, electrons are partially transferred from the metal atoms in the pigment to the oxygen molecules. Thus, there is a tendency toward oxidation of the pigment. However, the pigment is not truly oxidized in the simple sense. To emphasize this point, consider what happens when the iron atoms in molecules of hemoglobin *are* oxidized—in the full sense of the word—from their ordinary ferrous state to the ferric state: The hemoglobin molecules lose their ability to combine with oxygen and become useless for oxygen transport.

Hemoglobins

Chemical Properties   The basic molecular unit of hemoglobins consists of a **heme** group (Figure 13.1) bound to a protein—or **globin**—moiety. Heme is a particular metalloporphyrin, specifically ferrous protoporphyrin. (The complex ring structure surrounding the ferrous iron in Figure 13.1 is known as pro-

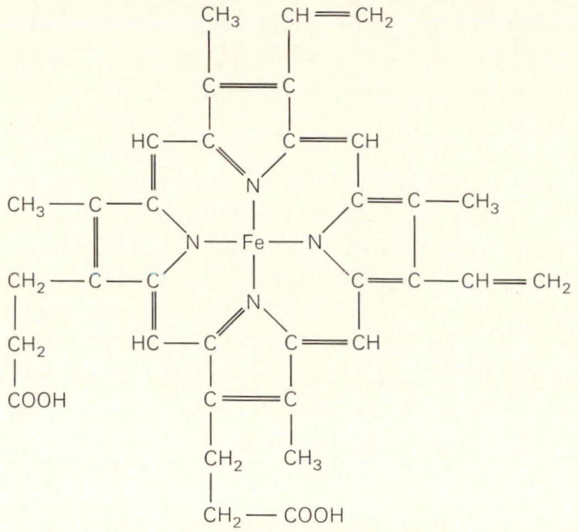

Figure 13.1 The chemical structure of heme. Ferrous iron is complexed with protoporphyrin. The positions assigned to double and single bonds in the porphyrin ring are arbitrary inasmuch as resonance occurs. Resonance also occurs in the central complex of iron with the four nitrogen atoms.

toporphyrin.) So far as is known, the heme groups of all hemoglobins are identical. Hemoglobins of different species differ in their chemical and physical properties, and several different hemoglobins may occur normally in any one species. These differences are attributable to differences in the globin moieties conjugated with heme. Indeed, only small differences in the globin moiety can cause highly significant alterations in the properties of hemoglobin. Consider, for example, the numerous (100+) mutant forms of human hemoglobin. Studies have revealed that most differ from the normal adult hemoglobin in only one of the over 140 amino acid residues conjugated with heme. These seemingly small changes in the amino acid composition can nonetheless substantially raise or lower the affinity of the hemoglobin for oxygen or critically change other properties such as solubility or structural stability.

Oxygen molecules combine with the heme loci in hemoglobin molecules. One oxygen molecule can combine with each heme group.

Unit molecules of hemoglobin, consisting of a heme group and associated globin, are frequently linked to form larger molecules. The blood hemoglobins of vertebrates, for example, are usually four-unit (tetrameric) molecules. The molecular weight of each unit molecule is typically about 16,000–17,000 in vertebrates. The four-unit blood hemoglobins thus have molecular weights of approximately 64,000–68,000. In the usual adult blood hemoglobin of humans there are two types of unit molecule, termed α and β, which differ in the composition of their globin moieties. Each molecule of the blood hemoglobin consists of two α units, each with 141 amino acid residues, and two β units, each with 146 amino acid residues. Because each molecule of the blood hemoglobin includes four heme groups, it can bind with four molecules of oxygen. Relatively huge hemoglobin molecules are found in some invertebrates. For example, in numerous annelids, including earthworms (*Lumbricus*), the blood hemoglobin has a molecular weight near 3 million. Such large molecules often contain around 100 or more oxygen-binding sites. In many of the invertebrate hemoglobins, there is one heme group per 14,000–17,000 units of molecular weight. In many others, however, a different ratio is found; for example, there is one heme per 22,000–30,000 units in a good number of the annelid hemoglobins of great molecular size.

Hemoglobins located in different parts of an animal often differ structurally and in their chemical and physical properties. Whereas the usual blood hemoglobin of adult humans, for example, is a four-unit molecule consisting of α and β chains, the hemoglobin found in human muscle tissue (myoglobin) is a single-unit molecule of molecular weight 17,800, consisting of one heme group and a globin moiety of different structure than the α or β chains in blood hemoglobin. Not uncommonly, the nature of hemoglobin changes over the life cycle. To again use hu-

mans as an example, the blood hemoglobin of the fetus is different from that of adults, being a four-unit molecule composed of two α chains (as in the adult) and two so-called γ chains.

In the adults of some species, the blood hemoglobin is of essentially uniform composition. It is increasingly clear, however, that in many species of poikilothermic vertebrates and invertebrates, the blood of adults normally contains considerable quantities of two, three, or more chemically different forms of hemoglobin. The hemoglobin of the sucker *Catostomus clarkii,* for example, consists of two major sets of components, one set accounting for about 80 percent of all hemoglobin and the other for 20 percent. When multiple chemical forms of hemoglobin occur in a species, the forms sometimes differ substantially in their oxygen-binding characteristics. Possession of multiple blood hemoglobins may thus permit a species to maintain adequate oxygen transport over a broader range of conditions than would be possible with only a single hemoglobin type (p. 323).

BOX 13.1 ABSORPTION SPECTRA

The hemoglobins and other respiratory pigments—in common with pigments in general—differentially absorb various wavelengths of light. The pattern of absorption—known as the absorption spectrum—can be used to determine the chemical class of a respiratory pigment. In addition, the amount of absorption at a given wavelength can be used to determine the pigment concentration in the blood. The absorption spectrum of hemoglobin and other respiratory pigments changes with oxygenation and deoxygenation (Figure 13.2). These changes are qualitatively evident to our eyes: We know, for example, that oxygenated hemoglobin is bright red, but deoxygenated hemoglobin is more purplish. The percentage of heme groups oxygenated in a blood sample can be determined by quantitative absorption measurements.

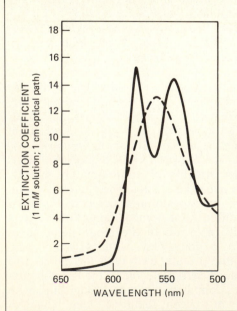

Figure 13.2 Absorption spectra for fully oxygenated (solid line) and fully deoxygenated (dashed line) human hemoglobin in the green through red parts of the spectrum. Spectra were determined by passing light through a hemoglobin solution and measuring the fraction of the light absorbed at each wavelength. The extinction coefficient varies directly with absorption: A high coefficient signifies a high degree of absorption. Additionally, strong absorption bands (not shown) occur in the violet–ultraviolet regions of the spectrum (peaking at 415 nm for oxyhemoglobin and 430 nm for deoxyhemoglobin). [After M.R. Waterman, Spectral characterization of human hemoglobin and its derivatives. In S. Fleischer and L. Packer (eds.), *Methods in Enzymology,* Vol. 52, pp. 456–463. Academic, New York, 1978.]

Distribution Hemoglobins are the most widely distributed of the respiratory pigments, being found in over 10 phyla (Figure 13.3). They are the only respiratory pigments of vertebrates, and with a few interesting exceptions (p. 316), all vertebrates have hemoglobin in their blood. The circulating hemoglobins of vertebrates are consistently contained in blood cells, the *erythrocytes*. Furthermore, as noted earlier, they are almost always four-unit (tetrameric) molecules. Muscle hemoglobins, termed *myoglobins,* are widespread in vertebrates and appear always to be single-unit (monomeric) molecules. Myoglobins tend to be especially concentrated in active muscles, such as the heart, and then impart a distinctly reddish color to the tissue (p. 524).

The distribution of hemoglobins among invertebrates is not only wide but sporadic. Hemoglobins may occur within certain subgroups of a phylum but not others and even within certain species but not other closely related species. Sometimes, out of all the members of a large assemblage of related species, just an isolated few possess hemoglobins; this is the case among insects, for example. The evolution of the wide but sporadic distribution of hemoglobins is difficult to understand. One possibility is that hemoglobins have evolved independently many times. This notion is not as farfetched as might at first appear. The cytochrome pigments of the electron-transport system, which are found universally in animals and undoubtedly evolved very early, consist of protein groups conjugated either with heme or with other closely similar iron porphyrins. Animals might have evolved hemoglobins on many separate occasions as an extension of their ancient ability to synthesize the chemically similar cytochromes.

The circulating hemoglobins of invertebrates—occasionally known as *erythrocruorins* (''red cruorins'')—may be found in blood or in other moving fluids, such as the fluids filling the coelomic or pseudocoelomic cavity. Sometimes, as in vertebrates, these hemoglobins are contained within cells. In other cases, however, the hemoglobins are simply dissolved in the body fluid. The hemoglobin in the blood of earthworms (*Lumbricus*) is dissolved, for example; and strikingly, when the blood is held to the light, it is wine red and *clear*—unlike vertebrate bloods, which are opaque because of their high concentration of red blood cells. The intracellular circulating hemoglobins of invertebrates are always of relatively low molecular weight ($\sim$14,000–70,000); structurally, they are generally one-, two-, or four-unit molecules. In sharp contrast, the extracellular, dissolved hemoglobins are almost always large, multiunit molecules having molecular weights of 200,000 to 12 million.

Circulating hemoglobins are widespread in annelid worms; although most commonly dissolved in the blood, they sometimes are found in coelomic cells or even dissolved in the coelomic fluid. Among arthropods, circulating hemoglobins—dissolved in the

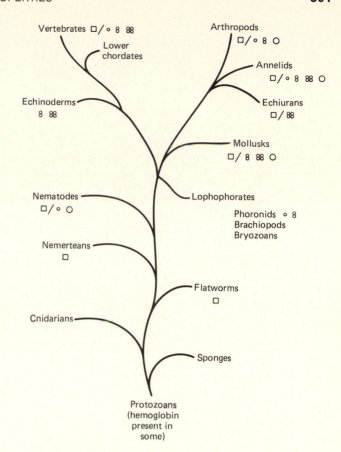

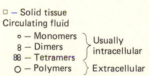

□ — Solid tissue
Circulating fluid
  ∘ — Monomers ⎫ Usually
  8 — Dimers ⎬ intracellular
  88 — Tetramers ⎭
  ○ — Polymers ⎫ Extracellular

Figure 13.3 Summary of the known phyletic distribution of hemoglobins (not exhaustive). A square indicates that hemoglobins occur in solid tissues such as muscle or nerve. Circles indicate the occurrence of hemoglobins in circulating body fluids. The small circle symbol (∘) corresponds to one unit molecule of heme plus globin: a hemoglobin monomer. If a single small circle is shown next to a phyletic group, monomeric molecules are found in circulating body fluids in some members of the group. A pair or foursome of small circles indicates, respectively, the occurrence in circulating body fluids of dimers or tetramers—molecules consisting of two or four joined unit molecules. A large circle (○) indicates the occurrence of polymeric hemoglobins of high molecular weight, consisting of many joined unit molecules. The polymeric hemoglobins are always extracellular: dissolved in the circulating fluid. With few exceptions, the monomeric, dimeric, and tetrameric forms are intracellular—contained within circulating cells (e.g., erythrocytes). Not all members of a phyletic group have all forms of hemoglobin shown. Most vertebrates, for example, have only tetrameric hemoglobin in their blood; the monomeric and dimeric forms are found just among cyclostome fish. Whereas some arthropods have hemoglobins, many others (e.g., most insects) lack them. [After R.C. Terwilliger, *Am. Zool.* **20**:53–67 (1980).]

blood—are found in a good number of the small, branchiopod crustaceans (e.g., *Daphnia, Artemia*) and a few insects but are absent from malacostracan crustaceans (e.g., crabs, crayfish). Numerous sea cucumbers (holothuroideans) have corpuscular hemoglobins in their blood and coelomic fluid; most other echinoderms lack respiratory pigments. Among mollusks, hemoglobin is found in the blood of some planorbid snails (dissolved) and a few clams (dissolved or corpuscular).

The concentration of blood hemoglobin in some branchiopod crustaceans, midge larvae (*Chironomus*), snails (*Planorbis*), and other invertebrates waxes and wanes dramatically with changes in the environmental availability of oxygen. Water fleas (*Daphnia*), for example, have little hemoglobin and are pale in appearance when they have been living in well-aerated waters. If ambient oxygen tensions become low, however, they increase their levels of hemoglobin markedly within a matter of days and become pink or red.

Hemoglobins are found widely in muscle and other solid tissues of invertebrates, usually as single-unit molecules. Muscle hemoglobins occur among mollusks, annelids, and parasitic nematodes. Some mollusks and annelids have nerve hemoglobins, which may impart a striking, pinkish color to their ganglia or nerves. A few bivalves have hemoglobins in the tissue of their gills. In a few insects, such as some of the backswimmer bugs, hemoglobins occur in tracheal organs, where they store oxygen for release to the tracheae during diving.

Hemocyanins

Hemocyanins are found in just two phyla, the arthropods and mollusks, but clearly rank as the second most common of the respiratory pigments. In turning to them, we encounter a minor problem common also to the chlorocruorins and hemerythrins: The names given to these compounds provide no clue to their chemical structure. Hemocyanins do not contain heme groups and, in fact, contain neither iron nor porphyrin. The metal they contain is copper, bound directly to the protein. Each oxygen-binding site contains two copper atoms. The binding ratio is thus one O_2 per two Cu. Hemocyanins are always found dissolved in the blood plasma and are always of large molecular size. Although colorless when deoxygenated, they turn a bright blue when oxygenated and then, if concentrated enough, impart a distinctly blue color to the blood.

Among mollusks, hemocyanins are found in the cephalopods, in many chitons and gastropods, and in some protobranch bivalves. They are not present in most bivalves; indeed, bivalves usually lack circulating respiratory pigments of any kind. The groups of arthropods in which hemocyanins occur are the malacostracan crustaceans (crabs, lobsters, shrimp, crayfish), the horseshoe crabs, and the spiders and scorpions. The molluscan and arthropod hemocyanins differ systematically in certain structural respects, suggesting that they may well be independently evolved. The molluscan pigments have one pair of copper atoms (one oxygen-binding site) per 50,000 units of molecular weight, whereas arthropod hemocyanins have a pair for every 70,000–75,000 units or so. The molecular weights of the molluscan pigments are commonly 4–9 million. Arthropod molecular weights are 0.4–3.3 million.

Hemocyanins are not found within muscle or other solid tissues. In those mollusks that have muscular respiratory pigments, they are hemoglobins.

When certain shrimps and lobsters are exposed to oxygen-poor environments, they evidently increase their blood concentration of hemocyanin, just as certain hemoglobin-bearing animals increase their blood content of hemoglobin under similar conditions.

Chlorocruorins

The chlorocruorins are found in just four families of polychaete annelids: the fanworms—Sabellidae and Serpulidae—and the Ampharetidae and Flabelligeridae. Chlorocruorins are always found in plasma solution and bear close chemical similarities to the ex-

---

BOX 13.2   WHY ARE DISSOLVED PIGMENTS LARGE?

When resiratory pigments are dissolved in the blood or other body fluids, they are almost always of large molecular size. This is true of not only the hemoglobins but also the hemocyanins and the chlorocruorins, and the question of why this should be so has repeatedly been raised. Firm answers cannot yet be given, but there are a number of considerations that seem likely to be important. Large molecular size may in part be an adaptation to confining dissolved pigments within the circulatory system; large dissolved molecules are less likely than small ones to diffuse out of the blood or be forced out by ultrafiltration (p. 151). Large size may also serve to prevent a dissolved pigment from imparting an excessive colloid osmotic pressure to the blood plasma (p. 148). Osmotic pressure depends on *molecular* concentration, and a relatively low molecular concentration of large molecules can provide as many oxygen-binding sites as a high molecular concentration of small molecules.

tracellular hemoglobins that are so commonly found dissolved in the blood of annelids. These chemical similarities are great enough that a close evolutionary affinity between the chlorocruorins and extracellular hemoglobins cannot be doubted. Like the extracellular hemoglobins, the chlorocruorins are large molecules, with molecular weights near 3 million, composed of iron porphyrin groups conjugated with protein. They bind one molecule of oxygen per iron-porphyrin group. With their many similarities to hemoglobins, the characteristic that has formed the basis for setting the chlorocruorins apart may emerge as being minor: The type of iron porphyrin they contain, although very similar to heme, differs in that one of the vinyl chains ($-CH=CH_2$) on the periphery of the protoporphyrin ring in heme (Figure 13.1) is replaced with a formyl group ($-CHO$). This difference gives the chlorocruorins a distinctive and striking color. In dilute solution, they are greenish. In more concentrated solution, they are deep red by transmitted light but greenish by reflected light. Chlorocruorin means "green cruorin."

Hemerythrins

The hemerythrins have a distribution that is puzzling for being simultaneously limited and farflung, encompassing four phyla. Hemerythrins occur in all known sipunculid worms, in some brachiopods (lamp shells), in both genera of the small phylum Priapulida, and in one genus only of polychaete annelids (*Magelona*). Despite their name, hemerythrins do not contain heme groups. They do contain ferrous iron, bound directly to the protein. Each oxygen-binding site contains two iron atoms, and there is one such site per 13,000–14,000 units of molecular weight. In some instances, single-unit hemerythrins, known as *myohemerythrins,* occur within muscle tissue. Better known are the circulating hemerythrins, which are always located intracellularly (e.g., in circulating coelomic or blood cells) and have molecular weights of about 40,000–110,000; many are octomeric, having eight oxygen-binding sites per molecule. Hemerythrins are colorless when deoxygenated but turn reddish-violet when oxygenated.

RESPIRATORY PIGMENTS: OXYGEN-BINDING CHARACTERISTICS

The combination of oxygen with each oxygen-binding locus of a respiratory pigment is stoichiometric. That is, one and only one molecule of oxygen can bind with each heme group of a hemoglobin or with each pair of copper atoms in a hemocyanin. In a body fluid containing a respiratory pigment, there is a large population of oxygen-binding loci. Human blood, for example, contains about 5.4×10^{20} heme groups per 100 mL. The fraction of these loci that is oxygenated depends in part on the oxygen tension.

If the tension is sufficiently high that all loci are oxygenated, the respiratory pigment is said to be **saturated**. At saturation, the heme groups in 100 mL of human blood collectively carry about 20 mL of oxygen, for example.

When we say that 20 mL (or any other volume) of oxygen is bound to a respiratory pigment, we do not mean that the oxygen actually occupies 20 mL in the blood. Rather, 20 mL is the volume that the oxygen will occupy, at STP, if released from the respiratory pigment into its gaseous state. Levels of oxygen in blood are often expressed in units of *volumes percent (vol %):* the volume of oxygen (at STP) carried per 100 volumes of blood. Blood carrying 20 mL of O_2 per 100 mL, for example, is said to have an oxygen content of 20 volumes percent.

When blood is equilibrated with oxygen tensions progressively lower than the tension required for saturation, the fraction of oxygenated binding sites falls and, with it, the amount of oxygen carried by the respiratory pigment. The degree of oxygenation at any given tension can be expressed as the percentage of binding sites oxygenated, as oxygen content in volumes percent, or as the percentage of saturation oxygen content (*percent saturation*). The first and third of these expressions are equal, and the second is proportional to them. For example, when 50 percent of the heme groups in human blood are oxygenated, the hemoglobin carries 50 percent as much oxygen as when saturated, or about 10 volumes percent.

The oxygen-binding characteristics of a respiratory pigment or of a pigment-containing body fluid are generally summarized by graphing volumes percent or percent saturation as a function of the partial pressure of oxygen with which the pigment or body fluid is at equilibrium. Such graphs are called *oxygen dissociation curves* or *oxygen equilibrium curves.* An example is provided in Figure 13.4 by the curve showing the total oxygen content of human arterial blood as a function of oxygen tension. The curve for total blood oxygen can be subdivided into two component curves, one for dissolved oxygen and one for hemoglobin-bound oxygen. The relation between the content of dissolved oxygen in human blood and the oxygen tension is shown in Figure 13.4 by the line near the bottom. The curve relating the content of hemoglobin-bound oxygen to the oxygen tension could be obtained by subtracting the content of dissolved oxygen from the total oxygen content at each oxygen tension. The amount dissolved is so small relative to the total, however, that clearly the curve for hemoglobin-bound oxygen in human arterial blood would differ only slightly from the curve shown in Figure 13.4 for total oxygen. A functionally important feature of the equilibrium curve for hemoglobin-bound oxygen is that it is sigmoid in shape; it imparts a strongly sigmoidal shape to the curve for total blood oxygen.

shifts in concentration probably have little bearing for normal physiology. In other animals, however, ion concentrations shift substantially during ordinary life-history events.

Many aquatic animals that live in estuaries or other places where fresh and marine waters mix experience wide ranges of ambient salinity during their lives. Often, their blood concentrations of inorganic ions rise or fall to some extent in parallel with the ups and downs in ambient salinity (p. 172). The effects of these shifts in blood ion concentrations are diverse and have only recently started to receive detailed study. As an example, data on the shore crab *Carcinus maenas* are presented in Figure 13.14. Each line in the figure depicts P_{50} as a function of pH at a particular overall concentration of blood ions, showing that a normal Bohr effect is present at each ion concentration. Comparison of the lines at any given pH reveals, in addition, that the P_{50} increases as the overall blood ion concentration decreases. That is, dilution of the inorganic ions in the crab's blood—as occurs in dilute environmental waters (Figure 8.5)—tends to lower the oxygen affinity of the hemocyanin in the blood.

The effects of inorganic ions are ion specific. When the crabs (*Carcinus maenas*) we have been discussing are transferred into dilute waters, for example, many of their blood ions become more dilute, but all are not equally responsible for the decrease in oxygen affinity observed. The concentrations of Ca^{2+} and Mg^{2+} in the blood are very influential, but those of Na^+ and Cl^- exert little effect.

Effects of Organophosphates and Other Organic Modulators; Mechanisms of Acclimatization and Acclimation of Affinity

One of the outstanding recent developments in the study of vertebrate oxygen transport has been the discovery that organophosphate compounds within the erythrocytes often play a key role in modulating the oxygen affinity of hemoglobin. Among mammals, the organophosphate of chief importance is 2,3-di-phosphoglycerate (2,3-DPG). In adult birds, inositol pentaphosphate is the major compound. Adenosine triphosphate (ATP) and guanosine triphosphate (GTP) assume key roles in fish; ATP is evidently also important in some amphibians and reptiles.

In cases where organophosphates act as modulators, increases in their intracellular concentration depress the oxygen affinity of hemoglobin, thus increasing P_{50} (Figure 13.15). The organophosphates are allosteric modulators; by combining with hemoglobin, they influence its molecular configuration in a way favoring deoxygenation. Some of their effect is also exerted by way of influences they have on intracellular pH.

In many species, substantial concentrations of red cell organophosphates are crucial for the maintenance of normal oxygen affinity. Human erythrocytes, for example, ordinarily contain about one molecule of 2,3-DPG per molecule of hemoglobin; if this 2,3-DPG is removed, the oxygen affinity of the hemoglobin increases drastically (P_{50} is reduced to about one-half its normal value), and thus the ability of the hemoglobin to unload is severely impaired. Many other mammals also possess hemoglobins that are of inherently high affinity and that therefore require substantial 2,3-DPG levels for their affinity to be shifted down into the normal, functional range. On the other hand, some ruminants, cats, civets, and related mammals have hemoglobins that inherently

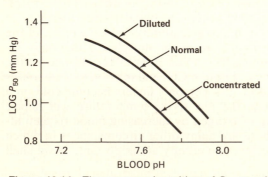

Figure 13.14 The common logarithm of P_{50} as a function of pH in blood serum of the crab *Carcinus maenas* at three levels of blood salts. The "normal" level of salts is that maintained when the crabs live in full-strength seawater. The diluted and concentrated bloods were obtained by dialyzing "normal" blood against dilute and concentrated physiological salt solutions. Hemocyanin concentration was held constant. [After J.P. Truchot, *Respir. Physiol.* **24**:173–189 (1975).]

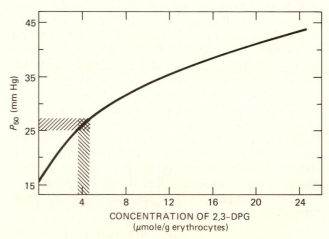

Figure 13.15 The P_{50} of human hemoglobin within erythrocytes as a function of the intracellular concentration of 2,3-DPG. The vertical dashed bar marks the range of 2,3-DPG concentrations observed during ordinary circumstances; the horizontal bar shows the P_{50} values associated with those 2,3-DPG concentrations (i.e., normal P_{50} values). Temperature (37°C), carbon dioxide tension (40 mm Hg), and extracellular pH (7.4) were held constant. [After J. Duhm, *Pflügers Arch.* **326**:341–356 (1971).]

display relatively low, functionally appropriate affinities; the red cells of these mammals contain little 2,3-DPG, and the affinity of their hemoglobin, in fact, exhibits little sensitivity to 2,3-DPG.

In many vertebrates, changes in the intracellular concentration of organophosphate modulators serve as a mechanism of adaptively altering oxygen affinity. Given that hours or days are required for such adjustments to take place, the adjustments qualify as mechanisms of acclimation or acclimatization. Humans suffering from anemia, to mention one example, often acclimate by raising the level of 2,3-DPG in their red cells; thus, the oxygen affinity of their hemoglobin is modestly reduced. The right shift in their oxygen equilibrium curve is not great enough to cause any substantial impairment of loading in the lungs, but it significantly increases unloading at any given oxygen tension in the systemic tissues (Figure 13.16). By aiding oxygen delivery in this way, the right shift tends to offset the disadvantage of having a reduced amount of hemoglobin and reduced oxygen-binding capacity per unit volume of blood.

Knowledge of organic modulators among invertebrates is in its infancy. Recently, the L-lactate ion has been shown to increase the oxygen affinity of

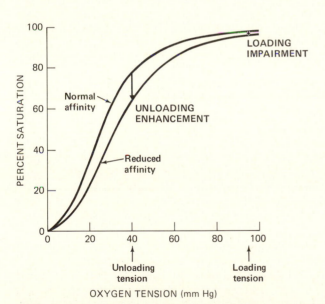

Figure 13.16 Schematic illustration of how an overall decrease in the oxygen affinity of hemoglobin can aid oxygen delivery in circumstances where the oxygen tension at the respiratory organs remains high. Two human equilibrium curves are shown. The loading tension in the lungs is assumed to be 95 mm Hg. The unloading tension in the systemic tissues is 40 mm Hg. Heavy vertical arrows depict changes in percent saturation at the two tensions caused by an overall shift from normal to reduced affinity. The shift enhances unloading of oxygen more than it impairs loading. (For simplicity, Bohr effects and other short-term effects occurring during the circulatory cycle are ignored. The reduction in affinity shown is exaggerated by comparison to reductions actually seen in anemia.)

hemocyanin in many decapod crustaceans. Urate may also serve as a modulator in these animals.

RESPIRATORY PIGMENTS: THEIR FUNCTION IN OXYGEN TRANSPORT

Oxygen-Carrying Capacity

A parameter of key importance in blood oxygen transport is the blood's *oxygen-carrying capacity:* the amount of oxygen carried by each unit volume of blood when the respiratory pigment in the blood is saturated. The carrying capacity sets an upper limit on the oxygen delivery per circulatory cycle.

In practice, the oxygen-carrying capacity of blood is usually determined by equilibrating the blood with air and then measuring its oxygen content. Most pigments are saturated under these conditions. The carrying capacity, being measured on whole blood, includes both dissolved oxygen and oxygen combined with the respiratory pigment. Since bloods equilibrated with air at any given temperature do not vary greatly in their content of dissolved oxygen, variations in the carrying capacity from animal to animal largely reflect variations in the amount of pigment present per unit volume of blood. As already seen in Figure 13.7, a wide range of carrying capacities is to be found. Capacities as high as 30–36 volumes percent are known, occurring, for example, in some of the diving marine mammals (see p. 373). At the other extreme, many crustaceans and mollusks with hemocyanins exhibit carrying capacities of just 1–1.5 volumes percent.

Birds and mammals generally have carrying capacities of 15–20 volumes percent. Capacities in fish, amphibians, and reptiles are typically lower, 6–15 volumes percent—in parallel with the lower oxygen-transport requirements of these animals. Active fish tend to have higher capacities than sluggish ones.

Among the invertebrates, oxygen-carrying capacities tend to be relatively low: usually well below the extreme values of 13–14 volumes percent reported in some of the hemoglobin-bearing annelid worms (e.g., the huge, meter-long earthworm, *Megascolex giganteus,* of South America). Capacities of 3 volumes percent or less are common in hemoglobin- and hemocyanin-bearing invertebrates. The highest of the capacities for hemocyanin-containing bloods occur among the cephalopod mollusks (squids and octopuses). Given the relatively large size and active way of life of many cephalopods, it is not surprising that they exhibit the highest capacities for their pigment type, but it is intriguing that their capacities are just 2–7 volumes percent: at or below the lower end of the range for fish.

Case Studies in Oxygen Transport

You will recall from the earlier discussion of circulatory oxygen transport in humans that our hemoglobin becomes nearly saturated as blood flows through our lungs. Furthermore, when we are at rest, our hemoglobin is far from completely desaturated in its passage through our systemic tissues. Our resting utilization of hemoglobin-bound oxygen is only about 25 percent, and thus we have a substantial *reserve* capacity for unloading of oxygen from our hemoglobin. We tap this reserve during exercise and other states in which our demand for oxygen is heightened (p. 305).

In its basic outlines, the pattern of oxygen transport seen in humans is found also in many other animals, both vertebrate and invertebrate. Figure 13.17 shows, for example, the similar pattern to be found in rainbow trout. In well-aerated water, the arterial blood of trout remains close to saturated regardless of swimming speed (note upper line). At rest, the trout use only about 30 percent of the oxygen in their arterial blood. Thus, they possess a large venous reserve of oxygen, which is used during exercise. Although the venous oxygen content is 7.1 volumes percent at rest, it is only 1.4 volumes percent at maximal swimming speed, and concomitantly the venous oxygen tension falls from 33 to 16 mm Hg. Another change that helps meet the heightened oxygen demand of peak swimming in trout is a tripling of their circulatory rate.

Certain crabs—to mention invertebrate examples—also exhibit a large venous reserve of oxygen at rest and make extensive use of the reserve during exercise. Data for Dungeness crabs (*Cancer magister*), in Figure 13.18, illustrate this pattern.

Because the crustaceans with hemocyanins tend to have low carrying capacities—usually below 3.5 volumes percent and often below 2 volumes per-

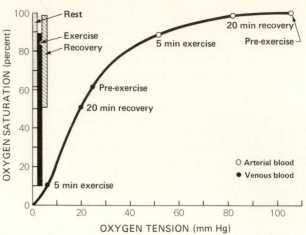

Figure 13.18 A representative example of results obtained on *Cancer magister* at 8–10°C. Blood was sampled using indwelling catheters. The animal was studied while at rest in its aquarium and then was provoked by external stimulation to 5 min of continuous activity. Points on the oxygen equilibrium curve depict oxygen tension and percent saturation in arterial and venous blood during rest ("pre-exercise"), during exercise, and after 20 min of recovery from exercise. Bars on the ordinate depict oxygen utilization. The mean carrying capacity in crabs studied was 3.4 volumes percent. The hemocyanin of *Cancer* exhibits a positive Bohr effect, but arteriovenous changes in pH were insufficient to shift the oxygen equilibrium curve appreciably. Arterial blood was sampled from blood spaces immediately efferent to the gills. Venous blood came from sinuses directly afferent to the gills. [From K. Johansen, C. Lenfant, and T.A. Mecklenburg, *Z. Vergl. Physiol.* **70**:1–19 (1970).]

cent—they have been a focus of attention for studies of whether respiratory pigments in fact contribute to oxygen transport when present in only meager amounts. By now the answer is clear. Even when crabs, crayfish, and lobsters are at rest in well-aerated water or air, their hemocyanin often accounts for 65–95 percent or more of their circulatory oxygen delivery (82 percent in resting Dungeness crabs, Figure 13.18). The adaptive value of having even a very modest concentration of respiratory pigment is thus established in these animals.

A noteworthy feature of cephalopod oxygen transport is the very small venous reserve typically observed. When *inactive,* the squids and octopuses that have received study use 80–90 percent of the oxygen available in their arterial blood (Figure 13.19). Thus, little room exists for enhancement of unloading, and the animals must meet the heightened oxygen demands of exercise largely by increasing their circulatory rate. For example, when octopuses (*Octopus vulgaris*) increase their rate of oxygen consumption by a factor of 2.8 during exercise, they increase unloading of their hemocyanin hardly at all and thus must increase their circulatory rate 2.5-fold. The small venous reserve of cephalopods signifies also that if they venture into poorly aerated waters and thereby suffer a decrease in arterial loading, they cannot compensate to any great degree by enhancing

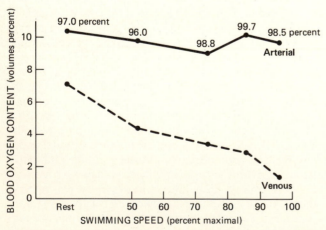

Figure 13.17 Average oxygen content of arterial and venous blood as functions of swimming speed in trout, *Salmo gairdneri,* exercising in well-aerated water. Numbers above arterial points show average arterial percent saturation. The arterial blood is nearly saturated at all speeds; some of the variation in its oxygen content is attributable to variations among test animals in carrying capacity. [Data from D.R. Jones and D.J. Randall, in W.S. Hoar and D.J. Randall (eds.), *Fish Physiology,* Vol. 7, pp. 425–501. Academic, New York, 1978.]

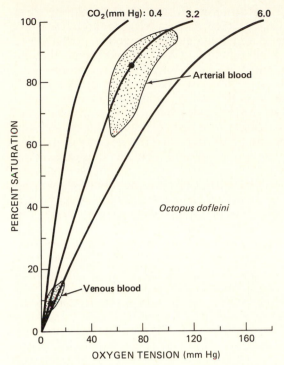

Figure 13.19 A summary of oxygen transport in *Octopus dofleini* during rest or moderate activity in well-aerated water. Oxygen equilibrium curves for hemocyanin at three CO_2 tensions are shown. Stippled areas encompass ranges of values (percent saturation, O_2 tension, CO_2 tension) prevailing in arterial and venous blood. Dots show approximate mean values. [Data from K. Johansen and C. Lenfant, *Am. J. Physiol.* **210**:910–918 (1966).]

unloading but again must rely mainly on increasing blood flow.

A final point worthy of note is that animals in well-aerated waters or air do not inevitably bring their arterial blood to near saturation. In most instances they do, but exceptions—largely unexplained—are known among spiders, octopuses (e.g., Figure 13.19), amphibians, and fish. In some toads, eels, and spiders in which arterial saturation is substantially less than complete at rest, the percent saturation of arterial blood *rises* during exercise. This unusual phenomenon signifies that the animals augment their oxygen delivery when exercising in part by using an *arterial* reserve (the mechanism is evidently that they increase the effectiveness of oxygen transfer across their respiratory organs).

Functions of Circulating Pigments of Very High Affinity

The circulating respiratory pigments of some invertebrates display remarkably high affinities for oxygen. For example, the blood hemoglobins of some of the annelid worms and small arthropods (e.g., *Daphnia*) exhibit P_{50} values of less than 5 mm Hg; in some instances, in fact, the P_{50} is below 1 mm Hg. Pigments of such high affinity load at low oxygen

tensions; in that respect they can help assure adequate oxygen delivery in low-oxygen environments. Such pigments, on the other hand, require such low tissue tensions for unloading that their precise role in oxygen delivery has long been subject to debate. Are tissue tensions always low enough for unloading? If so, the pigment in question could function in routine oxygen transport. Or do tissue tensions become low enough just in certain circumstances—such as during activity or exposure to low ambient oxygen levels? In that case, the pigment would function in transport only in special situations. Tissue tensions low enough for appreciable unloading might, in fact, arise only under circumstances when ambient oxygen levels are so low as to be inadequate for loading. Then the pigment would not really function in transport; that is, it would not at any one time become alternately loaded and unloaded in flowing between the sites of external respiration and tissues. Instead, its chief role would be *storage;* having taken up oxygen in times of oxygen availability, it would provide an internal source of oxygen at different times, times of external unavailability.

The pigment properties and life histories of animals with high-affinity pigments are so diverse that no single answer can be expected to the question of how such pigments function. With critical data available on only a few species, present answers must be tentative.

We noted earlier that some small arthropods, such as water fleas (*Daphnia*) and midge larvae (*Chironomus*), become rich in hemoglobin only when living in poorly aerated waters (p. 302). The hemoglobins they develop are typically of high affinity and, on these grounds, would appear well suited to aid oxygen transport when ambient oxygen levels are low. Experiments have revealed that if red, hemoglobin-rich *Daphnia* or *Chironomus* are placed in well-aerated waters, their rate of oxygen consumption is the same whether or not the possibility of oxygen transport by their hemoglobin has been blocked by exposure to carbon monoxide. In contrast, when the animals are in poorly aerated waters, their oxygen consumption is severely depressed by carbon monoxide poisoning. These and other findings indicate that the high-affinity pigments of these animals play little or no role in oxygen transport when ambient oxygen tensions are high but assume crucial roles in transport in poorly aerated waters.

Particularly in the last decade or so, it has become clear that the high-affinity pigments of some animals function in oxygen transport even when the animals are occupying high-oxygen environments. For example, the hemoglobins of marine lugworms (*Arenicola*) and the hemocyanin of certain tarantulas (*Eurypelma helluo*), despite having P_{50} values near 3 mm Hg, account for about half of blood oxygen transport at moderate temperatures (~22–25°C) when the animals are occupying well-aerated waters (worms) or air (spiders). It is correct to surmise from this statement that the tissues of these animals function at very low oxygen tensions even when ambient oxygen tensions are high (Figure 13.20). The reasons for this remarkable state of affairs remain to be fully understood.

In evaluating the role that a blood pigment might play in oxygen *storage,* a key consideration is to ask how long the oxygen bound to the pigment could meet the oxygen demands of the animal. The answer, although subject to substantial uncertainties unless all relevant parameters are firmly known, appears to be 20 min or less in the majority of invertebrates. Thus, pigment-bound oxygen seems generally to be capable of sustaining aerobic catabolism only during relatively short interruptions of oxygen input from the environment. Despite this limitation, there are important storage roles that blood pigments of high oxygen affinity sometimes play. For example, hemoglobin-bound oxygen contributes significantly to the length of time that air-breathing snails, *Planorbis,* can remain submerged during dives. Also, there is good reason to believe that the oxygen needs of some tube-dwelling polychaete worms and midge larvae are met in part by hemoglobin-bound oxygen during pauses in tube ventilation.

The Ice Fish: Vertebrates Without Hemoglobin

Given that most vertebrates have substantial amounts of hemoglobin in their blood, considerable

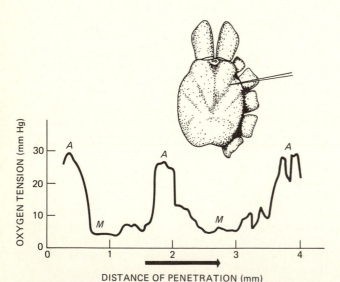

Figure 13.20 Recordings of oxygen tension measured by a microelectrode as it was forced through the head region of a tarantula (*Eurypelma helluo*) breathing air. The inset depicts the positioning of the electrode (tip diameter: 1–3 μm). The abscissa shows depth of penetration (0 mm corresponds to the dorsal body surface). As the electrode penetrated deeper and deeper, it passed successively through arterial blood channels (*A*) and through muscle or other solid tissue (*M*). Although the spiders were breathing air with an oxygen tension near 150 mm Hg, oxygen tensions within their muscles were very low: sometimes below 4 mm Hg. [Data from D. Angersbach, *J. Comp. Physiol.* **98:**133–145 (1975).]

interest has centered on the fish of the family Chaenichthyidae, which occupy frigid antarctic seas and lack circulating hemoglobin altogether (or have negligibly small amounts). Without red blood, they are whitish and translucent—thus their common name, *ice fish.* They are not small fish, and their lack of hemoglobin is thus all the more remarkable. *Chaenocephalus aceratus,* for example, often weighs over 2 kg and reaches a length of about 0.6 m. Without hemoglobin the oxygen-carrying capacity of its blood is limited to the amount that can dissolve in the plasma.

The habitat of these fish is undoubtedly important to understanding their respiratory physiology. Oxygen tensions in the antarctic seas are characteristically high, and temperatures typically remain low—near 0°C—the year round. The low temperatures depress the metabolism of the fish and also augment the solubility of oxygen not only in the ambient waters but in the fish's blood.

To elucidate the physiology of ice fish, comparisons have been made between the ice fish and other antarctic fish, such as species of the genus *Notothenia,* that possess more-usual piscine levels of hemoglobin. Available sets of data are in conflict concerning the metabolic levels of ice fish. Some data indicate that the resting rates of oxygen consumption of ice fish are similar to rates in red-blooded fish, whereas others indicate that they are only one-half or one-third as high. If the metabolic oxygen requirements of ice fish are indeed reduced, that reduction would certainly ease demands for oxygen transport by their circulatory system. However, even if the most extreme estimates of the metabolic reduction prove true, other factors must be involved in explaining how ice fish meet their oxygen requirements without hemoglobin: Consider that although the resting oxygen requirements of ice fish might be reduced to one-half or one-third of those in *Notothenia,* the carrying capacities of ice fish (about 0.7 volumes percent) are only about *one-tenth* as high as those of *Notothenia* (6–7 volumes percent). There is widespread agreement that one of the major adjustments in ice fish is an exceptionally high circulatory rate. The hearts of ice fish are dramatically larger than those of most fish of their body size and, it is estimated, pump at least 4–10 times as much blood per stroke.

The lack of hemoglobin in ice fish is hardly without its detractions. It probably limits the levels of activity that the fish can sustain, and it imposes a requirement that they remain in relatively cold and well-oxygenated waters.

Correlations of Oxygen Affinity with Habitat and Body Size

Having reviewed some of the diversity of respiratory transport physiology, we now turn to additional comparative information that can be derived from examination of oxygen equilibrium curves. Studies of transport in vivo are unfortunately limited in number, and for many animals the only information available is the oxygen equilibrium curve. It should now be obvious that deductions about the function of a pigment in vivo from the oxygen equilibrium curve alone must be made with caution.

Many instances are known among vertebrates in which species from high-oxygen environments tend to have blood pigments of lower oxygen affinity than related species from low-oxygen environments. Before looking at particular cases, it is important to consider why such a correlation between affinity and environmental oxygen levels might be expected. High affinities in species that regularly experience low ambient oxygen tensions are certainly not surprising: High-affinity pigments load more completely at low tensions than low-affinity ones. But why should affinity tend to decrease as the oxygen level in the environment increases? After all, a pigment of high affinity will load just as well at high tensions as at low, and with a decrease in affinity comes heightened vulnerability to inadequate loading should the ambient tension in a high-oxygen environment decrease from its usual high level. As already discussed in a different context (Figure 13.16), the advantage of low affinity is that it promotes *unloading*. A decrease in affinity is likely to decrease substantially the amount of oxygen that remains bound to pigment at any given tension in the systemic tissues; yet if ambient tensions are high, it may cause hardly any reduction in the amount of oxygen that becomes bound to the pigment in the respiratory organs (see Figure 13.16). At high ambient tensions, therefore, *delivery* of any particular amount of oxygen per circulatory cycle is likely to be attained at higher tissue tensions if the affinity is low rather than high (within limits).

Figure 13.21 depicts one way of summarizing these important ideas. At each arterial tension, the three plotted lines show the venous tensions required to attain a particular oxygen delivery (30 percent of carrying capacity) per circulatory cycle when three different affinities prevail. *When the arterial tension is high, low affinities enhance unloading more than they impair loading and thus permit higher venous tensions than high affinities. When the arterial tension is low, however, low affinities become so detrimental for loading that high affinities assume the advantage*, as judged by minimizing the arteriovenous drop in tension required for delivery. Quantitatively, Figure 13.21 applies only to systems meeting the specific assumptions made in the analysis. The basic patterns depicted in the figure are of wide applicability, however.

Fish that inhabit waters low in oxygen tend to have hemoglobins of higher affinity than those that occupy waters high in oxygen. For example, at roughly equivalent CO_2 tensions and temperatures, whereas the average P_{50} values of catfish and carp

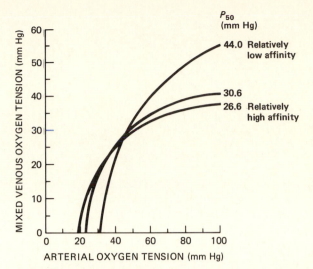

Figure 13.21 Over a range of arterial oxygen tensions, each curve shows the mixed venous tension required for the blood to deliver 30 percent of carrying capacity per circulatory cycle. The curves were calculated for three different values of P_{50}, assuming the blood characteristics to be those of mammals. The results are qualitatively applicable to many systems but apply quantitatively only to systems meeting the specific assumptions made in the calculations. To illustrate, if a higher delivery per circulatory cycle is stipulated, then high affinities become superior to low affinities at higher arterial tensions than shown. [After S. Lahiri, *Am. J. Physiol.* **229**:529–536 (1975).]

bloods are 1.4 and 5 mm Hg, respectively, those of rainbow trout and mackerel bloods are 18 and 16 mm Hg, respectively (see Figure 13.7). Correlations between affinity and habitat are also seen among mammals; species that live extensively in burrows, for example, tend to exhibit higher affinities than other mammals of their size (see also Box 13.4).

In some phyletic groups, both water- and air-breathing forms are found. Because the air provides much higher oxygen tensions than many aquatic habitats, a tendency for the air breathers in such groups to exhibit lower affinities than the water breathers might be expected. Among amphibians, this expectation seems to be fulfilled. Figure 13.22 provides, for example, a comparison of the adults of three species; it shows that mudpuppies (*Necturus*), which are fully aquatic as adults, exhibit a much higher blood oxygen affinity than bullfrogs, which are semiterrestrial. Interestingly, also, the tadpoles of semiterrestrial frogs have hemoglobins of higher affinity than the adults of their species. In some phyletic groups (e.g., crabs), the tendency of air breathers to have lower affinities than water breathers is less evident or absent.

Among mammals, and possibly some other vertebrate groups, there may exist a rough correlation between oxygen affinity and body size, small species tending to exhibit lower affinities than large. This correlation was once thought to be very clear but has become hazier as the matter has been reinvesti-

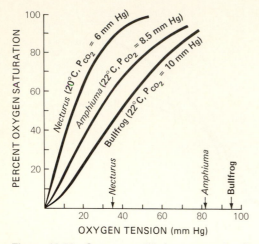

Figure 13.22 Oxygen equilibrium curves for three species of amphibians. For each species, carbon dioxide tension (P_{CO_2}) was held constant at a value about 2 mm Hg higher than the mean arterial carbon dioxide tension. Because none of the species showed a large Bohr effect, the differences in oxygen affinity of their hemoglobins are not to be explained by differences in experimental carbon dioxide tension. The mudpuppy, *Necturus maculosus*, though it possesses simple lungs, is chiefly an aquatically respiring animal, exchanging oxygen almost entirely across its gills and skin. *Amphiuma tridactylum* is a salamander that lives exclusively in water, like *Necturus*, but that lacks gills and depends strongly on air breathing. The bullfrog, *Rana catesbeiana*, is amphibious and spends much time out of water. There is a progressive drop in the oxygen affinity of the blood with increasing dependence on air breathing. The arrows on the abscissa indicate the mean oxygen tension of arterial blood of unrestrained animals living at 20–22°C in environments of high oxygen tension (water: 125 mm Hg; air: 152 mm Hg). The blood of all three species is at least 90 percent saturated at prevailing arterial tension. Note that *Necturus*, though having a far lower arterial tension than the other species, attains a comparable percent loading owing to the greater affinity of its hemoglobin for oxygen. [From C. Lenfant and K. Johansen, *Respir. Physiol.* **2:**247–260 (1967).]

gated using superior techniques. If small mammals do indeed tend to have lower affinities than large, the pattern might be rationalized in terms of their higher weight-specific oxygen demands. At the arterial oxygen tensions attained at sea level, relatively low affinities tend to favor relatively high oxygen tensions in the systemic blood capillaries (Figure 13.21). High systemic capillary tensions could aid animals with high weight-specific oxygen demands by favoring rapid diffusion of oxygen from the capillaries into the tissues.

As a general point, it is important to note in closing that interspecific differences in affinity are not necessarily attributable, in whole or even in part, to differences in their hemoglobins. Differences in the concentrations of organophosphate modulators in the red cells might be involved, for example.

Functions of Tissue Hemoglobins

As discussed earlier, many animals have hemoglobins in muscle or other solid tissues in addition to

their circulating respiratory pigment. It is notable that these solid-tissue hemoglobins are characteristically of higher affinity than the circulating pigment. In the large chiton *Cryptochiton stelleri*, for example, the P_{50} of the myoglobin of the radular (feeding) muscle is about 3 mm Hg, whereas that of the circulating hemocyanin is about 17 mm Hg (Figure 13.23). Similarly, human blood hemoglobin has a P_{50} of about 27 mm Hg, but human myoglobin has a P_{50} around 6 mm Hg (Figure 13.6). By virtue of their higher affinities, the myoglobins tend to load at the expense of unloading of the circulating pigment and thus draw oxygen from the blood.

The physiology of myoglobins is not well understood. However, their physiological importance has recently been firmly established by experiments showing that myoglobin-containing muscles sometimes suffer great impairment of their capacity for vigorous contractile activity when the ability of their myoglobin to combine with oxygen is chemically blocked. There is good reason to believe that myoglobins perform two functions: (1) facilitation of oxygen diffusion from the cell surface to the mitochondria and (2) intracellular storage of oxygen.

If two gas mixtures of different oxygen tension are separated by a thin layer of saline solution, the rate of oxygen flux through the layer turns out to be considerably greater if the solution contains myoglobin than if it does not. The mechanism of this ***facilitation of oxygen diffusion*** is believed to be that oxygen molecules are carried through the solution by diffusion of oxymyoglobin molecules when myoglobin is present. With or without myoglobin, free oxygen molecules diffuse through the solution. In the presence of myoglobin, the concentration of oxymyoglobin is higher on the side of the solution exposed to high oxygen tension than on the other side, and diffusion of oxymyoglobin molecules along this gradient provides a second "route" for the passage of oxygen through the solution, thereby increasing the overall rate at which oxygen appears on the low-tension side. For years, concern has existed that myoglobin in cells—as opposed to that dissolved in an artificial solution—might lack sufficient molecular mobility to facilitate oxygen diffusion in vivo as in vitro. Just recently, however, direct evidence for adequate intracellular mobility has been obtained.

By its very nature, oxymyoglobin in cells constitutes a store of oxygen that can be tapped if intracellular oxygen tensions fall low enough for unloading. In regard to storage, the questions at issue are not whether oxymyoglobin can act as a store but *when* and *to what extent*. In general, the myoglobin in muscle cells, if fully oxygenated, would appear to bear enough oxygen to meet oxygen demands for just relatively short periods. The average concentration of myoglobin in human skeletal muscle, for example, is about 0.5 μmole/g, meaning it can bind about 0.01 mL O_2/g; with the oxygen demand probably being at least 0.003–0.03 mL O_2/g·min over the

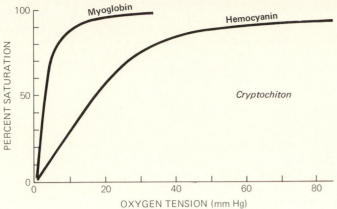

Figure 13.23 Oxygen equilibrium curves for circulating hemocyanin and radular myoglobin (hemoglobin) in a chiton, *Cryptochiton stelleri*. Both pigments were obtained from the same individual and were analyzed at 10°C. The radular muscles are colored deep red by their high content of myoglobin and are bathed by blood containing hemocyanin. The muscles are used to make a ribbon of teeth move back and forth in a rasping or scraping motion during feeding. [From C. Manwell, *J. Cell. Comp. Physiol.* **52**:341–352 (1958).]

usual range of contractile activity, myoglobin-bound oxygen could meet the oxygen needs of a muscle for anywhere from a few minutes to under 20 s. A short-term store such as this can play important roles. It can help sustain aerobic catabolism at the start of vigorous exercise, for example, by releasing oxygen when the amount supplied by the blood circulation is transiently inadequate to meet needs (p. 44).

Transfer of Oxygen from One Pigment to Another

In the exchange between circulating pigments and myoglobins, we have seen one example of oxygen transfer from one respiratory pigment to another. There are other prominent examples. These include the transfer from maternal to fetal blood hemoglobin in pregnant placental mammals. As maternal blood courses through the placenta, its oxygen tension drops—as a rough generalization—from an initial level of 70–100 mm Hg to a final level of 25–50 mm Hg. Given that the maternal blood is the source of oxygen for the fetal blood, it is evident that fetal blood must load at tensions much lower than those at which loading of maternal blood occurs in the lungs. Thus, it is not surprising to find that fetal blood typically exhibits a higher oxygen affinity than maternal blood; the fetal P_{50} is in most cases 3–17 mm Hg below maternal. In some species (e.g., dogs), the fetal and maternal hemoglobins are chemically similar, and the reason the fetal affinity is higher is that fetal red blood cells are lower in 2,3-DPG than adult red cells. In other species, the fetal hemoglobin is chemically distinct from adult; then the special structure of the fetal hemoglobin may itself endow it with higher affinity (e.g., ruminants), or the special structure may impart higher affinity wholly or partly by rendering the fetal hemoglobin less sensitive than adult hemoglobin to the affinity-lowering effects of red cell 2,3-DPG (e.g., primates).

The relatively high oxygen affinity of fetal blood in mammals is just one of several factors that can be involved in promoting oxygen transfer from the maternal to the fetal circulation. Another factor, for example, is that the loss of CO_2 from the fetal to the maternal blood in the placenta commonly induces—by virtue of the Bohr effect—a simultaneous rise in fetal oxygen affinity and fall in maternal oxygen affinity.

Some lower vertebrates possess placenta-type attachments between the young and mother. Garter snakes (*Thamnophis*) and spiny dogfish (*Squalus acanthias*) are two examples, and interestingly the young in each case exhibit a higher blood oxygen affinity than adults.

In the polychaete annelids that have hemoglobins in both their blood and coelomic fluids, the coelomic affinity typically exceeds the blood affinity. This difference evidently facilitates uptake of oxygen by the coelomic fluid from the blood, which in turn acquires oxygen from the environment across the gills.

Respiratory Pigments and Responses to Environmental Exigencies

In this section, we consider the properties of respiratory pigments as they influence the responses of animals to low ambient oxygen levels, high temperatures, and other environmental exigencies. Responses to environmental challenges can occur on three major time scales: the scale of immediate responses by individuals, that of acclimation or acclimatization, and that of genetic change through evolutionary time. To each environmental exigency discussed, responses on all three time scales are possible, if not already known. However, we shall not attempt here to discuss all three time scales in each instance.

Low Ambient Oxygen Levels ("Hypoxia")　The immediate challenge posed by low ambient oxygen levels is that they threaten the adequacy of loading of the circulating respiratory pigment in the gills or lungs. The magnitude of the threat depends in part on the pigment oxygen affinity a species has evolved, as already described (Figure 13.21). Thus, one of the reasons that carp (*Cyprinus*) are more tolerant of low ambient oxygen tensions than trout (*Salmo*) is that

---

BOX 13.4   MAMMALS AT HIGH ALTITUDE

In reviewing the oxygen cascade of mammals at high altitude in Chapter 12 (p. 282), we noted that the drop in oxygen tension from arterial to venous blood tends to be considerably smaller at altitude than at sea level. Indeed, Figure 12.31 reveals that the diminution of this drop is a prominent factor in limiting the decline in mixed venous oxygen tension at altitude.

The principal explanation for the reduced arteriovenous tension drop at altitude does not entail any special adaptations at altitude but rather rests simply on the *shape* of the mammalian oxygen equilibrium curve. Living at high altitude brings about a reduction in the arterial oxygen tension. Figure 13.5 (p. 305) illustrates the consequence: Moving the arterial tension off the plateau of the equilibrium curve sharply lessens the arteriovenous drop in tension required for the blood to yield any particular quantity of oxygen.

Historically, two hematological changes that occur in humans at altitude have been claimed to help limit the arteriovenous drop in tension: (1) an increase in the carrying capacity of the blood and (2) a drop in oxygen affinity. Not long ago, both of these changes were held up as shining examples of the adaptive process. Now the significance of both (and the existence of one) is sharply questioned.

When a number of lowland mammals—including humans—live at high altitude, their carrying capacity becomes elevated markedly above sea-level values by virtue of an increase in the concentration of red blood cells in the blood (polycythemia). For example, a rise in capacity from 20 to 28 volumes percent would not be unusual in people at 4000–5000 m. A reasonable argument can be made that this change helps prevent tissue hypoxia (p. 321). Skeptics have pointed to two contrary lines of thought, however. First, they have noted that many species of mammals (and birds) *endemic* to high altitudes do *not* have unusually high carrying capacities; and second, they have argued that a high carrying capacity could actually *interfere* with oxygen delivery because the increase in red cell density increases blood viscosity and thus not only places greater demands on the heart but also could impair flow through small blood vessels. Some of the skeptics have concluded that the increase in carrying capacity of lowland mammals at altitude is in fact *pathological*. By way of evidence they have presented (contested) data indicating that medical *removal* of red blood cells from humans at altitude either does not change or actually improves their performance.

With regard to oxygen affinity, there is at least equal debate. Humans and some other lowland species show increased red cell 2,3-DPG at altitude. This has been argued to help prevent tissue hypoxia by lowering the oxygen affinity of hemoglobin and thus promoting unloading at higher tissue tensions (p. 317). The counterarguments are complex. One concern has been whether the reasoning is even theoretically correct, for reduced affinity also has the potential to reduce loading, especially at the higher elevations (p. 317). There is also the notable observation that many species of mammals (and birds) *endemic* to high altitudes have particularly *high* affinities, just as fish from oxygen-poor waters tend to have high affinities (p. 317). Indeed, it has been argued that the human increase in 2,3-DPG at altitude is itself a pathologically misplaced response; according to this reasoning, humans have had so little contact with high altitudes in their evolutionary history that when they go to altitude, they mistakenly activate a response that they evolved to help deal with anemia at sea level (p. 313). From a practical viewpoint, an important consideration emphasized of late is that humans may well be at least slightly alkalotic at high altitude even after a long stay (p. 283); and their elevated pH can shift their affinity upward to at least as great an extent as the elevated 2,3-DPG can shift it downward. A recent careful study of people native to the Peruvian Andes revealed that with these *two* factors operative, their affinity was indistinguishable from that of people at sea level. Similarly, the P_{50} of members of the American medical expedition to Mount Everest in 1981 remained unaltered up to about 6000 m (and then fell).

The nature and significance of changes in carrying capacity and affinity in lowland species at altitude remain highly controversial. Just looking at the sides of the argument is valuable, however.

Another factor that could help reduce the arteriovenous drop in oxygen tension is an increase in blood flow. However, cardiac output is not systematically elevated in mammals at altitude except just after an ascent to higher elevation.

they have evolved much higher affinities than the trout (p. 317).

Individual animals exposed to reduced ambient oxygen levels are able to marshall defenses. Commonly, they increase the rate at which they pump water or air across the exchange surfaces of their gills or lungs. Increasing the amount of the medium passed across the exchange surfaces decreases the amount of oxygen that must be removed from each unit volume of medium and thus promotes loading of the respiratory pigment by keeping oxygen tensions at the exchange surfaces relatively high (e.g., see p. 282).

Among both vertebrates and invertebrates, another common response of individuals—occurring on a longer time scale than ventilatory adjustments—is an increase in the carrying capacity of the blood. Many fish, for example, increase the concentration of red cells in their blood when chronically exposed to poorly oxygenated waters (see also pp. 302 and 320). An increase in the carrying capacity of the blood can help protect the *tissues* from hypoxia because it means that at any given arterial oxygen tension, each unit volume of blood carries an increased quantity of oxygen, and thus the blood tension does not have to fall as much for the blood to yield a particular quantity of oxygen to the tissues.

Recent research has led also to a growing realization that individuals probably often modify the *functional properties* of their respiratory pigments in response to environmental exigencies. One possible mechanism of altering functional properties is to synthesize alternative molecular forms of the pigments. Another is to change the properties of preexisting pigment molecules by use of modulators. The former possibility is only beginning to be explored, but the latter process has been demonstrated to occur in a variety of animals, vertebrate and invertebrate. We know, for example, that several species of fish, when living in poorly oxygenated waters, increase the affinity of their hemoglobin for oxygen by decreasing red cell concentrations of their organophosphate modulators, ATP and GTP. Figure 13.24 illustrates the impressive advantages that eels in low-oxygen waters evidently gain by increasing their affinity in this way, along with their carrying capacity. In fish, a substantial change in red cell levels of ATP and GTP requires a day or more. Thus, this form of modulation operates on a relatively long time scale.

Just recently, evidence has been obtained that at least some fish possess a means of increasing the oxygen affinity of their hemoglobin on a much shorter time scale. The catecholamine hormones, adrenaline and noradrenaline, have been shown to stimulate red blood cells to actively transport H^+ outward into the plasma, rendering the insides of the cells more alkaline. This increase in intracellular alkalinity, in turn, increases the oxygen affinity of the hemoglobin by the Bohr effect. There is evidence that rainbow trout, for example, secrete catecholamines into their blood promptly when exposed to profound hypoxia, and the catecholamines induce a rapid pH-regulated increase in oxygen affinity which aids the fish until long-term adjustments in ATP and GTP levels are made.

Temperature   Previously (p. 249), we remarked that water-breathing animals can be caught in a two-pronged trap by rising temperatures: (1) Their oxygen demands tend to increase as temperature increases, and yet (2) the availability of oxygen in their environment tends at the same time to decrease because an elevated temperature suppresses oxygen solubility. Now it should be clear that the trap is potentially three-pronged, because increases in body temperature typically decrease the oxygen affinities of respiratory pigments and may in fact cause sufficient reductions in affinity for the loading of arterial blood to be impaired. Thermal effects on affinity also present potential perils at low temperatures: In cold environments, the affinity could become so great that unacceptably low tissue tensions would be required for adequate unloading.

It is always easy to list possibilities but more difficult to determine if they are borne out in reality. Just as we can say it is *possible* for changes in temperature to cause dangerous shifts in affinity, we can also say it is *possible* for animals to compensate in many ways. It is also possible for the consequences of temperature changes to be dependent on other simultaneous environmental changes; in shallow pools, for instance, at the same time the sun heats the water each day—reducing affinity—it may increase oxygen availability by promoting photosynthesis. The complexity of oxygen acquisition is so great and our knowledge so modest that we in fact know relatively little as yet about the impact of thermal affinity shifts in nature.

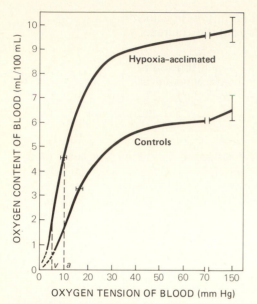

Figure 13.24 Oxygen dissociation curves for whole blood of eels (*Anguilla anguilla*). Controls were kept in water of high oxygen tension (140 mm Hg), whereas individuals acclimated to hypoxia were kept in water of low tension (15–40 mm Hg) for a week or two. The hypoxia-acclimated eels developed a much higher oxygen affinity (average P_{50} = 10.6 mm Hg) than control eels (P_{50} = 16.6 mm Hg), as shown by the horizontal bars, which for each group delimit the mean P_{50} ± twice the standard error of P_{50}. The increase in affinity was mediated by a decrease in red cell ATP and GTP. The hypoxia-acclimated eels also developed a higher carrying capacity (average: 9.8 mL O_2/100 mL) than the controls (6.6 mL O_2/100 mL), as shown by the vertical bars at the far right, which delimit mean carrying capacity ± 2 SE. Vertical dashed lines show mean arterial (*a*) and venous (*v*) oxygen tensions in eels living in water of low oxygen tension (30 mm Hg). Note that at the prevailing arterial tension, the blood of hypoxia-acclimated eels carries about 4.6 volumes percent, whereas that of controls carries only less than 2 volumes percent. Measurements were carried out at 20°C and at a pH of 7.8. [From S.C. Wood and K. Johansen, *Am. J. Physiol.* **225**:849–851 (1973).]

One evolutionary hypothesis is that among poikilotherms, species native to cold, oxygen-rich habitats (e.g., polar seas) might be expected to have evolved respiratory pigments exhibiting less severe enhancement of affinity at low temperatures than related warm-habitat species. Some dramatic examples of such a pattern are indeed known. Another popular hypothesis is that animals from thermally variable environments might have evolved especially low thermal sensitivity. This prediction has not been borne out by evidence as a general pattern, however.

Increasingly, it has become clear that blood oxygen affinity is subject to thermal acclimation in a variety of animals, including certain crabs, crayfish, fish, and turtles. The acclimation is typically compensatory. That is, after P_{50} has been altered by a sudden change of temperature, acclimation often returns it back toward its original level, as illustrated in Figure 13.25. This type of acclimation can be protective in many situations: It defends the animal against having its affinity rendered too high or low

by changes in temperature. The mechanism of acclimation has not yet been identified in most cases, though organophosphate modulation is known to be involved in some vertebrates.

Acidity and Carbon Dioxide Tension   Ordinarily, in animals with Root or Bohr effects, the pH and CO_2 tension of arterial blood permit ample loading, and yet the drop in pH and rise in CO_2 tension in the systemic tissues facilitate unloading. When the Root and Bohr effects are thus brought into play to enhance unloading *selectively* in the systemic tissues, the effects can have clear advantages, as already described (pp. 309 and 311). The Root and Bohr effects can potentially become disadvantages, however, if they are activated throughout the vascular system by an unusual acidification or elevation of CO_2 tension in the blood as a whole: If the blood at large is acidified or has its CO_2 tension raised, then the effects will suppress the carrying capacity or affinity of the blood even as it perfuses the respiratory organs, and thus they could interfere with adequate loading. In considering the threat these possibilities pose to animals in nature, the same caveat mentioned in connection with thermal effects deserves repetition: The possibilities for compensation are great, and we know more about pieces of the problem than the whole. Animals displaying large Bohr effects would appear to be more vulnerable to adverse consequences of generalized blood acidification than ones exhibiting small effects. Animals

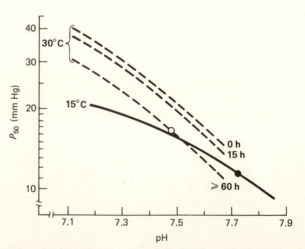

Figure 13.25   Blood oxygen affinity (P_{50}) as a function of pH in the shore crab *Carcinus maenas*. The solid curve is for animals acclimated to 15°C. Dashed curves are for animals at various times after transfer to 30°C from 15°C; note that the initial effect of transfer to 30°C (0 h) is to raise P_{50} at any given pH, but P_{50} then falls as acclimation occurs over the ensuing 60 h. Crabs have more-acid blood at high temperatures than low. Open and solid circles show pH and corresponding P_{50} after acclimation at 30 and 15°C, respectively. The P_{50} ultimately attained at 30°C is higher than at 15°C, but the elevation is attributable almost entirely to the pH change and is much smaller than it would be without thermal acclimation. [After J.P. Truchot, *Respir. Physiol.* **24**:173–189 (1975).]

having Root effects would seem to be especially threatened because the very carrying capacity of their blood is at risk.

One of the circumstances that can provoke increases in arterial acidity and CO_2 tension is a rise in the CO_2 tension of the environment (Chapter 12). Another circumstance that can increase acidity is intense exercise entailing production of lactic acid.

When some of the active species of fish undergo extreme exertion, it has seemed to certain investigators that the acidification of their blood sometimes leads to truly dire consequences. These investigators have argued that the acidification can impair oxygen loading by hemoglobin in the gills so severely as to compel an end to the exertion or even help cause the death of the fish by asphyxiation. Recently, on the other hand, it has become clear that both fish and crabs also have *defenses* against excessive acidification during exercise. Although they retain all the lactate they produce, they vigorously excrete H^+ into the environment. Furthermore, in crabs, the affinity-increasing effect of lactate helps offset the effects of lowered pH. In fish, there is recent evidence that some species increase blood levels of catecholamine hormones after severe exercise, and the catecholamines induce the red cells to transport H^+ outward into the plasma (p. 321). In this way, the pH within the red cells may be maintained well above that in the plasma, thereby reducing adverse pH effects on hemoglobin function.

Another relatively recent discovery is that in both fish and arthropods, when multiple molecular forms of hemoglobin are present in a species, the forms sometimes vary considerably in their pH sensitivity: Although certain molecular forms may exhibit strong Bohr effects, others may be nearly or completely immune to the effect. The sucker *Catostomus clarkii*—an inhabitant of fast-moving waters—provides one example. About 20 percent of its hemoglobin is insensitive to pH and retains a stable, relatively high affinity ($P_{50} \simeq 10$ mm Hg) as the blood is acidified. The remaining 80 percent of the hemoglobin, on the other hand, exhibits a Bohr effect so strong that its ability to bind oxygen is severely compromised at low pH values (P_{50} increases from 13 mm Hg at pH 7.2 to 76 mm Hg at pH 6.7!). The pH-insensitive hemoglobins in animals like this sucker are hypothesized to constitute a backup system, helping to assure that oxygen delivery will remain at least minimally adequate under circumstances when blood acidity is elevated. In a like vein, hemoglobin components that are relatively insensitive to thermal effects are sometimes found.

Salinity The effects of changes in ambient salinity on respiratory-pigment function vary a great deal from one animal group to another and are not yet well understood. One thing known emphatically is that the salt composition of the blood is often not the only influential blood parameter to be altered by changes in the ambient salinity. It is this important insight that we shall briefly explore here.

Earlier, in Figure 13.14, we saw that dilution of blood salts tends to reduce the oxygen affinity of hemocyanin in green crabs, *Carcinus maenas*. When such crabs are transferred from full-strength seawater (36 ppt) into diluted seawater (12 ppt), their blood salts are diluted enough to cause an increase of 3 mm Hg in their P_{50}. In fact, however, the *net* change in their P_{50} is a small *decrease* because the transfer of crabs to dilute seawater not only lowers their blood salt concentration but also provokes a rise in their blood pH, and the rise in pH overrides the effect of blood dilution on P_{50}. A similar process is seen also in some other crabs under like conditions. Why the blood pH rises in dilute waters is not definitively known. Possibly the increase in pH is a consequence of the increased production of alkaline ammonia that occurs in conjunction with cellular volume regulation (p. 176).

CARBON DIOXIDE TRANSPORT; ACID–BASE BALANCE

Carbon dioxide dissolves in blood as CO_2 molecules, but usually only a minor fraction of the carbon dioxide in the blood is present in this chemical form (e.g., about 5 percent in human arterial blood). One of the first steps in the study of carbon dioxide transport is to consider the other forms in which carbon dioxide may be carried. To set the stage for that undertaking, we need to look at the reactions of carbon dioxide with water.

When carbon dioxide is dissolved in aqueous solutions, it undergoes a series of reactions, the first of which is hydration to form carbonic acid,

$$CO_2 + H_2O \rightleftharpoons H_2CO_3 \qquad (2)$$

and the second, dissociation of the carbonic acid to yield bicarbonate and a hydrogen ion,

$$H_2CO_3 \rightleftharpoons H^+ + HCO_3^- \qquad (3)$$

Bicarbonate can then dissociate further, to yield carbonate (CO_3^{2-}) and an additional hydrogen ion. This final dissociation, however, occurs to only a small extent in most animals. Therefore, as a first approximation, we can disregard it and write the overall series of reactions as

$$CO_2 + H_2O \rightleftharpoons H_2CO_3 \rightleftharpoons H^+ + HCO_3^- \qquad (4)$$

Carbonic acid (H_2CO_3) is an important intermediate compound in these reactions, but it never accumulates to more than very slight concentrations (less than 1/200 the molar concentration of CO_2). Thus, carbonic acid can be disregarded as a *constituent* of the blood, and for purposes of *constituent* analysis, the reaction of CO_2 with water can be viewed simply as yielding HCO_3^- and protons:

$$CO_2 + H_2O \rightleftharpoons HCO_3^- + H^+ \qquad (5)$$

Equations (4) and (5) bring out two considerations of pivotal importance. First, carbon dioxide acts as an acid in aqueous systems (it reacts to yield H^+). Second, bicarbonate is potentially one of the important forms in which carbon dioxide exists in blood or other body fluids.

The Extent of Bicarbonate Formation; Buffers

Although almost no bicarbonate is generated when CO_2 is dissolved in distilled water or a NaCl solution, bicarbonate is typically the dominant form in which carbon dioxide exists in the bloods of animals. About 90 percent of the carbon dioxide in human arterial blood, for example, is in the form of bicarbonate. To understand why bicarbonate formation is negligible in a NaCl solution yet extensive in human blood—and why it is more extensive in some animal bloods than others—we need to examine its determinants.

Suppose that we bring a liter of aqueous solution—initially devoid of carbon dioxide—into contact with a source of CO_2 and that this source remains at a constant CO_2 tension regardless of how much CO_2 it donates to the solution. We know that the CO_2 *tension* of the solution will ultimately rise to equal that of the source. We need to consider, however, the *amount* of CO_2 the solution will take up in reaching this tension.

From Chapter 11, we know that the concentration of *dissolved* CO_2 is simply proportional to the CO_2 tension in a solution of any given ionic strength and temperature. Thus, the amount of CO_2 taken up in dissolved form by our liter of solution will depend straightforwardly on the laws of gas solubility and, at any particular tension, *will be identical regardless of the amount of bicarbonate formed.*

The extent of bicarbonate formation is governed, not by the laws of gas solubility, but by other considerations, one of which is the law of mass action as applied to Equations (4) and (5). Applying mass-action principles to Equation (5), we obtain

$$\frac{[HCO_3^-][H^+]}{[CO_2]} = K \qquad (6)$$

where the brackets signify concentrations and K is a constant (the apparent dissociation constant). Now we have already established that, in our solution, the concentration of CO_2, $[CO_2]$, is determined entirely by the CO_2 tension and thus is constant at any given tension. Given that K is also a constant, Equation (6) reveals that the amount of HCO_3^- formed at a given CO_2 tension depends inversely on the H^+ concentration. If $[H^+]$ is kept relatively low, a lot of HCO_3^- will be formed, but if it is allowed to rise to high levels, little HCO_3^- will be formed.

It should now be evident that, along with the principles of mass action, the buffering capacity of the solution is also a major determinant of the extent of bicarbonate formation. The very reactions that produce HCO_3^- also produce H^+. If the H^+ is allowed to accumulate freely in solution, raising the H^+ concentration, bicarbonate formation will be impaired. Buffers, however, can remove H^+ from solution, limit the rise in H^+ concentration, and thus enhance formation of HCO_3^-.

The extreme importance of buffering may be illustrated by contrasting the behavior of two NaCl solutions—each of the ionic strength and temperature of human blood plasma—one of which is completely unbuffered and the other of which is buffered so well that its H^+ concentration is held constant (an unrealistic but nonetheless instructive condition). In the absence of CO_2, both will have a pH of 6.8, the neutral pH at 37°C. We shall presume the volume of each to be a liter and the CO_2 tension to which they are exposed to be 40 mm Hg. By the laws of gas solubility, both solutions will take up 1.2 mmoles of CO_2 *in dissolved form*. In the unbuffered solution, as HCO_3^- is formed, the H^+ produced in tandem will simply accumulate in solution, driving the pH rapidly down. Indeed, formation of HCO_3^- from just 0.03 mmole of CO_2 will lower the pH to 4.5. At this pH, according to Equation (6), the HCO_3^- concentration at equilibrium with a CO_2 concentration of 1.2 mmoles/L is 0.03 mmole/L. Thus, in the end, the unbuffered solution will take up just 0.03 mmole of CO_2 as HCO_3^-—meaning that the total CO_2 absorbed into the solution will be 1.23 mmoles, and only 2.5 percent of it will exist as HCO_3^-. In the fully buffered solution, by contrast, when CO_2 reacts to form HCO_3^-, the H^+ formed will not lower the pH because it will be removed by the buffer. At a pH of 6.8, according to Equation (6), the HCO_3^- concentration in equilibrium with a CO_2 concentration of 1.2 mmoles/L is 6.0 mmoles/L. The buffered solution therefore will take up a total of 7.2 mmoles of CO_2—almost six times the amount absorbed by the unbuffered one—and over 80 percent of it will be present as HCO_3^-.

Buffering, in brief, augments formation of HCO_3^-, and consequently it enhances the total amount of CO_2 taken on by a solution when exposed to any given CO_2 tension. The way that buffering has these effects may be seen by reference to our original description of the chemical reactions involved, as summarized in Equation (4) or (5). By limiting the accumulation in solution of one reaction product, buffering shifts the reactions to the right, enhancing formation of the other product.

Principles of Buffering Of course, buffering is not an all-or-nothing phenomenon. The extent of buffering varies widely from one biological fluid to another, depending in part on both the *concentration* and *chemical properties* of available buffer compounds.

Buffer reactions can be represented by the general equation

$$HX \rightleftharpoons H^+ + X^- \qquad (7)$$

where X^- is a chemical group or compound that can combine with H^+. When H^+ is added to a buffered solution, the buffer reaction is shifted to the left, removing some of the H^+ from free solution; on the other hand, if H^+ is extracted from the buffered solution, the reaction shifts to the right, releasing free H^+ from compound HX. Together, HX and X^- are termed a **buffer pair**. The mass-action expression for the buffer reaction is

$$\frac{[H^+][X^-]}{[HX]} = K' \tag{8}$$

where K' is a constant (the apparent dissociation constant) that depends on the particular reaction under consideration and also on prevailing conditions, such as temperature. The negative of the common logarithm of K' is symbolized *pK'*, just as $-\log[H^+]$ is called pH. Now the effectiveness of any given reaction as a buffer is greatest when half of the X^- groups are combined with H^+ and half are not; that is, in a system buffered by one particular reaction, the change in pH caused by addition or removal of H^+ is minimized when $[HX] = [X^-]$. From the mass-action expression, it is clear that for [HX] and $[X^-]$ to be equal, $[H^+]$ must equal K'—or pH must equal pK'. Thus, buffering effectiveness is greatest when the prevailing pH matches the pK' of the buffer reaction. Buffering effectiveness falls off as the pH is shifted away from pK' and, as a rule of thumb, is seriously impaired at pH values that differ from pK' by over a unit. In general, then, *a particular buffer reaction will act effectively to limit changes in the free H^+ concentration caused by the addition or removal of acid over a span of about two pH units centered on the pK' of the reaction.*

Earlier, we said that the extent of buffering of H^+ generated from CO_2 in a biological fluid depends on not only the concentration but also the chemical properties of available buffer compounds. Now we can see that a property of central significance is pK': A fluid may contain an enormous variety of potential buffer pairs, but the ones that are functioning importantly will generally be those with pK' values within a unit of the pH of the fluid.

The Buffering of H^+ Generated from CO_2 In the blood of mammals, the compounds that are preeminent in buffering the H^+ generated from CO_2 are the blood proteins, *especially hemoglobin*. Proteins contain many types of chemical groups that are potentially able to contribute to buffer function. However, many of the types of groups have pK' values too different from blood pH to play much of a buffer role under physiological conditions. The two types of groups that have appropriate pK' values and contribute most to buffering are the terminal amino groups of the protein chains and the imidazole groups found in histidine residues. The imidazole groups are the dominant buffering groups. Their reaction with H^+

is shown in Figure 13.26. The buffering of human blood is so effective that when the carbon dioxide tension of the blood is increased within the physiological range, thus promoting increased formation of bicarbonate, the buffers remove from free solution over 99.999 percent of the H^+ produced.

Carbamino Compounds

About 5 percent of the carbon dioxide in human arterial blood is carried in the form of *carbamino groups*. Such groups are formed by the direct, reversible combination of CO_2 with free amino groups of proteins:

$$P\text{—}NH_2 + CO_2 \rightleftharpoons P\text{—}NHCOO^- + H^+ \tag{9}$$

where P represents protein. The CO_2 assimilated into carbamino groups is removed from the pool of dissolved CO_2 and no longer contributes directly to the partial pressure of CO_2. Thus, the formation of carbamino groups increases the total CO_2 content of the blood at given CO_2 tension. Although carbamino groups are formed in significant amounts in the blood of mammals, their formation under physiological conditions in other groups of animals—vertebrate and invertebrate—remains questionable.

Carbon Dioxide Equilibrium Curves; The Haldane Effect

Suppose that some blood is brought to equilibrium with an atmosphere containing no CO_2, so that the CO_2 tension of the blood is zero. Suppose then that the blood is exposed to an atmosphere containing CO_2 at some fixed, positive tension. And suppose that as the blood comes to equilibrium with the new atmosphere, we measure the total quantity of CO_2 it takes up, regardless of the chemical form assumed by the CO_2 once in the blood (CO_2, HCO_3^-, carbamino, etc.). This quantity—the total amount of CO_2 that must enter each unit volume of blood to raise the blood CO_2 tension from zero to any particular positive CO_2 tension—is termed the blood's *total carbon dioxide concentration* at the latter tension.

A plot of the total carbon dioxide concentration as a function of CO_2 tension is known as a *carbon dioxide equilibrium curve*. An example—the curve for oxygenated human blood—is shown in Figure 13.27.

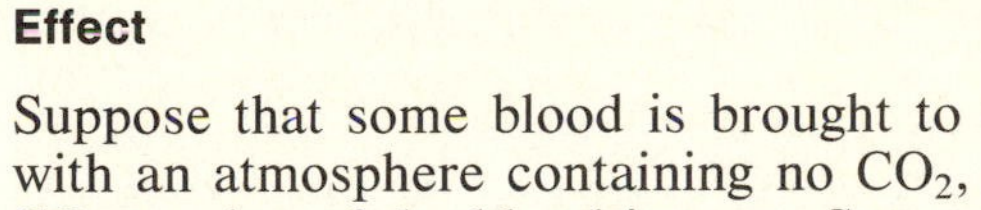

Shorthand expression:

$$Im + H^+ \rightleftharpoons ImH^+$$

Figure 13.26 The buffering reaction of the imidazole groups found in residues of the amino acid histidine within blood proteins, such as hemoglobin.

Carbon dioxide equilibrium curves—like oxygen equilibrium curves—are of great interpretive value because they permit estimation of the amount of gas the blood will take up or release in going from any one tension to another.

The shape of the human carbon dioxide equilibrium curve (Figure 13.27) is determined in large part by the kinetics of bicarbonate formation, because bicarbonate is the predominant form in which carbon dioxide exists in the blood. The shapes of the curves of many other species are similar for the same reason (Figure 13.31).

A critically important attribute of a carbon dioxide equilibrium curve is its slope in the physiological range of CO_2 tensions, for that slope indicates the extent to which the venous CO_2 tension must exceed arterial for the blood to pick up and discharge any particular amount of CO_2 per circulatory cycle. To illustrate, consider that in resting humans (1) the arterial CO_2 tension is about 40 mm Hg and (2) the rates of circulation and CO_2 production are such that each 100 mL of blood must carry about 4 mL of CO_2 from the systemic tissues to the lungs on each circulatory cycle. If CO_2 were carried only in dissolved form (see the "dissolved" line in Figure 13.27), the venous CO_2 tension would have to be about 100 mm Hg—60 mm Hg higher than arterial—for the venous CO_2 content to be 4 volumes percent higher than arterial. The actual CO_2 equilibrium curve seen in Figure 13.27 has a much greater slope than the "dissolved" curve. Indeed, its slope is so great that the venous tension needs to exceed the arterial tension by only about 8–9 mm Hg for the blood's CO_2 content to increase by 4 volumes percent in the tissues. The difference between the slopes of the dissolved and actual curves brings to light one of the crucially

important aspects of bicarbonate formation. The extensive reaction of CO_2 to form bicarbonate in the blood does not merely increase the blood's CO_2 content at any given tension, *it increases the amount of CO_2 the blood can take up when its tension is incremented by any given amount.* This important consequence, we should recall, can be attributed to the evolved properties of the blood buffers, for they are the chief determinants of the extent of bicarbonate formation.

The Haldane Effect   Commonly, the carbon dioxide equilibrium curve changes with the state of oxygenation of the blood, a phenomenon named the ***Haldane effect*** after one of its discoverers. If a blood exhibits the Haldane effect, the total CO_2 concentration at any given CO_2 tension is greater when the blood is deoxygenated than when it is oxygenated, as illustrated in Figure 13.28.

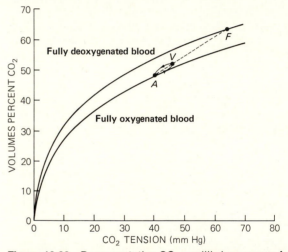

Figure 13.28   Representative CO_2 equilibrium curves for fully oxygenated and almost fully deoxygenated human blood. Because in the living organism changes in the CO_2 content of the blood are accompanied by contemporaneous changes in the state of oxygenation, neither equilibrium curve alone describes the true functional relation between CO_2 content and CO_2 tension. The dashed line indicates this true functional relation under the assumptions that fully oxygenated blood carries 20 volumes percent O_2 and the respiratory quotient is 0.80. If arterial blood carries 20 volumes percent O_2 and 48 volumes percent CO_2 (point *A*), complete utilization of the oxygen in the tissues would raise CO_2 content to 64 volumes percent on the curve for deoxygenated blood (point *F*). In resting humans blood is not fully deoxygenated in the tissues. The ellipsoid between *A* and *V* describes the approximate functional relation between CO_2 content and CO_2 tension as blood alternately becomes arterial (*A*) and venous (*V*) in resting humans. Note that an increase in CO_2 content of 4 volumes percent as the blood becomes venous is accompanied by an increase in CO_2 tension of about 6 mm Hg. If deoxygenation did not affect the CO_2 equilibrium curve and blood followed the curve for oxygenated blood throughout the circulatory cycle, an increase of 4 volumes percent in CO_2 content would raise the CO_2 tension by about 9 mm Hg. [From J.P. Peters and D.D. Van Slyke, *Quantitative Clinical Chemistry,* Vol. I. Williams & Wilkins, Baltimore, MD, 1932. Original figure © 1932 by The Williams & Wilkins Co., Baltimore.]

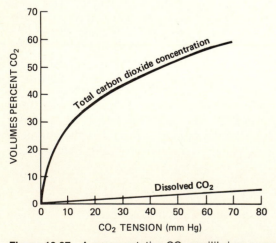

Figure 13.27   A representative CO_2 equilibrium curve for fully oxygenated human blood, showing total CO_2 content as a function of CO_2 tension. The portion of the total CO_2 content due to dissolved CO_2 is indicated at the bottom. Most of the excess of total CO_2 content over dissolved CO_2 content is attributable to bicarbonate; some is due to carbamino formation. [Equilibrium curve from A.V. Bock, H. Field, Jr., and G.S. Adair, *J. Biol. Chem.* **59:**353–378 (1924).]

The Haldane effect arises because the oxygen-transport pigments are intimately involved in determining the extent of CO_2 uptake by the blood. The most general role played by the respiratory pigments in CO_2 uptake is that of buffering, and in this role lies the most general explanation of Haldane effects. Deoxygenation of oxygen-transport pigments commonly increases their tendency to take up H^+ ions (this is the necessary obverse of the Bohr effect, as already established on p. 310). Thus, at any given CO_2 tension, deoxygenated pigments tend to buffer the blood at a more-alkaline pH than their oxygenated counterparts, an effect that, according to Equation (6), enhances formation of bicarbonate and thereby augments the total pool of CO_2 contained by the blood.

In mammals, deoxyhemoglobin binds more CO_2 in the form of carbamino groups than oxyhemoglobin. This also contributes to the Haldane effect.

The functional significance of the Haldane effect is illustrated in Figure 13.28, using data for humans. Measurements reveal that when people are at rest, their average mixed-venous CO_2 tension is about 46 mm Hg, whereas their average arterial CO_2 tension is about 40 mm Hg, as already mentioned. With this information, to determine the yield of CO_2 as the blood traverses the lungs, we must compare (1) the CO_2 concentration at 46 mm Hg of venous blood, which is only about 70 percent oxygenated, with (2) the CO_2 concentration at 40 mm Hg of arterial blood, which is essentially fully oxygenated. The ellipsoid between points A and V in Figure 13.28 illustrates this functional relation between CO_2 concentration and CO_2 tension in resting humans. The drop in CO_2 concentration between the venous and arterial blood is about 4 volumes percent, as we earlier said it must be. Previously, we determined that the venous CO_2 tension would have to exceed arterial by 8–9 mm Hg if the blood were to function on the arterial CO_2 equilibrium curve alone. Now we see that because of the Haldane effect, a smaller tension difference—just 6 mm Hg—suffices. The *functional slope* of the relation between CO_2 concentration and CO_2 tension is shown by the dashed line in Figure 13.28. In essence, the Haldane effect renders this functional slope greater than the slope of the curve for either oxygenated or deoxygenated blood alone. Deoxygenation promotes CO_2 uptake by the blood, and oxygenation promotes CO_2 offloading.

Previously, we stressed that the change in the buffer function of respiratory pigments with oxygenation and deoxygenation acts to limit variations in blood pH (p. 310). In humans, for example, whereas the arteriovenous drop in pH would be from about 7.41 to 7.32 without a change in hemoglobin buffer function, it is actually much smaller, from about 7.41 to 7.37, because hemoglobin tends to take up extra H^+ as it is deoxygenated. In total, the integration of performance by many of the respiratory pigments is nothing short of remarkable: As the blood courses

between the respiratory organs and systemic tissues, the pigments simultaneously promote oxygen delivery, CO_2 transport, and pH stabilization.

Thermal Effects Decreases in temperature typically enhance H^+ uptake by blood buffers, inducing them to buffer the pH at a more alkaline level. Thus, the extent of bicarbonate formation at given CO_2 tension is greater at low temperatures than high. Principally for this reason—but also because CO_2 solubility is enhanced at low temperatures—decreases in temperature tend to shift the CO_2 equilibrium curve toward higher CO_2 concentrations, as illustrated in Figure 13.29.

Carbon Dioxide Transport in Humans

Here, we summarize and examine additional details of carbon dioxide transport as presently understood in humans. Many of the principles developed are believed to apply to vertebrates in general.

An important attribute of the interconversion of CO_2 and H_2CO_3—Equation (2)—is that it occurs relatively slowly in the absence of catalysis. The native slowness of the reaction presents a potential bottleneck in the blood's uptake of CO_2 as bicarbonate in the systemic tissues and in the release of CO_2 from the bicarbonate pool in the lungs, for H_2CO_3 is a necessary intermediate in the interconversion of CO_2 and HCO_3^- [Equation (4)]. The transit time of blood through the systemic capillaries is so brief (on the order of 0.2–2 s) that bicarbonate formation must occur rapidly if it is to play the crucial role that it does in the uptake of CO_2 by the blood. Similarly, CO_2 must be able to exit the bicarbonate pool rapidly in the lungs. The bottleneck that could be posed by the native slowness of the interconversion of CO_2

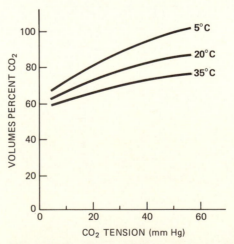

Figure 13.29 CO_2 equilibrium curves for blood of the turtle *Pseudemys scripta* at three temperatures, with the state of oxygenation held constant. [Data from R.B. Reeves and H. Rahn, in S.C. Wood and C. Lenfant (eds.), *Evolution of Respiratory Processes*, pp. 225–252. Dekker, New York, 1979.]

and H_2CO_3 has been circumvented by the evolution of an enzyme catalyst, *carbonic anhydrase*, that accelerates the reaction dramatically. The reaction is the only one known to be catalyzed in the entire reaction sequence associated with CO_2 transport.

Carbonic anhydrase is found at high concentrations in the red blood cells, where hemoglobin—the most important blood buffer—is immediately available to play its own critical role in bicarbonate formation. Blood plasma lacks carbonic anhydrase. This latter circumstance has long suggested that the interconversion of CO_2 and H_2CO_3 in the plasma would be entirely uncatalyzed. However, carbonic anhydrase has been found in lung tissue and certain systemic tissues, such as muscle; some investigators believe that catalysis of the interconversion of CO_2 and H_2CO_3 in the plasma is one of the functions of this tissue-bound enzyme.

The compartmentalization in the red cells of the blood's carbonic anhydrase and of its chief buffer (hemoglobin) gives rise to complexities that would not exist were the blood a homogeneous fluid. These will be evident as we now consider Figure 13.30, which summarizes the process of CO_2 uptake by the blood in the systemic tissues.

Most of the CO_2 diffusing into the blood from the systemic tissues is hydrated within the red cells because of the abundant presence there of carbonic anhydrase. This means, in turn, that most H^+ and HCO_3^- are generated within the red cells. The H^+ is buffered by hemoglobin, which develops a greater affinity for H^+ as it is deoxygenated. In mammals, however, much of the HCO_3^- diffuses out of the red cells into the plasma, following the concentration gradient created by the abundant formation of HCO_3^- within the cells. The plasma, in fact, ultimately carries most of the HCO_3^- added to the blood in the capillaries. Maintenance of electrical balance in the red cells demands that the outward diffusion

of HCO_3^- be accompanied by the outward diffusion of positive ions or the inward diffusion of other negative ions. Because the cell membranes of red cells are poorly permeable to the major cations in the red cells but rather freely permeable to chloride, the outward diffusion of HCO_3^- is in fact accompanied by inward diffusion of Cl^-: a phenomenon called the *chloride shift*. Some of the CO_2 entering the red cells is taken up in carbamino groups because hemoglobin increases its tendency to form such groups as it is deoxygenated. The entire sequence of events diagrammed in Figure 13.30 is reversed when the blood flows through the lungs.

With the rise in the blood CO_2 tension from 40 to 46 mm Hg in the systemic tissues, the blood's concentration of dissolved CO_2 increases by about 0.4 volumes percent. Recalling that the increase in total CO_2 concentration is about 4 volumes percent, we see that about 10 percent of the CO_2 carried from the systemic tissues to the lungs on each circulatory cycle is carried in dissolved form. Another 10 percent is transported in the form of carbamino groups when people are at rest. The remaining 80 percent of the CO_2 picked up and released per circulatory cycle is carried in the bicarbonate pool.

Comparative Aspects of Carbon Dioxide Transport

Figure 13.31 illustrates the diversity of CO_2 equilibrium curves found in various animals. Differences among the curves can be attributed for the most part to differences in the blood buffer systems that promote bicarbonate formation: differences in the concentrations of buffers, their pK' values, and their extent of titration by noncarbonic acids. Because the oxygen-transport pigments play important roles as buffers, there tends to be a rough correlation between the concentration of oxygen-transport pigment

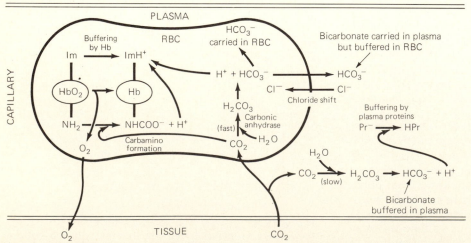

Figure 13.30 Schematic summary of processes of CO_2 uptake by the blood in the systemic capillaries of mammals. Processes occur in reverse in the lungs. RBC: red blood cell; Hb: hemoglobin; Im: imidazole groups; Pr: plasma proteins.

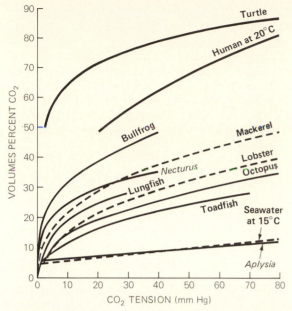

Figure 13.31 Typical CO_2 equilibrium curves of oxygenated blood in 10 species at 15–25°C. Because all curves were not determined at exactly the same temperature, some of the differences observed among curves may arise from temperature differences. There can be considerable variation in the curve among individuals of a single species. (1) River turtle (*Pseudemys floridana concinna*) at 25°C; (2) human blood at 20°C; (3) bullfrog (*Rana catesbeiana*) at 22°C; (4) mackerel (*Scomber scombrus*) at 20°C; (5) mudpuppy (*Necturus maculosus*) at 20°C; (6) lungfish (*Neoceratodus forsteri*) at 18°C; (7) lobster (*Palinurus vulgaris*) at 15°C; (8) octopus (*Octopus macropus*) at 15°C; (9) toadfish (*Opsanus tau*) at 20°C; and (10) sea hare (*Aplysia limacina*), a gastropod mollusk, at 15°C. [Sources of data: (1) F.C. Southworth, Jr. and A.C. Redfield, *J. Gen. Physiol.* **9**:387–403 (1926); (2) H. Harms and H. Bartels, *Pflügers Arch. Gesamte Physiol. Menschen Tiere* **272**:384–392 (1961); (3) and (5) C. Lenfant and K. Johansen, *Respir. Physiol.* **2**:247–260 (1967); (4) and (9) R.W. Root, *Biol. Bull.* (*Woods Hole*) **61**:427–456 (1931); (6) C. Lenfant, K. Johansen, and G.C. Grigg, *Respir. Physiol.* **2**:1–21 (1966); (7), (8), (10), and curve for seawater, T.R. Parsons and W. Parsons, *J. Gen. Physiol.* **6**:153–166 (1924).]

and the "CO_2 combining power" of the blood; animals with relatively high oxygen-carrying capacities (e.g., mackerel) tend to exhibit higher total CO_2 concentrations than related animals that have lower oxygen-carrying capacities (e.g., toadfish). The sea hare *Aplysia* (a gastropod mollusk) illustrates the extreme case of an animal whose blood lacks an oxygen-transport pigment and indeed contains little protein of any kind; without appreciable concentrations of protein buffers, its CO_2 equilibrium curve is not much different from that of seawater.

Thus far, attempts to identify adaptive rationales for interspecific differences in the CO_2 equilibrium curve have not borne much fruit. Interpretation of the differences is complex because the blood buffers that are such pivotal determinants of the CO_2 curves play multiple roles. They do not merely promote uptake of CO_2.

In analyzing the respiratory-gas exchange of ani-

mals with discrete respiratory organs in Chapter 12 (p. 249), we noted that air breathers characteristically raise the CO_2 tension of air in their lungs to well above the ambient CO_2 tension, whereas water breathers typically raise the tension in the water they breathe only slightly. Based on this observation and given that arterial blood exchanges gases with the air in the lungs or the water passing over the gills, it is not surprising to find that arterial CO_2 tensions tend to be much above ambient CO_2 tensions in air breathers but close to ambient in water breathers. Resting mammals and birds breathing atmospheric air, for example, exhibit arterial CO_2 tensions of 25 mm Hg or higher. Gill-breathing fish in well-aerated waters, by contrast, commonly have arterial CO_2 tensions of only 1–3 mm Hg. Studies of groups of related animals that include both water and air breathers—including fish, amphibians, and crabs—have revealed that a rise in the arterial CO_2 tension occurs regularly when animals make the transition from water to air breathing.

When arterial CO_2 tensions are high, as in mammals and many other air breathers, the functional range of CO_2 tensions is placed on the high, relatively flat portion of the CO_2 equilibrium curve. We have seen, for instance, that the blood of resting humans shifts back and forth between tensions of 40 and 46 mm Hg as it alternately flows to the lungs and systemic tissues. By contrast, when arterial tensions are low, as in water breathers occupying well-aerated waters, the functional range of tensions is placed on the steep, low-tension part of the CO_2 equilibrium curve; typically, even the venous CO_2 tensions of water breathers are below 10 mm Hg.

One of the generalizations that emerges from the study of CO_2 transport in both air and water breathers is that the rate of circulation and effective slope of the CO_2 equilibrium curve in the functional range of tensions are usually matched to CO_2 transport requirements in such a way that the arteriovenous difference in CO_2 tension is small: less—often much less—than 10 mm Hg.

Acid–Base Regulation

The topic of acid–base regulation could justly be discussed in a chapter of its own. The present introductory treatment is placed here, however, because of the important roles played in acid–base regulation by carbon dioxide transport and the CO_2–HCO_3^- buffer system.

The pH of the body fluids cannot vary far from normal levels without serious functional consequences. Among humans, for instance, the normal pH of arterial blood at deep-body temperature is about 7.4, and individuals are placed near death if their pH rises just to 7.7 or falls to 6.8. To a large extent, abnormal H^+ concentrations inflict their adverse effects by influencing the function of enzymes and other proteins.

According to a line of thought that has become ever more prominent in recent years, the state of molecular electrical charge is often the prime attribute of proteins to be immediately affected by abnormal deviations of pH. Thus, protection of a normal state of molecular electrical charge is often the key accomplishment of regulating pH within normal limits. The charge state of protein molecules is a critical parameter because it affects molecular conformation and other functionally important molecular properties.

The study of pH-induced variations in the charge state of proteins is restricted to just a limited number of types of chemical groups on the protein molecules, for only certain types of chemical groups are altered in charge by physiologically meaningful pH variations. These groups are the ones with pK' values within a unit or so of the prevailing pH, the very groups that we have already identified as being instrumental in blood buffering. Among them, imidazole groups (Figure 13.26) are by far the most important. By having pK' values near the prevailing pH, imidazole groups tend to take on H^+ ions as the blood H^+ concentration rises and release H^+ as $[H^+]$ falls. Earlier, we stressed the buffering effect of these reactions. Here, we stress that when imidazole groups on protein molecules take up or release H^+, the electrical charge of the protein molecules is rendered more positive or negative, as the case may be. For the molecular charge of a protein to remain within limits compatible with adequate protein function, the pH must remain within parallel limits.

Roles of Carbon Dioxide and Bicarbonate in Acid–Base Physiology   Let us consider again the interconversion of CO_2 and bicarbonate:

$$CO_2 + H_2O \rightleftharpoons HCO_3^- + H^+ \qquad (10)$$

In discussing CO_2 transport, we emphasized that these reactions are a *source* of H^+ and a *cause* of pH changes when forced to the right by the addition of CO_2 to the body fluids. This in fact is a very limited perspective, and now we must take a broader view of the roles that the reactions play.

To begin with, the CO_2—HCO_3^- system is itself an important *buffer* system, sometimes called the "bicarbonate buffer system." We have not mentioned this attribute so far because, obviously enough, H^+ generated from the reaction of CO_2 with water cannot be buffered by reacting with the other product of the selfsame reaction, HCO_3^-. That is, H^+ from the reaction of CO_2 and water cannot be buffered by the bicarbonate buffer system. Instead, H^+ of such origin must be buffered by proteins and other *nonbicarbonate* buffer systems. Suppose, however, that excess H^+ is added to the blood from some other source than the reaction of CO_2 with water (e.g., from lactic acid). Clearly, while some such H^+ would be taken up by the nonbicarbonate buffer sys-

tems, some also could be removed from solution by reacting with HCO_3^- to *form* CO_2 and water. In brief, *H^+ from noncarbonic acids can be buffered by the bicarbonate, as well as nonbicarbonate, systems*. Similarly, if some process other than formation of CO_2 from HCO_3^- were to remove H^+, the bicarbonate buffer system—by a shift toward the right—could participate along with nonbicarbonate buffers in limiting the pH change by replacing some of the H^+.

In discussing CO_2 equilibrium curves and the amount of CO_2 the blood is capable of taking up, we made much of the fact that accumulation of CO_2 tends to shift the pH downward. Now we must recognize that a certain extent of such pH shifting by CO_2 is in fact essential for normal acid–base regulation: *Without maintenance of an appropriately high blood CO_2 tension, the pH of the blood will be too alkaline.*

Because CO_2 is continuously manufactured by metabolism and voided across respiratory surfaces, its tension in the body fluids is a highly dynamic property, subject to rapid changes. Recognizing, in addition, that increases or decreases in the CO_2 tension titrate the body fluids in the acid or alkaline direction, physiological control of the CO_2 tension can be seen to be an important potential instrument of both short- and long-term pH regulation.

Mechanisms of Acid–Base Regulation   When a process acts to increase the amount of acid in the body of an animal, buffers provide the first line of defense (limiting the decline in pH). However, regulation of the animal's acid–base status ultimately demands that other processes be set in motion that will either export acid from the body or increase the body's content of base. Conversely, if the initial disturbance is a decrease in body acid, acid–base regulation requires a compensating uptake of acid or export of base from the body. An animal's regulatory exchanges of acid and base with the environment may be considered to be focused on CO_2, HCO_3^-, and H^+: the members of Equation (10).

The level of CO_2 in the body of an animal can be raised or lowered in service of acid–base regulation by the respiratory organs, especially in terrestrial organisms. Suppose, for example, that a person's blood becomes too acid. One possible compensatory response is for the person's rate of pulmonary ventilation to be increased. An increase in ventilation will accelerate export of CO_2, lower the blood CO_2 tension, pull reaction (10) to the left, and thus lower the H^+ concentration. Slowing of ventilation, on the other hand, can promote accumulation of CO_2 in the body fluids and thus counteract a shift of pH in the alkaline direction.

Organisms often have the ability to exchange H^+ itself with the environment. Of course, H^+ must be exchanged in solution. Among terrestrial animals, responsibility for its exchange rests with the kidneys.

Humans, for example, are regularly confronted with an excess of H^+ from their diet and void it in their urine; this urinary elimination of H^+ can be curtailed entirely, however, when appropriate. Among both fish and aquatic crustaceans, the gill epithelium is evidently the principal mediator of direct H^+ exchange with the environment (e.g., see Figure 7.6).

Bicarbonate ions are also exchanged with the environment in service of acid–base regulation. These exchanges are mediated by the kidneys in terrestrial animals but principally, it appears, by the gill epithelium in fish and crabs (Figure 7.6). *Bicarbonate functions as a base*. If retention or acquisition of HCO_3^- is enhanced, thus increasing the concentration of HCO_3^- in the body fluids, reaction (10) is shifted toward the left, tending to remove H^+ from solution and render the fluids more alkaline. Conversely, increased elimination of HCO_3^- tends to raise the H^+ concentration of the body fluids. One way to view this effect of HCO_3^- elimination is to recognize that HCO_3^- originates from CO_2 and H_2CO_3; when HCO_3^- is eliminated, just the H^+ of H_2CO_3 remains in the body fluids, acidifying them.

Temperature and Acid–Base Balance   The neutral pH—defined to be the pH of pure water—is a function of temperature, being higher at low temperatures than high ones. In recent decades, it has come to light that among poikilotherms, the normal blood pH varies with body temperature in parallel with the neutral pH—as illustrated in Figure 13.32. The blood pH is more alkaline than neutral. Within each species, the extent of this alkalinity tends to be consistent from one temperature to another. Thus, each species is said to maintain a consistent *relative alkalinity*.

Studies of humans and other large mammals have tended to engender the notion that pH is always regulated at a single, invariant level (e.g., 7.4 in human arterial blood). Now we recognize that this type of pH regulation is a special case, found only in organisms that maintain a consistent body-core temperature. A broader view of animals reveals that the pH defended by regulatory mechanisms is in fact a temperature-dependent *variable*.

In some air-breathing poikilotherms, such as certain turtles, the regulated changes of blood pH with temperature are effected almost entirely by modulation of pulmonary ventilation. In relation to their rate of CO_2 production, these animals breathe more rapidly at low temperatures than high temperatures, thus lowering their blood CO_2 tension at low temperatures and shifting their pH correspondingly upward (Figure 13.33). Aquatic animals are much more limited than terrestrial ones in the extent of control they can exert over their blood CO_2 tension (p. 271), and in certain fish it is now known that HCO_3^- regulation is the key element in adjusting pH along with temperature. In these fish, the blood CO_2 tension is about the same at all temperatures, but a higher blood concentration

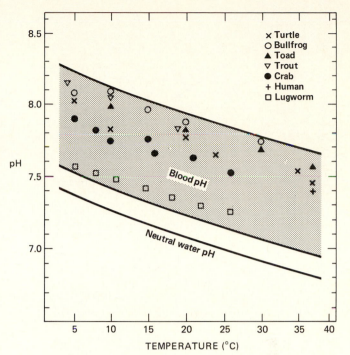

Figure 13.32   Blood pH as a function of body temperature in seven animals, including both vertebrates and invertebrates. Also shown is the neutral pH as a function of temperature. Note that within each particular species, the difference between blood pH and neutral pH is approximately constant regardless of temperature. This difference ("the constant of relative alkalinity") is usually about 0.6–0.8 pH units in vertebrates but is lower in some invertebrates. [Reprinted by permission from P. Dejours, *Principles of Comparative Respiratory Physiology*, 2nd ed. Elsevier/North-Holland, New York, 1981.]

of HCO_3^- is maintained at low temperatures than at high ones.

What advantage might animals realize by increasing their blood pH as temperatures fall? One prominent hypothesis—often called the *alphastat hypothesis*—is that the adjustment of pH is directed toward maintaining protein molecules at a consistent state of electrical charge (see p.330). As temperature falls, the pK' values of imidazole groups increase, meaning that the groups increase their inherent tendency to combine with H^+. If this change were unopposed, more groups would be combined with H^+—and thus positively charged—at low temperatures than at high temperatures. Decreasing the H^+ concentration opposes the heightened tendency of the imidazole groups to take up H^+ and thus prevents the proportion of positively charged groups from increasing.

Disturbances   When the pH of the body fluids is shifted abnormally to the acid or alkaline side of the usual pH at given body temperature, the disturbance of the pH is known as *acidosis* or *alkalosis*, respectively. Disturbances of pH are also classed as *respiratory* or *metabolic* according to their primary cause.

The *respiratory* disturbances are ones precipitated

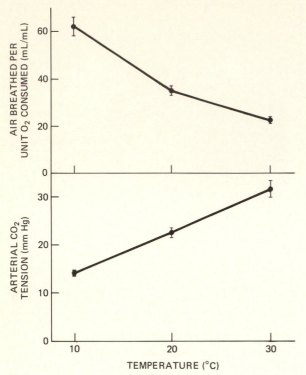

Figure 13.33 Aspects of CO₂ exchange as a function of body temperature in turtles, *Pseudemys scripta*. Upper plot shows that, as temperature falls, the ratio of pulmonary ventilation to metabolism increases. Consequently, as shown in the lower plot, the CO₂ tension maintained in arterial blood is lower at low temperatures than at high. Analysis of the trend in the upper plot reveals that when the temperature drops from 30 to 10°C, the rate of ventilation hardly changes at all even though the metabolic rate—and rate of CO₂ production—decrease by about 70 percent. Vertical lines through points delimit ±1 standard error. [Data from D.C. Jackson, S.E. Palmer, and W.L. Meadow, *Respir. Physiol.* **20**:131–146 (1974).]

ercise; the added H^+ reacts with HCO_3^- (buffering) and thus manifests itself as a drop in the concentration of HCO_3^-.

Compensation for a disturbance is the rule. For example, when accumulation of noncarbonic acid causes a fall in blood pH (metabolic acidosis), the problem must be corrected in the long term by the metabolic destruction or excretion of the acid. In the short term, however, compensation can be effected, especially in air breathers, by an increase in ventilation; heightening the rate of ventilation will lower the blood CO_2 tension and thereby tend to raise the pH back toward its original level (see also p. 323).

SELECTED READINGS

Angersbach, D. 1975. Oxygen pressures in haemolymph and various tissues of the tarantula, *Eurypelma helluo*. *J. Comp. Physiol.* **98**:133–145.

Angersbach, D. 1978. Oxygen transport in the blood of the tarantula *Eurypelma californicum:* pO₂ and pH during rest, activity, and recovery. *J. Comp. Physiol.* **123**:113–125.

Arp, A.J., J.J. Childress, and R.D. Vetter. 1987. The sulphide-binding protein in the blood of the vestimentiferan tube-worm, *Riftia pachyptila,* is the extracellular haemoglobin. *J. Exp. Biol.* **128**:139–158.

Bannister, J.V. (ed.). 1977. *Structure and Function of Haemocyanin.* Springer-Verlag, New York.

Bartels, H. 1970. *Prenatal Respiration.* North-Holland, Amsterdam.

*Bauer, C., G. Gros, and H. Bartels (eds.). 1980. *Biophysics and Physiology of Carbon Dioxide.* Springer-Verlag, New York.

Brendel, W. and R.A. Zink (eds.). 1982. *High Altitude Physiology and Medicine.* Springer-Verlag, New York.

Brix, O., G. Lykkeboe, and K. Johansen. 1979. Reversed Bohr and Root shifts in hemocyanin of the marine prosobranch, *Buccinum undatum:* Adaptations to a periodically hypoxic habitat. *J. Comp. Physiol.* **129**:97–103.

Cohen, J.J. and J.P. Kassirer. 1982. *Acid–Base.* Little, Brown, Boston.

Cole, R.P. 1982. Myoglobin function in exercising skeletal muscle. *Science* **216**:523–525.

*Davenport, H.W. 1974. *The ABC of Acid–Base Chemistry,* 6th ed. University of Chicago Press, Chicago.

*Dejours, P. 1981. *Principles of Comparative Respiratory Physiology,* 2nd ed. Elsevier/North-Holland, New York.

*Dickerson, R.E. and I. Geis. 1983. *Hemoglobin: Structure, Function, Evolution, and Pathology.* Benjamin-Cummings, Menlo Park, CA.

*Dickson, K.A. and G.N. Somero. 1987. Partial characterization of the buffering components of the red and white myotomal muscle of marine teleosts, with special emphasis on scombrid fishes. *Physiol. Zool.* **60**:699–706.

Edsall, J.T. and J. Wyman. 1958. *Biophysical Chemistry.* Academic, New York. (See for CO₂ chemistry.)

Fänge, R. 1983. Gas exchange in fish swim bladder. *Rev. Physiol. Biochem. Pharmacol.* **97**:111–158.

Ghiretti-Magaldi, A., F. Ghiretti, and B. Salvato. 1977. The evolution of haemocyanin. *Symp. Zool. Soc. London* **38**:513–523.

Heath, D. and D.R. Williams. 1981. *Man at High Altitude,* 2nd ed. Churchill Livingstone, London.

by modifications of CO_2 exchange with the environment across the respiratory organs. ***Respiratory alkalosis*** arises when ventilation is abnormally increased relative to CO_2 production, causing the CO_2 tension in the body fluids to be driven below the level consistent with a normal pH. Thermal panting, for example, sometimes causes this type of disturbance (p. 113). One major cause of ***respiratory acidosis*** is impairment of ventilation; another is exposure to a CO_2-rich environment. Both can lead to excessive CO_2 tensions in the body fluids.

Whereas the blood parameter initially altered in respiratory disturbances is the CO_2 tension, *metabolic* disturbances—by definition—initially alter the blood bicarbonate concentration. ***Metabolic acidosis*** and ***metabolic alkalosis*** both have numerous possible causes. Metabolic acidosis, for example, can be induced by excessive salivary or gastrointestinal loss of HCO_3^-, as during saliva spreading in rodents (p. 115) or diarrhea. Or it can result from excessive addition of noncarbonic acid to the body fluids, as when lactic acid is accumulated during vigorous ex-

Hebbel, R.P., J.W. Eaton, R.S. Kronenberg, E.D. Zanjani, L.G. Moore, and E.M. Berger. 1978. Human llamas. *J. Clin. Invest.* **62**:593–600.

Heisler, N. 1982. Transepithelial ion transfer processes as mechanisms of fish acid–base regulation in hypercapnia and lactacidosis. *Can. J. Zool.* **60**:1108–1122.

*Heisler, N. (ed.). 1986. *Acid–Base Regulation in Animals.* Elsevier, New York.

Hemmingsen, E.A. and E.L. Douglas. 1977. Respiratory and circulatory adaptations to the absence of hemoglobin in Chaenichthyid fishes. In G.A. Llano (ed.), *Adaptations within Antarctic Ecosystems,* pp. 479–487. Smithsonian Institution, Washington, DC. (See also: Holeton, G.F. 1970. *Comp. Biochem. Physiol.* **34**:457–471.)

Houlihan, D.F., G. Duthie, P.J. Smith, M.J. Wells, and J. Wells. 1986. Ventilation and circulation during exercise in *Octopus vulgaris. J. Comp. Physiol.* **156B**:683–689.

Ingermann, R.L. 1982. Physiological significance of Root effect hemoglobins in trout. *Respir. Physiol.* **49**:1–10.

Johansen, K. and C. Lenfant. 1966. Gas exchange in the cephalopod, *Octopus dofleini. Am. J. Physiol.* **210**:910–918.

Johansen, K., C. Lenfant, and T.A. Mecklenberg. 1970. Respiration in the crab, *Cancer magister. Z. vergl. Physiol.* **70**:1–19.

Johansen, K., G. Lykkeboe, R.E. Weber, and G.M.O. Maloiy. 1976. Respiratory properties of blood in awake and estivating lungfish, *Protopterus amphibius. Respir. Physiol.* **27**:335–345.

Jones, J.D. 1964. The role of hemoglobin in the aquatic pulmonate, *Planorbis corneus. Comp. Biochem. Physiol.* **12**:283–295.

Kagen, L.J. 1973. *Myoglobin. Biochemical, Physiological, and Clinical Aspects.* Columbia University Press, New York.

Kiceniuk, J.W. and D.R. Jones. 1977. The oxygen transport system in trout (*Salmo gairdneri*) during sustained exercise. *J. Exp. Biol.* **69**:247–260.

Kraus, D.W. and J.M. Colacino. 1986. Extended oxygen delivery from the nerve hemoglobin of *Tellina alternata* (Bivalvia). *Science* **232**:90–92.

Lahiri, S. 1975. Blood oxygen affinity and alveolar ventilation in relation to body weight in mammals. *Am. J. Physiol.* **229**:529–536.

Lenfant, C. 1973. High altitude adaptation in mammals. *Am. Zool.* **13**:447–456.

Linzen, B. (ed.). 1986. *Invertebrate Oxygen Carriers.* Springer-Verlag, New York.

Linzen, B. et al. 1985. The structure of arthropod hemocyanins. *Science* **229**:519–524.

*Lykkeboe, G. and K. Johansen. 1982. A cephalopod approach to rethinking about the importance of the Bohr and Haldane effects. *Pacific Sci.* **36**:305–313.

*Mangum, C.P. 1980. Respiratory functions of the hemocyanins. *Am. Zool.* **20**:19–38.

*Mangum, C.P. 1985. Oxygen transport in invertebrates. *Am. J. Physiol.* **248**:R505–R514.

Mangum, C. and D. Towle. 1977. Physiological adaptation to unstable environments. *Am. Sci.* **65**:67–75. (For update see: Mason, R.P., C.P. Mangum, and G. Godette. 1983. *Biol. Bull.* **164**:104–123.)

Mantel, L.H. 1983. *The Biology of Crustacea,* Vol. 5, *Internal Anatomy and Physiological Regulation.* Academic, New York.

McMahon, B.R., D.G. McDonald, and C.M. Wood. 1979. Ventilation, oxygen uptake and haemolymph oxygen transport, following enforced exhausting activity in the Dungeness crab *Cancer magister. J. Exp. Biol.* **80**:271–285.

Perutz, M.F. 1978. Hemoglobin structure and respiratory transport. *Sci. Am.* **239**(6):92–125.

Powers, D.A. 1980. Molecular ecology of teleost fish hemoglobins: Strategies for adapting to changing environments. *Am. Zool.* **20**:139–162.

Primmett, D.R.N., D.J. Randall, M. Mazeaud, and R.G. Boutilier. 1986. The role of catecholamines in erythrocyte pH regulation and oxygen transport in rainbow trout (*Salmo gairdneri*) during exercise. *J. Exp. Biol.* **122**:139–148.

*Prosser, C.L. 1973. Respiratory functions of blood. In C.L. Prosser (ed.), *Comparative Animal Physiology,* 3rd ed., Vol. I. Saunders, Philadelphia.

*Randall, D.J., W.W. Burggren, A.P. Farrell, and M.S. Haswell. 1981. *The Evolution of Air Breathing in Vertebrates.* Cambridge University Press, New York.

*Reeves, R.B. and H. Rahn. 1979. Patterns in vertebrate acid–base regulation. In S.C. Wood and C. Lenfant (eds.), *Evolution of Respiratory Processes,* pp. 225–252. Dekker, New York.

Riggs, A. 1979. Studies of the hemoglobins of Amazonian fishes: An overview. *Comp. Biochem. Physiol.* **62A**:257–272.

Roughton, F.J.W. 1964. Transport of oxygen and carbon dioxide. In W.O. Fenn and H. Rahn (eds.), *Handbook of Physiology. Section 3: Respiration,* Vol. I. American Physiological Society, Washington, DC.

Sanders, N.K., A.J. Arp, and J.J. Childress. 1988. Oxygen binding characteristics of the hemocyanins of two deep-sea hydrothermal vent crustaceans. *Respir. Physiol.* **71**:57–68.

Schmidt-Nielsen, K. and C.R. Taylor. 1968. Red blood cells: Why or why not? *Science* **162**:274–275.

Snyder, L.R.G., S. Born, and A.J. Lechner. 1982. Blood oxygen affinity in high- and low-altitude populations of the deer mouse. *Respir. Physiol.* **48**:89–105.

Steen, J.B. and A. Kruysse. 1964. The respiratory function of teleostean gills. *Comp. Biochem. Physiol.* **12**:127–142.

*Stewart, P.A. 1978. Independent and dependent variables of acid–base control. *Respir. Physiol.* **33**:9–26.

Stryer, L. 1981. *Biochemistry,* 2nd ed. Freeman, San Francisco. (Chapters 3–5.)

Truchot, J.P. 1975. Factors controlling the *in vitro* and *in vivo* oxygen affinity of the hemocyanin in the crab *Carcinus maenas* (L.). *Respir. Physiol.* **24**:173–189.

Truchot, J.P. 1987. *Comparative Aspects of Extracellular Acid–Base Balance.* Springer-Verlag, New York.

*Weber, R.E. 1980. Functions of invertebrate hemoglobins with special reference to adaptations to environmental hypoxia. *Am. Zool.* **20**:79–101.

West, J.B. 1984. Human physiology at extreme altitudes on Mount Everest. *Science* **223**:784–788.

West, J.B. and S. Lahiri (eds.). 1984. *High Altitude and Man.* American Physiological Society, Bethesda.

Wood, S.C. 1980. Adaptation of red blood cell function to hypoxia and temperature in ectothermic vertebrates. *Am. Zool.* **20**:163–172.

See also references in Appendix A.

chapter 14

Circulation

The mass flow of body fluids through the animal that we term *circulation* effects the transport of many commodities, including oxygen, carbon dioxide, nutrients, metabolic wastes, hormones, and heat. The circulatory system constitutes a vital link among the diverse specialized tissues of the organism. For example, tissues that do not immediately receive ingested food often gain nutrient molecules from digestive organs via the circulation, and tissues that lack immediate access to environmental oxygen often receive their oxygen from respiratory exchange membranes via the circulation.

Circulatory systems and the other systems that are dependent on them coevolve intimately. At any point in evolutionary time (such as the present), the capabilities of the extant circulatory system set limits on how rapidly commodities can be transported from one region of the body to another. These limits in turn can place constraints on the evolutionary development of the systems that are dependent on circulatory transport. Mutations that elevate tissue oxygen demands, for example, cannot be favored immediately in evolution if the demands are already as high as can be met by the extant circulatory system. Contrariwise, if mutations arise that heighten the transport capabilities of the circulatory system, their evolutionary fate depends on whether they are capitalized on by systems dependent on circulatory transport. Natural selection evidently cannot favor mutations unless they are put to use, and thus the features of the circulatory system come to be molded in part by the transport demands of the dependent systems.

As noted in Chapter 12, the exchanges of the respiratory gases between the animal and environment are typically the most immediately pressing of all exchanges. In turn, when the circulation is involved in the transport of the respiratory gases between sites of external respiration and the general tissues, the demands for gas transport are often the most immediately pressing of demands on the circulation. There is good evidence that requirements for gas transport have been of primary importance in the evolution of circulatory systems, and there seems to be little doubt that many animals could function with a far more sluggish internal circulation than they actually have were the need for gas transport to be eliminated. Because a discussion of the circulation must properly give emphasis to respiratory gas transport, the present chapter is in many respects a continuation of the treatment of oxygen and carbon dioxide exchange in Chapters 12 and 13.

BASIC ATTRIBUTES OF HEARTS

In some animals that have a blood–vascular system— such as many annelid worms—the blood is propelled through the system entirely by *peristaltic contractions of blood vessels*. In others, primary responsibility for propelling the blood is assumed by a discrete, localized pumping structure, or *heart*. Sometimes, as in arthropods, the heart is *single chambered*, consisting of a single muscular tube or sac. In other cases, including both vertebrates and mollusks, the heart is composed of two or more compartments through which blood passes in sequence and thus is classed as *multichambered*. Many animals, in addition to their principal heart, possess other hearts that assist with pumping blood through particular parts of the body; these are called *accessory* or *auxiliary hearts*.

The Initiation, Coordination, and Control of the Heartbeat

The contractile tissue of all hearts consists of muscle cells. Most hearts, in addition, are innervated. Given the presence of both types of excitable tissue—nerve and muscle—we must wonder what their respective roles are in initiating, coordinating, and controlling heart action. This has been one of the preeminent questions in cardiac physiology.

As described in Chapters 16 and 19, the cell membranes of nerve and muscle cells are strongly polarized electrically (inside negative relative to outside) but can become depolarized. Impulses in nerve cells are a type of depolarization. In muscle cells, depolarization is the immediate stimulus for contraction.

The rhythmic contraction of a heart reflects a rhythmic depolarization of its constituent muscle cells. One key question is, where does the impetus for this rhythmic depolarization originate? Do the contractile cells themselves spontaneously depolarize in a rhythmic manner? Or are they induced to depolarize by impulses arriving from other cells? If the latter is the case, which are the cells that spontaneously depolarize and thus originate the impulses? Whenever the initial, spontaneous rhythm of depolarization can be localized to particular cells, those cells are known as the heart's *pacemaker* (see p. 407 for the cellular physiology of pacemaker depolarization).

In some animals, the impulse to contract originates in nerve cells; the hearts of such animals are described as *neurogenic*. In other animals, the impulse to contract originates in muscle or modified muscle cells, and the heart is termed *myogenic*. We shall explore some of the attributes of neurogenic and myogenic hearts by looking at a classical example of each, but first we should note that the concept of classifying hearts into these categories should not be overdrawn. The categories in practice are not monolithic, and some hearts cannot be placed neatly in one or the other.

Neurogenicity The hearts of lobsters nicely exemplify neurogenicity. Their constituent muscle cells, according to present knowledge, are induced to depolarize and contract in much the same way as skeletal-muscle cells: Each cell is supplied with nerve fibers and typically contracts when and only when stimulated to do so by nerve impulses. As shown in Figure 14.1, a *cardiac ganglion* consisting of nine nerve cells is attached to the dorsal surface of the heart. The axonal processes of the five most-anterior cells (numbered 1–5) innervate the heart muscle. Those of the four posterior cells (numbered 6–9), on the other hand, are confined to the ganglion and make synaptic contact with the five anterior cells. One of the posterior cells ordinarily assumes the role of pacemaker; periodically and spontaneously, it produces a train of impulses, which in part excite the other posterior cells (see also p. 565). The impulses from the posterior cells activate the five anterior cells, which in turn send trains of impulses to the

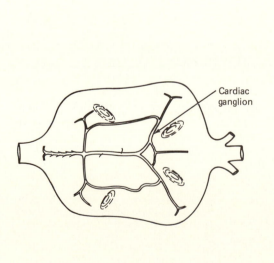

(A)

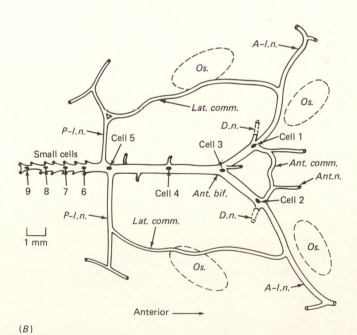

(B)

Figure 14.1 (*A*) A dorsal view of the heart of the lobster *Homarus americanus* showing the cardiac ganglion. (*B*) An enlarged view of the ganglion. The positions of the cell bodies of the nine nerve cells are identified by black dots. The posterior four cells (numbers 6–9) are small, whereas the anterior five are large. The nerve tracts are named: ant. comm. = anterior commissure; lat. comm. = lateral commissure; a-l. n. = anterolateral nerve; p-l. n. = posterolateral nerve; ant. n. = anterior nerve; d. n. = dorsal nerve; ant. bif. = anterior bifurcation. Os = ostium. [Part (*A*) reprinted by permission from D.K. Hartline, *Am. Zool.* **19**:53–65 (1979); part (*B*) reprinted by permission from D.K. Hartline, *J. Exp. Biol.* **47**:327–346 (1967).]

muscle cells, causing the latter to contract approximately in unison. If the ganglion and heart muscle are dissected apart, the ganglion continues to produce bursts of impulses periodically, but the muscle ordinarily ceases to contract. Other animals known or believed to have neurogenic hearts include most other crustaceans, horseshoe crabs (*Limulus*), and at least some spiders and scorpions.

Myogenicity The hearts of vertebrates are myogenic. They continue to beat even if stripped of their nervous connections. Other groups with myogenic hearts include the mollusks, tunicates, and many insects.

An important feature of vertebrate heart muscle is that adjacent muscle cells are electrically coupled (pp. 420 and 526). This coupling evidently occurs at gap junctions (p. 421), which in human and other mammalian heart muscles occur primarily at specialized regions of intercellular contact known as *intercalated discs*. Because adjacent cells are electrically coupled, depolarization of any one muscle cell *quickly* and *directly* provokes depolarization of other muscle cells neighboring it. In turn, the latter cells quickly induce their neighbors to depolarize and so on. Thus, within large regions of the heart muscle, once depolarization is initiated at any point, it rapidly spreads—muscle cell to muscle cell—to all cells in the region, leading all to contract together, as a unit. The depolarization is especially prolonged by comparison to that in most excitable tissues, an important attribute discussed in detail in Chapter 16 (p. 408).

Most or all of the muscle cells in a vertebrate heart possess an inherent capacity to depolarize and contract rhythmically. Thus, pieces of muscle cut from any part of the heart will beat. In the intact heart, of course, individual bits of heart muscle do not depolarize and contract on their own—at their own rhythm. Instead, they are all brought under the control of a particular group of specialized muscle cells, the pacemaker. In fish, amphibians, and reptiles, the pacemaker is located in the wall of the sinus venosus (the first heart chamber). In birds and mammals, which typically lack a sinus venosus, the pacemaker is found in the wall of the right atrium (Figure 14.2), where it is known as the *sinoatrial (S-A) node* (or *sinus node*). The cells of the pacemaker are modified in comparison to most heart muscle cells; for example, they have a relatively poorly developed contractile apparatus. Fundamentally, they are muscle cells, however, and the heart is thus myogenic. Critically, the pacemaker cells exhibit the highest frequency of spontaneous depolarization of all cells in the heart and therefore are normally the first to depolarize at each heartbeat. By thus *initiating* a wave of depolarization that spreads throughout the heart, they *impose* their rhythm of depolarization on the heart as a whole.

The process by which depolarization spreads through the vertebrate heart is known as *conduction*. Here, we look at conduction just in mammals, the group in which it is by far the best understood. To begin, some additional anatomical details need to be brought to light. The musculature of the two atria of the heart for the most part is separated from that of the two ventricles by fibrous connective tissue through which depolarization cannot pass. The one "electrical portal" through this fibrous layer is provided by a *conducting system* composed of specialized muscle cells. As shown in Figure 14.2, the conducting system starts with a group of cells in the right atrial wall known as the *atrioventricular (A-V) node*. Emanating from the node is a bundle of cells called the *atrioventricular bundle* (common bundle, bundle of His), which penetrates the fibrous layer and enters the interventricular septum (the tissue lying between the two ventricular chambers). Once in the septum, the bundle divides into right and left portions (bundle branches), which travel along the right and left surfaces of the septum and ultimately branch into the ventricular tissue on each side. Within the bundles and branches occur large cells of distinct histological appearance called *Purkinje cells*.

Functionally, the conducting system has two key properties: (1) Depolarization enters and traverses the atrioventricular node only relatively slowly, and (2) depolarization spreads down the bundles and their branches much more rapidly than it could travel

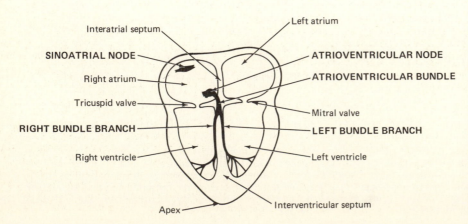

Interatrial septum — Left atrium
SINOATRIAL NODE — **ATRIOVENTRICULAR NODE**
Right atrium — **ATRIOVENTRICULAR BUNDLE**
Tricuspid valve — Mitral valve
RIGHT BUNDLE BRANCH — **LEFT BUNDLE BRANCH**
Right ventricle — Left ventricle
Apex — Interventricular septum

Figure 14.2 A semidiagrammatic representation of a mammalian heart with the sinoatrial node and specialized conducting system emphasized in black. The heart is viewed in section from its ventral side. The branches of the right and left bundle branches are in fact more elaborate than shown; traveling along the inner surfaces of the ventricles and across the ventricular cavities, they run to much of the *inner* wall of each ventricle. [After A.M. Scher and M.S. Spach, in R.M. Berne (ed.), *Handbook of Physiology*, Sec. 2, Vol. I, pp. 357–392. American Physiological Society, Bethesda, 1979.]

through ordinary ventricular muscle. The implications of these properties become apparent when we consider the sequence of events during a heartbeat, as diagrammed in Figure 14.3. Parts (A) and (B) show that once the sinoatrial node initiates the beat by depolarizing spontaneously, depolarization spreads rapidly throughout the muscle of both atria, leading to atrial contraction. Spread does not occur as rapidly into the ventricular muscle because it is dependent on activation of the conducting system, and the spread of depolarization into and through the initial member of the system—the atrioventricular node—is slow. This slowness of depolarization of the A-V node is responsible for the sequencing of contraction: atrial contraction distinctly first, ventricular distinctly second. Once the A-V bundle is activated, depolarization sweeps rapidly down the conducting system into the ventricles (C), precipitating wholesale ventricular depolarization and contraction (D). The rapid delivery of the depolarizing wave to far-flung parts of the ventricular tissue by the conducting system assures that all parts of the ventricles will contract approximately in unison.

Modulation of Heart Action Commonly, heart action is subject to nervous, hormonal, and "intrinsic" controls. We shall discuss each of these briefly, with emphasis on the lobster and mammalian hearts, as heretofore.

Hearts, even if myogenic, are usually innervated by *regulatory* nerves coming from the central nervous system. Typically, some of the regulatory nerve fibers evoke increased heart action, whereas others are inhibitory. In lobsters, for example, the cardiac ganglion receives two pairs of excitatory fibers and one of inhibitory fibers; these modulate both the frequency and intensity of the bursts of impulses generated by the ganglion and thus affect the rate and amplitude of the heartbeat. The mammalian heart is innervated by both the sympathetic and parasympathetic divisions of the autonomic nervous system. The sinoatrial node—the pacemaker—receives par-

ticularly profuse innervation; sympathetic impulses delivered to the node increase the frequency of spontaneous depolarization by the pacemaker cells and thus accelerate the heart, whereas parasympathetic impulses exert opposite effects. The heart muscle proper is also innervated. Sympathetic impulses arriving in the muscle markedly enhance the vigor of contraction, whereas parasympathetic impulses suppress contractile vigor. Exercise is one circumstance in which sympathetic stimulation of the heart is increased.

Hormonal controls are known among both invertebrates and vertebrates. In lobsters, for example, heart action is affected by neurohormones secreted by the pericardial organ, and in mammals it is enhanced by the adrenal medullary hormones, adrenaline and noradrenaline.

Intrinsic controls are adjustments in cardiac function that occur in response to changes in cardiac state without mediation of extrinsic nerves or hormones. One of the important elements of intrinsic control in the mammalian heart is known as the ***Frank–Starling relation:*** Stretching the cardiac muscle tends to increase the vigor of its contraction. This relation plays an important part in enabling the heart to match its output of blood to its input. Consider, for example, what happens if the rate of blood flow into the heart is increased. Because the heart chambers then tend to take on more blood in the time between beats, they become more stretched (distended), and thus, because of the Frank–Starling relation, they automatically contract more vigorously. This increased vigor of contraction increases the amount of blood ejected per beat, tending to match the increase in the amount of blood received by the heart. Lobster hearts function similarly in that they intrinsically increase both the vigor and rate of their contraction as they are stretched. The mechanism of this response in the lobsters is at least partly quite different from that in vertebrates, however, because the cardiac ganglion is involved; stretch induces the ganglion to fire more frequently and intensely.

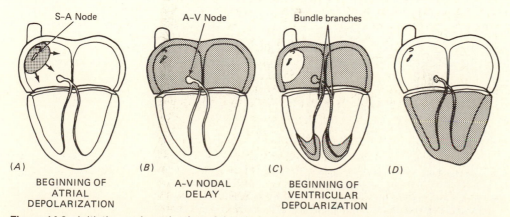

(A) BEGINNING OF ATRIAL DEPOLARIZATION

(B) A-V NODAL DELAY

(C) BEGINNING OF VENTRICULAR DEPOLARIZATION

(D)

S-A Node   A-V Node   Bundle branches

Figure 14.3 Initiation and conduction of depolarization in the mammalian heart. Darkened areas are depolarized. [Reprinted by permission from R.F. Rushmer, *Cardiovascular Dynamics,* 4th ed. Saunders, Philadelphia, 1976.]

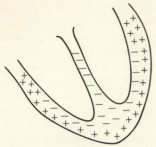

Figure 14.4 Relative electrical charges in the extracellular fluids of the ventricles at an instant during passage of a wave of depolarization. The part of the ventricular muscle lying nearest the ventricular chambers—known as the endocardium—depolarizes first because it is the part supplied immediately by the branches of the specialized conducting system (see Figures 14.2 and 14.3). The more-superficial part of the ventricular muscle—the epicardium—depolarizes later. At the instant captured here, much of the endocardium has depolarized, but the epicardium has not. Thus, the endocardium is negative with respect to the epicardium.

The Heart as a Source of Electrical Signals: The Electrocardiogram

When a mass of heart muscle is in the process of being depolarized, such that some regions of cells are depolarized already and yet others still await depolarization, a difference of electrical potential exists between the extracellular fluids in the depolar-

ized regions and those in the polarized regions (Figure 14.4). Such a potential difference in the heart sets up currents within the tissues and fluids surrounding the heart and in this way induces potential differences elsewhere in the body, even between various regions of the outer body surface. *Electrocardiograms (ECGs)* are measurements over time of potential differences of this sort. They are recorded using extracellular electrodes, which commonly are placed on the body surface but may in principle be placed anywhere in the body, even in contact with the heart itself.

Figure 14.5A shows electrocardiograms of two species, humans and octopuses. The waveforms in the human ECG are named with letters (Figure 14.5B). The *P wave* is produced by depolarization of the atrial muscle (= atrial contraction). The Q, R, and S waves, together known as the *QRS complex,* arise from ventricular depolarization. Repolarization of the ventricles generates the *T wave* (the waveform produced by repolarization of the atria, however, is typically not seen because it is obscured by the QRS complex).

The Heart as a Pump; Cardiac Output

During the heart cycle, the period of contraction is called *systole* (pronounced with a long *e:* systolē), whereas the period of relaxation is termed *diastole*

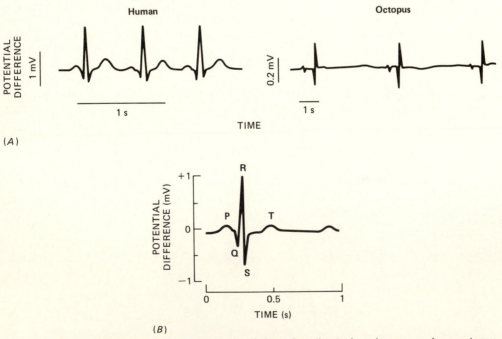

Figure 14.5 Electrocardiograms. (A) Records of three heartbeats in a human and an octopus (*Eledone cirrhosa*). The human record was obtained using electrodes placed on the skin surface on the right arm and left leg. Electrodes attached to the surface of the heart ventricle were used to obtain the octopus ECG. In both species, the sharp, high-amplitude waveform in each beat corresponds to ventricular depolarization. (B) A human ECG during one heartbeat with waveforms identified. [Octopus ECG from P.J.S. Smith, *Comp. Biochem. Physiol.* **70A**:103–105 (1981); human ECGs after A.C. Guyton, *Textbook of Medical Physiology*, 5th ed. Saunders, Philadelphia, 1976.]

(diastolē). The workings of the heart as a pump are revealed by analyzing pressure, flow, and volume during these periods.

As an example of the sort of analysis that is possible, let us consider the data on the mammalian left atrium and left ventricle shown in Figure 14.6. At the time marked by the arrow at the bottom, ventricular systole begins (note the QRS complex). The elevation of the pressure in the ventricle to above the pressure in the atrium causes the mitral valve between the two chambers to flip shut, yet for a time the ventricular pressure remains below the pressure in the systemic aorta, meaning that the aortic valve (leading into the aorta) is not forced open. With both the inflow and outflow valves shut, blood cannot leave the ventricle. Thus, although ventricular pressure rises sharply, the volume of blood in the ventricle remains constant. This is called the phase of *isovolumic contraction*. Once the ventricular pressure

rises high enough to exceed the aortic pressure, the aortic valve flips open, and the blood accelerates extremely rapidly, gushing out of the ventricle into the aorta (thus increasing aortic pressure). This marks the start of the phase of *ventricular ejection*. Even after the aortic pressure comes to slightly exceed the ventricular pressure, ejection continues for a while—at a rapidly falling rate—because of blood momentum. Ultimately, the ventricle starts to relax, and then the ventricular pressure falls rapidly away from the aortic pressure and the aortic valve shuts. A period of *isovolumic relaxation* follows, as ventricular pressure falls with both inflow and outflow valves shut. When the ventricular pressure drops below atrial pressure, the mitral valve is forced open and *ventricular filling* begins. Note that most filling occurs *before the atrial contraction,* the motive force being the pressure built up by accumulation of pulmonary venous blood in the atrium. Later, atrial sys-

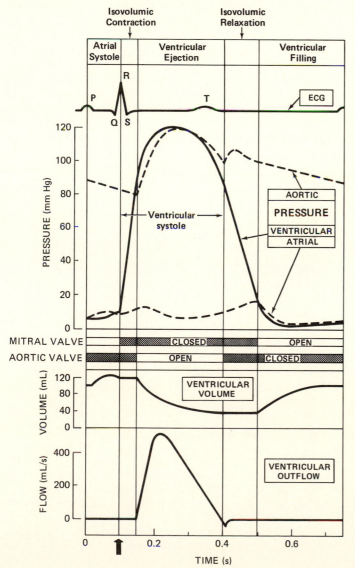

Figure 14.6 Dynamics of the human left heart: approximate values for pressure, volume, and flow in the left atrium, left ventricle, and systemic aorta during rest. The left atrium receives blood from the pulmonary veins. The mitral valve separates the atrium and ventricle, whereas the aortic valve separates the ventricle and aorta. Both are passive flap valves. The flaps of the mitral valve open into the ventricle; thus, the valve is forced open if atrial pressure exceeds ventricular, but the flaps are forced into the orifice—closing the valve—if ventricular pressure is greater than atrial. The flaps of the aortic valve open into the aorta, meaning that the valve is opened by ventricular pressures exceeding aortic. [After C.F. Rothe and E.E. Selkurt, in E.E. Selkurt (ed.), *Physiology,* 4th ed., pp. 289–309. Little, Brown, Boston, 1976.]

tole occurs (note the P wave), thus forcing some additional blood into the ventricle just before the next ventricular systole.

The volume of blood pumped by a heart per unit time into the artery or arteries emanating from the heart is known as the *cardiac output*. (In the case of the avian or mammalian heart, the term refers specifically to the output of the left ventricle into the systemic aorta unless stated otherwise.) The cardiac output is the product of the heart rate and the volume of blood pumped per heart cycle:

cardiac output (mL/min) = beats/min × mL/beat

The volume pumped per beat is called the *stroke volume*.

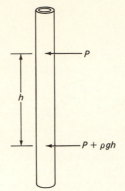

Figure 14.7 In a vertically positioned tube filled with non-moving fluid, if the pressure at one height is *P*, then the pressure at a height *h* units lower is *P* + ρ*gh*, where ρ is the density of the fluid and *g* is the acceleration attributable to gravity (about 980 cm/s²). The density of human blood is about 1.06 g/mL.

PRINCIPLES OF PRESSURE AND FLOW IN THE VASCULAR SYSTEM

Blood Pressure

The excess of blood pressure over ambient pressure is termed simply the *blood pressure* and is expressed in mm Hg or other pressure units (p. 136).

In arteries, the blood pressure waxes and wanes over the heart cycle. The highest pressure attained at the time of cardiac contraction is termed the *systolic pressure,* whereas the lowest pressure reached during cardiac relaxation is called the *diastolic pressure*. In resting young people, for example, the average aortic systolic pressure is about 120 mm Hg, and the average aortic diastolic pressure is about 75 mm Hg. These pressures are often expressed as a pseudo-ratio: for example, 120/75. The mean aortic pressure is obtained by averaging the pressure over the entire cardiac cycle and does not (except incidentally) correspond to the average of the two extreme pressures, systolic and diastolic. In resting young people, the mean aortic pressure is 90–100 mm Hg.

The principles of fluid statics tell us that a vertical column of fluid exerts pressure in proportion to its height (Figure 14.7). This phenomenon is important in the study of blood pressures because fluid columns are formed by the blood in the vessels of an animal.

If a person lies horizontally on a flat surface, the hydrostatic effects of fluid columns are virtually eliminated from consideration because different parts of the body do not differ greatly in height; all parts are at about the level of the heart. In this situation, the mean blood pressure is about the same in all major arteries. The mean pressure in the radial artery in the wrist, for example, is only about 3–4 mm Hg lower than the mean pressure in the systemic aorta. Thus, we see that *the head of pressure developed by the heart is diminished only slightly by the process of forcing blood to flow through the major arteries*.

When a person stands erect, the pattern of blood pressures in the major vessels becomes more complex because pressures are strongly affected by fluid-column effects. This is illustrated in Figure 14.8. In a vertical column of human blood, each 13 cm of height exerts about 10 mm Hg of pressure. At levels of the body below the heart, this static pressure is added to the head of pressure contributed by the heart; thus, in an erect person, average arterial pressures in the legs are several tens of mm Hg higher than the average pressure in the systemic aorta. At levels of the body above the heart, some of the head of pressure developed by the heart is lost in simply supporting the fluid column of blood; thus, the arterial blood pressure decreases by 10 mm Hg for every 13 cm of height. We defer a rigorous analysis of *flow* in the erect posture to a later section (p. 342). We may state here, however, that despite fluid-column effects on pressures in the erect posture, the force primarily responsible for driving blood through the arteries remains the head of pressure developed by contraction of the heart. Furthermore, just as in the horizontal posture, this head of pressure is diminished only slightly by the process of forcing blood to flow through the major arteries.

For many purposes, it is advantageous to simplify the analysis of pressures by considering animals in the horizontal posture. The usual convention in fact is to assume the prone posture unless otherwise stated.

Determinants of Flow Rate

When a simple liquid such as water flows steadily through a horizontal, rigid tube without turbulence, its rate of flow (mL/min) from one end of the tube to the other depends (Figure 14.9) on the pressure at the entry to the tube (P_{in}), the pressure at the exit (P_{out}), the internal radius of the tube (r), the length of the tube (l), and the viscosity of the fluid (η). The

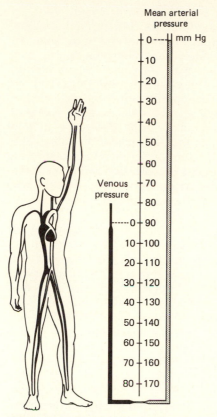

Figure 14.8 The effect of fluid columns on blood pressure in humans standing quietly erect. Below the heart, static, fluid-column effects increase both the venous and arterial pressures equally; thus, the arteriovenous difference in pressure is about the same at all levels of the body. Above the heart, although arterial pressure decreases with height, venous pressure outside the cranium is about zero (equal to ambient) regardless of height (negative pressures cannot develop in the extracranial veins because their walls are collapsible). [Reprinted by permission from R.F. Rushmer, *Cardiovascular Dynamics,* 4th ed. Saunders, Philadelphia, 1976.]

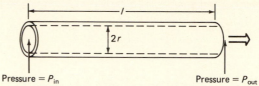

Figure 14.9 Parameters of importance in the Poiseuille equation.

inversely proportional to electrical resistance. An analogous formulation applies to steady fluid flow through a horizontal tube. The driving force for such flow is the difference in pressure from one end of the tube to the other ($P_{in} - P_{out}$), and it can be shown by calculation from Equation (1) that the **resistance to flow** is $8\eta l/\pi r^4$. The latter expression reveals that resistance is *inversely* proportional to the fourth power of the tube radius. Halving the radius increases the resistance to flow 16-fold.

The Poiseuille equation was derived to apply to flow of simple fluids like water through unbranched, rigid-walled tubes. Blood is not a simple fluid (e.g., it contains a high concentration of suspended "particles"), and blood vessels are not unbranched or rigid walled. Nonetheless, the equation has often proved to be a most-useful approximate model of the determinants of blood flow through animal vascular systems.

The Dissipation of Energy

The **linear velocity** of a bit of liquid in a stream flowing through a tube is defined to be the length of the tube traversed per unit time. When the flow of a liquid through a tube is steady and nonturbulent, the liquid ideally moves in a series of infinitesimally thin, concentric layers—or laminae—that differ in their linear velocities. This type of flow, called **laminar flow,** is illustrated in Figure 14.10. The outermost of the concentric layers of liquid—the layer immediately next to the wall of the tube—does not move at all. Layers closer and closer to the center move faster and faster.

Now, adjacent layers of fluid that are moving at different linear velocities do not slip effortlessly past each other. Instead, there is a resistance to their relative motion. **Viscosity** is a measure of this resistance, this *internal friction* within moving fluids. As a liquid flows through a tube, some of its energy of motion is steadily degraded to heat in overcoming viscosity. Sometimes, we hear it said that pumping the blood through vessels mainly involves overcoming friction between the blood and the vessel walls. This is entirely incorrect. In fact, there is no frictional force to be overcome between a laminarly flowing fluid and the walls of a tube because no relative motion occurs between the walls and the fluid layer next to them (that fluid layer is still). The frictional resistance to the flow of a fluid through a tube is entirely *internal* to the moving fluid.

formula relating these quantities was derived in the nineteenth century by Jean Poiseuille and thus is named the **Poiseuille equation:**

$$\text{flow rate} = (P_{in} - P_{out}) \left(\frac{\pi}{8}\right) \left(\frac{1}{\eta}\right) \left(\frac{r^4}{l}\right) \quad (1)$$

Note that increasing the difference in pressure between the ends of the tube will increase the flow rate. Raising the viscosity of the fluid will diminish flow. The final term in the equation is a geometrical term of great importance. Note that lengthening the tube without altering the pressure drop from end to end will decrease the rate of flow through the tube. Even more, note that the flow rate is a direct function of the *fourth power* of the radius. Because of this relation, the flow rate is very sensitive to changes in radius. Halving the radius (or diameter), for example, will reduce flow by a factor of 16!

In the study of electricity, Ohm's law states that the rate of current flow is directly proportional to the driving force for flow (the potential difference) and

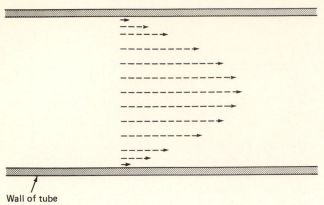

Wall of tube

Figure 14.10 Linear velocity as a function of distance from the wall as a simple fluid such as water flows laminarly through a tube. Lengths of dashed arrows are proportional to linear velocity. Velocity immediately next to wall is zero. The velocity profile for blood differs from that for a simple fluid.

At this point, the broad outlines of the energetics of flow through a tube become apparent. Pressure represents potential energy. In setting the blood in motion or maintaining it in motion, this potential energy is converted to kinetic energy. Then, along each millimeter of tube through which the fluid passes, some of the kinetic energy is lost as heat in overcoming viscous resistance to flow. Ultimately, therefore, pressure is converted to heat. In many situations, the drop in pressure from one point in a tube to another point downstream is in fact a good index of the heat evolved in overcoming opposing viscous forces.

The Whole Truth

It was Alan Burton (see Suggested Readings) who referred to the subject of this section as "the whole truth." As he pointed out, the proper place to introduce "the whole truth" is difficult to judge, for the material is essential and yet can be confusing at first.

Blood can possess three forms of high-grade energy (ignoring the bond energy of its chemical constituents): (1) its potential energy of pressure, (2) its kinetic energy (energy of motion), and (3) its potential energy of position—energy it possesses by virtue of its position in the earth's gravitational field. The *sum* of these three forms of energy has been termed the ***total fluid energy*** of the blood. Now, all three forms can contribute to blood flow, and it turns out that *the true driving force for blood flow from one place to another is not simply the difference in pressure between the two places but the difference in total fluid energy* (Figure 14.11).

Earlier, in our consideration of Figure 14.9, we said that the driving force for flow through a tube is the pressure difference from one end to the other (p. 341). We had stipulated there, however, that the tube under consideration was horizontal (so differences of potential energy of position could not arise) and the flow under consideration was steady (so the average kinetic energy was the same all along the tube). In that particular situation, the difference in total fluid energy between the two ends of the tube had to be precisely equal to the difference in pressure between the ends. A bit of reflection will reveal that situations like this are commonplace. That is why our "common sense" says that fluid flow is driven simply by pressure differences.

The importance of a fuller understanding of the driving force for flow is illustrated by two situations we have already encountered. (1) We have seen (Figure 14.6) that toward the end of left ventricular systole in humans, blood transiently continues to flow out of the left ventricle of the heart into the aorta even after the ventricular pressure has fallen below the aortic pressure. This flow occurs because, at that time in the heart cycle, the blood on the ventricular side of the aortic valve (being propelled by the ventricle) has a greater kinetic energy than that on the aortic side. Although the difference of *pressure* does not favor outflow of blood from the heart, the difference of *total fluid energy*—the true driving force—does. (2) We have seen that blood in the arteries of the legs of a standing person is at much higher pressures than that in the systemic aorta (Figure 14.8) because of the "fluid-column" effect (which is simply the obverse of gravitational potential energy). If pressure were the true driving force for flow, blood would flow from the legs toward the heart. However, blood in the aorta has a greater potential energy of position than that in the legs (it tends to "fall" into the legs), and when this is added in, the total fluid energy in the aorta proves to be higher than that in the leg arteries.

The instances just considered demonstrate that a full analysis of the total fluid energy is crucial. Nonetheless, just as we achieved simplifications in our analysis of the physical system of flow through a tube in Figure 14.9, simplifications are often possible in analyzing biological systems. Such simplifications can be achieved because differences of potential energy of position and differences of kinetic energy in numerous biological situations are small enough to be approximated as being equal to zero for many purposes. We now examine these simplifications to the equation for total fluid energy in Figure 14.11, recognizing that they are undertaken for conve-

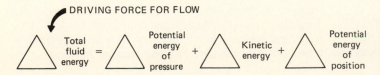

DRIVING FORCE FOR FLOW

Figure 14.11 The true driving force for flow, the difference in total fluid energy from one place to another, is equal to the sum of the differences in potential energy of pressure, kinetic energy, and potential energy of position.

nience. For full quantitative accuracy, it remains true that all factors affecting the total fluid energy must always be taken fully into account.

In analyzing animal blood flow, differences of potential energy of position can be approximated as being equal to zero if we assume the animal to be in a prone posture (for then all blood is roughly in one horizontal plane). As already noted, this approach is the usual one taken. It can be shown not to detract from the generality of conclusions drawn.

In some cases, differences of kinetic energy in the circulatory system are also small enough to be approximated as zero. This is the case in analyzing flow in the systemic vasculature of resting humans, for example. In most systemic human vessels, kinetic energy accounts for just a very small fraction of the total energy (e.g., only 1–3 percent in the aorta), and thus little accuracy is lost by disregarding differences of kinetic energy from place to place.

When differences of potential energy of position and those of kinetic energy can be approximated as being equal to zero, one of the useful simplifications achieved is that the difference of pressure from place to place can be treated as if it alone constitutes the driving force for blood flow (Figure 14.11). In regard to the analysis of energetics, to carry out a fully accurate assessment of the amount of energy degraded from high-grade form to heat as the blood flows from place to place, it is necessary to measure the drop in total fluid energy associated with the flow. However, *in cases where differences of kinetic energy and potential energy of position can be approximated as zero, the drop in blood pressure from place to place may be accepted as a measure of the energy degradation associated with flow.*

CIRCULATION IN MAMMALS AND BIRDS

Mammals and birds have evolved exceptionally high metabolic demands and in tandem have evolved circulatory systems clearly adapted to meeting such demands. Here, we emphasize the mammalian circulatory system, recognizing that many of its features are shared by birds.

The basic circulatory plan of mammals and birds is illustrated in Figure 14.12. Oxygen-depleted blood draining from the systemic tissues returns to the right heart via the great veins and is pumped by the right ventricle to the lungs, where oxygen is taken up and carbon dioxide released. The oxygenated blood from the lungs returns to the left heart and is then pumped by the left ventricle to the systemic aorta, which divides to supply all the systemic tissues. One of the key features of this circulatory plan is that it places the respiratory organs in series with the systemic tissues. The series arrangement, emphasized in Figure 14.12*B*, maximizes the efficiency of oxygen delivery to the tissues. All the blood pumped to the tissues by the heart is freshly oxygenated, and the tissues receive blood at the full oxygen tension established in the lungs. This feature is an important element in the ability of the circulatory system to meet high oxygen demands.

The Closed Nature of the Circulatory System

Circulatory systems are commonly classed as ***closed*** or ***open,*** depending on whether the entire circulatory path is enclosed in discrete vessels. In an open system, blood leaves discrete vessels to bathe at least some tissues directly, whereas in the ideal closed

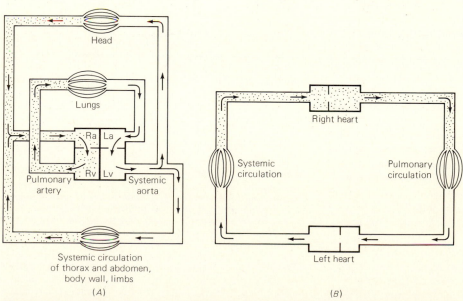

Figure 14.12 *(A)* Schematic representation of the circulatory plan in mammals and birds. Stippled parts carry relatively deoxygenated blood. Ra: right atrium of the heart; La: left atrium; Rv: right ventricle; Lv: left ventricle. *(B)* Another representation of the circulatory plan, serving to emphasize the arrangement of the pulmonary and systemic circuits in series with each other.

system, there is always at least a thin vessel wall separating the blood from the tissues. The distinction between closed and open systems is relative, for there are many intergradations. Birds and mammals are among the animals that have essentially closed circulatory systems.

The Systemic Vascular System: Its Anatomy and Functional Attributes

The vessels at various points in the circulatory path differ anatomically and functionally in important ways. We discuss the major vessel types in the order in which blood passes through them. In the course of this discussion, we shall start to use the term *perfusion*. *Perfusion* is the forced flow of blood (or other fluids) through vessels or tissues.

Arteries   The great arteries have thick walls heavily invested with smooth muscle and elastic tissue. They thus are equipped to convey blood under considerable pressure from the heart to the peripheral circulation. The elasticity of the great arteries allows for two important functions, the *damping of pressure oscillations* and the provision of a *pressure reservoir.* If the heart were to discharge into rigid, inelastic tubes, the head of pressure in the great arteries would oscillate violently upward and downward with each contraction and pause of the heart. By virtue of their elasticity, the great arteries in fact stretch when they receive blood discharged from the heart. Some of the energy developed by the heart is thus stored as elastic potential energy in the artery walls, and the increase in arterial pressure is limited to some extent. The energy stored elastically at the time of systole is released as the arteries contract again during diastole. In this way, some of the energy generated during the cardiac contraction is used to maintain the head of pressure in the great arteries between contractions. The end result is that variations in arterial pressure over the cardiac cycle are reduced (damping effect), and a substantial head of pressure is maintained in the arteries even when the heart is at rest (pressure reservoir effect).

The arteries become smaller as they branch outward toward the periphery. They also become more thinly walled, a fact that at first appears paradoxical when we recall that the mean blood pressure diminishes hardly at all over a considerable distance into the periphery. The paradox is reconciled by a principle identified by the Marquis de Laplace around 1820 and now known as *Laplace's law.* This law deals with the relation between wall tension and pressure in hollow structures. As applied to tubes, it says that when the pressure inside a tube exceeds that outside by any given amount, the tension developed in the walls of the tube is directly proportional to the radius of the tube: $T = r\Delta P$, where T is the tension, r is the radius, and ΔP is the pressure difference across the walls. Because of this relation, even

though a small artery may be exposed to the same blood pressure as a large one, the tension developed in its walls will be substantially lower than that developed in the walls of the large artery, and its walls accordingly need not be so well fortified as those of the large vessel to resist overexpansion. The same principle explains how capillaries can be exceedingly thin walled yet resist an appreciable pressure.

Microcirculatory Beds   Ultimately, the systemic arteries deliver blood to the microcirculatory beds of the organs and tissues (Figure 14.13). These beds consist of *arterioles, capillaries,* and *venules,* all microscopically small.

The walls of arterioles are so well endowed with muscular and fibrous elements that they are rather thick relative to the dimensions of the vessels themselves. Among the arterioles of humans, to illustrate, the mean diameter of the lumen is about 30 μm, and the mean thickness of the walls is about 20 μm. The smooth muscles in the walls of the arterioles play a

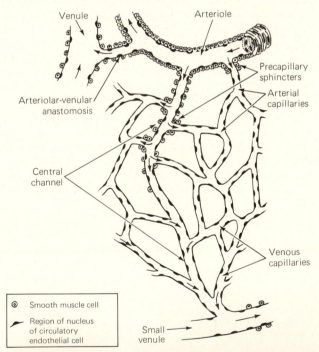

Figure 14.13   Diagram of a microcirculatory bed of a mammal. Capillaries form an anastomosing network between the arteriole and lower venule. The endothelial cells of the capillaries are thin and flat; they are thickened in the region of the cell nucleus. The arteriole is well invested with smooth muscle cells. The arteriolar end of the central thoroughfare channel contains dispersed smooth muscle cells and is known as a metarteriole. The venular end of the central channel lacks muscle cells and is structurally a capillary. Precapillary sphincters are located where capillaries arise from arterioles or metarterioles. "Arterial" and "venous" capillaries are distinguished by their position relative to arterioles and venules; they are not structurally different. [From W.M. Copenhaver, R.P. Bunge, and M.B. Bunge, *Bailey's Textbook of Histology,* 16th ed. Williams & Wilkins, Baltimore, 1971. © 1971 The Williams & Wilkins Company, Baltimore.]

pivotal role in the *vasomotor control of blood distribution*. The adjective **vasomotor** refers to changes in the luminal radius of blood vessels. In our earlier discussion of the Poiseuille equation (p. 341), we noted that the rate of flow through a tubular vessel is extremely dependent on the radius of the tubular lumen. By contracting and relaxing, the muscles of an arteriole control its luminal radius and thereby strongly affect the rate of blood flow into the capillary beds that the arteriole supplies. Control of the arteriolar muscles is mediated by the sympathetic division of the autonomic nervous system, by circulating hormones, and by incompletely understood chemical and physical effects at the local tissue level (p. 528). One familiar example of arteriolar control of tissue perfusion is provided by responses to cold. As noted in Chapter 6 (p. 105), mammals often curtail blood flow to large areas of their skin surface when exposed to low air temperatures. This curtailment is effected by vasoconstriction in arterioles and arteriolar-venular anastomoses (Figure 14.13) that control access of blood to the most superficial vascular beds of the skin. The vasconstriction is in part evoked by activity of sympathetic nerve fibers. Also, skin cooling exerts direct (local) vasoconstrictive effects. Among humans, blood flow through skeletal muscles can increase 10-fold or more during exercise; in part, this results from arteriolar vasodilation, which appears to be mediated principally by local effects of metabolites produced by the exercising muscles. Along with the arterioles, both precapillary sphincters (Figure 14.13) and smooth muscles in the walls of small terminal arteries also are involved in controlling blood flow to capillary beds.

The heart produces a head of pressure that is transmitted to all microcirculatory beds via the arteries. This driving force is always available and assures that vasodilation or vasoconstriction in the arterioles will result in immediate changes in tissue perfusion. Each microcirculatory bed has its own arteriolar valves and thus is readily controlled independently of other vascular beds. These important features, which allow for highly sensitive temporal and spatial control of blood distribution, depend anatomically on the fact that the entire circulatory system is enclosed in vessels. They are features that must be viewed as significant benefits of the closed type of circulation.

From the arterioles, the blood enters the capillaries, fine vessels that are often barely wide enough to allow the passage of red blood cells. The walls of capillaries consist of only a single layer of endothelial cells and contain no muscular or fibrous elements. Because of the thinness of their walls (only about 1 μm) capillaries are the *preeminent sites of exchange between the blood and tissues*. Even the relatively thin walls of arterioles are usually thick enough to pose an effective barrier to exchange by diffusion and ultrafiltration.

Capillary beds consist of many fine vessels that branch and anastomose among the tissue cells. The density of capillaries is different in different tissues. Fat, for example, has a relatively low capillary density, whereas skeletal and cardiac muscle and the brain have high densities. In muscle and other tissues rich in capillaries, the potential exchange surface provided by the capillary beds is truly enormous; cross sections of gastrocnemius muscles of various mammals, for example, reveal from 300 to over 600 capillaries per square *millimeter,* and the capillary surface area in various mammalian skeletal and cardiac muscles is from 150 to over 1000 cm^2/cm^3 of tissue. Only a fraction of the capillaries in skeletal muscle are open at rest, but during exercise the entire capillary bed can be brought into play.

The capillaries drain into small vessels, the walls of which contain fibrous and muscular elements but are nonetheless quite thin (2–5 μm in humans). These are the venules, and at least the smaller ones are sufficiently thin walled to permit a meaningful amount of exchange between blood and tissues.

Veins   The blood is led from the venules back to the heart through a series of veins of increasing diameter. Because the blood pressure has declined precipitously by the time the blood leaves the capillary beds, the walls of the veins need not be capable of resisting the high tensions associated with the high pressures in the arterial circulation. The walls of the veins are thus thin by arterial standards. They contain important elastic elements but are comparatively sparsely invested with muscle. One of the significant functions assumed by the venous vasculature is that of a *capacity reservoir*. At a given time, the circulatory system contains an essentially fixed volume of blood, which must always be accommodated somewhere. We have seen that the microcirculatory beds of the tissues can vary considerably in their extent of openness and thus in the volume of blood they contain. If the capacity of the capillary beds is reduced, the veins enlarge, thus providing a compensatory increase in venous capacity. Conversely, a reduction in venous capacity accompanies increases in capillary capacity.

Cardiac Performance and Body Size

Among mammals at rest, cardiac output per unit of body weight tends to be an inverse function of body size. Thus, small mammals meet their relatively high weight-specific demands for oxygen transport in part by maintaining relatively high weight-specific rates of blood flow. In relation to body weight, small mammals do not have larger hearts than large ones. The main way that the small species attain their relatively high cardiac outputs per unit weight is by maintaining relatively high heart rates. Whereas the resting heart rate of an elephant might be about 30 beats/min, that of a person is near 75 beats/min, and that of a house mouse is 500–600 beats/min. The average aortic

Table 14.1 REPRESENTATIVE AVERAGE SYSTOLIC AND DIASTOLIC BLOOD PRESSURES IN
LARGE ARTERIES LEAVING THE HEART OF SELECTED VERTEBRATES

Species	Pressure[a] (mm Hg)	
	Systolic	Diastolic
Human, male, 18 years old	120	74
Dog (*Canis familiaris*)	112	56
Rabbit (*Oryctolagus cuniculus*)	110	80
Monkey (*Macaca mulatta*)	134	76
Quail (*Coturnix coturnix*)	157	149
Starling (*Sturnus vulgaris*)	180	130
Frog (*Rana pipiens*)	31	21
Trout (*Salmo gairdneri*)	43	33

[a] Data given are results of particular studies carried out without general anesthesia. Often, different studies of a species
yield somewhat different results.

Source: P.L. Altman and D.S. Dittmer, *Biology Data Book,* 2nd ed., Vol. 3. Federation of American Societies for
Experimental Biology, Bethesda, 1972. Trout data from J.W. Kiceniuk and D.R. Jones, *J. Exp. Biol.* **69**:247–260 (1977).

blood pressure does not vary systematically with body size in mammals. Comparing birds and mammals, birds tend to have larger hearts, lower heart rates, and higher blood pressures than mammals of equivalent body size.

Pressure, Flow, and the Dissipation of Energy in the Systemic Circuit

Mammals and birds maintain the highest pressures in the great systemic arteries of any animals (Table 14.1). The mammals and birds require high flow rates because of their high oxygen-transport demands. Also, however, the resistance to flow through their systemic circuit is high. The maintenance of high flows in the face of this high resistance demands the high pressures seen. Why is the resistance high? The Poiseuille equation tells us that vessels of small diameter tend to make particularly great contributions to resistance (p. 341). In the closed circulatory system of birds and mammals, every tissue is densely invested with minute vessels. This property has its virtues, as already discussed (p. 345). Also, however, it imparts a high resistance to the systemic circuit as a whole.

As mentioned earlier, the blood makes its way from the systemic aorta through all the major arteries with little loss in pressure. The pressure drop is small in the major arteries because these vessels are large enough in diameter to pose only a small resistance.

As the blood moves into the terminal arteries, arterioles, and capillaries, the total cross-sectional area of the vasculature increases markedly (Figure 14.14). Since in steady state the volume passing through the vasculature must be the same at all levels of the vascular tree, the linear velocity of the blood falls sharply as it enters the smaller vessels (Figure 14.14). Given this fact and given the small diameters of the vessels, we see that the volume flowing through individual arterioles or capillaries per unit time is very markedly less than that flowing through individual arteries. Nonetheless, because resistance is so dramatically elevated by small vessel diameter, forcing blood through the small vessels in fact is responsible for dissipating most of the head of pressure (high-grade energy) developed by the heart. The mean blood pressure in resting humans, for example, declines from about 90 mm Hg to about 60 mm Hg across the terminal arteries, from about 60 to 30–35 mm Hg across the arterioles, and from about 30–35 to 15 mm Hg across the capillaries and initial venules. Whereas a pressure drop of only 2–3 mm Hg is sufficient to move blood at a relatively high flow rate from the shoulder to the wrist in the major arteries of the arm, a drop of perhaps 45 mm Hg is required to move blood at a relatively low flow rate over a distance of a few millimeters from the beginnings of arterioles to the initial venules in a microcirculatory bed. There is no more compelling evidence of the energy demand imposed by the type of closed circulatory system seen in birds and mammals.

The venous circulation is a low-pressure, low-resistance system. The average pressure at which blood is supplied to the venous vasculature (postcapillary pressure) in a resting, recumbent human is only around 10–15 mm Hg. The head of pressure developed by the left ventricle is essentially entirely dissipated by the time the blood reaches the entry of the great veins into the right atrium.

Exchange of Fluid Across the Walls of Systemic Capillaries

The blood pressure in systemic capillaries exceeds the hydrostatic pressure in tissue fluids by a good margin. This pressure difference favors outward ultrafiltration across the capillary walls (p. 151). On the other hand, the blood plasma has a higher osmotic pressure than the tissue fluids, meaning that forces of osmosis favor entry of fluid into the blood

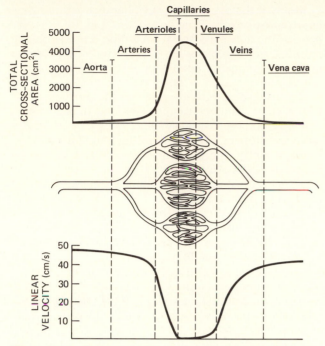

Figure 14.14 Estimated total cross-sectional area of the vascular tree in humans is shown at the top. Although the cross-sectional area of individual capillaries is minute, the capillaries are so numerous that their collective cross-sectional area greatly exceeds that of the aorta. In steady state, the volume of blood passing each dashed line per unit time must be identical. Thus, as shown at the bottom, the linear velocity of the blood varies inversely with the total cross-sectional area. Capillaries are so short (<1 mm) that the linear velocity must be low in them for the blood to remain long enough to exchange with the tissues. [Reprinted by permission from E.O. Feigl, in T.C. Ruch and H.D. Patton (eds.), *Physiology and Biophysics,* Vol. 2, 20th ed., pp. 10–22. Saunders, Philadelphia, 1974.]

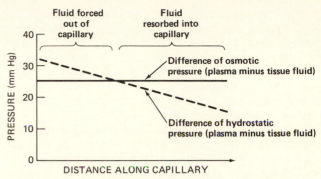

Figure 14.15 An outline of the Starling–Landis hypothesis of fluid exchange across capillary walls. The dashed line shows the extent to which blood pressure exceeds tissue-fluid hydrostatic pressure. The solid line shows the extent to which plasma osmotic pressure exceeds tissue-fluid osmotic pressure; losses and gains of water along the capillary are not great enough to affect this value substantially. Values are approximate.

from the tissue spaces. The interplay of the processes favoring efflux of fluid from the blood and those favoring influx has attracted considerable interest. Yet it still is not thoroughly understood, in part because critical parameters are difficult to measure at the microscopic level of capillaries.

Studies by E.H. Starling and E.M. Landis in the late nineteenth and early twentieth centuries gave rise to a model of fluid exchange across capillaries, diagrammed in Figure 14.15, that is still believed to summarize the essentials of exchange in many tissues. The osmotic pressure of the blood plasma exceeds that of the tissue fluids by about 25 mm Hg along the entire length of the capillaries. This difference, called the colloid osmotic pressure of the plasma, arises because the plasma is richer than tissue fluids in dissolved proteins (e.g., albumins) that do not pass freely through capillary walls (p. 148). At the arterial ends of the capillaries, the hydrostatic pressure of the blood might exceed that of the tissue fluids by somewhat over 30 mm Hg; because the hydrostatic-pressure gradient favoring exit of fluid

from the capillaries is thus greater than the osmotic-pressure gradient favoring entry, fluid is forced out of the capillaries in net fashion. Now, as the blood flows along the capillaries, its hydrostatic pressure falls precipitously, such that the blood pressure might exceed the tissue-fluid hydrostatic pressure by only about 15 mm Hg at the venous ends of capillaries. Thus, at the venous ends, the osmotic-pressure gradient favoring entry of fluid into the capillaries exceeds the hydrostatic-pressure gradient favoring exit, and a net entry of fluid into the capillaries occurs.

The *Starling–Landis hypothesis,* in brief, states that the blood tends to lose volume in the initial segments of systemic blood capillaries but regains it in the final segments. The overall effect is often a net loss: Fluid from the blood tends to be deposited in the intercellular tissue spaces. This fluid is picked up by lymphatic capillaries and ultimately returned to the blood by way of the lymphatic vasculature.

The Pulmonary Circulation

Discussion of the pulmonary circulation has been postponed to this point because an understanding of the forces affecting fluid movement across capillary walls is important to appreciating an interesting problem associated with perfusion of the lungs. The systemic circulation is a high-pressure, high-resistance system. We have seen that even in the presence of large precapillary resistances, pressures in the systemic capillaries are still sufficient to cause appreciable net losses of fluid into the systemic tissue spaces. Efficient gas exchange in the lungs requires that the pulmonary capillaries be in close apposition to the alveoli or air capillaries. Loss of fluid across the pulmonary capillaries at the rate seen in the systemic capillaries would flood these terminal air spaces with fluid, impairing gas exchange. Such flooding does not normally occur, however, because the pulmonary circuit differs dramatically from the systemic in its overall resistance and thus in the pressures required for its perfusion.

Various peculiarities of the pulmonary vasculature give it a far lower resistance than the systemic vasculature; the pulmonary circulatory path is much shorter than the systemic path, for example, and its vessels tend to have thinner, more distensible walls. Because the pulmonary circuit is connected in series with the systemic circuit (Figure 14.12B), the volume of blood passing through the pulmonary circuit per unit time must match that passing through the systemic circuit. However, because of the low resistance of the pulmonary circuit, equivalent flow can be maintained with much lower pressures than in the systemic circuit. The pulmonary circulation, in brief, is comparatively a low-pressure, low-resistance system.

In humans, the systolic pressure developed in the pulmonary arteries by the right ventricle averages about 22 mm Hg, and the diastolic averages about 9 mm Hg. The mean pulmonary aortic pressure is near 14 mm Hg. As every student of comparative anatomy knows, the right ventricle is far less muscular than the left; here, we see that it can be so because it needs to develop only much lower pressures. The mean blood pressure in pulmonary capillaries is near 6 mm Hg, as compared to a mean of perhaps 25 mm Hg in systemic capillaries. The low blood pressures prevailing in the pulmonary capillaries largely preclude net ultrafiltrational efflux of fluid from the plasma across the capillary walls.

The Circulation During Exercise

The circulatory system operates at a relatively leisurely pace when birds and mammals are at rest. Exercise heightens demands on the system and thus helps bring to light the full capabilities for transport that have been evolved.

We saw in Chapter 13 that an elevation of the cardiac output, along with an increase in the deoxygenation of the blood, is one of the major ways in which oxygen delivery to the tissues is increased during exercise (p. 306). Average young people can increase their cardiac output by at least a factor of 4, and outputs of six or seven times the resting level have been measured in trained athletes during heavy exertion. The increase in cardiac output is generally achieved by an increase in both heart rate and stroke volume.

Exercise is accompanied by a substantial drop in the total resistance presented by the systemic vasculature. Were it not for this, the aortic blood pressure would have to increase exorbitantly to drive blood through the circulation at rates that are three to four times (or more) greater than at rest. In fact, during whole-body exercise, the mean pressure in the human systemic aorta may rise by only about 20 mm Hg even when the level of exertion is high.

Much of the decrease in peripheral resistance during exercise results from vasodilation in the vascular beds of the active muscles, including the respiratory muscles. As noted earlier, there is a great increase in the percentage of open capillaries. The opening of the capillary beds not only reduces systemic resistance but also effectively reduces the average diffusion distance between capillaries and muscle fibers and permits the capillary beds to carry a greatly increased flow of blood without major changes in the linear velocity—or residence time of the blood—in individual capillaries.

The response of the peripheral vasculature during exercise is in fact a highly coordinated, adaptive one that brings about a preferential distribution of the increased cardiac output to the tissues that require increased perfusion. Whereas the skeletal muscles of people receive about 20 percent of the cardiac output at rest, they can receive 70–80 percent or more of the much-increased output during heavy, whole-body exertion. In contrast, vasoconstriction may decrease blood flow during exercise to well below resting rates in such organs as the intestines, kidneys, liver, and inactive muscles. The brain receives about the same amount of blood regardless of exercise state. Flow to the skin is often reduced at the start of exercise, helping to divert blood to the active muscles; but with continued exertion, the requirements of dissipating the heightened amounts of heat produced during exercise often elicit an increase in cutaneous blood flow.

Giraffes: The Problems of a Long Neck

We noted earlier that most mammals are rather similar in the systemic aortic blood pressures thay maintain at rest. Giraffes present an interesting exception, which most physiologists believe is related to their long neck. When people stand erect, their brain is about 35 cm above their heart. Given that a column of blood exerts 10 mm Hg of downward pressure for each 13 cm of height, this means that 27 mm Hg of the pressure developed by the left ventricle of the heart is used simply to support the column of arterial blood (Figure 14.8). If the mean aortic pressure is 90 mm Hg, then the mean arterial pressure available at the level of the brain to perfuse the vascular beds of the brain is about 63 mm Hg. For a giraffe, the brain may be over 160 cm higher than the heart. Thus, if a giraffe had the same aortic pressure as humans, blood would not even reach its brain; the highest column of blood that can be supported by 90–100 mm Hg of pressure is only about 120–130 cm. Giraffes in fact have especially well-developed left ventricles and maintain much higher aortic pressures than most mammals; their mean systemic aortic pressure when standing is about 220 mm Hg. This pressure is sufficient to lift the blood to the level of the brain and still provide an adequate perfusion pressure.

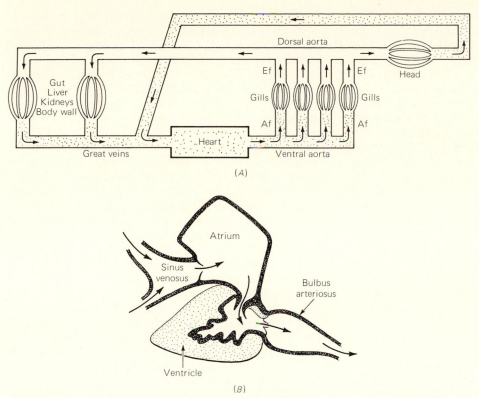

Figure 14.16 (*A*) The circulatory plan in teleost and elasmobranch fish. Stippled parts carry relatively deoxygenated blood. The heart, consisting of four chambers in series, pumps blood anteriorly into the ventral aorta, which gives off afferent branchial vessels (Af) to the gill arches. After perfusing the gills, blood is collected into efferent branchial vessels (Ef) that connect to the dorsal aorta. Blood is distributed to the major systemic circuits along the length of the body by the dorsal aorta. (*B*) The arrangement of the four heart chambers in a trout. [Part (*B*) from D.J. Randall, *Am. Zool.* **8**:179–189 (1968).]

CIRCULATION IN FISH

Like birds and mammals, the elasmobranch and teleost fish—and the amphibians and reptiles as well—possess closed circulatory systems. The circulatory plan of fish is illustrated in Figure 14.16*A*. Blood is pumped anteriorly by the heart into the ventral aorta, which distributes it to the afferent gill vessels. The blood then passes through the blood channels of the gills and is brought by the efferent gill vessels to the dorsal aorta, a large dorsal artery that distributes the blood to the systemic tissues. After perfusing the systemic capillaries, the blood returns in the veins to the heart. As in mammals and birds, the circulatory plan places the respiratory organs in series with the systemic tissues, assuring efficient oxygen transport. Unlike the case in mammals and birds, there is no heart between the respiratory circulation and systemic circulation to impart fresh energy to the blood before it leaves for the systemic tissues. The energy imparted by the heart to the blood entering the ventral aorta must be sufficient to assure an adequate rate of circulation through the resistances of both the branchial and systemic circuits.

The fish heart (Figure 14.16*B*) consists of four chambers arranged in series: a *sinus venosus* into which the great veins empty, an atrium, a ventricle, and a bulbus segment that empties into the ventral aorta. In elasmobranchs and dipnoans, the bulbus segment is endowed with cardiac muscle and is known as the *bulbus cordis* (or conus arteriosus). In teleosts, the segment consists of vascular smooth muscle plus elastic tissue and is called the *bulbus arteriosus*. An important morphological feature of the fish ventricle is that, unlike the ventricles of birds and mammals, it does not consist simply of a compact layer of muscle surrounding a capacious, relatively unobstructed central cavity. Instead, much of the ventricular muscle is of a spongy character, and the ventricular cavity is largely filled with interconnecting strands—or *trabeculae*—of this spongy muscle tissue. The outer shell of the fish ventricle is composed of a layer of compact muscle, and this muscle resembles mammalian ventricular muscle in receiving a vascular (coronary) blood supply. However, the inner, spongy muscle of the ventricle is not vascularized and must therefore obtain oxygen from the venous blood passing through the ventricular chamber. Amphibians—and reptiles to some extent—resemble fish in having the ventricular cavity rather fully occupied by muscle tissue of trabeculate, spongelike character.

The main propulsive force is developed by the ventricle. The bulbus arteriosus of teleosts does not contract in sequence with the other heart chambers but by virtue of its great elasticity plays important roles in smoothing pressure oscillations over the cardiac cycle and providing a pressure reservoir (see p. 344). The bulbus cordis of elasmobranchs is postulated to play similar roles but also contracts in sequence with the ventricle and helps pump the blood.

In keeping with their lower demands for oxygen transport, fish maintain far lower cardiac outputs than mammals of similar body size. They also maintain lower arterial pressures; mean pressures measured in the ventral aorta of teleosts and elasmobranchs have ranged from about 20 to 70 mm Hg, with most being toward the lower end of this range. Pressure drops as the blood perfuses the branchial circulation, and considerable interest has centered on the magnitude of this decline inasmuch as the pressure remaining in the dorsal aorta is that available for perfusion of the systemic circulation. In general, the mean dorsal aortic pressure is 20–45 percent lower than the mean ventral aortic pressure.

Cardiovascular responses to exercise can differ considerably from one species to another. Here, we look just at the responses of rainbow trout, which have been studied particularly thoroughly. Trout progressively increase their cardiac output as they swim faster; at high speeds they attain outputs about three times those seen at rest. Although the fish may raise their heart rate, their increases in cardiac output are effected primarily by augmentation of stroke volume. Ventral aortic pressures tend to increase with swimming speed; dorsal aortic pressures also increase, but much less. The resistance of the branchial circuit remains little changed; forcing the increased cardiac output of exercise through the branchial circulation therefore requires a commensurate increase in the pressure drop between the afferent and efferent branchial vessels, which is observed. The resistance of the systemic circuit drops markedly during exercise; thus, the increases in blood flow require only modest elevations of dorsal aortic pressure. Recent studies indicate that the reduction in systemic resistance is attributable in good measure to vasodilation in the swimming muscles, particularly the red muscles. A great deal remains to be learned about cardiovascular function in fish.

Air-Breathing Fish

Many fish are adapted to breathing from the atmosphere, as discussed in Chapter 12 (p. 271). In most cases, their aerial respiratory organs are derived from structures such as the pharyngeal membranes, gut, or swimbladder that primitively are serviced by the systemic circulation. An important attribute of these structures is that they typically continue to drain into the systemic venous vasculature—not the systemic arterial vasculature—even when adapted to respiration. Consider, for illustration, the electric eel, *Electrophorus,* which is an obligate air breather. Its gills are much reduced, and gas exchange is carried out largely across the wall of its mouth cavity, which is thrown into an elaborate system of well-vascularized papillae. The circulatory plan of *Electrophorus* is illustrated in Figure 14.17A. The afferent vessels to the buccal respiratory surfaces arise from the gill vasculature. The efferent vessels drain into the systemic venous vasculature. Thus, in contrast to the usual teleost pattern and to the avian and mammalian pattern, the respiratory circulation is placed in *parallel* with the systemic circulation. The implications of this arrangement are profound. Experiments confirm what anatomy suggests, namely, that oxygenated blood from the buccal respiratory surfaces freely *mixes* in the systemic veins and heart with deoxygenated blood from other tissues, and thus the heart of *Electrophorus* pumps a mixture of deoxygenated and oxygenated blood to both the respiratory surfaces and the systemic arteries (note Figure 14.17A). The level of oxygen saturation in the blood pumped by the heart never exceeds 60–65 percent even though that in blood draining the buccal respiratory surfaces may be over 90 percent. Clearly, the mixing of the oxygenated and deoxygenated blood reduces the efficiency of circulatory oxygen transport; blood deoxygenated in the systemic tissues is in part recycled directly back to those tissues, and blood oxygenated in the respiratory organ is in part recycled back to the respiratory surfaces. A similar anatomical and functional parallelism between the aerial respiratory circuit and the systemic circuit is typically observed also in fish that breathe using the gastrointestinal tract or swimbladder, as shown in Figure 14.17B–C.

Some of the air-breathing fish that suffer mixing of their freshly oxygenated blood with deoxygenated blood have evidently evolved compensations: traits that *promote* oxygen delivery just as mixing impairs it. *Electrophorus,* for example, has evolved an exceptionally high circulatory rate and an oxygen-carrying capacity well above that of most fish.

Although caution is necessary in drawing conclusions about the past from the condition of modern species, it appears that the early evolution of aerial respiration in piscine vertebrates, because it involved the respiratory adaptation of structures with a primitively systemic venous drainage, entailed the abandonment of a series circulation for a less-efficient parallel circulation that permitted free mixing of oxygenated and deoxygenated blood. Selective pressures favoring devices that would renew separation of the two types of blood then must have been enormous. In the evolutionary progression of the vertebrates, it is not until the level of birds and mammals that we again find the respiratory and systemic circuits connected inextricably in series by a circulatory anatomy that precludes mixing of oxygenated and deoxygenated blood. However, among some teleosts

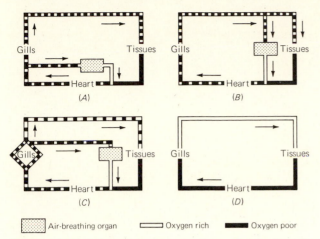

Air-breathing organ   Oxygen rich   Oxygen poor

Figure 14.17 Schematic representations of the circulatory plans of certain air-breathing fish. The general piscine arrangement is shown in (D) for comparison. The amounts of white and black in the circulatory tracts represent the approximate amounts of oxygenated and deoxygenated blood carried in the vessels. (A) The circulatory plan in *Electrophorus* and certain other fish with air-breathing organs derived from the pharyngeal and/or opercular mucosa. Afferent vessels to the air-breathing organ are derived from the afferent branchial vasculature. Efferent vessels from the air-breathing organ connect to systemic veins. (B) The circulatory plan in armored catfish (*Plecostomus*) and certain other fish in which the air-breathing organ is associated with the stomach or intestinal tract. Afferent vessels to the air-breathing organ are derived from the dorsal aorta; efferent vessels enter systemic veins. (C) The circulatory plan in the bowfin (*Amia*) and certain other fish that use the swimbladder as an air-breathing organ. Afferent vessels to the swimbladder are specialized, arising from the efferent branchial vasculature of the sixth aortic arch (similar to the condition in lungfish). Efferent vessels from the swimbladder enter systemic veins (unlike lungfish). The gill vasculature is represented in two tracts to emphasize that afferent vessels to the swimbladder arise from efferent vessels of only the posterior pair of gill arches. [From K. Johansen, Air breathing in fishes. In W.S. Hoar and D.J. Randall (eds.), *Fish Physiology,* Vol. IV. Academic, New York, 1970.]

(genus *Channa*) and the lungfish, amphibians, and reptiles, there are significant developments in anatomy and function that in many species assure a considerable separation of the oxygenated blood returning from the lungs and the deoxygenated blood returning from the systemic tissues. Furthermore, we find in these groups that the incomplete anatomical separation of the pulmonary and systemic circuits has often, in fact, been exploited to advantage.

Anatomical Features of Lungfish Some of the respiratory and circulatory specializations of lungfish are more fully developed in the African and South American genera, *Protopterus* and *Lepidosiren*—which are obligate air breathers—than in the Australian genus, *Neoceratodus* (p. 272). Here, we emphasize *Protopterus.*

The heart of lungfish is very different from that of other fish. For one thing, the atrial and ventricular chambers are partly divided into right and left halves

by septa. For another, the bulbus cordis—which takes the form of a sharply twisted tube—possesses two longitudinal ridges that project toward each other from opposite sides of its lumen, partially dividing the lumen into two channels.

The arteries supplying the lungs—which are homologous to the pulmonary arteries of tetrapod vertebrates—arise from the efferent vasculature of the posterior gill arches (Figure 14.18). The venous vasculature of the lungs in *Protopterus* and *Lepidosiren* is quite unlike the venous vasculature of other piscine air-breathing organs, for the veins lead directly into the left side of the atrium of the heart rather than connecting with the systemic venous vasculature. This is a very significant development, for on simple anatomical grounds, blood from the lungs is kept separate from systemic venous blood at least into the atrium. The sinus venosus, which now receives only systemic venous blood, connects to the right side of the atrium.

The ventral aorta, which primitively serves as a common conduit to all the afferent gill vessels (Figure 14.16), is virtually or entirely eliminated in lungfish. The four pairs of afferent branchial arteries thus arise immediately from the anterior end of the bulbus cordis, as do the homologous vessels of amphibians. Key features of the branchial vasculature of *Protopterus* are shown in Figure 14.18. Two of the pairs of afferent arteries—constituting aortic arches 3 and 4—emanate from the ventral bulbar channel and travel to anterior gill arches that have become devoid of gill lamellae; these arteries do not break up into capillaries but instead form direct through-connections to the dorsal aorta. The other two pairs of afferent branchial arteries—constituting aortic arches 5 and 6—lead from the dorsal bulbar channel to two pairs of posterior gill arches that retain rudimentary lamellae; these arteries break up to supply the lamellae, but the lamellar capillaries are of large diameter and can be bypassed to at least some extent by way of vascular shunts, meaning that blood often encounters only a relatively low resistance in flowing from the afferent to the efferent vessels of these gill arches.

Functional Features of *Protopterus* The numerous modifications of circulatory morphology in lungfish suggest that some separation of oxygenated and deoxygenated blood may be attained. However, the anatomy of lungfish does not preclude mixing of the two types of blood, and functional studies are therefore necessary to determine whether, and to what extent, separation is achieved.

A number of functional investigations have been carried out on *Protopterus.* Given that these studies have involved a variety of techniques and conditions, it is not surprising that they do not agree in all respects and leave important questions unanswered. They demonstrate dramatically, however, that oxygenated pulmonary venous blood and deoxygenated

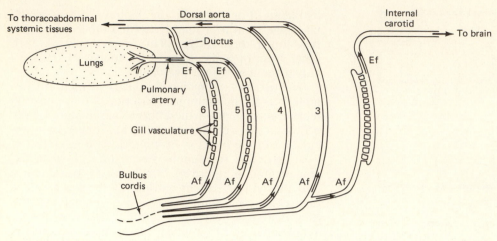

Figure 14.18 Schematic diagram of blood flow through the branchial arches on one side of the lungfish *Protopterus aethiopicus*. Vascular arches are numbered according to the aortic arches they represent (other numbering systems are sometimes used). The gill arches supplied by vascular arches 3 and 4 are devoid of gill lamellae; these vascular arches do not break up into capillaries but form direct through-connections to the dorsal aorta. Gill lamellae of a rudimentary sort are found on the gill arches supplied by vascular arches 5 and 6. Within these gill arches, the afferent branchial arteries (Af) break up to supply the gill lamellae, and blood is collected from the lamellae by efferent branchial vessels (Ef); direct shunts between the afferent and efferent vessels are possible, however, and capillaries in the lamellae are exceptionally large, thus posing only modest resistance to flow. A similar arrangement is found in the most anterior gill arch (shown) and in another posterior arch supplied by the vasculature of arch 6 (not shown). The efferent vessels of arches 5 and 6 form the pulmonary artery. Details of flow through the ductus connecting the pulmonary artery and dorsal aorta remain to be well elucidated. Arrows in vessels show direction of blood flow. [Based mostly on P. Laurent, R.G. DeLaney, and A.P. Fishman, *J. Morphol.* **156**:173–208 (1978).]

systemic venous blood follow substantially different paths through the central circulation.*

One method of studying flow patterns is to inject fluids that are opaque to x-radiation into selected vessels and, using x-rays, monitor where these fluids are carried by the blood. Studies of this sort show that pulmonary venous blood tends to follow a course through the left parts of the atrium and ventricle and is delivered preferentially to the ventral channel of the bulbus cordis and then to aortic arches 3 and 4, which provide direct through-channels to the dorsal aorta (Figures 14.18 and 14.19). In this way, the blood oxygenated in the lungs is directed preferentially to the systemic arterial circulation. Systemic venous blood tends to pass through the right parts of the atrium and ventricle. Then that blood appears either to be distributed about evenly to all four pairs of aortic arches or to be delivered preferentially to the dorsal bulbar channel and aortic arches 5 and 6, which lead in part to the pulmonary arteries.

Studies using radiopaque materials do not permit rigorous quantification of the extent of selective distribution of oxygenated and deoxygenated bloods. A technique that is better suited to quantification is monitoring of blood oxygen levels.

The difference in oxygen tension between sys-

temic arterial blood and pulmonary arterial blood can provide a useful index of the extent of selective blood distribution. If oxygenated blood from the lungs of a lungfish were to mix thoroughly with systemic venous blood in the heart, the blood pumped to the systemic circuit would be no higher in oxygen tension than that pumped to the lungs. On the other hand, to the extent that the distribution of the oxygenated and deoxygenated bloods is selective, the systemic arterial oxygen tension will exceed the pulmonary arterial tension. Several studies of *Protopterus* have found that the oxygen tension of systemic arterial blood averages 5–9 mm Hg higher than that of pulmonary arterial blood, reflecting a substantial degree of selective blood distribution. In one study, when the systemic venous oxygen tension averaged just 2 mm Hg and the pulmonary venous tension was 46 mm Hg, the blood pumped into aortic arches 3 and 4 for direct passage to the dorsal aorta had an average oxygen tension of 38 mm Hg, showing that it consisted mostly of oxygenated blood from the lungs. The blood pumped to aortic arches 5 and 6 had a much lower tension. Some of this less-oxygenated blood entered the dorsal aorta by way of the ductus (Figure 14.18) and, mixing with the well-oxygenated blood from aortic arches 3 and 4, lowered the tension in the blood delivered to thoracoabdominal arteries to 30 mm Hg. The pulmonary arteries received blood having a tension of 25 mm Hg.

The lungfish are intermittent breathers; having ventilated their lungs with air, they hold their breath

*The term *central circulation* is used throughout this chapter to refer to the heart and the veins and arteries leading to and from the heart.

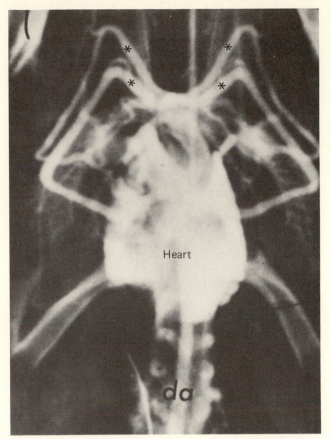

Figure 14.19 An x-ray image, in the horizontal plane, of the distribution of radiopaque material in the heart and aortic arches of a lungfish (*Protopterus aethiopicus*) following injection of the material into the pulmonary vein. White-colored parts have received radiopaque material. Note that of the four pairs of aortic arches, only two pairs—labeled with asterisks—have received appreciable amounts of material. The other two pairs, having not received much material, are not well visualized. The arches receiving material are numbers 3 and 4, and this image shows that blood from the pulmonary vein is distributed to these arches preferentially. The dorsal aorta is marked da. [Reprinted by permission from K. Johansen and R. Hol, *J. Morphol.* **126:**333–348 (1968).]

for a substantial time before ventilating again (p. 276). The circulatory function of *Protopterus* undergoes marked changes in parallel with the pulmonary ventilatory cycle. Immediately after an episode of ventilation—when pulmonary oxygen levels are at their highest—circulatory adjustments occur that favor rapid uptake of oxygen from the lungs and efficient distribution of the oxygen to the body. One such adjustment is that the rate of blood flow to the lungs is increased to about four times the rate observed just prior to ventilation. This increase is achieved in part by an increase in cardiac output. Also, however, it results from a *redistribution* of the cardiac output; the percentage of the heart's total outflow directed to the pulmonary arteries, while it may be as low as 20 percent just before ventilation, may rise to as high as 70 percent just after ventilation. Such redistribution of the heart's outflow is made possible by the incompletely divided nature of the central circulation, which permits flow to be shifted between the systemic and pulmonary circuits (under vasomotor control). The adjustments in flow pattern that occur in concert with the cycle of pulmonary ventilation have additional important consequences. One is that the extent of selective distribution of pulmonary venous blood to the systemic circuit is raised to a maximum immediately after the lungs have been freshly filled.

CIRCULATION IN AMPHIBIANS AND REPTILES

Anatomy and Selective Distribution in Amphibians

In the heart of typical lung-breathing amphibians, the atrium is completely divided into right and left halves by a septum. Pulmonary venous blood enters the left atrium, and systemic venous blood enters the right atrium via the sinus venosus. Separation of oxygenated and deoxygenated blood is assured on simple anatomical grounds until the blood enters the ventricle. The ventricle, however, lacks a septum altogether. An important feature of the ventricle, noted previously (p. 349), is that its inside does not consist of a wide-open cavity but is criss-crossed extensively with strands and cords of muscle—trabeculae—which give it a spongy consistency. The ventricle discharges into a contractile bulbus cordis (conus arteriosus), which in turn discharges typically into paired carotid, systemic, and pulmonary (pulmocutaneous) arteries. Running along the inside of the bulbus in most amphibians is a complexly twisted ridge of endothelial tissue—called the spiral fold or spiral "valve"—which incompletely divides the bulbar lumen.

Amphibians are so diverse in their respiratory physiology (p. 272) that considerable diversity in their circulatory attributes must be expected. This diversity is only gradually being elucidated. Already, however, it is clear that many amphibians—including certain frogs, toads, and salamanders—are capable of substantial selective distribution of blood in their central circulation. Furthermore, at least some can modulate the distribution adaptively (p. 355).

Recently, a study of blood oxygen levels in bullfrogs, *Rana catesbeiana,* has brought to light a particularly dramatic example of selective distribution. When blood entering the right atrium of the frogs from the systemic veins had an average oxygen concentration of 4.2 volumes percent and that entering the left atrium from the lungs had a concentration of 8.6 volumes percent, the blood pumped into the pulmonary arteries contained 4.4 volumes percent and that pumped into the arteries of the systemic circulation contained 8.0 volumes percent. Calculations indicated that 84 percent of the systemic venous blood arriving in the heart was directed into the pulmonary arteries, whereas 91 percent of pulmonary venous blood was channeled to the systemic circulation.

BOX 14.1   VENTILATION-PERFUSION MATCHING

An important tenet of respiratory physiology is that, ideally, the ability of the circulatory system to carry oxygen away from the gas-exchange organ should be closely matched to the ability of the ventilatory system to bring oxygen into the organ from the environment. The matching of the performance of the circulatory and ventilatory systems in this regard is often termed *ventilation-perfusion matching*. *Protopterus* provides a particularly graphic example. As we have seen, it increases perfusion of its lungs when breathing has made oxygen readily available within its lungs, and it decreases perfusion when the ventilatory supply of oxygen is interrupted.

Of course, when amphibians achieve such remarkable levels of selective distribution, they do so with a ventricle that entirely lacks a septum and a bulbus cordis that is only incompletely divided. The mechanism by which blood from the two atria is channeled along separate paths through the ventricle and bulbus remains only dimly understood. In the ventricle, the trabeculae are hypothesized to help guide the flows of left and right atrial blood; the manner in which they would accomplish this is by no means obvious, however. The spiral fold evidently separates and guides flows in the bulbus cordis.

Because the skin is typically an important respiratory site in amphibians (p. 274), its circulation deserves note. The pulmonary arteries, also known as the pulmocutaneous arteries, not only supply the lungs, but also give rise to the major cutaneous arteries. Thus, the skin receives blood of the same quality as the lungs: If the mechanisms of selective distribution send systemic venous blood preferentially to the lungs, they also send it to the respiratory surfaces of the skin. The venous vasculature of the skin is not so advantageously arranged, however, for it connects with the general systemic venous vasculature, placing the cutaneous circuit functionally in parallel with the systemic circuit at large. The effect of oxygenation of blood in the skin is to raise the level of oxygen in the systemic venous blood arriving in the right atrium. The modulation of cutaneous blood flow needs more study. However, we know, for example, that when frogs (*Rana esculenta*) are submerged in aerated water and must rely on cutaneous respiration, perfusion of the capillary beds of their skin is markedly increased. If the ambient water becomes oxygen-depleted, submerged bullfrogs curtail skin blood flow.

Anatomy and Selective Distribution in Noncrocodilian Reptiles

In all groups of reptiles, there is a complete atrial septum. Pulmonary venous blood returns, as usual, to the left atrium, and systemic venous blood returns to the right. The sinus venosus, which in mammals and birds is reduced into the right atrial wall, remains. The bulbus cordis, however, has become reduced to vestigial proportions in reptiles, with the important consequence that the systemic and pulmonary arteries arise directly and separately from the ventricle.

In the noncrocodilian reptiles—turtles, snakes, and lizards—although the ventricle is typically divided by muscular ridges and partial septa into three chambers, the physical barriers between the chambers are not complete. Anatomically, the ventricle is the one place where mixing of pulmonary and systemic venous blood can occur. As we examine what is known of ventricular anatomy and function, we shall not consider the varanid lizards, which differ from the other noncrocodilians.

The arrangement of the three ventricular chambers is so complex as to be difficult to depict. The cross section of the ventricle in Figure 14.20 shows, however, many of the key features. The chambers are termed the cavum arteriosum, cavum venosum, and cavum pulmonale. The cavum arteriosum and cavum venosum are separated entirely by a septum except for a single, localized fenestration that is shown in the section and termed the interventricular canal. This canal is positioned immediately between the two atrioventricular apertures. The cavum pulmonale is a small chamber separated incompletely from the cavum venosum by a muscular ridge. The cavum arteriosum is the chamber into which the left atrium opens. Note, however, that no arteries originate from the cavum arteriosum; the only exit from the chamber is the interventricular canal leading to the cavum venosum. The right atrium discharges into the cavum venosum, and all the systemic arteries arise from that chamber. The only entry to the cavum pulmonale is from the cavum venosum. The common pulmonary artery originates from the cavum pulmonale.

Numerous studies on turtles, snakes, and lizards have demonstrated that during periods of pulmonary ventilation, these animals typically display a high degree of selective blood distribution. The mechanism by which this selective distribution is achieved is not fully understood, however, and may be somewhat different in some groups than others. The *se-*

CIRC

BAS
SYST

The
open
syste
of th

A
have
charg
arter
vesse
cells.
vesse
small
thorc
of lac
to all
dissir
chann
are r
endot
space
nels 1
times
some
This 1
mean
and th
and ii
some

No
bound
a mer
with
know
there
and ti
certai
classe
theles
points
syster
cantly
worm:
to the
the de
the art
the ar
follow
and si
crete

CIRC

Some
blood
sively
sive fc

the systemic arteries then occurs (bypassing the lungs).

CONCLUDING COMMENTS ON VERTEBRATES

In the past, many texts have implied that there is an almost orthogenetic progression in the development of the central circulation from lungfish to mammals and have implied that the central circulation in "lower" air-breathing vertebrates is unequivocally defective because of its inability to guarantee selective blood distribution anatomically. These implications place an unwarranted and philosophically undesirable emphasis on the mammalian condition as a standard by which to judge other animals. True, a series connection between the pulmonary and systemic circuits has distinct advantages during times of pulmonary respiration. This does not mean, however, that an anatomically rigid series arrangement is always ideal. Each group of animals faces its own particular challenges; and it is evident that when animals fill their lungs only intermittently, incomplete anatomical division of the central circulation may have advantages over complete anatomical division because the incompletely divided state permits the rate of pulmonary blood flow to be modulated partly independently of systemic flow. The circulatory plans of the "lower" vertebrates, in brief, have their own potential virtues and are not simply defective versions of the mammalian plan.

Figure 14.23 provides a comparison of pressures and vascular resistances in a variety of vertebrates. One striking feature of the birds and mammals is that they maintain far higher *systemic* arterial pressures than most other vertebrates; because systemic resistance has undergone little, if any, systematic change in phylogeny, exceptionally high pressures are needed by the birds and mammals to maintain the relatively high rates of blood flow that their high

oxygen requirements demand. Of course, the birds and mammals must maintain pulmonary flow rates that are just as high as systemic. Notably, however, they have not evolved elevated *pulmonary* pressures but instead have facilitated pulmonary flow by evolving exceptionally low pulmonary resistances. Evidently, natural selection has favored lowering resistance rather than raising pressure in the pulmonary circuit because of the risks that elevated pressures would pose in regard to pulmonary flooding (p. 347).

CIRCULATION IN ANNELIDS

Most of the oligochaete and polychaete worms have a well-developed blood–vascular system modeled on a common plan, though exhibiting diverse specialized features in various species. Characteristically, there are two major longitudinal vessels that run the length of the body, a dorsal vessel above the gut and a ventral vessel below the gut. The connections between these longitudinal vessels reflect the strongly segmented body plan of these animals. In general, there are two lateral circulatory arcs in each body segment, which run between the dorsal and ventral longitudinal vessels on either side of the segment. These circulatory arcs vary greatly in complexity not only in different species but often in different body segments of a single species. Figure 14.24 depicts the lateral vascular system on one side of a body segment in a fairly generalized polychaete, *Nereis virens*. Note first the large longitudinal vessels above and below the gut. These are connected, quite typically for polychaetes and oligochaetes, via vessels in the gut wall. There is also a far-flung system of vessels, with connections to both longitudinal vessels, that supplies the parapodium, integument, nephridium, musculature of the body wall, and other structures on the side of the body segment shown. A similar

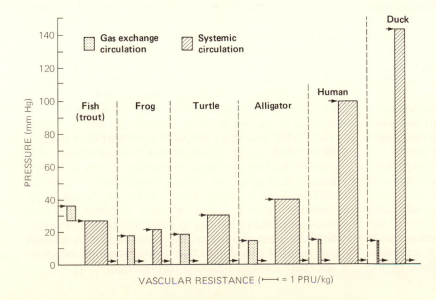

Figure 14.23 A summary of perfusion pressures and vascular resistances observed in representatives of the vertebrate classes. Arrows indicate pressures in major vessels going to and coming from the systemic circulation and gas exchange circulation (pulmonary circulation except in the fish). The height of each bar provides an indication of the pressure drop across the circulation. The width of each bar expresses vascular resistance measured in peripheral resistance units (PRU) per kilogram of body weight. One PRU is a resistance requiring a pressure gradient of 1 mm Hg to achieve a rate of flow of 1 mL/min. A summarizing diagram of this sort cannot possibly represent the range of variation observed even within a single species. This diagram is meant only to provide an overview of salient features as elucidated in selected species that have received study. [From K. Johansen, *Respir. Physiol.* **14**:193–210 (1972).]

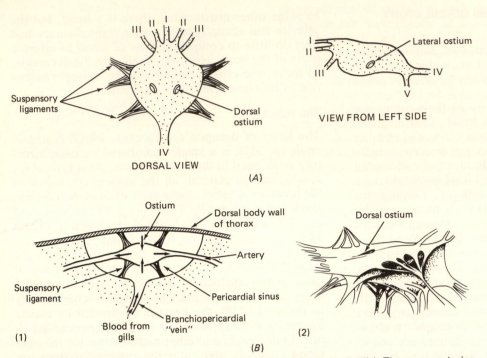

Figure 14.25 (A) Diagrams of the heart of a crayfish (*Astacus fluviatilis*). The seven arteries emanating from the heart are indicated by Roman numerals: I, one anterior aorta; II, two anterior lateral arteries; III, two hepatic arteries; IV, one posterior aorta; V, one sternal artery. There are six ostia: two dorsal, two lateral, and two ventral. The heart is suspended in a cavity, the pericardial sinus, by fibrous bands, the suspensory ligaments. (B) The suspension of the decapod heart. (B1) Highly schematic lateral view of heart suspended by suspensory ligaments in the pericardial sinus. The sinus is located in the dorsal thorax. Arrows in heart and vessels show the direction of blood flow. Blood enters the pericardial sinus from the gills through branchiopericardial "veins," is aspirated into the heart through the ostia during diastole, and is driven into the arteries during systole. (B2) Three-dimensional lateral view of the major suspensory ligaments of the heart of a lobster, showing the actual complexity of the system. [Part (A) from T.H. Huxley, *The Crayfish,* 4th ed. Kegan Paul, Trench & Company, London, 1884; part (B2) reprinted by permission from F. Plateau, *Arch. Biol.* **1**:595–695 (1880).]

capable of vasomotor control. As illustrated in Figure 14.26, the arterial network in decapods is extensive. Branching arteries lead the blood from the heart to most regions of the body. Sometimes the blood is discharged from the arteries directly to lacunar networks. Sometimes capillary beds intervene. Prominent capillary beds are observed, for example, in the brain and ventral ganglia. In many decapods and other malacostracans, an accessory heart, known as the cor frontale, is positioned along the anterior aorta just before the latter gives rise to the extensive vascular networks of the brain (Figure 14.26). Powered not by vascular musculature but evidently by specialized somatic muscles, the cor frontale is believed to help assure that adequate perfusion pressures are maintained in the brain at all times.

After blood has been delivered to lacunar networks throughout the body, it ultimately drains into a system of sinus thoroughfare channels located ventrally along the length of the animal. In these sinuses, blood from the posterior regions of the body flows anteriorly and that from the anterior regions of the body flows posteriorly, both flows converging on a thoracic sinus termed the infrabranchial sinus. The channels that carry blood into the gills (the afferent branchial channels) arise from this sinus. Blood flows into these channels, traverses systems of small sinuses or lacunae in the gill filaments or lamellae, and then exits the gills by way of efferent branchial channels. Importantly, these latter channels discharge into discrete channels, the branchiopericardial "veins," that lead directly to the pericardial sinus. Thus, after a period of seemingly poorly controlled flow following its discharge from the arterial system, the blood is channeled in a most significant way prior to its return to the heart. The gills are placed in series with the rest of the circulation, and oxygenation of the blood pumped by the heart is assured.

Pressure Drops

Like fish, decapods perfuse the branchial and systemic circuits in sequence, without repumping the blood between one circuit and the other. Unlike the case in fish, however, blood leaving the heart in decapods passes through the branchial circuit *after* trav-

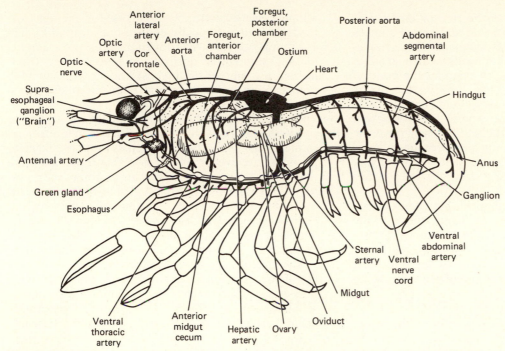

Figure 14.26 Diagrammatic representation of the heart and major arteries in a crayfish. [Reprinted by permission from P.A. McLaughlin, *Comparative Morphology of Recent Crustacea*. W.H. Freeman and Company, San Francisco, 1980. Copyright © 1980.]

ersing the systemic, not before. Thus, there has been some question about the source of the motive force for perfusion of the gills. Recent measures indicate that the branchial circulation of decapods typically poses a low enough resistance to flow that a pressure differential of only 1–2 mm Hg between the afferent and efferent branchial channels suffices for perfusion. In some species, at least, it appears that although the head of positive pressure developed in the arteries by the heart is substantially dissipated in perfusing the systemic circuit, enough of it remains— once the blood has reached the infrabranchial sinus— to force blood through the gills; that is, gill blood flow is driven largely by positive heart pressures. As noted earlier, forces of suction developed in the pericardial sinus during cardiac systole may also promote branchial flow.

The drops in pressure occurring in the systemic circuit of decapods have been quantified in only a few species but in general may be classed as moderately large. Perhaps the largest systemic pressure gradient has been observed in the spiny lobster (*Panulirus interruptus*), in which mean arterial pressure is about 10–15 mm Hg above mean infrabranchial-sinus pressure. The gradient is about 6–7 mm Hg in American lobsters (*Homarus*).

FUNCTIONAL PROPERTIES OF OPEN CIRCULATORY SYSTEMS

Relatively little is known about the functional implications of having an open circulatory system. None-

theless, some information is available; and given that the data on decapods and other crustaceans are as informative as can be found, a discussion at this point is appropriate.

Open circulatory systems, in comparison with closed, are commonly characterized by large blood volumes. The blood volume in vertebrates typically is equivalent to 2–10 percent of body weight (that is, 2–10 mL of blood per 100 g of body weight). Ranges of values observed in several groups of animals with open circulatory systems are: decapod crustaceans, 10–50 percent of body weight; insects, 1–45 percent; and noncephalopod mollusks, 35–80 percent. The higher blood volumes of animals with open circulatory systems are in part explained by the fact that their blood includes fluid that in animals with closed systems is distinguished as lymph and intercellular tissue fluid. In humans, for example, the blood volume is near 7 percent of body weight, but intercellular tissue fluid and lymph amount to about 12 percent of body weight, so that the total of blood, lymph, and intercellular fluid is near 19 percent of body weight, a value not unlike the blood volume of some decapods.

It is commonly observed in animals with open circulatory systems that body movements substantially affect the pressures in their blood spaces. When such movements create pressure *differences* from one part of the body to another, blood flow is promoted; in American lobsters, for example, quick flexion of the abdomen—an important locomotory movement—raises pressures in the abdomen above those in the thorax and can cause the rate of blood flow toward the thorax in the ventral abdominal sinuses to increase by a factor of 10 or more. Body movements and changes in body conformity can also affect the *overall* level of pressure throughout the body; if a change in posture, for instance, reduces the volume of an animal's blood spaces, it will tend to pressurize the blood throughout the body, much in the way that pushing on a water-filled rubber ball will raise the pressure of the water. Insofar as a pressure change is transmitted throughout the body, it will not in any immediate sense affect blood flow, for only differences in pressure cause flow.

In brief, we need to distinguish between two aspects of pressure. On the one hand, there is the *background pressure:* the level of pressure that prevails everywhere in the animal. On the other, there are *pressure differences,* responsible for flow, that are superimposed on the background pressure, as by the heart.

In the study of heart action, the distinction between background pressure and pressure differences is significant in all animals. However, the distinction is particularly important in analyzing the animals having open circulatory systems because in them the background pressure—and changes in it—are often large in comparison with the pressure differences the heart creates. In a decapod crustacean, for instance, it would not be unusual to find the lowest pressure anywhere in the body to be 12 mm Hg, with the infrabranchial-sinus pressure being slightly higher—14 mm Hg—and the mean arterial pressure being higher yet—22 mm Hg. Note that in this animal the *absolute pressure* in the arteries—22 mm Hg—would not, in itself, accurately represent either the pressure gradient required for blood flow or the heart's contribution to flow, for it includes a large component of pressure that is present everywhere. The *real magnitude of the head of pressure created by the heart* in the arteries is just 10 mm Hg (the extent to which arterial pressure exceeds background). Later, perhaps because of a change in posture, the background pressure in this animal might rise to 20 mm Hg, and the arterial pressure might then become 30 mm Hg: still 10 mm Hg above background. Often, in just this way, the pressure gradient created by the heart is superimposed on ups and downs of background pressure in animals with open circulatory systems.

When circulatory systems are classed as "high-pressure" or "low-pressure" systems, the terms should properly be used to describe the pressure *gradients* required for perfusion, not simply the absolute pressures that prevail. If a system functions with an overall pressure gradient of 4 mm Hg, for example, it is a "low-pressure" system whether the absolute pressures on its afferent and efferent sides are 6 and 2 mm Hg or 56 and 52 mm Hg.

Open circulatory systems typically are low-pressure or moderate-pressure systems. In bivalve and nonpulmonate gastropod mollusks, for example, arterial pressures typically exceed background by just 1–4 mm Hg or less. Arteriovenous pressure gradients of 14–15 mm Hg, as observed in spiny lobsters (*Panulirus*), are among the highest known to occur with regularity in animals having open circulatory systems.

The relatively low pressure gradients prevailing in open circulatory systems have often been interpreted to mean that blood flow in such systems is sluggish. By now, however, data are available to demonstrate that while flow may be sluggish in some open systems, it is vigorous in others. Table 14.2 shows, for example, that the cardiac output in certain crustaceans—including crabs (*Cancer*) having an arteriovenous pressure drop of just 3 mm Hg—is much greater than the cardiac output in fish of similar size.

It must always be remembered that the rate of blood flow through a system depends on not only the pressure gradient across the system but also the resistance to flow through the system. Among crustaceans, resistances across the systemic circuit are low (Table 14.2). These low resistances—which are probably a direct consequence of the open design of the circulation—permit relatively high rates of blood flow to be maintained with relatively low pressure gradients.

The high flow rates in decapods seem to represent a type of compensation for having low blood oxygen-carrying capacities (p. 314). By comparison to the fish in Table 14.2, the crustaceans transport less oxygen per unit volume of blood but circulate a greater volume per unit time and thus are able to sustain similar rates of oxygen consumption.

Inevitably, the question arises of how open and closed circulatory systems compare in performance. Data such as those in Table 14.2 would suggest that from many vantage points, although the two types of system are different, neither can be judged clearly or generically superior.

If open circulatory systems have inherent weaknesses relative to closed, the weaknesses may well be found in the realm of peripheral blood flow and its control. At first sight, flow through a system of lacunae and sinuses rather than discrete vessels would appear to be relatively haphazard. It is easy to imagine that flow through any particular region of the body could be erratic and that stagnant pools could develop. The extent to which these deleterious possibilities actually occur cannot be appraised fully at present, but there is evidence that peripheral flow in open circulatory systems is more defined and regular than might at first be supposed. Large sinuses are often subdivided by septa or body organs, such that more or less discrete channels for flow are established. Investigators who have observed systems as disparate as the large thoracoabdominal sinuses of lobsters and the minute sinus networks in the heads of insects have reported that remarkably well-defined patterns of blood flow can be recognized

Table 14.2 AVERAGE DATA ON CIRCULATORY FUNCTION IN TWO CRUSTACEANS, COMPARED WITH TWO FISH[a]

	Spiny lobster (*Panulirus interruptus*)	Rock crab (*Cancer productus*)	Starry flounder (*Platichthys stellatus*)	Rainbow trout (*Salmo gairdneri*)
Weight (g)	515	~370	684	~210
Temperature (°C)	16	12–16	8–11	9–15
Oxygen consumption (mL/kg·min)	0.80	0.60	0.46	0.65
Cardiac output (mL/kg·min)	128–148	125	39	18
Heart rate (beat/min)	65	101	35	63
Stroke volume (mL/kg·stroke)	2.1	1.2	1.2	0.3
Afferent systemic pressure (mm Hg)[b]	35	10	18	26
Efferent systemic pressure (mm Hg)[c]	21	7	2	4
Systemic resistance (mm Hg·kg·min/mL)[d]	0.1	0.03	0.4	1.2
Blood oxygen-carrying capacity (volumes percent)	2.0	1.3	5.7	7.8

[a] In some cases, some of the data given for a species were gathered under different conditions than others. Furthermore, the weights and conditions of study for the four species were different. Thus, conclusions drawn from this table should be viewed as first approximations only. The rock crab periodically ceases heart action; after resuming, its oxygen consumption, cardiac output, and stroke volume gradually fall. Values given are averages measured 5 min or more after a cardiac pause.

[b] Afferent systemic pressures were measured in systemic arteries, except for the crab pressure, which represents systolic ventricular pressure and thus probably overestimates arterial pressure (meaning that the calculated resistance is probably also an overestimate).

[c] Efferent systemic pressures were measured in systemic veins of fish and the infrabranchial sinus of crustaceans.

[d] Systemic resistance was calculated as the difference between the afferent and efferent pressures, divided by cardiac output.

Source: Lobster—B.W. Belman, *Mar. Biol.* **29**:295–305(1975), carrying capacity from J.R. Redmond, *J. Cell. Comp. Physiol.* **46**:209–247(1955); crab—B.R. McMahon and J.L. Wilkens, *J. Exp. Zool.* **202**:363–374(1977), pressures from B.W. Belman, *J. Exp. Zool.* **196**:71–78(1976); fish—C.M. Wood, B.R. McMahon, and D.G. McDonald, *J. Exp. Biol.* **78**:167–179(1979).

within them. A critical factor of which we know almost nothing is the modulation of blood flow and blood distribution in the periphery. In a closed system of continuous vessels, we have seen that flow through particular regions of the body can be controlled by vasomotor activity, permitting cardiac output to be redistributed readily according to tissue demands. Once blood pumped by the heart has exited the vessels in an open type of circulation, however, it is difficult to see how the distribution of the blood flow could be controlled in any general sense, though there can be little doubt that body movements, contractions of somatic muscles, and the activity of peripheral accessory hearts influence peripheral flow. Also, one is led to wonder from the design of the open circulation whether an increase in cardiac output, as during exercise, would result in a commensurate increase in flow through lacunar exchange networks or whether some of the increased cardiac output, following paths of least resistance, might simply skirt around lacunar networks and enter sinus channels rather directly. As these questions are probed, we must not forget that whatever weaknesses the open type of circulation might have, they have not been sufficient to prevent some crustaceans—even large ones—from assuming reasonably active and metabolically demanding ways of life.

CIRCULATION IN INSECTS

The circulatory system of insects follows the basic arthropodal plan. A dorsal vessel extends along most of the body (Figure 14.27) and in many insects is the only vessel. It is divisible, often indistinctly, into a posterior part, the heart—usually restricted to the abdomen—and an anterior part, the dorsal aorta, which leads forward into the thorax and head. The heart is perforated by 1–13 pairs of ostia, which are typically slit shaped and valved; the dorsal aorta, however, lacks ostia. The heart may be bound directly to the dorsal body wall or may be attached to it by threads of connective tissue. A septum of muscular and connective tissue, termed the dorsal or pericardial septum, divides the abdomen, to a greater or lesser extent, into a dorsal pericardial sinus and

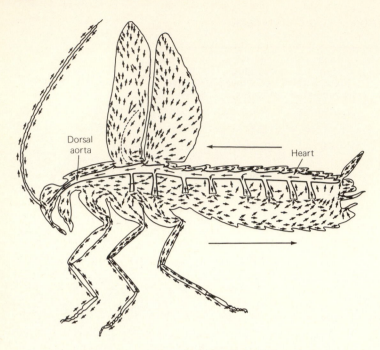

Dorsal aorta

Heart

Figure 14.27 Diagrammatic representation of blood flow in an insect. [From J.C. Jones, The circulatory system of insects. In M. Rockstein (ed.), *The Physiology of Insecta,* Vol. III. Academic, New York, 1964.]

ventral perivisceral sinus. This septum may be attached to the heart, extending laterally to the body wall on each side, or it may be separate, running across the abdomen underneath the heart and being attached to the heart by threads of connective tissue. The ostia usually open into the pericardial sinus in either case. The hearts of many insects are myogenic; others are neurogenic.

Refilling of the heart is accomplished much as in crustaceans. Cardiac contraction stretches the elastic elements to which the heart is attached, and rebound of these elements pulls the heart wall out again during diastole, causing aspiration of blood into the heart from the pericardial sinus through the ostia. Blood enters the pericardial sinus from the perivisceral sinus beneath by flowing around the posterior margin of the intervening septum or through holes in the septum.

The heart typically contracts in a peristaltic wave from posterior to anterior, forcing blood into the dorsal aorta and, in many species, into segmentally arranged lateral vessels (Figure 14.27). The dorsal aorta is itself contractile and commonly continues the peristaltic wave of the heart. The lateral vessels, if present, may also be vigorously contractile. The major vessels end abruptly with little branching, discharging blood directly to the lacunar circulation. The dorsal aorta, for example, runs to the head, where it sends off, at most, only a few branches before terminating. Once blood has exited the vessels, it is left to pass through lacunae and sinuses, ultimately to arrive in the perivisceral sinus, from which it returns to the pericardial sinus and heart. Figure 14.27 portrays the general course of the circulation.

Accessory pulsatile structures are widespread in insects, where they commonly serve to enhance the flow of blood through appendages: antennae, legs, and wings. Such pulsatile structures are sometimes positioned at the bases of the antennae or wings and can also occur within the wings or legs.

The state of the circulation in insects appears to offer important lessons. Many insects are active animals with relatively high oxygen demands. During sustained flight, strong fliers attain weight-specific rates of oxygen consumption that not only are many times higher than any reported for crustaceans but are among the highest in the entire animal kingdom. Within other phyla (notably chordates and mollusks), we are accustomed to finding that groups with a capacity for relatively intense aerobic catabolism show clear circulatory refinements in comparison to groups with a lower capacity for aerobic catabolism (compare mammals and fish, for example). This correlation between the metabolic and circulatory physiology of the animals is understandable because of the crucial importance of circulatory transport in meeting the oxygen demands of the tissues. When we compare insects and crustaceans, however, we find that the circulatory system in insects is not the more advanced either anatomically or physiologically. Indeed, in certain anatomical respects, the insect system appears less advanced: Witness the far less-extensive development of the arterial vasculature and the complete lack of capillary beds. Of course, it is highly relevant that the tracheal respiratory system of insects itself supplies oxygen to all tissues. The insects indicate then that the circulatory system can remain relatively simple in even highly active animals provided it is relieved of the burden of oxygen transport and needs only accomplish tasks of a less-pressing sort. In turn, the state of the circulation in insects supports the contention that selection for rapid respiratory-gas transport has been the primary force in the evolution of refined circulatory systems in other phyla (p. 334).

CIRCULATION IN ARACHNIDS

The arachnid heart resembles the insect heart in its basic form, being tubular, perforated by ostia, and suspended by elastic ligaments within a dorsal pericardial sinus. Typically, both posterior and anterior aortae emanate from the heart. The arachnids have received but little attention from circulatory physiologists, which is regrettable because some members of the group respire by book lungs whereas others are entirely tracheate, and there is thus opportunity for more or less direct comparative study of the function of circulatory systems that are responsible for respiratory gas transport and ones that are not. Notably, morphological investigations reveal that the spiders that have book lungs and the scorpions (all of which rely entirely on book lungs) possess substantially more-extensive arterial systems than insects.

In arachnids with book lungs, the lungs are characteristically drained by discrete vessels, the branchiopericardial or pulmonary veins, which lead directly to the pericardial sinus. The pattern of circulation in these animals is thus reminiscent of that in crustaceans: After the blood has been discharged to the lacunar circulation and made its way to ventral sinuses, it follows a channelized path back to the heart via the book lungs and pulmonary veins. Blood from the book lungs may be the only blood to enter the pericardial sinus, in which case the heart pumps only freshly oxygenated blood, or there may also be opportunity for direct entry of blood from systemic sinuses.

An interesting attribute of spiders is that extensor muscles are poorly developed or lacking at certain of the major joints of their walking legs. While flexion of their legs is accomplished using flexor muscles, extension of the legs is carried out hydraulically under force of elevated blood pressures. The blood of spiders therefore plays dual roles as a transport medium and hydraulic fluid.

The legs of spiders are attached to the prosoma (cephalothorax) and receive arteries (pedal arteries) from the heart, which is located in the abdomen. Studies of the tarantula *Dugesiella hentzi,* as well as other species, have revealed that when the animals walk, both their prosomal and abdominal cavities are compressed by contractions of somatic muscles. This compression, in *Dugesiella,* raises the prosomal background pressure to high levels: 40–60 mm Hg. Furthermore, superimposed on this background pressure, relatively modest additional pressures may be developed in the pedal arteries by the beating of the heart. The legs are extended under force of these high pressures whenever their flexor muscles are re-

laxed. When *Dugesiella* stops walking, it reduces its prosomal background pressure dramatically. Studies of fully quiescent animals have shown that pressures in the heart can be as low as 8–12 mm Hg, and pressure gradients of only a few mm Hg suffice to drive blood through the body.

CIRCULATION IN MOLLUSKS

Bivalves and Gastropods

The heart in bivalves and gastropods is myogenic and multichambered. It consists of a single, relatively muscular ventricle and one or two less-muscular atria that discharge into the ventricle. The atria receive blood from the gills, and usually there is a one-to-one relation between gills and atria; bivalves, for example, have two gills and correspondingly two atria. The ventricle gives rise to an anterior aorta and not uncommonly to a posterior aorta as well. The heart is enclosed in a pericardial cavity by a thin membrane, the pericardium. As in vertebrates, there are no direct connections between the heart lumen and pericardial cavity, and the pericardial cavity does not receive blood vessels. Usually in bivalves and occasionally in gastropods, the gut passes through the ventricle of the heart; the significance of this seemingly bizarre arrangement is unknown. The heart is positioned dorsally in the body with the atria in more or less close juxtaposition to the bases of the gills. Figure 14.28 depicts the central vascular system in a typical bivalve.

The arterial system is often extensive, branching throughout the body. Nonetheless, in bivalves and

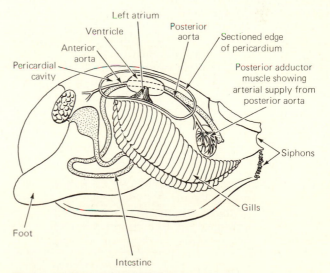

Figure 14.28 A freshwater mussel (*Anodonta*) with the left shell removed, semidiagrammatic. The pericardium has been sectioned to reveal the heart and aortae. The two atria (of which only the left is shown) are positioned laterally on either side of the ventricle. The gut passes through the ventricle. [After T.J. Parker and W.A. Haswell, *A Text-book of Zoology,* Vol. I, 6th ed. (as revised by O. Lowenstein). Macmillan and Company, Ltd., London, 1940.]

at least the nonpulmonate gastropods, the circulatory system is open, and the blood is ultimately discharged from the arteries into lacunar networks or into capillary or capillarylike beds that lead to lacunae. From the lacunar circulation, the blood enters a system of sinus thoroughfare channels, which sooner or later lead to discrete, lined vessels that are properly called veins. One of the striking commonalities in the circulatory plan of these animals is a tendency, then, for the blood to be returned to the heart via the kidneys and gills. In the simplest form of this path, blood from all the major regions of the body converges on vessels that lead to extensive lacunar networks in the walls of the kidneys. Then, after perfusing these networks, the blood is collected into afferent branchial vessels, passes through lacunae or capillaries in the gills, and travels to the atria of the heart via efferent branchial vessels. Thus, the gills are positioned just upstream from the heart. Many departures from this simple pathway are known. Sometimes, for example, part of the outflow from the kidneys is carried directly to the heart, bypassing the gills. And sometimes blood from the mantle gains direct access to the heart; this blood may be freshly oxygenated, because the mantle surfaces are possible sites of gas exchange with the ambient water.

In the terrestrial pulmonate snails, a lung derived from the mantle has become the principal site of external respiration (p. 262). In these gastropods, blood from the systemic sinuses is collected in vessels that supply the heavily vascularized mantle wall just prior to its return to the heart. The circulatory system may be more closed in pulmonates than in other gastropods.

The bivalves and nonpulmonate gastropods are typically characterized by low-pressure circulatory systems. The total pressure gradient between the major arteries leaving their heart and the veins entering it is just 4 mm Hg or less (Table 14.3).

An important recent discovery is that the circulatory system of mussels (*Mytilus* and *Geukensia*) plays little role in oxygen transport (Table 14.4). The mussels are constructed in such a way that they do not have many massive body parts in which cells are far removed from the body surface; instead, they consist substantially of thin sheets and cords of tissue, which are exposed to the stream of water they pump through their mantle cavity (p. 262). Evidently, much of their tissue obtains its oxygen directly by diffusion from one body surface or another. We know little about how widespread this pattern might be. The blood transports one-third of all oxygen consumed in certain scallops and could conceivably play an even larger role in oxygen delivery in some of the thick-bodied bivalves. A limit is placed on the role the blood can play in oxygen transport in bivalves by the fact that most species lack oxygen-transport pigments in their blood (p. 302). Gastropods commonly possess oxygen-transport pigments, and there is evidence for extensive delivery of oxygen by the circulation in some species.

Besides its transport functions, the blood frequently assumes hydraulic functions in bivalves and gastropods. Clams, for example, dig by hydraulically extending their foot into the sand or mud, anchoring its tip, and then contracting the foot, pulling their body toward the tip; by repetitions of this process, some species move through the substrate at amazingly rapid rates, as any weekend clam digger knows. Hydraulic displacements of blood are also intrumental in the extension of the siphons in clams and of the foot and penis in gastropods.

Cephalopods

Not only do the cephalopods include the most active of mollusks, but also some cephalopods rank among the most active of aquatic animals in general. The giant squids are the largest of all invertebrates. The circulatory system in cephalopods differs from that in most other mollusks in truly remarkable ways that seem clearly to be adaptive to meeting their heightened demands for internal transport, especially respiratory-gas transport. Our focus here shall be on the squids and octopuses.

The arterial and venous systems in squids and octopuses are very extensive and are joined in large part by capillary beds, giving the circulation a closed character. This is a most remarkable development in a phylum whose other major groups mostly have typical open circulatory systems. Some large blood sinuses are present in cephalopods, but they are lined, and the blood is confined to the circulatory system, distinct from intercellular tissue fluid. The basic circulatory plan remains typically molluscan: Blood enters the systemic heart from the gills, is pumped out to the body, and on its return to the heart passes first through the kidneys and then through the gills. The systemic heart consists of a powerful, muscular chamber that is valved at its inflows and outflows. The heart receives blood from two efferent branchial vessels, as shown in Figure 14.29 (sometimes the parts of these vessels nearest

Table 14.3 AVERAGE PRESSURES MEASURED IN THE AORTA AND IN THE VEINS LEADING TO THE ATRIA OF THE HEART (EFFERENT GILL VEINS) IN THE GASTROPOD MOLLUSK *Haliotis corrugata*: THE PINK ABALONE[a]

Location	Pressure (mm Hg)		
	Mean	Systolic	Diastolic
Aorta	5.4	6.5	4.3
Efferent gill veins[b]	1.4	1.5	0.8

[a] Of the total pressure drop from aorta to veins, 60 percent was incurred across the systemic circulation and 40 percent across the gills.

[b] The efferent gill veins lead away from the gills to the atria of the heart.

Source: G.B. Bourne and J.R. Redmond, *J. Exp. Zool.* **200**:9–16 (1977).

Table 14.4 OXYGEN CONSUMPTION IN THE MUSSEL *Mytilus edulis* AS A FUNCTION OF THE RATE OF PERFUSION OF THE CIRCULATORY SYSTEM AT VARIOUS ENVIRONMENTAL OXYGEN LEVELS[a]

Environmental oxygen tension (mm Hg)	Oxygen consumption (µL/g dry wt·h) at stated perfusion rate (mL/min)				
	0.0 mL/min	0.28 mL/min	0.73 mL/min	1.02 mL/min	1.36 mL/min
133	553	542	521	541	567
94	518	510	509	497	516
47	430	406	414	436	452
4	127	125	57	102	111

[a] The circulatory system was artificially perfused at blood flow rates near normal, 0.3–0.7 mL/min, and at rates well below and well above normal. The data show that oxygen uptake is virtually independent of circulatory rate, indicating that the tissues are not dependent on the circulation to acquire oxygen.

Source: P. Famme, *Comp. Biochem. Physiol.* **69A**:243–247 (1981).

the heart are called atria). It pumps blood into three major arteries (aortae).

As shown in Figure 14.29, blood returning from the systemic tissues in the major veins is split into two symmetrical paths to be directed to the gills. Much of the returning blood flows in the two lateral vena cavae through the kidneys before arriving at the bases of the gills, as described in Chapter 10 (Figure 10.17). Near the base of each gill is a bulbous accessory pulsatile organ, the branchial heart, which receives the systemic venous blood. This heart pumps the blood into an afferent branchial vessel, from which it passes through capillaries in the gill to arrive in the efferent branchial vessel and return to the systemic heart. The arrangement of hearts, you will note, is much "closer to home" than appears at first, being similar to that in mammals except that the respiratory pumps are anatomically separate from the systemic pump (compare Figure 14.12*B*).

Physiologically, the circulatory system of squids and octopuses resembles that of vertebrates far more than it does that of the other molluscan groups. To illustrate, let us look principally at data gathered on unrestrained *Octopus dofleini*, which are reasonably representative of other species that have been examined. The blood volume of *Octopus* is relatively small, about 6 percent of body weight: well within the range for vertebrates but radically lower than in noncephalopod mollusks (35–80 percent; p. 361). The systemic heart establishes a relatively large pressure gradient across the systemic circuit. In *O. dofleini,* the systolic pressure in the heart is approximately 33–52 mm Hg; the diastolic pressure is about 15 mm Hg lower. The pressure of the blood has fallen to 0–12 mm Hg by the time the blood has passed through the systemic tissues and reached the great systemic veins. In the highly active squid *Loligo peali,* a drop in pressure from a mean of 31 mm Hg in the anterior aorta to 2 mm Hg in the vena cava has been observed. The systemic pressure gradient in cephalopods is thus seen to be within the range for teleost fish (Table 14.2). The cardiac output of resting or mildly active *O. dofleini* is typically around 10–20 mL/kg·min; accordingly, it is in the lower end of the range for fish (e.g., Table 14.2). The cardiac output, considered together with the systemic pressure drop, indicates that the peripheral resistance of the systemic circuit is relatively high, as would be expected

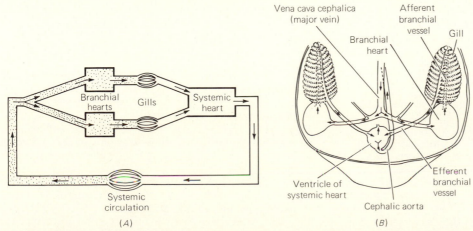

Figure 14.29 (*A*) Schematic representation of the circulatory plan in squids and octopuses. Stippled parts carry relatively deoxygenated blood. (*B*) Diagram of the major features of the central cardiovascular system in an octopus. [Part (*B*) from K. Johansen and C. Lenfant, *Am. J. Physiol.* **210**:910–918 (1966).]

from the closed character of the circulation. Overall, therefore, we find that the cephalopods resemble vertebrates in that they achieve reasonably rapid rates of flow through a relatively high-resistance systemic circuit by virtue of maintaining relatively high gradients of pressure. The branchial hearts of octopuses and squids boost the pressure of the blood at least modestly prior to its passage through the gills. The head of pressure produced by these hearts is important not only for perfusing the branchial circuit but also for producing ultrafiltration into the kidneys (p. 235).

The arteries and veins of cephalopods are muscular. Some veins undergo peristaltic contractions, which probably are important in moving the blood. Furthermore, the vascular muscles, which are richly innervated, are likely important in controlling blood distribution. These are matters needing more investigation.

In comparing the circulation of squids and octopuses with that in vertebrates, a striking degree of evolutionary convergence is evident. There is perhaps no more-compelling illustration of the interplay between the evolution of the circulation and that of a metabolically demanding way of life.

SELECTED READINGS

Anderson, M. and I.M. Cooke. 1971. Neural activation of the heart of the lobster *Homarus americanus*. *J. Exp. Biol*. **55**:449–468.

Belman, B.W. 1975. Some aspects of the circulatory physiology of the spiny lobster *Panulirus interruptus*. *Mar. Biol*. **29**:295–305.

Booth, C.E. and C.P. Mangum. 1978. Oxygen uptake and transport in the lamellibranch mollusc *Modiolus demissus*. *Physiol. Zool*. **51**:17–32.

Bourne, G.B. and J.R. Redmond. 1977. Hemodynamics in the pink abalone, *Haliotis corrugata* (Mollusca, Gastropoda). I. Pressure relations and pressure gradients in intact animals. *J. Exp. Zool*. **200**:9–16.

Boutilier, R.G., M.L. Glass, and N. Heisler. 1986. The relative distribution of pulmocutaneous blood flow in *Rana catesbeiana*: Effects of pulmonary or cutaneous hypoxia. *J. Exp. Biol*. **126**:33–39.

Burggren, W.W. 1982. Pulmonary blood plasma filtration in reptiles: A "wet" vertebrate lung? *Science* **215**:77–78.

*Burggren, W.W. 1987. Form and function in reptilian circulations. *Am. Zool*. **27**:5–19.

Burggren, W. and K. Johansen. 1982. Ventricular haemodynamics in the monitor lizard *Varanus exanthematicus*: Pulmonary and systemic pressure separation. *J. Exp. Biol*. **96**:343–354.

Burggren, W.W. and K. Johansen. 1986. Circulation and respiration in lungfishes (Dipnoi). *J. Morphol. Suppl*. **1**:217–236.

*Burton, A.C. 1972. *Physiology and Biophysics of the Circulation*, 2nd ed. Year Book Medical, Chicago.

Calder, W.A., III. 1981. Scaling of physiological processes in homeothermic animals. *Annu. Rev. Physiol*. **43**:301–322.

Cardiovascular Adaptation in Reptiles. (A Symposium.) 1987. *Am. Zool*. **27**:3–131.

Crone, C. 1981. Tight and leaky endothelia. In H.H. Ussing, N. Bindslev, N.A. Lassen, and O. Sten-Knudsen (eds.), *Water Transport Across Epithelia*, pp. 258–267. Munksgaard, Copenhagen.

Fretter, V. and A. Graham. 1962. *British Prosobranch Molluscs*. The Ray Society, London.

Functional Morphology of the Heart of Vertebrates. (A Symposium.) 1968. *Am. Zool*. **8**:177–229.

Greenberg, M.J. (ed.). 1979. Discussion: Common themes in cardiac control. *Am. Zool*. **19**:175–190.

*Grigg, G.C. and K. Johansen. 1987. Cardiovascular dynamics in *Crocodylus porosus* breathing air and during voluntary aerobic dives. *J. Comp. Physiol*. **157B**:381–392.

*Hartline, D.K. 1979. Integrative neurophysiology of the lobster cardiac ganglion. *Am. Zool*. **19**:53–65.

Hoppeler, H., O. Mathieu, E.R. Weibel, R. Krauer, S.L. Lindstedt, and C.R. Taylor. 1981. Design of the mammalian respiratory system. VII. Capillaries in skeletal muscles. *Respir. Physiol*. **44**:129–150.

Hudlická, O. 1985. Development and adaptability of microvasculature in skeletal muscle. *J. Exp. Biol*. **115**:215–228.

Johansen, K. 1970. Air breathing in fishes. In W.S. Hoar and D.J. Randall (eds.), *Fish Physiology*, Vol. IV, pp. 361–411. Academic, New York.

*Johansen, K. 1972. Heart and circulation in gill, skin and lung breathing. *Respir. Physiol*. **14**:193–210.

*Johansen, K. and W. Burggren. 1980. Cardiovascular functions in the lower vertebrates. In G.H. Bourne (ed.), *Hearts and Heart-like Organs*, Vol. 1, pp. 61–117. Academic, New York.

*Johansen, K. and W.W. Burggren (eds.). 1985. *Cardiovascular Shunts*. Munksgaard, Copenhagen.

Johansen, K., C. Lenfant, and D. Hanson. 1968. Cardiovascular dynamics in the lungfishes. *Z. vergl. Physiol*. **59**:157–186.

Johansen, K. and A.W. Martin. 1962. Circulation in the cephalopod, *Octopus dofleini*. *Comp. Biochem. Physiol*. **5**:161–176.

Johansen, K. and A.W. Martin. 1965. Circulation in a giant earthworm, *Glossoscolex giganteus*. *J. Exp. Biol*. **43**:333–347.

Johansen, K. and A.W. Martin. 1965. Comparative aspects of cardiovascular function in vertebrates. In W.F. Hamilton (ed.), *Handbook of Physiology. Section 2: Circulation*, Vol. III. American Physiological Society, Washington, DC.

Jones, D.R. and D.J. Randall. 1978. The respiratory and circulatory systems during exercise. In W.S. Hoar and D.J. Randall (eds.), *Fish Physiology*, Vol. VII, pp. 425–501. Academic, New York.

Jones, H.D. and D. Peggs. 1983. Hydrostatic and osmotic pressures in the heart and pericardium of *Mya arenaria* and *Anodonta cygnea*. *Comp. Biochem. Physiol*. **76A**:381–385.

Kiceniuk, J.W. and D.R. Jones. 1977. The oxygen transport system in trout (*Salmo gairdneri*) during sustained exercise. *J. Exp. Biol*. **69**:247–260.

Laughlin, M.H. 1987. Skeletal muscle blood flow capacity: Role of muscle pump in exercise hyperemia. *Am. J. Physiol*. **253**:H993–H1004.

Mantel, L.H. (ed.). 1983. *The Biology of Crustacea*, Vol. 5. Academic, New York. (See Chapters 1 and 6.)

*Martin, A.W. 1980. Some invertebrate myogenic hearts: The hearts of worms and molluscs. In G.H. Bourne (ed.), *Hearts and Heart-like Organs*, Vol. 1, pp. 1–39. Academic, New York.

Martin, A.W. and K. Johansen. 1965. Adaptations of the circulation in invertebrate animals. In W.F. Hamilton (ed.), *Handbook of Physiology. Section 2: Circulation*, Vol. III. American Physiological Society, Washington, DC.

Mayerson, H.S. 1963. The lymphatic system. *Sci. Am.* **208** (June):80–90.

Navaratnam, V. 1980. Anatomy of the mammalian heart. In G. Bourne (ed.), *Hearts and Heart-like Organs*, Vol. 1, pp. 349-374. Academic, New York.

*Randall, D.J., W.W. Burggren, A.P. Farrell, and M.S. Haswell. 1981. *The Evolution of Air Breathing in Vertebrates*. Cambridge University Press, New York.

Randall, D.J. and C. Daxboeck. 1982. Cardiovascular changes in the rainbow trout (*Salmo gairdneri* Richardson) during exercise. *Can. J. Zool.* **60**:1135–1140.

*Satchell, G.H. 1971. *Circulation in Fishes*. Cambridge University Press, London.

*Seymour, R.S. and K. Johansen. 1987. Blood flow uphill and downhill: Does a siphon facilitate circulation above the heart? *Comp. Biochem. Physiol.* **88A**:167–170.

Shelton, G. 1976. Gas exchange, pulmonary blood supply, and the partially divided amphibian heart. In P.S. Davies (ed.), *Perspectives in Experimental Biology*, Vol. 1, pp. 247–259. Pergamon, New York.

Stewart, D.M. and A.W. Martin. 1974. Blood pressure in the tarantula, *Dugesiella hentzi*. *J. Comp. Physiol.* **88**:141–172.

Tazawa, H., M. Mochizuki, and J. Piiper. 1979. Respiratory gas transport by the incompletely separated double circulation in the bullfrog, *Rana catesbeiana*. *Respir. Physiol.* **36**:77–95.

Trueman, E.R. and A.C. Brown. 1985. Dynamics of burrowing and pedal extension in *Donax serra* (Mollusca: Bivalvia). *J. Zool.* **207A**:345–355.

Van Citters, R.L., W.S. Kemper, and D.L. Franklin. 1968. Blood pressure responses of wild giraffes studied by radio telemetry. *Science* **152**:384–386.

Wells, M.J., G.G. Duthie, D.F. Houlihan, P.J.S. Smith, and J. Wells. 1987. Blood flow and pressure changes in exercising octopuses (*Octopus vulgaris*). *J. Exp. Biol.* **131**:175–187.

*Wiedeman, M.P., R.F. Tuma, and H.N. Mayrovitz. 1981. *An Introduction to Microcirculation*. Academic, New York.

See also references in Appendix A.

chapter 15

The Physiology of Diving in Birds and Mammals

This book is organized mostly around organ systems. That is a logical organization, and yet there is much to be said also for synthetic study of life-history phenomena. A synthetic study that cuts across many organ systems can make a unique contribution to understanding the intimate interplay of processes in the living animal. Choosing a life-history phenomenon for concerted study is difficult; flight, high-altitude life, and breeding cycles are just a few of the topics that would be rewarding. Here we look at diving in mammals and birds.* Diving involves an intriguing interplay of cellular catabolism, respiration, blood-gas transport, acid–base physiology, circulatory control, and behavior. Plus, in studying diving, you will obtain some vivid glimpses into the challenges we face when attempting to gain an accurate understanding of a natural phenomenon.

Many aquatic species of mammals and birds are capable of dives that are extraordinary by the standards of terrestrial species. These diving feats have been appreciated in a limited way for many years. For example, during the heroic era of whaling (chronicled most famously in *Moby Dick*), the whaling men became impressed with the depths to which wounded whales could dive, when sometimes a sperm whale had to be cut loose, having "sounded" so deep as to draw out two lengths of harpoon line, each over 370 m long. Only recently, however, have innovation and modern technology permitted the diving behavior of unfettered animals to be described in detail.

One species to have received thorough study is the Weddell seal (*Leptonychotes weddelli*), a large pinniped (400 kg) found principally along the coasts of Antarctica. By outfitting free-ranging animals with

* Diving in other groups of vertebrates is discussed on pp. 50–52, and that in insects is treated on pp. 293–295.

unobtrusive, self-powered devices that record the depths and durations of dives, investigators have obtained extensive, statistical descriptions of the diving the seals undertake for feeding and exploration. Figure 15.1, for example, summarizes data on the durations of over 5600 dives. A small percentage of dives lasted over an hour (the longest was 73 min), but most were considerably shorter: 20–25 min or less. Even many of the "short" dives were impressive; a 20-min dive, for instance, is astounding by human standards. Figure 15.2 summarizes data on depths attained. Whereas most dives are limited to 400 m (1312 ft) or less, the animals on rare occasions descend to near 600 m—over one-third of a mile down. A useful rule of thumb is that pressure increases by about an atmosphere for every 10 m of depth. Thus, even at 400 m, the seals expose themselves to about 40 atm.

The statistical pattern of diving behavior in other diving species is similar to that in Weddell seals in that the durations of most dives are appreciably shorter than the maximum duration of which each species is capable; dives of maximal or near-maximal duration are uncommon. Similarly, most dives are to depths substantially more shallow than the maximum a species can attain. These attributes of diving behavior will prove important later, but let us look for the moment at the maximal performances of a few additional species. The abilities of some aquatic mammals—albeit impressive—are modest by comparison to the Weddell seal's attainments. Among nearly 3000 dives recorded in northern fur seals (*Callorhinus ursinus*), for example, the longest was 5.6 min, and the deepest was 190 m. Emperor penguins (*Aptenodytes forsteri*)—which are perhaps the most accomplished of avian divers—are capable of remarkable excursions, having been observed to stay

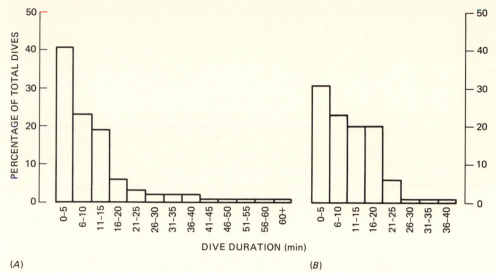

Figure 15.1 Summary of the durations of dives undertaken by unfettered Weddell seals (*Leptonychotes weddelli*). Results are expressed as the percentage of dives falling within various duration intervals. Parts (*A*) and (*B*) represent two different data sets, gathered under somewhat different conditions. Set (*A*) encompasses 1057 dives undertaken by 6 seals, whereas set (*B*) includes 4601 dives by 22 seals. [Reprinted by permission from G.L. Kooyman, E.A. Wahrenbrock, M.A. Castellini, R.W. Davis, and E.E. Sinnett, *J. Comp. Physiol.* **138B:**335–346 (1980).]

under for 18 min and to reach depths of over 265 m. The pinnacle of diving performance is attained by certain of the large whales. Sperm whales (*Physeter catodon*) occasionally remain submerged for at least an hour when unfettered and have been known to dive for at least 90 min after being harpooned. Free-ranging individuals have been detected by sonar at an astounding depth of 1140 m: 7/10 of a mile.

To put the performances of aquatic species into perspective, it is useful to look at the capabilities of a representative terrestrial species, *Homo sapiens*. Trained human divers are generally limited to about 3 min of submergence when at rest and about 90 s of submergence when swimming; some exceptional individuals can do slightly better. In Korea and Japan, there are people (mostly women) known as ama who earn their living by diving for shellfish and edible seaweeds. Their abilities are typical of those exhibited by other people (e.g., pearl and sponge divers) who have adopted diving as a career. The ama who dive in deep waters receive assistance during descent and ascent so their time at the bottom can be maximized. They carry weights to aid descent and are pulled back to the water's surface on a rope by an assistant in a boat. They routinely remain submerged for 60–80 s and reach depths of 15–25 m. They rest only about a minute between dives and thus average about 30 dives/h. The extremes recorded for human performance have been attained by competitive divers. The record holders have reached depths of 70–100 m (using assistance in descent and ascent) in dives lasting about 2 min. Dives of this length and depth tax humans to their limits but are routine and unremarkable for many marine mammals.

PROLOGUE

Our understanding of the physiology of diving remains in a state of flux. For several decades, it

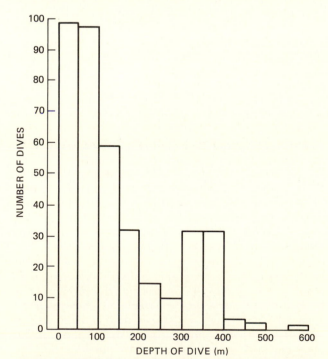

Figure 15.2 Summary of the depths of dives undertaken by unfettered Weddell seals. Data represent 381 dives undertaken by 27 individuals and are expressed as the number of dives falling within various depth intervals. [Reprinted by permission from G.L. Kooyman, *Science* **151:**1553–1554 (1966). Copyright 1966 by the American Association for the Advancement of Science.]

seemed logical to physiologists to study diving primarily in the laboratory because of the ease and technical sophistication of studies there. Most of this laboratory research was carried out on animals forcibly submerged. Commonly, for example, the animals were strapped to a movable platform and lowered under water whenever the investigator wished to elicit diving responses. From such work, a view of the physiology of diving developed that became virtually dogma by the 1960s. Then some pioneering investigators began to study animals such as Weddell seals in the wild. The techniques that could be applied in the wild were much more limited than those available in the laboratory, but quickly it became apparent that the physiology of voluntary diving in the wild sometimes differs from that of forced diving in the laboratory. At first, the differences were attributed largely to the distinction between forced and voluntary diving. More recently, however, investigators have pointed out that forced dives not only are forced but also involve little exercise, whereas voluntary dives may involve considerable swimming. The length of dives is also critical. Thus, by now, we realize we will ultimately need to understand not only the distinctions between *forced dives* (usually resting) and *voluntary dives* but also those between voluntary dives that are *short* or *long* and those that are *quiet* or *vigorously active*.

AEROBIC VERSUS ANAEROBIC METABOLISM: INTRODUCTION

Physiological studies of diving have revolved around two basic questions. One of these, that of how animals cope with high pressures at depth, is considered principally toward the end of this chapter. We turn now to the other major question: How do animals that are strictly dependent on the atmosphere for oxygen meet their metabolic energy demands during periods of submergence? It seems that most of the accomplished mammalian and avian divers resemble one another in their fundamental pattern of metabolism during dives. A review of this common pattern in this section will set the stage for a more thorough examination of its individual elements in subsequent sections.

When diving mammals and birds break contact with the atmosphere, they carry oxygen with them in three major internal stores: oxygen bound to blood hemoglobin, oxygen bound to muscle myoglobin, and oxygen contained in their lung air. These stores permit aerobic catabolism to continue to some extent during the dive.

In principle, the internal stores of oxygen are adequate to enable much or all of a diver's body to function aerobically for the duration of a relatively short dive. Available evidence increasingly indicates in fact that *voluntary* dives of *modest* duration are mostly or completely aerobic in many diving species

(i.e., there is little or no net accumulation of lactic acid during such dives).

On the other hand, the internal oxygen stores are manifestly inadequate to permit even an approximation of full aerobic function throughout the body over the course of protracted dives. Until recently, research on the energetics of diving has been preoccupied with this compelling problem. How are some species able to survive for 20, 40, 60—or even more—minutes without breathing? Our basic concept of how energy needs are met during protracted dives was first enunciated by Laurence Irving in 1934. He and Per Scholander then performed pioneering experiments that lent substantial support to the concept. Irving recognized that certain tissues—notably the central nervous system and heart—are strongly or exclusively dependent on aerobic catabolism for production of ATP and are quickly damaged by oxygen insufficiency; whereas other tissues—such as skeletal muscle—have a well-developed ability to meet their ATP demands anaerobically and thus are relatively tolerant of oxygen deprivation.* He then reasoned that dives could be prolonged if animals were to "reserve" certain of their oxygen supplies for the tissues that are dependent on oxygen. During a dive, if all tissues simply have equal access to all available oxygen, oxygen levels throughout the body will fall relatively quickly to a point that will impair the oxygen-dependent tissues; the animal will then be forced to surface even though many of its tissues could continue to function—anaerobically—for quite some time. On the other hand, if some oxygen is reserved for use by the oxygen-dependent tissues during a dive, then it will be possible for those tissues to continue to have adequate oxygen even while other parts of the body exhaust their oxygen supplies and resort to anaerobiosis—thus extending the time the animal can remain submerged.

The preferential distribution of some oxygen to the oxygen-dependent tissues is achieved, as Irving predicted, by adjustments of circulatory function. These adjustments were first elucidated in laboratory experiments. During forced dives, circulation to the appendages, trunk muscles, gut, kidneys, and certain other tissues is substantially curtailed by vasoconstriction in the vessels supplying those parts. The skeletal muscles, deprived of active blood flow, can make use of the hemoglobin-bound oxygen sequestered in their capillaries and of myoglobin-bound oxygen, but as these stores are depleted, they resort to anaerobic catabolism, and lactic acid accumulates in quantity. With the circulation to many parts of the body curtailed, the heart primarily pumps blood between itself and the lungs and head. The oxygen stores of this blood are thereby reserved primarily for the oxygen-dependent tissues, and whatever oxygen is extracted from the air in the lungs is likewise

* See Chapter 4 for a full discussion of the aerobic and anaerobic mechanisms of producing ATP.

delivered preferentially to these tissues. Consequently, adequate oxygen tensions can be maintained for a long period. Because circulation to the skeletal muscles is strongly limited during diving, lactic acid produced by the muscles tends to remain sequestered in the muscles during the dive. Once the animal surfaces, however, circulation to the muscles is restored, and there is a sudden rise of lactic acid in the circulating blood. The observation that circulating lactic acid increases principally *after* a forced dive was one of the early pieces of evidence that circulatory function is radically altered during submergence.

When animals are forcibly submerged, the adjustments in the pattern of blood flow just described tend to occur rapidly, consistently, and to a profound extent. Accordingly, during the era when diving was studied mainly in the laboratory, these responses came to be considered a "diving reflex." Now we realize from studies of voluntarily diving animals that the responses of the circulatory system are not nearly as inflexible and stereotyped as once was thought. Indeed, as suggested earlier, there is good reason to believe that, in certain species, relatively little redistribution of blood flow occurs during voluntary dives that are short enough for all catabolic needs to be met aerobically using oxygen stores. During protracted voluntary dives, a profound redistribution of blood flow does occur. It seems likely that forcibly submerged animals routinely exhibit a profound redistribution because they cannot predict with certainty how long they will be unable to breathe and thus consistently marshall responses that will maximize their underwater endurance.

THE OXYGEN STORES OF DIVERS

A critical question is whether diving species have greater capacities for oxygen storage than nondiving ones. In the discussion of this topic—as well as elsewhere in the chapter—it will be important to remember that different species of diving mammals and birds have different life histories and have been subject to different evolutionary selective pressures. Thus, we should not be surprised by the physiological diversity that can exist from species to species. Our emphasis will be on mammals.

The Blood Oxygen Store

The amount of oxygen available as oxyhemoglobin depends on three parameters: (1) the peak capacity of each unit volume of blood to bind with oxygen, termed the oxygen-carrying capacity (see p. 313), (2) the total volume of blood, and (3) the percentage to which the blood is fully loaded (saturated) with oxygen at the time of submergence. Whereas the former two parameters can be measured rather easily, the latter cannot and consequently remains poorly understood.

Some species of diving mammals and birds have exceptionally high oxygen-carrying capacities, but others have capacities that are well within the ordinary range for their terrestrial relatives. A high carrying capacity cannot be called typical of diving species. Bottle-nosed dolphins, northern fur seals, Steller sea lions, and sea otters, to cite some examples, have carrying capacities of 16.5–22 volumes percent—quite ordinary for mammals. On the other hand, species of diving mammals with especially high carrying capacities include the pygmy sperm whale (32 volumes percent), harbor seal (26–29 volumes percent), Weddell seal (29–36 volumes percent), and ribbon seal (34 volumes percent).* Among pinnipeds, a tendency is evident for species that undergo long dives—such as the three seals just mentioned—to have higher carrying capacities than those that perform shorter dives (e.g., northern fur seals).

Accurate determinations of blood volume have been performed on only a relatively few species of diving birds and mammals, but the evidence indicates a tendency toward high values. Humans, dogs, horses, and rabbits have average blood volumes of about 60–110 mL/kg of body weight. By contrast, such accomplished divers as the harbor seal, ribbon seal, and Weddell seal have volumes of 130–160 mL/kg. The northern fur seal, Steller sea lion, and sea otter have more-modest values of 90–110 mL/kg, but these are still relatively high.

By multiplying the blood volume by the oxygen-carrying capacity of the blood, the maximum possible oxyhemoglobin store is obtained. This figure is only tangentially relevant to normal physiology because the entire volume of blood is never fully oxygenated. However, it can be useful for interspecific comparisons: If two species are assumed to be similar in the average percentage saturation of their blood with oxygen at the time of submergence, then the one with the higher *maximal* oxyhemoglobin store will also—and in the same proportion—carry a higher *actual* oxyhemoglobin store during dives.

In diving mammals and birds, the maximal capacity to store oxygen as oxyhemoglobin tends to be high. Humans and horses have peak oxyhemoglobin storage capacities of about 14–15 mL O_2/kg of body weight. By contrast, very much higher storage capacities are found in those seals that combine the advantages of both high blood volume and high carrying capacity. Average reported values for the Weddell seal, harbor seal, and ribbon seal range between 39 and 47 mL O_2/kg—as much as three times higher than in humans and horses. In the northern fur seal and Steller sea lion, which have only modest carrying capacities but have blood volumes toward the high

---

* Recent studies indicate that in some seals the carrying capacity (and blood volume) may be lower between dives than during them (see p. 378). Values given here are believed to apply to animals during dives.

end of the range for terrestrial mammals, the maximum oxyhemoglobin storage capacity is 16–21 mL O_2/kg.

Oxymyoglobin

In oxymyoglobin, the muscles have a substantially private store of oxygen. Myoglobins have so much greater affinities for oxygen than hemoglobins that oxymyoglobin in the muscles will not yield appreciable amounts of oxygen to circulating blood for transport elsewhere until the oxygen tension in the circulating blood has itself fallen threateningly low (see p. 318). On the other hand, in skeletal muscles that are deprived of blood flow, oxymyoglobin will release its oxygen as local oxygen tensions plummet toward zero, thus staving off the time that the muscles will be forced to turn fully to anaerobic catabolism.

The amount of oxygen stored as oxymyoglobin at the time of submergence depends largely on how much myoglobin is present in each unit weight of muscle tissue. In the skeletal muscles of diving mammals and birds, myoglobin concentrations tend to be high—so high that the muscles of some proficient divers are reported to appear almost black. In humans and horses, skeletal muscles contain about 4–9 mg of myoglobin per gram of wet weight. By contrast, harbor seals have 55 mg/g, and ribbon seals have about 80 mg/g. Even such less-effective divers as the northern fur seal, Steller sea lion, and sea otter have 24–35 mg/g.

The Pulmonary Oxygen Store

At first it might appear that a large store of oxygen in the lungs would be of unquestioned advantage for diving mammals and birds. There are a number of considerations, however, that can militate against this conclusion. First, the density of air is so much lower than that of water that air contained in the lungs can strongly buoy the diving animal upward, thus increasing the effort of remaining submerged. Problems with the buoyant effect of air, if they occur, are particularly severe at shallow depths. They diminish as animals swim deeper because with increasing depth, the increasing ambient pressure compresses the air in the lungs into a smaller and smaller volume. This compression of the lung air at depth leads to another consideration. The alveoli, it is believed, are typically the first parts of a mammal's lungs to collapse as the air volume is reduced. That is, at depth the lung air comes to be contained mostly in the conducting airways—the trachea, bronchi, and nonrespiratory bronchioles (p. 278). Once this occurs, the oxygen in the lung air becomes largely unavailable, for the alveoli are the sites of oxygen transfer from the lung air to the blood. Thus, although diving mammals can make ready use of their pulmonary oxygen store during shallow diving, they

are not able to do so at depth, and in many species this again calls into question the utility of carrying a large amount of air in the lungs. A final concern is that a large pulmonary air store implies not only a large oxygen store but a fourfold greater nitrogen store; a large nitrogen store can be a disadvantage in the context of avoiding the bends, as discussed later (p. 381). With all these considerations, it is not surprising to find quite a bit of variation in the use of the lungs as an oxygen store in diving animals.

Lung capacity is defined to be the amount of air contained in the lungs at full inflation. Comparing the lung capacities of diving and terrestrial mammals is complicated by the fact that many of the diving species have exceptional amounts of body fat (e.g., blubber). Given this disparity in fat content, should we make comparisons by expressing lung capacities in relation to *total* body weight or *lean* weight? Depending on the approach taken, the lungs of diving species turn out generally to be similar in capacity to those of terrestrial species or just somewhat larger. One diving species, the sea otter, does have indisputably large lungs for its body size. There is good reason to believe, however, that sea otters have evolved these large lungs not as an adaptation for carrying lots of oxygen during dives but as one for providing lots of buoyancy between dives. Sea otters have little blubber and yet require considerable buoyancy because during feeding one of their means of opening shellfish and other prey is to float on their back, place a rock on their abdomen, and crack the prey against the rock.

The size of the pulmonary oxygen store carried upon submergence depends not only on the lung capacity but also on the extent to which the lungs are inflated. Some diving mammals dive following inhalation and thus may carry something approaching their full lung capacity. This appears to be true, for example, of sea lions, dolphins, and whales, as well as penguins. By contrast, deep-diving pinnipeds dive following exhalation; in a number of seals, the lungs are filled to just 20–60 percent of capacity upon submergence. Lung capacity itself is lower in many of the deep-diving seals and whales than in mammals of similar size that dive more shallowly; indeed, some of the deep-diving whales (e.g., fin whales) have exceptionally small lungs for their size. In all, it appears that little premium has been placed on carrying a large pulmonary oxygen store in many deep divers. It may be that they habitually dive to such great depths that alveolar collapse under the force of hydrostatic pressure largely voids the utility of a pulmonary oxygen store (see also p. 383).

To what extent do diving animals actually draw on the pulmonary oxygen they carry? This is a matter that is only partially understood, but clearly the answer depends not only on the length of a dive but also—because of the possibility of alveolar collapse—on the depth. One interesting experiment on the effects of depth was carried out on a trained male bottle-nosed dolphin (*Tursiops truncatus*) named Tuffy. When Tuffy held his breath quietly just below the water's surface, he would make thorough use of his pulmonary oxygen, reducing the oxygen concentration in his lung air from about 13 to 3 percent in 3–4 min. To dive to 300 m and return to the surface, he required this same length of time and more physical effort. Yet, during such dives, instead of reducing his pulmonary oxygen level to 3 percent, he reduced it just to about 5 percent. At least part of the explanation for his lessened use of pulmonary oxygen during deep dives appeared to lie in alveolar collapse: Although he could acquire oxygen from his lungs while traveling between the surface and about 100 m, at greater depths alveolar collapse was sufficient to prevent much oxygen uptake. Species differ in the depth at which oxygen uptake is blocked by alveolar collapse, but the basic phenomenon is widespread among deep divers.

Conclusions

Three major points emerge from the study of oxygen stores. Two of these are illustrated in Figure 15.3; namely, (1) some diving mammals have much greater oxygen stores per unit of body weight than terrestrial mammals, and (2) among the divers, less-accomplished species such as fur seals and sea lions tend to have smaller stores per unit of body weight than highly proficient species such as harbor and ribbon seals.

The third major point is that the oxygen stores of diving mammals and birds, despite their magnitude, are entirely inadequate to sustain a rate of oxygen consumption during maximally long submergence equivalent to that observed in these animals while they are at rest and breathing air. For example, Gerald Kooyman has estimated the total oxygen stores of a 450-kg Weddell seal to be 30 L. The resting rate of oxygen consumption of such a seal while breathing air is 1900–2300 mL/min. Thus, the seal could sustain its resting, aerial rate of oxygen consumption for 13–16 min during submergence if it could completely use all stores. Actually, as shown in Figure 15.1, Weddell seals sometimes remain under water for over an hour. Similar calculations have been performed for many other diving species, and even with generous assumptions about the magnitude and utilization of stores, dives of maximal length turn out to last from two to several times longer than would be predicted if the animal were to function aerobically at the rate seen during rest in air. Furthermore, the diving capabilities of proficient divers are disproportionate to their relative expansion of oxygen stores, as illustrated by comparing harbor seals and humans (Figure 15.3). Although the weight-specific oxygen stores of the seals are 2–2.5 times those of humans, the seals can remain submerged for over 12 times as long as humans when diving under comparable conditions.

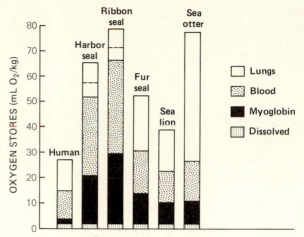

Figure 15.3 Total estimated oxygen stores of six species: human, harbor seal (*Phoca vitulina*), ribbon seal (*Histriophoca fasciata*), northern fur seal (*Callorhinus ursinus*), Steller sea lion (*Eumetopias jubata*), and sea otter (*Enhydra lutris*). Oxygen dissolved in tissues and body fluids other than blood was estimated as 2 mL/kg for all species. Oxygen present as oxymyoglobin was calculated from known myoglobin concentrations of skeletal muscle under the assumption that muscle constitutes three-tenths of body weight. Oxygen present in blood was calculated by assuming that one-third of the blood is arterial and 95 percent saturated and that two-thirds of the blood is venous and at 5 volumes percent lower oxygen content than arterial blood. Lung oxygen stores were calculated by assuming that the lungs are fully inflated and contain 15 percent oxygen by volume. The assumption of full inflation is probably reasonably realistic for humans and the sea lion, which dive on inspiration. The ribbon seal and harbor seal dive on expiration, and dashed lines show lung stores assuming the lungs to be 40 percent inflated. The sea otter also dives after partial expiration, although published data do not permit estimation of its percent inflation. The state of lung inflation during diving in fur seals is unknown. [Data for marine mammals from C. Lenfant, K. Johansen, and J.D. Torrance, *Respir. Physiol.* **9**:277–286 (1970).]

CIRCULATORY ADJUSTMENTS DURING DIVES

Adjustments in the Pattern of Blood Flow

The adjustments that take place in the pattern of blood flow during forced dives have been described extensively using a great range of techniques. Among the techniques, one that has yielded particularly striking results is based on the injection of radiopaque (contrast) materials into the bloodstream. Once administered, these materials flow with the blood, and because they can be visualized by x-ray, their distribution in the circulatory system can be monitored and used to determine the pattern of flow. Figure 15.4 shows the distribution of contrast material in the arteries of the posterior trunk of a harbor seal before and after a forced dive. Note that many of the arteries that received contrast material in the nondiving situation received little or none during submergence, demonstrating that blood flow to certain body regions is profoundly curtailed during a dive. This curtailment is achieved by vasoconstriction in the blood vessels supplying the affected body regions

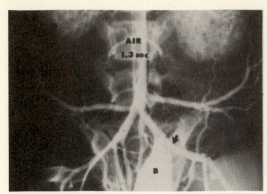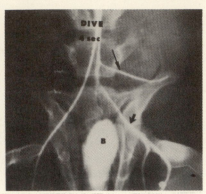

Figure 15.4 Angiograms of the posterior trunk of a harbor seal (*Phoca vitulina*) during air breathing and during forced submergence. Contrast material was injected into the aorta and arrived at the posterior trunk of the animal in the major artery positioned at the top-center of the photographs. The thin arrows point to an artery supplying the flanks, and the thick arrows point to an artery supplying the hind flippers. B marks the bladder. Note the severe restriction of peripheral blood flow during diving. Times indicate the interval between injection of contrast medium and exposure of the angiogram. The diving angiogram depicts the maximal amount of arterial filling observed during submergence. [From K.M. Bron, H.V. Murdaugh, Jr., J.E. Millen, R. Lenthall, P. Raskin, and E.D. Robin, *Science* **152**:540–543 (1966). Copyright 1966 by the American Association for the Advancement of Science.]

and is believed to be mediated at least partly by the sympathetic nervous system. Interestingly, although arterioles are the usual sites of vasomotor control in mammals and birds (p. 344), pronounced vasoconstriction occurs in sizable arteries in at least some diving species.

The parts of the body that receive little or no blood flow during forced dives commonly include the limbs (e.g., Figure 15.4), pectoral muscles, trunk muscles (e.g., Figure 15.4), skin, body wall, intestines, and kidneys. That is, flow to a considerable fraction of the body is curtailed, and the active circulation becomes substantially limited to the heart, lungs, and brain (blood travels from the right side of the heart to the lungs, then to the left side of the heart and brain, and back to the right heart). Quantitative studies of brain blood flow are in agreement that a considerable rate of flow is maintained during dives, although some studies indicate a decrease in brain blood flow relative to periods of air breathing, others an increase, and others no change. The rate of flow to the heart muscle (myocardium) is discussed in the next section.

We have little direct knowledge of patterns of blood flow during voluntary dives in the wild. However, new technical developments have just recently started to provide information. In Weddell seals, studies of the rates at which tracer chemicals disappear from the blood indicate that during *prolonged* voluntary dives, blood flow to much of the body is strongly curtailed. A flow pattern broadly similar to that in forced dives thus seems likely. During *short* voluntary dives, a more-modest curtailment of blood flow occurs, but the spatial and temporal pattern of this curtailment remains unknown. One of the biggest questions concerning voluntary dives is whether the

muscles used for swimming receive vigorous blood flow during the dives (allowing them to function aerobically but also permitting them to use up body oxygen) or whether they are denied blood flow as in forced dives. The only direct evidence comes from studies of ducks *(Aythya)*. During voluntary dives of short duration (when oxygen reserves are not greatly taxed), the swimming muscles of the ducks receive a vigorous blood flow. There is also indirect but persuasive evidence that the swimming muscles of Weddell seals function aerobically during short voluntary dives (see later), but their condition during long voluntary dives remains unknown.

Heart Rate

The earliest observation of cardiovascular involvement in diving physiology was the demonstration by Paul Bert in 1870 that the heart rate of ducks decreased from about 100 to 14 beats/min during forced dives. Such a slowing of the heart rate is termed *diving bradycardia*. Since Bert's time, research on other species has revealed that profound bradycardia is a general avian and mammalian response to forced or protracted submergence.

The existence of bradycardia during forced dives was established long before its significance became clear. Not until the 1930s and 1940s did it become known that blood flow to many regions of the body is curtailed during dives. Yet until that curtailment was understood, bradycardia could not be appreciated for what it is: simply one part of an integrated, far-flung reorganization of cardiovascular function.

Consider that during a forced or protracted dive the dimensions of the active circulatory path are greatly reduced by virtue of the peripheral vasoconstriction we have described. Within the vascular beds that *are* actively circulated (e.g., brain), an approximately normal rate of blood flow can be maintained with a much lower rate of blood output from the heart than is required to maintain normal flow throughout the entire body. Bradycardia represents a mechanism of reducing the heart's output—that is, of attuning cardiac output to the dimensions of the vascular path requiring perfusion. Studies of several species of birds and mammals have indicated that the stroke volume (the amount of blood pumped per ventricular beat) changes only modestly, if at all, during diving. Thus, in fact, cardiac output declines during a dive roughly in proportion to the decline in heart rate. Studies of blood pressure have demonstrated in a particularly direct way that the drop in cardiac output is highly integrated with the reduction in the dimensions of the active circulatory path during a dive. As discussed in Chapter 14 (p. 346), the pressure in the great systemic arteries is a function of both the rate of cardiac pumping and the resistance posed by the peripheral vascular system to through-flow of blood. A change in either factor without compensatory adjustments in the other could severely perturb the arterial pressure. The widespread vasoconstriction that occurs during a dive increases the overall resistance of the peripheral vascular system. However, pressures in the great systemic arteries remain unaltered or change only modestly. This stability of arterial pressure shows that the reduction of cardiac output during dives is closely matched to the increase in peripheral vascular resistance.

It might be supposed that the heart would have reduced demands for ATP and oxygen during dives because it is beating slower, pumping blood less rapidly, and doing less work per unit time. On this basis, the heart might be expected to require a lower rate of myocardial blood flow during dives than during air breathing. Recently, studies of Weddell and harbor seals have confirmed that in these species (others may be different) myocardial blood flow is reduced markedly during forced dives, probably because of vasoconstriction in the coronary arteries. It is noteworthy that the rate of myocardial flow is reduced in approximate proportion to the reduction in cardiac work. Weddell seals showing an 86 percent drop in cardiac output exhibited an 85 percent reduction in blood flow to the ventricular muscle.

Of all the potential cardiovascular responses to diving, changes in heart rate are the most easily studied. That is why, over a century ago, they were the first to be discovered. It is also why they were the focus for some of the first physiological studies to be carried out on the voluntary dives of free-ranging animals. These studies initiated the current reconsideration of views concerning cardiovascular responses to diving. When animals are forced under water, the cardiovascular responses typically proceed rapidly to a profound extent; a forcibly sub-merged harbor seal, for example, exhibits *within 10 seconds* a drop in heart rate to less than 10 percent of the nondiving rate and a drop in blood flow to near zero in the abdominal aorta and renal arteries. Decades of studies of forcibly submerged animals engendered the notion that such cardiovascular responses are obligatory and stereotyped: reflexive. Figure 15.5 presents the results of one of the first studies to challenge the applicability of this view to animals undergoing voluntary dives in their native habitat. When Weddell seals are forcibly submerged, their heart rate quickly falls from at least 50–60 beats/min to about 15 beats/min. As shown in the figure, however, voluntarily diving seals do not in general lower their heart rate to a similarly low level unless they anticipate remaining submerged for a long time. When undertaking their most common type of dive, a short excursion of 5 min or less (recall Figure 15.1), their bradycardia is much less profound: Their heart rate falls on average only to 30 beats/min (and their vasoconstriction is less profound as well). In some birds, such as Canada geese (*Branta*), double-crested cormorants (*Phalacrocorax*), and certain ducks (*Aythya*), voluntary dives typically last less than 30 s; and during such dives, except for brief transients in heart rate, there is no bradycardia at all. Such heart responses show in a particularly graphic way that responses to diving are flexible and that, in many species, the cardiovascular defenses seen in forced dives either are not invoked or are marshaled to a far from maximal extent during the short dives that dominate the voluntary diving repertoire.

The Significance of Vasoconstriction and Bradycardia during Dives

Whereas profound vasoconstriction and bradycardia do not occur during all dives, they are consistent features of forced or protracted dives, and their significance now deserves brief reconsideration. The intense vasoconstriction evoked by forced or protracted submergence can be blocked pharmacologically. When this is done, the length of time that animals can survive under water is markedly shortened. Forcibly submerged harbor seals, for example, can remain under water for only about 4 min when their vasoconstrictor response is blocked, whereas they have been known to endure for over 20 min with an intact cardiovascular response. By such results, the vasoconstrictor response is demonstrated to help enable protracted submergence, and as earlier described, we think we know how it does so: It isolates much of the body from access to certain oxygen stores, which then are substantially reserved for use by the brain and heart, which require oxygen. The cardiovascular responses to diving are seen in their basic form in terrestrial species of mammals and birds (see Box 15.1), but they occur in particularly pronounced and refined form in the diving spe-

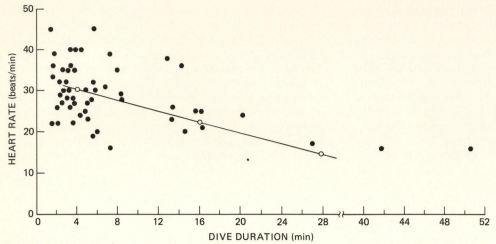

Figure 15.5 Heart rate in relation to dive duration in Weddell seals diving voluntarily in their native habitat. Closed circles depict individual data points. The open circles and line depict a statistical summary of the data. For their heart rates to be measured, seals trailed electrical leads during dives. When the leads were extended to their full length, they disconnected with ease. Leads often remained connected for the entire duration of dives lasting 4–5 min or less. During longer dives, which involved travel to deeper depths, the leads usually became detached within 30–45 s after submergence, meaning that heart rates were known only during the early parts of the dives. [Reprinted by permission from G.L. Kooyman and W.B. Campbell, *Comp. Biochem. Physiol.* **43A**:31–36 (1972). Copyright 1972, Pergamon Press.]

cies. Indeed, the vasoconstriction and bradycardia in the latter species during protracted submergence are probably the most profound cardiovascular adjustments observed anywhere in the animal kingdom during ordinary events of life history.

Sequestering and Release of Red Blood Cells

Innovations in technology have recently permitted monitoring of blood parameters in animals during voluntary diving in the wild. Studies of Weddell seals using this technology have provided a major new revelation: In the free-living seals, about a third of the red blood cells are removed from the circulation during the first 10 or so minutes following a dive and then are returned to the arterial blood during the first 10–15 min of the next dive (the effective oxygen-carrying capacity of the arterial blood rises from 21 volumes percent between dives to 30–34 volumes percent during dives). The place where the cells are sequestered between dives is unknown, although the spleen is a possibility. The sequestered cells are oxygenated; thus their addition to the blood during a dive represents an addition of oxygen. The added cells increase the blood's viscosity. Probably the advantage of sequestering them between dives is that the seals do not have to overcome this increased viscosity when merely resting on land. The details and full implications of this new discovery await further study. The need to extend investigations beyond the walls of the laboratory into the field is once more emphasized, however.

BOX 15.1   THE PHYLOGENETIC DISTRIBUTION OF CARDIOVASCULAR RESPONSES TO ASPHYXIA

Species of mammals and birds that habitually dive are far from alone in manifesting bradycardia and peripheral vasoconstriction when confronted with an interruption of respiratory oxygen intake. At least some terrestrial mammals and birds—including humans—exhibit similar, but less profound, cardiovascular responses when submerged under water. Certain fish, such as flying fish and grunion, show such responses when they emerge into the air, a situation that impairs gill breathing much as diving interrupts lung breathing. Fish also commonly develop bradycardia when exposed to poorly oxygenated water, and humans and other mammals do so during passage through the birth canal. The evidence suggests that bradycardia and other cardiovascular adjustments such as peripheral vasoconstriction are probably evolutionarily primitive responses to asphyxia in the vertebrates.

METABOLISM DURING DIVES

Protracted Dives

From the viewpoint of metabolism, the intense peripheral vasoconstrictor response that occurs during forced dives essentially divides the animal's body into two parts, as demonstrated by the classic data of Irving and Scholander on forcibly submerged harbor seals in Figure 15.6. The tissues that are denied blood flow during submergence, such as the skeletal muscles of the trunk, initially can continue to metabolize aerobically, using local oxygen stores. These tissues exhaust their oxygen supplies relatively rapidly, however—as seen in part (*A*) of the figure—and then turn to anaerobic catabolism, accumulating lactic acid—as seen in part (*B*). The data on the arterial blood in the central circulation, also shown in the figure, graphically illustrate the metabolic isolation of the actively circulated vascular bed from the peripheral tissues. Even as the trunk muscles are turning to anaerobic catabolism, the arterial oxygen content remains high enough to permit continued aerobic catabolism in the actively circulated parts of the body. Lactic acid accumulates to only a small extent in the arterial blood during submergence, showing that the acid produced in the muscles remains largely isolated there until the dive is over. During prolonged voluntary dives, metabolic patterns are thought to be broadly similar to those here described for forced dives. However, as already noted, we remain uncertain whether the swimming muscles become anaerobic during long voluntary dives as they do during forced dives.

Tissues are limited in the extent to which they can accumulate lactic acid (e.g., see p. 47). We often do not know, however, what sets the ultimate limits on dive length: whether it is excessive accumulation of anaerobic end products in anaerobic tissues, exhaustion of oxygen in the actively circulated part of the body, or other considerations. There are clearly species differences, and the answer may also depend on the type of dive.

Considerable research has been focused on the parameters that affect the limits of endurance of the oxygen-dependent tissues (e.g., the brain) that re-

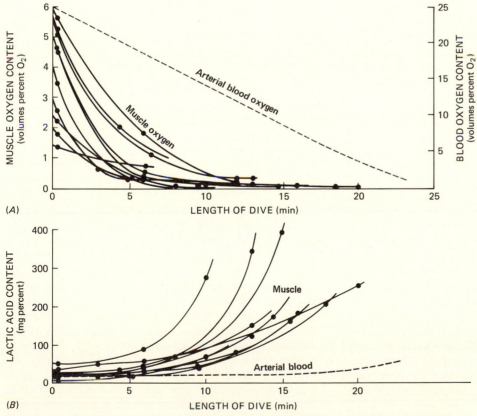

Figure 15.6 Oxygen and lactic acid content in the dorsal trunk muscles and circulating arterial blood of harbor seals (*Phoca vitulina*) during forced submergence in a bathtub, expressed as a function of time of submergence. (*A*) Changes in muscle oxygen content (see left ordinate) in each of 10 seals and the average course of oxygen depletion in circulating arterial blood (see right ordinate). (*B*) Changes in muscle lactic acid content in 10 seals and the average course of accumulation of lactic acid in circulating arterial blood during the dive. [From P.F. Scholander, L. Irving, and S.W. Grinnell, *J. Biol. Chem.* **142**:431–440 (1942).]

ceive an active circulation during the dive. Three parameters are important: the magnitude of the oxygen store available to the actively circulated part of the body, the rate of utilization of the store, and the extent to which the oxygen tension can fall before impairment of function sets in. Although there are interesting results and questions in all these realms, we shall note here just a few observations regarding the latter two parameters.

Recent studies of cardiac metabolism in harbor seals have experimentally confirmed what has long been expected (p. 377), namely, that the heart reduces its metabolic rate during a dive roughly in proportion to the decrease in heart rate. These same studies have made a less-expected revelation: Even though the heart is indeed believed to have a continuous requirement for oxygen, it nonetheless starts to employ anaerobic as well as aerobic catabolism to a significant extent after several minutes of submergence. Both the drop in cardiac metabolic rate and the partial resort to anaerobic catabolism reduce the heart's oxygen needs and thus should help postpone the time when oxygen levels in the circulating blood fall too low for the dive to continue. The brain of diving species is believed to be fully aerobic, as in nondiving ones. At least in Weddell seals, nonetheless, electroencephalographic patterns indicative of cerebral impairment appear at lower oxygen tensions than in terrestrial mammals. Such improved tissue tolerance to hypoxia should also help lengthen underwater endurance.

Short Dives

As discussed earlier, during voluntary dives of short duration, certain species either fail to develop bradycardia or exhibit a considerably smaller reduction in heart rate than during protracted dives. Inasmuch as the vasoconstrictor response is believed to operate in tandem with bradycardia, these observations on heart rate suggest that when the animals undertake short, voluntary dives, their circulation remains more open than during protracted dives, and there is now some direct evidence for this conclusion also. The maintenance of a relatively open circulation during short, voluntary dives would suggest that comparatively free access of much of the body to all available oxygen stores is not inimical to a short dive. Perhaps when a dive is short, oxygen stores are adequate for much or all of the body to function aerobically, thus obviating the need to employ cardiovascular adjustments to "wall off" certain oxygen stores from general access.

Using cardiovascular data to make deductions about the nature of metabolism is risky; thus, an important need at present is to test the deductions directly. One elegant test has been carried out on Weddell seals diving voluntarily in their native habitat. As shown in Figure 15.7, blood samples taken from the seals after their return from dives revealed

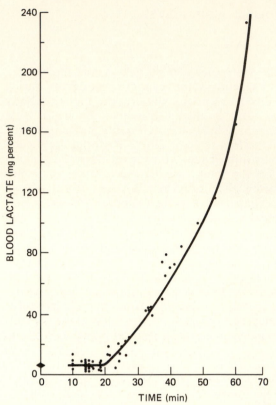

Figure 15.7 Maximum levels of lactic acid observed in arterial blood of freely diving Weddell seals after dives of various lengths. The diamond on the ordinate marks the average resting concentration of lactic acid. Recognize that the circulatory system is fully open after a dive. Thus, lactic acid produced anywhere in the body during a dive, even if temporarily sequestered while the animal remained submerged, may appear in the arterial blood after a dive, permitting whole-body production of lactic acid to be assessed using arterial blood samples. [Reprinted by permission from G.L. Kooyman, E.A. Wahrenbrock, M.A. Castellini, R.W. Davis, and E.E. Sinnett, *J. Comp. Physiol.* **138B:**335–346 (1980).]

no accumulation of lactic acid above resting levels unless the dives lasted longer than 18 min. For a Weddell seal, a dive of even 18 min is "short." Thus, short dives clearly prove to be fully aerobic. (Possibly, some tissues make lactic acid during such dives, but if so, the lactic acid is oxidized by other tissues during the dives. Why do Weddell seals exhibit any bradycardia and vasoconstriction at all during short, voluntary dives? We do not know.)

If an animal accumulates lactic acid during a dive, then after the dive, to recover its full diving capability, it needs to remain at the surface long enough to dissipate the lactic acid, which is a time-consuming process (see p. 46). On the other hand, an animal that has not had to resort to net anaerobic glycolysis during a dive can dive again, with its full diving capacity, after being at the surface only for the relatively brief period required to recharge its oxygen stores. These considerations probably help to explain why, as discussed at the start of the chapter, most dives are short by comparison to the maximum diving duration of which a species is capable. In essence, once recovery times are taken into account, it becomes clear that animals can spend a greater percentage of their time under water—for feeding or other activities—if they make many short, aerobic dives than if they undertake just a few lengthy and highly anaerobic ones. Consider, for instance, 115 min in the life of a Weddell seal. If the seal undertakes a 45-min dive, in the process accumulating a lot of lactic acid (Figure 15.7), it will require about 70 min for recovery; the dive and recovery together will take up the entire 115 min, but only 39 percent of the time will have been spent under water. By contrast, the seal could make six 15-min dives, each of which would entail no lactic acid buildup and would require just 4 min for recovery. Then, 90 min out of the total of 115 min would be spent under water, meaning that 78 percent of the time could be used for feeding or other underwater activities.

The comparatively high oxygen storage capabilities of diving species, it should be noted, make a particularly direct contribution to underwater endurance during aerobic dives.

Metabolic Rates While Diving

It is easy to envision in principle how diving animals would profit from maintaining relatively low metabolic rates during submergence. Whether we consider dives that are powered entirely aerobically or ones in which the body becomes subdivided into aerobic and anaerobic parts, one effect of maintaining a low metabolic rate during diving would be to extend the maximum time that an animal could remain submerged: The lower the metabolic rate, the slower oxygen supplies will be consumed and the slower lactic acid will accumulate.

While breathing air, many marine mammals exhibit resting metabolic rates that in fact are higher than those of terrestrial mammals of equivalent size. Many obstacles exist to quantifying the metabolic rates of mammals and birds during diving, and much of what we know about those rates is based on interpretation of indirect evidence. Nonetheless, the preponderance of available data indicates that going under water commonly brings about a suppression of metabolism in diving species; for example, the resting metabolic rates of animals while submerged tend to be below the resting rates of the animals while breathing air. Perhaps the least ambiguous evidence on the question of diving metabolic rates comes from recent studies of free-ranging Weddell seals that were trained to breathe consistently from a monitored source of air for hours at a time even as they engaged in repeated dives involving swimming and food capture (most of these dives were short enough to be fully aerobic). Remarkably, the average rate of oxygen consumption of these actively diving animals, if it differed at all from that of *resting* but nondiving

seals, was lower. The mechanisms of such metabolic suppression are little known.

Some Physiological Effects of Metabolic End Products

One important attribute shared by the anaerobic and aerobic catabolic pathways is that both produce acidic end products: lactic acid and carbon dioxide, respectively. The accumulation of these end products while animals are submerged often causes the pH of the body fluids to be reduced during a dive and the immediate postdive period. It is vital that this drop in pH be kept within tolerable limits. An important attribute of diving mammals in this regard is that their blood typically exhibits a particularly high buffering capacity in comparison with the blood of terrestrial mammals.

As discussed in Chapter 12 (p. 284), pulmonary ventilation is potently stimulated by a rise in blood carbon dioxide and/or fall in blood pH in terrestrial mammals and birds; indeed, the changes in these blood parameters that occur during submergence are believed to be largely responsible for eliciting the irresistible urge to breathe that we humans so commonly experience when diving. It has long been postulated that diving species might exhibit less ventilatory sensitivity to changes in blood CO_2 tension and pH than terrestrial species. By now, considerable evidence indicates that this is in fact the case. To illustrate, when free-ranging harbor seals are exposed to increasing concentrations of CO_2 in their breathing air, they do in fact increase the amount of air they breathe per unit time, especially by spending a greater fraction of their time breathing rather than diving; but for a given increase in alveolar CO_2 tension, they are only about half as sensitive as humans in the extent to which they increase ventilation.

DECOMPRESSION SICKNESS

In the annals of human diving, quite a number of "diver's diseases" have been described. In turn, considerable attention has been focused on how these diseases are avoided by marine mammals. Here, we consider just one of the diseases, that known as decompression sickness, caisson's disease, or the bends. An informative starting point is to examine the etiology of this disease in human divers.

Etiology in Humans

Undisputed cases of decompression sickness occur in humans when they are diving with a source of compressed air. During diving with a compressed-air source, the air pressure in the lungs is maintained equal to ambient pressure at depth. This prevents the lungs from collapsing under the force of the ambient pressure and allows continued breathing. The elevation of the total pressure in the lungs implies an increase in the partial pressure of each individual pulmonary gas. Of particular interest is the increase in alveolar nitrogen tension, which may be boosted from a normal level of about 570 mm Hg at sea level to many atmospheres at depth. When a person commences a dive, his or her tissues and body fluids are at equilibrium with the normal alveolar nitrogen tension; each contains dissolved nitrogen in accord with its absorption coefficient for a tension of 570 mm Hg (see Chapter 11). At depth, the tissues and body fluids take up nitrogen because their nitrogen tension is initially below the new, elevated alveolar tension. As nitrogen is extracted from the lungs, the alveolar tension does not fall because the person is breathing from a constantly renewed air source, and if the dive is continued for long enough, the tissues will dissolve sufficient nitrogen to come to a nitrogen tension as high as that maintained in the alveoli. Now if the person is then suddenly brought back to the surface (''decompressed''), the alveolar nitrogen tension will fall to its ordinary value, and the nitrogen-charged tissues will lose nitrogen across the lungs in a reversal of the processes that occurred at depth. Problems can arise because the tissue nitrogen tension, being as high as many atmospheres, exceeds the hydrostatic pressures prevailing in the body at sea level, meaning that bubbles of nitrogen can be formed within the body. This is not difficult to understand in principle. If the hydrostatic pressure in a particular tissue is 760 mm Hg (1 atm) and a minute gas space develops there, the pressure in the gas space will also be 760 mm Hg. If the body fluids surrounding the gas space contain nitrogen at a tension of, say, 5000 mm Hg, then they will lose nitrogen into the gas space even if the space is filled purely with nitrogen, for gases always move from regions of higher tension to regions of lower tension. Accordingly the minute gas space will grow into a bubble. Such bubbles are believed to be the primary agents causing decompression sickness. Most commonly, throbbing pains develop in the joints and muscles of the arms and legs (the bends). Additionally, the person may suffer neurological symptoms, such as paralysis, and severe breathing problems (the chokes). Exactly how the bubbles cause these symptoms remains a topic of active research. Some effects are physical; bubbles forming in the blood can occlude vascular beds, interrupting circulation, and bubbles in solid tissues can press on nerve endings. In addition, electrical phenomena at gas–water interfaces (i.e., bubble surfaces) are known to disturb the structures of proteins and can result in such phenomena as red cell clumping, altered blood enzyme function, and release of lipids from lipoproteins.

The factors that determine whether decompression sickness will occur are not completely known. The first and essential condition is that tissue gas tensions must be elevated. In addition, minute (microscopic) gas spaces must ''get started'' in the tissues; this process is subject to many influences. A final consideration is that excess gas is steadily eliminated across the lungs after a dive; if the gas overload is not too great, this process may lower tissue tensions sufficiently rapidly that even if bubbles do start to form, their proliferation will be halted before clinical symptoms are manifest. In general, humans can surface immediately without fear of the bends if their tissue nitrogen tension does not exceed 2 atm. At higher tensions, precautions against the bends are indicated.

When we consider humans undergoing *breath-hold* diving, a crucial difference from diving with compressed air is immediately apparent. The breath-hold diver carries only the limited amount of nitrogen contained within his or her lungs upon submergence; the nitrogen supply is not steadily renewed. During descent to depth, the lungs of the breath-hold diver are compressed under the force of increasing ambient pressure, and the pulmonary nitrogen tension increases initially to high levels, just as in diving with compressed air. This establishes a tension gradient favorable to the transfer of nitrogen from the lungs to the tissues and body fluids, but because the quantity of nitrogen in the lungs is limited, the pulmonary tension falls as the transfer occurs and ultimately the tension gradient is abolished. The increment in tissue nitrogen tension depends on (1) the total quantity of nitrogen transferred from the lungs, (2) the mass of the tissues to which it is distributed, and (3) the absorption coefficient of the tissues (which relates tension to concentration). In humans, it is clear that the nitrogen transfer during a single breath-hold dive is quite inadequate to cause decompression sickness.

What, however, if a person undergoes many *repeated* breath-hold dives? If the time between successive dives is insufficient for the tissues to release accumulated nitrogen after each dive, it is conceivable that the tissue nitrogen tension could be elevated *in increments* to a threatening level. This possibility seems to be far from theoretical. There are a number of reports of people who have developed the symptoms of decompression sickness after serial breath-hold dives. In a classic instance described by Paulev, for example, an individual complained of decompression symptoms after diving about 60 times to depths of 15–20 m over a period of 5 h.

Defenses in Marine Mammals

Marine mammals are, of course, breath-hold divers, and the first question to be asked is whether they face a potential problem of decompression sickness. Taking account of the serial nature of their diving, various calculations and experiments indicate that, indeed, if nitrogen is presumed able to move freely from their lungs into the rest of their body, there are many circumstances in which their tissue nitrogen tensions might rise threateningly high. One important factor that can increase the risk is that when a dive elicits a profound vasoconstrictor response, nitrogen absorbed from the lungs will be distributed by the circulation to just a modest fraction of the body.

During deep dives, defense against decompression sickness is believed to rest principally on features of the lungs that, in concert with the elevated pressures at depth, act to contain pulmonary nitrogen within the lungs, thus curtailing buildup of nitrogen in the blood and other tissues. This concept was first enunciated in its basic form by Scholander in 1940. As diagrammed in Figure 15.8, it is thought that as the pulmonary air is compressed to a smaller and smaller volume at depth, the respiratory air spaces (alveolar sacs) collapse preferentially and the air thus ultimately comes to be contained entirely in the conducting airways—anatomical dead space—of the lungs (p. 278). Consequently, below a certain depth, nitrogen invasion from the lungs into the blood and other tissues cannot occur because the nitrogen is safely sequestered from the respiratory exchange membranes. Oxygen, as already noted, is also sequestered, but that may be a price the animal must pay to avoid the bends.

The depth at which the alveolar sacs become completely collapsed is that where the pressure is sufficient to reduce the initial volume of pulmonary air to the volume of the pulmonary dead space. For example, if a species has a dead-space volume of 2 L and carries 30 L of pulmonary air on submergence, then alveolar collapse should be complete at about 150 m, for at that depth the pressure is near 15 atm—sufficient to reduce the initial volume by a factor of 15 (to 2 L). One reason that the deep-diving seals dive after exhaling may be to hasten alveolar collapse by reducing the initial lung volume relative to the dead-space volume. Studies of Weddell and northern elephant seals indicate that their alveolar exchange area is much reduced at depths as shallow as 30 m (see also p. 375).

Alveolar collapse is promoted in the marine mammals by special features of their thoracic and pulmonary anatomy. They have an exceptionally flexible, highly compressible thorax; thus, their thoracic walls are freely pushed inward as pressure increases at depth, and their lungs within are readily compressed. Another of their important traits is that their pulmonary airways are reinforced with cartilage or muscle right down to the openings to the alveolar sacs. In terrestrial mammals, such reinforcement stops in the terminal bronchioles, and thus the alveolar sacs are separated from the reinforced portions of the airways by several millimeters of delicate respiratory and terminal bronchioles (Figure 12.28). The significance of this difference is made clear by experiments comparing dogs and sea lions. When dog lungs were exposed to pressure, the delicate terminal and respiratory bronchioles collapsed shut before the alveolar sacs had emptied. Thus, a significant portion of the pulmonary air became trapped in the alveolar sacs, and during a dive this air would remain freely available for exchange with the blood. In the sea lions, however, the reinforced bronchioles remained open until the alveolar sacs had emptied. The more-extensive reinforcement of the airways in marine mammals thus assures that the alveolar sacs will be free to discharge their contents completely into the pulmonary dead space as pressure is increased.

In understanding how marine mammals avoid decompression sickness, alveolar collapse, despite its importance during deep dives, is probably not the only consideration. For alveolar collapse to prevent a buildup of nitrogen in the tissues, an animal must dive deep enough for effective collapse to occur. Certain marine mammals, however, sometimes dive extensively to depths that, while sufficient to elevate pulmonary nitrogen pressures to threatening levels, are too shallow to induce alveolar collapse. During such diving, nitrogen could invade the blood and other tissues to a great enough extent to pose a risk of decompression disease. Recent results on trained bottle-nosed dolphins undergoing repeated dives to modest depth suggest that this might in fact be a realistic concern. After an hour of diving, the animals had developed muscle nitrogen tensions—1.7–2.1 atm—that in a human would be only marginally safe. Possibly, unsafe tissue tensions are avoided behaviorally during natural diving, and possibly the animals can cope with higher tensions than humans. These considerations need study.

Increasing depth and pressure

Figure 15.8 Diagram of the hypothesis of preferential collapse of the alveolar sacs at depth. The heavy lines represent the trachea, bronchi, and bronchioles. The circles at the left represent the alveolar sacs. As the pulmonary air is compressed to a smaller volume at depth, it is postulated that the alveolar sacs collapse preferentially and that all the air thus comes to be contained in the anatomical dead spaces (conducting airways), where exchange with the blood is negligible.

SELECTED READINGS

Andersen, H.T. 1966. Physiological adaptations in diving vertebrates. *Physiol. Rev.* **46**:212–243.

Andersen, H.T. (ed.). 1969. *The Biology of Marine Mammals*. Academic, New York.

*Butler, P.J. and D.R. Jones. 1982. The comparative physiology of diving in vertebrates. In O. Lowenstein (ed.), *Advances in Comparative Physiology and Biochemistry*, Vol. 8, pp. 179–364. Academic, New York.

Comparative Physiology and Biochemistry of Cardiovascular, Respiratory, and Metabolic Responses to Hypoxia, Diving, and Hibernation. (A Symposium.) 1988. *Can. J. Zool.* **66**:1–190.

Castellini, M.A., G.N. Somero, and G.L. Kooyman. 1981. Glycolytic enzyme activities in tissues of marine and terrestrial mammals. *Physiol. Zool.* **54**:242–252.

Elsner, R., D.L. Franklin, R.L. Van Citters, and D.W. Kenney. 1966. Cardiovascular defense against asphyxia. *Science* **153**:941–949.

Falke, K.J., R.D. Hill, J. Qvist, R.C. Schneider, M. Guppy, G.C. Liggins, P.W. Hochachka, R.E. Elliott, and W.M. Zapol. 1985. Seal lungs collapse during free diving: Evidence from arterial nitrogen tensions. *Science* **229**:556–558.

Gallivan, G.J., J.W. Kanwisher, and R.C. Best. 1986. Heart rates and gas exchange in the Amazonian manatee (*Trichechus inunguis*) in relation to diving. *J. Comp. Physiol.* **156B**:415–423.

Guppy, M., R.D. Hill, R.C. Schneider, J. Qvist, G.C. Liggins, W.M. Zapol, and P.W. Hochachka. 1986. Microcomputer-assisted metabolic studies of voluntary diving of Weddell seals. *Am. J. Physiol.* **250**:R175–R187.

Hill, R.D., R.C. Schneider, G.C. Liggins, A.H. Schuette, R.L. Elliott, M. Guppy, P.W. Hochachka, J. Qvist, K.J. Falke, and W.M. Zapol. 1987. Heart rate and body temperature during free diving of Weddell seals. *Am. J. Physiol.* **253**:R344–R351.

Hong, S.K. and H. Rahn. 1967. The diving women of Korea and Japan. *Sci, Am.* **216**(5):34–43.

*Jones, D.R. 1987. The duck as a diver. In H. McLennan, J.R. Ledsome, C.H.S. McIntosh, and D.R. Jones (eds.), *Advances in Physiological Research,* pp. 397–409. Plenum, New York.

Kanwisher, J.W., G. Gabrielsen, and N. Kanwisher. 1981. Free and forced diving in birds. *Science* **211**:717–719.

Kjekshus, J.K., A.S. Blix, R. Elsner, R. Hol, and E. Amundsen. 1982. Myocardial blood flow and metabolism in the diving seal. *Am. J. Physiol.* **242**:R97–R104.

*Kooyman, G.L. 1981. *Weddell Seal, Consummate Diver.* Cambridge University Press, New York.

*Kooyman, G.L., M.A. Castellini, and R.W. Davis. 1981.

Physiology of diving in marine mammals. *Annu. Rev. Physiol.* **43**:343–356.

Kooyman, G.L., E.A. Wahrenbrock, M.A. Castellini, R.W. Davis, and E.E. Sinnett. 1980. Aerobic and anaerobic metabolism during voluntary diving in Weddell seals: Evidence of preferred pathways from blood chemistry and behavior. *J. Comp. Physiol.* **138B**:335–346.

Lenfant, C., K. Johansen, and J.D. Torrance. 1970. Gas transport and oxygen storage capacity in some pinnipeds and the sea otter. *Respir. Physiol.* **9**:277–286.

Qvist, J., R.D. Hill, R.C. Schneider, K.J. Falke, G.C. Liggins, M. Guppy, R.L. Elliot, P.W. Hochachka, and W.M. Zapol. 1986. Hemoglobin concentrations and blood gas tensions of free-diving Weddell seals. *J. Appl. Physiol.* **61**:1560–1569.

Ridgway, S.H. and R. Howard. 1979. Dolphin lung collapse and intramuscular circulation during free diving: Evidence from nitrogen washout. *Science* **206**:1182–1183.

Ridgway, S.H., B.L. Scronce, and J. Kanwisher. 1969. Respiration and deep diving in the bottlenose porpoise. *Science* **166**:1651–1654.

Schilling, C.W., C.B. Carlston, and R.A. Mathias (eds.). 1984. *The Physician's Guide to Diving Medicine.* Plenum, New York.

Scholander, P.F. 1964. Animals in aquatic environments: Diving mammals and birds. In D.B. Dill (ed.), *Handbook of Physiology. Section 4: Adaptation to the Environment.* American Physiological Society, Washington, DC.

Snyder, G.K. 1983. Respiratory adaptations in diving mammals. *Respir. Physiol.* **54**:269–294.

Zapol, W.M., G.C. Liggins, R.C. Schneider, J. Qvist, M.T. Snider, R.K. Creasy, and P.W. Hochachka. 1979. Regional blood flow during simulated diving in the conscious Weddell seal. *J. Appl. Physiol.* **47**:968–973.

chapter 16

Neurons and Nervous Systems:
The Basis of Excitability

INTRODUCTION: THE PHYSIOLOGICAL BASIS OF BEHAVIOR

As noted in Chapter 2, the functions of the organ systems within an animal are controlled and coordinated by the nervous system and the endocrine system. These two systems thus serve to integrate the functions of parts of an animal into coherent, adaptive processes of the whole organism. One aspect of this integration is the physiological regulation of organ and system function (e.g., regulation of respiration and circulation) already discussed in this book. A broader aspect of this integration is in the generation of animal behavior.

Until the last few centuries, human behavior was not considered to result from physical and chemical processes within the body tissues. Control of behavior was instead attributed to a nonmaterial *mind* that could will the body to function. Aspects of behavior difficult to explain by the workings of the mind, such as nightmares and epileptic seizures, were considered to be caused by demons inhabiting the body. Today, most scientists operate on the hypothesis that behavior, like all functions of the organism, results from complex interactions of physicochemical processes of body cells. This hypothesis, however, cannot be proved rigorously: We cannot prove that demons do not exist. Such uncertainty is the basis of the mind–body problem, which states that the physical basis of mental activity is unclear. For the meanwhile, we assume that relatively simple patterns of behavior of animals, consisting of movements and other externally recordable motor events, have in principal a physicochemical cellular basis.

Much of the progress in biology has resulted from our ability to explain function at a higher level of organization in terms of processes or mechanisms at a lower level of organization. Thus, functions of the liver are explained in terms of physiology of liver cells, and functions of the cells are explicated in terms of molecular mechanisms. Our primary goal in this and the following chapters is to apply this reductionist approach to the integrative systems that govern behavior. Thus, we want to understand the behavior of animals (including humans) in terms of actions of their nervous systems and endocrine systems. Moreover, we wish to understand nervous system function in terms of cellular neuronal properties, and neuron physiology in molecular terms. Likewise, we want similar explanations at lower levels of organization for endocrine functions. These reductionistic explanations are not sufficient to understand the integrated function of an animal, for which resynthesis of piecemeal information is required. Nevertheless, reductionist analysis is a good start. The following example illustrates the general form of a physiological analysis of behavior and defines some of the important terms in neurophysiology.

Suppose you walk into the kitchen and (horrors!) surprise a cockroach. The cockroach jumps. This simple behavioral act is mediated by electrical signals within the nervous system. As a preamble to discussing the physiology of *neurons* (nerve cells), let us examine what happens when a cockroach jumps. The jump is a *reflex,* a simple motor response to a distinct stimulus. In the cockroach, the stimulus is a vibration of hairs on the tail end of the animal by air currents or by the airborne vibration of sound (Figure 16.1). This stimulus triggers a brief series of *nerve impulses* or action potentials in sensory neurons located at the bases of the hairs. (Sensory processes are treated in Chapter 18). The long, conducting fibers (*axons*) of these sensory neurons are called *afferent* axons or fibers; by definition, an *afferent* fiber

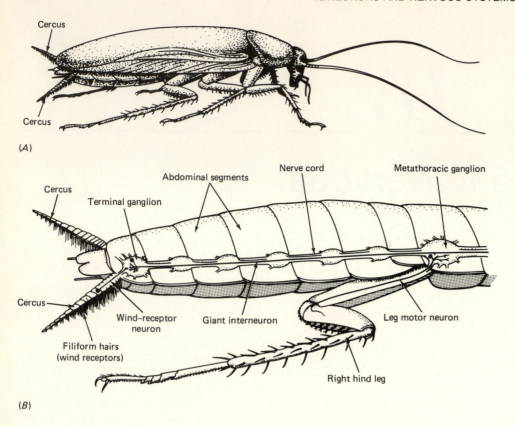

(A)

(B)

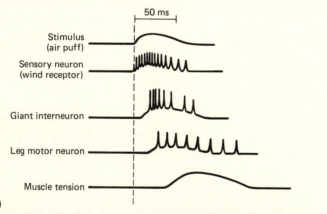

(C)

Figure 16.1   The neural circuit mediating escape in the cockroach. (A) The cockroach *Periplaneta* turns and runs away from a puff of air. The wind receptors that trigger this response are hairs located on the abdominal cerci. (B) Wind stimulation moves the hairs and generates nerve impulses in sensory neurons. The sensory neurons synapse with giant interneurons in the terminal abdominal ganglion, exciting the interneurons which extend up the central nerve cord. The interneurons in turn synaptically excite leg motor neurons, which activate leg muscles. (C) Responses of nerve and muscle cells in the escape circuit to a controlled puff of air of 50 msec duration. [A, B after J. Camhi, *Sci. Am.* **243**(6):158–172 (1980).]

is one that conveys impulses *toward* the central nervous system (CNS). The sensory afferent axons make contact with other neurons within the CNS. The contacts by which neurons interact are called *synapses,* and the interaction at a synapse is termed *synaptic transmission* (Chapter 17). In the cockroach, the sensory axons make synaptic contacts with a few large *interneurons* in the CNS; interneurons are neu-

rons that do not extend outside the CNS. These synapses are excitatory, so that the barrage of impulses in the several sensory axons excites the interneurons, which generate their own nerve impulses. The interneuron axons extend anteriorly in the ventral nerve cord (part of the CNS). They in turn make synaptic contact with *motor neurons* (or motoneurons), defined as neurons whose outgoing axons exit the CNS and innervate a muscle or another effector (Chapter 19); these axons are termed *efferent* because they carry impulses *away* from the CNS. The interneurons synaptically excite motor neurons that in turn excite the jump-producing extensor muscles of the legs. At the same time, the interneurons inhibit motor neurons that excite the antagonist flexor muscles of the legs. When you step on a tack, you withdraw the stricken foot by means of a similar reflex circuit: Pain receptors in the skin generate a barrage of nerve impulses that are propagated to the spinal cord (part of the CNS) along sensory afferent axons. Activity in these sensory axons synaptically excites both interneurons that excite motor neurons to (in this case) leg flexor muscles and other interneurons that inhibit antagonist extensor muscles. The details of these reflex circuits are considered more fully in Chapter 20. The important concepts for now are that nerve cells generate and transmit electrical signals along their elongated, cablelike axons and synaptically transmit signals from cell to cell. Information about the outside world is encoded into these electrical signals by receptors; this information is mixed (integrated) with signals arising within the animal and transmitted to effectors in such a way as to elicit a coordinated, adaptive response.

The functions of a nervous system described above—excitability, rapid transmission of signals, integration, coordination, and control—are present in all animals. Sometimes these functions are mediated by structures other than nerve cells: For example, protozoans and algal cells generate electrical impulses but clearly lack nerve cells; moreover, coelenterates such as *Hydra* may conduct impulses through epithelial tissues in addition to neural conducting systems. Control functions of endocrine systems may overlap extensively with neural control systems (Chapters 2 and 21). Nervous systems, nonetheless, play a primary role in excitability, coordination, and control in most animals, mediating physiological and behavioral interaction with the environment.

THE GENERAL ORGANIZATION OF NERVOUS SYSTEMS

Before we can explore the cellular basis of the function of nervous systems, we must briefly describe the gross organization of nervous systems, to provide a framework from which to proceed. Our aim is to provide necessary background information for a cellular analysis of neural functions, rather than to describe comprehensively the organization of invertebrate and vertebrate nervous systems.

To begin with, what exactly is a nervous system? Theodore Bullock (1977) has defined a *nervous system* as follows:

> An organized constellation of cells (neurons) specialized for the repeated conduction of an excited state from receptor sites or from any neuron to other neurons or to effectors. Higher nervous systems also

integrate the signals of excitation from converging neurons and generate new signals.

Protozoa, which are unicellular, clearly cannot have a nervous system; the most primitive of the multicellular phyla, the Mesozoa and Porifera, appear also to lack organized nervous systems.

Primitive Nervous Systems

The simplest and presumably most primitive form of nervous system is termed a *nerve net*. In a nerve net, neurons are dispersed in a thin layer, seemingly at random. The neurons make synaptic contacts where they cross, so that an excitation initiated in one neuron can spread in all directions along multiple, diffuse paths. The nervous system takes the form of a nerve net in some coelenterates, the most primitive phylum having a nervous system. Nerve nets are also present in peripheral parts of the body in many more-advanced intertebrate groups and in the intestines of vertebrates. Some coelenterates have evolved nervous systems more elaborately organized than simple nerve nets, with primitive integrative centers and directionally oriented through-conducting pathways (Figure 16.2).

There has been considerable speculation about the stages of evolution of primitive metazoan nervous systems. The dominant early theory was that of Parker, who in 1919 hypothesized three stages in nervous system evolution: (1) independent effectors (see p. 536) including epitheliomuscular cells; (2) a two-celled reflex arc consisting of a receptor cell–effector cell pair; and (3) a three-celled reflex arc of a receptor, a ganglion cell, and an effector. Connections between ganglion cells subsequently evolved to produce a primitive, coelenteratelike nerve net. Parker's theory was strongly influenced by Sherrington's

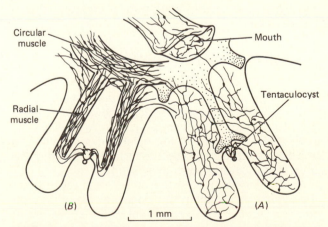

Figure 16.2 Nerve nets in an ephyra (immature stage) of the jellyfish *Aurelia*. (*A*). Unpolarized and presumably primitive nerve net, composed largely of multipolar neurons. (*B*). Superimposed net of bipolar neurons is oriented in relation to radial and circular muscles employed in swimming. This oriented net is thought to be a more recent evolutionary development. The tentaculocyst is a gravity receptor to which both nets are connected. [From G.A. Horridge, *Quart. J. Micr. Sci.* **97**:59–74 (1956).]

studies of the reflex organization of the vertebrate spinal cord and has not been supported by later studies. His hypothesized stage 2, for example, has not been found in any primitive nervous system. More recent theories about the evolution of nervous systems have started from the premise that a functional effector for movement is not a single muscle cell, but a connected field of muscle cells. The theories also recognize that many coelenterate coordination systems exhibit spontaneous activity of pacemakers such as those controlling jellyfish swimming. This spontaneity can be initiated or modulated by receptors, but receptor activity is not a necessary feature of coordinated movement. Finally, although we tend to think of coordinated movement as requiring nerves, animals possess nonnervous coordinating systems. Protozoa such as *Paramecium* generate action potentials and use them for intracellularly coordinated control of the direction of ciliary beating. Many coelenterates possess epithelial conducting systems as well as nervous systems. The epithelial cells generate action potentials that are propagated from cell to cell by low-resistance junctions (see Chapter 17). Thus, nervous systems may have evolved after preexisting epithelial conduction networks that also coordinated the activity of arrays of muscle cell effectors.

Coelenterates have radial symmetry, a body form with apparently limited potential for the evolution of nervous-system centralization. Echinoderms, which are evolutionarily more advanced but also radially symmetrical, also have relatively simple and uncentralized nervous systems. In contrast, even the most primitive groups with *bilateral symmetry* show evolutionary trends of increasing centralization and complexity of nervous-system organization. Apparently, the presence of a clear anterior end and a preferred direction of locomotion in bilateral animals have been important in the evolution of centralized nervous systems.

Evolution of Central Nervous Systems

There have been two major trends in the evolution of nervous systems in the bilaterally symmetrical metazoan phyla: *centralization* and *cephalization*. Both trends are already evident in flatworms (Platyhelminthes), considered the most ancient phylum to be characterized by three tissue layers, discrete organs, and a central nervous system.

In the flatworms and the more-advanced bilaterally symmetrical phyla, the trend toward *centralization* is evidenced by the advent and elaboration of *longitudinal nerve cords* that constitute a distinct **central nervous system** (CNS). Most neuronal cell bodies become located in the nerve cords, and neural control pathways become increasingly routed through the CNS. Motor neurons extend out from the CNS to effectors, and sensory neurons extend from the periphery of the body into the CNS. Increasing numbers of *interneurons*—neurons that are

neither sensory nor motor and are confined to the CNS—make their appearance and enhance the integrative capabilities of the nervous system. The **peripheral nervous system**—consisting of all the fibers of sensory and motor neurons that extend outside the CNS—also becomes increasingly consolidated. Instead of there being a random meshwork of fibers running in all directions in a nerve net, the peripheral sensory and motor fibers become condensed into **nerves**—discrete bundles of nerve fibers running between the CNS and the periphery. (Note that the term *nerve* always refers to a bundle of many nerve fibers located in the peripheral nervous system.) Nerves may contain only sensory axons or only motor axons, but usually contain both.

The other major evolutionary trend in nervous systems is *cephalization,* the progressively greater concentration of nervous structures and functions in the head. In the most primitive of centralized nervous systems, each region of the CNS largely controls just its own zone or segment of the body; indeed, elements of such segmental or regional organization persist in all higher phyla, including vertebrates. In most bilateral animals, however, the most anterior part of the CNS exerts some considerable degree of domination and control over other regions. This anterior part, typically larger and containing more neurons than other parts, is called the **brain**. A brain is a general term for any anterior enlargement of the CNS; the brains of different groups of animals are often not homologous. Cephalization is thought to have been an evolutionary adaptation resulting from the tendency of bilateral animals to move forward, so that information about newly encountered parts of the environment impinges first on the front of the animal. As a correlate of forward motion, most bilateral groups have evolved anterior placement of many of their major sense organs, and the brain that receives environmental information via these sense organs has become dominant over the rest of the CNS. Among vertebrates, a particular manifestation of cephalization is that the forebrain becomes successively enlarged in reptiles, birds, nonprimate mammals, and primates. Along with this development, functions formerly controlled by the spinal cord or brainstem come increasingly under forebrain control.

The evolutionary trends and the terms we have just described apply to all phyla above the coelenterates. Next, we consider certain important differences in the organization of complex nervous systems. There are *two major forms of organization* of central nervous systems in higher phyla: *ganglionic* central nervous systems characteristic of protostomes and *aganglionic* nervous systems characteristic of deuterostomes.

Ganglionic Nervous System of Arthropods

The basic organization of a ganglionic nervous system is present in all annelids and arthropods and to

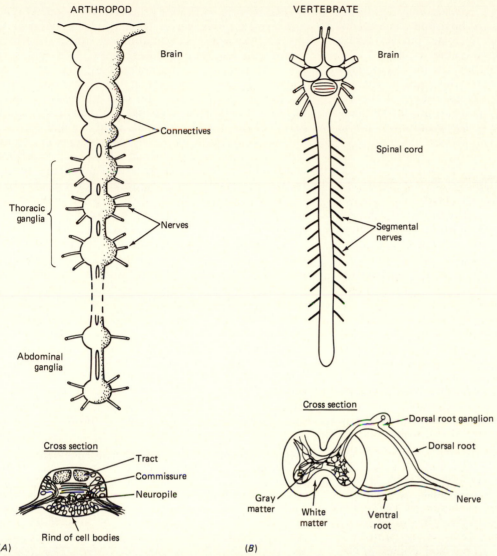

Figure 16.3 Organization of complex central nervous systems. (*A*) Arthropod central nervous systems consist of a chain of segmental ganglia linked by connectives. Each ganglion contains regions of cell bodies, of synaptic neuropile, and of axons (tracts and commissures), shown in cross section below. (*B*) Vertebrate central nervous systems consist of a continuous column of cells and fibers. Cross section of spinal cord shows the division of vertebrate central nervous system into *gray matter* (consisting of cell bodies, synapses, and unmyelinated processes) and *white matter* (consisting of tracts of myelinated axons).

some extent in mollusks. Specifically, in this type of organization, the *central* nervous system consists of a *chain of segmental ganglia* (a ganglion is a discrete collection of nerve cells), linked by paired bundles of fibers called **connectives.** We will use as an example the nervous system of an arthropod, the cockroach we startled several pages ago (Figure 16.1). The central nervous system of a cockroach (Figure 16.3*A*) consists of an anterior *brain* and a *ventral nerve cord* that is linked to the brain by circumesophageal connectives encircling the esophagus. The ventral nerve cord is a chain of ganglia, one ganglion for each thoracic and abdominal body segment. (In some arthropods there is secondary fusion of some of these segmental ganglia.) Each ganglion consists of an outer rind and an inner core. The rind consists mostly

of cell bodies of neurons and is devoid of axons and synapses. Indeed, nearly all neuron cell bodies of arthropods are confined to the outer rinds of the central ganglia, the major exceptions being cell bodies of sensory neurons, many of which are located in the periphery. The inner core of each central ganglion contains two kinds of regions: a region of synaptic contacts between neuron fibers that is termed **neuropile** and a region of **tracts** or bundles of nerve fibers within the ganglion. (Note that in an arthropod or other ganglionic nervous system, there are four terms for a bundle of nerve fibers, depending on where the bundle is located. In the peripheral nervous system a bundle is a **nerve,** between ganglia in the CNS it is a **connective,** within a ganglion it is a **tract,** and between right and left sides of a bilaterally symmetrical ganglion it is a *commissure*. The terms *nerve, tract,* and *commissure* have the same meanings for vertebrate nervous systems, but vertebrate nervous systems do not have connectives.)

Arthropod nervous systems are more compartmentalized than vertebrate nervous systems. This is evidenced by the compartmentalization of the arthropod CNS into discrete ganglia, and by the segregation of cell bodies and synaptic areas within the ganglia.

Vertebrate Nervous Systems

The central nervous system of vertebrates consists of a brain and spinal cord. It differs from the ganglionic central nervous systems of arthropods (and other protostomes) in several respects. The vertebrate central nervous system is dorsal, hollow, and single in embryological origin, developing from a neural tube that invaginates from the dorsal surface of the embryo. The nerve cords of arthropods, in contrast, are ventral, solid, and double in origin; each ganglion is a fusion of paired precursors, and the connectives typically remain paired. The vertebrate CNS, reflecting its origin as a continuous tube, is not clearly segmented into ganglia and connectives, as is the arthropod CNS (Figure 16.3*B*). The description of vertebrate nervous systems as *aganglionic* refers to the organization of the CNS as a continuous column of neuron cell bodies and fibers, rather than as discontinuous ganglia linked by (cell-body-free) connectives.

Terming the vertebrate central nervous system aganglionic can be confusing, because "ganglia" do exist in vertebrate nervous systems. However, the term "ganglion" has an entirely different meaning than in arthropod nervous systems. A *vertebrate ganglion* is a collection of neuronal cell bodies in the *peripheral* nervous system, such as an autonomic ganglion (pp. 14 and 440) or a dorsal root ganglion (see below). With rare exceptions such as the basal ganglia of the mammalian brain (p. 571), ganglia are not considered to exist in vertebrate *central* nervous systems.

In its basic histological organization, the vertebrate CNS consists of two types of tissue termed *gray matter* and *white matter*. Gray matter is composed of intermingled neuron cell bodies, fibers, and synaptic contacts. White matter, on the other hand, consists entirely of tracts of myelinated fibers; such fibers are sheathed with a material termed myelin (p. 411), and it is the myelin that imparts a distinctive white appearance to the area. In the spinal cord (Figure 16.3*B*) the white matter is superficial, and the gray matter is central. This arrangement does not hold in the brain, where white matter is often internal to gray matter, and the arrangement is considerably more complicated.

Peripheral nerves run between the CNS and the periphery; those connected to the brain are termed *cranial nerves,* and those attached to the spinal cord are termed *spinal nerves*. Spinal nerves connect to the cord via two sets of anatomical pathways, termed *dorsal roots* and *ventral roots* (Figure 16.3*B*). Sensory fibers (axons) enter the spinal cord via the dorsal roots, and motor axons exit via the ventral roots (with rare exceptions). The cell bodies of sensory neurons are located in *dorsal root ganglia,* enlargements of the dorsal roots outside the spinal cord.

The *organization of the vertebrate brain* is a major topic, but one we can only touch on. All vertebrate brains share a common structural organization, but in higher vertebrates, forebrain structures have evolved to a disproportionate size and complexity (the trend of cephalization discussed above). The vertebrate brain has three major divisions—the *forebrain, midbrain,* and *hindbrain* (Figure 16.4). The forebrain and hindbrain can be subdivided to yield five brain divisions: *telencephalon* and *diencephalon* (forebrain), *mesencephalon* (midbrain), and *metencephalon* and *myelencephalon* (hindbrain). Each of these divisions contains many tracts of nerve fibers and many clusters of cell bodies termed *nuclei*. There are two major outgrowths of the dorsal surface of the brain that become increasingly prominent in higher vertebrates: the *cerebellar cortex* of the metencephalon and the *cerebral cortex* of the telencephalon. Several of these brain regions will be discussed at appropriate places in Chapters 18 and 20.

THE CELLULAR ORGANIZATION OF NERVOUS SYSTEMS

As already suggested, nervous systems are composed of tissue, which in turn is composed of discrete cells. Nervous systems contain nerve cells or neurons, glial cells, and other nonneural cells of connective tissue and of the circulatory system. The concept of cellular organization of nervous systems is a part of the *cell theory,* which states that organisms are composed of cells, that these cells are the structural and functional units of organization of the organism, and that all cells come from preexisting cells as a result of cell division. The cell theory (cell doctrine) was formulated by Schleiden and Schwann in 1839 and gained widespread and rather rapid acceptance—except as applied to nervous systems. The dominant view of the organization of nervous systems in the latter half of the nineteenth century was instead the *reticular theory,* most strongly argued by Gerlach and Golgi. This theory held that nervous systems were composed of complex, continuous meshworks of cells and processes in protoplasmic continuity with each other. The reticular theory was supplanted only gradually, over the first third of the twentieth century, by an outgrowth of the cell theory known as the *neuron doctrine*. The main champion of the neuron doctrine was Ramon y Cajal, who used Golgi staining techniques to demonstrate convincingly that neurons were contiguous (in contact with each other)

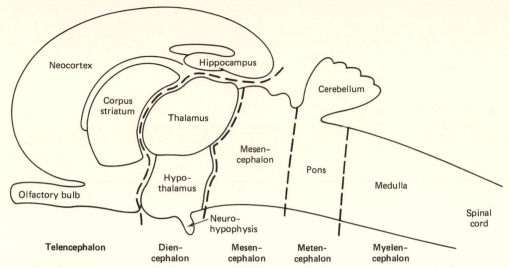

Figure 16.4 Schematic structure of the mammalian brain (side view; anterior is to left). The major structures of the five regions of the brain are shown. [Adapted from W.J.H. Nauta and M. Feirtag, *Sci. Am.* **241**(3):88–111 (1979).]

but were not continuous. The debate on continuity versus contiguity persisted until the 1950s, however, when electron microscopy permitted resolution of cell membranes and rigorously demonstrated the discontinuity of neurons in contact.

The neuron doctrine is a central cornerstone of our understanding of nervous systems. It states that neurons are morphologically independent, structural, functional, and developmental units of organization of nervous systems. Although each aspect of this statement can be debated in specific cases (e.g., squid giant axons arise from syncitial fusion of many embryonic neurons), the Neuron Doctrine represents an important generalization and starting place for examination of the cellular organization of nervous systems.

Neurons

It has been said that a neuron is a cell that is trying to be a telephone cable. This aphorism combines absurdity with the germ of a useful concept. Neurons are cells that are adapted to the function of generating electrical signals and transmitting these signals from place to place within the body, sometimes over considerable distances. As a major structural basis of this adaptation, nerve cells have long processes. Parenthetically, the elongated structure of neurons makes it hard to interpret histological sections of nerve tissue, since the long processes get sectioned into short segments. Historically, this difficulty obscured the structural relation of cells and fibers in histological sections and prolonged the debate between proponents of the reticular theory and the neuron doctrine. We have already made some brief allusions to the structure of neurons, but here we delve into this important matter in greater detail.

A neuron consists of a *cell body* region that con-

tains the nucleus and one or more *processes* arising from it. The cell body is often called the *soma* or *perikaryon* ("around the nucleus"). Neurons can be classified according to the number of processes emanating from the cell body. Neurons may be unipolar (one process), bipolar (two processes), or multipolar (three or more processes). Figure 16.5A shows some of the varieties in the form of neurons.

The soma region of a neuron is broadly similar in cytology to other nonneural cells. The soma contains a nucleus and most of the organelles familiar to cytologists: mitochondria, Golgi complex, microtubules, neurofilaments, smooth endoplasmic reticulum (ER), and rough ER. Neurons are very active secretory cells and thus have extensive, well-developed rough ER.

The processes of neurons are of great length and bewildering geometrical variety and complexity. Early anatomists attempted to bring order to this variety by classifying processes as axons and dendrites. Their classifications were usually based on vertebrate spinal motoneurons (Figure 16.5A) and are useful primarily for cells resembling spinal motoneurons in form. Spinal motoneurons have several relatively short *dendrites* that branch repeatedly; these have continuously varying diameters and lack myelin sheaths (see below). At the fine structural level, the broader dendritic trunks contain rough ER, mitochondria, microtubules, neurofilaments, and an occasional Golgi complex. In general, the fine structure of dendritic trunks resembles that of the soma. Finer dendritic branches lack Golgi complexes and rough ER. The dendrites of many vertebrate neurons bear specialized postsynaptic spines (Figure 16.5B). Dendrites are considered to be postsynaptic, receptive elements of neurons that convey information toward the cell body.

The *axon* of a neuron, in contrast, is classically

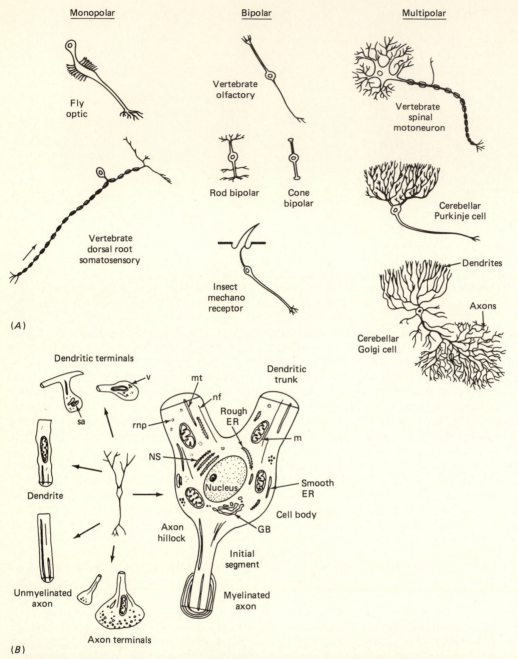

Figure 16.5 Structure of neurons. All neurons have a cell body (region around the nucleus, also termed soma or perikaryon) and processes usually classified as axons and dendrites. (A) Types of neurons. Neurons can be classified as monopolar, bipolar, or multipolar, based on whether one, two, or many processes connect to the soma. (B) Microscopic structure of neurons. A neuron is shown as it would appear with Golgi stain and light microscopy (center). Around it are diagrams of the electron microscopic structure of different parts of the neuron. ER, endoplasmic reticulum; GB, Golgi body; NS, Nissl substance; mt, microtubule; nf, neurofilament; rnp, ribonucleic particles; sa, spine apparatus; v, vesicles; m, mitochondria. [B from G.M. Shepherd, *The Synaptic Organization of the Brain,* 2nd ed. Oxford University Press, New York, 1979.]

single and long, with a relatively constant diameter and few branches (termed axon collaterals). The larger vertebrate axons are surrounded by myelin sheaths—multiple wrappings of insulating glial cell membranes that increase the speed of impulse trans-mission. Myelin is further discussed in the text accompanying Figure 16.25. Not all axons are myelinated; those of the smaller vertebrate neurons and nearly all invertebrate neurons lack myelin. At the fine structural level, axons contain microtubules, neurofilaments, elongate mitochondria, and sparse smooth endoplasmic reticulum. Rough ER and Golgi complexes are generally absent. Functionally, the axon is considered to be the portion of the neuron that supports propagated nerve impulses; it transmits information away from the cell body to the axon terminals.

The identifying criteria for axons and dendrites then include a mixture of anatomical, fine structural, and functional characteristics. Unfortunately, these do not always go together, so that there are exceptions to all the criteria listed. Cerebellar Golgi neurons (Figure 16.5A) have axons (as defined by fine structural criteria and direction of information transmission) that are shorter and more extensively branching than their dendrites. Retinal bipolar neurons (Figure 16.5A) have presumptive dendrites that are no more branched or variable in diameter than their presumptive axons. Moreover, these neurons (and some others) do not propagate nerve impulses, so that a propagative function cannot be used to define their axons. Finally, a vertebrate somatosensory neuron (Figure 16.5A) is myelinated over most of its length and has its cell body in a dorsal root ganglion adjacent to the spinal cord. Its myelinated distal process (i.e., the process away from the central nervous system) propagates impulses toward the soma. Most (but not all) anatomists class this distal process as an axon, making the direction of transmission unimportant. These examples should suffice to show that functional and gross morphological criteria are not sufficient to classify neuronal processes as axons and dendrites. Shepherd (1979; see selected readings), in his clear description of neuronal morphology, argues that definitions of axons and dendrites must ultimately rest on fine structural criteria and that even these may be ambiguous for the majority of neurons in the animal kingdom that are unmyelinated.

Glia

Neurons are not the only cellular constituents of central and peripheral nervous systems. Also present and surrounding the neurons are satellite cells called *glial cells* or *neuroglia* ("nerve glue"). Virchow discovered and named the neuroglial cells in 1846 and considered them to function primarily to bind the neurons together and maintain the form and structural organization of the nervous system. Our present ideas of glial function are only somewhat advanced from this idea, although we do know of some other types of function performed by glial cells. In vertebrates, certain glial cells called Schwann cells (peripheral nervous system) and oligodendroglia (central nervous system) form myelin sheaths around the axons of neurons and thus function to increase the velocity of nerve impulse propagation (see p. 411). Other glial cells called astrocytes line the surfaces of capillaries in the vertebrate CNS and have been

thought to help to maintain the blood–brain barrier (see below). Other functions of glial cells have been postulated, including metabolic nurturing and trophic maintenance of neurons, maintenance of ion balance, indirect electrical signaling, and even memory storage. Direct evidence concerning these functions is scant or absent, however, and glia remain the little understood, poor relations of the nervous system. Since glial cells are estimated to make up half the volume of the mammalian brain and to outnumber neurons by 10 : 1, clarification of their functional roles is necessary and overdue.

Other Cells

In addition to neurons and glia, nervous systems of higher invertebrates and vertebrates contain connective tissue and tissues associated with the circulatory system. One or more sheaths of connective tissue surround and protect both central and peripheral nervous systems. The mammalian brain is surrounded by three such sheaths: an outer dura mater, an arachnoid, and an inner pia mater. Between the arachnoid and pia mater is a subarachnoid space, filled with cerebrospinal fluid. This fluid-filled space around the brain provides it with an effective shock absorber, protecting against mechanical injury.

Nervous tissue requires an extensive blood circulation. In many invertebrates with open circulatory systems, the CNS may be suspended in a blood sinus and thus bathed in blood. Decapod crustaceans typically have a more extensive arterial and capillary supply to the brain than to the rest of the body. In the horseshoe crab *Limulus,* the sheath surrounding the nervous system is arterial, so that the entire nervous system is within an arterial sinus.

Vertebrate brains are very extensively vascularized; every neuron is within about 50 μm of a capillary. None of the neurons, however, directly contacts a capillary; instead, the capillaries are completely lined by astrocyte glia. The exchange of materials between blood and the extracellular fluid compartment of the brain appears to be tightly regulated, so that some materials (oxygen, glucose) exchange rapidly while others are excluded from the cerebrospinal fluid. For example, drugs such as curare injected into the circulation have rather rapid peripheral effects but no discernible effects on the brain. When injected into the cerebrospinal fluid, however, curare has potent effects on the brain. Such experiments have given rise to the concept of a *blood–brain barrier* that is permeable to many small molecules but relatively impermeable to others and to larger molecules such as proteins.

The term "blood–brain barrier" actually represents an oversimplification, since the differences between the composition of blood and of cerebrospinal fluid are maintained by active processes as well as by a passive barrier. Nevertheless, the expression is useful to indicate that the fluid of the vertebrate brain is more tightly homeostatically regulated than the blood and that many substances that may be toxic to neurons are excluded. Although astrocytes lining the capillaries have been postulated to create the barrier, recent evidence indicates that the barrier results predominantly from impermeable tight junctions between cells of the capillary endothelium in the brain.

IONIC BASIS OF MEMBRANE POTENTIALS

The major functional feature of neurons is their ability to generate and to transmit electrical signals. Neurons may have many other properties of functional importance, but we know little about them. In this section, we will explore the questions, What are the properties of the electrical signals of neurons, and how are these signals generated?

Let us begin with a brief review of basic electrical concepts. Protons and electrons have *electrical charge,* and ions bear a net charge because they have unequal numbers of protons and electrons. The net movement of charges constitutes an electric *current,* which is analogous to the hydraulic current flowing in a system of pipes. The separation of positive and negative electric charges constitutes a *voltage* or electrical *potential difference*. This potential difference can do work, causing charges to flow in a current. Voltage is analogous to a head of pressure in a hydraulic system, allowing water to flow downhill. A description of current flow in electrical circuits is beyond the scope of this book. The appendix in Kuffler et al. (1984) describes the properties of electrical circuits very clearly and simply; more abstract descriptions can be found in any physics text.

Both the inside and outside of cells are aqueous media, separated by a selectively permeable membrane (see Chapter 7). In these aqueous media, the electrical charges are ions rather than free electrons. In a radio, electric currents are carried by electron flow in metal wires. In a cell, all currents are carried by ions, and any voltage or potential difference results from local imbalances of ions. Another concept basic to the electrical activity of cells is that the only portion of a cell directly important in determining its electrical properties is its outer limiting cell membrane. This last principle is generally called the *membrane hypothesis*. According to the membrane hypothesis, the electrical activity of a nerve cell is a property of the cell membrane; the electrical potentials observed are transmembrane potentials. The only immediately important attribute of the rest of the cell is the concentration of ions in solution in the cytoplasm. Thus, in this context we can regard the cell as a membrane-bound bag of salt solution, the internal structure of which is unimportant.

All cells have transmembrane potentials, and all share certain responses to electrical current. These responses depend on properties of the cell that are not changed by the electrical current and therefore are termed the *passive electrical properties* of the cells. (This electrical use of the term ''passive'' differs from the thermodynamic use on p. 397.) To demonstrate the passive electrical properties of cells, let us examine a squid giant axon in some detail. The largest axons of a common squid may be 2 cm long and 700–1000 μm in diameter. It is relatively easy to cut out a length of the axon, ligate the ends, and penetrate the isolated axon with microelectrodes. Such an experiment is diagrammed in Figure 16.6. The microelectrode shown in the diagram consists of a glass capillary that has been heated and pulled to a fine tip (~0.5 μm), so that it can penetrate the cell without significant damage. The capillary is filled with a strong electrolyte such as 3 M KCl, to decrease the electrical resistance of its small tip as much as possible. When the tip of the microelectrode is outside the axon, there is no potential difference (no voltage) between it and the indifferent electrode in seawater outside the axon (Figure 16.6 A,C). This is because there is no source of potential and also negligible resistance between the two electrodes. When the electrode is advanced through the membrane into the axoplasm (cytoplasm of the axon), a potential difference (voltage) is recorded (Figure 16.6B,C). This potential difference recorded across the axon membrane in the absence of any stimulation is termed the *resting membrane potential* (E_m). (All known cells have a resting membrane potential with the inside negative with respect to the outside of the cell.) We also know that the axon membrane must have an electrical *resistance* (R_m) to current flow, or else a potential difference could not be maintained

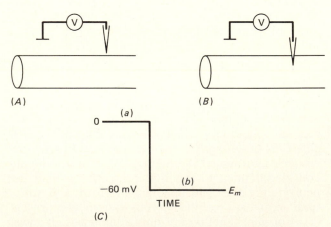

Figure 16.6 Recording the transmembrane potential of a squid giant axon. A voltmeter (Ⓥ) measures the potential difference between a glass capillary microelectrode and an indifferent electrode in the bath around the axon. The output of the voltmeter, recorded on a chartwriter or an oscilloscope, is shown in (C). In (A), the microelectrode is outside the axon, and there is no potential difference between the two electrodes. In (B), the microelectrode has been advanced through the axon membrane, and the transmembrane resting potential (E_m) is recorded.

in Figure 16.12 if we replaced half of the K^+ with Na^+? If the system were governed by passive forces alone and if both Na^+ and K^+ were permeable, they would come into an equilibrium in which the ratio $[Na^+]_{out} : [Na^+]_{in}$ would equal the ratio $[K^+]_{out} : [K^+]_{in}$ (Figure 16.13A). If the membrane were less permeable to Na^+ than to K^+, it would take longer for Na^+ to come into equilibrium, but the equilibrium concentrations for Na^+ would be the same as those for K^+. The identical concentration ratios for Na^+ and K^+ in our theoretical cell are in marked contrast to the great difference in these ratios in real cells (Figure 16.10), in which $[Na^+]_{out}$ is much greater than $[N^+]_{in}$. Thus, passive forces cannot explain the distributions of K^+ and especially of Na^+ that are found in real cells. Instead, these distributions result from *active transport of Na^+ and K^+* across the cell membrane. To clarify this point, let us return to our theoretical cell and add an active cation transport system to the membrane. This system, termed the sodium–potassium exchange pump (or more simply, and erroneously, the sodium pump), actively transports Na^+ out and K^+ into the cell. We shall assume that our theoretical pump transports Na^+ and K^+ in a 1 : 1 ratio, although many pumps are not 1 : 1. The exact ratio is not important here but will be later (see p. 400). If we start with the Na^+ and K^+ concentrations resulting from passive equilibrium (Figure 16.13A) and then turn on our Na^+–K^+ exchange pump, what happens? Na^+ is pumped out of the cell,

and K^+ is pumped in. The resulting ion concentrations, shown in Figure 16.13B, become similar to those found in real cells (Figure 16.10). Neither Na^+ nor K^+ ions are in equilibrium, since passive forces alone would produce net movement of Na^+ and K^+. Na^+ especially is far out of equilibrium, since both the concentration-diffusion gradient and the electrical gradient of the inside-negative membrane potential drive Na^+ inward. Since the membrane is only slightly permeable to Na^+ at rest, Na^+ enters only slowly and is pumped out as fast as it diffuses in. K^+ is closer to, but not at, equilibrium; the cell loses K^+ passively at a slow rate, since although K^+ permeability is large, the driving force is small. The slow passive loss of K^+ is also counteracted by the sodium–potassium exchange pump. Thus, the cation concentrations are maintained in a *steady state*, in which passive leaks are counteracted by active transport. This steady state is different from an equilibrium, in that metabolic energy is required to maintain concentrations different from the concentrations at equilibrium. Figure 16.13C summarizes the roles of active and passive forces in maintaining the steady-state concentrations of Na^+ and K^+ ions in intracellular and extracellular fluids.

Generation of Real-World Membrane Potentials As described above, a real cell is in a steady state with respect to ion concentrations, all ion concentrations being unequally distributed across the membrane and many being out of equilibrium. Because the membrane is more permeable to K^+ than to any other ion, the membrane potential is largely determined by K^+ concentrations. If the membrane were permeable only to K^+, then the membrane potential would be exactly equal to the K^+ equilibrium potential (i.e., $E_m = E_K$), as predicted by a Nernst equation employing the K^+ concentrations across the membrane. Since there is some permeability to other ions, however, they also contribute to the membrane potential. The contribution of each ion is weighted by its permeability, the more permeable ions having more effect. The value of the membrane potential predicted by the unequal contribution of more than one ion species is given by the *Goldman equation*:

$$E = \frac{RT}{F} \ln\left(\frac{P_K[K_o^+] + P_{Na}[Na_o^+] + P_{Cl}[Cl_i^-]}{P_K[K_i^+] + P_{Na}[Na_i^+] + P_{Cl}[Cl_o^-]}\right)$$

in which P_K is the permeability of the membrane to K^+, P_{Na} is the permeability to Na^+, and so on. (Note that the chloride term in the equation is inverted, to account for the valence of -1 for chloride ion.) In principle, it is necessary to add a term in the Goldman equation for every permeable ion species present, but in practice it is necessary to include terms only for Na^+, K^+, and Cl^-. The contribution of other ion species can be neglected, by reason of either low permeability (e.g., HCO_3^-) or low concentration (e.g., $[H^+] \simeq 10^{-7} M$).

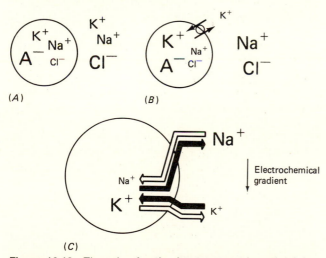

(A)   (B)

(C)

Figure 16.13 The role of active ion transport in maintaining concentration differences across the membrane. In (A), the theoretical cell from Figure 16.11 is shown, but half of the K^+ ions have been replaced by Na^+. The cell membrane is permeable to Na^+, so Na^+ and K^+ are treated indiscriminately by passive forces. In (B), an active sodium–potassium exchange pump has been added to the membrane and transports Na^+ out and K^+ in. The concentrations resulting from the combined passive and active processes resemble those of real cells (Figure 16.10). (C) The steady-state maintenance of Na^+ and K^+ concentrations across the membrane are shown. Na^+ enters slowly down its electrochemical gradient, and K^+ leaves slowly (open arrows). These slow passive leaks are counteracted by active transport (closed arrows).

Corrections to the Ionic Hypothesis As we have seen in the section above, the ionic hypothesis of the generation of membrane potentials states that the concentrations of ions inside and outside a cell are maintained in a steady state by a mixture of active forces (active ion transport) and passive forces (Gibbs–Donnan effects). The ionic hypothesis further asserts that the concentrations of ions inside and outside the cell, and the permeability of the cell membrane to these ions, determine the membrane resting potential (E_m) as described by the Goldman equation. The ionic hypothesis provides a good, approximate description of the factors giving rise to membrane potentials in real cells. This description is adequate for most purposes. For a complete explanation of the causes of membrane potentials, however, other factors must be considered. Two kinds of correction factor that may affect membrane potentials, but are not considered in the ionic hypothesis, are electrogenic pumps and the binding of ions.

There are two possible kinds of active ion transport mechanism: *electroneutral* pumps and *electrogenic* pumps. An electroneutral pump transports equal quantities of charge inward and outward across the membrane. Thus, an electroneutral pump transports no net charge and directly affects ion concentrations only. In Figure 16.13, we assumed that the sodium–potassium exchange pump in cell membranes was coupled 1 : 1, transporting equal numbers of Na^+ ions out and K^+ ions into the cell. Such a pump is electroneutral. An electrogenic pump is one that transports unequal numbers of charges inward and outward across the membrane. The best studied Na^+ – K^+ exchange pumps have a 3 : 2 ratio, trans-

porting 3 Na^+ out for each 2 K^+ transported into the cell. Other pumps may transport unidirectionally (1 : 0), pumping Na^+ or H^+ across a membrane without any ion transport in the reverse direction. Any pump that is not 1 : 1 in ratio of charges transported generates a net current (net movement of charge) across the membrane. This current, acting across the cell's membrane resistance, generates a potential directly, via Ohm's law. The resultant potential changes E_m from the value predicted by the Goldman equation to a different value. A 3 : 2 sodium–potassium pump generates an outward ionic current (outward movement of positive charge) that hyperpolarizes the cell to a level more inside negative than predicted by the Goldman equation.

To illustrate how an electrogenic pump contributes directly to a cell's membrane potential, let us examine a hypothetical experiment with molluscan neurons. We shall record intracellularly from a neuron of a gastropod mollusk such as *Aplysia* or *Helix*. The initial membrane potential is -60 mV (Figure 16.14). When the pump inhibitor ouabain is added to the bath, the cell is rapidly depolarized from -60 to -50 mV. This depolarization change is too rapid to be explained by the slow changes in ion concentration that would eventually result from pump inhibition. Rather, the depolarization represents the loss of that portion of E_m that results directly from the electrogenic pump current. Slowing the pump by cooling, or interfering with the cell's ATP supply by using metabolic inhibitors such as cyanide, also depolarizes the cell (Figure 16.14). If Na^+ ions are injected into a cell with a microelectrode, the cell hyperpolarizes (Figure 16.14) even though the Gold-

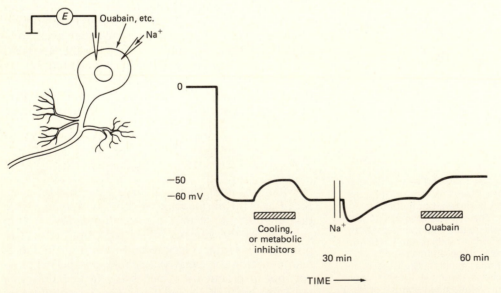

Figure 16.14 Demonstration of an electrogenic pump in the membrane of an *Aplysia* neuron. Cooling the preparation or addition of metabolic inhibitors depolarizes the cell with a time course too rapid to result from concentration changes. Increasing internal [Na^+], by iontophoric injection or by generating a train of nerve impulses, hyperpolarizes the cell by increasing pump rate. Ouabain, a rather specific pump inhibitor, depolarizes the cell irreversibly.

man equation predicts a depolarization when $[Na^+]_{in}$ is increased. The cell hyperpolarizes instead because the pump rate depends on internal $[Na^+]$; an increase in $[Na^+]_{in}$ speeds up the electrogenic pump, increasing the pump current.

A second correction to the ionic hypothesis concerns the *activities* of the ions contributing to the membrane potential. The activity of an ion is the concentration of the ion in the dissociated, freely diffusible form. Strictly speaking, the ion concentration values in the Nernst equation and Goldman equation should be ion activities rather than total concentrations; any ions contained in undissociated salts or bound to proteins would not contribute to transmembrane potentials. Measurements of ion activities in neuronal cytoplasm, using radioactively labeled ions or intracellular ion-specific electrodes, indicate that the activity coefficients of monovalent ions are fairly high and are close to the coefficients in extracellular fluids. Thus, only minor corrections are required to formulate the Goldman equation on the basis of activities rather than total concentrations. We should note, however, that divalent ions such as Ca^{2+} have low activity coefficients in cytoplasm, so that corrections for activity are important in considerations of Ca^{2+} and other divalent and polyvalent ions.

ACTION POTENTIAL OR THE NERVE IMPULSE

A major functional feature of neurons (and some other cells) is the ability to generate electrical signals and to transmit them from place to place. The main kind of electrical signal is the *nerve impulse* or *action potential*, sometimes called spike potential on account of its rapid time course. These terms will be considered synonymous and used interchangeably, although many muscle fibers generate action potentials. Because nerve and muscle cells generate action potentials, they are called excitable tissues. Action potentials then are the most important kind of electrical signals underlying the integrative activity of nervous systems, although as we shall see later, they are not the only kind.

General Features of Action Potentials

Action potentials result from *voltage-dependent* changes in membrane permeability. An action potential thus is initiated by a change in the membrane potential, namely, a sufficient depolarization. This voltage dependence underlies the general features of action potentials and makes impulses fundamentally different from potentials resulting from receptor and synaptic action.

To describe the properties of action potentials, let us perform a hypothetical experiment using a squid giant axon. As in Figure 16.7, we penetrate the axon with two glass capillary microelectrodes, one to record voltage and one to apply current (Figure 16.15).

Application of current in one direction (inward across the membrane and outward through the microelectrode) hyperpolarizes the membrane. The amount of hyperpolarization is proportional to the amount of current. This result is predicted by Ohm's law (ignoring capacitance and the time taken to reach a steady level of hyperpolarization), indicating that with hyperpolarization, the membrane resistance does not change. A small current in the opposite direction elicits a small depolarization that is nearly a mirror image of the corresponding hyperpolarization (Figure 16.15), again indicating little change in membrane resistance. Larger depolarizing currents, however, produce effects that are not mirror images of the corresponding hyperpolarizations. When the membrane potential approaches a critical level of depolarization (the *voltage threshold*) it triggers an action potential—a momentary reversal of membrane potential from about -60 mV (inside negative) to about $+40$ mV (inside positive). The action potential then is a voltage change of about 100 mV, lasting about 1 ms, followed by restoration of the original membrane potential. In the squid axon and in many other neurons, the action potential is followed by an *undershoot*—a transient hyperpolarization lasting a few milliseconds. A larger depolarizing current (above threshold) does not produce a larger action potential. Instead, action potentials are said to be *all-or-none*; a depolarization below threshold elicits no impulse, while all suprathreshold (above threshold) depolarizations produce complete impulses substantially alike in amplitude and duration. Immediately following an impulse, another impulse cannot be generated for about 1 ms (the *absolute refractory period*), and is harder to generate for a few ms longer (the *relative refractory period*. Because of the all-or-none nature of the nerve impulse and the refractory period after an impulse, impulses cannot summate with each other. Instead, a prolonged suprathreshold depolarizing current can elicit a train of discrete impulses (Figure 16.15). The frequency of impulses in such a train increases with increasing strength of depolarizing current up to a point.

An action potential, once initiated, *propagates* along the axon. It is conducted indefinitely without a decrease in amplitude and with a constant velocity that depends on the diameter of the axon. Thus, another voltage-measuring electrode (Figure 16.15, V_2), distant from the site of impulse initiation, records each impulse that the local V_1 electrode records, with no decrease in amplitude. Each impulse recorded at V_2 follows that at V_1 by a short latency that represents the time required for the impulse to propagate between the two electrodes. The subthreshold depolarizations and hyperpolarizations are not recorded by the distant V_2 electrode, since they are not propagated but instead spread decrementally (see Figure 16.9).

Thus, action potentials or nerve impulses can be viewed as all-or-none electrical signals that can be

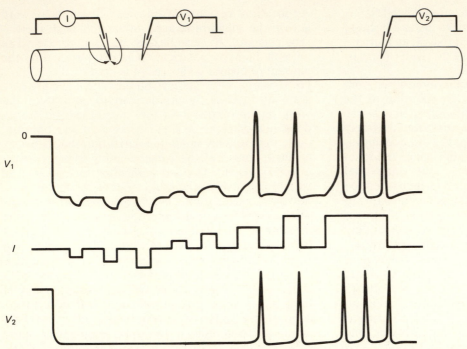

Figure 16.15 General features of action potentials. Pulses of current flowing inward across the membrane capacitance (arrows) hyperpolarize the membrane. Small pulses of outward current elicit depolarizations that are nearly the mirror image of the corresponding hyperpolarizations, but as a critical *voltage threshold* is reached, a brief, all-or-none reversal of membrane potential is generated. Larger pulses of depolarizing current produce identical impulses to these. A distant electrode V_2 records none of the subthreshold events, but records all the action potentials with no loss of amplitude.

transmitted over long distances rapidly and without degradation, by nondecremental impulse propagation. The evolution of large coordinated animals clearly required this ability to send signals over long distances rapidly and without distortion.

The Ionic Mechanism of the Action Potential

First, let us recall that the Goldman equation states that the value of the membrane potential can be predicted from the concentration ratios of ion species across the membrane and from the relative permeability of the membrane to different ion species. It is apparent from the Goldman equation that *any factor that changes the relative permeability of the membrane to one or more ion species also changes the value of the membrane potential*. The action potential results from such changes in ion permeabilities and from the fact that the permeability changes are voltage dependent.

In 1902, Bernstein hypothesized that neurons had an inside-negative membrane resting potential resulting from membrane permeability to potassium and that the nerve impulse resulted from a brief permeability increase to all ions, followed by restoration of the original, relatively low permeabilities. This hypothesis is partly correct and represented a major conceptual advance in understanding the action potential. Bernstein's hypothesis, however, predicted the E_m (the membrane potential) would go to zero during the action potential. When intracellular recording from squid axons was developed nearly 40 years later, it became possible to observe the inside-negative membrane potential and the action potential directly. The action potential, however, did not peak at zero. Instead, it overshot zero to a peak value of about $+40$ mV (inside positive). This overshoot could not be explained by an elevated and equal permeability to all ions. Direct measurements confirmed that the membrane resistance decreased during an action potential, indicating that permeability increases were somehow involved. Nevertheless, another hypothesis was clearly necessary to replace the nonspecific Bernstein hypothesis.

The Sodium Hypothesis

The newer hypothesis, usually called the *sodium hypothesis*, states that the rising phase of the action potential (depolarization and polarity reversal) results from a specific permeability increase to Na^+ ions. A key concept is that in the electrically excitable (impulse generating) membrane, the sodium permeability (P_{Na}) is *voltage dependent*. The sodium hypothesis states that (1) depolarization of the membrane increases its sodium permeability; (2) Na^+ ions flow into the cell as a result of the increased P_{Na}, since Na^+ ions are far out of equilibrium (Figure

16.12*B*); and (3) the inflow of Na$^+$ further depolarizes the membrane, since positive charges are moving from outside to inside. Membrane depolarization thus sets up a positive feedback loop, usually termed the *Hodgkin cycle* (Figure 16.16). The sodium hypothesis asserts then that changing E_m changes P_{Na} and changing P_{Na} changes E_m, as predicted by the Goldman equation. At rest, the membrane is 20–50 times as permeable to K$^+$ as to Na$^+$, so that the resting E_m is near E_K. The regenerative increase in P_{Na} in the Hodgkin cycle makes the membrane transiently much more permeable to Na$^+$ than to K$^+$, so that E_m approaches E_{Na} (+40–50 mV, inside positive; see Figure 16.10). The Hodgkin cycle explains only the rising phase of the action potential; if it alone were operating, the membrane potential would remain near E_{Na} indefinitely. Instead, the polarity reversal lasts only about a millisecond, because the membrane is rapidly repolarized by other factors described below.

The sodium hypothesis is an essentially correct explanation of the explosive rising phase of the action potential. But how would you prove it? The action potential is all over in a millisecond or two, too rapidly to test the starting assumption that depolarization increases P_{Na}. To measure the effect of depolarization on P_{Na} without having the experiment terminate itself in a millisecond, it is necessary to uncouple the positive feedback loop of the Hodgkin cycle. The feedback loop is uncoupled by means of a *voltage clamp*.

Voltage-Clamp Experiments A voltage clamp is an electronic device that allows the membrane potential to be set very rapidly to a predetermined value and held there. The strategy of a voltage-clamp experiment is to set the membrane potential to a given level, pass whatever current is necessary to keep it there, and measure that current. From the Hodgkin

cycle it is clear that any ion flow through the membrane constitutes an ionic current that tends to change the membrane potential. To keep the potential constant, the clamp circuit must generate a bucking current that is exactly equal and opposite to the net ionic current. Only when the ionic current and the bucking current are equal and opposite can the membrane potential remain constant. A voltage clamp senses any deviation of the membrane potential from the set level and instantaneously adjusts the bucking current to oppose the change. By measuring the bucking current, the experimenter has an accurate measure of the amplitude and time course of the net ionic current, since the two must be equal and opposite to each other. A voltage clamp then uncouples the feedback loop of the Hodgkin cycle at point three in Figure 16.16. Ionic currents resulting from permeability changes are prevented from changing the membrane potential.

In 1952, Alan Hodgkin and Andrew Huxley published a series of five papers that are of fundamental importance to our understanding of the ionic basis of excitability. Using voltage-clamped squid axons, they demonstrated and quantified the voltage-dependent permeability changes underlying the action potential. We shall study the ionic mechanisms of action potentials by examining the basic results of the Hodgkin–Huxley voltage-clamp experiments.

Figure 16.17 shows the results of clamping the membrane potential of a squid axon to a more hyperpolarized value (*A*) and to a more depolarized value (*B*) than at rest. When the membrane potential is clamped to a more hyperpolarized value, the current-measuring circuit shows only a brief blip of current required to set the membrane potential to a new level. This initial blip does not represent any ionic current and is not of interest to us. Following it, there is little current resulting from holding the membrane at a hyperpolarized level. (Actually, there is a slight leakage current, resulting from changing the membrane potential relative to ionic equilibrium potentials. It does not reflect any permeability changes and is also of no interest to us.) Hyperpolarization thus leads to no significant ionic current flow. Clamping the membrane potential to a value more depolarized than the resting potential has quite different effects (Figure 16.7*B*). Following the initial blip of current required to change the membrane potential, current is required to hold the membrane at the set value. The clamping current is first outward and then inward, corresponding to an early inward ionic current that is reversed in 1–2 ms to a later outward ionic current. Depolarization of the membrane thus induces permeability changes that (if the currents are carried by cations) result in first an inward movement of cations and then an outward movement of cations. If the membrane were not clamped, these ionic currents would produce first a depolarization and then a repolarization of the membrane, as in the action potential.

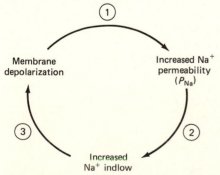

Figure 16.16 The Hodgkin cycle, which produces the rising phase of the action potential. The critical feature of the cycle is that Na$^+$ permeability is *voltage dependent,* with depolarization increasing P_{Na}. The increased Na$^+$ permeability allows inflow of Na$^+$ down its electrochemical gradient, which further depolarizes the membrane. The cycle is named after A.L. Hodgkin, who was a co-recipient of the Nobel Prize for his work clarifying the ionic mechanism of action potentials.

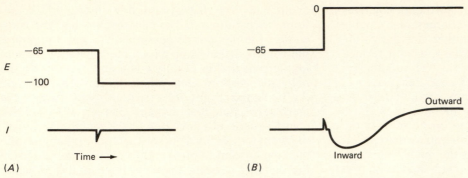

Figure 16.17 Results of a voltage-clamp experiment. In (A), the membrane potential is clamped to a level hyperpolarized relative to the resting potential (−65 mV). This hyperpolarization results in no significant ionic current. In (B), the membrane is clamped to a depolarized value, in this case zero. Depolarization induces first an early inward ionic current, followed by a later outward ionic current that persists as long as the depolarization is maintained.

According to the sodium hypothesis, the early inward ionic current (which generates the rising phase of the action potential in unclamped axons) is an influx of Na^+ ions. How could this prediction be tested? Hodgkin and Huxley replaced the Na^+ in the artificial seawater, with which they bathed the axon, with an impermeable cation. In the absence of extracellular Na^+, the early inward current was replaced by an early outward current (Figure 16.18). That is, depolarization induces a Na^+ permeability increase, which in the absence of extracellular Na^+ results in Na^+ diffusion outward down its concentration gradient. This interpretation predicts that if the Na^+ concentration were equal on both sides of the membrane, there would be no Na^+ concentration gradient and no early Na^+ current. Hodgkin and Huxley replaced about 90 percent of the extracellular Na^+ with impermeable ions so that $[Na^+]_{in} = [Na^+]_{out}$. When the membrane was clamped to 0 mV (so that there was no voltage gradient), there was no early current (Figure 16.18). Further evidence that the early inward current is carried by Na^+ was provided by experiments in which a squid axon in normal artificial seawater was clamped to the sodium equilibrium potential ($E_{Na} = +50$ mV). There was no resultant early current, since there was no driving force on Na^+ ions at E_{Na}. Clamping the membrane at a level beyond E_{Na} (more inside positive that E_{Na}) resulted in an early outward current, representing Na^+ efflux toward E_{Na}.

The experiments described above demonstrated that the early inward current during a voltage clamp

is carried by Na^+. What ion or ions carry the later outward current? Hodgkin and Huxley tested the hypothesis that the delayed outward current represented an efflux (outflow) of K^+ ions. After first clamping the membrane to a depolarized value to initiate the permeability increases, they reclamped it to E_K a few milliseconds later. At E_K there was no late current, a result indicating that the late current was carried by potassium. It was also possible to measure K^+ efflux directly. Since the late outward current persists as long as the membrane is depolarized by voltage clamp, potassium efflux (over the course of minutes) could be monitored directly by preloading a squid axon with the radioactive isotope ^{42}K. Such experiments showed that the late current was exactly equivalent to the K^+ efflux and was therefore carried by potassium ions.

In contrast to the late outward current, the early inward current lasts only 2–4 ms despite maintained depolarization. The major conclusion of the voltage-clamp experiments is that *depolarization has three effects on the excitable membrane* of the squid axon, each with its own time course. These effects are permeability changes, expressed in electrical units of **conductance** (inverse of resistance). These changes are diagrammed in Figure 16.19. First, depolarization increases the sodium conductance (or Na^+ permeability) of the membrane ①. Second, with a slower but overlapping time course, the sodium permeability is shut off ②. This process is called the inactivation of sodium conductance. Third, in the overlapping sequence, depolarization induces an increase in po-

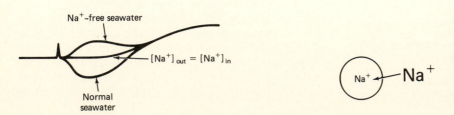

Figure 16.18 A demonstration that the early inward current is carried by sodium ions. Bathing the axon in Na^+-free seawater results in the reversal of the early inward current to an outward current. When $[Na^+]_{out}$ is made equal to $[Na^+]_{in}$, a voltage clamp to zero elicits no early current. Inset shows the relative Na^+ concentrations in normal seawater.

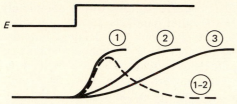

Figure 16.19 Time course of voltage-dependent permeability changes in a voltage-clamped squid axon membrane. ① The most rapid change is an increase in sodium permeability. ② Sodium permeability is turned off or inactivated by a process with a somewhat slower time course. ③ Still more slowly, potassium permeability is increased. The net sodium permeability is determined by the first two processes, that is, ① minus ②.

tassium conductance ③. In an unclamped axon, the ion fluxes resulting from these three permeability changes underlie the membrane action potential (Figure 16.20). Rapidly increased Na^+ permeability ① depolarizes the membrane via the Hodgkin cycle, but this permeability is quickly inactivated ②. The inactivation of Na^+ permeability and the slower increase of K^+ permeability both repolarize the membrane. Since the K^+ permeability increase results from depolarization, it is shut off by its own repolarizing effect. Both the inactivation of Na^+ conductance ② and the increase in K^+ conductance ③ outlast the nerve impulse, decreasing relatively slowly with repolarization just as they increase relatively slowly with depolarization. Their residual effects make the axon refractory to further stimulation for a few milliseconds after an impulse.

From their voltage-clamp data, Hodgkin and Huxley were able to quantify the voltage dependence and time dependence of the three parameters listed above. They developed a set of equations by which they showed that the magnitudes and time courses of these three voltage-dependent processes are sufficient to describe the behavior of action potentials in unclamped squid giant axons. These studies remain a cornerstone of our understanding of the physiology of excitable membranes.

Ion Movements in Action Potentials In the generation of an action potential, a neuron gains a small amount of Na^+ and loses a small amount of K^+. These amounts have been calculated to be $3–4 \times 10^{-12}$ mol/cm^2 of membrane per impulse. Like the slow passive leaks of Na^+ in and K^+ out across the resting membrane, the ions crossing the membrane during an impulse must be pumped back again by the Na^+–K^+ exchange pump. It is important to realize that the pumping process is slow relative to the time course of the action potential and serves only to keep the ion concentrations constant over minutes, hours, and days. The Na^+–K^+ exchange pump does not contribute directly to the generation of impulses, and the ion movements underlying impulse generation are very small relative to the concentrations of ions inside and outside the axon. If the Na^+–K^+ exchange pump of a squid giant axon is poisoned, the axon can still generate about 100,000 impulses before the internal Na^+ concentration is increased by 10 percent! Smaller axons, however, have a greater ratio of membrane surface to internal volume, so that the concentration changes produced by impulses are greater. Therefore, the smallest axons (~ 0.1 μm in

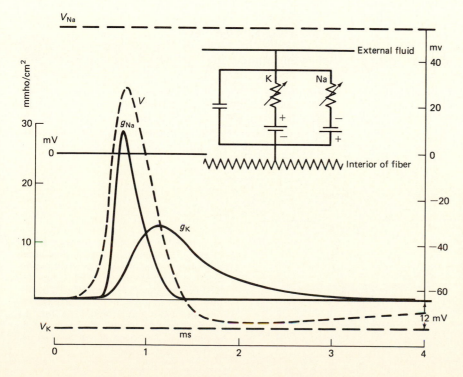

Figure 16.20 Theoretical action potential (V) and membrane conductance changes (g_{Na} and g_K) in an unclamped squid axon at 18.5°C. The theoretical curves are obtained from the Hodgkin-Huxley equations that describe the voltage and time dependence of the conductances. V_{Na} and V_K are the equilibrium potentials for sodium and potassium across the membrane. The inset shows an element of the excitable membrane of a nerve fiber. [From J.C. Eccles, *The Understanding of the Brain*, 2nd ed. McGraw-Hill, New York, 1977.]

diameter) presumably cannot generate impulses at a rate that greatly exceeds the ability of the Na^+-K^+ exchange pump to maintain normal concentrations.

Perfused Axons From our discussions above, it is apparent that, like the membrane resting potential, the action potential depends only on the properties of the axon membrane and on the ionic composition of the intracellular and extracellular fluids. This argument would predict that an axon filled with an isotonic solution of, say, K_2SO_4 would generate nearly normal resting potentials and action potentials (the membrane being impermeable to sulfate ions). Surprisingly, this experiment can be performed. One can place a squid axon on a rubber mat, gently squeeze out the axoplasm with a small rubber roller—rather like squeezing toothpaste from a tube—and refill the axon with K_2SO_4 solution or whatever one wishes. Axons perfused with the appropriate internal solutions generate normal resting potentials and action potentials (providing the membrane is not damaged), elegantly confirming the above predictions.

The Nature of the Voltage-Dependent Permeability Channels We have argued that there are two separate voltage-dependent channels in an axon membrane: One is selectively permeable to Na^+ and is rapidly opened and then closed as a result of membrane depolarization; the other is selectively permeable to K^+ and is less rapidly opened by depolarization. There are alternative interpretations to the Hodgkin–Huxley voltage-clamp experiments; for example, a single channel could become permeable first to Na^+ and then to K^+. What is the evidence that there are two types of channel, and what is known about the bases of their voltage dependence and their ion selectivity?

The best evidence for separate Na^+ and K^+ channels is pharmacological, from effects of drugs applied to the membrane. Tetrodotoxin (TTX), an extremely poisonous substance found in puffer fish, selectively blocks voltage-dependent Na^+ channels. If a squid axon is bathed in seawater containing TTX and is voltage clamped to a depolarized level such as 0 mV, the early inward Na^+ current is blocked. The de-

layed outward (K^+) current, however, is completely unaffected. On the other hand, tetraethylammonium (TEA) ions selectively block the delayed outward current flowing through the K^+ channel, when perfused into the internal axoplasm. TEA ions have no effect on the early inward current flowing through Na^+ channels. It is even possible to remove the Na^+ inactivation mechanism selectively and irreversibly, by perfusing a low concentration of the proteolytic enzyme pronase into an axon. The presumed sites of action of these pharmacological agents are shown in Figure 16.21. The independent effects of blockers of Na^+ channels and K^+ channels provide compelling evidence that the two channels are physically separate.

How can the permeabilities of these membrane channels be voltage dependent? Hodgkin and Huxley originally suggested that the part of a channel that opened or closed as a result of voltage changes (the gating subunit) must be charged in order to respond to changes in the electric field across the membrane. They further suggested that the movement of these charged subunits—opening the gates in the channels—would produce a very small current prior to the ionic currents underlying an action potential. This prediction has been confirmed: When the larger ionic currents are blocked by drugs or by substitution of impermeant ions, it is possible to measure a very small early *gating current* in voltage-clamped squid axons. This gating current corresponds to the synchronous opening of the Na^+ channels in the axon. Interestingly, tetrodotoxin does not block the Na^+ gating current, indicating that the gating subunit and the TTX binding site are at different places in the Na^+ channel (Figure 16.21). Usually, it is not possible to measure a potassium gating current, since the openings of membrane K^+ channels are much less tightly synchronized.

What is the basis of the ion selectivity of the channels underlying the nerve impulse? The ions Na^+ and K^+ are monovalent cations with similar properties. How can one kind of channel be selectively permeable to Na^+ and another be selectively permeable to K^+? Actually, neither channel is absolutely selective; each channel is slightly permeable to the "wrong" ion and is also permeable to other

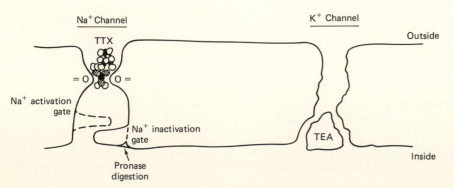

Figure 16.21 Schematic diagram of sodium and potassium voltage-dependent permeability channels in excitable membranes. The sodium channel is blocked by tetrodotoxin (TTX) from the outside. The potassium channel is blocked by tetraethylammonium ions (TEA) from the inside. The Na^+ inactivation gate can be irreversibly destroyed by pronase digestion from the inside.

ions that are not normally present in organisms. By studying the relative permeabilities of channels to different ions, one can infer something about the structure of the channels.

Ions in solution are normally *hydrated*; that is, each ion is surrounded by a shell of water molecules. The water molecules around a cation have their electronegative oxygen atoms facing in toward the ion, held to it by charge attraction. A naked Na^+ ion is smaller than a naked K^+ ion, but a fully hydrated K^+ ion is smaller than a fully hydrated Na^+ ion. The voltage-dependent Na^+ channel is thought to have a narrow selective region with a minimum pore size of 0.3×0.5 nm². This narrow pore is lined by oxygen atoms, which compete with the water molecules of the hydrated Na^+ ion, so that most of the water molecules are lost. It is calculated that the pore is just large enough for a Na^+ ion with one water molecule of hydration to pass through. A K^+ ion with one water molecule of hydration is larger, and thus K^+ is considerably less permeable. Obviously, the structural basis of selectivity of K^+ channels must be quite different. Apparently, cations pass through the potassium channel without losing their water molecules of hydration; because a fully hydrated K^+ ion is smaller than a fully hydrated Na^+ ion, it passes more easily through the K^+ channel.

Comparative Aspects of Action Potential Generation

How general are the ionic mechanisms of action potential generation? The mechanisms described above were first worked out for squid giant axons, which were large enough to permit use of multiple internal electrodes, voltage clamping, and internal perfusion of the axoplasm. Later studies have shown that the basic aspects of impulse generation apply to most excitable cells. The action potentials of vertebrate and invertebrate unmyelinated axons, amphibian myelinated axons (see p. 411), and vertebrate skeletal twitch muscle fibers (see Chapter 19) have ionic mechanisms qualitatively similar to those of squid axons. There are quantitative differences, however, among the potentials in these tissues: The amplitude of the membrane resting potential, the presence of after-hyperpolarization or after-depolarization following an impulse, and the degree, rate, and threshold of voltage-dependent P_{Na} and P_K may vary from one cell type to another. Recent reports, for example, indicate that in myelinated axons of fish and mammals (but not amphibians) it is difficult to demonstrate an outward potassium current repolarizing the membrane from the peak of the action potential. Thus, a voltage-dependent potassium permeability (P_K) may be reduced or absent in these axons. Such differences are relatively minor, however, and do not appreciably change the basic story.

On the other hand, there are cases where significant differences in ionic mechanism exist. These are modifications of or exceptions to the rule, some of which are functional adaptations to particular circumstances. A few of these modifications are considered here.

Calcium Spikes In some cells, Ca^{2+} ions, rather than Na^+ ions, carry the inward current that produces the rising phase of the action potential. Cells have very low internal Ca^{2+} concentrations (10^{-6}–10^{-7} M). Hence, the calcium equilibrium potential (E_{Ca}) is inside positive, and a voltage-dependent increase in Ca^{2+} permeability leads to an inward Ca^{2+} current. This calcium-dependent conductance is called **calcium electrogenesis**. Actually, there is some small degree of calcium conductance in most excitable cells, but in most cases such as squid axons, it is masked by a much larger sodium conductance. In many molluscan neurons and some arthropod muscle fibers, however, voltage-dependent calcium conductance accounts for the bulk of the early inward current underlying the action potential. The preponderance of voltage-dependent P_{Ca} and P_{Na} may vary from neuron to neuron even in the same organism; the central nervous system of mollusks such as *Aplysia* or *Helix* contains identified sodium-spiking neurons, calcium-spiking neurons, and mixed sodium–calcium spiking neurons. Calcium electrogenesis is also important in the endogenous generation of bursts of impulses in neurons (see Chapter 20, Figure 20.14).

Spontaneity and Pacemaker Potentials Many neurons are spontaneously active, generating action potentials at rather regular intervals without an external source of depolarization. The somata of some molluscan neurons, for example, generate action potentials in regular trains or even in repetitive bursts, in the absence of synaptic input. This activity (sometimes termed **endogenous activity**) can persist even after the soma has been tied off from the synaptic sites elsewhere on the neuron. Some other excitable cells, such as vertebrate cardiac muscle fibers, are also spontaneously active.

The membrane potential of a spontaneously active cell, instead of maintaining a fixed resting value, undergoes a continuous ramp of depolarization between action potentials, until it reaches threshold for the generation of the next action potential. The repolarizing phase of the action potential restores the membrane to a relatively hyperpolarized level, from which the next ramp of depolarization begins. These ramp depolarizations are termed **pacemaker potentials**, since they determine the rate of impulse generation of the cell. For example, in a cardiac muscle fiber in the pacemaker region of the vertebrate heart (see Chapter 14), the greater the rate of depolarization, the sooner the cell reaches threshold for the next action potential, and the greater the frequency of action potentials (and thus of heartbeats). For vertebrate cardiac muscle fibers, norepinephrine increases the rate of depolarization (and thus the heart rate) and acetylcholine decreases it.

The ionic basis of pacemaker potentials can be complex and may vary somewhat among cells. The simplest explanation for a pacemaker potential (such as that of cardiac muscle fibers) is that there is a relatively high permeability to Na$^+$ (and/or Ca^{2+}) in the cell, so that the inward sodium current is greater than the outward potassium current, leading to depolarization. After each impulse, the brief repolarizing potassium current (I_K) restores the membrane to a value near E_K; as the potassium current wanes, the cell begins the next ramp of depolarization.

In molluscan neurons and probably in some other spontaneously active cells, additional types of channel affect the characteristics of pacemaker potentials. One of these channel types produces an outward potassium current termed the *A current*, which slows the rate of pacemaker depolarization between impulses. These channels are activated by depolarization (e.g., to −65 mV) after a period of hyperpolarization and are subsequently inactivated by subthreshold depolarization (e.g., to −45 mV). The A current is not strong enough to repolarize the cell—only enough to slow the rate of its ramp depolarization. It may seem surprising that such a current is a significant component of some pacemaker potentials. The major functional attribute of the A current is that it allows repetitive impulse generation over a wide range of impulse frequencies. Axonal membranes, which lack A-current channels, can generate repetitive impulses (in response to a steady applied current) only over a narrow range of frequencies, if at all. Somata with A-current channels, in contrast, can generate repetitive impulse trains over a wide range of frequencies and conditions.

Other novel channels may be involved in pacemaker activity, as well as A-current channels. The details of function of these channels are found in Hille (1984). Recent studies reveal a wider range of ion channels in excitable membranes than had been supposed, but details of their operation and their functional significance are still being worked out.

Cardiac Muscle   The action potentials of heart muscle fibers demonstrate a significant departure from the common impulse-generating mechanisms seen in squid axons. In contrast to most action potentials, which have durations of about 0.4–3 ms, vertebrate cardiac muscle fibers have action potentials with typical durations of 100–500 ms (Figure 16.22). The long duration of cardiac action potentials is functionally important, since heart muscle must contract for about 100 ms to pump blood effectively (Chapter 14) and since depolarization is the necessary stimulus for the contraction. As shown in Figure 16.22, a cardiac muscle fiber action potential has a rapid upstroke and a rapid initial recovery to near zero but remains depolarized near zero for many milliseconds. This prolonged depolarization, termed the

plateau of the action potential, gradually decreases and is followed by a rather slow repolarization. The ionic basis of the cardiac action potential is more complex than that of most excitable cells. There are two separate inward currents underlying the cardiac action potential. The first is a sodium current (Figure 16.22) very similar to the sodium current in the squid axon. The sodium current produces the rapid upstroke of the cardiac action potential. Most of the sodium current is inactivated within a few milliseconds, allowing the potential to approach zero. The long plateau of the impulse results from two factors (Figure 16.22). The first of these is a second, slow inward current termed I_{slow}. This slow inward current is much smaller than the fast Na$^+$ current and takes at least 20 ms to develop. It is carried principally by Ca^{2+} ions, although with a variable secondary contribution of other ions. The slow current is slowly inactivated by the plateau depolarization. The other factor underlying the plateau is a *decrease* of the outward K$^+$ current from the resting level during the plateau depolarization (in contrast to the increased K$^+$ current in axons). Thus, the plateau represents a balance between two rather small currents, the slow inward current and the (decreased) outward potassium current. Membrane resistance of cardiac muscle fibers during the plateau is actually higher than the resistance between action potentials. Heart muscle fibers are typically depolarized as much as 30 percent of the time during the heartbeat cycle (Chapter 14). Thus, the ability to maintain prolonged depolarization without great increases in ion perme-

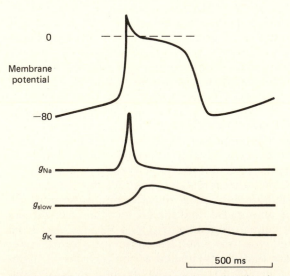

Figure 16.22   Action potentials of long duration in cardiac muscle fibers of vertebrates. The action potentials overshoot zero and then maintain a plateau of depolarization for hundreds of milliseconds. The initial rising phase of the action potential results from an increase in sodium conductance (g_{Na}) as in squid axons. A slow, prolonged conductance increase (g_{slow}) maintains the plateau depolarization. g_{slow} is largely an increased calcium conductance, although sodium is also implicated. The potassium conductance g_K is decreased during the action potential.

ability and consequent pump requirements must result in considerable energy savings.

Other Cells   Action potentials are not restricted to excitable tissues of metazoans. The protozoan *Paramecium* has an action-potential-generating mechanism, in which the impulse rising phase depends on regenerative changes in Ca^{2+} permeability. The unicellular algae *Chara* and *Nitella* generate impulses in which the rising phase results from a Cl^- outward current. These examples illustrate that the sodium hypothesis for the generation of action potentials is not universal; nevertheless, it is the most important generalization about the mechanism of action of electrically excitable cells.

Finally we should note that not all nerve cells generate action potentials (see Chapter 17, Reciprocal Synapses, and Chapter 18, Retina).

THE CONDUCTION OF ACTION POTENTIALS

General Mechanism of Impulse Propagation

So far, we have talked about stationary action potentials only, ignoring propagation. The major functional significance of action potentials, however, is their ability to be conducted over indefinitely long axonal distances without degradation. Consider that in large animals, single axons—such as those that control wiggling of your toes—can be at least a meter long. Now recall that a voltage change at one point on a membrane decreases exponentially with distance away from that point (Figure 16.9). Since the length constant describing this decrease is usually a fraction of a millimeter, it is clear that an electrical signal has to be amplified or refreshed to be conducted over any distance greater than a millimeter or two. We have seen that an action potential is all-or-none because the voltage-dependent, regenerative permeability increases bring the membrane potential toward a limiting value—the sodium equilibrium potential (E_{Na}). The action potential is propagated in an all-or-none, nondecremental fashion, simply because once it is initiated, it generates an action potential at successive locations along the axon.

To see how the propagation process works, let us look at the spatial distribution of voltage along the length of an axon (Figure 16.23). Figure 16.23*A*, for reference, shows an impulse recorded by an intracellular electrode at one point in space, over successive instants in time (V versus t). Since this impulse is propagated along the axon at constant velocity, the impulse can also be plotted as a voltage function of distance (V versus x) at any instant in time (Figure 16.23*B*). Given the constant velocity of propagation, these two pictures of an impulse are equivalent. (This can be a tricky point; try to assimilate it before continuing. Think of Figure 16.23*B* as a stop-action, flash picture of an impulse at an instant in its conduction along the axon. Which way is it moving?) Figure

16.23*C* shows the spatial distribution of charges along the axon at this instant and is a cartoon representation equivalent to Figure 16.23*B*. The impulse sets up *local circuits of current flow*, such that current flows away from the axoplasm and toward it in the extracellular fluid. These local currents depolarize the adjacent membrane regions, exactly as shown in Figure 16.9*D*.

Current must always flow in a complete circuit; it is useful to consider how the current is carried in completing its circuit. Current flow within the axoplasm and extracellular fluid (② and ④ in Figure 16.23*C*) is carried by small movements of many ions—cations moving in the direction of standard current flow and anions moving in the opposite direction. In the region of the impulse (① in Figure 16.23*C*), current is carried inward by movement of Na^+ ions through the membrane. Such current resulting from an ion flow through the membrane is called **ionic current**. On the other hand, the outward current flow (③ in Figure 16.23*C*) does not represent ion flow through the membrane. Rather, it results from a change in the distribution of charge (ions) across the membrane capacitance. Current flow resulting from such a change in charge distribution is called **capacitive current**. The longitudinal current flow inside the axon (② in Figure 16.23*C*) moves positive charges onto the inner surface of the membrane and takes away negative charges. Likewise, the extracellular current flow ④ moves negative charges onto the outer surface of the membrane and removes positive charges (Figure 16.23*D*). Thus, the inside becomes less negative and the outside becomes less positive (the membrane is depolarized). Even though no ions cross the membrane, current flows across the membrane capacitance and completes the circuit. The distinction between ionic current and capacitive current explains why both inward (ionic) current at ① and outward (capacitive) current at ③ can be depolarizing.

Thus, an action potential at one region of an axon tends to depolarize adjacent regions on both sides by setting up local circuits of current flow. What is the effect of this depolarization on the membrane region ahead of the impulse (to the left in Figure 16.23)? Sufficient depolarization activates the Hodgkin cycle (Figure 16.16); that is, each membrane region when depolarized to threshold undergoes a regenerative increase in Na^+ permeability and generates its own impulse. Because an impulse at one locus depolarizes the adjacent membrane to threshold for impulse generation at the next locus, the Hodgkin cycle provides the voltage amplification necessary for impulse propagation without diminution in amplitude. Thus, we can see that the undegraded propagation of impulses as well as their all-or-none amplitude results from the fact that the underlying ion permeabilities are *voltage dependent*. Without the voltage-dependent permeability changes of the Hodgkin cycle, neural electrical signals would be graded in amplitude and

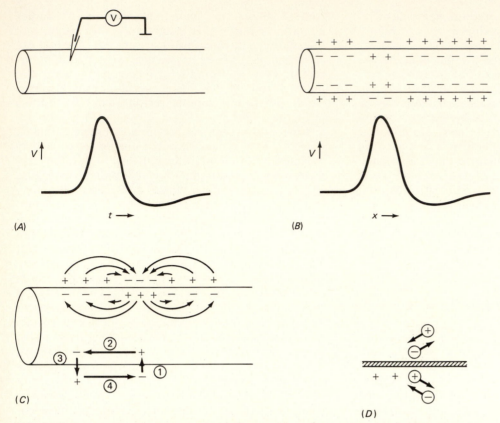

Figure 16.23 Propagation of the action potential. An action potential is recorded by a microelectrode at one place along the axon in (A); voltage changes as a function of time. In (B), a propagated action potential is viewed (conceptually) at one instant in time. The voltage change as a function of distance (x) at that instant is equivalent to the V versus t function in (A), since the impulse is propagated at constant velocity. (C) Local circuits of current flow ahead of an advancing action potential and behind it. The local circuit is divided into four components at the bottom of the axon. These components are described in the text. (D) This shows the way in which capacitative current at ③ in (C) depolarizes the membrane ahead of the advancing action potential. Further description is given in the text.

would spread over distance passively and decrementally, and conduction distances would be limited by the membrane length constant.

As shown in Figure 16.23C, the local circuits of current flow set up by an impulse also tend to depolarize the membrane region to the right, behind the advancing wavefront of the impulse, as well as ahead of it. What then prevents an impulse propagating from right to left in Figure 16.23 from triggering an impulse propagating backward from left to right? The axon itself is not polarized; it can in fact conduct equally well in either direction. An impulse artificially generated via an electrode in the middle of an axon will be propagated in both directions. The reason that the membrane behind an impulse is not automatically reexcited by the local currents is that the membrane is still in its refractory period. There are three aspects of the ionic mechanisms of action potentials that produce the absolute and relative refractory periods after the impulse and thereby prevent reexcitation. (1) The inactivation of sodium conductance (which turns off the voltage-dependent increase in Na$^+$ permeability) lasts for about a mil-

lisecond after the impulse, gradually decreasing to normal resting level following repolarization of the membrane. (2) The increased potassium permeability (the slowest of the three voltage-dependent processes in onset) gradually decreases to resting levels following repolarization. The residual P_K increase after an impulse may hyperpolarize the membrane toward E_K for a few milliseconds after the impulse (Figure 16.20). Thus, the membrane is hyperpolarized away from its voltage threshold. This after-hyperpolarization, called the undershoot, is prominent in squid axons and in some other neurons but is absent in some kinds of neurons and in vertebrate skeletal muscle fibers. (3) The increase in P_K referred to in (2) also renders the membrane refractory by its effect on membrane resistance. The increased K$^+$ permeability following an impulse corresponds to a decreased membrane resistance. Thus, by Ohm's law ($E = IR$), the local currents set up by the impulse have less of a depolarizing effect on the membrane, since they act on a decreased resistance. (If the resistance were zero, the currents would produce no depolarization at all.) These three effects, of which the first is probably the most important, explain why a membrane is refractory after an impulse. During the absolute refractory period (~ 1 ms), its voltage threshold is infinite; no amount of depolarization can open the inactivated Na^+ channels. During the relative refractory period, the voltage threshold, membrane potential, and membrane resistance gradually return to normal resting levels over a time course of a few milliseconds. The refractory effects outlast the backward spread of local currents and prevent reexcitation.

Conduction Velocity of Action Potentials

In the last section we saw that the active conduction of the action potential depends on local currents depolarizing the membrane ahead of the action potential, so that this membrane reaches its threshold and generates an impulse in turn. Several factors can affect the speed with which an impulse is propagated or actively conducted along an axon (or other membrane surface). In general, these factors affect one or both of two parameters of conduction: the spatial parameter and the temporal parameter. The greater the spatial spread of local currents along the axon, the greater the distance over which they can depolarize the membrane to threshold. Therefore, any factor that increases the membrane length constant (space constant, see Figure 16.9) tends to increase the conduction velocity of the action potential. Two major evolutionary adaptations for increased conduction velocity—increased axon diameter and myelination—derive much of their effect from increasing the length constant. The temporal parameter is the other general determinant of conduction velocity. The longer the time it takes the membrane to reach a threshold level of depolarization, the slower the conduction velocity. An increase in membrane capacity, for example, increases the time constant of the membrane (Figure 16.7C) so that it takes longer for a patch of membrane to be depolarized to threshold. Intrinsic membrane properties such as differences in membrane sodium–channel concentration may affect conduction velocity by influencing these spatial and temporal parameters. Such intrinsic membrane differences, however, appear to have relatively minor effects on conduction velocity. Rather, the three major variables employed by animals in the evolution of more rapid conduction velocity are axon diameter, myelination, and temperature.

Axon Diameter and Conduction Velocity   One of the major determinants of conduction velocity is axon diameter; larger diameter axons conduct more rapidly. This relation accounts for the repeated evolution of giant fiber systems (see Box 16.1). The increase in conduction velocity with diameter depends on the effect of diameter on the axon length constant. Referring to Figure 16.9D, it should be apparent that the length constant depends on the ratio of resistance across the membrane to resistance along the axon. Decreasing the membrane resistance (R_m) makes it easier for current to leak through the membrane and shortens the length constant (λ). On the other hand, decreasing the internal longitudinal resistance of the axoplasm (R_i) makes it easier for current to continue flowing along the axon and makes λ longer. In fact, the dependence of the length constant on the resistances in Figure 16.9 is

$$\lambda = \left(\frac{R_m}{R_i + R_o} \right)^{1/2}$$

In practice, the external longitudinal resistance R_o is considered to be (relatively) very small and is ignored. The membrane surface area increases proportionately with increasing axon diameter. This lowers R_m (by adding resistances in parallel). On the other hand, R_i decreases in proportion to an increase in cross-sectional area of the axoplasm, that is, in proportion to the square of the diameter. The net effect is that the ratio R_m/R_i increases linearly with increasing diameter. If other factors are equal, the velocity increases linearly with increasing λ and thus with the square root of the diameter. Figure 16.24 shows that this theoretical relation is approached for some unmyelinated axons. That the relation of velocity and diameter differs for different kinds of axons implies that other factors, including intrinsic membrane differences, are also involved.

Myelination   Myelinated axons of vertebrates represent a tremendous evolutionary advance, by allowing very high conduction velocities with relatively small axon diameters. A *myelinated axon* is wrapped with 200 or more concentric layers of glial membrane (Schwann cells in peripheral nervous systems, oligodendroglia in central nervous systems). The glial cytoplasm is extruded from between the glial membrane layers, so that the whole wrapping serves as an insulating layer (Figure 16.25). This multiply-wrapped insulating layer, termed *myelin*, is interrupted at intervals of a millimeter or so along the length of the axon by *nodes of Ranvier*, at which the glial wrappings are absent. The effect of myelination is to insulate the major part of the axon (the *internodes*) nearly completely, leaving only the nodes of Ranvier as loci of ion permeability and current flow.

The primary effect of the multiple wrappings of glial membrane is to increase membrane resistance of the internode by 1000–10,000-fold, by adding high-resistance membranes in series. Thus, when a node of Ranvier undergoes an action potential, the local currents cannot leak out through the high membrane resistance of the internode, but instead must flow to the next node of Ranvier. Myelination thus increases the axon length constant by increasing the effective membrane resistance.

A second, equally important function of myelin is to decrease membrane capacitance. If myelin only increased membrane resistance (without decreasing

---

BOX 16.1   GIANT AXONS

Since an increase in axon diameter increases the velocity of impulse propagation and since animals often face circumstances in which great speed of response is of advantage, very large axons have evolved in several animal groups. Such axons are called *giant axons,* a term meaning only that the axons are much larger than other axons in the same animal. Some axons are truly giant in cellular dimensions, such as the third-order giant axon in squid, which may be 1000 μm in diameter. At the other extreme, the giant axons in the fruit fly *Drosophila* are only about 4 μm in diameter but are still an order of magnitude larger than the other axons nearby.

Giant axons usually mediate activities such as escape movements, for which speed and short latency of response are paramount. The role of giant axons in squid escape locomotion provides an instructive example. Squid escape movements occur by jet propulsion. The animal contracts the muscles of its outer mantle to expel a jet of water from the mantle cavity through a moveable siphon. The giant axons serve to ensure a rapid and simultaneous contraction of all the mantle muscles, a necessary condition for fast, effective locomotion. There are actually three sets of giant axons, arranged in series. In the brain are two partially fused first-order giant axons, either of which can initiate activity of the entire propulsion system. Activation of a first-order giant axon excites second-order giant axons, which extend from the brain to paired stellate ganglia at the anterolateral margins of the mantle. Several third-order giant axons radiate out from the stellate ganglia to the mantle muscles, and are the motor axons that cause these muscles to contract. The muscles at the posterior end of the elongate mantle are much farther from the stellate ganglion than are the anterior muscles. As an adaptation ensuring simultaneous contraction of these widespread muscles, the radiating third-order giant axons differ greatly in size. The largest and most rapidly conducting third-order giant axon extends to the most distant, posterior portion of the mantle. (Consequently, this is the axon usually used in experiments!) It is this arrangement of the giant axons that enables the signal for muscle contraction to arrive simultaneously at all parts of the mantle, as rapidly as possible.

Giant axons have evolved independently in different groups of animals, and there have been different evolutionary paths to achieving axons of large diameter. Giant axons can be unicellular (with one soma) or multicellular (with several somata). Unicellular giant axons are found in cockroaches (see beginning of chapter), some annelids such as *Protula,* and crayfish (medial giant axons). Although most giant axons occur in invertebrates, the Mauthner neurons of fish and amphibians (see p. 557) are examples of unicellular, myelinated vertebrate giant axons.

Multicellular giant axons may be syncitial or segmented. Squid third-order giant axons develop by syncitial fusion of processes of 300–500 cells that retain discrete somata. Crayfish lateral giant axons and those of many annelids, in contrast, are made up of segmentally arranged cells that form end-to-end junctions. Imulses are electrically transmitted from cell to cell through these low-resistance junctions (see Chapter 17).

Apart from the evolutionary significance of giant axons to the animals possessing them, these axons are important experimental preparations in physiology. Their large size facilitates studies such as the voltage-clamping experiments described earlier in this chapter. Moreover, many giant axons are identifiable neurons that can be studied as unique, known entities in every member of a species. This ability to study identified neurons is an important tool in many aspects of contemporary neurobiology.

---

capacitance), its effect on the axon length constant would be largely offset by an increase in the membrane time constant (since $\tau = R_m C_m$). This increase in time constant would tend to slow conduction velocity, as noted above. Capacitance, however, is in-

versely proportional to the distance separating the charges on the "plates" of the capacitor, which in this case is the distance between the axoplasm and the extracellular fluid. This distance is increased in proportion to the number of glial membrane wrap-

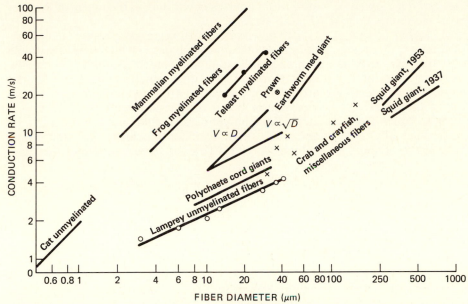

Figure 16.24 Velocity of nerve impulse conduction as a function of fiber diameter in a variety of animals. Points not connected by lines are fibers of different types. [From *Structure and Function in the Nervous System of Invertebrates* by Theodore H. Bullock and G. Adrian Horridge. Copyright © 1965. Reprinted with the permission of W.H. Freeman and Company.]

pings, so that capacitance is decreased about 1000-fold. Thus, the increase in R_m of the internode is offset by the decrease in C_m, and the membrane time constant is nearly unaffected. Capacitance is defined as the amount of charge stored per unit voltage ($C = Q/V$). It is calculated that a myelinated axon at

rest has as much charge stored at each node of Ranvier (2–3 μm in length) as at each internode (~1 mm in length)! Myelination then greatly increases conduction velocity by increasing the axon length constant without increasing the time constant. Currents are required to travel to the next node before cross-

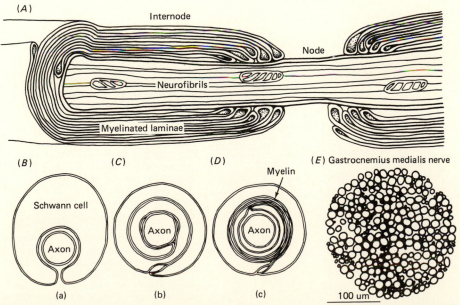

Figure 16.25 Myelinated nerve fibers. (*A*) diagrams a node of Ranvier in a longitidinal section. The layers of laminated myelin terminate leaving a short, bare region of axonal membrane at the node. (*B*), (*C*), and (*D*) show sequences in the formation of the myelin sheath by the Schwann cell. (*E*) shows a cross-section of myelinated fibers of a wide range in size in a muscle nerve of the cat. [From J.C. Eccles, *The Understanding of the Brain,* 2nd ed. McGraw-Hill, New York, 1977.]

ing the membrane, and the currents are not slowed by having to displace much charge in the internode.

As a consequence of myelination, action potentials can only occur at the nodes of Ranvier; there is no path for ions to carry current at the high-resistance internodes. Myelinated axons therefore exhibit *saltatory conduction*, in which the action potential jumps (saltates) from node to node without active propagation in the internode (Figure 16.26A). This hypothesis has been confirmed experimentally by demonstrating that transmembrane currents associ-

ated with impulses occur only at the nodes of Ranvier (Figure 16.26B).

The myelinated vertebrate axon represents an important evolutionary advance, which makes possible neural coordinating and control systems that employ large numbers of axons, which conduct rapidly but are relatively small. A frog myelinated axon 12 μm in diameter has a conduction velocity of 25 m/s at 20°C. An unmyelinated squid giant axon must be about 500 μm in diameter to achieve the same 25-m/s velocity at 20°C! Thus, myelination allows the same

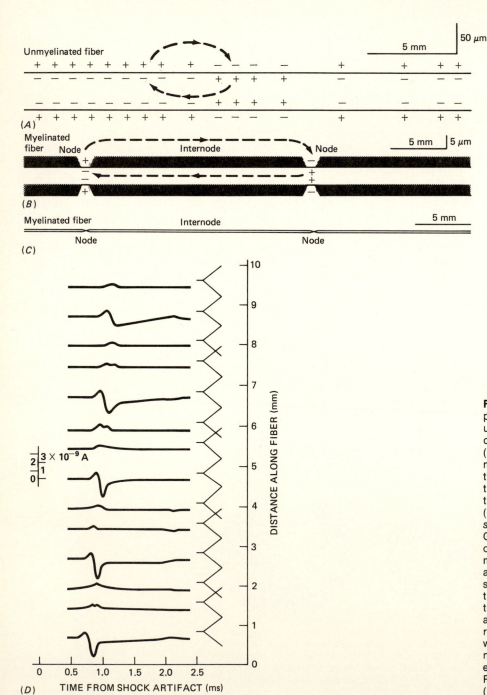

Figure 16.26 Propagation of an impulse along a nerve axon. (A) In an unmyelinated axon propagation is continuous, as shown in Fig. 16–23. (B) In a myelinated axon the transmembrane current flow is restricted to the nodes. The dimensions in (B) are transversely exaggerated as shown by the scale, but are correctly shown in (C). [From J.C. Eccles, *The Understanding of the Brain*, 2nd ed. Mc-Graw-Hill, New York, 1977.] (D). Demonstration of saltatory conduction in a myelinated axon. Membrane currents are measured along the length of a single frog myelinated axon. Each trace shows the difference between the currents recorded 0.75 mm apart, as indicated at the right. Inward current appears downward. Note that inward current is recorded only when a node occurs between the recording electrodes. [From A.F. Huxley and R. Stampfli, *J. Physiol.* **108:**315–339 (1949).]

velocity to be achieved with a 40-fold reduction in diameter and 1600-fold reduction in cross-sectional area and volume. The evolution of homeothermy in birds and mammals has further increased these savings (see discussion of temperature effects below).

Although myelin is usually considered to have evolved exclusively in vertebrates, some Crustacea have analogous sheaths with glial wrappings and increased conduction velocities similar to those of vertebrate compact myelin. The 30-μm myelinated axons of prawns conduct at 20 m/s at 17°C, a velocity comparable to that of a 350-μm squid axon. Myelinated shrimp axons 100–120 μm in diameter conduct at velocities exceeding 90 m/s at 20°C, rivaling the fastest mammalian myelinated axons. In contrast, unmyelinated lobster axons of the same 100–120-μm diameter have conduction velocities of only 8 m/s. Unlike vertebrate myelinated neurons, however, which are found in large numbers throughout vertebrate nervous systems, the convergently evolved arthropod myelinated axons are rare, specialized adaptations in high-velocity locomotory escape systems.

Temperature   The last factor that we shall consider, which affects conduction velocity, is temperature. The gating of the voltage-dependent ion channels and thus the time course of membrane depolarization to threshold are temperature dependent. The Q_{10} (see

p. 84) of the temperature effect on conduction velocity is about 1.8 for both myelinated and unmyelinated fibers. A 12–14-μm frog myelinated axon conducts at 25 m/s at 20°C, but a 3.5–4-μm cat myelinated axon conducts at the same 25 m/s at 37°C. Thus, the evolution of homeothermy (see p. 99) has allowed further miniaturization—and higher conduction velocities—in birds and mammals.

EXTRACELLULAR RECORDING OF NERVE IMPULSES

In our discussion of the properties of nerve impulses and the ionic mechanisms underlying their generation, we have depended on intracellular recording techniques to explain underlying concepts and causes. Intracellular recording with microelectrodes, however, is not always possible. Often the signs of impulse activity are recorded with pairs of extracellular electrodes. In this case, the potentials recorded are not true transmembrane potentials but instead result from the local currents flowing around the axon. Figure 16.27 shows the arrangement for such extracellular recording of impulse activity. You should recall that a recorded voltage is a *potential difference* between two points (two electrodes). First, let us assume that both electrodes are just external to the nerve axon and that the axon is inactive [Figure 16.27, (1)]. There is no potential dif-

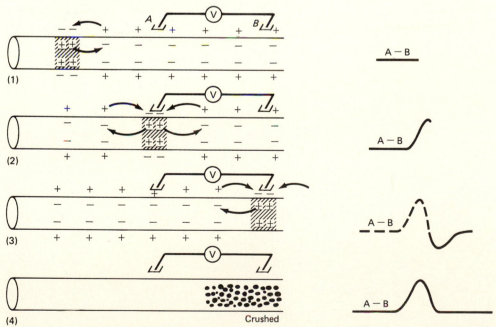

Figure 16.27 Extracellular recording of action potentials. Two extracellular electrodes are close to the outer surface of the axon membrane. Initially (1), there is no potential difference between them. (2) As the impulse passes electrode A, the electrode detects the resultant extracellular negativity, resulting in a potential difference between the electrodes (negativity at A is drawn as an upward deflection). (3) Electrode B becomes negative as the impulse passes it, so that the potential difference A − B is reversed. This is called *biphasic* recording. (4) If the nerve is crushed at electrode B or if the electrode is distant from the axon, a *monophasic* recording results.

ference between the electrodes (A − B = 0). As an impulse approaches electrode A, the outside of the membrane becomes negative relative to electrode B. (An impulse can be considered to be a wave of extracellular negativity sweeping along the axon surface.) If the oscilloscope is arranged so that negativity at electrode A is an upward displacement of the oscilloscope beam, then the beam is deflected upward [Figure 16.27, (2)]. An instant later, the impulse reaches electrode B [Figure 16.27, (3)]. At this time, the membrane near A is repolarized, while that near B is depolarized (outside negative). Electrode A is then positive relative to B, producing a downward deflection. So with both electrodes on active regions of the axon, the extracellularly recorded impulse is *biphasic* (two phases—up then down). If electrode B is placed on a region of the axon that has been crushed and is inactive, then a *monophasic* impulse is recorded [Figure 16.27, (4)]. Likewise, if electrode B is placed in a location distant from the axon, the impulse is recorded monophasically, since only electrode A is close enough to the axon to detect the local circuits of current flow. The waveform of the impulse recorded extracellularly thus depends on the

geometry of the electrodes recording it; it is a reflection of the true transmembrane potential changes but is not an accurate representation of them.

Compound Action Potentials

A *compound action potential* is the summated, extracellularly recorded activity of several axons that are active at nearly the same time. Compound action potentials are often recorded from *tracts* (bundles of nerve fibers within the central nervous system) or from *nerves* (bundles of nerve fibers in the peripheral nervous system). Such bundles can contain from a few to more than a million axons. Suppose that we place a whole peripheral nerve, such as a frog sciatic nerve or a *Limulus* leg nerve, across a set of wire recording electrodes (Figure 16.28). A single, strong electrical stimulus, applied by stimulating electrodes at one end of the nerve, initiates a single impulse in each of the axons. The impulses propagate down the axons past the recording electrodes. Since the largest axons have the highest conduction velocities, their impulses arrive earliest and are recorded first (Figure 16.28 *a,b*). Impulses in smaller and smaller axons are

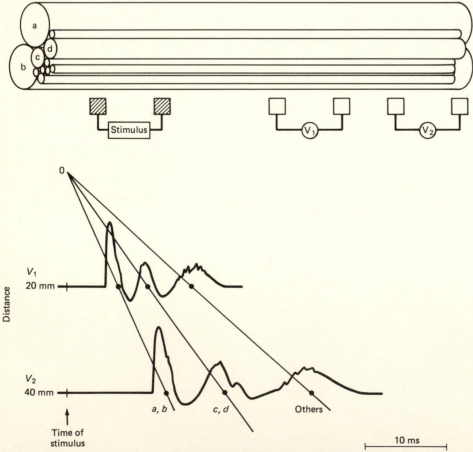

Figure 16.28 Compound action potential, recorded extracellularly from several axons of differing diameter. The impulses of the largest axons (*a, b*) have the fastest conduction velocity and pass the recording electrodes Ⓥ₁, Ⓥ₂ first. The slopes of the lines connecting the peaks in the compound action potential are a measure of conduction velocity.

recorded with increasing latencies after the stimulus. Each peak in this compound action potential represents the summated, extracellularly recorded activities of many axons with similar conduction velocities. Vertebrate peripheral nerve fibers are often classified according to the peak to which they contribute in a compound action potential. The largest, fastest myelinated fibers are designated A fibers, and the slow, small unmyelinated axons are called C fibers.

Unlike unitary action potentials, which are all-or-none, compound action potentials appear graded in amplitude. With increasing stimulus amplitude, the amplitude of a peak in the compound action potential increases, up to a maximum level. This is because the increasing stimulus excites more and more axons to threshold; each contributes an all-or-none potential, but the *summated* potential recorded from the whole group increases. This increase can appear smooth and continuous in a vertebrate nerve such as the frog sciatic nerve, because very many axons contribute to each peak of the compound action potential. Other nerves such as the *Limulus* leg nerve,

however, have few enough large axons that the all-or-none contribution of each can clearly be seen.

Amplitudes of Extracellularly Recorded Impulses

A final point about extracellularly recorded action potentials is that their amplitudes are much smaller than those of intracellularly recorded action potentials. An intracellular electrode can measure the true transmembrane potential changes of the action potential, which have typical amplitudes of about 100 mV (see Figure 16.15). Extracellular electrodes, however, measure only the local extracellular potential changes that result from local circuits of current flow. These are much smaller, ranging from over 1 mV to less than 10 μV. This amplitude depends on the distance of the electrode from the axon and, more importantly, on the size of the axon. Unlike intracellularly recorded action potentials, which have approximately the same amplitude regardless of axon diameter, extracellularly recorded impulses are larger for larger axons (which have larger local currents). Because of this size difference, it is possible to identify individual units (single neurons) in an extracellular record from several axons (Figure 16.29).

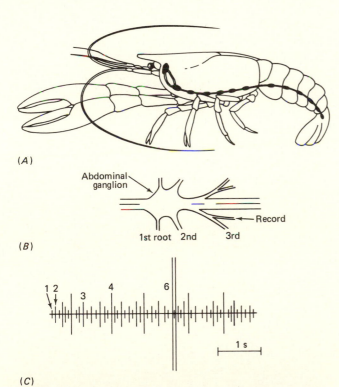

Figure 16.29 Extracellular recording of impulses from identified motor neurons in a crayfish. (*A*) Diagram showing the position of the ventral nerve cord of the crayfish CNS. Each abdominal segment contains an abdominal ganglion. (*B*) An abdominal ganglion (dorsal view). A branch of the third root contains only six motor neuron axons, which control postural flexion of the abdomen. (*C*) Extracellularly recorded activity of postural flexor motor axons. The amplitude of the extracellularly recorded impulse is porportional to axon diameter; hence axon 1 is smallest and axon 6 (which is characteristically active in impulse pairs) is the largest. Axon 5 is silent during flexion.

SELECTED READINGS

Abbott, N.J. 1985. Are glial cells excitable after all? *Trends Neurosci.* **8**: 141–144.

Adams, D.J., S.J. Smith, and S.H. Thompson. 1980. Ionic currents in molluscan soma. *Annu. Rev. Neurosci.* **3**: 141–167.

Adams, P. 1982. Voltage-dependent conductances of vertebrate neurons. *Trends Neurosci.* **5**: 116–119.

Aidley, D.J. 1978. *The Physiology of Excitable Cells*, 2nd ed. Cambridge University Press, New York.

Aldrich, R.W. 1986. Voltage-dependent gating of sodium channels: towards an integrated approach. *Trends Neurosci.* **9**: 82–86.

Armstrong, C.M. 1981. Sodium channels and gating currents. *Physiol. Rev.* **61**: 644–683.

Barchi, R.L. 1988. Probing the molecular structure of the voltage-dependent sodium channel. *Annu. Rev. Neurosci.* **11**: 455–495.

Bretscher, M.S. 1985. The molecules of the cell membrane. *Sci. Am.* **253**(4): 100–108.

Bullock, T.H., R. Orkand, and A. Grinnell. 1977. Introduction to Nervous Systems. Freeman, San Francisco.

Bullock, T.H. 1984. Comparative neuroscience holds promise for quiet revolutions. *Science* **225**: 473–478.

Camhi, J. 1980. The escape system of the cockroach. *Sci. Am.* **243**(6): 158–172.

Dorsett, D.A. 1980. Design and function of giant fiber systems. *Trends Neurosci.* **3**: 205–208.

Hagiwara, S. and L. Byerly. 1981. Calcium channel. *Annu. Rev. Neurosci.* **4**: 69–125.

Hille, B. 1984. *Ionic Channels of Excitable Membranes*. Sinauer, Sunderland, MA.

Hobbs, A.S. and R.W. Albers. 1980. The structure of proteins involved in active membrane transport. *Annu. Rev. Biophys. Bioeng.* **9**: 259–291.

Hodgkin, A.L. 1964. *The Conduction of the Nervous Impulse*. Liverpool University Press, Liverpool.

Hodgkin, A.L. and A.F. Huxley. 1952a. Currents carried by sodium and potassium ions through the membrane of the giant axon of *Loligo*. *J. Physiol.* **116:** 449–472.

Hodgkin, A.L. and A.F. Huxley. 1952b. The components of the membrane conductance in the giant axon of *Loligo*. *J. Physiol.* **116:** 473–496.

Hodgkin, A.L. and A.F. Huxley. 1952c. The dual effect of membrane potential on sodium conductance in the giant axon of *Loligo*. *J. Physiol.* **116:** 497–506.

Hodgkin, A.L. and A.F. Huxley. 1952d. A quantitative description of membrane current and its application to conduction and excitation in nerve. *J. Physiol.* **117:** 500–544.

Hodgkin, A.L., A.F. Huxley, and B. Katz. 1952. Measurement of current–voltage relations in the membrane of the giant axon of *Loligo*. *J. Physiol.* **116:** 424–448.

*Kandel, E.R. and J.A. Schwartz (eds.). 1985. *Principles of Neural Science*, 2nd ed. Elsevier, New York.

*Katz, B. 1966. *Nerve, Muscle, and Synapse*. McGraw-Hill, New York.

*Kuffler, S.W., J.G. Nicholls, and A.R. Martin. 1984. *From Neuron to Brain*, 2nd ed. Sinauer, Sunderland, MA.

Miller, R.J. 1987. Multiple calcium channels and neuronal function. *Science* **235:** 46–52.

McDonald, T.F. 1982. The slow inward calcium current of the heart. *Annu. Rev. Physiol.* **44:** 425–434.

Nauta, W.J.H. and M. Feirtag. 1979. The organization of the brain. *Sci. Am.* **241**(3): 88–111.

Rogart, R. 1981. Sodium channels in nerve and muscle membrane. *Annu. Rev. Physiol.* **43:** 711–725.

Poo, M.-M. 1985. Mobility and localization of proteins in excitable membranes. *Annu. Rev. Neurosci.* **8:** 369–406.

Ritchie, J.M. 1979. A pharmacological approach to the structure of sodium channels in myelinated axons. *Annu. Rev. Neurosci.* **2:** 341–362.

Rogawski, M.A. 1985. The A current: How ubiquitous a feature of excitable cells is it? *Trends Neurosci.* **8:** 214–219.

Roberts, A. and B.M.H. Bush (eds.). 1981. *Neurons Without Impulses: Their Significance for Vertebrate and Invertebrate Nervous Systems*. Cambridge University Press, Cambridge.

Rose, S.P.R. 1980. Can the neurosciences explain the mind? *Trends Neurosci.* **3**(5): I–IV.

Shepherd, G.M. 1979. *The Synaptic Organization of the Brain*, 2nd ed. Oxford University Press, New York.

*Shepherd, G.M. 1988. *Neurobiology*, 2nd ed. Oxford University Press, New York.

Strichartz, G., T. Rando, and G.K. Wang. 1987. An integrated view of the molecular toxinology of sodium channel gating in excitable cells. *Annu. Rev. Neurosci.* **10:** 237–267.

Thomas, R.C. 1972. Electrogenic sodium pump in nerve and muscle cells. *Physiol. Rev.* **52:** 563–594.

Tsien, R.W. 1983. Calcium channels in excitable cells. *Annu. Rev. Physiol.* **45:** 341–358.

Tsien, R.W. 1987. Calcium currents in heart cells and neurons. In L.K. Kaczmarek and I.B. Levitan (eds.), *Neuromodulation: The Biochemical Control of Neuronal Excitability*, pp. 206–242. Oxford University Press, New York.

Waxman, S.G. 1983. Action potential propagation and conduction velocity. *Trends Neurosci.* **6:** 157–161.

See also references in Appendix A.

chapter 17

Synapses

Neurons are not continuous with each other; in fact, they are rarely in direct physical contact with each other. Electron micrographs show that at the specialized regions termed **synapses** neurons are usually separated by a 20–30-nm space. At regions other than synapses, neurons are usually additionally separated from each other by intervening processes of glial cells. This anatomical discontinuity of neurons is a central tenet of the neuron doctrine (see Chapter 16), which has important functional implications. Much of the complexity of function of nervous systems is thought to result from the properties of the synaptic transmission of signals across this anatomical discontinuity.

To make clear the distinction between the functions of axonal transmission and synaptic transmission, let us again consider the example of the cockroach escape response discussed at the beginning of Chapter 16. Recall that the cockroach jumps in response to airborne vibrations—wind puffs or sound—that excite receptors on its cerci. In a simple view of this behavior, the receptor neurons excite a small group of giant interneurons, which excite leg motor neurons, which in turn excite leg muscles to contract (Figure 16.1). Thus, in this example there are two sets of *nerve–nerve* synapses and one set of *nerve–effector* synapses. The axons of each of the three groups of neurons have a single major functional role; their job is to convey a train of impulses over distance, rapidly and without degradation. For example, a single cercal receptor neuron generates a train of impulses that constitutes a signal or message about the environment (this aspect will be discussed further in Chapter 18). This signal (train of impulses) travels along the axon of the sensory neuron by impulse propagation, arriving at the axon terminals essentially unchanged after a delay of a millisecond or two. Synaptic transmission, in contrast, involves changing the signal rather than simply passing it on. A *postsynaptic* neuron (one of the giant interneurons in this example) typically does not generate an impulse as a result of each impulse in the *presynaptic* neuron (a cercal sensory neuron here). Instead, many presynaptic impulses may be required to generate one postsynaptic impulse. Moreover, the giant interneuron may receive synaptic input from hundreds of cercal sensory neurons and other kinds of neurons. The signal it generates is influenced by synaptic *convergence* from all the cells presynaptic to it, some of which make *excitatory* and some of which make *inhibitory* synaptic connections. Thus, synaptic transmission is rarely a simple relay in which each presynaptic impulse generates a postsynaptic impulse. Rather, the activity of a postsynaptic cell is a complex function of inputs from many presynaptic neurons that are added together and may mix with intrinsic activity of the postsynaptic cell. This complex blending of signals is termed **neuronal integration**.

The mechanism of synaptic transmission was a subject of much debate in the first half of the twentieth century. One group (the "sparks") argued that synaptic transmission was by direct electrical means. The other group (the "soups") postulated that transmission was by a chemical mechanism. It is now clear that most synaptic transmission is chemical; depolarization of the presynaptic terminal causes it to release a chemical transmitter which diffuses across the *synaptic cleft* to affect the postsynaptic cell. Direct electrical transmission also occurs but is much less common. We shall discuss electrical transmission first, to set the stage for treatment of the more common and more complex chemical transmission process.

ELECTRICALLY MEDIATED TRANSMISSION

Why Is Electrical Transmission Unusual?

Electrical synaptic transmission requires that the local circuits of current flow set up by an action potential in the presynaptic neuron be sufficient to affect the activity of the postsynaptic neuron. Figure 17.1 shows the paths of current flow at a synaptic terminal. In principle, some of the current flows through the postsynaptic cell and depolarizes it where the current exits. Quantitative considerations, however, show that this effect is so small as to be negligible at most synapses. The details of this argument are beyond our scope and are given in Katz (1966). A simplified version of the argument is sufficient to show that electrical transmission, far from being inevitable, is impossible except in special circumstances.

The amplitude of an action potential is about 100 mV. The local currents ahead of the action potential must depolarize the membrane by about 20 mV to reach threshold. Suppose a cell membrane is placed across an axon in the path of the internal longitudinal current. As shown in Figure 17.1, this membrane would act as a resistance in series in the current path and would decrease the current resulting from the 100-mV potential. The amount by which the current would decrease depends on the axon diameter. Katz calculated that for an axon 5 μm in diameter, the interposed membrane would decrease the current to less than 1 percent. Thus, the theoretical maximum depolarization past the membrane would be 1 mV (1 percent of 100 mV), and conduction of the action potential would certainly fail. At a synapse, however, the situation is much less favorable for electrical transmission. Instead of a single membrane, the two cells are separated by two membranes and a fluid-filled space (Figure 17.1). Current leaving the presynaptic ending may flow through the postsynaptic cell or through the fluid-filled space of the synaptic cleft. Most of the current flows through the low-resistance pathway of the cleft, rather than through the high-resistance pathway across two additional membranes. (The amount of current in any path is inversely proportional to the resistance in that path.) For a 5 μm axon interrupted by two membranes and a cleft (as at a synapse), the calculated attenuation of current is to 1/10,000. Therefore, the 100-mV action potential produces a 10-μV depolarization of the postsynaptic cell, an insignificant change that is below the limit of detectability with an intracellular electrode!

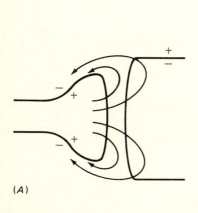

(A)

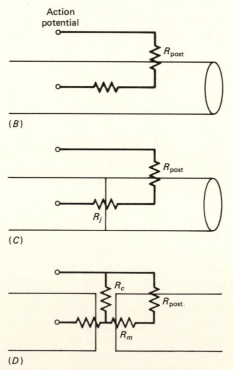

Figure 17.1 (A) Pattern of current flow at a synapse, showing how the currents resulting from a presynaptic impulse might be expected to depolarize a postsynaptic cell directly. (B) An action potential at one locus of an axon must depolarize the adjacent membrane at about 20 mV to trigger an action potential at the new location. (C) Addition of a membrane partition inside the axon adds a large resistance (R_j), which decreases the current available to depolarize the postjunctional membrane (R_{post}). (D) Two membranes and a cleft between them constitute a much greater barrier to current flow. Most of the current flows through the low-resistance cleft (R_c); very little current flows through the higher postjunctional resistances (R_m and R_{post}).

From the above argument, it is clear that electrical synaptic transmission is impossible under normal circumstances. What conditions are required to make electrical transmission possible? It is apparent from Figure 17.1 that current flow into the postsynaptic cell is increased by decreasing the resistances of the apposed membranes, or by increasing the cleft resistance R_c, or both. The membrane resistance can be decreased by changes in intrinsic membrane structure or by increasing the contact area of the pre- and postsynaptic cells. The cleft resistance can be increased by narrowing the cleft or by increasing the contact area, which increases the distance that current must flow through the narrow cleft. All these changes have been implicated in cases of electrical transmission.

Gap Junctions

The major structural specialization for electrical transmission is the *gap junction,* also called a nexus or electrotonic junction. A gap junction is a specialized locus where the pre- and postsynaptic membranes are only 2–3 nm apart instead of 20–30 nm. The resistance of the gap-junction membranes to current flow between cells is decreased by several orders of magnitude, so that the overall resistance of membranes between two joined cells is about 1 Ω · cm^2 rather than about 1000 Ω · cm^2. Gap junctions provide a low-resistance path for current flow, electrically coupling the cells they join. Thus, any electrical change in one cell is recorded in the other, with some attenuation but with negligible delay. Figure 17.2 shows the typical effects of this electrical coupling. Depolarization or hyperpolarization of cell A produces an attenuated corresponding change in cell B. The degree of attenuation (coupling ratio) is a measure of the strength of coupling of the cells. Most electrical junctions act as low-pass frequency filters; that is, they attenuate rapid potential changes (such as impulses) more than slow potentials. Strong electrotonic junctions act as short-latency relay synapses (each presynaptic spike triggering a postsynaptic spike) but in weaker junctions the depolarization of the postsynaptic cell may be subthreshold. Most electrically transmitting synapses are bidirectional, transmitting voltage changes roughly symmetrically in both directions. In the electrical synapse between the crayfish lateral giant axon and giant motor neuron, however, the electrical synapse is rectifying; that is, it allows current flow preferentially in one direction. Thus, an impulse in the lateral giant axon excites the giant motor neuron, but excitation of the motor neuron by other pathways cannot "backfire" to the lateral giant axon.

The structure of gap junctions has been examined by electron microscopy and x-ray diffraction studies. In the region of close membrane apposition, the gap separating the membranes is bridged by a regular array of channel structures termed connexons (Figure 17.3). Each connexon is thought to be composed of protein hexamers surrounding a 2-nm pore that connects the cytoplasm of the cells. These pores are the low-resistance pathways between the cells and are large enough to pass most ions, as well as dye and tracer molecules smaller than about 1000 daltons.

Cases of strong electrotonic coupling are relatively rare in nervous systems. Electrically mediated synaptic relays, in which a presynaptic spike nor-

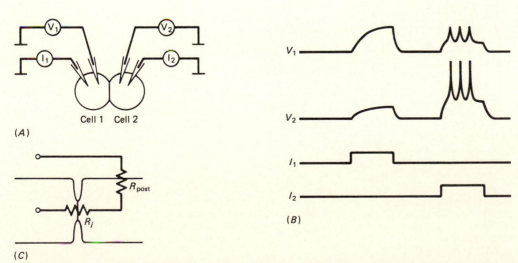

Figure 17.2 Electrical transmission between cells via a gap junction. (A) Electrodes are placed so that current can be passed through either cell and the resultant potential change of both cells is recorded. (B) The cells are electrically coupled so that depolarization of either cell produces a smaller depolarization of the other cell. Cell 2 generates impulses when depolarized, and attenuated impulses are recorded in cell 1 also. (C) Equivalent circuit for electrical transmission. The junctional resistance (R_j) of the gap junction is much lower than that of a normal membrane (cf. Figure 17.1).

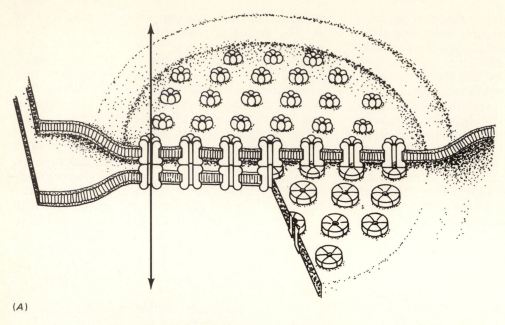

(A)

(B)

Figure 17.3 Models of gap junction structure. (A) The junction is a localized patch of close membrane apposition, the gaps being bridged by channels within protein hexamers termed connexons. (B) The structure of part of a gap junction, as inferred from electron microscopy and x-ray diffraction studies. [Reproduced from *The Journal of Cell Biology*, 1977, **74**:629–645 by copyright permission of The Rockefeller University Press.]

mally triggers a postsynaptic spike, are largely confined to escape systems employing giant fibers (Box 16.1), to groups of neurons that normally fire synchronously, and to systems controlling electric organ discharge in fish (Chapter 19). On the other hand, more and more cases of weak electrical coupling between neurons are being discovered. These weaker junctions may mediate important integrative functions in invertebrate and vertebrate nervous systems.

Gap junctions are not restricted to neurons. They are also found to couple many other cell types, including embryonic cells, vertebrate liver cells, and insect salivary gland cells. In these nonneural cases, the major function of gap junctions may be to allow intercellular passage of organic molecules and thus to mediate direct chemical communication, rather than to allow electrical interaction by serving as ionic current channels.

Electrical Transmission and Synaptic Contact Area

It is possible to have electrically mediated synaptic transmission between neurons without gap junctions, at least in principle. An alternative strategy for electrical transmission is via an increase in the contact area between pre- and postsynaptic cells. Such an increase can decrease the effective membrane resistance (by adding resistances in parallel) and increase the cleft resistance (by adding resistances in series, (Figure 17.1). The segmented giant axons of earthworms and crayfish (see Box 16.1) have electrically transmitting septal synapses with large contact areas, as well as a modest narrowing of the cleft. Electrical synapses in avian ciliary ganglia also have very large contact areas. In all these cases, however, gap junctions or similar membrane-bridging structures have been reported. Thus, it is not clear that a simple increase in contact area, without additional membrane specializations, is a sufficient mediator for electrical transmission in organisms.

Electrically Mediated Synaptic Inhibition

Before we leave the subject of electrically transmitting synapses, we must note that synaptic inhibition can be mediated by direct electrical interaction. An impulse in one neuron can generate local currents that directly hyperpolarize and inhibit an adjacent neuron. The proposed mechanism by which this electrical inhibition occurs is complex and beyond the scope of this book. Electrically mediated inhibition is known only for neurons in the Mauthner neuron circuit that mediates the startle response of fish. Thus, the phenomenon is a special case, too complex and specialized to explore in detail, but too curious and well documented to ignore.

CHEMICAL TRANSMISSION

As we noted in the introduction of Chapter 16, we take it as a working hypothesis that the behavior of organisms results from the integrated activity of interconnected central neurons. Much of this integrative interaction apparently results from chemically mediated synaptic excitation and inhibition. Thus, study of mechanisms of chemical synaptic transmission is central to our understanding of the physiology of nervous systems.

Although a major focus of our interest is on synapses between neurons in central nervous systems, most studies have concentrated on peripheral synapses. The reasons for this discrepancy are technical ones. Synapses in the vertebrate central nervous system, for example, are buried inside a rather large mass of tissue. They cannot be seen through a microscope, nor can their immediate environment be changed readily. Moreover, a single vertebrate central neuron may have tens of thousands of other neurons synapsing onto it, each producing its own synaptic effects. This complexity of organization and difficulty in visualizing and controlling the synaptic environment make it hard to study central synapses. Experimenters have turned to simpler *model systems,* in which detailed investigations of synaptic transmission are easier, and have later applied their findings to the more complex central nervous system. This strategy is analogous to studying squid giant axons to clarify mechanisms of generation of action potentials. The peripheral model system of chemical synaptic transmission that we shall consider is the one from which much of our basic knowledge of synaptic physiology is derived—the vertebrate skeletal neuromuscular junction.

The vertebrate neuromuscular junction has a number of advantages as a model synapse. First, it is anatomically a 1 : 1 synapse, in that each skeletal muscle fiber is innervated by only one motor neuron. Thus, all synaptic responses of a muscle fiber result from activity of a single presynaptic neuron. Second, the neuromuscular junction is physiologically 1 : 1, a relay synapse in which each presynaptic nerve action potential triggers one muscle fiber action potential. Third, the junction is anatomically open, accessible, and microscopically visible at the muscle surface. The experimenter can impale a muscle fiber with an electrode under visual control and can readily change the bathing fluid around the junction. Largely because of these experimental advantages, we know more about the vertebrate neuromuscular junction than about any other kind of synapse.

Overview of Events in Chemical Transmission

We shall first summarize the sequence of events of synaptic transmission at the neuromuscular junction (Figure 17.4). With only minor differences, these

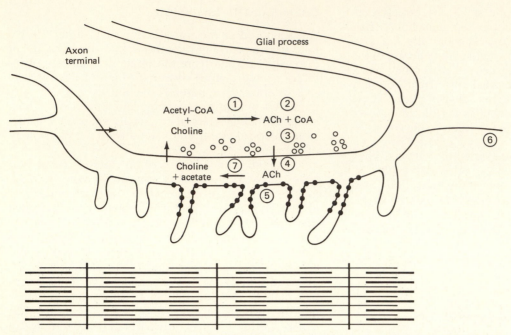

Figure 17.4   Summary of events in chemical synaptic transmission at the vertebrate neuromuscular junction. ① Acetylcholine (ACh) is synthesized from choline and acetyl CoA. ② The ACh is stored at the junction. ③ Depolarization of the terminal triggers ACh release. ④ ACh diffuses rapidly across the synaptic cleft. ⑤ ACh molecules bind to molecules of ACh receptor protein, eliciting a postsynaptic potential. ⑥ The resulting depolarization triggers a muscle fiber action potential (this step occurs at nonjunctional regions of the membrane). ⑦ The ACh is hydrolyzed by acetylcholinesterase.

events underlie all known chemically mediated synaptic transmissions.

1. *Synaptic transmitter* molecules are synthesized in the presynaptic neuron terminal. For the vertebrate neuromuscular junction, the synaptic transmitter is acetylcholine (ACh). ACh is just one of a number of transmitter compounds (see Synaptic Transmitters, p. 443). Acetylcholine is synthesized from choline and acetyl-CoA by the enzyme choline acetyltransferase.
2. The transmitter is stored in the presynaptic terminal. In a neuromuscular junction about half of the ACh is stored in *synaptic vesicles,* with the remainder in the cytoplasm of the terminal.
3. An action potential in the motor neuron depolarizes the terminal, causing release of ACh into the synaptic cleft.
4. Transmitter molecules diffuse across the synaptic cleft to the postsynaptic membrane. Because the diffusion path is so short (50 nm at the neuromuscular junction, 20–30 nm at central synapses) this diffusion is very rapid, accounting for only 20–50 μs of the synaptic delay.
5. Transmitter molecules bind to transmitter *receptor molecules* on the surface of the postsynaptic membrane. At the neuromuscular junction, ACh receptor molecules are located at the junctional folds of the muscle fiber membrane (Figure 17.4). The binding of ACh by ACh receptors triggers membrane changes that evoke an excitatory *postsynaptic potential.*
6. The excitatory postsynaptic potential at the neuromuscular junction depolarizes the muscle fiber membrane to its threshold and initiates a muscle fiber

action potential. The action potential propagates to the ends of the fiber, depolarizing the entire membrane and thereby initiating contraction of the fiber (Chapter 19).
7. The action of the synaptic transmitter is terminated by enzymatic degradation or by reuptake. At the neuromuscular junction, ACh is destroyed by acetylcholinesterase, an enzyme located within the synaptic cleft and at the postsynaptic membrane.

This sequence of events is fairly typical of chemical transmission in general. We shall first examine the nature of synaptic potentials in nerve–muscle and nerve–nerve synapses and then study the physiology of transmitter release and postsynaptic action, using information primarily from the neuromuscular junction and other peripheral model systems.

Synaptic Potentials

The central effect in chemical synaptic transmission is that the transmitter evokes a change in membrane potential of the postsynaptic cell. This effect can be studied best with intracellular microelectrodes in the postsynaptic cell. Consequently, the first direct studies of synaptic transmission were in the late 1940s and early 1950s, when microelectrode techniques were perfected. We will begin our discussion of chemical transmission by reconstructing such an experiment.

Figure 17.5 shows the basic plan and results of an

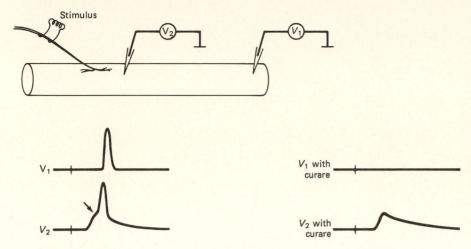

Figure 17.5 Demonstration of the excitatory postsynaptic potential (EPSP) at the vertebrate neuromuscular junction. A remote electrode (V_1) records only the muscle fiber action potential that results from junctional transmission. An electrode near the junction (V_2) also records an EPSP preceding the action potential (arrow). If curare is added, the EPSP is diminished below the threshold for action potential generation, allowing the EPSP to be recorded alone. Unlike the action potential, the EPSP is not propagated and does not reach the remote electrode V_1.

experiment on neuromuscular transmission. We impale a fiber in the sartorius muscle in the hind limb of a frog and arrange to stimulate the nerve to that muscle. If we place the electrode in a part of the fiber distant from the neuromuscular junction (Figure 17.5, V_1) and then stimulate the nerve, we record only a muscle fiber action potential. When we impale the fiber close to the neuromuscular junction (Figure 17.5, V_2), however, nerve stimulation evokes a more complex waveform V_2. The action potential is preceded by a depolarization, which triggers the impulse at its voltage threshold (arrow). It is apparent that another potential change underlies the impulse recorded near the junction. It is also apparent that although the impulse is propagated to the remote recording site (V_1), the other (underlying) potential is not propagated. To see only the underlying potential, we must prevent it from generating an impulse—but how? Bathing the preparation in tetrodotoxin would block the muscle fiber action potential, but it would also block the nerve impulse and prevent excitation of the terminal. Instead, we use curare, which interferes with transmitter action at the neuromuscular junction in a graded manner. When we add a low concentration of curare to the bathing fluid, the amplitude of the underlying potential is decreased but it is otherwise unchanged. By decreasing its amplitude below the threshold for impulse initiation, the underlying synaptic potential can be seen alone (Figure 17.5, V_2 + curare). This potential is termed an *excitatory postsynaptic potential,* or EPSP. The term originally applied to it was *end-plate potential* (EPP), after the term motor end plate, an alternative name for the neuromuscular junction of vertebrate twitch muscle. The term end-plate potential is still in common use, but we use the term excitatory postsynaptic potential to emphasize the now well-documented similarity of these potentials at nerve–nerve and nerve–muscle synapses.

Synaptic potentials, unlike action potentials, are not propagated. With curare added to the bath to prevent the generation of impulses by decreasing EPSP amplitude, the remote electrode (V_1) detects no

potential change (Figure 17.5). The amplitude of the EPSP decreases exponentially with distance from the neuromuscular junction. This exponential decay is described by the length constant of the muscle fiber (see Figure 16.9), a finding that illustrates that the decremental spread of the EPSP is a purely passive process. Recall that the propagation of the action potential results from the fact that the underlying permeability changes are voltage dependent. Therefore, we would predict that the changes underlying an EPSP are voltage independent. In fact, synaptic potentials result from ion permeability changes that are *transmitter dependent and voltage independent.* Now we will perform an experiment to show that these changes are transmitter dependent.

Figure 17.6 shows an experiment that compares the responses of a muscle fiber junctional membrane to nerve stimulation and to directly applied acetylcholine. We record intracellularly from the muscle fiber as before. We have two alternative ways of stimulating the muscle fiber: by electrical nerve stimulation and by direct application of ACh through a micropipette placed extracellularly as close to the nerve terminal as possible. Acetylcholine is a cation, so that when we pass current out through the tip of the pipette, ACh^+ is ejected near the postsynaptic membrane. This process is called *iontophoresis.* As shown in Figure 17.6, iontophoresis of ACh produces a depolarization of the muscle fiber membrane equivalent to the nerve-evoked EPSP. (The latency and time course of the iontophoretically evoked response are somewhat longer, because the diffusion path is longer from the pipette than from the neuron terminal.)

The fact that synaptic potentials evoked by presynaptic impulses can be mimicked by iontophoretic application of transmitter clearly demonstrates that the membrane permeability changes that underlie the EPSP are transmitter dependent.

The Ionic Basis of the EPSP   What are the permeability changes in the postsynaptic membrane that give rise to the EPSP? During the EPSP the resis-

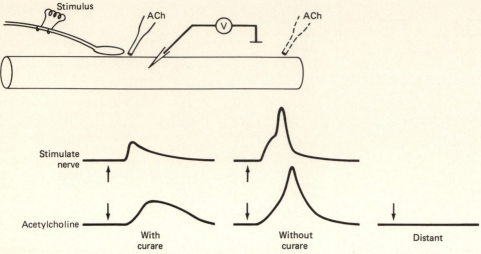

Figure 17.6 Demonstration that the neuromuscular EPSP results from acetylcholine action. Application of ACh through a micropipette very close to the junction produces a potential equivalent to the EPSP. Application of ACh to the muscle fiber membrane at a distance from the neuromuscular junction does not depolarize the membrane.

tance of the postsynaptic membrane is greatly decreased. This observation indicates that there is an *increased ion permeability* of the postsynaptic membrane and therefore an increase in ionic currents that generate the EPSP. To determine the ionic basis of the EPSP, we can vary two parameters of the preparation: membrane potential of the muscle fiber and ionic composition of the extracellular fluid.

Figure 17.7 shows an experiment to determine the effect of membrane potential changes on the EPSP.

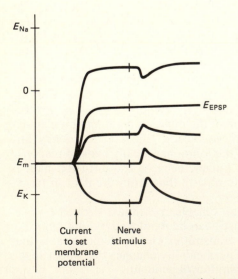

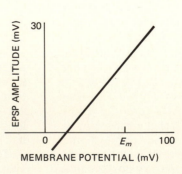

Figure 17.7 The action of the synaptic transmitter at the neuromuscular junction is to drive the muscle fiber membrane *toward* a potential (E_{EPSP}). A current-passing electrode is used to change the membrane potential prior to nerve stimulation. The membrane potential is displaced toward E_{EPSP} from any set value. Therefore E_{EPSP} is termed the *reversal potential* of the EPSP.

We impale a muscle fiber with electrodes, one to record potential and the other to deliver current to change the value of the membrane resting potential. We can then measure the amplitude of the EPSP (evoked by nerve stimulation) at varying values of "resting potential." As shown in Figure 17.7, hyperpolarization of the muscle fiber with the current-delivering electrode increases the amplitude of the EPSP. Conversely, depolarization decreases the amplitude of the EPSP. When we depolarize the membrane to -15 mV, the amplitude of the EPSP is decreased to zero; that is, nerve stimulation evokes no potential change. Still greater depolarization reverses the EPSP, so that it is hyperpolarizing. For this reason, the value -15 mV is said to be the *reversal potential* of the EPSP—the potential at which the EPSP reverses. We can abbreviate this reversal potential E_{EPSP}. The effect of nerve stimulation (and resultant release of ACh) is to drive the membrane potential of the postsynaptic membrane *toward* E_{EPSP}.

How can we interpret the results of this experiment in terms of ion permeabilities? Suppose for simplicity that the action of ACh on the postsynaptic membrane is to make it very permeable to Na$^+$ only. We can predict from the Goldman equation (see p. 399) that the application of ACh would drive the membrane potential toward the sodium equilibrium potential (E_{Na}). (Recall that this is what happens during the rising phase of the action potential.) During the EPSP, however, the membrane potential is driven toward -15 mV rather than toward E_{Na}. Since we know of no ion species with an equilibrium potential of -15 mV, we would guess that more than one ion species contributes to the generation of the EPSP.

To examine what ions contribute to the EPSP, we can vary the ion concentrations in the extracellular fluid. Changing the extracellular concentration of an ion species changes the equilibrium potential of that ion, as described by the Nernst equation. If the postsynaptic membrane becomes permeable to the ion species during an EPSP, the concentration change will alter the amplitude of the EPSP. For example, let us suppose that during an EPSP the membrane becomes roughly equally permeable to Na$^+$ and K$^+$. (This is in fact the case.) When we decrease the extracellular Na$^+$ concentration, E_{Na} is shifted toward zero from its normally positive value (Figure 17.8). This change in E_{Na} will shift the reversal potential (E_{EPSP}) downward if the EPSP reflects an increase in P_{Na}. Such a decrease in [Na$^+$]$_{out}$ decreases the amplitude of the EPSP, a result that indicates that Na$^+$ permeability is involved in generating the EPSP (Figure 17.8B). An increase [K$^+$]$_{out}$ shifts E_K toward zero from its normally negative value (Figure 17.8C) and increases the amplitude of the EPSP. This result indicates that K$^+$ permeability is also increased during an EPSP.* Changing the extracellular Cl$^-$ concentration, on the other hand, has no effect on the amplitude of the EPSP.

From the experiments described above, we can conclude that the membrane depolarization of an EPSP results from an increase in the permeabilities to Na$^+$ and K$^+$ but not to Cl$^-$. These increases in P_{Na} and P_K are *simultaneous*, rather than sequential as they were in the generation of the action potential. Therefore, they drive the membrane toward a value (E_{EPSP}) intermediate between E_{Na} and E_K. E_{EPSP} is slightly closer to E_{Na} than to E_K (Figure 17.8), a reflection of the fact that P_{Na} is slightly greater than P_K. Unlike the changes in permeability underlying the action potential, the *permeability changes that generate the EPSP are voltage independent*. EPSP amplitude changes with membrane potential (Figure 17.7) because the driving force ($E_m - E_{EPSP}$) is changed, not because the permeability is changed.

*The alert reader will object that [K$^+$]$_{out}$ also changes the membrane resting potential and confounds the analysis of the EPSP. For this reason, such experiments are best performed by voltage clamping the muscle fiber and measuring the ionic currents flowing during the EPSP. In this way, the effects of concentration changes on E_m can be eliminated.

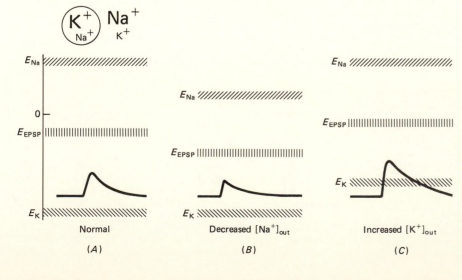

Figure 17.8 The amplitude of the EPSP depends on concentrations of Na$^+$ and K$^+$ ions. A decrease in [Na$^+$]$_{out}$ moves E_{Na} toward zero and decreases the amplitude of the EPSP. An increase in extracellular [K$^+$] moves E_K toward zero and increases the amplitude of the EPSP. These results indicate that the EPSP results from a simultaneous increase in permeability to both Na$^+$ and K$^+$.

Since the permeabilities are essentially unaffected by voltage, the depolarization of the EPSP itself is not regenerative; EPSPs are not all-or-none and are not propagated.

In summary, we have shown that neuromuscular transmission is mediated by an EPSP that depolarizes the postsynaptic membrane, as a result of a transmitter-dependent increase in permeability to Na^+ and K^+. The resulting ionic currents drive the membrane toward a value, E_{EPSP}, that is more depolarized than the threshold of the muscle. Normally, the amplitude of the EPSP at the junction is sufficient to exceed threshold and triggers a muscle fiber action potential. The EPSP itself is a nonregenerative, nonpropagated local response, because the transmitter-dependent permeability changes are not voltage dependent. Next we will examine postsynaptic potentials at nerve–nerve synapses.

Nerve–Nerve Synaptic Potentials Let us perform another experiment to record synaptic potentials from a spinal motoneuron of a cat. This experiment is technically more difficult, in part because we cannot see what we are doing in the spinal cord. Figure 17.9 shows the anatomy of the spinal cord and the design of the experiment. We impale a motoneuron with a microelectrode and stimulate various peripheral nerves that contain sensory neurons, some of which excite the motoneuron and some of which inhibit it.

First, how do we know that we have impaled a motoneuron? Our problem is like that of an airplane pilot in fog, flying by instruments alone. We advance the microelectrode into the ventral horn of the spinal cord, to the area and depth of motoneuron somata. We watch not the spinal cord, but the oscilloscope

screen. Cell penetration is signaled by the onset of the resting potential (Figure 17.9). The cell penetrated is probably a neuron, because the larger neurons are much more readily penetrated than the smaller glial cells—but it could be any neuron. To test whether we have impaled a motoneuron, we stimulate the ventral root exiting from the cord. We know the ventral root contains essentially only motoneuron axons; the incoming sensory axons all enter the cord through the dorsal root. Thus, a stimulus of the ventral root excites motoneuron axons only. If our cell is a motoneuron, our ventral root stimulus will generate an axonal impulse that is propagated back by the axon to the soma. This "wrong way" impulse is termed **antidromic,** as opposed to an **orthodromic** (normal direction) impulse. Only a motoneuron has an antidromic impulse in response to ventral root stimulation.

Now that we have confirmed that we are recording intracellularly from a motoneuron, we can examine its synaptic potentials. A cat spinal motoneuron receives about 10,000 synaptic terminals, many from sensory neurons and from spinal interneurons that are driven by sensory neurons. By electrical stimulation of various peripheral nerves, we can excite small groups of neurons that synapse onto the motoneuron. Some of these synapses are excitatory and some are inhibitory. Let us stimulate a group of sensory neurons that excite the motoneuron (Figure 17.10, A). We record an EPSP that has a brief depolarizing rising phase and an exponential decay over a time course of 10–20 ms. The motoneuron EPSP is similar to that recorded at the neuromuscular junction in all respects except amplitude. The amplitude of the neuromuscular EPSP is 40–50 mV, sufficient

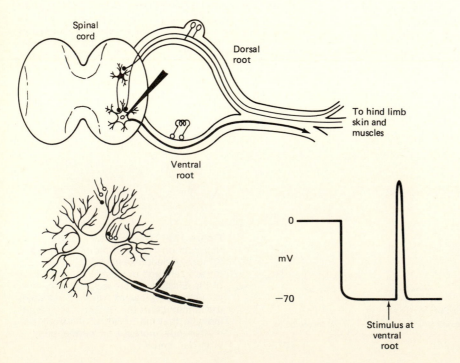

Spinal cord

Dorsal root

To hind limb skin and muscles

Ventral root

0

mV

−70

Stimulus at ventral root

Figure 17.9 Intracellular recording from a cat spinal motoneuron. To determine that the impaled cell is a motoneuron, the ventral root is stimulated, exciting motoneuron axons. A resulting antidromic action potential is propagated back up the axon and recorded at the motoneuron soma.

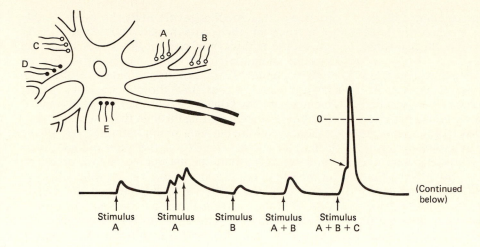

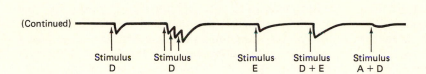

(Continued below)

Stimulus A   Stimulus A   Stimulus B   Stimulus A + B   Stimulus A + B + C

(Continued)

Stimulus D   Stimulus D   Stimulus E   Stimulus D + E   Stimulus A + D

Figure 17.10 EPSPs and IPSPs recorded intracellularly from a spinal motoneuron. Excitatory synapses are shown by open circles; inhibitory synapses are shown by closed circles. Inhibitory postsynaptic potentials (IPSPs) appear as near mirror-images of EPSPs. Both EPSPs and IPSPs exhibit temporal summation and spatial summation (see text).

to trigger a muscle fiber action potential under normal circumstances. The motoneuron EPSP is much smaller, even when several presynaptic neurons are activated synchronously. It is calculated that each presynaptic ending contributes only about 0.5 mV toward the depolarization of the motoneuron.

If we stimulate pathway A repeatedly, the resultant EPSPs summate with each other (Figure 17.10). This is called **temporal summation**. EPSPs from different sources can also summate with each other. In Figure 17.10, simultaneous stimulation of two excitatory pathways, A and B, leads to summation of their EPSPs, a process termed **spatial summation**. With sufficient presynaptic stimulation, EPSPs can summate to threshold and generate one or more impulses in the motoneuron (Figure 17.10, A + B + C).

Now let us stimulate some of the pathways that inhibit the motoneuron. As shown in Figure 17.10, stimulation of inhibitory pathway D evokes a synaptic potential that *hyperpolarizes* the motoneuron. This potential is termed an **inhibitory postsynaptic potential** or IPSP. It is a mirror image of an EPSP and can also exhibit temporal and spatial summation. Repeated stimulation of pathway D (Figure 17.10) evokes temporally summating IPSPs. IPSPs summate spatially with each other (Figure 17.10, D + E) and with EPSPs (A + D).

The ionic basis of EPSPs at nerve synapses is similar to that at neuromuscular junctions: The postsynaptic action of the synaptic transmitter is to open momentarily channels that increase permeability to Na^+ and K^+. What is the ionic basis of IPSPs? As shown in Figure 17.11, the IPSP of a spinal motoneuron has a reversal potential (E_{IPSP}) of about -80

mV. The potassium equilibrium potential (E_K) of a motoneuron is about -90 mV, and E_{Cl} is about -70 mV. It can be shown that permeabilities to both K^+ and Cl^- are increased during an IPSP, thereby driving the motoneuron membrane toward a point between E_K and E_{Cl}. Since it is difficult to estimate accurate values of E_K and E_{Cl} in central neurons, the relative contributions of changes in Cl^- permeability and K^+ permeability to the IPSP are not known.

The ionic mechanisms of motoneuron EPSPs and IPSPs described above are fairly typical of excitatory and inhibitory synapses in general. Most inhibitory synapses in arthropods depend on Cl^- permeability only, without an increase in K^+ permeability. On the other hand, there are IPSPs that are specifically K^+ dependent, for example, in the vertebrate heart. Neurons of mollusks such as *Aplysia* may have separate K^+-dependent and Cl^--dependent IPSPs. Al-

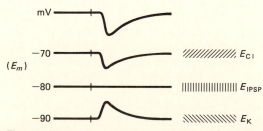

mV

-70 ////////// E_{Cl}

(E_m)

-80 |||||||||||||||| E_{IPSP}

-90 \\\\\\\\\\ E_K

Figure 17.11 An inhibitory postsynaptic potential (IPSP) drives the membrane potential toward a reversal potential (E_{IPSP}) of about -80 mV. If the membrane potential is set at different levels (see Figure 17.7), presynaptic stimulation will always drive it toward E_{IPSP}. The action of the inhibitory synaptic transmitter increases the permeability of the postsynaptic membrane to both Cl^- and K^+.

though most synaptic potentials are mediated by increases in ion permeability, other mechanisms have been discovered (see Other Synaptic Mechanisms, p. 440).

What determines whether a synapse will have an inhibitory effect or an excitatory effect on the postsynaptic cell? We can define *excitation* as an increase in the probability that a cell will generate an impulse, or if it is already generating impulses, an increase in impulse frequency. Similarly, we define *inhibition* as a decrease in the probability of impulse generation or a decrease in impulse frequency. Any process that drives the membrane toward a potential more depolarized than the potential of impulse threshold will be excitatory. Since the reversal potential of an EPSP is more depolarized than the threshold potential, EPSPs are excitatory. IPSPs, on the other hand, drive the membrane potential toward a point more hyperpolarized than the threshold potential—that is, toward E_{IPSP}. This inhibitory effect is most clear in spinal motoneurons and other cases in which the IPSP is a hyperpolarization from membrane resting potential. IPSPs mediated by increased Cl^- permeability, however, may not hyperpolarize the resting membrane. The chloride equilibrium potential (E_{Cl}) may be at the resting potential or even slightly depolarized relative to resting E_m. Thus, the activation of a Cl^--mediated inhibitory synapse may lead to no potential change at rest or even to a small depolarization. Such IPSPs are still functionally inhibitory; since E_{Cl} is more hyperpolarized than threshold, the IPSPs tend to "lock" the membrane potential at this subthreshold level. That is, if the membrane is depolarized to threshold by EPSPs, activation of a Cl^- inhibitory synapse hyperpolarizes the membrane from threshold toward E_{Cl}.

Integration of EPSPs and IPSPs If we recall that most neurons receive thousands of synaptic endings and that each synapse may contribute a synaptic potential of less than a millivolt, it is clear that the output of a neuron —its temporal sequence of action potentials—is a rather complex function of its synaptic input. The concept that a neuron's output is not the same as its input but is a function of that input is termed *neuronal integration*. The major process underlying neuronal integration is the spatial and temporal summation of EPSPs and IPSPs. The interaction of PSPs is more complex than a simple algebraic summation, however. To appreciate the complexity of this interaction, we must consider the geometric aspects of summation.

The site of action potential initiation of a spinal motoneuron is the axon hillock or initial segment of the axon. This region is devoid of synapses in spinal motoneurons and in most other vertebrate central neurons. The degree of excitation or inhibition that a synapse produces is a function of its effect on the generation of action potentials at the axon hillock. Since synaptic potentials are not propagated, the am-

plitude of this effect is governed by the passive cable properties of the postsynaptic cell. Figure 17.12 shows the pathways of current flow associated with synaptic potentials of a spinal motoneuron. For an EPSP (Figure 17.12A), the rising phase results from inward ionic current flow through the subsynaptic membrane (immediately under the synapse), and capacitative current exiting through the rest of the postsynaptic cell membrane, depolarizing the rest of the cell. The current density at the axon hillock determines the amplitude of the depolarization produced at the site of impulse initiation. The current pathways underlying an IPSP are diagrammed in Figure 17.12B. Ionic current flows out through the subsynaptic membrane, and capacitative current flows in across the rest of the cell membrane and hyperpolarizes it. The current density at the axon hillock determines the degree of hyperpolarization at the site of impulse initiation. If both the inhibitory and the excitatory synapse are activated simultaneously (Figure 17.12C), the ionic current entering the excitatory subsynaptic membrane tends to exit through the low-resistance path of the inhibitory subsynaptic membrane, rather than spreading to depolarize the rest of the cell. This "short-circuiting" effect occurs only during the initial 1–2 ms of the PSPs, the time during which the ion permeability channels are open. This resistance-decreasing effect of IPSPs can make their

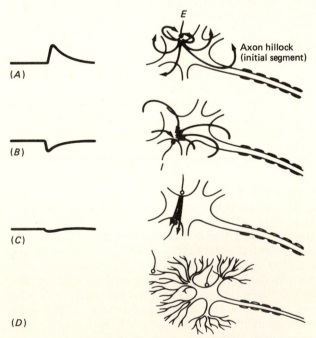

Figure 17.12 Integration of EPSPs and IPSPs. Action potentials are initiated at the axon hillock or initial segment of the motoneuron axon. Synaptic potentials spread decrementally from the synapse to distant locations, including the axon hillock. (A) Current flow underlying the spread of an EPSP (further explanation in text). (B) Current flow underlying the spread of an IPSP. (C) "Short-circuiting" of currents with simultaneous EPSP and IPSP. (D) Remote synapses have smaller effects on the axon hillock.

inhibitory effect stronger than would be produced by simple algebraic summation of the EPSP and IPSP amplitudes (Figure 17.12C).

So far we have considered synapses on the motoneuron soma. Synapses on the soma are relatively near the axon hillock in terms of *electrotonic distance;* that is, the amplitudes of their PSPs are little attenuated in their passive spread to the impulse initiation site. Synapses on dendrites, on the other hand, may be 200 μm or more distant from the axon hillock (Figure 17.12D). The electrotonic length of motoneuron dendrites is estimated to be 1–2λ (one or two times the membrane length constant), so that a synaptic potential at the dendrite tip is attenuated to 14–37 percent of its initial amplitude in its spread to the axon hillock. Although other factors may partially compensate for this attenuation (see Shepherd, 1979), it is clear that some synapses have more effect on the output of the postsynaptic cell than others. Thus, the summation of synaptic input is weighted by electrotonic distance of the synapses from the axon hillock.

Endogenous Activity of Neurons The above discussion of the neuron as an integrator of synaptic input is somewhat misleading because it implies that neurons are electrically silent in the absence of synaptic input. This view is in part an artifact of the experimental conditions used to record central neuronal activity in mammals. Such experiments usually employ general anesthetics that have in common the property of depressing the activity of the CNS. It is possible to implant electrodes chronically (over a long time period) in awake unanesthetized mammals. Under these conditions, many central neurons exhibit "spontaneous" activity, generating impulses in the absence of any apparent stimulation. In the CNS of vertebrates, it is impossible to determine whether this activity results from a background of undetected synaptic input or whether it represents endogenous impulse generation by the neuron in the absence of synaptic input. In mollusks and arthropods, however, it is possible to demonstrate unequivocally that single neurons may generate trains of impulses or even temporally patterned impulse bursts in the absence of any synaptic input (cf. pp. 407 and 563). Therefore, the impulse activity of a neuron should not be assumed to reflect summated synaptic input alone; such input may be integrated with endogenous impulse activity generated within the postsynaptic neuron.

Summary In this section, we have shown that in chemically mediated synaptic transmission, the transmitter released by the presynaptic terminal causes a potential change (the postsynaptic potential) at the postsynaptic membrane, usually by increasing membrane permeability to specific ions. (Exceptions to this generalization are treated under Other Synaptic Mechanisms, p. 440.) Excitatory postsynaptic potentials (EPSPs) result from increases in Na^+ and K^+ permeability and are similar at neuromuscular junctions and at nerve–nerve synapses. Inhibitory postsynaptic potentials (IPSPs) are typically hyperpolarizing mirror images of EPSPs and result from increased permeability to Cl^-, K^+, or both. A postsynaptic neuron integrates the synaptic potentials that result from presynaptic activity; the postsynaptic membrane potential and the resultant generation of impulses reflect a weighted summation of these synaptic potentials (as well as any endogenous impulse-generating capability of the neuron itself).

Presynaptic Mechanisms of Transmitter Synthesis, Storage, and Release

Now let us consider the processes by which synaptic transmitter molecules are synthesized and stored in presynaptic terminals and released by presynaptic impulses. As was the case earlier in the chapter, we depend on evidence from peripheral synapses such as the vertebrate neuromuscular junction. Having shown that the neuromuscular junction is a good model system for study of synaptic potentials, we will use it similarly to explore presynaptic mechanisms. Since the vertebrate neuromuscular transmitter is acetylcholine, we will consider primarily cholinergic (acetylcholine mediated) transmission here. Other transmitters are treated later (see Synaptic Transmitters, p. 443).

Synthesis and Storage of Acetylcholine Acetylcholine (ACh) is synthesized from choline and acetylcoenzyme A (Figure 17.13). The reaction is catalyzed by the specific enzyme choline acetyltransferase. Synthesis of ACh occurs in the cytoplasm of the presynaptic ending; choline acetyltransferase, like all proteins, is synthesized in the soma and must be transported the length of the axon. Choline is supplied by the circulation and by reuptake at the terminal. The availability of choline is the limiting factor in ACh synthesis.

Biochemical evidence indicates that nerve terminals contain several pools of ACh. About half of the stored ACh appears to be in synaptic vesicles and about half in the cytoplasm. Recently synthesized ACh is released more readily than the previously synthesized "depot storage" pool. About 15 percent of the ACh is not releasable. The relation of the vesicular and cytoplasmic fractions to the "readily releasable" and "depot store" pools is not clear.

Transmitter Release Is Voltage and Ca^{2+} Dependent How is synaptic transmitter released from the presynaptic terminal? This process is poorly understood, but it is clear that transmitter release requires both presynaptic depolarization and Ca^{2+} ions.

The normal trigger for transmitter release is the depolarization of the terminal by the presynaptic action potential. We can show that depolarization of

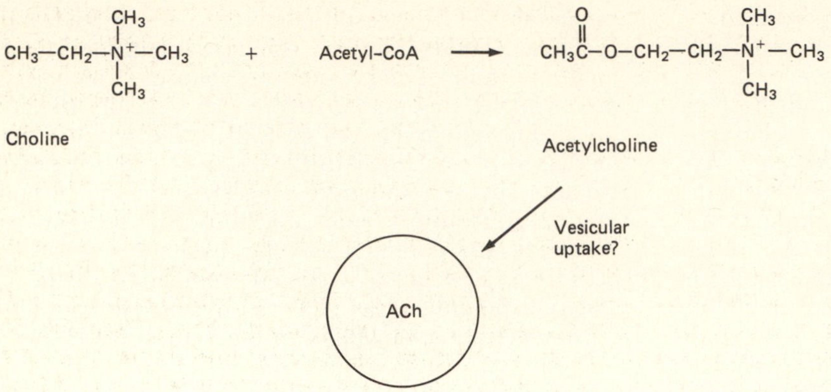

Figure 17.13 Synthesis of acetylcholine. The synthesis appears to occur in the cytoplasm of the presynaptic terminal. Synaptic vesicles presumably take up preformed acetylcholine, but the kinetics of uptake are poorly understood.

the terminal is a sufficient trigger by blocking action potentials with tetrodotoxin and TEA (Chapter 16, p. 406) and then depolarizing the terminal directly with a microelectrode. The clearest preparation for this demonstration is the squid *giant synapse* between second-order and third-order giant axons in the squid stellate ganglion. This synapse is unusual in that both the presynaptic neuron and the postsynaptic neuron can be impaled with microelectrodes. Thus, the relation of presynaptic depolarization to amount of transmitter released (measured as EPSP amplitude) can be determined directly. EPSP amplitude is a sigmoid function of depolarization (Figure 17.14), increasing steeply with increasing depolarization in the range of 60–100-mV depolarization. Similar experiments have been performed at the vertebrate neuromuscular junction, but only with an extracellular presynaptic electrode. There is a similar increase in EPSP amplitude with increasing depolarizing current, but the amplitude of presynaptic depolarization cannot be measured.

Transmitter release from the presynaptic terminal is also directly dependent on Ca^{2+} ions. If a synapse is bathed in a Ca^{2+}-free saline solution, presynaptic depolarization elicits no EPSP. In contrast, an EPSP evoked by direct iontophoretic application of transmitter is unaffected by the absence of Ca^{2+} ions, a result that demonstrates that Ca^{2+} ions are necessary for release of transmitter rather than for its postsynaptic action.

The dependence of transmitter release on Ca^{2+} ions appears to be true for all chemical synapses. The simplest explanation for the action of Ca^{2+} ions is that depolarization of the terminal membrane increases its permeability to Ca^{2+}, allowing Ca^{2+} to enter and somehow trigger transmitter release. (The intracellular concentration of free Ca^{2+} is much lower than the extracellular concentration, so that there is a strong inward driving force on Ca^{2+} ions.) This hypothesis has received experimental confirmation in the squid giant synapse, where it is possible to inject Ca^{2+} directly into the presynaptic terminal with a microelectrode. Pulsed Ca^{2+} injection triggers postsynaptic EPSPs, indicating that Ca^{2+} entry is sufficient to release transmitter. In other experi-

ments, the presynaptic terminal has been injected with aquorin, a protein that fluoresces in the presence of Ca^{2+}. Depolarization of the presynaptic terminal induces aquorin fluorescence, experimentally confirming that presynaptic depolarization leads to Ca^{2+} entry. The way in which intracellular Ca^{2+} mediates transmitter release, however, remains unclear.

Synaptic Transmitter Release Is Quantal Another general finding about the release process is that transmitter is released in multimolecular packets called *quanta*. That is, a synaptic transmitter such as acetylcholine appears to be released not a molecule at a time, but rather in quantal units of about 3000 molecules each.

The primary evidence for quantal release comes from the vertebrate neuromuscular junction. If we impale a muscle fiber near the neuromuscular junction with an intracellular microelectrode (Figure 17.15), we record a series of small depolarizations in the absence of any stimulation. These depolarizations have the same shape as a neuromuscular EPSP but are about 1/50 the amplitude; thus, they are termed *spontaneous miniature end-plate potentials* or *spontaneous miniature EPSPs* (mEPSPs). Each mEPSP is the postsynaptic response to a quantum. The mEPSPs typically have amplitudes of about 0.4 mV and occur nearly randomly over time. They do not represent responses to individual molecules of acetylcholine, because several thousand ACh molecules are required to produce an equivalent depolarization. These findings indicate that there is a spontaneous, low-frequency background release of quantal packets of transmitter in the absence of presynaptic stimulation.

Can we show that transmitter release evoked by presynaptic depolarization is also quantal? The neuromuscular EPSP evoked by a presynaptic impulse has an amplitude of 20–40 mV and would require the nearly simultaneous discharge of 100–300 quanta. This number is too large to determine whether the release is quantal or not, since the difference between the response to, say, 150 and 151 quanta is not detectable. To determine whether evoked release is quantal, it is necessary to *decrease the quantal*

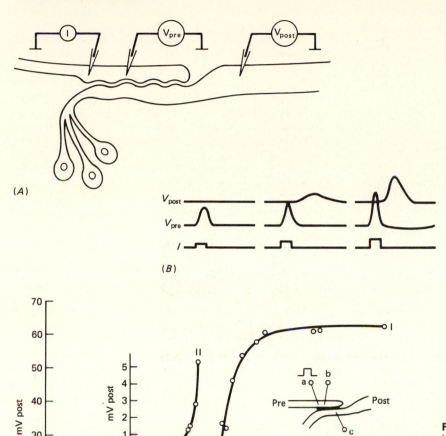

(A)

(B)

(C)

Figure 17.14 The amount of transmitter released and the amplitude of EPSPs depend on the amplitude of presynaptic depolarization. (*A*) Squid giant synapse. Impulses are blocked with TTX and TEA. The presynaptic terminal is depolarized with one electrode; two others record presynaptic and postsynaptic voltages. (*B*) EPSP amplitude (V_{post}) increases with increasing presynaptic depolarization. (*C*) Graph shows steep effect of presynaptic depolarization amplitude on EPSP amplitude [From B. Katz and R. Miledi, *J. Physiol.* **192**:407–436 (1967).]

content of evoked release, so that 1 or 2 quanta are released rather than hundreds. The quantal content is decreased by lowering extracellular [Ca^{2+}], since Ca^{2+} is necessary for transmitter release. Usually, the bath Mg^{2+} concentration is also increased, since Mg^{2+} competitively inhibits the action of Ca^{2+} on release. With an appropriately adjusted ratio of Ca^{2+} and Mg^{2+} concentrations, the amount of transmitter released by presynaptic stimulation can be decreased to correspond to only a few quanta. Are the resultant *evoked miniature EPSPs* quantal? As shown in Figure 17.16, evoked mEPSPs fall into size classes that are multiples of the amplitudes of spontaneous mEPSPs. That is, each presynaptic stimulus evokes the release of 0, 1, 2, or 3 quanta, but never $1\frac{1}{2}$ quanta. Since the amplitude of response to a single quantum is somewhat variable (Figure 17.16*B*), the peaks of quantal response in Figure 17.16*C* become indistinct when more than a few quanta are released. By these experiments, transmitter release evoked by a pre-

synaptic impulse can be shown to be quantal, at least under conditions of low quantal content. This result also has been found in other preparations (crustacean and insect neuromuscular junction, mammal and *Aplysia* CNS), and in some of these cases quantal release can be demonstrated without use of any ion manipulations or drugs. It is thought then that quantal release is the general rule for all chemical synapses.

Synaptic Vesicles and Transmitter Release   It has been shown that cholinergic vesicles contain 10^3–10^4 molecules of acetylcholine and that transmitter is released in quanta of 10^3–10^4 molecules; thus, the conclusion seems inescapable that a quantum corresponds to a synaptic vesicle. The vesicular release hypothesis states that vesicles fuse with the presynaptic membrane and discharge their content of transmitter, presumably by exocytosis. Such fusions occur sporadically in the resting terminal, producing

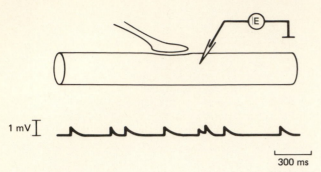

Figure 17.15 Spontaneous miniature excitatory postsynaptic potentials (spontaneous mEPSPs). The nerve terminal at a neuromuscular junction has a background low level of transmitter release in the absence of stimulation. The transmitter is released in multimolecular quantal packets, each quantum eliciting a miniature EPSP at the postsynaptic membrane.

spontaneous mEPSPs. Depolarization of the terminal by the presynaptic action potential greatly increases the probability of release of each of a number of vesicles so that at a neuromuscular junction about 150 vesicles discharge in a millisecond and produce a junctional EPSP. This *vesicular release* hypothesis is widely accepted for synapses in general.

There are, however, some objections to the vesicular release hypothesis that prevent us from accepting it as a foregone conclusion. These objections, mostly on biochemical grounds, are not fatal to the hypothesis, but they make it necessary to examine the experimental evidence on which the hypothesis rests. The major objections to vesicular release of ACh are as follows. First, studies of ACh metabolism show that newly synthesized ACh is released pref-

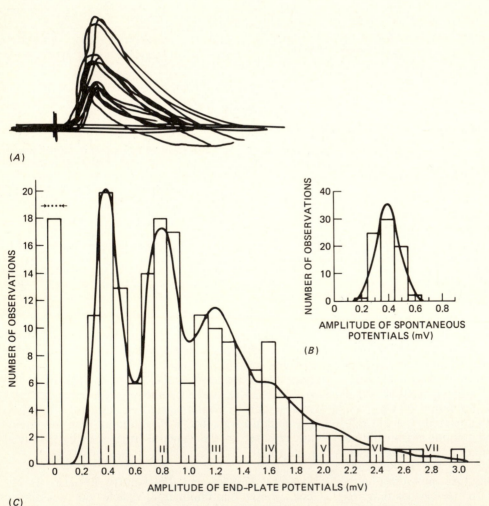

Figure 17.16 (*A*) Evoked mEPSPs, with amplitude decreased by low Ca^{2+}, high Mg^{2+} bath. Each response results from release of 0, 1, 2, or 3 quanta. About 20 responses have been superimposed. Amplitude histograms of spontaneous (*B*) and evoked (*C*), mEPSPs in mammalian muscle, recorded as above. Peaks in the mEPSP amplitude distribution occur at multiples (I, II, III, IV, etc.) of the spontaneous potential amplitude. The smooth curve is the expected distribution obtained by calculating the quantum, content, with the assumption that the amplitudes of the individual quanta are distributed like those of the spontaneous potentials. The arrows above 0 amplitude indicate that 19 failures were expected; 18 were actually counted. [From I.A. Boyd and A.R. Martin, *J. Physiol.* **132**:74–91 (1956).]

erentially to that previously synthesized. Since the enzyme choline acetyltransferase is located in the cytoplasm outside the synaptic vesicles, the most recently synthesized ACh is presumably extravesicular. Thus, it has been argued that the ACh outside the vesicles must be preferentially released. Other studies have shown that the proteins of vesicle membranes differ from the proteins of the plasma membrane. Moreover, the half-life of vesicular membrane proteins has been measured to be 19 days. On the other hand, the turnover of ACh in a stimulated superior cervical ganglion is 10 percent per minute; that is, the cholinergic synapses in the ganglion can release and resynthesize 10 percent of their ACh per minute, without any ACh depletion. If vesicles are fusing with the plasma membrane and discharging their ACh at such high rates, it is difficult to see how the protein differences and the 19-day half-life of vesicle membrane proteins can be maintained.

Another problem for the vesicular release hypothesis is that direct electron microscopic evidence of vesicle fusion and exocytosis is surprisingly difficult

to obtain. In electron micrographs of sections through the sites of presumed transmitter release, vesicles caught in the act of exocytosis (termed omega figures; see Figure 17.17) are only rarely seen. Moreover, it is difficult to demonstrate vesicle depletion with stimulation, unless the stimulation is abnormally intense or resynthesis is blocked with drugs such as black widow spider venom. Because of these objections, the validity of the vesicular release hypothesis has been questioned (see Cooper et al., 1982).

Many of these objections to the vesicular release hypothesis have been met in recent studies of Heuser and Reese. By rapid chemical fixation of neuromuscular junctions stimulated at modest rates, Heuser and Reese demonstrated that synaptic vesicles were depleted, but that within minutes, membrane was pinched off and returned to the inside of the terminal by endocytosis. This recycling of vesicular membrane is necessary both to allow reformation of synaptic vesicles and to prevent the plasma membrane from expanding by the addition of vesicular mem-

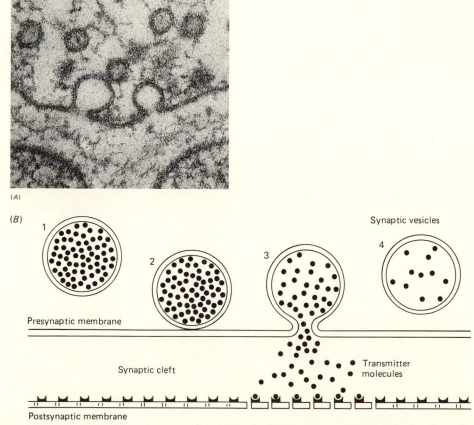

Figure 17.17 Exocytosis of synaptic vesicles at the presynaptic axonal membrane of a frog neuromuscular junction. In the electron micrograph above, vesicles appear in the act of fusion with the axon terminal membrane. Synaptic vesicles are clustered near the presynaptic membrane. The diagram below shows the probable steps in exocytosis. Vesicles filled with transmitter molecules move to the synaptic cleft, fuse with the membrane, discharge their contents, and are retrieved and refilled with transmitter. [Electron micrograph courtesy of J.E. Heuser.]

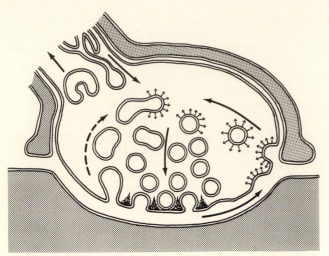

Figure 17.18 Summary of proposed mechanisms of synaptic-vesicle recycling at the frog neuromuscular junction, based on quick-freezing data. After synaptic vesicles undergo exocytosis and collapse into the presynaptic membrane, two different sorts of compensatory endocytosis can occur. The normal pathway is shown on the right. In it, coated vesicles selectively retrieve specific vesicle components from the presynaptic membrane and directly produce new synaptic vesicles. Under experimental conditions of high rates of transmitter secretion, the dotted pathway on the left may also occur, but may be abnormal. Random portions of the membrane may pinch off to form large internal vacuoles, which could be later converted to synaptic vesicles. [From J.E. Heuser and T.S. Reese, in F.O. Schmitt and F.G. Worden (eds.), *The Neurosciences: Fourth Study Program.* pp. 573–600. MIT Press, Cambridge, 1979.]

brane from exocytotic fusion. Figure 17.18 summarizes this process of recycling vesicular membrane.

Heuser and Reese have also obtained convincing evidence of the exocytotic process itself. It appears that fusion of synaptic vesicles with the plasma membrane is difficult to see in electron micrographs because the process is so brief. Heuser and Reese developed an ingenious device whereby the neuromuscular junction was instantaneously frozen a few milliseconds after the nerve was stimulated. The exact time interval between stimulation and freezing could be controlled precisely. The frozen tissue was then fractured, a procedure that selectively splits the lipid bilayer of cell membranes. A replica of the freeze-fractured junctional membrane was then prepared to show the ultrastructure of the membrane in surface view. The diagram in Figure 17.19 illustrates the structure of a neuromuscular junction as revealed by freeze-fracture electron microscopy. Figure 17.20 shows the results of freezing the junction 3 and 5 ms after stimulation. At 3 ms vesicles have not yet fused with the presynaptic membrane (under the conditions of the experiment), and all that is seen is a double row of membrane particles that delineate the active zone for transmitter release (compare Figures 17.20 and 17.19). These particles are probably the proteins that constitute the voltage-dependent Ca^{2+} channels

through which Ca^{2+} ions enter to trigger release. At 5 ms after stimulation (Figure 17.20*B*), the active zone contains holes and dimples that represent the junctions between the plasma membrane and the fused vesicles (see Figure 17.19 for interpretation). If the tissue is frozen as little as 50–100 ms after stimulation, vesicle fusion is already completed and the broken stalks of fusing vesicles are no longer seen.

Thus, there appears to be convincing microscopic evidence for the vesicular release hypothesis, at least at the neuromuscular junction. This mechanism of transmitter release may well be general, but it is also possible that there is more than one mechanism by which presynaptic depolarization leads to transmitter release.

Postsynaptic Mechanisms

Next, let us examine the ways in which transmitters act at postsynaptic membranes. All synaptic transmitters are thought to bind selectively to transmitter receptor proteins that are embedded in the postsynaptic membrane. This binding triggers changes in the receptor protein that lead directly or indirectly to the effects of the transmitter on the postsynaptic cell.

To examine these postsynaptic mechanisms, we will again use as our models the vertebrate neuromuscular junction and closely related cholinergic synapses.

Distribution of Acetylcholine Receptors Where on the surface of a vertebrate skeletal muscle fiber are acetylcholine receptor molecules located? The receptor distribution can be determined by recording responses to local iontophoretic application of ACh or by mapping the distribution of radioactively labeled α-bungarotoxin, a snake venom that binds specifically and nearly irreversibly to ACh receptors. Both techniques show that ACh receptors are present in high concentrations only at the neuromuscular junction region. The receptor density decreases abruptly at the edge of the junction, and the rest of the muscle fiber surface has a low concentration of receptors, 2 percent of that at the junction. During the development of the muscle fiber, ACh receptors first appear diffusely over the surface of the fiber. Only at or shortly before the time that the fiber is contacted by an outgrowing axon do receptors appear aggregated into localized "hot spots." Those "hot spots" that are not maintained by development of a neuromuscular junction soon disappear.

When a muscle is denervated by severing the nerve to it, the whole surface of each muscle fiber becomes very sensitive to ACh over the course of several days (Figure 17.21). This *denervation supersensitivity* results from synthesis or activation of new extrajunctional receptors, rather than from spread of junctional receptors over the fiber surface. Thus, ex-

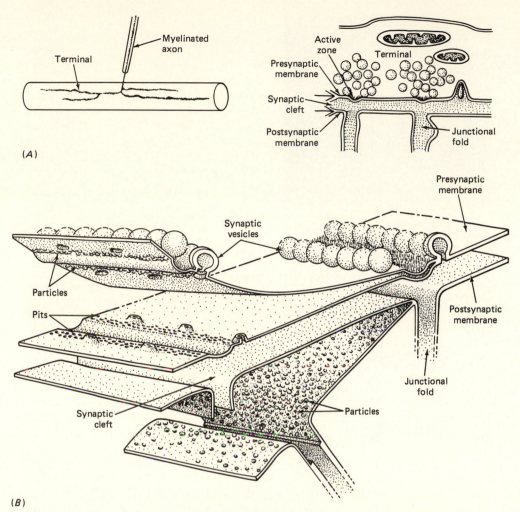

Figure 17.19 Synaptic membrane structure. (A) Entire frog neuromuscular junction (left) and longitudinal section through portion of nerve terminal (right). (B) Three-dimensional view of pre- and postsynaptic membranes with active zones and immediately adjacent rows of synaptic vesicles. The plasma membranes are split (at arrows in A) to illustrate structures observed upon freeze fracturing. The presynaptic membrane at the active zone shows protruding particles on the cytoplasmic half-membrane, and corresponding pits on the fracture face of the outer membrane leaflet. Vesicles that fuse with the presynaptic membrane give rise to characteristic protrusions and pores in the fracture faces. The fractured post-synaptic membrane in the region of the folds shows a high concentration of particles that are probably ACh receptors. [From S.W. Kuffler, J.G. Nicholls, and A.R. Martin, *From Neuron to Brain,* 2nd ed. Sinauer, Sunderland, MA, 1984.]

trajunctional receptors are suppressed by normal in-nervation of the muscle. This suppression of extra-junctional receptors appears to result from at least two factors: electrical activity of the muscle fiber and a trophic influence of the nerve. If a denervated mus-cle is stimulated directly by electrodes, the increase in extrajunctional receptors underlying ACh super-sensitivity is largely suppressed (Figure 17.21C). Moreover, supersensitivity can be induced by sur-rounding the nerve with a cuff containing tetrodo-toxin, as well as by cutting the nerve. The tetrodo-toxin cuff prevents nerve activity from stimulating the muscle but does not produce any degeneration

of the neuromuscular junction. These results indicate that electrical activity of the muscle fiber plays an important role in maintaining restricted distribution of ACh receptors.

It also appears that the very presence of the nerve helps to restrict the distribution of ACh receptors, perhaps by the release of trophic factors. The suppression of extrajunctional receptors in normal muscle is greater than the suppression with electrical stimulation of denervated muscle (Figure 17.21). Also, the increased density of extrajunctional ACh receptors is greater following severing of the nerve than following application of a tetrodotoxin cuff.

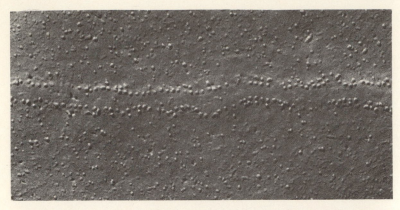

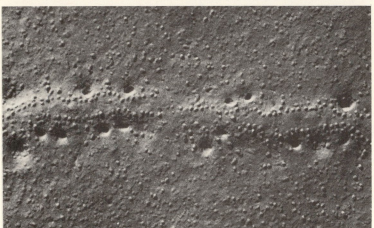

Figure 17.20 Freeze-fracture replicas of the presynaptic membrane of the frog neuromuscular junction, showing evidence of vesicular exocytosis. The upper micrograph shows the membrane 3 ms after the muscle had been stimulated, just before vesicle fusion. Running across the axon membrane surface is a double row of particles that may be calcium channels. The lower micrograph shows the membrane 5 ms after stimulation. The stimulation has caused synaptic vesicles to fuse with presynaptic membrane and form large pits or craters (see Figure 17.19) (Electron micrograph courtesy of J.E. Heuser).

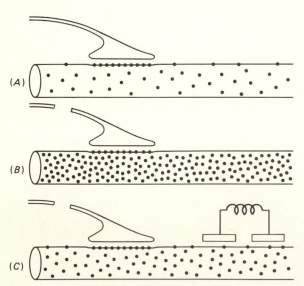

Figure 17.21 Acetylcholine receptor distribution of vertebrate skeletal muscles fibers. (*A*) With normal innervation, the acetylcholine receptors (dots) are largely confined to the immediate subsynaptic membrane of the neuromuscular junction. (*B*) Following denervation there is an increase in number of extrajunctional receptors and a corresponding supersensitivity to applied acetylcholine. (*C*) If the denervated muscle is electrically stimulated, the increase in extrajunctional receptors is largely (but not entirely) suppressed.

These results suggest that the nerve exerts other influences besides maintaining electrical activation of the muscle.

Both the restriction of transmitter receptors and denervation supersensitivity appear to be common features of chemical synapses. For example, both features are found in crustacean neuromuscular junctions that employ gamma-aminobutyric acid (GABA) as an inhibitory transmitter and (probably) glutamate as an excitatory transmitter. On the other hand, the somata of *Aplysia* neurons are synapse-free regions that nevertheless contain appreciable concentrations of membrane-bound transmitter receptors. Thus, the restriction of transmitter receptors to synaptic areas is not a universal property for all synapses.

Molecular Properties of Acetylcholine Receptors
Biochemists have made considerable progress in the isolation and characterization of ACh receptor molecules. Receptors were first isolated from the electric organs of marine electric fish (skates and rays). These electric organs consist almost entirely of modified neuromuscular junctions and therefore contain high concentrations of ACh receptors. The ACh receptor of the ray *Torpedo* is a glycoprotein of 250,000 daltons, composed of five subunits. Each molecule can bind two ACh molecules and, as a result of this

binding, is thought to change conformation to open a central channel large enough to pass Na^+ and K^+ ions (Figure 17.22). The ACh binding sites are located on two identical α subunits, and the ion permeability channel is a central hydrophilic region surrounded by the five subunits. The amino acid sequences of the subunits have been determined by molecular biological techniques; with this structural information, the function of the ACh receptor is becoming understood at the molecular level.

It is possible to record the current flowing through a single open ACh receptor, by a technique termed *patch clamp* or *single-channel* recording. In this technique, a patch of membrane containing a single channel is sealed by suction onto the smoothed tip of a fine glass micropipette electrode, so that the electrode records the single-channel current associated with channel opening (Figure 17.22). Both the amount and time course of the single-channel current can be measured. An ACh receptor channel remains open for a short, variable time (typically 1–3 ms), during which 1–5×10^4 ions pass through it. The synaptic current producing an EPSP is the sum of thousands of single-channel currents, since as noted above, acetylcholine is released in quanta, each composed of 10^3–10^4 molecules. A single quantum may transiently open as many as 2000 ACh receptors, and

a single presynaptic impulse causes release of many quanta.

The properties of the ACh receptor molecule can be compared to those of the voltage-dependent sodium channel that underlies the action potential. Both can be characterized and localized by specific binding agents: α-bungarotoxin in the case of the ACh receptor and tetrodotoxin (TTX) in the case of the sodium channel. The major difference is the nature of the factor that induces the permeability changes of the molecules. In the sodium channel, permeability depends directly on the membrane potential (Figures 16.16 and 16.21). Permeability of the ACh receptor is essentially voltage independent and depends instead on the binding of the transmitter ACh. As noted on p. 427, this difference in the control of ion permeability of the two molecules results in the difference in properties of the potentials they produce: Action potentials are all-or-none and propagated, while synaptic potentials are graded in amplitude and spread decrementally.

Termination of Transmitter Action: Enzymes and Reuptake   The action of synaptic transmitters is usually of short duration. Both the release and the receptor binding of transmitter molecules occur within a few milliseconds. For normal synaptic function,

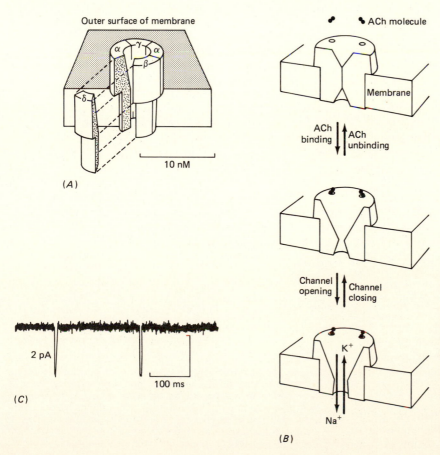

Figure 17.22 Acetylcholine receptor structure and conformation changes. (A) The receptor has five subunits; the two α subunits contain an ACh binding site. [From C.F. Stevens, *Trends Neurosci.* **8**:1–2 (1985).] (B) Two ACh molecules bind to the closed conformation of the receptor, which then changes conformation to open the channel to sodium and potassium ion flux. [Adapted from C.F. Stevens, *Sci. Am.* **241**(3):54–65 (1979).] (C) Single-channel currents resulting from two openings of an ACh receptor, recorded by patch clamping. [From E.R. Kandel and J.H. Schwartz (eds.), *Principles of Neural Science,* 2nd ed. Elsevier, New York, 1985.]

transmitter molecules must be prevented from "hanging around" in the synaptic cleft or from diffusing to other synapses. The temporal and spatial effects of transmitter action are limited in two ways: by enzymatic destruction of the transmitter molecules and by active reuptake of the native transmitter or its products.

At the neuromuscular junction, ACh is enzymatically digested by the enzyme acetylcholinesterase (AChE), located within the synaptic cleft. The action of AChE is very fast, the entire sequence of ACh release, diffusion, binding to ACh receptors, and digestion lasting perhaps 5 ms. Presumably, some ACh molecules are destroyed even before they can act postsynaptically. The products of ACh breakdown are choline and acetate. Choline is transported back into the nerve terminal by a specific high-affinity uptake system in the terminal membrane. Thus, the action of AChE both terminates the postsynaptic action of ACh and provides choline, the rate-limiting substrate for ACh synthesis, for reuptake and resynthesis of transmitter in the presynaptic terminal.

The termination of transmitter action by a localized enzyme is not a universal feature of all transmitter systems. For many transmitters (catecholamines, amino acids) the termination and reuptake processes are combined. For example, the transmitter noradrenaline, or norepinephrine, is itself transported back into the presynaptic cell by a high-affinity reuptake system. Although enzymes that catabolize noradrenaline are present, their action is slower than that of the reuptake system. Thus, the synaptic action of noradrenaline is terminated by reuptake of native transmitter rather than by enzymatic destruction.

Other Synaptic Mechanisms

Our discussion of synaptic action so far has been based on pioneering studies performed on synapses of vertebrate neuromuscular junctions and spinal motoneurons. In these synapses, the action of synaptic transmitter is to elicit an EPSP or IPSP by means of an increase in ion permeabilities. This permeability increase mechanism appears to be the dominant mode of synaptic action in all metazoan phyla. Other less widely distributed modes of synaptic action, however, have also been discovered. Two relatively well-characterized examples are PSPs resulting from a permeability decrease and presynaptic inhibition.

Synaptic Potentials Mediated by Permeability Decrease   It should be clear from the Goldman equation that any changes in ion permeabilities of a postsynaptic cell—decreases as well as increases—change the membrane potential. Thus, synaptic transmitter action that decreases ion permeability also produces a synaptic potential. Such a mechanism occurs in vertebrate sympathetic ganglia. As discussed in Chapter 2, the sympathetic nervous system is com-

posed of *preganglionic* neurons, with somata in the spinal cord and axon terminals in the peripheral sympathetic ganglia flanking the cord, and *postganglionic* neurons, with somata in the sympathetic ganglia and terminals at the visceral target organs (Figure 17.23). The synapses between preganglionic and postganglionic neurons in sympathetic ganglia were long thought to be simple relays of little integrative significance. Recent studies, however, reveal that synaptic interactions in sympathetic ganglia are more complex and interesting than had been suspected. For one thing, many sympathetic ganglia have been found to contain intrinsic interneurons that lack definite axons. These intrinsic neurons are termed SIF cells (small intensely fluorescent cells) because of their strong catecholamine-induced fluorescence.

The synaptic connections and interactions in a sympathetic ganglion are diagrammed in Figure 17.24. Recordings from the postganglionic neuron show three phases of synaptic response to stimulation of preganglionic axons (Figure 17.24A): a fast EPSP lasting about 30 ms, a slow IPSP lasting about 1 s, and a still slower EPSP of several seconds' duration. The synaptic connections thought to mediate these responses are shown in Figure 17.24B. The fast EPSP is mediated by increased ion permeability, but the slow IPSP and the slow EPSP are mediated by permeability decreases. This distinction is best shown by experiments in which standard current pulses are injected into the postganglionic neuron before and at various times after preganglionic axon stimulation (Figure 17.24C). The amplitude of depolarization produced by the standard current pulse is a measure of the membrane resistance, by Ohm's law. During the fast EPSP, the current pulse produces a smaller potential change than before stimulation, indicating a resistance decrease (Figure 17.24C₁). During the slow IPSP and the slow EPSP,

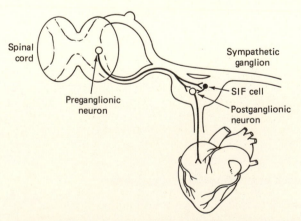

Figure 17.23 Anatomy of sympathetic neurons. Preganglionic neurons have somata in the spinal cord and axons terminating in a sympathetic ganglion adjacent to the segmental spinal nerve. They synapse upon postganglionic neurons that extend to visceral target organs such as the heart. Sympathetic ganglia also contain small, intensely fluorescent cells (SIF cells) that synapse upon the postganglionic neurons.

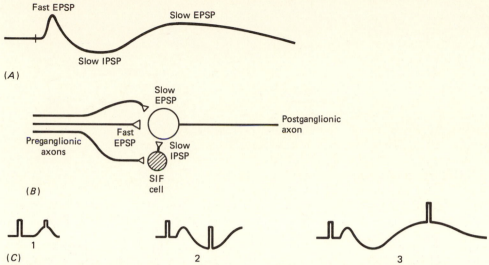

Figure 17.24 Synaptic potentials may be mediated by either a resistance decrease (permeability increase) or a resistance increase. (*A*) Intracellular recording of a sympathetic postganglionic neuron. Preganglionic stimulation elicits a sequence of a fast EPSP, a slow IPSP, and a slower EPSP. (*B*) Synaptic connections associated with the responses above. (*C*) Measurement of resistance changes during PSPs. Identical current pulses are applied to the postganglionic cell through a microelectrode. The voltage responses (vertical depolarization pulses) are a measure of resistance. ① During the fast EPSP, postganglionic neuronal resistance is decreased. ② and ③ In contrast, the resistance (voltage response to current) is *increased* during the slow IPSP and the slow EPSP.

however, the current pulse produces a larger potential change than a prestimulus pulse, demonstrating an increase in membrane resistance—that is, a decrease in conductance or decrease in ion permeability. Other results indicate that the slow IPSP results from a decrease in the resting membrane permeability to Na^+, and the slow EPSP results from a decrease in K^+ permeability.

Presynaptic Inhibition The nerve–nerve synapses we have considered so far have been *axodendritic* and *axosomatic* synapses in which the presynaptic axon terminal ends on a dendrite or soma of the postsynaptic cell. Axodendritic and axosomatic synapses are the most common types of synapse in nervous systems, but other types occur as well. Presynaptic inhibition is believed to be mediated by *axoaxonal* synapses as illustrated by the following example. Figure 17.25 shows the synapse of a sensory axon onto a cat spinal motoneuron. A presynaptic inhibitory (PSI) interneuron ends on the sensory axon in an axoaxonal synapse. The PSI interneuron exerts an inhibitory effect on the motoneuron only by decreasing the excitatory effect of the sensory neuron on the motoneuron. An impulse in the PSI neuron leads to no direct change in the membrane potential or membrane resistance of the motoneuron (Figure 17.25), showing that the inhibitory effect is indirect. The action of the PSI neuron is to *decrease the transmitter output* of the sensory neuron, thereby decreasing the amplitude of the sensory-neuron-generated EPSP. Evidence indicates that the PSI neuron depolarizes the sensory axon terminal and thus de-

creases the amplitude of the action potential in it, perhaps by partial inactivation of sodium conductance. Since the amount of transmitter released by the sensory terminal is a steep function of the amplitude of the action potential (Figure 17.14), a small decrease in action potential amplitude can considerably decrease the amount of transmitter liberated.

Presynaptic inhibition has functional properties quite different from those of synaptic inhibition mediated by IPSPs. Presynaptic inhibition is very specific, since a PSI neuron will oppose only those excitatory neurons it ends on and will have no effect on other excitatory neurons. This specificity stands in contrast to IPSP-mediated synaptic inhibition, which can oppose all excitatory synapses onto the postsynaptic cell (Figure 17.12). The time course of presynaptic inhibition also differs from that of IPSPs. Presynaptic inhibition lasts 100–150 ms, in contrast to 10–20 ms for IPSPs.

It is not clear how widespread presynaptic inhibition is in nervous systems. It is best known in crustacean neuromuscular junctions (see Chapter 19) and in mammalian spinal and brainstem sensory pathways. The selective presynaptic inhibition of sensory input by central and other sensory neurons is thought to underlie changes in how much attention we pay to sensory stimuli.

Reciprocal Synapses Another less common kind of synaptic structure is the *dendrodendritic* synapse. Synapses between dendrites are found in electron micrographs of mammalian brains, and some of them appear to be reciprocal. The mode of action and the

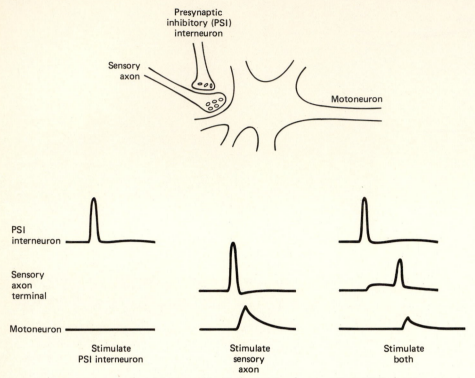

Figure 17.25 Presynaptic inhibition. Stimulation of the presynaptic inhibitory interneuron alone has no effect on the motoneuron. Rather, it decreases the amount of transmitter released by the sensory terminal, by decreasing the amplitude or the duration of the action potential at the sensory axon terminal.

functional significance of such synapses are unclear; many involve cells that lack a definable axon. Let us consider one example that illustrates the features of dendrodendritic synapses, reciprocal synapses, and neurons that do not generate action potentials.

The olfactory bulb, the most anterior structure in the vertebrate brain, is the major relay station for olfactory sensory neurons. Figure 17.26 shows a simplified circuit of the afferent (incoming) synaptic connections in the olfactory bulb. Olfactory receptor axons end on distal dendrites of mitral cells, which are the principal relay neurons to the rest of the brain. The horizontal secondary dendrites of the mitral cells form synapses with spine terminals of granule cell dendrites. The granule cells lack axons; the spines are their only synaptic output. Granule cell dendrites and mitral cell dendrites thus form *reciprocal synapses* in which each element is both presynaptic and postsynaptic (Figure 17.26, inset). The mitral-to-granule synapses are excitatory and the re-

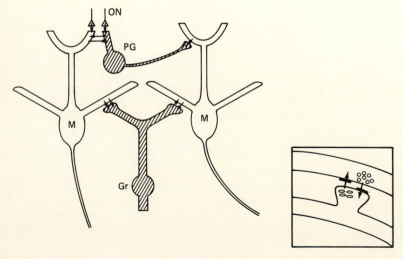

Figure 17.26 Reciprocal synapses in the mammalian olfactory bulb. Mitral cells (M) are excited by sensory input of the olfactory nerve (ON). The mitral cells form reciprocal synapses with axonless granule cells (Gr), in which both cells are both presynaptic and postsynaptic (inset). Additional reciprocal synapses occur between the distal dendrites of mitral cells and periglomerular cells (PG). [After G.M. Shepherd, *The Synaptic Organization of the Brain,* 2nd ed. Oxford University Press, New York, 1979.]

ciprocal granule-to-mitral synapses are inhibitory. Figure 17.27 shows how these synapses are thought to function. The granule cells, which do not generate action potentials, inhibit the mitral cells that excited them (after a brief delay) and also inhibit nearby dendrites of other mitral cells. These functions of self-inhibition with delay and of lateral inhibition of neighboring cells are presumably important features in the central integration of olfactory sensory information.

Reciprocal synapses also occur between periglomerular cells and mitral cell distal dendrites of the olfactory bulb (Figure 17.26), in the retina (Chapter 18), and elsewhere. Such local circuits, termed neural microcircuits, demonstrate that local synaptic interactions that need not involve action potentials may be an important mode of neural integration.

SYNAPTIC TRANSMITTERS

Perhaps two dozen chemical compounds have been implicated as possible synaptic neurotransmitters, and the list of candidates is increasing rapidly. So far, we have confined our attention to *cholinergic* synapses (defined previously as synapses in which the transmitter is acetylcholine). In so doing, we have run the risk of perpetuating the commonly held illusion that acetylcholine is *the* transmitter in nervous systems, or at least the most important and prevalent transmitter. In fact, our knowledge of different transmitter systems bears little relation to their prevalence. Acetylcholine-mediated transmission is well known because of the peripheral model system available for its study. Synapses in which the transmitter is norepinephrine or noradrenaline (termed *adrenergic* synapses) are the next best known, reflecting the existence of peripheral adrenergic synapses in the sympathetic nervous system, and of histofluorescent methods that illuminate their pathways. It is estimated, however, that fewer than 10 percent of the synapses in the vertebrate CNS are cholinergic and fewer than 1 percent employ catecholamines. In most central synapses, the identity of the transmitter is not known, but amino acids are thought to be the

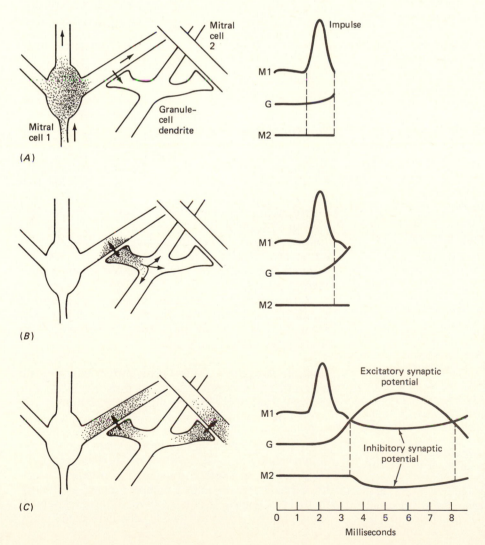

Figure 17.27 Hypothesis of reciprocal synaptic function in the olfactory bulb. The activity in the circuit during a physiological experiment is diagrammed. First the mitral-cell axon is electrically stimulated, triggering an impulse that is conducted backward up the fiber to the cell body and dendrites. The impulse activates excitatory synapses onto the dendrite spines of the granule cells (*A*). The excitation of the granule cell in turn activates inhibitory synapses from the granule-cell dendrite back onto the mitral-cell dendrite (*B*), inhibiting the initially excited mitral cell. The depolarization of the granule cell spreads passively, so that inhibitory synapses in other parts of the dendrite act to inhibit additional mitral cells (*C*). (Normally, the initial impulse in the mitral cell is triggered by input from the olfactory receptor cells, rather than by electrical stimulation.) [From Microcircuits in the nervous system, by G.M. Shepherd. Copyright © 1978, by Scientific American, Inc. All rights reserved.]

most abundant and widespread transmitters. This section discusses selected aspects of transmitter function, but it is beyond our scope to survey the subject comprehensively. We refer the reader to references at the end of the chapter that provide more extensive information.

Dale's Principle

The starting point for our consideration of synaptic transmitters is the idea, originally formulated by Sir Henry Dale in 1933, that a neuron releases the same transmitter at all its endings. This idea gained widespread support after the demonstration that mammalian spinal motoneurons release acetylcholine at the central synapses of their collaterals onto Renshaw cells (see Chapter 20) as well as at their peripheral synapses with muscle fibers. Dale's principle was informally rechristened Dale's law and was extended to assert that a differentiated neuron could synthesize and release only *one* transmitter, from all its terminals. This more general statement of Dale's law is presently under attack, since an increasing number of cases have been found in which a neuron appears to contain and perhaps release more than one putative transmitter.

Despite exceptions to Dale's principle (or ''law'' or ''hypothesis''), it remains an important generalization. Neurons appear to be metabolically specialized to synthesize and release a particular transmitter. This transmitter, however, may have a variety of postsynaptic effects (see Multiple Receptors, below). Moreover, a postsynaptic neuron typically receives synapses from many presynaptic neurons that release many different transmitters. Thus, it is relatively sound to characterize a neuron in terms of the transmitter it releases (e.g., as a cholinergic neuron), but it is usually unsound to characterize it as responding to a particular transmitter (e.g., as a cholinoceptive neuron), because it may respond to several transmitters.

Identification of Transmitters

What are the criteria used to determine whether a particular agent is in fact the transmitter at a particular synapse? There is general agreement that for acceptance of a candidate as the transmitter at a synapse, the five following experimental criteria should be met:

1. The candidate transmitter must be *present* in the presynaptic terminal, along with its synthetic machinery (enzymes, precursors).
2. The candidate transmitter is *released* upon presynaptic stimulation, in amounts sufficient to exert its postsynaptic action.
3. When applied exogenously (by iontophoresis or bath application) in moderate concentrations, the candidate transmitter should *mimic the effects* of pre-

synaptic stimulation. For example, it should induce the same ion permeability changes as the synaptic action.
4. A mechanism for *removal* of the candidate transmitter must exist. This removal can be by enzymatic inactivation or by reuptake into cells.
5. The *effects of drugs* on transmission at the synapse and on the candidate's action must be similar and consistent. That is, a drug that blocks, mimics, or potentiates synaptic transmission (if it acts postsynaptically) must similarly affect action of the applied candidate transmitter.

These criteria present imposing hurdles, and it is difficult to demonstrate all of them at a particular synapse, especially in the central nervous system. For example, the demonstration of release of a candidate transmitter requires that it be collected in sufficient quantity to be assayed, and further requires demonstration that the released transmitter came from the presynaptic terminal. This is rarely possible in the brain. Rigorous tests employing these criteria are important, however, to prevent uncritical acceptance of every synaptically active agent as a presumed transmitter. Some candidates can satisfy several criteria, but if they can be shown to fail any one, they should be rejected. Because these criteria are so difficult to satisfy experimentally, we have a long list of ''possible'' and ''probable'' transmitters and a short list of cases in which the transmitter at a particular synapse is clearly demonstrated.

Vertebrate Neurotransmitters

The principle synaptic transmitters of vertebrates, demonstrated and postulated, are summarized in Table 17.1. The list could be shorter or longer, depending on one's requirements for candidacy. The sites of probable transmitter distribution are particularly speculative and subject to change with further study. Just as the transmitter status for some candidates is better documented than for others, the evidence for a given transmitter is usually better at some sites than at others.

Rather than survey the metabolism and action of a number of transmitters, let us discuss a few conceptual issues that are important for understanding transmitter systems in general.

Multiple Receptors Many transmitters can mediate different postsynaptic effects at different sites. For example, acetylcholine excites skeletal muscle via EPSPs but inhibits the vertebrate heart via a hyperpolarization mediated by an increase in K^+ permeability. Since acetylcholine induces different permeability changes in the different postsynaptic cells (P_{Na} and P_K in skeletal muscle; P_K in heart), it follows that the postsynaptic receptor must be different in the two cases. Actually, two different kinds of acetylcholine receptor were characterized many

Table 17.1   NEUROTRANSMITTER CANDIDATES IN VERTEBRATE NERVOUS SYSTEMS

	Transmitter status	Usual synaptic action	Principal locations
Acetylcholine (nicotinic)	Demonstrated	+	Neuromuscular junctions, Renshaw cells, autonomic ganglia, rare in CNS
(muscarinic)	Demonstrated	− or +	Parasympathetic postganglionic, CNS, autonomic ganglia
Catecholamines			
Dopamine	Demonstrated	−	CNS epecially nigrostriatal, hypothalamus, SIF cells, retina
Norepinephrine	Demonstrated		Sympathetic postganglionic, locus coeruleus cells
Epinephrine	Probable		
Serotonin (5-HT)	Demonstrated	−	Brainstem raphe nuclei, pineal gland
Histamine	Probable		
Amino acids			
GABA	Demonstrated	−	Most brain IPSP; presynaptic inhibition
Glutamate	Probable	+	CNS EPSP, widespread
Aspartate	Probable	+	CNS, retina?
Glycine	Demonstrated	−	Spinal IPSP
Taurine	Possible	−	
Peptides			
Enkephalin	Probable	−	Brain and spinal; presynaptic inhibition
Endorphin	Possible		Pituitary, hypothalamus
Substance P	Probable	+	Spinal and brain sensory
TRF	Possible	−	Hypothalamus
Somatostatin	Possible	−	Hypothalamus, spinal cord
Neurotensin	Possible		
ATP	Possible		
Prostaglandins	Possible		

years ago, by their pharmacology (i.e., the effects of drugs). The ACh receptor of skeletal muscle is stimulated by nicotine and hence is termed *nicotinic*. As noted in Figure 17.5, it is blocked by curare. The ACh receptor of heart is stimulated by muscarine and hence is termed *muscarinic;* it is blocked by atropine. Muscarinic ACh receptors predominate over nicotinic receptors within the vertebrate CNS.

Several of the transmitters have more than one kind of postsynaptic receptor protein; some of these are listed in Table 17.2.

Permeability Effects and Metabolic Effects of Transmitter Action   In the examples we have considered, the action of transmitter has been to alter ion permeabilities in the postsynaptic membrane, thereby generating a postsynaptic potential. Permeability changes, however, are not the only kind of postsynaptic response to transmitter. Several transmitters have broad metabolic influences within their postsynaptic target cells, influences that are mediated by second messengers. A *second messenger* is an intracellular signaling molecule that alters activity of the cell in response to activation of outer membrane receptors.

To clarify the concept of an intracellular second messenger, let us anticipate Chapter 21 and consider an analogous example: How do hormones exert their effects on target cells? There are two broad chemical classes of hormones—steroids and peptides. Steroid hormones are lipid soluble and thus can penetrate the lipid bilayer of cell membranes to enter the target cell. Peptide hormones, on the other hand, are hydrophilic, relatively large molecules to which the cell membranes are impermeable. Thus, they cannot enter target cells to mediate their metabolic effects. Instead, peptide hormones bind to hormone receptor molecules on the outer cell membrane and trigger the production of an intracellular second messenger. The best-known second messenger is the cyclic nucleotide 3′,5′−cyclic adenosine monophosphate (cyclic AMP or cAMP). The hormone receptor protein is

Table 17.2 SOME TRANSMITTER RECEPTORS AND SECOND MESSENGERS

Neurotransmitter	Type of receptor	Candidate second messenger
Acetylcholine	Nicotinic	none
	Muscarinic	[a]
Dopamine	D-1	cAMP
	D-2	
Norepinephrine	α-Adrenergic	phosphoinositides
	β-Adrenergic	cAMP
Serotonin	5-HT$_1$	cAMP
	5-HT$_2$	phosphoinositides?
Histamine	H$_1$	phosphoinositides
	H$_2$	cAMP
Enkephalin	μ, κ, δ	cAMP

[a] Muscarinic ACh receptor stimulation can increase phosphoinositide turnover, increase cGMP levels, and decrease cAMP levels. Such multiple effects also may occur with other receptor types.

coupled to a membrane-bound enzyme, adenylate cyclase. Binding of hormone to the receptor activates the adenylate cyclase, which catalyzes the intracellular production of AMP from ATP.

AMP was first discovered to be an intracellular second messenger for peptide hormones, as outlined above. Subsequent experiments showed that several neurotransmitters also produced cAMP changes in neural tissue, both centrally and peripherally. The mechanism of action is thought to be similar to that for hormonal stimulation (Figure 17.28). cAMP is probably not the only second messenger for intracellular effects of neurotransmitters. Other candidate second messengers include Ca^{2+} ions, cyclic guanosine monophosphate (cGMP), and two products of phosphoinositide metabolism: inositol triphosphate and diacylglycerol. This area of research is rapidly changing, and is complicated by possible interactions between different second messenger systems. For example, various experiments have demonstrated that activation of muscarinic ACh receptors can in-

crease phosphoinositide metabolism, increase cGMP levels, and decrease cAMP levels, in the same or different kinds of cells. Table 17.2 lists some of the apparent second messenger systems implicated in the actions of neurotransmitters. Here we confine our attention to cAMP as a relatively well-studied example of second messenger action.

Intracellular second messengers such as cAMP can exert widespread metabolic effects. In general, cAMP exerts its effects by the activation of protein kinases, enzymes that phosphorylate proteins. An increase in intracellular cAMP concentration leads to increased activation of the protein kinases, which alter the structure and activity of various proteins by phosphorylating them. Depending on what kinases are present in the target cell, stimulation of a transmitter-dependent adenylate cyclase can trigger reactions that alter (by phosphorylation) membrane proteins involved in ion permeability changes, cytoplasmic proteins involved in control of cellular metabolism, or even nuclear proteins regulating gene

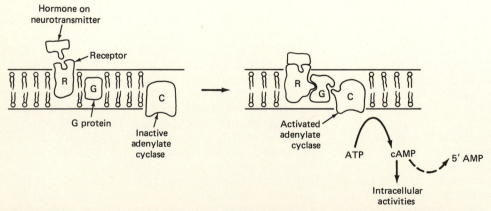

Figure 17.28 Formation of cAMP as an intracellular second messenger. The binding of a hormone or neurotransmitter to its membrane-bound receptor molecule induces activation of adenylate cyclase, an enzyme that catalyzes the synthesis of cAMP from intracellular ATP. The linkage of the receptor to adenylate cyclase is indirect, via an intermediate protein termed a G protein. Another enzyme (phosphodiesterase, not shown) catabolizes cAMP to 5'-AMP.

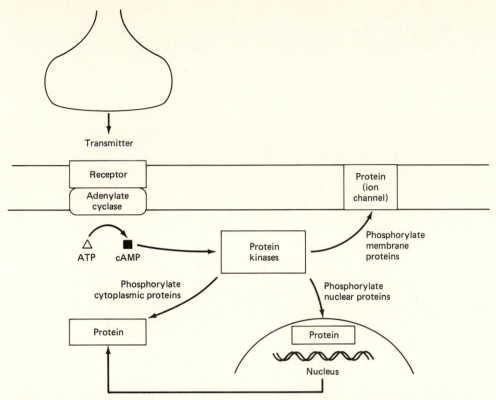

Figure 17.29 Cyclic AMP can exert a variety of intracellular effects, by activating various protein kinases. Some kinases may phosphorylate membrane proteins and thus alter membrane potentials. Others may phosphorylate cytoplasmic proteins to alter cell metabolism. Still others may phosphorylate nuclear proteins, controlling gene expression and synthesis of new proteins.

expression (Figure 17.29). The cAMP molecules are eventually catabolized to 5′-AMP, by action of the enzyme phosphodiesterase.

We shall consider the actions of cyclic nucleotides in greater detail in Chapter 21. The fact that transmitters can stimulate the production of second messengers is important, in part because it provides one possible mechanism whereby synaptic transmission can mediate relatively slow and long-lasting effects. The direct action of transmitters on ion permeability channels in membranes has a time course of milliseconds (Figure 17.5). The time course of effects mediated by second messengers may be as short as tens of milliseconds; for example, the slow PSPs mediated by decreased ion permeabilities in sympathetic ganglia (Figure 17.24) are cyclic nucleotide dependent. At the other extreme, second messenger actions may underlie synaptic changes involved in learning and memory, with a time course of days or years (see Synaptic Plasticity, p. 450).

McGeer et al. (see Selected Readings) have defined an *ionotropic* transmitter as one that directly changes ion permeabilities, while a *metabotropic* transmitter is one that affects cell metabolism via a second messenger. These terms, however, should be used with reference to the *effects* of transmitters rather than to the transmitters themselves, since many transmitters may have both ionotropic and metabotropic actions (Table 17.2). Moreover, metabotropic effects mediated by cAMP may affect ion permeabilities secondarily. Despite these objections and the cavil that the terms are not very euphonious, the distinction they represent is an important one.

Invertebrate Transmitters

The best known transmitters of vertebrates are also found in the nervous systems of the major invertebrate groups, and it is likely that the list of neurotransmitters is generally similar for most phyla. For example, there is evidence (of variable quality) supporting transmitter roles for acetylcholine, GABA, glutamate, dopamine, and 5-hydroxytryptamine (5-HT, serotonin), at least among annelids, arthropods, and mollusks. These transmitters are employed in different roles in different phyla, however. For example, most evidence indicates that in arthropods, glutamate is the major excitatory neuromuscular transmitter while acetylcholine is the major sensory neurotransmitter; their roles are reversed in the vertebrates (Figure 17.30).

Our knowledge of transmitter systems in invertebrates is less advanced than that for vertebrates, with the exception of a handful of biochemically charac-

BOX 17.1   PEPTIDES AND PAIN: ENKEPHALINS, ENDORPHINS, AND SUBSTANCE P

Morphine, heroin, and other opiate drugs have *analgesic* effects; that is, they decrease the strength and aversive quality of pain sensations. Opiates are thus presumed to inhibit pathways in the CNS that convey and interpret sensory information about pain. During the 1970s, several groups of investigators demonstrated that the opiates bound stereospecifically to anatomically discrete sites in the mammalian brain and that this specific binding was blocked competitively by opiate antagonists such as naloxone. These results showed that there are specific opiate receptor molecules in the brain. Other studies found that electrical stimulation of localized areas of the midbrain and brainstem produced analgesia that was also reversible by naloxone. The finding of specific opiate receptors implied the existence of endogenous opiates that bind to them, since the receptors were unlikely to have evolved to bind derivatives of opium poppies. Thus, the characterization of opiate receptors led to a search for endogenous opiates. Several peptides have been isolated that bind to opiate receptors in a naloxone-reversible manner. The most important of these are the pentapeptides methionine–enkephalin and leucine–enkephalin (*met-* and *leu-enkephalin*) and the 31-amino-acid peptide β-*endorphin*. β-Endorphin is one of several peptides derived from the cleavage of a larger precursor, pro-opiomelanocortin; other cleavage products include the pituitary hormones ACTH and αMSH (see Tables 21.3 and 21.4). Met-enkephalin and leu-enkephalin are derived from a different precursor and are localized in different neurons than β-endorphin. β-Endorphin is found in the pituitary and in hypothalamic neurons; it can exert a rather long-lasting analgesic effect and may act like a hormone. The enkephalins are found in neurons of many brain areas; they exert more transient effects and are thought to function more like neurotransmitters. More recent studies have found other endogenous opiates and have demonstrated the existence of several kinds of opiate receptors.

Another peptide, *substance P*, is involved in transmitting the pain signals that are inhibited by the endogenous opiates. Both the afferent sensory neurons and many of the central neurons that convey information about pain sensation contain substance P,

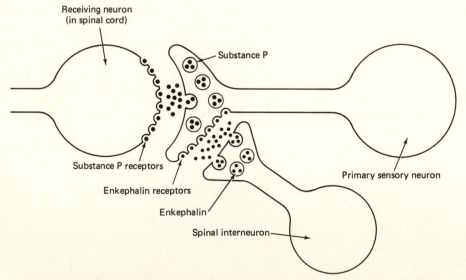

Diagrammatic representation of one way in which peptide transmitters may regulate the transmission of pain information from the peripheral pain receptors to the brain. Pain sensory neurons release substance P at their synapses in the dorsal horn of the spinal cord. Interneurons containing the peptide transmitter enkephalin make synapses onto the axon terminals of the pain neurons. Enkephalin released from the interneurons inhibits the release of substance P, so that the receiving neuron in the spinal cord receives less excitatory stimulation and hence sends fewer pain-related impulses to the brain. Opiate drugs such as morphine presumably bind to enkephalin receptors, mimicking the pain-suppressing effects of the enkephalin system. [Adapted from L.L. Iverson, *Sci. Am.* **241**(3):134–149 (1979).]

an undecapeptide (11 amino acids) thought to be a major transmitter of pain pathways. The opiates may act by regulating the activity of these substance-P-containing pain neurons. One idea is that enkephalin-producing neurons presynaptically inhibit the pain neurons, decreasing their release of substance P (see figure). This idea might explain how acupuncture and stimulation of nonpain neurons can inhibit pain sensations.

Study of the functional roles of peptides in nervous systems is one of the most rapidly expanding areas in physiology. We may find that a particular peptide such as an enkephalin acts in one place as a classical neurotransmitter, in another place as a neuromodulator that alters synaptic response to transmitter, and as a hormone somewhere else.

terized identifiable neurons. Thus, any speculation about the evolution of transmitter systems is probably premature. If the lists of major transmitter candidates are similar for several phyla, it may indicate that either many more transmitters remain to be discovered or that animals are using the same few in a variety of different ways. Certainly there is evidence for the latter view; molluscan neurons, for example, have at least six different pharmacologically separable 5-HT receptors and at least three ACh receptors. On the other hand, peptide transmitters have received little study in invertebrates and might be expected to show greater differences between phyla than the smaller amino acids and amines.

REGULATION AND PLASTICITY OF SYNAPTIC FUNCTION

The various aspects of synaptic transmission may change quantitatively over time. Rates of transmitter synthesis, storage, release, and reception may be increased or decreased in different circumstances. Even electrically mediated transmission can be altered by changes in pH and in $[Ca^{2+}]$. Changes in the parameters of synaptic transmission are important both for homeostatic regulation of synaptic functions and in synaptic plasticity—changes in synaptic efficacy over time. Because synaptic functions are more labile than other aspects of neuronal function such as axonal conduction, it is widely supposed that synaptic plasticity underlies changes in nervous system function over time. Thus, the synaptic bases of nervous system development, learning, and memory are subjects of active current investigation. We cannot survey this burgeoning field, but we will present examples that demonstrate some of the major issues.

Homeostatic Regulation of Transmitter Metabolism

The metabolism of transmitters must be regulated just as other aspects of metabolism are regulated. The problem is best illustrated by an example already mentioned (see Presynaptic Mechanisms, p. 435): A stimulated superior cervical ganglion releases 10 per-

cent of its acetylcholine content per minute, without a decrease in the total ACh content of the ganglion. The rate of stimulated release is 50 times greater than the resting release rate. Clearly, the rate of synthesis of ACh must also increase 50-fold if there is no depletion of ACh. Although the exact mechanism controlling the increased synthesis of ACh is not known, it probably involves an increased availability of free choline, since choline is normally rate limiting in ACh synthesis.

More complex but better understood ways of regulating transmitter synthesis have been discovered for the catecholamine transmitters dopamine and norepinephrine. Let us next examine catecholamine synthesis, to illustrate the various levels of regulation of transmitter metabolism.

Metabolism of Dopamine and Norepinephrine   Figure 17.31 shows the steps in the synthesis of dopamine and norepinephrine (noradrenaline). Both transmitters are synthesized by a common pathway starting with tyrosine. Some neurons, termed *dopaminergic,* accumulate dopamine as their transmitter product; other *noradrenergic* neurons synthesize norepinephrine as their transmitter. The enzyme tyrosine hydroxylase, present in both dopaminergic neurons and noradrenergic neurons, converts tyrosine to dihydroxyphenylalanine (DOPA). The tyrosine hydroxylase step is rate limiting in the synthesis of both transmitters. The DOPA is converted to dopamine by the enzyme DOPA decarboxylase. In noradrenergic neurons, dopamine is converted to norepinephrine by the enzyme dopamine β-hydroxylase; this enzyme is absent in dopaminergic neurons, and thus dopamine is the stored end product.

Dopaminergic and noradrenergic neurons are able to regulate their rates of transmitter synthesis with changes in activity, in at least three ways. First, both dopamine and noradrenaline inhibit the activity of tyrosine hydroxylase. This is an example of *end-product inhibition*, in which accumulation of the product of a series of reactions inhibits the rate-limiting enzyme of the series. When the neuron is active and releases its transmitter, the inhibition of tyrosine hydroxylase by the transmitter is presum-

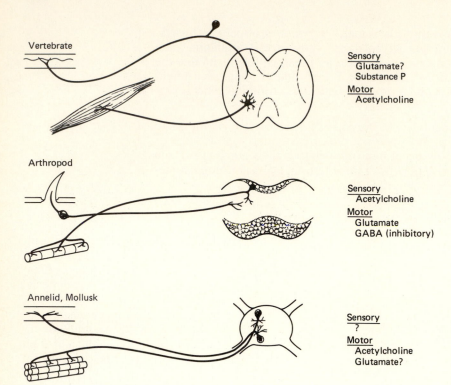

Vertebrate

Sensory
  Glutamate?
  Substance P
Motor
  Acetylcholine

Arthropod

Sensory
  Acetylcholine
Motor
  Glutamate
  GABA (inhibitory)

Annelid, Mollusk

Sensory
  ?
Motor
  Acetylcholine
  Glutamate?

Figure 17.30 Major candidate transmitters of peripheral neurons in the groups that have been most studied. The lists are incomplete, but indicate different functional roles of transmitters in the different phyla.

ably decreased and the rate of synthesis is increased. A second form of regulation involves change in the properties of tyrosine hydroxylase in the presynaptic terminal. With depolarization of dopaminergic and noradrenergic terminals, tyrosine hydroxylase undergoes a change termed activation; its kinetic properties change so that it has a higher affinity for tyrosine and for pteridine cofactor and is less susceptible to inhibition by dopamine or norepinephrine. This modification of the enzyme apparently involves phosphorylation by a mechanism that depends on cAMP and/or calcium. The third and slowest form of regulation involves changes in the concentration of tyrosine hydroxylase. Following prolonged activation of dopaminergic or noradrenergic neurons, their rate of tyrosine hydroxylase synthesis is somehow increased, eventually making more enzyme available at the terminals.

Transmitter synthesis is not the only aspect of synaptic function to be regulated. Both the release and the reception of transmitters can undergo changes that may be homeostatic. The terminals of dopaminergic, noradrenergic, and some other neurons bear *presynaptic receptors* (also termed *autoreceptors*) for the transmitter they release. The activation of these autoreceptors inhibits transmitter release, implying that the release process is somewhat self-limiting. The sensitivity of postsynaptic cells to transmitter can also change, as we have seen with denervation supersensitivity of muscle fibers (Figure 17.21). The roles of these forms or regulation in the normal functions of nervous systems, however, remain to be clarified.

Synaptic Plasticity

Synaptic potentials have time courses of milliseconds to seconds, long enough to have a transient effect on the excitability of postsynaptic cells, but ephemeral

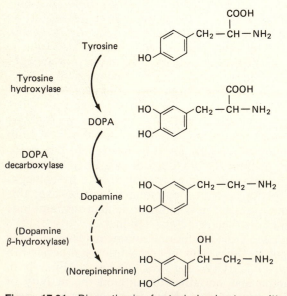

Tyrosine

Tyrosine
hydroxylase

DOPA

DOPA
decarboxylase

Dopamine

(Dopamine
β-hydroxylase)

(Norepinephrine)

Figure 17.31 Biosynthesis of catecholamine transmitters. Tyrosine is taken up from the blood. Noradrenergic neurons (but not dopaminergic neurons) contain the enzyme dopamine β-hydroxylase, which catalyzes the synthesis of norepinephrine. Adrenergic neurons contain an additional enzyme (phenylethanolamine N-methyl transferase) which converts norepinephrine to epinephrine.

Facilitation

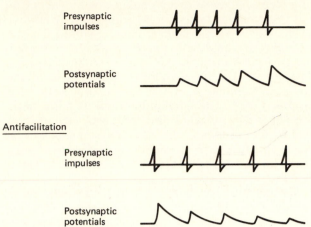

Figure 17.32 Synaptic facilitation and antifacilitation. In facilitation, the amplitudes of successive PSPs increase in response to repeated presynaptic action potentials. In antifacilitation, successive PSPs in a series decrease in amplitude.

nevertheless. If synapses are involved in the long-term behavioral changes of learning and memory (an assertion unproven at best), we would like to be able to demonstrate changes in synaptic efficacy—synaptic plasticity—that have a suitably long time course of minutes, days, or weeks. We examine next a few cases of such changes.

In many synapses, the amplitudes of individual synaptic potentials are not constant over time (Figure 17.32). *Synaptic facilitation* is defined as an increase in amplitude of postsynaptic potentials in response to successive presynaptic impulses (Figure 17.32, top). A decrease in amplitude of postsynaptic potentials with successive presynaptic impulses is termed *synaptic antifacilitation* or *synaptic depression*. Both synaptic facilitation and antifacilitation result from changes in the amount of transmitter liberated per presynaptic impulse. These changes are known to be calcium dependent, but their mechanisms are otherwise poorly understood.

Facilitation of synaptic transmission is often es-

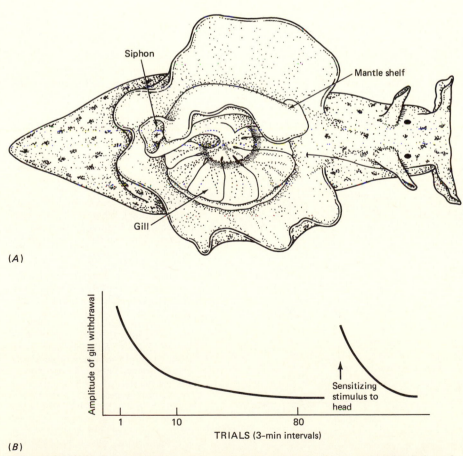

(A)

(B)

Figure 17.33 (A) Gill-withdrawal reflex of *Aplysia* occurs in response to stimulation of the siphon or the mantle shelf. The animal retracts the gill to the smaller indicated position. [Adapted from E.R. Kandel, *Sci. Am.* **241**(3):66–76 (1979).] (B) Habituation of the gill withdrawal reflex with repeated stimulation and its recovery following a sensitizing stimulus to the head.

pecially pronounced after tetanic stimulation of pre-synaptic neurons, e.g., by trains of stimuli at about 10–100/s for several seconds. The response to a single presynaptic impulse may be elevated several-fold after tetanic stimulation, and although the effect diminishes over time, it may persist for hours. This enhancement of synaptic response is termed *post-tetanic potentiation*. Post-tetanic potentiation indicates that synaptic efficacy can change with use, and these changes can be long lasting. Particularly long-term potentiation changes have been reported in the hippocampus of the vertebrate brain, a region implicated in learning and memory functions.

Habituation and Sensitization in *Aplysia*   Habituation and sensitization are two simple forms of learning that occur in nearly all kinds of animal. *Habituation* is defined as the decrease in intensity of a reflex response to a stimulus when the stimulus is presented

repeatedly. *Sensitization* is the prolonged enhancement of a reflex response to a stimulus which results from the presentation of a second stimulus that is novel or noxious. Several studies have explored the synaptic basis of habituation and sensitization of reflexive gill withdrawal in the marine mollusk *Aplysia*. The gill of *Aplysia* withdraws in response to mechanical stimulation of the siphon or mantle shelf (Figure 17.33). The amplitude of gill withdrawal decreases with repeated low-frequency stimulation; that is, the response habituates. After a shock to the head, the response to siphon stimulation is again large; that is, it is *sensitized* by the head shock (Figure 17.33).

Kandel and his colleagues have mapped the neural circuit of the gill withdrawal reflex and have determined the synaptic locus of the habituation and sensitization. The habituation of the gill withdrawal response results from a waning of synaptic excitation

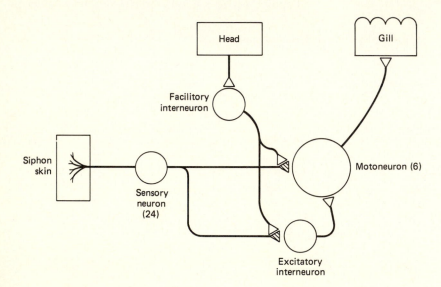

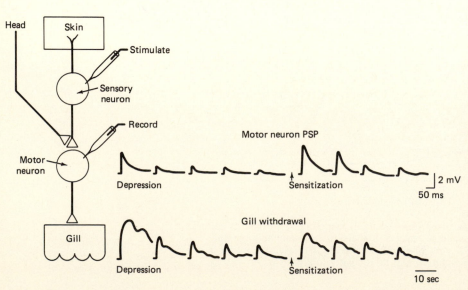

Figure 17.34 The gill-withdrawal reflex pathway and the behavioral and synaptic aspects of habituation and sensitization. The complete pathway is shown at the top. Tactile stimulation of the siphon skin activates sensory neurons which connect to gill motor neurons. The bottom of the figure compares responses of the whole animal with synaptic activity of a reduced preparation (shown at the left). Lower traces show gill withdrawal in the whole animal. When the skin is stimulated at low frequency, the gill-withdrawal response decrements, or habituates, while head shock causes the response to increase, or sensitize. Upper traces show responses of the reduced preparation, using intracellular stimulation of the sensory neurons instead of tactile stimulation and intracellular recording of motor neuron EPSPs instead of gill withdrawal. Stimulation of a nerve from the head replaces head shock. The decrement and enhancement of the motor neuron EPSPs mirror the behavioral habituation and sensitization of gill withdrawal. This result suggests that the sensory-to-motor neuron synapse is the primary site of the behavioral plasticity. [Adapted from M. Klein in J. Koester and J.H. Byrne, eds. *Molluscan Nerve Cells: From Biophysics to Behavior.* Cold Spring Harbor Laboratory, Cold Spring Harbor, NY, pp. 211–221 (1980).]

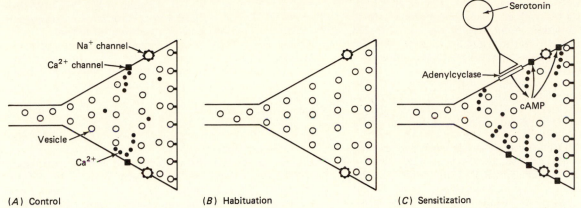

Figure 17.35 Model of short-term habituation and sensitization. (*A*). In the control state depolarization opens some Ca²⁺ channels. As a result, some Ca²⁺ flows into the terminals and allows a certain number of synaptic vesicles to bind to release sites and be released. (*B*) With habituation, fewer Ca²⁺ channels open in response to action potentials in the sensory terminal. The resulting decrease in Ca²⁺ influx functionally inactivates the synapse by preventing synaptic vesicles from binding. (*C*) Sensitization is produced by cells thought to be serotonergic. Serotonin acts on an adenyl cyclase in the terminals, which stimulates the synthesis of cAMP. cAMP in turn acts on a cAMP-dependent protein kinase to phosphorylate a membrane protein thought to be the K⁺ channel, which leads to a decrease in the repolarizing K⁺ current and a broadening of the action potential. The increase in the duration of the action potential increases the time during which Ca²⁺ channels can open, leading to a greater influx of Ca²⁺ and greater binding of vesicles to release sites, and therefore to an increased release. [Adapted from Klein and Kandel, *Proc. Nat. Acad. Sci. U.S.A.* **75**:3512–3516 (1978).]

of gill motor neurons by sensory neurons; the time course of diminution of sensory-to-motor EPSPs closely parallels the time course of behavioral habituation (Figure 17.34). This diminution of EPSPs does not result from any postsynaptic change, but rather from a decrease in the number of quanta of transmitter released by the sensory nerve endings. Thus, the synaptic basis of habituation in *Aplysia* is antifacilitation of the sensory synaptic terminals. (Habituation in crayfish and amphibians appears to have the same synaptic basis.)

Sensitization of the gill withdrawal response by head shock also occurs at the sensory-to-motor synapses. In contrast to habituation, a sensitizing stimulus increases the amount of transmitter released per impulse at the sensory neuron terminal. This facilitation apparently results from activation of synaptic endings of sensitizing interneurons that end on the sensory terminals in axoaxonal synapses (Figure 17.34). This process, termed presynaptic facilitation, is the synaptic basis of behavioral sensitization.

How is the amount of transmitter release diminished during habituation and increased by sensitization? Evidence indicates that the calcium current entering the presynaptic terminal during an impulse is depressed during habituation. This finding suggests that there is a progressive, long-lasting inactivation of Ca²⁺ channels with habituation, allowing less Ca²⁺ to enter and to trigger transmitter release. The presynaptic facilitation underlying sensitization is caused by an increased Ca²⁺ influx. According to the current evidence, the facilitating interneurons release 5-HT (serotonin), which acts to increase the amount of cAMP in the sensory terminals (Figure

17.35). The cAMP is thought to control the phosphorylation of the K⁺ channels in the terminal so as to terminate the K⁺ current that normally terminates the action potential. This K⁺ inactivation prolongs the action potential, leading to an increase in the Ca²⁺ influx and in resultant transmitter release.

The studies on *Aplysia* have determined the anatomical location of two forms of behavioral plasticity in specific, identifiable synapses and have made considerable progress in defining the synaptic mechanisms producing these changes. Short-term habituation in *Aplysia* lasts about an hour, but with repeated training sessions, habituation may persist for over 3 weeks. This long-term habituation depends on changes in the same sensory-to-motor synapses. The long-term synaptic changes may not have exactly the same underlying mechanisms as short-term habituation, but they are clearly related and both are reversed by the same sensitization mechanism. Classical conditioning in *Aplysia* involves mechanisms similar to those of sensitization; however, other mechanisms are suggested for learning in other kinds of animals. Nevertheless, the work described above and other studies of identified neurons in mollusks represent the greatest progress to date toward establishing a synaptic basis for learning and memory.

SELECTED READINGS

Abrams, T.W. and E.R. Kandel. 1988. Is contiguity detection in classical conditioning a system or a cellular property? Learning in *Aplysia* suggests a possible molecular site. *Trends Neurosci.* **11**:128–135.

Akil, H., S.J. Watson, E. Young, M.E. Lewis, H. Khach-aturian, and J.M. Walker, 1984. Endogenous opioids: Biology and function. *Annu. Rev. Neurosci.* **7**:223–255.

Augustine, G.J., M.P. Charlton, and S.J. Smith. 1987. Calcium action in synaptic transmitter release. *Annu. Rev. Neurosci.* **10**:633–693.

Barnard, E.A., M.G. Darlison, and P. Seeburg. 1987. Molecular biology of the GABAA receptor: the receptor/channel superfamily. *Trends Neurosci.* **10**:502–509.

Berridge, M.J. 1985. The molecular basis of communication within the cell. *Sci. Am.* **253**(4):142–152.

Byrne, J.H. 1987. Cellular analysis of associative learning. *Physiol. Rev.* **67**:329–439.

*Cooper, J.R., F. Bloom, and R.E. Roth. 1986. *The Biochemical Basis of Neuropharmacology,* 5th ed. Oxford University Press, New York.

Cotman, C.W., D.T. Monaghan, and A.H. Ganong. 1988. Excitatory amino acid neurotransmission: NMDA receptors and Hebb-type synaptic plasticity. *Annu. Rev. Neurosci.* **11**:61–80.

Dunant, Y. and M. Israël. 1985. The release of acetylcholine. *Sci. Am.* **252**(4):58–66.

Dunlap, K., G.G. Holz, and S.G. Rane. 1987. G proteins as regulators of ion channel function. *Trends Neurosci.* **10**:241–244.

Eccles, J.C. 1982. The synapse: From electrical to chemical transmission. *Annu. Rev. Neurosci.* **5**:325–339.

Fatt, P. and B. Katz. 1951. An analysis of the end-plate potential recorded with an intracellular microelectrode. *J. Physiol.* **115**:320–370.

Guy, H.R. and F. Hucho. 1987. The ion channel of the nicotinic acetylcholine receptor. *Trends Neurosci.* **10**:318–321.

Hertzberg, E.L., T.S. Lawrence, and N.B. Gilula. 1981. Gap junctional communication. *Annu. Rev. Physiol.* **43**:479–491.

Heuser, J.E. and T.S. Reese. 1977. Structure of the synapse. In E.R. Kandel (ed.), *Handbook of Physiology,* Section 1, Volume 1, *The Nervous System,* pp. 261–294. American Physiological Society, Bethesda, MD.

Heuser, J.E. and T.S. Reese. 1979. Synaptic-vesicle exocytosis captured by quick-freezing. In F.O. Schmitt and F.G. Worden (eds.), *The Neurosciences: Fourth Study Program,* pp. 573–600. MIT Press, Cambridge, MA.

Holden, A.V. and W. Winlow (eds.). 1984. *The Neurobiology of Pain.* Manchester University Press, Manchester.

*Iverson, L. 1979. Chemistry of the brain. *Sci. Am.* **241**(3):118–129.

Kaczmarek, L.K. and I.B. Levitan (eds.). 1987. *Neuromodulation: The Biochemical Control of Neuronal Excitability.* Oxford University Press, New York.

Kandel, E.R. 1979. Small systems of neurons. *Sci. Am.* **241**(3):66–76.

Kandel, E.R. and J.H. Schwartz. 1982. Molecular biology of learning: Modulation of transmitter release. *Science* **218**:433–443.

Kandel, E.R. and J.H. Schwartz (eds.). 1985. *Principles of Neural Science,* 2nd ed. Elsevier, New York.

Katz, B. 1966. *Nerve, Muscle, and Synapse.* McGraw-Hill, New York.

Krieger, D.T., M.J. Brownstein, and J.B. Martin (eds.). 1983. *Brain Peptides.* Wiley, New York.

*Kuffler, S.W., J.G. Nicholls, and A.R. Martin. 1984. *From Neuron to Brain,* 2nd ed. Sinauer, Sunderland, MA.

Kupferman, I. 1980. Role of cyclic nucleotides in excitable cells. *Annu. Rev. Physiol.* **42**:629–641.

Lester, H.A. 1977. The response to acetylcholine. *Sci. Am.* **236**(2):106–118.

Levitan, I.B. 1988. Modulation of ion channels in neurons and other cells. *Annu. Rev. Neurosci.* **11**:119–136.

Llinas, R. 1982. Calcium in synaptic transmission. *Sci. Am.* **247**(4):56–65.

Lundberg, J.M. and T. Hokfelt. 1983. Coexistence of peptides and classical transmitters. *Trends Neurosci.* **6**:325–333.

McCarthy, M.P., J.P. Earnest, E.F. Young, S. Choe, and R.M. Stroud. 1986. The molecular neurobiology of the acetylcholine receptor. *Annu. Rev. Neurosci.* **9**:383–413.

McGeer, P.L., J.C. Eccles, and E.G. McGeer. 1987. *Molecular Neurobiology of the Mammalian Brain,* 2nd ed. Plenum, New York.

Otsuka, M. and S. Konishi. 1983. Substance P—the first peptide neurotransmitter? *Trends Neurosci.* **6**:317–320.

Shepherd, G.M. 1978. Microcircuits in the nervous system. *Sci. Am.* **238**(2):92–103.

Shepherd, G.M. 1979. *The Synaptic Organization of the Brain,* 2nd ed. Oxford University Press, New York.

*Shepherd, G.M. 1988. *Neurobiology,* 2nd ed. Oxford University Press, New York.

The Synapse. 1976. Cold Spring Harbor Symposium on Quantitative Biology, Vol. 40. Cold Spring Harbor Laboratory, Cold Spring Harbor, NY.

See also References in Appendix A.

chapter *18*

Sensory Physiology

One of the major and evolutionarily important functions of nervous systems is to process information about the environment. The actions of the environment on organisms are mediated predominantly via specialized receptor cells. All cells are somewhat responsive to aspects of their environment and thus subserve some functions that can be considered sensory. For example, protozoa respond to light and bacteria to chemical gradients. Such cellular responses presumably preceded the evolution of specialized sensory neurons. With the evolution of multicellularity and the attendant specialization of different cell types, not all cells were exposed to environmental stimuli. It is likely that specialized sensory cells evolved in conjunction with neural coordinating systems (see Primitive Nervous Systems, p. 387). These sensory cells provided information about the external environment (exteroceptors) and also information about the internal environment (interoceptors).

Thus, a *receptor cell* is a cell that responds to a stimulus in its environment, whether internal or external to the animal. Immediately, we have set up at least two definitional problems. First, if all cells are somewhat able to respond to environmental perturbations, are all cells sensory receptors? The term would then be so general as to be meaningless. We are forced to redefine a receptor cell as a cell specialized to transform stimulus energy into a neural signal (usually but not always action potentials). Because most neurons are activated by external synaptic transmitter molecules, we should further clarify our definition of a receptor cell as an excitable cell normally activated by stimuli other than synaptic activity. The second problem is, what is a stimulus? We can define a *stimulus* as a form of external (to the cell) energy to which a receptor can respond. Our

definitions are now perfectly circular, receptors being defined in terms of stimuli and vice versa. We face the same problem we did in Chapter 1: Just as it is difficult to define *organism* and *environment* other than in terms of each other, it is impossible to define *receptor* and *stimulus* in noncircular terms. Since there is no satisfactory way out of the dilemma, we must tread carefully through this semantic minefield and hope for the best.

ORGANIZATION OF SENSORY SYSTEMS

We said above that a sensory receptor (or simply *receptor*) is a *cell* specialized to convert stimulus energy into a neural signal. This usage of the term "receptor" differs from that in Chapters 17 and 21, in which the term "receptor" refers to a transmitter receptor or a hormone receptor: *molecules* specialized to interact with molecules of synaptic transmitter or hormone, respectively. In this chapter, the term *receptor* refers to a sensory cell, unless noted otherwise. Receptors are commonly clustered together in *sense organs*. A sense organ is an organ specialized for the reception of a particular kind of stimulus energy. Usually, a sense organ contains many similar receptors, as well as several kinds of nonneural tissues. For example, the vertebrate eye is a sense organ; it contains photoreceptor cells (retinal rods and cones) as well as nonneural tissues such as those of the cornea, iris, and pigment layer. We can also speak of *sensory systems* such as the vertebrate visual system, which includes the eyes and the central areas in the brain that are primarily concerned with processing visual information.

Receptors can be classified in several ways. The most familiar of these, classification by *sensory modality,* is based on human subjective experience.

Aristotle distinguished five primary senses: vision, hearing, touch, smell, and taste. Within each of these sensory modalities there may be subdivisions termed sensory *qualities*, such as blue versus yellow light or salty versus sweet taste. Clearly, a classification based on the old "five senses" is inadequate even for humans. There are additional sensory modalities of which we are aware (e.g., balance, temperature), as well as many others (e.g., muscle length and tension, blood oxygen partial pressure) that do not normally impinge on consciousness. Early classifications of sensory modalities, moreover, are anthropocentric (human-centered) and thus inadequate for the range of receptors in different kinds of animals. Many animals possess receptors sensitive to modalities and qualities of stimuli not sensed by humans, including electric and magnetic fields and ultraviolet radiation (see p. 504).

Receptors can also be classified according to *location*; that is, location of the source of the stimulus energy relative to the body. We have already drawn the distinction between *exteroceptors* (responding to stimuli outside the body) and *interoceptors* (responding to internal stimuli). Exteroceptors can be subdivided into distance receptors, for which the stimulus source is at a distance from the body (vision, hearing, olfaction), and contact receptors, for which the stimulus contacts the body, as in touch and taste.

Probably the clearest classification of receptors is in terms of the *form of effective stimulus energy* at the receptor surface. Table 18.1 classifies receptors by forms of stimulus energy. Visual photoreceptors, electroreceptors, and magnetoreceptors all respond to different forms of electromagnetic energy; auditory receptors, gravity receptors such as statocysts, and touch receptors are all excited by mechanical stimuli, and so on. Although even these distinctions can be arbitrary, we follow the classification scheme of Table 18.1 in this chapter.

In practice, any form of stimulus energy will excite a particular receptor if there is enough of it. This observation leads us to one of the most important generalizations in sensory physiology: The sensory modality or quality of sensation associated with a stimulus depends solely on *which receptors are stimulated*, rather than on *how* they are stimulated. For example, any stimulus that excites photoreceptors is perceived as light, whether the stimulus is actually light, a poke in the eye ("seeing stars"), or electrical stimulation of the optic nerve. This principle was formulated 150 years ago by Johannes Muller, who gave it the rather misleading title, the *doctrine of specific nerve energies*. Why is it that any stimulation of a receptor is perceived as the same modality? Let us look at the question from the viewpoint of a central nervous system. Stimuli such as light, sound, and touch impinge on peripheral receptors, rather than on a brain. (There are central receptors, but they are not relevant here.) The brain receives *coded information* about stimuli, rather than the stimuli themselves. With rare exceptions, sensory information is coded as trains of action potentials (nerve impulses). All action potentials are rather similar in form, so that a brain cannot tell by the form of an action potential whether it was initiated by light, sound, or touch. Different populations of receptors code different kinds of stimuli into action potentials, and the central nervous system must *decode* the action potentials into information about stimulus quality. The central nervous system performs this decoding by the principle of **labeled lines** (Figure 18.1): Any action potentials of a particular axon are interpreted as a particular stimulus quality. Thus, any activity in

Table 18.1 **CLASSIFICATION OF SENSORY RECEPTORS, BASED PRIMARILY ON THE KIND OF STIMULUS ENERGY THAT EXCITES THEM**

Electromagnetic
  Light (vision)
  Ultraviolet
  Infrared
  Electric
  Magnetic
Thermal
Mechanical
  Touch, pressure
  Muscle length
  Muscle tension
  Joint position and movement
  Cuticular stress
  Acousticolateralis (vertebrate)
    Statocyst and vestibular
    Lateral line
    Auditory
Chemical
  Olfactory (distance chemoreception)
  Taste (contact chemoreception)
  Osmotic
  Oxygen, carbon dioxide, H^+

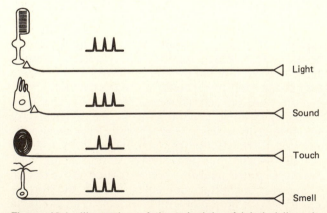

Figure 18.1 Illustration of the principle of labeled lines in sensory systems. Receptors sensitive to different kinds of stimuli send similar kinds of signals (action potentials) to the central nervous system (CNS). In the CNS, the signals are interpreted according to which lines (axons) convey the signals.

central axonal projections from photoreceptors is interpreted as light, and any activity in central auditory projections is interpreted as sound.

Because any activity emanating from a receptor is interpreted as a particular stimulus modality, receptors must have a high degree of specificity, so that they are normally excited by only one form of stimulus energy. The term for this "best" form of stimulus energy is the ***adequate stimulus*** of a receptor. We can define the adequate stimulus of a receptor as that form of stimulus energy to which the receptor is most sensitive, or that form to which it normally responds. This last phrase is a hedge to cover a few receptors such as the ampullae of Lorenzini (see Electroreceptors, p. 503) that are very sensitive to more than one stimulus modality but are normally stimulated by only one form.

The specificity of receptors for one stimulus modality is achieved in two ways. First, the receptors themselves are usually quite specific, responding to much smaller quantities of one form of stimulus energy than to other forms. Second, sense organs act as filters; they attenuate or filter out forms of stimulus energy other than the kind of energy for which the receptor is specialized. This ***peripheral filtering*** is an important function of the nonneural components of a sense organ. For example, vertebrate eyes are effectively "shock mounted," with the receptor cells in the back of a globe of viscous fluid that attenuates most mechanical stimulation to subthreshold levels. Sense organs in general are probably adapted to retain maximum sensitivity to one form of stimulus energy, while filtering other forms of energy as much as possible.

We have now presented enough background information to discuss how a generalized receptor works. Figure 18.2 is a block diagram of the functional elements in a sensory system. A sense organ mediates a series of operations, starting with an input of stimulus energy and ending with sensory output (usually trains of action potentials in particular sensory axons). The first operation, peripheral filtration, has already been discussed. Peripheral filtering mechanisms are typically nonneural, such as eyelids and the iris of the eye. They not only filter forms of stimulus energy other than the adequate stimulus, but they can also limit the amount of adequate stimulus energy that reaches the receptors. This limitation is under control of the central nervous system, as will be discussed later.

The critical and unique functions of sensory receptors take place in the box labeled ***sensory transducer*** in Figure 18.2. Actually, a series of processes occur in this box, so we could divide it into several boxes at the risk of the diagram becoming cumbersome. The processes in the transducer box are listed below.

1. *Absorption* of stimulus energy. In order for a stimulus to cause a change in a receptor, clearly the stimulus

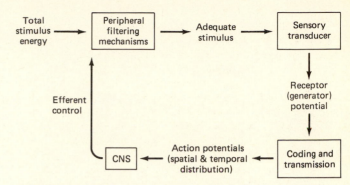

Figure 18.2 Functional elements or processes in a generalized sensory system. The functional operators are shown in boxes, the output of each box being the input of the next box.

energy must first be absorbed. A perfectly transparent photoreceptor, for example, could not function because it would absorb no light.

2. *Transduction* of the stimulus energy into an electrical event. The responses of all known receptors involve changes in electric currents and potentials, so any other form of stimulus energy must be transduced into electrical changes.

3. *Amplification* of energy. The output energy of a receptor—the energy required to generate action potentials—may be much greater than the energy of the stimulus. Thus, the stimulus energy merely triggers release of potential energy in the receptor. In the most sensitive receptors (e.g., human photoreceptors and auditory receptors, grasshopper tympanic organ), the amount of stimulus energy needed to elicit a sensory response is several orders of magnitude smaller than the energy of a single action potential.

4. *Integration and conduction* of the output potential of the transducer to a site of impulse initiation. In most receptors, the sites of sensory transduction are some distance away from the site of initiation of action potentials. We shall discuss these geometrical considerations and further illustrate other functions of sensory transducers below, by examining two well-studied receptors as illustrative examples.

Examples of Receptor Functions: Crayfish Stretch Receptor and Mammalian Pacinian Corpuscle

We want to find out how receptors work. Our overall strategy is to find a few receptors that are favorable examples for studying the mechanisms underlying the processes outlined in Figure 18.2. By exploring these mechanisms in the best examples (so-called "model" receptors) and by seeing what mechanisms are common to many of these, we can derive generalizations about the function of most receptors. Although such generalizations do not fit all receptor types, they do give us an idea of what to expect when we examine the diverse array of receptors in the animal kingdom.

What sorts of receptor should we use as models? Ideally, we want receptors that are large and individually isolated, to allow precise stimulation and ease

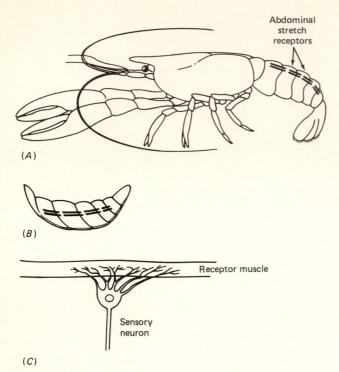

Figure 18.3 Location of abdominal stretch receptors in the crayfish. (A) There are two receptors on each side of each abdominal segment, spanning the joint. (B) View of dorsal abdominal carapace, showing receptors on the inner surface. (C) Each stretch receptor organ consists of a single sensory neuron and an associated specialized receptor muscle fiber, in which the sensory dendrites are imbedded.

of recording from a single receptor cell. Two mechanoreceptors that meet these criteria have been studied extensively as "model" receptors: the crustacean abdominal stretch receptor and the mammalian Pacinian corpuscle. We will discuss the crustacean stretch receptor first and use the Pacinian corpuscle for comparison and to illustrate specific points.

Crustacean abdominal stretch receptors are located on the dorsal side of each abdominal segment in certain decapod Crustacea such as crayfish and lobsters (Figure 18.3). There are two stretch receptor sensory neurons on each side of each abdominal segment; these two differ in the time course of their action (see below). Each of the stretch receptor sensory neurons is associated with a separate, specialized receptor muscle fiber, which spans an abdominal joint. When the abdomen is flexed ventrally, the receptor muscle is stretched, distorting the dendrites of the sensory cell that ramify within a central, noncontractile region of the muscle fiber. This muscle stretch constitutes the adequate stimulus of the stretch receptor neuron. The ends of the muscle fiber are contractile and receive motor innervation, and both the muscle and the sensory neuron receive inhibitory innervation. These complexities are considered below; at this point they serve to illustrate that each stretch receptor actually represents a miniature sense organ, termed a *muscle receptor organ* (MRO). The muscle receptor organ comprises one stretch

receptor sensory neuron, its associated receptor muscle, and the excitatory and inhibitory motor innervation of the organ.

Crustacean stretch receptor sensory neurons have large (up to 100 μm), exposed, visible somata that are readily impaled with microelectrodes. Because of this ease of intracellular recording, the stretch receptor has been an important model preparation for studies of receptor mechanisms. Figure 18.4 shows the results of such a study. The stretch receptor neuron has a membrane resting potential (E_m) of about −75 mV (inside negative). When the receptor muscle is stretched (either directly by the experimenter or by flexion of the abdomen), there is a depolarization of the membrane. This depolarization is graded in amplitude, increasing with increasing amounts of stretch, and lasts as long as the stretch is maintained (although with some decrease in amplitude). The graded depolarization of a receptor in response to stimulation is termed a *receptor potential*. (Because the depolarization of the receptor potential generates action potentials in response to a stimulus, it is sometimes also termed the *generator potential*.) If the receptor muscle is stretched enough to produce a suprathreshold depolarization of the stretch receptor neuron, the neuron generates action potentials (Figure 18.4). The greater the amount of stretch, the larger the amplitude of the receptor potential depolarization and the greater the frequency of the resultant action potentials. Figure 18.4 demonstrates two of the most important generalizations about receptor function: (1) *An adequate stimulus elicits a graded receptor potential,* the amplitude of which is a function of stimulus intensity; and (2) *the frequency of resultant action potentials in a receptor is a coded representation of the intensity of the adequate stimulus.* Although both these generalizations have exceptions, they apply to a great majority of receptors. Figure 18.5 shows the relation of stimulus intensity to receptor potential amplitude and to impulse frequency in a variety of receptors. The exact form of these functions differs for different receptors, but the overall effect is that the frequency of impulses in a receptor is an unambiguous code of stimulus intensity.

Ionic Basis of the Receptor Potential   The receptor potential of the crustacean stretch receptor (and of many other receptors) results from an inward ionic current carried primarily by sodium ions. (Replacement of sodium ions in the extracellular medium with impermeable cations decreases the amplitude of the receptor potential.) Presumably, muscle stretch distorts the membrane of embedded dendritic terminals, opening channels in the membrane by an unknown mechanism. The resulting receptor potential passively spreads from the dendrites to the soma and to the initial segment of the axon where action potentials are initiated. There, the receptor potential depolarization opens voltage-dependent sodium channels, generating action potentials.

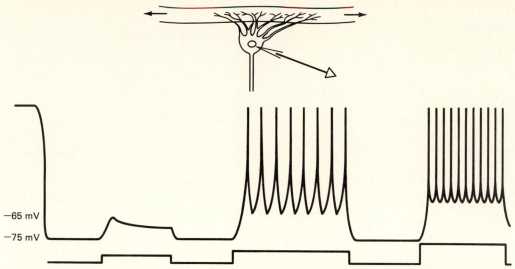

−65 mV
−75 mV

Figure 18.4 Excitation of an arthropod stretch receptor. Stretch of the receptor muscle elicits a graded, depolarizing *receptor potential,* the amplitude of which is proportional to the stimulus intensity. The receptor potential (if large enough) triggers action potentials at the axon hillock.

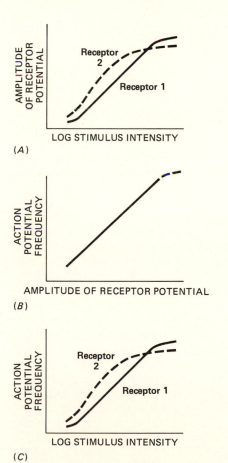

Figure 18.5 Sensory transfer functions. The output of a stage in the response of a receptor is plotted as a function of the input to that stage. (*A*) In receptor 1, the amplitude of the receptor potential increases linearly with the log of stimulus intensity over a wide range. (*B*) Action potential frequency is a linear function of receptor potential amplitude. so that frequency also increases linearly with the log of intensity (*C*). The input–output relation of receptor 2 is not linear with the log of intensity.

The channels responsible for the receptor potential are quite different from the voltage-dependent sodium channels that generate action potentials. Receptor potentials (and thus the channels that produce them) are not voltage dependent and are unaffected by tetrodotoxin (TTX), which blocks voltage-dependent sodium channels. Furthermore, the receptor potential channels are much less selective in their permeability: Large ions such as Tris and arginine can pass through them. As might be expected, there is a reversal potential for the receptor potential process (cf. Figure 17.7). In the stretch receptor, the value of the reversal potential is about + 15 mV (inside positive). This value is below the estimated sodium equilibrium potential, a difference suggesting that other unknown ions contribute to the receptor potential in addition to sodium ions.

Receptor Adaptation It is a common observation that our perception of a steady stimulus decreases over time. The smell of food cooking, the hum of a refrigerator, and even our awareness of the clothes we wear all decrease with time, particularly if the stimulus remains constant. Part of this decreased awareness is a higher brain function, but a large part of it results from *receptor adaptation.* We can define receptor adaptation as a decrease with time in the response of a receptor to a steadily maintained stimulus.* Receptors differ greatly in their rates of adaptation (Figure 18.6). *Tonic* receptors adapt relatively slowly and continue to discharge impulses throughout the duration of a prolonged stimulus, usually with some decline in frequency. *Phasic* receptors

*Note that this use of the term *adaptation* in a sensory context is very different from the use of the term in an evolutionary context in much of this book (see Chapter 1). Sensory adaptation is more akin to physiological acclimation than to evolutionary adaptation.

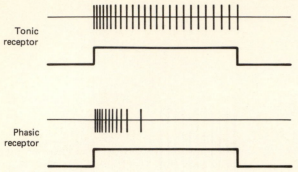

Figure 18.6 Tonic receptors adapt slowly and incompletely, so that the train of impulses continues with a prolonged stimulus. Phasic receptors adapt rapidly and completely, the impulse train ceasing during a prolonged stimulus. Phasic receptors thus convey information about change in stimulus intensity.

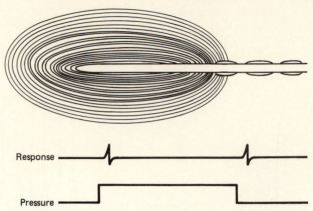

Figure 18.7 Pacinian corpuscle. The unmyelinated sensory terminal of the mechanoreceptor neuron is wrapped by concentric lamellar membrane layers. The response of a Pacinian corpuscle is a single impulse at the onset and cessation of a mechanical stimulus.

adapt more rapidly and cease impulse discharge during a prolonged stimulus. Tonic receptors provide information (in the form of impulse frequency) about the intensity of a prolonged stimulus. Phasic receptors provide information about the initial intensity of a stimulus but also provide information about the rate of change of the stimulus, impulse frequency increasing with increasing rate of change. The two types of crustacean abdominal stretch receptor, referred to above, differ in rate of adaptation. Of each pair of stretch receptors, one is tonic (slowly adapting) and the other is phasic (rapidly adapting). Thus, they convey different information to the CNS, the tonic receptor signaling abdominal position and the phasic receptor signaling abdominal movement.

The prime example of a very rapidly adapting, phasic receptor is the mammalian Pacinian corpuscle. The Pacinian corpuscle is an encapsulated nerve ending found in deep dermal layers of the skin and subcutaneously, as well as in mesenteries and other internal tissues. The sensory ending is devoid of myelin but is surrounded by many thin, concentric layers of epithelial cells (Figure 18.7). These multiple lamellae of nonneural cells that make up most of the volume of the Pacinian corpuscle are collectively called the onion skin, because of the resemblance of the corpuscle to a tiny onion. Pacinian corpuscles are mechanoreceptors, responding to mechanical deformation of the outer onion skin. They adapt extremely rapidly, normally generating only one impulse per stimulus, no matter how strong or prolonged (Figure 18.7). (Interestingly, they respond as vigorously to the sudden cessation of a prolonged stimulus as they do to its onset.)

How does receptor adaptation occur? Because the Pacinian corpuscle is such a rapidly adapting receptor, we begin our examination of mechanisms of adaptation with it. A review of Figure 18.2 shows that there are three possible sites of receptor adaptation: at the peripheral filtering mechanism, at the sensory transducer, and at the spike-encoding stage. Adap-

tation at either of the first two stages (peripheral filter, transducer) would produce a decrease in amplitude of the receptor potential with time. Adaptation at the spike-encoding stage would produce a decline in impulse frequency even with a constant receptor potential. For most receptors, adaptation occurs at more than one of these sites, the relative importance of the sites differing for different receptors. In the Pacinian corpuscle, the most important site of adaptation is the peripheral filtering mechanism, namely, the onion skin of epithelial cell lamellae that surrounds the transducer (membrane of the nerve ending). Because of the relatively large size of a Pacinian corpuscle (0.6×1 mm), it is possible to dissect away the onion-skin lamellae, leaving only a small inner core of lamellae around the nerve ending. The results of such an experiment are diagrammed in Figure 18.8. In the intact corpuscle (*A*), the amplitude of the receptor potential decreases to zero within 6 ms; therefore, adaptation occurs at or prior to the transduction step. In the decapsulated corpuscle (from which the onion skin has been dissected away), the receptor potential adapts much more slowly (Figure 18.8*B*). Thus, a major function of the onion skin is to act as a *frequency filter,* allowing sudden changes in mechanical displacement to pass through to the receptor membrane, but filtering out stationary forces and slow changes. Apparently as a result of the elasticity of the lamellar membranes and the viscosity of fluid between lamellar layers, maintained pressure is dissipated in the lamellae while rapid changes are transmitted to the core with little loss. The pressure at the core (the adequate stimulus) has been calculated to be proportional to the *velocity* of displacement (dx/dt) at the surface of the corpuscle.

Receptor adaptation, however, does not always depend on peripheral filtering mechanisms. For most receptors from which receptor potentials can be recorded, the amplitude of the receptor potential de-

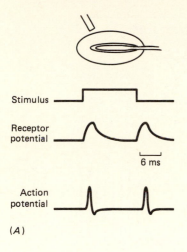

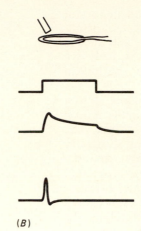

Stimulus

Receptor potential

|— 6 ms —|

Action potential

(A)

(B)

Figure 18.8 The rapid adaptation of a Pacinian corpuscle depends on peripheral filtration of the stimulus by the lamellae around the sensory terminal. (*A*) In the intact corpuscle, the receptor potential adapts within 6 ms. (*B*) In a dissected corpuscle from which most of the lamellae have been removed, the receptor potential persists with much less adaptation.

creases to a plateau during sustained stimulation. In some examples such as photoreceptors and chemoreceptors, this decrease is clearly not the result of peripheral filtration but appears to be part of the transduction process itself. We must be cautious in interpreting causes of adaptation, however. For example, the receptor potential of the tonic crustacean stretch receptor declines to a plateau during constant sustained stretch (Figure 18.4). If, however, the receptor muscle is clamped under constant *tension* rather than constant length, there is much less adaptation of receptor potential amplitude. This result indicates that most of the adaptation of the receptor potential results from mechanical slippage in the receptor muscle, a hidden form of peripheral filtration.

Finally, receptor adaptation may result from effects on action potential initiation (the spike encoding stage in Figure 18.2). An illustration of this form of adaptation is the contrast between the phasic and tonic crustacean stretch receptors. Intracellular recordings from the two neurons show that the receptor potentials of both cells adapt to the same degree (Figure 18.9). Despite this similarity, the spike trains of the two neurons differ greatly. The tonic receptor continues to generate action potentials for several minutes, provided the receptor potential remains above the receptor's original threshold for action potential initiation. In contrast, the phasic receptor ceases generation of action potentials in a few seconds, even if the receptor potential remains above the original threshold for impulse initiation. The same difference occurs if the cells are directly depolarized by a constant current: Action potentials always cease within a few seconds in the phasic receptor but persist in the tonic receptor. The reason for this difference is *accommodation* of the impulse-generating membrane of the phasic neuron. With prolonged depolarization, inactivation of sodium conductance (see p. 404) builds up to the point that the voltage-dependent sodium channels cannot open, no matter how great the depolarization. At their respective impulse-initiating sites, the phasic receptor has

considerable accommodation of its impulse-generating mechanism and the tonic receptor has little accommodation.

In summary then, receptor adaptation can occur at several stages: at the peripheral filtering stage (demonstrated by the Pacinian corpuscle), at the transduction stage, and at the impulse-generating stage. Both the degree of adaptation and the mechanism of adaptation differ for different kinds of receptors.

Efferent Control of Receptors   Receptors are commonly viewed as windows, through which aspects of the environment can influence an organism. This analogy, although useful, is inadequate because receptors are not merely passive windows. Instead, the sensitivity of many receptors can be modulated by the central nervous system. The ability of the central nervous system to adjust the information it receives is termed *centrifugal control* (from the center outward) or *efferent control* (efferent meaning outward from the CNS, as opposed to afferent, meaning

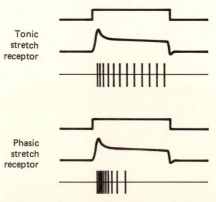

Tonic stretch receptor

Phasic stretch receptor

Figure 18.9 The difference in rate of adaptation between tonic and phasic crustacean stretch receptors reflects differences in the impulse-generating mechanisms. The receptor potentials of the tonic and phasic receptors have similar rates of adaptation.

toward the CNS). Efferent control is an important feature of many receptors and sense organs, including those of mammalian vision, hearing, and muscle sense. We will treat these examples later; here, we will describe efferent control of the crustacean stretch receptor.

The dendrites of a crustacean abdominal stretch receptor, as noted above, are imbedded in the central portion of a specialized receptor muscle fiber. The rest of the muscle fiber (but not the central sensory portion) is contractile and is innervated by one of the motor neurons that supply the adjacent muscles that extend the abdomen. As shown in Figure 18.10, stimulation of the excitatory motor neuron axon causes contraction of the receptor muscle, stretching the neuronal dendrites of the receptor cell and increasing the rate of impulse generation in the cell. Thus, the central nervous system, by activating the motor neuron to the receptor muscle, can increase the sensitivity of the stretch receptor. A similar form of efferent control occurs for the vertebrate muscle spindle (see Mechanoreception, p. 486). The behavioral significance of this control is treated in Chapter 19.

The crustacean stretch receptor receives inhibitory efferent control as well as excitatory efferent control. Each stretch receptor cell has one to three inhibitory axons associated with it. These efferent axons form inhibitory synapses on the dendrites and soma of the stretch receptor neuron (Figure 18.11A). If the receptor muscle is stretched to elicit a steady train of impulses in the stretch receptor neuron, stimulation of an inhibitory axon produces inhibitory postsynaptic potentials (IPSPs) in the receptor neuron, holding the neuron hyperpolarized below thresh-

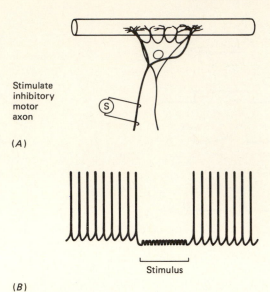

Stimulate inhibitory motor axon

(A)

Stimulus

(B)

Figure 18.11 Receptors are also subject to inhibitory efferent control. The crustacean stretch receptor receives inhibitory innervation (A) which, when stimulated, hyperpolarizes the receptor membrane, interrupting the train of action potentials resulting from passive stretch of the receptor muscle.

old (Figure 18.11B). Similar efferent inhibitory control mechanisms have been found for hair cells of the vertebrate auditory system and lateral line organs, to be considered below (see Mechanoreception, p. 488).

It is clear at this point that central nervous systems can set the level of activity of receptors, increasing their sensitivity by excitatory efferent control or decreasing sensitivity by inhibitory control. For example, an impulse frequency of 10/s in a crayfish stretch receptor does not necessarily signal a particular degree of abdominal flexion. For the crayfish CNS to have unambiguous information about abdominal position, it must "know" not only the frequency of impulses in the receptor but also its own efferent control signals to the receptor, sent via the excitatory and inhibitory motor neurons that modulate receptor activity. Thus, the CNS must perform more sophisticated sensory integration than would be the case if receptors were merely passive windows to the environment.

PHOTORECEPTION AND VISUAL SENSORY PROCESSING

In the next several portions of this chapter, we will examine different types of receptors, asking two sorts of questions: (1) How do they work? (2) What information do they provide to the CNS of the animals in question? We will consider photoreceptors and visual information processing first, because visual systems are rather better understood than other sensory modalities. There are several reasons for this relatively complete understanding of visual systems. First, an experimenter can control light stimuli more

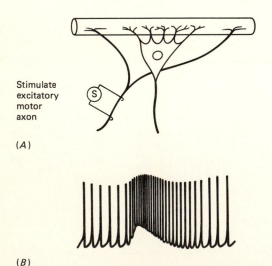

Stimulate excitatory motor axon

(A)

(B)

Figure 18.10 Sensory receptors are subject to excitatory efferent control from the CNS. The crustacean stretch receptor responds to active contraction (as well as passive stretch) of its receptor muscle fiber. Stimulation of the excitatory motor neuron contracts the ends of the muscle fiber, stretching the noncontractile central portion in which the dendrites of the receptor cell are imbedded.

precisely and more easily than other forms of stimulus energy, such as application of chemicals to a chemoreceptor. Second, humans are visual animals, relying more on sight than on other sensory modalities. Therefore, there has been more attention paid to visual systems because human investigators and students are interested in vision and tend to consider it more important than other senses.

Before discussing photoreception, we need to consider briefly the physical properties of light. *Light* is one of several kinds of electromagnetic radiation and like all electromagnetic radiation has both wave properties and quantal properties. Descriptions of light in terms of its wave properties and its quantal properties are equally valid and are useful to explain different attributes of electromagnetic radiation. The wave description of electromagnetic radiation states that waves of energy radiate out away from a source (an oscillating dipole in which a positive charge and a negative charge vibrate together and apart). The wave consists of an electrical force and a magnetic force, both oscillating in amplitude at the frequency of the oscillation of the source. Electromagnetic radiation travels at the velocity of light (*s*), approximately 3×10^8 m/s. Therefore, the radiation can be described either in terms of its frequency of oscillation *nu* (ν) or more commonly by its wavelength *lamda* (λ)—the distance between adjacent peaks of the electromagnetic wave. The relation between these two parameters is

$$\lambda \text{ (meters)} = \frac{s \text{ (meters/second)}}{\nu \text{ (second}^{-1})}$$

The portion of the electromagnetic spectrum that is *visible light* (i.e., exciting human visual photoreceptors) is 400–700 nanometers (nm) in terms of wavelength or roughly 7 to 14×10^{14} cycles/s in terms of frequency. The relation of visible light to other wavelengths of electromagnetic radiation is shown in Figure 18.12. The major source of electromagnetic energy on the earth is *solar* radiation (although as discussed on p. 79, all objects on earth radiate some infrared energy). Most of the sun's energy is contained in a wavelength band of 300–1100 nm. Moreover, wavelengths much longer or shorter than visible light are largely blocked by the earth's atmosphere, so that only visible light and the nearer portions of ultraviolet and infrared radiation reach the earth's surface in appreciable amounts.

Electromagnetic radiation has particle properties as well as wave properties. For some issues such as the absorption of light, it is better to consider radiation not as a wave but as a stream of particles called *quanta*. Quanta of visible light are termed *photons*. The energy of a quantum increases linearly with increasing frequency of the radiation. For a molecule to be photosensitive, it must absorb a light quantum and be chemically changed by it (usually by raising an electron to a higher energy state). Quanta of infrared radiation lack sufficient energy to trigger photochemical reactions. Quanta of ultraviolet radiation create the opposite problem: They contain too much energy, enough to damage nucleic acids and proteins. Therefore, most sensory systems are shielded from ultraviolet radiation.

Photochemistry

Light must be absorbed as a necessary first step in the excitation of photoreceptors. The light-absorbing molecules of these receptors are termed *photopigments*. Several types of photopigment occur in photoresponsive protists, single-celled organisms in which locomotion is influenced by light. In the multicellular Metazoa, however, all photoreceptors employ carotenoid photopigments termed *rhodopsins*. The term rhodopsin originally referred only to the photopigment of vertebrate retinal rods, but with increased appreciation of the chemical similarity of all metazoan photopigments, the term is now applied to all of them.

Rhodopsin is a conjugate of two moities: *retinal,* the aldehyde of vitamin A, and the protein *opsin*. The structures of vitamin A and of retinal are shown in Figure 18.13. Vitamin A (also known as *retinol*) is an alcohol derivative of β-carotene, a yellow pigment of carrots and a parent compound of the carotenoids. Animals cannot synthesize carotenoids such as vitamin A and so must obtain them in their diets. Two molecular forms of vitamin A have been found in eyes: vitamin A_1 has one double bond in its ring portion, and vitamin A_2 (3-dehydroretinol) has two (Figure 18.13). These are oxidized to two distinct retinals: $retinal_1$ with one double bond and $retinal_2$ (3-dehydroretinal) with two double bonds in the ring. $Retinal_1$ is the chromophore (light-absorbing part) of most rhodopsins, including those of mammals, birds, most marine fishes, and all invertebrate Metazoa. $Retinal_2$ is found in many freshwater fishes and freshwater stages of amphibians. There is presumably some evolutionary advantage favoring $retinal_2$ (and its associated carotenoprotein, which has been termed porphyropsin) in freshwater vertebrates, but this association is more complex than once suspected. We will confine our attention to the more widespread and better-studied pigment system based on $retinal_1$. Mammalian rhodopsins have been studied the most and are considered to be fairly typical of the structures and reactions of other rhodopsins.

Retinal is conjugated to the protein opsin by a linkage of the aldehyde end to a lysine residue of opsin. Many rhodopsins with different light-absorption spectra have been characterized, all containing the same carotenoids ($retinal_1$ or $retinal_2$). Thus, the differences among rhodopsins must result from differences in the opsin moiety and possibly in the linkage between retinal and opsin. Opsins have a molecular weight of 38,000–49,000 daltons and contain two short oligosaccharide chains and 30 or more attached phospholipids that help to embed the protein in the

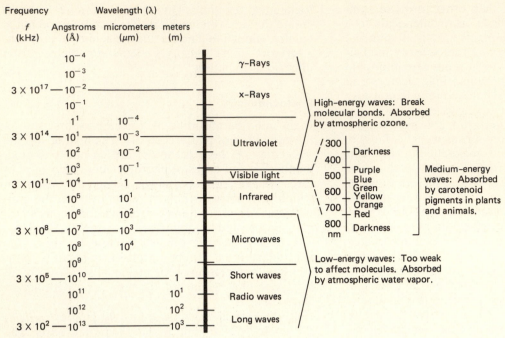

Frequency

Wavelength (λ)

f (kHz)   Angstroms (Å)   micrometers (µm)   meters (m)

3×10^{17}	10^{-4} 10^{-3} 10^{-2} 10^{-1} 1^{1}	10^{-4}	γ-Rays / x-Rays
3×10^{14}	10^{1} 10^{2} 10^{3}	10^{-3} 10^{-2} 10^{-1}	Ultraviolet
3×10^{11}	10^{4} 10^{5} 10^{6}	1 10^{1} 10^{2}	Visible light / Infrared
3×10^{8}	10^{7} 10^{8} 10^{9}	10^{3} 10^{4}	Microwaves
3×10^{5}	10^{10} 10^{11} 10^{12}	1 10^{1} 10^{2}	Short waves / Radio waves
3×10^{2}	10^{13}	10^{3}	Long waves

High-energy waves: Break molecular bonds. Absorbed by atmospheric ozone.

300 — Darkness
400
500 — Purple, Blue, Green
600 — Yellow, Orange
700 — Red
800 nm — Darkness

Medium-energy waves: Absorbed by carotenoid pigments in plants and animals.

Low-energy waves: Too weak to affect molecules. Absorbed by atmospheric water vapor.

Figure 18.12   The electromagnetic spectrum. [Adapted from G.M. Shepherd, *Neurobiology.* Oxford University Press, New York, 1983.]

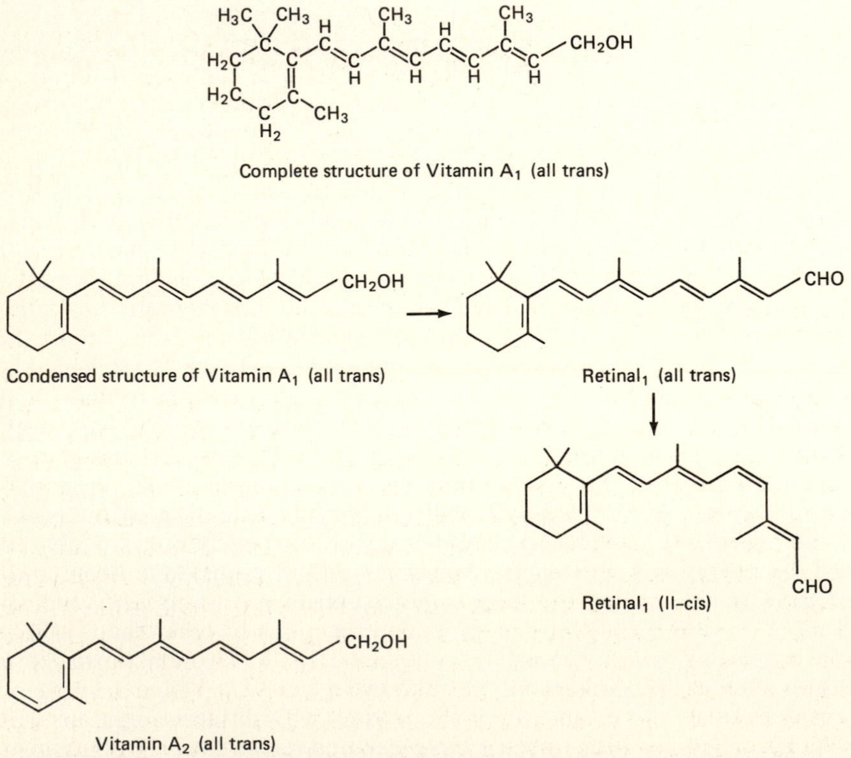

Complete structure of Vitamin A₁ (all trans)

Condensed structure of Vitamin A₁ (all trans)   Retinal₁ (all trans)

Retinal₁ (II–cis)

Vitamin A₂ (all trans)

Figure 18.13   Chemical structures of vitamin A and retinal. There are two forms of vitamin A: Vitamin A₁ and Vitamin A₂. Vitamin A₁ is shown both as a complete structure (top) and as a skeleton structure (middle). Vitamin A₁ is converted to retinal₁; Vitamin A₂ is converted to retinal₂ (not shown).

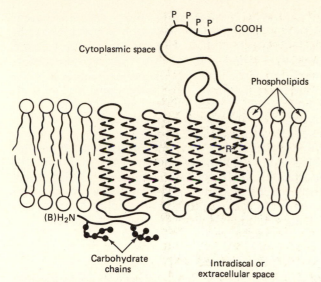

Figure 18.14 A model of the structure of vertebrate rhodopsin. The probable location of retinal is indicated by R. Seven α-helical regions of the protein span the membrane; P indicates amino acid residues that become phosphorylated by a light-induced process. [From A. Fein and E.Z. Szuts, *Photoreceptors: Their Role in Vision.* Cambridge University Press, Cambridge, MA, 1982.]

photoreceptor membrane. Figure 18.14 shows one proposed three-dimensional structure of rhodopsin and of its incorporation into the membrane.

Photochemical Reactions The photochemical reactions of rhodopsin depend primarily on geometrical isomerization of retinal and secondarily on conformational changes in the whole molecule. Retinal con-

tains five double bonds and thus can exist in many geometrical (cis–trans) isomers. The retinal in rhodopsin is in the *11-cis* configuration, and the action of light is to photoisomerize this bent form to the straight, *all-trans* configuration (Figure 18.13). Because retinal is intimately associated with opsin, the photoconversion of retinal to all-trans is followed by a series of spontaneous changes in conformation of opsin, ultimately leading to dissociation of retinal from opsin. Figure 18.15 shows the sequence of chemical reactions following photoactivation of rhodopsin. The intermediate compounds are characterized by their different absorption spectra. By exposing rhodopsin to light at successively lower temperatures, it is possible to stop the spontaneous reactions progressively earlier in the sequence. For example, even at temperatures below −140°C rhodopsin is converted to bathorhodopsin (also termed prelumirhodopsin). At temperatures greater than −140°C, bathorhodopsin is spontaneously converted to lumirhodopsin, and so on. The photoconversion of rhodopsin and the subsequent spontaneous conversions leading to metarhodopsin II are all very rapid, so that metarhodopsin II is produced in about 1 ms. The splitting of metarhodopsin II into *trans*-retinal and opsin (termed **bleaching**) proceeds spontaneously but very slowly (tens of seconds). Since latencies of photoreceptor excitation are usually greater than 10 ms, any of the photochemical changes up to the formation of metarhodopsin II could act as the trigger for receptor excitation.

How do changes in rhodopsin molecules excite photoreceptors? Despite considerable effort, we still do not know how the photochemical events summarized in Figure 18.15 lead to receptor excitation.

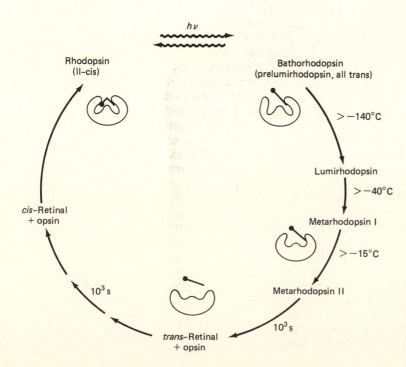

Figure 18.15 The rhodopsin cycle. Light (hμ) acts to convert ll-cis rhodopsin to bathorhodopsin, which undergoes rapid, spontaneous transformations to Metarhodopsin II. Metarhodopsin II slowly dissociates to trans retinal + opsin, which can be reconverted to ll-cis rhodopsin.

As will be discussed below, the excitation of photo-receptors involves electrical events resulting from ion permeability changes. We shall see that the links between rhodopsin changes and excitation may be complex and indirect, as well as diverse.

Regeneration of Rhodopsin   After photoactivation of rhodopsin, its products must be regenerated to the original state, with retinal in the 11-cis configuration. Regeneration can occur in either of two ways: en-zymatically or photochemically. For vertebrate re-ceptors, most regeneration is by enzymatic means. After retinal and opsin split, the all-*trans*-retinal is reconverted to 11-*cis*-retinal by an isomerase. This process, however, is neither direct nor simple. Much of the enzymatic regeneration occurs not in the pho-toreceptors but in the adjacent pigment epithelium. A typical pathway for regeneration is thought to be as follows: All-*trans*-retinal is converted to all-*trans*-retinol (vitamin A) by retinal oxidoreductase. Retinol migrates to the pigment epithelium (the pigment of which is melanin, not rhodopsin). Vitamin A (retinol) is converted from all-trans to 11-cis in the pigment epithelium by an isomerase. Vitamin A (11-cis) mi-grates back to the photoreceptors and is converted to 11-*cis*-retinal by retinal oxidoreductase. The 11-*cis*-retinal recombines with opsin, which remains sta-tionary as an intrinsic protein of the photoreceptor membranes. It is not clear what functional advantage results from the major role of the pigment epithelium in regeneration of vertebrate rhodopsin.

Rhodopsin also can be photoregenerated directly from its products, rather than enzymatically regen-erated. Bathorhodopsin, lumirhodopsin, and meta-rhodopsin I and II can absorb light quanta and re-convert directly to 11-*cis*-rhodopsin without requiring metabolic energy. Such direct photoregen-eration is a minor pathway for vertebrate rhodop-sin, accounting for perhaps 6 percent of all regener-ation under normal light conditions. In contrast, in the well-studied invertebrate examples (the fly *As-calaphus* and the octopus *Eledone*), most regenera-tion of rhodopsin is by direct photoregeneration of products similar to vertebrate metarhodopsins.

Early Receptor Potentials   In many photoreceptors it is possible to record electrical events that cor-respond directly to the photochemical conversions of rhodopsin to metarhodopsin. These electrical changes are termed *early receptor potentials,* because they occur with latencies of microseconds, in con-trast to latencies of "ordinary" receptor potentials, which exceed 10 ms. The reason for this latency difference is that ordinary receptor potentials result from ionic currents that depend on membrane perme-ability changes; early receptor potentials, in contrast, result from the conformation changes in rhodopsin molecules (which later lead to the permeability changes). Thus, early receptor potentials are an elec-trical measure of photochemical events, rather than

of photoreceptor excitation. Early receptor poten-tials can be recorded only with very bright light flashes that simultaneously activate a large fraction of all the rhodopsin in the eye. The resultant confor-mation changes in activated rhodopsin molecules dis-place fixed charges within the rhodopsins. The sum-mated result of these charge displacements is an electrical signal, analogous to the gating currents in excitable membranes (Chapter 16).

As noted above, early receptor potentials are un-measurably small unless much of the photopigment is simultaneously activated, and they neither reflect nor produce receptor excitation. They are useful as another way of measuring photochemical changes besides photometric changes in light absorption.

Evolution of Visual Systems

Most organisms are light sensitive in some fashion. Plants grow toward light (phototropism) and most animals make simple directed movements in re-sponse to light stimuli (phototaxes; see p. 501). Prim-itive organisms such as *Amoeba* undergo phototactic behavior without any apparent photoreceptive struc-tural specialization. Most photoreceptors, however, are associated with one of two sorts of organized intracellular structure, both of which serve to in-crease the surface area of photopigment-bearing membranes. The two kinds of photoreceptor struc-tures are termed *ciliary* and *rhabdomeric* (Figure 18.16). Many photoreceptor specializations represent modified cilia, often recognizable only by the persis-tence of the ciliary basal body, as in vertebrate retinal rods. The association of sensory cilia with receptor specializations is quite widespread, occurring not only in photoreceptors of several phyla but also in mechanoreceptors and chemoreceptors of both ver-tebrates and invertebrates. The other major struc-tural specialization of photoreceptor evolution is an elaboration of microvilli that greatly increases the membrane surface area of the photoreceptor, Mic-rovillar photoreceptors not associated with cilia are found in the protostome phyla (e.g., Annelida, Ar-thropoda, and Mollusca, Figure 18.16) and are most elaborately developed in arthropods. Because the stacked microvillar projections of arthropod photo-receptors are collectively termed the rhabdomere, microvillar photoreceptors are termed rhabdomeric. As shown in Figure 18.16, photoreceptors appear to have evolved in two lines, a ciliary line (protists, coelenterates, deuterostomes) and a rhabdomeric line (protostomes). Photoreceptor evolution must have been more complicated than this hypothesis suggests, however, because photoreceptors of both the ciliary and the rhabdomeric types have been found in the same phylum and even in the same animal. The eyes of the scallop *Pecten,* for example, contain two layers of photoreceptors. The receptors of the proximal retina are rhabdomeric and those of the distal retina are ciliary. Such exceptions show

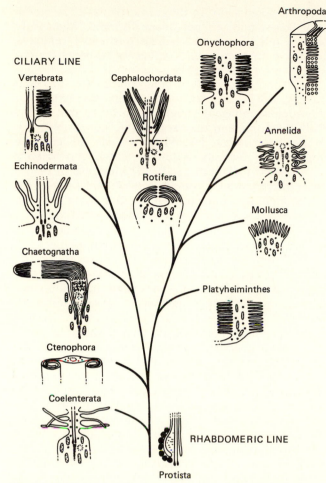

Figure 18.16 A simplified evolutionary relation of photoreceptive structures. Two major lines of photoreceptor evolution have been proposed, one employing modified cilia and the other a rhabdomeric arrangement of microvilli. [From R.M. Eakin, in T. Dobzhansky, M.K. Hecht, and W.C. Steere (eds.) *Evolutionary Biology,* Vol. 2, pp. 194–242. Appleton-Century-Crofts, New York, 1968.]

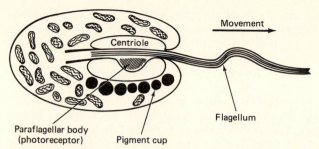

Figure 18.17 Photoreceptive structure associated with the flagellum of *Euglena,* a photosynthetic unicellular protist. [From J.J. Wolken, *Photoprocesses, Photoreceptors, and Evolution.* Academic, New York, 1975.]

Euglena swims through the water with a twisting spiral motion, the paraflagellar body is alternately illuminated by and shaded from a directional light source. The eyespot can then respond to directional light by sequential sampling.

The first stage in the evolution of multicellular eyes is thought to have been a flat plate of photoreceptors on a pigmented base (Figure 18.18*A*). The receptors are shielded from light on one side but can provide only general directional information. Such retinal plates are found in primitive members of all major phyla. The presumed next evolutionary stage is a lensless pigmented eyecup (Figure 18.18*B*), also found in many phyla. The cupped shape of the screening pigment enhances the directional selectivity of the photoreceptors. If the opening in the pigment cup is small enough, the eye can be truly image forming, acting like a pinhole camera. Such a pinhole eye, of course, is not very light sensitive. To combine the image-forming qualities of a pinhole eye with

that any phyletic generalization about lines of photoreceptor evolution is probably oversimplified.

The evolution of *eyes* (secialized photoreceptive organs) is an equally speculative issue. The major selective advantage of localizing photoreceptors into eyes is to allow directionality of photoreceptor stimulation. Photoreceptors bounded by pigment on one side respond to light from only one direction, permitting oriented movements by the animal. A good starting point in considering the evolution of visual systems is the photosynthetic protist *Euglena*. The photoreceptive organelle (eyespot) of a *Euglena* is a swelling on the side of the base of the flagellum, termed a paraflagellar body (Figure 18.17). It is tempting to speculate that the association of this apparently primitive photoreceptive organelle with a flagellum is the evolutionary precursor of the ciliary line of photoreceptor cells. The paraflagellar body is shielded by pigment from light on one side. As a

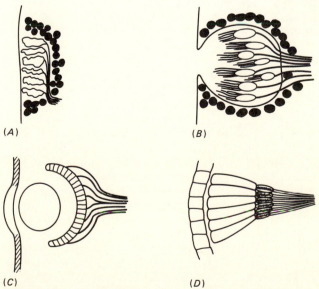

Figure 18.18 Postulated evolution of eyes: (*A*) pigmented photoreceptor plate, (*B*) eyecup, (*C*) camera eye, and (*D*) compound eye.

greater light-gathering power, a lens must be added. The resulting *camera eye* (Figure 18.18C) is most familiar as the eye type of vertebrates but has evolved independently in many phyla. The other major evolutionary trend is the *compound eye* (Figure 18.18D), in which each facet or *ommatidium* has its own lens and pigment-shielded photoreceptors. A compound eye is image forming in a different way than a camera eye. Each ommatidium "sees" a small portion of the visual field, focusing light from this small area onto its photoreceptor membranes. Adjacent ommatidia receive light from adjacent parts of the visual field, with relatively little overlap. The result is an erect (noninverted) image termed a *mosaic image,* by analogy to a mosaic picture made up of many small tiles; each ommatidium "sees" the equivalent of one tile. In reality, of course, all image-forming eyes produce a mosaic image at the level of the photoreceptors, even if not at the level of the optical system. The rods and cones of the human retina, for example, are themselves a mosaic, each "seeing" light from a small part of the visual field. Thus, application of the term mosaic image to a compound eye merely stresses that the mosaic character is a property of the optical system of many separate lenses arranged as facets. Other aspects of the optics of compound eyes are considered in the section below.

Physiology of Some Invertebrate Visual Systems

In this section, we will consider the distinctive features of a few examples of invertebrate visual systems. The examples selected are those that are relatively well studied and that illustrate salient features of a variety of visual systems.

Simple Eyes of Barnacles   Barnacles have three simple eyes, termed *ocelli,* that are particularly advantageous for study. The two lateral ocelli each have three photoreceptor cells; the median ocellus has

four to seven cells in different species. The ocelli lack lenses and are not image forming. Their only known function is to mediate a shadow reflex, a closing of the opercular valves in response to the shadow of a passing object.

The photoreceptors are large (30–100 μm) and have large (10–30 μm), visible axons that extend to the supraesophageal ganglion, up to 10 mm away. Both the soma and the axon can be penetrated with intracellular electrodes. The response to light is a depolarizing receptor potential that is graded with light intensity (Figure 18.19). The waveform of the receptor potential is simple at low light intensities, but at high light intensities it is more complex, with an initial peak (that may overshoot zero potential) and a rapid diminution to a smaller plateau. The inward current producing the receptor potential is carried primarily by Na^+ and to a lesser extent by Ca^{2+}. The decay in receptor potential amplitude at high light intensities is due to several factors, including a decrease in Na^+ current by a rise in intracellular Ca^{2+} concentration and a delayed K^+ permeability increase that depends on voltage, $[Ca^{2+}]$, or both. Like most arthropod photoreceptors, barnacle photoreceptors do not generate action potentials. The absence of action potentials reflects the fact that the inward currents of the receptor potential are predominantly light dependent rather than voltage dependent. The total picture of the ionic basis of the receptor potential is nevertheless complex, with several interdependent components.

If barnacle photoreceptors do not generate action potentials, how are their sensory responses conveyed the considerable distance to the CNS (about 10 mm for large barnacles)? The answer is that the receptor potential spreads passively along the axon to its end, in accordance with the electrotonic cable properties of membranes described in Chapter 16. This answer is somewhat surprising, particularly considering the distance involved. We are not used

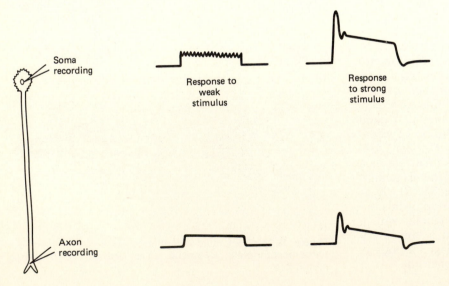

Figure 18.19   Responses of a medial photoreceptor of the barnacle *Balanus* to light. Responses are recorded from the photoreceptive soma and from the axon in the brain, 13 mm away. Graded receptor potentials spread from soma to axon terminals directly because of the long membrane length constant of the axon.

to thinking of neurons that do not generate action potentials, although recent studies show that they are quite common (e.g., in the vertebrate retina). Intracellular recordings from the axon terminal in the CNS (Figure 18.19) show that the receptor potential waveform is rather faithfully preserved, with a very modest decrease in amplitude. Clearly, the membrane length constant (λ, cf. Figure 16.8) is longer than the approximate 10-mm length of the axon. This unusually long length constant results in part from the large diameter of the axon and in part from an unusually high specific resistance of the axon membrane ($\sim$300 k$\Omega \cdot$ cm^2 versus $\sim$1 k$\Omega \cdot$ cm^2 for most membranes).

The physiological properties of barnacle photoreceptors are fairly typical of those of most invertebrate photoreceptors that have been studied. The depolarizing, sodium-dependent receptor potential and the absence of action potentials are common, but not universal, features. Most invertebrate photoreceptors have much shorter axons than those of barnacles (or lack axons entirely), so that receptor potentials can spread to the receptor terminal with relative ease. Several crustacean mechanoreceptors share with barnacle photoreceptors the property of electrotonic conduction of receptor potentials over relatively large distances (see Mechanoreception, p. 487).

Compound Eyes of Arthropods   Compound eyes are usually associated with the Arthropoda, although they also occur in a few annelids and mollusks and are absent in ostracod Crustacea. As noted above, compound eyes are made up of many individual optical units termed ***ommatidia***. Each ommatidium has its own lens system and cluster of ***retinular*** cells (photoreceptors) (Figure 18.20). The eight or more retinular cells of an ommatidium are arranged in a circle, like sections of an orange but more elongate. The sides of the ommatidium are lined with pigment cells, which are capable of shielding against light leakage between ommatidia. The light-sensitive photopigment of a retinular cell is localized in the membranes of microvilli. Typically, all the microvilli of a retinular cell are arranged along one edge of the retinular cell; this array is termed a ***rhabdomere***. In most eyes the rhabdomeres of all eight or more retinular cells are tightly packed along the central axis of an ommatidium (Figure 18.20). The resulting elongated zone of photosensitive microvilli is termed the ***rhabdom*** of the ommatidium. The rhabdom thus corresponds to the aggregated rhabdomeres of all the retinular cells of the ommatidium. The eyes of Diptera (flies) and Hemiptera (true bugs) are unusual in that the rhabdomeres of an ommatidium remain separate rather than being joined into a single central rhabdom.

There are two types of compound eye that differ in their means of image formation. In ***apposition eyes,***

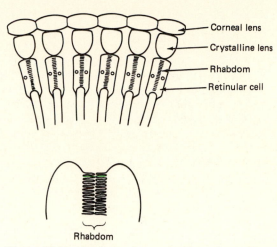

Figure 18.20   Compound eye structure. A compound eye is made up of many *ommatidia,* or faceted optical elements. Each ommatidium has a corneal lens and a crystalline lens, which focus light from a narrow acceptance angle onto photosensitive microvilli of sensory retinular cells. The closely packed microvilli are collectively termed the rhabdom.

each ommatidium is optically isolated from other ommatidia and receives light from a narrow region of the visual field (Figure 18.21A). Light from outside this narrow *acceptance angle* is absorbed by lateral pigment cells and does not reach the rhabdom. In apposition eyes, the distal end of the rhabdom is directly adjacent to the inner end of the crystalline lens. In ***superposition eyes,*** in contrast, the rhabdoms are separated from the crystalline lenses by a wide optically clear zone. Light can pass between ommatidia in this zone, so that light can reach a given rhabdom from many ommatidial lenses (Figure 18.21B). The lenses of superposition eyes are thick cylinders that refract eccentric light beams back at the same angle at which they enter the lens. Figure 18.21B shows the resultant convergence of light onto a rhabdom from many ommatidial lenses. Crayfish and lobsters achieve the same effect by prismatic reflection of light by the sides of squared crystalline cones.

What is the functional significance of a superposition eye? It certainly does not allow as great visual acuity as an apposition eye (which in turn lacks the acuity of a camera eye such as that of vertebrates). The adaptive value of a superposition eye is its great increase in light-gathering power. Light striking much of the eye surface can be brought to focus on a rhabdom. In contrast, in an apposition eye only the light falling on a single ommatidial lens is focused on the rhabdom. Superposition eyes therefore provide increased sensitivity to dim light (at some cost in resolving power), rather like opening the aperture of a camera lens. Most arthropods with superposition eyes are nocturnal (e.g., fireflies and moths), while most with apposition eyes are diurnal (e.g., brachyuran crabs, diurnal insects). In superposition eyes,

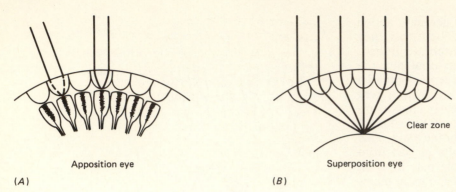

Apposition eye

(A)

Clear zone

Superposition eye

(B)

Figure 18.21 Apposition and superposition compound eyes. *(A)* In an apposition eye, each ommatidium focuses light in its optical axis onto the rhabdom. Light not in the optical axis misses the rhabdom and is absorbed by screening pigment. *(B)* In a dark-adapted superposition eye, parallel light rays are focused by many ommatidial lenses onto a single rhabdom, increasing the light-gathering power. When the same eye is light adapted (not shown), migrating screening pigment prevents light passage between ommatidia. Therefore, light-adapted superposition eyes function similarly to apposition eyes.

the degree of light passage between ommatidia is controlled by light-induced pigment migration. In the dark-adapted eye (e.g., at night), pigment is withdrawn from the clear zone (Figure 18.21*B*), allowing light passage and image superposition. In the light-adapted state, pigment migrates between the ommatidia into the clear zone and blocks light passage, so that image formation becomes similar to that of an apposition eye. Thus, superposition eyes function as such only when dark adapted, under conditions where their increased light-gathering power is important.

Physiology of Compound Eyes We shall now concentrate on the compound eye of the horseshoe crab *Limulus,* a primitive arthropod, which has been studied most extensively. The compound eyes of insects and crustacea have complex optic ganglia—outgrowths of the brain—immediately behind the eye. Because of this proximity of higher-order interneurons, photoreceptor physiology has been harder to study in these groups. For this reason, investigators led by H.K. Hartline have concentrated on the more accessible eye of *Limulus,* in which the optic ganglia are in the brain, several centimeters away from the

eye. In 1932, long before the advent of intracellular recording techniques, Hartline recorded light-induced activity of single receptor axons in the *Limulus* optic nerve. The photoreceptor physiology even in the Limulus compound eye is complicated by integrative interactions. Nevertheless, the *Limulus* eye is an important model system for studies of visual physiology and of sensory processes in general.

The anatomy of an ommatidium in the *Limulus* eye is shown in Figure 18.22. The eye is of the apposition type, with optically isolated ommatidia and a rhabdom close to the proximal end of the crystalline lens. There are about 12 retinular cells and a single modified retinular cell termed the **eccentric cell.** The soma of the eccentric cell lies to the side of the ommatidium, and a narrow dendrite extends from the soma through the center of the rhabdom, making intimate contact with the microvilli of the retinular cells. Eccentric cells are an unusual feature of *Limulus* eyes that are not found in other arthropods.

Now let us examine the results of an experimental demonstration of photoreception in the *Limulus* compound eye (Figure 18.23). Early studies employed extracellular recording from axons in the optic nerve, so that is a good place for us to begin. The response

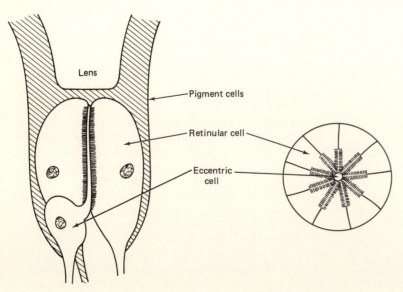

Lens

Pigment cells

Retinular cell

Eccentric cell

Figure 18.22 Structure of an ommatidium in the compound eye of the horseshoe crab *Limulus.*

to a light stimulus is a train of action potentials (nerve impulses) in the axon of the eccentric cell of the illuminated ommatidium (Figure 18.23, bottom record). With more intense light stimuli, the frequency of action potentials is increased, the impulse frequency being approximately proportional to the logarithm of the intensity (Figure 18.5). Because of this logarithmic relation, the photoreceptors are able to encode information about light intensity over a wide range (about five orders of magnitude) of intensity.

Let us now examine the intracellularly recorded responses of a retinular cell and an eccentric cell to light stimulation. A microelectrode in a retinular cell (Figure 18.23, top record) records a light-induced, depolarizing receptor potential, the amplitude of which is graded with light intensity. Microelectrode recordings from the eccentric cell (Figure 18.23, middle record) reveal a graded depolarization on which are superimposed action potentials corresponding to the train of action potentials recorded from the axon. Experiments have demonstrated electrical junctions between the eccentric cell and each of the retinular cells of the ommatidium. These electrical junctions selectively pass slow potentials, allowing electrotonic spread of the receptor potentials from retinular cells to the eccentric cell, but severely attenuating spread of impulses from the eccentric cell to retinular cells. Action potentials (other than the attenuated spikes from eccentric cells) are never recorded from retinular cells or from their axons. Evidently, the receptor potentials of the retinular cells spread to and summate at the eccentric cell, and only action potentials of the eccentric cell are transmitted to the brain. There are also isolated ventral photoreceptor cells in *Limulus,* not part of organized eyes, that have been studied extensively because they lack the complication of electrical coupling. The ventral photoreceptors also lack impulses, as do photoreceptors of barnacles (Figure 18.19), and other Crustacea, insects, and most other invertebrates. In most of these cases, the receptor potential must spread electrotonically only a short distance to reach second-order cells. The eccentric cell of the *Limulus* compound eye may have evolved as a specialization to generate impulses for long-range transmission to the distant brain.

The receptor potential of a *Limulus* photoreceptor appears to represent a summation of small electrical responses to individual light quanta. With very weak light stimulation, discrete individual depolarizations termed **quantum bumps** are recorded (Figure 18.24). Evidence suggests that each quantum bump represents the absorption of a single light quantum by a photopigment (rhodopsin) molecule, leading to photochemical changes similar to those in vertebrate photoreceptors (to be described below). Each photon capture leads to a receptor current of 10^{-9} A. With increasing intensities of light stimulation, the frequency of quantum bumps increases (Figure 18.24), so that they summate into a continuous receptor potential. The amplitude of the individual quantum bumps, however, decreases at higher light intensities and with prior exposure to light. This decrease in depolarizing current produced by individual quanta has two important consequences. First, it is the basis of the wide dynamic range of the photoreceptors, in which the amplitude of the receptor potential is proportional to the logarithm of stimulus intensity. Second, the decrease is a major component of light adaptation—the decrease in responsiveness of receptors after prior exposure to light.

The receptor potential of *Limulus* photoreceptors results primarily from a permeability increase to Na^+ ions, as well as a smaller and slower contribution of Ca^{2+} ions. The absorption of a single photon leads

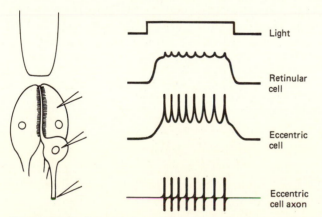

Figure 18.23 Response of *Limulus* photoreceptor cells to light. Retinular cells undergo a graded depolarization (receptor potential), which spreads to the electrically coupled eccentric cell. The eccentric cell generates a train of action potentials, which propagate down the eccentric cell axon. Attenuated spikes from the eccentric cell appear in the electrically coupled retinular cell, which does not itself generate action potentials.

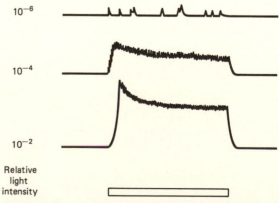

Figure 18.24 Responses to individual light quanta (quantum bumps) in a *Limulus* photoreceptor. At low light intensities, individual quantal depolarizations are clear. With higher light intensities, the frequency of quantal responses increases, and the amplitude of individual quantal responses decreases.

to the opening of as many as 1000 ionic channels to elicit a quantum bump, a calculation that suggests that one or more amplifying steps must intervene between photon capture by photopigment and channel opening to produce the receptor current. The light-dependent increase in Na^+ permeability that underlies the receptor potential is itself inhibited by intracellular Ca^{2+} ions. Thus, with prolonged or repeated light stimulation, Ca^{2+} entry and accumulation in the photoreceptor lead to a decrease in the light-dependent Na^+ current and a diminution in amplitude of the receptor potential. This action of Ca^{2+} ions is thought to play an important role in light adaptation of the photoreceptors.

The eccentric cells of *Limulus* ommatidia interact with each other by *lateral inhibition,* a simple form of sensory integration of widespread occurrence. Lateral inhibition, although not confined to sensory systems, is an important aspect of information processing in many sense organs. The *Limulus* eye has served as the major model system for studying the mechanism and function of this process. Eccentric cells have axon collaterals that form a plexus beneath the receptor cells. These collaterals make inhibitory synaptic contacts with other eccentric cells, exerting their greatest inhibitory influence on eccentric cells 3–5 ommatidia away. Figure 18.25 demonstrates lateral inhibition between two eccentric cells in a *Limulus* eye. It is clear that the generation of impulses in cell A decreases impulse frequency in cell B and vice versa. What is the functional significance of lateral inhibition? Clearly, strongly activated cells suppress their weakly activated neighbors more than the converse, since the strongly activated cells generate more effectively summating IPSPs (cf. Figure 17.10). The major functional role of this differential suppression in sensory systems is *enhancement of contrast* of a spatially distributed stimulus. To see how lateral inhibition enhances visual contrast, let us consider a single row of ommatidia in a *Limulus* eye (Figure 18.26). We can imagine a light stimulus above the eye (A) that is bright on the left side and

dimmer on the right. The ommatidia on the left then receive strong light stimulation, and those on the right (22–40) receive weaker light stimulation. A few ommatidia at the light–dark edge receive an intermediate amount of light, since the edge falls within their acceptance angle. Figure 18.26*B* shows the expected response of the photoreceptors *if there were no lateral inhibition*. The receptor potential amplitude and the resultant frequency of eccentric cell impulses would then depend only on light level for each individual ommatidium. Now let us add the effects of lateral inhibition to our conceptual experiment. On the left side of the row of ommatidia (e.g., ommatidia 5–10 in Figure 18.26*C*), each eccentric cell is strongly excited by light stimulation but is also strongly inhibited by the lateral inhibitory effects of all the other stimulated eccentric cells in the neighborhood. On the right side of the row (e.g., ommatidia 30–35), eccentric cells are less excited by light but also less inhibited by their neighbors. Near the light–dark edge, the eccentric cells receive different degrees of inhibition from the two sides. Cells 22–26 (on the darker side of the boundary) receive little photoexcitation but are strongly inhibited by the excited cells on their left. Thus, the net activities of the cells 22–26 are considerably less than for cells 30–35. Conversely, cells 14–18 are strongly excited by light but receive strong inhibition from only one side. Thus, their net activities are greater than those of cells 5–10 (which are strongly inhibited from both sides). The overall effect of lateral inhibition is therefore to enhance contrast at edges so that the light side of a boundary appears lighter and the dark side appears darker. At the same time, lateral inhibition diminishes the apparent difference in brightness between areas that are uniformly illuminated (such as between ommatidia 5–10 and 30–35 in Figure 18.26). The enhancement of contrast at edges by lateral inhibition also occurs in the human visual system, as demonstrated in Figure 18.27. Although each band in the figure is uniform gray (shown by covering the other bands with white cards), each band appears

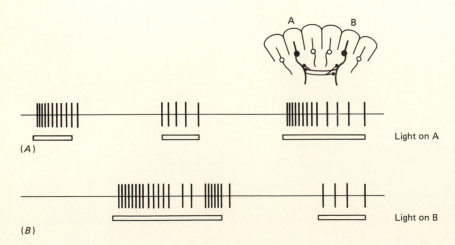

(A)

(B)

Light on A

Light on B

Figure 18.25 Lateral inhibition in the *Limulus* eye. (A) and (B) are recordings from the axons of two eccentric cells in nearby ommatidia (see inset). Illumination of ommatidium A leads to a spike train in A and also decreases the response of B to light stimulation. Conversely, activity in B inhibits the response of A to illumination.

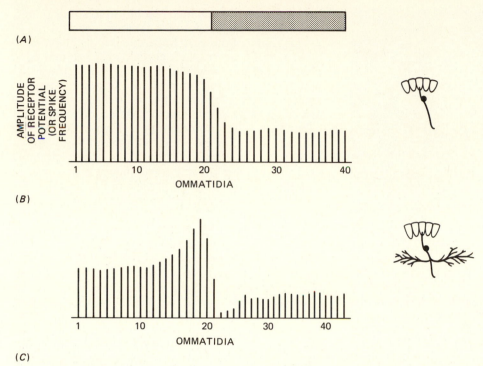

Figure 18.26 Contrast enhancement at a light–dark edge of light shone on the *Limulus* eye. The first 20 ommatidia in a row are directed toward the light; the next 20 point to the dark side of the edge (*A*). (*B*) Activity of the photoreceptors if there were *no* lateral inhibition. (*C*) Activity of the photoreceptors *with* lateral inhibition. Contrast at the edge is enhanced.

darker on the side facing the lighter band, just as described for the *Limulus* eye.

Limulus eyes are subject to *efferent control* of their sensitivity. Efferent neurons from the brain, under the control of a circadian clock, act to increase receptor sensitivity at night by as much as five orders of magnitude over daytime sensitivity. There are several components of this sensitivity increase. (1) Retinal structure is modified so that more light reaches the rhabdom at night. The aperture of pigment cells between the lens and rhabdom is broadened and the rhabdom broadens and shortens toward the lens, increasing acceptance angle and quantum capture. (2) There is an increased amount of depolarization per quantum of light (quantum gain). (3) Photoreceptor noise (fluctuations similar to quantum bumps in the absence of light) is suppressed. Thus, spontaneous activity of the receptors is suppressed while photosensitivity is enhanced. These three changes are distinct from light/dark adaptation. Two additional changes that are induced by light require "priming" by *prior* efferent activity. (4) Migration of pigment granules partially screens the rhabdom during light adaptation. (5) Microvilli of the rhabdom disperse into membrane whorls and then reform during the first half-hour of daylight. These effects do not occur in the absence of efferent activity the previous night. The effects of efferent activity are prevented by severing the optic nerves and can be produced at any time by optic nerve stimulation to activate the effer-

ent axons. The major transmitter mediating efferent control appears to be octopamine, but one or more peptides, possibly related to substance P and/or ranatensin, may also be involved.

Vertebrate Visual System

As already noted, the visual system is an exemplary one for studies of sensory processes. Vertebrate vision has been studied most extensively and is in many respects the best-known sensory system. Consequently, we shall explore it in greater depth than other systems. We will consider several questions: How does the vertebrate eye function as an optical device? How do vertebrate photoreceptors function? How does the central nervous system integrate and process visual information?

Structure and Optics of the Vertebrate Eye   The vertebrate eye is an example of a camera eye, as noted previously. The lens system focuses an inverted image on the **retina,** the photoreceptor-containing layer at the back of the eye (Figure 18.28). For terrestrial vertebrates, the greatest refraction occurs at the air-cornea interface, because the difference in refractive index (which governs the bending of light rays) is greater at that interface than between the lens and the watery fluid within the eye. For aquatic vertebrates, the cornea does little refraction, because its refractive index is similar to water; most refraction

Figure 18.27 Demonstration of the effect of lateral inhibition in the human visual system. Each vertical rectangle is a uniform gray but appears of uneven brightness; the side facing a lighter border appears darker than the side facing a darker border. If the adjacent rectangles are covered with white cards, the rectangle appears uniformly gray, a demonstration that the apparent nonuniformity results from an enhancement of contrast with the surround.

light entry and to enhance resolution to some extent, like the lens diaphragm of a camera.

The retina of the vertebrate eye is a developmental outgrowth of the brain. It contains the rod and cone photoreceptors, as well as neurons that perform the first stages of visual integration (horizontal, bipolar, amacrine, and ganglion cells, to be discussed below) and a pigmented epithelium that absorbs light not captured by the photoreceptors (Figure 18.29). The retina is said to be inverted, with the photoreceptors in the outermost layer, farthest away from incoming light. This inversion is a consequence of the way in which the retina develops in the embryo, as the outer layer of a two-layered optic cup. Although light must pass through all the retinal layers to reach the outer segments of the rods and cones, the inverted retina does not degrade the image greatly, because the retinal layers are quite transparent. There is some light scattering, however, and many retinas have a high-acuity region in which the intervening cells and blood vessels are displaced to the side. In humans this area is the *fovea,* a depression 1.5 mm in diameter (5° of visual angle) containing tightly packed cones to the exclusion of other neurons. Rods are absent in the human fovea but outnumber cones elsewhere in the retina. Primates and some birds have well-developed foveas, and many other vertebrates have a less elaborate and broader area of relatively high acuity termed the area centralis.

Another consequence of the inverted retina is that the axons of retinal ganglion cells, which form the optic nerve, come off the inner side of the retina, facing the lens. The axons exit through the retina at the optic disc, producing a blind spot in the visual field. We are normally unaware of this blind spot because it falls in the binocular field and thus the missing information is supplied by the other eye. Moreover, we depend on the fovea for much of our detailed pattern vision, and we make unconscious rapid eye movements, further decreasing any visual deficit resulting from the blind spot.

Photoreception in Vertebrates   Retinal photoreceptors are classified as *rods* and *cones*. In mammals, cones are used in color vision and for high acuity in the fovea; the more sensitive rods are used in dim light and do not contribute to color vision. Nocturnal animals tend to have retinas in which most or all

is done by a nearly spherical lens, the refractive index of which is greater at the center than at the edges. Most vertebrate lenses are focused by changing the shape of the lens, although in fish the lens is moved forward and back like a camera lens. The iris diaphragm constricts the opening of the pupil to limit

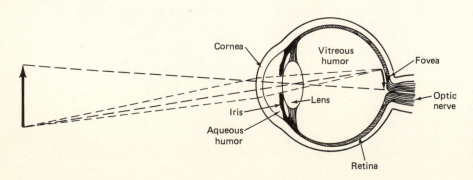

Figure 18.28 Structure of the mammalian eye. The cornea and the lens focus an inverted image on the retina at the back of the eye.

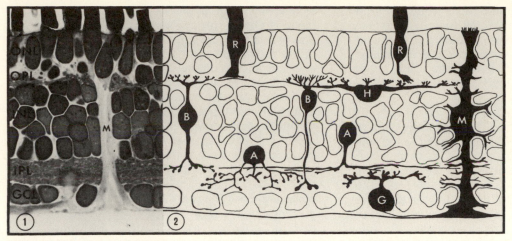

Figure 18.29 Structure of the vertebrate retina. *Left:* Light micrograph of the retina of *Necturus,* showing its layered structure. ONL, outer nuclear layer; OPL, outer plexiform layer; INL, inner nuclear layer; IPL, inner plexiform layer; GCL, ganglion cell layer; M, Müller (glial) cell. *Right:* Principal cell types of the vertebrate retina, drawn from Golgi-stained *Necturus* cells. R, receptors; H, horizontal cells; B, bipolar cells; A, amacrine cells; G, ganglion cells; M, Müller (glial) cells. [From J.E. Dowling, in F.O. Schmitt and F.G. Worden (eds.), *Neurosciences: Fourth Study Program.* MIT Press, Cambridge, 1979, pp. 163–182. Courtesy of J.E. Dowling.]

receptors are rods, while cones predominate in retinas of strongly diurnal animals. Because rods are larger and less diverse than cones, they have been studied more extensively.

Figure 18.30 shows the structure of retinal rods and cones. Both kinds of photoreceptors consist of an ***outer segment,*** containing the photosensitive membranes, and an ***inner segment,*** containing the nucleus, mitochondria, and other cell organelles, and the synaptic terminal. The inner and outer segments are

connected by a short ciliary stalk, the outer segment being derived from a modified cilium. The photoreceptor outer segments contain many flattened lamellae of membranes derived from the plasma membrane. In the cones of mammals and of some other vertebrates, these lamellae retain continuity with the outer membrane, so that the lumen of each lamella is continuous with extracellular space. In rods, in contrast, the lamellae become separated from the outer membrane and form internalized flattened ***discs***

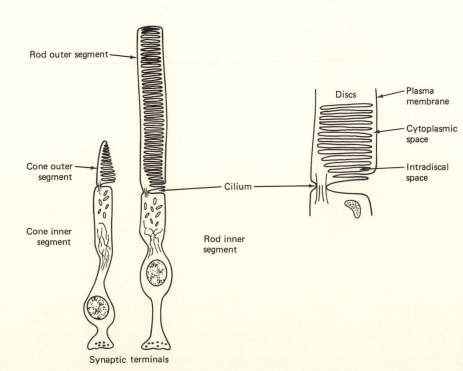

Figure 18.30 Vertebrate photoreceptors. Both rods and cones have an inner segment, containing the nucleus and synaptic terminal, and an outer segment that contains ordered lamellae bearing photopigment molecules. Cone lamellae are continuous with the outer plasma membrane. Rod lamellae are internal membrane discs, discontinuous with the outer plasma membrane.

(Figure 18.30). Several hundred to a few thousand discs, stacked like pancakes, fill the rod outer segment. Because rhodopsin is contained in the membranes of rod discs and cone invaginated lamellae, these membranes are the sites of the primary photochemical events that trigger photoreceptor excitation.

To describe the excitation of a retinal rod, it is advantageous to record from the rod intracellularly. This is possible although difficult, owing to the small size of the cells (a toad rod outer segment may be 50 μm long but only 5 μm in diameter). Before impaling a rod, let us ask: What is our expectation about the response of the cell to light? Recalling the responses to sensory stimulation of receptors we have already considered (crustacean stretch receptor, Pacinian corpuscle, barnacle and *Limulus* photoreceptors), we might expect vertebrate photoreceptors to depolarize in response to light, with or without impulses. In fact, their response is just the opposite. Figure 18.31*A* shows the effect of light stimulation on a retinal rod, recorded intracellularly. Rods (and cones) have an unusually small resting potential in the dark. The response to light is a *hyperpolarization* that is graded according to the intensity of the light flash. This hyperpolarization is accompanied by an increase in membrane resistance (decrease in conductance). From this conductance decrease, we can conclude that the membrane becomes *less* permeable

to one or more ion species as a result of light stimulation. Other sorts of experiment show that light absorption leads to a *decrease in Na^+ permeability* of the rod outer segment. In the dark, the rod outer segment has a relatively high Na^+ permeability. As a result, the membrane potential in the dark is relatively depolarized because (as you recall from the Goldman equation in Chapter 16) the membrane potential depends on the ratio of Na^+ and K^+ permeabilities. Also, because of Na^+ entry in the outer segment, rods have a *dark current* flowing into the outer segment and exiting the rod inner segment (Figure 18.31*B*). The light-induced decrease in Na^+ permeability of the outer segment decreases the dark current (Figure 18.31*C*) and hyperpolarizes the rod toward E_K (by increasing the $P_K : P_{Na}$ ratio). Thus, we have a paradox: An illuminated rod acts like a "resting" neuron (with E_m close to E_K), while a rod in the dark is like an "excited" neuron or other receptor, depolarized as a result of a high Na^+ permeability. The contrast between light responses of vertebrate and invertebrate photoreceptors is shown in Figure 18.32. In terms of consistency of response with other cells, a retinal rod (or cone) acts as if it were "excited" (depolarized) by dark and "inhibited" (hyperpolarized) by light. This semantic trap indicates that in using terms such as "excitation" of photoreceptors, we must be careful not to equate excitation with depolarization.

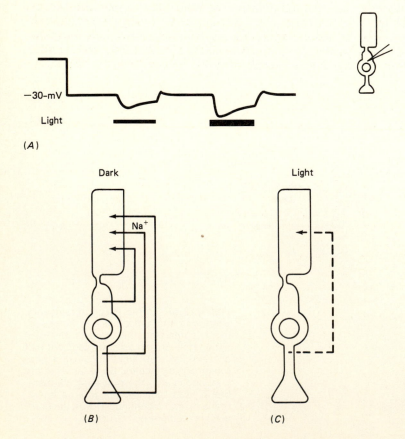

−30-mV

Light

(*A*)

Dark

Na⁺

Light

(*B*) (*C*)

Figure 18.31 (*A*) Light hyperpolarizes vertebrate photoreceptors. (*B*) A *dark current* enters the rod outer segment in the dark, carried by Na^+ ions. (*C*) Light acts to decrease the dark current.

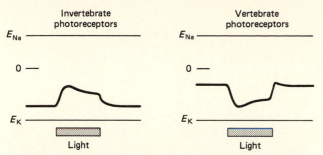

Figure 18.32 Receptor potentials of invertebrate and vertebrate photoreceptors. Invertebrate photoreceptors are depolarized in response to light, by an increase in permeability to Na^+. Vertebrate photoreceptors are unusual; they hyperpolarize in light as a result of a decrease in permeability to Na^+.

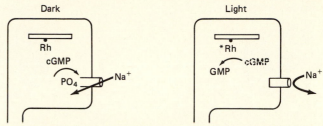

Figure 18.33 Cyclic GMP hypothesis of intracellular messenger action in vertebrate photoreceptors. In the *dark*, cGMP phosphorylates sodium channels in the outer membrane of the photoreceptor outer segment, keeping the sodium channels open. *Light* acts on rhodopsin (Rh) at the inner lamellar membranes, indirectly activating a phosphodiesterase that catalyzes the conversion of cGMP to GMP. In the absence of cGMP, the sodium channel dephosphorylates and closes.

The light-induced decrease in Na^+ permeability occurs at the outer plasma membrane of the rod outer segment—a necessity if it is to affect membrane potential. We have already noted, however, that rhodopsin is localized at the internal membranes of the discs. Because the disc membranes and the outer plasma membrane are discontinuous, we face a complication in our understanding of rod function: How do light-induced changes in rhodopsin at the disc membranes affect Na^+ permeability of the outer membrane? An electrical change in the disc membrane does not affect the surface membrane, since the two are separate. Instead, there must be an *intracellular messenger* that conveys a change from the discs to the outer surface membrane. There have been two major candidates for this intracellular messenger: Ca^{2+} ions and cGMP (cyclic guanosine monophosphate). We will discuss only the cGMP hypothesis, which is supported by most recent evidence. This hypothesis postulates that the high Na^+ permeability in the dark depends on a cGMP-dependent phosphorylation of the Na^+ channel protein (Figure 18.33). Light-induced change in rhodopsin activates a series of reactions at the disc membrane that results in an enzymatic degradation of cGMP to GMP. The resultant decrease in cGMP concentration leads to dephosphorylation of the Na^+ channel protein and closing of the channel. It is calculated that each photon absorbed by rhodopsin leads to approximately 10^6 Na^+ ions *not* entering a rod outer segment. One functional attribute of an intracellular messenger is to provide the necessary amplification for this process. Suppose that the activation of one rhodopsin molecule causes the enzymatic degradation of 10^3 cGMP molecules. Each missing cGMP molecule could let close one Na^+ channel, blocking entry of 10^3 Na^+ ions for a fraction of a second, providing the necessary amplification.

Processing of Visual Information in the Retina The aspects of visual stimuli that are most important for the behavior of animals (including humans) are *patterns* of light, dark, and color (see below), rather

than the overall light level. In behavioral terms, the significant features of the visual world are spatial patterns of visual stimuli that represent objects in the world, and temporal patterns that indicate movements. To a frog, for example, a small dark area of the visual field may have great behavioral importance (a fly for lunch)—particularly if it moves relative to the rest of the visual field. Changes in overall illumination, in contrast, may be less important, merely indicating that a cloud has passed in front of the sun. The photoreceptors themselves respond only to light level at one point in space. Therefore, we would expect that visual systems might integrate signals from receptors in ways that abstract the behaviorally significant spatial patterns and movements of stimuli. This integration has been studied most extensively in the vertebrate visual system, where information about visual pattern and movement is abstracted first in the retina and then in central visual areas. In this section, we consider how neural circuits in the retina perform the early stages of this visual integration.

We begin our examination of retinal integration by looking at the output of the retina: the response properties of the retinal ganglion cells, the axons of which form the optic nerve. The properties of mammalian retinal ganglion cells were first assessed in the 1950s, by Steven Kuffler and others. In contrast to photoreceptors, ganglion cells respond to stimulation over a relatively large visual area. This brings us to an important concept—that of the *receptive field* of a sensory neuron. For the vertebrate visual system, the receptive field of a neuron is defined as the area of the retina from which the impulse activity of that neuron can be influenced by light. For a retinal rod or cone, the receptive field essentially corresponds to the retinal area occupied by the receptor itself. For visual interneurons such as retinal ganglion cells, in contrast, the receptive field is typically much larger, embracing an area containing many receptors.

Let us examine the receptive fields of two ganglion cells in the retina of a cat. We can record from the ganglion cells with an extracellular microelectrode

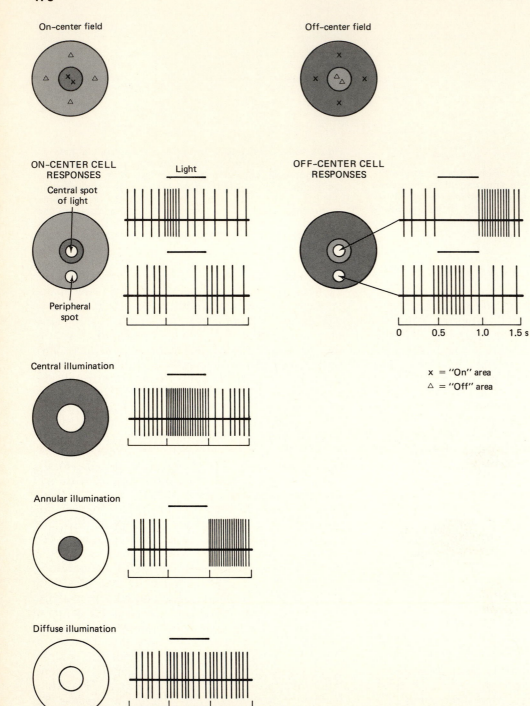

Figure 18.34 Receptive fields and responses of ganglion cells in the retina of a cat. The parts of the figure each illustrate a light stimulus within a cell's receptive field, and the response of the cell to that stimulus. The receptive fields of cats and monkeys are grouped into two main classes: on-center and off-center fields. The activity of an *on-center* cell increases in response to illumination within the center of the cell's receptive field; the cell's activity decreases in response to illumination in the surround of its receptive field, with some increased activity when the light is turned off. The receptive field and the activity of an *off-center* cell are analogous but reversed: light in the center of the receptive field decreases activity of the cell, and light in the surround increases activity. [From S.W. Kuffler, J.G. Nicholls, and A.R. Martin, *From Neuron to Brain,* 2nd ed. Sinauer, Sunderland, MA, 1984.]

made of tungsten or stainless steel wire, etched to a fine tip and insulated with resin to a few micrometers of the tip. Such a microelectrode records impulses of neurons in the immediate vicinity. We can map the receptive fields by shining small spots of light on a screen in front of the eye. Figure 18.34 shows the responses of two retinal ganglion cells to different kinds of light stimulation. The cells have receptive fields divided into two areas: the first cell, termed an **on-center cell,** increases its rate of impulse discharge when the center of its receptive field is illuminated by the spot of light. The same spot of light, however, suppresses activity when it is presented in the larger surrounding part of the receptive field. The on-center cell is maximally stimulated when the entire center of its receptive field is illuminated, but none of its surround. The cell's activity is maximally inhibited or suppressed when the surround, but not the center, is illuminated by an annulus of light. Such suppression is followed by an increased discharge when the light is turned off (an "off-response"). Diffuse light stimulation, covering the entire receptive field of the ganglion cell, has little effect on the cell's activity, because the excitatory effect of light at the center and the inhibitory effect of light in the surround are antagonistic, canceling each other out. The second cell, termed an **off-center cell,** also has a receptive field with a concentric, antagonistic center and surround (Figure 18.34). The off-center cell, however, is inhibited by light in its center and excited by light in its surround. Its receptive field is thus the converse of the field of the on-center cell.

The two retinal ganglion cells described above are fairly typical of ganglion cells in cat and monkey retinas, in which there are roughly equal numbers of the two types. There are other classifying features of ganglion cell organization, notably the classification of X, Y, and W cells (described in the section Central Integration of Visual Information, p. 483). Furthermore, other vertebrates such as rabbits and frogs have ganglion cells with more complex properties than those of cats. Nevertheless, the above description of response properties of on-center and off-center ganglion cells is a sufficient starting place for analysis of retinal integration.

To see how the response properties and receptive fields of ganglion cells are derived from neural circuits in the retina, we need to examine the synaptic connections and properties of the other retinal neurons. Figure 18.35 shows the basic synaptic organization of the vertebrate retina. The major retinal synaptic contacts can be described as follows: Rods and cones synapse on bipolar cells and horizontal cells. Horizontal cells synapse on bipolar cells. Bipolar cells synapse on amacrine cells and ganglion cells. Amacrine cells synapse on ganglion cells. (In some retinas there are additional synapses, for example, between photoreceptor cells. These are not considered here.) We can distinguish between two sorts of retinal pathway: *straight-through pathways*

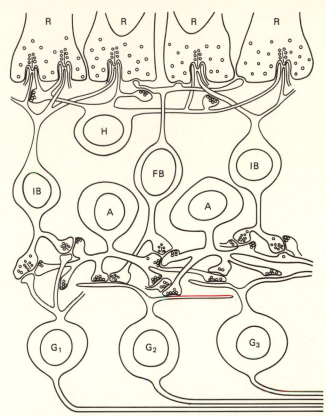

Figure 18.35 Major synaptic contacts found in vertebrate retinas. The diagram summarizes synaptic relationships seen in electron micrographs. Receptor terminals (R) form specialized ribbon synapses with invaginating bipolar cells (IB) and horizontal cells (H). Other, so-called flat bipolar cells (FB) contact the receptor terminals without invaginations. Horizontal cells make conventional synaptic contacts onto bipolar dendrites. These synapses are all in the outer plexiform layer. In the inner plexiform layer, bipolar terminals usually form ribbon synapses with one ganglion cell (G) and one amacrine cell (A) process (left), or with two amacrine cell processes (right). Amacrine cells make reciprocal synapses back onto bipolar terminals. Amacrine cells may form serial synapses with other amacrine terminals. Ganglion cells G_1, G_2, and G_3 show differences in proportions of bipolar and amacrine input. [From J.E. Dowling, in F.O. Schmitt and F.G. Worden (eds.), *The Neurosciences: Fourth Study Program.* MIT Press, Cambridge, 1979, pp. 163–182.]

that project through the retina at right angles to its surface (photoreceptors → bipolar cells → ganglion cells); and *lateral pathways* that extend along the retinal sheet, via horizontal cells and amacrine cells. The straight-through pathways give rise to the properties of the center of a ganglion cell's receptive field, while the lateral pathways give rise to the properties of the antagonistic surround.

To clarify further the retinal circuits, we need to record intracellularly from retinal neurons. For technical reasons, it is difficult to achieve intracellular recordings from mammalian retinal cells, which are very small and not readily penetrated with microelectrodes. For this reason, most of our information about response properties of retinal neurons has

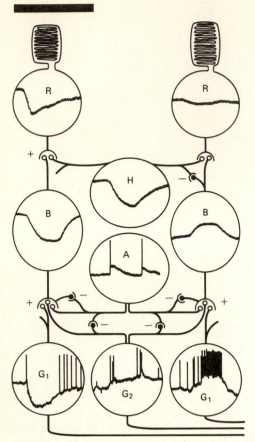

Figure 18.36 Diagram summarizing the intracellularly recorded responses of retinal ganglion cells of *Necturus*. The responses are superimposed on the synaptic organization of the retina, in an attempt to show how the synaptic connections determine the receptive-field properties of hyperpolarizing bipolar cells and off-center ganglion cells. A spot of light (indicated by bar) falls on the retinal area containing the receptor (R) at left. This light spot is therefore in the *center* of the receptive fields of cells on the left, and in the *surround* of the receptive fields of cells on the right. On the left, the hyperpolarizing bipolar cell (HB) and the off-center ganglion cell (G₁) respond to central illumination by hyperpolarizing. On the right, the corresponding cells respond to illumination in the surround by depolarizing, as a result of lateral pathways via a horizontal cell (Ho) and an amacrine cell (A); these cells have sign-inverting synaptic contacts with the bipolar cell. Cell G₂ is an on-off ganglion cell. Synapses labeled +, with open circles designating the postsynaptic element, are "excitatory" (non sign-inverting); synapses labeled −, with closed circles, are "inhibitory" (sign-inverting). Further discussion in the text. [From J.E. Dowling, in F.O. Schmitt and F.G. Worden (eds.), *The Neurosciences: Fourth Study Program.* MIT Press, Cambridge, 1979, pp. 163–182.]

come from fish, amphibians, and turtles. Figure 18.36 summarizes the results of studies by Werblin and Dowling on the retina of the mudpuppy *Necturus,* an amphibian with relatively large retinal cells. Imagine that a spot of light is directed onto the region of the retina containing the left column of cells in the figure. The light spot then falls in the center of the receptive fields of cells on the left and in the surround of the receptive fields of the cells on the right.

The light spot excites the rod on the left, which

as we saw above, corresponds to a hyperpolarization of that rod. Since the light does not fall on the rod on the right, there is little electrical effect on it. The illuminated rod synapses onto bipolar and horizontal cells. Like rods and cones, bipolar and horizontal cells do not generate impulses. There are two types of bipolar cell: *H-type* bipolar cells, in which light in the center of the receptive field hyperpolarizes the cell, and *D-type* bipolar cells, in which light in the center of the receptive field depolarizes the cell. Only H-type bipolar cells are considered in Figure 18.36; D-type cells are mentioned below. Hyperpolarization of the left rod by light leads to hyperpolarization of the H-type bipolar cell, as well as hyperpolarization of the horizontal cell. The horizontal cell synapses on the right bipolar cell and produces a light-induced depolarization of the right (H-type) bipolar cell. Thus, we see that in the *Necturus* retina, bipolar cells have receptive fields with an antagonistic center and surround. Light in the center of the field of an H-type bipolar cell (at left) hyperpolarizes it, and light in the surround of an H-type bipolar cell (at right) depolarizes it. The center–surround antagonism is mediated by horizontal cells, which act in a fashion resembling lateral inhibition in the *Limulus* eye. The amacrine cells provide a second layer of lateral inhibitory type of interaction in the retina, enhancing the center–surround antagonism and in some cases imparting sensitivity of ganglion cells to moving light stimuli. The roles of amacrine cells are more diverse and complex than those of horizontal cells and are incompletely understood. For our purposes it is sufficient to show in general how the receptive-field properties of retinal ganglion cells emanate from synaptic interactions of the retina, the center of the receptive field resulting from straight-through pathways (as exemplified by the left G₁ ganglion cell) and the surround resulting from lateral pathways through horizontal and amacrine cells (exemplified by the right G₁ ganglion cell). As indicated in Figure 18.36, evidence suggests that off-center ganglion cells (G₁) receive their central synaptic input from H-type bipolar cells, while on-center ganglion cells (not shown) receive their central input from D-type ganglion cells.

Investigators have recorded cellular activities in the retinas of a number of vertebrate species. The kinds of synaptic interactions described above for *Necturus* appear fairly typical, but some aspects of retinal integration differ for different vertebrates. In mammals, horizontal cells are less important and amacrine cells are more important than in *Necturus* in determining the properties of the surround of ganglion cell receptive fields. In birds, there is a significant efferent control of the retina, analogous to the efferent control of the *Limulus* eye. Additional complexities and differences among retinas are involved in color vision (see section on Color Vision, p. 483). Finally, in some species there is more elaborate retinal integration than we have indicated, with correspondingly more complex properties of retinal ganglion cells. The ganglion cells of frogs, for example, may respond only to quite specific features of a visual stimulus. Figure 18.37 shows one such type of ganglion cell that responds only to a small, dark, convex edge that moves relative to the background. The optimum size of a dark, stimulating object (1°) is about the size of a fly at striking distance, and stimuli that activate these ganglion cells tend to trigger feeding strike movements (oriented jump and ejection of the sticky tongue) of frogs. It is plausible therefore to interpret ganglion cells with these properties as "fly detectors" adapted to respond to a specific, behaviorally significant feature of the visual world. We discuss the concept of feature detectors further in the next section.

The greater complexity of visual integration in the frog retina is also found in several other lower vertebrates, in birds, and in rabbits. The degree of complexity of ganglion cell properties correlates poorly with phylogeny and much better with the capacity for binocular vision. Species with laterally placed eyes and limited binocular overlap of the fields of the two eyes (e.g., frogs, rabbits) tend to have ganglion cells with *more* complex properties. Species with

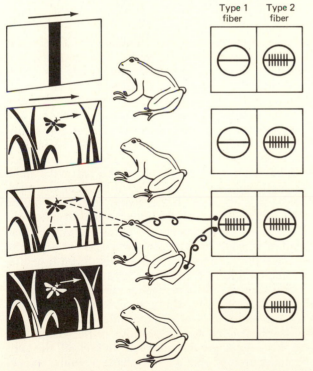

Figure 18.37 Abstraction of sensory pattern in the frog retina. One type of fiber (Type 1) in the optic nerve responds only to a small, dark object moving relative to a lighter background, and does not respond to other moving stimuli. The kind of stimulus to which Type 1 fibers respond is the kind that elicits prey-capture strike movements of frogs. Other fibers (Type 2) in the optic nerve respond to a wide variety of moving stimuli in the visual field. [After T.H. Bullock and G.A. Horridge, *Structure and Function in the Nervous Systems of Invertebrates*, Vol. 1. Freeman, San Francisco, 1965.]

frontally placed eyes and extensive binocular overlap of fields (e.g., cats, primates) tend to have ganglion cells with simpler properties, as in Figure 18.34. Presumably it is adaptive for strongly binocular species to delay complex visual integration, performing it in areas (e.g., visual cortex) that receive binocular input, rather than in the (monocular) retina.

Central Integration of Visual Information   In the preceding discussion of the retina, we have seen that the activities of ganglion cells (the retinal output) convey information about visual *pattern* rather than overall level of illumination. Activity in an on-center ganglion cell signals that a region of visual space (corresponding to the center of its receptive field) is *brighter* than the surround. Corresponding activity in an off-center ganglion cell indicates that the center of its receptive field is *darker* than the surround. This responsiveness to stimulus pattern or contrast is continued in central visual projections in the brain. We will examine central integration of visual information as an example from which we can derive some general principles of the way in which a brain processes sensory information.

Visual information is conveyed over several different central pathways in the vertebrate brain. Vision is an important sensory component of many kinds of animal behavior, and different aspects or features of the visual world are important for different kinds of behavior. Thus, it is perhaps not surprising that different pathways convey information about different aspects of a complex visual stimulus, such as color, fine details of form, and stimulus movements that elicit responsive eye movement.

In lower vertebrates, the major visual projection of the optic nerve is to the optic tectum of the midbrain. The optic nerves cross the brain midline at the optic chiasm and connect to neurons in the contralateral (opposite) tectum. In higher vertebrates, the region homologous to the optic tectum is the superior colliculus. Visual projections to the superior colliculus are important in visually guided movements such as those in prey capture and predator avoidance.

The major visual projection of higher vertebrates, however, is the geniculostriate system. The axons of retinal ganglion cells that form the optic nerve synapse in the *lateral geniculate nucleus* of the thalamus. Neurons of the lateral geniculate nucleus project to the *visual cortex* at the posterior end of the cerebrum (Figure 18.38). Unlike the optic projections of lower vertebrates, the projections of most mammals are only partially crossed at the optic chiasm. In cats and primates, the projections of the nasal (inner) half of the retina cross to the contralateral side, while those of the temporal (outer) half go to the ipsilateral (same side) lateral geniculate nucleus. This mixing of input from the two eyes allows mammals with forward-facing eyes to merge binocular input for depth perception at the visual cortex.

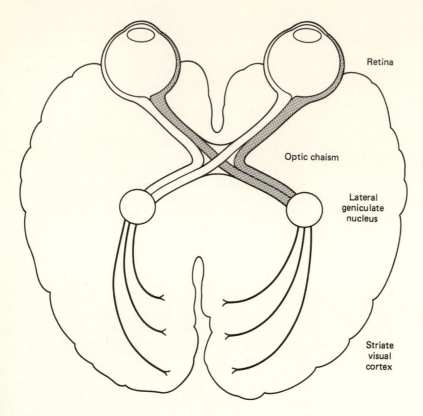

Retina

Optic chaism

Lateral
geniculate
nucleus

Striate
visual
cortex

Figure 18.38 Central visual projections of a mammal, as seen from above in cutaway view. Note that only part of the optic tract crosses the midline at the optic chiasm. Therefore, a stimulus in the *left* visual field projects to the *right* lateral geniculate nucleus (LGN) via both eyes. Conversely, a stimulus in the *right* visual field projects to the left LGN and visual cortex.

Our understanding of the central processing of visual information is based largely on studies by Hubel and Wiesel on the response properties of visual neurons in cats and monkeys. They recorded from visual cells of anesthetized animals while projecting patterned stimuli (light and dark bars, edges, and spots) on a screen in front of the animals. They found that lateral geniculate neurons had response properties similar to those of retinal ganglion cells: They responded to stimuli to only one eye and had concentric receptive fields with an antagonistic center and surround, either on-center or off-center. (The lateral geniculate nucleus is a layered structure; each layer receives input from either the contralateral eye or the ipsilateral eye, but the whole nucleus receives input from both eyes.)

The receptive fields of neurons in the visual cortex are quite different from those of lower levels of the visual pathway. We can differentiate two major kinds of neurons in the visual cortex: *simple cells* and *complex cells*. Both types are binocular: They can respond to visual stimuli presented to either eye, although one eye may predominate. Both kinds of cells have a preferred *axis of orientation* of visual stimuli, a term best explained by describing the receptive fields of a few simple cells in the visual cortex of a cat. First, let us recall that a receptive field is the area on the retina in which light stimuli influence the activity of a particular neuron. Figure 18.39 shows the receptive fields of three simple cells, as they might be mapped with small spots of light. Plus signs (+) denote areas in which a light spot excites a

simple cell, increasing the frequency of its discharge of action potentials. Minus signs (−) denote areas in which the frequency of action potentials decreases when the light spot is on and increases when the spot is turned off ("off-response"). Each of the receptive fields has a line drawn through it that defines its axis of orientation. To be most effective, a stimulus must be aligned with this axis of orientation. The optimal stimuli for these three cells are as follows: for (*A*), a dark bar in the central band of the receptive field; for (*B*), a light bar in the central band of the field; and for (*C*), an edge on the center line, light on the upper left and dark on the lower right. In all cases, the stimulus pattern must be precisely aligned and oriented; in (*A*), for example, a vertical bar is a weak or ineffective stimulus, since it falls on both excitatory and inhibitory areas. For the same reason, visual cortical cells are quite insensitive to changes in overall illumination. The three receptive fields shown have the same axis of orientation, which is to say that the best edge or bar stimulus in the visual field is at the same angle for each cell. In fact, all the cells in a small area of visual cortex have a similar axis of orientation. All possible axes of orientation, however, are represented in the cortex. If the recording electrode is advanced sideways through the cortex, it records from cells with progressively changing axes of orientation, a finding that suggests that the cortex is organized in columns with an orderly arrangement of axes. (There is also an orderly arrangement of bands of ocular dominance, so that a region of the cortex about 1 mm × 1 mm—termed a hyper-

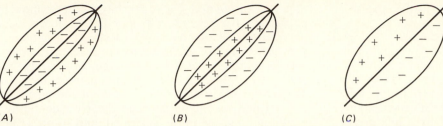

(A)   (B)   (C)

Figure 18.39 Receptive fields of simple cells in the visual cortex of a cat. Each receptive field is an area of the retina that affects the activity of a single cortical cell. All three cells have the same axis of stimulus orientation; other cell fields have different axes of orientation. Cortical simple cells typically respond best to a dark bar (A), a light bar (B), or a light–dark edge (C). The + signs indicate a region of receptive field in which light excites the cell; the − signs indicate a region of receptive field in which light inhibits the cell.

column—contains cells responding to all axes of orientation, driven by both eyes.)

Complex cells, like simple cells, are responsive to bars and edges and have receptive fields with a preferred axis of orientation. Unlike simple cells, however, complex cells are insensitive to the *position* of a stimulatory bar or edge within the receptive field. Figure 18.40 shows this defining feature of a complex cell. The figure shows the receptive field of a complex cell that responds to a horizontal bar anywhere within a certain area of the retina. The same bar stimulus presented at an angle different from the preferred axis of orientation is much less stimulatory. No part of the receptive field of a complex cell can be defined as excitatory or inhibitory; instead, the stimulus *pattern* (a dark horizontal bar) is excitatory. Complex cells then have strict requirements about stimulus form and orientation, but less strict requirements about stimulus position within the field.

It is not clear how the receptive field properties of cortical simple and complex cells result from their synaptic input from other cells such as lateral geniculate neurons. Early studies suggested a *hierarchical* organization in which many similar geniculate cells, the receptive field centers of which were aligned in a row, converged onto a cortical cell. Complex cells

were envisioned to receive convergent synaptic input from several simple cells and in turn to synapse upon still higher-order cells. Aspects of this hierarchical organization are no doubt correct, but complex cells have been shown to receive some synaptic input directly from geniculate axons. These and other studies suggest a *parallel* organization of central visual projections. We mentioned in passing (p. 479) that there are three types of retinal ganglion cell in the cat, termed *W, X,* and *Y cells.* X cells are smaller and more responsive to fine detail of stationary visual patterns, while Y cells are larger, more rapidly conducting, and more responsive to stimulus changes and movements. (W cells are small and have complex and varied receptive field properties that do not concern us here.) The separation of X- and Y-type pathways is preserved in the lateral geniculate nucleus, and some studies suggest that simple cells receive primarily X-type input while Y-type input predominates for complex cells. Central visual pathways thus have elements of both parallel and hierarchical organization, the details of which are complicated and somewhat controversial.

The ways in which central neurons function in visual perception are also not clear. Simple and complex cells can be envisioned as *feature detectors* (cf. Figure 18.37), that are responsive to line segments and contrast edges at particular orientations. We might see the world as a sort of line drawing composed of the activities of a number of such feature detectors. The response properties of known cortical cells, however, are also consistent with other theories of visual perception. Thus, at this point there remains a gulf between our understanding of visual cell physiology and our questions about mechanisms of visual perception.

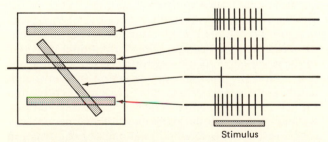

Stimulus

Figure 18.40 Receptive field of a complex cell in the visual cortex. Complex cells respond best to a bar (or edge) at the correct angle of orientation, but unlike simple cells, the stimulatory bar may be anywhere within the receptive field. A bar at an angle of orientation different from the preferred angle is much less stimulatory.

Color Vision The ability to distinguish colors depends on differential sensitivities of photopigments to different wavelengths of light. Although many animals are color blind, many animals with well-developed diurnal visual systems have evolved color vision. Examples include several orders of insects, teleost fish, frogs, turtles, lizards, birds, and primates.

Theories of color vision are strongly based on human perceptual studies. In 1801, T. Young proposed that human color perception was based on separate receptor classes sensitive to red, green, and blue light. This theory (termed the trichromaticity theory) was supported by perceptual observations that any color could be duplicated by a mixture of three primary colors. The physiological basis of the trichromaticity theory has been clarified in the last 25 years. In humans and other primates, there are three populations of cone photoreceptors, which are sensitive to different wavelengths of light. Figure 18.41 shows the spectral characteristics of the three classes of primate cones, which can be determined by measuring for different wavelengths of light, either light absorption (absorption spectrum) or stimulating effectiveness (action spectrum). The three types of cones are termed red cones, green cones, and blue cones, approximating the color of light to which they are most sensitive. The spectral sensitivities of the three types of cone are rather broad and overlapping, so that perception of color must be based on the ratio of excitation of different cone populations. For example, we might expect that perception of long-wavelength red light might depend on an analyzer that is excited by red cones but inhibited by green cones. Just this sort of integration occurs in the retina.

There are two kinds of *color opponent* process that occur in the mammalian retina: red–green opponency and blue–yellow opponency. These opponent processes explain why we can perceive a color as bluish-green or as reddish-yellow (orange), but we cannot perceive a color as reddish-green or as bluish-yellow. Many of the ganglion cells of a primate retina

that are stimulated by cones have color-opponent properties. (Cones and rods stimulate different populations of cells in the retina.) Red–green opponent ganglion cells, for example, have concentric antagonistic receptive fields that are color opponent. One cell may be inhibited by red light in its receptive field center and excited by green light in its surround; another may be excited by red light in the center and inhibited by green in the surround, and so on. Other classes of ganglion cell may be excited by blue light and inhibited by yellow (a sum of red and green cone input); such cells lack center–surround antagonism.

As we might expect, central processing of color information becomes complicated at the level of the visual cortex. We merely note that color information appears to be integrated by clusters of cells in separate districts in the cortex from cells that are not color selective but process information about brightness contrasts. This apparent segregation of color channels and so-called achromatic channels is another example of parallel organization in the mammalian visual system.

Mechanisms for color vision in animals other than mammals are always based on several populations of receptor cells with different spectral sensitivities, although these receptors are not always cones. Fish such as carp have three populations of cones analogous to those of primates, with absorption maxima of 455, 530, and 625 nm. In contrast, frogs have two or more classes of rod photopigment; their color vision involves input from both rods and cones. Although evidence for color vision is lacking in most invertebrate groups, insects have well-developed mechanisms for color vision, including receptors sensitive to ultraviolet radiation that is invisible to vertebrates. The honeybee, for example, has photoreceptors with spectral maxima of 350, 450, and 550 nm and readily distinguishes both ultraviolet radiation and colors of visible light.

MECHANORECEPTION

Earlier in this chapter we raised two kinds of questions that physiologists ask about a sensory receptor (or indeed about any part of an organism). These questions are: (1) How does it work? What is the *mechanism* of action of a receptor? (2) What is its *function* for the organism? Of what functional significance is the receptor and the information it provides to the organism? At the beginning of the chapter, we considered the mechanisms of action of two mechanoreceptors—the crustacean stretch receptor and the mammalian Pacinian corpuscle. In this section, we take a more functional view of mechanoreceptors. We examine selected examples of the ways in which mechanoreceptors are used to provide different sorts of functionally significant information to the animal. The examples chosen are proprioceptors, receptors for equilibrium, and hearing. In this section and the following discussion of other types of recep-

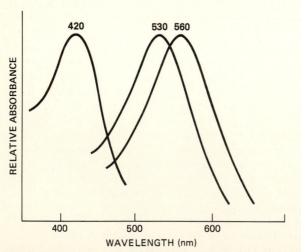

Figure 18.41 Spectral sensitivities of human retinal cones, determined by measuring absorption spectra of single cones. The three types of cone have photopigments with different absorption maxima: 420 nm (blue cones), 530 nm (green cones), and 560 nm (red cones).

tors, treatment is not as detailed as in the preceding section on photoreception and vision.

Proprioception

At the beginning of the chapter, we drew the distinction between exteroceptors (responding to stimuli outside the body) and interoceptors (responding to internal stimuli). This distinction applies to many stimulus modalities; for example, mammals contain internal carotid body chemoreceptors sensitive to the partial pressure of oxygen in the blood, as well as external chemoreceptors. Nowhere else, however, is the distinction as useful as for mechanoreceptors. All animals have exteroceptive mechanoreceptors sensitive to touch, pressure, water flow, or other surface stimulation. Probably all animals also have interoceptive mechanoreceptors that monitor movements, position, stresses, and tensions of parts of the body. Such internal mechanoreceptors are termed *proprioceptors* (from Latin *proprius*—one's own). Strictly speaking, proprioceptors are mechanoreceptors associated with the musculoskeletal system; they provide most of the information about muscle contraction, position, and movement of parts of the body, although other receptors such as skin mechanoreceptors may make a secondary contribution.

Proprioceptors are particularly important (and well studied) in animals with articulated joints and rigid skeletons, that is, arthropods and vertebrates. Figure 18.42 diagrams some of the kinds of proprioceptors and other mechanoreceptors in a crayfish and a mammal. The figure illustrates a general principle that there has been considerable parallel evolution in sensory systems, sense organs quite different in structure having evolved independently in different groups to provide the same kinds of sensory information. For example, tetrapod vertebrates sense position of limbs with muscle receptors and joint receptors (see below), while arthropods may use elastic

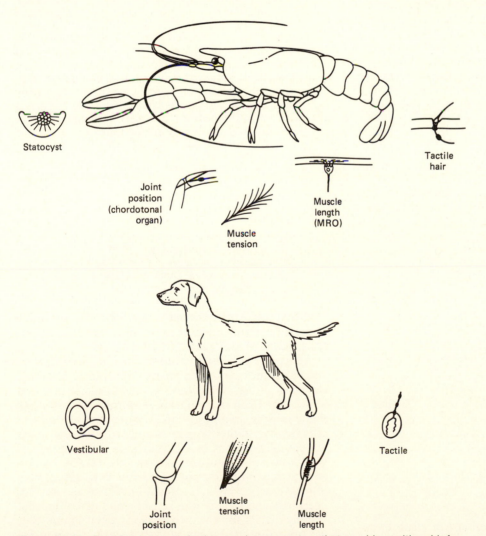

Figure 18.42 Proprioceptors and other mechanoreceptors that provide positional information in a crustacean and a mammal. Both groups have receptor types that are analogous but not homologous.

sensory strands termed chordotonal organs (Figure 18.42) or arrays of hairs (hair plates) that are stimulated by contact with the adjacent joint segment (Figure 18.43).

The best known proprioceptive organ is the vertebrate *muscle spindle* organ that monitors length of a skeletal muscle. Most skeletal muscles of vertebrates contain small groups of such organs. One component of each organ is a specialized muscle fiber termed an *intrafusal* fiber. The ends of the fiber are contractile, but the central sensory portion is noncontractile (Figure 18.44). When an intrafusal fiber is stretched by elongation of the muscle, it activates sensory neurons—the endings of which are associated with the central sensory portion of the fiber. Figure 18.44 diagrams the response of a muscle spindle sensory axon to stretch imposed on the muscle.

The functional organization of a muscle spindle is similar in several important respects to that of the crustacean abdominal stretch receptor organ (muscle receptor organ, MRO) discussed early in this chapter. The three most important similarities are as follows:

1. Both are proprioceptive organs arranged *in parallel* to the force-producing muscle fibers. Like the receptor muscle of the MRO, muscle spindle intrafusal fibers are too few and weak to generate appreciable tension themselves. Because they are located in parallel to the tension-producing extrafusal fibers, muscle spindles do not sense muscle tension. Instead, their activity is related to *muscle length*. Contraction of extrafusal muscle fibers tends to decrease spindle sensory activity, by shortening the muscle (Figure 18.45).

2. In both sense organs there is a distinction between *phasic* and *tonic* response properties. For the crus-

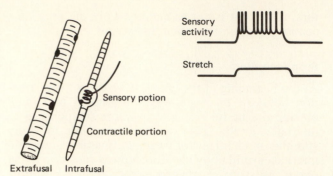

Figure 18.44 Vertebrate skeletal muscles contain muscle spindle stretch receptors located in the central region of specialized intrafusal muscle fibers. Each muscle contains relatively few intrafusal fibers as well as many tension-producing extrafusal fibers. Stretch of the intrafusal fiber produces a depolarizing receptor potential and a train of action potentials in the afferent sensory neuron of the muscle spindle.

tacean MRO, there are separate phasic and tonic receptor neurons that differ markedly in rate of adaptation of their response (Figure 18.9). Mammalian muscle spindles also have two kinds of sensory neurons, termed primary and secondary endings. The smaller secondary endings have tonic responses that provide information about muscle length but are rather insensitive to movement. The larger primary endings signal muscle length but also have a rapidly adapting, phasic component of their response that is particularly sensitive to *changes* in length. Their activity thus provides information about both muscle length (position) and rate of change of length (velocity of movement).

3. Both the crustacean MRO and the vertebrate muscle spindle are under efferent control of the central nervous system. The receptor muscle of the crustacean stretch receptor contracts when its motor neuron is stimulated, stretching the dendrites of the receptor cell and evoking a train of impulses (Figure 18.10). Similarly, the intrafusal fibers of muscle spindles receive motor innervation; in birds and mammals this innervation is by a separate class of small motor neurons termed gamma (γ) motoneurons.

The functional roles of muscle spindles, crustacean stretch receptors, and their respective forms of efferent control are treated in Chapter 20.

There are also several differences between muscle spindles and crustacean stretch receptors, although they are not as important as the similarities listed above. (1) Each muscle of a vertebrate typically has many muscle spindles, with considerable anatomical complexity of their organization. In contrast, each abdominal segment of a crayfish has a single pair of stretch receptor neurons on each side. (2) Crustacean stretch receptor neurons are fairly easy targets for intracellular recording, a technique that has not yet been possible for the fine sensory endings of muscle spindles. It is for the two reasons cited above that we used the crustacean stretch receptor as our example with which to illustrate sensory responses at the beginning of the chapter. (3) Crustacean stretch

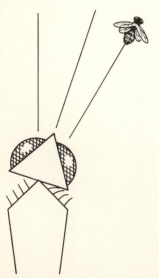

Figure 18.43 Proprioceptive function of hair plates in the neck of the praying mantis. The mantis turns its head toward its target before striking at it with its forelegs. Cervical hair plates monitor the angle of the head relative to the thorax.

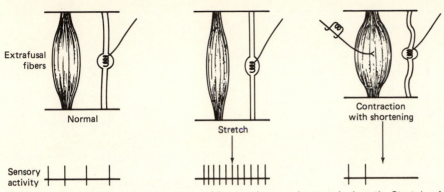

Figure 18.45 Muscle spindles are sensitive to changes in muscle *length*. Stretch of the muscle increases sensory activity. Contraction of the muscle (by activating extrafusal muscle fibers) shortens the muscle and thus *decreases* sensory activity.

receptors receive efferent inhibitory control (Figure 18.11). In contrast, there is no inhibition of vertebrate muscle spindles.

A final note on proprioceptors is that in at least three cases in crustaceans, proprioceptors send information to the central nervous system by passive spread of a receptor potential, without the generation of nerve impulses. In these cases (thoracico-coxal receptors at the bases of the legs of crabs, tailfan receptors, and gill-bailer receptors), the receptive endings are relatively close to the CNS and have large-diameter axons that increase the membrane length constant. Their mode of operation is thus similar to that of barnacle photoreceptors (Figure 18.19).

Equilibrium Reception

All animals tend to maintain an equilibrium of their body position with respect to gravity. This orientation to gravity may involve several sources of sensory information, including vision, proprioceptive and tactile receptors, and specialized organs of balance or equilibrium. Equilibrium receptors are thought to have been important in the early stages of evolution of nervous systems. Sense organs for equilibrium reception appear to be particularly important in aquatic swimmers of many phyla, such as pelagic crustaceans and fish. These animals must orient in a three-dimensional world in which they are nearly neutrally buoyant and thus lack information about gravity from other sensory sources.

The most common and widespread organ of equilibrium sense is the *statocyst,* some version of which is present in members of all metazoan phyla. Many subphyletic groups (e.g., insects), however, lack statocysts. The essential feature of a statocyst is that a dense, mineral structure termed a *statolith* contacts and stimulates mechanosensory cells that surround it (Figure 18.46). The statolith, which is usually composed of sand or of a calcareous concretion, has a greater specific gravity than the surrounding fluid. As a result, gravitational force presses the statolith against the underlying receptors. Statocyst receptor

cells have hairlike projections, typically derived from cilia, which transduce bending or shearing forces into a depolarizing generator potential.

Statocysts provide unambiguous information about gravity only if the animal is stationary. Whenever the animal moves or is moved, the statolith presses against receptors by a force proportional to acceleration. A simple statocyst of the sort shown in Figure 18.46A thus provides information about both gravity and acceleration, but these two kinds of information are mixed. One way to resolve this ambiguity in the response of a statocyst is to have a

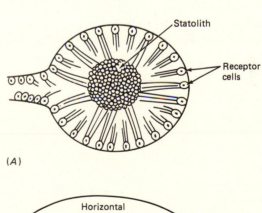

(A)

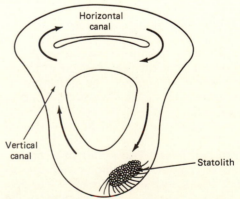

(B)

Figure 18.46 (A) Statocyst of *Pecten* (Mollusca). (B) Statocyst of a crab.

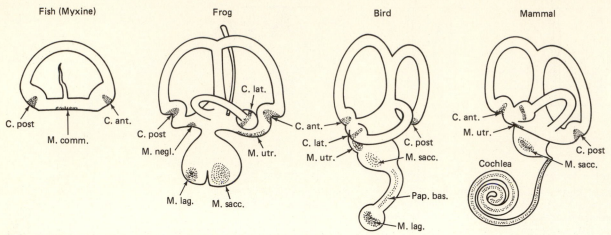

Figure 18.47 Evolution of the vertebrate labyrinth, as suggested by the labyrinths of living vertebrates. C. ant., anterior crista; C. lat., lateral crista; C. post., posterior crista; M. comm., macula communis; M. lag., macula lagenae; M. negl., macula neglecta; M. sacc., macula sacculi; M. utr., macula utricula; Pap. bas., papilla basilaris. [From G.M. Shepherd, *Neurobiology.* Oxford University Press, New York, 1983.]

compound organ in which one part detects only acceleration. The statocyst of lobsters and crabs, located at the base of the antennule, is one such compound organ (Figure 18.46*B*). The statocyst walls are convoluted to form a horizontal canal and a vertical canal, within which fluid moves when the animal is rotated. The structure of these canals is analogous to the semicircular canals of the vertebrate vestibular organ (see below). Sensory hairs (termed thread hairs and free hook hairs) are displaced by the moving fluid and signal rotational acceleration. The statolith and its associated hairs are located at the bottom of the statocyst. The statolith hairs do not adapt to constant stimulation, unlike thread hairs, which adapt with constant displacement and thus signal acceleration.

Vertebrate Vestibular Organ Vertebrate organs of equilibrium are part of a sensory system termed the acousticolateralis system. The sensory cells within the acousticolateralis system are *hair cells* derived from the epithelium; they make synaptic contact with sensory neurons. The acousticolateralis system includes the vestibular organs of equilibrium, the lateral line system of surface receptors in fish and amphibians, and the mammalian cochlea, an auditory organ to be considered below. The vestibular organs and cochlea together are termed the membranous labyrinth, which develops from the anterior end of the lateral line system.

Figure 18.47 diagrams stages in the evolution of the vertebrate labyrinth. The basic structure of a vestibular organ consists of two parts: an *otolith organ* that functions as a gravity detector and *semicircular canals* that detect angular acceleration. The otolith organ is divided into two bony chambers, the saccule and the utricle; the utricle connects to the semicircular canals. Sensory hair cells are located in discrete patches in the saccule, utricle, and enlarge-

ments at the ends of the semicircular canals termed ampullae (Figure 18.47). The hair cell groups of the saccule and utricle, termed maculae, are attached to overlying otoliths by a gelatinous matrix. The sensory hairs of the ampullar receptors (cristae) are embedded in a gelatinous flap termed a *cupula*. As the head is accelerated in one of three rotational planes, the endolymph fluid in one of the three semicircular planes lags behind the motion by static inertia, like water in a rotated glass. The movement of endolymph fluid relative to the cupula displaces the cupula, stimulating the sense hairs embedded in it.

Transduction in Hair Cells The hair cells of the vertebrate acousticolateralis system are so named because they bear projecting cilia that impart to the cells exquisite sensitivity to minute mechanical stimuli. For each hair cell there is a single *kinocilium* and 20 or more *stereocilia*. The kinocilium has the 9 + 2 arrangement of microtubules characteristic of motile cilia (Figure 19.35). Stereocilia are not true cilia; they have the appearance of microvilli, contain actin microfilaments, and are developmentally unrelated to the kinocilium. Unlike invertebrate mechanoreceptors, hair cells lack axons; they synaptically excite sensory axons that project to the central nervous system (Figure 18.48).

What is the site of sensory transduction in hair cells? It is the projecting stereocilia and kinocilium that are displaced by mechanical stimuli. Presumably, the transduction of displacement into a change in membrane potential involves effects of bending and shearing forces on membranes of the cilia or of the adjacent apical surface of the cell. The cilia of all hair cells are arranged in a polarized array, with the shortest stereocilia at one side of the cluster and the longest at the other side. The kinocilium is always located at the edge with the longest stereocilia (Figure 18.48). Displacement of the cilia in opposite di-

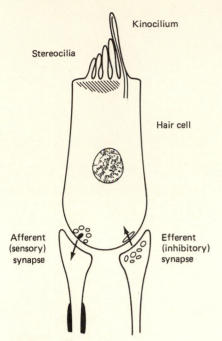

Figure 18.48 Structure of a hair cell of the vertebrate acousticolateralis system.

olfactory and invertebrate chemoreceptors, and photoreceptors of vertebrates and of many invertebrates (Figure 18.16). However, the kinocilium appears not to be necessary for transduction in hair cells. Mammalian auditory hair cells lack a kinocilium in their mature form, although a kinocilium is present transiently in development and its basal body persists in the mature cell. In some reptilian and avian hair cells, the kinocilium is present but appears uninvolved in transduction. Moreover, Hudspeth and Jacobs have microdissected the kinocilium away from the stereocilia in bullfrog saccular hair cells (see Hudspeth, 1983). Stimulation of the kinocilium alone produced no response in the hair cell, and removal of the kinocilium did not alter the cellular response to stimulation of the stereocilia. Thus, the stereocilia appear to be the site of transduction. Evidence suggests that mechanical stimulation opens conductance channels near the outer ends of the stereocilia, thereby depolarizing the cell.

Auditory Reception

One important use of mechanoreceptors, especially by vertebrates and insects, is for sound reception. Sound receptors are termed *auditory receptors* or phonoreceptors, and the sense of hearing is termed *audition*. *Sound* consists of waves of compression of air or water, which propagate away from a vibrating source. Before considering the reception of sound, we need to consider briefly the physical characteristics of sound itself.

Suppose we have a loudspeaker, the cone or membrane of which is vibrating in and out (Figure 18.50). When the membrane pushes out, it compresses air molecules, momentarily increasing the air pressure. When the membrane vibrates back into the speaker, there is a momentary rarefaction of air molecules (they occupy more volume) and the pressure de-

rections produces opposing effects (Figure 18.49). Displacement from the stereocilia toward the kinocilium depolarizes the hair cell and increases the action potential frequency in the sensory axon with which it synapses. Displacement from the kinocilium toward the stereocilia hyperpolarizes the cell and decreases sensory neuron activity.

Initial attention was focused on the kinocilium as a potential site of sensory transduction, largely because many other sensory transducers appear to be derived from true cilia. Examples of sensory derivatives of cilia include most invertebrate mechanoreceptors (including statocyst receptors), vertebrate

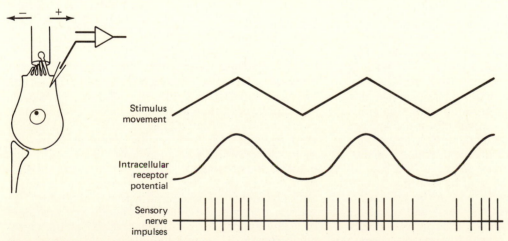

Figure 18.49 Hair cells are depolarized and excited by movements of the stereocilia toward the kinocilium. Displacement of stereocilia away from the kinocilium hyperpolarizes the hair cell, decreasing impulse frequency in the sensory axon.

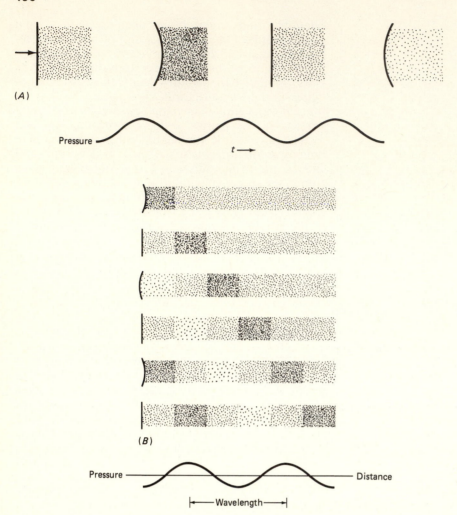

Pressure

$t \longrightarrow$

(A)

(B)

Pressure ————————————— Distance

|← Wavelength →|

Figure 18.50 Generation of a sound pressure wave by a vibrating source such as a loudspeaker. [Adapted from J.M. Camhi, *Neuroethology.* Sinauer, Sunderland, MA, 1984.]

creases. Therefore, as shown in Figure 18.50*A*, the vibrating speaker cone causes a repeating pressure wave in the air, the frequency of which is identical to the frequency of movement of the speaker cone. This wave of sound pressure propagates away from its source (the speaker cone) at (obviously) the speed of sound. The propagation results from collisions between air molecules. Compressing the air molecules increases the frequency of collisions between them; in fact, you may recall from physics that it is the increased collision frequency that produces the increase in pressure. As a result of these molecular collisions, the wave of compression (and high collision frequency) propagates outward from the source, even though individual air molecules may move only a very short distance. A mechanical analogy for the propagation of a sound pressure wave is a row of balls, each connected to the next by a spring. If you press together two balls at the end of the row, compressing the spring connecting them, and let go, a wave of oscillation spreads down the row as the balls are successively pushed together and the springs push the balls back apart.

Because it is propagated, a sound pressure wave

has a wavelength as well as a frequency. Figure 18.50*B* diagrammatically relates the wavelength to sound frequency; the quantative relation is

$$\lambda \text{ (meters)} = \frac{s \text{ (meters/second)}}{f \text{ (seconds}^{-1})}$$

where λ is the wavelength, f the frequency, and s the speed of propagation in the medium. For air at sea level and 0°C, $s = 334$ m/s.

The pressure component of sound, discussed above, is the aspect of sound particularly important for most auditory systems. We should note in passing, however, that sound also has a displacement component: A vibrating source such as the speaker cone in Figure 18.50 physically displaces nearby air molecules toward and away from itself. This displacement component, unlike the pressure component, loses energy very rapidly with distance and is only significant within about one wavelength of the source. Behavioral evidence indicates that receptors of a few animals (e.g., cuticular hairs of the caterpillar *Barathra,* lateral line hair cells of fish) can respond to this near-field displacement component of a sound source.

Insect Auditory Receptors Many kinds of animals can detect low-frequency vibrations in air, water, or within the substrate, presumably using sensory hairs or other undifferentiated mechanoreceptors. For example, cockroach cercal wind receptors (Figure 16.1) respond to loud sounds at frequencies below 1 kHz. Specialized organs for sound detection, however, are largely confined to insects and vertebrates. The most common form of auditory organ in insects is a **tympanal organ** in which a thin cuticular tympanum is displaced by sound. Mechanosensory cells are attached to the tympanum and are stimulated by its movement. Tympanal organs are air-filled cavities derived from some aspect of the tracheal respiratory system. (The functional importance of an air-filled chamber is considered below for vertebrates.) Tympanal organs may occur at any of several locations in the body, including the thorax (e.g., in noctuid moths), abdomen (locusts, cicadas), legs (crickets, katydids), or labial palps (sphyngid moths). This diversity, as well as the fact that nontympanic organs are also used for sound reception, shows that auditory organs have evolved repeatedly in different insect groups.

One of the simplest and best studied tympanal organs is that of noctuid moths. Each of the paired thoracic organs contains only two neurons that respond to sound. The range of frequency sensitivity is from 3 to 150 kHz, with maximum sensitivity at 50–70 kHz. Most of this range is **ultrasonic**—above the frequency range of sound audible to humans (20 Hz to 20 kHz). Moth auditory organs are most sensitive to the ultrasonic cries of echolocating bats (see Echolocation, p. 502). The two auditory cells (A_1 and A_2) respond similarly to ultrasonic pulses, but the threshold of A_1 is 20 dB lower than the threshold of A_2. The cells convey no information about sound frequency; moths and most insects cannot discriminate frequencies of stimulating sound. Sound intensity is coded by the impulse frequency in each receptor, by a decreased response latency with increasing intensity, and by recruitment of the high-threshold A_2 cell.

The behavioral significance of auditory responses in moths has been well studied by Roeder and others (as have central neuronal responses not considered here). Moth tympanal organs provide good directional information about a source of ultrasound, in part because the sound wavelength is short relative to the size of the moth's body. If a bat emits ultrasonic pulses to the left of a moth, the response of the left "ear" (tympanal organ) is greater, the moth's body creating an effective sound shadow for short-wavelength ultrasound. The left A_1 cell responds with a shorter latency and a higher frequency than the right A_1 cell. As a result of this asymmetry, the moth turns away from the sound source and flies away. This response is effective for predator avoidance if the bat is distant, but not if the bat is close enough to detect the moth (since bats are stronger

fliers). A nearby bat emits an ultrasonic cry loud enough to stimulate the A_2 cell, which triggers a quite different response from the moth. Instead of turning away, the moth flies erratically or dives to the ground, responses presumably more adaptive than an attempt to "outrun" the bat once detected.

Most insect tympanal organs are somewhat similar to those of moths, in that they are sensitive detectors and encoders of sound intensity over a certain range of frequencies but are poor at detecting frequency differences. Next, we examine two refinements of tympanal organs, one that improves localization of low-frequency sound and one that provides some information about sound frequency.

In crickets and several other kinds of insects, the tympanal organs function as *pressure difference* receivers, a feature that enhances their ability to localize sound sources. The noctuid moth organs described above function as simple pressure receivers: They respond to sound pressure peaks and troughs only on the outer tympanal surface, the inner chamber being a completely enclosed space. Such a pressure receiver is insensitive to the direction of a sound source if the wavelength of the sound is large relative to the size of the receiver. This lack of directional sensitivity is because pressure acts in all directions, so that the force on the membrane is independent of the direction of the sound. (Noctuids can localize ultrasound, however, because the ultrasound wavelength is small relative to the moth's size and thus is diffracted by the moth's body, producing a "sound shadow.") Crickets and other Orthoptera use audition primarily for intraspecies communication, for which sound frequencies are lower. For a cricket song with a carrier frequency of 4–5 kHz, the sound wavelength is about 8 cm, larger than the cricket. Tympanal organs on the two sides of the cricket thus receive similar amplitudes of sound pressure, regardless of source. The cricket tympanum, however, is a pressure difference receiver that makes use of a difference in phase of the sound wave with different path lengths. Sound pressure can reach the tympanum from both the outside and the inside, because the tympanic cavities of the two sides of the animal are connected to each other and to the outside by tracheal tubes. Sound reaches a tympanum by an external path and an internal path; the difference in length of these paths differs according to the location of the sound source. If the path-length difference results in a phase difference between sound pressure waves on opposite sides of the tympanum, the tympanum is vibrated by the pressure difference. Many orthopteran tympanal organs function as pressure difference receivers, and in the tested cases their directional selectivity is lost if the pathways for sound to reach the inner tympanal surface are blocked.

Most insects are thought to be tone deaf, as noted above. Their auditory organs may be selectively responsive to certain sound frequencies, but their receptors all respond to the same frequency range and so provide no frequency-specific information (just as mammalian rods provide no wavelength-specific visual information about color). Two adaptations allow some frequency differentiation in insect auditory organs. In crickets, individual receptor cells are more sharply tuned than is the tympanum; that is, the receptors are sensitive to a narrower range of frequencies than the tympanum, and different receptors have different frequencies of greatest sensitivity. The receptors are attached not directly to the tympanum, but rather onto tracheal branches indirectly linked to the tympanum. Presumably the tracheal system acts as a secondary frequency filter that narrows the frequency range of vibration reaching the receptor. In locusts, in contrast, frequency tuning is a property of the tympanum itself. The locust tympanum has a thin and a thick region, with different resonant properties. Four groups of receptor cells make complex contact with specialized attachment bodies on different parts of the tympanum, which have different resonant frequencies (rather like a Caribbean steel drum). As shown in Figure 18.51, cells in the four groups have differential frequency responses and thus can analyze sound frequency.

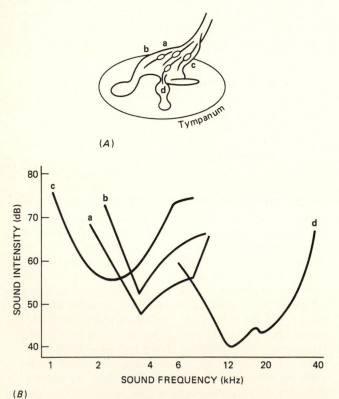

Figure 18.51 Locust tympanal auditory receptors can discriminate sound frequency because different populations of receptors are coupled to regions of the tympanum that have different resonant properties. (A) Anatomy of the tympanal organ. (B) Threshold curves of the four groups of receptors shown in (A). [After A. Michelsen, *Am. Scientist* **67**:696–706 (1979).]

Finally, we cannot leave the subject of insect hearing without mention of Johnston's organ, a complex chordotonal organ at the base of the insect antenna. (Chordotonal organs have mechanoreceptors with ciliary distal processes associated with specialized attachment structures termed scolopidia. They include elastic strand proprioceptors and tympanal receptors.) Johnston's organ is a nontympanal proprioceptive organ that is also used as a wind receptor and gravity receptor in different insect groups. In small insects, Johnston's organ can function as a displacement auditory receptor. In male mosquitoes, for example, the antenna is vibrated by sound at a resonant frequency of about 500 Hz, the flight frequency of female mosquitoes. Antennal vibration is detected by the receptors in Johnston's organ. Male mosquitoes will approach and molest an artificial sound source humming at 500 Hz. This example clearly indicates that insect auditory receptors have had multiple and opportunistic evolutionary origins.

Vertebrate Hearing   The auditory organs of all vertebrates are derived from largely homologous structures of the acousticolateralis system. The auditory receptors are hair cells similar to those of the related vestibular system and lateral line organs. In fact, as noted above, some lateral line receptors of fish may be sensitive to the near-field displacement component of sound.

Early vertebrates evolved in an aquatic environment, an environment that presents a problem for sensitive sound detection. The body of a fish is largely aqueous and therefore quite transparent to waterborne sound pressure waves. Most fish have limited auditory sensitivity, achieved by their otolith organs (saccule and lagena). Because the otoliths are more dense than the surrounding tissues, they are displaced relative to the tissues by sound pressure waves, exciting the attached hair cells. In many fish, the swim bladder functions as a pressure-to-displacement transformer of sound waves, increasing auditory sensitivity. Some fishes have evolved bony conduction links from the swim bladder to the saccule and lagena of the inner ear, an adaptation providing further increases in high-frequency responsiveness and overall sensitivity. Anuran amphibians, reptiles, and birds have an external tympanum or eardrum connected to the inner ear by a single bony ossicle, the columella. The function of this ossicle linkage is similar to that in mammals (see below). There is progressive elongation of the lagena cavity in reptiles and birds, forming a tube homologous to the mammalian cochlea and containing a basilar membrane.

The structure of the *mammalian* ear is shown in Figure 18.52. The ear consists of three parts: an *external ear* distal to the eardrum, an air-filled *middle ear*, and a liquid-filled *inner ear* or *cochlea*. Sound pressure waves vibrate the eardrum, and this vibration is transmitted to the membranous *oval window* of the inner ear by three middle-ear ossicles—the

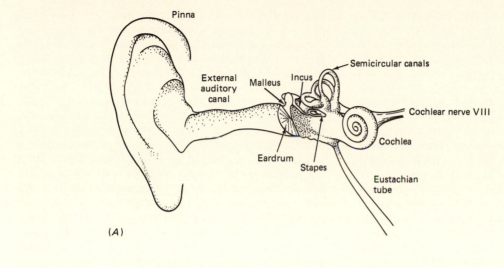

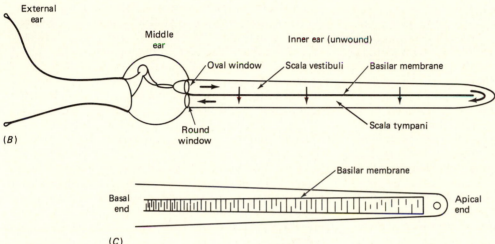

Figure 18.52 Anatomy of the human ear (A) and diagrammatic representation of how the ear would appear if the inner ear cochlea were unwound (B). (C) View of the surface of the basilar membrane (unwound). The basilar membrane is narrower and stiffer at its basal end than at its apical end.

malleus, incus, and stapes. Movements of the ossicles can be damped by two middle-ear muscles—the tensor tympani and the stapedius—protecting the membranes from damage by loud sounds. The eustachian tube connects the middle ear with the pharynx, equalizing pressure in the middle ear with environmental pressure. The cochlea or inner ear of mammals is a coiled tube that develops as an outpouching of the vestibular labyrinth (Figure 18.47). The structure of the cochlea is more easily understood if we consider it to be uncoiled, as in Figure 18.52*B*. The cochlea is longitudinally divided into two major compartments: the scala vestibuli and scala tympani (there is a third compartment, which we consider below). These longitudinal compartments are separated by the ***basilar membrane,*** which bears the auditory hair cells. Sound pressure waves are transformed into vibrations of a series of structures: eardrum, ossicles, oval window, basilar membrane, and round window (Figure 18.52*B*). This long succes-

sion of mechanical displacements points out the anatomical complexity of the ear. We can then ask: Why so complex? Of what functional significance are the various stages of transmission of mechanical displacement from the outside to the hair cells? We will now examine two aspects of auditory function that help to make sense (no pun intended) of the anatomical complexity of the ear. These functional aspects are impedance matching and sound frequency analysis.

The major function of the middle-ear ossicles is the efficient transfer of sound energy from air to the liquid of the inner ear. This process is termed ***impedance matching***. Airborne sound striking a liquid surface is almost all reflected; only about $\frac{1}{30}$ of the sound energy is transferred to the liquid. (If you stand on the shore and talk to a fish, it cannot hear you very well.) The energy transfer is poor because liquids are incompressible and so have a low volume of movement in response to sound pressure. Therefore, there must be a corresponding increase in pressure for significant transfer of energy from the eardrum (vibrating in air) to the liquid medium of the inner ear. The middle-ear ossicles apply forces from a relatively large area (the eardrum) onto a much smaller area (the footplate of the stapes, which covers the oval window). The concentration of force on a small area provides the necessary increase in pressure, allowing efficient transfer of sound energy.

Our second functional question is, How can auditory receptors respond differently to different sound frequencies? We will consider below the way in which movement of the basilar membrane stimulates the auditory hair cells located on it. First, let us examine the effects of sound waves on the basilar membrane itself. The basilar membrane varies in width and thickness along its length. It is narrow and rigid at the base of the cochlea near the oval window, and wider and more flexible at the apex, farthest from the oval window (Figure 18.52C). (The rest of the width of the cochlea is spanned by a rigid bony shelf.) In the nineteenth century, Helmholtz formulated a resonance theory of cochlear mechanics: Each portion of the basilar membrane was thought to be under tension, "tuned" like strings in a piano. Sound of a particular frequency would then induce resonant vibration of just the one part of the basilar membrane that was "tuned" to that frequency. When von Bekesy later tested this theory, however, he found aspects of it to be untrue. Using cochleas obtained from cadavers, he found that the basilar membrane was not under tension and so lacked strict resonant properties. Moreover, he found that sound induced a traveling wave that covered much of the length of the cochlea (Figure 18.53). Low-frequency sounds vibrated the entire length of the basilar membrane, with a maximum amplitude near the apex. High-frequency sounds vibrated only the region near the

base, and for every frequency there was a different place of maximum amplitude of the traveling wave along the length of the basilar membrane. This observation forms the basis of the modern *place theory* of sound frequency discrimination: The pitch or frequency of a sound is encoded by the *place* of maximum vibration (and thus maximal hair cell stimulation) over the length of the basilar membrane. The place theory is somewhat similar to Helmholtz's resonance theory in that both argue that the cochlea is *tonotopically* organized so that different places along the length of the basilar membrane respond to different frequencies. The theories differ in that the place theory postulates an apparently much broader responsive region than Helmholtz envisaged. More-recent experiments have used the Mösbauer effect to measure basilar membrane movements for lower sound intensities in living cochleas. These experiments have largely confirmed von Bekesy's observations, although some studies suggest that the basilar membrane movements are more sharply localized than indicated in his measurements.

How do movements of the basilar membrane excite the overlying auditory hair cells? Let us first examine the structure of the part of the cochlea that contains the hair cells, a region termed the *organ of Corti*. Figure 18.54 shows a cross section of the cochlea, divided by the horizontal basilar membrane. The hair cells and various accessory structures of the organ of Corti sit on the basilar membrane and vibrate up and down with it. Stereocilia of the hair cells project into a separate fluid compartment of the cochlea, the scala media. The scala media is separated from the scala vestibuli by Reissner's membrane, a thin layer of cells that is transparent to sound pressure but separates the endolymph fluid of the scala media from the perilymph of the scala vestibuli and scala tympani. The hair cells of the organ of Corti are of two kinds: Typically there are three rows of *outer hair cells* and a single row of *inner hair cells* (Figure 18.54B). Auditory hair cells lose their kinocilium during development and thus have only stereocilia. In other respects, they are similar to the vestibular hair cells described earlier. The hair cells are covered by a flap of tissue termed the tectorial membrane. The stereocilia of the outer hair cells contact the tectorial membrane, but stereocilia of the inner hair cells are coupled to the membrane only by intervening viscous fluid. Displacement of the basilar membrane is thought to generate shearing forces of the tectorial membrane on the stereocilia. The transduction mechanism of auditory hair cells is unclear but is probably similar to that of vestibular hair cells.

Intracellular responses of auditory hair cells have been recorded, a feat made difficult by the small size and inaccessibility of the cells. Inner hair cells depolarize in response to tone bursts, the amount of depolarization depending on sound intensity and frequency. Surprisingly, the hair cells respond over a narrower range of frequencies than does the under-

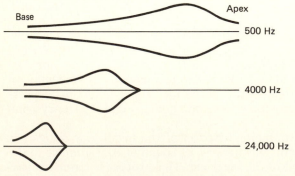

Figure 18.53 Amplitude of movement of the basilar membrane at different sound frequencies. Low-frequency sounds displace the whole length of the membrane, with a maximum near the apex. High-frequency sounds move only the basal portion of the membrane; the position of maximal movement is thus a function of sound frequency. [Adapted from G. von Bekesy, *Experiments in Hearing.* McGraw-Hill, New York, 1960.]

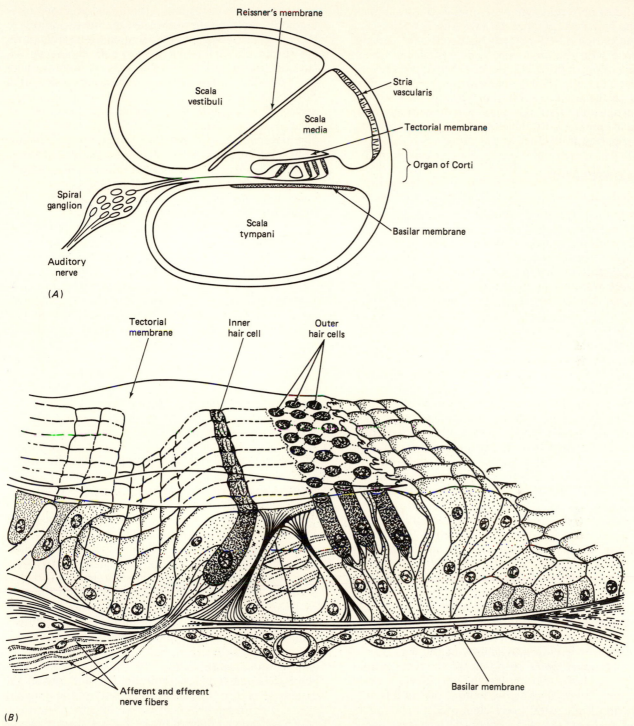

Figure 18.54 Structure of the mammalian cochlea: (A) cross section through one turn of the cochlea and (B) detail of the organ of Corti. [From ANIMAL PHYSIOLOGY, 2nd ed. by Roger Eckert and David Randall. Copyright © 1978, 1983 W.H. Freeman and Company. Reprinted with permission.]

lying basilar membrane, a finding that suggests either that the hair cells themselves have differences in preferred frequency, or that an unknown second frequency filter sharpens the hair cell response. Hair cells of reptiles achieve narrow frequency responses without an overlying tectorial membrane, as do those

of amphibians in which even a basilar membrane is lacking. Thus, von Bekesy's place theory appears to be only a partial explanation of auditory frequency selectivity.

Auditory hair cells make synaptic contact with afferent nerve fibers of the auditory (cochlear) nerve and receive efferent synapses similar to those of vestibular hair cells (Figure 18.48). Another surprising finding is that 80–95 percent of the afferent neurons synapse with the *inner* hair cells, which in humans represent only about 20 percent of the approximately 24,000 hair cells. Thus, the major source of auditory input to the brain is from inner hair cells, the more numerous outer hair cells making a numerically more modest contribution. The functional significance of this apparent imbalance is not clear.

CHEMORECEPTION

The behavior of humans is less dependent on the chemical senses than is that of many other kinds of animals. Consequently, it is easy for us to underestimate the importance of chemoreceptors as a major source of sensory input for a variety of animals. Popular accounts of bloodhounds tracking escaped convicts, moths locating mates, and salmon returning to their "home" rivers—all by means of chemoreceptors—seem fantastic to us. Our lesser reliance than other animals on chemical senses in part may result from our lesser sensitivity to chemical stimuli, and in part it may reflect Marcel Proust's observation that our behavioral use of chemoreceptors is only partly conscious.

Sensory chemoreception is only one kind of chemical sensitivity of cells. Many animal cells have evolved sensitivities to hormones, synaptic transmitters, and growth factors. Moreover, nonmetazoan organisms such as bacteria and protozoa exhibit responses to chemicals in their environment. Such cells and organisms may serve as important model systems for studying chemosensory mechanisms, but we do not consider them because they cannot be classed as differentiated chemoreceptor cells. If we exclude these nonsensory cells, we can divide chemoreceptors into four loose categories. (1) *General chemical sensors* are relatively insensitive, nondiscriminating receptors, which when stimulated elicit protective responses of the organism. For example, stimulation of the skin of a frog with salt or acid solutions activates general chemical sensory endings that elicit avoidance or wiping movements of the legs. (2) *Internal chemoreceptors* respond to chemical stimuli within the body. Examples include blood glucose receptors, internal chemoreceptors of the digestive tract, and the carotid body chemoreceptors that respond to blood oxygen concentration (see Chapter 12). (3) *Contact chemoreceptors* have relatively high thresholds and respond to dissolved chemicals from a nearby source. Contact chemoreceptors usually play a role in feeding behavior, as in the familiar example of vertebrate taste receptors. (4) *Distance chemoreceptors* or *olfactory receptors* are more sensitive than contact chemoreceptors and are adapted to respond to external chemicals from a distant source. The distinction between contact and distance chemoreception is relatively clear for animals that live

in air. In these cases, the stimuli of olfactory (distance) receptors are airborne chemicals, and the stimuli of taste or gustatory contact chemoreceptors are dissolved in liquid that contacts the receptive surface. Because stimuli of distance receptors also must dissolve in liquid at the receptive surface, this distinction is not as clear as we would like. For animals that live in water, the distinction is even less clear, since all chemical stimuli are dissolved in the aquatic environment, whether their source is distant or nearby. Nevertheless, the distinction remains useful, because many aquatic animals respond quite differently to dilute or distant stimuli than to more concentrated or proximate stimuli. Lobsters, for example, orient and search in response to chemicals at low concentrations (and presumably distant sources) that stimulate distance chemoreceptors on the antennules; high concentrations of chemicals may trigger feeding movements by stimulating "contact" chemoreceptors on the mouthparts. We now consider contact and distance chemoreceptors of insects and of terrestrial vertebrates, as examples in which the distinction is relatively clear.

Insect Contact Chemoreceptors

The contact chemoreceptors of flies are among the most fully investigated sensory cells of any animal. Insect surface receptors are organized in *sensilla,* specialized cuticular structures such as sensory hairs, pegs, plates, and pits. In flies such as the blowfly *Phormia,* contact chemoreceptors are localized in sensory hairs of the *tarsus* (terminal segment of the leg) and of the *labellum* (tip of the extensible proboscis used for feeding). Stimulation of even a single tarsal sensory hair with sugar solution elicits extension of the proboscis. If increasing concentrations of salt or of quinine are added to the sugar solution, there is increasing inhibition of proboscis extension. Stimulation of a labellar hair with sugar solution elicits drinking behavior, in which the proboscis works as a suction pump. As with tarsal hair stimulation, addition of salt or quinine to the solution stimulating the labellar hairs inhibits the behavioral effects of sugar stimulation. Blowfly contact chemoreceptors are appealing to study because they mediate these clear behavioral choices, and because it is easy to record the electrical activity of the relevant sensory neurons.

The contact chemosensory hairs ("taste hairs") of flies contain five sensory neurons. One neuron is a mechanoreceptor, the dendrite of which ends at the base of the hair. The other neurons are chemosensory and have single dendrites extending through the lumen of the hair to a pore at the hair tip. The dendrites are modified cilia, another example of sensory function of a ciliary derivative. A number of investigators, most notably Vincent Dethier, have explored the chemical responsiveness of the sensory neurons. Chemical stimuli are applied by placing a small, fluid-filled pipette over the tip of a hair, and the resultant train of impulses is recorded extracellularly via an electrode in the stimulating pipette or by a second electrode at a crack made in the side wall of the hair. Figure 18.55 shows the responses of neurons in a tarsal taste hair to different stimuli. Different sensory cells in a hair can be identified by differences in size and waveform of their extracellularly recorded impulses or spikes (see Figure 16.28). One of the sensory cells responds best to water, the response diminishing with increasing concentration of dissolved substances. Activity of this cell (termed the water cell) elicits proboscis extension and subsequent drinking if the fly is thirsty but does not induce proboscis extension in a water-sated fly. A second cell (termed the sugar cell) responds most strongly to solutions of sugars. The frequency of spikes in the sugar cell increases with increasing concentration of sugar and thus encodes information about sugar concentration. Any stimulus that elicits proboscis extension in a water-sated fly is found to activate the sugar cell above a certain minimum spike frequency. A different cell (the salt cell) responds preferentially to a range of salts, particularly to monovalent cations. Salt cell activity is correlated with

inhibition of proboscis extension, a finding that suggests that the CNS of a water-sated fly controls proboscis extension by assessing the logical function: sugar cell activity minus salt cell activity. Not all feeding deterrents activate the salt cell, however. Some may inhibit sugar cell activity directly, while some anions may stimulate the less-fully characterized fifth cell of taste hairs. In tarsal hairs of another dipteran, the apple maggot fly *Rhagoletis,* a cell that may correspond to the blowfly fifth cell responds to an oviposition-deterrent pheromone. This and other examples suggest that some insect contact chemoreceptor neurons may convey information about specific pheromones or food plant substances, rather than simply signaling nonspecific categories such as sugars and salts.

Insect contact chemoreceptors demonstrate several important features of sensory coding. The impulse frequency of a receptor response codes stimulus magnitude (chemical concentration). Each cell has a characteristic spectrum of sensitivity and acts (to some degree) as a *labeled line* for a stimulus quality (e.g., sugar) in the same way that different visual cones are labeled lines signaling colors. Chemoreceptor response spectra can be broad or narrow, and a stimulus may excite several cells, the activities of which must be integrated centrally. Most chemosensory systems are thought to code stimulus quality and magnitude in ways similar to this, but in other systems coding is more complicated and less clear, as we shall see below.

Insect Olfactory Receptors

The distance chemoreceptors of insects can be termed olfactory receptors, since they respond to airborne odorants at low concentrations. The distinction between olfaction and taste is an equally valid one for insects and for terrestrial vertebrates, although the receptors of the two groups are not homologous. Insect olfactory receptors are localized in a variety of sensilla and are concentrated on the antennae. The internal structures of olfactory sensilla are similar to those of taste hairs, although olfactory sensilla may contain up to 50 or more sensory neurons. The ciliary dendrites of olfactory neurons may branch within the sensillum to end at or under pores in the cuticular wall of a peg. Odorant molecules are thought to adsorb to the hair surface, diffuse through the pores, and bind to sites on the dendrite membrane to stimulate a receptor cell.

Olfactory sensory processes have primary roles in many aspects of insect behavior, especially among the social insects. One important kind of use of olfactory receptors is in response to pheromones. A **pheromone** is a metabolically produced chemical that is released to the outside world by one individual and elicits specific behavioral or systemic responses when detected by other individuals of the same species. Various insects have sex attractant (and deter-

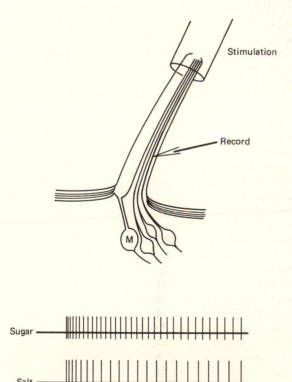

Figure 18.55 Responses of a blowfly contact chemoreceptor sensillum to different stimulatory solutions. One of the three chemosensory neurons is most responsive to one of the three kinds of stimulation shown.

rent) pheromones, oviposition-deterrent pheromones, trail-making pheromones, alarm pheromones, and colony-recognition pheromones. The
best studied examples of olfaction of pheromones
concern sex attractants of moths. Female moths
release pheromones, which, when detected by
males, induce the males to fly upwind to find the
females. Males are extraordinarily sensitive to the
female attractant and can be attracted over distances
of miles. A female gypsy moth produces 1 mg of
disparlure, its sex attractant. This amount is sufficient to attract one billion males if efficiently distributed.

The receptors for sex pheromones and other odors
have been studied in many insects. We will discuss
the silkworm moth, *Bombyx mori,* the first insect for
which a sex attractant pheromone was characterized.
Male *Bombyx* have large pinnate or comblike antennae with up to 50,000 hair sensilla responsive to the
sex attractant pheromone, while females have simple
antennae that lack hair sensilla (although they have
other types of olfactory sensilla—see below). Because of the elaborate branching and the density of
pheromone-sensitive olfactory hairs, the antennae of
male *Bombyx* are estimated to catch up to one-third
of the odor molecules in the air passing through
them. The pheromone receptor cells are exquisitely
sensitive and very specific in their responses. The
major sex attractant pheromone in *Bombyx* is bombykol, a 16-carbon unsaturated alcohol (Figure
18.56). The hair sensilla contain a sensory neuron
specifically sensitive to bombykol. Electrophysiological recordings indicate that binding of a single molecule of bombykol elicits sufficient depolarization to
generate an impulse. The threshold for a behavioral
response of a male moth occurs with activation of
about 200 receptor cells per second, which occurs at
a phenomenally low concentration of 1000 molecules/
cm^3 of air! Females, in contrast, completely lack
receptors sensitive to bombykol.

Female *Bombyx* release small amounts of bombykal (the corresponding aldehyde), along with bom-

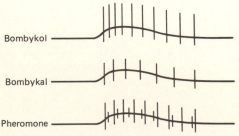

Figure 18.56 Responses of olfactory neurons in the moth,
Bombyx mori, to bombykol and bombykal. Separate neurons respond to the two compounds, both of which are
released by female moths (see text). Bombykol, the specific
attractant pheromone, has the chemical formula
$CH_3CH_2CH_2CH=CH—CH=CH(CH_2)_8CH_2OH$ (*trans*-10, *cis*-12-
hexadecadien-1-o1). Bombykal has the chemical formula
$CH_3CH_2CH_2CH=CH—CH=CH(CH_2)_8CHO$.

bykol. Bombykal activates separate receptor cells in
males from those that respond to bombykol (Figure
18.56), a finding that demonstrates the specificity of
response of pheromone receptors. Surprisingly, activation of bombykal receptors centrally inhibits the
moths' behavioral response to bombykol. What is
the adaptive value of such an inhibitory response to
a component of the female's emission? When sex
attractant pheromones were first discovered, it was
supposed that each species would employ a single,
species-specific kind of molecule. The reality is more
complicated, however. Females of many species release a mixture of two or more compounds, and the
same compound may be released in more than one
species. Bombykal is a putative sex attractant in
another moth species (the tobacco hornworm moth
Manduca sexta). Thus, the inhibitory response to
bombykal in *Bombyx* may have evolved to suppress
attraction of *Bombyx* males to females of other species.

Not all olfactory receptors of insects, of course,
respond to pheromones. Moths of both sexes possess
olfactory sensilla sensitive to other odorants, and
honeybees can behaviorally discriminate hundreds of
kinds of odorants. Although the response spectrum
of each species is different, a few generalities emerge
from a wide range of studies of insect olfaction. The
most important generalization is that olfactory receptor cells can be either *odor specialists,* responding
to a narrow spectrum of kinds of chemicals, or *odor
generalists,* responding to a broad spectrum. The
bombykol and bombykal receptors we considered
above are good examples of odor specialist cells. In
addition to such pheromone specialist cells, food
odor specialist cells have also been described. Odor
generalist cells typically respond to a broad spectrum
of plant and food odorants, different receptors having
different relative sensitivities. For example, two generalist receptors of a moth may respond to geraniol,
citral, and other odorants, but one may be most sensitive to geraniol and the other most sensitive to
citral.

Odor specialist and odor generalist receptors
clearly provide different kinds of information to the
central nervous system. Odor specialists act as *labeled lines;* that is, any activity in, say, a bombykol
receptor neuron can be reliably interpreted as indicating the presence of bombykol, since the receptor
is largely insensitive to anything else. Odor generalist
neurons, in contrast, are not specifically labeled
lines, because any of a number of odorants can elicit
their activity. For the central nervous system to interpret sensory activity correctly as the presence of,
say, geraniol, it must compare activities of many
receptors rather than just monitoring a specifically
labeled line. Generalist receptors are said to convey
information about stimulus quality (e.g., the kind of
odorant) by the *across-fiber pattern* of sensory activity. The distinction between labeled lines and across-
fiber pattern is of general importance in sensory cod-

ing, but it is a difference of degree. In color vision, a red cone can be viewed as a labeled line, but it has a rather broad response spectrum. A particular ratio of activation of red, green, and blue cones (an across-fiber pattern) is interpreted as orange light. Across-fiber patterning is especially important in chemoreception, in which the range of stimulus qualities (e.g., kinds of odorants) is too great to allow narrow-spectrum specialist receptors and labeled lines for each kind. We shall see other examples of this principle in chemosensory systems of vertebrates.

Vertebrate Taste

When we speak of the taste of food, most of the complex perception of flavor to which we refer is actually mediated by olfactory processes, rather than by taste receptors. We have little awareness of the distinctness of olfactory and taste sensory systems. It is probably for this reason that even the central organizing principles of taste physiology are products of this century. The most fundamental of these organizing principles, derived from psychophysical studies, is that there are *four basic taste qualities—sweet, sour, salty, and bitter*. All true taste perception (as opposed to olfactory perception of flavors) in humans is thought to involve these four taste qualities, singly or in combination. The sensory physiology of taste in other vertebrates is thought to be broadly similar to that in humans.

Vertebrate taste receptor cells are grouped together in **taste buds**. In higher vertebrates, taste buds are confined to the tongue and back of the mouth, but some fish have taste buds over much of the body surface. The structure of a typical taste bud is shown in Figure 18.57. There are at least three types of cell in a taste bud, all derived from epithelial cells. The taste receptor cells surround a shallow lumen or taste

pore and have extensions into it that are presumably the sites of sensory transduction. Supporting cells are interspersed between the taste cells and appear to secrete fluid into the lumen. Basal cells, the third type, are derived from the epithelium adjacent to the taste bud and are precursors of the taste cells. Studies with radioactive tracers have shown that individual taste receptor cells have lifetimes of only about 5–10 days and are constantly being replaced by new receptor cells derived from the basal cells. Because the receptor cells are epithelial rather than neural, they have no axons. Instead, the receptors make synaptic contact with sensory neurons in the taste bud, in the same way that hair cells of the acousticolateralis system synapse with sensory neurons. As a result of the high rate of turnover of taste receptor cells, the synapses between these cells and their sensory neurons must be continuously broken and reformed.

Taste buds of the human tongue have different distributions for different taste qualities, as determined by tests with small drops of solutions confined to localized regions of the tongue. The front of the tongue is most sensitive to sweet substances, with successively posterior zones of greatest sensitivity to salty, sour, and bitter tasting substances. From these and other psychophysical studies, we would expect that individual taste receptor cells should respond specifically to stimuli that evoke only one of the four perceptual taste qualities. This expectation, however, is not completely met. Recordings from individual taste receptor cells and sensory axons show that each receptor responds preferentially to stimuli of one of the taste qualities but also responds to other kinds of taste stimuli at a lower sensitivity. Figure 18.57 shows the responses of a single receptor cell, recorded intracellularly. The cell responds to stimulation with a graded depolarizing receptor potential, which excites sensory axons that synapse with the cell. The cell responds most strongly to salt but also depolarizes in response to other stimuli. We can characterize the cell as a ''salt-best'' cell, but any information it provides to the central nervous system is ambiguous. This finding is typical of most studies of vertebrate taste receptors. Most receptor cells (and sensory axons) respond preferentially to stimuli of one of the taste qualities, but with some response to the others. Therefore, the coding of taste quality must be by the across-fiber pattern of incoming signals; the brain must integrate the *relative* activities of salt-best cells, sugar-best cells, and so on to assess quality.

Taste receptors have much higher thresholds than do olfactory receptors. In humans, NaCl solutions must be at least 0.05 M to taste salty, and sucrose must be at least 0.17 M to taste sweet. Thresholds are lower for acids (0.001 M) and much lower for quinine (3 × 10^{-5} M). These values are still much higher than for olfaction, although some fishes have taste bud responses to amino acids at very low

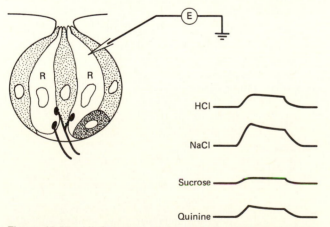

Figure 18.57 (*A*) Structure of a mammalian taste bud, showing receptor cells (R), supporting cells (shaded), and a basal cell (stippled). (*B*) Intracellular recordings from a single receptor cell. The cell responds most strongly to NaCl but also responds to other taste stimulants.

thresholds $(10^{-8}-10^{-10}\ M)$ that are comparable to their olfactory thresholds. The mechanism of transduction in taste receptors is unknown, as is the case for all chemoreceptors. Molecules or ions are presumed to bind to membrane-bound proteins and cause the opening of ion channels, depolarizing the cell in a manner analogous to the action of acetylcholine receptor proteins (Chapter 17). Characterization of taste receptor mechanisms may be aided by the discovery of specific ligands for the receptor sites. For example, gymnemic acid specifically blocks sweetness, while miraculin (a glycoprotein of the "miracle fruit" *Synsepalum*) makes normally sour substances taste sweet, without affecting responses to salty, bitter, or sweet substances.

Vertebrate Olfaction

Vertebrates can employ olfaction not only in feeding behavior but also for orientation and for social behavior. Vertebrate pheromones, although not as well characterized as those of insects, are implicated in some species in reproductive and maternal behavior, territoriality, and dominance. Olfactory systems are characterized by great sensitivity and by large numbers of qualitatively distinguishable odors. It is difficult to determine just how many kinds of odors humans and other animals can distinguish, but most estimates are in the hundreds. Any conversation with a perfumer or a wine connoisseur will convince you of both the range of discriminability of human olfaction and the difficulty in talking about it. The sensitivity of human olfaction is also quite impressive, but not nearly as much so as that of some other vertebrates and invertebrates. For example, the human behavioral threshold for ethyl mercaptan in inhaled

air is $7 \times 10^{-1}\ M,$ or about 10^8 molecules/mL. For butyric acid, thresholds in molecules/mL are: for human—7×10^9; for *Triturus* (amphibian)—4×10^9; for dog—9×10^3; and for bee—1.1×10^3. The physiological bases of this extraordinary sensitivity and of the discrimination of so many odor qualities are very poorly understood.

The anatomy of the olfactory system is rather similar for all vertebrates. The olfactory receptive surface, termed the olfactory epithelium, lines part of the internal nasal cavity. The area of the nasal mucosa that constitutes the olfactory epithelium varies greatly among species, being only 2–4 cm^2 in humans but 18 cm^2 in dogs and 21 cm^2 in cats. Humans have an estimated 10^7 olfactory receptor neurons and dogs have 2×10^8. In contrast, the mammalian auditory nerve contains only 3×10^4 axons and the optic nerve only 10^5 axons.

Olfactory receptors are neurons, unlike epithelial taste receptors. Each olfactory receptor is a bipolar sensory neuron with a perikaryon in the olfactory epithelium (Figure 18.58). A single, narrow dendrite extends from the perikaryon to the mucus-covered epithelial surface and ends in a knob that projects into the overlying layer of mucus (secreted by supporting cells and by underlying glands). From this knob project a number of olfactory cilia (typically 8–20 in mammals) that extend as an intermeshed feltwork in the mucous layer. The membranes of these cilia are thought to be the site of receptor transduction. The receptor neurons have fine, unmyelinated axons that extend a short distance to the olfactory bulb of the forebrain, where they synapse with second-order cells (see Figure 17.26). The receptor axons are typically only 0.2 μm in diameter, among the smallest axons in the nervous system. Olfactory neu-

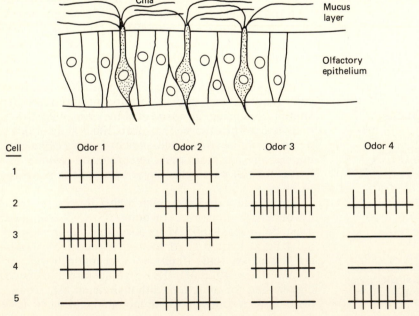

Figure 18.58 Vertebrate olfactory receptors. The olfactory receptors are small bipolar neurons, the sensory cilia of which extend into the overlying mucus layer. The neurons appear to respond as odor generalists, each cell responding to more than one kind of odor. Thus, the central nervous system may have to integrate a complex across-fiber pattern of receptor activities to discriminate odor quality.

rons undergo continuous turnover throughout adult life and are the only mammalian neurons known to do so. The receptor neurons differentiate from basal cells in the epithelium, develop dendrites and axons, and have life spans of about 60 days before they degenerate.

The excitation of olfactory receptors is poorly understood. Inhaled air is warmed and moistened by passage along a respiratory pathway that can be quite elaborate in some species. (In aquatic vertebrates, of course, the incoming medium is water.) The air (or water) passes across the olfactory epithelium, and odorant molecules are absorbed into the mucous layer, where they are presumed to bind to acceptor sites on the surface membranes of olfactory cilia in the outer mucous layer. Intracellular recordings of receptor activity are nearly impossible, because of the very small size of the cells. Consequently, we are dependent on extracellular recordings for any attempt to make sense of olfactory responses. Most of these extracellular recordings are not very helpful; because of the same small size of the somata and their axons, one usually records average activities of many cells. The few studies that allow us to distinguish responses of individual olfactory neurons all indicate that the cells act as odor generalists, each cell responding to many kinds of stimulating odors (Figure 18.58). Thus, the brain must integrate relative activities of very many receptor neurons to discriminate odors.

Theories of Coding of Olfactory Quality The psychophysical view of vertebrate olfaction is that vertebrates, including humans, can distinguish hundreds of kinds of odors and can make fine discriminations among similar odor mixtures, all at high levels of sensitivity. Physiological studies provide little evidence about how such discriminations can occur. Let us examine two kinds of theories about how information about qualities or kinds of odors may be encoded in the olfactory system. The first is termed the *stereochemical theory*. This theory originally postulated that odors fall into seven odor types (musky, floral, camphoraceous, minty, ethereal, pungent, and putrid). Molecules in each category tend to have similar sizes and shapes and were envisioned to interact stereochemically with corresponding membrane binding sites, in a manner analogous to substrate binding at an enzyme active site. More recent studies suggest that there may be 27 or more odor types, rather than seven. Moreover, if receptor cells are odor generalists with each responding to a wide odor spectrum, each cell presumably has several types of binding sites for different odor types. *Place theories* suggest that different *regions* of the olfactory epithelium may respond preferentially to different odor types, because of receptor differences between regions and/or because the patterns of air flow across the epithelium may separate odorants, as in a gas chromatograph. There are other theories of olfactory function, but none seems completely satisfactory at present. All are limited by the difficulty of recording activity of individual receptors. Moreover, the available evidence of broad generalist response properties of the receptors suggests that the receptors place an enormous burden on the brain to decode the properties of an olfactory stimulus by integrating a very large across-fiber pattern of activities of nonspecific receptors. One cautionary note to consider in contemplating this unsatisfying conclusion is that of sampling bias: Extracellular recording techniques select for larger cells. It is possible that some olfactory neurons are more specific in their response properties than are the units recorded but are missed because of their small size. In any case, it is fair to conclude that we know less about the function of vertebrate olfactory systems than about nearly any other aspect of nervous system activity.

SENSORY GUIDING OF BEHAVIOR: ORIENTATION AND NAVIGATION

In this chapter, we have examined two kinds of questions about sensory processes: *mechanism* questions about how receptors work and *function* questions about the roles of sensory input in the natural behavior of animals. We have discussed the major sensory modalities, but the variety of kinds of receptors in different animals is enormous. Many animals, for example, respond to changes in environmental temperature, osmotic pressure, or humidity. Rather than survey these and other remaining sensory processes, we will instead examine the roles of different senses in animal orientation. Some of these functional roles depend on sensation of kinds of environmental stimuli that are undetected and unfamiliar to humans, such as polarized and ultraviolet light, ultrasound, and electric and magnetic fields. We shall see that there is great variety both in the degree of complexity of orientation behavior and in the kinds of sensory cues that animals use for orientation and navigation.

Simple Reflexive Orientation: Taxes and Kineses

All animals perform simple, reflexive movements that orient them in their environments. A reflexive orienting movement toward or away from a form of stimulus energy is termed a *taxis*. The water flea *Daphnia,* for example, swims in ponds and lakes and feeds on algae. It tends to swim upward against the force of gravity (negative geotaxis) and also toward light (positive phototaxis). Such taxes are common in planktonic animals, keeping them near the water surface in the daytime. In some taxes, the animal acts to equalize stimulation of right and left receptors; for example, grayling butterflies escape predators by flying directly toward the sun, but if one eye is experimentally blinded, they fly in circles because they cannot match the light input to the two eyes.

Animals can also reflexively achieve the effects of

orientation without making directionally orienting movements relative to a stimulus. Such responses, termed *kineses,* consist of changes in activity and random, unoriented changes in direction with stimulation. Paramecia congregate in zones of mild acidity in water (e.g., around a bubble of CO_2), as a result of swimming more slowly when in their optimal pH zone and turning in random directions when they swim into a region in which the pH is higher or lower than the optimum. Similarly, sow bugs (isopod crustaceans that live in humid, dark conditions under logs) increase their rate of undirected movement in dry and light conditions but slow or stop if they encounter a moist, dark spot.

Kineses and taxes are characteristic of simple organisms but are not necessarily confined to them. Elements of these simple reflexive behaviors are described in complex animals including vertebrates. For example, a mouse tends to move along a wall, particularly when exploring unfamiliar territory. This tendency has been termed positive thigmotaxis or orientation to tactile contact. Most such cases in complex animals, however, involve greater behavioral complexity than true taxes and kineses. Older descriptions of vertebrate taxes may thus represent overzealous applications of the reductionist approach to animal behavior that was popular early in the twentieth century.

Spatial Orientation

Animals employ a wide variety of sensory information for orientation within the environment. We have already mentioned orientation with respect to gravity and orientation upwind by insects in response to pheromone stimulation. Many animals possess an alerting response, turning toward a sudden visual, auditory, or other stimulus. The tendency of animals to track a moving visual stimulus, termed *optokinetic nystagmus,* has widely been used as a behavioral measure of visual perception. In this section, we will consider two examples of spatial orientation that involve sensory abilities foreign to humans: echolocation in bats and orientation to electric fields in fish.

Echolocation Many kinds of bats have poor vision and yet fly well at night, avoiding obstacles and catching insects at high rates in profound darkness. They orient by emitting ultrasonic pulses (i.e., sound at frequencies too high to be audible to humans) and detecting echoes reflected by objects around them.

The Italian naturalist Lazzaro Spallanzani provided the first evidence that "bats see with their ears" in 1793. Spallanzani placed hoods over the heads of bats and found that they were disoriented, flying into walls. When the bats were blinded, however, they flew with normal orientation and were still able to catch insects. In an ingenious set of experiments, Spallanzani plugged the ears of bats with either hollow brass tubes or tubes filled with wax. The bats with wax-filled tubes were disoriented, but those with hollow tubes (through which sound could pass) flew normally. Spallanzani correctly concluded that the ears were necessary for oriented flight in the dark, but their eyes were not. Although Spallanzani's results were confirmed and extended by Jurine in 1794, they were disbelieved and became buried in obscurity. Why were Spallanzani's elegant and surprisingly "modern" experiments given so little credence? We must realize that in the eighteenth century the wave nature of sound was not understood, and thus the concept of sound too high in frequency for humans to hear was inconceivable. In fact, there was no general understanding of the possibility that animals had sensory capabilities that humans lacked. "Spallanzani's bat problem" remained unexplained until 1938, when Donald Griffin, using newly developed ultrasound detectors, showed that bats emitted high-intensity ultrasound in the frequency range of 30–100 kHz. Griffin subsequently independently confirmed Spallanzani's experiments and demonstrated that bats oriented by detecting the echoes of their ultrasonic cries.

Even with our present, relatively good understanding of the phenomenon, bat echolocation is an amazing display of sensitivity and precision. Most species of bats emit ultrasonic cries, typically as pulses of either constant frequency (CF bats) or decreasing frequency within each pulse (frequency-modulating or FM bats). The little brown bat *Myotis lucifugus* is an example of an FM bat. A cruising *Myotis* emits pulses that sweep in frequency from 80 to 40 kHz. The sound energy of these pulses is an enormous 120 dB, equivalent to a jet plane taking off 100 m away. Bat cries would be nearly deafening to humans if their sound energy was within the human auditory frequency range. Despite the intensity of the cries, the echoes that bats receive are extremely faint. Echolocating bats thus must detect and orient to faint echoes that arrive within 20 ms after emitting a potentially deafening cry. Several physiological and anatomical specializations contribute to this ability. Because the inner ear is mechanically isolated from the rest of the skull, bone conduction of the cry to the ear is decreased. Auditory sensitivity is effectively decreased during a cry by contraction of tensor tympani muscles (p. 493); the recovery of sensitivity after the cry is extremely rapid. Some bats have a selective short-term enhancement of auditory sensitivity during a period 2–20 ms after calling, appropriate to detect echoes returning from an object 34–340 cm away. Constant-frequency (CF) bats make use of the Doppler shift of sound to enhance echo detection. If the bat is approaching an object, the echoes returning from the object are shifted to a higher frequency by the Doppler effect—the same phenomenon that causes the familiar drop in frequency when an approaching train passes and the sound source begins to recede. The ears of CF bats are sharply tuned to a frequency several kilohertz higher than their cries. Thus, their auditory sensitivity is much greater to Doppler-shifted echoes (from an object being approached) than to the cry itself.

Bats, of course, must not only detect echoes but also determine their direction with great accuracy. A number of auditory specializations contribute to this localization. For example, contralateral inhibition is well developed in bat auditory centers. Many central auditory neurons are excited by sound stimulation of one ear and inhibited by sound stimulation of the other ear. These neurons are sensitive to differences in latency of stimulation of the two ears as small as 0.1 ms. This ability to determine minute differences in response latency of the two ears plays a major role in sound localization. Several other auditory mechanisms contributing to echolocation have been explored. No matter how much we learn about the mechanisms of bat echolocation, however, the performance itself remains amazing. Many bats can echolocate insects as small as mosquitoes and fruit-flies and catch them at rates of two captures per second!

Echolocation is not confined to bats. Central American oilbirds orient through caves by echolocation, and dolphins employ sound pulses for echolocation as well as for communication.

Electro-orientation   Several kinds of fish possess electric organs (see p. 540). The strong electric organs of electric eels, rays, and catfish produce a powerful discharge used to stun prey and potential predators. Other less well-known groups of fish have weak electric organs, which, together with *electroreceptors* on the body surface, are used for orientation in the murky tropical waters that these fish inhabit. The mode of action of electro-orientation in weakly electric fish was first suggested by Lissmann's behavioral studies in 1958 and was later confirmed physiologically. Three groups of teleosts (gymnotids, mormyrids, and gymnarchids) can orient by producing electric fields around themselves (by discharges of their electric organs) and by detecting the perturbations of these fields by nearby objects. Figure 18.59 shows the electric field set up by electric organ discharge in *Gymnarchus*. An electrical conductor such as a metal ball or plate concentrates the lines of electric force, increasing the field density at the nearby surface of the fish. Poor conductors such as glass or plastic conversely spread the lines of force and decrease the electric field density. These perturbations are sensed by electroreceptors in tuberous organs that are part of the lateral line system on the sides of the fish. Although this system would appear relatively insensitive, the fish can use it to make impressive discriminations. For example, a *Gymnarchus* was trained to discriminate between two porous clay flowerpots, one containing a glass rod 2 mm wide and the other containing a glass rod 0.8 mm wide!

Electroreceptors and electric organs of fish are also used in other ways than the electric field orien-

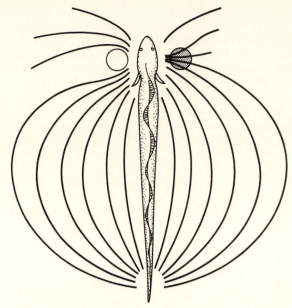

Figure 18.59 Electric field around a weakly electric fish (*Gymnotus*). The electric field is set up by the electric organ in the tail and is sensed by anterior electroreceptors. A conductile object (right side) concentrates lines of current through it, increasing the field density.

tation described above. Weakly electric fish also employ their electric organs and electroreceptors for social communication. Moreover, some other fishes that lack electric organs nevertheless have electroreceptors, which they use in locating prey and perhaps in other ways. For example, elasmobranch fish such as sharks and skates possess sense organs termed ampullae of Lorenzini, which are associated with their lateral line systems. Ampullae of Lorenzini are sensitive to a wide variety of stimuli (touch, temperature, salinity, and electrical stimuli). Behavioral and physiological tests indicate that the adequate stimulus is electrical and that sharks can detect voltage gradients as small as 0.005 V/cm. Calculations suggest that this electrical sensitivity would allow a shark to detect a camouflaged or buried flounder at 30 cm, by sensing the electrical activity of its respiratory muscles. Sharks will attack buried electrodes rather than ground flounder extracts, a finding that suggests that electroreception is in fact used for prey detection. Other calculations suggest that a shark swimming east or west through the earth's magnetic field will generate an induced current strong enough to stimulate the ampullary electroreceptors, but it is not clear whether sharks use this potential information for directional navigation.

Animal Navigation

Many animals employ their senses not only to orient within their immediate environment but also to navigate over long distances with impressive precision. Familiar examples include the return of foraging honeybees to their hives, the long-distance migration of many birds between summer and winter ranges, and homing in pigeons. In many circumstances, familiarity with the terrain can be ruled out as the orienting mechanism, so that the performance requires true navigation—calculation of present position, goal position, and the correct direction of movement, all from nonlocal sensory cues. We will confine our attention to the ways in which animals use sensory information to determine the correct direction of orientation, and we will pay only lip service to the intricacy and behavioral complexity of the navigational performance itself.

Landmark Recognition and Random Search   Aspects of animal homing behavior depend on sensory recognition of local landmarks, but dependence on local sensory cues is often only part of a more complex navigation system. A classic example of the use of visual landmarks comes from tests by Tinbergen on the female digger wasp (*Philanthus triangulum*). Tinbergen placed a ring of pinecones around the nest entrance, a simple hole in the ground. When the wasp emerged she flew around the nest for about 6 s before disappearing to hunt food. Tinbergen then moved the ring of pinecones about 30 cm while the wasp was away. The returning wasp flew to the ring of pinecones and was completely unable to find the nest. If, however, the ring of pinecones was moved farther than 1–2 m from the nest, the wasp was unable to locate the ring. This latter observation clearly demonstrates that only the final localization of the nest depends on landmark orientation; other cues must be used to return to the correct neighborhood.

Sensory recognition of landmarks, of course, need not be visual. One impressive example of homing that involves chemosensory cues is the spawning migration of salmon. Salmon are anadromous (see p. 177). Young salmon hatch in freshwater streams and migrate to the ocean to grow to adulthood. The mature salmon always return to spawn in the same stream from which they hatched. In one study, nearly 11,000 tagged salmon were recovered from the same stream they had left several years before, while none of the tagged salmon was found in any other stream. Much of the sensory basis of locating the home stream appears to be olfactory. Salmon can be conditioned to discriminate water from different streams and containing different plant products. Moreover, young salmon at the stage they begin to migrate downriver can be imprinted to chemicals such as morpholine that are not normally present in streams, learning to recognize and respond to the chemicals as sensory cues. Adults that were previously exposed to morpholine (but not untreated controls) will enter a stream to which morpholine has been added experimentally.

Other studies stress the roles of different chemical cues, such as pheromone differences between geographically isolated races of salmon. In any case,

local chemical stimuli clearly play a major role in the stream–selection phase of the migration.

Navigation in Bees   Many animals use the position of the sun as a compass to guide their oriented movements. The sun moves across the sky from east to west at 15° per hour. Therefore, animals that know the time of day can determine an appropriate heading from the sun's position. (Actually, this calculation is quite complex and also requires information concerning latitude and season.) One of the first demonstrations of this kind of orientation (termed **sun-compass orientation**) was for foraging bees. Bees that find a source of nectar can fly directly and accurately back to the hive, where they perform a dance that tells other bees the approximate distance of the nectar source and its direction as an angle relative to the sun's position. Other bees, using this information, can fly at the correct angle to the sun, directly to the nectar source. The bees accurately compensate for the change in solar position over time, by using their internal circadian clock (see Biological Clocks in Chapter 2). Even when the sun's position is obscured by clouds, bees can correctly indicate the direction of the nectar source as long as a patch of blue sky remains. Bees and other insects are sensitive to ultraviolet light and can also detect the direction of polarization of light diffracted by particles in the air. This direction of polarization is always at right angles to a plane that passes through the sun's position (as well as the positions of the observer and the point in the sky being observed). Bees can use the direction of polarization of ultraviolet light to determine the sun's position (for sun-compass navigation) when the sun itself is obscured. Thus, bees possess a backup navigational system that depends on two aspects of electromagnetic radiation (ultraviolet light and polarization) to which humans are insensitive.

It should be clear from the foregoing discussion that bees can perform complex navigational tasks even under conditions in which we would be quite unable to orient. The navigational performance of bees can be even more sophisticated, involving aspects that are beyond our scope but are discussed in many textbooks on animal behavior.

Bird Navigation: Migration and Homing   Many kinds of birds migrate long distances between summer breeding grounds and winter ranges, often returning to the same area year after year. For many species, the migratory routes are along geographic features that serve as learned landmarks or are genetically preprogrammed. Other species, however, require true navigational abilities of surprising precision. For example, the Pacific golden plover flies nonstop from Alaskan summer breeding grounds to winter grounds in the Hawaiian islands, a distance of several thousand kilometers. The migratory route is nearly entirely over featureless open ocean, and the birds cannot land on water. Consequently, there is strong selective pressure for accurate navigation! (cf. p. 69)

Such bird migrations and the homing ability of pigeons require true navigation, based on both a map sense that tells the animal its location and the location of the goal, and a compass sense that tells the animal in which direction to travel. The basis of the map sense is poorly understood; we will concentrate on experiments that explore the sensory basis of the compass.

The primary means of determining direction in birds is a sun compass like that used by bees. Many experimenters have examined birds' choices of direction, either by tracking the movements of migrating or homing birds or by measuring the strong orientation preferences of caged birds in migratory phase. Most birds use the sun's position and their internal clock to determine which direction is (say) south. For caged birds, changing the apparent direction of the sun with mirrors causes an appropriate change in orientation. A more rigorous test of sun-compass orientation is to reset the bird's circadian clock by imposing an artificial light–dark cycle different from natural sunrise and sunset (see Chapter 2). A bird whose clock is set ahead 6 h, when again exposed to the natural sun, will misinterpret the sun's position as (say) the 2 p.m. position when in fact it is 8 a.m. Therefore the bird will orient in a direction that deviates from the correct direction by 90° (Figure 18.60A). Such experiments show that homing pigeons and virtually all diurnal migrants use sun-compass orientation as the primary mechanism of navigation.

Sun-compass orientation is not the only directional cue, however. Figure 18.60B shows that on an overcast day when the sun is not visible, control birds still orient correctly and clock-shifted birds *also* orient correctly. Thus, the birds have a backup system *not* based on solar position but appear to use it only when the sun is not available. We will return to this backup system in a minute, after considering nocturnal migrants.

Many bird species migrate at night, when solar position is obviously unavailable as a source of directional information. Experiments have convincingly demonstrated that nocturnal migrants such as indigo buntings use the pattern of stars in the night sky as a star compass to determine their heading. Caged birds able to see the night sky orient well only on clear nights when the northern constellations are visible. Moreover, caged birds in a planetarium orient relative to the projected position of the northern constellations even when the stars are projected in an incorrect direction. For example, if the northern sky is projected in the south side of the planetarium, the birds' orientation is reversed by 180°. The young birds must learn the constellations within 35° of Polaris (the Northern Star) during a critical period prior to their first migration in order to orient successfully.

What is the backup system that birds use when

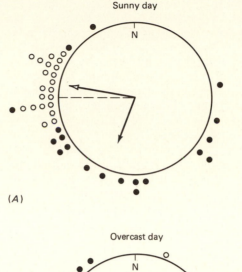

Sunny day

(A)

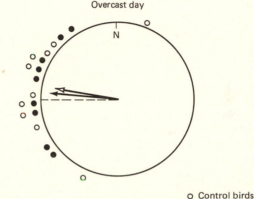

Overcast day

(B)

○ Control birds
● Clock–shifted birds

Figure 18.60 Demonstration that birds use sun-compass orientation as a primary orienting mechanism, but also have a backup system to use when the sun is hidden. (A) On sunny days, pigeons depart from a release point in the homeward direction (dotted line). Birds with their internal clocks shifted by 6 h misinterpret the sun's position and depart 90° left of the homeward direction. (B) On overcast days, both control birds and clock-shifted birds depart in the homeward direction, a demonstration that they use a backup mechanism rather than sun-compass orientation. Arrows indicate the mean direction of departure of the group; arrow length is a measure of statistical significance of the orientation. [From W.T. Keeton, *Science* **165**:922–928 (1969). Copyright 1969 by the AAAS.]

solar or stellar cues are obscured? Recent studies indicate that birds and other animals can sense the earth's magnetic field and orient to it when other cues are lacking. Pigeons with magnets attached to their backs or heads can home normally on sunny days but are disoriented on overcast days; control pigeons with brass bars instead of magnets home normally on both sunny and overcast days. Presumably, the magnets disrupt sensing of the earth's magnetic field. This view was confirmed by attaching small Helmholtz coils to the pigeons, thereby inducing a magnetic field that overrides the earth's field. Reversing the direction of the artificial field (by reversing the battery) reversed the direction of orientation. Similar experiments with large Helmholtz coils around the cages of European robins demonstrated that the robins' orientations were shifted by changing the direction of the artificial field.

Other organisms including bees and bacteria have been shown to orient to magnetic fields. The sensory basis of this performance is not clear, but crystals of the ferromagnetic mineral *magnetite* have been found in the bacteria, in the honeybee thorax, and in the pigeon head. These internal magnets are thought to be involved in magnetic orientation in some way that remains to be clarified.

From the foregoing examples, it is clear that sensory processes play many important roles in animal behavior, often in ways that seem wondrous to our own limited human sensory repertoire. It is clear that behavioral studies are important for further expansion of the range of our knowledge of animals' sensory capacities.

SELECTED READINGS

Ache, B. and C.D. Derby. 1985. Functional organization of olfaction in crustaceans. *Trends Neurosci.* **8**:356–360.

Besson, J.-M. and A. Chaouch. 1987. Peripheral and spinal mechanisms of nociception. *Physiol. Rev.* **67**:67–186.

Bolis, L., R.D. Keynes, and S.H.P. Maddrell (eds.). 1984. *Comparative Physiology of Sensory Systems.* Cambridge University Press, Cambridge.

Bullock, T.H. 1982. Electroreception. *Annu. Rev. Neurosci.* **5**:121–170.

*Camhi, J.M. 1984. *Neuroethology.* Sinauer, Sunderland, MA.

Cronin, T.W. 1986. Photoreception in marine invertebrates. *Am. Zool.* **26**:403–415.

Dethier, V.G. 1976. *The Hungry Fly.* Harvard University Press, Cambridge, MA.

Dowling, J.E. 1979. Information processing by local circuits: The vertebrate retina as a model system. In F.O. Schmitt and F.G. Worden (eds.), *The Neurosciences: Fourth Study Program*, pp. 163–182. MIT Press, Cambridge, MA.

*Fein, A. and E.Z. Szuts. 1982. *Photoreceptors: Their Role in Vision.* Cambridge University Press, Cambridge.

Finger, T.E. and W.L. Silver (eds). 1987. *Neurobiology of Taste and Smell.* Wiley, New York.

Getchell, T.V. 1986. Functional properties of vertebrate olfactory reception neurons. *Physiol. Rev.* **66**:772–818.

Gould, J.L. 1984. Magnetic field sensitivity in animals. *Annu. Rev. Physiol.* **46**:585–598.

Hubel, D.H. and T.N. Wiesel. 1962. Receptive fields, binocular interaction, and functional architecture in the cat's visual cortex. *J. Physiol.* **160**:106–154.

Hubel, D.H. and T.N. Wiesel. 1979. Brain mechanisms of vision. *Sci. Am.* **241**(3):150–162.

Hudspeth, J. 1983. Mechanoelectrical transduction by hair cells in the acousticolateralis system. *Annu. Rev. Neurosci.* **6**:187–215.

Kaissling, K.-E. 1986. Chemo-electrical transduction in insect olfactory receptors. *Annu. Rev. Neurosci.* **9**:121–145.

Kuffler, S.W. 1953. Discharge patterns and functional organization of the mammalian retina. *J. Neurophysiol.* **16**:37–68.

*Kuffler, S.W., J.G. Nicholls, and A.R. Martin. 1984. *From Neuron to Brain*, 2nd ed. Sinauer, Sunderland, MA.

Land, M.F. 1981. Optics and vision in invertebrates. In H. Autrum (ed.), *Handbook of Sensory Physiology,* Vol. VII 6B, *Comparative Physiology and Evolution of Vision in Invertebrates. B: Invertebrate Visual Centers and Behavior I,* pp. 471–594. Springer, New York.

Lennie, P. 1984. Recent developments in the physiology of color vision. *Trends Neurosci.* **7**:243–248.

Liebman, P.A., K.R. Parker, and E.A. Dratz. 1987. The molecular mechanism of visual excitation and its relation to the structure and composition of the rod outer segment. *Annu. Rev Physiol.* **49**:765–791.

Livingstone, M. and D. Hubel. 1988. Segregation of form, color, movement, and depth: Anatomy, physiology, and perception. *Science* **240**:740–749.

Loewenstein, W.R. 1971. Mechano-electric transduction in the Pacinian corpuscle. Initiation of sensory impulses in mechano-receptors. In W.R. Loewenstein (ed.), *Handbook of Sensory Physiology.* Vol 1, *Principles of Receptor Physiology,* pp. 269–290. Springer, New York.

Matthews, P.B.C. 1981. Evolving views on the internal operation and functional role of the muscle spindle. *J. Physiol.* **320**:1–30.

Michelson, A. 1979. Insect ears as mechanical systems. *Am. Sci.* **67**:696–706.

Pugh, E.N., Jr. 1987. The nature and identity of the internal excitational transmitter of vertebrate phototransduction. *Annu. Rev. Physiol.* **49**:715–741.

Rhode, W.S. 1984. Cochlear mechanics. *Annu. Rev. Physiol.* **46**:231–246.

Roeder, K.D. 1967. *Nerve Cells and Insect Behavior.* Harvard University Press, Cambridge, MA.

*Schmidt, R.E. (ed.). 1978. *Fundamentals of Sensory Physiology.* Springer, New York.

Schmidt-Koening, K. and W.T. Keeton (eds.). 1978. *Animal Migration, Navigation, and Homing.* Springer, New York.

Shapley, R. and P. Lennie. 1985. Spatial frequency analysis in the visual system. *Annu. Rev. Neurosci.* **8**:547–583.

*Shepherd, G.M. 1988. *Neurobiology 8,* 2nd ed. Oxford University Press, New York.

Sinclair, D. 1981. *Mechanisms of Cutaneous Sensation.* Oxford University Press, New York.

Stryer, L. 1987. The molecules of visual excitation. *Sci. Am.* **257**(1):42–50.

*Uttal. W.R. 1973. *The Psychobiology of Sensory Coding.* Harper & Row, New York.

Werblin, F.S. and J.E. Dowling. 1969. Organization of the retina of the mudpuppy, *Necturus maculosus.* II. Intracellular recording. *J. Neurophysiol.* **32**:339–355.

Yost, W.A. and D.W. Nielson. 1985. *Fundamentals of Hearing*, 2nd ed. Holt, Rinehart, & Winston, New York.

See also references in Appendix A.

chapter *19*

Muscle and Other Effectors

The generation of movement is characteristic of all living things. Movements of or within cells and organisms require the generation of forces. Plants are capable of slow movements produced by differential rates of growth, which are under hormonal control. Some plants, such as the Venus flytrap and the sensitive plant *Mimosa,* can produce rapid movements by sudden changes in the turgor pressure of cells. Animals, on the other hand, are capable of much less restricted and more rapid movements, and most are able to move from place to place. Thus, we usually think of active movement as a characteristic property of animals.

For present purposes, an *effector* is a part of an animal that carries out some action relative to the environment and to behavior. Many effectors such as muscle, cilia, and flagella generate forces that produce movement of some part of the animal. Other effectors have other sorts of actions, such as glandular secretion, modulation of color (by chromatophores), light generation (by bioluminescent organs), and electrical discharge (by electric organs). Effectors' actions are usually subject to extrinsic control by the nervous system, the endocrine system, or both. In fact, the control of effectors is a major function of nervous systems, and all externally observable behavior is an expression of effector actions (muscle contractions, glandular secretions, etc.), which as a group are under predominantly nervous control.

In this chapter, we will examine muscle and other effectors, and aspects of their control. In the next chapter we will concentrate on neural control of muscle and on the organizational features of nervous systems that enable animals to produce coordinated, behaviorally significant patterns of movement such as walking, swimming, and flight.

MUSCLE

All animal phyla above the Porifera have clearly differentiated muscle cells. The muscles of different organisms are diverse in many respects, but there is evidence that all muscles share many underlying similarities. We will first consider vertebrate muscles, which have been the most studied. Vertebrate muscle is divided into three types, based on differences in structure, contractile properties, and mode of control. *Skeletal* muscle is typically attached to a part of the skeleton and produces skeletal movement. Skeletal muscle is *striated,* exhibiting transverse bands visible under the light microscope. Skeletal muscle is innervated by motor neurons of the somatic nervous system (see Chapter 2) and is said to be under voluntary control. The term ''voluntary'' does not mean that skeletal muscles must be activated by conscious acts of will; in fact, we are normally unaware of many movements brought about by skeletal muscle, such as breathing and eye movement. The term merely means that such movements (at least in humans) can be initiated or controlled voluntarily. *Smooth* or *nonstriated* muscle lacks observable transverse bands. Smooth muscle is found in the walls of hollow internal organs of the body such as the intestine, the uterus, and blood vessels. Smooth muscle contraction is controlled primarily by the autonomic nervous system. This control is unconscious (involuntary); we are usually quite unaware of the activity of smooth muscle in our blood vessels and intestines. *Cardiac* muscle, the third type, is the muscle of the vertebrate heart. It is somewhat intermediate in its properties between skeletal and smooth muscle. Cardiac muscle is striated like skeletal muscle but is regulated by the autonomic nervous system and is not normally under voluntary control. Unlike skeletal muscle, cardiac muscle contracts spontaneously and rhythmically in the absence of innervation. Because of this spontaneous contractility, the beating of the vertebrate heart is classed as *myogenic* (originating in the muscle), and only the rate of the heartbeat is controlled by nerves (see Chapter 14).

In this section, we will consider two sorts of questions about muscle. The first is a ''how does it work?'' question of physiochemical mechanism: How do molecules act to generate and regulate large-scale forces and movements of an animal? Many aspects of this question can now be answered rather completely, making muscle contraction a physiological process relatively well understood at the molecular level. The second question is a ''what does it do for you?'' question of functional significance: How are different types of vertebrate and invertebrate muscles specialized to perform different functions? The general mechanisms of muscle contraction appear to be broadly similar for all types of muscle, but there are functionally significant differences. To address these two questions, we will first discuss the structure and function of vertebrate skeletal muscle and then consider some other specialized types of muscle.

Structure of Vertebrate Skeletal Muscle

Muscles are composed of long, cylindrical cells called *muscle fibers*. Small muscles may contain only a few hundred muscle fibers, while large limb muscles of mammals contain hundreds of thousands of fibers. An individual muscle fiber is 10–100 μm in diameter and as much as 0.3 m (1 ft) long. Muscle fibers are unusual cells in that they are multinucleate (containing many nuclei). Nevertheless, the terms muscle fiber and muscle cell are used interchangeably. The muscle fiber is surrounded by a cell membrane termed the sarcolemma.

The internal structure of a muscle fiber is shown in Figure 19.1. Each muscle fiber contains hundreds of parallel, cylindrical *myofibrils*. (The prefixes *myo-* and *sarco-* both denote muscle.) The myofibrils are 1 or 2 μm in diameter, as long as the muscle fiber, and are not membrane bound. A myofibril has regularly repeating, transverse bands (Figure 19.1). The major bands are the dark *A bands* and the lighter *I bands*. In the middle of each I band is a narrow, dense *Z line* (Figure 19.1). The functional unit of a myofibril is called a *sarcomere*. A sarcomere is defined as the portion of a myofibril between one Z line and the next Z line (Figure 19.1). Thus, a myofibril consists of longitudinally repeating sarcomeres. The Z lines of adjacent myofibrils are lined up with each other, so that the pattern of alternating A bands and I bands

appears continuous for all the fibrils of a muscle fiber. It is this alignment of banding within a skeletal muscle fiber that gives the fiber its striated appearance.

The structures of muscle described above are visible by light microscopy. When suitably thin sections of muscle are examined by electron microscopy, the myofibrils are seen to contain two kinds of *myofilaments*. The *thick filaments* (15-nm diameter) are composed of the protein *myosin* and are confined to the A band (Figures 19.1 and 19.2). The *thin filaments* (7-nm diameter) contain the protein *actin* and the regulatory proteins *troponin* and *tropomysin* (see p. 513). Thin filaments extend from the Z line part way into the A band, where they interdigitate with thick filaments. Electron micrographs of cross sections of a myofibril show the relation of thick and thin filaments in a sarcomere (Figure 19.2). The I bands contain only thin filaments. In much of the A band, thick and thin filaments overlap in a very regular array, each thick filament being surrounded by six thin filaments. The central region of the A band, which contains only thick filaments and which appears lighter than the rest of the A band in longitudinal sections, is called the *H zone*. In the center of the H zone (which is also the center of the A band) is a narrow dense region, the *M line*. At the M line, adjacent thick filaments are interconnected, maintaining their regular spacing.

In the region of overlap between thick and thin filaments, the filaments interact by means of *cross-bridges* (Figure 19.3). The cross-bridges are projections of myosin molecules in the thick filaments that span the distance between adjacent thick and thin filaments. The interactions of these myosin cross-bridges with actin molecules generate the forces for muscle contraction, as will be described below.

Muscle fibers contain other structures in addition to the arrays of myofilaments within the myofibrils. These other structures, particularly the internal membrane systems involved in initiating contraction, will be described later, after consideration of the mechanism of contraction.

Mechanism of Muscle Contraction

The Sliding Filament Basis of Contraction   When a muscle fiber shortens, the width of the A bands remains constant (Figure 19.4). Thus, the length of the thick filaments does not change during shortening. The widths of the I bands and of the H zones both decrease, but the distance from the edge of one H zone through the I band to the edge of the next H zone also remains constant. Since the H-to-H distance represents the length of the thin filaments (Figure 19.4), it follows that the length of the thin filaments also does not change during muscle fiber shortening. Therefore, muscle shortens in contraction *without either of the kinds of protein filaments changing length*. Instead, the filaments slide past

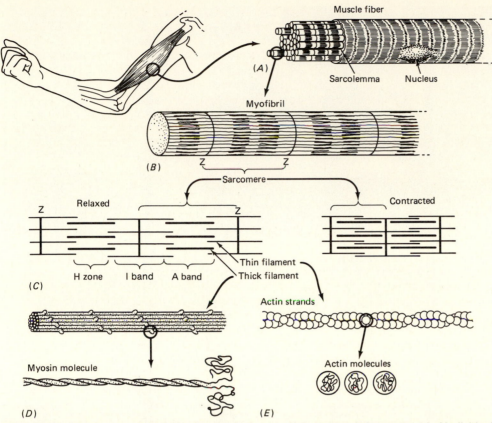

Figure 19.1 Structural organization of skeletal muscle. (*A*) Each muscle is composed of individual muscle cells termed muscle fibers. (*B*) Each muscle fiber is made up of many cylindrical subunits termed myofibrils, that extend the length of the muscle fiber. (*C*) A myofibril is divided into sarcomeres, which are longitudinal segments separated by Z lines. (*D*) Each sarcomere contains many thick filaments and thin filaments. Thick filaments consist of bundles of molecules of the protein myosin. (*E*) Thin filaments are made up of actin and other proteins. Each thin filament contains two strands of actin, each strand composed of globular actin molecules. [Adapted from H. Curtis, *Biology*, 4th ed. Worth, New York, 1983.]

each other in their region of overlap. These observations were first made in 1954 by two independent teams, H.E. Huxley and Hanson, and A.F. Huxley and Niedergerke. They formed the basis for the *sliding-filament model* of muscle contraction, which has since been amply confirmed. According to the sliding-filament model, the force of contraction is generated by the process that actively moves thick and thin filaments past each other. This process must involve the cross-bridges that extend from the thick filaments to contact the thin filaments. To understand how these cross-bridges generate the forces of contraction, we must next examine the molecular structure of the thick and thin myofilaments.

Molecular Basis of Muscle Contraction As noted above, the thick filaments are composed of the protein *myosin,* and the thin filaments contain primarily the protein *actin*. Thus, thick filaments are also called myosin filaments and thin filaments are called actin filaments. Myosin molecules are very large proteins [500 kilodaltons (kDa)], each consisting of a double-headed globular region joined to a long rod, or tail.

A myosin molecule contains two *heavy* chains; part of a heavy chain makes up one head and part extends the length of the tail (Figure 19.5). The myosin heads also contain four smaller *light* chains, so that each myosin molecule consists of six polypeptide chains, two heavy and four light. The tails of many myosin molecules together make up the thick filament, while the globular heads project to the side, forming the cross-bridges (Figure 19.5). The arrangement of actin molecules in thin filaments is rather different. Each actin molecule is a globular protein (42 kDa). Thin filaments contain two chains of actin molecules wound around each other in a double helix (Figure 19.6).

Myosin and actin molecules are incorporated into filaments with specific orientations. The myosin molecules in a thick filament all have their head ends oriented toward the ends of the filament and their tails pointing toward the middle (Figure 19.5). As a result of this orientation, there is a short bare zone, devoid of cross-bridges, at the middle of the thick filament. Actin molecules in thin filaments are also oriented; all the molecules on one side of the Z line

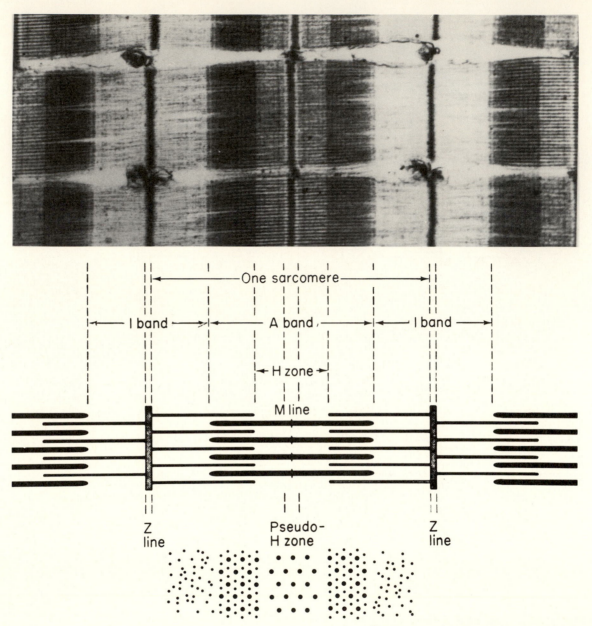

Figure 19.2 Structure of a sarcomere, which is the contractile unit of muscle fibers. The electron micrograph shows portions of three myofibrils in longitudinal section. Longitudinal and cross-section diagrams (below) interpret the structures in the micrograph. The A band contains thick filaments; the I band contains only thin filaments, which are anchored at the Z line. Thick and thin filaments overlap in the lateral parts of the A band. The central portion of the A band (termed the H zone) contains only thick filaments. [courtesy of Dr. H.E. Huxley.]

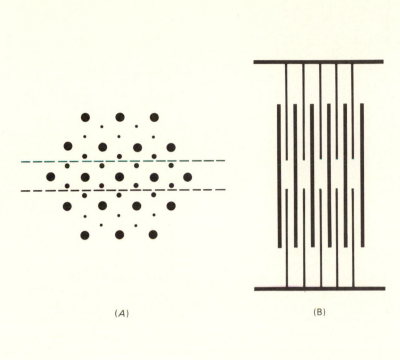

(A)   (B)

(C)

Figure 19.3 Diagram and electron micrograph showing interdigitation and cross-bridges between thick and thin filaments in striated muscle. (A) End view of thick and thin filaments. Dotted lines indicate the outline of a thin longitudinal section that would have the appearance shown in diagram (B) and micrograph (C). Note that thick and thin filaments appear connected by cross-bridges in (C). [From H.E. Huxley, *J. Biophys. Biochem. Cytol.* **3**:631–648 (1957); micrograph courtesy of H.E. Huxley.]

have one orientation, and all those on the other side of the Z line have the opposite polarity. Thus, the polarities of both the actin and myosin molecules are reversed on opposite sides of the middle of a sarcomere.

A muscle requires ATP to contract. The energy required to move myosin and actin filaments past each other comes from the binding and splitting of ATP by the globular heads of myosin molecules.

These heads (which form the cross-bridges) cyclically attach to actin molecules and then swivel, acting as oars that pull the actin and myosin filaments past each other. The cycle of molecular interactions underlying contraction is shown in Figure 19.7. The globular subunit of myosin has two active sites: one for actin and one for ATP. In the cross-bridging cycle, the globular head binds ATP and splits it to ADP + P_i (this process requiring Mg^{2+} ions) but

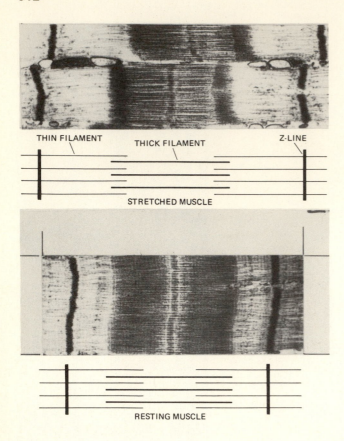

Figure 19.4 Demonstration of sliding of thick and thin filaments in muscle contraction. Stretch or contraction of muscle fibers change the width of the I band and the H zone, but do not change the width of the A band. [courtesy of Dr. H.E. Huxley.]

does not release the ADP and P_i. The energy released by hydrolysis of ATP is stored in the myosin–ADP–complex. This complex then binds actin, forming an actin–myosin–ADP–P_i complex. In the next step, ADP and P_i are released by myosin, and the myosin head changes conformation (swivels), pulling the attached actin toward the middle of the myosin filament. The myosin head then binds a new ATP, triggering its release from actin. Subsequently, the new ATP is hydrolyzed, "cocking" the myosin head in position to bind another actin molecule.

During contraction, this cycle of attachment, swiveling, and detachment recurs 30–100 times a second. With each cycle, the actin filaments are pulled a short distance toward the center of the sarcomere. The cross-bridges work independently and asynchronously, so that at any instant during a contraction about 50 percent of the cross-bridges are bound to actin, the rest being in other phases of the cycle. The summated effect of all the cross-bridge cycles is to pull the thin filaments toward the middle of the sarcomere, like a team of people in a rope-pulling contest.

Control of Contraction by Calcium and Regulatory Proteins Since an intact skeletal muscle fiber contains the elements necessary to complete the cross-bridges cycle—actin, myosin, and ATP—why

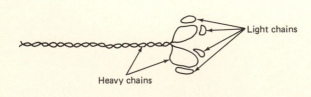

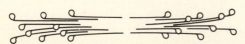

Figure 19.5 Myosin molecule (above) consists of two heavy chains and four light chains; the arrangement of myosin molecules in a thick filament is illustrated below. Note the polarity of the molecules, all pointing away from the center of the filament.

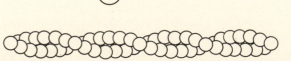

Figure 19.6 Actin molecule (above) is a globular protein monomer. Thin filament (below) contains two helical strands of actin monomers.

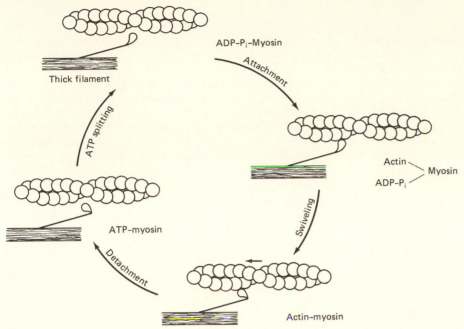

Figure 19.7 Proposed molecular events of a single cross-bridge cycle. The thick and thin filaments move relative to each other by the swiveling action of the myosin head. The initial state in a relaxed muscle fiber is at the top of the figure. [After L. Streyer, *Biochemistry*. Freeman, San Francisco, 1981.]

doesn't contraction occur all the time? The answer is that, at rest, contraction of skeletal muscle fibers is prevented by the regulatory proteins *troponin* and *tropomyosin*. These proteins reside on the thin fila-

ments in association with actin (Figure 19.8). Tropomyosin is a long protein that is coiled along the groove between the two actin chains of the thin filament. Troponin (actually a complex of three sub-

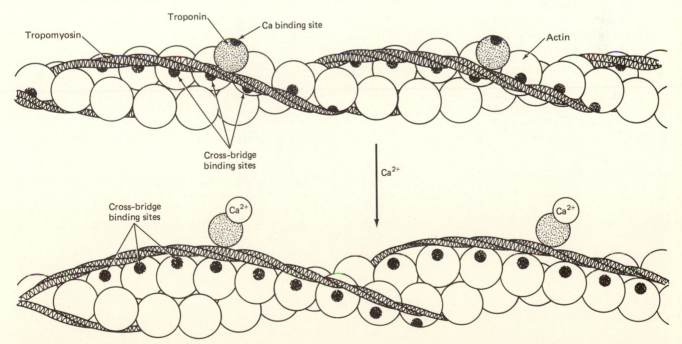

Figure 19.8 Actin-linked regulation of contraction in vertebrate skeletal muscle. In the absence of calcium ions, tropomyosin blocks the cross-bridge binding sites of actin molecules. Calcium ions bind to troponin, inducing the movement of tropomyosin to uncover the binding sites and allowing cross bridges to bind to the thin filaments. [From A.J. Vander, J.H. Sherman, and D.S. Luciano, *Human Physiology,* 3rd ed. McGraw-Hill, New York, 1980.]

units) is located on the thin filaments at intervals of 38 nm, approximately the half-period of the actin helix. In the resting state, the tropomyosin molecule blocks the myosin-binding sites of the adjacent actin molecules and so prevents cross-bridge formation between thin and thick filaments. Since in resting muscle tropomyosin prevents contraction by sterically inhibiting cross-bridge formation, the activation of contraction must counteract this inhibition. *The physiological regulator of muscle contraction is* Ca^{2+}. In skeletal and cardiac muscle, Ca^{2+} acts via troponin and tropomyosin to uncover binding sites on actin molecules. Therefore, skeletal muscle contraction is said to be *actin regulated*. The Ca^{2+} ions bind to troponin, triggering conformational changes in both troponin and tropomyosin (Figure 19.8). The effect of these changes is to uncover the myosin-binding sites of actin, permitting the cross-bridging cycle to proceed until the Ca^{2+} is removed. The

muscle will therefore contract when and only when Ca^{2+} ions are available to bind troponin. As described in the next section, excitation of muscle triggers the release of Ca^{2+} ions to permit contraction.

Excitation–Contraction Coupling   Skeletal muscle contracts when it is excited by a motor neuron via neuromuscular synaptic transmission (p. 423; Box 19.1). Neuromuscular transmission produces a depolarization of the sarcolemma, or outer surface membrane of the muscle fiber. Muscle fiber depolarization is referred to as *excitation* of the fiber. Somehow the excitation of the muscle fiber surface membrane must activate the contractile machinery of filaments and cross-bridges, tens of micrometers deep in the muscle fiber, within a few milliseconds. This relation is called *excitation–contraction coupling*.

We have seen that contraction is triggered by the availability of Ca^{2+} ions that bind troponin. What is

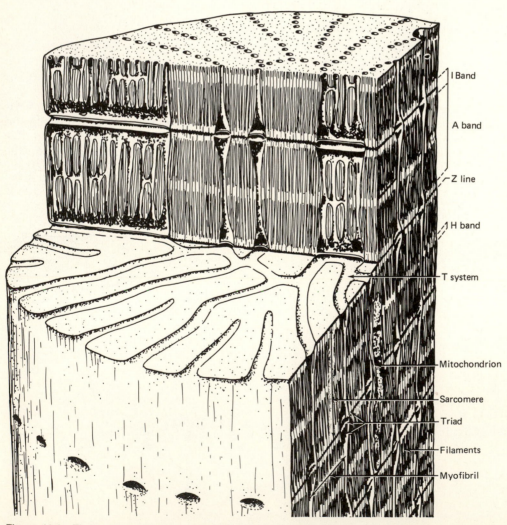

Figure 19.9   Three-dimensional view of part of an amphibian skeletal muscle fiber, showing the T system of transverse extensions of the plasma membrane into the interior of the fiber, and triad junctions of the T system with tubules of the sarcoplasmic reticulum. [From The sarcoplasmic reticulum. Copyright © 1965, by Scientific American, Inc. All rights reserved.]

the source of these Ca^{2+} ions? They do not diffuse in across the sarcolemma from outside the muscle fiber; this path is too long and the resulting diffusion would be too slow to produce the known time course of excitation–contraction coupling. Instead, the external depolarization is brought inward to the center of the muscle fiber and triggers Ca^{2+} release from an intracellular storage compartment.

A muscle fiber contains two separate but intimately joined internal membrane systems. The first of these is the system of *transverse tubules,* or *t tubules*. As their name implies, the transverse tubules cross the muscle fiber at each sarcomere in a network that branches between the myofibrils (Figure 19.9). The position of the transverse tubules is somewhat variable between phyletic groups but is usually at the level of the Z lines (e.g., amphibian muscle) or at the junction of the A and I bands (e.g., muscles of crabs, reptiles, and humans). The t tubule membrane is continuous with the outer sarcolemma, and the tubule lumen is continuous with extracellular space. Thus, t tubules represent invaginations of the muscle fiber membrane surface into the fiber interior.

The other internal membrane system is the *sarcoplasmic reticulum,* a highly organized smooth endoplasmic reticulum of muscle fibers. Each compartment of the sarcoplasmic reticulum forms a hollow sleeve around each myofibril and extends longitudinally between adjacent t tubules (Figure 19.9). An individual compartment of the sarcoplasmic reticulum has enlarged lateral sacs at its ends, which oppose the t tubules. Many smaller longitudinal tubules join the two lateral sacs. The sarcoplasmic reticulum appears to be a completely internal membrane system, not continuous with either the t tubules or the sarcolemma. It does form close contacts, however, with the t tubules along most of their length. In vertebrate muscle fibers these regions of close contact are called *triads* because of their characteristic association of a t tubule with the two adjacent lateral sacs of the sarcoplasmic reticulum (Figure 19.10). Most invertebrate muscles have similar regions of close contact between t tubules and one element of the sarcoplasmic reticulum; such regions are termed *dyads*.

As noted above, a motor neuron excites a muscle fiber by depolarizing it. For most vertebrate skeletal muscle fibers, neuromuscular transmission elicits a muscle fiber action potential that propagates along the surface of the entire fiber. Most invertebrate and a few vertebrate muscle fibers do not generate action potentials (see Neural Control of Muscle, p. 533), but surface depolarization is still the trigger for contraction. The t tubules convey the surface depolarization to the interior of the muscle fiber, either by passive spread (Figure 16.6) or by t-tubular action potentials. There is indirect evidence that the t tubules of vertebrate skeletal muscle fibers can generate action potentials, but the small size of the tubules (~20 nm) makes it hard to test this ability directly.

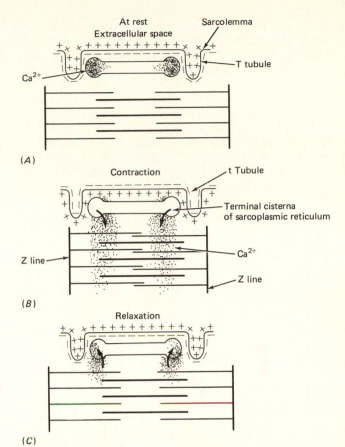

Figure 19.10 Diagram showing how calcium release and recovery control the sliding of actin and myosin filaments. (*A*) T tubules continuous with the sarcolemma (plasma membrane) penetrate the interior of the muscle fiber. (*B*) When action potentials pass down the T tubules, calcium ions are released from terminal cisternae of the sarcoplasmic reticulum. Calcium permits cross-bridge interaction and sliding of actin and myosin filaments. (*C*) Relaxation results from the recovery of calcium ions by active transport back into the terminal cisternae.

The depolarization of the t tubules triggers the release of Ca^{2+} ions from the sarcoplasmic reticulum, which serves as the intracellular Ca^{2+} store. In resting muscle, Ca^{2+} is largely confined to the lateral sacs of the sarcoplasmic reticulum. Ca^{2+} is actively transported into the reticulum by an ATP-dependent Ca^{2+} pump and may be stored in association with a calcium-binding protein, calsequestrin. The sequestration of Ca^{2+} in the sarcoplasmic reticulum is sufficiently effective to keep the sarcoplasmic $[Ca^{2+}]$ below $10^{-7} M$, below the Ca^{2+} dissociation constant of troponin. With activation of the muscle fiber, t-tubular depolarization somehow produces Ca^{2+} release from the sarcoplasmic reticulum. Ca^{2+} rapidly diffuses the short distance to the adjacent myofilaments, binds to troponin, and triggers contraction. The exact mechanism by which t-tubular depolarization causes Ca^{2+} release from the separate membrane system of the sarcoplasmic reticulum is not clear, but indirect evidence indicates that the two membrane systems may be electrically coupled by low-resistance junctions at the triad. If so, the membranes of the terminal sacs would become depolarized, perhaps activating voltage-dependent calcium channels in the membranes of the reticulum.

Box 19.1 provides an overview of the sequence of events by which an action potential in a vertebrate spinal motor neuron elicits contraction of a muscle fiber on which it ends. Part of this sequence was described in Chapter 17, and the rest is treated above.

Mechanics of Muscle Contraction

Muscle contraction is defined as the active generation of force in a muscle. The force exerted by a muscle on an external load is called the muscle *tension.* If the tension developed by a muscle is greater than the external force exerted on it by a load, the muscle shortens. If the force exerted by the load (e.g., the weight of the load) is greater than or equal to the muscle tension, the muscle does not shorten its overall length. A muscle can thus contract with or without shortening, since it develops tension in either case.

Because of the distinction between tension development and shortening, there are two different ways of measuring the mechanical responses of muscle contraction. In one of these, termed *isotonic contraction,* the muscle is allowed to shorten against a load that is less than the muscle tension (Figure 19.11*A*). During the period in which the muscle lifts the load, the tension is constant and is equal to the force exerted by the load; hence, the contraction is isotonic (constant tension). In isotonic contraction, the *length* of the muscle is measured. The second method of measuring muscle contraction is to measure the *tension* developed by a muscle that is contracting without shortening. Such a contraction is termed *isometric* (constant length) (Figure 19.11*B*). Isometric contraction is usually not measured by hanging from the muscle a weight that is too heavy to be lifted, since this would stretch the muscle. Instead, a very stiff force transducer (such as a silicon strain gauge) is used to measure force with negligible changes in length.

Muscle Twitch   The mechanical response of a muscle to a single electrical stimulus is called a *twitch.* Muscles differ considerably in the time course and the degree of their response to a single stimulus; these differences are considered in later sections.

BOX 19.1   SEQUENCE OF EVENTS IN THE CONTROL OF VERTEBRATE SKELETAL MUSCLE CONTRACTION

---

1. Motoneuron action potential propagates down the axon and depolarizes the presynaptic terminal.
2. Ca^{2+} ions enter the axon terminal and cause release of acetylcholine (ACh) from synaptic vesicles.
3. Acetylcholine diffuses across the synaptic cleft.
4. Acetylcholine binds to ACh receptors of the postsynaptic membrane of the muscle fiber, opening transmitter-dependent membrane channels.
5. The resulting Na^+ and K^+ movements give rise to an excitatory postsynaptic potential.
6. The EPSP spreads passively to adjacent electrically excitable membrane regions and triggers a muscle fiber action potential.
7. The muscle fiber action potential propagates rapidly over the entire surface of the fiber and also propagates along t-tubular membranes into the fiber interior.
8. Depolarization of the t-tubular membranes triggers release of Ca^{2+} from the adjacent lateral sacs of the sarcoplasmic reticulum.
9. Ca^{2+} ions diffuse to the thin filaments and bind to troponin.
10. Troponin and tropomyosin molecules change conformation, exposing the myosin-binding sites of adjacent actin molecules on the thin filaments.
11. Myosin heads bind to the actin sites, forming cross-bridges and continuing the cross-bridge cycle as long as sufficient Ca^{2+} is present. The cross-bridge cycle moves thick and thin filaments relative to each other, shortening the sarcomeres.
12. The Ca^{2+} concentration of the sarcoplasm is decreased by active Ca^{2+} reuptake into the sarcoplasmic reticulum. Ca^{2+} dissociates from troponin, and the troponin–tropomyosin complex again inhibits contraction. The muscle relaxes.

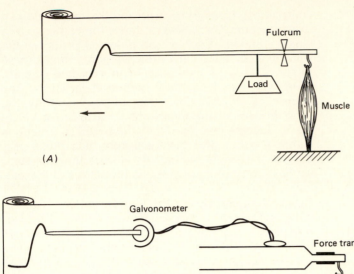

(A)

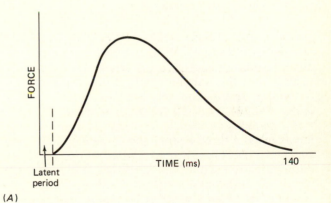

(B)

Figure 19.11 (A) Apparatus for measuring isotonic muscle contraction. The muscle is allowed to shorten at constant tension. (B) Apparatus for measuring isometric muscle contraction. The muscle length is held constant and the force is measured.

Figure 19.12 shows the twitch response of a mammalian muscle recorded under isometric and under isotonic conditions. For the isometric contraction, there is a brief latent period before any tension is recorded; this latent period largely reflects the time required for excitation–contraction coupling to occur. The latent period is followed by a rather rapid increase in tension and a slower tension decay. For isotonic contraction, the latent period is longer than for isometric contraction, and it is followed by a nearly linear rate of shortening. The duration of isotonic contraction is considerably shorter than the duration of isometric contraction.

If a muscle is stimulated to contract isotonically and the load is varied, the result is shown in Figure 19.13. With increasingly heavy loads, the latency prior to shortening increases and the duration of shortening and the velocity of shortening (the rising slope in Figure 19.13A) decrease. Figure 19.13B shows the relation of the velocity of shortening to load on the muscle.

Summation of Contractions When a muscle is stimulated more than once within a brief period of time, the contractions produced by the stimuli may summate (Figure 19.14). The electrical events triggering the contractions—action potentials in the nerve axon and muscle fibers—are all-or-none and do not summate, but since a muscle twitch lasts many milliseconds, several action potentials can be initiated during a twitch. The amplitude of the resultant contraction depends on the time interval between the

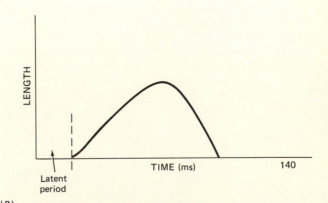

Figure 19.12 Form and time course of a muscle twitch, measured by (A) isometric and (B) isotonic recording. [After A.J. Vander, J.H. Sherman, and D.S. Luciano, *Human Physiology,* 3rd ed. McGraw-Hill, New York, 1980.]

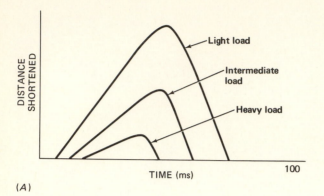

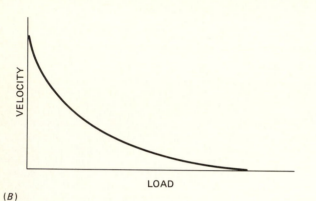

(A)

(B)

Figure 19.13 (A) Time course and (B) velocity of isotonic muscle contractions at different loads.

contraction, it may be surprising that a muscle can exert a tetanic force several times the response to a single stimulus. Each action potential produces the same degree of depolarization, so there is no increase in the degree of excitation with repeated stimuli. Moreover, each action potential triggers the release of enough Ca^{2+} ions to activate fully the cross-bridges of the muscle fiber, at least for vertebrate skeletal muscle. How then can contractions summate to greater tension levels than those produced by a single twitch? To answer this question, we have to examine the elastic properties of muscle.

Elastic Components of Muscle   Muscle does not behave as if it were a purely contractile tissue. Instead, it acts as if there are elastic elements both in series and in parallel with the contractile elements. The *parallel elastic component* presumably results from connective tissue, the sarcolemma, and sarcoplasm. The *series elastic component* includes tendons and associated connective tissue, Z-line material, and elasticity associated with the cross-bridges themselves. These components can be schematically represented by a mechanical model, as shown in Figure 19.15. The elastic elements of muscle can serve to store energy of contraction, smoothing the development of tension and broadening the range of tension that can be produced.

Because of the presence of the elastic elements, externally recordable measurements of muscle contraction do not accurately represent the state of the contractile process at the cross-bridges. For example, Figure 19.15 shows an isotonic contraction in which a muscle lifts a load. In the first stage of contraction, the cross-bridge cycle is actively generating force but the initial contractile force is largely expended in stretching the series elastic component, in effect taking up internal slack in the muscle. The externally recorded tension increases only gradually, and a considerable latent period elapses before the external tension exceeds the weight of the load and begins to lift it. This illustration explains why isotonic contractions have a longer latency and shorter duration than isometric contractions (Figure 19.12) and why isotonically recorded latency increases and duration decreases with increasing load (Figure 19.13).

impulses. As shown in Figure 19.14, impulses at relatively long intervals (e.g., 50–200 ms) produce contractions that summate but are not fused. Higher-frequency stimulation (e.g., 10-ms intervals) produces a smoothly fused contraction called a *tetanus*. For a mammalian muscle, the amplitude of the tetanus is usually three or four times the amplitude of a single twitch. This *twitch : tetanus ratio*, however, differs greatly for different kinds of muscle, many of which may not give a twitch response to a single stimulus. Muscles also differ greatly in the length of time they can maintain tetanic contraction without depleting their energy stores (see later, Vertebrate Muscle Fiber Types).

For the reader who has mastered the above discussion of the mechanisms of muscle excitation and

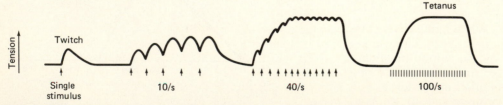

Figure 19.14 Isometric contractions in twitch muscle, evoked by repetitive stimulation at different stimulus frequencies. At low stimulus frequency (10/s), contractions summate but remain unfused. At high stimulus frequency (100/s), contractions summate in a smoothly fused *tetanus*. Note that the amplitude of the tetanic contraction is about three times the amplitude of a single twitch in this example.

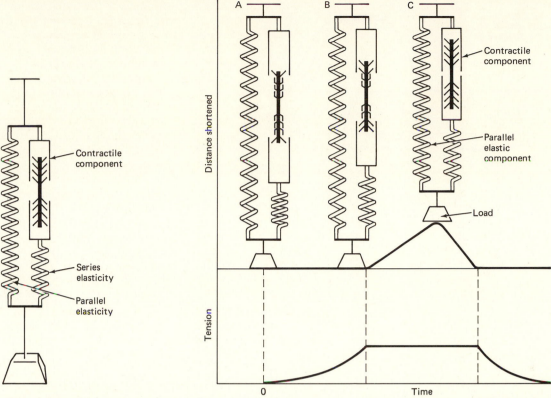

Figure 19.15 Diagram illustrating series and parallel elastic components in muscle. (*Left*) A mechanical representation of muscle, with a contractile component in series and in parallel with elastic components. (*Right*) At the beginning of a contraction (A) the weight rests on the substratum. Initial shortening of the contractile component stretches the series elasticity. Tension increases until point (B); contraction is isometric before this point. At point (C), tension equals the weight of the load; the load is lifted and the contraction becomes isotonic. [After A.J. Vander, J.H. Sherman, and D.S. Luciano, *Human Physiology,* 1st ed. McGraw-Hill, New York, 1970.]

The relation between the tension produced by an isometric twitch and by a tetanus can also be explained in terms of elastic components of muscle. In an isometric twitch, much of the force generated by the cross-bridges is dissipated in stretching the series elastic elements and is not recorded externally. We can define a parameter, the ***active state,*** as the amount of tension produced by the contractile elements when the elastic elements are neither lengthening or shortening. Thus, the active state represents the tension that the contractile elements would produce if there were no elastic elements. The active state cannot be measured directly but can be approximated by quick stretches of a muscle during its contraction, which take up the slack in the series elastic component. As shown in Figure 19.16, the amplitude of the active state is considerably greater than the tension of an externally recorded twitch. Moreover, the time course of development of the active state is much more rapid than the force increase of the external twitch and is probably similar to the time course of increased sarcoplasmic $[Ca^{2+}]$.

During a tetanic contraction, the active state is maintained for a longer period by repeated excitation

(Figure 19.16). Therefore, the series elastic component can become fully stretched, unlike the situation in a single twitch in which the active state decays before the muscle can fully develop (external) tension. After the series elastic component is fully stretched, the external muscle tension is equal to the active-state tension of the contractile elements, and is considerably greater than the tension produced during a single twitch.

Length–Tension Relation The amount of tension that a muscle can develop in isometric contraction depends on the length at which the muscle is held. For most muscles, the maximum tension is developed at lengths close to the resting length of the muscle (Figure 19.17). For whole muscle (inset), the amount of tension is a smooth function of length, since not all fibers are stretched or shortened to the same degree. With careful measurements on single amphibian muscle fibers, however, the length–tension relation consists of four straight-line segments (Figure 19.17). According to the sliding-filament theory of contraction, the amount of tension generated should be proportional to the number of active cross-bridges be-

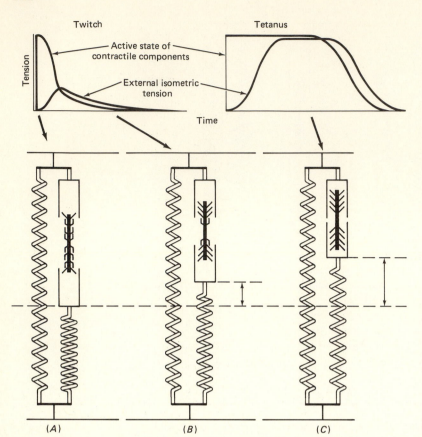

Figure 19.16 The *active state* of muscle contractile components is the force the contractile components generate when they are neither lengthening nor shortening. During a twitch contraction, much of the contractile force of the active state is dissipated in stretching the series elastic component. During tetanic contraction, the active state is prolonged so that the whole-muscle tension increases to the force of the active state. [After A.J. Vander, J.H. Sherman, and D.S. Luciano, *Human Physiology,* 1st ed. McGraw-Hill, New York, 1970.]

tween thick and thin filaments, which number depends on the degree of filament overlap (and hence on sarcomere length). As shown in Figure 19.17, the tension increases linearly with increasing overlap, from a sarcomere length of 3.65 μm (*1,* zero overlap) to a length of 2.25 μm (*2,* complete overlap of cross-bridge areas). With further shortening to *3,* there is no increase in cross-bridge area, since the central bare zone of the thick filament is devoid of cross-bridges. With still further shortening, several factors occur that presumably disrupt tension development. First (*4*), the thin filaments encounter "wrong way" cross-bridges on the opposite side of the bare zone. These thin filaments might actually bind to the "wrong way" cross-bridges or might simply interfere with cross-bridge binding to a "right way" thin filament. Next (*5*), the thick filaments reach the Z line and appear to crumple at the sarcomere ends. Finally (*6*), the thin filaments reach the opposite Z line. The length–tension curve of single muscle fibers agrees strikingly with the area of filament overlap for effective cross-bridge formation. This agreement strongly implies that each cross-bridge contributes an independent increment of tension, and provides powerful support for the sliding filament theory of muscle contraction.

Whole-Muscle Force, Velocity, and Work   The velocity with which a muscle can contract is a function of muscle length. If each sarcomere contracts, say, 10 percent of its length over 100 ms, then the absolute velocity of shortening over that time depends on the number of sarcomeres *in series.* That is, the ends of a long muscle approach each other at a faster rate than do the ends of a short muscle. The distance a muscle can shorten clearly depends on length, in the same way.

The amount of tension that a muscle can develop, on the other hand, does not increase with increasing length. A longer chain cannot support any more load than a shorter length of the same chain; the force on all the links is equal. Like the chain, a muscle is only as strong as its weakest sarcomere. The maximum force of a muscle depends on the number of contractile elements *in parallel* and hence on the cross-sectional area of the muscle, regardless of length. Thus, a muscle adapted to exert maximal force is relatively short and thick, while a muscle functioning to exert maximal velocity is relatively long and thin. Different muscles may differ greatly in their intrinsic velocity of shortening, as will be considered later. The relation of maximum force and cross-sectional area, however, is nearly constant among muscles of a wide variety of animals.

The amount of work that a muscle performs in lifting a load is the product of the force exerted (which equals the weight of the load) and the distance the load is lifted. In a purely isometric contraction,

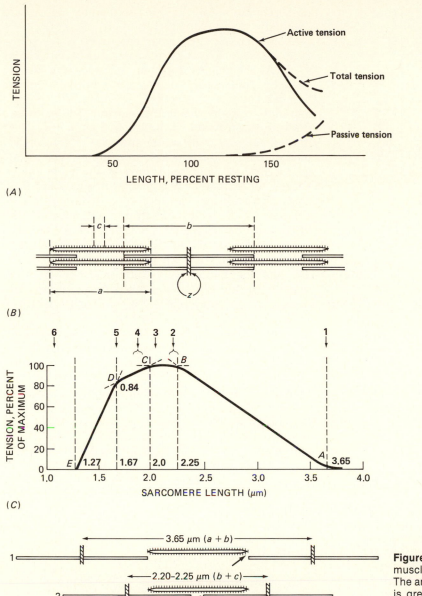

(B)

(C)

(D)

Figure 19.17 Length-tension diagram of whole muscle (A) and of a single muscle fiber (B to D). (A) The amount of active tension that a muscle can exert is greatest at lengths near its resting length. With stretches much beyond resting length, active tension decreases but passive tension increases because of stretch of elastic components. The length-tension relationship of a single muscle fiber (C) is related to the geometry of thick and thin filaments in a sarcomere (B, D). Between 1 and 2, the amount of tension depends on the degree of filament overlap and the resultant number of functional cross bridges. At shorter lengths, several factors decrease tension: thin filaments meet at mid-sarcomere (3) and begin to interact with cross-bridges that pull the wrong way (4); thick filaments (5) and thin filaments (6) begin to crumple against the Z lines. [After G. Gordon, A.F. Huxley, and F.J. Julian, *J. Physiol.* **184**:170–192 (1966).]

no physical work is done, since the distance moved is zero. Likewise, the isotonic contraction of a completely unloaded muscle does no work, since negligible force is exerted in the absence of a load. As shown in Figure 19.18, a muscle performs the max-

imal amount of work at an intermediate load, typically about 40 percent of the maximum load that can be lifted. As a simplified generalization, the amount of work that a muscle can do depends on its volume, since volume is the product of length and cross-

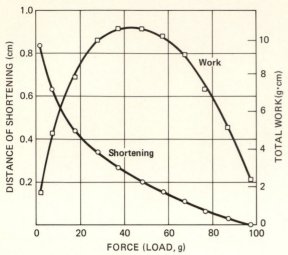

Figure 19.18 Amplitude of isotonic contraction and work (force times distance) as a function of load in muscle. The amplitude of shortening decreases to zero with increasing load. The work performed in the contraction first increases as the load is increased, and then decreases at larger loads (and smaller amplitudes of shortening). [From K. Schmidt-Nielson, *Animal Physiology,* 3rd ed. Cambridge University Press, Cambridge, 1983.]

sectional area, and since force is proportional to cross-sectional area and distance is proportional to length.

Energetics of Muscle Contraction

The immediate source of energy for muscle contraction is ATP, which is required to perform at least three functions in the contraction–relaxation cycle: (1) ATP splitting activates the myosin cross-bridge preparatory to binding to actin and swiveling (Figure 19.7), (2) ATP binding to the cross-bridge is necessary for detachment of myosin from actin, (3) ATP energy drives the calcium pump that transports Ca^{2+} ions into the sarcoplasmic reticulum. Muscle, however, contains only enough ATP (2–4 mM) to sustain contraction for a few seconds. Ultimately, the ATP is regenerated by glycolysis and oxidative phosphorylation, but these multistep pathways do not increase their rates immediately. The most immediately available path for regeneration of ATP is from one of two energy-rich compounds termed phosphagens (Chapter 4). In vertebrate muscle, the phosphagen is *creatine phosphate* (CP), which rephosphorylates ADP in a reversible reaction

$$CP + ADP \rightleftharpoons C + ATP$$

In resting muscle, ATP is abundant and drives the reaction to the left by mass action. As ATP is used in contraction, the reaction proceeds to the right. ATP levels decline only slightly until creatine phosphate is largely depleted. In the muscle of most invertebrates, *arginine phosphate* acts as the phosphagen instead of creatine phosphate. The sources

of regeneration of ATP in vertebrate muscle are summarized in Figure 19.19.

A contracting muscle can be viewed as being in a race to resupply its ATP as fast as it is used (Chapter 4). If ATP becomes depleted, the muscle is unable to keep contracting, a process known as *muscle fatigue.* Tetanic contraction, in which ATP is used especially rapidly, leads to more rapid fatigue than does repeated twitch contraction (Figure 19.20). At modest rates of contraction and of ATP hydrolysis, oxidative metabolism is able to resupply ATP as fast as it is used. In mammalian muscle, the major energy source for oxidative phosphorylation is from fatty acids, with glucose playing a secondary role (see Chapter 4). At levels of activity and rates of ATP breakdown exceeding about 50 percent of the maximum aerobic levels possible, oxidative metabolism becomes unable to keep up with ATP breakdown. The limiting factors in oxidative phosphorylation may be the rate of O_2 delivery to the muscle, the rate of delivery of metabolic substrates, or the intrinsic reaction rates of oxidative pathways. At activity levels above about 50 percent of the aerobic maximum, vertebrate skeletal muscle becomes increasingly dependent on anaerobic glycolysis to resupply ATP. Glycolysis can generate ATP at a faster absolute rate than can oxidative phosphorylation but requires much greater quantities of glucose to do so because it yields relatively little ATP per glucose (Table 19.1; see also Chapter 4). The primary substrate for anaerobic glycolysis is glycogen stored in the muscle. Thus, under extreme conditions, muscle can generate ATP rapidly without requiring extrinsic energy substrates or oxygen. The duration of anaerobic activity, however, is limited by the amount of stored glycogen and by lactic acid accumulation. As discussed in Chapter 4, the use of anaerobic glycolysis incurs an oxygen deficit.

Vertebrate Muscle Fiber Types   The fibers of vertebrate skeletal muscles are heterogeneous in their speed of contraction, fatigue resistance, and metabolism. By far, the most common types are varieties of *twitch* fibers, in which the muscle fibers sustain action potentials and each action potential gives rise to a muscle twitch. *Tonic* muscle fibers (discussed under Neural Control of Muscle, p. 533) do not sustain action potentials and many do not twitch in response to a single stimulus. Tonic fibers are far less common among vertebrate skeletal muscles than are twitch fibers.

Some twitch muscle fibers are capable of contracting faster than others. Therefore, fibers have long been classified as either fast twitch or slow twitch fibers, some muscles having a preponderance of fast, and others having mostly slow twitch fibers. More recently, it has been found that many biochemical and physiological differences are correlated with speed of contraction. Table 19.2 compares the three major types of mammalian twitch fiber. These are

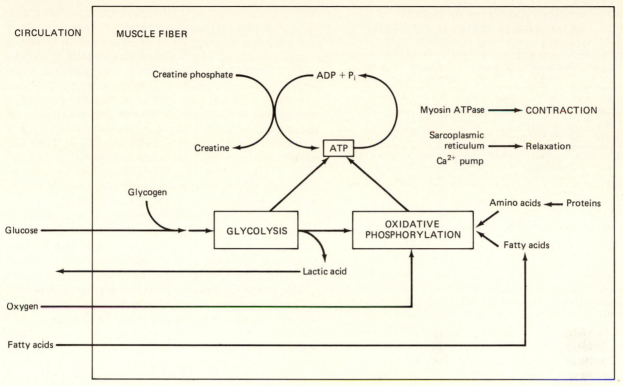

Figure 19.19 Biochemical pathways producing ATP utilized during vertebrate muscle contraction. [From A.J. Vander, J.H. Sherman and D.S. Luciano, *Human Physiology,* 3rd ed. McGraw-Hill, New York, 1980.]

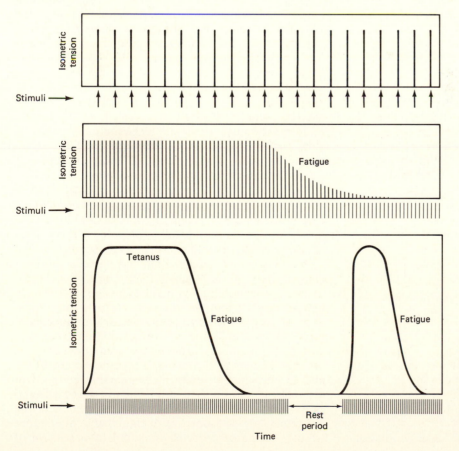

Figure 19.20 Muscle fatigue resulting from prolonged stimulation. The muscle recovers its ability to contract after a rest period, but subsequent fatigue is more rapid if the recovery is incomplete. [From A.J. Vander, J.H. Sherman, and D.S. Luciano, *Human Physiology,* 3rd ed. McGraw-Hill, New York, 1980.]

Table 19.1 SOURCES AND RATES OF ENERGY METABOLISM IN VERTEBRATE SKELETAL MUSCLE

	Relative rate	Glucose used	Lactate produced	O_2 needed?
Oxidative metabolism	36-38 ATP	1	0	Yes
Glycolytic metabolism[a]	64 ATP	32	64	No

[a] Glycolytic pathways make less ATP per glucose than oxidative pathways but can be active at a higher rate of ATP production (cf. Table 4.1 for slightly different estimates of rates of ATP production).

usually termed (a) FG (fast, glycolytic) or FF (fast, fatiguing), (b) FOG (fast, oxidative, glycolytic) or FR (fast, resistant to fatigue), and (c) SO (slow, oxidative) or S (slow, resistant to fatigue). The primary differences between the fiber types are in their rates of splitting ATP and in their metabolic sources for regenerating ATP. Fast twitch fibers (FG and FOG) contain myosin with a higher intrinsic ATPase activity than the myosin of slow twitch fibers. Since the rate of ATP hydrolysis governs the rate of cross-bridge cycling, the higher myosin ATPase activity of fast twitch fibers allows more rapid contraction. The amount of tension developed per cross-bridge, however, is the same for the different fiber types. The regeneration of ATP can be from primarily oxidative or primarily glycolytic sources. Slow twitch fibers (SO) are adapted to regenerate ATP by oxidative processes. They contain many mitochondria and a dense network of capillaries to supply oxygen and substrates. Moreover, they contain appreciable quantities of myoglobin, a protein related to hemoglobin (see Chapter 13), which facilitates oxygen entry into muscle fibers and serves as a modest internal

Table 19.2 PHYSIOLOGICAL AND BIOCHEMICAL PROPERTIES OF MAMMALIAN MUSCLE FIBER TYPES

	FG	FOG	SO
Biochemical classification	FG	FOG	SO
Physiological classification	FF	FR	S
Speed of contraction	Fast	Fast	Slow
Myosin ATPase activity	High	High	Low
Source of ATP production	Glycolysis	Oxidative to mixed	Oxidative
Myoglobin content	Low	—	High
Mitochondria	Few	Many	Many
Capillaries	Few	—	Many
Glycogen content	High	Intermediate	Low
Rate of fatigue	Fast	Intermediate	Slow
Fiber diameter + amount of tension	Large	Intermediate	Small

oxygen store. Because of their high myoglobin content and extensive vascularization, slow muscle fibers have a distinctly red appearance. In contrast, some fast twitch fibers are predominantly glycolytic (FG). These fibers contain fewer mitochondria, a less extensive capillary network, and less myoglobin than do slow twitch fibers. Because of their paucity of myoglobin and sparse circulation, FG fibers are pale in appearance and are termed white muscle. The third muscle fiber type, FOG or FR, is generally intermediate in its properties between FG and SO fibers. FOG fibers have high myosin–ATPase activities and rapid contractions and also have high oxidative capacity and extensive vascularization.

The different muscle fiber types vary greatly in their resistance to fatigue. Slow oxidative (SO) fibers fatigue very slowly or not at all. Their high rates of oxidative metabolism and extensive circulation can keep pace with their modest rates of ATP hydrolysis, maintaining high ATP levels for prolonged periods. FOG fibers also have a high oxidative capacity and fatigue rather slowly despite their high rates of ATP hydrolysis. FG fibers fatigue rapidly, being largely dependent on anaerobic glycolysis to regenerate their rapidly split ATP.

The functional significance of the differences in properties of muscle fiber types is that different functions can be subserved by muscle fibers that are specifically adapted to fulfill these functions. Slow twitch, oxidative fibers (and tonic fibers when present) are adapted for isometric postural functions and for small, slow movements. Slow fibers operate efficiently and without fatigue under these conditions but do not generate much force. Fast oxidative (FOG) fibers generate more force and more rapid contraction yet are rather fatigue resistant. They are adapted for, and are used for, repeated movements such as locomotion. Fast glycolytic (FG) fibers generate rapid contractions and large increments of tension but fatigue rapidly. They are used for occasional, forceful, fast movements such as bursts of speed or leaps in escape or prey capture.

Let us examine a few examples of the distribution and use of different fiber types of different animals. The ankle extensor group of the cat hind limb consists of three muscle heads: the soleus, the medial gastrocnemius, and the lateral gastrocnemius. The soleus is a slowly contracting, red muscle that consists entirely of slow twitch (SO) fibers. It is active in postural standing, with little further increase in activity during locomotion. In contrast, the medial and lateral gastrocnemii are faster, white muscles of mixed fiber composition. Most of the fibers are fast twitch, the medial gastrocnemius containing approximately 45 percent type FG, 25 percent type FOG, and 25 percent SO fibers. (The remaining 5 percent are intermediate in their properties between FG and FOG fibers.) Because of the greater diameter of the FG fibers, their 45 percent of the muscle fibers contributes 75 percent of the maximal total tension of the medial gastrocnemius. Experimental evidence indicates that only about 25 percent of the maximal tension of the medial gastrocnemius is exerted during walking and most running. This is approximately the amount of tension contributed by the FOG and SO fibers, indicating that these fatigue-resistant fibers are sufficient for most locomotion. The large force contribution of the FG fibers is presumably reserved for short bursts of contraction in galloping and jumping.

In fish, the myotomal (body trunk) muscles are divided into separate regions of red slow muscle and white fast muscle (Figure 19.21). The muscle fibers in these regions have many histological, biochemical, and physiological similarities to mammalian SO and FG fibers, respectively. The slow red muscle is less than 10 percent of the total myotomal muscle in most fish species and never exceeds 25 percent. Nevertheless, experimental recordings indicate that only slow muscle is used for swimming at all speeds up to cruising. The fast muscle that constitutes the bulk of the muscle mass is used only for high-speed swimming bursts and fatigues very rapidly.

A similar distinction between rapidly and slowly contracting muscle fibers occurs in arthropods. Although the metabolic basis of the distinction is poorly understood, in arthropods there is a correlation between sarcomere length and speed of contraction (Figure 19.22) that is not found in vertebrates. Rapidly contracting arthropod muscle fibers have short (1.5–3 μm) sarcomeres and relatively low ratios of thin to thick filaments. Slowly contracting muscle

fibers have long (6–15 μm) sarcomeres and high ratios of thin to thick filaments. All degrees of intermediates occur, often mixed within the same muscle. Less commonly, fast and slow muscles are separate, for example, in the muscles that flex and extend the abdomens of crayfish and lobsters (Figure 19.22). When you eat lobster tail, the bulk of the muscle mass consists of short-sarcomere, rapidly contracting flexors and extensors. Separate thin sheets of long-sarcomere, slow muscle fibers lie just under the carapace, the slow flexors ventrally and the slow extensors dorsally. The slow muscles are used for slow movements and postural adjustments, while the massive fast muscles are used only in rapid movements such as swimming and escape tail flips (see Chapter 20).

In other studied examples, both vertebrate and invertebrate, the specialization of separate fast and slow muscles is less complete than in the examples just given. Most muscles have a mixture of fiber types, like the cat medial gastrocnemius. Whether the fiber types are arranged in separate muscles or intermingled, the specialization of distinct fiber types is an important functional feature of muscle.

Specialized Muscles

The above sections have described the structure and function of vertebrate skeletal muscle. To a large extent, these descriptions also apply to most kinds of invertebrate muscles, which have similar patterns of sarcomere organization and similar mechanisms

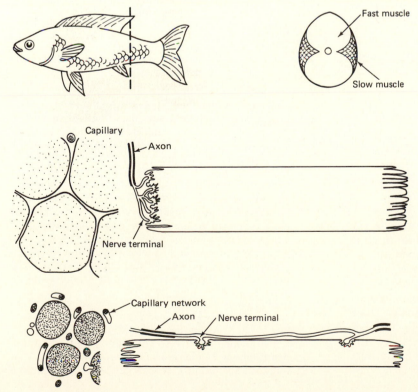

Figure 19.21 Fast and slow muscle fibers in fish. (*A*) Most of the trunk musculature consists of fast muscle, slow muscle being confined to relatively narrow lateral bands. (*B*) Fast muscle fibers (above) are relatively large, with sparse mitochondria and capillaries and with innervation at the fiber end. Slow muscle fibers (below) have a denser capillary network and more mitochondria, and have multiterminal innervation along the length of the fiber. [After Q. Bone in W.S. Hoar and D.J. Randall (eds.), *Fish Physiology,* Vol. VII, pp. 361–424. Academic Press, New York, 1978.]

SARCOMERES OF CRUSTACEAN MUSCLE FIBERS

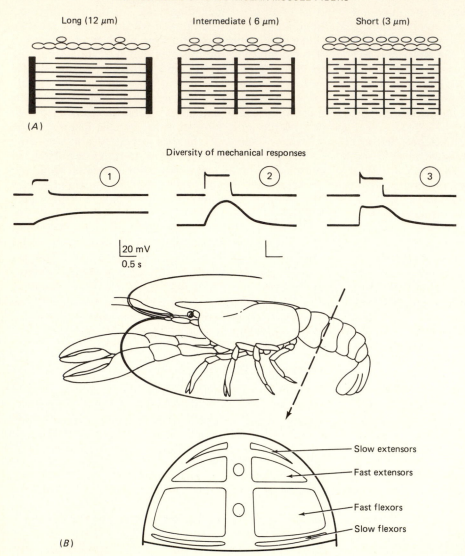

(A)

Diversity of mechanical responses

|20 mV
0.5 s

(B)

Slow extensors
Fast extensors
Fast flexors
Slow flexors

Figure 19.22 Crustacean slow and fast muscle fibers. (*A*) Diversity in structure and mechanical response of crustacean muscle fibers. Fibers with long sarcomeres and widely-spaced transverse tubules (above) tend to have slow tension responses (below). The mechanical responses (1, 2, 3, lower traces) differ in speed, although the membrane responses to direct depolarization (upper traces) are similar in these non-impulse-generating muscle fibers. [From H.L. Atwood in G. Hoyle (ed.), *Identified Neurons and Behavior of Arthropods.* pp. 9–29. Plenum, New York, 1977.] (*B*) Separate slow and fast muscles in the crayfish abdomen. Arthropod muscle fibers may be segregated in different muscles by fiber type, as in this example, or may co-occur within a single muscle.

of action. Nevertheless, there is considerable diversity among different kinds of muscle, in both vertebrate and invertebrate groups. In this section, we discuss vertebrate cardiac muscle and smooth muscle, as well as some of the diverse kinds of invertebrate muscle that do not fit easily into the above picture.

Vertebrate Cardiac Muscle Cardiac muscle is classed as striated, since cardiac myofibrils have a cross-banded microscopic appearance similar to skeletal myofibrils. The striations are not aligned from fibril to fibril in cardiac muscle, however, so that the striated appearance is less obvious. Cardiac muscle fibers are mononucleate and smaller than skeletal fibers. Sarcoplasmic reticulum and t tubules may be well developed or absent. There are several important functional properties of cardiac fibers that differ from skeletal muscle.

1. Cardiac muscle fibers generate endogenous action potentials at periodic intervals. If individual cardiac fibers are dissociated and grown in tissue culture, each one generates action potentials and beats at its own endogenous frequency.
2. Cardiac muscle fibers are electrically coupled by gap junctions (see Figure 17.2) located in the longitudinal portions of specialized contact regions termed *intercalated discs*. As a result, cardiac fibers that contact each other (in tissue culture or in the heart) beat nearly synchronously.
3. The action potentials of vertebrate cardiac fibers are greatly prolonged, with typical durations of 100–500 ms (Figure 16.19) versus about 1 ms for most muscle and nerve. This long duration and the rather slow excitation–contraction coupling of cardiac fibers ensure a smoothly prolonged contraction rather than a brief twitch. Moreover, since the cardiac action potential is long relative to conduction time within the heart, all ventricular muscle fibers contract at essentially the same time (see Chapter 14). These features

are essential for efficient pumping of blood by heart muscle contraction.

4. Each cardiac action potential is followed by a refractory period of several hundred milliseconds. This long refractory period prevents the heart from contracting tetanically, allowing it to fill with blood between contractions.

Vertebrate Smooth Muscle The terms smooth muscle and nonstriated muscle are applied to any muscle that lacks transverse striations when viewed by light microscopy. As might be expected, this definition embraces considerable diversity among different kinds of smooth muscle. Most vertebrate smooth muscle lines hollow internal organs and blood vessels, including the digestive tract, uterus, urinary bladder, ureters, arteries, and arterioles. Smooth muscles also occur in the iris and in association with hairs in the skin.

Smooth muscle cells are small and spindle-shaped, typically 2–10 μm long. Each has a single central nucleus. Often cells are electrically coupled via gap junctions, so that the cells in an area may contract as a single functional unit.

The internal structure of smooth muscle cells appears less highly organized than in skeletal and cardiac muscles. Much of the cytoplasm is filled with longitudinally arrayed filaments of three kinds: thick myosin-containing filaments, thin actin-containing filaments, and intermediate filaments that are thought to act as an internal cytoskeleton but may not function in active contraction. Thick filaments are less numerous and thin filaments more numerous than in skeletal muscle, a reflection of the lower myosin content and higher actin content of smooth muscle. The arrangement of thick and thin filaments in smooth muscle appears nearly random (Figure 19.23). Sarcomeric organization and Z bands are absent. Instead, thin filaments are attached to dense bodies in the cytoplasm and at the plasma membrane. Despite the apparent lack of internal organization, there is electron microscopic evidence of cross-bridging between thick and thin filaments (Figure 19.23), and the contraction mechanism in smooth muscle is considered to be similar to the sliding filament model of contraction of striated muscle.

Contraction of smooth muscle depends on the intracellular concentration of Ca^{2+} ions, as in skeletal and cardiac muscle. The mechanism by which Ca^{2+} regulates contraction, however, is different in smooth muscle. Recall that contraction of skeletal muscle is actin regulated by Ca^{2+} interaction with troponin (p. 514). The thin filaments of smooth muscle contain actin and tropomyosin but apparently lack troponin. Instead, Ca^{2+} activates smooth muscle by *myosin-linked regulation*. Smooth muscle myosin can interact with actin filaments only when the myosin light chains are phosphorylated. Ca^{2+} ions regulate the phosphorylation of myosin light chains indirectly, by combining with the Ca^{2+}-binding protein *calmodulin*. The Ca^{2+}-calmodulin complex activates the enzyme *myosin light-chain kinase*, which phosphorylates the myosin light chains, initiating contraction and allowing continued cross-bridge cycling as long as Ca^{2+} is present. Myosin-linked regulation of contraction (rather than troponin–tropomyosin–actin-linked regulation, as in skeletal muscle) also occurs in muscle of mollusks and several other invertebrate groups and in nonmuscular actin–myosin contractile systems (see Other Cellular Movements, p. 538).

Excitation–contraction coupling is also rather different in smooth muscle. An organized system of t tubules is absent, and the sarcoplasmic reticulum is usually much less extensive. Some of the Ca^{2+} ions that activate contraction may come from the sarcoplasmic reticulum, but much of the calcium enters the smooth muscle cell directly across the outer plasma membrane. Because smooth muscle cells are so small, no part of the cell is more than a few micrometers from the membrane surface. This short diffusion path and the slow contraction time of smooth muscle make unnecessary the elaborate system of t tubules and sarcoplasmic reticulum of skeletal muscle. Depolarization of the surface membrane makes it permeable to Ca^{2+} ions, allowing Ca^{2+} entry down its concentration gradient. Both the plasma membrane and the sarcoplasmic reticulum contain Ca^{2+} pumps that maintain a very low resting internal Ca^{2+} concentration. Thus, depolarization allows Ca^{2+} entry to trigger contractions, and contractions are terminated by pumping Ca^{2+} back out. In some smooth muscle cells, the sarcoplasmic reticulum forms junctions with the outer membrane that may be analogous to dyads, triggering additional release of Ca^{2+} from the sarcoplasmic reticulum.

Unlike skeletal muscle, smooth muscle can be activated in a confusing variety of ways. Skeletal muscle is excited only via action potentials in the motor neurons innervating it. Cardiac muscle is activated by electrotonic spread of depolarization from pacemaker regions, which are rhythmically spontaneously active; the rate of pacemaker activity may be modulated by regulatory nerves and hormones (see Chapter 14). Smooth muscle can be excited spontaneously, by nerves, by hormones, and in some cases by stretch of the muscle. All these sources of excitation act ultimately by increasing the concentration of free intracellular Ca^{2+} ions.

Some smooth muscle cells undergo changes in membrane potential and tension in the apparent absence of external stimulations. This *endogenous activity* (also termed spontaneous activity) often consists of slow depolarizations that may trigger action potentials if they exceed the voltage threshold of the fiber (Figure 19.24). Even subthreshold depolarizations may produce measurable tension in the muscle, but the amount of tension is greater when action potentials are generated. In smooth muscle, the in-

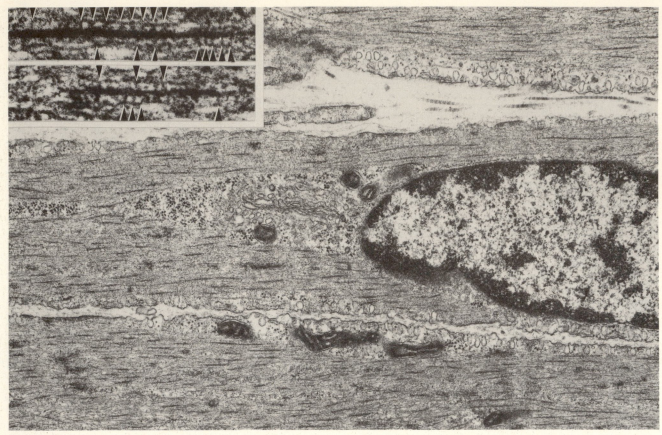

Figure 19.23 Electron micrograph of portions of three smooth muscle fibers. Thick and thin filaments are present but are not aligned in sarcomeres. (*Insert*) Higher magnification showing thick filaments with projections (arrows) interpreted as cross-bridges that connect to adjacent thin filaments. [From A.P. Somlyo, C.E. Devine, A.V. Somlyo, and R.V. Rice, *Philos. Trans. R. Soc. London* B ser. **265**:223–229 (1973); micrograph courtesy of A.P. Somlyo.]

ward current of the action potential is a Ca^{2+} influx, so action potentials lead to a direct increase in intracellular Ca^{2+} concentration.

Smooth muscle contraction is controlled in varying degrees by nerves and by hormones. A few smooth muscles lack innervation, but most are innervated by the autonomic nervous system (Chapter 2). Sympathetic postganglionic axons, parasympathetic postganglionic axons, or both branch and ramify among the muscle fibers. The axons have repeated varicosities, giving them a beaded appearance. Nerve activity releases transmitter molecules from the varicosities. Unlike vertebrate skeletal muscle, smooth muscle can be both excited and inhibited by transmitter actions. Moreover, a given transmitter can be either excitatory or inhibitory. For example, most vascular smooth muscle is excited to contract by norepinephrine from its sympathetic innervation and inhibited (relaxed) by parasympathetic acetylcholine. Intestinal smooth muscle, on the other hand, is inhibited by sympathetic norepinephrine. The differences in action of norepinephrine reflect differences in the type of postsynaptic transmitter receptor present in the smooth muscles: Vascular smooth muscle contains predominantly α-adrenergic recep-

tors and intestinal smooth muscle has primarily β-adrenergic receptors (see Table 17.2).

Smooth muscle contraction can also be controlled or modulated by a variety of other factors. Hormones such as circulating epinephrine and oxytocin can directly control contraction. Local factors such as acidity, oxygen and ion concentrations, prostaglandins (see Chapter 21), and osmolarity can also affect contraction. For example, during exercise the arterioles and precapillary sphincters dilate in regions of intense metabolic activity, in response to locally decreased O_2 concentration and increased concentrations of CO_2, H^+, and other metabolites (see Chapter 14). Finally, some smooth muscle can be activated mechanically, depolarizing and contracting in response to sudden stretching.

Smooth muscle is a diverse tissue that tends to defy generalization, and many of the above descriptions have exceptions. One useful, simplifying classification scheme differentiates two major types of vertebrate smooth muscle: *single-unit* and *multiunit* smooth muscles. In single-unit smooth muscle, the muscle cells are tightly electrically coupled by numerous gap junctions. Because of this tight electrical coupling, groups of muscle fibers are depolarized and

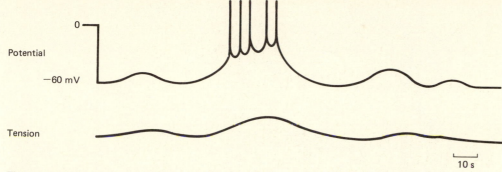

Figure 19.24 Endogenous activity in smooth muscle. Smooth muscle fibers may undergo apparently spontaneous changes in membrane potential and tension. The tension changes with fluctuations in membrane potential, larger depolarizations triggering action potentials and larger increases in tension.

contract together, functioning as a single unit. Single-unit smooth muscle is often spontaneously active, action potentials propagating from cell to cell via the gap junctions. Neural and hormonal controls may modulate the endogenous activity to varying degrees, but some single-unit smooth muscles are not innervated. The smooth muscles of the uterus, gastrointestinal tract, and small-diameter blood vessels are examples of the single-unit types.

Multiunit smooth muscles are less extensively interconnected by gap junctions, so that the muscle cells function more nearly like separate and independent units. They are extensively innervated by autonomic nerves and are under more direct neural control than are single-unit smooth muscles. Multiunit smooth muscles may or may not generate action potentials and may be activated hormonally as well as neurally. They are not stretch sensitive. Mus-

cles of the iris, skin hairs, large arteries, and large respiratory airways are examples of multiunit smooth muscles.

Table 19.3 compares the properties of the major types of vertebrate muscle.

Asynchronous Insect Flight Muscles   There are two kinds of muscular mechanisms for generating flight in insects. In grasshoppers, moths, dragonflies, and butterflies, the flight muscles attach directly to the wings (Figure 19.25A). Each motor neuron impulse triggers contraction of the muscle, so that the contractions that raise and lower the wing are synchronous with the respective motor neuron impulses. This type of flight muscle is called *synchronous* or *direct flight* muscle. In contrast, in Diptera (flies), Hymenoptera (bees and wasps), Coleoptera (beetles), and Hemiptera (true bugs), the flight muscles are *asyn-*

Table 19.3   CHARACTERISTICS OF VERTEBRATE MUSCLE FIBERS

Characteristic	Skeletal muscle	Smooth muscle Single-unit	Smooth muscle Multiunit	Cardiac muscle
Thick and thin filaments	Yes	Yes	Yes	Yes
Sarcomeres—banding pattern	Yes	No	No	Yes
Transverse tubules	Yes	No	No	Yes
Sarcoplasmic reticulum (SR)	+ + +	+	+	+ +
Source of activating calcium	SR	SR and extra-cellular	SR and extra-cellular	SR and extra-cellular
Site of calcium regulation	Troponin (thin filaments)	Myosin (thick filaments)	Myosin (thick filaments)	Troponin (thin filaments)
Speed of contraction	Fast–slow	Very slow	Very slow	Slow
Inherent tone (low levels of maintained tension)	No	Yes	No	No
Spontaneous production of action potentials by pacemakers	No	Yes	No	Yes
Gap junctions between fibers	No	Yes	Few	Yes
Effect of nerve stimulation	Excitation	Excitation or inhibition	Excitation or inhibition	Excitation or inhibition
Physiological effects of hormones on excitability and contraction	No	Yes	Yes	Yes
Quick stretch of fiber produces contraction	No	Yes	No	Yes
Slow wave potentials may be present	No	Yes	No	No

Source: A.J. Vander, J.H. Sherman, and D.S. Luciano, *Human Physiology,* 3rd ed. McGraw-Hill, New York, 1980.

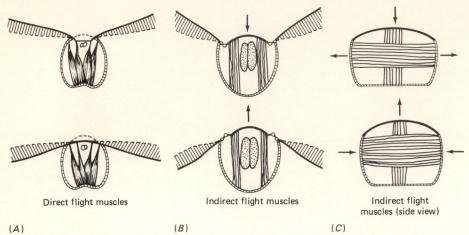

Direct flight muscles   Indirect flight muscles   Indirect flight muscles (side view)

(A)   (B)   (C)

Figure 19.25 Mechanics of direct and indirect flight muscles in insects. (A) In insects with direct flight muscles, the muscles are connected to the wings. (B) In insects with indirect flight muscles, the downstroke is produced by raising the roof of the thorax, a result of contraction of dorsal longitudinal muscles (shown in side view in C). The movement of the roof of the thorax moves the wings via a skeletal click mechanism that involves two hinges. [After G. Goldspink, in R. McN. Alexander and G. Goldspink (eds.), *Mechanics and Energetics of Animal Locomotion*. Chapman and Hall, London, 1977.]

chronous. In these groups, the frequency of wing movements is too great to be controlled directly by neuronal activation. Wing beat frequencies (and frequencies of wing muscle contractions) of 100–300/s are common, and a frequency of 1046/s has been recorded for a midge. The frequency of motor neuron impulses is much lower and is asynchronous with muscle contraction. In these insects, the flight muscles attach to the walls of the thorax rather than directly to the wings (Figure 19.25B). The muscles ''pop'' the thorax into one of two stable positions: one with the roof of the thorax elevated and the wings depressed, and the other with the roof of the thorax depressed and the wings elevated. Such flight muscles are termed asynchronous (because contraction is asynchronous with nerve excitation) or *indirect* flight muscles. Because asynchronous flight muscles have large fibers with prominent myofibrils, they are also called fibrillar flight muscles.

The most striking functional attribute of asynchronous flight muscle is that its contraction is normally triggered by stretch. When the wing depressors contract, they elevate the roof of the thorax (Figure 19.25B), stretching the wing elevators and triggering elevator contraction. Since the intermediate position of the thorax is unstable, the thorax ''pops'' into its elevated position, aiding the stretch of the wing elevators and suddenly shortening the wing depressors. This sudden shortening terminates the active state in the depressors, which relax until they are stretched again by depression of the thorax (= elevation of the wings). The depressors and elevators undergo the same cycle (but out of phase with each other): stretch → contraction → shortening → relaxation → stretch . . . . The frequency of the cycle depends on the resonant mechanical properties of the thorax and wings, not on the frequency of nerve impulses. If the wings are clipped, the flight frequency increases because of decreased wing inertia and air resistance, but the nerve impulse frequency is unchanged.

The stretch sensitivity of asynchronous flight muscle is a property of the contractile machinery, rather than of the fiber surface membrane. Membrane potential does not change during flight oscillation. Moreover, flight muscle can be extracted with glycerol, which destroys membrane systems but leaves the contractile machinery intact. Glycerol-extracted flight muscle will undergo oscillatory contraction if provided with a resonant mechanical system to permit it to shorten and stretch alternately.

Asynchronous flight muscle has sarcomeres with an extremely regular arrangement of myofilaments. The thick myosin filaments extend nearly to the Z line, so that the I bands are extremely short. The length–tension relation of contraction is very steep, no tension being generated at lengths less than 90 percent of the peak-tension length (cf. Figure 19.17). The chemical composition of the filaments and the mechanism of contraction appear to be similar to those of vertebrate skeletal muscles. The mechanism of stretch sensitivity, on the other hand, is not well understood. Ca^{2+} is necessary for stretch-sensitive contraction, and there is evidence that stretch increases the activity of Ca^{2+}-activated myosin ATPase. Stretch thus acts like an increase in Ca^{2+} concentration. The thick myosin filaments have tapered ends and have been reported to attach to the Z line by fine extensions termed C filaments. Applied stretch might directly stress the thick filaments via the C filaments, somehow increasing Ca^{2+} binding.

The functional role of motor nerves in asynchronous flight muscles is reduced to one of enabling stretch to trigger contraction. Flight is terminated by cessation of nerve activity, permitting muscle fiber membrane repolarization and Ca^{2+} reuptake.

Catch Muscle   Many bivalve mollusks have non-striated adductor muscles with unusual catch properties. The term *catch* means that the contracted state is greatly prolonged. A brief train of nerve impulses can initiate a prolonged catch contraction that can persist for hours or even days, without further nerve impulses and with little expenditure of energy. The muscle in the relaxed state is pliable and easily stretched, but in catch it is relatively rigid and resistant to stretch. Bivalve mollusks are relatively sedentary and ill adapted to "outrun" predators (although some have escape swimming responses). Shell-closing muscles that can remain contracted for long periods without fatigue are therefore of obvious adaptive significance.

Muscle in catch remains contracted until a relaxation mechanism is activated. In intact bivalves, the maintenance of catch is associated with low-frequency nerve activity, but catch persists after the nerves to the muscle are cut. Both the development and the termination of the catch state are neurally mediated, by separate neurons. Acetylcholine induces contraction, and serotonin (5-HT) and dopamine both induce relaxation from catch. The motor neurons for contraction are apparently cholinergic and the relaxation-inducing motor neurons are probably serotonergic (and possibly dopaminergic). Relaxation of muscle fiber catch is not an inhibition. Muscle fibers are not hyperpolarized during relaxation, and relaxing agents such as serotonin actually increase the amplitude of phasic contraction (in addition to preventing catch).

The mechanism of the catch property is not known but is thought to result from a delay in breaking actin–myosin cross-bridges. Catch muscle fibers are 1–2 mm long and 4–5 μm in diameter and are interconnected by gap junctions. They lack sarcomeres and differ from other smooth muscles by containing large amounts of the protein *paramyosin*. The thick filaments are much thicker than in other muscles (40–120 nm versus 15 nm) and consist of a paramyosin core surrounded by myosin molecules. Thin filaments are numerous (24 thin filaments per thick filament) and contain actin and tropomyosin but lack troponin. They attach to dense bodies in the cytoplasm, as in vertebrate smooth muscle. Contraction requires Ca^{2+} ions, which act via myosin-linked regulation (also like vertebrate smooth muscle). Since other aspects of catch muscles appear similar to other smooth muscles, attention has focused on paramyosin to explain the catch property. There is no clear evidence for conformational changes in paramyosin during catch, however, and paramyosin's effect on cross-bridges would have to be indirect, since the paramyosin is in the interior of the thick filaments. The relaxing effect of serotonin is thought to be mediated by cAMP, which increases fourfold before relaxation occurs. Phosphorylated paramyosin inhibits actin-activated myosin ATPase in vitro. One hypothesis is that Ca^{2+} remains bound to contractile proteins and blocks the breaking of cross-bridges, until it is released by a cAMP-dependent process.

Finally, it should be noted that many invertebrate muscles that lack the catch property contain paramyosin, although in concentrations an order of magnitude lower. Paramyosin must therefore have other functions in muscle, and its presence is not a sufficient condition for catch.

NEURAL CONTROL OF MUSCLE

In order for muscle to be physiologically and behaviorally useful to an animal, its contraction must be precisely controlled. Tension must be developed and maintained smoothly, and the degree of tension must be under continuously graded control, rather than appearing as a series of twitchlike, episodic jerks. In this section, we discuss the ways in which the amount of tension in muscle is controlled by nervous systems. The generation of temporal patterns of motor output that underlie coordinated movements is considered in Chapter 20. This section considers neural control of skeletal muscle only. As discussed above, in smooth muscle, neural control is only one of several ways of controlling tension development, and myogenic cardiac muscle is not directly controlled by nerves, which only modulate intrinsic contractions.

There are two contrasting evolutionary approaches ("strategies") for the gradation of amount of tension of a muscle, one exemplified by vertebrates and the other by arthropods. We can call these approaches the *vertebrate plan* and the *arthropod plan* (Figure 19.26). In most of the other well-studied invertebrate groups, muscle tension is controlled using variations on the arthropod plan.

Vertebrate Twitch Muscle

A vertebrate skeletal muscle is innervated typically by about 100–1000 motor neurons. Each muscle fiber receives endings of only one motor neuron, and each motor neuron typically innervates a few hundred muscle fibers. A motor neuron and all the muscle fibers it innervates are termed a **motor unit**. When the motor neuron generates an impulse, all the muscle fibers in the motor unit generate impulses and contract. As discussed earlier (Mechanics of Muscle Contraction), a single motor neuron impulse elicits a single twitch contraction. Repeated impulses at low frequencies elicit a series of partially summating, unfused twitches, and repeated impulses at higher frequencies elicit a fused tetanic contraction (Figure 19.14). Fusion occurs at less than 100 impulses/s in mammalian slow twitch, oxidative muscle fibers and

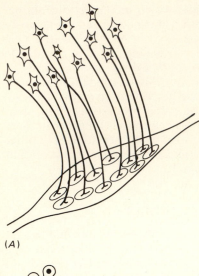

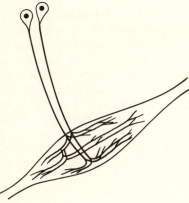

Figure 19.26 Organization of neural control of muscle. (A) Vertebrate plan in which there are many motor neurons, each innervating a separate array of muscle fiber (nonoverlapping motor units). (B) Arthropod plan in which there are a few motor neurons, each innervating many or most of the muscle fibers (overlapping motor units).

at about 300 impulses/s in fast twitch, glycolytic fibers. For most vertebrate muscles, the amount of tetanic tension is only two to five times the twitch tension. Since there is no possible gradation of tension between inactivity and a twitch, vertebrate twitch muscle has only a very limited ability to gradate tension *within* a motor unit.

Above the level of the individual motor unit, in contrast, there must be a large degree of tension control, since vertebrates can control the amount of tension in a whole muscle over a range of several hundredfold. This control is exerted by *recruitment* of additional motor units of the muscle. Individual motor neurons in the pool innervating a muscle require different amounts of excitatory synaptic input to activate them (see Chapter 17, section entitled Integration of EPSPs and IPSPs). An increase in muscle tension is effected by an increased excitatory "synaptic drive," which has two effects on the motor neuron pool: (1) It increases the frequency of action

potentials in motor neurons that are already active, increasing the tension *within* motor units; and (2) it brings more motor neurons to threshold, activating more motor units. This latter mechanism of recruitment of additional motor units is the dominant means of control of amount of tension in vertebrate twitch muscle. Because most vertebrate muscles have several hundred motor units, recruitment provides a finely graded control over the amount of whole-muscle tension.

The recruitment of additional motor units does not occur randomly. Rather, motor units tend to be recruited in a sequence from the smallest to the largest motor units. This tendency is called the *size principle*. A number of studies have shown that many aspects of motor unit size are positively correlated: Motor neurons with larger somata have larger, more rapidly conducting axons than motor neurons with smaller somata, and they innervate larger numbers of muscle fibers. The muscle fibers of these large motor units are fast twitch, glycolytic, rapidly fatiguing fibers, whereas the muscle fibers of smaller motor units are slow twitch, oxidative, and resistant to fatigue (see Table 19.2). Large motor neurons have relatively high thresholds of activation by synaptic input. When central neural circuits excite the motor neurons to a muscle, the smaller motor neurons have the lowest thresholds and are recruited first. The smaller motor neurons each activate a relatively small number of slow oxidative (SO) muscle fibers, and each contributes a relatively small increment of tension. With greater central nervous excitation, successively larger motor neurons are recruited, each representing larger increments of tension. Thus, there is a tendency to recruit not constant increments of tension but a constant *percentage* increment. For example, a small motor neuron might be recruited at a tension level of 50 g and contribute a tension increment of 5 g. A large motor neuron might contribute 50 g of additional tension but not be recruited until the tension level is 500 g. The size principle also indicates that slower, fatigue-resistant muscle fibers are recruited at low tension levels and thus are used much of the time, for example, for maintenance of posture and slow movements. The fast, readily fatiguing muscle fibers of large motor units are recruited only when nearly maximal tension levels are needed, levels that cannot be sustained for more than a few seconds.

The above discussion of the size principle is oversimplified, since each motor unit does not contribute a fixed increment of tension; its tension can be varied a fewfold by tetanic summation as shown in Figure 19.14. The size principle reflects not a requirement but a *tendency* for motor neuron recruitment to occur in order of increasing size. In rapid movements of some mammalian muscles, larger motor neurons may precede smaller ones in order of activity. Nevertheless, the size principle is a useful generalization about the ordering of motor units within the pool that controls tension in a muscle.

Vertebrate Tonic Muscle

Although most vertebrate muscle fibers are twitch fibers, several vertebrate postural muscles contain some *tonic* muscle fibers that do not develop all-or-none action potentials. Tonic muscle fibers contract more slowly than the slowest twitch fibers, and the mode of their neural control is quite different. Figure 19.27 shows the pattern of innervation of a vertebrate twitch muscle fiber and a tonic muscle fiber. The twitch fiber has a single neuromuscular junction near the middle of the fiber; its activation results in a now familiar muscle fiber action potential that propagates over the length of the fiber. In contrast, a tonic muscle fiber has many neuromuscular junctions distributed over its length. This pattern is termed *multiterminal* innervation. Each motor neuron action potential results in an excitatory postsynaptic potential (EPSP) at each of the distributed junctions. The muscle fiber has little or no ability to generate action potentials; the amount of tension generated depends directly on the amount of depolarization produced by the EPSPs (Figure 19.28).

Tonic muscle fibers usually make up only a small proportion of the fibers of postural muscles. In some muscles of lower vertebrates (e.g., frog rectus abdominis and iliofibularis), tonic fibers may have a significant postural role. In mammals, however, tonic fibers occur only in extraocular muscles and as intrafusal fibers of muscle spindles (see Chapter 18). Most of the muscle fibers involved in postural maintenance of vertebrates are slow twitch fibers rather than tonic fibers.

The innervation pattern of vertebrate tonic muscle fibers is intermediate between that of vertebrate twitch muscle and that of arthropod muscle, to be described next. Thus, the distinction between a vertebrate pattern and an arthropod pattern of muscle innervation is an oversimplification of the diversity of modes of neuromuscular control.

Arthropod Muscle

A typical arthropod muscle is innervated by 1–10 motor neurons, in contrast to the hundreds or thousands of motor neurons that innervate a vertebrate muscle. Most arthropod muscle fibers are innervated by more than one motor neuron, a pattern termed *polyneuronal* innervation. Polyneuronal innervation is rare in vertebrates. Arthropod muscle fibers are multiterminally innervated (Figure 19.27), and most do not generate all-or-none action potentials.

Because arthropod muscle fibers are polyneuronally innervated, the motor units of arthropods are overlapping; each muscle fiber is part of several motor units. Thus, arthropods have few overlapping motor units per muscle, while vertebrates have many nonoverlapping motor units. Another feature of neural control of arthropod muscle that is absent for vertebrate muscle is *peripheral inhibition,* in which muscle fiber depolarization is directly opposed by action of inhibitory efferent neurons (see later section).

The control of amount of tension in arthropod muscles is very different from that in twitch muscles

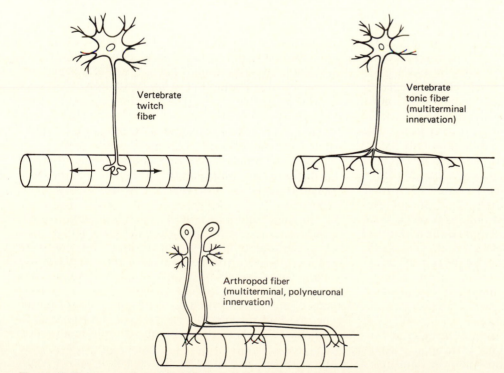

Figure 19.27 Innervation of vertebrate twitch and tonic fibers and arthropod fibers.

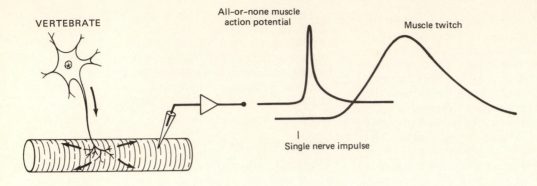

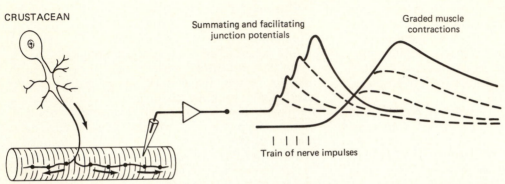

Figure 19.28 Comparison of vertebrate and arthropod muscle excitation. (*Top*) Vertebrate twitch muscle produces an all-or-none twitch in response to each all-or-none muscle action potential, which is propagated over the entire fiber. (*Bottom*) Most muscle fibers of crustaceans and insects produce graded contractions in response to graded, summating EPSPs. Distributed (multiterminal) neuromuscular junctions insure that all regions of the muscle fiber are depolarized. [From ANIMAL PHYSIOLOGY, 2nd ed. by Roger Eckert and David Randall. Copyright © 1978, 1983 W.H. Freeman and Company. Reprinted with permission.]

of vertebrates. Arthropod muscle tension is controlled largely or entirely by gradation of contraction within a motor unit, and recruitment of motor units is minimal. Since arthropod muscle fibers usually do not generate impulses, neuromuscular EPSPs must directly depolarize the membrane to trigger contraction (Figure 19.28). A single motor neuron impulse generates a single EPSP, which may elicit only a very small twitchlike contraction or none at all. Repeated motor neuron impulses produce sequential EPSPs that summate and facilitate (cf. Figure 17.32), their greater depolarizations eliciting stronger contractions (Figure 19.29). Thus, the dominant mechanism of tension control in arthropod muscles is by *control of the degree of depolarization of muscle fibers,* which in turn depends on the frequency of action potentials in a motor neuron. This mechanism is unavailable to vertebrate twitch muscle fibers, in which an EPSP triggers an all-or-none action potential. In contrast, recruitment of additional motor units is at most a minor strategy of tension control in arthropods, for several reasons. Many arthropod muscles have only one excitatory motor neuron, and many muscles innervated by two or more excitatory motor neurons may use only one in executing a contraction

in nature (see examples below). At most, arthropod muscles have only a handful of tension increments available by recruitment; fine control of muscle tension must depend largely on gradation of effects within motor units.

Examples of Neuromuscular Control in Arthropods
The neuromuscular organization of arthropods is sufficiently diverse to resist generalization. We shall describe a few representative examples that provide a (nonexhaustive) overview of control of arthropod muscles.

In one common pattern of muscle control, the muscle is innervated by two excitatory motor neurons and one inhibitory neuron. Many limb muscles of insects and crustaceans fit into this category, the claw closer muscle of crabs being a well-studied example (Figure 19.30). The excitatory motor neurons are often differentiated as a phasic ("fast") motor neuron and a tonic ("slow") motor neuron. Stimulation of the phasic axon produces rapid contraction of muscle, while tonic axon stimulation elicits slow contraction. The tonic motor neuron is active much of the time and is used for slow movements and postural functions; the phasic axon is normally silent

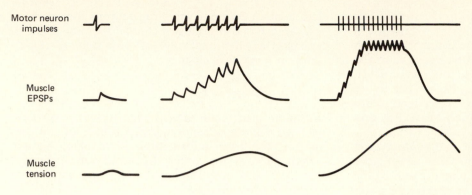

Figure 19.29 Effect of frequency of stimulation on EPSP summation and on tension in arthropod muscle. Higher frequencies of motor neuron impulses produce greater depolarizations by summation of EPSPs, and thus greater muscle tensions.

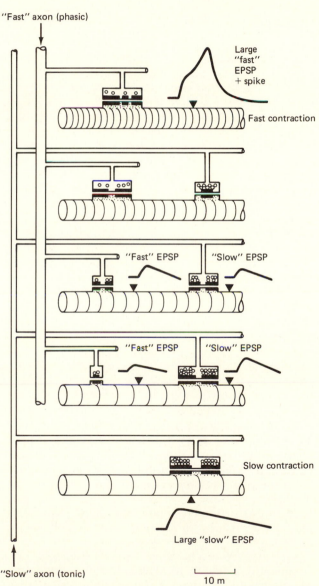

"Fast" axon (phasic)

Large "fast" EPSP + spike

Fast contraction

"Fast" EPSP   "Slow" EPSP

"Fast" EPSP   "Slow" EPSP

Slow contraction

"Slow" axon (tonic)

10 m

Large "slow" EPSP

Figure 19.30 Crustacean neuromuscular innervation. The diagram illustrates the matching of motor axons with muscle fibers of different types (indicated by sarcomere length) in a crab muscle. The phasic ("fast") axon preferentially innervates short-sarcomere fibers that produce fast contractions. The tonic ("slow") axon preferentially innervates long-sarcomere fibers that produce slow contractions. Some short-sarcomere fibers in this muscle can produce regenerative spike responses. [From H.L. Atwood, *Am. Zool.* **13**:357–378 (1973).]

and is activated in short bursts to produce rapid movements. The claw closer muscle contains fibers of a variety of sarcomere lengths and speeds of contraction (cf. Figure 19.22). The most rapidly contracting, short-sarcomere fibers are innervated preferentially or exclusively by the phasic axon, and the slowest, long-sarcomere fibers are only supplied by the tonic axon (Figure 19.30). Fibers with intermediate contraction times are innervated by both axons, but there is still a matching of motor neuron and muscle fiber properties, longer-sarcomere fibers receiving stronger excitation from the "slow" axon, and shorter-sarcomere fibers being more excited by the fast axon. Thus, the muscle can contract over a range of speeds and durations, as a result of matching of nerve and muscle fiber properties and connections.

Many arthropod muscles are innervated by only a single excitatory motor neuron, with or without peripheral inhibition. An extreme example is that of the crustacean claw opener muscle and the stretcher (extensor) of the next more proximal leg segment. These two muscles share a *single* excitatory motor neuron between them. Each muscle is supplied with a separate inhibitor, so that with coactivation of the excitatory motor neuron and one or the other inhibitor, the two muscles can be activated somewhat independently. The ways in which peripheral inhibitory neurons act are discussed in the following section.

Most arthropod muscles are mixed muscles, containing fibers with a variety of sarcomere lengths and contraction speeds. Some muscles, however, are composed of all long-sarcomere slow fibers or all short-sarcomere fast fibers. One example of such a division, previously mentioned, is the set of abdominal flexor and extensor muscles of crayfish and lobsters (Figure 19.22). The slow flexor and extensor muscles each receive one inhibitory and five excitatory tonic motor neurons, although most fibers are innervated by two to four axons rather than all six. Many of the phasic muscles receive three excitatory axons and an inhibitor. Chelicerate arthropods such as *Limulus* have leg muscles with denser innervation, with as many as 11 excitatory axons to a muscle and as many as six to a single fiber.

Peripheral Inhibition   There are two modes of peripheral inhibition of arthropod muscle: postsynaptic inhibition and presynaptic inhibition (Figure 19.31). In *postsynaptic* inhibition the inhibitory axon ends in multiterminal synapses on the muscle fiber. Each impulse in the inhibitory axon causes release of the transmitter GABA (gamma-aminobutyric acid), which produces an IPSP at the muscle fiber by an increase in Cl^- conductance. The IPSPs tend to oppose both membrane depolarization by EPSPs and resultant contraction.

At some muscles the endings of an inhibitory axon also terminate on excitatory axon terminals (Figure 19.31) and exert *presynaptic inhibition,* decreasing the amount of excitatory transmitter released as a result of excitatory motor neuron impulses (cf. Figure 17.25). For presynaptic inhibition to be effective, the inhibitory neuron impulse must arrive at the terminals shortly before the excitatory neuron impulse. In a few cases where presynaptic inhibition is relatively strong, impulse generation in the inhibitory neuron is timed within the central nervous system to arrive at the terminal shortly before excitatory impulses, maximizing the presynaptic inhibitory effect.

Of the two mechanisms of peripheral inhibition, postsynaptic inhibition is much more widespread. All peripheral inhibitory axons exert a postsynaptic effect. Presynaptic inhibition, in contrast, is significant only in some limb muscles of Crustacea (although it has been reported in an insect). The functions of peripheral inhibition are not clear. Certainly it provides another variable in the control of muscle tension, and in the opener and stretcher muscles of crustacean claws, it provides a measure of independent control of the two muscles that share a single excitatory axon. In several cases, the inhibitor neuron is active when the excitatory neurons to a muscle are silent and vice versa. In such cases, the inhibitor cannot serve to grade the amount of tension and may serve instead to speed the rate of relaxation of a contracted muscle. Finally, crustacean and insect

legs contain common inhibitor neurons that each innervate several muscles, including functional antagonists. The functional roles of these common inhibitors are even less clear.

OTHER EFFECTORS

Effectors other than muscle tend to be overlooked, because of the widespread importance of muscles and the wealth of information about them. We will briefly survey some other effectors, more to remind the reader of their existence than to convey any great depth of knowledge.

Independent Effectors

Effector cells that function without requiring external control are called *independent effectors*. The term has implied that such effector cells are themselves sensitive to stimuli that elicit the effector response; that is, independent effectors act as their own receptors. The term would also apply, however, to a hypothetical effector cell that was autonomously rhythmically active in the absence of stimulation. Independent effectors may have been important starting elements in the evolution of nervous systems (cf. p. 387).

Examples of independent effectors include several primitive kinds of muscle cells. Sponges contain porocytes—conical cells with a channel through the middle through which water enters the interior of the sponge. The porocytes can contract to close the channel, for example, to prevent sediments in turbid water from entering. Epitheliomuscular cells of coelenterates and smooth muscle precapillary sphincters of vertebrate circulatory systems (see Smooth Muscle section above and Chapter 14) may also function as independent effectors.

Nematocysts   Nematocysts are examples of more complex independent effectors that are not related to muscle. Many coelenterates possess specialized epithelial cells called nematocytes or cnidoblasts. A nematocyte produces a nematocyst as an intracellular organelle (Figure 19.32). The nematocyst contains a coiled, introverted thread that can be explosively discharged when an appropriate stimulus contacts the nematocyte surface. The thread is everted by the discharge and may wrap around, stick to, or penetrate the unfortunate stimulating object. Coelenterates use nematocysts for prey capture and for defense.

Nematocyst discharge is usually triggered by a combination of mechanical and chemical stimulation. Many nematocysts have a triggerlike protrusion, the cnidocil, which is thought to act as the sensor initiating discharge. Mechanical stimulation by a clean glass rod is a relatively ineffective stimulus by itself; chemical priming appears to be more important. If the glass rod is smeared with foodstuffs or if extracts are added to the bath, light mechanical stimulation

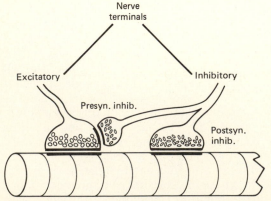

Figure 19.31 Structural basis of presynaptic and postsynaptic inhibition of arthropod muscle. Inhibitory nerve terminals form neuromuscular synapses (postsynaptic inhibition) and axo-axonal synapses upon excitatory terminals (presynaptic inhibition). [From F. Lang and H.L. Atwood, *Am. Zool.* **13**:337–355 (1973).]

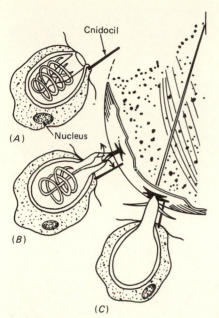

Figure 19.32 Structure of a nematocyst of a coelenterate, (A) before discharge, (B) during discharge, and (C) after discharge. [From A.C. Hardy, *The Open Sea, Its Natural History, part 1, The World of Plankton.* Collins, London, 1956.]

elicits a vigorous discharge. Proteins and especially lipids are effective in lowering the threshold to mechanical stimulation.

The mechanism of nematocyst discharge is poorly understood. Presumably, a pressure increase within the nematocyst drives the explosive eversion of the thread. The pressure increase may result from inflow of water by osmotic forces or by hydration of a colloid within the nematocyst or may result from contraction of the capsule by contractile fibers in the nematocyte.

Many Protozoa contain explosive organelles termed *trichocysts* that are analogous in function to the nematocysts of coelenterates.

Cilia and Flagella Cilia and flagella are hairlike cell organelles that act as motile cellular appendages. They are similar in diameter (0.2 μm) and internal structure but differ in length. Cilia are generally less than 15 μm long while flagella may be as long as 100–200 μm. Cilia and flagella or their derivatives occur in all animal phyla. They constitute the primary locomotor structures of many Protozoa and of several kinds of metazoan cells (e.g., spermatozoa, ciliated epithelia of flatworms, mollusks, and vertebrate trachea and oviduct). Moreover, ciliary derivatives occur in a wide variety of photoreceptor, mechanoreceptor, and chemoreceptor cells (Chapter 18), including those of arthropods and nematodes, two groups that are sometimes described as lacking cilia.

The internal structure and the molecular composition of cilia and flagella have been studied by electron microscopy and biochemistry. Since the major

features are comparable for all eukaryotic cilia and flagella, we shall describe a "standard" cilium without considering variants. A cilium is bounded by an evagination of the plasma membrane. At the base of the cilium is a **basal body,** from which a ciliary rootlet projects into the cytoplasm of the cell (Figure 19.33). The major internal structures of a cilium are **microtubules,** which extend continuously from the base to the tip. The microtubules are arranged in a so-called 9 + 2 configuration, consisting of nine outer doublets surrounding two single central microtubules (Figure 19.33). Each microtubule is a hollow cylinder composed of polymers of the globular proteins α- and β-tubulin. The outer doublets each consist of a complete tubule (the A tubule) and an attached incomplete B tubule (Figure 19.34). Each A tubule bears two **dynein arms** that project laterally toward the B tubule of the next doublet. Less distinct radial links (radial spokes) connect the outer A tubule to the region of the two inner microtubules and nexin links connect adjacent outer doublets (Figure 19.34). The entire array of microtubules and associated arms and links is called the **axoneme.**

The beating movement of a cilium can be divided into two phases (Figure 19.35): a power stroke or effective stroke, in which the cilium is held rigid and bends at the base, and a recovery stroke, in which a wave of bending proceeds from the base to the tip, the cilium being held relatively close to the cell surface. Flagellar beating is more complex and more variable.

The force-generating mechanism that produces ciliary and flagellar movement must act to bend the organelle. In principle, there are two ways in which this bending could occur: (1) Microtubules could change length, those on the inside of the bend shortening or those on the outside of the bend lengthening; or (2) microtubules could slide past each other without themselves changing length. There is considerable evidence favoring the latter mechanism. First, in electron micrographs of the tips of cilia, the microtubules at the inside of the bend always project farther than the microtubules at the outside of the bend (Figure 19.36). This difference is a necessary condition of microtubular sliding and is inconsistent with bending resulting from microtubular shortening. Second, no structural changes have been seen in the microtubules themselves during bending, even in cases in which a 40–50 percent local length change would be required to produce the observed degree of bending. Finally, there is direct evidence that microtubules slide past each other by an ATP-dependent mechanism. It is possible to remove the membranes from isolated cilia or flagella with detergent. If the resultant axonemes are treated with trypsin, the radial spokes and nexin links are digested away. With addition of ATP (in the presence of Mg^{2+}), the doublet tubules slide relative to each other (Figure 19.37). Since the isolated protein dynein possesses Mg^{2+}-dependent ATPase activity, it is inferred that

(A)

(B)

Figure 19.33 Structure of cilia, shown in electron micrographs. (A) Transverse section; (B) Longitudinal section. Rootlets (R) attach to basal bodies (B) of the cilia. [Electron micrographs courtesy of P. Satir.]

the dynein arms "walk" along the next adjacent B tubule in a fashion analogous to the cross-bridge cycle of muscle contraction. In the intact cilium, this sliding of tubules is converted to a bending of the cilium, presumably by the attachments of radial spokes to the central microtubules to constrain the sliding. The roles of these radial spokes and the nexin links are complex and incompletely understood.

Cilia are independent effectors, but their beating is usually coordinated. The cilia of protozoans and of epithelia beat in *metachronal* waves in which all adjacent cilia are in phase with each other (Figure 19.38). In most cases, the cilia are entrained by viscous forces of the water, but in ctenophores the co-

ordination appears to be maintained by underlying nerves. Neural stimulation can increase or decrease the frequency of ciliary beating in mollusks and vertebrates. Thus, the capacity for independent ciliary action does not preclude exogenous modulation of the activity.

Other Cellular Movements Many cells possess the ability to move in whole or in part. The amoeboid movement of protozoans, leucocytes, and developing cells is well known, but other less dramatic examples are becoming appreciated only recently. It is beyond our scope to discuss general cell movement, and the interested reader should consult the refer-

(A)

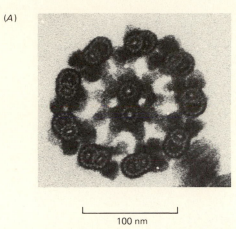

(B)

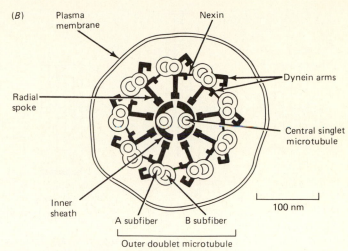

Plasma
membrane

Nexin

Radial
spoke

Dynein arms

Central singlet
microtubule

Inner
sheath

A subfiber   B subfiber

Outer doublet microtubule

100 nm

100 nm

Figure 19.34 (A) Electron micrograph of a cilium in cross section. The distinctive arrangement of 9 doublet outer microtubules and 2 singlet inner microtubules is characteristic of cilia and eukaryotic flagella. (B) Diagram of a cilium in cross section, showing projections from the microtubules. These projections are thought to mediate interactions between microtubules in ciliary movement. [(A) courtesy of R. Link; (B) from B. Alberts, D. Bray, J. Lewis, M. Raff, K. Roberts, and J.D. Watson, *Molecular Biology of the Cell.* Garland, New York, 1983.]

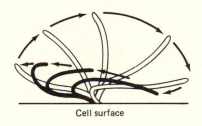

Cell surface

(A)

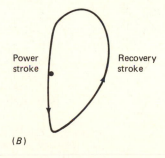

Power
stroke

Recovery
stroke

(B)

Figure 19.35 Ciliary beat cycle. (A) Seen from the side; power stroke shown in white, recovery stroke in black. (B) Seen from above; spot indicates the position of the base of the cilium, and arrow traces the path of the ciliary tip.

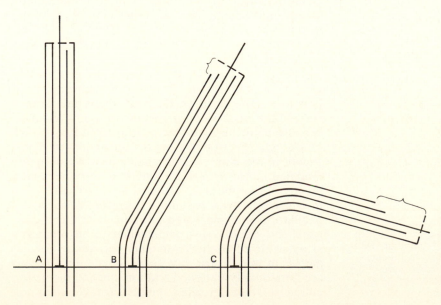

Figure 19.36 Diagram illustrating the sliding of microtubules during ciliary bending. The shorter lines represent the outer doublet microtubules; the longer middle line represents the central microtubules. In the bent position (A, B), the doublets on the inside of the bend extend farther than those on the outside. This observation suggests that the microtubules slide relative to each other, without themselves changing in length. The bracket indicates the extent of sliding between the doublets on the inside and outside of the bend. [From H. Stebbings and J.S. Hyams, *Cell Motility.* Longmans, London, 1979.]

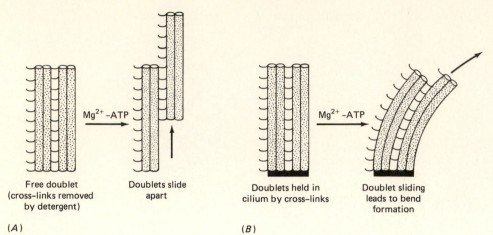

Mg²⁺–ATP

Mg²⁺–ATP

Free doublet
(cross-links removed
by detergent)

Doublets slide
apart

Doublets held in
cilium by cross-links

Doublet sliding
leads to bend
formation

(A)

(B)

Figure 19.37 Diagram illustrating (A) the experimental demonstration of sliding of outer microtubule doublets *in vitro,* and (B) the bending of the doublets if they are bound together at one end, as in the cilium. [From B. Alberts, D. Bray, J. Lewis, M. Raff, K. Roberts, and J.D. Watson, *Molecular Biology of the Cell.* Garland, New York, 1983.]

ences at the end of the chapter. We note only that most cells contain actin and myosin, as well as microtubules (Figure 19.39). The degrees to which these proteins function as a skeletal support or as a force-generating mechanism for movement, however, remain to be clarified.

Electric Organs

Electric organs are effector organs specialized for the production of an electric field outside the body.

Figure 19.38 Coordination of ciliary beating in the ciliate protozoan *Paramecium.* Adjacent cilia beat together, with metachronal waves travelling over the surface. In this scanning electron micrograph the metachronal waves were "frozen" by rapid fixation. [From S.L. Tamm, *J. Cell Biol.* **55:**250–255 (1972). Micrograph courtesy of S.L. Tamm.]

They occur only in fish but apparently have evolved independently in at least six groups of elasmobranchs and teleosts.

There are two types of electric fish: strongly and weakly electric. Strongly electric fish emit a powerful, painful discharge to defend against predators and to stun prey. Strongly electric fish have been known since ancient times, but weakly electric fish have been recognized only rather recently. Charles Darwin considered the evolution of electric organs to be a "case of special difficulty." The strongly electric fish that he knew of had large, specialized electric organs of clear adaptive value, but he could not imagine how natural selection could have favored hypothetical ancestral fish with primitive electric organs too weak to stun anything. One hundred years after publication of *The Origin of Species,* experiments demonstrated that other fish (now classed as weakly electric) generated weak electric fields around themselves and oriented by detecting changes in these fields with electroreceptors (see Chapter 18). This demonstration of electro-orientation resolved Darwin's paradox.

The electric organs of both strongly electric and weakly electric fish are derived from neuromuscular junctions. The cells of electric organs are termed *electrocytes* (also called myocytes or electroplaques, because many are flattened cells arrayed like a stack of plates). Electrocytes are modified muscle cells, except in the family Stenarchidae in which they are modified motor neuron terminals. Electrocytes vary greatly in morphology, but most are flattened, asymmetrical cells with a smooth innervated face and a papilliform noninnervated face. In most electric organs, the electrocytes are stacked in an organized array, with all the cells similarly oriented. For example, in the electric ray *Torpedo,* the electrocytes are horizontally flattened and are stacked in vertical columns. There are over 1000 electrocytes in a col-

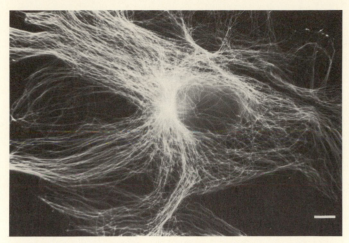

Figure 19.39 Proteins associated with cellular movements are revealed in a wide variety of cells by immunofluorescence microscopy. These proteins include actin, myosin, and (in this micrograph) tubulin. The cell is a cultured lung epithelium cell of a newt (*Taricha granulosa*), labelled with a monochonal antibody to β-tubulin. The marker bar is 10 μm. [Micrograph courtesy of P. Wadsworth.]

umn and 500–1000 columns in an organ. In the electric eel *Electrophorus,* the columns are longitudinal, parallel to the spinal cord. There are about 60 columns on each side, with 6000–10,000 electrocytes per column. The electric organs of *Electrophorus* can generate a discharge of over 500 V, while *Torpedo* can generate 60 V and several amperes (1 kilowatt).

How are modified muscle cells able to generate such a large electrical output? All electric organs share the same basic principles of operation. (1) The two faces of an activated electrocyte generate different membrane potentials, so that a potential difference of tens of millivolts occurs across the whole electrocyte (Figure 19.40A). (2) The electrocytes in a column are activated at the same time. Thus, the small potential differences across each electrocyte are added to produce a large voltage across the entire

column, in the same way that four 1.5-V batteries connected in series produce 6 V (Figure 19.40B). (3) Columns of electrocytes are arranged in parallel to each other. The simultaneous activation of many columns does not further increase the voltage but instead increases the current (just as adding batteries in parallel increases the current; Figure 19.40C). (4) The electrical resistance of myocyte membranes (and of skin and other structures in series with the myocytes) is typically low, tending to maximize current flow. (5) The arrangement of accessory structures such as connective tissue tends to channel current flow through the electric organ.

There are two ways in which electrocytes can generate electric currents: by action potentials and by neuromuscular synaptic potentials. In the electric organs of freshwater teleosts, myocytes generate ac-

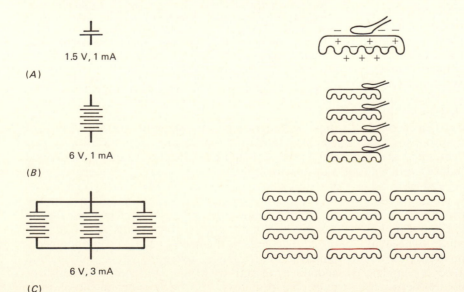

(A) 1.5 V, 1 mA

(B) 6 V, 1 mA

(C) 6 V, 3 mA

Figure 19.40 Addition of electrocytes in series increases the resultant voltage; addition of electrocytes in parallel increases the current. (*Left*) The 1.5-V battery cells are connected together. (*Right*) The analogous arrangements of electrocytes is depicted. Each electrocyte generates 100–150 mV rather than 1.5 V, but the effect of their arangement in columns is similar to adding plates in a battery.

tion potentials, whereas in the organs of marine elasmobranchs and teleosts only synaptic potentials are
generated. Let us further consider a case in which
myocytes generate action potentials, in the electric
eel *Electrophorus*. Only the innervated face of the
eel myocyte generates an action potential. This is
shown in Figure 19.41, which depicts the effect of
repeatedly depolarizing the innervated face of an
electrocyte while slowly advancing one of two microelectrodes through the electrocyte. When both
electrodes are outside the innervated face (Figure
19.41A), no potential difference appears in response
to depolarization with external electrodes. With one
microelectrode inside the electrocyte (B), a resting
potential and an action potential in response to depolarization are recorded. When the microelectrode
is advanced through the uninnervated face of the
electrocyte (C), the resting potential disappears
(showing that both faces have the same resting potential), but the depolarization-induced action potential is essentially unchanged. Thus, the full amplitude
of the action potential appears across the entire electrocyte, since only the innervated face is activated
(D). [The fact that the action potential amplitude is
undiminished in (C) indicates that the resistance of
the uninnervated face is very low and that accessory
structures act as a high-resistance seal preventing
current flow around the edges of the electrocyte.]
Since the electric organs of *Electrophorus* contain
thousands of simultaneously activated electrocytes
in series, and since each active electrocyte can generate a 140-mV pulse, the summated output of the
organ can be hundreds of volts. The high voltage is
thought to be necessary to overcome the high resistance of the freshwater medium.

The electric organs of the marine elasmobranch
ray *Torpedo,* in contrast, generate a discharge of
unusually high current. *Torpedo* electrocytes do not
generate action potentials and are not directly electrically excitable. Instead, the entire innervated face
of the electrocyte consists of a modified neuromuscular junction. *Torpedo* electrocytes contain the
highest overall concentration of acetylcholine receptors known and in fact are the major source of acetylcholine receptors used for biochemical studies.
When the motor axon terminal is depolarized and
releases acetylcholine, the resulting activation of
these acetylcholine receptors produces a large EPSP
with an extraordinary synaptic current. Since only
the innervated face of each electrocyte produces the
EPSP, the EPSPs of simultaneously activated electrocytes in a column sum to produce a large voltage.
The large number of columns in parallel and the large
synaptic currents of each electrocyte maximize the
current output of the *Torpedo* electric organ. This
high current output renders the organ effective in the
ray's high-conductivity seawater medium. The marine teleost *Astroscopus,* unlike freshwater teleosts,
also has non-impulse-generating electrocytes similar
to those of *Torpedo*. Thus, the distinction between

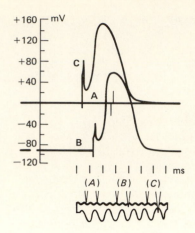

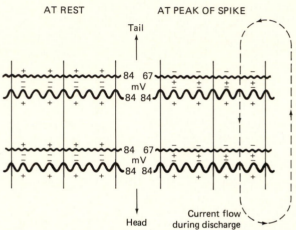

Figure 19.41 Responses of electrocytes in the electric eel.
One electrode is advanced through an electrocyte as shown
in the diagram, and separate external electrodes depolarize
the innervated face of the cell. (A) Both electrodes external to
the innervated face; no response is seen. (B) One electrode is
advanced into the cell. The inside-negative resting potential
and an overshooting action potential of about 140 mV are
recorded. (C) The electrode is advanced to outside the uninnervated face of the cell. The resting potential disappears, but
the spike is essentially unchanged. (D) Mechanism of additive
discharge in eel electrocytes. At rest, there is no net potential
across the cells. At the peak of the spike, all the potentials are
in series (see Fig. 19.40), and the head of the eel becomes
positive with respect to the tail. [From R.D. Keynes and H.
Martins-Ferreira, *J. Physiol* **119**:315–351 (1953).]

impulse-generating and synaptic-current types of
electrocyte is an evolutionary adaptation to the medium, rather than a phyletic separation of teleost and
elasmobranch lines.

The electric organs of weakly electric fish are all
of the impulse-generating type. In contrast to the
rarity of discharge of strongly electric fish, weakly
electric fish emit pulses continually, often at high
rates. The pulses themselves may be monophasic,
diphasic, or triphasic, reflecting more complicated
sequences of activity of the two faces of the electrocyte. Nevertheless, the basic mechanism of their

pulse generation is broadly similar to that described for *Electrophorus* (Figure 19.41). The role of electric organs in electro-orientation of these fish is considered in Chapter 18.

Bioluminescence

Many organisms are able to produce light by means of catalyzed chemiluminescent reactions in which a substrate is oxidized to a light-emitting product. Bioluminescence occurs in bacteria, algae, fungi, and in members of all major phyla of animals. Because the occurrence of bioluminescence is very sporadic among representatives of different groups, the phenomenon is thought to have evolved independently many times. In many organisms, the functional significance of bioluminescence is not clear. Primitive bacteria, for example, may have first evolved bioluminescent reactions as an adaptation to rid cells of free oxygen during early stages of biochemical evolution, when an accumulation of free oxygen may have been toxic to anaerobic organisms. Bioluminescence in present-day aerobic bacteria might thus be a mere evolutionary vestige. In several animal groups, however, bioluminescent organs are highly developed and play clear roles in mating, defense, or prey capture. We will illustrate some aspects of bioluminescence by considering one example—that of the fireflies or lightning bugs (actually beetles of the order Coleoptera).

Biochemistry of Luminescence In chemiluminescent reactions, the change in free energy associated with the reaction raises electrons to an excited state of high energy level. When the excited electrons return to the ground state, they give off photons (quanta of light). In bioluminescent reactions, the oxidation of a *luciferin* substrate by molecular oxygen is catalyzed by a *luciferase* enzyme. Different organisms have different molecular species of luciferin and luciferase. The general overall reactions of bioluminescence are:

$$\text{luciferin (L)} + O_2 + \text{luciferase (E)} \rightarrow \text{E–L*}$$

$$\text{E–L*} \rightarrow \text{E} + \text{L} + h\upsilon \text{ (light)}$$

The symbol L* indicates an electron-excited state. Bioluminescent reactions usually require a source of chemical energy to raise electrons to the excited state. In the firefly example we are considering, this energy is supplied in the form of ATP. The structures of firefly luciferin and of its products are shown in Figure 19.42. The first reaction in the firefly is the formation of a luciferin–luciferase–AMP complex, the AMP complexing with the carboxyl group of reduced luciferin (LH_2):

$$\text{E} + LH_2 + \text{ATP} \xrightarrow{\text{Mg}^{2+}} \text{E–}LH_2\text{—AMP} + \text{PP}$$

This enzyme–luciferyl–adenylate complex then reacts with oxygen to yield a series of intermediates

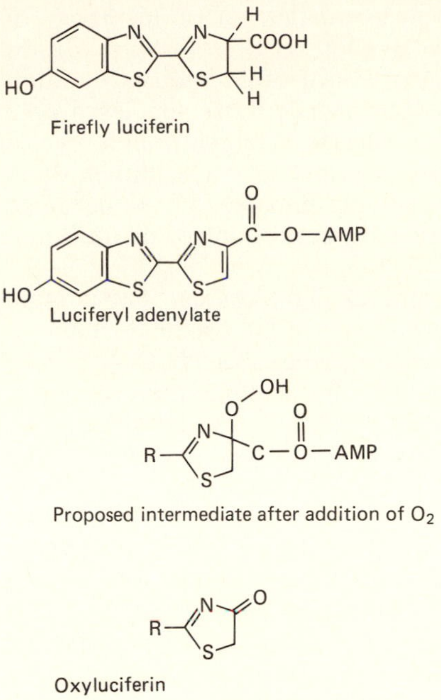

Figure 19.42 Molecular structures of firefly luceferin and of its derivatives in firefly luminescence.

with loss of AMP and CO_2, finally forming an excited oxyluciferin (L*), which emits a photon on return to the ground state:

$$\text{E–}LH_2\text{–AMP} + O_2 \rightarrow \text{L} + CO_2 + \text{AMP} + \text{light}$$

Bioluminescent reactions in other organisms are broadly similar but involve different luciferins, luciferases, and cofactors. In the marine ostracod crustacean *Cypridina,* bioluminescent reactions are extracellular, unlike the intracellular reactions of fireflies. *Cypridina* luciferin and luciferase are stored in separate glands. When stimulated the ostracod secretes luciferin and luciferase into the surrounding seawater, where they react with molecular oxygen to produce blue light. This reaction does not require ATP or Mg^{2+}. Another variant on the reactions of bioluminescence occurs in the luminescent jellyfish *Aequorea*. A photoprotein, isolated from this coelenterate and termed aequorin, binds Ca^{2+} ions and produces light in a reaction that requires no exogenous oxygen. Aequorin has been shown to consist of a stable complex of luciferin and luciferase that contains its own oxygen within the photoprotein. The action of Ca^{2+} ions is to release the stored oxygen, triggering light production.

Control of Firefly Luminescent Organs The behavioral functions of firefly bioluminescence are under neural control. In *Photinus pyralis*, the common firefly of the eastern United States, flying males flash at intervals of about 7 s; nonflying females flash back with a latency of 2 s at 25°C. Neither sex responds to flashes at inappropriate intervals, which are presumably interpreted as responses of members of other species. Females of some *Photinus* species may attract and eat males of other species by mimicking the flashes of females of those species. In the genus *Pteroptyx* of southeast Asia, males of a species congregate in dense swarms in trees and flash in synchrony to attract flying females. These behavior patterns require precise neural control of the luminescent organs.

Fireflies have complex light organs that contain photocytes (light-producing cells), tracheae, and motor nerves. In *Photuris*, rosettes of photocytes surround cylindrical spaces that each contain a trachea and the nerve supply. Tracheoles fan out between each of the photocytes. Synapses end not on the photocytes but on tracheal end cells that enclose the junctions between tracheae and tracheoles. There are two kinds of theories about how photocytes flash in response to a neural signal. In the first kind of theory, control involves the admission of oxygen to otherwise anoxic photocytes, perhaps by transmitter action on the tracheal end cells. The other type of theory postulates that axon endings release a transmitter that directly stimulates light production in photocytes. There is evidence that the transmitter is octopamine (an amine related to norepinephrine), which induces flashing by a cAMP-dependent mechanism. The second theory would explain control of light organs of a wider variety of species, many of which lack tracheal end cells or tracheae themselves. It is possible, however, that the varieties of control mechanisms of bioluminescence may turn out to be nearly as diverse as the varieties of bioluminescence itself.

Chromatophores

Chromatophores are pigmented effector cells. Changes in the form or the number of pigmented cells underlie changes in color and color pattern of animals. Many color patterns change only slowly, if at all—with developmental stage, sexual maturation, or season. These gradual changes, termed ***morphological color changes,*** typically involve changes in amount of pigment and numbers of pigmented cells. They are not considered further here. Several groups of animals, in contrast, employ movement of pigments within specialized chromatophores to produce relatively rapid color changes. These more rapid changes, termed ***physiological color changes,*** are typically controlled by nerves, hormones, or both. The chromatophores mediating physiological color changes are thus true effectors.

The capacity for physiological color change is most highly developed among cephalopod mollusks, crustaceans, and lower vertebrates (but not birds and mammals). Members of other groups, including some annelids, echinoderms, and insects, can also control their color changes. The most common function of physiological color change is protective, tending to

match the animal's coloration to the background. Other functions may include aggressive and sexual displays, thermoregulation, and protection of tissues from intense light.

Structure and Mode of Action of Chromatophores There are two basic types of chromatophore. In one type, found only in mollusks, a central baglike pigment cell is attached to a radiating array of muscle fibers (Figure 19.43*A*). The pigment cell contains an elastic sac filled with pigment granules. Contraction of the radial muscles pulls outward on the membrane of the pigment cell, stretching the elastic sac into a flattened sheet and spreading the pigment over a large area. When the muscles relax, the elasticity of the pigment sac restores it to a small ball. In the relaxed, concentrated state of the chromatophore, the pigment cell membrane is highly convoluted. Thus, a molluscan chromatophore is actually a tiny organ. The neuromuscular control of molluscan chromatophores provides them with a rapidity and a precision of control of color change unequaled by other animals.

The second, more common type of chromatophore is a very irregularly shaped cell in which pigment granules can be concentrated in a small central region or dispersed throughout the cytoplasm (Figure 19.43*B*). The size and shape of the pigment cell itself apparently does not change; the only change is in the degree of dispersal of pigment granules within it. The term chromatophore usually refers to this latter type unless otherwise specified. Such chromatophores change more slowly than do molluscan chromatophores, but there are considerable differences within the type in rapidity of change and in the degree of concentration possible. Chromatophores are named by color: melanophores are black or brown, xanthophores are yellow, erythrophores are red, and iridophores are silvery white.

The mechanism of pigment migration in classical (nonmolluscan) chromatophores is only recently becoming understood. Studies by Porter and his associates have shown that the cytoplasm of cells is enmeshed by a microtrabecular lattice of microtubules, microfilaments, and intermediate filaments. In chromatophores, the movement of pigment granules is associated with changes in the microtrabecular lattice. The aggregation of pigment granules is correlated with the shrinking and collapse of the microtrabecular lattice. Pigment dispersal is associated with the reformation of the lattice, which appears to carry pigment granules toward the periphery. The dispersal requires ATP while concentration is triggered by an increase in cytoplasmic Ca^{2+} concentration. The role of ATP in dispersal could be in microtrabecular lattice assembly, in pumping Ca^{2+} out of the cytoplasm, or both.

Control of Chromatophores The two different types of chromatophore—the muscle-associated chromato-

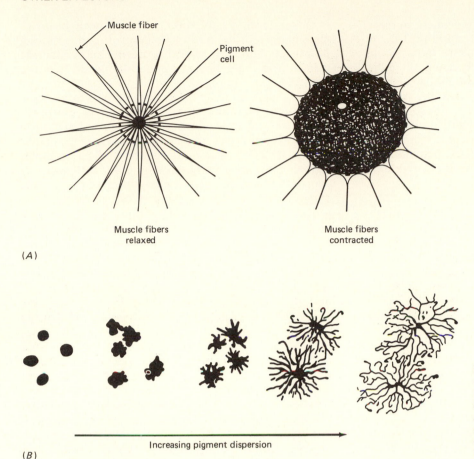

Muscle fiber

Pigment cell

Muscle fibers relaxed

Muscle fibers contracted

(A)

Increasing pigment dispersion

(B)

Figure 19.43 Chromatophores of a cephalopod mollusk (A) and of an amphibian (B). Cephalopod chromatophores are expanded by contraction of radially arranged muscle fibers; in amphibian (and other) chromatophores pigment is dispersed or contracted by intracellular movements of pigment granules.

phores of cephalopod mollusks and the more common chromatophores in which pigment granules migrate in the cytoplasm—clearly have different modes of control. Cephalopod chromatophores are under rapid and precise neuromuscular control. A chromatophore motor neuron innervates the radiating muscle fibers of several chromatophores. Most chromatophores are controlled by a single motor neuron, but some appear to be controlled by two or more motor neurons. The motor neurons controlling cephalopod chromatophore muscles originate in color control centers in the subesophageal ganglion. The coordination of complex waves of color change that sweep over the animal and the rapidity of individual chromatophore responses (0.2–1 s) must require complex central neural integration.

The single-celled chromatophores, best studied in crustaceans and vertebrates, may be under hormonal control, nervous control, or both. Crustacean chromatophores are entirely under hormonal control (p. 613). Vertebrate chromatophores are controlled both by hormones and by the autonomic nervous system, separately or in combination. Hormonal control occurs in all the lower vertebrates, while neural control mechanisms are largely confined to teleosts and reptiles. The major hormonal regulators are melanocyte-stimulating hormones (MSH) from the pituitary in-

termediate lobe, the catecholamines norepinephrine and epinephrine from the adrenal medulla, and melatonin from the pineal organ. In melanocytes, MSH causes pigment dispersal via an increase in levels of intracellular cAMP (see Chapters 17 and 21). The catecholamines can induce either pigment dispersal (by activating β-adrenergic receptors and increasing cAMP) or pigment concentration (by activating α-adrenergic receptors and apparently decreasing cAMP). A single chromatophore may have either or both kinds of adrenoreceptor; in the latter case, the α receptors are usually dominant. Other hormonal effects may also occur; for example, a melanophore-concentrating hormone may be released from the posterior pituitary of teleosts.

Neural control of chromatophores, when present, is predominantly via the sympathetic nervous system. Norepinephrine, released by sympathetic postganglionic neurons, can concentrate pigment in melanophores by activating α receptors or can disperse pigment by activating β receptors. Acetylcholine can also produce pigment dispersal in some melanophores and may be released in nerve-induced dispersal in teleosts.

Finally, some vertebrate chromatophores can respond to changes in local light intensity, either directly as photosensitive independent effectors or via local reflexes mediated by dermal photoreceptors.

Xenopus chromatophores disperse pigment in direct response to local light, even after total destruction of the spinal cord or after maintenance in tissue culture. The cells thus act as independent effectors. Chromatophores of blinded chameleons also disperse with increasing local light intensity, but this response ceases when the skin is denervated. These and further experiments have demonstrated that chameleon chromatophores are not independent effectors, but rather have coordinated nonvisual responses in which dermal photoreceptors in the skin activate the chromatophores via their motor nerves.

Glands

The most widespread effector organs of animals are muscles and glands. In fact, experimental analysts of animal behavior have argued that at least in mammals, all observable behavior consists of muscle contractions and glandular secretions. Thus, glands comprise the second major kind of effectors in animals. There are two types of gland: *endocrine* glands (Chapter 21) that secrete hormones into the bloodstream and *exocrine* glands that secrete material to a body surface, often via a duct. Exocrine glands are subdivided into *external* glands that deliver secretions to the outer body surface (e.g., sweat glands, glands secreting pheromones, toxins, inks, adhesives, or mucus) and *internal* glands such as salivary glands, mucus glands, and other digestive glands. Many other cells secrete substances but are not normally considered to be gland cells. For example, probably all neurons secrete synaptic transmitter substances but are not termed gland cells. Only those neurons that are specialized to secrete hormones into the blood are considered glandular and are termed ***neurosecretory*** or ***neuroendocrine*** cells (see Chapter 21).

In this section, we consider the basic mechanisms of glandular secretion and of its control, concentrating on aspects common to both exocrine and endocrine glands. Specific aspects of endocrine gland function are treated in Chapter 21.

Most secreted substances are peptides or proteins or protein-containing complexes such as glycoproteins. We will examine the synthesis and secretion of digestive enzymes in mammalian pancreatic acinar cells as a well-studied and probably typical example. Acinar cells are exocrine gland cells that are specialized to secrete digestive enzymes and their precursors into the digestive tract. The movements of protein secretory products within acinar cells were first studied by incubating the cells with a brief pulse of [³H] amino acids, fixing cells at various times after the pulse labeling and examining the distribution of labeled, newly synthesized proteins by electron microscopic autoradiography, which showed the distribution of labeled protein as developed silver grains in the thin photographic emulsion coating the tissue sections. As shown in Figure 19.44, the first appearance of labeled protein was at the rough endoplasmic reticulum (ER). (In fact, such studies were one of the first demonstrations that proteins are synthesized at the rough ER.) About 20 min after labeling, proteins were found to be associated with the Golgi complex, and at later times the label was localized in membrane-bounded *secretory vesicles* or *secretory granules*. Studies indicate that secretory proteins are synthesized across the rough ER membrane and enter the lumen of the rough ER, and are then transported to the Golgi complex by shuttle vesicles. Within the Golgi complex, material destined for secretion is thought to be segregated from other cellular components, modified (such as by addition of carbohydrate or lipid components), and concentrated into

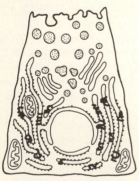
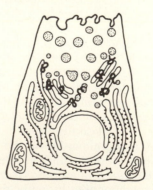
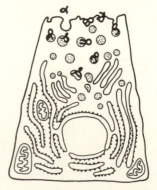

3 min:
Silver grains over the ER

20 min:
Silver grains over the Golgi apparatus

90 min:
Silver grains over secretory vesicles

Figure 19.44 Experimental demonstration of the sites of protein synthesis and packaging in a mammalian pancreatic acinar cell. Cells are briefly pulse labeled with [³H] amino acids, fixed at various times after the label, and examined by electron microscopic autoradiography. Developed silver grains, showing the locations of newly synthesized protein molecules, are localized first at the rough ER, then at the Golgi complex, and finally at secretory vesicles. [From B. Alberts, D. Bray, J. Lewis, M. Raff, and J.D. Watson, *Molecular Biology of the Cell.* Garland, New York, 1983.]

secretory vesicles that pinch off from the outer cisternae of the releasing face of the Golgi complex. The secretory vesicles remain in the cytoplasm until released by exocytosis, a process that requires energy and free calcium ions (see below). Cells that synthesize and secrete lipids (such as steroid hormones) have an extensive elaboration of tubular smooth ER, steroids being synthesized in smooth rather than rough ER.

Control of Glandular Secretion   The secretory activity of exocrine and endocrine glands can be under endocrine control, neural control, or both. Gland cells bear membrane receptors for hormones and neurotransmitters, and the activation of these receptors depolarizes the gland cells. Some gland cells such as pituitary cells and insulin-secreting pancreatic islet cells can generate action potentials as a result of depolarization. The depolarization (with or without action potentials) elicits secretion, a process termed *excitation–secretion coupling*. Glandular secretion results from an increased concentration of intracellular free Ca^{2+} and Ca^{2+}-dependent exocytosis of secretory vesicles, factors similar to the control of neurotransmitter release at chemical synapses. From this similarity, it has been suggested that excitation–secretion coupling in gland cells and in synaptic terminals may have a common mechanism and possibly a common evolutionary origin. The similarity of these two forms of secretion is shown in Figure 19.45.

How are gland cells excited by hormones or neurotransmitters? It is tempting to hypothesize that the actions of hormones and transmitters that stimulate

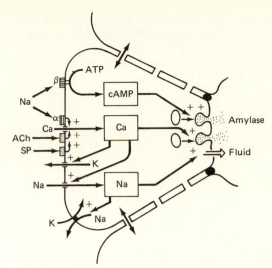

Figure 19.46   Schematic diagram of a gland cell, summarizing the mechanisms of excitation-secretion coupling. Several external agents can affect secretion, acting via one or more second messengers. Note the polarity of the cell: excitation occurs at the outer surface, and secretion at the inner, lumenal surface. [From O.H. Petersen, *The Electrophysiology of Gland Cells.* Academic Press, London, 1980.]

secretion (termed *secretagogues*) are similar to the actions of transmitters at nerve–nerve and nerve–muscle synapses. Recent microelectrode studies of gland cells, however, suggest a degree of diversity in the stimulation of secretion. For example, in mammalian salivary gland cells, acetylcholine, norepinephrine, and substance P all depolarize the cell, decrease membrane resistance, and stimulate secretion. Experimentally induced increases in intracellular Ca^{2+} concentration mimic all three effects, a finding suggesting that receptor activation leads to increased intracellular Ca^{2+} (both from membrane-bound intracellular stores and from entry of extracellular Ca^{2+}), which in turn increases Na^+ and K^+ permeability (Figure 19.46). Experimental activation of noradrenergic β receptors, in contrast, stimulates enzyme secretion via a cAMP-dependent mechanism, with relatively little change in membrane resistance. Thus, as indicated in Figure 19.46, some secretagogues exert convergent effects on a common mechanism (ACh, substance P, α-receptor action of NE on intracellular Ca^{2+}) while other actions may involve a parallel mechanism (β-receptor action of NE on cAMP). The situation is similar in pancreatic acinar cells, which are stimulated by acetylcholine and by several gut hormones. Acetylcholine, cholecystokinin, and bombesin all act to increase intracellular Ca^{2+} concentration; Ca^{2+} triggers secretion and may also increase Na^+, Cl^-, and K^+ permeabilities, producing depolarization. Other hormones (secretin, vasoactive intestinal polypeptide—VIP) stimulate secretion by increasing cAMP, with little effect on membrane potential or resistance. It is likely that further studies will uncover greater diver-

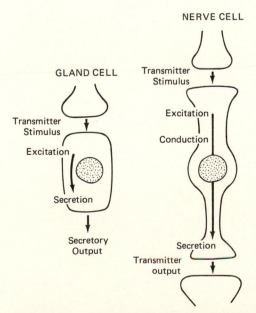

Figure 19.45   Diagrams showing the similarities between the processes of excitation-secretion coupling in gland cells and in nerve cells [From G.M. Shepherd, *Neurobiology.* Oxford University Press, New York, 1983.]

sity in stimulus–secretion coupling, as is the case with synaptic transmission mechanisms.

SELECTED READINGS

Alberts, B., D. Bray, J. Lewis, M. Raff, K. Roberts, and J.D. Watson. 1983. *Molecular Biology of the Cell.* Garland, New York.

Amos, W.B. and J.G. Duckett (eds.). 1982. *Prokaryotic and Eukaryotic Flagella.* Society of Experimental Biology Symposium 35. Cambridge University Press, Cambridge.

Bulbring, E., A.F. Brading, A.W. Jones, and T. Tomita. 1981. *Smooth Muscle: An Assessment of Current Knowledge.* Arnold, London.

Burke, R.E. 1981. Motor units: Anatomy, physiology and functional organization. In V. Brooks (ed.), *Handbook of Physiology. Section 1: The Nervous System,* Vol. 2, Part 1, pp. 345–422. American Physiological Society, Bethesda, MD.

Darnell, J., H. Lodish, and D. Baltimore. 1986. *Molecular Cell Biology.* Scientific American Books, New York.

Design and Performance of Muscular Systems (Review Volume). 1985. *J. Exp. Biol.* **115.**

Enoka, R.M. 1984. Henneman's "size principle": Current issues. *Trends Neurosci.* **7**:226–228.

Gauthier, G.F. 1987. Vertebrate muscle fiber types and neuronal regulation of gene expression. *Am. Zool.* **27**:1033–1042.

*Goldspink, G. 1977. Design of muscles in relation to locomotion. In R.McN. Alexander and G. Goldspink (eds.), *Mechanics and Energetics of Animal Locomotion,* pp. 1–22. Chapman and Hall, London.

Goldspink, G. 1981. Design of muscle for locomotion and the maintenance of posture. *Trends Neurosci.* **4**:218–221.

Govind, C.K. and H.L. Atwood. 1982. Organization of neuromuscular systems. In H.L. Atwood and D.C. Sandeman (eds.), *Biology of Crustacea,* Vol. 3. Academic, New York.

Hartshorne, D.J. and R.F. Siemankowski. 1981. Regulation of smooth muscle actomyosin. *Annu. Rev. Physiol.* **43**:519–530.

Henneman, E. and L.M. Mendell. 1981. Functional organization of motoneuron pool and its inputs. In V. Brooks (ed.), *Handbook of Physiology. Section 1: The Nervous System,* Vol. 2, Part 1, pp. 423–507. American Physiological Society, Bethesda, MD.

Hopkins, C.D. 1988. Neuroethology of electric communication. *Annu. Rev. Neurosci.* **11**:497–535.

Hopkins, C.R. and C.J. Duncan (eds.). 1979. *Secretory Mechanisms.* Society of Experimental Biology Symposium 33. Cambridge University Press, Cambridge.

Hoyle, G. 1982. *Muscles and Their Nervous Control.* Wiley, New York.

Huxley, A. 1988. Prefatory chapter: Muscular contraction. *Annu. Rev. Physiol.* **50**:1–16.

Kelly, R.B. 1985. Pathways of protein secretion in eucaryotes. *Science* **230**:25–32.

Martonosi, A.N. 1984. Mechanism of Ca^{2+} release from sarcoplasmic reticulum of skeletal muscle. *Physiol. Rev.* **64**:1240–1320.

McElroy, W.D. and M. DeLuca. 1978. Chemistry of firefly luminescence. In P.J. Herring (ed.), *Bioluminescence in Action,* pp. 109–127. Academic, New York.

Petersen, O.H. and I. Findlay. 1987. Electrophysiology of the pancreas. *Physiol. Rev.* **67**:1054–1116.

Poisner, A.M. and J.M. Trifaro. 1985. *The Electrophysiology of the Secretory Cell.* Elsevier, Amsterdam.

Porter, K.R. and J.B. Tucker. 1981. The ground substance of the living cell. *Sci. Am.* **244**(3):57–67.

Rüegg, J.C. 1986. *Calcium in Muscle Activation.* Springer, Berlin.

Schliwa, M. and U. Euteneuer. 1983. Comparative ultrastructure and physiology of chromatophores, with emphasis on changes associated with intracellular transport. *Am. Zool.* **23**:479–494.

Silverman, H., W.J. Costello, and D.L. Mykles. 1987. Morphological fiber type correlates of physiological and biochemical properties of crustacean muscle. *Am. Zool.* **27**:1011–1019.

Squire, J. M. 1981. *The Structural Basis of Muscle Contraction.* Plenum, New York.

*Squire, J.M. 1986. *Muscle: Design, Diversity, and Disease.* Benjamin/Cummings, Menlo Park, CA.

*Stebbings, H. and J.S. Hyams. 1979. *Cell Motility.* Longmans, London.

Twarog, B.M., R.J.C. Levine, and M.M. Dewey. 1982. *Basic Biology of Muscles: A Comparative Approach* (Society of General Physiologists Series V. 37). Raven, New York.

*Vander, A.J., J.H. Sherman, and D.S. Luciano. 1985. *Human Physiology,* 4th ed. McGraw-Hill, New York.

White, D.C.S. 1977. Muscle mechanics. In R.McN. Alexander and G. Goldspink (eds.), *Mechanics and Energetics of Animal Locomotion,* pp. 23–56. Chapman and Hall, London.

White, D.C.S. 1987. Muscle mechanics. In H. McLennan, J.R. Ledsome, C.H.S. McIntosh, and D.R. Jones (eds.), *Advances in Physiological Research,* pp. 271–293. Plenum, New York.

See also references in Appendix A.

chapter *20*

Control of Movement: Examples of the Organization of Nervous Function

All externally observable behavior is a direct result of activation of effectors. That is, all behavior that can be seen by an observer is a series of movements (usually resulting from muscle contractions), production of sounds (also from muscle contractions), gland secretions, color changes, and other miscellaneous effector responses. Thus, the mechanisms by which animals generate patterns of behavior involve the control of muscles and other effectors to produce patterned movements, secretions, and so on. There is a school of psychology (behaviorism) that also views human behavior in terms of effector activity, ignoring mental states. We can leave such issues to the psychologists (but see introduction to Chapter 16); for our purposes, it is sufficient to assert that animal behavior results from the coordinated activation of effectors. The nervous system exerts the major and immediate control of effectors to generate behavior, although hormonal and other controls may also be important. In this chapter we will consider issues in the neural control and coordination of movement, particularly including behaviorally significant patterns of movement such as walking, swimming, and flying.

BEHAVIORAL BACKGROUND: REFLEXES AND FIXED ACTION PATTERNS

Kandel has provided a useful classification of behavior (see Selected Readings list at end of chapter). *Complex behavior* consists of patterns or sequences of responses (e.g., locomotion or feeding) that may involve several effector organs. *Elementary behavior* consists of an isolated response of a single effector organ. Complex behavior is considered to be made up of sequences of elementary behavior (see Table 20.1). Both the patterns of complex behavior and the acts of elementary behavior are classified as lower-order behavior; higher-order behavior consists of still more elaborate patterns or sequences of activity such as courtship, nest building, or communication. In general, higher-order behavior is too complicated to be analyzed readily in terms of neural control of effectors and is not considered further here.

Lower-order behavior, both elementary and complex, can be categorized further as either reflex or fixed (Table 20.1). *Reflex* acts and patterns are responses to an eliciting stimulus; reflex acts are graded, being greater or more complete with a stronger eliciting stimulus. A familiar example of a spinal reflex is the flexion reflex of a dog, in which a leg is lifted (flexed) in response to a sharp or painful stimulus. Reflex patterns of behavior require continuous sensory stimulation, each movement producing sensory feedback that stimulates or modifies further movements in the sequence (see below, Neural Generation of Rhythmic Behavior). Reflexive behavior can be learned and may be subject to considerable learned modification. A familiar example is that of Pavlov's classical conditioning experiments, in which dogs were trained to salivate in response to a bell by repeated pairing of the bell and a food stimulus.

The terms *fixed acts* and *fixed action patterns* stem from the ethological studies of Lorenz and Tinbergen, prior to and during the 1950s. Fixed action patterns are stereotyped, species-specific, and little modified by learning. A fixed action pattern is typically triggered by a specific stimulus (the sign stimulus), but once initiated it proceeds in an all-or-nothing way. For example, a greylag goose retrieves an egg that has rolled out of the nest by a fixed action pattern of bill movement; once this response is triggered, it goes to completion even if the egg is re-

Table 20.1 CLASSIFICATION OF TYPES OF BEHAVIOR

Lower-order behavior	*Higher-order behavior*
Elementary (acts)	Courtship
Reflex acts	Nest-building
Fixed acts	Communication
Complex (patterns)	
Reflex patterns	
Fixed action patterns	

Source: E.R. Kandel, *Cellular Basis of Behavior*. Freeman, San Francisco, 1976.

moved. Fixed action patterns may occur in the apparent absence of a stimulus, a phenomenon known as vacuum activity. Single fixed acts are usually triggered by stimuli, but unlike reflex acts (which are smoothly graded responses that depend on the amount and form of the stimulus), fixed acts are all-or-none responses the form of which is relatively independent of the amount or form of stimuli that trigger them (Figure 20.1). The distinctions among reflex acts, fixed acts, and action patterns are not absolute, but rather represent attempts to categorize behaviors that in fact fall along a continuum. Nevertheless, the distinction between stereotyped, fixed patterns of behavior and stimulus-dependent, modifiable reflex patterns remains an important organizing principle for studies of the physiological basis of behavior.

A major question in the analysis of the neural basis of behavior is: What are the neural circuits—the patterns of synaptic interconnection of neurons—by which particular patterns of behavioral movements are generated? In this chapter, we examine several neural circuits and relate their actions to the behavior they mediate.

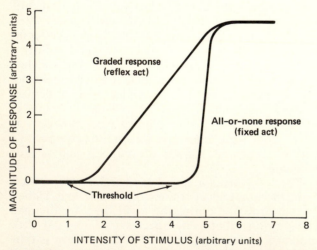

Figure 20.1 Comparison of graded reflex acts and all-or-none fixed acts. The stimulus–response curve of the reflex act has a gradual slope, while that of the fixed act has a steep slope. [From CELLULAR BASIS OF BEHAVIOR: An Introduction to Behavioral Neurobiology, by Eric R. Kandel. Copyright © 1976 W.H. Freeman and Company. Reprinted with permission.]

NEURAL CIRCUITS MEDIATING REFLEXES AND FIXED ACTS

Vertebrate Spinal Reflexes

As a result of the pioneering studies of Sherrington and Pavlov at the turn of the century, the analysis of behavior in terms of reflexes dominated studies of neural circuits until at least the 1960s. Therefore, it is appropriate to begin an examination of motor circuits with vertebrate spinal reflexes. In spinal reflexes, somatosensory input (from receptors of the skin, muscles, tendons and joints—see Chapter 18) enters the spinal cord through the dorsal roots. This sensory input, via intervening synapses in the spinal cord, excites some motor neurons and inhibits others. The sensory inputs from different populations of receptors have different connections in the spinal cord and thereby initiate different reflexes. We shall examine three of the many reflexes of the mammalian hindlimb that have been studied extensively in the last century.

Stretch Reflex The first spinal reflex that we consider is the *stretch reflex* or *myotactic reflex*. A familiar example of the stretch reflex is the knee-jerk response to a tap on the patellar tendon, a test that is a staple of routine medical examinations. (This procedure is the simplest test for major damage to the spinal cord, such as degeneration resulting from tertiary syphilis.) When a physician taps you on the patellar tendon, the tap stretches the extensor muscle in the thigh. This stretch stimulates *muscle spindles* (see Chapter 18), which contain stretch-sensitive receptor endings located in a noncontractile portion of specialized *intrafusal* muscle fibers. The sensory axons associated with muscle spindles are known as *1a afferent* fibers—afferent meaning conducting toward the central nervous system, and 1a because they are the largest and most rapidly conducting sensory fibers in the body. (Not all muscle spindle sensory neurons are of the 1a class, but we will simplify the discussion by considering only the 1a sensory neurons here.) The 1a axons from muscle spindles enter the spinal cord and make direct, excitatory synaptic contact with motor neurons to the same muscle (Figure 20.2). This direct synaptic excitation is unusual; most vertebrate sensory neurons synapse directly only onto interneurons (intrinsic neurons that do not leave the central nervous system). The simplest manifestation of a stretch reflex then involves only two kinds of neurons—1a sensory neurons and motor neurons. When a muscle spindle is stretched, its 1a afferent neuron generates a train of nerve impulses. These inpulses elicit EPSPs in motor neurons, leading to motor neuron impulses and ultimately to contraction of the stretched muscle. Muscle spindles are said to be in parallel to the other fibers of a muscle (termed *extrafusal* fibers), because the spindles act beside the extrafusal fibers (Figure 20.3A). Because muscle

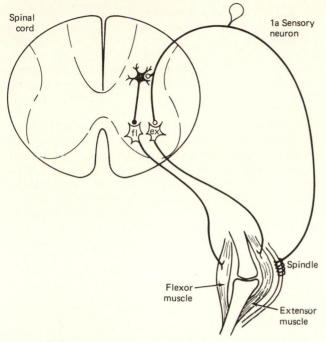

Figure 20.2 Stretch reflex (myotactic reflex): Stretch of the extensor muscle activates a muscle spindle of the extensor. The resultant activity of the 1a sensory neuron excites motor neurons to the extensor muscle and also excites inhibitory interneurons (shaded) that inhibit motor neurons to the flexor muscle.

spindles are in parallel to the force-producing extrafusal fibers, they are sensitive to muscle *length* (in contrast to Golgi tendon organs—treated later—that are in series with extrafusal fibers and therefore are sensitive to muscle tension, Figure 20.2*B*).

The stretch reflex is one of the simplest spinal reflexes, but it is not as simple as the above description suggests. For one thing, the 1a afferent axons do not synapse only on motor neurons. They also make excitatory synapses on inhibitory interneurons that inhibit motor neurons to the antagonist (opposing) muscle. Thus, a tap on the patellar tendon not only excites motor neurons to the extensor muscle to produce the familiar knee jerk but also inhibits motor neurons to the antagonist flexor muscle. These synaptic connections illustrate one of the most basic features of reflexes and of organization of motor systems: the principle of *reciprocity* or *reciprocal innervation*. Muscles tend to be arranged in antagonist pairs that oppose each other, such as the flexor that bends the knee and the extensor that straightens it. In general, any signal that activates movements, whether it is the sensory input to a reflex or a command of the central nervous system, is coordinated to contract one set of opposed muscles (the *agonists*) and relax the opposite (*antagonist*) set. This reciprocal activation of muscles ensures that two mutually antagonistic muscles do not counteract each other. The circuits of spinal reflexes (and motor circuits in general) are organized by reciprocal innervation, so that reciprocity of muscle action is largely assured by neural connections. The stretch reflex provides an example of reciprocal innervation, because each 1a sensory axon excites motor neurons to the agonist (stretched) muscle and also excites inhibitory interneurons that inhibit antagonist motor neurons, ensuring relaxation of the antagonist muscle (Figure 20.2).

Another added complexity in the stretch reflex (and in motor circuits in general) is the number of neurons involved in even the simplest behavioral act.

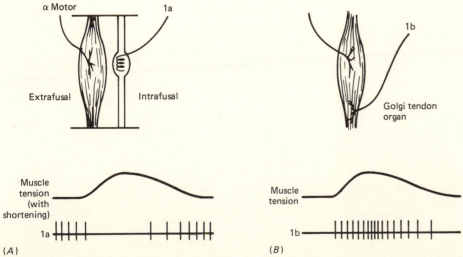

Figure 20.3 Mammalian muscle spindle and Golgi tendon organ. (*A*) A muscle spindle senses muscle *length;* it is arranged *in parallel* with extrafusal muscle fibers. As shown in the recording, muscle spindle activity decreases with contraction of extrafusal fibers. (*B*) A Golgi tendon organ senses muscle *tension;* it is arranged *in series* with extrafusal muscle fibers. As shown in the recording, Golgi tendon organ activity increases with contraction of extrafusal fibers.

Figure 20.2 shows one 1a sensory neuron and one extensor motor neuron, but these only represent larger populations of these neuron types. A muscle such as the knee extensor contains many muscle spindles and its stretch thus activates many sensory neurons. Moreover, the muscle is supplied by at least 300 motor neurons (see Chapter 19, the section entitled Neural Control of Muscle). Each 1a sensory neuron synapses with most, and probably all, of the extensor (α) motor neurons, as well as with a large number of interneurons of different types. This example illustrates the principle of *divergence* of central neural connections: Each presynaptic neuron usually contacts many postsynaptic neurons. The converse principle of *convergence* is also true; each postsynaptic neuron is contacted by many presynaptic neurons. For example, recall from Chapter 17 that each extensor motor neuron receives approximately 10,000 synapses, representing many 1a sensory neurons and many more excitatory and inhibitory interneurons. Thus, the cartoon view of a stretch reflex circuit in Figure 20.2 is a great oversimplification.

A third aspect of the stretch reflex that increases its organizational complexity is the motor innervation of the stretch receptor organs by *gamma motor neurons*. Recall that in muscle spindles the stretch-sensitive 1a sensory neurons are associated with intrafusal muscle fibers. The intrafusal muscle fibers are innervated by a separate population of small motor neurons, the gamma (γ) motor neurons. The extrafusal muscle fibers (i.e., all the fibers that are not part of muscle spindles) are innervated by alpha (α) motor neurons. (When the term motor neuron is used without the greek-letter prefix, it denotes an alpha motor neuron. Hence, the previous discussions of vertebrate motor neurons in Chapter 17 and in this chapter refer to alpha motor neurons.) Activation of γ motor neurons excites the 1a sensory neurons by contracting the contractile ends of intrafusal fibers and thereby stretching the noncontractile central sensory portion of the spindle. Therefore, there are two ways to increase muscle spindle receptor activity: by passive (external) stretch of the muscle and by γ motor neuron activity. Note also that the activity of γ neurons and the activity of α motor neurons have opposite effects on muscle spindle sensory activity.

What is the function of the stretch reflex? Surely it has not evolved to mediate a sudden extension of the leg when the knee is struck with a rubber mallet. We can illustrate one aspect of stretch reflex function with the following theoretical example: Suppose that while you are standing, a large monkey or a small person jumps on your back. The added weight will cause your knees to start to buckle, stretching the extensor muscles and activating muscle spindle sensory neurons. This sensory activity will reflexively excite motor neurons to the extensor muscles, generating more muscle force to counteract the increased load and maintain upright posture. This scenario, although scarcely more plausible than the

doctor's mallet as a selective force, illustrates a functional role of the stretch reflex in postural maintenance, counteracting changes in load, muscle fatigue, or other factors. In a later section (Proprioceptive Feedback and Load Compensation), we consider other aspects of stretch reflex function, including the role of γ motor neurons.

Golgi Tendon Reflex   There are several different kinds of vertebrate spinal reflexes, because different groups of spinal sensory neurons make different synaptic connections onto the populations of central neurons in the spinal cord. We will consider two further examples. The *Golgi tendon reflex* is mediated by muscle tension receptors of Golgi tendon organs. These sense organs consist of arborized sensory endings of 1b sensory neurons, imbedded in the tendons at the ends of skeletal muscles. (The designation 1b denotes that the axons of these neurons are the second most rapidly conducting group in a muscle nerve, after the 1a axons of muscle spindles.) Because Golgi tendon organs are in series with the force-generating extrafusal muscle fibers, they are sensitive to muscle tension (Figure 20.3). Thus, vertebrate muscles have two separate kinds of receptors, one (muscle spindles) sensing muscle length and the other (Golgi tendon organs) sensing muscle tension. The 1b sensory axons of Golgi tendon organs indirectly *inhibit* motor neurons to the same muscle, by making excitatory synapses onto inhibitory interneurons in the spinal cord (Figure 20.4). This Golgi tendon reflex then opposes the effect of the stretch reflex, which increases the force output of the same muscle. The functional significance of the Golgi tendon reflex is not clear. Originally, it was thought that Golgi tendon organs were activated only at very large muscle tensions and served to protect muscles from damage when overloaded. The reflex may subserve this function, but recent studies have demonstrated that Golgi tendon organs are activated at lower tension levels when the muscle actively generates its own tension. These studies suggest that Golgi tendon reflexes may act as a tension-compensating system in postural maintenance. For example, if muscle tension decreases over time because of fatigue or decreased synaptic transmission, the Golgi tendon organ discharge would decrease, stabilizing the muscle tension level.

Flexion Reflex   The last vertebrate spinal reflex we examine is the *flexion reflex*. When you step on a tack, you reflexively withdraw your foot from the offending stimulus. The neural circuit mediating this flexion reflex is shown in Figure 20.5. Sensory neurons known as flexion reflex afferents have endings in the skin which are sensitive to painful and noxious stimuli. The central endings of the flexion reflex afferents make excitatory synaptic contacts on interneurons which in turn excite flexor motor neurons and (via inhibitory interneurons) inhibit extensor mo-

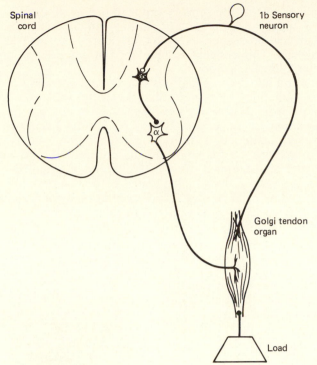

Figure 20.4 *Golgi tendon reflex.* An increase in muscle tension (from either contraction or an external load) activates 1b sensory neurons, which act via inhibitory interneurons (shaded) to inhibit α motor neurons to the same muscle.

tor neurons. Thus, as in the stretch reflex (and in spinal reflexes in general), synaptic interactions in the spinal cord maintain reciprocity of action between antagonist pools of flexor and extensor motor neurons. Unlike the stretch reflex, however, flexion reflex afferents make only indirect connections to motor neurons, via at least one layer of intervening interneurons. The obvious function of the flexion reflex is a protective one; the offended limb is flexed, lifted, and withdrawn from a painful and potentially damaging stimulus. The reflex circuit is relatively short, local, and rapid. Of course, flexion reflex afferents also connect to other interneurons that ascend the spinal column to the brain, so that you become aware of the painful stimulus. This slower process occurs while the reflex flexion is taking place, so that in most cases the foot is lifted (or the hand is withdrawn from the hot stove) before you are aware of the stimulus triggering the withdrawal.

If you stepped on a tack with your left foot while your right foot was lifted off the ground, it would be a good idea to extend your right foot while flexing your left foot. In fact, one component of the flexion reflex ensures this. As shown in Figure 20.5, flexion reflex afferents synapse onto interneurons that cross the midline of the spinal cord and indirectly excite extensor motor neurons of the contralateral (opposite side) leg. Thus, the right leg is extended (by exciting extensor motor neurons and inhibiting flexor motor neurons) while the stimulated left leg is flexed (by exciting flexor motor neurons and inhibiting extensor motor neurons). The reflex extension of the contralateral leg has been given a separate name (the crossed extensor reflex), but it is a set part of the flexion reflex, a product of the synaptic connections "wired in" to the spinal cord. This example illustrates that reflexes do not operate in a vacuum, influencing only a single antagonist pair of muscles. Instead, reflexes may have diverse and widespread

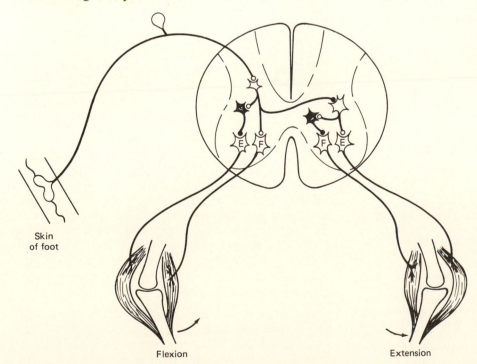

Flexion   Extension

Figure 20.5 Neural circuit of the flexion reflex (left) and the crossed extension reflex (right). Noxious stimulation of the skin activates pain sensory neurons that (via interneurons) excite flexor motor neurons and inhibit extensor motor neurons on the stimulated side (left). On the opposite side (right), extensor motor neurons are excited and flexors are inhibited.

effects and must also interact with all other synaptic influences on motor neurons.

Limitations of the Spinal Reflex Viewpoint   Having described several spinal reflexes, we must now address the more difficult question of just how fruitful is such analysis for an understanding of motor function of the spinal cord. Neurophysiologists have approached spinal circuitry from the viewpoint of reflexes for largely technical reasons that go back to Sherrington's work at the turn of the century. Shepherd and others have pointed out, however, that it is more useful to consider the primary input to the motor circuitry of the spinal cord to be the descending input from higher centers of the central nervous system and the sensory fibers mediating spinal reflexes to be a secondary input. This viewpoint is shown in Figure 20.6, in which the primary descending input enters from the top and sensory input enters from the bottom. Such an orientation implies that a

major role of the sensory input to the spinal cord is to supply *sensory feedback* that can modulate or correct the responses of motor neurons to central signals. This concept is developed in more detail in the following sections.

Additional complications to our understanding of the functions of neural networks in the spinal cord arise from the presence of other types of neurons (in addition to those already described) and from the sheer numbers of neurons in spinal circuits. To illustrate these complexities, let us consider Renshaw cells, an additional class of inhibitory interneurons in the ventral horn of the spinal cord (Figure 20.6). Renshaw cells receive direct excitatory synaptic input from axon collaterals of spinal motor neurons; these collaterals branch off the motor axons before the axons exit the ventral root. Renshaw cells inhibit the motor neurons that excite them, as well as other motor neurons to the same muscle. They also inhibit interneurons, including the 1a inhibitory interneurons that inhibit antagonist motor neurons. The functional significance of the recurrent (feedback) inhibition mediated by Renshaw cells is not clear. Suggested possibilities include a decrease in the duration of motor neuron discharge, an inhibition of weakly active motor neurons by their more strongly active neighbors, and a disabling of reciprocal inhibition (by inhibiting 1a inhibitory interneurons). The difficulty in clarifying Renshaw cell function results from the inability to identify individual neurons within the motor pool innervating a muscle. Thus, one does not know which motor neurons in a pool excite a given Renshaw cell and which neurons the Renshaw cell inhibits. Is Renshaw inhibition of the motor neurons that excited it (i.e., self-inhibition) stronger or weaker than inhibition of other motor neurons in the pool (i.e., lateral inhibition)? In most studies, vertebrate central neurons cannot be identified as unique individuals but instead can only be identified as members of a class (such as an alpha motor neuron to the gastrocnemius muscle). Because the neurons can be recognized only as anonymous members of a motor pool, questions that require more precise identification of cells (*which* motor neuron?) are difficult to answer.

One result of our inability to study identified neurons in the vertebrate CNS is that investigators have turned to numerically simpler nervous systems—especially those of arthropods and mollusks—in which individual neurons can be identified uniquely. Next, we examine several such networks, considering principles or features of organization that may be of general importance.

Crayfish Escape Behavior

The tail flip escape response of crayfish to mechanical stimulation is one of the most completely analyzed behavioral acts, in terms of the roles of individually identified neurons in a neural circuit. The

Spinal cord: ventral horn

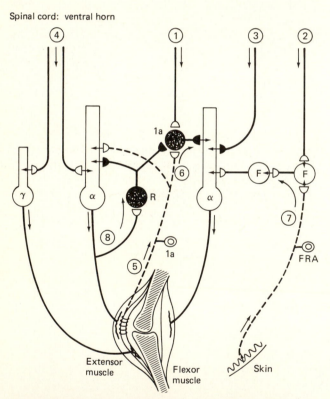

Figure 20.6 Basic circuit diagram of the ventral horn of the mammalian spinal cord. *Input pathways:* central descending pathways from the brain (1–4); the 1a afferent excitatory (5) and inhibitory (6) pathways from muscle spindles; the flexion reflex afferent pathway (7); and the Renshaw recurrent inhibitory pathway (8). *Principal output neurons:* alpha (α) motor neurons to flexor and extensor muscles; gamma (γ) motor neurons to intrafusal fibers of muscle spindles. *Local interneurons:* 1a inhibitory interneuron; (F) interneuron in the flexion reflex pathway; (R) Renshaw cell in the recurrent inhibitory pathway. Neurons and terminals that are excitatory are in open profiles; those that are inhibitory are shaded. [From G.M. Shepherd, *Synaptic Organization of the Brain,* 2nd ed. Oxford University Press, New York, 1979.]

escape response has two components: a single rapid flexion of the abdomen (tail flip) and a subsequent swimming sequence consisting of repeated alternate flexions and extensions of the abdomen, propelling the crayfish away from the source of stimulation. We consider only the initial single flexion, which has received the most study. There are three different kinds of initial tail flip, each triggered by a separate pathway. The first kind is triggered by a sudden tap or water jet stimulus to the abdomen, resulting in an abdominal flexion mediated by the *lateral giant* interneurons (Figure 20.7*A*). This flexion is greatest in the anterior abdominal segments and pitches the crayfish upward and foreward, as in a somersault. The second kind is triggered by a sudden stimulus to the anterior portion of the animal, causing a flexion mediated by the *medial giant* interneurons (Figure 20.7*B*). This flexion is greatest in the posterior ab-

dominal segments and propels the animal backward. Weaker or less abrupt stimulation may induce the third kind of response, a less stereotyped abdominal flexion that is mediated by separate, nongiant neurons. The abdominal flexions that are mediated by giant interneurons have properties of a fixed act: They occur in an all-or-none manner (cf. Figure 20.1) and involve the coordinated activation of muscles in many segments of the body.

The neural circuit for escape abdominal flexion mediated by lateral giant interneurons (LGIs) has been studied most extensively and is the focus of our attention. The neural circuit by which LGI-mediated escape occurs is shown schematically in Figure 20.8. There are two lateral giant interneurons, one on each side. Each is actually a compound series of cells electrically coupled by septate junctions (see Box 16.1). For simplicity, we show only the circuit of one

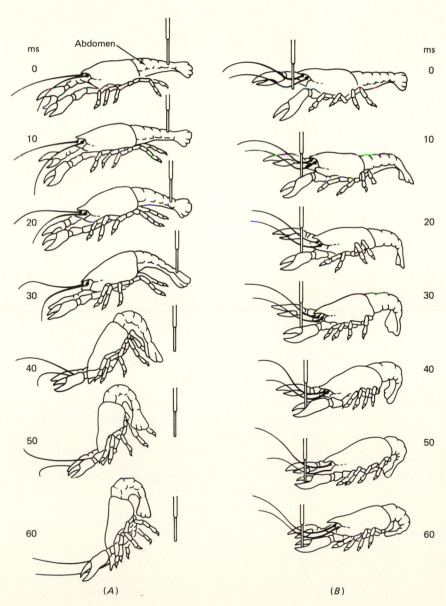

(A)   (B)

Figure 20.7 Excape responses of the crayfish *Procamabrus* to tactile stimuli. (*A*) Stimulation of the abdomen (top) evokes an abdominal flexion that moves the crayfish up and forward. This response is mediated by the lateral giant interneurons (LGIs). (*B*) Anterior stimulation (e.g. to the antenna, top) evokes an abdominal flexion of a different form that propels the animal backward. This response is mediated by the medial giant interneurons (MGIs). In both cases, the movement is away from the source of stimulation. [From J.M. Camhi, *Neuroethology.* Sinauer, Sunderland, MA, 1984.]

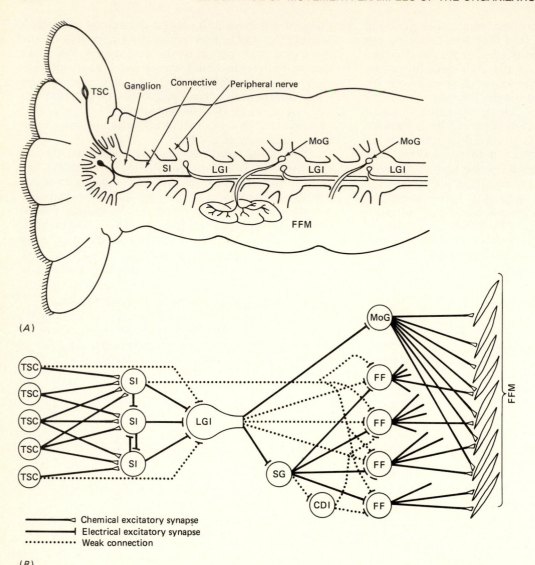

Figure 20.8 The neural circuit for the escape response of a crayfish to an abdominal tactile stimulus. (*A*) Crayfish abdomen, showing the arrangement of some of the cells in the escape circuit. Two more anterior abdominal ganglia are not shown. TSC, tactile sensory cell; SI, sensory interneuron; LGI, lateral giant interneuron; MoG, giant motor neuron; FFM, fast flexor muscle. (*B*) The escape circuit shown in detail, for one side of a single abdominal ganglion. Three additional cell types are included: the fast flexor motor neurons (FF), the segmental giant cell (SG), and one of a group of cells called CDIs. The strongest synaptic connections are indicated by the solid lines. Most synapses in the circuit are electrical, exceptions being the neuromuscular junctions and the TSC-to-SI connections. [From J.M. Camhi, *Neuroethology.* Sinauer, Sunderland, MA, 1984.]

side and consider the LGIs as a single neuron. The major sensory input to the escape circuit is from tactile sensory neurons associated with cuticular sensory hairs that cover the surface of the animal. The surface of the abdomen contains about 1000 tactile hairs, the sensory neurons of which are excited by touch and by waterborne vibration. The axons of these sensory neurons make excitatory chemical synapses with a number of sensory interneurons within the abdominal portion of the central nervous system (Figure 20.8*B*). The sensory interneurons in turn make electrical excitatory synapses with one or both of the LGIs. In addition, many of the tactile sensory

neurons make direct, weak electrical synapses onto the LGIs (Figure 20.8*B*). The LGIs synapse onto segmental motor giant (MoG) neurons (see Chapter 17, Electrical Synapses), the axons of which excite the fast flexor muscles (FFMs) that produce the escape flexion. The LGIs also excite a segmental giant (SG) neuron, which in turn excites the nongiant FF motor neurons to the fast flexor muscles. All these cells are uniquely identified, individual neurons, rather than populations as in the vertebrate spinal cord. (Exceptions: There are at least 20 sensory interneurons, only some of which are uniquely identified; tactile sensory cells are not clearly individually identified. Also, other cells, identified and unidentified, participate in the full circuit in addition to those described.) This escape tail-flip circuit then involves many tactile sensory neurons relaying convergent excitatory synaptic input, largely via sensory interneurons, onto the LGIs; if this synaptic input is great enough to trigger an action potential in the LGIs, the action potential powerfully excites the motor giant and (via the segmental giant neuron) the other fast flexor motor neurons. There is thus considerable convergence of input onto the LGIs and considerable divergence of LGI output to many motor neurons, primarily of the first three abdominal segments. (The LGIs also excite inhibitory neurons, not shown, which exert widespread inhibitory effects to suppress competing behavioral activities.) The extensive convergence onto and divergence from the LGIs magnify the importance of the LGIs in the escape circuit. Next, we consider the functional consequences of the uniquely important position of the LGIs.

One principle exemplified by the crayfish escape flexion circuit is that of a **command neuron**. A command neuron is a neuron whose activity is *sufficient* to elicit a behavioral act or action pattern. In the crayfish, a single action potential in a LGI (e.g., as a result of direct electrical stimulation) elicits a tail flip indistinguishable from one in response to a tap on the abdomen. Moreover, a tap normally adequate to elicit a tail flip will not do so if the LGI is hyperpolarized and thus prevented from generating an action potential. This experiment shows that the LGI is necessary for the generation of one kind of escape tail flip. Other examples of command neurons are considered below.

The lateral giant interneuron can also be thought of as mediating a decision—the decision to escape. Any activity of tactile sensory neurons and sensory interneurons that produces a synaptic potential in the LGI that is above threshold for generating an action potential results in an escape response. In contrast, any activity producing a subthreshold synaptic potential in the LGI does not lead to escape. Thus, the initiation of an action potential in a LGI is the cellular equivalent of a decision to escape. The LGI therefore functions as a decision unit as well as a command neuron (although these two functions are not exact logical equivalents). Recall that an important distinction between a fixed act and a reflex act is that a fixed act (such as the escape response) has an all-or-nothing character, while a reflex response is graded with the intensity of stimulation (Figure 20.1). The decision character of impulse generation in the LGI provides one kind of cellular basis for the all-or-nothing expression of a fixed act, in this case of an escape flexion response. An analogous fixed act, the rapid turn of a fish away from a vibrational stimulus, has a similar cellular basis. A pair of giant *Mauthner neurons* in the brainstem of the fish act as decision and command neurons, commanding a flexion to the side opposite the stimulus. (The two

Mauthner neurons generate flexions in opposite directions, and inhibitory circuitry prevents the two from acting together.) Mauthner neurons are the only known command neurons in a vertebrate. Another case somewhat similar to the crayfish escape response is the escape response of cockroaches, discussed in Chapter 16 (Figure 16.1). The cockroach also employs giant interneurons to mediate escape, but there are several sets of giant interneurons and their effects on motor neurons appear more labile than in the crayfish. The escape startle response of the cockroach may therefore have properties intermediate between those of a fixed act and of a reflex act.

The crayfish escape response circuit is unusual in that most of the synapses in the circuit are electrical (Figure 20.8*B*). With the exceptions of the neuromuscular synapses (which are chemically transmitting in all known animals) and the synapses of tactile sensory neurons onto sensory interneurons, all other synapses in the circuit appear to be electrical. As discussed in Box 16.1, both electrical synapses and giant neurons minimize latency and maximize speed of response, functions of obvious evolutionary importance in an escape system in which natural selection may well operate in millisecond time frames. The crayfish escape circuit appears surprisingly complex, with numerous parallel pathways and reinforcing synapses (many of which we have omitted for clarity in this discussion). Wine and Krasne (see Selected Readings at end of chapter) have suggested that the giant neuron escape pathways evolved over older, preexisting nongiant pathways and that all synapses were retained unless they were maladaptive. This hypothesis is difficult to test but would tend to explain much of the apparent redundancy found in the circuit.

Gill Withdrawal in *Aplysia*

A second well-studied neural circuit is that controlling gill withdrawal in the marine mollusk *Aplysia* (see Figure 17.33). Defensive gill withdrawal is a reflex act, in contrast to the fixed act of the crayfish escape tail flip. Mechanical stimulation of the siphon or of the gill itself elicits withdrawal; stronger stimulation evokes proportionately more vigorous and rapid withdrawal responses. The neural circuit mediating the gill-withdrawal reflex was described in Chapter 17 (Figure 17.34) and is shown in more detail in Figure 20.9. The circuit consists of about 24 siphon sensory neurons (of which four are shown), six motor neurons to the gill, and at least three interneurons. All these neurons are identified, although the sensory neuron identifications are provisional. The most important synaptic connections for the gill-withdrawal reflex appear to be the direct, monosynaptic connections from sensory to motor neurons, with the interneurons playing a subsidiary role.

The circuit for the gill-withdrawal reflex elucidates

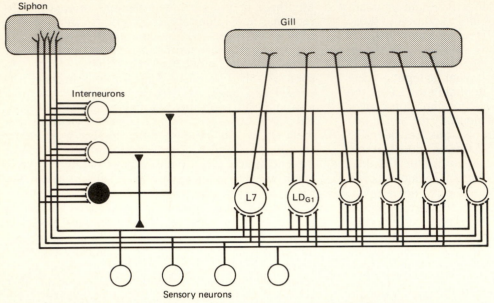

Figure 20.9 Neural circuit mediating the gill-withdrawal reflex of *Aplysia*. Mechanical stimulation of the siphon excites about 24 sensory neurons, four of which are shown. The sensory neurons make direct, monosynaptic contacts onto six identified gill motor neurons, two of which are labeled (L7 and LD_{G1}). The sensory neurons also synaptically excite at least two excitatory interneurons and one inhibitory interneuron (black). [After E.R. Kandel, Sci. Am. **241**(3):66–76 (1979).]

the neural basis for an important distinction between reflexes and fixed acts, namely, the graded and proportionate nature of reflexes versus the all-or-nothing nature of a fixed act (see Figure 20.1). Increasing the strength of siphon stimulation elicits more impulses in a sensory neuron and recruits more sensory neurons. (A weak stimulus may excite eight sensory neurons with overlapping receptive fields on the siphon.) The increased number and frequency of impulses lead to larger summated EPSPs and more impulses in the motor neurons, producing a larger contraction of the gill. In contrast to the crayfish escape tail-flip circuit, at no point in the gill-withdrawal circuit is a single impulse in a single neuron necessary or sufficient for complete expression of the behavior. This analysis should not imply, however, that all fixed acts critically depend on the activity of a single command neuron. Two fixed acts in *Aplysia*—defensive release of ink and egg laying—are all-or-nothing responses mediated by populations of electrically coupled cells. When sufficiently excited by a stimulus, such cells can mutually excite each other in an accelerating crescendo of impulses, via their electrical connections. In these cases, it is the mutually excitatory effect of electrically coupled neurons that is responsible for the all-or-nothing nature of the response.

Proprioceptive Feedback and Load Compensation

In the discussion of vertebrate spinal circuits above, we suggested that a major function of reflex circuits was to provide sensory feedback to modify a centrally initiated movement or posture. One function of sensory feedback is to compensate for a load that acts as a resistance to the execution of an intended movement. A neural circuit mediating such load compensation is most clearly exemplified for postural changes in the crayfish abdomen.

Recall from Figure 19.22 that in crayfish the muscles and control systems for postures and slow movements are separate from those for rapid abdominal movements such as swimming and escape tail flips. The thin, superficial slow flexor and extensor muscles are used for all slow movements and postures. Only these slow muscles are considered in this section. In each abdominal segment, there are six bilateral pairs of motor neurons innervating the slow extensor muscle and six pairs of motor neurons innervating the slow flexor muscle. These motor neurons are individually identifiable and are well characterized. Each abdominal segment also contains a bilateral pair of slowly adapting stretch receptor organs (MROs; see Chapter 18), as well as a pair of rapidly adapting MROs, which are not considered. The stretch receptor sensory neuron is associated with a receptor muscle fiber that is in parallel with the main, "working" slow extensor muscle. The sensory neuron is activated either by passive flexion of the abdomen or by activation of the receptor muscle fiber. The receptor muscle fiber is too weak to affect abdominal position and is analogous to an intrafusal muscle fiber in a vertebrate muscle.

Figure 20.10 shows the neural circuit by which a

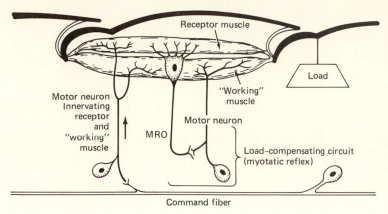

Figure 20.10 The muscle receptor organ (MRO) in the crayfish abdomen is arranged in a circuit that can mediate load compensation in abdominal extension. The receptor muscle is arranged in parallel to the "working" slow abdominal extensor muscle, and shares innervation with the working muscle. The MRO excites an unshared motor neuron that innervates only the working muscle. Commands to the shared motor pathway produce shortening of both muscles at once; but if there is a load, proportional excitation develops in the MRO because the (unshortened) receptor muscle produces tension. This load-dependent excitation activates the unshared motor neuron and adds proportional excitation to the working muscle. [From D. Kennedy in J.C. Fentress (ed.), *Simpler Networks and Behavior*. Sinauer, Sunderland, MA, 1976.]

crayfish is thought to execute a centrally generated abdominal extension. Although it is not known exactly how a crayfish commands abdominal extension, a number of command interneurons for abdominal posture have been identified in the ventral nerve cord. Repeated electrical stimulation of one of these command neurons elicits an abdominal posture (extension or flexion) that lasts as long as the stimulation. As shown in Figure 20.10, an extension command neuron activates shared motor neurons that innervate both the working slow extensor muscle and the receptor muscle. The stretch receptor neuron, in contrast, excites unshared motor neurons that innervate only the working muscle. When the crayfish commands an extension and there is *no load* on the abdomen, the shared motor neuron (Figure 20.10) activates contraction of both the receptor muscle and the working muscle. In the absence of a load, the working extensor muscle shortens to extend the abdomen, allowing the receptor muscle to shorten as it contracts. The shortening of the receptor muscle prevents it from generating tension and activating the stretch receptor. Thus, in an abdominal extension with no load, the stretch receptor does not discharge, even though its receptor muscle is activated. In contrast, *if there is a load* or resistance to abdominal extension, the working muscle is prevented by the load from shortening when it is activated. In the absence of shortening of the abdominal segment, contraction of the receptor muscle activates the stretch receptor (since the receptor muscle is not allowed to shorten). Thus, the discharge of the stretch receptor neuron is an error signal—a measure of the degree to which the abdominal segment failed to shorten when commanded to do so. The stretch receptor axon makes excitatory synaptic contact with an unshared motor neuron (Figure 20.10), which innervates only the working extensor muscle. Activity in the stretch receptor neuron (the error signal) excites proportionate activity in the unshared motor neuron, generating additional tension in the working muscle to overcome the load. The stretch receptor–unshared motor neuron pair is a reflex circuit that functions as a *load-compensating servo loop*, de-

tecting and counteracting an error (failure to shorten) in the centrally commanded extension.

There are clear parallels between the crayfish postural control system and the vertebrate stretch reflex spinal circuit. Figure 20.11 emphasizes the similarities of the two circuits. (Note, however, that for the crayfish the neurons shown are single, identified elements, whereas the vertebrate neurons represent tens or hundreds of similar cells.) One major difference between the two circuits is the presence of shared motor neurons in the crayfish, in contrast to separate α and γ motor neurons (to extrafusal and intrafusal muscle fibers, respectively) in the spinal cord. How could the vertebrate spinal circuit function like the crayfish circuit, when it lacks shared motor neurons? One way would be to have the central command for a movement *coactivate* α and γ motor neurons, exciting both groups together so that they act like shared motor neurons (Figure 20.11*B*). That is exactly what is found in mammals: In centrally generated ("volitional") movements, descending commands coactivate α and γ motor neurons to co-contract extrafusal and intrafusal muscle fibers. Muscle spindle sensory neurons (1a in Figure 20.11*B*) then function as error detectors (like the crayfish stretch receptor), signaling the degree to which the muscle failed to shorten. The stretch reflex circuit thus may function as a load-compensating servo loop, functionally equivalent to the load-compensating circuit in the crayfish.

NEURAL GENERATION OF RHYTHMIC BEHAVIOR

Most animal behavior consists not just of isolated single acts of the sorts discussed above, but rather of action patterns: *sequences* of effector actions that result from sequences of motor output of the nervous system. These sequences of motor activity are patterned in space and time. For example, consider the activity of your nervous system required to pick up a pencil. First you extend your arm by contracting extensor muscles at the shoulder and upper arm, then you flex your fingers to oppose your thumb, then you elevate and flex the arm to lift the pencil. This motor

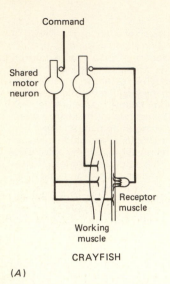

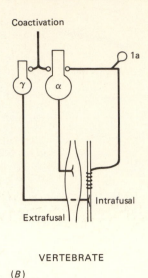

Figure 20.11 Neuronal circuits mediating load compensation in the crayfish abdomen and the vertebrate limb. In the vertebrates, there are no shared motor neurons innervating both working muscle and receptor muscle. The same effect is achieved, however, by coactivating α and γ motor neurons.

performance may involve varying amounts of visual, tactile, and proprioceptive sensory input, and, moreover, the temporal and spatial pattern of the sequence of contractions may differ considerably from one time to the next. Because of this variability, neurophysiological studies of action patterns have concentrated on *rhythmic behavior:* stereotyped, repetitive sequences of movement such as walking, swimming, and flying in which the motor output is stable, repeatable, and predictable from cycle to cycle of the activity. Next we examine several examples of neurophysiological analysis of rhythmic behavior, attempting to extract principles that may be of general importance in motor control systems.

Central and Peripheral Control: Locust Flight

Let us begin our exploration of the control of rhythmic behavior by asking the question: How does a locust fly? As shown in Figure 20.12*A,* the movement of a single wing of a flying locust can be viewed as a simple up-and-down oscillation, generated by a set of elevator and depressor muscles. The electrical activity of these muscles can be recorded from a tethered locust flying in a windstream. This activity consists of alternate bursts of muscle potentials, the depressors being activated when the wings are up and the levators being activated when the wings are down. Since locusts have direct flight muscles (Figure 19.25*A*), each muscle potential results from an action potential in a motor neuron to that muscle, and thus it is clear that flight results from the generation in the CNS of alternate bursts of action potentials in levator and depressor motor neurons. This kind of pattern—alternating bursts of activity in motor neurons to antagonist muscles—underlies most forms of rhythmic behavior.

How are the motor neurons to antagonist muscles activated in alternation to produce a rhythmic movement such as that of a locust wing? Historically, two kinds of hypotheses have been advanced to explain the neural basis of rhythmic movements: peripheral control and central control. According to the *peripheral control* hypothesis, each movement activates receptors that trigger the next movement in the sequence. The position of a locust wing is monitored by several proprioceptors: a single wing-hinge stretch receptor that generates a train of impulses when the wing is elevated and several other receptors that are activated when the wing is depressed. Locust flight could (in principle) operate by peripheral control by having sensory feedback from wing receptors activate the motor neurons for the next movement (Figure 20.12*B*). Thus, elevation of the wing would excite the wing-hinge stretch receptor, which would synaptically excite depressor motor neurons, lowering the wing. The lowered wing would terminate excitation of the wing-hinge stretch receptor and would excite the depression-sensitive receptors, which would synaptically excite levator motor neurons, elevating the wing and completing the cycle. The peripheral control hypothesis is also called a *chained-reflex* hypothesis, since each movement is a reflex response to sensory feedback resulting from the last movement.

According to a *central control* hypothesis, locust flight is sustained by a **central pattern generator**—a neural circuit in the CNS that can generate the sequential, patterned activation of levator and depressor motor neurons without requiring sensory feedback to trigger the next movement. Thus, in central control of locust flight, the basic pattern of alternation of motor neurons would result from an intrinsic central pattern generator (CPG) rather than from a chained reflex (Figure 20.12*C*).

How would one determine whether peripheral control or central control is responsible for the patterned motor activity underlying locust flight? The

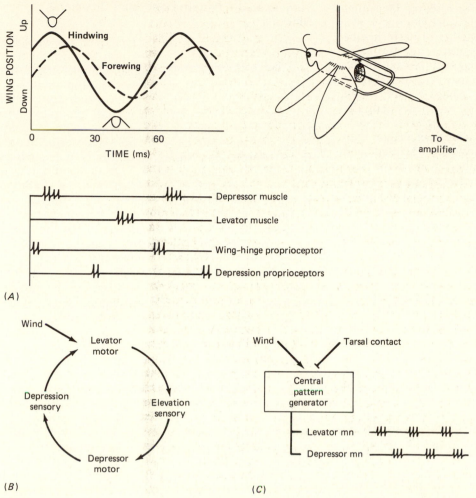

Figure 20.12 Control of flight in the locust. (A) Cyclic wing movements and associated temporal pattern of motor and sensory activity, recorded from a tethered locust. Two sorts of hypotheses could explain the generation of the motor pattern of wing muscle excitation: a peripheral control hypothesis (B), in which sensory feedback resulting from a movement triggers the next movement; and a central control hypothesis (C), in which a central pattern generator produces the motor pattern without requiring moment-to-moment sensory timing.

obvious answer is to remove the relevant sensory input, a process termed *deafferentation* (afferent meaning toward the CNS). In the locust, most if not all wing sensory input can be removed by cutting nerves to the wing-hinge area. Donald Wilson, who pioneered analysis of locust flight control, found that tethered locusts could maintain flight in the absence of wing sensory feedback, although the flight frequency was slower than normal. Normal flight frequency could be restored by providing temporally unpatterned stimulation of the cut sensory nerve stumps or of the ventral nerve cord, a finding that suggested that the sensory input provided general excitation to the CNS but was not necessary to supply timing information for pattern generation itself. These experiments demonstrated the existence of a central pattern generator for locust flight.

Subsequent experiments in other animals have shown that many patterns of rhythmic behavior are under central control. These rhythmic activities (some of which are discussed below) include walking, swimming, breathing or ventilation, and feeding in a variety of invertebrates and vertebrates. Thus, the concept of central pattern generation is a generally important aspect of the control of coordinated behavior.

Does the demonstration of a central pattern generator for a behavior pattern (such as locust flight) mean that sensory input is unimportant? The hypotheses of central control and peripheral control appear to be logical alternatives, but they are not mutually exclusive. Thus, sensory feedback can play significant roles in a centrally controlled behavior. This statement may seem paradoxical, but suppose you are walking down the sidewalk. If you suddenly lost all sensation to your legs, you would probably still be able to generate the motor output sequence of walking. That is, evidence from cats (see later) suggests that you have a CPG for walking. Does that mean that sensory input is irrelevant? Of course not. Sensory input may affect the quality of performance of walking and is essential for correcting the basic pattern, as when walking over uneven terrain. In the locust, several functions of sensory feedback have been found. First, as already noted, sensory input has a generally stimulatory effect of speeding up the flight rhythm. This is termed a *tonic* effect, because it increases the "tone" of the system rather than providing specific timing information. Second, and more surprisingly, sensory feedback can also provide timing information, adding an element of peripheral control to the system. Electrophysiological studies have shown that wing proprioceptors *do* have the synaptic effects diagrammed in Figure 20.12*B:* The stretch receptor monitoring wing elevation excites depressor motor neurons, and depression-sensitive receptors excite levator motor neurons. Thus, the synaptic connections necessary for a chained reflex are present, and these reflexes operate with latencies appropriate to reinforce the flight rhythm. In other experiments, one wing of a tethered, flying locust was moved up and down at a set frequency by a penmotor, the forced cyclic movement overriding normal flight-generated sensory feedback. When the forced movement of the one wing was at a rate close to the normal flight frequency, the flight frequency (recorded from muscles to all four wings) changed to match the driving frequency of the penmotor! Therefore, sensory information from the driven wing can *entrain* the pattern generator to the driven frequency. Our conclusions at this point are (1) there is a central pattern generator for flight that can maintain the flight pattern in the absence of sensory timing information, and (2) sensory timing information (when present) can reset the central pattern generator, entraining it to a slightly different driven frequency. The roles of the central pattern generator and the sensory timing information in this case are analogous to those in a circadian (about 24 h) endogenous activity rhythm of animals. As discussed in Chapter 2, many animals if kept in constant light conditions have an activity rhythm with a period near (but not exactly) 24 h. If a light–dark cycle (such as 12 h light–12 h dark) is added, it provides timing information (the Zeitgeber) that entrains the endogenous circadian rhythm to an exactly 24-h period (Figure 2.9). In the same manner, sensory timing information in locust flight can entrain the central pattern generator. Thus, the original hypotheses of central and peripheral (reflex) control, although they at first seem to be logical alternatives, are not mutually exclusive. The central pattern generator is *sufficient* to maintain flight, but this sufficiency does not rule out a contribution of peripheral control. The relative contributions of central and peripheral control can be expected to differ for different rhythmic behavior patterns. It is likely however, that they interact in most cases.

Turning on the Pattern Generator: Command Neurons

In the above discussion of locust flight, we have deliberately omitted consideration of how flight is started and stopped. The sensory inputs triggering and terminating flight are known, but the circuits by which they affect the flight pattern generator remain obscure. One way in which pattern generators are thought to be turned on and off is by means of command neurons. We have previously described command neurons controlling nonrhythmic acts such as abdominal posture and the escape tail flip of crayfish, but they have also been shown to drive rhythmic patterns of activity in a number of arthropods and mollusks. In fact, the first clear demonstration of command neurons, as well as one of the first demonstrations of a central pattern generator, was for the control of swimmeret beating in crayfish.

Swimmerets are paired, oarlike abdominal appendages of many crustaceans, used in swimming, righting responses, and incubating eggs. They beat in a metachronal, back-to-front sequence (Figure 20.13*A*), at a frequency of about 1.4 beats/s in crayfish. Wiersma and Ikeda demonstrated that following deafferentation and complete isolation of the crayfish abdominal ventral nerve cord, normally coordinated rhythmic motor output of the swimmeret motor neurons in abdominal ganglia persisted in the absence of sensory feedback. Thus, the control and coordination of swimmeret beating involves central pattern generators. Moreover, they found about five bilateral pairs of interneuron axons ("command fibers") in the abdominal connectives, which when individually stimulated at about 30 stimuli/s produced normally coordinated rhythmic activity in the swimmeret motor neurons (Figure 20.13*B*). Thus, the continued activity of a command neuron is sufficient to generate the expression of swimmeret beating, presumably by activating the central pattern generator (Figure 20.13*C*). Command neurons have been described for a number of rhythmic and nonrhythmic activities in arthropods and mollusks, including walking, feeding, and ventilatory patterns. Although it has been difficult to show how individual command neurons or sets of command neurons function in normal animal behavior, they may have rather widespread roles in controlling the expression of patterned behavior.

Mechanisms of Central Pattern Generation

What is the cellular nature of a central pattern generator? We have seen that central nervous systems must contain neurons or networks of neurons that determine the spatiotemporal pattern of motor output (the "motor program") appropriate to generate rhythmic fixed action patterns. Since many of the rhythmic behavior patterns studied are oscillatory,

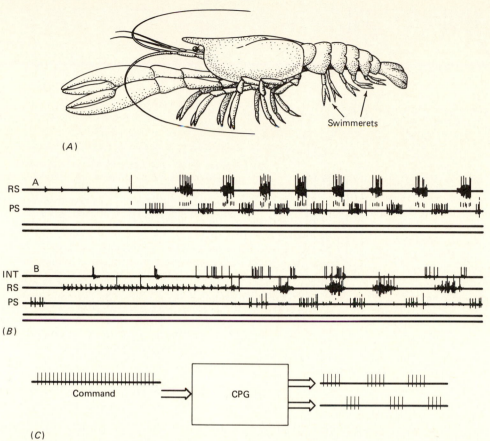

Swimmerets

(A)

(B)

(C)

Figure 20.13 Central pattern generation and central commands for swimmeret beating in lobsters and crayfish. (A) Position of the swimmerets. (B) Properties of a swimmeret command interneuron. A—Electrical stimulation of the interneuron causes rhythmic, alternating activity of motor neurons to return stroke (RS) and power stroke (PS) muscles. B—tactile stimulation of the abdomen activates the interneuron (INT) and also elicits rhythmic swimmeret output. The bottom two traces are a stimulus monitor and a time base (100 marks/s). [From W.J. Davis and D. Kennedy, *J. Neurophysiol.* **35**:1–12 (1972).] (C) Relationship of command neurons to a central pattern generator (CPG). Unpatterned input from a command neuron activates a CPG to produce temporally patterned motor output for a rhythmic behavior pattern.

the central pattern generators underlying them have been termed *oscillators,* analogous to the timing oscillator of a clock. In theory, there are two logical categories of oscillators: cellular oscillators and network oscillators. The central pattern generators that have been studied appear to employ one or both kinds of oscillatory mechanism in differing degree.

A *cellular oscillator* is a neuron that generates temporally patterned activity by itself, without depending on synaptic interaction with other cells. Such cells may generate endogenous bursts of impulses (Figure 20.14A) or may show oscillations of membrane potential without generating any impulses (Figure 20.14B). The underlying mechanism of oscillation is thought to be similar for both types, since cells that generate impulse bursts will continue to oscillate after impulse generation is blocked with tetrodotoxin (TTX, see Chapter 16). A proposed mechanism of cellular oscillation is shown in Figure 20.14C.

Cellular oscillators are thought to play a role in central pattern generation in several cases studied, including those controlling cockroach walking and crustacean heartbeat and stomach contractions (see later). One example of the function of a cellular oscillator in central pattern generation is in the control of crustacean scaphognathites, or gill bailers. Scaphognathites are paired, oarlike appendages located at the anterior openings of the gill chambers. Their paddling action draws water across the gills and pumps it out the anterior openings. Each scaphognathite is controlled by motor neurons in the subesophageal ganglion of the CNS. The control circuit in the subesophageal ganglion contains at least one nonspiking oscillator cell for each scaphognathite. The intracellularly recorded activity of one such cell is shown in Figure 20.15. The membrane potential of the cell oscillates in temporal correlation with the motor neuron activity (recorded extracellularly from

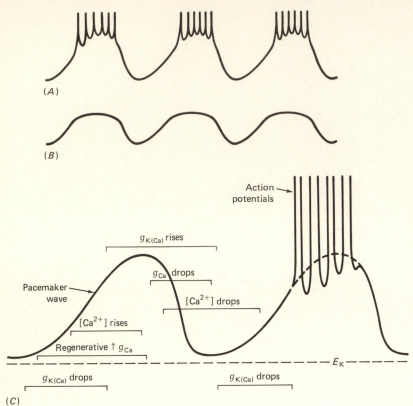

(A)

(B)

Action potentials

$g_{K(Ca)}$ rises

g_{Ca} drops

$[Ca^{2+}]$ drops

Pacemaker wave

$[Ca^{2+}]$ rises

Regenerative ↑ g_{Ca}

E_K

$g_{K(Ca)}$ drops $g_{K(Ca)}$ drops

(C)

Figure 20.14 Cellular oscillators. *(A)* Neuron generating bursts of impulses (e.g., in *Aplysia*). *(B)* Neuron with membrane-potential oscillation but without impulses (e.g., cockroach interneuron in walking). *(C)* A proposed basis for endogenous oscillations of membrane potential. The slow depolarization of the pacemaker wave results from a calcium conductance (g_{Ca}). The regenerative g_{Ca} allows Ca^{2+} influx; as the $[Ca^{2+}]$ rises, it activates a calcium-dependent potassium conductance, $g_{K(Ca)}$. This potassium conductance repolarizes the membrane toward E_K, and the repolarization terminates g_{Ca}. Intracellular $[Ca^{2+}]$ drops (as a result of Ca^{2+} transport) decreasing $g_{K(Ca)}$ and allowing initiation of the next cycle of depolarization. This pacemaker wave can occur with action potentials (right) or without them (left). [Adapted from R. Eckert and D. Randall, *Animal Physiology,* 2nd ed. Freeman, San Francisco, 1983.]

the nerve roots leaving the ganglion). This correlation by itself, however, does not indicate that the cell's oscillation *causes* the motor neuron output pattern. More significantly, when the cell is depolarized by passing current through the intracellular microelectrode, its potential ceases to oscillate, and the motor neurons activated during the depolarized phase of the oscillation remain active throughout the depolarization. When the cell is hyperpolarized via the microelectrode, it again ceases oscillation and the motor neurons active during the hyperpolarizing phase remain active. The finding that the cell oscillates only over a restricted range of membrane potentials suggests that the oscillation is an endogenous property of the cell rather than depending on synaptic input. The finding that controlling the membrane potential of the oscillating cell also controls activity of the motor neurons demonstrates that the cellular oscillations are causally related to the patterned activity of the motor neurons. Many aspects of the neural circuitry controlling scaphognathite beating remain to be worked out; for example, there may be several oscillator cells, coupled by electrical or chemical synapses, rather than only one. Nevertheless, the central pattern generator for the scaphognathite rhythm appears to depend in large part on cellular oscillator properties.

A ***network oscillator*** is a network of neurons that interact in such a way that the output of the network is temporally patterned, although no neuron in the

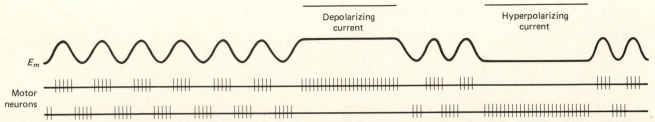

Depolarizing current

Hyperpolarizing current

E_m

Motor neurons

Figure 20.15 Oscillation of cellular membrane potential (*Em*) underlying scaphognathite rhythm in the subesophageal ganglion of a hermit crab. At left, *Em* undergoes rhythmic oscillatory depolarization and repolarization in synchrony with the alternate bursts of impulse in antagonistic motor neurons. Depolarization of the cell (center) through the intracellular electrode prevents oscillation; one set of motorneurons (normally active during the depolarizing phase of cellular oscillation) remains active throughout experimental depolarization. Experimental hyperpolarization of the cell (right) also prevents oscillation and activates the motor neurons normally active during cell repolarization. [Drawing is based on experiments performed by M. Mendelson, *Science* **171:**1170–1173 (1971).]

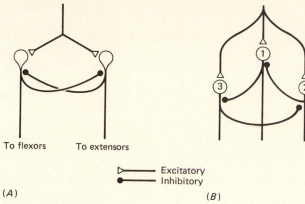

To flexors   To extensors

▷——— Excitatory
●——— Inhibitory

(A)   (B)

Figure 20.16 Network oscillator models. (A) Reciprocal inhibition or inhibitory half-center mode. (B) Cyclic inhibition or inhibitory closed-loop model.

network functions as a cellular oscillator. Thus, the oscillatory or pattern-generating property is said to be an *emergent property* of the network, resulting from cellular interactions in the network rather than from intrinsic cellular properties. The simplest model of an oscillatory network is shown in Figure 20.16A. Two neurons (or pools of neurons) synaptically inhibit each other so that when one stops generating impulses the other is released from inhibition and generates a train of impulses, inhibiting the first. This model appears straightforward but is actually rather unstable. Unless some additional time-dependent property is added (such as "fatigue" in the impulse-generating capability of the cells, or antifacilitation of the inhibitory synapses—see Figure 17.32), the first neuron to reach threshold will tend to remain active and perpetually inhibit the other. This instability is greatly lessened if three or more neurons are connected in a cyclic inhibitory loop. Figure 20.16B shows a closed-loop model that would produce a stable network oscillation. All three cells receive unpatterned excitatory input or are spontaneously active. Suppose that cell 1 is active first. Its activity inhibits cell 3, but this inhibition prevents cell 3 from inhibiting cell 2. Cell 2 can now be active, inhibiting cell 1 and thus releasing cell 3 from inhibition. Cell 3 can then be active, inhibiting cell 2 and releasing cell 1 from inhibition and so on. Such a model produces a stable pattern of bursts in the sequence 1–2–3–1–2–3··· without any cell possessing endogenous oscillator properties. An elaboration of this cyclic inhibition model has been proposed to generate the motor output pattern of swimming in leeches (Figure 20.17).

A central pattern generator can combine the properties of both cellular oscillators and network oscillators. Such a *hybrid oscillator* might have one or more oscillatory cells acting within a network that stabilizes and reinforces the oscillation. Several examples of hybrid oscillator circuits have been studied, including the cardiac ganglia of several arthro-

Intrasegmental Interactions

Phases

−10° to 50°   130° to 170°   220° to 260°

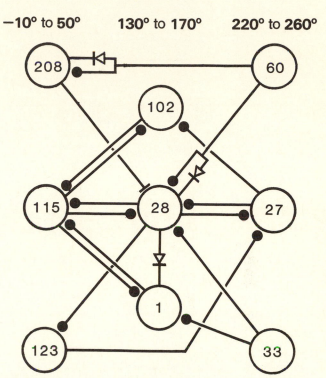

Figure 20.17 Network oscillator proposed as a central pattern generator for swimming motor output in one segmental ganglion of a leech. Interneurons in inhibitory closed loops (e.g., 27, 28, 123) generate rhythmic patterns of activity and rhythmically drive motor neurons (not shown) that excite swimming muscles. Closed circles indicate inhibitory synapses; diode symbols indicate rectifying electrical junctions. [From W.O. Friesen in J.W. Jacklet, ed. *Cellular and Neuronal Oscillators*. Marcel Dekker, New York, *in press*.]

pods and the circuit in the lobster stomatogastric ganglion that controls the pyloric rhythm of stomach activity. Figure 20.18 shows a slightly simplified diagram of the stomatogastric pyloric circuit. The circuit contains three oscillator (pacemaker) cells (one AB and two PD) that are electrically coupled. These cells give oscillatory bursting activity even when synaptically disconnected from the rest of the circuit by high Mg^{2+}, low Ca^{2+} perfusion. These cells synaptically inhibit 11 follower cells, of which the LP cell and eight smaller PY cells are shown. The follower cells (which appear not to have endogenous oscillatory properties) synaptically inhibit the pacemaker cells and many also inhibit each other. When the pacemaker cells (PD + AB) burst, the followers are inhibited. At the end of the PD/AB burst, the LP cell recovers from inhibition faster than the PY cells; therefore, the LP cell bursts next and prolongs PY inhibition. PY cells eventually escape from (relatively weak) LP inhibition and burst, inhibiting LP

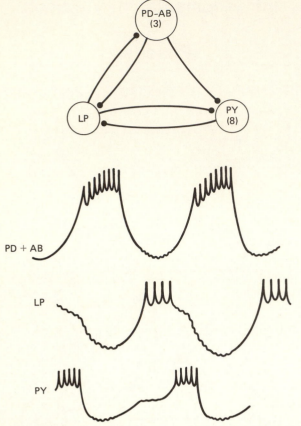

Figure 20.18 Simplified network and cyclic neuronal activity in the pyloric rhythm of the crustacean stomatogastric ganglion. The circuit is a hybrid central pattern generator, the PD and AB cells being cellular oscillators that are electrically coupled.

cells. The next PD–AB pacemaker burst completes the cycle. The pyloric circuit thus has both cellular oscillator and network oscillator properties. The period of the triphasic cycle (PD/AB–LP–PY–··) depends both on the oscillatory properties of the PD and AB cells and on the timing and strength of inhibitory synapses from the follower cells.

Hybrid oscillators such as the pyloric circuit of the stomatogastric ganglion may turn out to be the most common type of central pattern generator. A hybrid network would be expected to be more stable than a "pure" network oscillator (one without oscillator cells) and would rely less on a single neuron and perhaps be more flexible than a "pure" cellular oscillator. Moreover, cellular oscillator properties are being discovered in an ever increasing number of preparations and thus appear to be fairly common.

Central Pattern Generators and Complex Behavior

How elaborate are the behavioral performances that can be carried out under the control of a central pattern generation mechanism? There is a great difference between a short-term rhythmic activity such

as swimmeret beating or locust flight and a complex behavior pattern such as mating behavior in the three-spined stickleback fish. Are such complex patterns as the mating behavior simply elaborations and chains of centrally programmed acts? For technical reasons, it is difficult to explore the neurophysiological bases of increasingly complex fixed action patterns, but some progress is being made.

Central pattern-generating mechanisms have been shown to be sufficient for several behavior patterns that are significantly more complex and long lasting than the simple cyclic patterns described above. One of the more elaborate of these long-term sequences involves gill movements of the horseshoe crab *Limulus*. Gill ventilation in *Limulus* is a simple, cyclic activity analogous to swimmeret beating in crayfish. Gill ventilation, however, is intermittent: Periods of rhythmic ventilation a few minutes long may alternate with approximately 1-min periods of quiescence (apnea) or of gill cleaning. Gill cleaning is a relatively complex fixed action pattern in which the gill plates are adducted across the midline and make rhythmic flicking movements between the lamellae of the opposite book gill. There are two mirror-image patterns of gill cleaning, since a particular pair of gill plates can cross right-over-left (right-leading) or left-over-right (left-leading). An undisturbed *Limulus* may spend several hours performing a given long-term sequence such as ventilation/gill cleaning, employing both left-leading and right-leading patterns of cleaning without any strict alternation. When the abdominal ventral nerve cord is dissected out of a *Limulus* and its unstimulated motor activity is recorded in isolation, the motor output pattern underlying all the above behavior persists (Figure 20.19): Periods of ventilatory motor output rhythm alternate with periods of gill-cleaning motor pattern. Moreover, the rough alternation between left-leading and right-leading gill cleaning can also be expressed in isolation. Thus, relatively elaborate and long-lasting sequences of behaviorally significant patterns of motor output can persist in isolation, without muscles, movement, or sensory feedback.

Still more complex fixed action patterns have been analyzed in insects, including molting activity of crickets and moths and reproductive behavior in several insects. These stereotyped behaviors appear to have centrally patterned components but also have stages at which appropriate sensory feedback is necessary to proceed to the next stage. It is likely that more complex behavior patterns will have increasingly elaborate interactions between sensory components and central motor programs.

CONTROL AND COORDINATION OF VERTEBRATE MOVEMENT

The principles of central pattern generation and of the interaction of central and peripheral control of movement have been developed largely from inver-

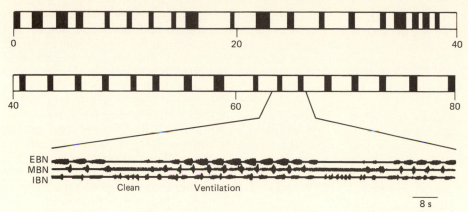

Figure 20.19 Pattern of long-term sequential alternation of gill-ventilation and gill-cleaning motor programs in the isolated abdominal nerve cord of *Limulus*. (*Bottom*) An approximately 140 s record of left branchial nerves of the first abdominal ganglion (EBN, MBN, IBN). Three bouts of ventilation rhythm alternate with two bouts of gill-cleaning pattern. (*Top*) An 80-min period of recorded activity in the same isolated cord, during which ventilation bouts (white) and gill-cleaning bouts (black) alternated fairly regularly. Such stable alternation is common in intact animals. [From G.A. Wyse, D.H. Sanes, and W.H. Watson, *J. Comp. Physiol.* **141**:87–92 (1980).]

tebrate studies, principally with arthropods and mollusks. In this section, we consider the degree to which these principles also apply to vertebrates. We can start with the question: How does a cat walk? For the moment, let us consider the cat nervous system as simply composed of three compartments: brain, spinal cord, and sensory input (Figure 20.20). The immediate generators of walking movements in a cat, of course, are the spinal motor neurons that control the limb muscles. The spinal circuitry associated with these motor neurons was introduced earlier (Vertebrate Spinal Reflexes, p. 554). The motor neurons receive direct or indirect synaptic input from three kinds of sources: descending input from the brain, sensory input from proprioceptors and other receptors in the periphery, and local input from intrinsic spinal circuits. If the spinal motor neurons are to be activated in the right spatiotemporal pattern to produce walking, what are the roles of these three compartments in generating this pattern?

Locomotion in Spinally Transected and Deafferented Cats

In the arthropod systems discussed earlier, the compartments of neural control could be experimentally isolated with relative ease. For example, it is technically easy to isolate the abdominal ganglia of a crayfish or a *Limulus* and to ask the question: What do abdominal ganglia do by themselves, without the brain and without sensory feedback? Such questions are harder to treat experimentally in vertebrates; if you take out the spinal cord of most vertebrates, both the cord and the animal die. Fortunately, with refinements of technique in the 1970s, it has become possible to perform experiments analogous to those in invertebrates.

To begin to determine the role of the three com-

partments (Figure 20.20) in walking in cats, we can consider the result of an experiment that removes the influence of the brain by transection of the spinal cord. Under most circumstances, a cat with a transected spinal cord can maintain a standing posture but cannot walk. This result, however, does not indicate that a spinally transected cat is incapable of walking, as shown by two sorts of experiments. First, in chronic (long-term) experiments, cats with the spinal cord transected 1–2 weeks after birth recover the ability to walk on a treadmill at a speed dependent on the treadmill speed. Second, in acute (short-term) experiments, spinally transected cats can walk on a treadmill if given the norepinephrine precursor DOPA (Figure 17.31) or the norepinephrine receptor stimulator clonidine. These experiments show that the brain is not needed to provide timing information for walking. Noradrenergic fibers descending from the brain presumably command or enable the expression of the walking pattern by spinal circuits but are not necessary to time the stepping cycle of a limb; certainly, injected DOPA does not provide timing information. In other experiments, decerebrate cats can walk on a treadmill when given unpatterned electrical stimulation to a mesen-

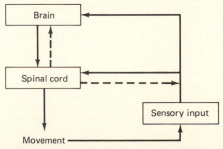

Figure 20.20 The major components of the control of movement in a vertebrate.

cephalic locomotor command area (Figure 20.21). With increasing strength of stimulation, the rate of locomotion increases and the gait changes to a trot and finally to a gallop. Thus, the brain may initiate locomotion and modulate it subject to conditions, but the brain is not necessary to generate the locomotor pattern.

Sensory feedback from the hindlimbs is also unnecessary for hindlimb stepping movements, as can be shown by experiments similar to those described earlier. Cats with or without spinal transection can make normally alternating stepping sequences of the hindlimbs after hindlimb deafferentation by sectioning of the dorsal roots that contain the sensory afferent fibers (see Figures 20.4 and 20.5). (For the spinally transected cats, walking is initiated with DOPA or clonidine.) These experiments indicate that the cat spinal cord contains a central pattern generator for walking movements. Similar experiments suggest that fish, salamanders, toads, and turtles also have spinal locomotor central pattern generators.

Sensory feedback, of course, can still have important functional roles in locomotion of intact vertebrates. Spinal reflexes stabilize and modulate the effects of centrally patterned locomotor output, but spinal reflexes may themselves be modulated by the central pattern generator. For example, the effect of mechanical stimulation of the top of the foot of a walking cat depends on the position of the foot in the stepping cycle. If the foot is off the ground and swinging forward, it is lifted higher when stimulated ("exaggerated flexion"). If the foot is on the ground and bearing the cat's weight, the same stimulation produces a more forceful extension. This reversal of a spinal reflex (which is clearly adaptive for stable walking) shows that the central events of the stepping cycle can strongly modulate reflex function.

The experiments described in this section demonstrate that the mechanisms of control of rhythmic locomotor movements are fundamentally similar in many invertebrates and vertebrates. For example, Pearson (see Selected Readings at end of chapter) describes impressive similarities in the control of walking in cats and cockroaches. Although the cellular aspects may vary (e.g., different cellular mechanisms of central pattern generation), the functional

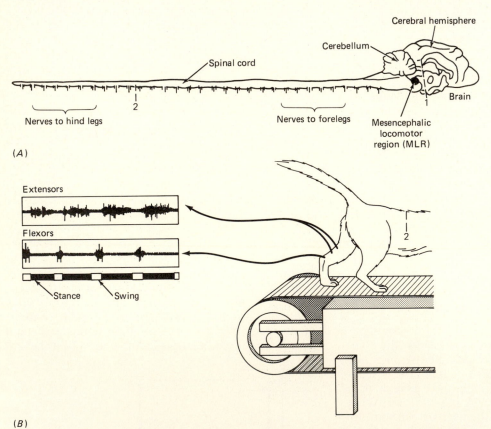

Figure 20.21 Spinal and brain control of mammalian locomotion. (A) Spinal cord and lower brain stem of a cat isolated from cerebral hemispheres by transection at point ①. Electrical stimulation of the mesencephalic locomotor region can produce locomotion in this preparation. Transection of the spinal cord at point ② isolates the hind limb segments of the cord. The hind limbs are still able to walk on a treadmill after recovery from surgery. (B) Locomotion on a treadmill of a cat with a spinal transection at point ②. Reciprocal bursts of electrical activity are recorded from flexors during the swing phase and from extensors during the stance phase of walking. [From E.R. Kandel and J.H. Schwartz, *Principles of Neural Science*, 2nd ed. Elsevier, New York, 1985.]

roles of central and reflex aspects of control appear to be generally similar in many cases.

Vertebrate Brain Areas and Generation of Movement

The vertebrate brain is profoundly important in the control of movement. We have discussed experiments showing that patterned locomotor movements can persist in spinally transected vertebrates and hence do not require the brain. This finding, however, does not contradict the importance of the brain in initiation, coordination, and regulation of normal movements. Next, we consider the ways in which brain areas interact with sensory input and spinal centers in movement control. We must admit at the outset, however, that the roles of brain areas in motor control are very incompletely understood and are difficult to separate from sensory, motivational, and other aspects of brain function. It is well beyond our scope to survey brain function in any comprehensive fashion. We only attempt to provide some ideas concerning the interaction of brain areas in movement control, and we refer the reader to works in neurobiology and physiological psychology for more comprehensive coverage.

Cerebral Cortex   We will begin our examination of the execution of a voluntary movement with the *cerebral motor cortex*. This area of the cerebral cortex (sometimes termed the *precentral gyrus*) lies just anterior to the central fissure, a prominent fold in the convoluted cortical surface of most mammals (Figure 20.22). Early studies demonstrated that electrical stimulation of areas of the motor cortex elicited movements of particular parts of the body, with a point-to-point correspondence between the area stimulated and the muscles activated. Thus, the body

regions are represented on the surface of the motor cortex by a map (Figure 20.23), which for humans is distorted by the disproportionately large areas of cortex serving regions such as the hand and the mouth. The neurons of the motor cortex that mediate these actions are pyramidal cells (neurons with pyramid-shaped somata), the axons of which end on brainstem motor nuclei and also continue down the spinal cord as a major component of the corticospinal tract. (This tract is also known as the pyramidal tract, because the axons funnel through a pyramid-shaped structure on the ventral surface of the brainstem—not because the cells are pyramidal neurons.) The corticospinal axons end primarily on interneurons in the spinal cord, although in primates they also end directly on spinal motor neurons. The neurons of the motor cortex therefore activate spinal motor circuits directly via the corticospinal tract and indirectly via brainstem motor nuclei such as the reticular nuclei, vestibular nucleus, and red nucleus (Figure 20.22).

It seems reasonable to hypothesize that activation of motor cortex pyramidal cells initiates a voluntary movement, although definitive evidence is lacking. The corticospinal tract appears essential for voluntary movement, but the brainstem motor nuclei are also important. Activity of neurons in the motor cor-

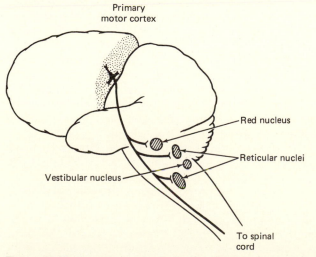

Figure 20.22 Location of the primary motor cortex (shaded area) in the mammalian cerebrum. Neurons from the motor cortex descend to activate motor areas of the brainstem and spinal cord.

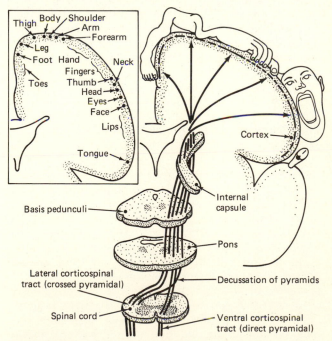

Figure 20.23 Locations of large pyramidal cells in a cross-section of the primary motor cortex. The cells are arranged somatotopically, so that stimulation of cells in a particular area causes movement of a corresponding part of the body (indicated in the inset and by the "motor homunculus" drawn over the cortical surface). Tracts of pyramidal cell axons descend through the internal capsule, brainstem, and spinal cord. [From P.L. McGeer, J.C. Eccles, and E.G. McGeer, *Molecular Neurobiology of the Mammalian Brain.* Plenum, New York, 1978.]

tex precedes and correlates well with voluntary movements, a finding that suggests a control function but is by no means conclusive. One argument against an initiating function of the motor cortex is that non-mammalian vertebrates lack a corticospinal tract and a clearly organized motor cortex. Thus, it is argued that certainly for nonmammalian vertebrates and perhaps also for mammals the generation of movement is controlled subcortically, with an evolutionarily recent cortical overlay in mammals that may modulate and correct the movement. This view, although not widely held, demonstrates our uncertainty about even basic issues in motor control.

Even if we accept the hypothesis that the motor cortex initiates voluntary movements, we have just brushed the tip of the iceberg. How does a decision to initiate a voluntary movement initiate the neural events that lead to activation of motor cortex pyramidal cells? How does the motor cortex interact with other brain areas, such as the cerebellum and basal ganglia, to produce smoothly coordinated, skilled movements without our having to expend continuous conscious effort? For the first question we have little semblance of an answer. If we place surface electroencephalogram (EEG) electrodes on the skull of a subject and ask him or her to move one finger whenever he or she wishes, we record a small but consistent rising wave of activity that precedes the movement by about 800 ms—from much of the entire cortical surface. This activity, termed a readiness potential, becomes localized to the relevant portion of the motor cortex only in the last 50–80 ms preceding the movement. It appears that the decision to initiate a voluntary movement involves many areas of the cortex, including the so-called association cortex, and that this decision is passed to a specific motor cortical site for initiation of the movement. The readiness potential itself is too widespread and diffuse to allow further analysis.

Cerebellum and Basal Ganglia Many studies suggest that both the initial preprogramming of a movement and its modification once initiated involve interaction of the motor cortex with other brain areas. Much attention has been focused on two of these areas: the cerebellum and the basal ganglia. The *cerebellum* is a rather large, highly convoluted structure at the dorsal side of the metencephalon of the hindbrain. It is present in all vertebrates. The cerebellum is clearly involved in the coordination of movement, as demonstrated by the effects of cerebellar lesions in various animals including humans. Voluntary movements are still possible following cerebellar lesions but are clumsy and disordered, lacking the smooth and effortless precision of normal movements. Movements are accompanied by tremor, and patients report they have to concentrate on each part of a movement, joint by joint. It is as if the cerebellum serves to compile instructions for the smooth and coordinated execution of complex movements.

The cellular architecture and synaptic interactions of the cerebellar cortex (the outer portion of the cerebellum) are elegantly simple and are as well known as those of any area of the brain. As shown in Figure 20.24, the cerebellar cortex contains five types of neurons (granule cells, Golgi cells, basket cells, stellate cells, and Purkinje cells), and three types of input fibers (mossy fibers, climbing fibers, and sparse noradrenergic fibers). Moreover, there is only one kind of output from the cerebellar cortex: the axons of Purkinje cells, which end in the deep cerebellar nuclei below the cortical surface. The major synaptic interactions of the cerebellar cortex are summarized in Figure 20.25. Climbing fibers make powerful excitatory 1:1 synaptic contacts with Purkinje cells. Mossy fibers, in contrast, provide divergent excitatory input to about 400 granule cells. The axons of granule cells ascend to the surface layer of the cortex and branch in opposite directions; the continuations of granule cell axons parallel to the cortical surface are termed parallel fibers. Parallel fibers make excitatory synaptic contacts with the four other types of cerebellar cortical cell: Purkinje, basket, stellate, and Golgi cells. The most anatomically impressive of these is the synaptic interaction of parallel fibers and Purkinje cells. Purkinje cells have large, flattened or planar dendritic trees extending across a ridge (folium) of the convoluted cerebellar surface (Figure 20.24). The parallel fibers run along the folium and pass through the planar dendrites at right angles. Each Purkinje cell receives excitatory synapses from about 100,000 parallel fibers (in addition to one climbing fiber). Thus, the climbing fiber and mossy fiber inputs differ greatly in the degree of divergence and convergence of their synaptic effects.

With the exception of the granule cells, which exert excitatory synaptic effects, all other cell types of the cerebellar cortex are *inhibitory* in their effects. In Figure 20.25, the inhibitory cells are shaded black, emphasizing the importance of inhibitory interactions. Basket cells and stellate cells send axons across a folium and thus inhibit Purkinje cells flanking the row of Purkinje cells excited by a beam of parallel fibers. Golgi cells mediate local feedback inhibition onto granule cells. Finally, the Purkinje axons that are the sole output of the cerebellar cortex are inhibitory in their effects on the deep cerebellar nuclei.

Despite extensive studies of the circuitry of the cerebellum (only superficially described above) we have little clear idea how it serves to coordinate movements. The preponderance of inhibitory operations in the cerebellar cortex is noteworthy, all input being converted to inhibitory effects after a maximum of two synapses. Eccles has used the term "inhibitory sculpturing" to describe one idea of how the cerebellum may function, refining a movement by inhibiting unwanted portions. Thus, a cerebral command to flex the index finger could be initially

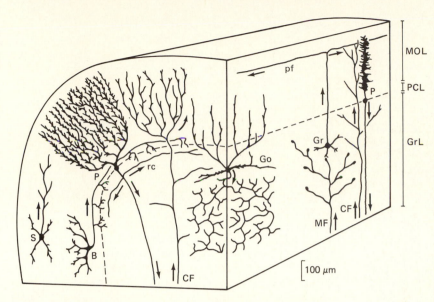

Figure 20.24 Neuronal organization of the mammalian cerebellar cortex. Inputs: mossy fibers (MF) and climbing fibers (CF). Output neuron: Purkinje cell (P), with recurrent collaterals (rc). Local interneurons: granule cell (Gr), stellate cell (S), basket cell (B), Golgi cell (Go). Histological layers: molecular layer (MOL), Purkinje cell body layer (PCL), granule layer (GrL). [From G.M. Shepherd, *Synaptic Organization of the Brain*, 2nd ed. Oxford University Press, New York, 1979.]

and roughly expressed as "flex fingers," which the cerebellum might refine by adding "inhibit all but the index finger." There are other theories about the role of inhibition, but it is clear that inhibitory operations are predominant. Another functional generalization we can make is that cerebellar coordination probably involves extensive interaction with other brain areas. The cerebellar cortex lacks intrinsic long-range connections of the sort that might be expected to coordinate, say, an arm *and* a leg. Moreover, the predominance of inhibitory operations means that all effects of a cerebellar input are over within a half-second. These spatial and temporal restrictions of effects of a cerebellar input imply that the cerebellum must interact with other areas for any smooth, large-scale coordination. We will consider some of these interactions after examination of the basal ganglia.

The term **basal ganglia** is applied to a set of nuclei (brain areas) located in the telencephalon of the fore-

brain, under the cerebral hemispheres. The most important areas (in terms of motor control) are the *neostriatum* (or simply *striatum*) and the *globus pallidus* (Figure 20.26). The neostriatum is subdivided into the more medial *caudate nucleus* and the more lateral *putamen*. Although not part of the forebrain, the mesencephalic *substantia nigra* is considered part of the basal ganglia because of its important interconnections with the striatum. Other forebrain areas of the basal ganglia are the subthalamic nucleus, amygdala, and nucleus accumbens. Of these, only the subthalamic nucleus is primarily motor in function.

The basal ganglia are important in initiating movements, but little is known about how they act in motor control. Most evidence about their role comes from lesions and diseases that affect movement and from pharmacological manipulations of synaptic transmitter systems. We will consider two examples

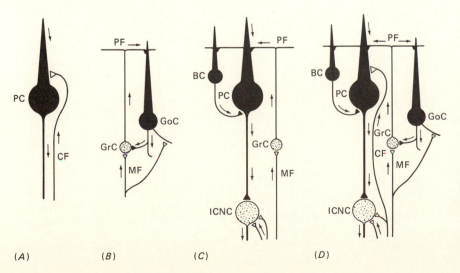

Figure 20.25 Major neural circuits in the cerebellar cortex. The component circuits of A, B, and C are assembled together in D. Arrows show directions of impulse travel. Inhibitory cells are shown in black. BC, basket cell; CF, climbing fiber; GoC, Golgi cell; GrC, granule cell; ICNC, intracerebellar nuclear cell; MF, mossy fiber; PC, Purkinje cell; PF, parallel fiber. [From J.C. Eccles, *The Understanding of the Brain,* 2nd ed. McGraw-Hill, New York, 1977.]

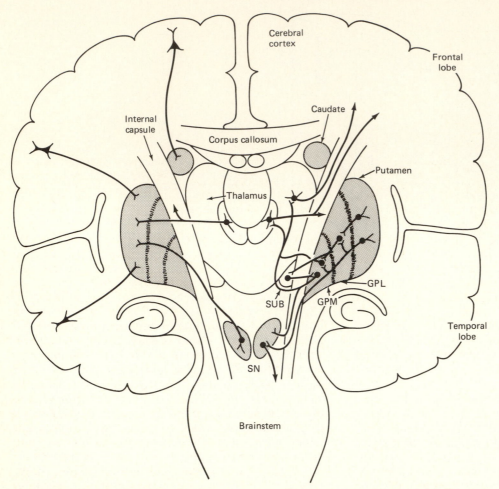

Figure 20.26 Locations of the basal ganglia (shaded) shown in a frontal section of the human brain. The globus pallidus contains a lateral segment (GPL) and a medial segment (GPM). SUB, subthalamic nucleus; SN, substantia nigra. Connections of the candate are similar to those shown for the putamen. [After G.M. Shepherd, *Neurobiology*. Oxford University Press, New York, 1983.]

of abnormal function of the basal ganglia: Parkinson's disease and Huntington's chorea. In *Parkinson's disease* there is difficulty in initiating movements (akinesia), so that a simple task such as climbing stairs or getting up from a chair becomes almost impossible to carry out. Akinesia is often accompanied by postural rigidity and by tremors in limbs at rest. *Huntington's chorea* represents the opposite problem from Parkinsonism: Movements occur uncontrolledly and are difficult to stop. Both chorea (uncontrolled but coordinated jerky movements) and athetosis (slow writhing movements) are associated with damage to the striatum.

Both Parkinson's disease and Huntington's chorea appear to involve neuronal degeneration and altered neurotransmitter activity in the basal ganglia. In Parkinson's disease, dopaminergic neurons in the substantia nigra degenerate. The synaptic endings of these dopamine neurons are in the striatum, so that degeneration of the neurons deprives the striatum of dopaminergic input. Dopamine replacement therapy

(by providing the dopamine precursor L-DOPA, which crosses the blood–brain barrier) alleviates many of the symptoms of Parkinson's disease. In Huntington's chorea, there is loss of neurons in the striatum and globus pallidus; many of these neurons appear to be cholinergic. Pharmacological evidence suggests that normal function of the basal ganglia requires a balance between effects of acetylcholine (ACh) and dopamine (DA). As diagrammed in Figure 20.27, the functional levels of the two transmitters are normally in balance. Any factor that increases ACh or decreases DA, including excess ACh, lowered DA, or blockade of DA receptors, leads to Parkinsonian symptoms. (Some antipsychotic drugs such as phenothiazines block DA receptors and can produce Parkinsonian symptoms.) In contrast, any factor that decreases ACh or increases DA, including cholinergic blockade and excess L-DOPA, leads to chorea.

What then is the role of the basal ganglia in normal motor function? The implications of the above stud-

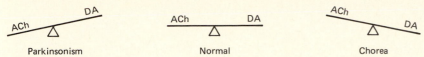

Figure 20.27 Function of the basal ganglia, viewed as balance between functional effects of acetylcholine (ACh) and dopamine (DA). [After P.L. McGeer, J.C. Eccles, and E.G. McGeer, *Molecular Neurobiology of the Mammalian Brain*. Plenum, New York, 1978.]

ies of abnormal function are suggestive but hardly conclusive. One hypothesis is that the function of the basal ganglia is to select and maintain motor behavior patterns, suppressing conflicting motor activity while reinforcing the ongoing behavior. Chorea results from a decreased ability to maintain ongoing behavior and suppress unwanted activity. Parkinsonian symptoms, in contrast, reflect an inability to stop one behavioral activity (e.g., walking or sitting) and start another. In any case, the basal ganglia appear more involved in initiation and gross control of movements, in contrast to the cerebellum, which appears to "fine tune" movement, smoothly modifying the execution of a movement to match a command.

Interaction of Brain Areas in Movement Control We will now attempt to integrate the hypothesized roles of the cerebral cortex, cerebellum, and basal ganglia in the control of voluntary movement. As shown in Figure 20.28, the planning and programming of a movement can be thought of as separate from the execution of the movement. We can suppose that the decision to move starts in the association cortex (cortex that is not linked to any particular sensory or motor system), since the readiness potentials recorded prior to a movement are not localized to any specific cerebral area. Two loops from the association cortex are thought to be involved in preprogramming a movement: one loop through the basal ganglia (selection and initiation) and another through the lateral cerebellum (initial programming). Both loops feed back to the motor cortex via the ventrolateral nucleus of the thalamus (VL thal). The motor cortex

then generates the appropriate pattern of activity in the pyramidal tract to initiate the movement. Information about the command is sent to the intermediate lobe of the cerebellum, via several subcortical nuclei. This process, termed command monitoring, "informs" the cerebellum of the intended movement. The cerebellar intermediate lobe also receives ascending information, both sensory information about joint position and muscle tension and also central information from spinal and brainstem motor centers (Figures 20.28 and 20.29). The intermediate lobe of the cerebellum is thought to integrate this feedback information about the state of lower motor centers (internal feedback) and about the periphery (external feedback) with the monitored cerebral command (Figure 20.29). The cerebellar output can then modify and correct the command on a continuous basis, before and during the generation of the movement as well as after sensory feedback returns to the brain. Figure 20.29 diagrams this proposed action of the cerebellum: to integrate all relevant information (command, motor state, and sensory feedback) on a moment-to-moment basis and to correct a movement continuously as the movement evolves. This continuous correction, based on command monitoring and internal and external feedback loops, is presumably faster and smoother than, say, a correction system based on sensory feedback alone.

The aim of Figures 20.28 and 20.29 is to provide a conceptual framework for thinking about higher motor control in mammals. The ideas presented are still hypothetical and may not prove strictly accurate. Nevertheless, they may illustrate some important

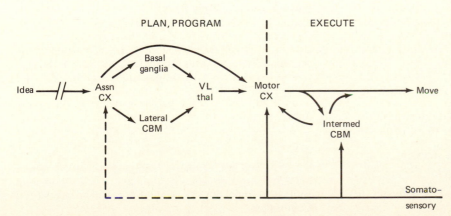

Figure 20.28 Diagram showing pathways concerned in the planning, execution, and control of voluntary movement. ASSN CX, association cortex; lateral CBM, cerebellar hemisphere; intermediate CBM, pars intermedia of cerebellum. [From J.C. Eccles, *The Understanding of the Brain,* 2nd ed. McGraw-Hill, New York, 1977.]

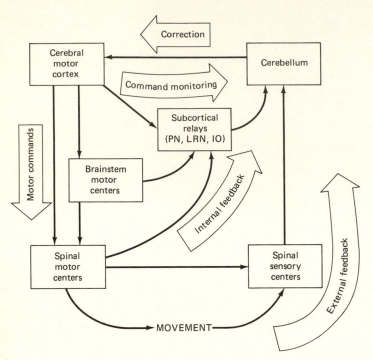

Figure 20.29 Hypothetical functional interactions of brain areas in the control of mammalian voluntary movement. Descending motor commands are modified by cerebellar correction influenced by three feedback loops: command monitoring within the brain, internal feedback from brainstem and spinal motor centers, and external feedback from receptors affected by the evolving movement.

principles, and they serve to bring our discussion of motor control to a closure point.

SELECTED READINGS

Arshavsky, Y.I., I.M. Gelfand, and G.N. Orlovsky. 1983. The cerebellum and control of rhythmical movements. *Trends Neurosci.* **6:**417–422.

*Bentley, D. and M. Konishi. 1978. Neural control of behavior. *Annu. Rev. Neurosci.* **1:**35–59.

Brooks, V.B. 1986. *The Neural Basis of Motor Control.* Oxford University Press, New York.

*Camhi, J.M. 1984. *Neuroethology.* Sinauer, Sunderland, MA.

Delcomyn, F. 1980. Neural basis of rhythmic behavior in animals. *Science* **210:**492–498.

Eccles, J.C. 1977. *The Understanding of the Brain,* 2nd ed. McGraw-Hill, New York.

Evarts, E.V. 1979. Brain mechanisms of movement. *Sci. Am.* **241**(3):164–179.

Evarts, E.V., M. Kimura, R.H. Wurtz, and O. Hikosaka. 1984. Behavioral correlates of activity in basal ganglia neurons. *Trends Neurosci.* **7:**447–453.

Fentress, J.C. (ed.). 1976. *Simpler Networks and Behavior.* Sinauer, Sunderland, MA.

Grillner, S. 1985. Neurobiological bases of rhythmic motor acts in vertebrates. *Science* **228:**143–149.

Grillner, S., P.S.G. Stein, D.G. Stuart, H. Forssberg, and R.M. Herman (eds.). 1986. *Neurobiology of Vertebrate Locomotion.* Macmillan, London.

Grillner, S. and P. Wallén. 1985. Central pattern generators for locomotion, with special reference to vertebrates. *Annu. Rev. Neurosci.* **8:**233–261.

Hasan, Z. and D.G. Stuart. 1988. Animal solutions to the problems of movement control: The role of proprioceptors. *Annu. Rev. Neurosci.* **11:**199–223.

Hoyle, G. (ed.). 1977. *Identified Neurons and Behavior in Arthropods.* Plenum, New York.

Huber, F. and H. Markl (eds.). 1983. *Neuroethology and Behavioral Physiology: Roots and Growing Points.* Springer, Berlin.

Ito, M. 1986. Neural systems controlling movement. *Trends Neurosci.* **9:**515–518.

Kandel, E.R. and J.H. Schwartz. 1985. *Principles of Neural Science,* 2nd ed. Elsevier, New York.

Kandel, E.R. 1976. *Cellular Basis of Behavior.* Freeman, San Francisco.

Kennedy, D. and W. J. Davis. 1977. Organization of invertebrate motor systems. In E.R. Kandel (ed.), *Handbook of Physiology,* Sec. 1, Vol. 1, *Cellular Biology of Neurons,* pp. 1023–1088. American Physiological Society, Bethesda, MD.

Kupfermann, I. and K.R. Weiss. 1978. The command neuron concept. *Behav. Brain Sci.* **1:**3–39.

Lent, C.M. and M.H. Dickinson. 1988. The neurobiology of feeding in leeches. *Sci. Am.* **258**(6):98–103.

Marsden, C.D. 1980. The enigma of the basal ganglia and movement. *Trends Neurosci.* **3:**284–287.

Marsden, C.D., J.C. Rothwell, and B.L. Day. 1984. The use of peripheral feedback in the control of movement. *Trends Neurosci.* **7:**253–257.

*Pearson, K.G. 1976. The control of walking. *Sci. Am.* **253** (6):72–86.

Pearson, K.G. 1985. Neuronal circuits for patterning motor activity in invertebrates. In M.J. Cohen and F. Strumwasser (eds.), *Comparative Neurobiology,* pp. 225–244. Wiley, New York.

Pearson, K.G. 1987. Central pattern generation: A concept under scrutiny. In H. McLennan, J.R. Ledsome, C.H.S. McIntosh, and D.R. Jones (eds.). *Advances in Physiological Research,* pp. 167–185. Plenum, New York.

Roberts, A. and B.L. Roberts (eds.). 1984. *Neural Origin of Rhythmic Movements.* Society of Experimental Biology Symposium 37. Cambridge University Press, Cambridge.

Selverston, A.I. and M. Moulins, 1985. Oscillatory neural networks. *Annu. Rev. Physiol.* **47**:29–48.

Selverston, A.I. (ed.). 1985. *Model Neural Networks and Behavior.* Plenum, New York.

Shepherd, G.M. (1979. *The Synaptic Organization of the Brain,* 2nd ed. Oxford University Press, New York.

*Shepherd, G.M. 1988. *Neurobiology,* 2nd ed. Oxford University Press, New York.

Stein, P.S.G. 1978. Motor systems, with specific reference to the control of locomotion. *Annu. Rev. Neurosci.* **1**:61–81.

Stein, P.S.G. 1985. Neural control of the vertebrate limb: Multipartite pattern generators in the spinal cord. In M.J. Cohen and F. Strumwasser (eds.), *Comparative Neurobiology,* pp. 245–253. Wiley, New York.

Stent, G.S., W.B. Kristan, Jr., W.O. Friesen, C.A. Ort, M. Poon, and R. Calabrese. 1978. Neural generation of the leech swimming movement. *Science* **200**:1348–1356.

Walters, E.T., J.H. Byrne, T.J. Carew, and E.R. Kandel. 1985. A comparison of simple defensive reflexes in *Aplysia:* Implications for general mechanisms of integration and plasticity. In M.J. Cohen and F. Strumwasser (eds.), *Comparative Neurobiology,* pp. 181–205. Wiley, New York.

Wilson, D.M. 1968. The flight-control system of the locust. *Sci. Am.* **218**(5):83–90.

Wine, J.J. and F.B. Krasne. 1982. The cellular organization of crayfish escape behavior. In D.C. Sandeman and H. Atwood (eds.), *The Biology of Crustacea,* Vol. 4, *Neural Integration and Behavior,* pp. 241–292. Academic, New York.

Wise, S.P. and E.V. Evarts. 1981. The role of the cerebral cortex in movement. *Trends Neurosci.* **4**:297–300.

See also references in Appendix A.

chapter 21

Endocrine and Neuroendocrine Physiology

The endocrine glands release their hormonal secretions into the blood, and thus hormones are carried throughout the body, potentially bathing every organ and tissue. When a tissue is exposed to a particular hormone, its response depends on its differentiated properties: It may fail to respond, or it may alter its function in any number of ways. Moreover, the response of one tissue may be quite different from that of another. Thus, endocrine communication systems are well suited for the economical coordination of various processes throughout the body. A single chemical released at a single point into the bloodstream can signal widespread tissues to alter their function in diverse ways that, through the molding of natural selection, form a coordinated and adaptive overall response.

An important contrast between nervous and endocrine control is that individual endocrine signals typically operate on much longer time scales than individual nervous signals (p. 13). Several seconds to several minutes are likely to pass between the time a hormone is initially released by a gland and the time it begins to affect the function of other tissues, because before the hormone can exert its influences, circulation of the blood must transport it to target tissues and it must diffuse from the blood to target cells in effective concentrations. On the other hand, a hormone once secreted can continue to exert its effects for as long as it remains in the blood at a sufficient concentration; thus, a single release of hormone may be effective for hours or days. In brief, individual hormonal signals characteristically are slower to take effect than individual nervous signals but are more persistent.

Temporally and spatially, nervous systems are capable of an extremely fine resolution of signaling that seems to be well outside the realm of possibility for

endocrine systems (p. 13). Therefore, processes that require fine spatiotemporal resolution for their control—such as the activation of the locomotory muscles in an insect or mammal—are always principally under nervous control. Such fine resolution is unnecessary for many other processes, however. Many processes require only long-term control, and often their control involves the simultaneous and similar modulation of extensive populations of cells. Although processes of this sort certainly can be controlled by nervous mechanisms (and sometimes are), in many animals a predominantly endocrine control of such processes has evolved. Growth, reproductive cycles, nutrient metabolism, and mechanisms of mineral and water balance are all examples of processes that are often controlled primarily by hormones. As discussed on p. 13, it is evidently more economical to control such processes by endocrine rather than nervous means. Presumably, this economy explains the adaptive advantage to modern animals of having both endocrine and nervous systems.

We have already discussed some basic aspects of endocrine systems on pp. 12–13. We have also described some examples of endocrine control in previous chapters. For example, the hormonal control of salt and water balance was discussed on pp. 193–196, and the control of specific aspects of renal function was discussed on pp. 214–217 and in many other parts of Chapter 10. In this chapter, we first examine the *principles* of endocrine function, using appropriate vertebrate and invertebrate examples. Later, we examine how the integrated interactions of hormonal systems control particular physiological functions. Our aim is to be selective rather than comprehensive; that is, we seek to illustrate concepts of endocrine regulation rather than survey all the hormones and endocrine glands known in all animals.

INTRODUCTION TO ENDOCRINE PRINCIPLES

Endocrinology is the study of hormones. It is not always easy, however, to define what is or is not a hormone. Conceptually, a **hormone** is a metabolically produced chemical substance that is released directly into the blood by specialized secretory cells and that, in small amounts, exerts regulatory influences on the function of other distant cells that it reaches via the circulation of the blood. The secretory cells that produce hormones often—but not always—are organized into discrete organs termed **endocrine glands**. Endocrine glands, unlike exocrine ones, lack outflow ducts (they are often called **ductless glands**). The hormones they secrete pass directly from each of their constituent cells into the blood that perfuses them.

The conceptual definition of a hormone stated in the previous paragraph is valuable, but it does not always suffice as an *operational* definition. An operational definition of a hormone is one that would allow us to determine experimentally whether or not any particular chemical substance we might name functions as a hormone. Some substances are unambiguously hormones; examples include thyroxine—a secretion of the thyroid gland—and testosterone and estrogen—secretions of the gonads. Such compounds epitomize what we mean by the word *hormone,* and indeed the conceptual definition of hormones is phrased to describe just such examples. As a contrasting case, consider carbon dioxide. Is it a hormone? Although it is a metabolically produced chemical compound that enters the blood directly from cells and exerts regulatory effects on distant cells (e.g., cells of the mammalian respiratory control center, p. 284), biologists have little difficulty agreeing that it is *not* an example of what we mean when we speak of hormones. The reference to release by "specialized secretory cells" in the conceptual definition is in fact designed to make clear that substances that emanate from most or all tissues are not hormones. The criterion that hormones exert their effects "in small amounts" helps to sharpen this distinction. Not all compounds, however, are as easily categorized as thyroxine and carbon dioxide. We know now, for example, of a large array of compounds that are released by certain cells and affect the function of other *nearby* cells located in the same organ or tissue mass, evidently making their way from cell to cell by diffusion in the tissue fluids but, in some cases, probably being carried by the tissue capillary circulation as well. Are some or all of these compounds true hormones? And how would we refine the definition to make clear which are included or excluded? How distant must the point of action be from the point of release for a compound to qualify as a hormone, and to what extent must transport of the compound be by the blood? We shall talk more about these substances in a later section (p. 629). Here we stress the general point that regulatory or communicatory chemicals do not fall naturally into distinct categories, and therefore any arbitrary division of them into hormones and nonhormones is bound to be partly artificial.

To develop further some of the basic principles of endocrinology, let us now focus on a specific example: the vertebrate thyroid gland and its hormonal secretion thyroxine. Thyroid gland cells *synthesize* thyroxine molecules from precursors taken up from the blood. The specific precursors of thyroxine are iodine and the amino acid tyrosine. (One of the earliest discoveries of clinical endocrinology was that goiter, an enlargement of the thyroid, could be prevented by small amounts of iodine in the diet.) Thyroxine is in fact synthesized as part of a larger molecule, which is stored. Later, the thyroxine is released and then secreted into the blood, where it circulates to the tissues of the body. Typically, only certain kinds of cells undergo responses to a particular hormone; those distant cells on which a hormone exerts its effects are termed **target cells** (Figure 21.1). This term "target cell" has been criticized, because it could imply that hormones are directed toward, or seek out, specific targets. We know that such "seeking" behavior is not the case. All cells are exposed to a hormone, but only those that contain *receptors* for the hormone are target cells. Thyroxine exerts a wide range of metabolic, structural, and developmental effects on many or most tissues of the body; that is, many (but not all) body tissues are targets of thyroxine action. In contrast, thyrotropic hormone (to be considered below) has just one major target: the cells of the thyroid gland (Figure 21.1).

Thyroxine and other hormone molecules secreted into the blood do not circulate indefinitely. Instead, they are metabolically broken down at their targets or by organs such as the liver and kidneys. Thus, individual hormone molecules have discrete life spans, from synthesis and release to action and eventual metabolic destruction. We shall return to this concept later (p. 579).

For hormones to function in physiological regulation, the rate of their synthesis and secretion into the blood must be controlled. In general, the higher the rate of secretion of a hormone such as thyroxine, the higher its concentration is in the blood and thus the greater its effect is on target cells. *Control* of hormone synthesis and secretion is commonly effected by other hormones. For example, the rates of thyroxine synthesis and secretion are controlled primarily by a pituitary hormone called thyrotropic hormone or thyroid-stimulating hormone (TSH) (Figure 21.1). The elucidation of these control pathways has been an important goal in endocrinology, and the pathways are discussed in considerable detail later (p. 598).

Finally, we must appreciate that all aspects of endocrine function have *evolved*. Thyroid glands are discrete structures in most vertebrate groups but are patchy clusters of cells in cyclostomes and most teleosts. In protochordates and cyclostome larvae, ap-

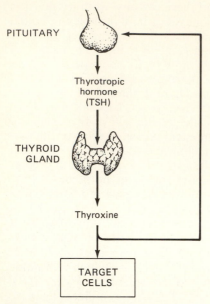

Figure 21.1 Diagram illustrating some basic organizational features common in endocrine systems. The vertebrate thyroid gland produces thyroid hormones—principally thyroxine—that are carried in the blood and that exert regulatory effects on other cells of the body (termed target cells). The blood concentration of thyroxine is controlled principally by the action of another hormone, called thyrotropic hormone or thyroid-stimulating hormone (TSH), which is secreted by the anterior pituitary; TSH regulates the rates of synthesis and release of thyroxine by the thyroid gland. The rate of secretion of TSH is itself partly under control of blood thyroxine levels (feedback).

parent homologs of the thyroid are part of the ventral wall of the pharynx. Thus, we can infer an evolutionary sequence in the thyroid structure of chordates, from cells in the pharyngeal wall to patches of free-standing cells to discrete glands. The evolution of thyroid *hormones* appears to have been more conservative; thyroxine is the major thyroid hormone in all vertebrates. In contrast, there has been considerable evolutionary change in *target cell responses* to thyroxine, as discussed on p. 596.

It is neither desirable nor practical to illustrate all endocrine principles using one example such as the thyroid. Now that we have introduced some of the important principles with this example, we shall illustrate more fully these and other principles with further examples. Before doing so, however, we need to introduce the major hormones and endocrine glands of vertebrates. To the unfamiliar reader, a discussion of endocrinology can read like a Russian novel, with a large, strangely named cast of interacting characters. To aid the reader, we provide here a list and diagram of the cast of mammalian hormones and glands and some of their major effects (Table 21.1; Figure 21.2). This list is not exhaustive and will be expanded later in the chapter. It is presented at this point to help you to keep track of the examples we use to illustrate principles of endocrine

action. We suggest you review the table and illustration, now and from time to time, to keep track of the cast of vertebrate endocrine characters.

THE CHEMICAL NATURE OF HORMONES

Hormones are identified as such by the physiological roles they play, not by their chemical composition. A particular chemical compound that is present in several groups of animals may function as a hormone in some but not in others. Similarly, a compound may be used in *both* hormonal and other roles in the same organism; among vertebrates, for example, norepinephrine is not only secreted as a hormone by the medullae of the adrenal glands but also functions as a synaptic transmitter substance (p. 443).

Although the chemical identity of a substance does not in itself determine hormonal status, the chemical nature of hormones remains a matter of great interest and importance. The compounds now commonly recognized as hormones fall into several chemical classes, four of which we mention here:

1. *Steroids.* The steroid hormones are synthesized from cholesterol. Prominent in their structure is a series of carbon rings: typically three six-membered rings conjugated with a five-membered one (Figure 21.3*A*). Among vertebrates, the gonadal sex hormones and the hormones of the adrenal cortex are steroid hormones. Arthropod molting hormones (e.g., ecdysone) are also steroids.
2. *Peptides and proteins* and derivatives thereof (e.g., glycoproteins). Hormones structured from chains of amino acids are numerous. In vertebrates, for example, they include the antidiuretic hormones, insulin, and the various anterior-pituitary hormones such as growth hormone and ovarian follicle-stimulating hormone. The gamete-shedding hormone of starfish and the diuretic hormones of insects are examples of peptide/protein hormones in invertebrates. The peptide/protein hormones vary enormously in molecular size, from tripeptides (consisting of just three amino acid residues) to proteins containing over 200 residues. Often all such hormonal compounds are simply called "peptide hormones" (blurring the size distinction), and we shall frequently follow that practice. Figure 21.3*B* shows the structure of one example of a peptide hormone, mammalian gonadotropin releasing hormone (a decapeptide).
3. *Catecholamines.* The catecholamines, so named because they are amine derivatives of the six-carbon ring structure catechol, are found widely in both invertebrates and vertebrates and commonly function as synaptic transmitter substances. There are two well-established catecholamine hormones in vertebrates: epinephrine (Figure 21.3*C*) and norepinephrine—also called adrenaline and noradrenaline. Dopamine, another catecholamine, may also act as a hormone.
4. *Iodothyronines.* Although evidence exists for the synthesis of iodothyronines by protochordates, it is only in vertebrates that these compounds are known to be hormones. Secreted by the thyroid gland, they

Table 21.1 PRINCIPAL MAMMALIAN ENDOCRINE GLANDS AND HORMONES[a]

Gland or tissue	Hormone	Some general effects (Effects are stimulated by hormone unless noted.)
Endocrine hypothalamus	Thyrotropin releasing hormone (TRH)	TSH release from pituitary
	Corticotropin releasing hormone (CRH)	ACTH release from pituitary
	Gonadotropin releasing hormone (GnRH)	LH and FSH release from pituitary
	Somatostatin (GHRIH)	Inhibits growth-hormone release
	Growth-hormone releasing hormone	Growth-hormone release from pituitary
Neurohypophysis (posterior pituitary)	Oxytocin	Uterine contraction, milk ejection from mammary gland
	Vasopressin (ADH)	Water reabsorption by kidney tubules
Adenohypophysis (anterior pituitary)	Thyroid-stimulating hormone (TSH)	Thyroid hormone secretion
	Adrenocorticotropic hormone (ACTH)	Adrenal cortex hormone secretion
	Follicle-stimulating hormone (FSH)	Ovarian follicle development and secretion, sperm production by testes
	Luteinizing hormone (LH)	Secretion of ovarian/testicular hormones, luteinization of ovarian follicles
	Growth hormone (GH or STH)	Growth, protein synthesis
	Prolactin	Milk secretion
Pineal	Melatonin	Affects LH and FSH production
Thyroid	Thyroxine and triiodothyronine	Oxidative metabolism, affect growth and development
	Calcitonin	Lowers blood calcium
Parathyroid	Parathyroid hormone	Regulates calcium metabolism
Adrenal cortex	Mineralocorticoids (e.g., aldosterone)	Ion transport by kidney tubules
	Glucocorticoids (e.g., cortisol)	Carbohydrate formation from protein, fat catabolism
Adrenal medulla	Epinephrine (adrenaline)	Elevates blood glucose, "fight or flight" responses
	Norepinephrine (noradrenaline)	Vasoconstriction, "fight or flight" responses
Endocrine pancreas (islets of Langerhans)	Insulin	Synthesis of glycogen, protein, and fat; glucose and amino acid uptake from blood
	Glucagon	Glycogen conversion to glucose
Stomach	Gastrin	Gastric juice secretion
Duodenum	Secretin	Pancreatic juice flow, inhibits gut motility and HCl secretion
	Cholecystokinin (CCK)	Release of bile from gallbladder, pancreatic enzyme secretion
Ovary	Estrogens	Development and maintenance of female reproductive system
	Progesterone	Uterine changes necessary for pregnancy
Testis	Testosterone	Development and maintenance of male reproductive system

[a] See later tables for greater detail.

have the unique property of being rich in iodine. One of the iodothyronines is thyroxine. Thyroxine is a four-iodine compound and thus is sometimes known as tetraiodothyronine, or T_4. Thyroid secretion also includes a three-iodine compound called triiodothyronine or T_3 (Figure 21.3D).

THE LIFE HISTORIES OF HORMONES

A hormone molecule has a definite physiological life span in terms of its ability to act as a messenger affecting body functions. Hormone molecules are synthesized, stored, and released by endocrine cells,

travel through the circulating blood, exert their effects on target tissues, and are metabolically destroyed or excreted from the body. Thus, a hormone molecule can be said (metaphorically) to be born, travel, do its job, and die. Bentley has used the phrase "the life history of hormones" to describe these stages in the life cycle of hormone molecules.

Hormones typically exert *proportionate effects,* the magnitude of effect depending on the concentration of the hormone. The blood concentration represents a balance between the *rate of addition* of hormone to the blood (by endocrine cell secretion)

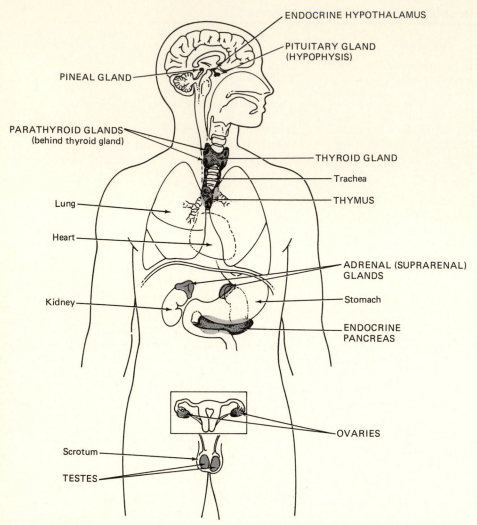

Figure 21.2 The discrete glands of mammals, as illustrated by the human example. The status of the thymus is uncertain; it may or may not be endocrine in function. [From G.J. Tortora and N.P. Anagnostakos, *Principles of Anatomy and Physiology,* 5th ed. Harper & Row, New York, 1987.]

and the *rate of removal* of hormone from the blood (by metabolic destruction and excretion) (Figure 21.4). The control of hormone concentrations—and thus the major control of hormone effects—rests primarily on the rate of hormone addition to the blood. The blood concentration of insulin, for example, depends primarily on the rate of insulin synthesis and secretion and much less on the relatively invariant rate of removal of insulin from the blood.

Synthesis, Storage, and Release

The major functions of endocrine gland cells are to synthesize their characteristic hormonal product, store it, and release it into the bloodstream. We examine these processes using the peptide hormone insulin as our primary example. Then we consider steroid hormones as secondary examples of the syn-

thesis and release of a different chemical class of hormones.

Insulin and the Pancreas   Insulin is the major vertebrate hormone that lowers the concentration of blood glucose (hypoglycemic action) and promotes the incorporation of foodstuff molecules into storage compounds (see Table 21.6 and p. 615). Insulin is synthesized in the *endocrine pancreas,* which in most vertebrates consists of islets of endocrine tissue—the *islets of Langerhans*—embedded in the exocrine tissue that constitutes the bulk of the pancreas and that secretes digestive enzymes. Mammalian islets of Langerhans contain four different types of endocrine cell that synthesize different peptide hormones: insulin (B cells), glucagon (A cells), somatostatin (D cells), and pancreatic polypeptide (PP cells). We shall consider only the most numerous insulin-producing

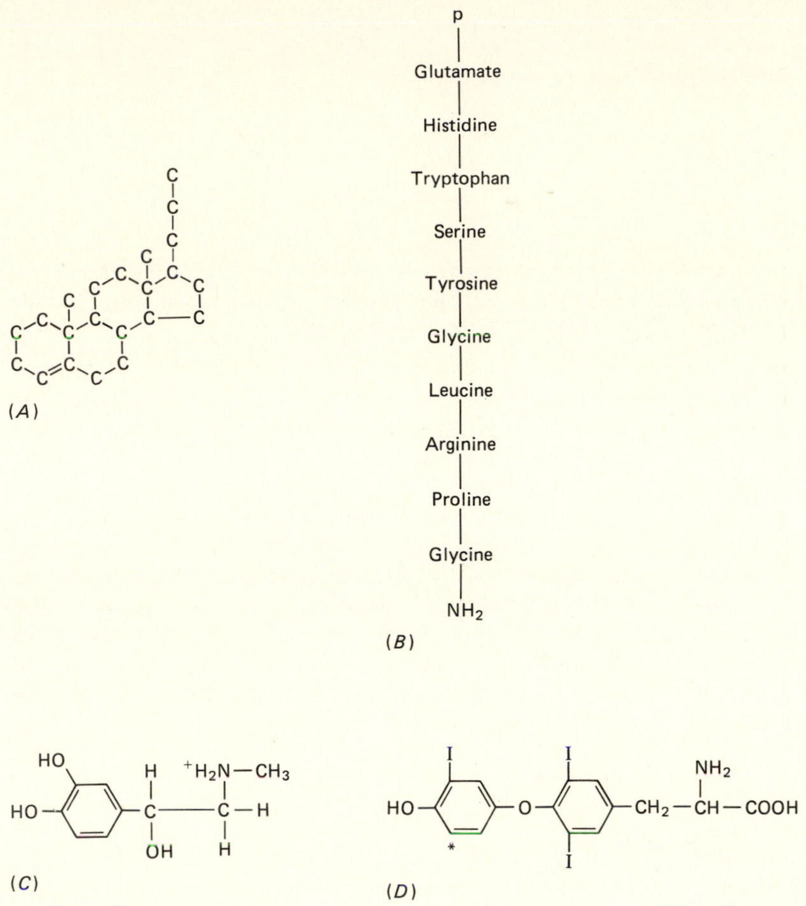

p
Glutamate
|
Histidine
|
Tryptophan
|
Serine
|
Tyrosine
|
Glycine
|
Leucine
|
Arginine
|
Proline
|
Glycine
|
NH₂

(A)

(B)

(C)

(D)

Figure 21.3 Chemical structures of some hormones. (A) The carbon structure of a representative steroid hormone. Various steroids differ, one to another, in the side chains attached to the ring structure and sometimes in the detailed structure of the rings. (B) The sequence of amino acid residues in mammalian gonadotropin releasing hormone, a peptide hormone secreted by the hypothalamus. (C) Epinephrine (adrenaline). (D) Triiodothyronine. The asterisk in (D) marks the position of the fourth iodine atom present in tetraiodothyronine (thyroxine). In both (C) and (D), the angles of the rings are occupied by carbon atoms.

B cells as an example of production of peptide hormones in general.

Insulin Synthesis and Secretion The general processes of synthesis and packaging of peptides and proteins for export are thought to be similar for all kinds of secreted peptides and proteins, including peptide hormones, peptide transmitters, and exocrine proteins such as digestive enzymes. The processes of synthesis and packaging were considered in Chapter 19 (p. 546). Figure 19.44 outlines the sequence of cellular activities underlying these processes. Here, we consider the example of insulin synthesis in greater detail.

The insulin molecule consists of two peptide

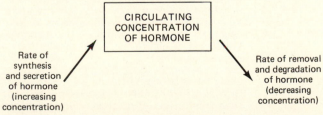

Figure 21.4 Control of hormone concentration in the blood. The blood concentration determines the amount of effect.

chains (termed A and B chains) connected by disulfide bonds (Figure 21.5). There is some variation among species in the amino acid sequences of the two chains (p. 593), but the general structure, the amino acid sequences of certain regions, and the positions of the disulfide bonds are highly conserved. The information determining the amino acid sequences of the chains is encoded in the DNA of the cell nucleus, as is the case for all peptides. It has been possible to clone the DNA for human insulin, by inserting it in a bacterial genome. This powerful technique has helped considerably to clarify the way in which insulin is synthesized and processed.

Figure 21.6 summarizes the steps in insulin synthesis. First, the DNA nucleotide sequence of the insulin gene is *transcribed* into the nucleotide sequence of messenger RNA, which leaves the nucleus and interacts with ribosomes of the granular endoplasmic reticulum (ER). The nucleotide sequence of messenger RNA is then *translated* to determine the amino acid sequence of the polypeptide that is synthesized at the ribosomes.

The two peptide chains of insulin are not synthesized independently but instead are formed as two parts of a single, larger molecule termed *pre-proinsulin*. Pre-proinsulin is a *precursor* of insulin, syn-

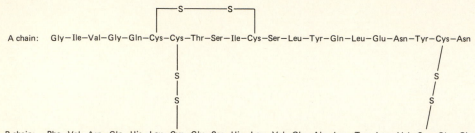

A chain: Gly—Ile—Val—Gly—Gln—Cys—Cys—Thr—Ser—Ile—Cys—Ser—Leu—Tyr—Gln—Leu—Glu—Asn—Tyr—Cys—Asn

B chain: Phe—Val—Asn—Gln—His—Leu—Cys—Gly—Ser—His—Leu—Val—Glu—Ala—Leu—Tyr—Leu—Val—Cys—Gly—Glu—Arg—Gly—Phe—Phe—Tyr—Thr—Pro—Lys—Thr

Figure 21.5 Structure of human insulin. Two polypeptide chains, termed the A chain and B chain, are held together by disulfide bonds. The three-letter codes refer to amino acid residues (e.g., "Gly" symbolizes a glycine residue).

thesized at the ribosomes. The processes involved in altering such a precursor polypeptide to make the final product are collectively termed *post-translational processing*. Pre-proinsulin is a single polypeptide consisting of four regions termed P, B, C, and A. Regions A and B become the A and B chains of the insulin molecule, C is the segment connecting these chains, and P is the "pre" segment that is the first part of pre-proinsulin to be synthesized. As earlier noted, the ribosomes involved in synthesis of pre-proinsulin are associated with membranes of the ER. The P segment is thought to function as a leader sequence

that facilitates passage of the nascent polypeptide chain through the ER membrane to the lumen of the ER. The P segment is then enzymatically cleaved off the polypeptide, even before synthesis is completed. The resulting polypeptide (minus the P segment) is termed *proinsulin*. Proinsulin initially contains sulfhydryl (—SH) groups that are subsequently reduced to the three disulfide (—S—S—) bridges that fold the molecule (Figure 21.6). After formation of these disulfide bridges, the C segment is cleaved away, leaving the A and B chains linked by the disulfide bonds. These stages of post-translational pro-

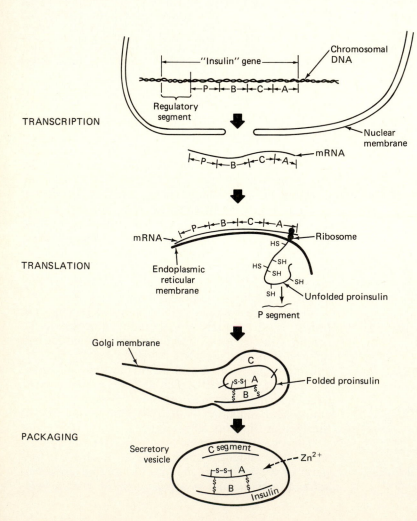

Figure 21.6 An outline of insulin biosynthesis. At the top, messenger RNA (mRNA), produced by transcription of the "insulin" gene, exits the nucleus. The mRNA is then translated by ribosomes associated with the endoplasmic reticulum to form a precursor polypeptide known as pre-proinsulin. Part of this precursor (the P segment) is cleaved away even before translation is complete, leaving proinsulin. The proinsulin then becomes folded, and part of it (the C segment) is cleaved away, leaving insulin. The insulin and C segment are packaged into secretory vesicles. Pre-proinsulin consists of four segments: the A and B segments destined to be the A and B chains in insulin, the P ("pre") segment, and the C ("connecting") segment interconnecting the A and B segments. The letters along the DNA and mRNA mark the parts of the DNA and mRNA that are translated into these segments of the polypeptide. Zinc is required for normal aggregation of insulin molecules. [From A. Gorbman, W.W. Dickhoff, S.R. Vigna, N.B. Clark, and C.L. Ralph, *Comparative Endocrinology.* Wiley, New York, 1983.]

cessing occur as the proinsulin molecules are first transported to the Golgi apparatus in shuttle vesicles and then packaged into secretory vesicles that bud off from the Golgi complex. The C-peptide fragments are retained in the secretory vesicles and secreted along with the "mature" insulin molecules. The general sequence of synthesis of a pre-prohormone and post-translational cleaving to produce a mature hormone is thought to be similar for all peptide hormones, and in fact for all secreted peptides.

Secretion of insulin is by exocytosis of secretory granules, a process that appears similar to the secretion of synaptic transmitters (Chapter 17, p. 433) and of digestive enzymes (Chapter 19, p. 546). Most endocrine cells synthesize and release some hormone all the time. As noted earlier (Figure 21.4), the *rate* of release of a hormone is variable and is one of the major determinants of the hormone concentration in the circulation and thus of hormone effects on targets. Rates of hormone synthesis and release are controlled by a variety of means, to be discussed later (p. 598). The rate of insulin secretion (and thus the level of circulating insulin) is modulated by several factors, the most important of which are the levels of glucose and of certain amino acids in the

blood. These agents stimulate secretion via an effect on Ca^{2+} ions; like all cellular secretion, insulin secretion is Ca^{2+} dependent. Stimulants may directly increase intracellular Ca^{2+} concentrations or may do so indirectly by increasing levels of cyclic AMP in the cell, which in turn increases the concentration of Ca^{2+}. The B cells of the islets can generate action potentials that promote Ca^{2+}-dependent insulin release. This finding suggests that the later stages in the biochemical control of peptide hormone release may be quite similar to the control of release of synaptic transmitters (p. 431; see also p. 547).

Steroid Hormone Synthesis   Hormones other than peptide hormones are made by enzymatic means, rather than by transcription and translation. Enzymatically synthesized hormones include steroids, catecholamines such as epinephrine, and iodothyronines such as thyroxine (see p. 578 and Figure 21.3). We shall briefly consider steroid synthesis as an example (not necessarily typical) of the synthesis of a nonpeptide hormone.

Steroid hormones are synthesized from cholesterol, a 27-carbon cyclic lipid (Figure 21.7). In vertebrates, cholesterol is in part obtained from the diet

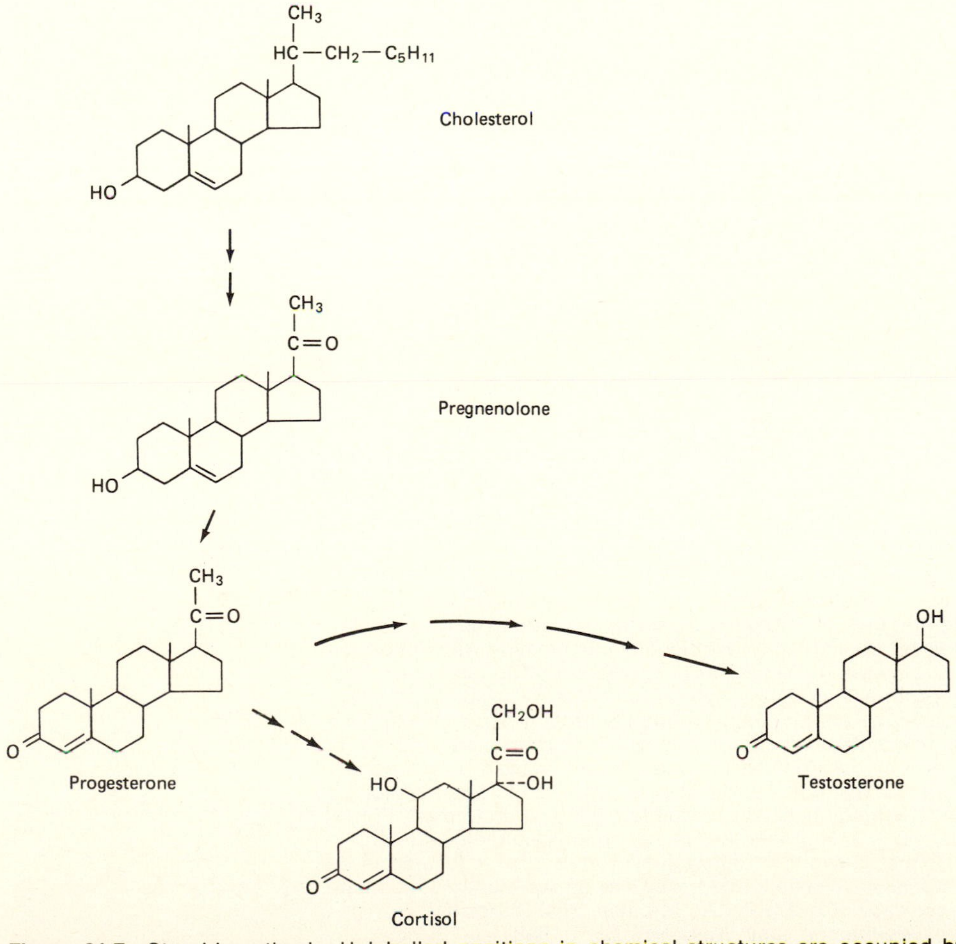

Figure 21.7 Steroid synthesis. Unlabelled positions in chemical structures are occupied by carbon atoms.

and can also be synthesized by steroid-producing cells of the adrenal cortex and gonads, by the liver, and by other cells. Cholesterol from the diet and from synthesis in the liver is carried in the blood as cholesteryl esters in association with low-density lipoprotein (LDL) and high-density lipoprotein (HDL). The lipoproteins are taken into cells by endocytosis, after combining with surface receptors such as LDL receptors.

The conversion of cholesterol to steroid hormones occurs in the endocrine cells of the adrenal cortex and gonads. The six-carbon side chain is cleaved from cholesterol to form pregnenolone (Figure 21.7), which is then converted via one of several parallel biochemical pathways to one of the estrogens, androgens, or adrenal cortical secretions. Different cell types contain the enzymes for different pathways and

major end products; cells of the testis, for example, contain a preponderance of the enzymes for synthesis of testosterone.

Steroid synthesis occurs at two sites within the relevant cells: smooth (or agranular) ER and mitochondria. Figure 21.8 summarizes the process. Intracellular synthesis of cholesterol occurs at smooth ER tubules (so termed because they lack ribosomes). Synthesized cholesterol and cholesterol derived from LDL uptake then enter mitochondria and are converted to pregnenolone. Most subsequent conversions in the biosynthesis of steroids occur at the smooth ER, although synthesis of some steroids (such as cortisol) requires reentry into mitochondria for at least one step of the synthetic sequence.

Steroids are stored as lipid droplets and granules in the cytoplasm. Secretion may occur by simple

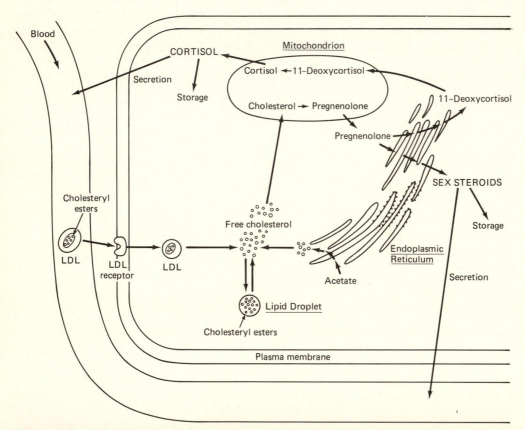

Figure 21.8 A diagrammatic outline of the biosynthesis of steroid hormones. Synthesis of both cortisol and sex steroids is shown although both would not ordinarily be made equally by any one steroid-producing cell. Cholesteryl esters are carried in the blood in combination with lipoprotein particles: low-density lipoprotein (LDL, shown) or high-density lipoprotein (HDL, not shown). The lipoprotein particles combine with receptors (e.g., LDL receptors) and enter the cell by endocytosis. The cholesteryl esters are then released by an interaction of the lipoprotein particles with lysosomes. Cholesterol and cholesteryl esters may be stored in the cell in lipid droplets. Cholesteryl esters may be converted to cholesterol by deesterification. Cholesterol may also be synthesized from acetate by the steroidogenic cell. Cholesterol is enzymatically converted to pregnenolone in the mitochondria. Pregnenolone is then enzymatically converted to the steroid hormones. For the most part these latter conversions take place in or near the smooth endoplasmic reticulum. In the adrenal cortex, however, the enzyme that converts 11-deoxycortisol to cortisol is located in the mitochondria. Once the steroid hormones have been formed, they may be secreted into the blood, stored within the steroidogenic cell, or catabolized. [After A. Gorbman, W.W. Dickhoff, S.R. Vigna, N.B. Clark, and C.L. Ralph, *Comparative Endocrinology.* Wiley, New York, 1983.]

diffusion of steroids through the lipid of the plasma membrane, although release by exocytosis of granules has also been suggested.

Transport in Blood

As noted above, the concentration of a hormone in the blood reflects a balance between the rate of hormone release by endocrine gland cells and the rate of hormone removal from the circulation by cellular uptake, metabolic destruction, and excretion. Concentrations of circulating hormones are very low, typically 10^{-12} to 10^{-9} M for peptide hormones and 10^{-11} to 10^{-5} M for steroid hormones. Hormone concentrations, of course, change with physiological circumstances; for example, in the rat, the vasopressin (ADH) concentration increases 25-fold with dehydration, from 10^{-11} to 2.5×10^{-10} M.

Some hormones are transported in the blood primarily in a bound form, in association with specific binding proteins. Protein binding is a reversible (noncovalent) interaction, with free and bound forms of the hormone in equilibrium. Only the free form of the hormone is thought to be physiologically active. The hormones for which protein binding is most important are thyroid hormones and steroids. Since these hormones are relatively insoluble in water, binding to soluble proteins may increase the hormone-carrying capacity of the blood and provide a relatively stable reserve supply of the hormone. Protein binding also protects hormones from rapid inactivation (e.g., by the liver) and/or excretion.

Some hormones may be converted to a more active form after release from the endocrine glands, a process termed *peripheral activation*. For example, some secreted thyroxine (T_4) is converted to T_3 (a more active form in many tissues) in liver, pituitary, and kidneys. The peptide hormone angiotensin I (p. 611) and some steroids are also converted to more active forms in the circulation or at their targets.

Mechanisms of Hormone Action

The range of processes influenced by hormones is broad, including metabolism of carbohydrates, lipids, and proteins; growth, development, and reproduction; modulation of membrane permeabilities; and muscle contraction. The mechanisms of hormone action, by which hormones exert all these effects, involve the binding of hormone molecules to specific receptor molecules at the target cells and subsequent biochemical changes in the cells that result from this binding.

Hormone Receptors Different hormones typically affect different target cells and cause different effects even on the same cells. This *specificity of hormone action* depends on specific hormone receptor molecules of the target cells; only cells with insulin re-

ceptor molecules, for example, are affected directly by circulating insulin.

The binding interaction of a hormone and its receptor is thought to be analogous to the binding between an enzyme and a substrate: The two molecules fit together in a shape-dependent, lock-and-key fashion. Thus, binding of hormone and receptor is highly *specific*. Other molecules, even those chemically related to the hormone, lack distinctive details of structure and shape that allow a specific fit to the receptor site. A related characteristic of hormone–receptor binding is the *high affinity* of receptors for their hormones. Because of this affinity, a hormone can bind to a receptor and exert its effects even at very low circulating concentrations.

There are two different classes of hormone receptors: cell surface receptors and intracellular receptors. Lipid-soluble hormones (steroids and iodothyronines) can readily enter target cells by passing through the lipid bilayer of the cell membrane. Receptors for these lipid-soluble hormones are located inside the cell, in the cytoplasm or the nucleus. Peptide and catecholamine hormones, in contrast, are water soluble and do not diffuse into cells. Their receptors are located at the cell membrane and have external hormone-binding sites. These membrane-bound surface receptors mediate hormone actions by activating an intracellular *second-messenger* system. We first consider cyclic AMP as an example of a second messenger that mediates hormone actions and then examine hormone actions mediated by intracellular receptors for lipid-soluble hormones.

Cyclic AMP and Other Second Messengers It is thought that all intracellular actions of nonpermeating, water-soluble hormones are mediated by second messengers. (Another possible way in which a hormone such as a peptide might act is by receptor-mediated endocytosis, a process in which receptor–hormone complexes are taken into the cell by endocytosis of coated vesicles. It has been shown that insulin, for example, can enter cells by receptor-mediated endocytosis, but it is not clear whether this entry is involved in hormone action. Here, we consider only the second-messenger mechanism of peptide hormone action.)

Cyclic adenosine-3′,5′-monophosphate (cyclic AMP or cAMP, Figure 21.9) is the best-studied and one of the most widespread hormonal second messengers. It mediates most effects of peptide and catecholamine hormones (Table 21.2), as well as many metabolic actions of synaptic transmitters (p. 445 and Table 17.2). We have previously considered the role of cAMP in synaptic transmission in Chapter 17. Here, we describe an example of its similar function as an endocrine second messenger.

Epinephrine promotes the breakdown of glycogen to glucose in skeletal muscle and liver, among other actions (see Table 21.6). This effect was the first

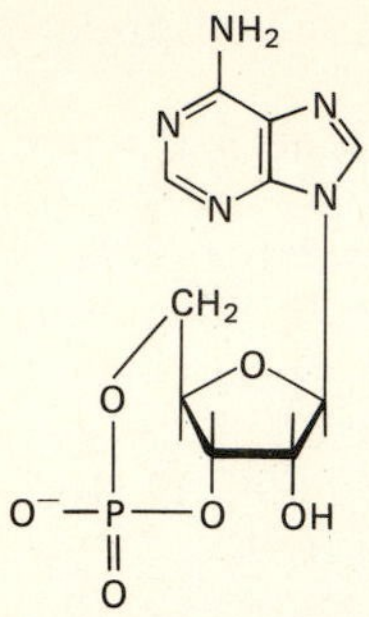

Figure 21.9 Structure of cAMP. Unlabelled positions in the structure are occupied by carbon atoms.

hormonal action shown to involve cAMP and will serve as an example of the ways in which a second messenger regulates a metabolic process. The cAMP-mediated action of epinephrine on glycogen metabolism involves the following sequence of steps (Figure 21.10):

1. *Epinephrine in the extracellular fluid binds to receptors at the cell membrane surface, activating the enzyme adenylate cyclase.* Adenylate cyclase is a membrane-bound enzyme that catalyzes the conversion of ATP to cAMP. The epinephrine–receptor complex activates adenylate cyclase by linking to it indirectly, via an intermediate G-protein (see Figure 17.28).

2. *The activation of adenylate cyclase allows that enzyme to catalyze the conversion of ATP to cAMP* at the cytoplasmic side of the membrane. It is the *concentration* of cAMP that determines the magnitude of its effect. The cAMP is metabolized to $5'$-AMP by the intracellular enzyme phosphodiesterase. The concentration of cAMP at any given time reflects a balance between the rate of its synthesis (which is under hormonal control) and the rate of its degradation (typically not under hormonal control). The

balance of processes that determines the intracellular concentration of cAMP is thus analogous to the balance of synthetic and degradative processes that determines the circulating concentration of a hormone (Figure 21.4).

3. *The cAMP binds to protein kinase,* an intracellular regulatory enzyme, *and activates it* by removal of an inhibitory subunit. Different types of animal cell may have different cAMP-dependent protein kinases.

4a. *Protein kinases catalyze the phosphorylation of regulatory proteins.* In skeletal muscle and liver cells, cAMP-dependent protein kinase indirectly regulates the rate of glycogen breakdown by this means. The protein kinase catalyzes the phosphorylation of a second regulatory protein, phosphorylase kinase (Figure 21.10). Active phosphorylase kinase in turn activates the enzyme glycogen phosphorylase, which converts glycogen to glucose-1-phosphate, at last completing the action of epinephrine on glycogen breakdown.

4b. The cAMP also leads to inhibition of glycogen synthesis, in addition to stimulating glycogen breakdown. This inhibition occurs because the cAMP-dependent protein kinase phosphorylates glycogen synthetase (the enzyme mediating the last step in glycogen synthesis) and thereby *inactivates* it. Thus, a single second messenger may direct several coordinated metabolic effects in a target cell.

5. When blood levels of epinephrine decrease, intracellular concentrations of cAMP also fall (as a result of phosphodiesterase action). Protein kinase then reverts to its inactive form, and the various enzymes affected by it become dephosphorylated.

The role of cAMP in the action of epinephrine on liver and muscle cells is fairly typical of cAMP function in the mediation of effects of other hormones. All actions of cAMP are thought to involve protein phosphorylation via protein kinases. The proteins phosphorylated, however, differ in different target cells. A wide variety of proteins appear subject to

Table 21.2 SOME HORMONE-INDUCED EFFECTS MEDIATED BY cAMP

Hormone	Target tissue	Effect mediated by cAMP
Epinephrine	Muscle, liver	Glycogen breakdown
Epinephrine	Fat cell	Lipid breakdown
Epinephrine	Heart	Increased rate and force of contraction
Adrenocorticotropic hormone (ACTH)	Adrenal cortex	Steroid synthesis and release
Luteinizing hormone (LH)	Corpus luteum (ovary)	Progesterone synthesis and release
Thyroid-stimulating hormone (TSH)	Thyroid	Thyroid hormone synthesis and release
Vasopressin (ADH)	Kidney	Increased water reabsorption
Melanocyte-stimulating hormone (MSH)	Melanocyte	Dispersion of melanin
Parathyroid hormone	Bone	Calcium mobilization

Sources: P.J. Bentley, *Comparative Vertebrate Endocrinology*. Cambridge University Press, New York, 1982; B. Alberts, D. Bray, J. Lewis, M. Raff, K. Roberts, and J.D. Watson, *Molecular Biology of the Cell*. Garland, New York, 1983.

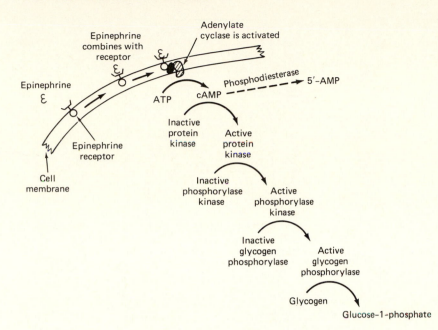

Figure 21.10 The cAMP-dependent effects of epinephrine on glycogen breakdown in cells of skeletal muscle and liver.

control by cAMP (see Figure 17.29). The actions of cAMP are typically indirect and, as in the control of glycogen breakdown, can involve several intervening steps. These intervening steps provide great *amplification* of the cellular response to a hormone: A single hormone–receptor complex can lead to the production of many cAMP molecules, each of which can activate a protein kinase molecule that then can phosphorylate many protein molecules. The greater the number of steps in this cascade, the greater the possible amplification of the original hormonal signal.

Other chemical agents besides cAMP may act as intracellular second messengers in mediating hormone action. These include Ca^{2+}, cyclic guanosine monophosphate (cGMP), and inositol phospholipids. Clarification of the roles of these agents is complicated by evidence that they may often interact with each other and with cAMP. For example, cGMP may act as an intracellular messenger by activating cGMP-dependent protein kinases. However, the enzyme guanylate cyclase—which catalyzes formation of cGMP—appears to differ from adenylate cyclase in being cytoplasmic rather than membrane-bound; thus, it could not directly carry hormonal signals into the cell. Instead, changes in cGMP may result from receptor-stimulated increases in intracellular Ca^{2+} concentration.

Calcium ions appear to play widespread roles as second messengers. Intracellular Ca^{2+} concentrations regulate muscle contraction (pp. 512 and 527), cellular secretion (pp. 431 and 547), and activities of several enzymes. Most effects of intracellular Ca^{2+} are mediated by **calmodulin,** a calcium-binding regulatory protein. Calmodulin is related to troponin C, the Ca^{2+}-dependent protein that regulates contraction of skeletal and cardiac muscle. When activated by binding Ca^{2+}, calmodulin can activate several pro-

teins including adenylate cyclase and phosphorylase kinase. Thus, Ca^{2+} and cAMP may co-act as second messengers.

Lipid-Soluble Hormones Act Via Intracellular Receptors (Figure 21.11)   Steroid hormones and iodothyronines are lipid-soluble and can diffuse through the cell membrane, as noted earlier. Once in the cell, the hormone molecules bind to intracellular receptor proteins, which activate specific genes of the target cell. Early studies suggested that unoccupied steroid receptors were mostly cytoplasmic; after binding to steroids the receptors were thought to move into the nucleus and to bind to DNA. Recent studies indicate that (at least for estrogen receptors) the unoccupied receptors remain in the nucleus. Estrogen moves into the nucleus and binds to the estrogen receptors, inducing activation of the receptors. The activated steroid receptors can then bind to chromatin at specific sites and activate RNA polymerase, the enzyme responsible for synthesis of messenger RNA. Thus, as shown in Figure 21.11, the hormone initiates transcription of messenger RNA (mRNA) associated with specific genes. The mRNA, in turn, codes for the synthesis of specific proteins, many of which are regulatory proteins (e.g., enzymes). The mechanism of action of iodothyronine hormones is similar; the hormones bind to nuclear receptors and control gene expression.

Steroid hormone–receptor complexes apparently directly regulate the transcription of only a small number of genes in a target cell, an action termed the *primary response*. Some gene products of the primary response, however, may activate other genes, producing a delayed *secondary response*. The same hormone–receptor complex can activate different genes in different cells. For example, the diverse

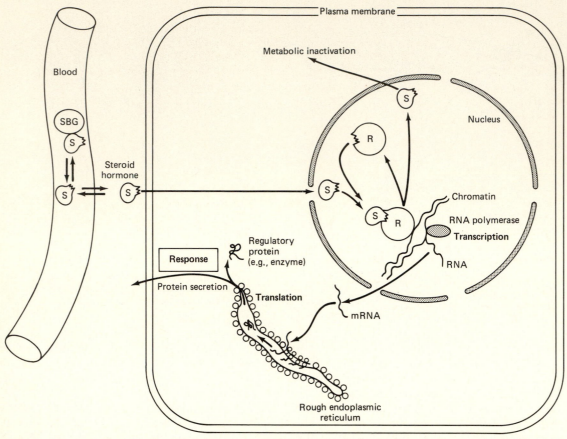

Figure 21.11 A diagram of the mechanism of action of steroid hormones as understood from recent studies of estrogens. Steroid hormone (S) which has disassociated from steroid-binding globulin (SBG) in the blood enters the cell cytoplasm and then the cell nucleus. In the nucleus, it binds with a receptor (R), and the hormone-receptor complex activates transcription to form a specific mRNA which is then translated into protein. [After A. Gorbman, W.W. Dickhoff, S.R. Vigna, N.B. Clark, and C.L. Ralph, *Comparative Endocrinology*. Wiley, New York, 1983.]

cellular responses to testosterone (see p. 619) may all result from differential gene activations by a single testosterone receptor protein. The basis of this differential gene action remains to be clarified.

Termination of Hormone Actions

Most hormones persist in the circulation for only a rather short period of time. Hormones are enzymatically degraded (usually by the liver or kidneys in vertebrates) and their products are excreted in urine or bile. Hormones have different average lifetimes in the circulation. In humans, for example, the biological half-life (time required to decrease the concentration of a circulating hormone by one-half) of vasopressin is 15 min; the half-life of cortisol is about 1 h, and that of thyroxine is nearly a week. These differences in half-life reflect both the activities of degrading enzymes and the protection of some hormones (e.g., thyroxine and cortisol) by circulating hormone-binding proteins.

Dilution by the blood volume can effectively terminate the action of some hormones. For example,

the hypothalamo–hypophysial portal system (discussed later, Figure 21.25) carries releasing hormones directly from the hypothalamus to the pituitary before their concentrations are diluted in the general circulation. By the time these hormones are in the general circulation, they are typically so dilute as to be ineffective.

TYPES OF ENDOCRINE GLAND AND CELL

Often, hormone-producing cells are grouped together into discrete, grossly observable endocrine glands (Figure 21.2). At another extreme, however, they may be scattered within the tissues of organs that are otherwise nonendocrine in function; these scattered populations of cells are sometimes termed *diffuse glands*. Discrete and diffuse endocrine glands are ends of a continuous spectrum of tissue organization. Several endocrine tissues, such as the islets of Langerhans of the pancreas and the endocrine tissues of the testes (see Figure 21.34), have structural organizations intermediate between discrete and diffuse glands. Moreover, many discrete glands such as the adrenal glands appear to have evolved from more-diffuse glands of ancestral animals (see pp. 593–595).

The largest and best-known site of diffuse hormone production in mammals is the gut. The walls of the small intestine, for example, harbor a population of distinctive cells known as S cells that produce a hormone called *secretin* when exposed to stomach acid, and the walls of both the stomach and upper small intestine contain cells of another type, I cells, that produce *cholecystokinin* when digestive products are encountered. Secretin and cholecystokinin stimulate release of juice and digestive enzymes by the exocrine part of the pancreas, thus aiding digestion. Recent work indicates that the heart (see p. 195), lungs, and other tissues are also sites of diffuse hormone production.

Endocrine cells themselves commonly have been

divided into two major classes, as mentioned in Chapter 2 (p. 12). On the one hand are cells that have structural and physiological resemblances to neurons, known as *neurosecretory cells*. On the other hand are cells lacking such affinities, known as *epithelial* or *nonneural endocrine cells*. When cells of these types are organized into discrete glands, the glands themselves may be described as *neurosecretory* or *epithelial* (nonneural) *glands,* respectively. The endocrine secretions of either type of cell or gland are properly termed hormones, but those of neurosecretory structures are often distinguished by being called *neurohormones* or *neurosecretions*.

Secretion by nonneural endocrine cells is usually controlled by hormones that reach them from other endocrine structures via the blood, as diagrammed in Figure 21.12*B*. Nonneural endocrine cells often appear not to be innervated. Some invertebrate non-

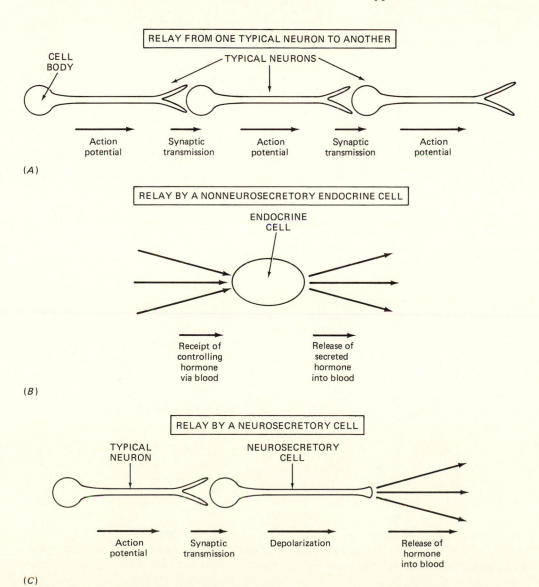

Figure 21.12   Three types of signal-relay system: (*A*) relay from one typical neuron to another; (*B*) relay by a nonneurosecretory endocrine cell; and (*C*) relay by a neurosecretory cell.

neural cells, however, have long been known to be innervated, and recent studies indicate that vertebrate nonneural cells (e.g., the pancreatic islets and ovarian corpus luteum) may be more commonly innervated than has been realized.

Neurosecretory endocrine cells are always in synaptic contact with typical neurons. Thus, they are *endocrine cells that directly interface with the nervous system.* As diagrammed in Figure 21.12*C,* neurosecretory cells synaptically *receive controlling impulses* from the typical neurons with which they are in contact. Instead of synaptically sending impulses to other cells, however, they release chemicals into the blood when stimulated. They thus are capable of transducing neural signals into endocrine ones.

The functional similarity between typical neurons and neurosecretory cells should not escape notice. Both types of cell release communicatory chemicals. In the one case, the chemicals are released into an exceedingly narrow gap between two cells—a synapse—and, traveling no further, are almost immediately destroyed or resorbed; these chemicals we call synaptic transmitter substances. In the other case, the chemicals are released into the blood and, persisting for a time, travel some appreciable distance before encountering the cells they control; these we call neurohormones.

The fundamental similarity between neurons and neurosecretory cells [and the presence of intergrades between them (e.g., see p. 629)] suggests evolutionary

---

BOX 21.1   SIMILARITIES BETWEEN VERTEBRATE NEUROSECRETORY AND NONNEURAL ENDOCRINE CELLS: THE APUD HYPOTHESIS

The distinction between vertebrate neurosecretory cells and nonneural endocrine cells is not absolute. The cells of the adrenal medulla, for example, lack axons and neuronal cytological features. Embryologically, however, they develop as modified sympathetic ganglionic neurons. Thus, in their developmental origin and their mode of control (Figure 21.12*C*), adrenal medullary cells are closely related to neurons.

Many peptide-secreting vertebrate endocrine cells, although not neurosecretory, also have certain similarities to neurons. One theory, termed the *APUD hypothesis,* suggests that peptide- and amine-secreting endocrine cells consistently are embryologically derived from neuroectoderm and thus are all related to neurons. The cells claimed to be embryologically related in this way include those of the adrenal medulla and the neurosecretory cells of the hypothalamus (p. 600), as well as many cells that have been classed as nonneural. Examples of this latter group include cells of the anterior pituitary, the pancreatic islet cells that produce insulin and glucagon, the calcitonin-secreting cells of the thyroid, and the several sets of diffuse-gland cells in the gut that produce secretin, gastrin, cholecystokinin, and a variety of other peptides.

The evidence for the claimed relation among such diverse endocrine cells rests principally on their biochemical and histological similarities. One similarity is that the cells contain amines and/or synthesize amines by decarboxylation of precursors. The name "APUD" hypothesis is an acronym for a description of this property: "*a*mine content and/or *p*recursor *u*ptake and *d*ecarboxylation." The significance of the association of peptides and amines in peptide-secreting cells is not clear.

In further support of the APUD hypothesis, there is immunological evidence that several peptide hormones occur within discrete groups of neurons in vertebrate brains. For example, various groups of neurons contain peptides of the gastrin, cholecystokinin, and secretin families (see p. 589). In the brain, these peptides do not act as hormones but instead may serve as neurotransmitters. Whatever the functional significance to the brain of neurons containing agents such as gastrinlike peptide, the presence of the same or homologous peptides in certain neurons and in nonneural endocrine cells of the gut suggests that the two cell types are related.

The APUD hypothesis, if correct, suggests that the most fundamental division of the endocrine cells of vertebrates is not between neurosecretory and epithelial cells but rather between cells of ectodermal origin (synthesizing amines and peptides) and cells of mesodermal origin (synthesizing steroids). Because aspects of the hypothesis are controversial and continue to undergo revision, we shall continue to employ the more-traditional distinction of neural and nonneural endocrine cells.

continuity between the neural and the endocrine control systems. We do not know, however, whether neurosecretory cells evolved from neurons, or vice versa, or whether both types of cell trace a common ancestry.

Neurohemal Organs

Neurosecretory cells typically resemble neurons in being highly elongate. Commonly, each consists of a cell body and a long axon. The cell bodies are ordinarily located within the central nervous system, but the axons often extend well outside the central nervous system. Neurohormones are synthesized in the cell bodies. They are released, however, at the ends of the axons, which they reach by traveling down the axons, sometimes in combination with carrier molecules. It should be evident from all these considerations that the sites of release of neurohormones are often distant from their sites of synthesis.

Often, release occurs in a neurohemal organ. A *neurohemal organ* is an anatomically distinct site of release of neurohormones from the axons of neurosecretory cells into the blood. It consists of one or more clusters of axon ends and associated blood vessels or other circulatory specializations. Neurohemal organs are often anatomically more prominent than the clusters of cell bodies that actually produce the neurohormones that are released within them.

Neurohemal organs occur in both vertebrates and invertebrates. As we shall see, the posterior pituitary gland is one prominent example of a vertebrate neurohemal organ (p. 600). For our first detailed look at neurohemal organs, however, we shall examine an invertebrate case. Earlier, we summarized the glands of vertebrates (Figure 21.2). Now, we summarize those of insects and in the process encounter several neurohemal structures.

The Endocrine Organization of Insects: A Case Study

Figure 21.13 depicts many of the important endocrine structures of insects. Some insect glands are principally or entirely of the epithelial (nonneural) sort. Prominent among these are the *corpora allata* (singular: corpus allatum) and the *prothoracic glands* (or their homologs). These glands are innervated (though most of their innervation is not shown in the figure). The corpora allata receive nerve fibers from the brain (discussed later) and sometimes from the subesophageal ganglion. The prothoracic glands are innervated variously from the subesophageal and thoracic ganglia.

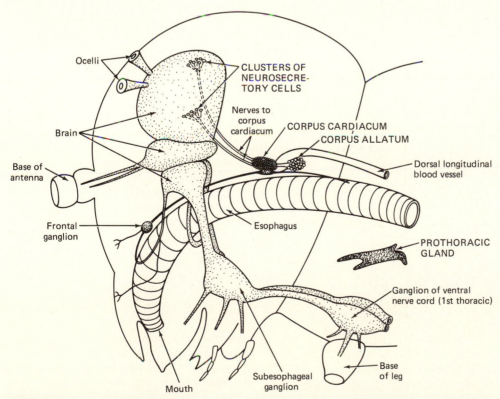

Figure 21.13 A lateral view of the anterior end of an insect, showing some of the major endocrine structures. Not all the neurosecretory centers in the brain are shown. The ganglia of the ventral nerve cord also contain neurosecretory cells. Some endocrine structures (e.g., the ovaries) are not included in this view. [Adapted by permission from P.M. Jenkin, *Animal Hormones*. Pergamon, New York, 1962.]

Neurosecretory cells are also of great importance in the endocrine systems of insects, as they are in the systems of all invertebrates. The brain contains clusters of **brain neurosecretory cells,** and neurosecretory cells are found in the *subesophageal ganglia* and *ganglia of the ventral nerve cord.*

Among the most prominent of endocrine structures in insects are the *corpora cardiaca,* which take their name from their close anatomical association with the dorsal longitudinal blood vessel ("heart"). Primitively, insects have two separate corpora cardiaca. Often, however, these have become fused to form a single structure, which is then called by the singular name, corpus cardiacum. The corpora cardiaca produce some hormones intrinsically. However, *they are principally neurohemal organs.* They are sites of release for neurosecretions synthesized in the brain.

As shown in Figure 21.13, the clusters of brain neurosecretory cells send axons via nerves to the corpora cardiaca. The axon terminals are located in the corpora cardiaca and are bathed there by the blood. The neurohormones produced within the cell bodies in the brain enter the blood in the corpora cardiaca.

Small neurohemal organs are also associated with neurosecretory cells of the ventral ganglia. Furthermore, it has recently been discovered that the corpora allata may serve in part as neurohemal organs for certain brain neurosecretory cells, the axons of which reach the corpora allata by traversing the corpora cardiaca (Figure 21.13).

THE EVOLUTION OF ENDOCRINE SYSTEMS

Endocrine systems evolve at all levels of their organization. Hormones exhibit molecular evolution; glands evolve in their histological structure and location; and tissues throughout the body change evolutionarily in their responses to hormones. In this section, we use case studies to examine these and other aspects of endocrine evolution.

Hormone Evolution

The posterior pituitary gland of tetrapod vertebrates—a part of the neurohypophysis—releases two peptide hormones. These *neurohypophysial peptides* provide one of the most interesting chapters in the study of hormonal molecular evolution.

In most mammals, the two peptides of the posterior pituitary are arginine vasopressin and oxytocin. The former is often called antidiuretic hormone (ADH), its most visible effect being limitation of urinary volume (p. 195). Oxytocin, on the other hand, functions in reproduction, causing contraction of the uterus and ejection of milk by the mammary glands. Structurally, arginine vasopressin and oxytocin bear close resemblances. Both are octapeptides, and both consist of a ring of five amino acid residues and a side chain of three residues. Indeed, as shown in Figure 21.14, the two hormones differ in just two residues; where vasopressin has a phenylalanine residue, oxytocin has isoleucine, and where vasopressin contains arginine, oxytocin has leucine. These close chemical similarities strongly suggest evolutionary affinity: Arginine vasopressin and oxytocin are probably descended from a single ancestral octapeptide. The divergence in their chemical structures, although involving just two amino acid residues, has been accompanied by a sharp divergence of roles. Oxytocin has little effect on urine production; vasopressin, on the other hand, has little effect on uterine contraction or milk ejection. Such functional divergence of hormones increases the capability of the endocrine system to control different processes independently.

Several mammalian species (e.g., pigs, peccaries, and hippopotamuses) produce lysine vasopressin as their antidiuretic hormone—either along with or to the exclusion of arginine vasopressin. Lysine vasopressin differs chemically from arginine vasopressin only in having the arginine residue replaced with a lysine residue. Presumably, the gene for lysine vasopressin in the piglike mammals arose by mutation from the gene for arginine vasopressin. Some marsupials produce both lysine vasopressin and another compound, phenypressin, which also differs from arginine vasopressin in just one amino acid residue. Lysine vasopressin and phenypressin diverge only a little from arginine vasopressin in their antidiuretic effects. Thus, although they provide a striking illustration of molecular evolution, we remain uncertain why they have replaced arginine vasopressin as the antidiuretic hormone in some mammalian groups.

In birds, reptiles, and amphibians, the usual neurohypophysial peptides are arginine vasotocin (the antidiuretic hormone) and mesotocin; both are octapeptides of the same basic sort as seen in mammals. In fish, additional octapeptides are found.

Figure 21.15 summarizes the composition of some of the compounds we have been discussing, emphasizing their close chemical similarity. We cannot say in all cases which compounds have evolved from which during the course of vertebrate evolution. There can be little doubt, however, that all these compounds are related by descent. Many biochemists have concluded that arginine vasotocin is the ancestral compound from which all the others have evolved.

Sets of hormones that are related by descent are called hormone *families*. The nomenclature of the members of families follows mixed principles and deserves brief mention. In the family we have been discussing, the neurohypophysial peptides, each chemically distinct substance has been assigned a distinctive common name. This practice promotes clarity, but particularly in regard to the peptide hormones of high molecular weight, it has not by any

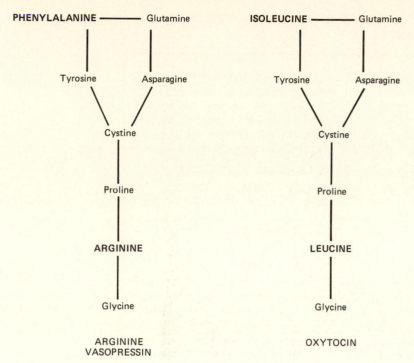

Figure 21.14 The chemical structures of arginine vasopressin and oxytocin, emphasizing the two loci at which the molecules differ.

means been uniformly followed. Consider, for example, insulin. Humans, horses, rats, chickens, and codfish are all said to produce "insulin." In fact, though, they do not all produce the same substance. All the vertebrate insulins bear a family resemblance, being polypeptides consisting of two linked chains and a usual total of 51 amino acid residues. Evolutionary divergence, however, has occurred within this family, so that, for example, the insulin of horses differs from that of humans in two residues and that of codfish differs from the human in 16. Vertebrates

produce *insulins*, not insulin. Prolactin, parathyroid hormone, corticotropin, and other hormones are actually all hormone families, not single compounds. It is important to note that the chemical differences among compounds in a family are sometimes accompanied by significant functional differences.

Evolution of Glands

Glandular tissues are known to have undergone evolutionary changes in their location and tissue orga-

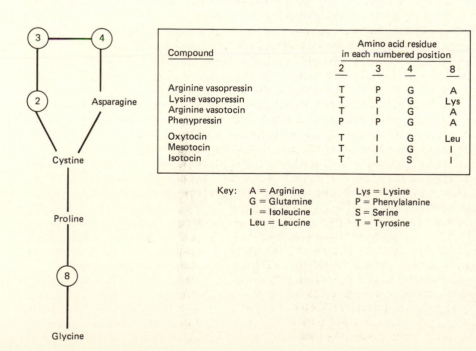

Compound	Amino acid residue in each numbered position			
	2	3	4	8
Arginine vasopressin	T	P	G	A
Lysine vasopressin	T	P	G	Lys
Arginine vasotocin	T	I	G	A
Phenypressin	P	P	G	A
Oxytocin	T	I	G	Leu
Mesotocin	T	I	G	I
Isotocin	T	I	S	I

Key: A = Arginine   Lys = Lysine
      G = Glutamine   P = Phenylalanine
      I = Isoleucine   S = Serine
      Leu = Leucine   T = Tyrosine

Figure 21.15 All the known neurohypophysial peptides are octapeptides that are identical in four of their eight amino acid residues. Their differences occur at just four positions, which are conventionally numbered 2, 3, 4, and 8, as shown at the left. The table at the right specifies the amino acids found at each of these four positions in seven hormones. Isotocin is a hormone of bony fish. Some of the known neurohypophysial peptides of fish are omitted.

BOX 21.2   CROSS-PHYLETIC RESEMBLANCES

Recent studies using immunological techniques have led to the striking discovery that compounds immunologically similar to vertebrate hormones exist within invertebrate tissues. Molecules immunologically similar to insulin, for example, are located in the guts of some mollusks and bees. Molecules having close affinities to gastrin, cholecystokinin, and arginine vasopressin have been identified in the nervous systems of insects, and *Hydra* neurons are immunoreactive for neurotensin and substance P.

Immunological similarity does not necessarily imply chemical identity. Thus, the molecules that have been found in invertebrate tissues may or may not prove to be exactly the same as vertebrate hormones. Nonetheless, even close similarity is evolutionarily revealing. We do not yet know the roles played by such compounds in invertebrates. Possibly, they function as synaptic transmitter substances, perhaps even as hormones. Clearly, in any case, molecules of these sorts occupy a much broader place—and have a far longer evolutionary history—than heretofore thought.

Compounds similar in their biological activity to certain vertebrate hormones are known even in plants. Prominent among these are the phytoestrogens: molecules that typically are chemically different from vertebrate estrogens but that share certain active subgroups with the vertebrate molecules and thus have estrogenic activity. Mammals that eat plants rich in phytoestrogens sometimes experience reproductive effects. In some instances, for example, phytoestrogens in fresh plant growth seem to promote reproduction in herbivores. At another extreme, phytoestrogens in red clover have been known to disturb profoundly the reproduction of grazing sheep, even causing abortion.

nization. An outstanding example is provided by the adrenal glands of vertebrates.

In mammals, the adrenals take the form of two discrete glands, usually located at or near the anterior ends of the kidneys (Figure 21.16*A*). Each mammalian adrenal gland consists of two major types of tissue *organized into two discrete regions* (Figure 21.16*B*). One type of tissue occupies the core or *adrenal medulla* and thus is known as *adrenomedullary* tissue. The other forms an outer layer, or *adrenal cortex,* surrounding the medulla and is called *adrenocortical* tissue. The adrenomedullary tissue arises embryologically from neural crest cells and is homologous with the sympathetic ganglia. It is extensively innervated by the sympathetic nervous system and produces two major *catecholamine* hormones, epinephrine and norepinephrine. The adrenocortical tissue is of mesodermal origin. It receives no, or very little, innervation and produces two classes of *steroid* hormones: *glucocorticoids* that affect the metabolism of foodstuffs and *mineralocorticoids* that affect ion balance.

The two basic types of adrenal tissue possessed by mammals are found also in the other groups of vertebrates. However, in no other group are they organized into a discrete cortex and medulla, and thus other vertebrates present a problem of nomenclature. Sometimes, the tissues of nonmammalian vertebrates that are homologous with mammalian cortical and medullary tissues are simply termed ad-

renocortical and adrenomedullary tissues, despite the anatomical inaccuracy. Often, the adrenal medullae of mammals and the homologous tissues of other vertebrates are known collectively as **chromaffin tissues.** Sometimes, the tissues of nonmammals that are homologous to mammalian adrenal cortex are called **interrenal tissues.** Here, we refer to the homologs of the mammalian adrenal medulla as **catecholamine-producing adrenal tissues** and to those of the adrenal cortex as **steroid-producing adrenal tissues.**

The evidence provided by modern fish suggests that, in the phylogeny of the vertebrates, the catecholamine- and steroid-producing tissues were initially separate—or at least were not at all as intimately associated as in mammals. There can be little doubt as well that the tissues originally were distributed more diffusely in the body than they are in mammals. Figure 21.17 shows, for example, that in a dogfish shark, the steroid-producing tissue is positioned apart from the catecholamine-producing tissue, and the latter exhibits a particularly diffuse distribution. In some bony fish, the two types of tissue are also anatomically separate, but in others much of the catecholamine-producing tissue has become associated with the steroid-producing tissue.

In reptiles, birds, and many amphibians, the two types of tissue have become intimately associated but not in the way seen in mammals. Instead of forming two discrete zones, the tissues commonly *intermingle* in the adrenal glands, as illustrated by

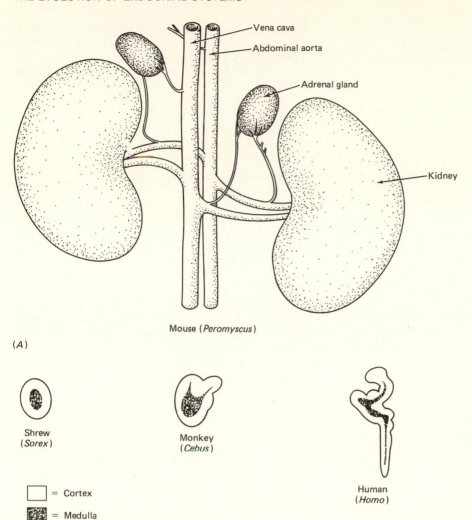

Mouse (*Peromyscus*)

(*A*)

Shrew
(*Sorex*)

Monkey
(*Cebus*)

Human
(*Homo*)

☐ = Cortex

▨ = Medulla

(*B*)

Figure 21.16 Mammalian adrenal anatomy. (*A*) The position and some of the vascular connections of the adrenals and kidneys in the New World mouse *Peromyscus leucopus.* (*B*) Sections through the adrenals of three mammals showing the positions of cortical and medullary tissue. [Reprinted by permission from F.A. Hartman and K.A. Brownell, *The Adrenal Gland.* Lea & Febiger, Philadelphia, 1949.]

Figure 21.18. The adrenal tissue of amphibians is diffusely distributed (Figure 21.19), but in birds and many reptiles—as in mammals—a transition to compact glands has been made.

Clearly, extensive reorganization of adrenal tissue has occurred in the evolution of the vertebrates. However, we do not understand much about the adaptive implications of this reorganization. Adrenal steroids have been shown to increase the activity of the enzyme (phenylethanolamine-*N*-methyl transferase or PNMT) that converts norepinephrine to epinephrine. Thus, the association of steroid-producing and catecholamine-producing tissues may result in an increased tendency to produce epinephrine rather than norepinephrine. In this regard, in the mammalian layered adrenal the blood supply of the adrenal medulla comes directly from the cortex, so that the two tissues are strongly associated through their circulation.

A final point to note is that the mammalian adrenal cortex is actually more elaborately organized than mentioned heretofore. It is not simply a homogeneous mass of tissue but instead is usually composed of three histologically distinct layers that are distinguished by their cell shape, tissue organization, and other properties: an outer layer called the *zona glomerulosa,* an inner *zona reticularis,* and in between a *zona fasciculata.* The layers differ in the mix of steroids they produce; typically, for example, the major mineralocorticoid hormone aldosterone is made chiefly by the zona glomerulosa, while the other layers produce glucocorticoids. The steroid-producing tissue of nonmammalian vertebrates is not subdivided into histologically distinct zones although there is some evidence for cellular specialization in its production of various hormones. The adaptive significance of the zonal organization in mammals remains speculative.

Evolution of Tissue Responses to Hormones

Hormones undergo molecular evolution, and glands evolve in their location and structure, but if we were compelled to specify which aspect of endocrine evo-

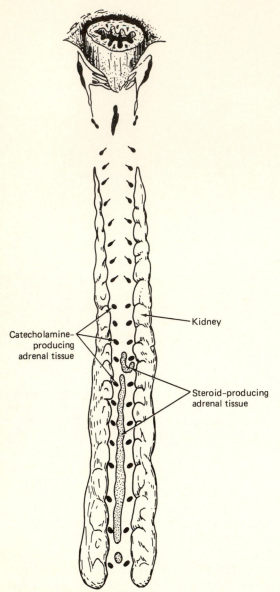

Catecholamine-
producing
adrenal tissue

Kidney

Steroid-producing
adrenal tissue

Figure 21.17 The adrenal tissues of a dogfish shark, *Mustelus canis.* The catecholamine-producing tissue is arrayed in a long series of paired bodies (black). The steroid-producing tissue (stippled) is centered between the kidneys. The kidneys have been turned outward to reveal the adrenal tissues. In life, the caudal adrenal tissues (toward bottom of figure) are in fact imbedded in kidney tissue. [Reprinted by permission from F.A. Hartman and K.A. Brownell, *The Adrenal Gland.* Lea & Febiger, Philadelphia, 1949.]

lution has been most significant, probably we would have to select the evolution of *responses* to hormones. Two species can secrete the same chemical compound as a hormone and yet exhibit different responses to it, thus employing it in different regulatory roles. Considering the number of tissues in an animal's body and the number of potential responses of each, it is apparent that the possibilities for endocrine evolution through response evolution are extensive.

From fish to mammals, all the major groups of vertebrates possess thyroid tissue, and in all the groups this tissue produces identical hormonal compounds: thyroxine (T_4) and often triiodothyronine (T_3). In fact, all the vertebrate groups maintain roughly similar concentrations of thyroxine in their blood. It would appear that thyroxine has been *available* as a hormone throughout the evolution of the vertebrates. Thyroxine's regulatory roles, however, have been far from static.

Consider, for example, the control of metamorphosis in amphibians. Secretion of thyroid hormones increases markedly at the time of metamorphosis, as indicated indirectly by the hypertrophy of the thyroid cells (Figure 21.20) and directly by the elevation of blood hormone concentrations (Figure 21.21). Experiments show that the heightened levels of thyroid hormones play a pivotal role in precipitating metamorphosis. If elevated amounts of thyroxine are experimentally administered to frog tadpoles before the usual age of metamorphosis, the tadpoles metamorphose prematurely into exceptionally small young frogs. Conversely, if the thyroid gland is removed from tadpoles, metamorphosis does not occur, and the animals—continuing to grow—become tadpole giants. Tissues throughout the bodies of tadpoles respond to elevated thyroxine by undergoing metamorphic changes. It is possible to test the responses of individual tissues—independently of other tissues—by inserting a small thyroxine-containing pellet into each tissue of interest; thyroxine diffusing from the pellet markedly raises the local thyroxine concentration without causing great changes in concentration in the body at large. A pellet placed in the tail of a frog tadpole will cause local resorption of tail tissue (Figure 21.22*A*), and one placed in a hindlimb will promote both growth and skeletal ossification in that particular limb (Figure 21.22*B*). The thyroid hormones exert major effects on growth and development in other vertebrates as well. However, their metamorphic effects in amphibians—involving dramatic, specific responses of tissues throughout the body—are exceptional. Neither in fish nor in any other group of tetrapod vertebrates is such a concerted alteration of body plan so strongly under thyroid control. The presence of thyroid hormones in modern fish strongly suggests that the hormones were already present in vertebrates before amphibians or the process of amphibian metamorphosis evolved. Thus, the evolution of thyroid control of metamorphosis involved the appropriation of a preexisting hormone to a new role, and an important part of that appropriation was the evolution of new responses to thyroid hormone by many tissues.

Amphibian metamorphosis presents other interesting evidence for the occurrence of evolutionary changes in the responses of tissues to hormones. Some species of amphibians do not ordinarily undergo metamorphosis but retain gills and other larval characteristics throughout life. Certain of these will

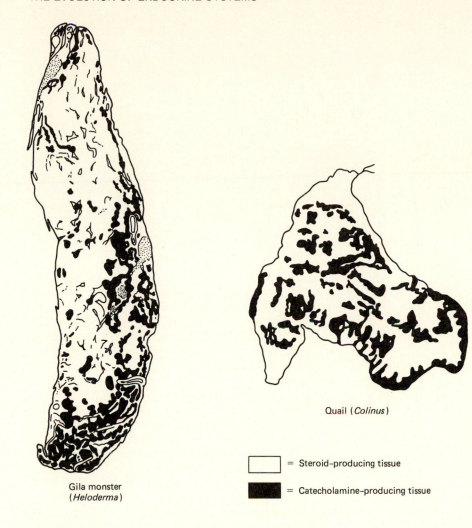

Quail (*Colinus*)

☐ = Steroid-producing tissue

■ = Catecholamine-producing tissue

Gila monster
(*Heloderma*)

Figure 21.18 Sectioned adrenals of a lizard and bird, showing the intermingling of steroid- and catecholamine-producing tissues. Stippled parts in lizard section are ganglia. [Reprinted by permission from F.A. Hartman and K.A. Brownell, *The Adrenal Gland*. Lea & Febiger, Philadelphia, 1949.]

metamorphose if thyroid hormones are artificially administered. They therefore resemble other amphibians in their tissue responses to thyroid hormones but differ in failing to exhibit an intrinsic increase in their rate of thyroid secretion. Others of the nonmetamorphosing amphibians—including mudpuppies (*Necturus*), for example—undergo little or no metamorphic change even when experimentally administered thyroid hormones. These species are of great interest in the present context because they have achieved their nonmetamorphic life cycle in part by the evolutionary loss of the tissue responses characteristic of most amphibians. The responses of tissues to thyroid hormones can also change during the ontogeny of individual amphibians.

The first physiological role of thyroid hormones to be recognized was the *calorigenic effect* of the hormones in mammals and birds. Suppression of thyroid secretion in such animals considerably lowers the metabolic rate (rate of metabolic heat production). Conversely, the elevation of hormone levels above normal evokes marked increases in metabolism. These effects are well known in clinical medicine, as illustrated in Figure 21.23. A lot of uncertainty remains about possible *short-term* calorigenic

responses mediated by thyroid hormones when a bird or mammal is exposed to cold or heat. However, the *long-term* actions of the hormones are well established and consistent from species to species (note the time scale in Figure 21.23): A proper *overall,* or *background,* level of metabolism is dependent on proper chronic levels of thyroid hormones, and often the steady maintenance of appropriate thyroid activity seems to play an essential *permissive* role, allowing for other, more volatile modulators of metabolism to be fully effective.

A positive correlation between the background rate of metabolic heat production and the rate of thyroid secretion is found uniformly in birds and mammals. A very different situation prevails among the poikilothermic vertebrates, however. In the latter groups, even if a species exhibits a calorigenic response to thyroid hormones, the response is likely to be inconsistent or highly dependent on test conditions (e.g., body temperature); more often than not, calorigenesis is in fact entirely unaffected by thyroid secretion. The consistent, dramatic calorigenic effect of the thyroid hormones on the tissues of birds and mammals appears to provide another striking example of a hormonal action that emerged

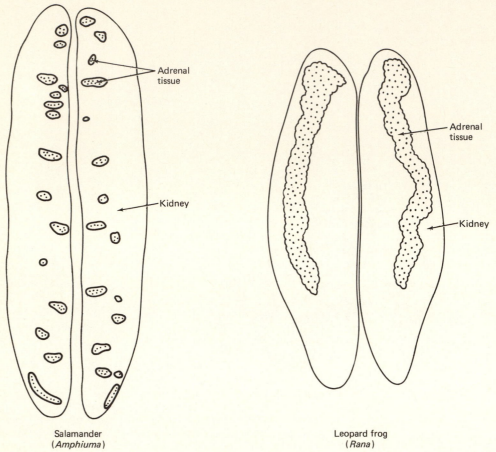

Salamander
(*Amphiuma*)   Leopard frog
                                                   (*Rana*)

Figure 21.19 Ventral views of the kidneys of two amphibians showing the diffusely distributed masses of adrenal tissue (stippled). The masses of adrenal tissue are embedded in the ventral renal surfaces and contain mingled steroid- and catecholamine-producing tissues. [Salamander reprinted by permission from F.A. Hartman and K.A. Brownell, *The Adrenal Gland.* Lea & Febiger, Philadelphia, 1949; frog adapted from same source.]

through the evolution of altered tissue responses to preexisting hormonal compounds.

CONTROL OF ENDOCRINE SYSTEMS

We have already noted that hormones exert proportional effects on their targets; in general, the higher the blood concentration of a hormone, the greater its effect is. For hormonal regulation of targets to be physiologically useful, animals must therefore be able to regulate the blood concentrations of their hormones. We stated earlier (p. 580) that this regulation is usually exerted by controlling the rates of release of hormones, and we briefly discussed some of the immediate biochemical aspects of hormone release. In this section, we consider in a broader sense how animals control the circulating concentration of a hormone, often via the effects of either other hormones or nervous activity on hormone release.

Some of the control pathways for endocrine cells are simple and direct. When the B cells of the pancreatic islets are exposed to high blood glucose con-

centrations, for example, they increase their secretion of insulin, a hormone that both promotes the transfer of glucose out of the blood and stimulates glucose use by tissues (p. 614). In this case, a very immediate regulatory relation has evolved: The B cells are controlled (in part) by detecting the level of one of the very substances that they themselves control.

Another relatively straightforward type of control pathway is *direct control of a gland by the nervous system,* as illustrated by the mammalian adrenal medulla. The medulla of each adrenal gland is homologous to a sympathetic ganglion. Preganglionic sympathetic neurons make synaptic contact with the secretory cells of the medulla and control their secretion, using acetylcholine as a synaptic transmitter substance.

At another extreme, control pathways for many types of endocrine cell are lengthy and indirect. Some of the most intricate of all control pathways involve the vertebrate pituitary gland. Let us now examine the basic characteristics of the pituitary and

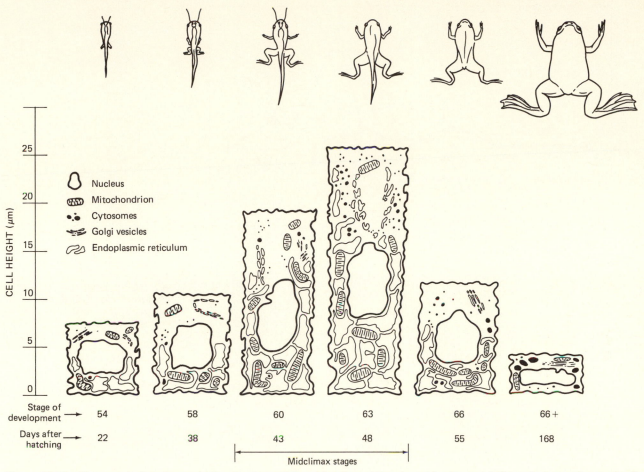

Figure 21.20 Changes in the size and ultrastructure of thyroid cells and in external body morphology during metamorphosis in the clawed toad *Xenopus laevis* reared at 22°C. The numbers (54-66) labelled "stage of development" refer to standardized stages in the course of development from tadpole to adult; the morphology of the animal at each stage is pictured at the top. Climax is the period of development when concerted metamorphic changes occur. During metamorphosis, dramatic increases occur in the size of the thyroid cells and in their content of cellular constituents—such as rough endoplasmic reticulum—that are involved in thyroid-hormone synthesis. These changes are reversed as metamorphosis ends. [Reprinted by permission from R. Coleman, P.J. Evennett, and J.M. Dodd, *Gen. Comp. Endocrinol.* **10**:34–46 (1968).]

then progress to a full examination of these complex endocrine control pathways.

The Vertebrate Pituitary

The vertebrate pituitary is a preeminent example of an endocrine gland that controls other endocrine glands—so much so that it used to be fashionable to call the pituitary "the master gland." The pituitary is a complex gland; although many of its functions involve control of endocrine activities, it has other functions as well. Here, we present an integrated overview of pituitary structure and function, concentrating on three questions: (1) What are the attributes of the pituitary? (2) What sorts of endocrine and other activities are controlled by the pituitary? (3) How are the various functions of the pituitary *themselves* controlled?

The vertebrate pituitary consists of two major divisions, which differ in embryological origin and in their functions. The divisions are the *adenohypophysis* (less formally known as the *anterior pituitary*) and the *neurohypophysis* (which is in part called the *posterior pituitary*).

The Neurohypophysis The neurohypophysis forms embryologically as a downgrowth of the brain and actually constitutes part of the brain floor. As shown in Figure 21.24, the neurohypophysis consists of three segments: (1) an anterior section, the *median eminence,* (2) a posterior extension, the *pars nervosa,* and (3) an interconnecting segment, the *infundibular stalk.* Together, the anterior pituitary and pars nervosa are suspended from the base of the brain by the infundibular stalk. The immediately adjacent brain region is the hypothalamus; the median eminence, in fact, constitutes part of the floor of the hypothalamus. In discussing the neurohypophysis, our initial

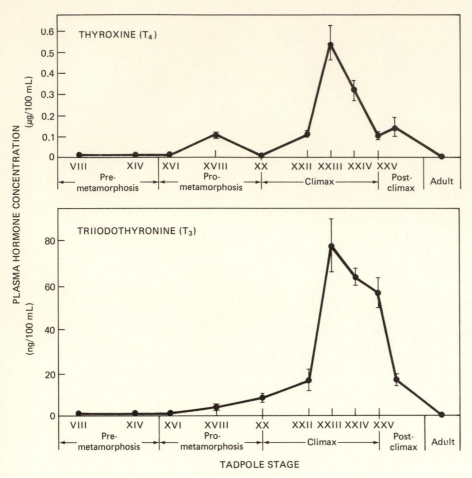

Figure 21.21 Plasma concentrations of thyroid hormones during development in bullfrogs, *Rana catesbeiana*. Both hormones peak during metamorphic climax. See legend to Figure 21.20 for notes on staging and climax; a different system of stage numbers is used for bullfrogs than for clawed toads. Vertical bars on graphs delimit ±1 standard deviation. [Reprinted by permission from E. Regard, A. Taurog, and T. Nakashima, *Endocrinology* **102**:674–683 (1978). Copyright 1978, The Endocrine Society, per The Williams and Wilkins Company (agent).]

focus shall be on the pars nervosa. *Pars* means "part." The pars nervosa is the "nervous part" of the tissue suspended beneath the brain. Sometimes called the *neural lobe* of the pituitary and sometimes called the *posterior lobe,* it is anatomically the most prominent segment of the neurohypophysis.

Typically, two peptide hormones are introduced into the blood in the pars nervosa. The exact nature of these hormones varies with the animal group; in most tetrapods, as already noted (p. 592), one of the hormones is *antidiuretic hormone,* whereas the other is *oxytocin* (mammals) or *mesotocin* (other tetrapods). Endocrinologists originally believed that the hormones of the pars nervosa were synthesized there. However, starting in the 1930s, research revealed that these hormones are actually neurosecretory and synthesized within the hypothalamus. Among mammals, as diagrammed in Figure 21.24, two paired clusters of cell bodies in the hypothalamus, known as the *paraventricular* and *supraoptic nuclei,* constitute the principal sites of production for the posterior-lobe hormones. Axons extend in well-defined tracts from these nuclei to the pars nervosa, where they release their hormones into the blood. Therefore, the pars nervosa is a neurohemal organ, analogous in

this respect to the insect corpora cardiaca described on p. 592.

The control of neurohormone release from the pars nervosa of the pituitary illustrates one relatively simple form of control of endocrine function: *neural control of neurosecretory cells*. Recall that the major effect of antidiuretic hormone (ADH) is to increase water reabsorption from the urine in the kidney tubules (p. 216). The release of ADH is controlled by hypothalamic receptors of osmotic concentration and other receptors (p. 195). The osmotic receptors directly or indirectly sense the osmotic concentration of body fluids and activate neurons that excite or inhibit the hypothalamic neurosecretory cells that secrete ADH in the pars nervosa. A high osmotic concentration of the body fluids leads to increased secretion of ADH and increased retention of water in the body. Neuronal sensors that provide information on blood volume are also involved in the control of ADH secretion. These include sensors of arterial blood pressure in the carotid sinus and sensors of stretch in the left atrial wall (low blood volume tends to cause low blood pressure and diminished atrial filling, thus diminished stretch). Information from the sensors is relayed neuronally

(A)

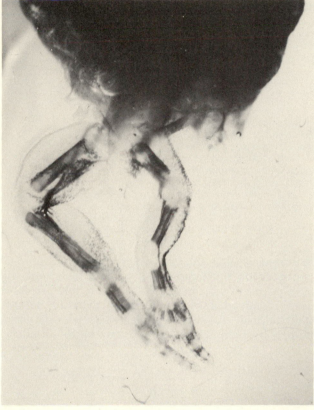

(B)

Figure 21.22 Developing frogs (*Rana pipiens*) implanted with pellets containing thyroxine, showing localized metamorphic changes. (A) A tadpole that had received a pellet in the dorsal tail fin 10 days earlier. The pellet has induced early tail resorption in its vicinity. (B) Hindlimbs of a tadpole that had received a pellet in its left leg 22 days earlier. Note that the left limb has developed faster than the right. [Photographs courtesy of Jane Kaltenbach Townsend.]

to the hypothalamus and is used there to modulate neurosecretion. Low blood volume acts by this pathway to increase ADH secretion.

Oxytocin in mammals promotes contraction of smooth muscles of the uterus and mammary glands. The neurosecretory release of oxytocin is also under neural control. One distinctive mode of neural control of neurosecretory cells is the neuroendocrine reflex. A **neuroendocrine reflex** is a *reflex* (stimulus-elicited response mediated by the nervous system,

see p. 549) in which the effector response is secretion of a hormone.* The secretion of oxytocin from the neurohypophysis to effect milk ejection from the mammary glands is a classic example of a neuroendocrine reflex. (Milk production or *lactation* is a distinct phenomenon from *milk ejection* or *"let-down."* Lactation is controlled by adenohypophysial and steroid hormones and is not considered here.) Oxytocin acts on smooth muscle cells of the alveolar sacs and ducts in the mammary gland, contracting the sacs and forcing stored milk through the ducts to the nipple. When an infant sucks on the nipple, the resultant sensory stimulation of receptors in the nipple initiates nerve impulses, which are relayed via the spinal cord to the mother's hypothalamus. This sensory activity excites hypothalamic neurosecretory neurons, the secretory endings of which are in the pars nervosa. Thus, the neuroendocrine reflex response to nipple stimulation is release of oxytocin from the posterior pituitary. Blood levels of oxytocin begin to increase in a few seconds after the start of suckling, and milk ejection normally occurs within a minute.

The Adenohypophysis; Control of Peripheral Glands by the Hypothalamo–Hypophysial System The *adenohypophysis* originates from mouth ectoderm (reptiles, birds, and mammals) and in the adult forms the anterior portion of the pituitary gland (Figure 21.25). In gross morphology, it is subdivided into parts called the pars distalis, pars intermedia, and pars tuberalis. The exact positions and relative sizes of these parts vary considerably from one animal group to another, and in some groups not all parts are present. The adenohypophysis manufactures its hormones intrinsically. All the hormones are proteins, glycoproteins, or polypeptides, and all or most are made by the pars distalis. When the pars intermedia is present, it makes melanocyte-stimulating hormone (MSH), but when it is absent (as in birds and many adult mammals), MSH may be made by the pars distalis.

In terms of function, adenohypophysial hormones may be roughly categorized into two groups. The hormones of one group, although they sometimes affect other glands, exert their *principal* effects on *nonendocrine* tissues. These hormones are listed in Table 21.3. They include for example, *melanocyte-stimulating hormone,* which in lower vertebrates acts mainly on cutaneous pigment cells. The second group is composed of a set of hormones that principally control *other endocrine glands*. Included—as summarized in Table 21.4—are *thyrotropic hormone* (*TSH*), *adrenocorticotropic hormone* (*ACTH*), and the two *gonadotropic hormones* (*FSH* and *LH*). The suffix

* In principle, a system in which the initial stimulus is internal could be called a neuroendocrine reflex. However, in ordinary usage the concept of neuroendocrine reflex is applied only to systems in which the initial stimulus comes from the external environment.

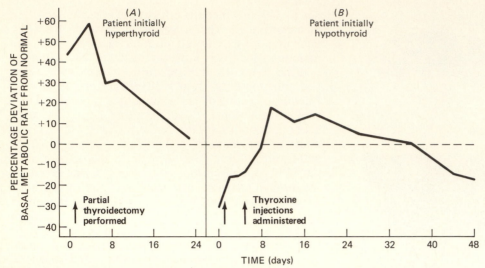

Figure 21.23 Metabolic effects of thyroid hormones as illustrated by case records of two human patients suffering thyroid disorders. (*A*) This patient initially suffered from an excess of thyroid secretion (hyperthyroidism) attributable to a thyroid tumor. The patient's BMR was initially over 40 percent above normal, but partial thyroid removal on day 1 led ultimately to a fall of BMR toward normal (0 = normal). (*B*) This patient suffered at first from too little thyroid secretion (hypothyroidism). Two injections of thyroxine on days 1 and 5 provoked a rise in metabolism that lasted over a month. Data are from one of the first studies to quantify the effects of thyroxine in humans. [After W.M. Boothby and I. Sandiford, *Physiol. Rev.* **4**:69–161 (1924).]

tropic in the name of each of these latter hormones is derived from the Greek "to turn." Etymologically, thyrotropic hormone, for example, is "thyroid-turning" hormone, meaning that it alters or guides the functioning of the thyroid. Sometimes these hormones are called *tropins.* For example, thyrotropic hormone is also known as *thyrotropin,* and adrenal corticotropic hormone is also called *corticotropin.*

A moment's reflection will reveal that we have now identified several endocrine glands—the thyroid, adrenal cortex, and gonads—that are controlled in part by secretions of another gland, the pituitary.

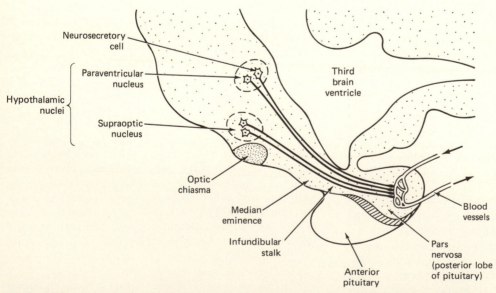

Figure 21.24 A schematic representation of a longitudinal section through the hypothalamus and pituitary gland of a mammal, emphasizing the neurohypophysis. Stippling demarks the brain and tissue derived from it. Arrows at blood vessels depict blood flow. [After A. Gorbman, W.W. Dickhoff, S.R. Vigna, N.B. Clark, and C.L. Ralph, *Comparative Endocrinology.* Wiley, New York, 1983.]

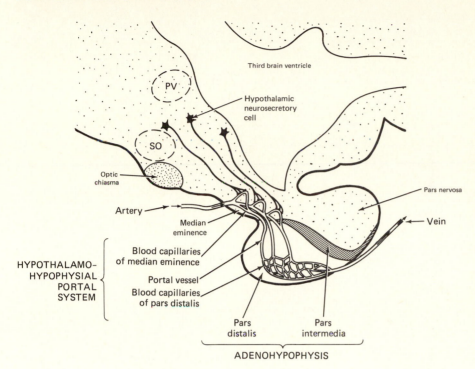

Figure 21.25 Schematic longitudinal section through the hypothalamus and pituitary gland of a mammal, emphasizing the median eminence and adenohypophysis. Stippling demarks the brain and tissue derived from it. Arrows in blood vessels show blood flow. The venous drainage of the pars distalis (shown) connects with that of the pars nervosa (not shown). The pars tuberalis, which is not shown, is generally small. Compare with Figure 21.24. SO, supraoptic nucleus; PV, paraventricular nucleus. [After A. Gorbman, W.W. Dickhoff, S.R. Vigna, N.B. Clark, and C.L. Ralph, *Comparative Endocrinology.* Wiley, New York, 1983.]

In fact, the control of these glands involves even further hormonal "levels of command."

One of the great achievements of modern endocrinology has been the recognition that *secretion of the adenohypophysial hormones is substantially under control of a part of the brain, the endocrine hypothalamus.* An important constituent of this control pathway is the vascular connection that exists between the adenohypophysis and the median eminence. As diagrammed in Figure 21.25, the capillaries (and other minute blood channels) of the median eminence do not empty directly to the venous drainage of the brain but instead coalesce into *portal vessels* that travel the short distance to the adenohypophysis and there break up to form the microcirculatory beds of the adenohypophysis. This whole system is termed the **hypothalamo–hypophysial portal system.** Now, as also shown in Figure 21.25, the median eminence is a neurohemal organ: the terminus of the axons of substantial populations of hypothalamic neurosecretory cells. These cells produce hormones (or "factors"), known as **hypophysiotropic hormones,** that control adenohypophysial secretion. Some of the hypothalamic neurohormones stimulate release of particular hormones by the adenohypophysis. These are called **releasing hormones (RHs).** Other hypothalamic neurohormones inhibit release of adenohypophysial hormones and are termed **release-inhibiting hormones (RIHs).** Each neurohormone is specific in its actions. For example, in mammals there is a particular neurohormone that causes the adenohypophysis to secrete thyrotropic hormone and another that causes it to reduce secretion of growth hormone. These are known, respectively, as

*thyrotropic-hormone releasing hormone (**TRH**)* and *growth-hormone release-inhibiting hormone (**GHRIH** or somatostatin).* The hypothalamic neurosecretory cells secrete their RHs and RIHs into the microcirculation of the median eminence, and the hormones are then carried via the portal system to the adenohypophysis where they exert their effects on their target cells.

Knowledge of the available array of releasing and release-inhibiting hormones remains far from complete. It now seems clear that at least one adenohypophysial hormone is controlled by both a RH and a RIH acting antagonistically; a RIH for mammalian growth hormone has been known for years, and recently a RH for it has been reported. More commonly, it appears that a given adenohypophysial hormone is under the control of either a RH or a RIH. The structures of several of the neurohormones have been worked out, and all are polypeptides. Mammalian TRH, for example, is a tripeptide, and GHRIH is a 14-amino-acid peptide.

We see now that the control of certain hormones of the peripheral glands—notably the thyroid and gonadal hormones and the glucocorticoid hormones of the adrenal cortex—commonly involves *two* superimposed levels of hormonal control arrayed hierarchically: the neurohormones of the endocrine hypothalamus plus the tropic hormones of the pituitary (Figure 21.26). When one gland acts on another, endocrinologists speak of the whole system as an *axis.* Thus, we would say that secretion of the thyroid hormones is mediated by the hypothalamus–pituitary–thyroid axis, and that of glucocorticoids by the hypothalamus–pituitary–adrenocortex axis. Secre-

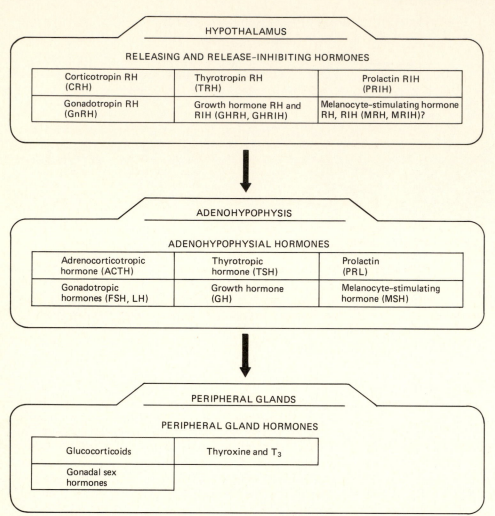

Figure 21.26 Sequential hormonal controls in the hypothalamo–adenohypophysial system. The upper array of boxes lists major known and postulated releasing hormones (RHs) and release-inhibiting hormones (RIHs). The names and abbreviations given are among those in common use but are not in all cases highly standardized. Only one RH is known for both gonadotropins; this RH is called gonadotropin RH (GnRH) or, sometimes, luteinizing hormone RH (LHRH). Each box in the middle array of boxes lists adenohypophysial hormones controlled by the RHs and RIHs listed in the *similarly positioned* box in the upper array. Likewise, each hormone in a box in the lower array is controlled by the tropic hormone(s) listed in the similarly positioned box of the middle array. Not all relationships among hormones are shown; for example, evidence exists that GHRIH can inhibit secretion of thyrotropic hormone as well as growth hormone.

tion of the adenohypophysial hormones prolactin and melanocyte-stimulating hormone (Table 21.3) involves a two-gland axis (composed of hypothalamus and pituitary).

An example of multiglandular, hierarchical control of hormone secretion is provided by the adrenocortical response to stress. Many sorts of conditions that represent a challenge to the organism cause a rapid increase in secretion of cortisol and other adrenal glucocorticoid steroids in mammals and other vertebrates (see Table 21.4 for some of the characteristics of glucocorticoids). Increased glucocorticoid secretion occurs, for example, when mammals are wounded, exposed to thermal extremes, or forced to

exercise vigorously. It also can be elicited by high emotion, such as the anxiety students may feel before an important examination. *Stress* has traditionally been used as a generalized term for the circumstances that evoke glucocorticoid secretion, and we shall not break with that custom here. It should be noted, however, that circumstances producing the stress response (as we use the term) are not always *subjectively* stressful. One of the key ways that stress induces glucocorticoid secretion is that brain neural activity associated with the stress brings about a rapid increase in hypothalamic output of *corticotropin releasing hormone, CRH*. Secreted at the median eminence or other nearby structures, the CRH is carried

Table 21.3 ADENOHYPOPHYSIAL HORMONES THAT PRINCIPALLY CONTROL NONENDOCRINE TISSUES AND SOME OF THEIR MAJOR ACTIONS

Adenohypophysial hormone	Principal actions
Growth hormone, also called somatotropic hormone (GH or STH)[a]	Stimulates growth of skeleton and somatic soft tissues; promotes formation of proteins, elevation of blood glucose, fat breakdown, and other metabolic effects (see also Table 21.6); stimulates release of insulin
Melanocyte-stimulating hormone (MSH)	Particularly in fish, amphibians, and reptiles, MSH stimulates dispersion of pigment in cutaneous melanocytes (dark-pigment-containing cells), thus causing skin darkening; stimulates production of more melanocytes and pigment acquisition by melanocytes. MSH possibly has effects on steroid-producing glands, learning, and development
Prolactin (PRL)	PRL is involved in an extraordinary range of functions, varying in major ways from one animal group to another. Examples include the promotion of growth of the mammary glands and milk synthesis in mammals; stimulation of the production of crop milk by pigeons; stimulation of nestbuilding and incubating behavior by some mammals; promotion of survival of certain fish in fresh water by exerting favorable effects on ion and water exchange in gills, kidneys, and other organs; stimulation of salt-gland secretion in birds; promotion of the "water-drive" effect in some newts (causes them to seek water in preparation for breeding)

[a] In recent years, it has become increasingly clear that many effects of GH are mediated via diffusely produced hormones called somatomedins made, for example, in liver. Thus, although GH has traditionally been thought to act mainly on nonendocrine cells, it may ultimately be classed with the hormones in Table 21.4.

promptly to the adenohypophysis by the portal vascular system, and there it precipitates a rise in the secretion of **adrenocorticotropic hormone, ACTH**. The ACTH then is carried to the adrenal cortex, where it stimulates glucocorticoid secretion (Figure 21.27). There is considerable evidence that elevated levels of glucocorticoids aid animals in coping with stresses in the short term, by mechanisms largely unknown; prolonged stress, in contrast, can lead to decreased target responsiveness to glucocorticoids, with deleterious consequences.

One enormously important attribute of the hypothalamo–hypophysial control apparatus is clear: *It provides an interface between much of the vertebrate body's endocrine system and the brain*. By virtue of the diversity of adenohypophysial hormones produced, the anterior pituitary holds power over many other tissues, including several glands. Now we recognize, however, that the pituitary itself is substantially under control of the brain—as exemplified by the stress-related responses we have discussed. The hypothalamo–hypophysial system provides a mechanism by which the brain exerts control over many endocrines and by which the sophisticated integra-

Table 21.4 HORMONES OF THE ADENOHYPOPHYSIS THAT ACT PRINCIPALLY TO CONTROL ACTIONS OF OTHER ENDOCRINE GLANDS

Adenohypophysial hormone	Gland affected	Principal actions
Thyrotropic hormone (TSH) (also called thyroid-stimulating hormone, thyrotropin)	Thyroid	TSH promotes hypertrophy and hyperplasia of thyroid cells, stimulates iodide uptake by the thyroid, and promotes synthesis and release of thyroid hormones
Gonadotropic hormones (also called gonadotropins) (1) Follicle-stimulating hormone (FSH) (2) Luteinizing hormone (LH)	Gonads (ovaries or testes)	Although named for actions in the female, FSH and LH control gonad function in the male as well. They prompt gamete (egg or sperm) development (FSH being involved in this more than LH). They also control gonadal endocrine function: the secretion of sex steroids such as testosterone and estradiol (LH being more involved in this role than FSH). Their exact functions are complex and vary with species
Adrenal corticotropic hormone (ACTH) (also called corticotropin)	Adrenal cortex	ACTH principally stimulates the synthesis and release of the set of adrenocortical steroids known as glucocorticoids (e.g., cortisol, cortisone). The glucocorticoids control metabolism of nutrient molecules (e.g., stimulate glucose synthesis), control certain aspects of development and calcium metabolism, and suppress the inflammatory response (including suppression of the release of histamine and certain kinins and prostaglandins)

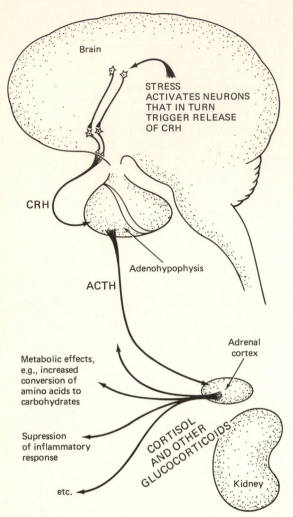

Figure 21.27 Summary of the activation of the hypothalamus–hypophysis–adrenocortex axis by stress. Corticotropin releasing hormone, CRH, has not yet been fully characterized chemically and may in fact consist of more than one chemical constituent. ACTH, adrenocorticotropic hormone (corticotropin).

tive capabilities of the central nervous system can be brought to bear on endocrine integration.

Control of peripheral endocrine glands by hormonal secretions of the central nervous system is by no means limited to vertebrates. We noted earlier (e.g., p. 592) that neurosecretory structures are prominent in many invertebrates. Cases of neurosecretory control of hormone secretion in invertebrate endocrine systems (e.g., the control of molting in insects) are treated later (p. 626).

Modulation of Control Pathways

The rate of secretion of a hormone is governed by the modulation of its control pathways. When the control pathways are lengthy, the possibilities for their modulation are extensive. To illustrate this important point, we shall examine modulation of con-

trol pathways involving the adenohypophysis. First, in this section, we take a conceptual point of view. Then we turn to some informative case studies in subsequent sections.

Consider the diagram in Figure 21.28, representing the major steps leading to secretion of a particular end hormone under control of the hypothalamo–hypophysial system. A change in the rate of firing of the brain neurons at the top may be considered to represent an *input signal* to this control pathway. This signal is translated to a new physical form three times as it is relayed through the pathway:

1. First, the changed rate of neuronal firing is translated to a changed rate of hypothalamic neuroendocrine secretion.
2. Second, the changed output of hypothalamic hormone provokes a changed rate of adenohypophysial tropic-hormone secretion.
3. Finally, the altered tropic-hormone output is translated to an altered output of the end hormone.

In this simplified model, we consider the functional relation between each successive pair of signal forms to be linear. This is what is symbolized by the graphs to the right in Figure 21.28; the uppermost graph indicates, for example, that the hypothalamus' rate of releasing-hormone secretion R is a positive, linear function of the rate N of firing in the input neurons. Now, if the control pathway were isolated from outside influences and each of the three functional relations between signal forms were immutable, the process of relaying a signal through the pathway would be stereotyped: A given change in neuronal firing rate—despite the several translations along the pathway—would always elicit a precisely stereotyped change in the rate of end-hormone secretion. In fact, however, control pathways have never evolved as isolated systems. Rather, each set of cells that transforms and translates the control signal in Figure 21.28 can *modulate* the translation, depending on one or more conditions of its physiological environment. Consider, for example, the production of an adenohypophysial tropic hormone. As shown in the figure, the functional relation between the pituitary's output of tropic hormone (T) and its input of releasing hormone (R) may depend strongly on blood levels of some third hormone, X: When the concentration of X is high, a particular rate of neuronal firing in the brain elicits a greater output of tropic hormone *and end hormone* than when it is low. Similarly, although not shown, the level of X or some other hormone (or some nonhormonal agency) could affect the hypothalamic translation of N to R or the translation of T to E. Possibilities for adaptive modulation of end-hormone output are presented at all levels of the control pathway and have permitted the evolution of intricately sophisticated controls overall.

Although Figure 21.28 may seem complicated at

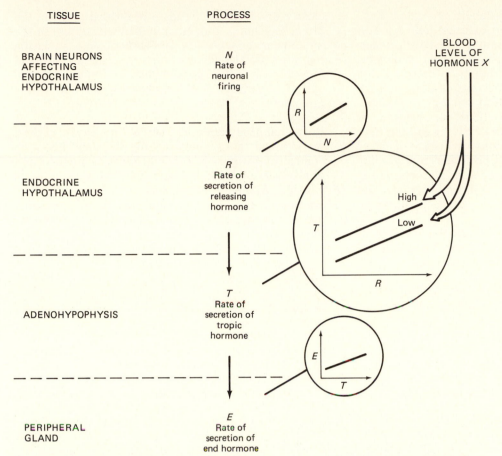

Figure 21.28 The sequence of signal forms in a vertebrate control pathway involving the hypothalamo–adenohypophysial system. The plot in the large circle illustrates that the translation of signal-form *R* to signal-form *T* by the adenohypophysis is subject to modulation, as by the level of a hormone *X*. Similarly, although not shown, the translations of *N* to *R* and of *T* to *E* are susceptible to modulation. In this way, control of the secretion of end hormone can be exerted at all levels of the control pathway. For simplicity, linear relations between successive variables are arbitrarily assumed.

first glance, it actually represents a simplified view of the modulation of the control pathways of the hypothalamus and pituitary. In fact, the secretion of many hypothalamic and pituitary hormones is now understood to be *in pulses,* so that there are intermittent brief periods of high hormone concentration in the blood. Modulation of the control pathways is expressed in part by a change in the *frequency* of the pulsatile release of the hypothalamic hormones and therefore of the pituitary hormones. Thus, the implication in Figure 21.28 that hormone concentrations change in a smooth, continuous manner is a simplification of what really happens.

Modulation by the Central Nervous System, Including Rhythms We have already discussed some cases in which a major cause of a change in hormone output is a change in neural activity in the central nervous system, or CNS (i.e., a change in *N* in Figure 21.28). Examples include the effects of emotional stress on glucocorticoid secretion (p. 604) and those of suckling

on oxytocin secretion and milk ejection (p. 601). In this section, we discuss some additional examples; in the next section, we examine modulation at other levels of control pathways.

Hormone outputs can be under the influence of virtually any neuronal sensory modality. Sometimes *internal* parameters are important. We have seen, for instance, that the CNS uses information on blood pressure, atrial stretch, and the osmotic concentration of the body fluids to control the secretion of antidiuretic hormone (p. 600).

Sensation of *external* parameters is also important. The amount of light per day, for example, is often a prime determinant of reproductive condition in birds and mammals. Among many species that live in the temperate zones and mate in the spring, long days elicit growth and activation of the gonads. The system responsible for this reaction receives information on external light from the eyes in mammals, but in birds the relevant light receptors are in the hypothalamus (where they receive light penetrating through the thin skull). Day length is "calculated" from the raw information on external light by neuronal mechanisms, just beginning to be well understood, that make intriguing use of circadian biological clocks. Then, if a long day length is detected, hypothalamic neurosecretory cells are stimulated to increase the frequency of pulsed releases of gonadotropin releasing hormone (GnRH), leading to enhanced secretion of LH and FSH by the pituitary and then to gonadal development, encompassing both secretion of gonadal sex hormones and production of eggs or sperm (Figure 21.29). Among mammals, the pineal gland and its hormone melatonin help mediate the control of releasing-hormone secretion.

An important aspect of control by the CNS is that sometimes neuronal biological clocks are employed to drive rhythmic hormone secretion (see p. 17). Perhaps the best-known clock-driven hormonal rhythm is that in the secretion of glucocorticoids by the adrenal cortex. Among humans, blood levels of glucocorticoids regularly rise and fall on a daily basis, being highest in early morning hours (3:00–8:00 a.m.) and lowest during the evening (6:00 p.m.–midnight). Rhythms of glucocorticoid secretion are also known in a number of other mammals as well as some other vertebrates. The mammalian rhythms are clearly driven by a circadian biological clock in the brain. Evidently, this clock induces a rhythm in the hypothalamic secretion of corticotropin releasing hormone (CRH), which in turn affects glucocorticoid secretion via the hypothalamus–pituitary–adrenal axis. The sensitivity of the pituitary to CRH varies in tandem.

Hormonal Modulation of Control Pathways Earlier, we stressed that signals passing through endocrine control pathways—whether originating in the CNS or elsewhere—are likely to be modulated along their passage. Much of this modulation is mediated by hormones. The most widespread type of hormonal modulation of control pathways is *negative feedback;* that is, a hormone controlled via a particular pathway often modulates that pathway in ways that tend to suppress its own production. As diagrammed in Figure 21.30, for example, glucocorticoids are believed to exert negative feedback at two levels of the hypothalamus–pituitary–adrenocortex axis. They tend to suppress secretion of CRH by the hypothalamus, and they reduce the responsiveness of the pituitary to CRH. Both effects tend to decrease the output of ACTH by the pituitary and thus reduce the stimulus for glucocorticoid secretion.

Negative feedback is very common. Our example

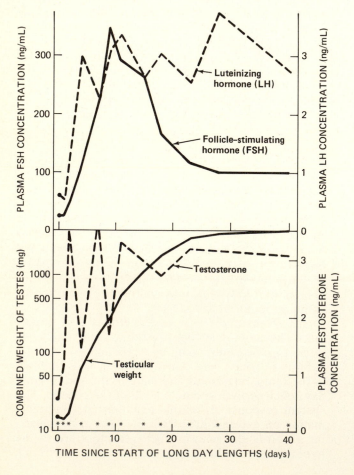

Figure 21.29 Effect of long day lengths on the reproductive condition of young male quail. The animals, which were growing, were transferred at time zero from a daily light-dark cycle of 8 h light/16 h dark to one of 20 h light/4 h dark. The change in day length provoked precipitous increases in blood levels of pituitary gonadotropic hormones (upper frame) and in testicular size and testosterone secretion (lower frame). Note that testicular size is expressed on a logarithmic scale. The testes were over 35 times heavier on day 11 than day zero. Only a small fraction of this marked increase could be attributed to mere growth of the birds. Asterisks along abscissa mark times data were collected. [Data from B.K. Follett, *J. Endocrinology* **69**:117–126 (1976).]

609

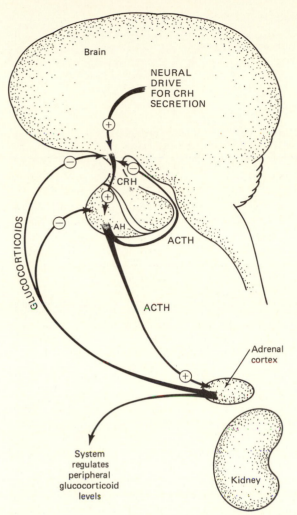

Figure 21.30 A summary of the regulation of glucocorticoid secretion, including negative feedback. Each arrow is labeled + or − to signify, respectively, whether the effect depicted stimulates or inhibits hormone secretion. Compare Figure 21.27. AH, adenohypophysis; CRH, corticotropin releasing hormone; ACTH, adrenocorticotropic hormone.

of thyroid function at the beginning of the chapter (Figure 21.1) provides a similar case to that of glucocorticoids; thyroid hormones tend to inhibit adenohypophysial secretion of thyrotropic hormone (TSH).

Mechanisms of negative feedback do *not* tend to reduce hormone outputs to zero, as is sometimes supposed. They are only one element in control pathways, and their function is to *stabilize* blood concentrations of end hormones at levels determined by the interaction of all control elements. Witness the fact, for example, that with all the negative-feedback relations in the hypothalamus–pituitary–adrenocortex axis, steady-state glucocorticoid levels in the blood still *rise* in humans between evening and morning, as central-nervous drive for CRH secretion is heightened.

The negative-feedback properties of control path-

ways account at times for pathological conditions. If dietary intake of iodine is too low, adequate amounts of thyroid hormones cannot be made, and the ordinary thyroid-hormone inhibition of thyrotropic-hormone (TSH) secretion can become so diminished that the pituitary's output of TSH is excessive. Continued over a long time, these high levels of TSH will stimulate an abnormal growth of the thyroid gland, termed goiter. Drugs with glucocorticoid activity are currently used extensively in clinical medicine. When administered to a patient, they can suppress ACTH secretion to the point that function of the adrenal cortex is not sustained. Over a long time, if due caution is not exercised, cortical function may become permanently depressed, unable to return to normal even when the exogenous glucocorticoids are discontinued.

Mammalian Ovulation: An Example of Interaction of Nervous and Endocrine Control   So far, we have discussed the modulation of control pathways by the nervous system and by hormones in isolation. Many cases of modulation of endocrine control, however, involve interaction of both neural and endocrine sources of modulating influence. Here, we discuss the control of mammalian ovulation as an example of a hormonally mediated event orchestrated by such mixed modulation of the control path.

Ovulation refers to the release of mature eggs from the ovaries into the fallopian tubes, where fertilization can occur if sperm are present. Ovulation is controlled by the hypothalamo–hypophysial endocrine axis, much like the control of glucocorticoid secretion (p. 604). Characteristically, a key stimulus for ovulation is a marked rise in blood levels of luteinizing hormone (LH), secreted by the adenohypophysis. In most mammals, ovulation-provoking surges of LH are generated largely or entirely endogenously by the female, resulting in *spontaneous ovulation* (see next paragraph). In domestic cats, rabbits, and quite a few others, however, these surges are induced by mating, leading to *induced ovulation*. That is, once eggs have matured in the ovary, their release awaits copulation (or other intimate association with a male)—an elegant mechanism for assuring a high likelihood that sperm will be available for fertilization of the ovulated eggs. The evidence regarding the exact control of induced ovulation is not definitive but points strongly to the following sequence. First, the copulatory act provides sensory inputs that are relayed by nerve pathways to the brain. Second, neuronal processes in the brain induce hypothalamic neurosecretory cells to secrete gonadotropin releasing hormone (GnRH, also termed LH releasing hormone, LHRH) in the median eminence. Finally, the GnRH stimulates release of LH by the pituitary, yielding the LH surge in the blood (Figure 21.31). Induced ovulation is another example of a neuroendocrine reflex, analogous to the mammalian milk ejection response (p. 601).

Spontaneously ovulating mammals employ the same hypothalamo–hypophysial control pathway but use different modulators than induced ovulators. Spontaneously ovulating rodents such as rats or hamsters in reproductive season undergo ovulation every 4–5 days. As in induced ovulators, the immediate stimulus for ovulation is a surge of luteinizing hormone (LH), which precedes ovulation by several hours. In the spontaneous ovulators, however, this surge is initiated endogenously, rather than in response to copulation or some other extrinsic stimulus. A notable feature of rodent spontaneous ovulation is that it occurs reliably at a certain time of day: early morning. There is strong evidence that the *time of day of the LH surge* (and thus the timing of ovulation) is determined by a circadian clock in the CNS. This clock probably initiates an appropriately timed neural signal every day. The signal precipitates a modest rise in hypothalamic secretion of GnRH on at least some nonovulatory days and a marked rise in GnRH secretion on the day prior to ovulation. (This rise is appropriately timed to stimulate ovulation about 8 h later, in the early morning hours of the next day.) The responsiveness of the pituitary to GnRH is particularly great on the day of the marked surge in GnRH secretion. The *differences between days* in both the extent of the GnRH increase and the pituitary responsiveness to GnRH result from *hormonal* modulation of the control pathway. A key feature of the control pathway for LH secretion is that the responsiveness of the LH-producing pituitary cells to GnRH has evolved to be highly sensitive to blood levels of estradiol (an estrogenic hormone). Estradiol is secreted in the ovaries by ovarian follicles, the structures that nurture eggs to maturity (p. 619). After a rodent has ovulated, only immature follicles are present in its ovaries. Over the next several days, a set of follicles matures, bringing about an increase in estradiol secretion as it also prepares a new set of eggs for ovulation. High estradiol levels render the LH-producing cells of the rat pituitary about 50 times more responsive to GnRH than they are when estradiol levels are low. Accordingly, when estradiol has become abundant, the pituitary responds particularly vigorously with LH secretion after the circadian clock discussed earlier emits its daily signal inducing GnRH secretion by the hypothalamus. The *positive* effect of estradiol on LH secretion thus contributes to the ovulation-inducing LH surge, and *it does so in synchrony with the availability of mature eggs*. Certain investigators have recently postulated that estradiol also exerts favorable modulating effects on either the neurons of the biological clock or on the hypothalamic GnRH-producing cells. Heightened stimulation of the GnRH-producing cells by the clock in the presence of high estradiol levels—or heightened responsiveness of the GnRH-producing cells to the clock cells—could explain why especially large amounts of GnRH are secreted on the day before ovulation.

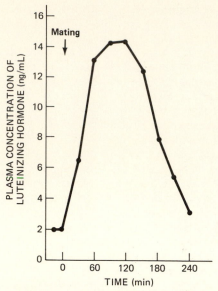

Figure 21.31   The LH surge in the peripheral blood plasma of rabbits following copulation. Copulation occurred at time zero. Data are averages for nine animals. [After R.C. Tsou, R.A. Dailey, C.S. McLanahan, A.D. Parent, G.T. Tindall, and J.D. Neill, *Endocrinology* **101**:534–539 (1977).]

Thus, we see that ovulation is orchestrated by a complex interaction of neural signals (circadian clock signals in spontaneous ovulators, copulation-induced sensory signals in induced ovulators) with levels of modulating hormones such as estradiol. Such complexity of endocrine and neural interaction is the rule rather than the exception in the control of hormone secretion.

Control of the Nervous System by Hormones

The controls exercised by the nervous system over the endocrine system deserve the emphasis they have received, but we need to emphasize equally that the overall function of physiological control is performed by the two systems interacting mutually, not by one unilaterally controlling the other. Many instances of endocrine control of the nervous system are known.

One elegant example is provided by the motions that insects undergo when emerging as adults from their pupal encasement. (Hormonal control of insect molting is discussed in greater detail on p. 626.) In saturniid moths, experiments reveal that these highly stereotyped emergence movements are intrinsically programmed in certain of the abdominal ganglia, which neuronally stimulate appropriate muscles in the appropriate order. The ganglia, however, do not execute their program normally unless stimulated to do so hormonally. The activating hormone (termed eclosion hormone) is a neurosecretion of the brain, released by the corpora cardiaca at a certain time on the day of emergence from the pupa, under control of a biological clock. Without the hormone, emergence from the pupa is aberrant.

Among vertebrates, it is well established that hormones help control the expression of a variety of sexual, nurturing, and aggressive behaviors by the brain. Levels of sex hormones experienced during fetal life, for example, can affect behavioral development, as is illustrated in a particularly interesting way by the intrauterine position phenomenon in mice and rats. Male fetuses produce much more testosterone than female ones, and in some way the high levels of testosterone in a particular male act to raise testosterone levels in the male's immediate neighbors. Female mouse fetuses that undergo their intrauterine development between two male fetuses have 30 percent higher testosterone levels than ones whose intrauterine position is between two other females. Once the fetuses have matured to adulthood, the females that developed between males prove to be much less prone to assume the female mating posture (lordosis) than ones that developed between females, and various types of evidence indicate that this difference in mating behavior is attributable to the different endocrine environments of the animals during development.

Many types of adult mammal (unlike humans) exhibit strong cycles of mating drive, which are mediated by endocrine effects. For example, cyclic increases in ovarian steroids (e.g., estradiol, progesterone) or even in adrenal steroids are responsible in various species for eliciting female receptivity to the male (estrus, heat). Nestbuilding and other nurturing behaviors are also under endocrine control.

The Renin–Angiotensin System; Control of a Diffusely Generated Hormone

So far, we have focused on controls of hormone secretion mediated via the pituitary and hypothalamus. This focus reflects the dominant role of the hypothalamus and pituitary in control of many endocrine glands. It is important to note, however, that for certain hormones secretion is controlled in very different ways. Here we discuss the renin–angiotensin system as an example of a hormonal system not under control of the pituitary or hypothalamus.

Angiotensin II is a hormone that in mammals and other vertebrates exerts important influences on blood pressure, sodium balance, water balance, and other physiological systems. It has regulatory effects on a number of nonendocrine tissues. For example, it is the most potent vasoconstrictive substance known; by inducing arteriolar vasoconstriction, it tends to raise blood pressure, and it is implicated in the etiology of some forms of hypertension. Angiotensin is also a major controller of the secretion of **aldosterone** and other mineralocorticoids by the adrenal cortex. The mineralocorticoids are the adrenocortical steroids that exert their primary effects on salt and water balance. Aldosterone, secreted in response to elevated angiotensin II, enhances resorption of sodium from the urinary fluid by the kidney tubules, thus aiding sodium retention in the body (p. 194).

A peculiarity of angiotensin II is that *it is generated diffusely in the blood*. The chief actor in modulating this generation is a substance named **renin** (pronounced rē-nin) secreted by the **juxtaglomerular cells** in the walls of the afferent glomerular arterioles of the kidneys. Renin is sometimes called a hormone, sometimes an enzyme. When secreted, it catalyzes breakdown of a blood glycoprotein—**angiotensinogen**—to release angiotensin I, a decapeptide. Angiotensin I is then quickly cleaved to angiotensin II, an octapeptide, by an enzyme called **converting enzyme** located in the vasculature of the lungs and other parts of the body. Angiotensinogen (made by the liver) and converting enzyme are ordinarily present in sufficient amounts to permit ample synthesis of angiotensin II. Thus, the rate-limiting factor in this synthesis is renin secretion. Two of the stimuli that promote renin secretion are low blood pressure and signs of low body sodium (conditions that angiotensin II tends to correct). Figure 21.32 illustrates the major elements of this important diffuse hormone system. (Although this system is usually considered the primary angiotensin system, we should note that angiotensin II is also probably made in certain solid tissues, such as the kidneys, where it may play local regulatory roles.)

ARENAS OF ENDOCRINE CONTROL

Among both vertebrates and invertebrates, there are certain physiological processes that are often under predominantly endocrine control. These arenas of endocrine control include:

A. Nutrient and energy metabolism
B. Reproduction
C. Growth and maturation
D. Salt and water balance
E. Calcium (including skeletal) metabolism
F. Molting
G. Color change

The processes in this list are usually, but by no means always, controlled by hormones. Integumentary color change, for example, is hormonally controlled in a wide array of invertebrates (e.g., crustaceans, insects) and vertebrates (e.g., amphibians, Figure 19.43*B*); but in cephalopod mollusks (Figure 19.43*A*), color change is mediated by muscle cells contracting and relaxing under nervous control.

As we turn now to a synoptic examination of endocrine functions in vertebrates, we shall largely limit our focus to nutrient and energy metabolism (arena A) and reproduction (arena B). These arenas will also be our principal concern when we later look at invertebrate endocrine functions. Some examples

Table 21.5 EXAMPLES OF HORMONAL CONTROL IN FIVE PHYSIOLOGICAL ARENAS[a]

Process	Major hormones involved	Some major effects of hormones
Growth and maturation		
Mammalian growth	Growth hormone	Promotes growth of skeleton and soft tissues, promotes protein synthesis (At least some effects are mediated by *somatomedins* liberated by liver and muscle in response to GH.)
	Thyroid hormones	Potentiate effects of growth hormone; necessary for proper growth and maturation of some tissues, e.g., CNS, skeleton
	Gonadal or adrenal androgens ("male hormones")	Promote growth of muscle, cartilage, kidneys
Amphibian metamorphosis	Thyroid hormones	Precipitate metamorphic changes in many tissues (p. 596)
	Prolactin	Impedes metamorphosis prior to time of onset
	Glucocorticoids	Induce changes in nutrient metabolism during metamorphosis; reinforce morphological effects of thyroid hormones
Insect metamorphosis	20-Hydroxyecdysone	Induces molting (p. 627)
	Juvenile hormone	Decreased secretion during final larval stage permits tissues to metamorphose at next molts (p. 627)
	Various brain neurohormones:	
	(A) Anterior retraction hormone	Promotes assumption of pupal form by retraction of anterior body segments (flies)
	(B) Tanning hormones	Promote tanning and hardening of pupal and adult exoskeletons
	(C) Eclosion hormone	Activates neuronally programmed motions effecting exit of adult from pupal case (p. 610)
Salt and water balance		
Mammalian salt–water balance	Antidiuretic hormone (vasopressin)	Promotes renal water resorption by effects on renal tubular permeability (p. 232)
	Aldosterone (and other adrenal mineralocorticoids)	Promotes active resorption of Na^+ from urinary fluid in kidneys; lowers Na^+ concentration in sweat, saliva, milk; promotes renal K^+ excretion (p. 194)
	Natriuretic hormone(s)	Promote loss of Na^+ in urine (p. 195)
	Angiotensin II	Promotes thirst and drinking; increases appetite for Na^+
Insect water balance	Diuretic hormones (neurohormones produced by brain or thoracic ganglia)	Stimulate fluid secretion by Malpighian tubules; inhibit fluid resorption by rectum
	Antidiuretic hormones (also neurohormones)	Promote water retention
Calcium metabolism		
Mammalian calcium metabolism	Parathyroid hormone	Increases blood Ca^{2+}; stimulates resorption of Ca^{2+} from bone (an intricate process, not well understood); enhances resorption of Ca^{2+} from urinary fluid
	Calcitonin—secreted by C cells of thyroid	Decreases blood Ca^{2+}; inhibits resorption of bone Ca^{2+}; influences renal Ca^{2+} excretion
	Vitamin D metabolites (vitamin D is formed in skin and altered in liver and kidneys, resulting metabolites now considered hormones)	Increase blood Ca^{2+}; promote resorption of Ca^{2+} from bone; essential for adequate active absorption of Ca^{2+} from gut contents (stimulate absorption)
Decapod crustacean calcium metabolism	Neurohormones produced by brain and peripheral ganglia and released at neurohemal organs in eyestalks (e.g., sinus glands)	Promote resorption of Ca^{2+} from old exoskeleton at start of molting process; promote deposition of Ca^{2+} in new exoskeleton (p. 157)

Table 21.5   EXAMPLES OF HORMONAL CONTROL IN FIVE PHYSIOLOGICAL ARENAS^a (continued)

Process	Major hormones involved	Some major effects of hormones
Molting		
Shedding of hair/feathers in mammals/birds	Thyroxine (prolactin and gonadal steroids are also involved)	Often stimulates molting
Molting in decapod crustaceans	Molting hormones (e.g., 20-hydroxyecdysone)—secreted by epithelial glands called Y organs	Activate molting process (p. 628); promote growth
	Various neurohormones produced by brain and peripheral ganglia and released at neurohemal organs in eyestalks:	
	(A) Molt-inhibiting hormone	Prevents secretion of molting hormones by Y organs during periods between molts (p. 628)
	(B) Water-balance hormone	Controls the body enlargement brought about by water uptake between shedding of old exoskeleton and hardening of new (the enlargement assures that the new exoskeleton will be larger than the old)
	(C) Hormones controlling Ca^{2+}	See preceding section of this table
Integumentary color change		
Color change in amphibians	Melanocyte-stimulating hormone (from pars intermedia)	Promotes dispersion of pigment in melanocytes, thus darkening skin (p. 545)
	Melatonin (from pineal)	In tadpoles only, promotes aggregation of pigment and thus skin lightening; mediates daily rhythm of lightening and darkening
Color change in decapod crustaceans	Multiple neurohormones (chromatophorotropins)—released in eyestalks and elsewhere	Some promote pigment dispersion in chromatophores; some promote pigment aggregation (crustaceans typically possess several colors of controllable pigments) (p. 629)

^{a} Although the hormonal effects listed are believed to occur commonly in the animal groups specified, they are not necessarily operative in all species. Only end hormones are listed, not hormones that primarily control the secretion of other hormones (e.g., thyroxine is listed, but TRH and TSH are not). Lists of hormones and hormonal effects are not necessarily exhaustive.

of hormonal control in arenas C–G are outlined in Table 21.5. This table provides an overview of some of the vast diversity of vertebrate and invertebrate processes that are controlled endocrinologically. Some aspects of endocrine control of these processes have been discussed in other chapters (particularly salt and water balance in Chapters 8 and 10), and some will be discussed further in this chapter. One important point to note is that multiple hormonal control is very common. In crustaceans, for example, molting of the exoskeleton is a major, recurrent event in the life cycle, and no fewer than four hormones are sequentially utilized in its regulation.

VERTEBRATE ENDOCRINE SYSTEMS

Control of Nutrient Metabolism

Animals use the major classes of nutrients—carbohydrates, lipids, and proteins—in a great variety of processes in tissues scattered throughout their bodies. They acquire nutrients when they eat, but their cells need nutrients at all times. Moreover, the cells may require the three major classes of nutrients in very different proportions than are to be found in digested foods. Thus, the body's management of nutrients involves not only their dietary acquisition and metabolic use but also their transport, temporary storage, removal from storage, and interconversion. Interconversions and storage are not carried out in all tissues equally but are specialties of some, adding to the complexity of the nutrient-management task.

Our focus here will be on the hormones that control the transport, storage, mobilization, and transformation of nutrients in mammals. The number of these hormones is difficult to specify because it depends on exactly where the boundaries of "nutrient metabolism" are drawn. Certainly, no fewer than eight or nine end hormones play pivotal roles in one context or another. Table 21.6 lists these hormones

Table 21.6   HORMONES INVOLVED IN MAMMALIAN NUTRIENT MANAGEMENT

Hormone (source)	Some major actions of hormone
Insulin (endocrine pancreas)	*Overall*, insulin promotes the transfer of glucose, fatty acids, and amino acids out of the blood into solid tissues and their incorporation into glycogen, lipids, and proteins. It exerts the following effects: • Decreases plasma concentration of glucose ("hypoglycemic effect"); it is the only hormone with this effect • Increases rate of glucose uptake by certain tissues, notably heart, skeletal muscle, and adipose tissue • Promotes formation of glycogen from glucose, particularly in liver but also in muscle • Promotes formation of free fatty acids (FFAs) from glucose and their deposition as lipids in adipose tissue • Promotes catabolism of glucose in muscle • Inhibits formation of glucose (gluconeogenesis) from amino acids and proteins • Decreases plasma concentrations of FFAs • Increases uptake of FFAs by adipose tissue • Inhibits breakdown of lipids to form FFAs • Increases uptake of amino acids by muscle and liver and promotes protein synthesis
Glucagon (endocrine pancreas)	*Overall*, glucagon acts primarily on the liver to increase formation of glucose, which enters the blood. It exerts the following effects: • Increases plasma glucose concentration ("hyperglycemic effect") • Increases formation of glucose from glycogen in liver (but not in muscle) • Increases formation of glucose (gluconeogenesis) from amino acids and proteins (glucose formation occurs in liver) • May promote breakdown of lipids and increase plasma concentrations of FFAs
Epinephrine (adrenal medulla)	• Increases plasma glucose concentration • Promotes formation of glucose from glycogen in liver *and* muscle • Often promotes breakdown of lipids and increases plasma concentrations of FFAs • Inhibits release of insulin and antagonizes the positive effects of insulin on cellular glucose uptake; stimulates release of glucagon and ACTH (note that these effects on other hormones reinforce the other effects of epinephrine)
Cortisol, cortisone, and other glucocorticoids (adrenal cortex)	• Increase plasma glucose concentration • Antagonize positive effects of insulin on cellular glucose uptake • Promote formation of glucose (gluconeogenesis) from amino acids and proteins; enhance like actions of glucagon • Promote formation of liver glycogen from glucose • Reduce glucose oxidation • Promote breakdown of lipids to form FFAs; enhance like actions of epinephrine and glucagon
Thyroxine and T_3 (thyroid)	• Promote oxidation of nutrients (raise metabolic rate) • Increase plasma concentrations of glucose and FFAs • Promote actions of epinephrine, glucocorticoids, and growth hormone • Affect protein metabolism in complex ways, as by acting synergistically with growth hormone to promote growth and deposition of new protein in growing animals
Growth hormone (adenohypophysis)	• Promotes growth and formation of new protein • Increases uptake of amino acids by liver and muscle • Increases lipid oxidation and breakdown of lipids to form FFAs in long term; enhances like action of epinephrine • Decreases glucose utilization and increases blood glucose concentration in long term
Androgens (gonads and adrenal cortex)	• Promote protein formation and growth of muscle

and their major effects. As we discuss the substantial roles of the hormones, it will be important to keep in mind that the regulation of nutrient metabolism is by no means entirely endocrinological. Important influences are exerted also, for example, by the regulatory properties (e.g., allosteric ones) of enzymes of intermediary metabolism.

In humans and numerous other mammals, bouts of feeding are regularly separated by many hours of not feeding. This alternation between feeding and fasting is one important realm in which nutrient metabolism must be regulated. Without appropriate regulation, nutrient molecules absorbed from the gut might be excreted or otherwise wasted during the nutrient surfeit immediately following a meal, and yet cells might be starved for nutrients hours later when the gut has been empty for some time.

Such alternations of feast and famine at the cellular level are prevented by mechanisms that favor storage of nutrient molecules in the immediate aftermath of a meal and then promote mobilization of nutrients from stores as the hours pass until the next meal. There is evidence that these mechanisms are partly nonendocrine; the elevation of blood glucose following a meal, for example, evidently acts in and of itself to promote storage of glucose in the form of glycogen in the liver (by positively affecting the ac-

tivity of glycogen-synthesizing enzymes). Endocrine mechanisms are also of great significance; and of all the hormones, *insulin* is the one that stands out as being most important in the management of short-term fluctuations of food availability. During the digestion of a meal, a variety of factors—including rising concentrations of glucose and amino acids in the blood and the secretion of hormones by the gut—induce the B cells in the pancreatic islets of Langerhans to increase their secretion of insulin. A review of insulin's actions in Table 21.6 reveals that it favors the *storage* of all three major classes of nutrients. It promotes the uptake of glucose, fatty acids, and amino acids from the blood into tissues like muscle and fat, and it promotes the conversion of glucose to glycogen and lipids, that of fatty acids to lipids, and that of amino acids to proteins. It also inhibits the breakdown of glycogen, lipids, and proteins. Overall, the elevated levels of insulin after a meal act to remove digestive products from the blood and deposit them in storage depots. Insulin secretion subsides as digestion comes to an end, and most evidence suggests that the ebbing of insulin levels is often the only endocrine change necessary for a shift to net *mobilization* of nutrients from stores. Breakdown of stored glycogen and lipids—to yield blood glucose and free fatty acids—is enhanced by a de-

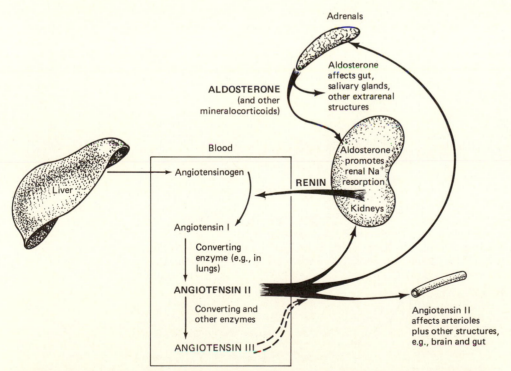

Figure 21.32 Summary of the renin–angiotensin system and its control of aldosterone secretion. Aldosterone and angiotensin II have both renal and extrarenal effects (in the kidneys, aldosterone promotes sodium resorption, and angiotensin II exerts, for example, feedback inhibition on the secretion of renin). Angiotensin III, a heptapeptide made enzymatically from angiotensin II, has many of the same activities as angiotensin II. Aldosterone secretion is under control of other factors besides the angiotensins (e.g., blood potassium concentration).

cline in insulin and also by the drop in blood glucose that tends to occur as glucose from digested food ceases to be available.

Figure 21.33 shows the average variations in plasma insulin over the course of a day in four individuals consuming a mixed diet. The rise in insulin after each meal and its fall between meals are evident. An important overall effect of the pattern of insulin secretion is that it tends to *stabilize blood concentrations of nutrients*—(1) by promoting the removal of nutrient molecules from the blood to stores when blood nutrient levels are tending to rise because of abundant inputs from the gut and (2) by promoting the entry of nutrients into the blood from stores when blood levels are tending to fall because of low gut inputs. In brief, the pattern of insulin secretion helps to mediate a negative-feedback relation between blood nutrient levels and the direction of net nutrient exchange between the blood and storage depots. As illustrated by the data on glucose in Figure 21.33, complete stability of blood nutrient concentrations is not attained (see p. 11). However, concentrations remain far *more* stable than they would without the negative-feedback mediated by insulin. This point is made dramatically clear by the plight of people who are diabetic. Diabetics secrete abnormally low amounts of insulin or suffer from diminished tissue responsiveness to insulin. After a carbohydrate meal, untreated diabetics experience far higher blood glucose concentrations than the normal population. Their blood glucose levels become so high that their kidneys are unable to recover all glucose in the process of urine formation (p. 232), and thus glucose is excreted in their urine and wasted.

Second to insulin, the hormone most involved in managing short-term fluctuations of food availability is **glucagon**. Glucagon is secreted by the A cells of the pancreatic islets and affects primarily the carbohydrate metabolism of the liver. Within its range of action (which is narrower than that of insulin), glucagon exerts effects that are the mirror image of insulin's effects (Table 21.6). It promotes *formation* of glucose from both glycogen and proteins and thus raises blood glucose levels.

The pattern of glucagon secretion depends on the nutrient composition of a meal. When humans consume a mixed diet, their glucagon secretion may remain virtually steady, as shown in Figure 21.33. Carbohydrate meals, however, provoke a fall in glucagon secretion, a response that reinforces the actions of insulin and favors transfer of glucose out of the blood into tissue stores. After a protein meal, *both* insulin and glucagon rise. The rise in insulin is important in promoting the incorporation of absorbed amino acids into body protein. The adaptive value of a rise in glucagon under these circumstances seems to hinge

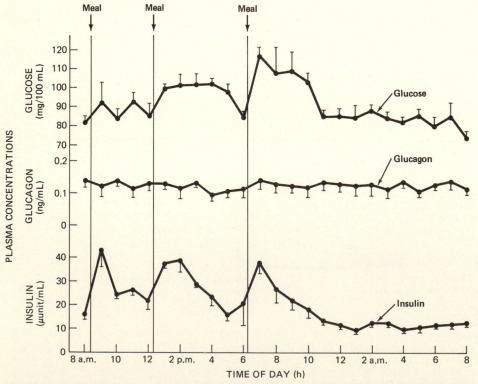

Figure 21.33 Variations over the course of a day (8:00 a.m.–8:00 a.m.) in plasma concentrations of glucose, glucagon, and insulin in four normal human adults eating meals of mixed composition. Vertical bars show the magnitude of 1 standard error. [After Y. Tasaka, M. Sekine, M. Wakatsuki, H. Ohgawara, and K. Shizume, *Horm. Metab. Res.* **7:**205–206 (1975).]

on the fact that a high-protein meal in itself supplies little glucose and yet the brain requires glucose, meaning that an output of glucose from body stores must be maintained in the face of the high insulin levels. The high glucagon levels counteract the effects of insulin on liver carbohydrate metabolism and assure an adequate mobilization of glucose from hepatic glycogen.

Of the other hormones listed in Table 21.6, only growth hormone is reported to help modulate meal-to-meal responses. These other hormones find their chief roles in different contexts. A review of their actions reveals at the outset two interesting properties: (1) In some respects, certain of these hormones are not only similar in their actions but also *enhance* each other's effects. Thyroxine, for example, acts synergistically with growth hormone to promote growth in young animals, and both growth hormone and glucocorticoids enhance the effects of epinephrine on lipid breakdown. (2) As similar as some of these hormones are in certain of their actions and as frequently as they in fact act in concert, each exerts its own unique *suite* of effects. Thus, each hormone is positioned to play a unique regulatory role.

Background levels of some of these hormones have evolved to play an essential *permissive role* in nutrient metabolism. Without glucocorticoids, for example, metabolism is seriously deranged. Fasting and certain other stresses, for instance, tend to precipitate excessively low blood glucose levels in the absence of glucocorticoids, not only because the direct positive actions of glucocorticoids on glucose formation are lacking but also because glucocorticoids are required for glucagon and epinephrine to exert their own positive effects.

Several hormones—notably growth hormone and the thyroid hormones and androgens—play key roles in the *growth of young animals,* in part because they promote protein formation singly and synergistically. Testicular androgen, principally testosterone, is responsible during puberty for the greater muscular development that occurs in boys as compared to girls.

Exercise and *fasting* are two other circumstances in which features of nutrient metabolism are altered by hormones. Both circumstances require a mobilization of metabolic fuels from stores. Glucose must be liberated from glycogen and free fatty acids from lipids. Additionally, prolonged fasting or exercise demands heavy reliance on gluconeogenesis: the formation of glucose from noncarbohydrate precursors, mainly amino acids. This dependency on gluconeogenesis arises because certain tissues—notably the brain—ordinarily require glucose as a fuel, but body stores of glycogen are sufficient to supply glucose for only limited periods of time (e.g., only a day or two of fasting). Once glycogen stores are depleted, sustenance of the brain and other glucose-requiring tissues demands that glucose be made de novo. Gluconeogenesis entails both the mobilization of amino acids from body proteins and their conversion to glucose. The latter occurs in the liver.

Both exercise and fasting elicit complex, multihormonal responses that are not fully understood. Two prominent changes that occur in both circumstances are a decline in insulin secretion and rise in glucagon secretion. In prolonged exercise or fasting, the decline in insulin secretion becomes particularly marked. Low insulin levels allow the processes of glycogen breakdown, lipid breakdown, protein breakdown, and gluconeogenesis to occur at heightened rates (insulin suppresses all of them). Increased glucagon levels stimulate both liver glycogen breakdown and gluconeogenesis. Secretion of growth hormone, epinephrine, glucocorticoids, and thyroid hormones may also be affected by exercise or fasting, and in respect to these hormones, the two circumstances do not always elicit similar responses. An increase in epinephrine secretion, for example, occurs commonly in exercise but is not so consistent a feature of fasting. Epinephrine (unlike glucagon) has the advantage during exercise of stimulating breakdown of glycogen in muscle as well as liver. A response seen often in prolonged fasting, but not in exercise, is a decline in thyroid-hormone levels; this decline serves to lower metabolic demands and conserve fuels.

Control of Reproductive Physiology

Endocrine controls of reproduction are exceedingly widespread and are the most commonly recognized endocrine function among invertebrates. We shall discuss invertebrate examples later (p. 625). Here we consider vertebrate reproduction, with particular focus on mammals. Different mammalian species can differ markedly in their reproductive endocrinology (e.g., recall the contrast between induced and spontaneous ovulators, p. 609). Our emphasis is on humans.

Mammalian sex steroids are placed in three loosely defined categories:

1. *Androgens*—hormones, like testosterone and dihydrotestosterone, that promote the characteristics of maleness (these are 19-carbon steroids).
2. *Estrogens*—hormones, such as estradiol and estrone, that promote the characteristics of femaleness (these are 18-carbon steroids).
3. *Progestogens* (progestins)—hormones, like progesterone, that act principally to prepare and sustain a state of pregnancy (these are among the 21-carbon steroids).

The testes and ovaries are the major sources of sex steroids, but they are by no means the only sources. During pregnancy, for example, the placenta acts as an endocrine gland, producing progesterone, other steroids, and several protein hormones. The adrenal cortex is also a source of sex steroids. *In adults,* the adrenal cortex produces significant amounts of androstenedione, which is readily converted to testos-terone in peripheral tissues. This testosterone of adrenal origin is overshadowed by testicular tes-tosterone in men, but in women it is the principal component of androgenic hormonal activity (the role of androgens in women is poorly understood). Some estrogens also originate by peripheral conversion of adrenal androstenedione; indeed, in women follow-ing menopause, the adrenals are the main source of estrogens. *In the fetuses* of a variety of mammals, including humans, the adrenal cortex is dominated by a special "fetal zone" of tissue, which manufac-tures copious quantities of steroids that are con-verted to estrogens by the placenta. These are the principal estrogens of mid- to late pregnancy.

Endocrinology of the Male   The testicle consists of two types of tissue of principal interest here. First, approximately 85 percent of its volume is occupied by the *seminiferous tubules,* which are lined with germ cells and Sertoli cells and are responsible for the production of sperm. Second, in the spaces between the seminiferous tubules (Figure 21.34) are well-vas-cularized clusters of cells called *interstitial cells* or *Leydig cells.*

The principal androgen of human males is testos-terone, and its principal source is the Leydig cells. These cells are under control of the hypothalamo–adenohypophysial axis. The hormones of this axis

that are of chief importance in the gonadal physiol-ogy of males are the same as in females: the hypo-thalamic neurohormone called gonadotropin releas-ing hormone (GnRH), and the two adenohypophysial gonadotropic hormones, luteinizing hormone (LH) and follicle-stimulating hormone (FSH). Production of testosterone by the Leydig cells is controlled chiefly by LH, although FSH seems also to play a role. Testosterone, FSH, and LH are all required for normal sperm production.

In the life of a normal human male, there are three periods of heightened testosterone production by the testes (Figure 21.35). The first occurs toward the end of the first trimester of fetal life. This episode of testosterone production is essential for differentia-tion of the male genital organs, including not only the external genitalia (penis, scrotum) but also such internal structures as the vas deferens, seminal ves-icles, and ejaculatory duct. Without testicular andro-gen secretion at the proper period of fetal life, the genital primordia intrinsically differentiate to the fe-male form. Some uncertainty exists about the tropic hormones that elicit the episode of testosterone se-cretion in male fetuses. Fetal LH and FSH may be involved, particularly toward the later stages of the episode, but also a *placental* glycoprotein, human chorionic gonadotropin (hCG), plays a role in stim-ulating the fetal testes. By the 14th week after con-ception, the form of the genitalia (whether male or

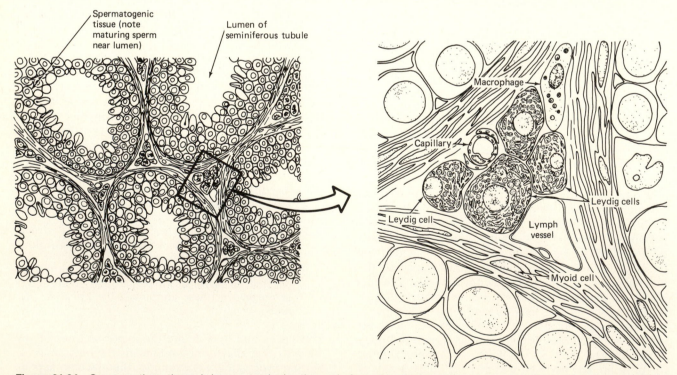

Figure 21.34   Cross sections through human testicular tissue. Left section depicts parts of six seminiferous tubules and shows positions of interstitial or Leydig cells. Right section shows Leydig cells at greater magnification; they are rich in endoplasmic reticulum and are the main source of testosterone. [Reprinted by permission from A.K. Christensen, Leydig cells. In D.W. Hamilton and R.O. Greep (eds.), *Handbook of Physiology,* Sec. 7, Vol. 5, *Male Reproductive System,* pp. 57–94. American Physiological Society, Washington, DC, 1975.]

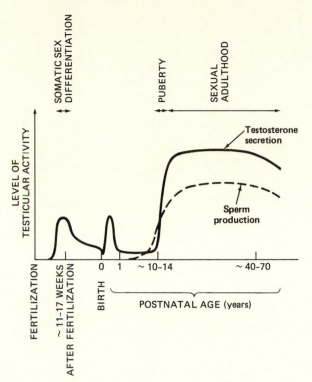

Figure 21.35 Testosterone secretion and spermatogenesis over the life span of a human male. Vertical scale is qualitative. Note that the time scale is not even. [After P. Troen and H. Oshima, in P. Felig, J.D. Baxter, A.E. Broadus, and L.A. Frohman (eds.), *Endocrinology and Metabolism*, pp. 627–668. McGraw-Hill, New York, 1981.]

female) is determined, and shortly thereafter the testes of the human male enter a quiescent phase for the remainder of uterine life. Prenatal androgens affect brain differentiation in several groups of mammals (e.g., p. 611).

The second episode of testicular androgen secretion in the life of a human male occurs shortly after birth. Elicited by heightened LH and FSH, this period of testosterone production peaks at about 1 month of postnatal age. Its function presently remains enigmatic.

The final period of high testosterone output begins at puberty and lasts for the rest of life. It is set in motion by a variety of factors, not all well understood. Included are changes in the CNS mechanisms controlling GnRH secretion, leading to heightened GnRH output, and increases in the sensitivity of the pituitary to GnRH; both the elevation of GnRH secretion and the heightened sensitivity to GnRH lead to increased secretion of LH and FSH, which act on the testes to promote secretion of testosterone. The initial major action of the increase in testosterone is to bring about sexual maturation. It promotes growth of both the penis and such other reproductive structures as the prostate and seminal vesicles. Along with FSH, the testosterone stimulates testicular growth and the onset of sperm production. Pubertal testos-

terone secretion also leads to the appearance of numerous secondary sexual traits; for example, it causes deepening of the voice, development of sexual drive (libido), and growth of facial hair. It contributes too to muscular development. In many mammals, testosterone secretion and breeding condition vary markedly from season to season during adult life. Human males, however, maintain a relatively steady state of sexual readiness and exhibit steady testosterone levels until a slow decline commences in middle age.

Endocrinology of the Female   During the first half of the fetal life of a human female, the germ cells of the ovaries multiply mitotically to produce some 7 million primordial eggs. These grow and mature to the completion of prophase of the first meiotic division, at which point their maturation arrests and they are known as *oocytes.* Each oocyte becomes surrounded with a single layer of *granulosa cells,* forming a unit called a *primordial ovarian follicle* (Figure 21.36). Each follicle is destined to undergo further development at some point in the woman's life. No additional oocytes are ever made.

Individual follicles undertake their further development at various ages, starting during fetal life and ending at menopause. Figure 21.36 shows the early stages of a follicle's development. First, the oocyte enlarges and the granulosa cells multiply, forming a *primary follicle.* Subsequently, a fluid-filled space, or *antrum,* opens up within the mass of granulosa cells, and the follicle becomes an early *antral* (or *secondary*) *follicle.* Development to the late primary and early antral stages requires some promotion by gonadotropins. Development beyond those stages demands even more endocrine stimulation, however, and cannot occur until after puberty, as described later. Most follicles fail to complete their development, for one reason or another. When a developing follicle arrests its progress at any stage prior to full maturation, both the oocyte and the other follicular constituents undergo a process of degeneration called *atresia.*

By mid-fetal life, some primordial follicles are embarking on their developmental path, and some of these reach the primary or early antral stages. But the fetal endocrine environment is inadequate for further maturation, and all become atretic. The second half of fetal life is in fact a period of profound oocyte degeneration, such that only 1–2 million oocytes (out of the initial population of 7 million) remain at birth. Then, from birth to puberty, the process continues. Many thousands of follicles initiate their development only to undergo atresia, leaving perhaps 300,000 primordial follicles (and oocytes) at the onset of sexual maturity.

As in males, puberty is ordinarily initiated by an increase in the average rate of pituitary gonadotropin secretion, mediated by the hypothalamo–hypophysial system. In females, as we shall shortly see, this heightened gonadotropin secretion is highly cyclic

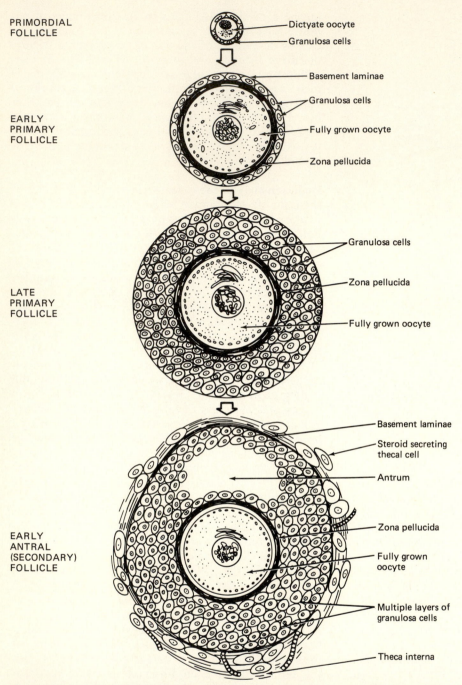

PRIMORDIAL FOLLICLE
— Dictyate oocyte
— Granulosa cells

EARLY PRIMARY FOLLICLE
— Basement laminae
— Granulosa cells
— Fully grown oocyte
— Zona pellucida

LATE PRIMARY FOLLICLE
— Granulosa cells
— Zona pellucida
— Fully grown oocyte

EARLY ANTRAL (SECONDARY) FOLLICLE
— Basement laminae
— Steroid secreting thecal cell
— Antrum
— Zona pellucida
— Fully grown oocyte
— Multiple layers of granulosa cells
— Theca interna

Figure 21.36 Early follicular development. The zona pellucida is an amorphous layer of carbohydrate and protein material that accumulates around the oocyte as the primordial follicle matures into a primary one. The mass of granulosa cells becomes surrounded with a concentric layer of cells, the *theca*, toward the end of the primary stage and particularly during the antral stage. Although the theca is well vascularized, the follicular granulosa tissue is not. [Reprinted by permission from G.F. Erickson, *Clin. Obstet. Gynecol.* **21**(1):31–52 (1978).]

and interacts with cyclic endocrine responses of the ovary. Proper integration of these cycles, once attained, creates a hormonal environment that can sustain the full development of follicles. Thereafter, a set of primordial follicles undertakes development each month, and although most fail to complete de-

velopment (and become atretic), one follicle attains full maturity and releases its oocyte in the process termed *ovulation* (p. 609).

Figure 21.37 illustrates the morphological changes that occur in a follicle when its development is sustained beyond the early antral stage. Although the

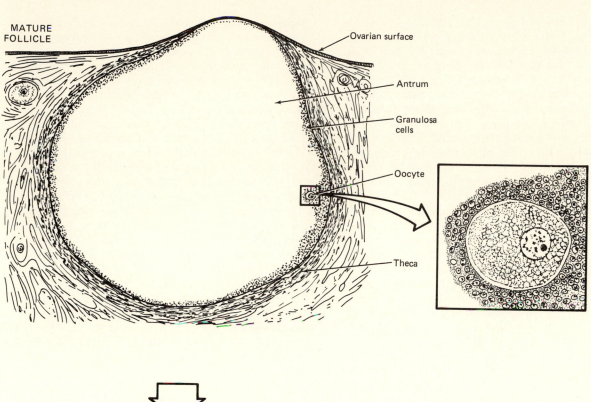

MATURE
FOLLICLE

Ovarian surface

Antrum

Granulosa
cells

Oocyte

Theca

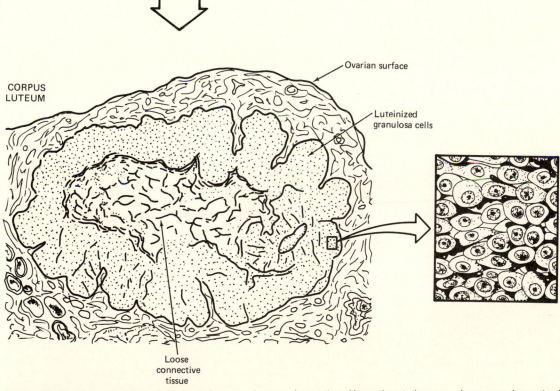

CORPUS
LUTEUM

Ovarian surface

Luteinized
granulosa cells

Loose
connective
tissue

Figure 21.37 Human mature follicle and corpus luteum, in section. Upper inset shows region around oocyte; lower shows luteinized granulosa cells (granulosa lutein). [Follicle reprinted by permission from B.M. Patten, *Human Embryology*. Blakiston, Philadelphia, 1946; corpus luteum after M.S.H. DiFiore, R.E. Mancini, and E.D.P. DeRobertis, *New Atlas of Histology*. Lea & Febiger, Philadelphia, 1977; granulosa lutein after G.S. Richardson, *Ovarian Physiology*. Little Brown, Boston, 1967.]

oocyte itself does not grow, the follicle as a whole grows considerably. Whereas a late primary or early antral human follicle measures about 0.2 mm in diameter, a fully mature follicle attains a diameter of approximately 15–17 mm (over 1/2 in.) and the follicles in a monthly cohort that fail to reach maturity attain, on average, 7 mm. The fully mature follicle is known as a *mature, Graafian,* or *preovulatory* follicle. Its development from the early antral stage involves a substantial proliferation of its granulosa cells and a large increase in its volume of follicular fluid and the size of its antrum. Having assumed a position just under the outer epithelium of the ovary, the mature follicle forms a pronounced bulge on the ovarian surface. As the day of ovulation approaches, the layers of cells and other membranes separating the follicular antrum from the peritoneal cavity become thin and degenerative. Additionally, the oocyte completes its first meiotic division (the second is completed after fertilization). Finally, after a period of swelling, the follicle ruptures, and the oocyte (with a layer of granulosa cells) is discharged from the ovary, to be picked up by the fallopian tube for transport to the uterus.

After ovulation, the empty mature follicle undergoes changes entirely different from the atresia that is the fate of most follicles. The granulosa cells proliferate, enlarge, become vacuolated, and in the long term accumulate a yellowish pigment known as lutein. Furthermore, the granulosa tissue is rapidly infiltrated by blood vessels (the granulosa tissue is nonvascular in the follicular stage, although the theca surrounding it is well vascularized). The structure formed by all these changes is called a *corpus luteum* (Figure 21.37). If pregnancy fails to occur, the corpus luteum lasts for only about 14 days, attaining a peak diameter of 15–20 mm, and is known as a *corpus luteum of the* (menstrual) *cycle.* After its 2 weeks of life, it degenerates, often forming a fibrous remnant structure called a corpus albicans. If pregnancy occurs, the corpus luteum not only persists but grows, reaching perhaps 40–50 mm in diameter, and is termed a *corpus luteum of pregnancy.*

Endocrinologically, the chief objectives in the study of female reproductive physiology have been to understand the hormonal bases for both the monthly reproductive cycle and the maintenance and termination of pregnancy. A crucial finding in both realms is that *the ovarian follicle, corpus luteum, and placenta are all glands.* Moreover, a central theme that continues to emerge with greater force and clarity is that these structures, in interaction with the hypothalamo–hypophysial axis, *exert substantial endocrinological control over their own maturation and destiny.*

We often say that "the ovary" is a source of hormones. It is more precise and revealing, however, to recognize that these hormones are produced by particular ovarian constituents, principally the follicles and corpora lutea. *Follicles produce chiefly estrogens* although they do at times secrete progesterone and even make androgens. An interesting and potentially very important area of investigation concerns the comparative functional properties of the granulosa cells and the cells of the outer sheath, or theca, of the follicle. When the granulosa cells produce estrogens (such as estradiol), they do so chiefly using precursors made by the theca; thus, there is heterogeneity of function and cooperation *within* the follicle. The theca itself often makes little estrogen directly. In most species, it primarily produces the estrogen precursors testosterone and androstenedione. These precursors then in part diffuse to the adjacent granulosa cells, which contain the aromatase enzyme necessary to convert them to estrogen. *Corpora lutea secrete chiefly progesterone* and also secrete some other progestogens as well as substantial amounts of estrogens. The progesterone is made intrinsically by the luteinized granulosa cells.

Figure 21.38 summarizes the changes in the blood levels of four key hormones over the course of a *nonfertile* monthly cycle. We first review the major *effects* of these changes on the reproductive organs and later address the question of how the hormonal cycles are controlled.

In discussing effects, five major processes in the reproductive organs deserve mention:

1. *Follicular growth and maturation.* The relatively early stages of follicular development are stimulated by FSH, which, as can be seen in Figure 21.38, reaches high levels a couple of weeks before ovulation (recall: FSH = follicle-stimulating hormone). Later stages of follicular development are promoted by LH and estrogens as well as FSH. Without appropriate hormonal support all through development, a follicle will lapse into atresia.
2. *Ovulation.* Ovulation is induced by a high level of LH maintained over an adequate length of time (p. 609). The monthly surge of LH that causes ovulation is dramatically evident in Figure 21.38. The mechanism by which the LH signal is translated into the act of ovulation is not understood, although prostaglandins are known to be involved.
3. *Luteinization.* The formation and later persistence of the corpus luteum are dependent on stimulation by LH (luteinizing hormone), which is abundant at the time of follicular rupture.
4. *Uterine wall development.* As diagrammed in Figure 21.39, the inner wall of the uterus—the endometrium—is potently affected by ovarian hormones. The estrogens secreted in quantity by the follicle prior to ovulation induce extensive multiplication of endometrial cells—including gland cells and blood vessels—and resultant endometrial thickening. The progesterone and estrogens secreted by the corpus luteum provoke even greater thickening by stimulating further cell proliferation and also by inducing enlargement of cells and formation of intercellular edema; furthermore, at this stage the endometrial exocrine glands are stimulated to become secretory, and other endometrial cells accumulate nutrients (e.g., glycogen). All these changes prepare the uter-

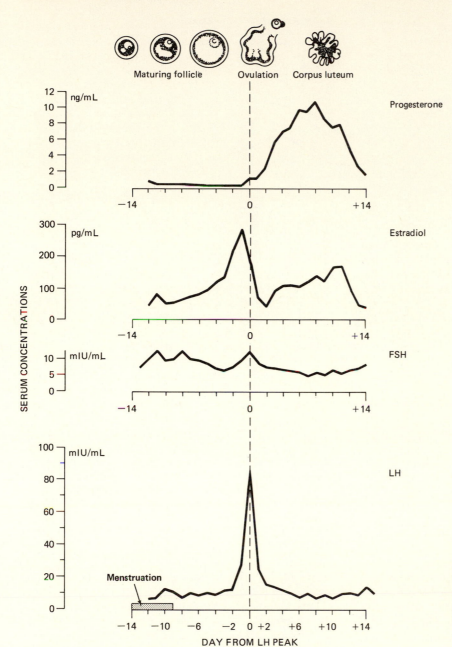

Figure 21.38 Mean serum concentrations of progesterone, estradiol (a prominent estrogen), FSH, and LH over the course of a normal nonfertile human menstrual cycle. Time is reckoned relative to the LH peak (labeled as time zero). Shaded bar just above lowest time axis marks approximate period of menstruation. IU, international unit. [After G. Leyendecker, K. Hinckers, W. Nocke, and E.J. Plotz, *Arch. Gynaekol.* **218**:47–64 (1975); drawings after C.P. Channing, F.W. Schaerf, L.D. Anderson, and A. Tsafriri, *Int. Rev. Physiol.* **22**:117–201 (1980).]

ine wall to be a favorable site for embryonic development should the ovulated ovum be fertilized and implant.

5. *Luteolysis and menstruation.* Levels of LH fall off markedly after ovulation (Figure 21.38) and are not sufficient to maintain the corpus luteum beyond 12–14 days. If pregnancy occurs, tropic-hormone stimulation above and beyond that of LH is provided (see later). If pregnancy does not occur, however, the corpus luteum reaches peak development at 7–10 days after ovulation, then ebbs, and finally undergoes degeneration or *luteolysis*. One view of the cause of this degeneration is that the corpus luteum requires steady tropic-hormone support and does not receive it in an infertile cycle because of the slackening of LH levels after ovulation. Another view is

that the corpus luteum is programmed to persist only 12–14 days unless it receives a large tropic-hormone stimulus relatively early in its existence, one larger than LH alone provides. Just as the corpus luteum requires tropic-hormone support for persistence, the uterine endometrium requires progesterone and estrogen to maintain its thickened, glandular state. Thus, as the corpus luteum—the source of the steroids—fails, endometrial regression sets in, culminating in an episode of bleeding and sloughing of cells known as *menstruation* (Figure 21.39).

The *mechanisms that control the monthly cycling of hormones* are complex and incompletely understood. We shall discuss some of the salient features

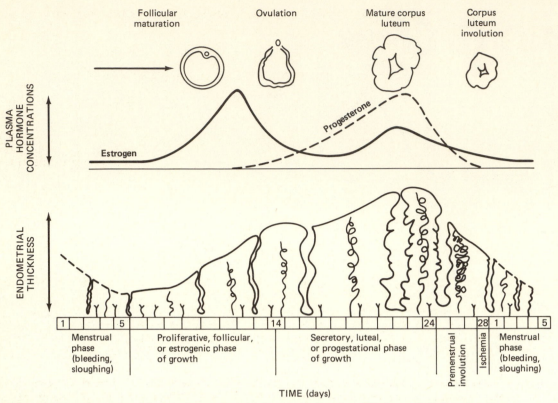

Figure 21.39 Summary of the uterine endometrial cycle (bottom) and its immediate hormonal controls (top) in humans. Note that the days (1–5) of the menstrual phase (characterized by bleeding and sloughing) are duplicated on both sides of the diagram. At least half of the tissue loss during the premenstrual and menstrual phases is by involution and resorption rather than sloughing. Compare with Figure 21.38, noting that the time of ovulation is labeled day 14 here but about day +1 there. [After S.T. Shaw, Jr. and P.C. Rouche, *Oxford Rev. Reprod. Biol.* **2**:41–96 (1980). Used by permission of Oxford University Press.]

of these control mechanisms and at the same time summarize the extraordinary physiological integration that results from the hormonal cycles.

The changes in hormone concentrations shown in Figure 21.38 result from interactions between the ovary, pituitary, and brain. We first consider the *follicular phase* of the cycle: the phase when follicles are developing. This phase begins during menstruation. At that time, progesterone and estrogen levels have dropped, and the pulsed secretion of GnRH by the hypothalamus (which prior to menstruation is inhibited by high levels of progesterone) rises again. As a result of GnRH stimulation, the adenohypophysis increases release of LH and particularly FSH, thereby promoting renewed development of new follicles in the ovary. The developing follicles secrete increasing amounts of estrogen; this secretion is supported by LH, FSH, and the estrogens themselves, which increase ovarian responsiveness to FSH. The increasing levels of ovarian hormones during the follicular phase exert feedback effects on the hypothalamus and adenohypophysis. FSH levels gradually decrease (day −8 to day −3), probably as a result of negative feedback exerted by estrogen and by inhibin, a gonadal peptide hormone that appears

to act specifically on secretion of FSH at the pituitary level. The feedback effects of estrogen on LH are more complex. During the follicular phase, estrogen and GnRH both promote synthesis of LH in pituitary cells, but estrogen somewhat inhibits LH release. These effects increase the store of LH for the ovulation phase.

The *ovulation phase* of the cycle is characterized by a surge of LH release that triggers ovulation of the mature follicle. The LH surge results from two influences: (1) a midcycle episode of large pulsed releases of GnRH from the brain and (2) the high estrogen levels prevailing at midcycle (Figure 21.38), which increase pituitary responsiveness to GnRH. Recalling our earlier discussion of the follicular phase, we see that the effects of estrogen on the adenohypophysis are complex, having aspects of both negative feedback and positive feedback. Note the important feature that it is the mature follicle that secretes the high estrogen levels involved in inducing the LH surge; thus, the follicle helps control the timing of its own ovulation. By secreting estrogen, the follicle also initiates preparation of the uterus to receive its ovum, should fertilization occur.

A small FSH surge accompanies the LH surge,

but overall there is a surprising degree of independence of release of LH and FSH, considering that the two pituitary hormones share one common hypothalamic releasing hormone—GnRH. There appear to be several reasons for this independence. GnRH stimulates LH release more than FSH release, and the *frequency* of the pulsed releases of GnRH affects the ratio of LH to FSH in the blood. Moreover, inhibin from the ovary specifically inhibits pituitary secretion of FSH, further damping the effects of GnRH on FSH release.

After ovulation, the conversion of the ruptured follicle to a corpus luteum begins the **luteal phase** of the cycle. The high LH levels that prevail before and after ovulation provide the primary stimulus for luteinization, but follicle rupture is also necessary. Prior to ovulation, the granulosa cells are in an avascular compartment; at ovulation, they become exposed to factors in the blood plasma that are necessary for luteinization. Within a few days after ovulation, the newly transformed corpus luteum secretes increasing amounts of progesterone and estrogen. Progesterone and estrogen cause continued preparation of the uterus, as already noted. High levels of progesterone combined with estrogen also strongly inhibit GnRH secretion by the hypothalamus and thereby inhibit secretion of LH and FSH. Since FSH and LH are necessary to permit maturation of new follicles, the presence of a corpus luteum thus inhibits follicular maturation; that is, the corpus luteum associated with the presence of one fertilizable ovum prevents release of other ova. (The inhibition of follicle development by use of progesterone combined with estrogen is the basic rationale of the birth-control pill.)

If fertilization fails to occur, uterine preparation has been fruitless and new follicles are needed. The evolved adaptive response to these realities is regression of the corpus luteum, effected (as described already) either by some type of programmed self-destruction or by an immediate reaction to the diminishing levels of LH (the drop in LH being attributable to the inhibitory effects of the corpus luteum's own progesterone on GnRH secretion). With the regression of the corpus luteum and ebbing of its steroid output, the uterus is returned to its initial state, and outputs of GnRH and FSH are allowed to rise, initiating and sustaining development of new follicles.

If fertilization does occur, the corpus luteum is rescued by a hormonal signal emitted by the early embryo–placenta unit. The hormone involved is **chorionic gonadotropin,** known specifically in women as **human chorionic gonadotropin (hCG).** This is a glycoprotein that not only has close chemical similarities to LH but also resembles LH physiologically in that it stimulates and supports the corpus luteum. The embryo resulting from fertilization starts to implant in the uterine wall 6–8 days following ovulation, and by 7–9 days after ovulation, markedly elevated levels of hCG are detected in maternal blood. Chorionic gonadotropin is known to be synthesized by certain placental cells, and the traditional view is that these cells, activated following implantation, are the exclusive source of the hCG in maternal blood. It is not known if the embryo also contributes to this hCG, perhaps even before it implants. In any case, the hCG furnishes sufficient tropic-hormone support to rescue the corpus luteum from degeneration and sustain it to become a corpus luteum of pregnancy. Just as a rising level of hCG is the maternal body's first signal from the uterus that conception has occurred, the presence of hCG in the mother's blood or urine is what clinical pregnancy tests use as their first definitive signal.

The crucial significance of luteal rescue is that luteal progesterone and estrogen are required to sustain the uterine endometrium in a thickened, secretory condition that will nurture the implanted embryo. A consequence of this uterine sustenance is that menstruation does not occur. Its absence is sometimes a woman's first indication of pregnancy.

In humans, luteal production of progesterone is essential for maintenance of the uterine wall in a pregnant state for about 6 weeks after conception; destruction of the corpus luteum prior to this time results in physiological abortion. Gradually, the placenta itself produces and secretes greater and greater amounts of progesterone and estrogens, thus providing for its own endocrine support. After about 6–7 weeks of pregnancy, the placenta can be self-sustaining, and although evidence exists that the corpus luteum remains functional through most of pregnancy, destruction of the corpus luteum will not interrupt the pregnancy.

During the course of pregnancy, the fetal–placental unit repeatedly exercises "self-interested" endocrine control over the physiology of the mother. We have already noted, for instance, that it hormonally rescues the corpus luteum. The fetal–placental unit also secretes hormones that favorably influence maternal nutrient metabolism and mammary-gland development, and toward the end of pregnancy, it helps to orchestrate the events leading to labor and exit of the fetus to its new environment in the outside world.

INVERTEBRATE ENDOCRINE SYSTEMS

Because invertebrates comprise about 95 percent of the species in the animal kingdom, we might expect a correspondingly great diversity of invertebrate endocrine mechanisms. While this expectation may ultimately be confirmed, our present knowledge of the endocrine systems of many invertebrate groups is still highly incomplete. Here, we describe only a handful of relatively well-studied examples of invertebrate endocrine systems. The invertebrate systems are grouped in a separate section for the convenience of the reader, rather than because of any fundamental difference in organization or action between invertebrate and vertebrate endocrine systems. In fact, the basic principles presented in this chapter apply equally well to endocrine systems of vertebrates and invertebrates. The only general difference is that in most groups of invertebrates, unlike vertebrates, neurosecretory systems are distinctly more prominent than nonneural endocrine glands; indeed, in many invertebrate groups, nonneural endocrine glands appear to be absent.

Annelids

All annelids have brain neurosecretory cells that control reproduction, growth and maturation, and probably other functions. Polychaete worms such as *Nereis* have been studied most extensively. Many polychaetes have a distinct reproductive body form, termed an *epitoke,* that arises from the asexual body form by metamorphosis or by budding. This transformation occurs only once in the lives of some species but recurs cyclically in others. In *Nereis,* maturation of eggs and sperm and transformation to the reproductive epitoke form are both controlled by one or more brain neurohormones. Removal of the brain from a young worm results in rapid maturation of gametes and premature epitokal metamorphosis, but both processes may be abnormal and incomplete. Normal reproductive development appears to depend on the gradual withdrawal of brain neurohormones with increasing age. These hormones are inhibitory in high concentrations, but necessary in low concentrations for completion of the reproductive changes.

Many annelids possess considerable abilities to regenerate lost parts. If a worm is cut into two pieces, the anterior piece will commonly regenerate a new tail; less commonly, the posterior piece will regenerate a new head. Regeneration is probably in general controlled by neurohormones. In *Nereis,* for example, tail regeneration requires a hormone from the brain of a young animal.

Echinoderms

In starfish and sea urchins, spawning of eggs is precipitated by secretion of a "shedding hormone," which is a polypeptide neurohormone emitted by the radial nerves. The hormone induces two responses: freeing of the oocytes from surrounding follicle cells and completion of meiosis of the maturing oocytes. Both effects of shedding hormone are mediated indirectly. The hormone induces follicle cells to produce 1-methyladenine, which acts to dissolve the adhesion of follicle cells to the oocyte and to each other; 1-methyladenine also acts on the underlying oocyte to trigger the completion of meiosis.

Mollusks

Many mollusks have neurons that have the histological appearance of neurosecretory cells and that change their apparent secretory activity with conditions such as reproductive state. In a few of these cases, there is experimental evidence for neurosecretory control of reproduction, water balance, or heart function. We consider as an example the endocrine control of reproduction in *Aplysia.*

The sea hare *Aplysia* is an opisthobranch gastropod. It is a simultaneous hermaphrodite, each individual being simultaneously male and female. After copulation, an *Aplysia* lays a long string of more than a million eggs. This egg laying is triggered by a neurohormone secreted by distinct clusters of *bag cells* in the parietovisceral (abdominal) ganglion. Copulation presumably leads to activation of the bag cells, an effect that can be mimicked experimentally by electrical stimulation of connectives to the ganglion. Following strong stimulation, all the bag cells of a cluster undergo electrical discharge synchronously and repetitively for several minutes to half an hour, in a nearly all-or-none manner. This prolonged, repetitive discharge produces secretion of the hormone, which contracts muscles around the ovotestis and reproductive duct, first inducing ovulation and then ejection of strings of eggs.

The egg-laying hormone (ELH) is well characterized. It is a peptide consisting of 36 amino acid residues. In common with other peptide hormones and transmitters, such as β-endorphin (p. 448) and insulin (p. 581), ELH is formed by post-translational processing of a larger protein that is a gene product. The ELH is not the only fragment to be split off from the larger precursor protein; at least three other fragments are released with it. Interestingly, ELH and two of the other fragments have local effects on specific neurons in the abdominal ganglion, in addition to the hormonal effect of ELH on contraction of the reproductive duct.

Insects

Probably the most elaborate and certainly the best studied invertebrate endocrine system is that of insects. Insects possess discrete clusters of neurosecretory cells, well-developed neurohemal organs, and (unlike many invertebrates) nonneural endocrine glands. We have already taken an overview of insect endocrine organization (p. 591). Here, we discuss specifically the functions of hormones in insect molting and development, and we also list hormonal roles in some other physiological processes.

Control of Molting and Development As insects develop from larvae to adults, they are compelled periodically to shed their exoskeleton and synthesize a new, larger one. This replacement of the exoskeleton is necessary both for growth and for changes in body form (such as acquisition of wings in adults). The process of replacing one exoskeleton with another is termed *molting;* the shedding of the old exoskeleton is termed *ecdysis.* Molting is a complex series of events involving several major steps and quite a few hormones. Here, we look at only the most salient aspects of endocrine control.

Molting is precipitated by steroid hormones called *ecdysteroids* that are secreted by peripheral, epithelial glands—mainly the *prothoracic glands* or their homologs (Figure 21.13). The secretion of the ecdysteroids is substantially under control of the brain. Brain control is mediated in major part by a tropic neurohormone, usually called *prothoracicotropic hormone* *(PTTH),* which was the first of all insect hormones to be discovered. PTTH is synthesized by neurosecretory cells within the brain but is released into the blood predominantly at peripheral neurohemal organs. Traditionally, the corpora cardiaca (see Figure 21.13) were thought to be the exclusive sites of release, but now there is evidence that the corpora allata are release sites also and may in fact be the preeminent or sole sites in some insects. At each time of molting (Figure 21.40), neural processes, which may involve biological clocks as well as neural sensory inputs, activate secretion of PTTH, which then travels in the blood to the prothoracic glands and stimulates the latter to secrete a particular ecdysteroid called *ecdysone.* Traditionally, ecdysone itself was believed to be the true molt-inducing hormone. Now it is clear, however, that a variety of tissues (e.g., fat bodies) convert circulating ecdysone into another ecdysteroid, *20-hydroxyecdysone,* and this latter hormone appears to be the preeminent

molt inducer (this conversion is an example of peripheral activation). The action of 20-hydroxyecdysone on target cells is similar to the action of steroids in vertebrates (Figure 21.11): The steroid binds to intracellular receptors, inducing transcription of specific genes. The transcription process can sometimes be visualized directly in insects, as hormone-dependent enlargements (or puffs) within the giant chromosomes of the salivary glands and other tissues.

Another hormone prominent in controlling developmental molts is a terpenoid termed *juvenile hormone* (which exists in several forms). Juvenile hormone is secreted by the corpora allata (Figure 21.13) and determines whether molting will yield a larva or adult. Specifically, it promotes retention of larval characteristics. In insects such as beetles, butterflies, and moths that undergo a sudden, wholesale metamorphosis from larva to pupa and adult, juvenile hormone is secreted at high levels during each larval stage (instar) except the last; this secretion of juvenile hormone assures that each molt will yield an animal that, although bigger, is larval in form. Then, during the final larval stage, secretion of juvenile hormone is curtailed. At the time of the next molt following this curtailment, the tissues are exposed (for the first time ever) to a pulse of ecdysteroids in the absence or near absence of juvenile hormone. This set of circumstances reprograms the tissues to progress to pupal form. Premature cessation of juvenile hormone secretion, as by removal of the cor-

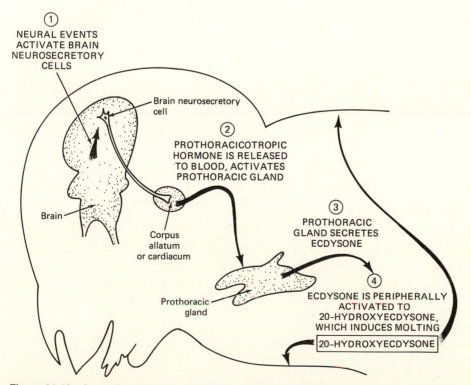

Figure 21.40 An outline of major steps in the activation of the molting process in insects. Compare with Figure 21.13.

pora allata, induces a premature metamorphosis, into an unusually small pupa and adult.

Prothoracicotropic hormone, ecdysteroids, and juvenile hormone exert principal control over insect molting, as just described. They are not, however, the only hormones involved. For example, the peptide hormone *bursicon* is important for tanning (hardening and darkening) of the new cuticle after a molt, and *eclosion hormone* (p. 610) is necessary for the neural generation of movements by which the adult emerges from the pupa.

Reproduction In insects, hormones play permissive roles in the development of the reproductive system but are not normally involved in the sex determination of reproductive structures. The major reproductive role of hormones is in guiding normal development of eggs. Juvenile hormone is necessary for egg development. It acts on the fat body to induce synthesis of vitellogenins (yolk proteins) for the eggs and may also act on the ovary to promote vitellogenin uptake and oocyte maturation. In the males of most species, in contrast, spermatogenesis proceeds without juvenile hormone.

Ecdysone may be present in adult insects, despite degeneration of the prothoracic gland. Adult ecdysone, synthesized in the ovary, is also implicated in oocyte development.

Other Hormones Myriad other insect hormones have been described or suggested. We mention only a few of these, which are involved in nutrient metabolism and in salt and water balance.

Two hormones synthesized by cells that are endogenous to the corpora cardiaca affect lipid metabolism and carbohydrate metabolism, respectively. *Adipokinetic* (lipid mobilizing) *hormone* stimulates lipid release from the fat body. Lipid may be released, for example, during flight, for which it may serve as an important fuel. *Hypertrehalosemic hormone* serves a similar function for carbohydrates. It mobilizes trehalose (the major insect blood sugar) from stored glycogen in the fat body, raising trehalose levels in the blood.

Several insects also produce *diuretic hormones* that increase rates of water excretion. *Antidiuretic hormones* also occur in some species. The functional roles of these hormones were discussed previously (pp. 196 and 241).

Crustaceans

Decapod crustaceans such as crabs, shrimps, lobsters, and crayfish have well-developed endocrine systems that are second only to insects in complexity among invertebrates. We shall focus on two prominent neurosecretory systems that (in some cases along with nonneural glands) control molting, reproduction, mineral and water balance, color changes, and heart function.

The X Organ–Sinus Gland Neurosecretory System and Control of Molting In decapod crustaceans, a sizable part of the brain is located in the eyestalks; the major neurosecretory system in decapods is in this eyestalk portion of the brain. The *X organ* is a cluster of cell bodies of neurosecretory neurons within the ganglia of each eyestalk. These cell bodies send a bundle of axons to the *sinus gland,* a neurohemal organ also located in the eyestalk. The crustacean X organ–sinus gland complex is analogous in many ways to the system in insects composed of the brain neurosecretory cells and corpus cardiacum. The anatomical arrangement of the crustacean system, however, has advantages for experimentation, because eyestalk removal functionally removes the neurosecretory system.

Neurohormones released at the sinus gland control several processes, of which *molting* has been studied most extensively. Crustaceans, like insects, must periodically molt the exoskeleton in order to grow. (Crustacean exoskeletons, unlike those of most insects, are heavily calcified. This difference in exoskeletal composition requires some differences in hormonal control of molting.)

Neurons of the X organ–sinus gland complex produce and secrete a *molt-inhibiting hormone.* Removal of the eyestalks of a decapod thus induces an early molt. The molt-inhibiting hormone acts on paired nonneural glands, called *Y organs,* located in the maxillae or bases of the antennae. The Y organs, sometimes termed *ecdysial glands,* are analogous and perhaps homologous to insect prothoracic glands. They secrete ecdysteroids, principally ecdysone and 20-hydroxyecdysone, which directly control the various processes involved in molting. Thus, the control of molting in crustaceans is strikingly parallel to that in insects, with the major exception that in crustaceans the neurosecretory control of steroid secretion by the ecdysial glands is *inhibitory,* in contrast to the *excitatory* neurosecretory control of the prothoracic glands in insects (Figure 21.41).

Other hormones participate in control of molting besides molt-inhibiting hormone and ecdysteroids. For example, hormones from the sinus gland and elsewhere regulate molt-related water balance and calcium metabolism (Table 21.5).

Reproduction and Sex Determination Hormones play an important role in sex determination in crustaceans. Young genetic males develop an *androgenic gland* associated with the genital tract. The androgenic gland, which does not develop in genetic females, masculinizes the differentiating gonads into testes and also induces male secondary sexual characteristics. Implantation of an androgenic gland into a young genetic female causes transformation of the ovaries into testes. Thus, as in mammals, the crustacean reproductive system becomes male if provided with an appropriate hormonal signal and becomes female if this signal is absent.

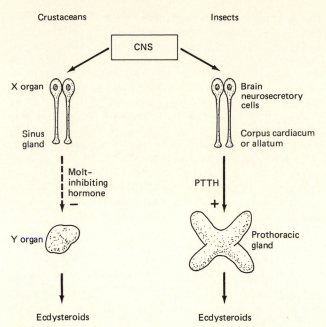

Figure 21.41 Endocrine control of molting in crustaceans and insects. In crustaceans, a brain neurohormone *inhibits* secretion of ecdysteroids by the Y organ. In insects, a brain neurohormone *stimulates* secretion of ecdysteroids by the prothoracic gland.

Chromatophores and Color Change Many crustaceans are capable of pronounced changes in color, mediated by chromatophores (p. 544). Each chromatophore is a highly branched cell within which pigment can be concentrated or dispersed. Many crustaceans have several color classes of chromatophores and can control them rather independently. The control of crustacean chromatophores is hormonal, mediated by **chromatophorotropins** from the eyestalks and other neurosecretory centers. Each color class of chromatophores is under control of antagonistic concentrating and dispersing hormones, which have been given names such as red pigment concentrating hormone and black pigment dispersing hormone. Similar hormones control the movement of screening pigments during light and dark adaptation of the eyes. The various pigment concentrating hormones are quite specific and independent in their actions, but recent evidence suggests that a single hormone mediates light adaptation and pigment dispersal in all chromatophore classes.

Pericardial Organs and Cardioregulation *Pericardial organs* are the second most prominent neurohemal organs of higher Crustacea. They lie in the dorsal venous cavity surrounding the heart and exert their clearest effects on heart function. The cell bodies of most of the neurosecretory neurons that have endings in pericardial organs lie in or near thoracic ganglia of the central nervous system; their axons reach the pericardial organs through segmental nerves. Some cell bodies, however, are located in the peri-

cardial organs themselves. The major hormones released from the pericardial organs are the biogenic amines octopamine, serotonin, and dopamine, plus at least one peptide. All have excitatory effects on the heart and also facilitate neurally evoked contraction of peripheral skeletal muscles. The hormonal release of cardiostimulatory amines by the pericardial organs is analogous in certain ways to the release of catecholamines (epinephrine and norepinephrine) by the vertebrate adrenal medulla (p. 337). It also serves to remind us that although most known neurohormones are peptides, not all are.

HORMONELIKE SUBSTANCES

In this section, we briefly examine a variety of regulatory chemicals that in their modes of action bear partial—but *only* partial—resemblances to classical hormones. Knowledge of such chemicals has been escalating dramatically in recent years. Earlier (p. 577), we noted that because regulatory chemicals fall along a continuum in their properties, they cannot neatly be classified into distinct categories. The substances now to be discussed are among those that do not "fit" into the traditional scheme of classification, a fact that is reflected in their nomenclature. All are substances that are emitted by certain cells and affect the activity of other cells. But some, for example, not only are released but also exert their effects within the confines of a single tissue mass, making their way from their cells of origin to the cells they affect by strictly local mechanisms (e.g., diffusion and tissue capillary flow) rather than by transport in the general circulation. Certain investigators have described these particular types of substance as **local hormones** or **tissue hormones**. Others have wished to avoid calling them "hormones" at all and have described them with novel terms like **paracrines** or **autacoids**. Some textbooks of endocrinology describe these substances at length, whereas others hardly mention them, such is their ambiguous status.

Neuromodulators

Many recent studies strongly suggest that neurons may exert chemical control at distances (and time courses) intermediate between the extremes of classical neuroendocrine and neural synaptic control. Thus, a neuron may release a regulatory chemical that diffuses within a tissue, acting on many cells in the immediate area without entering the circulation in appreciable quantity. Such agents are usually termed **neuromodulators**. Most candidate neuromodulators are peptides or amines.

The "gray area" between neurohormones and synaptic transmitters is exemplified by the genesis of the "late slow EPSP" in frog sympathetic ganglia. We have already discussed synaptic transmission in sympathetic ganglia (Figures 17.23 and 17.24). In frog sympathetic ganglia, there is a late slow excitatory postsynaptic potential (EPSP) in addition to the fast EPSP and slow EPSP shown in Figure 17.24. The late slow EPSP is mediated by release of a peptide closely related to gonadotropin releasing hormone (GnRH). The GnRH-like peptide is released from presynaptic neurons and can act on postsynaptic neurons *several tens of micrometers away* (versus 20–30 nm for conventional synapses). The resulting late slow EPSP has a duration of several minutes. This kind of neuromodulator action is suggested in several other vertebrate and arthropod preparations and may prove to be a common, intermediate form of chemical regulation of target cells.

Prostaglandins and Prostaglandinlike Substances

The *prostaglandins* are a family of chemically related, 20-carbon fatty acid derivatives (Figure 21.42). They were first discovered in the seminal fluid of men and initially were believed to be synthesized by the prostate gland. This is how the class of compounds got its name. Research ultimately revealed that the seminal vesicles, not the prostate, are the principal source of the compounds in semen, but by then the name had become so established as to endure. Usually, specific prostaglandins are described by names like "prostaglandin E_1" (PGE_1), "prostaglandin F_2" (PGF_2), and so on. The letters and subscripted numerals in these names, such as *E* and *F* and *1* and *2*, refer to specific structural features of each compound. Recently, three additional types of compound have been described that differ in certain regular structural respects from the classical prostaglandins and yet resemble them chemically and are synthesized by related biochemical pathways. These have been named the *leukotrienes,* the *thromboxanes,* and *prostacyclin.* Some authorities refer to these classes of compounds as prostaglandins, whereas others distinguish them as "prostaglandinlike" substances.

By now, prostaglandins of one sort or another have been found to be synthesized in virtually every mammalian tissue as well as in tissues of other vertebrates and some invertebrates. At least in mammals, the prostaglandins are believed to function largely as tissue hormones, exerting their effects within the very tissue masses where they are made. One reason they do not exert more-long-range effects is that many tissues possess catabolic pathways that destroy prostaglandin molecules before they enter the general bloodstream (also, some prostaglandins rapidly break down spontaneously). In addition, the lungs are extremely efficient at inactivating prostaglandins in the blood. Thus, prostaglandin molecules that do get into the bloodstream are generally rendered ineffective before they complete even a single circulatory circuit. In some exceptional cases, prostaglandins do function like classical hormones. For example, among sheep and some other mammals (but not humans), it seems that during nonfertile reproductive cycles, prostaglandins from the (nonpregnant) uterus make their way by a special vascular arrangement to the ovaries, where they help provoke involution of corpora lutea (p. 625). The normal role of the prostaglandins in semen has yet to be definitively established, but many investigators believe them to exert fertility-promoting influences on the contractility of muscles in the female reproductive tract following copulation, and if this is true, these prostaglandins also operate in an exceptional way; the name *exohormone* has been proposed for them.

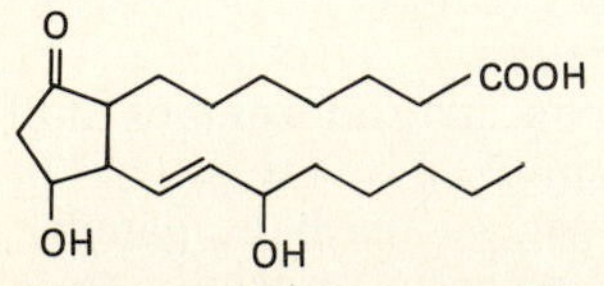

The thousands of research reports on prostaglandins identify a huge, almost bewildering array of physiological effects that these substances *can* exert. In general, however, considerable ignorance remains about the roles the prostaglandins normally play; some of their known effects are undoubtedly products of unusual experimental conditions rather than being reflective of normal function (it is difficult to mimic in the laboratory the normal life history of substances that are released within local tissues and have only a fleeting existence there before being destroyed). A very abbreviated list of actions of a few prostaglandins is given in Table 21.7. There is good evidence that the "normal" repertoire of some prostaglandins (as well as histamine and some kinins) includes pathological effects as well as regulatory ones. Some leukotrienes, for example, are implicated in asthmatic and other hypersensitive reactions.

Aspirin inhibits one of the enzymes involved in prostaglandin synthesis. Its well-known actions are believed to result at least in part from lowering of the concentrations of certain prostaglandins.

Histamine and the Kallikrein–Kinin System

Histamine, an amine derivative of histidine, was one of the first tissue hormones identified. Among vertebrates, it is synthesized for the most part in mast cells within the body's various tissues. It is capable of exerting numerous effects, but considerable ignorance remains concerning which of the effects occur in normal physiology. Probably the best-established role of histamine in mammals is in provoking parts of the inflammatory response: When released from mast cells in response to one or another sort of tissue insult, histamine causes dilation of microcirculatory beds (thus tissue redness) and promotes exit of fluid from capillaries into the intercellular spaces

Figure 21.42 The structure of prostaglandin E_1. Angles are occupied by carbon atoms.

Table 21.7 CERTAIN ACTIONS OF SOME OF THE PROSTAGLANDINS

Prostaglandin	Action
Prostaglandin E_1	Inhibition of platelet aggregation Vasodilatation Fever induction Inflammatory response, including tissue swelling as well as vasodilatation
Prostaglandin E_2	Vasodilatation Contraction of uterine smooth muscle Gut motility increased Stimulation of renal Na^+ excretion Inhibition of renal antidiuretic-hormone effect Bronchodilatation Inhibition of gastric acid secretion Stimulation of erythropoietin secretion Sensitization to pain stimuli Fever induction
Prostaglandin $F_{2\alpha}$	Uterine and fallopian-tube contraction Involution of corpora lutea (nonprimates) Bronchoconstriction Gut motility increased
Prostacyclin	Vasodilatation Inhibition of platelet aggregation and of thrombus formation Renin release from kidneys Relaxation of uterine smooth muscle Bronchodilatation
Thromboxane A_2	Platelet aggregation Vasoconstriction Bronchoconstriction
Leukotrienes C_4 and D_4	Bronchoconstriction Increased cutaneous vascular permeability Vasoconstriction Impaired sputum clearance

Source: Expanded from J.C. Froelich and H.S. Margolius, in P. Felig, J.D. Baxter, A.E. Broadus, and L.A. Frohman (eds.), *Endocrinology and Metabolism*, pp. 1247–1273. McGraw-Hill, New York, 1981.

of the tissue (thus swelling). Histamine occurs in some invertebrates; for instance, it is a prominent component of bee venom.

The *kinins,* such as *bradykinin* and *lysylbradykinin* (kallidin), are polypeptides that act as local hormones. Rather than being secreted directly from cells, they are often formed enzymatically from extracellular proteins (called kininogens) that are found, for example, in blood plasma. The enzymes that catalyze the release of kinins from kininogens are called *kallikreins*. Essentially, kinin formation is controlled by the mechanisms that control kallikrein formation. Again, as in the case of prostaglandins and histamine, it has been far easier to demonstrate laboratory effects of the kinins and kallikreins than to determine which effects represent normal functions, and the list of known effects is long. Bradykinin in mammals, for example, dilates arterial vessels, lowers blood pressure, increases vascular permeability, causes bronchoconstriction, increases renal excretion of sodium and water, and promotes release of both histamine and certain prostaglandins

(these latter may be immediately responsible for some of bradykinin's effects).

The effects of histamine and kinins are localized, as are those of most prostaglandins, because the agents are rapidly destroyed once released. Kinins, for example, are destroyed locally by enzymes found in blood and various other tissues, and any kinins that escape local destruction are largely destroyed on their first pass through the lungs.

Pheromones

Whereas hormones are employed for communication between tissues within a single animal's body, pheromones serve communication between animals. As discussed previously (p. 497), pheromones are metabolically produced chemicals released by an organism that elicit specific behavioral or systemic responses in other individuals of the same species. Animals employ an extraordinary variety of chemicals as pheromones, including proteins, hydrocarbons, alcohols, and esters. Often, pheromones are used as sexual signals, serving, for example, to attract one sex to the other. They are also employed in other roles, such as in mediating orientation and spacing behavior.

CODA

In this chapter we have tried to show some of the principles of endocrine function, some of the diversity of endocrine action, and the widespread power of endocrine control. It is becoming increasingly difficult to draw lines between endocrine control and other aspects of physiological regulation. Furthermore, we have needed to omit many examples and much richness of detail, in the interest of keeping our treatment manageable. (This last disclaimer, of course, applies to the entire book.) We hope you will be stimulated to learn more.

SELECTED READINGS

Ball, J.N. 1981. Hypothalamic control of the pars distalis in fishes, amphibians, and reptiles. *Gen. Comp. Endocrinol.* **44:**135–170.

*Balthazart, J., E. Pröve, and R. Gilles (eds.). 1983. *Hormones and Behaviour in Higher Vertebrates.* Springer-Verlag, New York.

Barrington, E.J.W. 1979. Chemical coordination. In E.J.W. Barrington, *Invertebrate Structure and Function,* 2nd ed., pp. 492–541. Nelson, Sudbury-on-Thames.

*Barrington, E.J.W. 1979. Introduction. In E.J.W. Barrington (ed.), *Hormones and Evolution,* pp. vii–xxi. Academic, New York.

*Barrington, E.J.W. (ed.). 1979. *Hormones and Evolution.* Academic, New York.

Behavioral Endocrinology. Special section of *BioScience.* 1983. *BioScience* **33:**545–582.

Bennett, K.L. and J.W. Truman. 1985. Steroid-dependent survival of identifiable neurons in cultured ganglia of the moth *Manduca sexta. Science* **229:**58–60.

*Bentley, P.J. 1982. *Comparative Vertebrate Endocrinology.* Cambridge University Press, New York.

*Cameron, J.N. 1985. Molting in the blue crab. *Sci. Am.* **252**(5):102–109.

Cantin, M. and J. Genest. 1986. The heart as an endocrine gland. *Sci. Am.* **254**(2):76–81.

*Carmichael, S.W. and H. Winkler. 1985. The adrenal chromaffin cell. *Sci. Am.* **253**(2):40–49.

Chester Jones, I. and I.W. Henderson (eds.). 1976–1980. *General, Comparative and Clinical Endocrinology of the Adrenal Cortex,* 3 vols. Academic, New York.

Chester Jones, I., P.M. Ingleton, and J.G. Phillips (eds.). 1987. *Fundamentals of Comparative Vertebrate Endocrinology.* Plenum, New York.

Clarke, I.J. 1982. Prenatal sexual development. *Oxford Rev. Reprod. Biol.* **4:**101–147.

Dahl, K.D., T.A. Bicsak, and A.J.W. Hsueh. 1988. Naturally occurring antihormones: Secretion of FSH antagonists by women treated with a GnRH analog. *Science* **239:**72–74.

Darras, V.M. and E.R. Kuehn. 1982. Increased plasma levels of thyroid hormone in a frog *Rana ridibunda* following intravenous administration of TRH. *Gen. Comp. Endocrinol.* **48:**469–475.

Dodd, M.H.I. and J.M. Dodd. 1976. The biology of metamorphosis. In B. Lofts (ed.), *Physiology of the Amphibia,* Vol. 3, pp. 467–599. Academic, New York.

Dusting, G.T., S. Moncada, and J.R. Vane. 1982. Prostacyclin: Its biosynthesis, actions, and clinical potential. *Adv. Prostaglandin Thromboxane Leukotriene Res.* **10:**59–106.

Farner, D.S. and B.K. Follett. 1979. Reproductive periodicity in birds. In E.J.W. Barrington (ed.), *Hormones and Evolution,* pp. 827–872. Academic, New York.

*Felig, P., J.D. Baxter, A.E. Broadus, and L.A. Frohman (eds.). 1981. *Endocrinology and Metabolism.* McGraw-Hill, New York.

Feuerstein, G. and J.M. Hallenbeck. 1987. Leukotrienes in health and disease. *FASEB J.* **1:**186–192.

Fink, G. 1979. Feedback actions of target hormones on hypothalamus and pituitary with special reference to gonadal steroids. *Annu. Rev. Physiol.* **41:**571–585.

Forsyth, I.A. 1982. Growth and differentiation of mammary glands. *Oxford Rev. Reprod. Biol.* **4:**47–85.

Friedman, W.F., M.P. Printz, R.A. Skidgel, L.N. Benson, and M. Zednikova. 1982. Prostaglandins and the ductus arteriosus. *Adv. Prostaglandin Thromboxane Leukotriene Res.* **10:**277–302.

Froelich, J.L. and H.S. Margolius. 1981. Prostaglandins, the kallikrein–kinin system, Bartter's syndrome, and the carcinoid syndrome. In P. Felig, J.D. Baxter, A.E. Broadus, and L.A. Frohman (eds.), *Endocrinology and Metabolism,* pp. 1247–1273. McGraw-Hill, New York.

Fuchs, F. and A. Klopper (eds.). 1983. *Endocrinology of Pregnancy.* Harper & Row, New York.

Golding, D.W. 1983. Endocrine programmed development and reproduction in *Nereis. Gen. Comp. Endocrinol.* **52:**456–466.

Goldsworthy, G.J., J. Robinson, and W. Mordue. 1981. *Endocrinology.* Wiley, New York.

*Gorbman, A., W.W. Dickhoff, S.R. Vigna, N.B. Clark, and C.L. Ralph. 1983. *Comparative Endocrinology.* Wiley, New York.

*Greep, R.O. and E.B. Astwood (eds.). 1972–1976. *Handbook of Physiology, Sec. 7: Endocrinology,* 7 vols. American Physiological Society, Washington, DC.

*Highnam, K.C. and L. Hill. 1977. *The Comparative Endocrinology of the Invertebrates,* 2nd ed. Arnold, London.

Hoffmann, J. and M. Porchet (eds.). 1984. *Biosynthesis, Metabolism, and Mode of Action of Invertebrate Hormones.* Springer-Verlag, New York.

Houck, J.C. (ed.). 1979. *Chemical Messengers of the Inflammatory Process.* Elsevier/North-Holland, New York.

Jackson, I.M.D. 1986. Phylogenetic distribution and significance of the hypothalamic releasing hormones. *Am. Zool.* **26:**927–938.

James, V.H.T. (ed.). 1979. *The Adrenal Gland.* Raven, New York.

Jan, Y.N. and L.Y. Jan. 1983. LHRH—A new neuromodulator capable of "action at a distance" in autonomic ganglia. *Trends Neurosci.* **6:**320–325.

Jones, M.T. 1979. Control of adrenocortical hormone secretion. In V.H.T. James (ed.), *The Adrenal Gland,* pp. 93–102. Raven, New York.

*Knobil, E. 1980. The neuroendocrine control of the menstrual cycle. *Recent Prog. Horm. Res.* **36:**53–88.

Kolata, G. 1984. Steroid hormone systems found in yeast. *Science* **225:**913–914.

Lee, J.B. (ed.). 1982. *Prostaglandins*. Elsevier, New York.

Levine, J.E. and V.D. Ramirez. 1982. Luteinizing hormone-releasing hormone release during the rat estrous cycle and after ovariectomy, as estimated with push–pull cannulae. *Endocrinology* **111**:1439–1448.

Licht, P. 1979. Reproductive endocrinology of reptiles and amphibians: Gonadotropins. *Annu. Rev. Physiol.* **41**:337–351.

Milligan, S.R. 1982. Induced ovulation in mammals. *Oxford Rev. Reprod. Biol.* **4**:1–46.

Newsholme, E.A. and C. Start. 1973. *Regulation in Metabolism*. Wiley, New York.

Nijhout, H.F. 1981. Physiological control of molting in insects. *Am. Zool.* **21**:631–640.

*Norman, R.L., J.E. Levine, and H.G. Spies. 1983. Control of gonadotropin secretion in primates: Observations in stalk-sectioned rhesus macaques. In R. Norman (ed.), *Neuroendocrine Aspects of Reproduction,* pp. 263–285. Academic, New York.

Norris, D.O. and R.E. Jones. 1987. *Hormones and Reproduction in Fishes, Amphibians, and Reptiles*. Plenum, New York.

Novák, V.J.A. 1975. *Insect Hormones,* 4th ed. Chapman and Hall, London.

Pang, P.K.T. and M.P. Schreibman (eds.). 1986–1987. *Vertebrate Endocrinology. Fundamentals and Biomedical Implications*, 2 vols. Academic, New York.

Piper, P.J. (ed.). 1986. *The Leukotrienes. Their Biological Significance*. Raven, New York.

*Ralph, C.L. (ed.). 1986. *Comparative Endocrinology. Developments and Directions*. (Progr. Clin. Biol. Res., Vol. 205). Liss, New York.

Regoli, D. and J. Barabe. 1980. Pharmacology of bradykinin and related kinins. *Pharmacol. Rev.* **32**:1–46.

Samuelsson, B. and R. Paoletti (eds.). 1982. *Leukotrienes and Other Lipoxygenase Products*. Raven, New York.

Segerson, T.P., J. Kauer, H.C. Wolfe, H. Mobtaker, P. Wu, I.M.D. Jackson, and R.M. Lechan. 1987. Thyroid hormone regulates TRH biosynthesis in the paraventricular nucleus of the rat hypothalamus. *Science* **238**:78–80.

Serra, G.B. (ed.). 1982. *The Ovary*. Raven, New York.

Shaw, S.T., Jr. and P.C. Roche. 1980. Menstruation. *Oxford Rev. Reprod. Biol.* **2**:41–96.

Shorey, H.H. 1976. *Animal Communication by Pheromones*. Academic, New York.

Tepperman, J. and H.M. Tepperman. 1987. *Metabolic and Endocrine Physiology,* 5th ed. Year Book Medical, Chicago.

Truman, J.W. and L.M. Riddiford. 1973. Hormonal mechanisms underlying insect behavior. *Adv. Insect Physiol.* **10**:297–352.

Turek, F.W., J. Swann, and D.J. Earnest. 1984. Role of the circadian system in reproductive phenomena. *Recent Prog. Horm. Res.* **40**:143–183.

Van Cauter, E. and E. Honinckx. 1985. Pulsatility of pituitary hormones. In H. Schulz and P. Lavie (eds.), *Ultradian Rhythms in Physiology and Behavior,* pp. 41–60. Springer-Verlag, New York.

VomSaal, F.S. 1983. The interaction of circulating oestrogens and androgens in regulating mammalian sexual differentiation. In J. Balthazart, E. Pröve, and R. Gilles (eds.), *Hormones and Behaviour in Higher Vertebrates,* pp. 157–177. Springer-Verlag, New York.

See also references in Appendix A.

appendix A

References

Each of the works listed here covers a variety of aspects of physiology. These are good reference sources. They have generally not been listed with the individual chapters of this text inasmuch as many of them would have to be listed under virtually all chapters, such is the breadth of their coverage.

GENERAL WORKS

Barrington, E.J.W. 1979. _Invertebrate Structure and Function_, 2nd ed. Nelson, Sudbury-on-Thames.

Dill, D.B. (ed.). 1964. _Handbook of Physiology,_ Sec. 4, _Adaptation to the Environment._ American Physiological Society, Washington, DC.

Florkin, M. and B.T. Scheer (eds.). 1967–1979. _Chemical Zoology_, 13 vols. Academic, New York.

Fretter, V. and A. Graham. 1976. _A Functional Anatomy of Invertebrates._ Academic, New York.

Hochachka, P.W. and G.N. Somero. 1984. _Biochemical Adaptation._ Princeton University Press, Princeton.

Newell, R.C. (ed.). 1976. _Adaptation to Environment. Essays on the Physiology of Marine Animals._ Butterworths, London.

Newell, R.C. 1979. _Biology of Intertidal Animals._ Marine Ecological Surveys Ltd., Faversham.

Prosser, C.L. (ed.). 1973. _Comparative Animal Physiology_, 3rd ed. Saunders, Philadelphia.

Prosser, C.L. 1986. _Adaptational Biology. Molecules to Organisms._ Wiley, New York.

Ruch, T.C. and H.D. Patton (eds.). 1973. _Physiology and Biophysics_, 3 vols. Saunders, Philadelphia.

Schmidt-Nielsen, K. 1972. _How Animals Work._ Cambridge University Press, London.

Seymour, R.S. (ed.). 1984. _Respiration and Metabolism of Embryonic Vertebrates._ Junk, Dordrecht.

Vernberg, F.J. and W.B. Vernberg (eds.). 1981. _Functional Adaptations of Marine Organisms._ Academic, New York.

WORKS DEALING WITH SPECIFIC GROUPS OF INVERTEBRATES

Anderson, O.R. 1988. _Comparative Protozoology. Ecology, Physiology, Life History._ Springer-Verlag, New York.

Bell, W.J. and K.G. Adiyodi (eds.). 1981. _The American Cockroach._ Chapman and Hall, London.

Berquist, P.R. 1978. _Sponges._ University of California Press, Berkeley.

Binyon, J. 1972. _Physiology of Echinoderms._ Pergamon, New York.

Bliss, D.E. (ed.). 1982–1985. _The Biology of Crustacea_, 10 vols. Academic, New York.

Blum, M.S. (ed.). 1985. _Fundamentals of Insect Physiology._ Wiley, New York.

Boolootian, R.A. (ed.). 1966. _Physiology of Echinodermata._ Wiley, New York.

Hughes, R.N. 1986. _A Functional Biology of Marine Gastropods._ Johns Hopkins University Press, Baltimore.

Kerkut, G.A. and L.I. Gilbert (eds.). 1985. _Comprehensive Insect Physiology, Biochemistry and Pharmacology_, 13 vols. Pergamon, New York.

Laverack, M.S. 1963. _The Physiology of Earthworms._ Pergamon, Oxford.

Lawrence, J. 1987. _A Functional Biology of Echinoderms._ Johns Hopkins University Press, Baltimore.

Lee, D.L. and H.J. Atkinson. 1977. _Physiology of Nematodes_, 2nd ed. Columbia University Press, New York.

Mill, P.J. (ed.). 1978. _Physiology of Annelids._ Academic, New York.

Nentwig, W. (ed.). 1987. _Ecophysiology of Spiders._ Springer-Verlag, New York.

Obenchain, F.D. and R. Galun (eds.). 1982. _Physiology of Ticks._ Pergamon, New York.

Rockstein, M. (ed.). 1973–1974. _The Physiology of Insecta_, 2nd ed., 6 vols. Academic, New York.

Sawyer, R.T. 1986. _Leech Biology and Behaviour._ Clarendon, Oxford.

Sutton, S.L. and D.M. Holdrich (eds.). 1984. *The Biology of Terrestrial Isopods* (Symp. Zool. Soc. London No. 53). Clarendon, Oxford.

Wells, M.J. 1978. *Octopus*. Chapman and Hall, London.

Wigglesworth, V.B. 1972. *The Principles of Insect Physiology*, 7th ed. Chapman and Hall, London.

Wilbur, K.M. (ed.). 1983–1987. *The Mollusca*, 12 vols. Academic, New York.

WORKS DEALING WITH SPECIFIC GROUPS OF VERTEBRATES

Bemis, W.E., W.W. Burggren, and N.E. Kemp (eds.). 1987. *The Biology and Evolution of Lungfishes*. Liss, New York.

Farner, D.S. and J.R. King (eds.). 1971–1975. *Avian Biology*, Vols. 1–5. Academic, New York.

Farner, D.S., J.R. King, and K.C. Parkes (eds.). 1982–1985. *Avian Biology*, Vols. 6–8. Academic, New York.

Foreman, R.E., A. Gorbman, J.M. Dodd, and R. Olsson (eds.). 1985. *Evolutionary Biology of Primitive Fishes*. Plenum, New York.

Gans, C. (ed.). 1969–1987. *Biology of the Reptilia,* 16 vols. Academic, New York.

Handbook of Physiology. 1960–1987. American Physiological Society, Bethesda. (A comprehensive mammalian physiology; numerous volumes and editors.)

Hoar, W.S. and D.J. Randall (eds.). 1969–1987. *Fish Physiology*, 11 vols. Academic, New York.

King, A.S. and J. McLelland (eds.). 1979–1985. *Form and Function in Birds*, 3 vols. Academic, New York.

Lofts, B. (ed.). 1974–1976. *Physiology of the Amphibia*, Vols. 2 and 3. Academic, New York.

Moore, J.A. (ed.). 1964. *Physiology of the Amphibia*, Vol. 1. Academic, New York.

Mountcastle, V.B. (ed.). 1980. *Medical Physiology*, 14th ed. Mosby, St. Louis.

Sturkie, P.D. (ed.). 1986. *Avian Physiology*, 4th ed. Springer-Verlag, New York.

Tytler, P. and P. Calow (eds.). 1985. *Fish Energetics. New Perspectives*. Johns Hopkins University Press, Baltimore.

appendix **B**

Units of Measure

This table presents units of measure in the Système International (SI) and shows their relations to various traditional units of measure. The table is divided into two sections: base SI units and derived SI units. The SI recognizes just seven base units. All other units in the SI are derived from these seven (e.g., the unit of measure for velocity is the m/s, a derivative of the base units for length and time). The column labeled "SI unit" gives the unit of measure in the SI. Sometimes, derived units in the SI are given special names, and those are indicated (e.g., the unit for energy, the $kg \cdot m^2/s^2$, is termed the joule). Prefixes indicating decimal multiples of units are acceptable in the SI; thus 1000 joules (J) may be called a kilo-joule (kJ). Relations between SI units and selected traditional units are shown in the column labeled "relations among units." Relations between different sets of traditional units are also occasionally listed there. To obtain relations converse to those shown, divide both sides of the relevant equation by the number to the right of the equal symbol. For example, the table states "1 m = 3.28 ft." Dividing both sides by 3.28 yields "1 ft = 0.305 m." For more detail on unit conversions, consult a recent edition of a standard reference such as the *Handbook of Chemistry and Physics* (CRC Press, Boca Raton, FL) or *Lange's Handbook of Chemistry* (McGraw-Hill, New York).

Quantity	SI unit	Relations among units
Base SI units		
Length	meter (m)	1 m = 3.28 feet (ft)
		1 inch (in) = 25.4 millimeter (mm)
		1 statute mile (mi) = 1609.3 m
Mass	kilogram (kg)	1 kg = 2.20 pound, avoirdupois (lb)
		1 ounce, avoirdupois (oz) = 28.3 gram (g)
Temperature	degree Celsius (°C) = 1 kelvin (K)	A difference of 1°C = a difference of 1.8 degree Fahrenheit (°F)
Time	second (s)	—
Electric current	ampere (A)	—
Amount of substance	mole (mol)	—
Luminous intensity	candela (cd)	1 cd ≃ 1 candle (pentane)
Derived SI units		
Area	m^2	1 m^2 = 10000 square centimeters (cm^2) = 10.8 ft^2
Volume	m^3	1 m^3 = 1 × 10^6 cm^3 = 1000 liter (L)
		1 cm^3 = 1 milliliter (mL)
		1 U.S. gallon = 3.785 L
		1 U.S. fluid ounce = 29.6 mL
Density	kg/m^3	1 kg/m^3 = 0.001 g/mL
Velocity	m/s	1 m/s = 3.28 ft/s = 2.24 statute mile/hour
Acceleration	m/s^2	1 m/s^2 = 3.28 ft/s^2
Force	$kg \cdot m/s^2$ = 1 newton (N)	1 N = 0.102 kilogram of force = 0.225 pound of force = 1 × 10^5 dyne
Energy, work	$kg \cdot m^2/s^2$ = 1 N·m = 1 joule (J)	1 J = 0.239 calorie (cal) = 1 × 10^7 erg = 0.000948 British thermal unit = 0.738 foot-pound
Power	$kg \cdot m^2/s^3$ = 1 J/s = 1 watt (W)	1 W = 0.239 cal/s = 0.0013 horsepower = 3.41 British thermal unit/h
Pressure	$kg/(s^2 \cdot m)$ = 1 N/m^2 = 1 pascal (Pa)	1 Pa = 0.0075 mm of mercury (mm Hg)
		1 kilopascal (kPa) = 1000 Pa
		1 atmosphere = 101.3 kPa
		1 atmosphere = 760 mm Hg
		1 mm Hg = 1 torr
		1 lb/in^2 = 51.7 mm Hg
Frequency	1/s = 1 hertz (Hz)	1 Hz = 1 cycle/s
Electric potential	$kg \cdot m^2/(s^3 \cdot A)$ = 1 W/A = 1 volt (V)	—
Electric resistance	$kg \cdot m^2/(s^3 \cdot A^2)$ = 1 V/A = 1 ohm (Ω)	—
Electric charge	s·A = 1 coulomb (C)	1 C = 0.00028 ampere-hour

Index

Boldfaced page numbers indicate the definition of a
word or concept.